S0-ATC-027

Structural Concrete

Theory and Design

Second Edition

M. Nadim Hassoun
South Dakota State University

Prentice Hall, Upper Saddle River, NJ 07458

Library of Congress Cataloging-in-Publication Data

Hassoun, M. Nadim.
Structural concrete: theory and design/M. Nadim Hassoun.
p. cm.
Includes index.
ISBN 0-13-042171-5
1. Reinforced concrete construction. I. Title.

TA683.2 .H365 2002
624.1'8341—dc21

2001021993

Vice President and Editorial Director, ECS: Marcia J. Horton
Acquisitions Editor: Laura Curless
Editorial Assistant: Erin Katchmar
Vice President and Director of Production and Manufacturing, ESM: David W. Riccardi
Executive Managing Editor: Vince O'Brien
Managing Editor: David A. George
Production Editor: Barbara A. Till
Director of Creative Services: Paul Belfanti
Creative Director: Carole Anson
Art Director: Jayne Conte
Art Editor: Adam Velthaus
Cover Designer: Bruce Kenselaar
Manufacturing Manager: Trudy Pisciotti
Manufacturing Buyer: Lisa Mc Dowell
Marketing Manager: Holly Stark
Marketing Assistant: Karen Moon

Upper Saddle River, New Jersey 07458

Printed in the United States of America
10 9 8 7 6 5 4 3 2 1

ISBN 0-13-042171-5

Prentice-Hall International (UK) Limited, *London*
Prentice-Hall of Australia Pty. Limited, *Sydney*
Prentice-Hall Canada Inc., *Toronto*
Prentice-Hall Hispanoamericana, S.A., *Mexico City*
Prentice-Hall of India Private Limited, *New Delhi*
Prentice-Hall of Japan, Inc., *Tokyo*
Pearson Education Asia Pte. Ltd., *Singapore*
Editora Prentice-Hall do Brasil, Ltda., *Rio de Janeiro*

CONTENTS

PREFACE TO THE FIRST EDITION

The main objective of a course on reinforced concrete design is to develop, in the engineering student, the ability to analyze and design a reinforced concrete member subjected to different types of forces in a simple and logical manner using the basic principles of statics and some empirical formulas based on experimental results. Once the analysis and design procedure is fully understood, its application to different types of structures becomes simple and direct, provided that the student has a good background in structural analysis.

The material presented in this book is based on the requirements of the American Concrete Institute (ACI) Building Code (318-95). Also, information has been presented on material properties, including volume changes of concrete, stress-strain behavior, creep, and elastic and nonlinear behavior of reinforced concrete.

Concrete structures are widely used in the United States and almost all over the world. The progress in the design concept has increased in the last few decades, emphasizing safety, serviceability, and economy. To achieve economical design of a reinforced concrete member, specific restrictions, rules, and formulas are presented in the codes to ensure both safety and reliability of the structure. Engineering firms expect civil engineering graduates to understand the code rules and, consequently, to be able to design a concrete structure effectively and economically with minimum training period or overhead costs. Taking this into consideration, this book is written to achieve the following objectives:

1. To present the material for the design of reinforced concrete members in a simple and logical approach.
2. To arrange the sequence of chapters in a way compatible with the design procedure of actual structures.
3. To provide a large number of examples in each chapter in clear steps to explain the analysis and design of each type of structural member.
4. To provide an adequate number of practical problems at the end of each chapter to achieve a high level of comprehension.
5. To explain the failure mechanism of a reinforced concrete beam due to flexure and to develop the necessary relationships and formulas for design.
6. To explain *why* the code used specific equations and specific restrictions on the design approach based either on a mathematical model or experimental results. This approach will improve the design ability of the student.

7. To provide adequate number of design aids to help the student in reducing the repetitive computations of specific commonly used values.
8. To enhance the student's ability to use a total quality and economical approach in the design of concrete structures and to help the student to design reinforced concrete members with confidence.
9. To explain the nonlinear behavior and the development of plastic hinges and plastic rotations in continuous reinforced concrete structures.
10. To provide flowcharts to aid the students in writing their own programs, because computers are needed in the design of concrete structures. Flowcharts and computer programs are discussed in Chapter 21.
11. To provide a summary at the end of each chapter to help the student to review the materials of each chapter separately.
12. To provide new information on the design of some members, such as beams with variable depth (Chapter 8), stairs (Chapter 18); and curved beams (Chapter 19), that are not covered in other books on concrete.
13. To present information on the design of reinforced concrete frames, principles of limit design, and moment redistribution in continuous reinforced concrete structures.
14. To provide examples in SI units in all chapters of the book. Equivalent conversion factors from customary units to SI units are also presented. Design tables in SI units are given in Appendix B.
15. References are presented at the end of most chapters.

The book is an outgrowth of the author's lecture notes, which represent his teaching and industrial experience over the past 30 years. The industrial experience of the author includes the design and construction supervision and management of many reinforced, prestressed, and precast concrete structures. This is in addition to the consulting work he performed to international design and construction firms, professional registration in U.K. and other countries, and a comprehensive knowledge of other European codes on the design of concrete structures.

The book is written to cover two courses in reinforced concrete design. Depending on the proficiency required, the first course may cover Chapters l through 11 and part of Chapter 13, whereas the second course may cover the remaining chapters. Parts of the late chapters may also be taught in the first course as needed. A number of optional sections have been included in various chapters. These sections are indicated by an asterisk (*) in the Table of Contents, and may easily be distinguished from those which form the basic requirements of the first course. The optional sections may be covered in the second course or relagated to a reading assignment. Brief descriptions of the chapters are given below.

The first chapter of the book presents information on the historical development of concrete, codes of practice, loads and safety provisions, and design philosophy and concepts. The second chapter deals with the properties of concrete as well as steel reinforcement used in the design of reinforced concrete structures, including stress–strain relationships, modulus of elasticity and shear modulus of concrete, shrinkage, creep, fire resistance, high-performance concrete, and fibrous concrete. Because the current ACI Code gives emphasis to the strength design method, this approach has been adopted throughout the text, except in Chapter 5, where the analysis of reinforced concrete sections by the working stress design is explained in order to enable the student and designer to check the deflection of flexural members under service loads. Chapters 3 and 4 cover the analysis and design of reinforced concrete sections based on strength design concept. The behavior of reinforced concrete beams loaded to failure, the types of flexural failure, and failure mechanisms are explained in a way that differs from other textbooks. It is essential for the student to understand the failure concept and the inherent reserve strength before using the necessary design formulas.

Chapters 5 and 6 deal with the elastic behavior and serviceability of concrete beams, including deflection and control of cracking. Chapters 7 and 8 cover the bond, development length, shear, and diagonal tension. In Chapter 8, expressions are presented for designing members of variable depth in addition to prismatic sections. It is quite common to design members of variable depth in actual structures. Chapter 9 covers the design of one-way slabs, including joist-floor systems. Distribution of loads from slabs to beams and columns are also presented in this chapter to enhance the student's understanding of the design loads on each structural component. Chapters 10, 11, and 12 cover the design of axially loaded, eccentrically loaded, and long columns, respectively. Chapter 10 allows the student to understand the behavior of columns, failure conditions, ties and spirals, and other code limitations. Absorbing basic information, the student is introduced in Chapter 11 to the design of columns subjected to compression and bending. New mathematical models are introduced to analyze column sections controlled by compression or tension stresses. Biaxial bending for rectangular and circular columns are introduced using Bresler, PCA, and Hsu methods. Design of long columns is presented in Chapter 12 using the ACI moment magnifier method.

Chapters 13 and 14 cover the design of footings and retaining walls, whereas Chapter 15 covers the design of reinforced concrete sections for shear and torsion. Torsional theories as well as ACI Code design procedure are explained. Chapter 16 deals with continuous beams and frames. A unique feature of this chapter is the introduction of the design of frames, frame hinges, limit state design collapse mechanism, rotation and plastic hinges, and moment redistribution. Adequate examples are presented to explain these concepts.

Design of two-way slabs is introduced in Chapter 17. All types of two-way slabs, including waffle slabs, are presented with adequate examples. Summary of the design procedure is introduced with tables and diagrams. Chapter 18 covers the design of reinforced concrete stairs. Slab-type as well as stepped-type stairs are explained. The second type, although quite common, has not been covered in any text. Chapter 19 deals with the design of curved beams. In actual structures, curved beams are used frequently. These beams are subjected to flexure, shear, and torsion. Design coefficients are presented in this chapter. Chapter 20 covers an introduction to prestressed concrete. Methods of prestressing, fully and partially prestressed concrete design, losses, and shear design are presented with examples. Chapter 21 introduces computer programs as well as flowcharts.

The unified design method (UDM) for the design of reinforced and prestressed concrete flexural and compression members is presented in Chapter 22. This new approach introduces some basic changes in the design limits. Provisions for this method are introduced in the ACI Code, Appendix B. The author suggests that the concept of UDM be explained to the students with Chapters 3, 4, and 11. Examples of Chapter 22 can be presented with these chapter.

In the Appendix of this book, design tables using customary units and SI units are presented.

All the photos shown in this book were taken by the author.

My sincere thanks to the reviewers of the manuscript for their constructive comments and valuable suggestions:

Jack D. Bakos, Jr., Youngstown State University; Kassim Tarhini, Valparaiso University; Hany J. Farran, California Polytechnic University, Pomona; Jerry R. Bayless, University of Missouri, Rolla; J. Michael Stallings, Auburn University; Steven D. Leftwich, West Virginia Tech; Amir Mirmiran, University of Central Florida; Jean-Paul Pinelli, Florida Tech University; Richard N. White, Cornell University; Gajanan M. Sabnis, Howard University. Special thanks are due to the civil engineering students at South Dakota State University for their feedback while using the manuscript. Special thanks to Dr. Moayyad Al-Nasra for arranging the software programs in an executable form.

PREFACE TO THE SECOND EDITION

The second edition of this book revises the previous text to conform to the latest American Concrete Institute (ACI) Code 318-99. It also includes additional sections, revisions, and editing of various chapters of the book. In Chapter 1, Section 1.11 has been added to help the student understand the accuracy of calculations in engineering design. Additional examples are introduced in Chapters 3 and 4 to elaborate on the behavior of reinforced concrete beams at failure and to combine structural analysis with concrete design. In Chapter 6, the section on crack control has been revised to conform to the ACI Code limits on distribution of flexural reinforcement. The code also made some changes in the shear for circular sections and spiral lap splices and introduced a limit on column slenderness ratio. These changes are covered in Chapters 8, 10 and 12, respectively. An additional section on Coulomb theory for soil pressure has been introduced in Chapter 13. Revisions are also made in the design for torsion (Chapter 15) and in bars layout and extensions in two-way slabs without beams (Chapter 17). A new section on partially prestressed concrete has been added to Chapter 20. Structural aid tables are added as Appendix C to help the student and reader to determine the moment, shear, and deflection for simple and continuous beams. The book also contains numerous examples in International System (SI) units, summaries at the end of each chapter, and flow charts (in Chapter 21).

I would like to extend my sincere thanks to the reviewers for their constructive comments to revise the book in this final second edition.

Finally, the book is written to provide basic and reference materials on the analysis and design of reinforced concrete members in a simple, practical, and logical approach. Because this is a required course for seniors in civil engineering, I believe it will be accepted by reinforced concrete instructors at different universities as well as designers who can make use of the information in this book in their practical design of reinforced concrete structures. A solutions manual is provided. Software for the design of different reinforced concrete members is also available on a ftp site maintained by the publisher. Instructors should have received login instructions for themselves and their students from their Prentice Hall Engineering sales representative. All other users of the text should e-mail publisher at Prentice Hall at **www.prenhall.com/hassoun** to request login information.

M. Nadim Hassoun

NOTATION

a	Depth of the equivalent rectangular concrete stress block
a_b	Value of a for a balanced condition
A	Effective tension area of concrete surrounding one bar (This value is used for control of cracking.)
A_b	Area of individual bar
A_c	Area of core of spirally reinforced column
A_{cp}	Gross area enclosed by outside perimeter of cross section
ACI	American Concrete Institute
A_g	Gross (total) area of cross section
A_ℓ	Total area of longitudinal torsion steel
A_o	Gross area enclosed by shear flow $0.85A_{oh}$
A_{oh}	$x_1 y_1$
A_{ps}	Area of prestressed reinforcement in the tension zone
A_s	Area of flexural tension steel
A'_s	Area of compression steel
A_{sb}	Area of balanced steel
A_{st}	Total steel area in the section (column)
A_{sf}	Area of reinforcement to develop compressive strength of overhanging flanges in T- or L-sections
A_t	Area of one leg of closed stirrups used to resist torsion
A_{tc}	Transformed concrete area
A_v	Total area of shear reinforcement within a spacing S
A_1	Loaded area
A_2	Maximum area of supporting surface geometrically similar and concentric with the loaded area
b	Width of compression zone at extreme fiber
b_e	Effective width of flange
b_o	Perimeter of critical section for punching shear
b_w	Width of beam web
c	Distance from extreme compression fiber to neutral axis
c_b	c for a balanced section
c_1	Side of a rectangular column measured in the direction of span

c_2	Side of rectangular column measured transverse to the span
C	Cross-sectional constant $= \Sigma(1 - 0.63x/y)x^3y/3$; compression force
C_c	Compression force in a concrete section with a depth equal to a
C_m	Correction factor applied to the maximum end moment in columns
C_r	Creep coefficient = creep strain per unit stress per unit length
C_s	Force in compression steel
C_t	Factor relating shear and torsional stress properties $= b_w d/\Sigma x^2 y$
C_w	Compression force in web
C_1	Force in the compression steel
d	Distance from extreme compression fiber to centroid of tension steel
d'	Distance from extreme compression fiber to centroid of compression steel
d_b	Nominal diameter of reinforcing bar
d_c	Distance from tension extreme fiber to center of bar closest to that fiber, used for crack control
D	Dead load, diameter of a circular section
e	Eccentricity of load
e'	Eccentricity of load with respect to centroid of tension steel
E	Modulus of elasticity, force created by earthquake
E_c	Modulus of elasticity of concrete $= 33W^{1.5}\sqrt{f'_c}$
E_{cb}	Modulus of elasticity of beam concrete
E_{cc}	Modulus of elasticity of column concrete
E_{cs}	Modulus of elasticity of slab concrete
EI	Flexural stiffness of compression member
E_s	Modulus of elasticity of steel $= 29 \times 10^6$ psi $= 2 \times 10^5$ MPa
f	Flexural stress
f_c	Maximum flexural compressive stress in concrete due to service loads
f_{ca}	Allowable compressive stress in concrete (alternate design method)
f'_c	28-day compressive strength of concrete (standard cylinder strength)
f_{ci}	Compressive strength of concrete at transfer (initial prestress)
f_{pc}	Compressive stress in concrete due to prestress after all losses
f_{pe}	Compressive stress in concrete at extreme fiber due to the effective prestressing force after all losses
f_{ps}	Stress in prestress steel at nominal strength
f_{pu}	Tensile strength of prestressing tendons
f_{py}	Yield strength of prestressing tendons
f_r	Modulus of rupture of concrete $= 7.5\sqrt{f'_c}$ psi
f_s	Stress in tension steel due to service load
f'_s	Stress in the compression steel due to service load
f_{se}	Effective stress in prestressing steel after all losses
f_t	Tensile stress in concrete
f_y	Yield strength of steel reinforcement
F	Lateral pressure of liquids
G	Shear modulus of concrete (in torsion) $= 0.45E_c$
h	Total depth of beam or slab or column
h_f	Depth of flange in flanged sections
h_v	Total depth of shearhead cross section
H	Lateral earth pressure
I	Moment of inertia
I_b	Moment of inertia of gross section of beam about its centroidal axis
I_c	Moment of inertia of gross section of column
I_{cr}	Moment of inertia of cracked transformed section

I_e	Effective moment of inertia, used in deflection
I_g	Moment of inertia of gross section neglecting steel
I_s	Moment of inertia of gross section of slab
I_{se}	Moment of inertia of steel reinforcement about centroidal axis of section
j	A factor relating internal couple arm to d in working stress analysis
J	Polar moment of inertia
k	A factor relating the position of the neutral axis with respect to d in working stress analysis, a constant
K	Kip = 1000 lb, a factor used to calculate effective column length
K_b	Flexural stiffness of beam
K_c	Flexural stiffness of column
K_{ec}	Flexural stiffness of equivalent column
K_s	Flexural stiffness of slab
K_t	Torsional stiffness of torsional member
KN	Kilonewton
Ksi	Kip per square inch
ℓ_n	Clear span
ℓ_u	Unsupported length of column
L	Live load, span length
l_d	Development length
l_{dh}	l_{hb} times the applicable modification factor
l_{hb}	Basic development length of a standard hook
l_n	Clear span
l_u	Unsupported length of compression member
l_v	Length of shearhead arm
l_1	Span length in the direction of moment
l_2	Span length in direction transverse to span l_1
M	Bending moment
M_1	Smaller end moment at end of column
M_2	Larger end moment at end of column
M_a	Maximum service load moment
M_b	Balanced moment in columns, used with P_b
M_{cr}	Cracking moment
M_m	Modified moment
M_n	Nominal ultimate moment $= M_u/\phi$
M'_n	Nominal ultimate moment using an eccentricity e'
M_o	Total factored moment
M_p	Plastic moment
M_u	Ultimate moment, moment due to factored loads
M_{u1}	Part of M_u when calculated as singly reinforced
M_{u2}	Part of M_u due to compression reinforcement or overhanging flanges in T- or L-sections
M'_u	Ultimate moment using an eccentricity e'
M_v	Shearhead moment resistance
n	Modular ratio $= E_s/E_c$
N	Normal force
N_u	Factored normal load
N_1	Normal force in bearing at base of column
NA	Neutral axis
psi	Pounds per square inch
P_{cp}	Outside perimeter of gross area $= 2(x_0 + y_0)$

P_o	Perimeter of shear flow in area A_0
P	Unfactored concentrated load
P_b	Balanced load in column (at failure)
P_c	Euler buckling load
P_n	Nominal axial strength of column for a given e
P_o	Axial strength of a concentrically loaded column
P_s	Prestressing force in the tendon at the jacking end
P_u	Factored load $= \phi P_n$
P_x	Prestressing force in the tendon at any point x
q	Soil-bearing capacity
q_a	Allowable bearing capacity of soil
q_u	Ultimate bearing capacity of soil using factored loads
r	Radius of gyration, radius of a circle
R	Resultant of force system, reduction factor for long columns, or $R = R_u/\phi$
R_u	A factor $= M_u/bd^2$
s	Spacing between bars, stirrups, or ties
SI	International system of units
t	Thickness of a slab
T	Torque, tension force
T_c	Nominal torsional strength provided by concrete
T_{cr}	Cracking torsional moment
T_n	Nominal torsional strength provided by concrete and steel
T_s	Nominal torsional strength provided by reinforcement
T_u	Torque provided by factored load $= \phi T_n$
u	Bond stress
U	Design strength required to resist factored loads
V	Shear stress produced by working loads
v_c	Shear stress of concrete
v_{cr}	Shear stress at which diagonal cracks develop
v_h	Horizontal shear stress
v_t	Shear stress produced by a torque
v_u	Shear stress produced by factored loads
V	Unfactored shear force
V_c	Shear strength of concrete
V_{ci}	Nominal shear strength of concrete when diagonal cracking results from combined shear and moment
V_{cw}	Nominal shear strength of concrete when diagonal cracking results from excessive principal tensile stress in web
V_d	Shear force at section due to unfactored dead load (d = distance from the face of support)
V_n	Nominal shear strength $= V_c + V_s$
V_p	Vertical component of effective prestress force at section
V_s	Shear strength carried by reinforcement
V_u	Shear force due to factored loads
w	Width of crack at the extreme tension fiber, unit weight of concrete
w_u	Factored load per unit length of beam or per unit area of slab
W	Wind load or total load
x_o	Length of the short side of a rectangular section
x_1	Length of the short side of a rectangular closed stirrup
y_b	Same as y_t, except to extreme bottom fibers
y_o	Length of the long side of a rectangular section

y_t	Distance from centroidal axis of gross section, neglecting reinforcement, to extreme top fiber
y_1	Length of the long side of a rectangular closed stirrup
z	Factor related to width of crack $= f_s\sqrt[3]{Ad_c}$
α	Angle of inclined stirrups with respect to longitudinal axis of beam, ratio of stiffness of beam to that of slab at a joint
α_c	Ratio of flexural stiffness of columns to combined flexural stiffness of the slabs and beams at a joint; $(\Sigma K_c)/\Sigma(K_s + K_b)$
α_{ec}	Ratio of flexural stiffness of equivalent column to combined flexural stiffness of the slabs and beams at a joint: $(K_{ec})/\Sigma(K_s + K_b)$
α_m	Average value of a for all beams on edges of a panel
α_v	Ratio of stiffness of shearhead arm to surrounding composite slab section
β	Ratio of long to short side of rectangular footing, measure of curvature in biaxial bending
β_1	Ratio of a/c, where a = depth of stress block and c = distance between neutral axis and extreme compression fibers (This factor is 0.85 for $f'_c \leq 4000$ psi and decreases by 0.05 for each 1000 psi in excess of 4000 psi but is at least 0.65.)
β_a	Ratio of unfactored dead load to unfactored live load per unit area
β_c	Ratio of long to short sides of column or loaded area
β_d	Ratio of maximum factored dead load moment to maximum factored total moment
β_t	Ratio of torsional stiffness of edge beam section to flexural stiffness of slab: $E_{cb}C/2E_{cs}I_s$
γ	Distance between rows of reinforcement on opposite sides of columns to total depth of column h
γ_f	Fraction of unbalanced moment transferred by flexure at slab-column connections
γ_p	Factor for type of prestressing tendon (0.4 or 0.28)
γ_v	Fraction of unbalanced moment transferred by eccentricity of shear at slab-column connections
δ	Magnification factor
δ_{ns}	Moment magnification factor for frames braced against sidesway
δ_s	Moment magnification factor for frames not braced against sidesway
Δ	Deflection
ε	Strain
ε_c	Strain in concrete
ε_s	Strain in steel
ε'_s	Strain in compression steel
ε_y	Yield strain $= f_y/E_s$
θ	Slope angle
λ	Multiplier for additional long-time deflection
μ	Poisson's ratio; coefficient of friction
ζ	Parameter for evaluating capacity of standard hook
π	A constant equal to approximately 3.1416
ρ	Ratio of the tension steel area to the effective concrete area $= A_s/bd$
ρ'	Ratio of compression steel area to effective concrete area $= A'_s/bd$
ρ_1	$(\rho - \rho')$
ρ_b	Balanced steel ratio
ρ_g	Ratio of total steel area to total concrete area
ρ_p	Ratio of prestressed reinforcement A_{ps}/bd
ρ_s	Ratio of volume of spiral steel to volume of core

ρ_w	$A_s/b_w d$
ϕ	Strength-reduction factor
ω	$\rho f_y/f'c$
ω'	$\rho' f_y/f'_c$
ω_p	$\rho_p f_{ps}/f'_c$
$\omega_{w'}$	Reinforcement indices for flanged sections comuputed as for ω, ω_p, and ω'
$\omega_{pw'}$	except that b shall be the web width, and reinforcement area shall be that
ω'_w	required to develop compressive strength of web only

CONVERSION FACTORS

To Convert	To	Multiply By
1. Length		
Inch	Millimeter	25.4
Foot	Millimeter	304.8
Yard	Meter	0.9144
Meter	Foot	3.281
Meter	Inch	39.37
2. Area		
Square inch	Square millimeter	645
Square foot	Square meter	0.0929
Square yard	Square meter	0.836
Square meter	Square foot	10.76
3. Volume		
Cubic inch	Cubic millimeter	16390
Cubic foot	Cubic meter	0.02832
Cubic yard	Cubic meter	0.765
Cubic foot	Liter	28.3
Cubic meter	Cubic foot	35.31
Cubic meter	Cubic yard	1.308
4. Mass		
Ounce	Gram	28.35
Pound (lb)	Kilogram	0.454
Pound	Gallon	0.12
Short ton (2000 lb)	Kilogram	907
Long ton (2240 lb)	Kilogram	1016
Kilogram	Pound (lb)	2.205
Slug	Kilogram	14.59
5. Density		
Pound/cubic foot	Kilogram/cubic meter	16.02
Kilogram/cubic meter	Pound/cubic foot	0.06243

To Convert	To	Multiply By
6. Force		
Pound (Ib)	Newton (N)	4.448
Kip (1000 lb)	Kilonewton (kN)	4.448
Newton (N)	Pound	0.2248
Kilonewton (kN)	Kip (K)	0.225
7. Force/length		
Kip/foot	Kilonewton/meter	14.59
Kilonewton/meter	Pound/foot	68.52
Kilonewton/meter	Kip/foot	0.06852
8. Force/area (stress)		
Pound/square inch (psi)	Newton/square centimeter	0.6895
Pound/square inch (psi)	Newton/square millimeter (MPa)	0.0069
Kip/square inch (Ksi)	Meganewton/square meter	6.895
Kip/square inch (Ksi)	Newton/square millimeter	6.895
Pounds/square foot	Kilonewton/square meter	0.04788
Pound/square foot	Newton/square meter	47.88
Kip/square foot	Kilonewton/square meter	47.88
Newton/square millimeter	Kip/square inch (Ksi)	0.145
Kilonewton/square meter	Kip/square foot	0.0208
Kilonewton/square meter	Pound,/square foot	20.8
9. Moments		
Foot · Kip	Kilonewton · meter	1.356
Inch · Kip	Kilonewton · meter	0.113
Inch · Kip	Kilogram force · meter	11.52
Kilonewton · meter	Foot · Kip	0.7375

Structural Concrete

Theory and Design

Second Edition

1 INTRODUCTION

Water Tower Place, 74 Stories, U.S. Tallest Concrete Building, Chicago.

1.1 STRUCTURAL CONCRETE

The design of different structures is achieved by performing, in general, two main steps: (1) determining the different forces acting on the structure using proper methods of structural analysis, and (2) proportioning all structural members economically, considering the safety, stability, serviceability, and functionality of the structure. Structural concrete is one of the materials commonly used to design all types of buildings. Its two component materials, concrete and steel, work together to form structural members that can resist many types of loadings. The key to its performance lies in strengths that are complementary: Concrete resists compression and steel reinforcement resists tension forces.

The term *structural concrete* indicates all types of concrete used in structural applications. Structural concrete may be plain, reinforced, prestressed, or partially prestressed concrete; in addition, concrete is used in composite design. Composite design is used for any structural member, such as beams or columns, when the member contains a combination of concrete and steel shapes.

1.2 HISTORICAL BACKGROUND

Concrete has its first modern record as early as 1760, when John Smeaton used it in Britain in the first lock on the river Calder [1]. The walls of the lock were made of stones filled in with concrete. In 1796 J. Parker discovered Roman natural cement, and 15 years later Vicat burned a mixture of clay and lime to produce cement. In 1824 Joseph Aspdin manufactured portland cement in Wakefield, Britain. It was called portland cement because when it hardened, it resembled stone from the quarries of the Isle of Portland.

In France, François Marte Le Brun built a concrete house in 1832 in Moissac, in which he used concrete arches of 18-ft spans. He used concrete to build a school in St. Aignan in 1834 and a church in Corbarièce in 1835. Joseph Louis Lambot [2] exhibited a small rowboat made of reinforced concrete at the Paris Exposition in 1854. In the same year, W. B. Wilkinson of England obtained a patent for a concrete floor reinforced by twisted cables. The Frenchman François Cignet obtained his first patent in 1855 for the system he used of iron bars, embedded in concrete floors, that extended to the supports. One year later, he added nuts at the screw ends of the bars, and in 1869 he published a book describing the applications of reinforced concrete.

Joseph Monier, who obtained his patent in Paris on July 16, 1867, was given credit for the invention of reinforced concrete [3]. He made garden tubs and pots of concrete reinforced with iron mesh, which he exhibited in Paris in 1867. In 1873, he registered a patent to use reinforced concrete in tanks and bridges, and four years later, he registered another patent to use it in beams and columns [1].

In the United States, Thaddeus Hyatt conducted flexural tests on 50 beams that contained iron bars as tension reinforcement and published the results in 1877. He found that both concrete and steel can be assumed to behave in a homogeneous manner for all practical purposes. This assumption was important for the design of reinforced concrete members using elastic theory. He used prefabricated slabs in his experiments and considered that prefabricated units were best cast of T-sections placed side by side to form a floor slab. Hyatt is generally credited with developing the principles upon which the analysis and design of reinforced concrete is now based.

A reinforced concrete house was built by W. E. Ward near Port Chester, New York, in 1875. It used reinforced concrete for walls, beams, slabs, and staircases. P. B. Write wrote in the *American Architect and Building News* in 1877, describing the applications of reinforced concrete in Ward's house as a new method in building construction.

E. L. Ransome, head of the Concrete Steel Company in San Francisco, used reinforced concrete in 1879 and deformed bars for the first time in 1884. During 1889–1891, he built the two-story Leland Stanford Museum in San Francisco using reinforced concrete. He also built a reinforced concrete bridge in San Francisco. In 1900, after Ransome introduced the reinforced concrete skeleton, the thick wall system started to disappear in construction. He registered the skeleton type of structure in 1902; using spiral reinforcement in the columns as was suggested by Armand Considére of France.

A. N. Talbot, of the University of Illinois, and F. E. Turneaure and M. O. Withney, of the University of Wisconsin, conducted extensive tests on concrete to determine its behavior, compressive strength, and modulus of elasticity.

In Germany, G. A. Wayass bought the French Monier patent in 1879 and published his book on Monier methods of construction in 1887. Rudolph Schuster bought the patent rights in Austria, and the name of Monier spread throughout Europe, which is the main reason for crediting Monier as the inventor of reinforced concrete.

In 1900 the Ministry of Public Works in France called for a committee headed by Armand Considére, chief engineer of roads and bridges, to establish specifications for reinforced concrete, which were published in 1906.

The Barwick House, a three-story concrete building built in 1905 Montreal, Canada.

Reinforced concrete was further refined by introducing some precompression in the tension zone to decrease the excessive cracks. This refinement was the preliminary introduction of partial and full prestressing. In 1928 Eugene Freyssinet established the practical technique of using prestressed concrete [4].

From 1915 to 1935, research was conducted on axially loaded columns and creep effects on concrete; in 1940, eccentrically loaded columns were investigated. Ultimate-strength design started to receive special attention, in addition to diagonal tension and prestressed concrete. The American Concrete Institute Code (ACI Code) specified the use of ultimate-strength design in 1963 and included this method in all later codes. Building codes and specifications for the design of reinforced concrete structures are established in most countries, and research continues on developing new applications and more economical designs.

1.3 ADVANTAGES AND DISADVANTAGES OF REINFORCED CONCRETE

Reinforced concrete, as a structural material, is widely used in many types of structures. It is competitive with steel if economically designed and executed.

The advantages of reinforced concrete can be summarized as follows:

1. It has a relatively high compressive strength.
2. It has better resistance to fire than steel.
3. It has a long service life with low maintenance cost.
4. In some types of structures, such as dams, piers, and footings, it is the most economical structural material.
5. It can be cast to take the shape required, making it widely used in precast structural components. It yields rigid members with minimum apparent deflection.

The disadvantages of reinforced concrete can be summarized as follows:

1. It has a low tensile strength of about one-tenth of its compressive strength.
2. It needs mixing, casting, and curing, all of which affect the final strength of concrete.
3. The cost of the forms used to cast concrete is relatively high. The cost of form material and artisanry may equal the cost of concrete placed in the forms.
4. It has a low compressive strength as compared to steel (the ratio is about 1:10, depending on materials), which leads to large sections in columns of multistory buildings.
5. Cracks develop in concrete due to shrinkage and the application of live loads.

1.4 CODES OF PRACTICE

The designer engineer is usually guided by specifications called the *codes of practice.* Engineering specifications are set up by various organizations to represent the minimum requirements necessary for the safety of the public, although they are not necessarily for the purpose of restricting engineers.

Most codes specify design loads, allowable stresses, material quality, construction types, and other requirements to building construction. The most significant code for structural concrete design in the United States is the Building Code Requirements for Structural Concrete, **ACI-318,** or the ACI Code. Most of the design examples of this book are based on this code. Other codes of practice and material specifications in the United States include the Uniform Building Code, Standard Building Code, National Building Code, Basic Building Code, South Florida Building Code, American Association of State Highway and Transportation Officials (AASHTO) specifications, and specifications issued by the American Society for Testing and Materials (ASTM), American Railway Engineering Association (AREA), and Bureau of Reclamation, Department of the Interior.

Different codes other than those of the United States include the British Standard (B.S.) Code of Practice for Reinforced Concrete, CP-110 and BS 8110; the National Building Code of Canada; the German Code of Practice for Reinforced Concrete, DIN 1045; Specifications for Steel Reinforcement (Russia); and Technical Specifications for the Theory and Design of Reinforced Concrete Structures, CC-BA (France), the CEB Code (Comitè European Du Beton), and EuroCodes. EuroCodes are unified codes developed by the EC (European Community) countries for adoption throughout the member states. EuroCode 2 (EC2) deals with the design of concrete structures.

1.5 DESIGN PHILOSOPHY AND CONCEPTS

The design of a structure may be regarded as the process of selecting the proper materials and proportioning the different elements of the structure according to state-of-the-art engineering science and technology. In order to fulfill its purpose, the structure must meet the conditions of safety, serviceability, economy, and functionality. This can be achieved using the strength design method.

The strength design method (SDM) is based on the ultimate strength of structural members assuming a failure condition, whether due to the crushing of the concrete or to the yield of the reinforcing steel bars. Although there is some additional strength in the bars after yielding (due to strain hardening), this additional strength is not considered in the analysis of reinforced concrete members. In the strength design method, the actual loads, or working loads, are multiplied by load factors to obtain the ultimate design loads. The load factors represent a high percentage of the factor for safety required in the design. Details of this method are presented in Chapters 3 and 4. The ACI Code emphasizes this method of design, and its provisions are presented in the body of the code.

A second method, which is not commonly used, is called the alternate design method, also called the Working Stress Design (WSD) or the Elastic Design Method. The design concept is based on elastic theory, assuming a straight-line stress distribution along the depth of the concrete section. The actual loads or working loads acting on the structure are estimated, and members are proportioned on the basis of certain allowable stresses in concrete and steel. The allowable stresses are fractions of the crushing strength of concrete, f'_c, and the yield strength of steel, f_y. This method is presented in Chapter 5, and it is introduced in Appendix A of the ACI Code.

A third approach for the design of reinforced and prestressed concrete flexural and compression members is called the unified design method (UDM). This new approach introduces substantial changes in design for flexure and axial loads. Provisions for this method were introduced in the ACI Code, Appendix B. The reason for introducing this approach relates to the fact that different design methods were developed for reinforced and prestressed concrete beams and columns. Also, design procedures for prestressed concrete were different from those of reinforced concrete. The purpose of the code in the UDM approach is to simplify and unify the design requirements for reinforced and prestressed flexural and compression members. When the UDM method is used in design, the designer adheres to all sections of Appendix B and substitutes accordingly for the corresponding sections in the code. Reinforcement limits, strength reduction factor, ϕ, and moment redistribution are affected. The provisions of the unified design method satisfy the Code and are equally acceptable. See Chapter 22.

To provide guidance to the designers of buildings containing combinations of structural steel and reinforced concrete elements (composite design), the ACI Code in Appendix C introduced new provisions for this type of design, called alternative load and strength-reduction factors. The method is based on the provisions of the load factor combinations in the ASCE 7-88 publication and the strength-reduction factors in Appendix C.

Limit state design is a further step in the strength design method. It indicates the state of the member in which it ceases to meet the service requirements, such as losing its ability to withstand external loads or suffering excessive deformation, cracking, or local damage. According to the limit state design, reinforced concrete members have to be analyzed with regard to three limiting states:

1. Load-carrying capacity (safety, stability, and durability)
2. Deformation (deflections, vibrations, and impact)
3. The formation of cracks

The aim of this analysis is to ensure that no limiting state will appear in the structural member during its service life.

1.6 UNITS OF MEASUREMENT

Two units of measurement are commonly used in the design of structural concrete. The first is the U.S. customary system (lying mostly in its human scale and its ingenious use of simple numerical proportions), and the second is the SI (Le Système International d'Unités), or metric, measurements.

The metric system is planned to be in universal use within the coming few years. The United States is committed to change to SI units. Great Britain, Canada, Australia, and other countries have been using SI units for several years.

The base units in the SI system are the units of length, mass, and time, which are the meter (m), the kilogram (kg), and the second (s), respectively. The unit of force, a derived unit called the newton (N), is defined as the force that gives the acceleration of 1 meter per second per second ($1\ \text{m/s}^2$) to a mass of 1 kg, or $1\ \text{N} = 1\ \text{kg} \cdot \text{m/s}^2$.

The weight of a body, W, which is equal to the mass m multiplied by the local gravitational acceleration $g(9.81\ \text{m/s}^2)$, is expressed in newtons (N). The weight of a body of 1 kg mass is $W = mg = 1\ \text{kg} \times 9.81\ \text{m/s}^2 = 9.81\ \text{N}$.

Multiples and submultiples of the base SI units can be expressed through the use of prefixes. The prefixes most frequently used in structural calculations are the kilo (k), mega (M), milli (m), and micro (μ). For example,

$$1\ \text{km} = 1000\ \text{m}, \qquad 1\ \text{mm} = 0.001\ \text{m} \qquad 1\ \mu\text{m} = 10^{-6}\ \text{m}$$

$$1\ \text{kN} = 1000\ \text{N}, \qquad 1\ \text{Mg} = 1000\ \text{kg} = 10^6\ \text{g}$$

1.7 LOADS

Structural members must be designed to support specific loads.

Loads are those forces for which a given structure should be proportioned. In general, loads may be classified as dead or live.

Dead loads include the weight of the structure (its self-weight) and any permanent material placed on the structure, such as tiles, roofing materials, and walls. Dead loads can be determined with a high degree of accuracy from the dimensions of the elements and the unit weight of materials.

Live loads are all other loads that are not dead loads. They may be steady or unsteady or movable or moving; they may be applied slowly, suddenly, vertically, or laterally; and their magnitudes may fluctuate with time. In general, live loads include the following:

- Occupancy loads caused by the weight of the people, furniture, and goods
- Forces resulting from wind action and temperature changes
- The weight of snow if accumulation is probable
- The pressure of liquids or earth on retaining structures
- The weight of traffic on a bridge
- Dynamic forces resulting from moving loads (impact), earthquakes, or blast loading

The ACI Code does not specify loads on structures; however, occupancy loads on different types of buildings are prescribed by the American National Standards Institute (ANSI) [5]. Some typical values are shown in Table 1.1. Table 1.2 shows weights and specific gravity of various materials.

AASHTO and AREA specifications prescribe vehicle loadings on highway and railway bridges, respectively. These loads are given in [6] and [7].

Snow loads on structures may vary between 10 and 40 lb/ft^2 (0.5 and 2 kN/m^2), depending on the local climate.

Wind loads may vary between 15 and 30 lb/ft^2, depending on the velocity of wind. The wind pressure of a structure, F, can be estimated from the following equation:

$$F = 0.00256 C_s V^2 \tag{1.1}$$

where V = velocity of air (mi/h)

C_s = shape factor of the structure

F = the dynamic wind pressure (lb/ft^2)

Table 1.1 Typical Uniformly Distributed Design Loads

Occupancy	Contents	Design Live Load lb/ft²	kN/m²
Assembly hall	Fixed seats	60	2.9
	Movable seats	100	4.8
Hospital	Operating rooms	60	2.9
	Private rooms	40	1.9
Hotel	Guest rooms	40	1.9
	Public rooms	100	4.8
	Balconies	100	4.8
Housing	Private houses and apartments	40	1.9
	Public rooms	100	4.8
Institution	Classrooms	40	1.9
	Corridors	100	4.8
Library	Reading rooms	60	2.9
	Stack rooms	150	7.2
Office building	Offices	50	2.4
	Lobbies	100	4.8
Stairs (or balconies)		100	4.8
Storage warehouses	Light	100	4.8
	Heavy	250	12.0
Yards and terraces		100	4.8

Table 1.2 Density and Specific Gravity of Various Materials

Material	Density lb/ft³	kg/m³	Specific Gravity
Building materials			
Bricks	120	1,924	1.8–2.0
Cement, portland, loose	90	1,443	—
Cement, portland, set	183	2,933	2.7–3.2
Earth, dry, packed	95	1,523	—
Sand or gravel, dry, packed	100–120	1,600–1,924	—
Sand or gravel, wet	118–120	1,892–1,924	—
Liquids			
Oils	58	930	0.9–0.94
Water (at 4°C)	62.4	1,000	1.0
Ice	56	898	0.88–0.92
Metals and minerals			
Aluminum	165	2,645	2.55–2.75
Copper	556	8,913	9.0
Iron	450	7,214	7.2
Lead	710	11,380	11.38
Steel, rolled	490	7,855	7.85
Limestone or marble	165	2,645	2.5–2.8
Sandstone	147	2,356	2.2–2.5
Shale or slate	175	2,805	2.7–2.9
Normal weight concrete			
Plain	145	2,324	2.2–2.4
Reinforced or prestressed	150	2,405	2.3–2.5

As an example, for a wind of 100 mi/h with $C_s = 1$, the wind pressure is equal to 25.6 lb/ft^2. It is sometimes necessary to consider the effect of gusts in computing the wind pressure by multiplying the wind velocity in equation (1.1) by a gust factor, which generally varies between 1.1 and 1.3.

The shape factor, C_s, varies with the horizontal angle of incidence of the wind. On vertical surfaces of rectangular buildings, C_s may vary between 1.2 and 1.3. Detailed information on wind loads can be found in [5].

1.8 SAFETY PROVISIONS

Structural members must always be proportioned to resist loads greater than the service or actual load in order to provide proper safety against failure. In the strength design method, the member is designed to resist factored loads, which are obtained by multiplying the service loads by load factors. Different factors are used for different loadings. Because dead loads can be estimated quite accurately, their load factors are smaller than those of live loads, which have a high degree of uncertainty. Several load combinations must be considered in the design to compute the maximum and minimum design forces. Reduction factors are used for some combinations of loads to reflect the low probability of their simultaneous occurrences. The ACI Code presents specific values of load factors to be used in the design of concrete structures (see Section 3.5).

In addition to load factors, the ACI Code specifies another factor to allow an additional reserve in the capacity of the structural member. The nominal strength is generally calculated using accepted analytical procedure based on statistics and equilibrium; however, in order to account for the degree of accuracy within which the nominal strength can be calculated and for adverse variations in materials and dimensions, a strength-reduction factor ϕ should be used in the strength design method. Values of the strength-reduction factors are given in Section 3.6.

To summarize this discussion, the ACI Code has separated the safety provision into an overload or load factor and to an undercapacity (or strength-reduction) factor ϕ. A safe design is achieved when the structure's strength, obtained by multiplying the nominal strength by the reduction factor ϕ, exceeds or equals the strength needed to withstand the factored loadings (service loads times their load factors). For example,

$$M_u \leq \phi M_n \quad \text{and} \quad V_u \leq \phi V_n \tag{1.2}$$

where M_u and V_u = external factored moment and shear forces
M_n and V_n = nominal ultimate moment and shear capacity of the member, respectively

Given a load factor of 1.4 for dead load and a load factor of 1.7 for live load, the overall safety factor for a structure loaded by a dead load D and a live load L is

$$\text{Factor of safety} = \frac{1.4D + 1.7L}{D + L}\left(\frac{1}{\phi}\right) = \frac{1.4 + 1.7(L/D)}{1 + (L/D)}\left(\frac{1}{\phi}\right) \tag{1.3}$$

The factors of safety for the various values of ϕ and L/D ratios are shown in the following table.

ϕ	0.9				0.8				0.7			
L/D	0	1	2	3	0	1	2	3	0	1	2	3
Safety factor	1.56	1.72	1.78	1.81	1.75	1.94	2.00	2.03	2.00	2.21	2.29	2.32

For members subjected to flexure (beams), $\phi = 0.9$, and the factor of safety ranges between 1.56 for $L/D = 0$ and 1.81 for $L/D = 3$.

For members subjected to axial forces (columns), $\phi = 0.7$, and the factor of safety ranges between 2.00 for $L/D = 0$ and 2.32 for $L/D = 3$. The increase in the factor of safety in columns reflects the greater overall safety requirements of these critical building elements.

A general format of equation (1.2) is [8]

$$\phi R_n \geq \nu_o \sum (\nu_i Q_i) \tag{1.4}$$

where R_n = nominal strength of the structural number

ϕ = undercapacity factor (<1.0)

$\sum Q_i$ = sum of load effectss

ν_i = overload factor

ν_o = analysis factor (>1.0)

The subscript i indicates the load type, such as dead load, live load, and wind load. The analysis factor, ν_o, is greater than 1.0 and is introduced to account for uncertainties in structural analysis. The overload factor, ν_i, is introduced to account for several factors, such as an increase in live load due to a change in the use of the structure and variations in erection procedures. The design concept is referred to as *load and resistance factor design* (LRFD) [8,9].

1.9 STRUCTURAL CONCRETE ELEMENTS

Structural concrete can be used for almost all buildings, whether single story or multistory. The concrete building may contain some or all of the following main structural elements, which are explained in detail in other chapters of the book:

- *Slabs* are horizontal plate elements in building floors and roofs. They may carry gravity loads as well as lateral loads. The depth of the slab is usually very small relative to its length or width (Chapters 9 and 17).
- *Beams* are long horizontal or inclined members with limited width and depth. Their main function is to support loads from slabs (Chapters 3 and 4).
- *Columns* are critical members that support loads from beams or slabs. They may be subjected to axial loads or axial loads and moments (Chapters 10 and 11).
- *Frames* are structural members that consist of a combination of beams and columns or slabs, beams, and columns. They may be statically determinate or statically indeterminate frames (Chapter 16).
- *Footings* are pads or strips that support columns and spread their loads directly to the soil (Chapter 13).
- *Walls* are vertical plate elements resisting gravity as well as lateral loads as in the case of basement walls (Chapter 14).

1.10 STRUCTURAL CONCRETE DESIGN

The first step in the design of a building is the general planning carried out by the architect to determine the layout of each floor of the building to meet the owner's requirements. Once the architectural plans are approved, the structural engineer then determines the most adequate structural system to ensure the safety and stability of the building. Different structural options must be considered to determine the most economical solution based on the materials available and the soil condition. This result is normally achieved by

1. Idealizing the building into a structural model of load-bearing frames and elements;
2. Estimating the different types of loads acting on the building;
3. Performing the structural analysis using computer or manual calculations to determine the maximum moments, shear, torsional forces, axial loads, and other forces;
4. Proportioning the different structural elements and calculating the reinforcement needed;
5. Producing structural drawings and specifications with enough details to enable the contractor to construct the building properly.

1.11 ACCURACY OF CALCULATIONS

In the design of concrete structures, exact calculations to determine the size of the concrete elements are not needed. Calculators and computers can give an answer to many figures after the decimal point. For a practical size of a beam, slab, or column, each dimension should be approximated to the nearest one or one-half of an inch. Moreover, the steel bars available in the market are limited to specific diameters and areas, as shown in Table A.12 (Appendix A). The designer should choose a group of bars from the table with an area equal or greater than the area obtained from calculations. Also, the design equations in this book based on the ACI Code are approximate. Therefore, for a practical and economical design it is adequate to use four figures (or the full number with no fractions if it is greater than four figures) for the calculation of forces, stresses, moments, or dimensions like length or width of section. For strains, use five or six figures because strains are very small quantities measured in a millionth of an inch (for example, a strain of 0.000358 in/in.). Stresses are obtained by multiplying the strains by the modulus of elasticity of the material, which has a high magnitude (for example, 29,000,000 lb/sq in.) for steel. Any figures less than five or six figures in strains will produce quite a change in stresses.

Examples

For forces, use 28.45 K, 2845 lb, 567.8 K (four figures).

For force/length, use 2.451 K/ft or 2451 lb/ft.

For length or width, use 14.63 in., 1.219 ft (or 1.22 ft).

For areas, use 7.537 in.^{2}, and for volumes, use 48.72 in.^{3}.

For strains, use 0.002078.

1.12 CONCRETE HIGH-RISE BUILDINGS

High-rise buildings are becoming the dominant feature of many U.S. cities; a great number of these buildings are designed and constructed in structural concrete.

Table 1.3 Examples of Reinforced Concrete Skyscrapers

Year	Structure	Location	Stories	Height, ft (m)
1965	Lake Point Tower	Chicago	70	645 (197)
1969	One Shell Plaza	Houston	52	714 (218)
1975	Peachtree Center Plaza Hotel	Atlanta	71	723 (220)
1976	Water Tower Place	Chicago	74	859 (262)
1976	CN Tower	Toronto	–	1465 (447)
1977	Renaissance Center Westin Hotel	Detroit	73	740 (226)
1983	City Center	Minneapolis	40	528 (158)

Although at the beginning of the century the properties of concrete and joint behavior of steel and concrete were not fully understood, a 16-story building, the Ingalls Building, was constructed in Cincinnati in 1902 with a total height of 210 ft (64 m). In 1922, the Medical Arts Building, with a height of 230 ft (70 m) was constructed in Dallas, Texas. The design of concrete buildings was based on elastic theory concepts and a high factor of safety, resulting in large concrete sections in beams and columns. After extensive research, high-strength concrete and high-strength steel were allowed in the design of reinforced concrete members. Consequently, small concrete sections as well as savings in materials were achieved, and new concepts of structural design were possible.

To visualize how high concrete buildings can be built, some structural concrete skyscrapers are listed in Table 1.3 [16]. The CN Tower is the world's tallest free-standing concrete structure.

The reader should realize that most concrete buildings are relatively low and range from 1 to 5 stories. Skyscrapers and high-rise buildings constitute less than 10% of all concrete buildings.

The photos of different concrete buildings and structures are shown.

Renaissance Center, Detroit, Michigan.

Marina City Towers, Chicago, Illinois.

City Center, Minneapolis, Minnesota.

CN Tower, Toronto, Canada (height 1465 ft, or 447 m).

Concrete bridge for the city transit system, Washington, D.C.

Concrete bridge, Knoxville, Tennessee.

Reinforced concrete grain silo using the slip form system, Brookings, South Dakota.

REFERENCES

1. Ali Ra'afat. *The Art of Architecture and Reinforced Concrete.* Cairo: Halabi, 1970.
2. R. S. Kirby, S. Withington, A. B. Darling, and F. G. Kilgour. *Engineering in History.* New York: McGraw-Hill, 1956.
3. Hans Straub. *A History of Civil Engineering.* London: Leonard Hill, 1952.
4. E. Freyssinet. "The Birth of Prestressing." Cement and Concrete Association Translation No. 29. London, 1956.
5. American National Standards Institute. *ANSI A58.1.* 1992.
6. American Association of State Highway and Transportation Officials (ASSHTO). *Standard Specifications for Highway Bridges,* 12th ed. Washington, D.C., 1994.
7. American Railway Engineering Association (AREA). *Specifications for Steel Railway Bridges.* Chicago, 1992.
8. C. W. Pinkham and W. C. Hansell. "An Introduction to Load and Resistance Factor Design for Steel Buildings." *Engineering Journal AISC* 15 (1978 (first quarter)).
9. M. K. Ravindra and T. V. Galambos. "Load and Resistance Factor Design for Steel." *Journal of Structural Division ASCE* 104 (September 1978).
10. National Research Council of Canada. *National Building Code of Canada.* Ottawa, 1980.
11. Comitè Euro-International du Bèton: *CEB-FIP Model Code for Concrete Structures,* 3d ed., Paris, 1990.
12. B. Ellingwood, T. V. Galambos, J. G. MacGregor, and C. A. Cornell. "Development of a Probability-based Load Criterion for American National Standards, A58." National Bureau of Standards Special Publication 577, 1980.
13. T. V. Galambos, B. Ellingwood, J. G. MacGregor, and C. A. Cornell. "Probability-based Load Criteria: Assessment of Current Design Practice." *ASCE Proceedings* 108 (May 1982).
14. B. Ellingwood, J. G. MacGregor, T. V. Galambos, and C. A. Cornell. "Probability-based Load Criteria: Load Factors and Load Combinations." *ASCE Proceedings* 108 (May 1982).
15. J. G. MacGregor. "Load and Resistance Factors for Concrete Design." *ACI Journal,* no. 4 (July--August 1983).
16. *Concrete Construction* 28, no. 2 (February 1983).
17. American Concrete Institute. "Building Code Requirements for Structural Concrete." ACI (318–99). Detroit, Michigan, 1999.

2

PROPERTIES OF REINFORCED CONCRETE

IBM Building, Montreal, Canada (the highest concrete building in Montreal, with 50 stories).

2.1 FACTORS AFFECTING THE STRENGTH OF CONCRETE

In general, concrete consists of coarse and fine aggregate, cement, water, and—in many cases—some kind of admixtures. The materials are mixed together until a cement paste is developed, filling most of the voids in the aggregates and producing a uniform dense concrete. The plastic concrete is then placed in a mold and left to set, harden, and develop adequate strength. For the design of concrete mixtures, as well as composition and properties of concrete materials, the reader is referred to [1], [5], and [7].

The strength of concrete depends upon many factors and may vary within wide limits with the same production method. The main factors that affect the strength of concrete are described next.

Water-cement ratio The water-cement ratio is one of the most important factors affecting the strength of concrete. For complete hydration of a given amount of cement, a water-cement ratio (by weight) equal to 0.25 is needed. A water-cement ratio of about 0.35 or higher is needed for the concrete to be reasonably workable, without additives. This ratio corresponds to 4 gal of water per sack of cement (94 lb) (or 17.5 l per 50 kg of cement). Based on this cement ratio, a concrete strength of about 6000 psi may be achieved. A water-cement ratio of 0.5 and 0.7 may produce a concrete strength of about 5000 psi and 3000 psi, respectively.

Properties and proportions of concrete constituents Concrete is a mixture of cement, aggregate, and water. An increase in the cement content in the mix and the use of well-graded aggregate increase the strength of concrete. Special admixtures are usually added to the mix to produce the desired quality and strength of concrete.

Method of mixing and curing The use of mechanical concrete mixers and the proper time of mixing both have favorable effects on strength of concrete. Also, the use of vibrators produces dense concrete with a minimum percentage of voids. A void ratio of 5% may reduce the concrete strength about 30%.

The curing conditions exercise an important influence on the strength of concrete. Both moisture and temperature have a direct effect on the hydration of cement. The longer the period of moist storage, the greater the strength. If the curing temperature is higher than the initial temperature of casting, the resulting 28-day strength of concrete is reached earlier than 28 days.

Age of the concrete The strength of concrete increases appreciably with age, and hydration of cement continues for months. In practice, the strength of concrete is determined from cylinders or cubes tested at the age of 7 days and 28 days. As a practical assumption, concrete at 28 days is 1.5 times as strong as at 7 days: The range varies between 1.3 and 1.7. The British code of practice [2] accepts concrete if the strength at 7 days is not less than two-thirds of the required 28-day strength. For a normal portland cement, the increase of strength with time, relative to 28-day strength, may be assumed as follows:

Age	7 days	14 days	28 days	3 months	6 months	1 year	2 years	5 years
Strength ratio	0.67	0.86	1.0	1.17	1.23	1.27	1.31	1.35

Loading conditions The compressive strength of concrete is estimated by testing a cylinder or cube to failure in a few minutes. Under sustained loads for years, the ultimate strength of concrete is reduced by about 30%. Under 1-day sustained loading, concrete may lose about 10% of its compressive strength. Sustained loads and creep effect as well as dynamic and impact effect, if they occur on the structure, should be considered in the design of reinforced concrete members.

Shape and dimensions of the tested specimen The common sizes of concrete specimens used to predict the compressive strength are either 6- by 12-in. (150- by 300-mm) cylinders or 6-in. (150-mm) cubes. When a given concrete is tested in compression by means of cylinders of like shape but of different sizes, the larger specimens give lower strength indexes. Table 2.1 [4] gives the relative strength, for various sizes of cylinders, as a percentage of the strength of the standard cylinder; the heights of all cylinders are twice the diameters.

Table 2.1 Effect of Size of Compression Specimen on Strength of Concrete

Size of Cylinder		Relative Compressive Strength
(.in)	(mm)	
2×4	50×100	1.09
3×6	75×150	1.06
6×12	150×300	1.00
8×16	200×400	0.96
12×24	300×600	0.91
18×36	450×900	0.86
24×48	600×1200	0.84
36×72	900×1800	0.82

Table 2.2 Strength Correction Factor for Cylinders of Different Height-Diameter Ratios

Ratio	2.0	1.75	1.50	1.25	1.10	1.00	0.75	0.50
Strength correction factor	1.00	0.98	0.96	0.94	0.90	0.85	0.70	0.50
Strength relative to standard cylinder	1.00	1.02	1.04	1.06	1.11	1.18	1.43	2.00

Table 2.3 Relative Strength of Cylinder vs. Cube [7]

Compressive strength											
Compressive	(psi)	1000	2200	2900	3500	3800	4900	5300	5900	6400	7300
strength	(N/mm^2)	7.0	15.5	20.0	24.5	27.0	24.5	37.0	41.5	45.0	51.5
Strength ratio of cylinder to cube		0.77	0.76	0.81	0.87	0.91	0.93	0.94	0.95	0.96	0.96

Sometimes concrete cylinders of nonstandard shape are tested. The greater the ratio of specimen height to diameter, the lower the strength indicated by the compression test. To compute the equivalent strength of the standard shape, the results must be multiplied by a correction factor. Approximate values of the correction factor are given in Table 2.2, extracted from ASTM C42-57. The relative strengths of a cylinder and a cube for different compressive strengths are shown in Table 2.3.

2.2 COMPRESSIVE STRENGTH

In designing structural members, it is assumed that the concrete resists compressive stresses and not tensile stresses; therefore, compressive strength is the criterion of quality concrete. The other concrete stresses can be taken as a percentage of the compressive strength, which can be easily and accurately determined from tests. Specimens used to determine compressive strength may be cylindrical, cubical, or prismatic.

Test specimens in the form of a cube 6 in. (150 mm) or 8 in. (200 mm) are used in Great Britain, Germany, and other parts of Europe.

Prism specimens are used in France, Russia, and other countries and are usually 70 by 70 by 350 mm or 100 by 100 by 500 mm. They are cast with their longer sides horizontal and are tested, like cubes, in a position normal to the position of cast.

Before testing, the specimens are moist-cured and then tested at the age of 28 days by gradually applying a static load until rupture occurs. The rupture of the concrete specimen may be caused by the applied tensile stress (failure in cohesion), the applied shearing stress (sliding failure), the compressive stress (crushing failure), or combinations of these stresses.

The failure of the concrete specimen can be in one of three modes [5], as shown in Figure 2.1. First, under axial compression, the specimen may fail in shear, as in Figure 2.1(a). Resistance to failure is due to both cohesion and internal friction.

The second type of failure (Figure 2.1(b)) results in the separation of the specimen into columnar pieces by what is known as splitting, or columnar, fracture. This failure occurs when the strength of concrete is high, and lateral expansion at the end bearing surfaces is relatively unrestrained.

The third type of failure (Figure 2.1(c)) is seen when a combination of shear and splitting failure occurs.

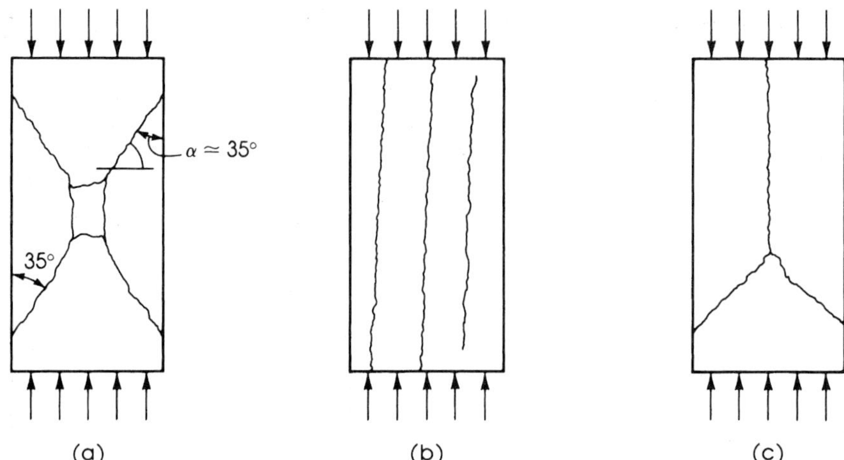

Figure 2.1 Modes of failure of standard concrete cylinders.

2.3 STRESS-STRAIN CURVES OF CONCRETE

The performance of a reinforced concrete member under load depends, to a great extent, on the stress-strain relationship of concrete and steel and on the type of stress applied to the member. Stress-strain curves for concrete are obtained by testing a concrete cylinder to rupture at the age of 28 days and recording the strains at different load increments.

Figure 2.2 shows typical stress-strain curves for concretes of different strengths. All curves consist of an initial relatively straight elastic portion, reaching maximum stress at a strain of about 0.002; then rupture occurs at a strain of about 0.003.

Concrete having a compressive strength between 3000 and 6000 psi (21 and 42 N/mm^2) may be adopted. High-strength concrete with a compressive strength greater than 6000 psi (6000–15,000 psi) is becoming an important building material for the design of concrete structures.

2.4 TENSILE STRENGTH OF CONCRETE

Concrete is a brittle material, and it cannot resist the high tensile stresses that are important when considering cracking, shear, and torsional problems. The low tensile capacity can be attributed to the high stress concentrations in concrete under load, so that a very high stress is reached in some portions of the specimen, causing microscopic cracks, while the other parts of the specimen are subjected to low stress.

Direct tension tests are not reliable for predicting the tensile strength of concrete, due to minor misalignment and stress concentrations in the gripping devices. An indirect tension test in the form of splitting a 6- by 12-in. (150- by 300-mm) cylinder was suggested by the Brazilian Fernando Carneiro. The test is usually called the *splitting test.* In this test, the concrete cylinder is placed with its axis horizontal in a compression testing machine. The load is applied uniformly along two opposite lines on the surface of the cylinder through two plywood pads, as shown in Figure 2.3. Considering an element on the vertical diameter and at a distance y from the top fibers, the element is subjected to a compressive stress

$$f_c = \frac{2P}{\pi LD}\left(\frac{D^2}{y(D - y)} - 1\right) \tag{2.1}$$

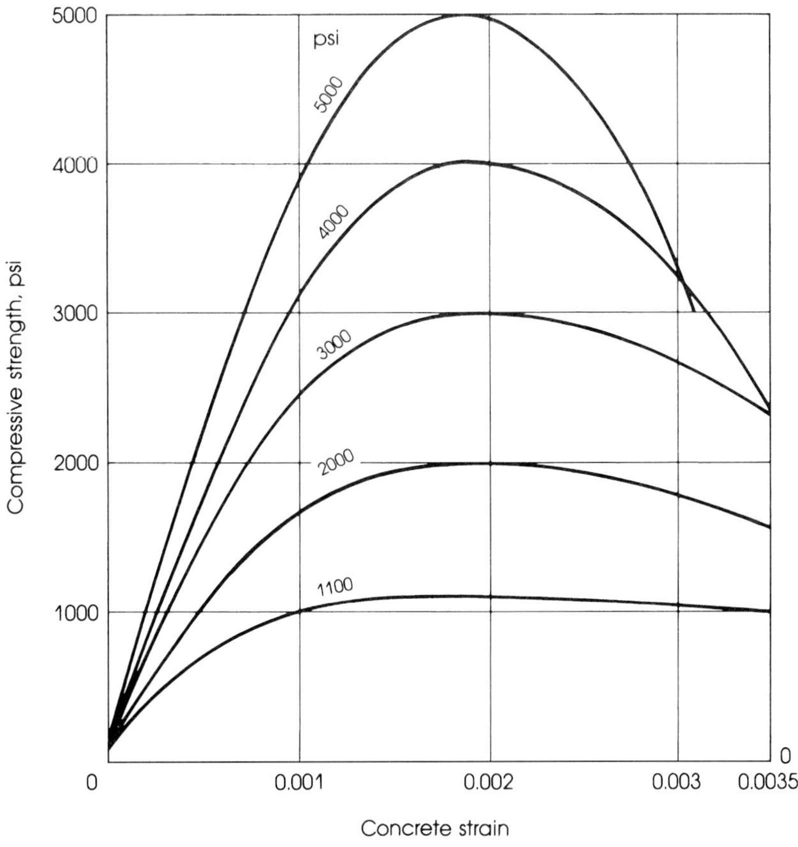

Figure 2.2 Typical stress-strain curves of concrete.

Standard cylinders ready for testing.

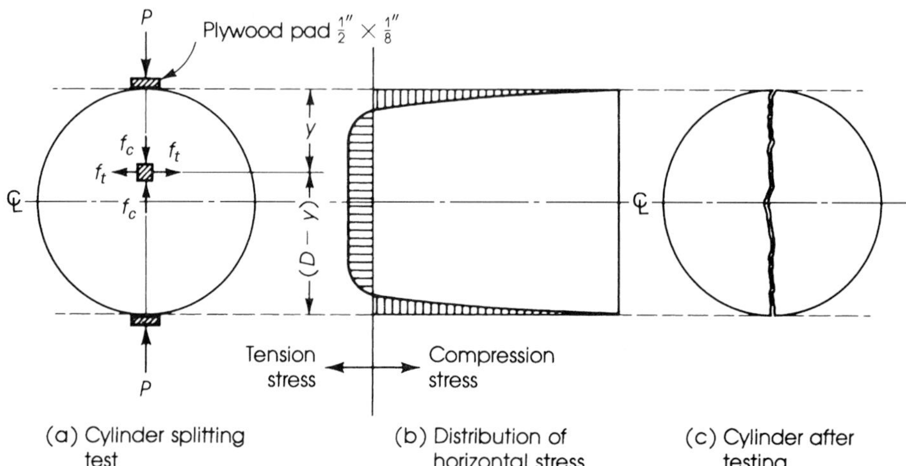

Figure 2.3 Cylinder splitting test [7]: (a) configuration of test, (b) distribution of horizontal stress, and (c) cylinder after testing.

and a tensile stress

$$f'_{\text{sp}} = \frac{2P}{\pi LD} \tag{2.2}$$

where P is the compressive load on the cylinder and D and L are the diameter and length of the cylinder. For a 6- by 12-in. (150- by 300-mm) cylinder and at a distance $y = D/2$, the compression strength is $f_c = 0.0265P$, and the tensile strength is $f'_{\text{sp}} = 0.0088P = f_c/3$.

The splitting strength of f'_{sp} can be related to the compressive strength of concrete in that it varies between 6 and 7 times $\sqrt{f'_c}$ for normal concrete and between 4 and 5 times $\sqrt{f'_c}$ for lightweight concrete. The direct tensile stress, f'_t, can also be estimated from the split test: its value varies between $0.5\,f'_{\text{sp}}$ and $0.7\,f'_{\text{sp}}$. The smaller of these values applies to higher-strength concrete. The splitting strength, f'_{sp}, can be estimated as 10% of the compressive strength up to $f'_c = 6000$ psi (42 N/mm^2). For higher values of compressive strength, f'_{sp} can be taken as 9% of f'_c.

In general, the tensile strength of concrete ranges from 7% to 11% of its compressive strength, with an average of 10%. The lower the compressive strength, the higher the relative tensile strength.

2.5 FLEXURAL STRENGTH (MODULUS OF RUPTURE) OF CONCRETE

Experiments on concrete beams have shown that ultimate tensile strength in bending is greater than the tensile stress obtained by direct or splitting tests. Flexural strength is expressed in terms of the modulus of rupture of concrete (f_r), which is the maximum tensile stress in concrete in bending. The modulus of rupture can be caluculated from the flexural formula used for elastic materials, $f_r = Mc/I$, by testing a plain concrete beam. The beam, 6 by 6 by 28 in. (150 by 150 by 700 mm), is supported on a 24-in. (600-mm) span and loaded to rupture by two loads, 4 in. (100 mm) on either side of the center. A smaller beam of 4 by 4 by 20 in. (100 by 100 by 500 mm) on a 16-in. (400-mm) span may also be used.

(a)

(b)

Concrete cylinder splitting test.

The modulus of rupture of concrete ranges between 11% and 23% of the compressive strength. A value of 15% can be assumed for strengths of about 4000 psi (28 $\mathrm{N/mm^2}$). The ACI Code prescribes the value of the modulus of rupture as

$$f_r = 7.5\sqrt{f'_c}\ (\text{psi}) = 0.62\sqrt{f'_c}\ (\mathrm{N/mm^2}) \tag{2.3}$$

The modulus of rupture as related to the strength obtained from the split test on cylinders may be taken as $f_r = (1.25 \text{ to } 1.50)f'_{sp}$.

2.6 SHEAR STRENGTH

Pure shear is seldom encountered in reinforced concrete members, because it is usually accompanied by the action of normal forces. An element subjected to pure shear breaks transversely into two parts. Therefore, the concrete element must be strong enough to resist the applied shear forces.

Shear strength may be considered as 20% to 30% greater than the tensile strength of concrete, or about 12% of its compressive strength. The ACI Code allows an ultimate shear strength of $2\sqrt{f'_c}$ psi $(0.17\sqrt{f'_c}\ \mathrm{N/mm^2})$ on plain concrete sections. For more information, refer to Chapter 8.

2.7 MODULUS OF ELASTICITY OF CONCRETE

One of the most important elastic properties of concrete is its modulus of elasticity, which can be obtained from a compressive test on concrete cylinders. The modulus of elasticity, E_c, can be defined as the change of stress with respect to strain in the elastic range:

$$E_c = \frac{\text{unit stress}}{\text{unit strain}} \tag{2.4}$$

The modulus of elasticity is a measure of stiffness, or the resistance of the material to deformation. In concrete, as in any elastoplastic material, the stress is not proportional to the strain, and the stress-strain relationship is a curved line. The actual stress-strain curve of concrete can be obtained by measuring the strains under increments of loading on a standard cylinder.

The *initial tangent modulus* (Figure 2.4) is represented by the slope of the tangent to the curve at the origin under elastic deformation. This modulus is of limited value and

$$E_c = \frac{df_c}{d\varepsilon_c}$$

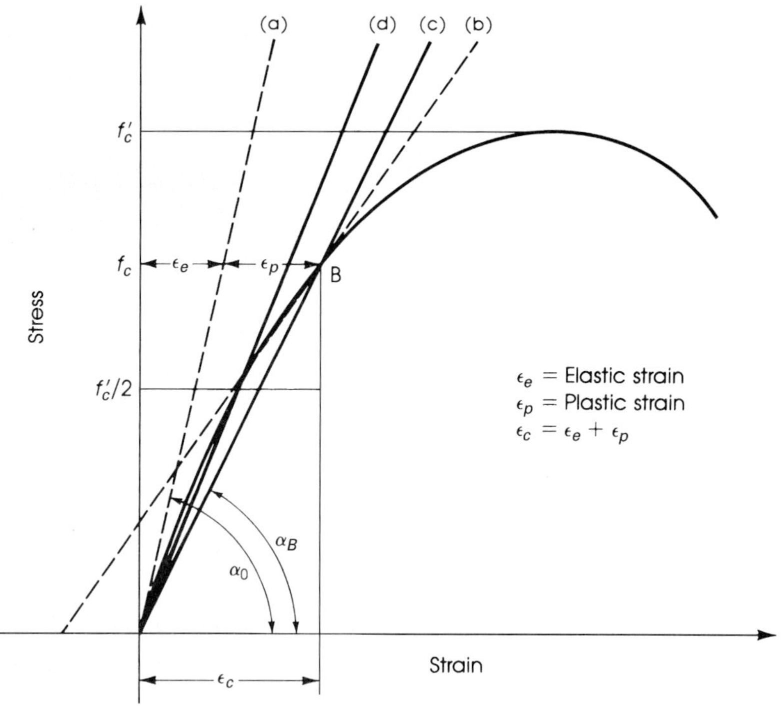

Figure 2.4 Stress-strain curve and modulus of elasticity of concrete. Lines a–d represent (a) initial tangent modulus, (b) tangent modulus at a stress f_c, (c) secant modulus at a stress f_c, and (d) secant modulus at a stress $f'_c/2$.

cannot be determined with accuracy. Geometrically, the tangent modulus of elasticity of concrete, E_c, is the slope of the tangent to the stress-strain curve at a given stress. Under long-time action of load and due to the development of plastic deformation, the stress-to-total-strain ratio becomes a variable nonlinear quantity.

For practical applications, the *secant modulus* can be used. The secant modulus is represented by the slope of a line drawn from the origin to a specific point of stress (B) on the stress-strain curve (Figure 2.4). Point B is normally located at $f'_c/2$.

The ACI Code gives a simple formula for calculatng the modulus of elasticity of normal and lightweight concrete considering the secant modulus at a level of stress f_c equal to half the ultimate concrete strength, f'_c:

$$E_c = 33w^{1.5}\sqrt{f'_c}\ \text{psi}\ (w \text{ in pcf}) = 0.043w^{1.5}\sqrt{f'_c}\ \text{N/mm}^2 \tag{2.5}$$

where w = unit weight of concrete (between 90 and 155 lb/ft^3 (pcf) or 1400 to 2500 kg/m^3) and f'_c = ultimate strength of a standard concrete cylinder. For normal-weight concrete, w is approximately 145 pcf (2320 kg/m^3); thus,

$$E_c = 57{,}600\sqrt{f'_c}\ \text{psi} = 4780\sqrt{f'_c}\ \text{MPa} \tag{2.6}$$

The ACI Code allows the use of $E_c = 57{,}000\sqrt{f'_c}$ (psi) $= 4700\sqrt{f'_c}$ MPa. The moduli of elasticity, E_c, for different values of f'_c are shown in Table A.10.

Test on a standard concrete cylinder to determine the modulus of elasticity of concrete.

2.8 POISSON'S RATIO

Poisson's ratio, μ, is the ratio of the transverse to the longitudinal strains under axial stress within the elastic range. This ratio varies between 0.15 and 0.20 for both normal and lightweight concrete. Poisson's ratio is used in structural analysis of flat slabs, tunnels, tanks, arch dams, and other statically indeterminate structures. For isotropic elastic materials, Poisson's ratio is equal to 0.25. An average value of 0.18 can be used for concrete.

2.9 SHEAR MODULUS

The modulus of elasticity of concrete in shear ranges from about 0.4 to 0.6 of the corresponding modulus in compression. From the theory of elasticity, the shear modulus is taken as follows:

$$G_c = \frac{E_c}{2(l + \mu)} \tag{2.7}$$

where μ = Poisson's ratio of concrete. If μ is taken equal to $\frac{1}{6}$, then $G_c = 0.43E_c = 24{,}500\sqrt{f'_c}$.

2.10 MODULAR RATIO

The modular ratio, n, is the ratio of the modulus of elasticity of steel to the modulus of elasticity of concrete: $n = E_s/E_c$.

Because the modulus of elasticity of steel is considered constant and is equal to 29×10^6 psi and $E_c = 33w^{1.5}\sqrt{f'_c}$,

$$n = \frac{29 \times 10^6}{33w^{1.5}\sqrt{f'_c}} \tag{2.8}$$

For normal-weight concrete, $E_c = 57{,}400\sqrt{f'_c}$; hence n can be taken as

$$n = \frac{500}{\sqrt{f'_c}}\,(f'_c \text{ in psi}) = \frac{42}{\sqrt{f'_c}}\,(f'_c \text{ in N/mm}^2) \tag{2.9}$$

The significance and the use of the modular ratio are explained in Chapter 5.

2.11 VOLUME CHANGES OF CONCRETE

Concrete undergoes volume changes during hardening. If it loses moisture by evaporation, it shrinks, but if the concrete hardens in water, it expands. The causes of the volume changes in concrete can be attributed to changes in moisture content, chemical reaction of the cement with water, variation in temperature, and applied loads.

2.11.1 Shrinkage

The change in the volume of drying concrete is not equal to the volume of water removed [7]. The evaporation of free water causes little or no shrinkage. As concrete continues to dry, water evaporates and the volume of the restrained cement paste changes, causing concrete to shrink, probably due to the capillary tension that develops in the water remaining in concrete. Emptying of the capillaries causes a loss of water without shrinkage. But once the absorbed water is removed, shrinkage occurs.

Many factors influence the shrinkage of concrete caused by the variations in moisture conditions [5]:

1. *Cement and water content.* The more cement or water content in the concrete mix, the greater the shrinkage.
2. *Composition and fineness of cement.* High-early-strength and low-heat cements show more shrinkage than normal portland cement. The finer the cement, the greater is the expansion under moist conditions.
3. *Type, amount, and gradation of aggregate.* The smaller the size of aggregate particles, the greater is the shrinkage. The greater the aggregate content, the smaller is the shrinkage [14].
4. *Ambient conditions, moisture, and temperature.* Concrete specimens subjected to moist conditions undergo an expansion of 200 to 300×10^{-6}, but if they are left to dry in air, they shrink. High temperature speeds the evaporation of water and, consequently, increases shrinkage.
5. *Admixtures.* Admixtures that increase the water requirement of concrete increase the shrinkage value.
6. *Size and shape of specimen.* As shrinkage takes place in a reinforced concrete member, tension stresses develop in the concrete, and equal compressive stresses develop in the steel. These stresses are added to those developed by the loading action. Therefore, cracks may develop in concrete when a high percentage of steel is used. Proper distribution of reinforcement, by producing better distribution of tensile stresses in concrete, can reduce differential internal stresses.

The values of final shrinkage for ordinary concrete vary between 200 and 700 $\times$ 10^{-6}. For normal-weight concrete, a value of 300 $\times$ 10^{-6} may be used. The British Code CP110 [12] gives a value of 500 $\times$ 10^{-6}, which represents an unrestrained shrinkage of 1.5 mm in 3 m length in thin, plain concrete sections. If the member is restrained, a tensile stress of about 10 N/mm^2 (1400 psi) arises. If concrete is kept moist for a certain period after setting, shrinkage is reduced; therefore, it is important to cure the concrete for a period of no fewer than 7 days.

Exposure of concrete to wind increases the shrinkage rate on the upwind side. Shrinkage causes an increase in the deflection of structural members, which in turn increases with time. Symmetrical reinforcement in the concrete section may prevent curvature and deflection due to shrinkage.

Generally, concrete shrinks at a high rate during the initial period of hardening, but at later stages the rate diminishes gradually. It can be said that 15% to 30% of the shrinkage value occurs in 2 weeks, 40% to 80% occurs in 1 month, and 70% to 85% occurs in 1 year.

2.11.2 Expansion Due to Rise in Temperature

Concrete expands with increasing temperature and contracts with decreasing temperature. The coefficient of thermal expansion of concrete varies between 4 and 7 $\times$ 10^{-6} per degree Fahrenheit. An average value of 5.5 $\times$ 10^{-6} per degree Fahrenheit (12 $\times$ 10^{-6} per degree Celsius) can be used for ordinary concrete. The B.S. Code [12] suggests a value of 10^{-5} per degree Celsius. This value represents a change of length of 10 mm in a 30-m member subjected to a change in temperature of 33°C. If the member is restrained and unreinforced, a stress of about 7 N/mm^2 (1000 psi) may develop.

In long reinforced concrete structures, expansion joints must be provided at lengths of 100 to 200 ft (30 to 60 m). The width of the expansion joint is about 1 in. (25 mm). Concrete is not a good conductor of heat, whereas steel is a good one. The ability of concrete to carry load is not much affected by temperature.

2.12 CREEP

Concrete is an elastoplastic material, and beginning with small stresses, plastic strains develop in addition to elastic ones. Under sustained load, plastic deformation continues to develop over a period that may last for years. Such deformation increases at a high rate during the first 4 months after application of the load. This slow plastic deformation under constant stress is called *creep.*

Figure 2.5 shows a concrete cylinder that is loaded. The instantaneous deformation is ε_1, which is equal to the stress divided by the modulus of elasticity. If the same stress is

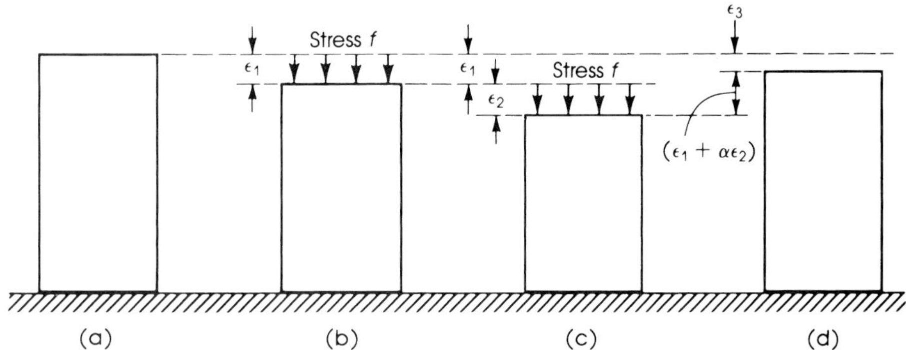

Figure 2.5 Deformation in a loaded concrete cylinder: (a) specimen unloaded, (b) elastic deformation, (c) elastic plus creep deformation, (d) permanent deformation after release of load.

kept for a period of time, an additional strain ε_2, due to creep effect, can be recorded. If load is then released, the elastic strain, ε_1, will be recovered, in addition to some creep strain. The final permanent plastic strain, ε_3, will be left, as shown in Figure 2.5. In this case $\varepsilon_3 = (1 - \alpha)\varepsilon_2$, where α is the ratio of the recovered creep strain to the total creep strain. The ratio α ranges between 0.1 and 0.2. The magnitude of creep recovery varies with the previous creep and depends appreciably upon the period of the sustained load. Creep recovery rate will be less if the loading period is increased, probably due to the hardening of oncrete while in a deformed condition.

The ultimate magnitude of creep varies between 0.2×10^{-6} and 2×10^{-6} per unit stress (lb/in.^{2}) per unit length. A value of 1×10^{-6} can be used in practice. The ratio of creep strain to elastic strain may be as high as 4.

Creep takes place in the hardened cement matrix around the strong aggregate. It may be attributed to slippage along planes within the crystal lattice, internal stresses caused by changes in the crystal lattice, and gradual loss of water from the cement gel in the concrete.

The different factors that affect the creep of concrete can be summarized as follows [15]:

1. *The level of stress.* Creep increases with an increase of stress in specimens made from concrete of the same strength and with the same duration of load.
2. *Duration of loading.* Creep increases with the loading period. About 80% of the creep occurs within the first 4 months; 90% occurs after about 2 years.
3. *Strength and age of concrete.* Creep tends to be smaller if concrete is loaded at a late age. Also, creep of 2000 psi- (14 N/mm^2-) strength concrete is about 1.41×10^{-6} whereas that of 4000 psi- (28 N/mm^2-) strength concrete is about 0.8×10^{-6} per unit stress and length of time.
4. *Ambient conditions.* Creep is reduced with an increase in the humidity of the ambient air.
5. *Rate of loading.* Creep increases with an increase in the rate of loading when followed by prolonged loading.
6. *Percentage and distribution of steel reinforcement in a reinforced concrete member.* Creep tends to be smaller for higher proportion or better distribution of steel.
7. *Size of the concrete mass.* Creep decreases with an increase in the size of the tested specimen.
8. *Type, fineness, and content of cement.* The amount of cement greatly affects the final creep of concrete, as cement creeps about 15 times as much as concrete.
9. *Water-cement ratio.* Creep increases with an increase in the water-cement ratio.
10. *Type and grading of aggregate.* Well-graded aggregate will produce dense concrete and consequently a reduction in creep.
11. *Type of curing.* High-temperature steam curing of concrete as well as the proper use of a plasticizer will reduce the amount of creep.

Creep develops not only in compression, but also in tension, bending, and torsion.

The ratio of the rate of creep in tension to that in compression will be greater than 1 in the first 2 weeks, but this ratio decreases over longer periods [5].

Creep in concrete under compression has been tested by many investigators. Troxell, Davis, and Raphael [16] measured creep strains periodically for up to 20 years and estimated that of the total creep after 20 years, 18% to 35% occurred in 2 weeks, 30% to 70% occurred in 3 months, and 64% to 83% occurred in 1 year.

For normal concrete loaded after 28 days, $C_r = 0.13\sqrt[3]{t}$, where C_r = creep strain per unit stress per unit length. Creep augments the deflection of reinforced concrete beams appreciably with time. In the design of reinforced concrete members, long-term deflection may be critical and has to be considered in proper design. Extensive deformation may influence the stability of the structure.

Sustained loads affect the strength as well as the deformation of concrete. A reduction of up to 30% of the strength of unreinforced concrete may be expected when concrete is subjected to a concentric sustained load for 1 year.

The fatigue strength of concrete is much smaller than its static strength. Repeated loading and unloading cycles in compression lead to a gradual accumulation of plastic deformations. If concrete in compression is subjected to about 2 million cycles, its fatigue limit is about 50% to 60% of the static compression strength. In beams, the fatigue limit of concrete is about 55% of its static strength [17].

2.13 UNIT WEIGHT OF CONCRETE

The unit weight, w, of hardened normal concrete ordinarily used in buildings and similar structures depends on the concrete mix, maximum size and grading of aggregates, water-cement ratio, and strength of concrete. The following values of the unit weight of concrete may be used:

1. Unit weight of plain concrete using maximum aggregate size of $\frac{3}{4}$ in. (20 mm) varies between 145 and 150 lb/ft^3 (2320 to 2400 kg/m^3). For concrete of strength less than 4000 psi (280 kg/cm^2), a value of 145 lb/ft^3 (2320 kg/m^3) can be used, whereas for higher-strength concretes, w can be assumed equal to 150 lb/ft^3 (2400 kg/m^3).
2. Unit weight of plain mass concrete of maximum aggregate size of 4 to 6 in. (100 to 150 mm) varies between 150 and 160 lb/ft^3 (2400 to 2560 kg/m^3). An average value of 155 lb/ft^3 (2500 kg/m^3) may be used.
3. Unit weight of reinforced concrete, using about 0.7% to 1.5% of steel in the concrete section, may be taken as 150 lb/ft^3 (2400 kg/m^3). For higher percentages of steel, the unit weight, w, can be assumed to be 155 lb/ft^3 (2500 kg/m^3).
4. Unit weight of lightweight concrete used for fireproofing, masonry, or insulation purposes varies between 20 and 90 lb/ft^3 (320 and 1440 kg/m^3). Concrete of upper values of 90 pcf or greater may be used for load-bearing concrete members.

The unit weight of heavy concrete varies between 200 and 270 lb/ft^3 (3200 and 4300 kg/m^3). Heavy concrete made with natural barite aggregate of $1\frac{1}{2}$ in. maximum size (38 mm) weighs about 225 lb/ft^3 (3600 kg/m^3). Iron ore sand and steel-punchings aggregate produce a unit weight of 270 lb/ft^3 (4320 kg/m^3) [19].

2.14 FIRE RESISTANCE

Fire resistance of a material is its ability to resist fire for a certain time without serious loss of strength, distortion, or collapse [20]. In the case of concrete, fire resistance depends on the thickness, type of construction, type and size of aggregates, and cement content. It is important to consider the effect of fire on tall buildings more than low or single-story buildings, because occupants need more time to escape.

Reinforced concrete is a much better fire-resistant material than steel. Steelwork heats rapidly, and its strength drops appreciably in a short time. Concrete itself has low thermal conductivity. The effect of temperatures below 250°C is small on concrete, but definite loss is expected at higher temperatures.

(a)

(b)

Casting and finishing precast concrete wall panels.

2.15 HIGH-PERFORMANCE CONCRETE

High-performance concrete may be assumed to imply that the concrete exhibits combined properties of strength, toughness, energy absorption, durability, stiffness, and a relatively higher ductility than normal concrete. This improvement in concrete quality may be achieved by using a new generation of additives and superplasticizers, which improves the workability of concrete and, consequently, its strength. Also, the use of active microfillers such as silica fume, fly ash, and polymer improves the strength, porosity, and durability of concrete. The addition of different types of fiber to the concrete mix enhances many of its properties, including ductility, strength, toughness, and many other properties.

Because it is difficult to set a limit to measure high-performance concrete, one approach is to define a lower-bound limit based on the shape of its stress-strain response in tension [23]. If the stress-strain relationship curve shows a quasi strain-hardening behavior—or, in other words, a postcracking strength larger than the cracking strength with an elastic-plastic behavior—then high performance is achieved [23]. In this behavior, multicracking stage is reached with high energy-absorption capacity. Substantial progress has been made recently in understanding the behavior and practical application of high-performance concrete.

2.16 LIGHTWEIGHT CONCRETE

Lightweight concrete is a concrete that has been made lighter than conventional normal-weight concrete and, consequently, it has a relatively lower density. Basically, reducing the density requires the inclusion of air in the concrete composition. This, however, can be achieved in four distinct ways:

1. By omitting the finer sizes from the aggregate grading, thereby creating what is called *no-fines* concrete. It is a mixture of cement, water, and coarse aggregate only ($\frac{3}{4}$–$\frac{3}{8}$ in.), mixed to produce concrete with many uniformly distributed voids.
2. By replacing the gravel or crushed rock aggregate by a hollow cellular or porous aggregate, which includes air in the mix. This type is called *lightweight aggregate concrete.* Lightweight aggregate may be natural, such as pumice, pozzolans, and volcanic slags; artificial (from industrial by-products), such as furnace clinker and foamed slag; or industrially produced, such as perlite, vermiculite, expanded clay, shale, or slate.
3. By creating gas bubbles in a cement slurry, which, when it sets, leaves a spongelike structure. This type is called *aerated concrete.*
4. By forming air cells in the slurry by chemical reaction or by vigorous mixing of the slurry with a preformed stable foam, which is produced by using special foam concentrate in a high-speed mixer. This type is called *cellular concrete.*

Structural lightweight concrete has a unit weight that ranges from 90 to 120 lb/ft^3, compared with 145 lb/ft^3 for normal-weight concrete. It is used in the design of floor slabs in buildings and other structural members where high-strength concrete is not required. Structural lightweight concrete can be produced with a compressive strength of 2500 to 5000 psi for practical applications.

2.17 FIBROUS CONCRETE

Fibrous concrete is made primarily of concrete constituents and discrete reinforcing fibers. The brittle nature of concrete and its low flexural tensile strength are major reasons for the growing interest in the performance of fibers in concrete technology. Various types of fibers—mainly steel, glass, and organic polymers—have been used in fibrous concrete. Generally, the length and diameter of the fibers do not exceed 3 in. (75 mm) and 0.04 in. (1 mm), respectively. The addition of fibers to concrete improves its mechanical properties, such as ductility, toughness, shear, flexural strength, impact resistance, and crack control. A convenient numerical parameter describing a fiber is its aspect ratio, which is the fiber length divided by an equivalent fiber diameter. Typical aspect ratios range from about 30 to 150, with the most common ratio being about 100. More details on fibrous concrete are given in [24].

2.18 STEEL REINFORCEMENT

Reinforcement, usually in the form of steel bars, is placed in the concrete member, mainly in the tension zone, to resist the tensile forces resulting from external load on the member. Reinforcement is also used to increase the member's compression resistance. Steel costs more than concrete, but it has a yield strength about 10 times the compressive strength of concrete. The function and behavior of both steel and concrete in a reinforced concrete member are discussed in Chapter 3.

Longitudinal bars taking either tensile or compression forces in a concrete member are called *main reinforcement.* Additional reinforcement in slabs, in a direction perpendicular to the main reinforcement, is called *secondary,* or *distribution, reinforcement.* In reinforced concrete beams, another type of steel reinforcement is used, transverse to the direction of the main steel and bent in a box or U shape. These are called *stirrups.* Similar reinforcements are used in columns, where they are called *ties.* Refer to Figure 8.8 and Figure 10.3.

2.18.1 Types of Steel Reinforcement

Different types of steel reinforcement are used in various reinforced concrete members. These types can be classified as follows:

Round bars Round bars are used most widely for reinforced concrete. Round bars are available in a large range of diameters, from $\frac{1}{4}$ in. (6 mm) to $1\frac{3}{8}$ in. (36 mm), plus two special types, $1\frac{3}{4}$ in. (45 mm) and $2\frac{1}{4}$ in. (57 mm). Round bars, depending on their surfaces, are either plain or deformed bars. Plain bars are used mainly for secondary reinforcement or in stirrups and ties. Deformed bars have projections or deformations on the surface for the purpose of improving the bond with concrete and reducing the width of cracks opening in the tension zone.

The diameter of a plain bar can be measured easily, but for a deformed bar, a nominal diameter is used that is the diameter of a circular surface with the same area as the section of the deformed bar. Requirements of surface projections on bars are specified by ASTM Specification A305, or A615. The bar sizes are designated by numbers 3 through 11, corresponding to the diameter in eighths of an inch. For instance, a no. 7 bar has a nominal diameter of $\frac{7}{8}$ in. and a no. 4 bar has a nominal diameter of $\frac{1}{2}$ in. The two largest sizes are designated no. 14 and no. 18, respectively. American standard bar marks are shown on the steel reinforcement to indicate the initial of the producing mill, the bar size, and the type of steel (Figure 2.6). The grade of the reinforcement is indicated on the bars by either the continuous-line system or the number system. In the first system, one longitudinal line is added to the bar, in addition to the main ribs, to indicate the high-strength grade of 60 ksi (420 N/mm^2), according to ASTM Specification A617. If only the main ribs are shown on the bar, without any additional lines, the steel is of the ordinary grade according to the ASTM A615 for the structural grade ($f_y = 40$ ksi, or 280 N/mm^2). In the number system, the yield strength of the high-strength grades is marked clearly on every bar. For ordinary grades, no strength marks are indicated. The two types are shown in Figure 2.6.

Welded fabrics and mats Welded fabrics and mats consist of a series of longitudinal and transverse cold-drawn steel wires, generally at right angles and welded together at all points of intersections. Steel reinforcement may be built up into three-dimensional cages before being placed in the forms.

Prestressed concrete wires and strands Prestressed concrete wires and strands use special high-strength steel (see Chapter 20). High-tensile steel wires of diameters 0.192 in. (5 mm) and 0.276 in. (7 mm) are used to form the prestressing cables by winding six steel wires around a seventh wire of slightly larger diameter. The ultimate strength of prestressed strands is 250 ksi or 270 ksi.

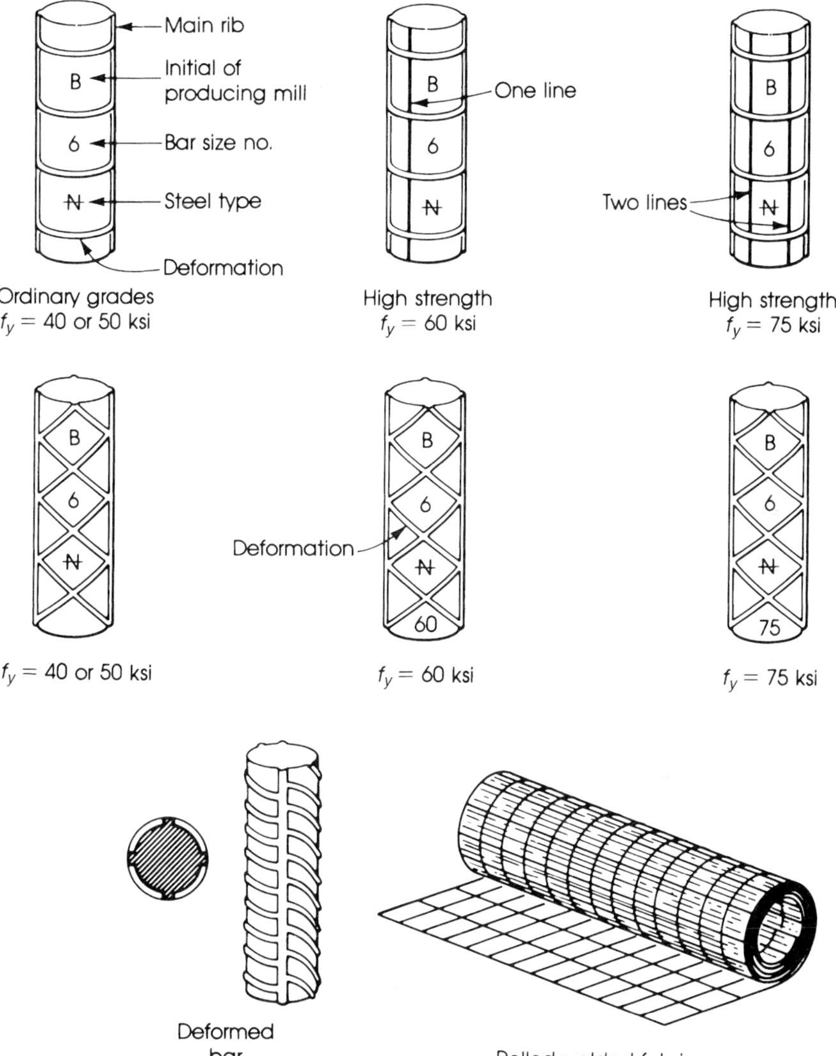

Figure 2.6 Some types of deformed bars and American standard bar marks.

2.18.2 Grades and Strength

Different grades of steel are used in reinforced concrete. Limitations on the minimum yield strength, ultimate strength, and elongation are explained in ASTM specifications for reinforcing steel bars (Table 2.4). The properties and grades of metric reinforcing steel are shown in Tables 2.5 and 2.6.

2.18.3 Stress-Strain Curves

The most important factor affecting the mechanical properties and stress-strain curve of the steel is its chemical composition. The introduction of carbon and alloying additives in steel increases its strength but reduces its ductility. Commercial steel rarely contains more than 1.2% carbon; the proportion of carbon used in structural steels varies between 0.2% and 0.3%.

Two other properties are of interest in the design of reinforced concrete structures; the first is the modulus of elasticity E_s. It has been shown that the modulus of elasticity is

Table 2.4 Grade of ASTM Reinforcing Steel Bars

Steel	Minimum Yield Strength f_y		Ultimate Strength f_{su}	
	ksi	MPa	ksi	MPa
Billet steel				
Grade 40	40	276	70	483
60	60	414	90	621
75	75	518	100	690
Rail steel				
Grade 50	50	345	80	551
60	60	414	90	621
Deformed wire				
Reinforcing	75	518	85	586
Fabric	70	483	80	551
Cold-drawn wire				
Reinforcing	70	483	80	551
Fabric	65	448	75	518
Fabric	56	386	70	483

Table 2.5 ASTM 615M (Metric) for Reinforcing Steel Bars

Bar no.	Diameter mm	Area mm^2	Weight kg/m
10M	11.3	100	0.785
15M	16.0	200	1.570
20M	19.5	300	2.355
25M	25.2	500	3.925
30M	29.9	700	5.495
35M	35.7	1000	7.850
45M	43.7	1500	11.770
55M	56.4	2500	19.600

Table 2.6 ASTM Metric Specifications

ASTM	Bar Size no.	Grade	
		MPa	ksi
A 615 M	10, 15, 20	300	43.5
billet steel	10–55	400	58.0
	35, 45, 55	500	72.5
A 616 M	10–35	350	50.75
rail steel	10–35	400	58.0
A 617 M	10–35	300	43.5
axle steel	10–35	400	58.0
A 706	10–55	400	58.0
low alloy			

constant for all types of steel. The ACI Code has adopted a value of $E_s = 29 \times 10^6$ psi $(2.0 \times 10^5$ MPa$)$. The modulus of elasticity is the slope of the stress-strain curve in the elastic range up to the proportional limit; $E_s =$ stress/strain. Second is the yield strength f_y. Typical stress-strain curves for some steel bars are shown in Figure 2.7. In high-tensile steel, a definite yield point may not show on the stress-strain curve. In this case, ultimate strength is reached gradually under an increase of stress (Figure 2.7). The yield strength or proof stress is considered as the stress that leaves a residual strain of 0.2% on the release of load, or a total strain of 0.5% to 0.6% under load.

SUMMARY

Section 2.1

The main factors that affect the strength of concrete are the water-cement ratio, properties and proportions of materials, age of concrete, loading conditions, and shape of tested specimen.

$$f'_c\,(\text{cylinder}) = 0.85\, f'_c\,(\text{cube}) = 1.10\, f'_c\,(\text{prism})$$

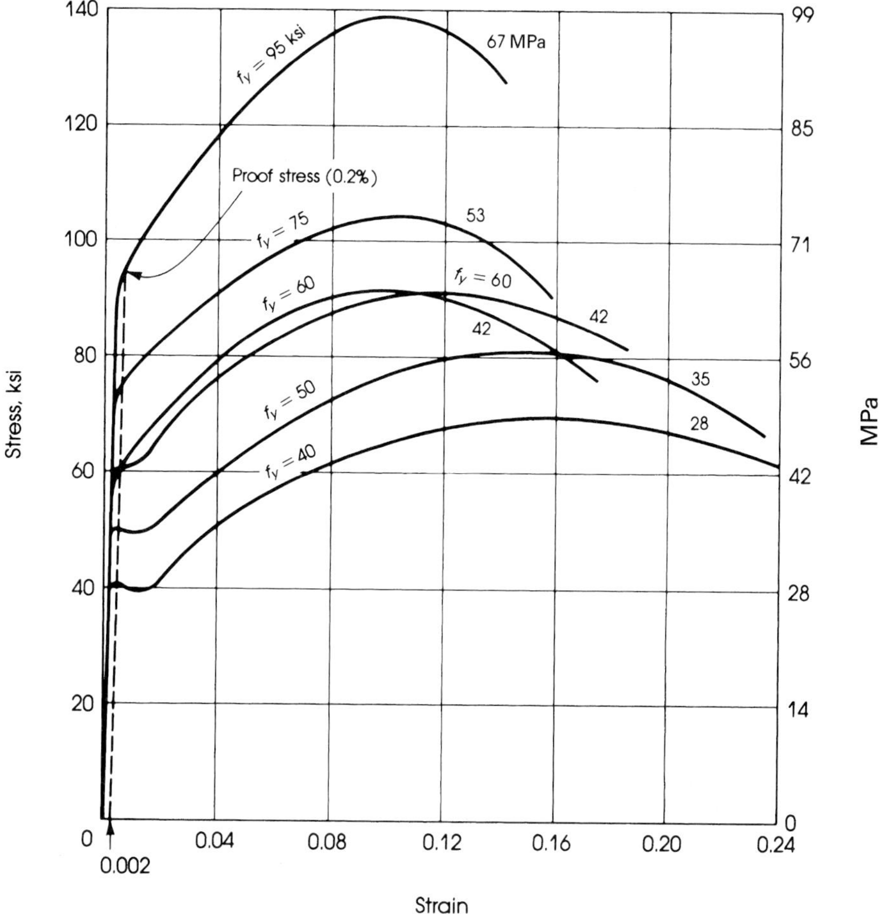

Figure 2.7 Typical stress-strain curves for some reinforcing steel bars of different grades. Note that 60-ksi steel may or may not show a definite yield point.

Sections 2.2–2.6

1. The usual specimen used to determine the compressive strength of concrete at 28 days is a 6- by 12-in. (150- by 300-mm) cylinder. Compressive strength between 3000 and 6000 psi is usually specified for reinforced concrete structures. Maximum stress, f'_c, is reached at an estimated strain of 0.002, whereas rupture occurs at a strain of about 0.003.
2. Tensile strength of concrete is measured indirectly by a splitting test performed on a standard cylinder using formula $f'_{sp} = 2P/\pi LD$. Tensile strength of concrete is approximately $0.1f'_c$.
3. Flexural strength (modulus of rupture, f_r) of concrete is calculated by testing a 6- by 6- by 28-in. plain concrete beam, $f_r = 7.5\sqrt{f'_c}$ (psi).
4. Ultimate shear strength is $2\sqrt{f'_c}$ (psi).

Sections 2.7–2.9

The modulus of elasticity of concrete E_c, for unit weight w between 90 and 155 pcf, is $E_c = 33w^{1.5}\sqrt{f'_c}$ (psi) $= 0.043w^{1.5}\sqrt{f'_c}$ MPa.

For normal-weight concrete, $w = 145$ pcf.

$$E_c = 57{,}600\sqrt{f'_c} \text{ or } E_c = 57{,}000\sqrt{f'_c} = 4700\sqrt{f'_c}\text{ MPa}$$

The shear modulus of concrete is $G_c = E_c/2(1 + \mu) = 0.43\,E_c$ for a Poisson's ratio $\mu = \frac{1}{6}$.

Poisson's ratio μ varies between 0.15 and 0.20, with an average value of 0.18.

Section 2.10

Modular ratio is $n = E_s/E_c = 500/\sqrt{f'_c}$, where f'_c is in psi.

Section 2.11

1. Values of shrinkage for normal concrete fall between 200×10^{-6} and 700×10^{-6}. An average value of 300×10^{-6} may be used.
2. The coefficient of expansion of concrete falls between 4×10^{-6} and $7 \times 10^{-6}/°\text{F}$.

Section 2.12

The ultimate magnitude of creep varies between 0.2×10^{-6} and 2×10^{-6} per unit stress per unit length. An average value of 1×10^{-6} may be adopted in practical problems. Of the ultimate (20-year) creep, 18% to 35% occurs in 2 weeks, 30% to 70% occurs in 3 months, and 64% to 83% occurs in 1 year.

Section 2.13

The unit weight of normal concrete is 145 pcf for plain concrete and 150 pcf for reinforced concrete.

Section 2.14

Reinforced concrete is a much better fire-resistant material than steel. Concrete itself has a low thermal conductivity. An increase in concrete cover in structural members such as walls, columns, beams, and floor slabs will increase the fire resistance of these members.

Sections 2.15–2.17

1. High-performance concrete implies that concrete exhibits properties of strength, toughness, energy absorption, durability, stiffness, and ductility higher than normal concrete.
2. Concrete is made lighter than normal-weight concrete by inclusion of air in the concrete composition. Types of lightweight concrete are no-fines concrete, lightweight aggregate concrete, aerated concrete, and cellular concrete.
3. Fibrous concrete is made of concrete constituents and discrete reinforcing fibers such as steel, glass, and organic polymers.

Section 2.18

The grades of steel mainly used are grade 60 ($f_y = 60$ ksi) and grade 40 ($f_y = 40$ ksi). The modulus of elasticity of steel is $E_s = 29 \times 10^6$ psi (2×10^{-5} MPa).

REFERENCES

1. Portland Cement Association. *Design and Control of Concrete Mixtures.* Chicago, 1992.
2. British Standard Institution. *B.S. Code of Practice For Reinforced Concrete,* CP 114. 1973.
3. H. E. Davis, G. E. Troxell, and C. T. Wiskocil. *The Testing and Inspection of Engineering Materials.* New York: McGraw-Hill, 1964.

4. United States Bureau of Reclamation. *Concrete Manual,* 7th ed. 1963.
5. G. E. Troxell and H. E. Davis. *Composition and Properties of Concrete.* New York: McGraw-Hill, 1956.
6. V. Murashev and V. B. Segalov. *The Design of Reinforced Concrete Structures.* Moscow: Mir Publications, 1968.
7. A. M. Neville. *Properties of Concrete.* London: Pitman and Sons, 1985.
8. H. Gonnerman and E. C. Shuman. "Compression, Flexural and Tension Tests of Plain Concrete." *ASTM Proceedings* 28 (Part II, 1928).
9. S. Musa. "The Effect of Steel Fibers on the Behavior of High Strength Reinforced Concrete Beams." M. S. Thesis, South Dakota State University, August 1982.
10. American Concrete Institute and the Cement and Concrete Association. *Recommendations for an International Code of Practice for Reinforced Concrete.* London, 1963.
11. T. C. Powers. "Measuring Young's Modulus of Elasticity by Means of Sonic Vibrations." *ASTM Proceedings 40* (1940).
12. British Standard Institution. *B.S. Code of Practice For Structural Use of Concrete.* BS8110. London, 1985.
13. L. M. Legatski. *Cellular Concrete.* Ann Arbor, Mich.: Elastizell Corp. of America, 1980.
14. G. Pickett. "Effect of Aggregate on Shrinkage of Concrete and Hypothesis Concerning Shrinkage," *ACI Journal 52* (Jan. 1956).
15. "Symposium on Shrinkage and Creep of Concrete." *ACI Journal 53* (Dec. 1957).
16. G. E. Troxell, J. M. Raphale, and R. E. Davis. "Long Time Creep and Shrinkage Tests of Plain and Reinforced Concrete." *ASTM Proceedings 58* (1958).
17. "Fatigue of Concrete—Reviews of Research." *ACI Journal 58* (1958).
18. A. Ruettgers, E. N. Vidal, and S. P. Wing. "An Investigation of the Permeability of Mass Concrete with Particular Reference to Boulder Dam." *ACI Journal 31* (1935).
19. E. J. Callan. "Concrete for Radiation Shielding." *ACI Journal 50* (1954).
20. J. Faber and F. Mead. *Reinforced Concrete.* London: Spon Ltd., 1967.
21. F. Walley and S. C. Bate. *A Guide to the B. S. Code of Practice for Prestressed Concrete.* London: Concrete Publications, 1959.
22. American Concrete Institute. "Building Code Requirements for Structural Concrete." ACI 318–99, Detroit, Michigan, 1999.
23. A. E. Newman and H. W. Reinhardt. "High Performance Fiber Reinforced Cement Composites." *Proceedings*, Vol. 2. University of Michigan, Ann Arbor, Michigan, June 1995.
24. American Concrete Institute. "State-of-the-Art Report on Fiber Reinforced Concrete." ACI Committee 544 Report, 1994.

PROBLEMS

2.1 Explain the modulus of elasticity of concrete in compression and the shear modulus.

2.2 Determine the modulus of elasticity of concrete by the ACI formula for a concrete cylinder that has a unit weight of 120 pcf (1920 kg/m^3) and a compressive strength of 3000 psi (21 MPa).

2.3 Estimate the modulus of elasticity and the shear modulus of a concrete specimen with a dry density of 150 pcf (2400 kg/m^3) and compressive strength of 4500 psi (31 MPa) using Poisson's ratio, $\mu = 0.18$.

2.4 What is meant by the modular ratio and Poisson's ratio? Give approximate values for concrete.

2.5 What factors influence the shrinkage of concrete?

2.6 What factors influence the creep of concrete?

2.7 What are the types and grades of the steel reinforcement used in reinforced concrete?

2.8 On the stress-strain diagram of a steel bar, show and explain the following: proportional limit, yield stress, ultimate stress, yield strain, and modulus of elasticity.

2.9 Calculate the modulus of elasticity of concrete E_c for the following types of concrete.

Density	Strength f'_c
160 pcf	5000 psi
45 pcf	4000 psi
25 pcf	2500 psi
2400 kg/m^3	35 MPa
2300 kg/m^3	30 MPa
2100 kg/m^3	25 MPa

$$E_c = 33\, W^{1.5}\sqrt{f'_c}\ (\text{ft}),$$
$$E_c = 0.043\, W^{1.5}\sqrt{f'_c}\ (\text{SI})$$

2.10 Determine the modular ratio n and the modulus of rupture for each case of Problem 2.9. Tabulate your results.

$$f_r = 7.5\sqrt{f'_c}\ (\text{psi}), \quad f_r = 0.62\sqrt{f'_c}\ (\text{MPa})$$

2.11 A standard normal concrete cylinder 6 × 12 in. was tested to failure, and the following loads and strains were recorded.

Load, kips	Strain × 10^{-4}	Load, kips	Strain × 10^{-4}
0.0	0.0	72	10.0
12	1.2	84	13.6
24	2.0	96	18.0
36	3.2	108	30.0
48	5.2	95	39.0
60	7.2	82	42.0

a. Draw the stress-strain diagram of concrete and determine the maximum stress and corresponding strain.

b. Determine the initial modulus and secant modulus.

c. Calculate the modulus of elasticity of concrete using the ACI formula for normal-weight concrete and compare results.

$$E_c = 57{,}000\sqrt{f'_c}\ \text{psi},$$
$$E_c = 4730\sqrt{f'_c}\ \text{MPa}$$

3

Apartment building, Fort Lauderdale, Florida.

STRENGTH DESIGN METHOD: FLEXURAL ANALYSIS OF REINFORCED CONCRETE BEAMS

3.1 INTRODUCTION

Strength design is a method of determining the dimensions of a structural member based on ultimate strength of sections. Ultimate loads are determined by multiplying the working loads, the dead load, the assumed live load, and other loads—such as wind load—by load factors. Ultimate strength of sections is reached by the yielding of steel or by the crushing of concrete.

Loads cause external forces, such as bending moments, shear, or thrust, depending on how these loads are applied to the structure. The section of the member is designed in such a way that its internal ultimate capacity is equal to or greater than the external ultimate forces acting on the member.

In proportioning reinforced concrete structural members, we investigate three main items:

1. The safety of the structure, which is maintained by providing adequate internal ultimate strength capacity.
2. Deflection of the structural member under working loads. The maximum value of this must be limited and is usually specified as a factor of the span, to preserve the appearance of the structure.

3. Width of cracks under working loads. Visible cracks spoil the appearance of the structure and also permit humidity to penetrate the concrete, causing corrosion of steel and weakening the reinforced concrete member. The ACI Code implicitly limits crack widths for interior and exterior exposure to 0.016 in. (0.40 mm) and 0.013 in. (0.33 mm), respectively, which is achieved by adopting and limiting the spacings of the tension bars (see Chapter 6).

It is worth mentioning that the ultimate-strength design method was first permitted in Britain in 1957 and in the United States in 1956. The ACI Code of 1963 put equal emphasis on both ultimate-strength design and working-stress design methods, whereas the later ACI codes emphasized the ultimate-strength concept.

3.2 ASSUMPTIONS

Reinforced concrete sections are heterogeneous (nonhomogeneous), because they are made of two different materials, concrete and steel. Therefore, proportioning structural members by ultimate-strength design is based on the following assumptions:

1. Strain in concrete is the same as in reinforcing bars at the same level, provided that the bond between the steel and concrete is adequate.
2. Strain in concrete is linearly proportional to the distance from the neutral axis.
3. The modulus of elasticity of all grades of steel is taken as $E_s = 29 \times 10^6$ lb/in.^{2} (200,000 MPa or N/mm^2). The stress in the elastic range is equal to the strain multiplied by E_s.
4. Plane cross sections continue to be plane after bending.
5. Tensile strength of concrete is neglected because (a) concrete's tensile strength is about 10% of its compressive strength, (b) cracked concrete is assumed to be not effective, and (c) before cracking, the entire concrete section is effective in resisting the external moment.
6. The method of elastic analysis, assuming an ideal behavior at all levels of stress, is not valid. At high stresses, nonelastic behavior is assumed, which is in close agreement with the actual behavior of concrete and steel.
7. At ultimate strength, the maximum strain at the extreme compression fibers is assumed equal to 0.003, by the ACI Code provision.
8. At ultimate strength, the shape of the compressive concrete stress distribution may be assumed to be rectangular, parabolic, or trapezoidal. In this text, a rectangular shape will be assumed.

3.3 BEHAVIOR OF A SIMPLY SUPPORTED REINFORCED CONCRETE BEAM LOADED TO FAILURE

Concrete being weakest in tension, a concrete beam under an assumed working load will definitely crack at the tension side, and the beam will collapse if tensile reinforcement is not provided. Concrete cracks occur at a loading stage when its maximum tensile stress reaches the modulus of rupture of concrete. Therefore, steel bars are used to increase the moment capacity of the beam; the steel bars resist the tensile force, and the concrete resists the compressive force.

To study the behavior of a reinforced concrete beam under increasing load, let us examine how two beams were tested to failure. Details of the beams are shown in Figure 3.1. Both beams had a section of 4.5 in. by 8 in. (110 mm by 200 mm), reinforced only on the

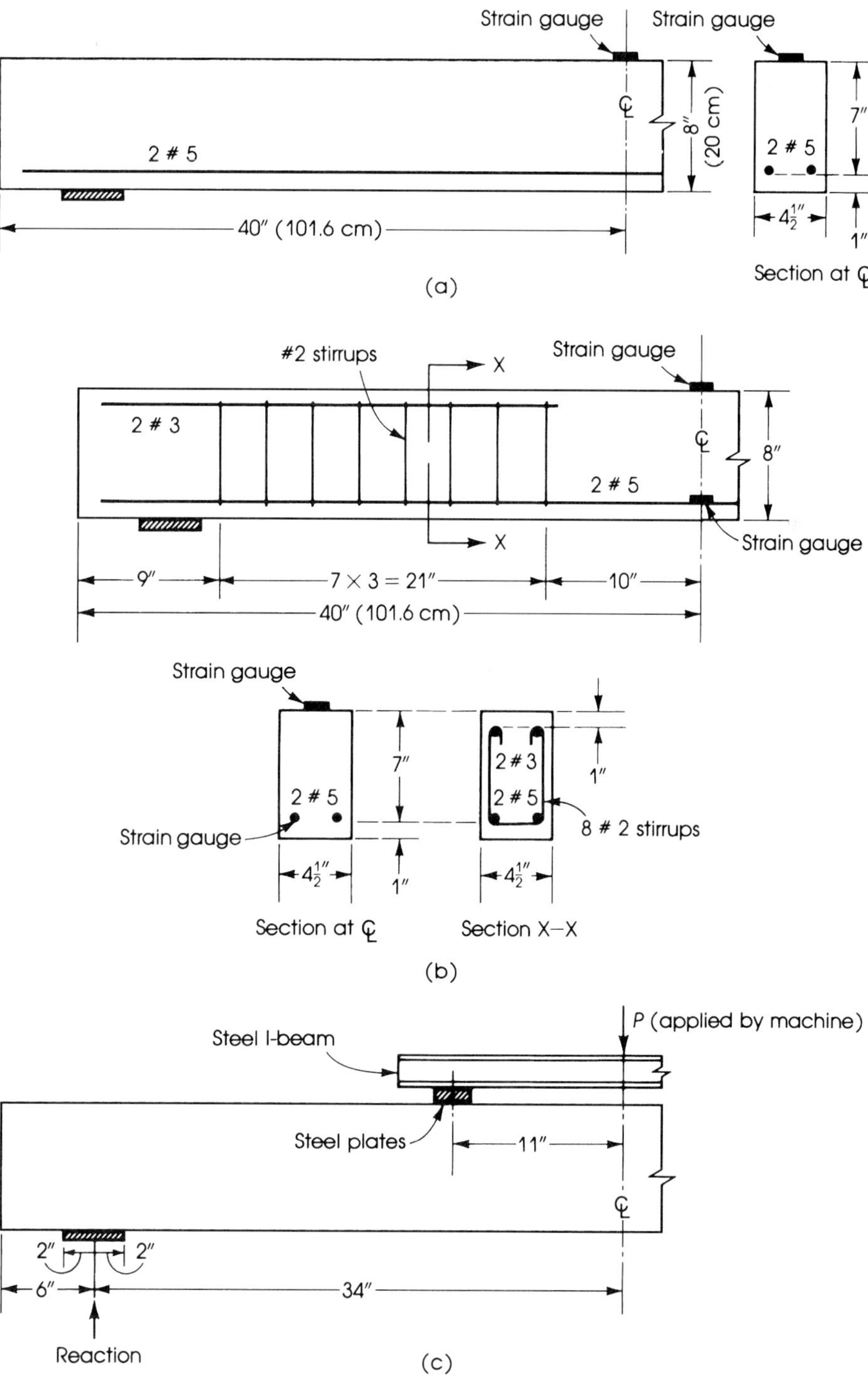

Figure 3.1 Details of tested beams: (a) beam 1, (b) beam 2, and (c) loading system. All beams are symmetrical about the centerline.

tension side by two no. 5 bars. They were made of the same concrete mix. Beam 1 had no stirrups, whereas beam 2 was provided with no. 3 stirrups spaced at 3 in. The loading system and testing procedure were the same for both beams. To determine the compressive strength of the concrete and its modulus of elasticity, E_c, a standard concrete cylinder was

(a)

(b)

Tests on a simply supported beam and a two-span continuous beam loaded to failure.

tested, and strain was measured at different load increments. The following observations were noted at different distinguishable stages of loading.

Stage 1: At zero external load, each beam carried its own weight in addition to that of the loading system, which consisted of an I-beam and some plates. Both beams behaved similarly at this stage. At any section, the entire concrete section, in addition to the steel reinforcement, resisted the bending moment and shearing forces. Maximum stress occurred at the section of maximum bending moment—that is, at midspan. Maximum tension stress at the bottom fibers was much less than the modulus of rupture of concrete. Compressive stress at the top fibers was much less than the ultimate concrete compressive stress f'_c. No cracks were observed at this stage.

Stage 2: This stage was reached when the external load P was increased from zero to P_1, which produced tensile stresses at the bottom fibers equal to the modulus of rupture of concrete. At this stage the entire concrete section was effective, with the steel bars at the tension side sustaining a strain equal to that of the surrounding concrete.

Stress in the steel bars was equal to the stress in the adjacent concrete multiplied by the modular ratio n, the ratio of the modulus of elasticity of steel to that of concrete. The compressive stress of concrete at the top fibers was still very small as compared with the compressive strength, f'_c. The behavior of beams was elastic within this stage of loading.

Stage 3: When the load was increased beyond P_1, tensile stresses in concrete at the tension zone increased until they were greater than the modulus of rupture, f_r, and cracks developed. The neutral axis shifted upward, and cracks extended close to the level of the shifted neutral axis. Concrete in the tension zone lost its tensile strength, and the steel bars started to work effectively and to resist the entire tensile force. Between cracks, the concrete bottom fibers had tensile stresses, but they were of negligible value. It can be assumed that concrete below the neutral axis did not participate in resisting external moments.

In general, the development of cracks and the spacing and maximum width of cracks depend on many factors, such as the level of stress in the steel bars, distribution of steel bars in the section, concrete cover, and grade of steel used.

At this stage, the deflection of the beams increased clearly, because the moment of inertia of the cracked section was less than that of the uncracked section. Cracks started about the midspan of the beam, but other parts along the length of the beam did not crack. When load was again increased, new cracks developed, extending toward the supports. The spacing of these cracks depends on the concrete cover and the level of steel stress. The width of cracks also increased. One or two of the central cracks were most affected by the load, and their crack widths increased appreciably, whereas the other crack widths increased much less. It is more important to investigate those wide cracks than to consider the larger number of small cracks.

If the load were released within this stage of loading, it would be observed that permanent fine cracks, of no significant magnitude, were left. On reloading, cracks would open quickly, because the tensile strength of concrete had already been lost. Therefore, it can be stated that the second stage, once passed, does not happen again in the life of the beam. When cracks develop under working loads, the resistance of the entire concrete section and gross moment of inertia are no longer valid.

At high compressive stresses, the strain of the concrete increased rapidly, and the stress of concrete at any strain level was estimated from a stress-strain graph obtained by testing a standard cylinder to failure for the same concrete. As for the steel, the stresses were still below the yield stress, and the stress at any level of strain was obtained by multiplying the strain of steel, ε_s, by the modulus of elasticity of steel, E_s.

Stage 4: In beam 1, at a load value of 9500 lb (42.75 kN), shear stress at a distance of about the depth of the beam from the support increased and caused diagonal cracks at approximately 45° from horizontal in the direction of principal stresses resulting from the combined action of bending moment and shearing force. The diagonal crack extended downward to the level of the steel bars and then extended horizontally at that level toward the support. When the crack, which had been widening gradually, reached the end of the beam, a concrete piece broke off and failure occurred suddenly (Figure 3.2). The failure load was 13,600 lb (61.2 kN). Stresses in concrete and steel at the midspan section did not reach their failure stresses. (The shear behavior of beams is discussed in Chapter 8.)

In beam 2, at a load of 11,000 lb (49.5 kN), a diagonal crack developed similar to that of beam 1; then other parallel diagonal cracks appeared, and the stirrups started to take an effective part in resisting the principal stresses. Cracks did not extend along the horizontal main steel bars, as in beam 1. On increasing the load, diagonal cracks on the other end of the beam developed at a load of 13,250 lb (59.6 kN). Failure did not occur at this stage because of the presence of stirrups.

Stage 5: When the load on beam 2 was further increased, strains increased rapidly until the maximum carrying capacity of the beam was reached at ultimate load $P_u = 16{,}200$ lb (72.9 kN).

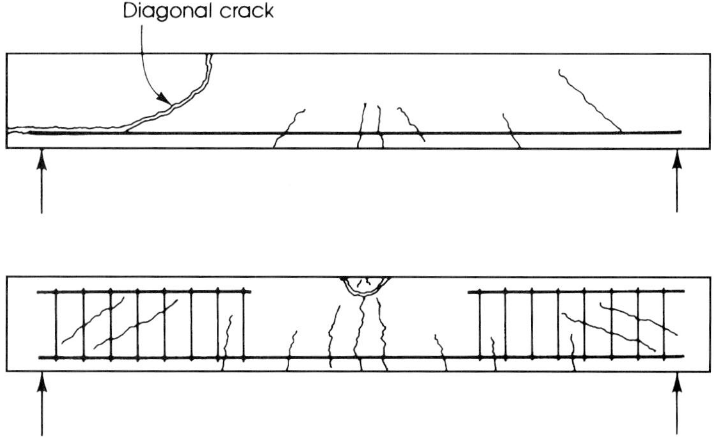

Figure 3.2 Shape of beam 1 at shear failure (top) and beam 2 at bending moment failure (bottom).

Two-span continuous reinforced concrete beam loaded to failure.

In beam 2, the amount of steel reinforcement used was relatively small. When the strain in the steel reached the yield strain, which can be considered equal to yield stress divided by the modulus of elasticity of steel, $\varepsilon_y = f_y/E_s$, the strain in the concrete, ε_c, was less than the strain at maximum compressive stress, f'_c. The steel bars yielded, and the strain in steel increased to about 12 times that of the yield strain without increase in load. Cracks widened sharply, deflection of the beam increased greatly, and the compressive strain on the concrete increased. After another very small increase of load, steel strain hardening occurred, and concrete reached its maximum strain, ε'_c, and it started to crush under load; then the beam collapsed. Figure 3.2 shows the failure shapes of the two beams.

3.4 TYPES OF FLEXURAL FAILURE

Three types of flexural failure of a structural member can be expected, depending on the percentage of steel used in the section.

1. Steel may reach its yield strength before the concrete reaches its maximum strength (Figure 3.3(a)). In this case, failure is due to yielding of steel. The section contains a relatively small amount of steel and is called an *under-reinforced* section.
2. Steel may reach its yield strength at the same time as concrete reaches its ultimate strength (Figure 3.3(b)). The section is called a *balanced* section. Steel and concrete fail simultaneously.
3. Concrete may fail before the yield of steel (Figure 3.3(c)) due to the presence of a high percentage of steel in the section. In this case the concrete strength, f'_c, and maximum strain, ε'_c, are reached, but the steel stress is less than the yield strength: $f_s < f_y$. The section is called an *over-reinforced* section.

It can be assumed that concrete fails in compression when the concrete strain reaches 0.003. A range of 0.0025 to 0.004 has been obtained from tests; therefore, ε'_c can be taken as 0.003.

In a structure based on under-reinforced sections, steel yields before the crushing of concrete. Cracks widen extensively, giving a warning before the concrete crushes and the

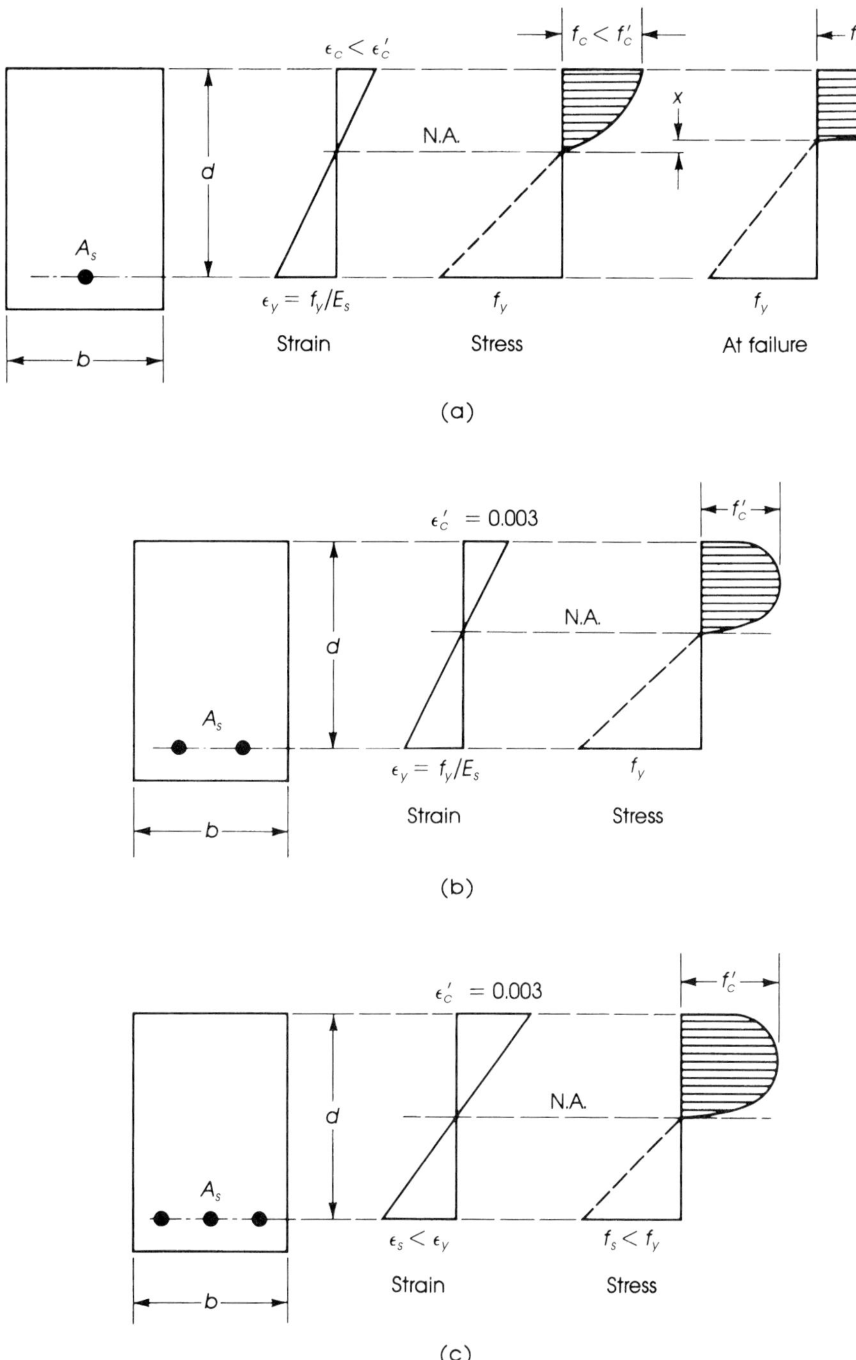

Figure 3.3 Stress and strain diagrams for (a) under-reinforced, (b) balanced, and (c) over-reinforced sections.

structure collapses. This type of design is adopted by the ACI Code. In a structure designed with balanced or over-reinforced conditions, the concrete fails suddenly, and collapse occurs immediately with no warning. This type of design is not allowed by the ACI Code. Consequently, the ACI Code limits the maximum steel percentage to 75% of the balanced steel ratio.

(a)

(b)

Failure conditions at the positive- and negative-moment sections in a continuous reinforced concrete beam.

3.5 LOAD FACTORS

The types of load and the safety provisions were explained in Sections 1.7 and 1.8.

In the strength design method, the factored design load is obtained by multiplying the dead load by a load factor and multiplying the specified live load by another load factor. The magnitude of the load factor must be adequate to limit the probability of failure and to permit an economical structure. The choice of a proper load factor or, in general, a proper factor of safety depends mainly on the importance of the structure (whether a courthouse or a warehouse), the degree of warning needed prior to collapse, the importance of each structural member (whether a beam or a column), the expectation of overload, the accuracy of artisanry, and the accuracy of calculations.

Based on historical studies of various structures, experience, and the principles of probability, the ACI Code adopts a load factor of 1.4 for dead loads and 1.7 for live loads. The dead-load factor is smaller, because the dead load can be computed with a greater degree of certainty than the live load. Moreover, the choice of the factors reflects the degree of economical design as well as the degree of safety and serviceability of the structure. It is also based on the fact that the performance of the structure under actual loads must be satisfactory within specific limits.

If the ultimate load is denoted by U and those due to wind load and earthquake are W and E, respectively, then according to the ACI Code, the ultimate required strength, U, shall be the most critical of the following:

1. In the case of dead, live, and wind loads,

$$U = 1.4D + 1.7L \tag{3.1a}$$

$$U = 0.75(1.4D + 1.7L + 1.7W) \tag{3.1b}$$

$$U = 0.9D + 1.3W \tag{3.1c}$$

2. In cases when earthquake forces, E, must be included in the design, $1.1E$ shall be substituted for W in the preceding equations.
3. In cases when earth pressure load, H, must be included in the design,

$$U = 1.4D + 1.7L + 1.7H \tag{3.2a}$$

Where dead load, D, and live load, L, reduce the effect of H, U shall be checked for

$$U = 0.9D + 1.7H \tag{3.2b}$$

For any combination of D, L, or H,

$$U = 1.4D + 1.7L$$

4. If weight and pressure loads from liquids, F, must be included in the design,

$$U = 1.4D + 1.7L + 1.4F \tag{3.3a}$$

Where dead load, D, and live load, L, reduce the effect of F,

$$U = 0.9D + 1.4F \tag{3.3b}$$

For any combination of D, L, or F,

$$U = 1.4D + 1.7L$$

The vertical pressure of liquids shall be considered as dead load with due regard to variation in liquid depth.

5. When impact effects are taken into account, it shall be included in the live load. The ACI Code does not specify a value, but AASHTO specifications give the impact effect, I, as a percentage of the live load, L, as follows:

$$I = \frac{50}{125 + S} \leq 30\% \tag{3.4}$$

where

I = percentage of impact (with a maximum of 30%)
S = part of the span loaded in ft

When a better estimation is known from experiments or experience, the actual value shall be used:

$$\text{Live load including impact to be used} = L(1 + I)$$

6. Where the structural effects of differential settlement, creep shrinkage, or temperature change may be significant, they shall be included with the dead load D:

$$U = 0.75(1.4D + 1.7L) \tag{3.1d}$$

Equation (3.1a) is most generally used. The dead-load factor is equal to 1.4, whereas the live-load factor is equal to 1.7. These values are less than those specified by ACI Code of 1963 of 1.5 for the dead load and 1.8 for the live load. The decrease was suggested because of the more comprehensive code provisions, additional research and experience, and improved concrete and steel control.

For applied concentrated dead and live loads, P_D and P_L, the ultimate concentrated load is $P_U = 1.4P_D + 1.7P_L$; also $M_U = 1.4M_D + 1.7M_L$, where M_D and M_L are the actual dead-load and live-load moments, respectively.

3.6 CAPACITY-REDUCTION FACTOR

The nominal strength of a section is reduced by a factor ϕ to account for small adverse variations in material strengths, artisanry, dimensions, control, and degree of supervision. The factor ϕ constitutes a portion of the factor of safety, as discussed in Section 1.8.

The ACI Code, Section 9.3.2, specifies the following values to be used:

- Bending, axial tension, bending and axial tension $\phi = 0.90$
- Shear and torsion $\phi = 0.85$
- Bearing on concrete $\phi = 0.70$
- Bending in plain concrete or in concrete with minimum reinforcement of $200/f_y$ $\phi = 0.65$
- Axial compression, with or without bending:
  - Members with spiral reinforcement $\phi = 0.75$
  - Other reinforced members $\phi = 0.70$

Limitations of the axial-compression case are explained in detail in Chapter 10.

3.7 SIGNIFICANCE OF ANALYSIS AND DESIGN EXPRESSIONS

Two approaches for the investigations of a reinforced concrete member will be used in this book:

Analysis of a section implies that the dimensions and steel used in the section (in addition to concrete strength and steel yield strength) are given, and it is required to calculate the internal ultimate moment capacity of the section so that it can be compared with the applied external ultimate moment. It may also be necessary to check maximum stresses under external loads.

Design of a section implies that the external ultimate moment is known from structural analysis, and it is required to compute the dimensions of an adequate concrete section and the amount of steel reinforcement. Concrete strength and yield strength of steel used are given.

3.8 EQUIVALENT COMPRESSIVE STRESS DISTRIBUTION

The distribution of compressive concrete stresses at failure may be assumed to be a rectangle, trapezoid, parabola, or any other shape that is in good agreement with test results.

When a beam is about to fail, the steel will yield first if the section is under-reinforced, and in this case the steel is equal to the yield stress. If the section is over-reinforced, concrete crushes first and the strain is assumed equal to 0.003, which agrees with many tests of beams and columns. A compressive force C develops in the compression zone and a tension force T develops in the tension zone at the level of the steel bars. The position of the force T is known, because its line of application coincides with the center of gravity of the steel bars. The position of the compressive force C is not known unless the compressive volume is known and its center of gravity is located. If that is done, the moment arm, which is the vertical distance between C and T, will consequently be known.

In Figure 3.4, if concrete fails, $\varepsilon_c' = 0.003$, and if steel yields, as in the case of a balanced section, $f_s = f_y$.

The compression force C is represented by the volume of the stress block, which has the nonuniform shape of stress over the rectangular hatched area of bc. This volume may be considered equal to $C = bc(\alpha_1 f_c')$ where $(\alpha_1 f_c')$ is an assumed average stress of the nonuniform stress block.

The position of the compression force C is at a distance z from the top fibers, which can be considered as a fraction of the distance c (the distance from the top fibers to the neutral axis), and z can be assumed equal to $\alpha_2 c$, where $\alpha_2 < 1$. The values of α_1 and α_2 have

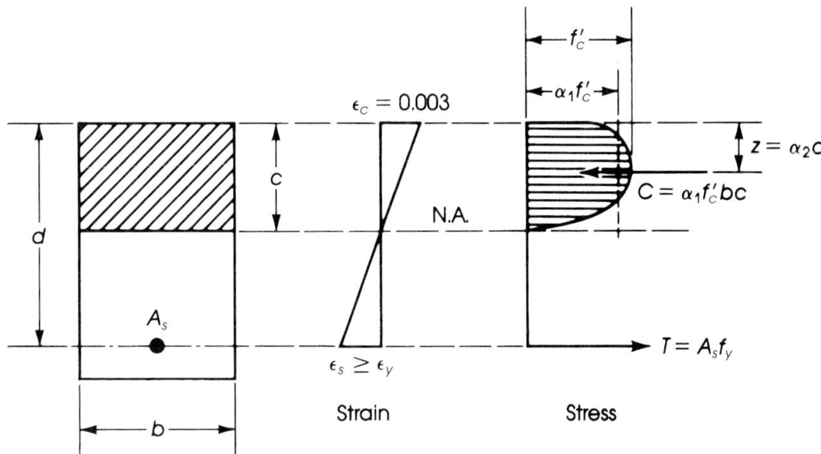

Figure 3.4 Ultimate forces in a rectangular section.

been estimated from many tests, and their values, as suggested by Mattock, Kriz, and Hognestad [3], are as follows:

α_1 = 0.72 for $f'_c \leq 4000$ psi (27.6 MPa); it decreases linearly by 0.04 for every 1000 psi (6.9 MPa) greater than 4000 psi

α_2 = 0.425 for $f'_c < 4000$ psi (27.6 MPa); it decreases linearly by 0.025 for every 1000 psi (6.9 MPa) greater than 4000 psi

The decrease in the value of α_1 and α_2 is related to the fact that high-strength concretes show more brittleness than low-strength concretes [2].

To derive a simple rational approach for calculations of the internal forces of a section, the ACI Code adopted an equivalent rectangular concrete stress distribution, which was first proposed by C. S. Whitney and checked by Mattock and others [3]. A concrete stress of $0.85f'_c$ is assumed to be uniformly distributed over an equivalent compression zone bounded by the edges of the cross section and a line parallel to the neutral axis at a distance $a = \beta_1 c$ from the fiber of maximum compressive strain, where c is the distance between the top of the compressive section and the neutral axis (Figure 3.5). The fraction β_1 is 0.85 for concrete strengths $f'_c \leq 4000$ psi (27.6 MPa) and is reduced linearly at a rate of 0.05 for each 1000 psi (6.9 MPa) of stress greater than 4000 psi (Figure 3.6), with a minimum value of 0.65.

The preceding discussion applies in general to any section, and it is not confined to a rectangular shape. In the rectangular section, the area of the compressive zone is equal to ba, and every unit area is acted on by a uniform stress equal to $0.85f'_c$, giving a total stress volume equal to $0.85f'_c ab$ that corresponds to the compressive force C. For any other shape, the force C is equal to the area of the compressive zone multiplied by a constant stress equal to $0.85f'_c$.

For example, in the section shown in Figure 3.7, the force C is equal to the shaded area of the cross section multiplied by $0.85f'_c$:

$$C = 0.85f'_c(6 \times 3 + 10 \times 2) = 32.3f'_c \text{ lb}$$

The position of the force C is at a distance z from the top fibers, at the position of the resultant force of all small-element forces of the section. As in this case when the stress is

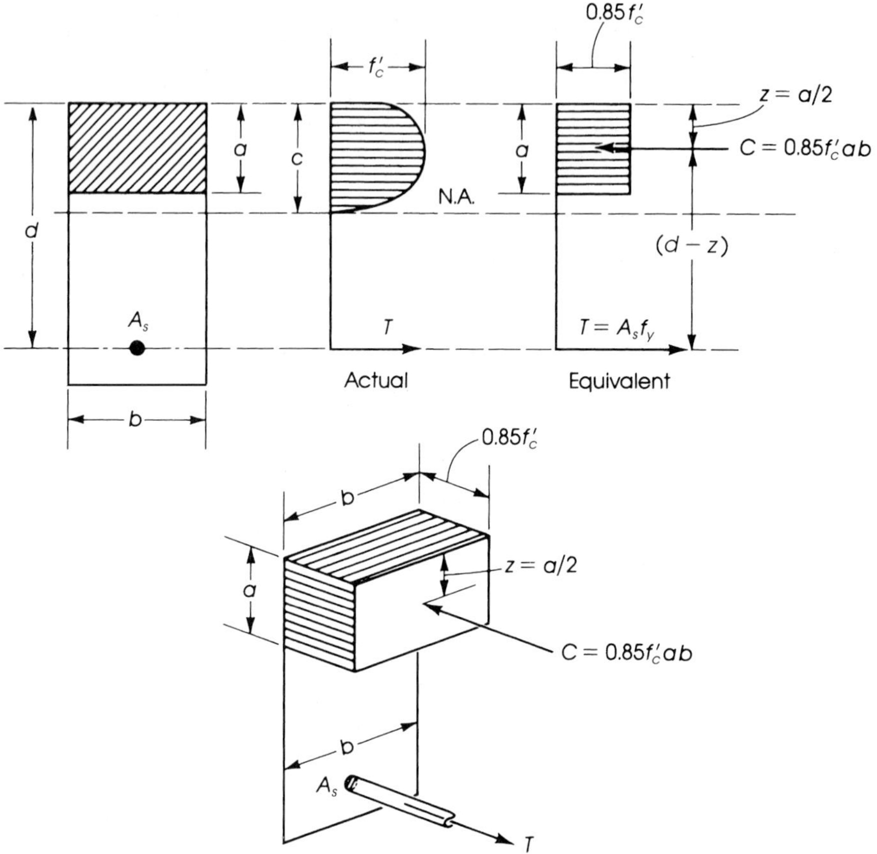

Figure 3.5 Actual and equivalent stress distributions at failure.

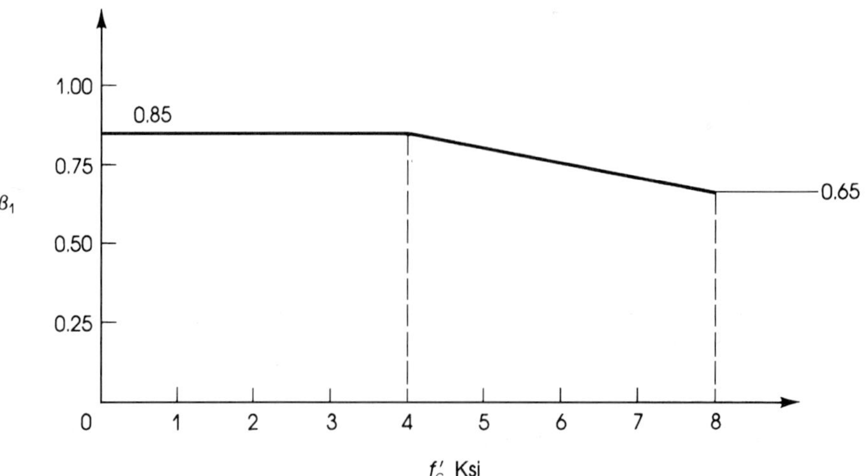

Figure 3.6 Values of β_1 for different compressive strengths of concrete, f'_c.

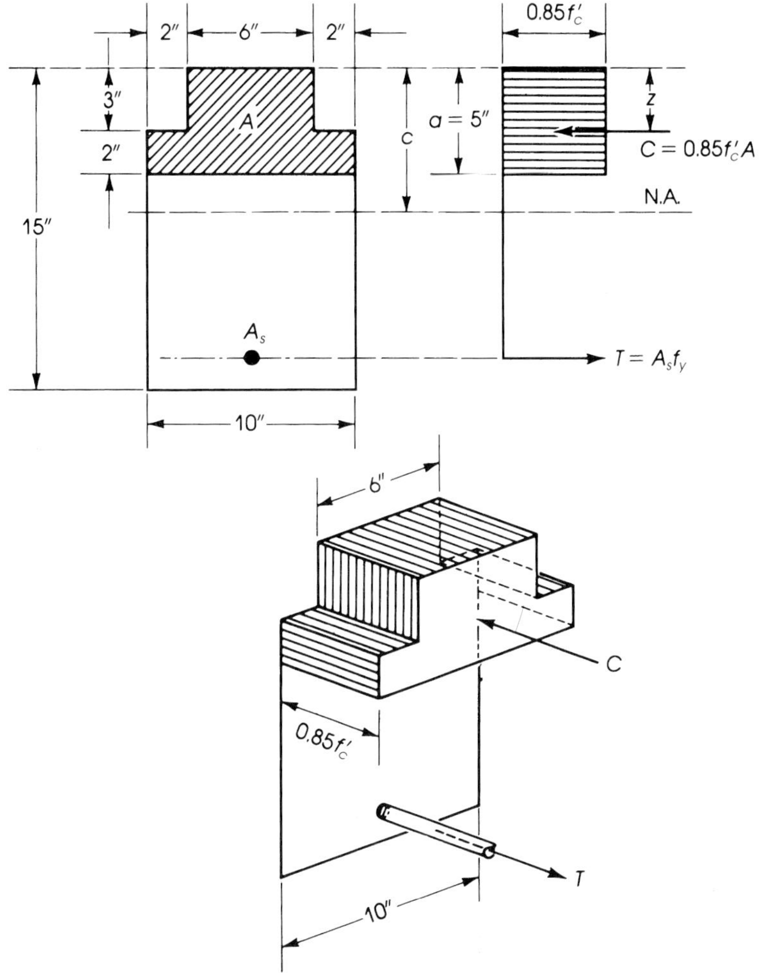

Figure 3.7 Ultimate forces in a nonrectangular section.

uniform and equals $0.85f'_c$, the resultant force C is located at the center of gravity of the compressive zone, which has a depth of a.

In this example z is calculated by taking moments about the top fibers:

$$z = \frac{(6 \times 3 \times \frac{3}{2}) + 10 \times 2(1 + 3)}{6 \times 3 + 10 \times 2} = \frac{107}{38} = 2.82 \text{ in.}$$

3.9 SINGLY REINFORCED RECTANGULAR SECTION IN BENDING

We explained previously that a balanced condition is achieved when steel yields at the same time as the concrete fails, and that failure usually happens suddenly. This implies that the yield strain in the steel is reached $(\varepsilon_y = f_y/E_s)$ and that the concrete has reached its maximum strain of 0.003. The percentage of reinforcement used to produce a balanced condition is called the *balanced steel ratio,* ρ_b. This value is equal to the area of steel, A_s, divided by the effective cross section, bd:

$$\rho_b = \frac{A_s(\text{balanced})}{bd} \tag{3.5}$$

where

b = width of the compression face of the member

d = distance from the extreme compression fiber to the centroid of the tension reinforcement

If a smaller percentage of reinforcement is used, the section is under-reinforced, and steel bars yield before the crushing of concrete, which avoids sudden failure. To ensure this, the ACI Code limits the maximum tension reinforcement to 0.75 of that producing a balanced condition at ultimate loading:

$$\rho_{\max} = 0.75\rho_b \tag{3.6}$$

If a greater percentage of reinforcement is used, the section is over-reinforced, and concrete fails suddenly before steel yields. This case is not permitted by the ACI Code.

Two basic equations for the analysis and design of structural members are the two equations of equilibrium that are valid for any load and any section:

1. The compression force should be equal to the tension force; otherwise a section will have linear displacement plus rotation:

$$C = T \tag{3.7}$$

2. The internal nominal ultimate bending moment M_n is equal to either the compressive force C multiplied by its arm or the tension force T multiplied by the same arm:

$$M_n = C(d - z) = T(d - z)$$
$$(M_u = \phi M_n \text{ after reduction by the factor } \phi) \tag{3.8}$$

The use of these equations can be explained by considering the case of a rectangular section with tension reinforcement (Figure 3.5). The section may be balanced, under-reinforced, or over-reinforced, depending on the percentage of steel reinforcement used.

3.9.1 The Balanced Section

Let us consider the case of a balanced section, which implies that at ultimate load the strain in concrete equals 0.003 and that of steel equals the yield stress divided by the modulus of elasticity of steel, f_y/E_s. This case is explained by the following steps.

Step 1: From the strain diagram of Figure 3.8,

$$\frac{c_b}{d - c_b} = \frac{0.003}{f_y/E_s}$$

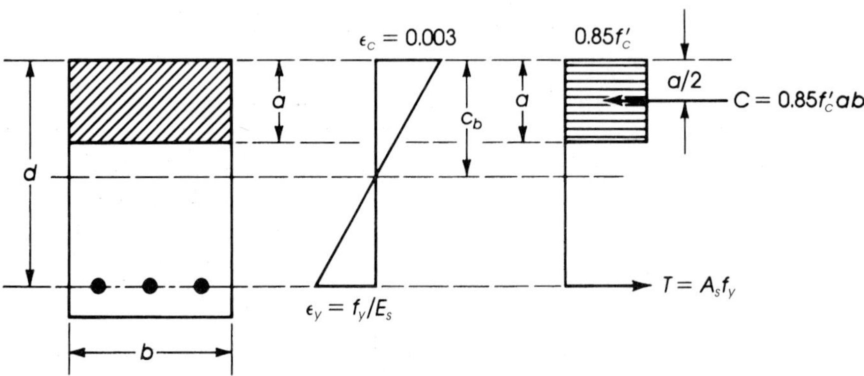

Figure 3.8 Rectangular balanced section.

From triangular relationships (where c_b is c for a balanced section) and by adding the numerator to the denominator,

$$\frac{c_b}{d} = \frac{0.003}{0.003 + f_y/E_s}$$

Substituting $E_s = 29 \times 10^3$ Ksi,

$$c_b = \left(\frac{87}{87 + f_y}\right) d \tag{3.9}$$

where f_y is in Ksi.

Step 2: From the equilibrium equation,

$$C = T \tag{3.10}$$

$$0.85f'_c ab = A_s f_y$$

$$a = \frac{A_s f_y}{0.85f'_c b} \tag{3.11}$$

Here, a is the depth of the compressive block, equal to $\beta_1 c$, where $\beta_1 = 0.85$ for $f'_c \leq 4000$ psi (27.6 MPa) and decreases linearly by 0.05 per 1000 psi (6.9 MPa) for higher concrete strengths (Figure 3.6). Because the balanced steel reinforcement ratio is used,

$$\rho_b = \frac{A_s(\text{balanced})}{bd} = \frac{A_{sb}}{bd} \tag{3.12}$$

and substituting the value of A_{sb} in equation (3.10),

$$0.85f'_c ab = f_y \rho_b bd$$

Therefore,

$$\rho_b = \frac{0.85f'_c}{df_y} a = \frac{0.85f'_c}{df_y}(\beta_1 c_b)$$

Substituting the value of c_b from equation (3.9), the general equation of the balanced steel ratio becomes

$$\rho_b = 0.85\beta_1 \frac{f'_c}{f_y}\left(\frac{87}{87 + f_y}\right) \tag{3.13}$$

where f'_c and f_y are in Ksi. The ACI Code limits the maximum percentage of reinforcement used to $0.75\rho_b$; therefore,

$$\rho_{\max} = 0.75\rho_b = 0.6375\beta_1 \frac{f'_c}{f_y}\left(\frac{87}{87 + f_y}\right) \tag{3.14}$$

For concrete strength of 4000 psi (28 N/mm^2) or less, $\beta_1 = 0.85$, and then

$$\rho_{\max} = 0.542 \frac{f'_c}{f_y}\left(\frac{87}{87 + f_y}\right) \tag{3.15}$$

Step 3: The internal ultimate nominal moment, M_n, is calculated by multiplying either C or T by the distance between them:

$$M_n = C(d - z) = T(d - z) \tag{3.8}$$

For a rectangular section the distance $z = a/2$ as the line of application of the force C lies at the center of gravity of the area ab, where

$$a = \frac{A_s f_y}{0.85 f'_c b}$$

$$M_n = C\left(d - \frac{a}{2}\right) = T\left(d - \frac{a}{2}\right)$$

For a balanced or an under-reinforced section, $T = A_s f_y$. Then

$$M_n = A_s f_y\left(d - \frac{a}{2}\right) \tag{3.16}$$

To get the usable ultimate moment ϕM_n, the previously calculated M_n must be reduced by the capacity-reduction factor, ϕ. In bending, $\phi = 0.9$.

$$\phi M_n = \phi A_s f_y\left(d - \frac{a}{2}\right) = \phi A_s f_y\left(d - \frac{A_s f_y}{1.7 f'_c b}\right) \tag{3.16a}$$

Equation (3.16a) can be written in terms of the steel percentage ρ:

$$\rho = \frac{A_s}{bd}, \qquad A_s = \rho bd$$

$$\phi M_n = \phi f_y \rho bd\left(d - \frac{\rho bd f_y}{1.7 f'_c b}\right) = \phi \rho f_y bd^2\left(1 - \frac{\rho f_y}{1.7 f'_c}\right) \tag{3.17}$$

Equation (3.17) can be written as

$$\phi M_n = R_u bd^2 \tag{3.18}$$

where

$$R_u = \phi \rho f_y\left(1 - \frac{\rho f_y}{1.7 f'_c}\right) \tag{3.19}$$

It is clear that R_u is a function of ρ for a given concrete strength and steel yield strength; R_u is maximum when ρ is maximum. Therefore, using ρ_{max},

$$(R_u)_{max} = \phi \rho_{max} f_y\left(1 - \frac{\rho_{max} f_y}{1.7 f'_c}\right) \tag{3.20}$$

Similarly,

$$a_{max} = \frac{A_{s\,max} f_y}{0.85 f'_c b} \tag{3.11}$$

The ratio of the equivalent compressive stress block depth a to the effective depth of the section d can be found from equation (3.10):

$$0.85 f'_c ab = \rho bd f_y \tag{3.21}$$

$$\frac{a}{d} = \frac{\rho f_y}{0.85 f'_c}$$

The maximum ratio (a_{max}/d) is obtained by substituting the value of ρ_{max} (equation (3.14)) in equation (3.21).

Example 3.1

For the section shown in Figure 3.9, calculate

a. The balanced steel reinforcement;

b. The maximum reinforcement area allowed by the ACI Code;

c. The position of the neutral axis and the depth of the equivalent compressive stress block for case (b).

Given: $f'_c = 4$ Ksi, $f_y = 60$ Ksi.

Solution

a. $$\rho_b = 0.85\beta_1 \frac{f'_c}{f_y}\left(\frac{87}{87 + f_y}\right) \tag{3.13}$$

Because $f'_c = 4000$ psi, $\beta_1 = 0.85$:

$$\rho_b = (0.85)^2\left(\frac{4}{60}\right)\left(\frac{87}{87 + 60}\right) = 0.0285$$

The area of steel reinforcement to provide a balanced condition is

$$A_{sb} = \rho_b bd = 0.0285 \times 16 \times 25.5 = 11.63 \text{ in.}^2$$

b. $\rho_{\max} = 0.75\rho_b = 0.75 \times 0.0285 = 0.02138$ or, from equation (3.15),

$$\rho_{\max} = 0.542 \times \frac{f'_c}{f_y} \times \frac{87}{87 + f_y}$$

$$= 0.542 \times \frac{4}{60} \times \frac{87}{147} = 0.02138$$

$$A_{s\max} = \rho_{\max} bd = 0.02138 \times 16 \times 25.5 = 8.72 \text{ in.}^2$$

c. The depth of the equivalent compressive block using $A_{s\max}$ is

$$a_{\max} = \frac{A_{s(\max)}f_y}{0.85 f'_c b} = \frac{8.72 \times 60}{0.85 \times 4 \times 16} = 9.62 \text{ in.}$$

The distance from the top fibers to the neutral axis is $c = a/\beta_1$, as defined by equation (3.11). Because $f'_c = 4000$ psi, $\beta_1 = 0.85$; thus,

$$c = \frac{9.62}{0.85} = 11.32 \text{ in.}$$

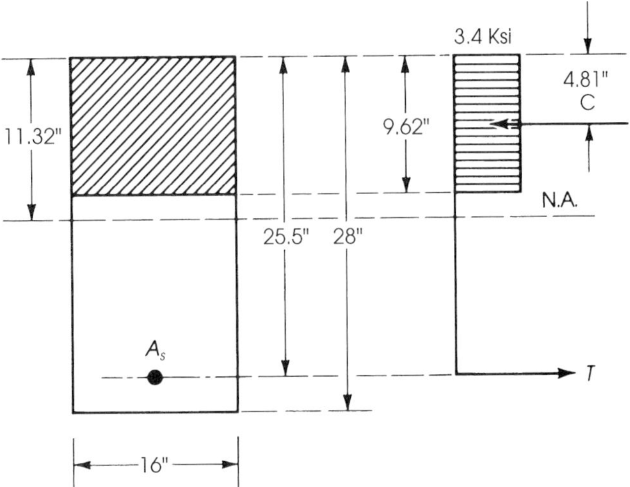

Figure 3.9 Example 3.1.

If the balanced steel area is used, then the position of the neutral axis will be lower than that shown, and c can be calculated by equation (3.9), which is valid only for a balanced section.

$$c_b = \left(\frac{87}{87 + 60}\right)25.5 = 15.1 \text{ in.}$$

Example 3.2

Determine the allowable ultimate moment and the position of the neutral axis of the rectangular section shown in Figure 3.10 if the reinforcement used is three no. 9 bars. Given: $f'_c = 3$ Ksi, $f_y = 60$ Ksi.

Solution

1. The area of three no. 9 bars is 3.0 in.^{2}

$$\rho = \frac{A_s}{bd} = \frac{3.0}{18 \times 12} = 0.01389$$

2. $$\rho_{\max} = 0.542\frac{f'_c}{f_y} \times \frac{87}{87 + f_y}$$

$$= 0.542 \times \frac{3}{60} \times \frac{87}{(87 + 60)} = 0.016 \quad (3.15)$$

$\rho \leq \rho_{\max} \leq \rho_b$

Therefore, the section is under-reinforced. Because $\rho < \rho_{\max}$, the ratio ρ should be used to calculate ϕM_n:

3. $$\phi M_n = \phi A_s f_y\left(d - \frac{a}{2}\right)$$

$$a = \frac{A_s f_y}{0.85 f'_c b} = \frac{3.0 \times 60}{0.85 \times 3 \times 12} = 5.88 \text{ in.} \quad (3.16)$$

$$\phi M_n = 0.9 \times 3.0 \times 60\left(18 - \frac{5.88}{2}\right) = 2440 \text{ K}\cdot\text{in.} = 203.3 \text{ K}\cdot\text{ft}$$

4. The distance of the neutral axis from the top fibers is $c = a/\beta_1$. Because $f'_c < 4000$ psi, $\beta_1 = 0.85$:

$$c = \frac{5.88}{0.85} = 6.92 \text{ in.}$$

In this example, the section is under-reinforced, which implies that the steel will yield before the concrete reaches its ultimate strain. A simple check can be made from the strain diagram (Figure 3.10). From similar triangles,

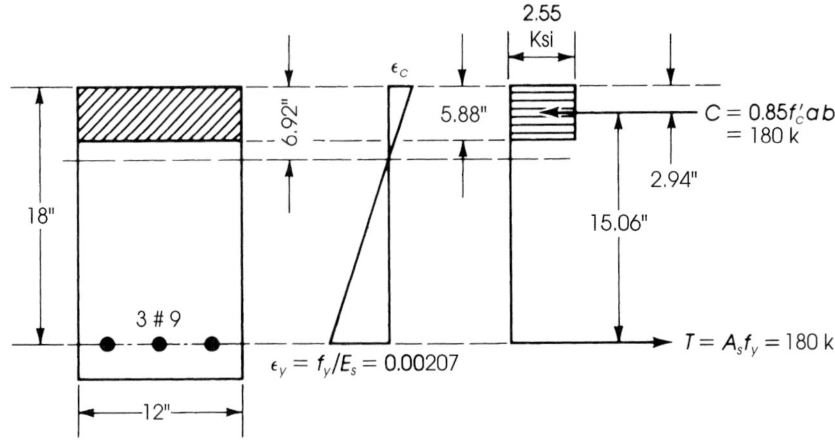

Figure 3.10 Example 3.2.

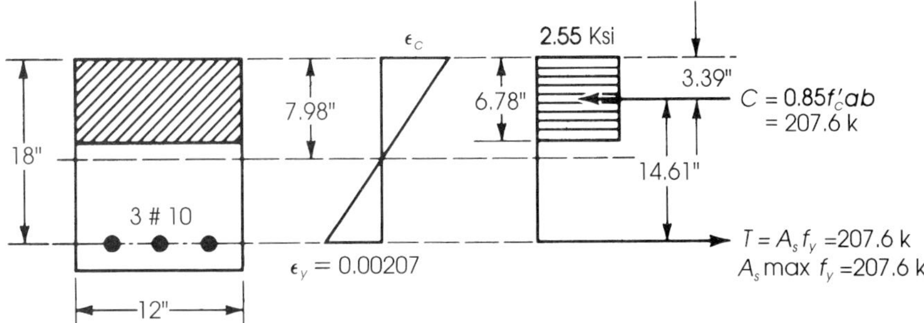

Figure 3.11 Example 3.2 (using $A_{s(\max)} = 3.46$ in.^{2}).

$$\frac{\varepsilon_c}{\varepsilon_y} = \frac{c}{(d - c)} \quad \text{and} \quad \varepsilon_y = \frac{f_y}{E_s} = \frac{60}{29000} = 0.00207$$

$$\varepsilon_c = \frac{6.92}{(18 - 6.92)} \times 0.00207 = 0.00129$$

which is much less than 0.003. Therefore, steel yields before concrete reaches its limiting strain of 0.003.

Discussion:

If three no. 10 bars are used (see Figure 3.11), then $A_s = 3.81$ and $\rho = 3.81/(12 \text{ in.} \times 18 \text{ in.}) = 0.0176$, which is greater than $\rho_{\max} = 0.016$ and less than the balanced ratio $\rho_b = 0.0214$. The section is still considered under-reinforced, but the steel ratio used is greater than that allowed by the ACI Code, as compared with the previous example, when the A_s of three no. 9 bars meets the ACI Code limit. The maximum A_s that can be used in this section to calculate its ultimate moment capacity is equal to $\rho_{\max} bd = 0.016(12)(18) = 3.46$ in.^{2}, and it produces a maximum allowable ultimate moment $\phi M_n = 227.4$ K · ft. Actually, the section can resist a higher ultimate moment, and collapse will occur due to the yield of steel, but little warning is to be expected.

If ρ is greater than the balanced ratio, then the section is considered over-reinforced, and the beam will collapse due to the crushing of concrete and without warning. The tension steel in this case will not yield. This type of section must be avoided in the design of concrete beams.

3.9.2 Under-Reinforced Sections

A section is under-reinforced when the steel reinforcement percentage is less than that of a balanced condition ($\rho < \rho_b$). The steel reaches its yield strength, f_y, before the crushing of concrete. The previous equations for ϕM_n and R_u (3.16), (3.17), and (3.18) are still valid.

At the moment when steel starts to yield, its strain equals f_y/E_s, and the strain in concrete, ε_c, is less than 0.003. Due to the increase of strain in steel after yielding, $\varepsilon_s > f_y/E_s$ (Figure 3.12), the deflection of the structural member will increase rapidly, causing excessive compression strain in the top fibers until concrete reaches its assumed maximum strain of 0.003. The steel bars will not fail because they undergo strain hardening; collapse will instead be the result of concrete crushing. For a given b, d, and ρ, the distance of the neutral axis from the top fibers in an under-reinforced section, c_u, is less than that for the balanced case; that is, $c_u < c_b$. Consequently, the depth of the equivalent stress block is less than that at the balanced condition: $a < a_b$. Then $C_u = (0.85f'_c b)a$ is less than $C_b = (0.85f'_c b)a_b$. If the steel percentage ρ is increased slightly to ρ_1 such that $\rho_1 < \rho_b$, then tension T will increase, which indicates that compression force $C = 0.85f'_c ab$ has increased ($C = T$ at all times), and consequently a is increased, indicating that the neutral axis has moved downward.

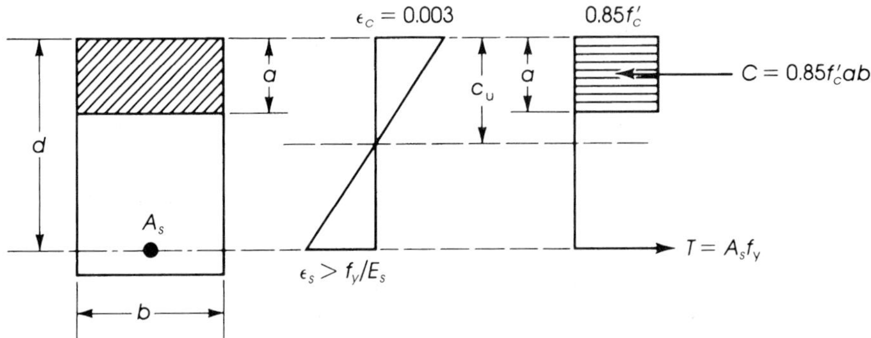

Figure 3.12 Under-reinforced rectangular section.

3.10 ADEQUACY OF SECTIONS

A given section is said to be *adequate* if the internal moment capacity of the section is equal to or greater than the externally applied ultimate moment M_u, or $\phi M_n \geq M_u$. The procedure can be summarized as follows:

1. Calculate the external applied ultimate moment, M_u.

$$M_u = 1.4\, M_D + 1.7\, M_L$$

2. Calculate ϕM_n for the basic singly reinforced section:
   a. Check that $\rho_{\min} \leq \rho \leq \rho_{\max}$.
   b. Calculate $a = A_s f_y/(0.85\, f'_c b)$.
   c. Calculate $\phi M_n = \phi A_s f_y(d - a/2)$.
3. If $\phi M_n \geq M_u$, then the section is adequate; see Example 3.3.

Example 3.3

An 8-ft-span cantilever beam has a rectangular section and reinforcement as shown in Figure 3.13. The beam carries a dead load, including its own weight, of 1.5 K/ft and a live load of 0.7 K/ft. Using $f'_c = 4$ Ksi and $f_y = 60$ Ksi, check if the beam is safe to carry the above loads.

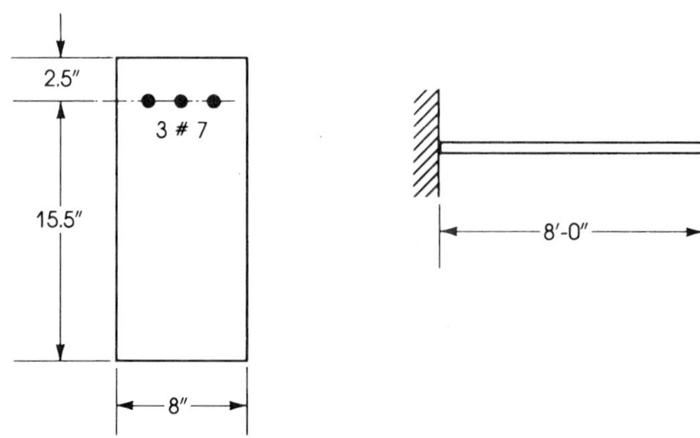

Figure 3.13 Example 3.3.

Solution

1. Calculate the external ultimate moment:

$$M_u = 1.4M_D + 1.7M_L$$
$$= 1.4\left(\frac{W_D L^2}{2}\right) + 1.7\left(\frac{W_L L^2}{2}\right)$$
$$= 1.4\left(\frac{1.5}{2} \times 8^2\right) + 1.7\left(\frac{0.7}{2} \times 8^2\right) = 105.3 \text{ K}\cdot\text{ft} = 1263 \text{ K}\cdot\text{in.}$$

2. Calculate the internal ultimate moment capacity:

$$A_s = 3 \times 0.6 = 1.8 \text{ in.}^2, \qquad \rho = \frac{1.8}{8 \times 15.5} = 0.0145 \tag{3.15}$$
$$\rho_{\max} = 0.542\frac{f'_c}{f_y} \times \frac{87}{87 + f_y} = 0.02138$$

Therefore, the section is under-reinforced; hence, use $\rho < \rho_{\max}$. Also, $\rho > \rho_{\min} = 200/f_y = 0.00333$.

3. $\phi M_n = \phi A_s f_y\left(d - \dfrac{A_s f_y}{1.7 f'_c b}\right)$

$$= 0.9 \times 1.8 \times 60\left(15.5 - \frac{1.8 \times 60}{1.7 \times 4 \times 8}\right) = 1314 \text{ K}\cdot\text{in.} = 109.5 \text{ K}\cdot\text{ft} \tag{3.16}$$

4. Or

$$R_u = \phi\rho f_y\left(1 - \frac{\rho f_y}{1.7 f'_c}\right)$$
$$= 0.9 \times 0.0145 \times 60\left(1 - \frac{0.0145 \times 60}{1.7 \times 4}\right) = 0.684 \text{ Ksi} \tag{3.19}$$
$$\phi M_n = R_u b d^2$$
$$= 0.684 \times 8 \times (15.5)^2 = 1314 \text{ K}\cdot\text{in.}$$

5. The internal ultimate moment of 1314 K · in. is greater than the external applied ultimate moment of 1263 K · in.; therefore, the section is adequate.

Example 3.4

A simply supported beam has a span of 20 ft. If the cross section of the beam is as shown in Figure 3.14, f'_c = 3 Ksi, and f_y = 60 Ksi, determine the allowable uniformly distributed service

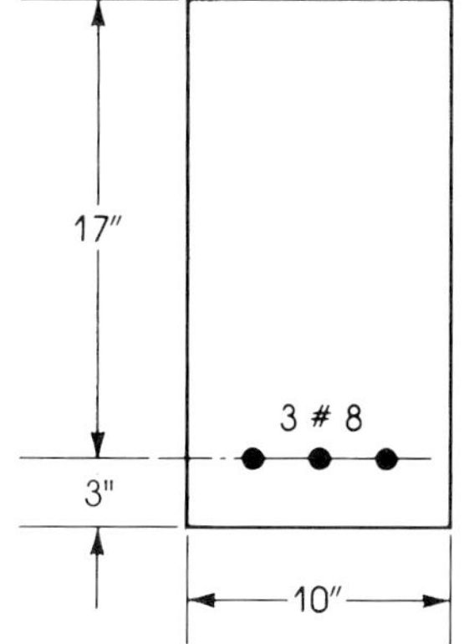

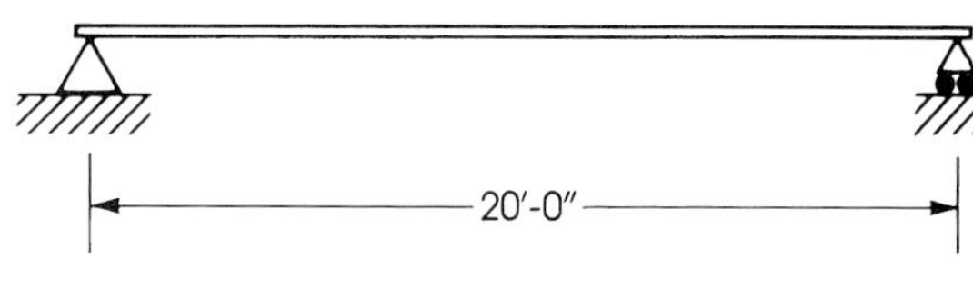

Figure 3.14 Example 3.4.

live load on the beam. Given: $b = 10$ in., $d = 17$ in., total depth $h = 20$ in., and reinforced with three no. 8 bars $(A_s = 2.37 \text{ in.}^2)$.

Solution

1. Determine the internal ultimate moment capacity:

$$\rho = \frac{A_s}{bd} = \frac{3 \times 0.79}{10 \times 17} = 0.0139$$

$$\rho_{\max} = 0.542 \frac{f'_c}{f_y} \times \frac{87}{87 + f_y} = 0.54\left(\frac{3}{60}\right)\left(\frac{87}{87 + 60}\right) = 0.016$$

$$\rho < \rho_{\max}, \qquad \text{therefore, use } \rho = 0.0139$$

Also, $\rho > \rho_{\min} = 200/f_y = 0.00333$.

2. $\phi M_n = \phi A_s f_y \left(d - \dfrac{A_s f_y}{1.7 f'_c b}\right)$

$$= 0.9 \times 2.37 \times 60\left(17 - \frac{2.37 \times 60}{1.7 \times 3 \times 10}\right) = 1819 \text{ K}\cdot\text{in.} = 151.6 \text{ K}\cdot\text{ft}$$

3. The dead load acting on the beam is self-weight (assumed):

$$W_D = \frac{10 \times 20}{144} \times 150 = 208 \text{ lb/ft} = 0.208 \text{ K/ft}$$

where 150 is the weight of reinforced concrete in pounds per cubic foot.

4. The external ultimate moment is

$$M_u = 1.4M_D + 1.7M_L$$

$$= 1.4\left(\frac{0.208}{8} \times 20^2\right) + 1.7\left(\frac{W_L}{8} \times 20^2\right) = 14.56 + 85W_L$$

where W_L = uniform service live load on the beam in K/ft.

5. Internal ultimate moment equals the external ultimate moment:

$$151.6 = 14.56 + 85W_L \quad \text{and} \quad W_L = 1.61 \text{ K/ft}$$

The allowable uniform service live load on the beam is 1.61 K/ft.

3.11 MINIMUM PERCENTAGE OF STEEL

If the external ultimate moment applied on a beam is very small and the dimensions of the section are specified (as is sometimes required architecturally) and are larger than needed to resist the external ultimate moment, the calculation may show that very small or no steel reinforcement is required. In this case, the maximum tensile stress due to bending moment may be equal to or less than the modulus of rupture of concrete: $f_r = 7.5\sqrt{f'_c}$. If no reinforcement is provided, sudden failure will be expected when the first crack occurs, thus giving no warning. The ACI Code, Section 10.5.1, specifies a minimum steel area A_s,

$$A_{s\,\min} = \left(\frac{3\sqrt{f'_c}}{f_y}\right) b_w d \geq \left(\frac{200}{f_y}\right) b_w d$$

or the minimum steel ratio, $\rho_{\min} = (3\sqrt{f'_c}/f_y) \geq 200/f_y$, where the units of f'_c and f_y are in psi. The first term of the preceding equation was specified to accommodate a concrete strength higher than 5 Ksi. The two minimum ratios are equal when $f'_c = 4440$ psi. This indicates that

$$\rho_{\min} = 200/f_y \qquad \text{when } f'_c < 4500 \text{ psi}$$

$$\rho_{\min} = \frac{3\sqrt{f'_c}}{f_y} \qquad \text{when } f'_c \geq 4500 \text{ psi}$$

For example, if $f_y = 60$ Ksi, $\rho_{\min} = 0.00333$ when $f'_c < 4500$ psi, whereas $\rho_{\min} = 0.00353$ when $f'_c = 5000$ psi and 0.00387 when $f'_c = 6000$ psi.

In the case of a rectangular section, use $b = b_w$ in the preceding expressions. For statically determinate T-sections with the flange in tension, as in the case of cantilever beams, the value of $A_{s\,\min}$ should be equal or greater than the *smaller* of (a) and (b):

(a) $$A_{s\,\min} = \left(\frac{6\sqrt{f'_c}}{f_y}\right)b_w d$$

(b) $$A_{s\,\min} = \left(\frac{3\sqrt{f'_c}}{f_y}\right)bd \geq \left(\frac{200}{f_y}\right)bd$$

where b_w and b are the width of the beam web and flange, respectively, and f'_c and f_y are in psi. For example, if $b = 48$ in., $b_w = 16$ in., $f'_c = 4000$ psi, and $f_y = 60{,}000$ psi, then $A_{s\,\min} = 2.02$ in.^{2} in (a) controls, which is smaller than the value of $A_{s\,\min}$ in (b) (3.2 in.^{2}).

Example 3.5

Check the design adequacy of the section shown in Figure 3.15 to resist a moment $M_u = 30$ K·ft, using $f'_c = 3$ Ksi and $f_y = 40$ Ksi.

Solution

1. Check ρ provided in the section:

$$\rho = \frac{A_s}{bd} = \frac{3 \times 0.2}{10 \times 18} = 0.00333$$

2. Check $\rho_{\min}$ required according to the ACI Code:

$$\rho_{\min} = \frac{200}{f_y} = 0.005 > \rho = 0.00333$$

Therefore, use $\rho = \rho_{\min} = 0.005 < \rho_{\max} = 0.0278$:

$$A_{s\,\min} = \rho_{\min} bd = 0.005 \times 10 \times 18 = 0.90 \text{ in.}^2$$

Use three no. 5 bars $(A_s = 0.91 \text{ in.}^2)$, because three no. 4 bars are less than the minimum specified by the code.

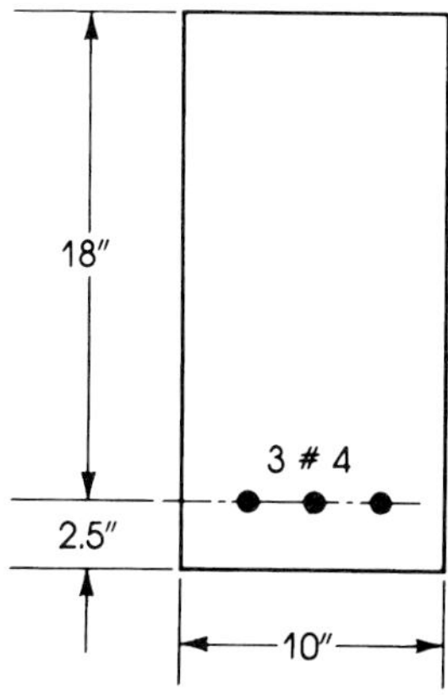

Figure 3.15 Example 3.5.

3. Check ultimate moment capacity: $\phi M_n = \phi A_s f_y(d - a/2)$.

$$a = \frac{A_s f_y}{0.85 f'_c b} = \frac{0.91 \times 40}{0.85 \times 3 \times 10} = 1.43 \text{ in.}$$

$$\phi M_n = 0.9 \times 0.91 \times 40\left(18 - \frac{1.43}{2}\right) = 566 \text{ K}\cdot\text{in.} = 47.2 \text{ K}\cdot\text{ft}$$

4. An alternative solution, according to the ACI Code, Section 10.5: For three no. 4 bars, $A_s = 0.6 \text{ in.}^2$:

$$a = \frac{A_s f_y}{0.85 f'_c b} = \frac{0.6 \times 40}{0.85 \times 3 \times 10} = 0.94 \text{ in.}$$

$$\phi M_n = \frac{0.9}{12} \times 0.6 \times 40\left(18 - \frac{0.94}{2}\right) = 31.55 \text{ K}\cdot\text{ft}$$

$$A_s \text{ required for } 30 \text{ K}\cdot\text{ft} = \frac{30}{31.55} \times 0.6 = 0.57 \text{ in.}^2$$

The minimum A_s required according to the ACI Code, Section 10.5.2, is at least one-third greater than 0.57 in.^2:

$$\text{Minimum } A_s \text{ required} = 1.33 \times 0.57 = 0.76 \text{ in.}^2$$

which exceeds the 0.6 in.^2 provided by the no. 4 bars. Use three no. 5 bars, because $A_s = 0.91 \text{ in.}^2$ is greater than the 0.76 in.^2 required.

3.12 BUNDLED BARS

When the design of a section requires the use of a large amount of steel—for example, when ρ_{max} is used—it may be difficult to fit all bars within the cross section. The ACI Code, Section 7.6.6, allows the use of parallel bars placed in a bundled form of two, three, or four bars, as shown in Figure 3.16. Up to four bars (no. 11 or smaller) can be bundled when they are enclosed by stirrups.

The same bundled bars can be used in columns, provided that they are enclosed by ties. All bundled bars may be treated as a single bar for checking the spacing and concrete cover requirements. The single bar diameter shall be derived from the equivalent total area of the bundled bars.

Summary: Singly Reinforced Rectangular Section

The procedure for determining the allowable ultimate moment capacity of a singly reinforced rectangular section according to the ACI Code limitations can be summarized as follows:

1. Calculate the steel ratio in the section, $\rho = A_s/bd$.
2. Calculate

$$\rho_{max} = 0.75\rho_b = 0.75(0.85)\beta_1\left(\frac{f'_c}{f_y}\right)\left(\frac{87}{87 + f_y}\right)$$

Figure 3.16 Bundled bar arrangement.

Also, calculate $\rho_{min} = 200/f_y$ when $f'_c < 4500$ psi (f'_c and f_y are in psi units) and $\rho_{min} = 3\sqrt{f'_c}/f_y$ when $f'_c \geq 4500$ psi.

3. If $\rho_{min} \leq \rho \leq \rho_{max}$, then the section meets the ACI Code limitations. If $\rho \leq \rho_{min}$, the section is not acceptable (unless a steel ratio $\rho \geq \rho_{min}$ is used). If $\rho \geq \rho_{max}$, only ρ_{max} can be used to calculate the allowable ultimate moment capacity of the section. It is not recommended to use $\rho \geq \rho_{max}$ in singly reinforced concrete sections.

4. Calculate $a = \dfrac{A_s f_y}{0.85 f'_c b}$.

5. Calculate $\phi M_n = \phi A_s f_y \left(d - \dfrac{a}{2}\right)$.

Flow charts representing this section and other sections are given in Chapter 21.

3.13 RECTANGULAR SECTIONS WITH COMPRESSION REINFORCEMENT

In concrete sections proportioned to resist the bending moments resulting from external loading on a structural member, the internal moment is equal to or greater than the external moment. But a concrete section of a given width and effective depth has a maximum capacity when $\rho_{max}(=0.75\,\rho_b)$ is used. If external moment is greater than the internal moment capacity, more compressive and tensile reinforcement must be added.

Compression reinforcement is used when a section is limited to specific dimensions due to architectural reasons, such as a need for limited headroom in multistory buildings. Another advantage of compressive reinforcement is that long-time deflection is reduced, as is explained in Chapter 6. A third use of bars in the compressive zone is to hold stirrups, which are used to resist shear forces.

Two cases of doubly reinforced concrete sections will be considered, depending upon whether compression steel yields or does not yield.

3.13.1 When Compression Steel Yields

Internal moment can be divided into two moments, as shown in Figure 3.17. M_{u1} is the moment produced by the concrete compressive force and an equivalent tension force in steel A_{s1}, acting as a basic section. M_{u2} is the additional moment produced by the compressive force in compression steel A'_s and the tension force in the additional tensile steel A_{s2}, acting as a steel section.

The moment M_{u1} is that of a singly reinforced concrete basic section,

$$T_1 = C_c \tag{3.22}$$

$$A_{s1} f_y = C_c = 0.85 f'_c a b \tag{3.23}$$

$$a = \frac{A_{s1} f_y}{0.85 f'_c b} \tag{3.24}$$

$$M_{u1} = \phi A_{s1} f_y \left(d - \frac{a}{2}\right) \tag{3.25}$$

The restriction for M_{u1} is that $\rho_1 < A_{s1}/bd$ shall be equal to or less than $\rho_{max}(=0.75\,\rho_b)$ for singly reinforced concrete basic sections given in equation (3.14):

$$\rho_1 \leq \rho_{max} = 0.6375\beta_1 \left(\frac{f'_c}{f_y}\right)\left(\frac{87}{87 + f_y}\right) \tag{3.26}$$

where f_y is in Ksi.

Considering the moment M_{u2} and assuming that the compression steel designated as A'_s yields,

$$M_{u2} = \phi A_{s2} f_y (d - d') \tag{3.27a}$$

$$M_{u2} = \phi A'_s f_y (d - d') \tag{3.27b}$$

In this case $A_{s2} = A'_s$, producing equal and opposite forces, as shown in Figure 3.17. The total resisting moment M_u is then the sum of the two moments M_{u1} and M_{u2}:

$$\phi M_n = M_{u1} + M_{u2} = \phi\left[A_{s1} f_y \left(d - \frac{a}{2}\right) + A'_s f_y (d - d')\right] \tag{3.28}$$

The total steel reinforcement used in tension is the sum of the two steel amounts A_{s1} and A_{s2}. Therefore,

$$A_s = A_{s1} + A_{s2} = A_{s1} + A'_s \tag{3.29}$$

and

$$A_{s1} = A_s - A'_s$$

Then, substituting $(A_s - A'_s)$ for A_{s1} in equations (3.24), (3.25), and (3.28),

$$a = \frac{(A_s - A'_s) f_y}{0.85 f'_c b} \tag{3.30}$$

$$\phi M_n = \phi\left[(A_s - A'_s) f_y \left(d - \frac{a}{2}\right) + A'_s f_y (d - d')\right] \tag{3.31}$$

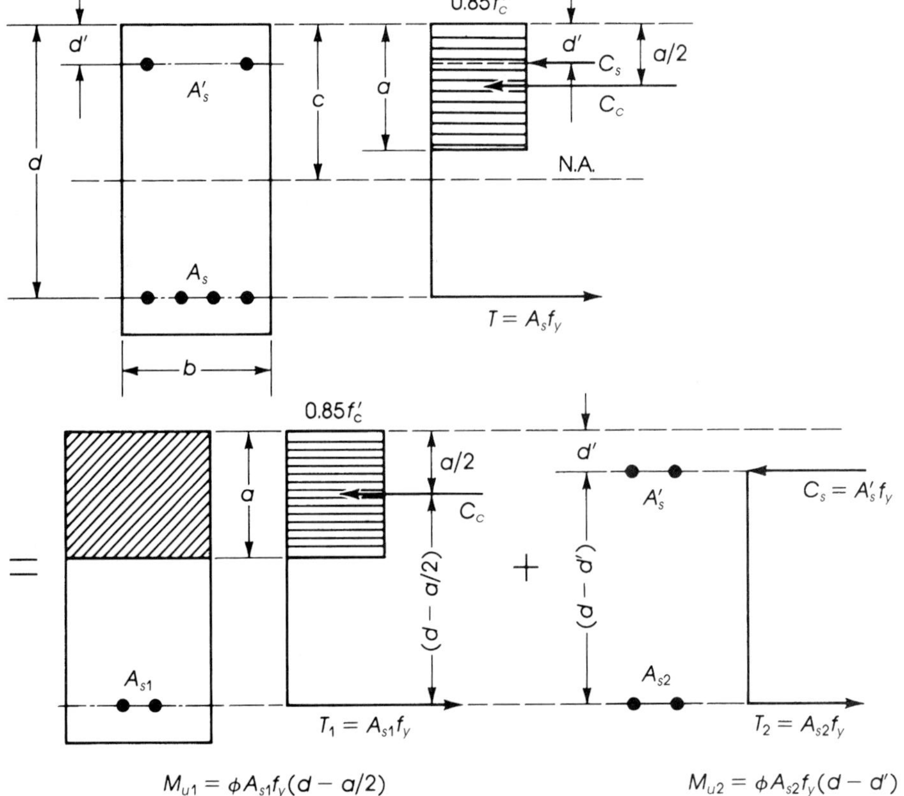

Figure 3.17 Rectangular section with compression reinforcement.

and

$$(\rho - \rho') \leq \rho_{\max} = 0.6375\beta_1\left(\frac{f'_c}{f_y}\right)\left(\frac{87}{87 + f_y}\right) \tag{3.32}$$

Equation (3.32) must be fulfilled in doubly reinforced concrete sections, which indicates that the difference between total tension steel and the compression steel should never exceed the maximum steel allowed for singly reinforced concrete sections. Failure due to the yielding of the total tensile steel will then be expected, and sudden failure of concrete is avoided.

In the compression zone, the force in the compression steel is $C_s = A'_s(f_y - 0.85f'_c)$, taking into account the area of concrete displaced by A'_s. In this case,

$$T = A_s f_y = C_c + C_s = 0.85f'_c ab + A'_s(f_y - 0.85f'_c)$$

$$A_s f_y - A'_s f_y + 0.85f'_c A'_s = 0.85f'_c ab = C_c = A_{s1} f_y \qquad \text{(for the basic section)}$$

Dividing by (bdf_y),

$$\rho - \rho'\left(1 - 0.85\frac{f'_c}{f_y}\right) = \rho_1, \qquad \text{where } \rho_1 = \frac{A_{s1}}{bd} \leq \rho_{\max}$$

Therefore,

$$\rho - \rho'\left(1 - 0.85\frac{f'_c}{f_y}\right) \leq \rho_{\max} = 0.6375\beta_1\left(\frac{f'_c}{f_y}\right)\left(\frac{87}{87 + f_y}\right) \tag{3.33}$$

Although equation (3.33) is more accurate than equation (3.32), it is quite practical to use both equations to check the condition for maximum steel ratio in rectangular sections when compression steel yields.

For example, if $f'_c = 3$ Ksi and $f_y = 60$ Ksi, equation (3.33) becomes $(\rho - 0.9575\rho') \leq 0.016$; if $f'_c = 4$ Ksi and $f_y = 60$ Ksi, then $(\rho - 0.9575\rho') \leq 0.02138$.

The maximum total tensile steel ratio ρ that can be used in a rectangular section when compression steel yields is as follows:

$$\text{Max } \rho = (\rho_{\max} + \rho') \tag{3.34}$$

where $\rho_{\max}$ is maximum tensile steel ratio for the basic singly reinforced concrete section. This means that maximum total tensile steel area that can be used in a rectangular section when compression steel yields is

$$\text{Max } A_s = bd(\rho_{\max} + \rho') \tag{3.34a}$$

In the preceding equations, it is assumed that compression steel yields. To investigate this condition, refer to the strain diagram in Figure 3.18. If compression steel yields, then

$$\varepsilon'_s \geq \varepsilon_y = \frac{f_y}{E_s}$$

From the two triangles above the neutral axis, substitute $E_s = 29{,}000$ Ksi and let f_y be in Ksi. Then

$$\frac{c}{d'} = \frac{0.003}{0.003 - f_y/E_s} = \frac{87}{87 - f_y}$$

$$c = \left(\frac{87}{87 - f_y}\right)d' \tag{3.35}$$

From equation (3.23),

$$A_{s1} f_y = 0.85f'_c ab \tag{3.23}$$

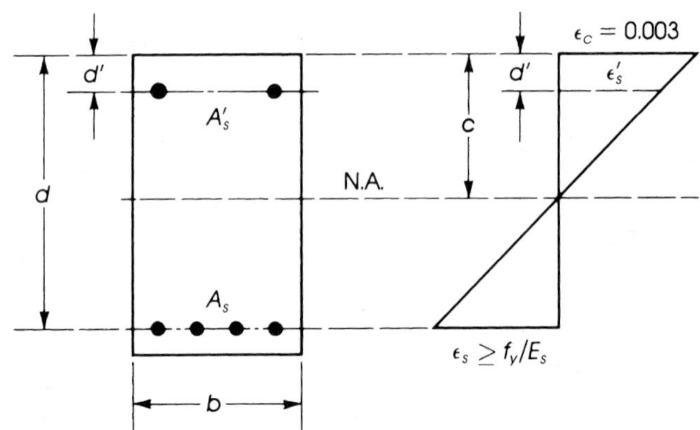

Figure 3.18 Strain diagram in doubly reinforced section.

but

$$A_{s1} = A_s - A'_s \quad \text{and} \quad \rho_1 = (\rho - \rho')$$

Therefore, equation (3.23) becomes $(A_s - A'_s)f_y = 0.85f'_c ab$:

$$(\rho - \rho')bdf_y = 0.85f'_c ab$$

$$(\rho - \rho') = 0.85\left(\frac{f'_c}{f_y}\right)\left(\frac{a}{d}\right)$$

Also,

$$a = \beta_1 c = \beta_1\left(\frac{87}{87 - f_y}\right)d'$$

Therefore,

$$(\rho - \rho') = 0.85\beta_1\left(\frac{f'_c}{f_y}\right)\left(\frac{d'}{d}\right)\left(\frac{87}{87 - f_y}\right) = K \tag{3.36}$$

The quantity $(\rho - \rho')$ is the steel ratio, or $(A_s - A'_s)/bd = A_{s1}/bd = \rho_1$ for the singly reinforced basic section.

If $(\rho - \rho')$ is greater than the value of the right-hand side in equation (3.36), then compression steel will also yield. In Figure 3.19 we can see that if A_{s1} is increased, T_1 and, consequently, C_1 will be greater and the neutral axis will shift downward, increasing the strain in the compression steel and ensuring its yield condition. If the tension steel used (A_{s1}) is less than the right-hand side of equation (3.36), then T_1 and C_1 will consequently be smaller, and the strain in compression steel ε'_s will be less than or equal to ε_y, because the neutral axis will shift upward, as shown in Figure (3.19(c)), and compression steel will not yield.

Therefore, equation (3.36) can be written

$$(\rho - \rho') \geq 0.85\beta_1 \frac{f'_c}{f_y} \times \frac{d'}{d} \times \frac{87}{87 - f_y} = K \tag{3.36a}$$

where f_y is in Ksi, and this is the condition for compression steel to yield.

For example, the values of K for different values of f'_c and f_y are as follows (Table 3.1).

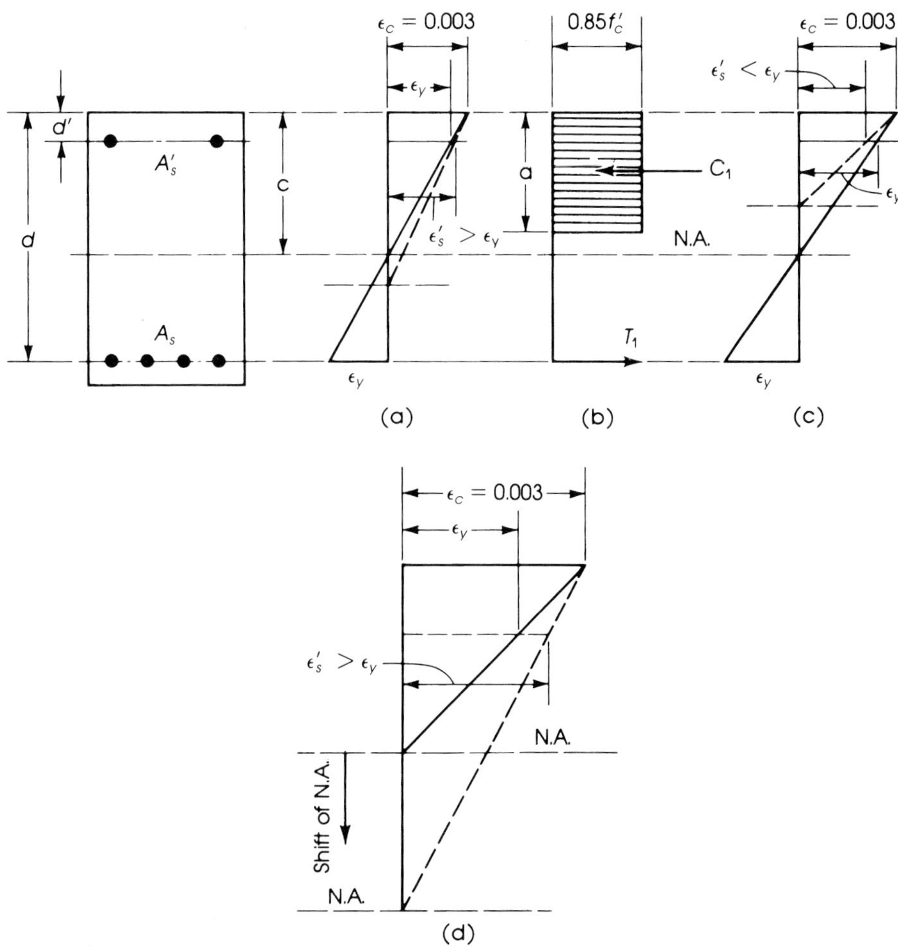

Figure 3.19 Yielding and nonyielding cases of compression reinforcement. Diagram (d), a close-up of (a), shows how the neutral axis responds to an increase in A_{s1}.

Table 3.1

f'_c (Ksi)	f_y (Ksi)	K	K (for d' = 2.5 in.)
3	40	$0.1003d'/d$	$0.251/d$
	60	$0.1164d'/d$	$0.291/d$
4	60	$0.1552d'/d$	$0.388/d$
5	60	$0.1826d'/d$	$0.456/d$

Example 3.6

A rectangular beam has a width of 12 in. and an effective depth, d = 21.5 in., to the centroid of tension steel bars. Tension reinforcement consists of six no. 9 bars in two rows; compression reinforcement consists of two no. 7 bars placed as shown in Figure 3.20. Calculate the ultimate moment capacity of the beam if f'_c = 4 Ksi and f_y = 60 Ksi.

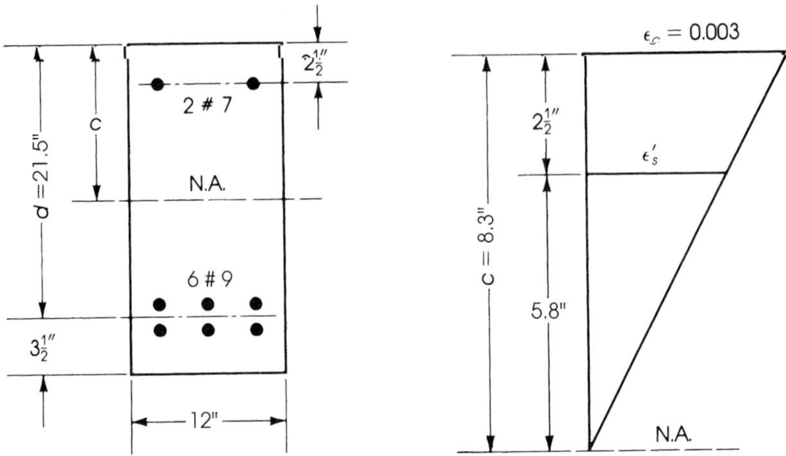

Figure 3.20 Example 3.6.

Solution

1. Check if compression steel yields:

$$A_s = 6.0 \text{ in.}^2 \qquad \rho = \frac{A_s}{bd} = \frac{6.0}{12 \times 21.5} = 0.02325$$

$$A_s' = 1.2 \text{ in.}^2 \qquad \rho' = \frac{A_s'}{bd} = \frac{1.2}{12 \times 21.5} = 0.00465$$

$$A_s - A_s' = 4.8 \text{ in.}^2, \qquad \rho - \rho' = 0.0186$$

For compression steel to yield,

$$(\rho - \rho') \geq 0.85\,\beta_1 \frac{f_c'}{f_y} \times \frac{d'}{d} \times \frac{87}{87 - f_y} = \text{K}$$

β_1 is 0.85 because $f_c' = 4000$ psi; therefore,

$$K = (0.85)^2\left(\frac{4}{60}\right)\left(\frac{2.5}{21.5}\right)\left(\frac{87}{87 - 60}\right) = 0.01805$$

$$(\rho - \rho') = 0.0186 > 0.01805$$

Therefore, compression steel yields.

2. Check that $(\rho - \rho') \leq \rho_{\max}$ (equation (3.15)):

$$\rho_{\max} = 0.542\frac{f_c'}{f_y}\left(\frac{87}{87 + f_y}\right) = 0.542\left(\frac{4}{60}\right)\left(\frac{87}{147}\right) = 0.02138$$

$$(\rho - \rho') = 0.0186 < \rho_{\max}$$

3. ϕM_n can be calculated by equation (3.31):

$$\phi M_n = \phi\left[(A_s - A_s')f_y\left(d - \frac{a}{2}\right) + A_s'f_y(d - d')\right]$$

$$a = \frac{(A_s - A_s')f_y}{0.85 f_c' b} = \frac{4.8 \times 60}{0.85 \times 4 \times 12} = 7.06 \text{ in.}$$

$$\phi M_n = (0.9)\left[4.8 \times 60\left(21.5 - \frac{7.06}{2}\right) + 1.2 \times 60(21.5 - 2.5)\right]$$

$$= 5889 \text{ K}\cdot\text{in.} = 490.8 \text{ K}\cdot\text{ft}$$

4. An alternative approach for checking if compression steel yields can be made as follows: From the strain diagram,

$$c = \frac{a}{0.85} = \frac{7.06}{0.85} = 8.3 \text{ in.}$$

$$\varepsilon_s' = \frac{5.8}{8.3} \times 0.003 = 0.0021, \qquad \varepsilon_y = \frac{f_y}{E_s} = \frac{60}{29{,}000} = 0.00207$$

Because ε_s' exceeds ε_y, compression steel yields.

5. The maximum total tension steel allowed in this section, max A_s, is equal to

$$\text{Max } A_s = bd(\rho_{\max} + \rho') = 12 \times 21.5(0.02138 + 0.00465)$$
$$= 6.7 \text{ in.}^2 > A_s = 6.0 \text{ in.}^2 \quad \text{used in the section.}$$

3.13.2 When Compression Steel Does Not Yield

As was explained earlier, if

$$(\rho - \rho') < \left(0.85\beta_1 \times \frac{f_c'}{f_y} \times \frac{d'}{d} \times \frac{87}{87 - f_y}\right) = K \tag{3.37}$$

then compression steel does not yield. This indicates that if $(\rho - \rho') < K$, the tension steel will yield before concrete can reach its maximum strain of 0.003, and the strain in compression steel, ε_s', will not reach ε_y at failure (Figure 3.19). Yielding of compression steel will also depend on its position relative to the extreme compressive fibers d'. A higher ratio of d'/c will decrease the strain in the compressive steel, ε_s', as it places compression steel A_s' nearer to the neutral axis.

If compression steel does not yield, a general solution can be performed by analysis based on statics. Also, a solution can be made as follows: Referring to Figures 3.17 and 3.18,

$$\varepsilon_s' = 0.003\left(\frac{c - d'}{c}\right), \quad \text{and} \quad f_s' = E_s\varepsilon_s' = 29{,}000(0.003)\left(\frac{c - d'}{c}\right) = 87\left(\frac{c - d'}{c}\right)$$

Let $C_c = 0.85f_c'\beta_1 cb$:

$$C_s = A_s'(f_s' - 0.85f_c') = A_s'\left[87\frac{(c - d')}{c} - 0.85f_c'\right]$$

Because $T = A_s f_y = C_c + C_s$, then

$$A_s f_y = (0.85f_c'\beta_1 cb) + A_s'\left[87\left(\frac{c - d'}{c}\right) - 0.85f_c'\right]$$

Rearranging terms yields

$$(0.85f_c'\beta_1 b)c^2 + [(87A_s') - (0.85f_c'A_s') - A_s f_y]c - 87A_s'd' = 0$$

This is similar to $A_1c^2 + A_2c + A_3 = 0$, where

$$A_1 = 0.85f_c'\beta_1 b$$
$$A_2 = A_s'(87 - 0.85f_c') - A_s f_y$$
$$A_3 = -87A_s'd'$$

Solve for c:

$$c = \frac{1}{2A_1}\left[-A_2 \pm \sqrt{A_2^2 - 4A_1A_3}\right] \tag{3.38}$$

Once c is determined, then calculate f'_s, a, C_c, and C_s. $f'_s = 87[(c - d')/c]$; $a = \beta_1 c$; $C_c = 0.85 f'_c ab$; and $C_s = A'_s(f'_s - 0.85 f'_c)$.

$$\phi M_n = \phi\left[C_c\left(d - \frac{a}{2}\right) + C_s(d - d')\right] \tag{3.39}$$

When compression steel does not yield, $f'_s < f_y$, and the maximum total tensile steel reinforcement needed for a rectangular section is

$$\text{Max } A_s = (0.75\rho_b)bd + A'_s\frac{f'_s}{f_y} = bd\left(\rho_{\max} + \frac{\rho' f'_s}{f_y}\right) \tag{3.40}$$

Using steel ratios and dividing by bd:

$$\text{Max } \rho = \frac{\max A_s}{bd} \leq \rho_{\max} + \rho'\frac{f'_s}{f_y} \tag{3.41}$$

or

$$\left(\rho - \rho'\frac{f'_s}{f_y}\right) \leq \rho_{\max} \tag{3.42}$$

where $\rho_{\max}$ is the maximum steel ratio for the basic singly reinforced rectangular section, equation (3.14).

In this case,

$$a = \frac{A_s f_y - A'_s f'_s}{0.85 f'_c b} \tag{3.43}$$

$$\phi M_n = \phi\left[(A_s f_y - A'_s f'_s)\left(d - \frac{a}{2}\right) + A'_s f'_s(d - d')\right] \tag{3.44}$$

In summary, the procedure for analyzing sections with compression steel is as follows:

1. Calculate ρ, ρ', and $(\rho - \rho')$. Also, calculate $\rho_{\max}$ and $\rho_{\min}$.

2. Calculate

$$K = 0.85\beta_1\left(\frac{f'_c}{f_y}\right)\left(\frac{87}{87 - f_y}\right)\left(\frac{d'}{d}\right)$$

Use Ksi units.

3. If $(\rho - \rho') \geq K$, then compression steel yields, and $f'_s = f_y$; if $(\rho - \rho') < K$, then compression steel does not yield, and $f'_s < f_y$.

4. If compression steel yields, then

a. Check that $\rho_{\max} \geq (\rho - \rho') \geq \rho_{\min}$;

b. Calculate $a = \dfrac{(A_s - A'_s)f_y}{0.85 f'_c b}$;

c. Calculate $\phi M_n = \phi\left[(A_s - A'_s)f_y\left(d - \frac{a}{2}\right) + A'_s f_y(d - d')\right]$;

d. The maximum A_s that can be used in the section is

$$\text{Max } A_s = bd(\rho_{\max} + \rho') \geq A_s \qquad \text{(used)}$$

5. If compression steel does *not* yield, then

a. Calculate the distance to the neutral axis c by using analysis (Example 3.7) or by using the quadratic equation (3.38);

b. Calculate $f'_s = 87\left(\frac{c - d'}{c}\right)$(Ksi);

c. Check that $(\rho - \rho f'_s/f_y) \le \rho_{max}$ or max A_s that can be used in the section is greater than or equal to the A_s used.

$$\text{Max } A_s = bd\left(\rho_{max} + \frac{\rho' f'_s}{f_y}\right) \ge A_s \qquad \text{(used)}$$

d. Calculate $a = \frac{(A_s f_y - A'_s f'_s)}{0.85 f'_c b}$; or $a = \beta_1 c$

e. Calculate $\phi M_n = \phi\left[(A_s f_y - A'_s f'_s)\left(d - \frac{a}{2}\right) + A'_s f'_s (d - d')\right]$.

For flow charts, refer to Chapter 21.

Example 3.7

Determine the ultimate moment capacity of the section shown in Figure 3.21 using $f'_c = 5$ Ksi, $f_y = 60$ Ksi, $A'_s = 2.37$ in.^{2} (three no. 8 bars), and $A_s = 7.59$ in.^{2} (six no. 10 bars).

Solution

1. Calculate ρ and ρ':

$$\rho = \frac{A_s}{bd} = \frac{7.59}{14 \times 22.5} = 0.0241 \qquad \rho' = \frac{A'_s}{bd} = \frac{2.37}{14 \times 22.5} = 0.00754$$

$$\rho - \rho' = 0.01656$$

2. Apply equation (3.37), assuming $\beta_1 = 0.8$ for $f'_c = 5000$ psi.

$$K = 0.85\beta_1 \times \frac{f'_c}{f_y} \times \frac{d'}{d} \times \frac{87}{87 - f_y} = 0.85 \times 0.8\left(\frac{5}{60}\right)\left(\frac{2.5}{22.5}\right)\left(\frac{87}{87 - 60}\right) = 0.0203$$

(or from Table 3.1, $K = 0.456/d = 0.0203$).

$$(\rho - \rho') = 0.01656 < 0.0203$$

Therefore, compression steel does not yield, $f'_s < 60$ Ksi

$$\rho_{max} = 0.75\rho_b = \frac{0.75(0.85)(0.8)(\frac{5}{60})87}{87 + 60} = 0.02515$$

$(\rho - \rho') < \rho_{max}$ (for the basic section)

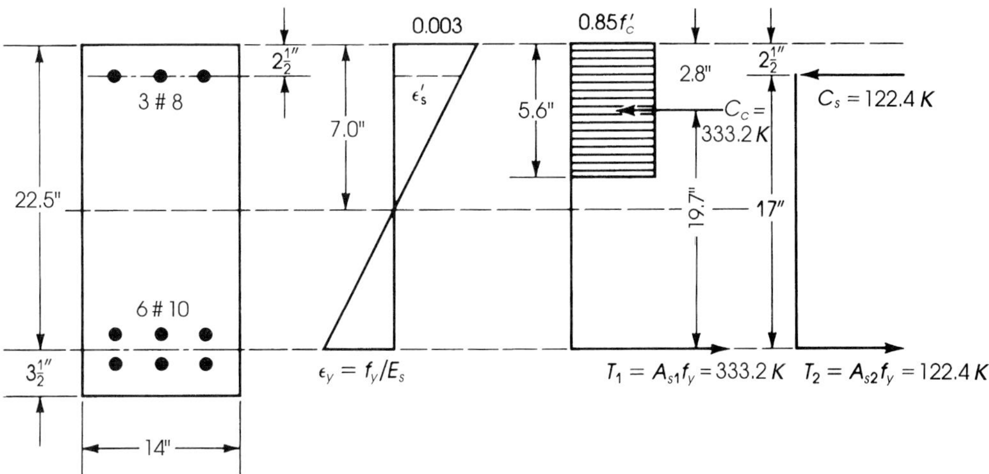

Figure 3.21 Example 3.7 analysis solution.

3. Calculate ϕM_n by analysis.

Internal forces:

$$C_c = 0.85 f'_c ab, \qquad a = \beta_1 c = 0.8c$$

$$C_c = 0.85 \times 5(0.8c) \times 14 = 47.6c$$

$$\begin{aligned} C_s &= \text{the force in compression steel} \\ &= A'_s f'_s - \text{force in displaced concrete} \\ &= A'_s(f'_s - 0.85 f'_c) \end{aligned}$$

From strain triangles,

$$\varepsilon'_s = 0.003 \frac{(c - d')}{c}$$

$$f'_s = E_s \varepsilon'_s \qquad \text{(since steel is in the elastic range)}$$

$$= 29{,}000\left[\frac{0.003(c - d')}{c}\right] = \frac{87(c - d')}{c} \qquad \text{(Ksi)}$$

Therefore,

$$C_s = 2.37[87(c - d')/c - (0.85 \times 5)] \text{ kips} = \left[\frac{206.2(c - 2.5)}{c}\right] - 10.07$$

$$T = T_1 + T_2 = (A_{s1} + A_{s2})f_y = A_s f_y = 7.59(60) = 455.4 \text{ kips}$$

4. Equate internal forces to determine the position of the neutral axis (the distance c):

$$T = C = C_c + C_s$$

$$455.4 = 47.6c + \frac{206.2(c - 2.5)}{c} - 10.07$$

$$c^2 - 5.447c - 10.83 = 0$$

$$c = 7.0 \text{ in.} \quad \text{and} \quad a = 0.8c = 5.6 \text{ in.}$$

5. Calculate f'_s, C_c, and C_s:

$$f'_s = \frac{87(c - 2.5)}{c} = \frac{87(7.0 - 2.5)}{7} = 55.9 \text{ Ksi}$$

which confirms that compression steel does not yield.

$$C_c = 47.6c = 47.6(7.0) = 333.2 \text{ kips}$$

$$C_s = (A'_s f'_s - 10.07) = 2.37(55.90) - 10.07 = 122.40 \text{ kips}$$

6. To calculate ϕM_n, take moments about the tension steel A_s:

$$\phi M_n = \phi\left[C_c\left(d - \frac{a}{2}\right) + C_s(d - d')\right] = 0.9[333.2(22.5 - 2.8) + 122.40(22.5 - 2.5)]$$

$$= 8110.8 \text{ K}\cdot\text{in.} = 675.9 \text{ K}\cdot\text{ft}$$

7. Check that $(\rho - \rho' f'_s/f_y) \le \rho_{\max}$, equation (3.42):

$$0.0241 - 0.00754\left(\frac{55.9}{60}\right) = 0.0171 < \rho_{\max} = 0.02515$$

The maximum total tension steel that can be used in this section is calculated by equation (3.40):

$$\text{Max } A_s = bd\left(\rho_{\max} + \frac{\rho' f'_s}{f_y}\right)$$

$$= 14(22.5)\left(0.02515 + \frac{0.00754 \times 55.9}{60}\right) = 10.13 \text{ in.}^2$$

This value is greater than $A_s = 7.59$ in.^{2} used in the section.

8. *Note:* A faster solution can be made using equation (3.38), which is developed from the preceding analysis. In this case, $A_1 = 4.76$, $A_2 = -259.3$, $A_3 = -515.5$, and $c = 7.0$ in.

3.14 ANALYSIS OF T- AND I-SECTIONS

3.14.1 Description

It is normal to cast concrete slabs and beams together, producing a monolithic structure. Slabs have smaller thicknesses than beams. Under bending stresses, those parts of the slab on either side of the beam will be subjected to compressive stresses, depending on the position of these parts relative to the top fibers and relative to their distances from the beam. The part of the slab acting with the beam is called the flange, and it is indicated in Figure 3.22(a) by area bt. The rest of the section confining the area $(h - t)b_w$ is called the *stem*, or *web*.

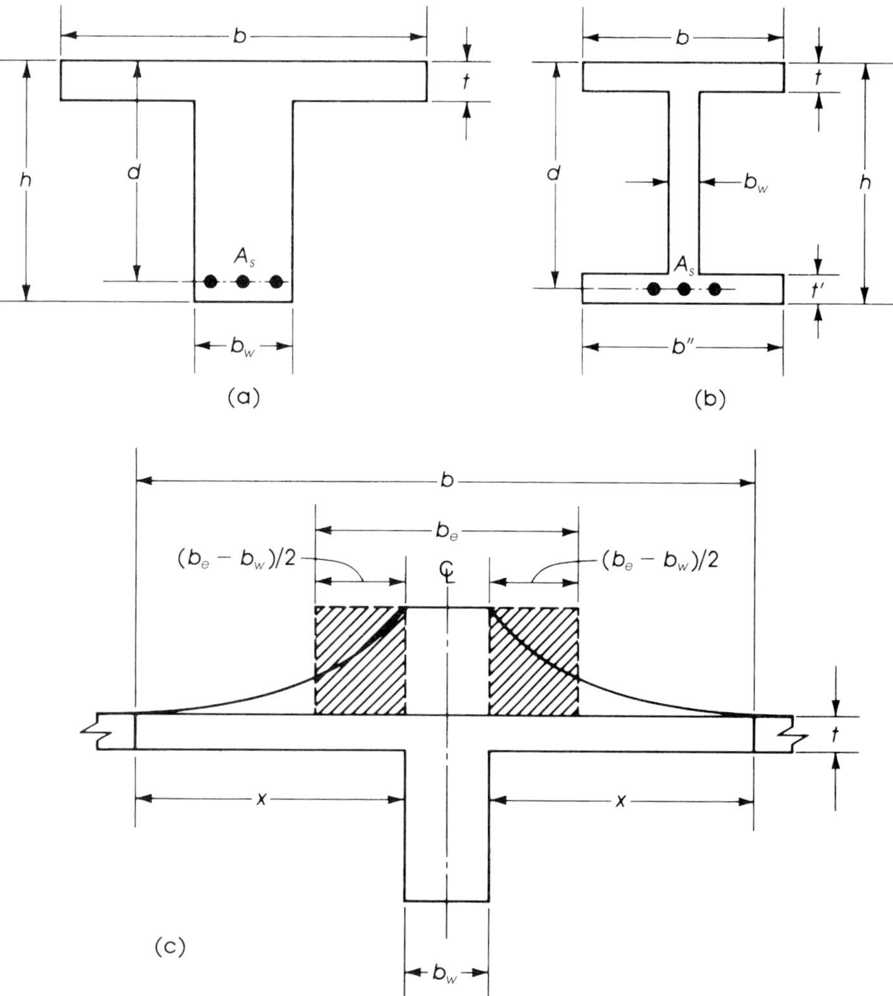

Figure 3.22 (a) T-section and (b) I-section, with (c) illustration of effective flange width b_e.

In an I-section there are two flanges, a compression flange, which is actually effective, and a tension flange, which is ineffective, because it lies below the neutral axis and is thus neglected completely. Therefore, the analysis and design of an I-beam is similar to that of a T-beam.

3.14.2 Effective Width

In a T-section, if the flange is very wide, the compressive stresses are at a maximum value at points adjacent to the beam and decrease approximately in a parabolic form to almost zero at a distance x from the face of the beam. Stresses also vary vertically from a maximum at the top fibers of the flange to a minimum at the lower fibers of the flange. This variation depends on the position of the neutral axis and the change from elastic to inelastic deformation of the flange along its vertical axis.

An equivalent stress area can be assumed to represent the stress distribution on the width b of the flange, producing an equivalent flange width, b_e, of uniform stress (Figure 3.22(c)).

Analysis of equivalent flange widths for actual T-beams indicate that b_e is a function of span length of the beam [7]. Other variables that affect the effective width b_e are (Figure 3.23)

- Spacing of beams;
- Width of stem (web) of beam b_w;
- Relative thickness of slab with respect to the total beam depth;
- End conditions of the beam (simply supported or continuous);
- The way in which the load is applied (distributed load or point load);
- The ratio of the length of the beam between points of zero moment to the width of the web and the distance between webs.

The ACI Code, Section 8.10.2, prescribes the following limitations on the effective flange width b_e, considering that the span of the beam is equal to L:

1. $b_e = L/4$
2. $b_e = 16t + b_w$
3. $b_e = b$, where b is the distance between centerlines of adjacent slabs

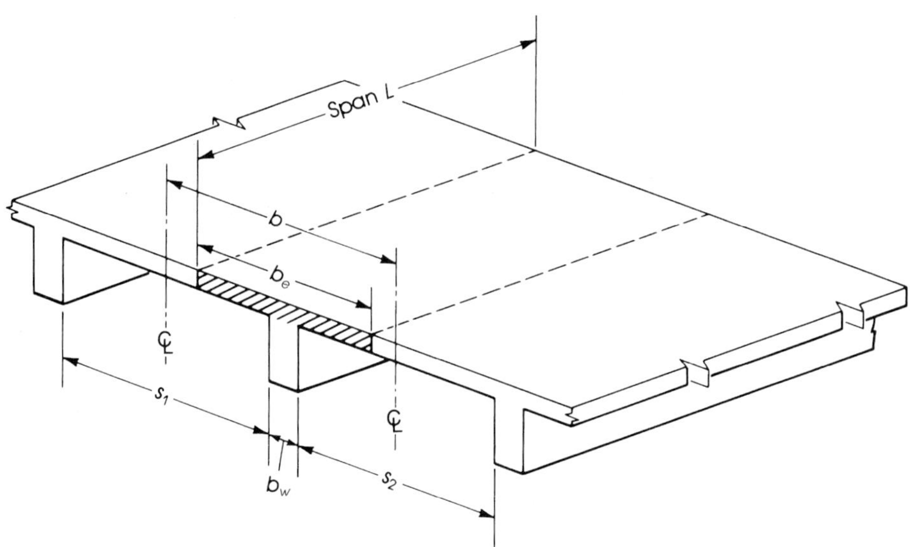

Figure 3.23 Effective flange width of T-beams.

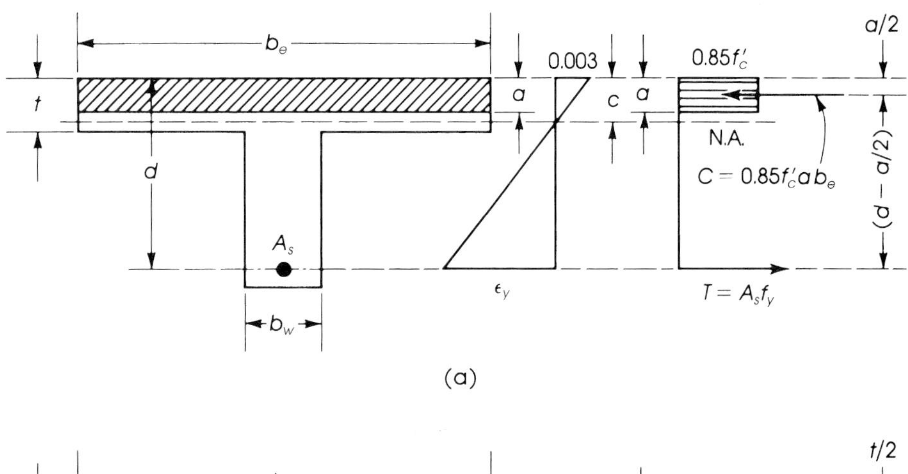

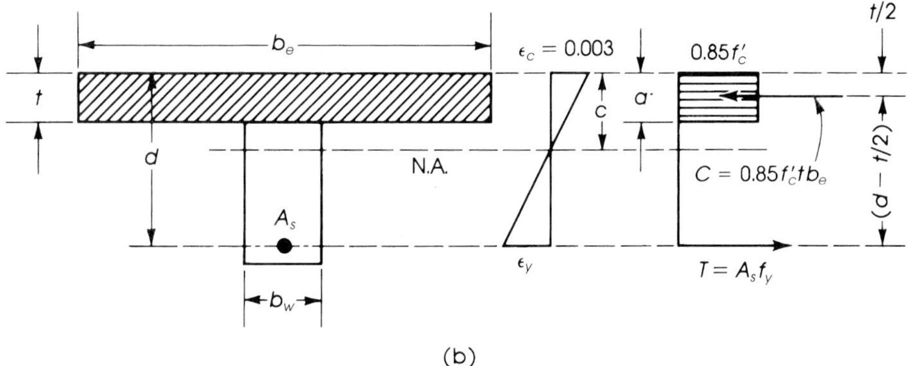

Figure 3.24 Rectangular section behavior (a) when the neutral axis lies within the flange and (b) when the stress distribution depth equals the slab thickness.

The *smallest* of the aforementioned three values must be used.

These values are conservative for some cases of loading and are adequate for other cases. A similar effective width of flange can be adopted for I-beam sections.

Investigations indicate that the effective compression flange increases as load is increased toward the ultimate value [6]. Under working loads, stress in the flange is within the elastic range. The restrictions on b_e just mentioned are used for both elastic and ultimate-strength methods.

A T-shaped or I-shaped section may behave as a rectangular section or a T-section. The two cases are investigated as follows.

3.14.3 T-Sections Behaving As Rectangular Sections

In this case, the depth of the equivalent stress block a lies within the flange, with extreme position at the level of the bottom fibers of the compression flange ($a \leq t$).

When the neutral axis lies within the flange (Figure 3.24(a)), the depth of the equivalent compressive distribution stress lies within the flange, producing a compressed area equal to $b_e a$. The concrete below the neutral axis is assumed ineffective, and the section is considered as singly reinforced, as explained earlier, with b_e replaced by b. Therefore,

$$a = \frac{A_s f_y}{0.85 f'_c b_e} \tag{3.45}$$

and

$$\phi M_n = \phi A_s f_y \left(d - \frac{a}{2} \right) \tag{3.46}$$

If the depth a is increased such that $a = t$, then the ultimate moment capacity is that of a singly reinforced concrete section:

$$\phi M_n = \phi A_s f_y \left(d - \frac{t}{2} \right) \tag{3.47}$$

In this case

$$t = \frac{A_s f_y}{0.85 f'_c b_e} \quad \text{or} \quad A_s = \frac{0.85 f'_c b_e t}{f_y} \tag{3.48}$$

In this analysis, the limit of the steel area in the section should apply: $A_s \le A_{s\,\max}$.

3.14.4 Analysis of a T-section

In this case the depth of the equivalent compressive distribution stress lies below the flange; consequently, the neutral axis also lies in the web. This is due to an amount of tension steel A_s more than that calculated by equation (3.48). Part of the concrete in the web will now be effective in resisting the external moment. In Figure 3.25, the compressive force C is equal to the compression area of the flange and web multiplied by the uniform stress of $0.85 f'_c$:

$$C = 0.85 f'_c \left[b_e t + b_w (a - t) \right]$$

The position of C is at the centroid of the T-shaped compressive area at a distance z from top fibers.

The analysis of a T-section is similar to that of a doubly reinforced concrete section, considering an area of concrete $(b_e - b_w)t$ as equivalent to the compression steel area A'_s. The analysis is divided into two parts, as shown in Figure 3.26:

1. A singly reinforced rectangular basic section $b_w d$ and steel reinforcement A_{s1}. The compressive force C_1 is equal to $(0.85 f'_c a b_w)$, the tensile force T_1 is equal to $A_{s1} f_y$, and the moment arm is equal to $(d - a/2)$.
2. A section that consists of the concrete overhanging flange sides $2 \times \left[(b_e - b_w)t \right] / 2$ developing the additional compressive force (when multiplied by $0.85 f'_c$) and a moment arm equal to $(d - t/2)$. If A_{sf} is the area of tension steel that will develop a force equal to the compressive strength of the overhanging flanges, then

$$A_{sf} f_y = 0.85 f'_c (b_e - b_w) t$$

$$A_{sf} = \frac{0.85 f'_c t (b_e - b_w)}{f_y} \tag{3.49}$$

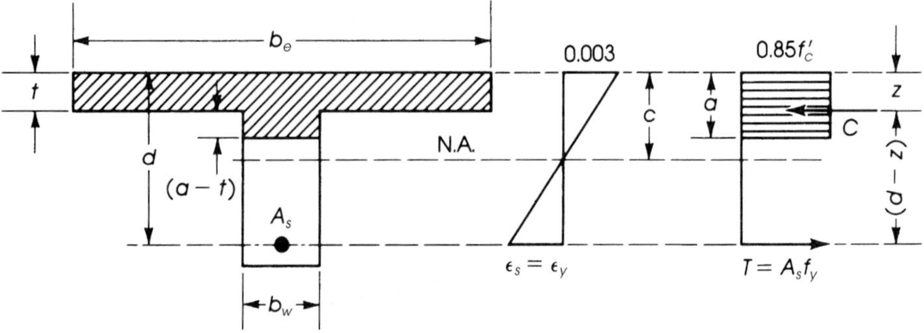

Figure 3.25 T-section behavior.

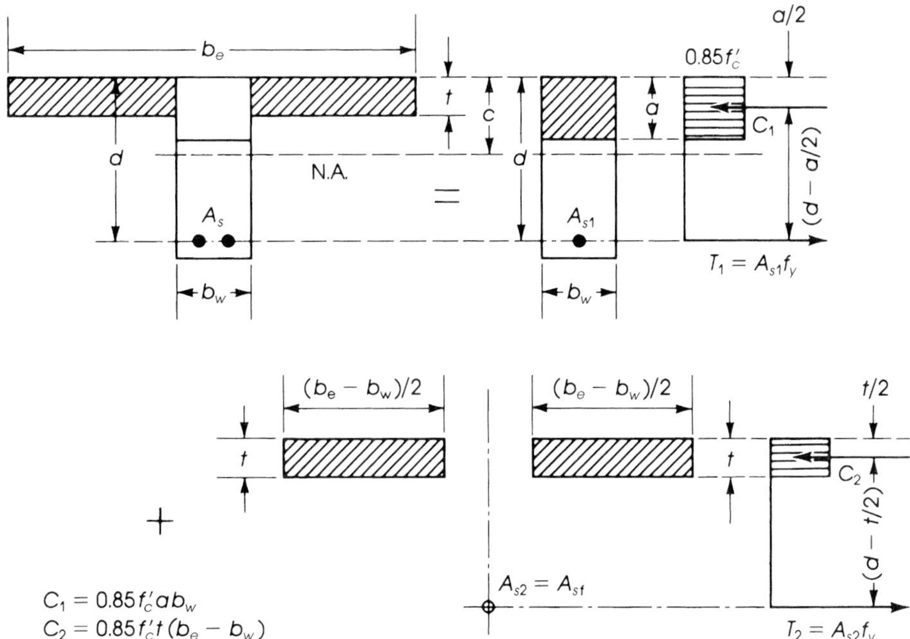

Figure 3.26 T-section analysis.

The total steel used in the T-section A_s is equal to $A_{s1} + A_{sf}$, or

$$A_{s1} = A_s - A_{sf} \tag{3.50}$$

The T-section is in equilibrium, so $C_1 = T_1$, $C_2 = T_2$, and $C = C_1 + C_2 = T_1 + T_2 = T$. Considering equation $C_1 = T_1$ for the basic section, then $A_{s1}f_y = 0.85f'_c ab_w$ or $(A_s - A_{sf})f_y = 0.85f'_c ab_w$; therefore,

$$a = \frac{(A_s - A_{sf})f_y}{0.85f'_c b_w} \tag{3.51}$$

Note that b_w is used to calculate a. The ultimate moment capacity of the section is the sum of the two moments M_{u1} and M_{u2}.

$$\phi M_n = M_{u1} + M_{u2}$$

$$M_{u1} = \phi A_{s1}f_y\left(d - \frac{a}{2}\right) = \phi(A_s - A_{sf})f_y\left(d - \frac{a}{2}\right)$$

where

$$A_{s1} = (A_s - A_{sf}) \quad \text{and} \quad a = \frac{(A_s - A_{sf})f_y}{0.85f'_c b_w}$$

$$M_{u2} = \phi A_{sf}f_y\left(d - \frac{t}{2}\right) \tag{3.52}$$

$$\phi M_n = \phi\left[(A_s - A_{sf})f_y\left(d - \frac{a}{2}\right) + A_{sf}f_y\left(d - \frac{t}{2}\right)\right]$$

The ultimate moment of a T-section or I-section can be calculated from the preceding equation. It is always necessary to check the following:

1. The total tension steel ratio relative to the web effective area is equal to or greater than $\rho_{\min}$, i.e.,

$$\rho_w = \frac{A_s}{b_w d} \geq \rho_{\min}$$

$$\rho_{\min} = \frac{3\sqrt{f'_c}}{f_y} \geq 200/f_y \qquad \text{(psi units)} \tag{3.53}$$

2. Also, check that A_s used in the T-(web) section is less than max A_s, where max $A_s = 0.75A_{sb}$ (balanced) for the T-section.

$$A_{sb} \quad \text{(balanced)} = A_{sf} \quad \text{(flange)} + \rho_b(b_w d) \quad \text{(web)}$$

$$\text{Max } A_s = 0.75A_{sf} \quad \text{(flange)} + \rho_{\max}(b_w d) \quad \text{(web)} \tag{3.54}$$

$$\text{Max } A_s = \left(\frac{1}{f_y}\right)[0.75(0.85f'_c)t(b - b_w)] + (\rho_{\max} b_w d) \tag{3.55}$$

The total tension steel used in the section should be less than or equal max A_s.
In steel ratios, relative to the web only, divide equation (3.54) by $b_w d$:

$$\rho_w = \frac{A_s}{b_w d} \leq \left(\frac{0.75A_{sf}}{b_w d} + \rho_{\max}\right) \tag{3.56}$$

or

$$(\rho_w - 0.75\rho_f) \leq \rho_{\max} \quad \text{(web)} \tag{3.57}$$

where $\rho_{\max}$ is the maximum steel ratio for the basic singly reinforced web section and $\rho_f = A_{sf}/b_w d$.

A general equation for calculating (max A_s) in a T-section when $a > t$ can be developed as follows:

$$C = 0.85f'_c[b_e t + b_w(a - t)]$$

For a balanced section,

$$a = a_b = \beta_1 d\left(\frac{87}{87 + f_y}\right) \quad \text{and} \quad C_b = 0.85f'_c[b_e t + b_w(a - t)]$$

The balanced steel area is $A_{sb} = C_b/f_y$, and max $A_s = 0.75A_{sb} = 0.75C_b/f_y$:

$$\text{Max } A_s = (0.75)(0.58)\frac{f'_c}{f_y}\left[b_e t + b_w\left(\frac{87\beta_1 d}{87 + f_y} - t\right)\right] \tag{3.58}$$

where max A_s is the total maximum tension steel area that can be used in a T-section when $a > t$.

For example, for $f'_c = 3$ Ksi and $f_y = 60$ Ksi, the preceding equation is reduced to

$$\text{Max } A_s = 0.0319[b_e t + b_w(0.5d - t)] \tag{3.59}$$

For $f'_c = 4$ Ksi and $f_y = 60$ Ksi,

$$\text{Max } A_s = 0.0425[b_e t + b_w(0.5d - t)] \tag{3.60}$$

In summary, the procedure to analyze a T- or inverted L-section is as follows:

1. Determine the effective width of the flange b_e (refer to Section 3.14.2). Calculate $\rho_{\max}$ and $\rho_{\min}$.

2. Check if $a \le t$ as follows: let

$$a' = \frac{A_s f_y}{0.85 f'_c b_e}$$

3. If $a' \le t$, it is a rectangular section analysis.
   a. Calculate $\phi M_n = \phi A_s f_y \left(d - \frac{a}{2} \right)$, $a = a'$.
   b. Check that $\rho_w = A_s / b_w d \ge \rho_{\min}$.
   c. Check that A_s used is less than or equal to max A_s. Max A_s is given by equation (3.58). When $a \le t$, normally this condition is already met.
4. If $a' > t$, it is a T-section analysis:
   a. Calculate $A_{sf} = 0.85 f'_c t (b - b_w) / f_y$.
   b. Check that $(\rho_w - 0.75\rho_f) \le \rho_{\max}$ (relative to the web area), where

$$\rho_w = \frac{A_s}{b_w d}, \qquad \rho_f = \frac{A_{sf}}{b_w d}$$

   or calculate max A_s that can be used in this section, and this value must be greater than or equal to A_s (used):

$$\text{Max } A_s = 0.75 A_{sf} + \rho_{\max}(b_w d) \ge A_s \qquad \text{used}$$

   c. Check that $\rho_w = A_s / b_w d \ge \rho_{\min}$. This condition is normally met when $a' > t$.
   d. Calculate $a = (A_s - A_{sf}) f_y / 0.85 f'_c b_w$ (for the web section).
   e. Calculate $\phi M_n = \phi[(A_s - A_{sf}) f_y (d - a/2) + A_{sf} f_y (d - t/2)]$.

Example 3.8

A series of reinforced concrete beams spaced at 7 ft 10 in. on centers have a simply supported span of 15 ft. The beams support a reinforced concrete floor slab 4 in. thick. The dimensions and reinforcement of beams are shown in Figure 3.27; $f'_c = 3$ Ksi and $f_y = 60$ Ksi. Determine the moment capacity of a typical intermediate beam.

Solution

1. Determine the effective flange width. Effective flange width b_e is the smallest of

$$16t + b_w = (16 \times 4) + 10 = 74 \text{ in.}$$

$$\frac{\text{Span}}{4} = \frac{15 \times 12}{4} = 45 \text{ in.}$$

$$\text{Center to center of beams} = (7 \times 12) + 10 = 94 \text{ in.}$$

   Therefore, take $b_e = 45$ in. $= b$.

2. Check the depth of the stress block. If the section behaves as a rectangular one, then the stress block lies within the flange (Figure 3.24(a)). In this case, the width of beam used is equal to 45 in.

$$a' = \frac{A_s f_y}{0.85 f'_c b} = \frac{2.37 \times 60}{0.85 \times 3 \times 45} = 1.24 \text{ in.} < t \qquad (A_s = 2.37 \text{ in.}^2)$$

   Therefore, the stress block lies within the flange and the section is considered rectangular, or $a = a'$.

3. Check that $\rho_w = A_s / b_w d \ge \rho_{\min} = 0.00333$.

$$\rho_w = \frac{2.37}{10(16)} = 0.0148 > \rho_{\min}$$

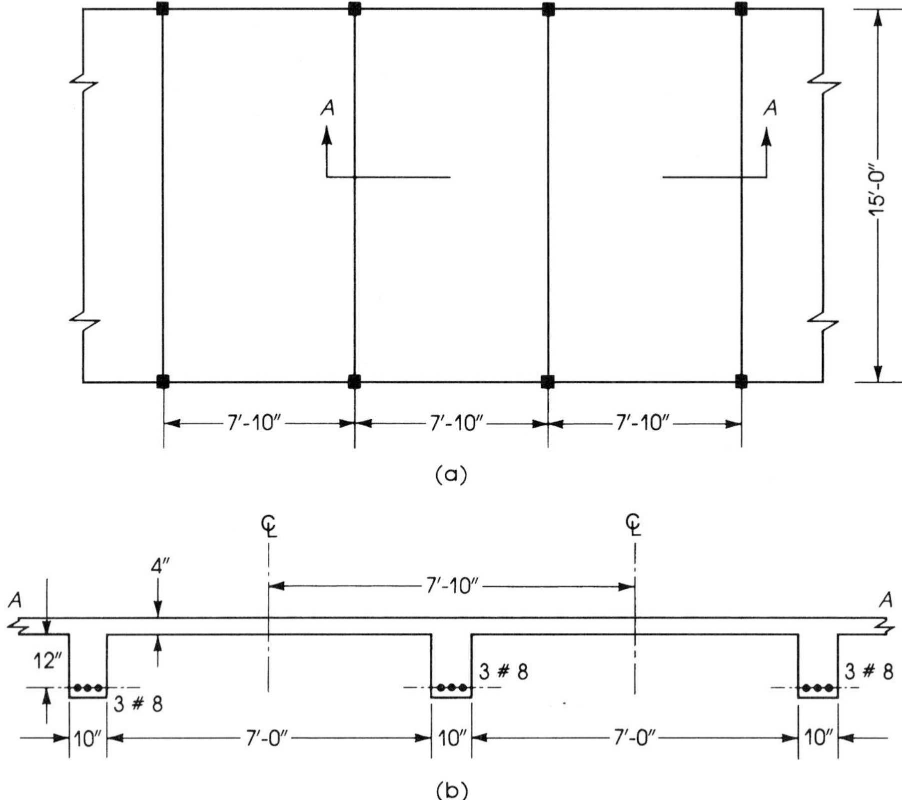

Figure 3.27 Example 3.8: (a) plan of slab-beam roof and (b) section *A-A*.

4. Calculate

$$\phi M_n = \phi A_s f_y\left(d - \frac{a}{2}\right)$$

$$= 0.9(2.37)(60)\left(16 - \frac{1.24}{2}\right) = 1968 \text{ K}\cdot\text{in.} = 164 \text{ K}\cdot\text{ft}$$

5. Check that the A_s used is less than or equal to max A_s (equation (3.59)):

$$\text{Max } A_s = 0.0319[(45 \times 4) + (0.5 \times 16 - 4)] = 7.02 \text{ in.}^2$$

$$A_s \text{ used} = 2.37 \text{ in.}^2 < 7.02 \text{ in.}^2$$

This condition is met whenever $a \leq t$.

Example 3.9

Calculate the ultimate moment capacity of the T-section shown in Figure 3.28 using $f'_c = 3.5$ Ksi and $f_y = 60$ Ksi.

Solution

1. Given $b = b_e = 36$ in., $b_w = 10$ in., $t = 3$ in., $d = 17$ in., and $A_s = 6.0$ in.^{2} Check if $a \leq t$:

$$a' = \frac{A_s f_y}{0.85 f'_c b} = \frac{6(60)}{0.85(3.5)(36)} = 3.36 \text{ in.}$$

Because $a' > t$, it is a T-section analysis.

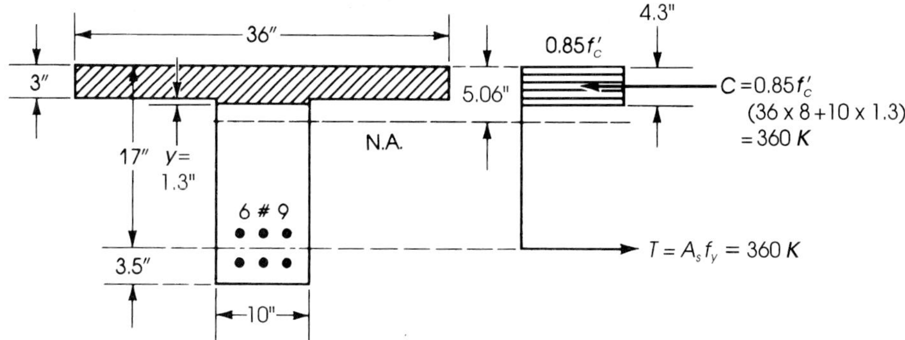

Figure 3.28 Example 3.9.

2. $A_{sf} = \dfrac{0.85f'_c(b - b_w)t}{f_y} = \dfrac{0.85 \times 3.5(36 - 10) \times 3}{60} = 3.87 \text{ in.}^2$

3. a. Check that $A_s \leq \max A_s$ (from equation (3.58)).

$$\text{Max } A_s = (0.75)(0.85)\left(\frac{3.5}{60}\right)\left[(36 \times 3) + 10\left(\frac{87}{147}(0.85)(17) - 3\right)\right]$$

$$= 6.09 \text{ in.}^2 > A_s = 6.0 \text{ in.}^2 \qquad \text{(used)}$$

b. Check that $A_s \geq A_{s\,\min}, \rho_{\min} = 0.00333, A_{s\,\min} = \rho_{\min} b_w d$:

$$A_{s\,\min} = 0.00333(10)(17) = 0.57 \text{ in.}^2 < A_s = 6.0 \text{ in.}^2 \qquad \text{(used)}$$

4. Calculate ϕM_n using equation (3.52):

$$\phi M_n = \phi\left[(A_s - A_{sf})f_y\left(d - \frac{a}{2}\right) + A_{sf}f_y\left(d - \frac{t}{2}\right)\right]$$

$$a = \frac{(A_s - A_{sf})f_y}{0.85f'_c b_w} = \frac{2.13 \times 60}{0.85 \times 3.5 \times 10} = 4.3 \text{ in.}$$

$$\phi M_n = 0.9\left[2.13 \times 60\left(17 - \frac{4.3}{2}\right) + 3.87 \times 60\left(17 - \frac{3}{2}\right)\right]$$

$$= 4947 \text{ K} \cdot \text{in.} = 412.3 \text{ K} \cdot \text{ft}$$

Another approach to checking whether $a \leq t$ is to calculate the tension force $T = A_s f_y$ and compare it to the compressive force in the total flange (Figure 3.28).

$$T = A_s f_y = 6.0 \times 60 = 360 \text{ kips}$$

$$C = 0.85f'_c t b_e = 0.85 \times 3.5 \times 3 \times 36 = 321.3 \text{ kips}$$

Because T exceeds C, then $a > t$, and the section behaves as a T-section.

An additional area of concrete should be used to provide for the difference, that is, a compressive force of $(360 - 321.3) = 38.7$ kips. This area has a width equal to 10 in. and a depth equal to y. Therefore,

$$b_w y(0.85f'_c) = 38.7 \text{ kips}$$

$$10(y)(0.85 \times 3.5) = 38.7, \qquad y = \frac{38.7}{10 \times 0.85 \times 3.5} = 1.3 \text{ in.}$$

Thus

$$a = y + t = 1.3 + 3 = 4.3 \text{ in.}$$

as calculated earlier.

3.15 DIMENSIONS OF ISOLATED T-SHAPED SECTIONS

In some cases, isolated beams with the shape of a T-section are used in which additional compression area is provided to increase the compression force capacity of sections. These sections are commonly used as prefabricated units.

The ACI Code, Section 8.10.4, specifies the size of isolated T-shaped sections as follows:

1. Flange thickness, t, shall be equal to or greater than one-half of the width of the web, b_w.
2. Total flange width b shall be equal to or less than four times the width of the web, b_w (Figure 3.29).

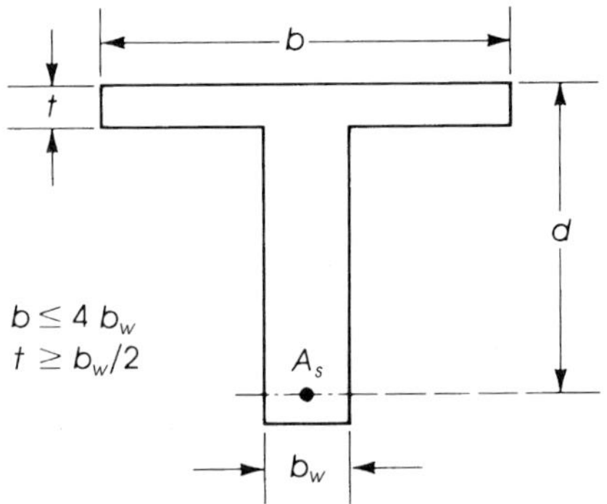

Figure 3.29 Isolated T-shaped sections.

3.16 INVERTED L-SHAPED SECTIONS

In slab-beam girder floors, the end beam is called a *spandrel beam.* This type of floor has part of the slab on one side of the beam and is cast monolithically with the beam. The section is unsymmetrical under vertical loading (Figure 3.30(a)). The loads on slab S_1 cause torsional moment uniformly distributed on the spandrel beam B_1. Design for torsion is explained later. The overhanging flange width $(b - b_w)$ of a beam with the flange on one side only is limited by the ACI Code, Section 8.10.2, to the smallest of

1. One-twelfth of the span of the beam;
2. Less than or equal to six times the thickness of the slab;
3. Less than or equal to one-half the clear distance to the next beam.

If this is applied to the spandrel beam in Figure 3.30(b), then

1. $(b - 12) \leq (20 \times 12)/12 = 20$ in. (controls)
2. $(b - 12) \leq 6 \times 6 = 36$ in.
3. $(b - 12) \leq 3.5 \times 12 = 42$ in.

Therefore, the effective flange width is $b = 20 + 12 = 32$ in., and the effective dimensions of the spandrel beam are as shown in Figure 3.30(d).

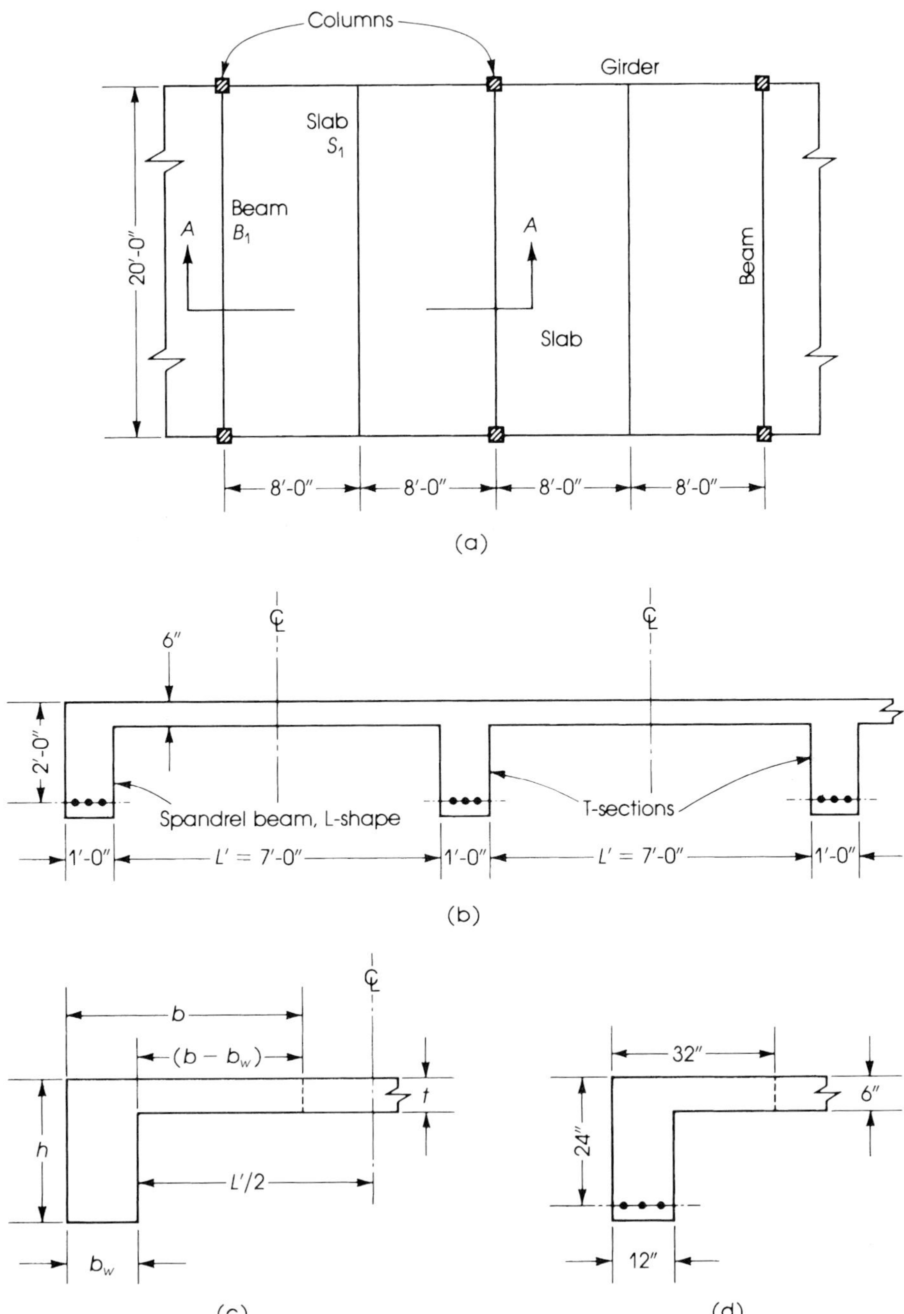

Figure 3.30 Slab-beam-girder floor, showing (a) plan, (b) section including spandrel beam, (c) dimensions of the spandrel beam, and (d) its effective flange width.

3.17 SECTIONS OF OTHER SHAPES

Sometimes a section different from the previously defined sections is needed in special requirements of structural members. For instance, sections such as those shown in Figure 3.31 may be used in the precast concrete industry. The analysis of such sections is similar to that of a rectangular section, taking into consideration the area of removed or added concrete. The best way to explain this analysis is by giving an example.

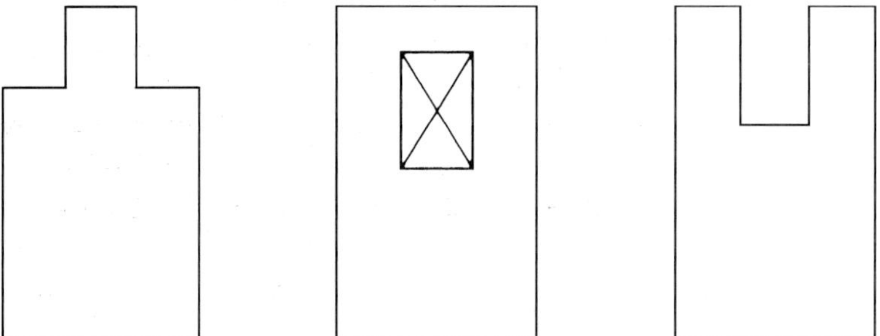

Figure 3.31 Sections of other shapes.

Example 3.10

The section shown in Figure 3.32 represents a beam in a structure containing prefabricated elements. The total width and total depth are limited to 12 in. and 19 in., respectively. Tension re-

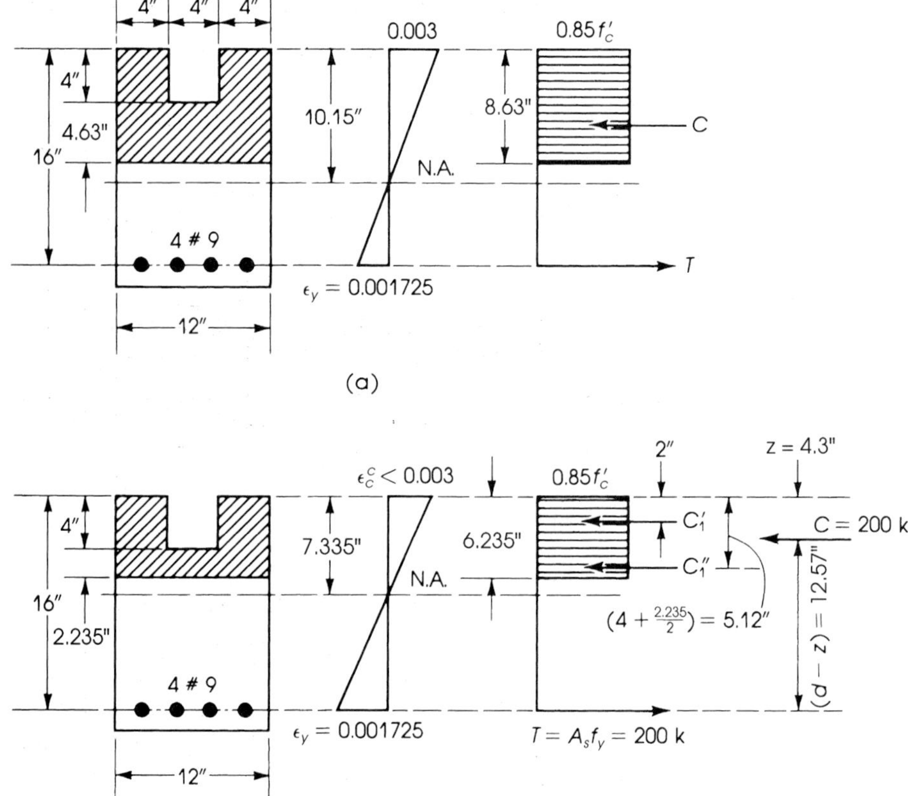

Figure 3.32 Example 3.10: (a) balanced and (b) under-reinforced sections.

inforcement used is four no. 9 bars; $f'_c = 4$ Ksi and $f_y = 50$ Ksi. Determine the ultimate moment capacity of the section.

Solution

1. Check whether the section is balanced, under-reinforced, or over-reinforced. If the section is balanced, then the maximum strain of concrete is 0.003 and steel yields; that is,

$$\varepsilon_y = \frac{f_y}{E_s} = \frac{50}{29{,}000} = 0.001725$$

From similar triangles, find the position of the neutral axis:

$$c = \frac{0.003}{0.003 + 0.001725} \times 16 = 10.15 \text{ in.}$$

$$a = 0.85c = 0.85 \times 10.15 = 8.63 \text{ in.}$$

$$\begin{aligned} C &= \text{total compressive force} \\ &= 0.85f'_c \times \text{area of section from top fibers to a depth equal to } a \\ &= 0.85 \times 4[2(4 \times 4) + 12(8.63 - 4)] = 298 \text{ kips} \end{aligned}$$

$$T = A_s f_y = 4 \times 1.0 \times 50 = 200K$$

$T < C$; thus, tension steel will yield first, and the section is under-reinforced.

2. Check that the steel used is equal to or less than the maximum steel specified by the ACI Code $(0.75\,A_{sb})$.

$$A_s\,(\text{balanced}) = \frac{C_b}{f_y} = \frac{298}{50} = 5.96 \text{ in.}^2$$

$$A_{s\,\max} = 0.75A_s\,(\text{balanced}) = 0.75 \times 5.96 = 4.47 \text{ in.}^2$$

$$A_s\,(\text{used}) = 4 \text{ in.}^2 < 4.47 \text{ in.}^2$$

Use the actual steel area of 4 in.^{2} and $T = 200$ k. Note that if $A_s > A_{s\,\max}$, then $A_{s\,\max}$ should be used, and the tension force is $T = A_{s\,\max} f_y$.

3. Determine the position of the neutral axis based on $T = 200$ kips. Because $T = C$,

$$200 = 0.85f'_c[2(4 \times 4) + 12(a - 4)]$$

where a = depth of equivalent compressive block needed to produce a total compressive force of 200 k.

$$200 = 0.85 \times 4(32 + 12a - 48)$$

$$a = 6.235 \text{ in.} \qquad c = \frac{6.235}{0.85} = 7.335 \text{ in.}$$

4. Calculate ϕM_n by taking moments of the two parts of the compressive forces (each by its arm) about tension steel.

$$\begin{aligned} C'_1 &= \text{compressive force on two small areas } 4 \times 4 \text{ in.} \\ &= 0.85 \times 4[2 \times 4 \times 4] = 108.8 \text{ kips} \end{aligned}$$

$$\begin{aligned} C''_1 &= \text{compressive force on area as } 12 \times 2.25 \text{ in.} \\ &= 0.85 \times 4 \times 12 \times 2.235 = 91.2 \text{ kips} \end{aligned}$$

$$\begin{aligned} \phi M_n &= \phi[C'_1(d - 2) + C''_1(d - 5.12)] \\ &= 0.9[108.8(14) + 91.2(10.88)] = 2269 \text{ K}\cdot\text{in.} = 189 \text{ K}\cdot\text{ft} \end{aligned}$$

5. Alternatively, the resultant of C'_1 and C''_1 can be found:

$$C'_1 + C''_1 = 108.8 + 91.2 = 200 \text{ kips}$$

The position of the resultant z from the top fibers is

$$z = \frac{108.8 \times 2 + 91.2 \times 5.12}{200} = 3.43 \text{ in.}$$

$$\text{Arm} = (d - z) = 16 - 3.43 = 12.57 \text{ in.}$$

$$\phi M_n = 0.9 \times 200 \times 12.57 = 2263 \text{ K}\cdot\text{in.} = 189 \text{ K}\cdot\text{ft}$$

3.18 MORE THAN ONE ROW OF STEEL BARS IN THE SECTION

In the case where more than one layer of tension steel bars are present in the section (Figure 3.33), the strain in the steel is located at the centroid of the area of steel bars. Actually, because in an under-reinforced concrete section, for instance, tension steel yields, so the strain in the bottom layer will reach the yield strain ε_y, whereas that in the top layer will not. As the load is increased, the top layer will reach its yield strain ε_y, whereas the strain in the bottom layer will be slightly greater than the yield strain, but the stress will still be f_y. Therefore, the preceding assumption that the strain in the steel is located at the centroid of the area of the steel bars is valid and can be assumed in the analysis of reinforced concrete sections with more than one layer of steel. If, in some cases, the steel is placed at different depths of the section, which may not cause the yielding of the steel, the layer closest to the extreme tension fibers will be assumed to have a strain equal to $\varepsilon_y = f_y/E_s$, whereas the strain in the other layers can be calculated from the strain triangles of Figure 3.33. The preceding assumption is valid as long as the strain in concrete ε_c is less than or equal to 0.003.

3.19 ANALYSIS OF SECTIONS USING TABLES

Reinforced concrete sections can be analyzed and designed using the tables shown in Appendix A (for U.S. customary units) and Appendix B (for SI units). The tables give the values of R_u as related to the steel ratio, ρ, in addition to the maximum and minimum values of ρ and R_u. When the section dimensions are known, R_u is calculated; then ρ and A_s are determined from tables.

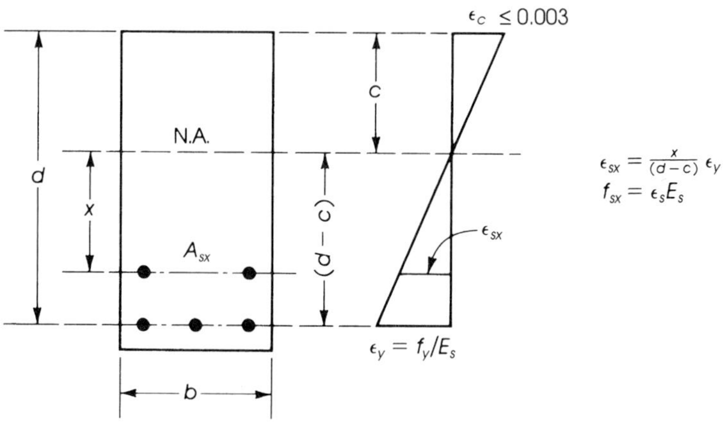

Figure 3.33 Strain and stress in steel bars of different layers.

$$\phi M_n = R_u bd^2 \quad \text{or} \quad R_u = \frac{M_u}{bd^2}$$

$$R_u = \phi \rho f_y \left(1 - \frac{\rho f_y}{1.7f'_c}\right)$$

$$A_s = \rho bd \quad \text{or} \quad \rho = \frac{A_s}{bd}$$

For any given value of ρ, R_u can be determined from tables; then ϕM_n can be calculated. The values of ρ and R_u range between a minimum value of R_u(min) when $\rho = \rho_{\min}$ to a maximum value as limited by the ACI Code, when $\rho = \rho_{\max}$.

The use of the tables will reduce the manual calculation time appreciably. Example 3.11 explains the use of tables.

Example 3.11

Calculate the ultimate moment capacity of the section shown in Example 3.2 (Figure 3.10) using tables. Given $b = 12$ in., $d = 18$ in., $f'_c = 3$ Ksi, $f_y = 60$ Ksi, and three no. 9 bars.

Solution

1. Using three no. 9 bars,

$$A_s = 3.0 \text{ in.}, \quad \rho = \frac{A_s}{bd} = \frac{3}{18 \times 12} = 0.01389$$

From Table A.4, $\rho_{\max} = 0.016$. Because $\rho = \rho_{\max}$, the section is under-reinforced. From Table A.1, for $\rho = 0.01389$, $f'_c = 3$ Ksi and $f_y = 60$ Ksi, get $R_u = 628$ psi (by interpolation).

2. $\phi M_n = R_u bd^2$

$$\text{For } b = 12 \text{ in. and } d = 18 \text{ in.}, bd^2 = 3888 \text{ in.}^3$$

$$\phi M_n = 0.628 \times 3888 = 2440 \text{ K}\cdot\text{in.} = 203.3 \text{ K}\cdot\text{ft}$$

3.20 ADDITIONAL EXAMPLES

The following examples are introduced to enhance the understanding of the strength design concept and applications.

Example 3.12

Calculate the factored moment capacity of the precast concrete section shown in Figure 3.35 using $f'_c = 4$ Ksi and $f_y = 60$ Ksi.

Solution

1. The section behaves as a rectangular section with $b = 14$ in. and $d = 21.5$ in. The width b is the width of the section on the compression side.
2. Check the steel ratio: $\rho = \dfrac{A_s}{(bd)} = \dfrac{5}{(14 \times 21.5)} = 0.01661$. From Table A.4 in Appendix A, $\rho_{\max} = 0.0214$ and $\rho_{\min} = 0.00333$; therefore, ρ lies within the limits.
3. $a = \dfrac{A_s \cdot f_y}{(0.85 f'_c b)} = \dfrac{5 \times 60}{(0.85 \times 4 \times 14)} = 6.3$ in.

$$\phi M_n = \phi A_s f_y \left(d - \frac{a}{2}\right) = 0.9 \times 5 \times 60 \times \left(\frac{21.5 - \frac{6.3}{2}}{12}\right) = 412.9 \text{ K}\cdot\text{ft.}$$

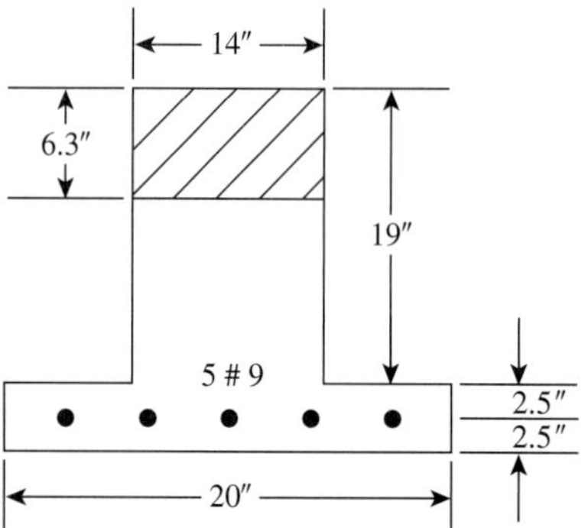

Figure 3.35 Example 3.12

Example 3.13

A reinforced concrete beam was tested to failure and had a rectangular section, $b = 14$ in., and $d = 18.5$ in. At ultimate moment (failure), the strain in the tension steel was recorded and was equal to 0.002613. The strain in the concrete at failure may be assumed $= 0.003$. If $f'_c = 3$ Ksi and $f_y = 60$ Ksi, it is required to

1. Check if the section is balanced, over-reinforced, or under-reinforced.
2. Calculate the steel area provided in the section to develop the above strains. Then calculate the ultimate applied moment.
3. Calculate the maximum factored moment allowed by the ACI Code.

Solution

1. Check the strain in the steel relative to the yield strain. The yield strain $\varepsilon_y = \dfrac{f_y}{E_s} = \dfrac{60}{29{,}000} = 0.00207$. The measured strain in the steel bars $= 0.002613$, which is much greater than 0.00207, indicating that the steel bars have yielded and in the elastoplastic range. The concrete strain was 0.003, indicating that the concrete has failed and started to crush. Therefore, the steel yielded before the crushing of concrete, which indicates an under-reinforced section.

2. Calculate the depth of the neutral axis c from the strain diagram (Figure 3.36). From the triangles:

$$\frac{c}{d} = \frac{0.003}{(0.003 + 0.002613)}, \quad \text{and} \quad c = 18.5\left(\frac{3}{5.613}\right) = 9.887 \text{ in.}$$

Also, $a = \beta_1 c = 0.85(9.887) = 8.4$ in.

$$C_c = 0.85 f'_c ab = 0.85(3)(8.4)(14) = 299.9 \text{ K}$$

$$A_s = \frac{C_c}{f_y} = \frac{299.9}{60} = 5.0 \text{ in.}^2 \text{ (section has 5 no. 9 bars)}$$

$$M_n = A_s f_y\left(d - \frac{a}{2}\right) = 5 \times 60\left(\frac{18.5 - \dfrac{8.4}{2}}{12}\right) = 357.5 \text{ K}\cdot\text{ft}$$

This is the applied moment to develop the strains mentioned in the problem. $\phi M_n = 321.75 \text{ K}\cdot\text{ft}$.

3. According to the ACI Code limitations, $\rho_{\text{max}} = 0.00161$ and $\rho_{\text{min}} = 0.00333$. Therefore, maximum $A_s = 0.0161bd = 0.0161(14)(18.5) = 4.17$ in.^{2} The balanced steel area $= \dfrac{4.17}{0.75} = 5.56$ in.^{2}, and $a = \dfrac{A_s \cdot f_y}{(0.85 f'_c b)} = \dfrac{4.17 \times 60}{(0.85 \times 3 \times 14)} = 7.0$ in.

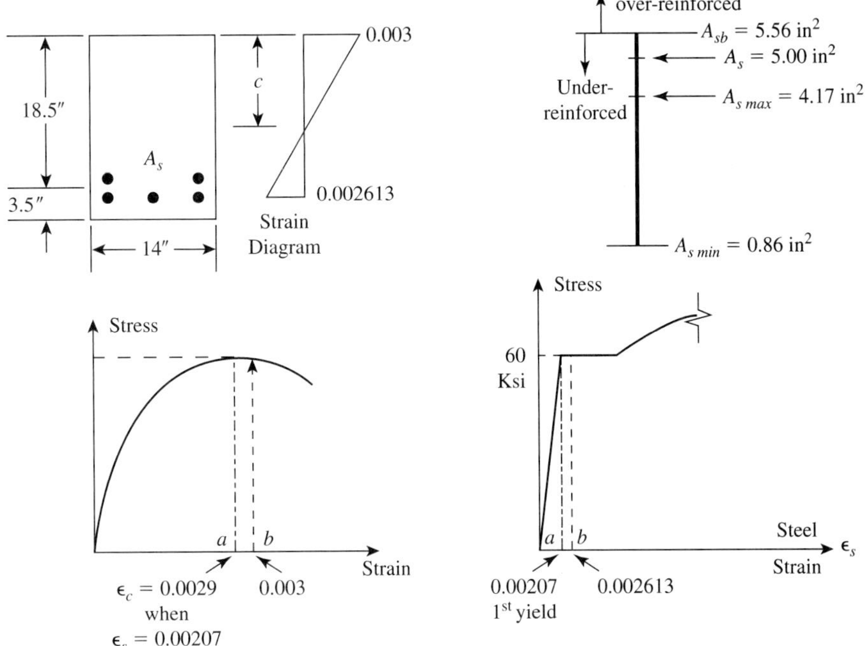

Figure 3.36 Example 3.13

The allowable factored moment is

$$\phi M_n = \phi A_s f_y\left(d - \frac{a}{2}\right) = \frac{0.9 \times 4.17 \times 60(18.5 - 3.5)}{12} = 281.5 \text{ K}\cdot\text{ft}$$

The section is under-reinforced, but the steel provided exceeds the maximum allowed by the Code. Minimum $A_{s\,\min} = 0.00333 \times 14 \times 18.5 = 0.863$ in.^{2} (see Figure 3.36).

Example 3.14

A reinforced concrete section has a width $b = 15$ in., and an effective depth $d = 24.5$ in. If $f'_c = 4$ Ksi and $f_y = 60$ Ksi, determine

a. The maximum factored moment capacity of the section according to the ACI Code requirements

b. The factored moment capacity of the section if the strain in the row of the tension steel bars is limited to 0.005 for tension to control (refer to Figure 22.1, Chapter 22)

Solution

a. For $f'_c = 4$ Ksi and $f_y = 60$ Ksi, $\rho_{\max} = 0.0214$, $R_u(\max) = 936$ psi, and $\rho_{\min} = 0.00333$.

The factored moment $\phi M_n = R_u b d^2 = \dfrac{0.936(15)(24.5)^2}{12} = 702.3$ K·ft.

$A_{s\,\max} = 0.0214(15)(24.5) = 7.865$ in.^{2} Use 8 no. 9 bars in two rows. Total depth $h = 24.5 + 3.5 = 28$ in. (Figure 3.37).

b. If the strain in the lower row of the tension bars is 0.005, as required by the Unified Design Method (ACI Code, Appendix B, and Chapter 22 in this book), for tension controlled sections, then the depth to the neutral axis c' can be calculated from the strain diagram as follows:

$$\left(\frac{c'}{d}\right) = \frac{0.003}{(0.003 + 0.005)} = \frac{3}{8}$$

Then $c' = \left(\dfrac{3}{8}\right)(25.5) = 9.562$ in. (assuming 1 in. spacing between the two rows of bars), and $a = 0.85c' = 8.128$ in.

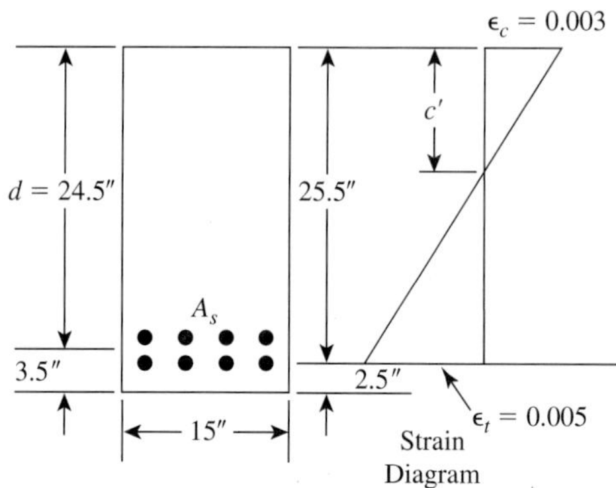

Figure 3.37 Example 3.14

Calculate the steel area from $a = \dfrac{A_s f_y}{(0.85 f'_c b)}$.

$$8.128 = \frac{A_s(60)}{(0.85 \times 4 \times 15)}, \quad \text{and} \quad A_s = 6.91 \text{ in.}^2$$

Choose 7 no. 9 bars with an area of 7.0 in.^{2}.

$$\phi M_n = \phi A_s f_y \left(d - \frac{a}{2} \right) = 0.9 \times 6.91 \times 60 \frac{\left(24.5 - \frac{8.128}{2} \right)}{12} = 635.5 \text{ K} \cdot \text{ft}$$

The maximum steel ratio in the Unified Design Method, part b, is less than what the Code allows of 0.75 the balanced ratio, part a.

3.21 EXAMPLES USING SI UNITS

The following equations are some of those mentioned in this chapter but converted to SI units. The other equations, which are not listed here, can be used for both U.S. customary and SI units. Note that f'_c and f_y are in MPa (N/mm^2).

$$\rho_b = 0.85\beta_1 \frac{f'_c}{f_y} \left(\frac{600}{600 + f_y} \right) \tag{3.13}$$

$$\rho_{\max} = 0.6375\beta_1 \frac{f'_c}{f_y} \left(\frac{600}{600 + f_y} \right) \tag{3.14}$$

$$\rho_{\max} = 0.542 \frac{f'_c}{f_y} \left(\frac{600}{600 + f_y} \right) \quad \text{when } \beta_1 = 0.85 \text{ and } f'_c \leq 30 \text{ MPa} \tag{3.15}$$

$$(\rho - \rho') \leq 0.85\beta_1 \frac{f'_c}{f_y} \times \frac{d'}{d} \times \left(\frac{600}{600 - f_y} \right) = K \tag{3.36}$$

Example 3.15

Determine the ultimate moment capacity and the position of the neutral axis of a rectangular section that has $b = 300$ mm and $d = 500$ mm and is reinforced with five bars 20 mm in diameter. Given $f'_c = 20$ MPa and $f_y = 400$ MPa.

Solution

1. Area of five bars 20 mm in diameter is 1570 mm^2.

$$\rho = \frac{A_s}{bd} = \frac{1570}{300 \times 500} = 0.01047, \qquad \rho_{\min} = \frac{1.4}{f_y} = 0.0035$$

$$\rho_{\max} = 0.542 \frac{f'_c}{f_y}\left(\frac{600}{600 + f_y}\right) = 0.542\left(\frac{20}{400}\right)\left(\frac{600}{600 + 400}\right) = 0.01626$$

Because $\rho < \rho_{\max}$, the ratio ρ should bc used to calculate ϕM_n; the section is under-reinforced. Also, $\rho > \rho_{\min}$.

2. Calculate the moment capacity:

$$\phi M_n = \phi A_s f_y\left(d - \frac{a}{2}\right)$$

$$a = \frac{A_s f_y}{0.85 f'_c b} = \frac{1570 \times 400}{0.85 \times 20 \times 300} = 123 \text{ mm}$$

$$\phi M_n = 0.9 \times 1570 \times 400\left(500 - \frac{123}{2}\right) \times 10^{-6} = 247.8 \text{ kN}\cdot\text{m}$$

(ϕM_n was multiplied by 10^{-6} to get the answer in kN·m.) The distance of the neutral axis from the compression fibers $c = a/\beta_1$. Because $f'_c < 30$ MPa, then $\beta_1 = 0.85$. Therefore,

$$c = \frac{123}{0.85} = 145 \text{ mm}$$

Example 3.16

A 2.4-m-span cantilever beam has a rectangular section of $b = 300$ mm, $d = 490$ mm and is reinforced with three bars 25 mm in diameter. The beam carries a uniform dead load (including its own weight) of 22 kN/m and a uniform live load of 30 kN/m. Check the adequacy of the section if $f'_c = 30$ MPa, and $f_y = 400$ MPa.

Solution

1. $U = 1.4D + 1.7L = 1.4 \times 22 + 1.7 \times 30 = 81.8$ kN·m.

External ultimate moment $= M_u = 1.4M_D + 1.7M_L$

$$M_u = 1.4\left(\frac{22 \times 2.4^2}{2}\right) + 1.7\left(\frac{30 \times 2.4^2}{2}\right) = 235.6 \text{ kN}\cdot\text{m}$$

2. Calculate the internal ultimate moment:

$$A_s = 1470 \text{ mm}^2 \quad \text{and} \quad \rho = \frac{A_s}{bd} = \frac{1470}{300 \times 490} = 0.01$$

$$\rho_{\max} = 0.542 \frac{f'_c}{f_y}\left(\frac{600}{600 + f_y}\right) = 0.542\left(\frac{30}{400}\right)\left(\frac{600}{600 + 400}\right) = 0.0244$$

$$\rho_{\min} = \frac{1.4}{f_y} = \frac{1.4}{400} = 0.0035$$

Because $\rho < \rho_{\max}$ and ρ is greater than $\rho_{\min}$, the section is under-reinforced, and actual ρ must be used to calculate ϕM_n.

$$\phi M_n = \phi A_s f_y\left(d - \frac{a}{2}\right)$$

$$a = \frac{A_s f_y}{0.85 f'_c b} = \frac{1470 \times 400}{0.85 \times 30 \times 300} = 77 \text{ mm}$$

$$\phi M_n = 0.9 \times 1470 \times 400\left(490 - \frac{77}{2}\right) \times 10^{-6} = 238.9 \text{ kN}\cdot\text{m}$$

3. The internal ultimate moment is greater than the external applied moment; therefore, the section is adequate.

Example 3.17

Calculate the ultimate moment capacity of a rectangular section with the following details: $b = 250$ mm, $d = 440$ mm, $d' = 60$ mm; tension steel is six bars 25 mm in diameter, compression steel is three bars 20 mm in diameter; $f'_c = 20$ MPa and $f_y = 350$ MPa.

Solution

1. Check if compression steel yields

$$A_s = 490 \times 6 = 2940 \text{ mm}^2, \quad A'_s = 314 \times 3 = 942 \text{ mm}^2, \quad A_s - A'_s = 1998 \text{ mm}^2$$

$$\rho = \frac{A_s}{bd} = \frac{2940}{250 \times 440} = 0.0267$$

$$\rho' = \frac{A'_s}{bd} = \frac{942}{250 \times 440} = 0.00856$$

$$\rho' - \rho' = 0.01814$$

For compression steel to yield,

$$(\rho - \rho') \geq 0.85 \times 0.85 \times \frac{20}{350} \times \frac{60}{440} \times \frac{600}{600 - 350} = 0.01351$$

$\rho - \rho'$ (provided) > 0.01351. Therefore, compression steel yields.

2. Check if $\rho - \rho'$ (provided) $< \rho_{max}$ as singly reinforced:

$$\rho_{max} = 0.542\frac{f'_c}{f_y}\left(\frac{600}{600 + f_y}\right) = 0.542\left(\frac{20}{350}\right)\left(\frac{600}{950}\right) = 0.01956$$

$\rho - \rho'$ (provided) $< \rho_{max}$.

3. Calculate:

$$\phi M_n = \phi\left[(A_s - A'_s)f_y\left(d - \frac{a}{2}\right) + A'_s f_y(d - d')\right]$$

$$a = \frac{(A_s - A'_s)f_y}{0.85 f'_c b} = \frac{1998 \times 350}{0.85 \times 20 \times 250} = 164 \text{ mm}$$

$$\phi M_n = 0.9\left[1998 \times 350\left(440 - \frac{164}{2}\right) + 942 \times 350(440 - 60)\right] \times 10^{-6}$$

$$= 375.6 \text{ kN}\cdot\text{m}$$

SUMMARY

Flow charts for the analysis of sections are given in Chapter 21.

Sections 3.1–3.8

a. Three types of failure are expected in a reinforced concrete flexural member based on the amount of tension steel used, A_s.

b. Load factors for dead and live loads are $U = 1.4D + 1.7L$. Other values are given in text.

c. The reduction capacity factor for beams, ϕ, is equal to 0.9.

d. An equivalent rectangular stress block can be assumed to calculate the ultimate moment capacity of the beam section, ϕM_n.

Sections 3.9–3.12 Analysis of a Singly Reinforced Rectangular Section

Given: f'_c, f_y, b, d, and A_s. Required: ϕM_n.

To determine the ultimate moment capacity of a singly reinforced rectangular section,

1. Calculate the compressive force, $C = 0.85f'_c \cdot a \cdot b$.
   Calculate the tensile force, $T = A_s \cdot f_y$

$$a = \frac{A_s \cdot f_y}{0.85f'_c \cdot b}$$

   Ultimate moment capacity $= \phi M_n$:

$$\phi M_n = \phi C\left(d - \frac{a}{2}\right) = \phi T\left(d - \frac{a}{2}\right) = \phi A_s f_y\left(d - \frac{a}{2}\right)$$

2. Calculate the maximum and minimum steel ratios:

$$\rho_{\max} = 0.75\rho_b \qquad \text{(balanced)} = 0.75(0.85)\beta_1\left(\frac{f'_c}{f_y}\right)\left(\frac{87}{87 + f_y}\right)$$

$$\rho_{\min} = \frac{0.2}{f_y} \qquad (f'_c \text{ and } f_y \text{ are in Ksi}), \text{ for } f'_c < 4.5 \text{ Ksi}$$

   (See Section 3.11.) The steel ratio in the section is $\rho = A_s/bd$. Check that $\rho_{\min} \leq \rho \leq \rho_{\max}$.
3. The allowable ultimate moment in the section according to the ACI Code is ϕM_n, where $\phi = 0.9$.
4. Another form of M_n is $M_n = \rho f_y(bd^2)\left(1 - \frac{\rho f_y}{1.7f'_c}\right) = R_n(bd^2)$.

$$R_n = \rho f_y\left(1 - \frac{\rho f_y}{1.7f'_c}\right) \quad \text{and} \quad R_u = \phi R_n$$

5. For $f_y = 60$ Ksi and $f'_c = 3$ Ksi:

   $\rho_{\max} = 0.0161$, $\rho_{\min} = 0.00333$, $R_n = 782$ psi, $R_u = 704$ psi

   For $f_y = 60$ Ksi and $f'_c = 4$ Ksi:

   $\rho_{\max} = 0.0214$, $\rho_{\min} = 0.00333$, $R_n = 1040$ psi, $R_u = 936$ psi

   For example, for $f_y = 60$ Ksi, $f'_c = 4$ Ksi, $b = 14$ in., $d = 26.5$ in., and $A_s = 5.0$ in.^{2} (five no. 9): $\rho = 0.01348$, $R_n = 712$ psi, and $\phi M_n = 525.3$ K·ft.

Section 3.13 Analysis of Rectangular Sections with Compression Steel

Given: $b, d, d', A_s, A'_s, f'_c, f_y$, and $\phi = 0.9$
Required: Ultimate moment capacity, ϕM_n
In general,

1. Calculate $\rho = \frac{A_s}{bd}$, $\rho' = \frac{A'_s}{bd}$, and $(\rho - \rho')$.
2. Calculate $\rho_{\max} = 0.75(0.85\beta_1)\left(\frac{f'_c}{f_y}\right)\left(\frac{87}{87 + f_y}\right)$ and $\rho_{\min} = \frac{0.2}{f_y}$ for $f'_c < 4.5$ Ksi (f'_c and f_y are in Ksi). (See Section 3.11.)
3. Calculate $K = 0.85\beta_1\left(\frac{f'_c}{f_y}\right)\left(\frac{d'}{d}\right)\left(\frac{87}{87 - f_y}\right)$ (f'_c and f_y are in Ksi).

   Case 1: When compression steel yields:
   a. Check that $\rho > \rho_{\min}$.
   b. Check that $(\rho - \rho') \geq K$.
      If $(\rho - \rho') \geq K$, then compression steel yields.
      If $(\rho - \rho') < K$, then compression steel *does not* yield; see Case 2.

c. If compression steel yields, then $f'_s = f_y$.
d. Check that $\rho \le (\rho_{max} + \rho')$ or $(\rho - \rho') \le \rho_{max}$.
e. Calculate $a = \dfrac{(A_s - A'_s)f_y}{0.85f'_c b}$.
f. Calculate $M_u = \phi M_n = \phi[(A_s - A'_s)f_y(d - a/2) + A'_s f_y(d - d')]$.

Case 2: When compression steel does *not* yield:
a. If $(\rho - \rho') < K$, then compression steel does not yield, or $f'_s < f_y$. The magnitude of f'_s is not known.
b. Calculate the distance to the neutral axis c from the following quadratic equation:

$$A_1 c^2 + A_2 c + A_3 = 0$$

$$A_1 = 0.85f'_c \beta_1 b$$

$$A_2 = A'_s(87 - 0.85f'_c) - A_s f_y$$

$$A_3 = -87A'_s d'$$

or solve for c from the following equation (which is similar to the equation in (b)):

$$C + C' = T$$

where

$$C = 0.85f'_c(\beta_1 cb - A'_s)$$

$$C' = A'_s\left[87\left(\frac{c - d'}{c}\right)\right] - 0.85f'_c A'_s$$

$$T = A_s \cdot f_y$$

c. Calculate $f'_s = 87\left(\dfrac{c - d'}{c}\right)$ (in Ksi), must be $\le f_y$.
d. Check that

$$\rho \le \left[\rho_{max} + \rho'\left(\frac{f'_s}{f_y}\right)\right] \quad \text{or} \quad A_s \le \left[\rho_{max}(bd) + A'_s\left(\frac{f'_s}{f_y}\right)\right]$$

e. Calculate $a = \dfrac{(A_s f_y - A'_s f'_s)}{0.85f'_c b}$ or $a = \beta_1 c$.
f. Calculate $M_u = \phi M_n = \phi[(A_s f_y - A'_s f'_s)(d - a/2) + A'_s f'_s(d - d')]$ or $M_u = \phi M_n = \phi[A_{s1} f_y(d - a/2) + A'_s f'_s(d - d')]$, where $A_{s1} = A_s - A_{s2} = A_s - (A'_s f'_s / f_y)$ and $A_{s2} f_y = A'_s f'_s$. Also, $a = A_{s1} f_y / 0.85f'_c b$ (same as in (e)).

Sections 3.14–3.16 Analysis of T-Sections

Given: f'_c, f_y, section dimensions, and A_s
Required: ultimate moment capacity, ϕM_n
Two possible cases exist (determine b_e first).

Case 1:

a. If $a \le t$ (slab thickness), then it is a T-section shape but acts as a singly reinforced rectangular section using b (flange width) as the width of compression top fibers (flange):

$$a' = \frac{A_s \cdot f_y}{0.85f'_c b_e} \le t$$

Or, check that A_c (area of concrete compression area) $\leq bt$, $A_c = A_s f_y / 0.85 f'_c \leq bt$.
Note: If $A_c > bt$, then it is Case 2, T-section analysis.

b. If $a' \leq t$ or $A_c \leq bt$, then $a' = a$.

$$\phi M_n = \phi A_s f_y (d - a/2), \text{ or}$$

$$\phi M_n = \phi R_n b d^2 = R_u b d^2 \qquad R_u = \phi \rho f_y \left(1 - \frac{\rho f_y}{1.7 f'_c}\right)$$

c. Check that $\rho_w = A_s / b_w d \geq \rho_{\min}$ (ρ_w = steel ratio in web) or $A_s > A_{s\,\min}$, where $A_{s\,\min} = \rho_{\min}(b_w d)$.

d. Check that $A_s \leq A_{s\,\max}$.

$$A_{s\max} = (0.75)\frac{(0.85 f'_c)}{f_y}\left[bt + b_w \beta_1 d \left(\frac{87}{87 + f_y}\right) - b_w t\right]$$

$$= 0.6375\left(\frac{f'_c}{f_y}\right)\left[t(b - b_w) + b_w \beta_1 d \left(\frac{87}{87 + f_y}\right)\right]$$

Note: b (flange width) is the *smallest* of

1. Span/4;
2. Center to center of adjacent slabs;
3. $(b_w + 16t)$, t = slab thickness.

Case 2: When $a > t$ or $A_c > bt$, it is a T-section analysis.

a. Flange: $C_f(\text{flange}) = 0.85 f'_c\, t(b - b_w) = A_{sf} \cdot f_y$. Calculate $A_{sf} = C_f / f_y = 0.85 f'_c\, t(b - b_w)/f_y$.

b. Web: (A_{sw} = tension steel in web):

$$A_{sw} = A_s - A_{sf}$$

$$a = \left(\frac{A_s - A_{sf}}{0.85 f'_c b_w}\right) f_y$$

$$C_w \text{ (web)} = 0.85 f'_c a b_w = A_{sw} \cdot f_y$$

c. $$\phi M_n = \phi\left[C_w\left(d - \frac{a}{2}\right) + C_f\left(d - \frac{t}{2}\right)\right]$$

$$= \phi[M_w(\text{web}) + M_f(\text{flange})]$$

$$M_u = \phi M_n = \phi\left[0.85 f'_c a b_w \left(d - \frac{a}{2}\right) + 0.85 f'_c t(b - b_w)\left(d - \frac{t}{2}\right)\right]$$

$$= \phi\left[(A_s - A_{sf}) f_y \left(d - \frac{a}{2}\right) + A_{sf} \cdot f_y \left(d - \frac{t}{2}\right)\right]$$

d. Check that $A_{s\,\min} \leq A_s \leq A_{s\,\max}$. See Case 1.

Sections 3.17–3.19

a. Analysis of nonuniform sections is explained using Example 3.10.

b. More than one row of steel bars can be used in reinforced concrete sections.

c. Tables in Appendix A may be used for the analysis of rectangular sections.

REFERENCES

1. E. Hognestad, N. W. Hanson, and D. McHenry. "Concrete Distribution in Ultimate Strength Design." *ACI Journal* 52 (December 1955): 455–79.
2. J. R. Janney, E. Hognestad, and D. McHenry. "Ultimate Flexural Strength of Prestressed and Conventionally Reinforced Concrete Beams." *ACI Journal* (February 1956): 601–20.
3. A. H. Mattock, L. B. Kriz, and E. Hognestad. "Rectangular Concrete Stress Distribution in Ultimate Strength Design." *ACI Journal* (February 1961): 875–929.
4. A. H. Mattock and L. B. Kriz. "Ultimate Strength of Nonrectangular Structural Concrete Members." *ACI Journal* 57 (January 1961): 737–66.
5. American Concrete Institute. "Building Code Requirements for Structural Concrete." ACI Code 318-99, American Concrete Institute, Detroit, 1999.
6. Franco Levi. "Work of European Concrete Committee." *ACI Journal* 57 (March 1961): 1049–54.
7. UNESCO. *Reinforced Concrete,* An International Manual. Butterworth, London, 1971.
8. M. N. Hassoun. "Ultimate-Load Design of Reinforced Concrete," *View Point Publication.* Cement and Concrete Association, London, 1981, 2nd ed.

PROBLEMS

3.1 Singly reinforced rectangular sections Determine the moment capacity of the sections given in the following table, knowing that $f'_c = 4$ Ksi and $f_y = 60$ Ksi. (Answers are given in the right column.)

No.	b (in.)	d (in.)	A_s (in.^{2})	ϕM_n (K·ft)
a	14	22.5	5.08 (4 no. 10)	441.2
b	18	28.5	7.62 (6 no. 10)	849.1
c	12	23.5	4.00 (4 no. 9)	370.1
d	12	18.5	3.16 (4 no. 8)	230.0
e	16	24.5	6.35 (5 no. 10)	600.0
f	14	26.5	5.00 (5 no. 9)	525.3
g	10	17.5	3.00 (3 no. 9)	200.5
h	20	31.5	4.00 (4 no. 9)	535.2

For problems in SI units: 1 in. = 25.4 mm
1 in.^{2} = 645 mm^2
1 Ksi = 6.9 MPa (N/mm^2)
1 M_u (K·ft) = 1.356 kN·m

3.2 Rectangular section with compression steel Determine the moment capacity of the sections given in the following table, knowing that $f'_c = 4$ Ksi, and $f_y = 60$ Ksi, and $d' = 2.5$ in. (Answers are given in the right column. In the first four problems, $f'_s = f_y$.)

No.	b (in.)	d (in.)	A_s (in.^{2})	A_s' (in.^{2})	ϕM_n (K·ft)
a	14	22.5	8.0 (8 no. 9)	2.0 (2 no. 9)	685.4
b	16	24.5	10.08 (8 no. 10)	2.54 (2 no. 10)	941.6
c	12	21.5	7.00 (7 no. 9)	1.8 (3 no. 7)	567.5
d	10	18.5	5.08 (4 no. 10)	1.2 (2 no. 7)	348.6
e	14	20.5	7.62 (6 no. 10)	2.54 (2 no. 10)	597.9
f	16	20.5	9.0 (9 no. 9)	4.0 (4 no. 9)	716.3
g	20	18.0	12.0 (12 no. 9)	6.0 (6 no. 9)	820.3
h	18	20.5	10.16 (8 no. 10)	5.08 (4 no. 10)	813.7

For problems in SI units: 1 in. = 25.4 mm
1 in.^{2} = 645 mm^2
1 Ksi = 6.9 MPa (N/mm^2)
1 M_u (K·ft) = 1.356 kN·m

3.3 T-sections Determine the moment capacity of the T-sections given in the following table, knowing that $f'_c = 3$ Ksi and $f_y = 60$ Ksi. (Answers are given in the right column. In the first three problems, $a < t$.)

No.	b (in.)	b_w (in.)	t (in.)	d (in.)	A_s (in.^{2})	ϕM_n (K·ft)
a	54	14	3	17.5	5.08 (4 no. 10)	374.8
b	48	14	4	16.5	4.0 (4 no. 9)	279.4
c	72	16	4	18.5	10.16 (8 no. 10)	769.9
*d	32	16	3	15.5	6.0 (6 no. 9)	N.G.
e	44	12	4	20.5	8.0 (8 no. 9)	660.1
f	50	14	3	16.5	7.0 (7 no. 9)	466.8
g	40	16	3	16.5	6.35 (5 no. 10)	415.0
h	42	12	3	17.5	6.0 (6 no. 9)	425.8

For problems in SI units: 1 in. = 25.4 mm
1 in.^{2} = 645 mm^2
1 Ksi = 6.9 MPa (N/mm^2)
1 M_u (K·ft) = 1.356 kN·m

*Answer = 308.6 K·ft if ρ_{max} is used.

3.4 Calculate ρ_b, ρ_{max}, $R_u(\max)$, R_u, a/d, and max (a/d) for a rectangular section that has a width $b = 12$ in. (300 mm) and an effective depth $d = 20$ in. (500 mm) for the following cases:

a. $f'_c = 3$ Ksi, $f_y = 40$ Ksi, $A_s = 4$ no. 8 bars

b. $f'_c = 4$ Ksi, $f_y = 60$ Ksi, $A_s = 4$ no. 7 bars

c. $f'_c = 4$ Ksi, $f_y = 75$ Ksi, $A_s = 4$ no. 9 bars

d. $f'_c = 5$ Ksi, $f_y = 60$ Ksi, $A_s = 4$ no. 9 bars

e. $f'_c = 30$ MPa, $f_y = 400$ MPa, $A_s = 3 \times 30$ mm

f. $f'_c = 20$ MPa, $f_y = 300$ MPa, $A_s = 3 \times 25$ mm

g. $f'_c = 30$ MPa, $f_y = 500$ MPa, $A_s = 4 \times 25$ mm

h. $f'_c = 25$ MPa, $f_y = 300$ MPa, $A_s = 4 \times 20$ mm

3.5 Using the ACI Code requirements, calculate the ultimate moment capacity of a rectangular section that has a width $b = 250$ mm (10 in.) and an effective depth $d = 550$ mm (22 in.) when $f'_c = 20$ MPa (3 Ksi), $f_y = 420$ MPa (60 Ksi), and the steel used is as follows.

a. 4×20 mm **b.** 4×25 mm **c.** 4×30 mm

d. 3 no. 9 bars **e.** 6 no. 9 bars

3.6 A reinforced concrete simple beam has a rectangular section with width $b = 8$ in. (200 mm) and effective depth $d = 18$ in. (450 mm). At ultimate moment (failure), the strain in the steel was recorded and was equal to 0.0015. (The strain in concrete at failure may be assumed to be 0.003.)

a. Check if the section is balanced, under-reinforced, or over-reinforced.

b. Determine the steel area that will make the section balanced.

c. Calculate the steel area provided in the section to produce the aforementioned strains, and then calculate its ultimate moment. Compare this value with the ultimate moment allowed by the ACI Code using ρ_{max}.

d. Calculate the ultimate moment capacity of the section if the steel percentage used is $\rho = 1.4\%$. Use $f'_c = 3$ Ksi (20 MPa) and $f_y = 50$ Ksi (350 MPa) for all parts.

3.7 A 10-ft.- (3-m-) span cantilever beam has an effective cross section (bd) of 12 in. by 24 in. (300 by 600 mm) and is reinforced with five no. 8 (5×25 mm) bars. If the uniform load due to its own weight and the dead load are equal to 685 lb/ft (10 kN/m), determine the allowable uniform live load on the beam using the strength design method and ACI load factors. Given: $f'_c = 3$ Ksi (20 MPa) and $f_y = 60$ Ksi (400 MPa).

3.8 The cross section of a 17-ft.- (5-m-) span simply supported beam is 10 by 28 in. (250 by 700 mm), and it is reinforced symmetrically with eight no. 6 bars (8×20 mm) in two rows. Determine the allowable concentrated live load at midspan considering the total acting dead load (including self-weight) is equal to 2.55 K/ft (37 kN/m). Given: $f'_c = 3$ Ksi (20 MPa) and $f_y = 40$ Ksi (300 MPa).

3.9 Determine the ultimate moment capacity of the sections shown in Figure 3.34. Neglect the lack of symmetry in (b). Given: $f'_c = 4$ Ksi (30 MPa) and $f_y = 60$ Ksi (400 MPa).

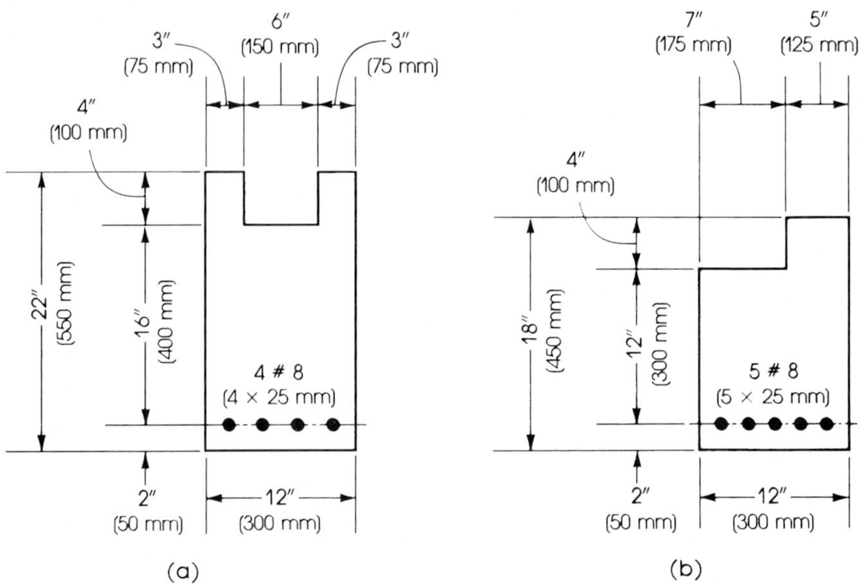

Figure 3.34 Problem 3.9.

3.10 A rectangular concrete section has a width $b = 12$ in. (300 mm), an effective depth $d = 18$ in. (450 mm), and $d' = 2.5$ in. (60 mm). If compression steel consisting of two no. 7 bars (2×20 mm) is used, calculate the allowable ultimate moment that can be applied on the section if the tensile steel A_s is as follows.

a. Four no. 7 (4×20 mm) bars **b.** Eight no. 8 (8×25 mm) bars

Given: $f'_c = 3$ Ksi (20 MPa) and $f_y = 40$ Ksi (300 MPa).

3.11 A 16-ft.- (4.8-m-) span simply supported beam has width $b = 12$ in. (300 mm), $d = 22$ in. (500 mm), $d' = 2.5$ in. (60 mm), and $A'_s =$ three no. 6 bars (3×20 mm). The beam carries a uniform dead load of 2 K/ft (30 kN/m), including its own weight. Calculate the allowable uniform live load that can be safely applied on the beam. Given: $f'_c = 3$ Ksi (20 MPa) and $f_y = 50$ Ksi (350 MPa). (Hint: Use ρ_{max} for the basic section to calculate M_u.)

3.12 Check the adequacy of a 10-ft- (3-m-) span cantilever beam, assuming a concrete strength $f'_c = 4$ Ksi (30 MPa) and a steel yield strength $f_y = 60$ Ksi (400 MPa) are used. The dimensions of the beam section are $b = 10$ in. (250 mm), $d = 20$ in. (500 mm), $d' = 2.5$ in. (60 mm), $A_s =$ six no. 7 bars (6×20 mm), and $A'_s =$ two no. 5 bars (2×15 mm). The dead load on the beam, excluding its own weight, is equal to 2 K/ft (30 kN/m), and the live load equals 1.25 K/ft (20 kN/m). (Compare the internal M_u with the external ultimate moment.)

3.13 A series of reinforced concrete beams spaced at 9 ft (2.7 m) on centers are acting on a simply supported span of 18 ft (5.4 m). The beam supports a reinforced concrete floor slab 4 in. (100 mm) thick. If the width of the web is $b_w = 10$ in. (250 mm), $d = 18$ in. (450 mm), and the beam is reinforced with three no. 9 bars (3×30 mm), determine the moment capacity of a typical interior beam. Given: $f'_c = 4$ Ksi (30 MPa) and $f_y = 60$ Ksi (400 MPa).

3.14 Calculate the ultimate moment capacity of a T-section that has the following dimensions:

- Flange width = 30 in. (750 mm)
- Flange thickness = 3 in. (75 mm)
- Web width = 10 in. (250 mm)
- Effective depth $d = 18$ in. (450 mm)
- Tension reinforcement: six no. 8 bars (6×25 mm)
- $f'_c = 3$ Ksi (20 MPa)
- $f_y = 60$ Ksi (400 MPa)

3.15 Repeat Problem 3.14 if $d = 24$ in. (600 mm).

3.16 Repeat Problem 3.14 if the flange is an inverted L-shape with the same flange width projecting from one side only. (Neglect lack of symmetry.)

Reinforced concrete office building. Amman, Jordan.

STRENGTH DESIGN METHOD: FLEXURAL DESIGN OF REINFORCED CONCRETE BEAMS

4.1 INTRODUCTION

In the previous chapter, the analysis of different reinforced concrete sections was explained: Details of the section were given, and we had to determine the ultimate moment capacity of the section. In this chapter, the process is reversed: The external moment is given, and we must find safe, economic, and practical dimensions of the concrete section and the area of reinforcing steel that provide adequate internal moment capacity.

4.2 RECTANGULAR SECTIONS WITH TENSION REINFORCEMENT ONLY

From the analysis of rectangular singly reinforced sections (Section 3.9), the following equations were derived, where f'_c and f_y are in Ksi:

$$\rho_b = 0.85\beta_1 \frac{f'_c}{f_y}\left(\frac{87}{87 + f_y}\right) \tag{3.13}$$

$$\rho_{\max} = 0.75\rho_b = 0.6375\beta_1 \frac{f'_c}{f_y}\left(\frac{87}{87 + f_y}\right) \tag{3.14}$$

For $f'_c \leq 4000$ psi,

$$\rho_{\max} = 0.542 \frac{f'_c}{f_y}\left(\frac{87}{87 + f_y}\right) \tag{3.15}$$

The value of β_1 is 0.85 when $f'_c \leq 4000$ psi (30 N/mm^2) and decreases by 0.05 for every increase of 1000 psi (7 N/mm^2) in concrete strength. The steel percentage of a balanced section, ρ_b, and the maximum allowable steel percentage, $\rho_{\max}$, can be calculated for different values of f'_c and f_y, as shown in Table A.4 in Appendix A and Table 4.1.

Table 4.1 Suggested Design Steel Ratios, ρ_s

f'_c (Ksi)	f_y (Ksi)	$\%\rho_b$	$\%\rho_{\max}$	$\%\rho_s$	RATIO ρ_s/ρ_b	RATIO $\rho_s/\rho_{\max}$	R_u (psi)	$R_{u\max}$ (psi)
3	40	3.71	2.78	1.4	0.38	0.50	450	783
	60	2.15	1.61	1.2	0.56	0.75	556	702
4	60	2.85	2.14	1.4	0.49	0.65	662	936
	75	2.07	1.55	1.2	0.58	0.77	702	867
5	60	3.36	2.52	1.4	0.42	0.56	681	1120
	75	2.44	1.83	1.2	0.49	0.66	722	1033

The ultimate-moment equations were derived in the previous chapter in the following forms:

$$M_u = R_u bd^2 \tag{3.18}$$

where

$$R_u = \phi \rho f_y\left(1 - \frac{\rho f_y}{1.7 f'_c}\right) = \phi R_n \tag{3.19}$$

and $\phi = 0.9$, or

$$\phi M_n = M_u = \phi A_s f_y\left(d - \frac{A_s f_y}{1.7 f'_c b}\right) \tag{3.16a}$$

Also,

$$\phi M_n = M_u = \phi \rho f_y bd^2\left(1 - \frac{\rho f_y}{1.7 f'_c}\right) \tag{3.17}$$

We can see that for a given ultimate moment and known f'_c and f_y, there are three unknowns in these equations: the width, b, the effective depth of the section, d, and the steel ratio, ρ. A unique solution is not possible unless values of two of these three unknowns are assumed. Usually ρ is assumed (using $\rho_{\max}$, for instance), and b can also be assumed.

Based on the preceding discussion, the following cases may develop for a given M_u, f'_c, and f_y:

1. If ρ is assumed, then R_u can be calculated from equation (3.19), giving $bd^2 = M_u/R_u$. The ratio of d/b usually varies between 1 and 3, with a practical ratio of 2. Consequently, b and d can be determined, and $A_s = \rho bd$. The ratio ρ for a singly reinforced rectangular section must be equal to or less than $\rho_{\max}$, as given in equation (3.14). It is a common practice to assume a value of ρ that ranges between $\frac{1}{2}\rho_{\max}$ and $\frac{1}{2}\rho_b$. Table 4.1 gives suggested values of the steel ratio ρ to be used in singly reinforced sections when ρ is not assigned. For example, if $f_y = 60$ Ksi, then $\rho_s = 1.4\%$ is suggested for $f'_c = 4$ Ksi or 5 Ksi and $\rho_s = 1.2\%$ for $f'_c = 3$ Ksi. The designer may use ρ up to $\rho_{\max}$,

which produces the minimum size of the singly reinforced concrete section. Using $\rho_{\min}$ will produce the maximum concrete section. If b is assumed in addition to ρ, then d can be determined as follows:

$$d = \sqrt{\frac{M_u}{R_u b}} \tag{4.1}$$

If $d/b = 2$, then $d = \sqrt[3]{(2M_u/R_u)}$ and $b = d/2$, rounded to nearest higher inch.

2. If b and d are given, then the required reinforcement ratio ρ can be determined by rearranging equation (3.17) to obtain

$$\rho = \frac{0.85f'_c}{f_y}\left[1 - \sqrt{1 - \frac{4M_u}{1.7\phi f'_c bd^2}}\right] \tag{4.2}$$

$$= \frac{0.85f'_c}{f_y}\left[1 - \sqrt{1 - \frac{2R_n}{0.85f'_c}}\right] \tag{4.2a}$$

or

$$\rho = \frac{f'_c}{f_y}\left[0.85 - \sqrt{(0.85)^2 - Q}\right]$$

where

$$Q = \left(\frac{1.7}{\phi f'_c}\right)\frac{M_u}{bd^2} = \left(\frac{1.7}{\phi f'_c}\right)R_u \tag{4.3}$$

$$A_s = \rho bd = \left(\frac{f'_c}{f_y}\right)bd\left[0.85 - \sqrt{(0.85)^2 - Q}\right] \tag{4.4}$$

where all units are in kips (or pounds) and inches, and Q is dimensionless. For example, if $M_u = 2440$ K·in., $b = 12$ in., $d = 18$ in., $f'_c = 3$ Ksi, and $f_y = 60$ Ksi, then $\rho = 0.01389$ (from equation (4.2)) and $A_s = \rho bd = 0.01389(12)(18) = 3.0$ in.^{2}, or directly from equation (4.4), $Q = 0.395$ and $A_s = 3.0$ in^2. When b and d are given, it is better to check if compression steel is or is not required because of a small d. This can be achieved as follows:

a. Calculate $\rho_{\max}$ and $R_u(\max) = \phi\rho_{\max}[1 - (\rho_{\max} f_y/1.7f'_c)]f_y$.

b. Calculate $\phi M_n(\max) = R_u(\max)bd^2 =$ the maximum moment capacity of a singly reinforced concrete section.

c. If $M_u < \phi M_n(\max)$, then no compression reinforcement is needed. Calculate ρ and A_s from the preceding equations.

d. If $M_u > \phi M_n(\max)$, then compression steel is needed. In this case, the design procedure is explained in Section 4.4.

3. If ρ and b are given, calculate R_u:

$$R_u = \phi\rho f_y\left(1 - \frac{\rho f_y}{1.7f'_c}\right)$$

Then calculate d from equation (4.1):

$$d = \sqrt{\frac{M_u}{R_u b}} \quad \text{and} \quad A_s = \rho bd$$

4.3 SPACING OF REINFORCEMENT AND CONCRETE COVER

4.3.1 Specifications

Figure 4.1 shows two reinforced concrete sections. The bars are placed such that the clear spacings shall be at least equal to nominal bar diameter D but not less than 1 in. (25 mm). Vertical clear spacings between bars in more than one layer shall not be less than 1 in. (25 mm), according to the ACI Code, Section 7.6.

The width of the section depends on the number, n, and diameter of bars used. Stirrups are placed at intervals; their diameters and spacings depend on shear requirements, to be explained later. At this stage, stirrups $\frac{3}{8}$ in. (10 mm) diameter can be assumed to calculate the width of the section. There is no need to adjust the width b if different diameters of stirrups are used. The specified minimum concrete cover for cast-in-place and precast concrete is given in the ACI Code, Section 7.7. Concrete cover for beams and girders is equal to $\frac{3}{2}$ in. (38 mm), and that for slabs is equal to $\frac{3}{4}$ in. (20 mm), when concrete is not exposed to weather or in contact with ground.

4.3.2 Minimum width of concrete sections

The general equation for the minimum width of a concrete section can be written in the following form:

$$b_{\min} = nD + (n - 1)s + 2(\text{stirrup's diameter}) + 2(\text{concrete cover}) \qquad (4.5a)$$

where

n = number of bars
D = diameter of the largest bar used
s = spacing between bars (equal to D or 1 in., whichever is larger)

If the stirrup's diameter is taken equal to $\frac{3}{8}$ in. (10 mm) and concrete cover equals $\frac{3}{2}$ in. (38 mm), then

$$b_{\min} = nD + (n - 1)s + 3.75 \text{ in.} \quad (95 \text{ mm}) \qquad (4.5b)$$

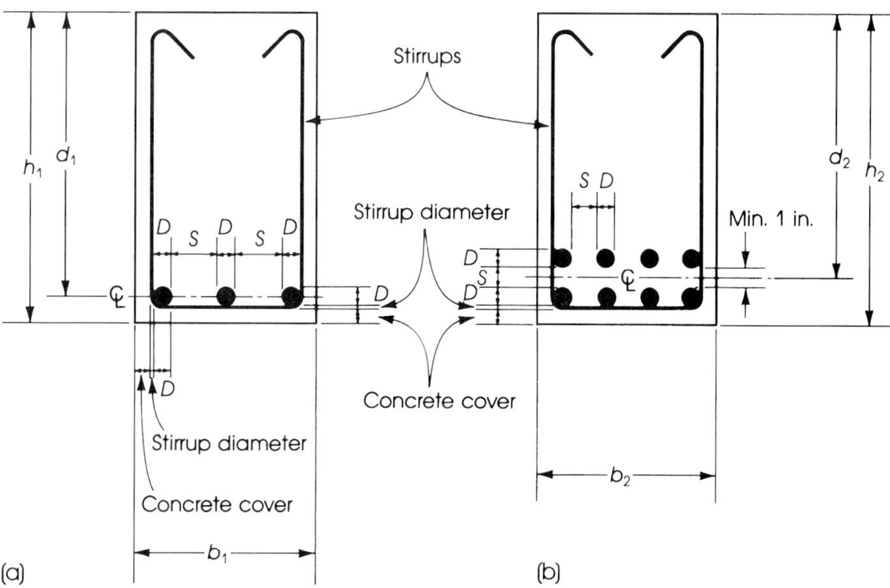

Figure 4.1 Spacing of steel bars (a) in one row or (b) two rows.

This equation, if applied to the concrete sections in Figure 4.1, becomes

$$b_1 = 3D + 2S + 3.75 \text{ in.} \quad (95 \text{ mm})$$

$$b_2 = 4D + 3S + 3.75 \text{ in.} \quad (95 \text{ mm})$$

To clarify the use of equation (4.5), let the bars used in sections of Figure 4.5 be no. 10 (32-mm) bars. Then

$$b_1 = 5 \times 1.27 + 3.75 = 10.10 \text{ in.} \quad (s = D), \qquad \text{say, 11 in.}$$

$$b_1 = 5 \times 32 + 95 = 225 \text{ mm}, \qquad \text{say, 250 mm}$$

$$b_2 = 7 \times 1.27 + 3.75 = 12.64 \text{ in.}, \qquad \text{say, 13 in.}$$

$$b_2 = 7 \times 32 + 95 = 319 \text{ mm}, \qquad \text{say, 320 mm}$$

If the bars used are no. 6 (20 mm), the minimum widths become

$$b_1 = 3 \times 0.75 + 2 \times 1 + 3.75 = 8.0 \text{ in.} \quad s = 1.0 \text{ in.}$$

$$b_1 = 3 \times 20 + 2 \times 25 + 95 = 205 \text{ mm}, \qquad \text{say, 210 mm}$$

$$b_2 = 4 \times 0.75 + 3 \times 1 + 3.75 = 9.75 \text{ in.}, \qquad \text{say, 10 in.}$$

$$b_2 = 4 \times 20 + 3 \times 25 + 95 = 250 \text{ mm}$$

The width of the concrete section shall be increased to the nearest inch. Table A.7 in Appendix A gives the minimum beam width for different numbers of bars in the section.

4.3.3 Minimum overall depth of concrete sections

The effective depth, d, is the distance between the extreme compressive fibers of the concrete section and the centroid of the tension reinforcement. The minimum total depth is equal to d plus the distance from the centroid of the tension reinforcement to the extreme tension concrete fibers, which depends on the number of layers of the steel bars. In application to the sections shown in Figure 4.1,

$$\begin{aligned} h_1 &= d_1 + \frac{D}{2} + \tfrac{3}{8} \text{ in.} + \text{concrete cover} \\ &= d_1 + \frac{D}{2} + 1.857 \text{ in.} \quad (50 \text{ mm}) \end{aligned} \tag{4.6a}$$

for one row of steel bars and

$$\begin{aligned} h_2 &= d_2 + 0.5 + D + \tfrac{3}{8} \text{ in.} + \text{concrete cover} \\ &= d_2 + D + 2.375 \text{ in.} \quad (60 \text{ mm}) \end{aligned} \tag{4.6b}$$

for two layers of steel bars. The overall depth, h, shall be increased to the nearest half inch (10 mm) or, better, to the nearest inch (20 mm in SI). For example, if $D = 1$ in. (25 mm), $d_1 = 18.9$ in. (475 mm), and $d_2 = 20.1$ in. (502 mm),

$$\text{Minimum } h_1 = 18.9 + 0.5 + 1.875 = 21.275 \text{ in.}$$

say, 21.5 in. or 22 in.,

$$h_1 = 475 + 13 + 50 = 538 \text{ mm}$$

say, 540 mm or 550 mm, and

$$\text{Minimum } h_2 = 20.1 + 1.0 + 2.375 = 23.475 \text{ in.}$$

say, 23.5 in. or 24 in.,

$$h_2 = 502 + 25 + 60 = 587 \text{ mm},$$

say, 590 mm or 600 mm.

If no. 9 or smaller bars are used, a practical estimate of the total depth, h, can be made as follows:

$$h = d + 2.5 \text{ in.} \quad (65 \text{ mm}), \qquad \text{for one layer of steel bars}$$

$$h = d + 3.5 \text{ in.} \quad (90 \text{ mm}), \qquad \text{for two layers of steel bars}$$

For more than two layers of steel bars, a similar approach may be used.

It should be mentioned that the minimum spacing between bars depends on the maximum size of the coarse aggregate used in concrete. The nominal maximum size of the coarse aggregate shall not be larger than one-fifth of the narrowest dimension between sides of forms, nor one-third of the depth of slabs, nor three-fourths of the minimum clear spacing between individual reinforcing bars or bundles of bars (ACI Code, Section 3.3).

Example 4.1

Design a singly reinforced rectangular section to resist a factored moment of 412 K · ft using the maximum allowable steel percentage ρ_{max}. Given: $f'_c = 3$ Ksi and $f_y = 60$ Ksi.

Solution

For $f'_c = 3$ Ksi, $f_y = 60$ Ksi, $\beta_1 = 0.85$ and ρ_{max} is calculated as follows:

$$\rho_{max} = (0.75)(0.85)\beta_1(f'_c/f_y)[87/(87 + f_y)],$$

$$\rho_{max} = (0.75)(0.85)^2\left(\frac{3}{60}\right)\left(\frac{87}{147}\right) = 0.016$$

$$(R_u)_{max} = \phi\rho_{max}f_y\left(1 - \frac{\rho_{max}f_y}{1.7f'_c}\right)$$

$$= 0.9 \times 0.016 \times 60 \times \left(1 - \frac{0.016 \times 60}{1.7 \times 3}\right) = 0.702 \text{ Ksi}$$

(Or, use the tables in Appendix A.)
Since $M_u = R_u bd^2$,

$$bd^2 = \frac{M_u}{R_u} = \left(\frac{412 \times 12}{0.702}\right) = \frac{4944}{0.702} = 7043 \text{ in.}^3$$

Thus, for the following assumed b, calculate d and $A_s = \rho bd$:

$b = 10$ in., $d = 26.5$ in., $A_s = 4.24$ in.^{2},

$b = 12$ in., $d = 24.2$ in., $A_s = 4.65$ in.^{2}, 6 no. 8 bars ($A_s = 4.71$ in.^{2})

$b = 14$ in., $d = 22.4$ in., $A_s = 5.01$ in.^{2}, 5 no. 9 bars ($A_s = 5.0$ in.^{2})

$b = 16$ in., $d = 21.0$ in., $A_s = 5.37$ in.^{2}

The choice of the effective depth d depends on three factors:

1. The width b required. A small width will result in a deep beam that decreases the headroom available. Furthermore, a deep narrow beam may lower the ultimate moment capacity of the structural member due to possible lateral deformation.
2. The amount and distribution of reinforcing steel. A narrow beam may need more than one row of steel bars, thus increasing the total depth of the section.
3. The wall thickness. If cement block walls are used, the width b is chosen to be equal to the wall thickness. Exterior walls in buildings in most cases are thicker than interior walls. The architectural plan of the structure will show the different thicknesses.

A reasonable choice of d/b varies between 1 and 3, with a practical value of about 2. It can be seen from the previous calculations that the deeper the section, the more economical it is, as far as the quantity of concrete used, expressed by the area bd of a 1-ft length of the beam. Alternatively, calculate $bd^2 = M_u/R_u$ and then choose adequate b and d.

The area of the steel reinforcement, A_s, is equal to ρbd. The area of steel needed for the different choices of b and d for this example was shown earlier. Because the steel percentage required is constant $(\rho_{max} = 0.016)$, A_s is proportional to bd. For a choice of a 12×24.2-in. section, the required A_s is 4.65 in.^{2} Choose six no. 8 bars in two rows (actual $A_s = 4.71$ in.^{2}). The minimum b required for three no. 8 in one row is 8.9 in. < 12 in., and total $h = 24.2 + 3.5 = 27.7$ in., say, 28 in. (actual $d = 24.6$ in.). Another choice is a 14×22.4-in. section with a total depth $h = 25$ in., and five no. 9 bars in one row. The choice of bars depends on

1. Adequate placement of bars in the section, normally in one or two rows, fulfilling the restrictions of the ACI Code for minimum spacing between bars;
2. The area of steel bars chosen closest to the required calculated steel area.

The final section is shown in Figure 4.2.

Example 4.2

Solve Example 4.1 using a steel percentage ρ of about 1% and $b = 14$ in.

Solution

1. For $f'_c = 3$ Ksi and $f_y = 60$ Ksi, $\rho_{max} = 0.016$ and $\rho_{min} = 0.00333$.

$$R_u = \phi \rho f_y \left(1 - \frac{\rho f_y}{1.7 f'_c}\right)$$

$$= 0.9 \times 0.01 \times 60 \left(1 - \frac{0.01 \times 60}{1.7 \times 3}\right) = 0.476 \text{ Ksi}$$

(From the tables in Appendix A, for $\rho = 0.01$, $R_u = 476$ psi.)

2. $bd^2 = \dfrac{M_u}{R_u} = \dfrac{4944}{0.476} = 10{,}387$ in.^{3}

Choosing $b = 14$ in. and $d = 27.24$ in.,

$$A_s = \rho db = 0.01 \times 14 \times 27.24 = 3.81 \text{ in.}^2$$

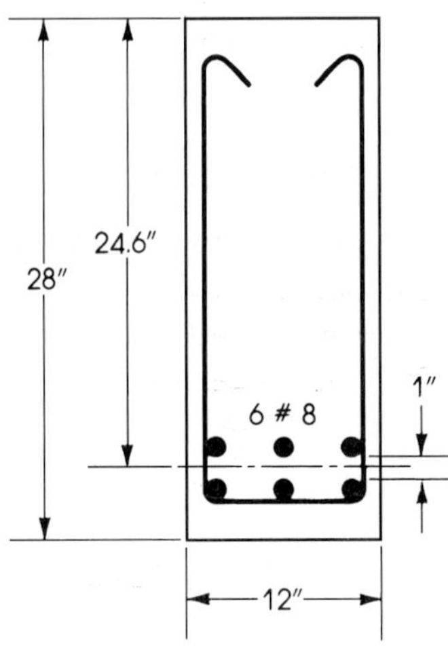

Figure 4.2 Example 4.1.

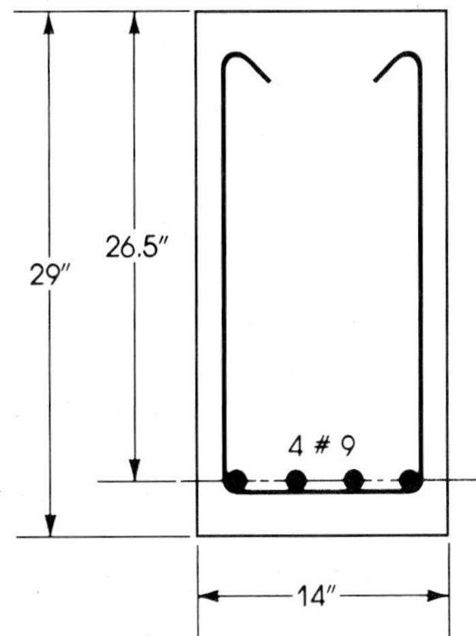

Figure 4.3 Example 4.2.

Choose four no. 9 bars in one layer; $A_s = 4.00 \text{ in.}^2$

$$b_{\min} = nD + (n-1)s + 3.75$$
$$= 7 \times 1.128 + 3.75 = 11.7 \text{ in.} < 14 \text{ in.}$$
$$h_{\min} = d + \frac{D}{2} + 1.875$$
$$= 27.24 + 1.138/2 + 1.875 = 29.7 \text{ in.}, \qquad \text{say, 30 in.}$$

3. Because the actual A_s used is greater than the calculated A_s, a smaller depth can be adopted. Therefore, take $h = 29$ in. Then $d = 29 - 1.138/2 - 1.875 = 26.5$ in.

For small variation in depth, $A_s = 3.81(27.5/26.5) = 3.95 \text{ in.}^2$, which is less than the 4.00 in.² used (Figure 4.3). A check of the ultimate moment capacity of the section can be made.

$$a = \frac{A_s f_y}{0.85 f'_c b} = \frac{4.0 \times 60}{0.85 \times 3 \times 14} = 6.72 \text{ in.}$$
$$\phi M_n = \phi A_s f_y\left(d - \frac{a}{2}\right)$$
$$= 0.9 \times 4 \times 60\left(26.5 - \frac{6.72}{2}\right) = 4998 \text{ K}\cdot\text{in.} > 4944 \text{ K}\cdot\text{in.}$$

which is acceptable.

The final section is shown in Figure 4.3.

Example 4.3

Find the necessary reinforcement for a given section 10 in. wide and 28 in. total depth (Figure 4.4) if it is subjected to an external factored moment of 245 K · ft. Given: $f'_c = 4$ Ksi and $f_y = 60$ Ksi.

Solution

1. Assuming one layer of no. 8 steel bars (to be checked later), $d = 28 - 0.5 - 1.875 = 25.625$ in. (or $d = 28 - 2.5 \text{ in.} = 25.5$ in.).
2. Check if the section is adequate without compression reinforcement. Compare ultimate moment capacity of the section (using $\rho_{\max}$) with the design moment.

   For $f'_c = 4$ Ksi and $f_y = 60$ Ksi, $\rho_{\max} = 0.02138$.

$$R_{u(\max)} = \phi \rho_{\max} f_y\left(1 - \frac{\rho_{\max} f_y}{1.7 f'_c}\right) = 937 \text{ psi}$$

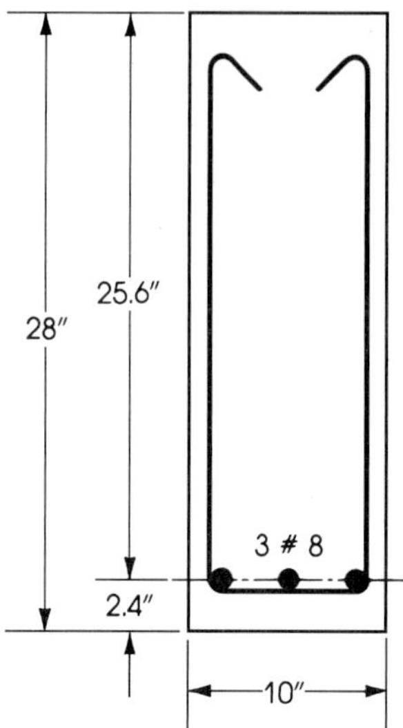

Figure 4.4 Example 4.3.

The maximum moment capacity of a singly reinforced basic section is

$$\phi M_{n(\max)} = R_{u(\max)} bd^2 = 0.937(10)(25.6)^2$$

$$= 6141 \text{ K}\cdot\text{in.} > 245 \times 12 = 2940 \text{ K}\cdot\text{in.}$$

Therefore, $\rho < \rho_{\max}$ and the section is singly reinforced.

3. Calculate ρ from equation (4.2) or (4.3):

$$Q = \left(\frac{1.7}{\phi f'_c}\right) \times \frac{M_u}{bd^2} = \left(\frac{1.7}{0.9 \times 4}\right) \times \left(\frac{2940}{10 \times 25.6^2}\right) = 0.2118$$

$$\rho = \frac{f'_c}{f_y}\left(0.85 - \sqrt{(0.85)^2 - Q}\right) = 0.009$$

$A_s = \rho bd = 0.009(10)(25.6) = 2.31 \text{ in.}^2$ Use three no. 8 bars $\left(A_s = 2.35 \text{ in.}^2\right)$ in one row, $b_{\min} < 10$ in. The final section is shown in Figure 4.4.

4.4 RECTANGULAR SECTIONS WITH COMPRESSION REINFORCEMENT

A singly reinforced section has a maximum internal ultimate moment capacity when $\rho_{\max}$ of steel is used. If the applied ultimate moment is greater than the internal ultimate moment capacity, as in the case of a limited cross section, a doubly reinforced section may be used, adding steel bars in both the compression and the tension zones. Compression steel will provide compressive force in addition to the compressive force in the concrete area.

The procedure for designing a rectangular section with compression steel when M_u, f'_c, f_y, b, d, and d' are given can be summarized as follows:

1. Calculate the maximum steel ratio, $\rho_{\max}$, using equation (3.14).

$$\rho_{\max} = 0.6375\beta_1 \frac{f'_c}{f_y}\left(\frac{87}{87 + f_y}\right)$$

Calculate $A_{s\,\max} = A_{s1} = \rho_{\max} bd$ (maximum steel area as singly reinforced).

2. Calculate $R_{u\,max}$ using ρ_{max}:

$$R_{u\,max} = \phi\rho_{max}f_y\left(1 - \frac{\rho_{max}f_y}{1.7f'_c}\right)$$

($R_{u\,max}$ can be obtained from the tables in Appendix A.)

3. Calculate the maximum moment capacity of the section M_{u1}, as singly reinforced, using ρ_{max} and $R_{u\,max}$.

$$M_{u1} = R_{u\,max}bd^2$$

If $M_{u1} < M_u$ (the applied moment), then compression steel is needed. Go to the next step. If $M_{u1} > M_u$, then compression steel is not needed. Use equation (4.2) to calculate ρ and $A_s = \rho bd$, as explained earlier.

4. Calculate $M_{u2} = M_u - M_{u1}$ = the moment to be resisted by compression steel.
5. Calculate A_{s2} from $M_{u2} = \phi A_{s2}f_y(d - d')$.
   Then calculate the total tension reinforcement A_s:

$$A_s = A_{s1} + A_{s2}$$

6. Calculate the stress in the compression steel as follows:
   a. Calculate $a = A_{s1}f_y/(0.85f'_cb)$ and $c = a/\beta_1$. (A_{s1} was calculated in Step 1.)
   b. Calculate $f'_s = 87[(c - d')/c]$ Ksi $\leq f_y$. (f'_s cannot exceed f_y.)
   c. Or, ε'_s can be calculated from the strain diagram, and $f'_s = (\varepsilon'_s \cdot E_s)$. If $\varepsilon'_s \geq \varepsilon_y$, then compression steel yields and $f'_s = f_y$.
   d. Calculate A'_s from $M_{u2} = \phi A'_s f'_s(d - d')$. If $f'_s = f_y$, then $A'_s = A_{s2}$. If $f'_s < f_y$, then $A'_s > A_{s2}$, and $A'_s = A_{s2}(f_y/f'_s)$.
7. Choose bars for A_s and A'_s to fit within the section width b. In most cases, A_s bars will be placed in two rows, whereas A'_s bars are placed in one row.
8. Calculate $h = d + 2.5$ in. for one row of tension bars and $h = d + 3.5$ in. for two rows of tension steel. Round h to the next higher inch. Now check that $[\rho - \rho'(f'_s/f_y)] \leq \rho_{max}$ using the new d, or check that $A_{s\,max} = bd[\rho_{max} + \rho'(f'_s/f_y)] \geq A_s$ (used). Actual d = final h − 2.5 in. (or 3.5 in. for two rows of bars),

$$\rho = A_s/(bd) \text{ and } \rho' = A'_s/(bd)$$

This check may not be needed if ρ_{max} is used in the basic section.

9. If desired, the ultimate moment capacity of the final section, ϕM_n, can be calculated and compared with the applied moment M_u: $\phi M_n \geq M_u$. Note that a steel ratio ρ smaller than ρ_{max} can be assumed in Step 1, say, $\rho = 0.7\rho_b$ or $\rho = 0.9\rho_{max}$, so that the final tension bars can be chosen to meet the given ρ_{max} limitation.

For flow charts, refer to Chapter 21.

Example 4.4

A beam section is limited to a width $b = 10$ in. and a total depth $h = 22$ in. and has to resist a factored moment of 3150 K · in. Calculate the required reinforcement. Given: $f'_c = 3$ Ksi and $f_y = 50$ Ksi.

Solution

1. Determine the ultimate moment capacity that is allowed for the section as singly reinforced. This is done by starting with ρ_{max}. For $f'_c = 3$ Ksi and $f_y = 50$ Ksi,

$$\rho_{max} = 0.0206, \qquad R_{u\,max} = 740 \text{ psi}$$

$$M_u = R_u bd^2, \qquad b = 10 \text{ in.}, \qquad d = 22 - 3.5 = 18.5 \text{ in.}$$

(This calculation assumes two rows of steel, to be checked later.) $M_{u1} = 0.740 \times 10 \times (1.85)^2 = 2533\ \text{K}\cdot\text{in.} = \max \phi M_n$, as singly reinforced. Design $M_u = 3150\ \text{K}\cdot\text{in.} > 2533\ \text{K}\cdot\text{in.}$ Therefore, compression steel is needed to carry the difference.

2. Compute A_{s1}, M_{u1}, and M_{u2}:

$$A_{s1} = \rho_{\max} bd = 0.0206 \times 10 \times 18.5 = 3.81\ \text{in.}^2$$

$$M_{u1} = 2533\ \text{K}\cdot\text{in.}$$

$$M_{u2} = M_u - M_{u1} = 3150 - 2533 = 617\ \text{K}\cdot\text{in.}$$

3. Calculate A_{s2} and A'_s, the additional tension and compression steel due to M_{u2}. Assume $d' = 2.5$ in.; $M_{u2} = \phi A_{s2} f_y (d - d')$.

$$A_{s2} = \frac{M_{u2}}{\phi f_y (d - d')} = \frac{617}{0.9 \times 50(18.5 - 2.5)} = 0.86\ \text{in.}^2$$

Total tension steel is equal to A_s.

$$A_s = A_{s1} + A_{s2} = 3.81 + 0.86 = 4.67\ \text{in.}^2$$

The compression steel has $A'_s = 0.86\ \text{in.}^2$ (if A'_s yields).

4. Check if compression steel yields:

$$\varepsilon_y = \frac{f_y}{29{,}000} = \frac{50}{29{,}000} = 0.00172$$

Let $a = (A_{s1} f_y)/(0.85 f'_c b) = (3.81 \times 50)/(0.85 \times 3 \times 10) = 7.47$ in.

$$c(\text{distance to neutral axis}) = \frac{a}{\beta_1} = \frac{7.47}{0.85} = 8.80\ \text{in.}$$

$$\varepsilon'_s = \text{strain in compression steel (from strain triangles)}$$

$$= 0.003 \times \left(\frac{8.8 - 2.5}{8.8}\right) = 0.00215$$

Because $\varepsilon'_s > \varepsilon_y$, compression steel yields. $A'_s = A_{s2} = 0.86\ \text{in.}^2$.

5. Choose steel bars as follows: $A_s = 4.67\ \text{in.}^2$. Choose six no. 8 bars $(A_s = 4.71\ \text{in.}^2)$ in two rows, as assumed. $A'_s = 0.86\ \text{in.}^2$ Choose two no. 6 bars $(A'_s = 0.88\ \text{in.})$.

6. Check actual d:

$$\text{Actual } d = 22 - (1.5 + 0.375 + 1.5) = 18.625\ \text{in.}$$

It is equal approximately to the assumed depth. The final section is shown in Figure 4.5.

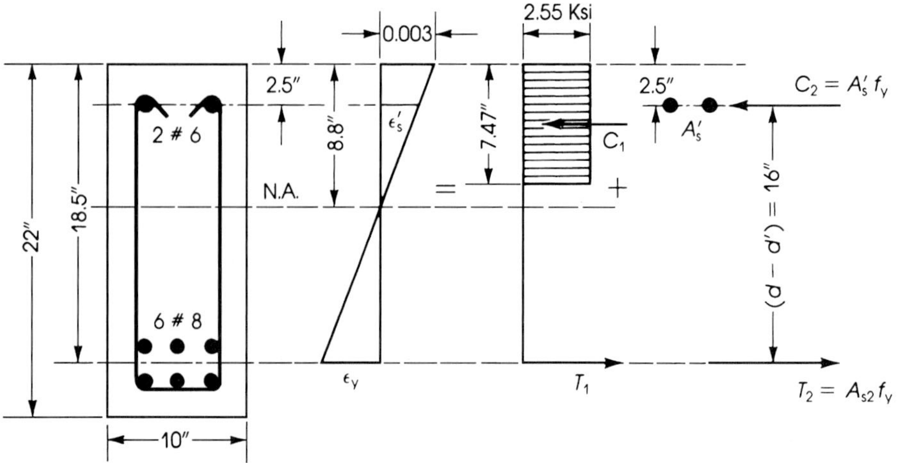

Figure 4.5 Example 4.4, doubly reinforced concrete section.

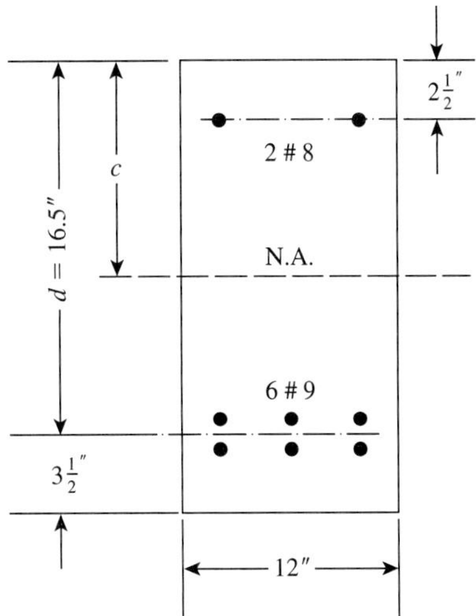

Figure 4.6 Example 4.5.

Example 4.5

A beam section is limited to $b = 12$ in. and to a total depth $h = 20$ in. and is subjected to a factored moment $M_u = 330$ K·ft. Determine the necessary reinforcement using $f'_c = 4$ Ksi and $f_y = 60$ Ksi. (Refer to Figure 4.6.)

Solution

1. Determine the maximum moment capacity of the section as singly reinforced. For $f'_c = 4$ Ksi and $f_y = 60$ Ksi, $\rho_{\max} = 0.0214$ and $R_{u\max} = 936$ psi. Assuming two rows of bars, $d = 20 - 3.5 = 16.5$ in.

$$\text{Max}\, M_{u1} = R_{u\max} bd^2 = 0.936(12)(16.5)^2 = 3058 \text{ K}\cdot\text{in.}$$

The design moment is $M_u = 330 \times 12 = 3960$ K·in. $> M_{u1}$; therefore, compression steel is needed.

2. Calculate A_{s1}, M_{u2}, A_{s2}, and total A_s.

$$A_{s1} = \rho_{\max} bd = 0.0214(12)(16.5) = 4.24 \text{ in.}$$

$$M_{u2} = M_u - M_{u1} = 3960 - 3058 = 902 \text{ K}\cdot\text{in.}$$

$$M_{u2} = \phi A_{s2} f_y (d - d'), \quad \text{assume } d' = 2.5 \text{ in.}$$

$$902 = 0.9A_{s2}(60)(16.5 - 2.5), \quad A_{s2} = 1.19 \text{ in.}^2$$

$$\text{Total } A_s = A_{s1} + A_{s2} = 4.24 + 1.19 = 5.43 \text{ in.}^2 \qquad \text{(6 no. 9 bars)}$$

3. Check if compression steel yields by equation (3.37). Compression steel yields if

$$\rho - \rho' \geq K = 0.85\beta_1 \frac{f'_c}{f_y}\left(\frac{d'}{d}\right)\left(\frac{87}{87 - f_y}\right)$$

$$K = (0.85)^2\left(\frac{4}{60}\right)\left(\frac{2.5}{16.5}\right)\left(\frac{87}{27}\right) = 0.0235$$

$$\rho - \rho' = \frac{A_{s1}}{bd} = \frac{4.24}{(12)(16.5)} = 0.0214 \leq K$$

Therefore, compression steel does not yield: $f'_s < f_y$.

4. Calculate f'_s: $f'_s = 87[(c - d')/c] \le f_y$. Determine c from A_{s1}: $A_{s1} = 4.24$ in.^{2},

$$a = \frac{A_{s1}f_y}{0.85f'_c b}$$

$$= \frac{4.24 \times 60}{0.85 \times 4 \times 12} = 6.235 \text{ in.}$$

$$c = \frac{a}{\beta_1} = \frac{6.235}{0.85} = 7.336 \text{ in.}$$

$$f'_s = 87 \times \left(\frac{7.336 - 2.5}{7.336}\right) = 57.3 \text{ Ksi} < 60 \text{ Ksi}$$

5. Calculate A'_s from $M_{u2} = \phi A'_s f'_s(d - d')$:

$$902 = 0.9A'_s(57.3)(16.5 - 2.5)$$

so $A'_s = 1.25$ in.^{2}, or calculate A'_s from $A'_s = A_{s2}(f_y/f'_s) = 1.25$ in.^{2} (2 no. 8 bars).

Note that the condition $[\rho - \rho'(f'_s/f_y)] = (\rho - \rho') \le \rho_{max}$ is already met.

$$\left(\rho - \rho'\frac{f'_s}{f_y}\right) = \frac{1}{bd}(A_s - A_{s2}) = \frac{4.24}{12 \times 16.5} = 0.0214$$

as assumed in the solution.

4.5 DESIGN OF T-SECTIONS

In slab-beam-girder construction, the slab dimensions as well as the spacing and position of beams are established first. The next step is to design the supporting beams, namely, the dimensions of the web and the steel reinforcement. Referring to the analysis of a T-section in the previous chapter, we can see that a large area of the compression flange, forming a part of the slab, is effective in resisting a great part or all of the compressive force due to bending. If the section is designed on this basis, the depth of the web will be small; consequently, the moment arm is small, resulting in a large amount of tension steel, which is not favorable. Shear requirements should be met, and this usually requires quite a deep section.

In many cases web dimensions can be known based on the flexural design of the section at the support in a continuous beam. The section at the support is subjected to a negative moment, the slab being under tension and considered not effective, and the beam width is that of the web.

In the design of a T-section for a given ultimate moment, M_u, the flange thickness t and width b would have been already established from the design of the slab and the ACI Code limitations for the effective flange width b, as given in Section 3.14. The web thickness, b_w, can be assumed to vary between 8 in. and 20 in., with a practical width of 12 to 16 in. Two more unknowns still need to be determined, d and A_s. Knowing that M_u, f'_c, and f_y are always given, two cases may develop as follows:

1. When d is given and we must calculate A_s:

a. Check if the section acts as a rectangular or T-section by assuming $a = t$ and calculating the moment capacity of the whole flange:

$$\phi M_{nf}(\text{flange}) = \phi(0.85f'_c)bt\left(d - \frac{t}{2}\right) \tag{4.7}$$

If $M_u > \phi M_{nf}$, then $a > t$; if $M_u < \phi M_{nf}$, then $a < t$, and the section behaves as a rectangular section.

b. If $a < t$, then calculate ρ using equation (4.2), and $A_s = \rho bd$. Check that $\rho_w \geq \rho_{min}$.

c. If $a > t$, determine A_{sf} for the overhanging portions of the flange, as explained in Section 3.14.4.

$$A_{sf} = 0.85f'_c(b - b_w)t/f_y \tag{4.8}$$

$$M_{u2} = \phi A_{sf} f_y \left(d - \frac{t}{2}\right) \tag{4.9}$$

The moment resisted by the web is

$$M_{u1} = M_u - M_{u2}$$

Calculate ρ_1 using M_{u1}, b_w, and d in equation (4.2) and determine $A_{s1} = \rho_1 b_w d$.

$$\text{Total } A_s = A_{s1} + A_{sf}$$

Then check that $A_s \leq A_{s\,max}$ or $\rho_w - 0.75\rho_f \leq \rho_{max}$ (web), as explained in Section 3.14. Also check that $\rho_w = A_s/(b_w d) \geq \rho_{min}$.

d. If $a = t$, then $A_s = \phi(0.85f'_c)bt/f_y$.

2. When d and A_s are not known, the design may proceed as follows:

a. Assume $a = t$ and calculate the amount of total steel, A_{sft}, needed to resist the compression force in the whole flange, bt.

$$A_{sft} = \frac{(0.85f'_c)bt}{f_y} \tag{4.10}$$

b. Calculate d based on A_{sft} and $a = t$ from the following equation:

$$M_u = \phi A_{sft} f_y \left(d - \frac{t}{2}\right) \tag{4.11}$$

If the depth d is acceptable, then $A_s = A_{sft}$ and $h = d + 2.5$ in. for one row of bars or $h = d + 3.5$ in. for two rows of bars.

c. If a new d_1 is adopted greater than the calculated d, then the section behaves as a rectangular section, and ρ can be calculated using equation (4.2); $A_s = \rho bd < A_{sft}$.

d. If a new d_2 is adopted that is smaller than the calculated d, then the section will act as a T-section, and the final A_s will be greater than A_{sft}. In this case proceed as in Step 1(c) to calculate A_s.

Example 4.6

The T-beam section shown in Figure 4.7 has a web width, b_w, of 10 in., a flange width, b, of 40 in., a flange thickness of 4 in., and an effective depth, d, of 14.5 in. Determine the necessary reinforcement if the applied factored moment is 3800 K · in. Given: $f'_c = 3$ Ksi and $f_y = 60$ Ksi.

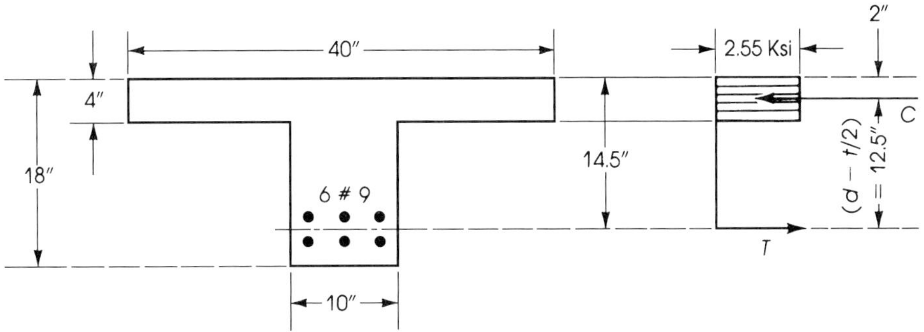

Figure 4.7 Example 4.6: T-section.

Solution

1. Check the position of the neutral axis; the section may be rectangular. Assume the depth of compression block a is 4 in.; that is, $a = t = 4$ in. Then

$$\phi M_n = \phi(0.85f'_c)bt(d - t/2) = 4590 \text{ K}\cdot\text{in.} > M_u = 3800 \text{ K}\cdot\text{in.}$$

   The ultimate moment that the concrete flange can resist is greater than the ultimate moment applied. Therefore, the section behaves as a rectangular section.

2. Determine the area of tension steel, considering a rectangular section, $b = 40$ in.

$$R_u = \phi M_n/(bd^2) = \frac{3{,}800{,}000}{40 \times 14.5^2} = 452 \text{ psi}$$

   From the tables in Appendix A, for $R_u = 452$ psi, $\rho = 0.0094$. Or, from equation (4.2),

$$A_s = \rho bd = 0.0094 \times 40 \times 14.5 = 5.46 \text{ in.}^2$$

   Use six no. 9 bars, $A_s = 6.00$ in.^{2} (in two rows).

3. Check that $\rho_w = A_s/b_w d \geq \rho_{\min}$; $\rho_w = 5.46/(10 \times 14.5) = 0.0377 > \rho_{\min} = 0.00333$. Note that A_s used is less than max A_s of 6.15 in.^{2} calculated by equation (3.59).

Example 4.7

The floor system shown in Figure 4.8 consists of 3-in. slabs supported by 18-ft-span beams spaced at 10 ft on centers. The beams have a web width, b_w, of 12 in. and an effective depth, d, of 17.5 in. ($h = 20$ in.). Calculate the necessary reinforcement for a typical interior beam if the factored applied moment is 6200 K·in. Use $f'_c = 3$ Ksi and $f_y = 60$ Ksi.

Solution

1. Find the beam flange width: Flange width is the smallest of

$$b = 16t + b_w = 3 \times 16 + 12 = 60 \text{ in.}$$

$$b = \frac{\text{span}}{4} = \frac{18 \times 12}{4} = 54 \text{ in.}$$

   Center-to-center of adjacent slabs is $10 \times 12 = 120$ in. Use $b = 54$ in.

2. Check the position of the neutral axis, assuming $a = t$.

$$\phi M_u(\text{based on flange}) = \phi \times 0.85f'_c bt\left(d - \frac{t}{2}\right)$$

$$= 0.9 \times 0.85 \times 3 \times 54 \times 3(17.5 - 1.5) = 5949 \text{ K}\cdot\text{in.}$$

   The applied moment is $M_u = 6200$ K·in. > 5949 K·in.; the beam acts as a T-section, so $a > t$.

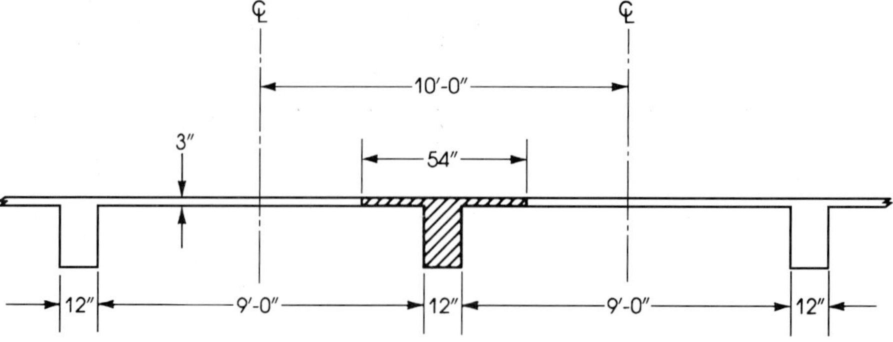

Figure 4.8 Example 4.7: effective flange width.

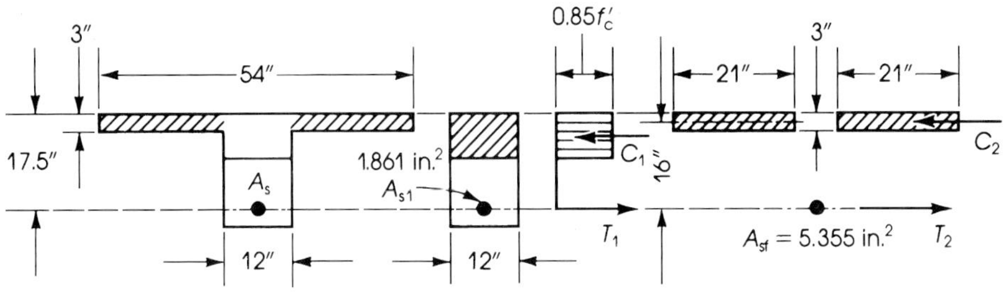

Figure 4.9 Analysis of Example 4.7.

3. Find the portion of the ultimate moment taken by the overhanging portions of the flange (Figure 4.9). First calculate the area of steel required to develop a tension force balancing the compressive force in the projecting portions of the flange:

$$A_{sf} = \frac{0.85f'_c(b - b_w)t}{f_y} = \frac{0.85 \times 3 \times (54 - 12) \times 3}{60} = 5.355 \text{ in.}$$

$\phi M_n = M_{u1} + M_{u2}$, that is, the sum of the ultimate moment of the web and the ultimate moment of the flanges.

$$M_{u2} = \phi A_{sf} f_y \left(d - \frac{t}{2}\right)$$

$$= 0.9 \times 5.355 \times 60\left(17.5 - \frac{3}{2}\right) = 4627 \text{ K}\cdot\text{in.}$$

4. Calculate the ultimate moment of the web (as a singly reinforced rectangular section):

$$M_{u1} = M_u - M_{u2} = 6200 - 4627 = 1573 \text{ K}\cdot\text{in.}$$

$$R_u = \frac{M_{u1}}{(b_w d^2)} = \frac{1{,}573{,}000}{12 \times 17.5^2} = 428 \text{ psi}$$

From equation (4.2) or the tables in Appendix A, for $R_u = 428$ psi, $\rho_1 = 0.00886$.

$$A_{s1} = \rho_1 b_w d = 0.00886(12)(17.5) = 1.861 \text{ in.}^2$$

$$\text{Total } A_s = A_{sf} + A_{s1} = 5.355 + 1.861 = 7.216 \text{ in.}$$

5. Check that $(\rho - 0.75\rho_f) \leq \rho_{\max}$ or calculate $A_{s\max}$ for T-sections using equation (3.58) or (3.59):

$$\text{Max } A_s = 0.0319\left[bt + b_w\left(\frac{d}{2} - t\right)\right] = 7.37 \text{ in.}^2 > 7.216 \text{ in.}^2$$

Example 4.8

In a slab-beam floor system, the flange width was determined to be 48 in., the web width was $b_w = 12$ in., and the slab thickness was $t = 4$ in. (Figure 4.10). Design a T-section to resist an external factored moment $M_u = 812$ K·ft. Use $f'_c = 3$ Ksi and $f_y = 60$ Ksi.

Solution

1. Because the effective depth is not given, let $a = t$ and calculate A_{sft} for the whole flange.

$$A_{sft} = \frac{0.85f'_c bt}{f_y} = \frac{0.85(3)(48)(4)}{60} = 8.16 \text{ in.}^2$$

Let $M_u = \phi A_{sft} f_y(d - t/2)$ and calculate d:

$$812 \times 12 = 0.9(8.16)(60)\left(d - \frac{4}{2}\right), \qquad d = 24.1 \text{ in.}$$

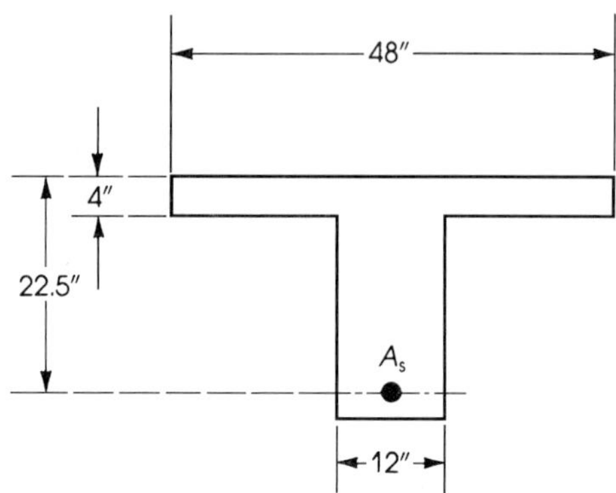

Figure 4.10 Example 4.8.

Now, if an effective $d = 24.1$ in. is chosen, then

$$A_s = A_{sft} = 8.16 \text{ in.}^2$$

2. If a depth $d > 24.1$ is chosen, say, 26.5 in., then $a < t$ and it is a rectangular section analysis. The steel ratio can be calculated from equation (4.2) with $\rho = 0.00574$ and $A_s = \rho bd$:

$$0.00574 \times 48 \times 26.5 = 7.3 \text{ in.}^2$$

3. If a depth $d < 24.1$ in. is chosen, say, 22.5 in., then $a > t$, and the section behaves as a T-section. Calculate $A_{sf} = 0.85 f'_c t(b - b_w)/f_y = 0.85(3)(4)(48 - 12)/60 = 6.12 \text{ in}^2$.

$$M_{u2} = \phi A_{sf} f_y \left(d - \frac{t}{2}\right) = 0.9(6.12)(60)\left(22.5 - \frac{4}{2}\right) = 6774.8 \text{ K}\cdot\text{in.}$$

$$M_{u1} = 812 \times 12 - 6774.8 = 2969.2 \text{ K}\cdot\text{in.}$$

4. For the basic singly reinforced section, $b_w = 12$ in., $d = 22.5$ in., and $M_{u1} = 2969.2 \text{ K}\cdot\text{in.}$ Calculate ρ_1 from equation (4.2) to get $\rho_1 = 0.0103$.

$$A_{s1} = \rho_1 b_w d = 0.0103(12)(22.5) = 2.78 \text{ in.}^2$$

$$\text{Total } A_s = A_{sf} + A_{s1} = 6.12 + 2.78 = 8.90 \text{ in.}^2$$

5. Calculate the total max A_s that can be used for the T-section by equation (3.58) or (3.59).

$$\text{Max } A_s = 0.0319\left[bt + b_w\left(\frac{d}{2} - t\right)\right]$$

$$= 0.0319\left[48 \times 4 + 12\left(\frac{22.5}{2} - 4\right)\right] = 8.91 \text{ in.}^2$$

$$A_s \text{ (used)} \le \max A_s$$

6. *Note:* If there are no restrictions on the total depth of the beam, it is a common practice to adopt the case when $a \le t$ (Step 2). This is because an increase in d produces a small increase in concrete in the web only, while decreasing the quantity of A_s required.

4.6 ADDITIONAL EXAMPLES

The following design examples give some practical applications and combine structural analysis with concrete design of beams and frames.

Example 4.9

For the precast concrete I-section shown in Figure 4.11, calculate the reinforcement needed to support a factored moment of 360 K · ft. Use $f'_c = 4$ Ksi and $f_y = 60$ Ksi.

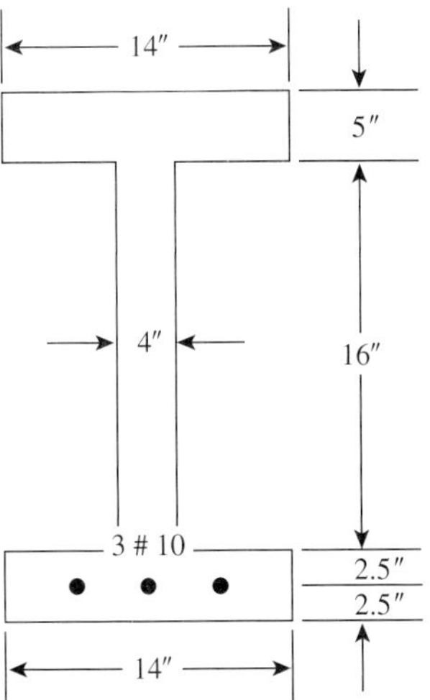

Figure 4.11 Example 4.9.

Solution

Determine if the force in the flange area 14 × 5 in. will be sufficient to resist a factored moment of 360 K · ft. Let $d = 23.5$ in.
Force in flange $C_c = 0.85 \times f'_c$ (flange area) $= 0.85 \times 4 \times (14 \times 5) = 238$ K, located at 2.5 in. from the top fibers, and $a = 5$ in.

$$\phi M_n = 0.9C_c(d - a/2) = 0.9 \times 238(23.5 - 2.5)/12 = 374.9 \text{ K} \cdot \text{ft}$$

which is greater than the applied moment of 360 K · ft. Therefore, a is less than 5 in.

$$\phi M_n = \phi A_s f_y(d - a/2), \text{where } a = A_s f_y/(0.85 f'_c b)$$

$$360 \times 12 = 0.9A_s(60)(23.5 - 60A_s/1.7 \times 4 \times 14)$$

Solve to get $A_s = 3.79$ in.^{2} Or use equation (4.2) to get $\rho = 0.01152$ and $A_s = 0.01152 \times 14 \times 23.5 = 3.79$ in.^{2} Use three no. 10 bars in one row, as shown in Figure 4.11.

For similar T-sections or I-sections, it is better to adopt a section with a flange size to accommodate the compression force C_c. In this case a is less than or equal to the flange depth. The bottom flange is in tension and not effective.

Example 4.10

The simply supported beam shown in Figure 4.12 carries a uniform service load of 2.4 K/ft (including self-weight) in addition to a service live load of 1.5 K/ft. Also, the beam supports a concentrated dead load of 12 K and a concentrated live load of 8 K at C, 10 ft from support A.

a. Determine the maximum factored moment and its location on the beam.

b. Design a rectangular section to carry the loads safely using a steel percentage of 2%, $b = 20$ in., $f'_c = 4$ Ksi, and $f_y = 60$ Ksi.

Solution

a. Calculate the uniform factored load: $w_u = 1.4(2.4) + 1.7(1.5) = 5.91$ K/ft. Calculate the concentrated factored load: $P_u = 1.4(12) + 1.7(8) = 30.4$ K. Calculate the reaction at A by taking moments about B:

$$R_A = 5.91(30)(30/2)/30 + 30.4(20)/30 = 108.92 \text{ K},$$

$$R_B = 5.91(30) + 30.4 - 108.92 = 98.78 \text{ K}$$

Maximum moment in the beam occurs at zero shear. Starting from B,

$$V = 0 = 98.78 - 5.91x \text{ and } x = 16.71 \text{ ft from } B \text{ at } D$$

$$M_u \text{ (at } D) = 98.72(16.71) - 5.91(16.71)(16.71/2) = 825.5 \text{ K} \cdot \text{ft. (Design moment)}$$

$$M_u \text{ (at } C) = 98.78(20) - 5.91(20)(20/2) = 793.6 \text{ K} \cdot \text{ft.}$$

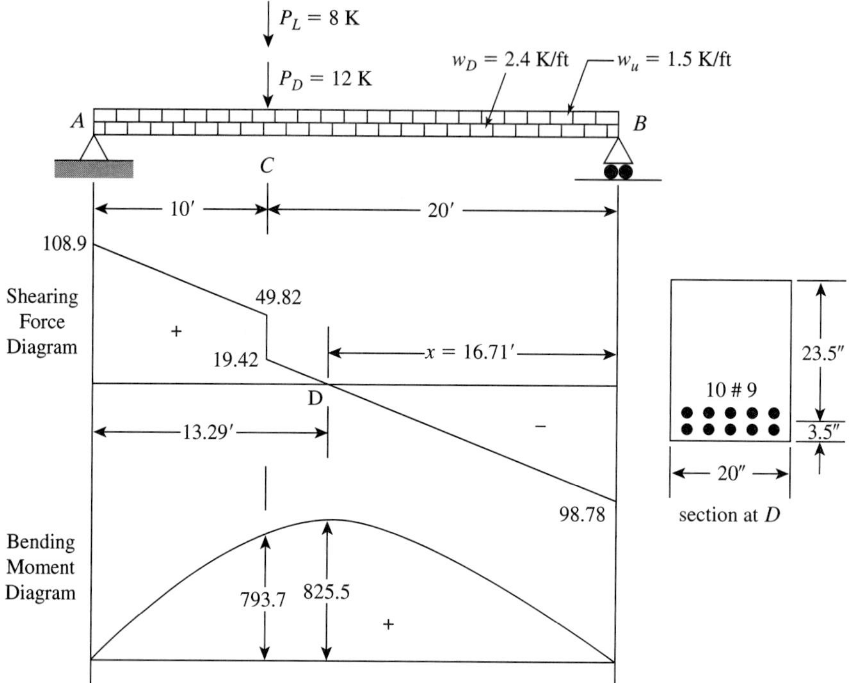

Figure 4.12 Example 4.10.

b. Design of the section at D: For $f'_c = 4$ Ksi and $f_y = 60$ Ksi, $\rho_{max} = 0.0214$ and $\rho_{min} = 0.00333$, and the design steel ratio of 2% is within the limits. For $\rho = 0.002$, $R_u = 890$ psi (from Table A.2) or from equation (3.19).

$$M_u = R_u bd^2, \text{or } 825.5 \times 12 = 0.89(20)d^2$$

Solve to get $d = 23.6$ in.

$$A_s = 0.002 \times 20 \times 23.6 = 9.44 \text{ in.}^2$$

Choose 10 no. 9 bars in two rows (area = 10 in.^{2}) or 8 no. 10 bars in two rows (area = 10.12 in.^{2}). Minimum b for 5 no. 9 bars in one row is 14 in. (Table A.7). Total depth $h = 23.6 + 3.5 = 27.1$ in. Use $h = 27$ in. because the steel area used is greater than the required area. Actual $d = 27 - 3.5 = 23.5$ in. Check the moment capacity of the section, $a = 10 \times 60/(0.85 \times 4 \times 20) = 8.82$ in.

$$\phi M_n = 0.9 \times 10 \times 60(23.5 - 8.84/2)/12 = 859 \text{ K}\cdot\text{ft}$$

which is greater than 825.5 K · ft. Check that $A_s = 10$ in.^{2} is less than $A_{s(max)}$.

$$A_{s(max)} = 0.0214 \times 20 \times 23.5 = 10.06 \text{ in.}^2$$

which exceeds 10 in.^{2} The final section is shown in Figure 4.12.

Example 4.11

The two-hinged frame shown in Figure 4.13 carries a uniform service dead load (including estimated self-weight) of 2.4 K/ft and a uniform service live load of 1.1 K/ft on frame beam BC. The moment at the corner B (or C) can be evaluated for this frame dimension, $M_b = M_c = -wL^2/18.4$ and the reaction at A or $D = wL/2$. A typical section of beam BC is shown, the column section is 16 × 21 in. It is required to:

a. Draw the bending moment and shear diagrams for the frame $ABCD$ showing all critical values.

b. Design the beam BC for the factored moments, positive and negative, using $f'_c = 4$ Ksi and $f_y = 60$ Ksi. Show reinforcement details.

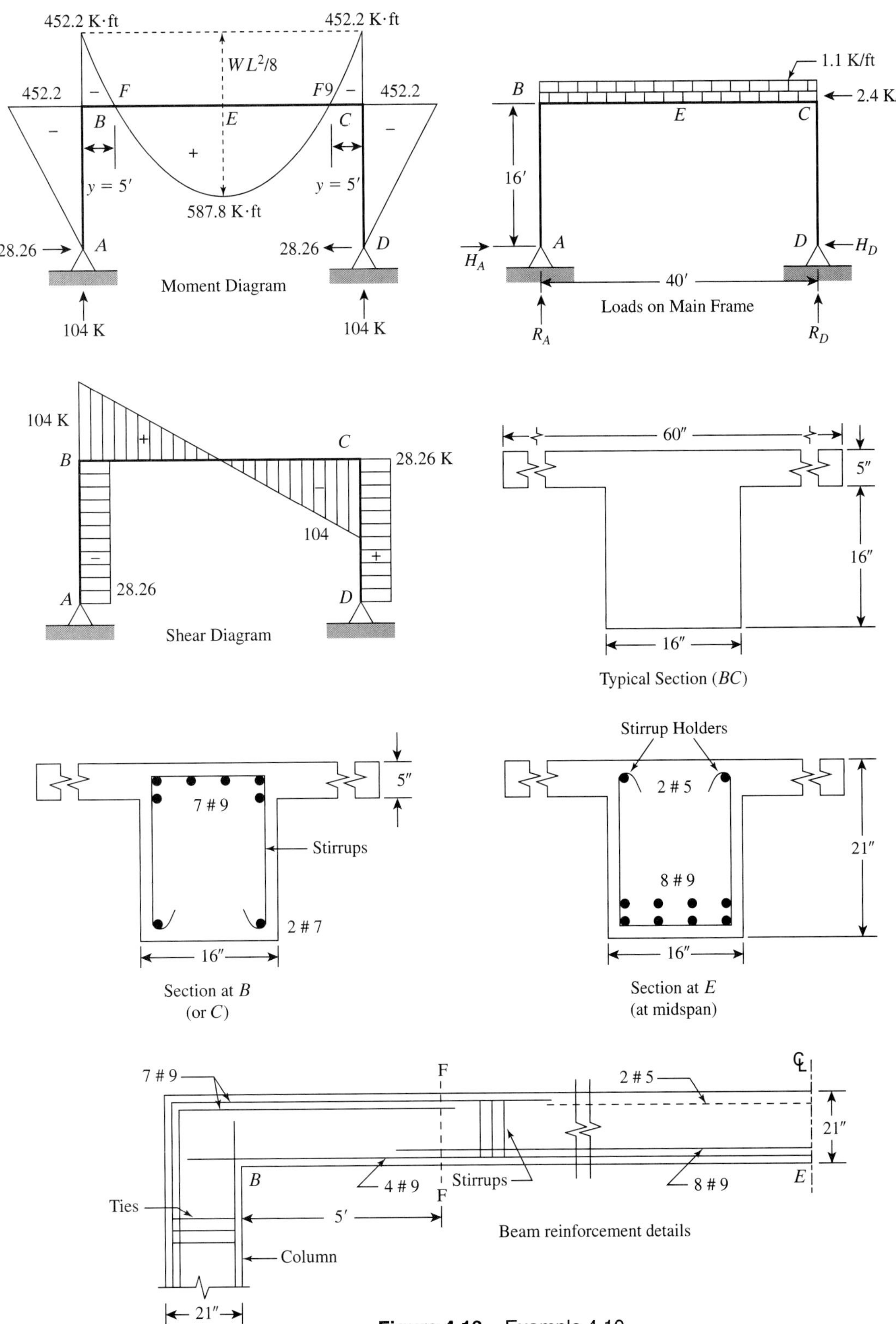

Figure 4.13 Example 4.10.

Solution

a. Calculate the forces acting on the frame using a computer program or the values mentioned previously. Factored load, $w_u = 1.4(2.4) + 1.7(1.1) = 5.2$ K/ft. Because of symmetry, $M_B = M_C = -w_u L^2/18.4 = -5.2(40)^2/18.4 = -452.2$ K·ft. Positive moment at midspan $E = w_u L^2/8 + M_B = 5.2(40)^2/8 - 452.2 = 587.8$ K·ft. Vertical reaction at $A = R_A = R_D = w_u L/2 = 5.2(40)/2 = 104$ K. Horizontal reaction at $A = H_A = H_D = M_B/h = 452.2/16 = 28.26$ K. The moment and shear diagrams are shown in Figure 4.13.

Determine the location of zero moment at section F on beam BC by taking moments about F:

$$104(y) - 28.26(16) - 5.2(y)^2/2 = 0, \qquad y = 4.963 \text{ ft, say, 5 ft from joint } B.$$

b. Design of beam BC:

1. Design of section E at midspan: $M_u = +587.8$ K·ft. Assuming two rows of bars, $d = 21 - 3.5 = 17.5$ in. Calculate the moment capacity of the flange using $a = 5.0$ in.

$$\phi M_n \text{ (flange)} = \phi(0.85 f'_c)ab(d - 5/2)$$

$$= 0.9(0.85 \times 4) \times (5 \times 60) \times (17.5 - 1.0)/12 = 1147.5 \text{ K·ft}$$

which is greater than the applied moment; therefore, a is less than 5.0 in.

Assume $a = 2.0$ in. and calculate A_s.

$$M_u = \phi A_s f_y (d - a/2)$$

$$587.8 \times 12 = 0.9 \times 60 A_s (17.5 - 1.0), \quad \text{and} \quad A_S = 7.92 \text{ in.}^2$$

Check assumed $a = A_s f_y/(0.85 f'_c b) = 7.92 \times 60/(0.85 \times 4 \times 60) = 2.33$ in. Revised $A_s = 587.8 \times 12/(0.9 \times 60 \times 16.33) = 7.99$ in.^{2} Check revised a: $a = 7.99 \times 2.33/7.92 = 2.35$ in, very close to 2.33 in.

Alternatively, equation (4.2) can be used to get ρ and A_s. Choose eight no. 9 bars in two rows (area $= 8.0$ in.^{2}), ($b_{\min} = 11.8$ in.). Extend four no. 9 bars on both sides to the columns. The other four bars can terminate where they are not needed, beyond section F; see the longitudinal section in Figure 4.13.

2. Design of section at B: $M_u = -452.2$ K·ft:

The section acts as a rectangular section, $b = 16$ in. and $d = 17.5$ in. The main tension reinforcement lies in the flange.

$$\rho_{\max} = 0.0214 \text{ and } R_{u(\max)} = 936 \text{ psi (Table A.2)}$$

Check the maximum moment capacity of the section as singly reinforced.

$$\phi M_{n(\max)} = R_{u(\max)} bd^2 = 0.936(16)(17.5)^2/12 = 382.2 \text{ K·ft}$$

which is less than the applied moment. Compression steel is needed.

$$A_{s1} = 0.0214(16)(17.5) = 5.99 \text{ in.}^2$$

$$M_{u2} = 452.2 - 382.2 = 70 \text{ K·ft}$$

$$M_{u2} = \phi A_{s2} f_y (d - d'); \text{ assume } d' = 2.5 \text{ in.}$$

$$70 \times 12 = 0.9 A_{s2}(60)(17.5 - 2.5) \quad \text{and} \quad A_{s2} = 1.04 \text{ in.}^2$$

Total tension steel $= 5.99 + 1.04 = 7.03$ in.^{2} Use seven no. 9 bars in two rows (area used $= 7.0$ in.^{2}, which is adequate). For compression steel, use two no. 9 bars (area $= 2.0$ in.^{2}), extended from the positive moment reinforcement to the column. Actually, four no. 9 bars are available; see the longitudinal section in Figure 4.13.

The seven no. 9 bars must extend in the beam BC beyond section F into the compression zone, and also must extend into the column BA to resist the column moment of 452.2 K·ft without any splices at joints B or C.

Stirrups are shown in the beam to resist shear (refer to Chapter 8), and two no. 5 bars were placed at the top of the beam to hold the stirrups in position. Ties are used in the column to hold the vertical bars (refer to Chapter 10). For the development length of bars to determine their extension in beams or columns, refer to Chapter 7.

4.7 EXAMPLES USING SI UNITS

Example 4.12

Design a singly reinforced rectangular section to resist an factored moment of 280 kN·m using the maximum allowable steel percentage. Given: $f'_c = 20\ \text{N/mm}^2$, $f_y = 400\ \text{N/mm}^2$, and $b = 250$ mm.

Solution

$$\rho_{\max} = (0.75)(0.85)\beta_1\left[\frac{f'_c}{f_y}\right]\left(\frac{600}{600 + f_y}\right)$$

$$= 0.6375 \times 0.85 \times \frac{20}{400} \times \left(\frac{600}{600 + 400}\right) = 0.01626$$

$$R_{u\max} = \phi\rho_{\max}f_y\left(1 - \frac{\rho_{\max}f_y}{1.7f'_c}\right)$$

$$= 0.9 \times 0.01626 \times 400\left(1 - \frac{0.01626 \times 400}{1.7 \times 20}\right) = 4.73\ \text{N/mm}^2 \quad (\text{MPa})$$

$$M_u = R_u bd^2,$$

$$d = \sqrt{\frac{M_u}{R_u b}} = \sqrt{\frac{280 \times 10^6}{4.73 \times 250}} = 487\ \text{mm}$$

$$A_s = \rho bd = 0.01626 \times 250 \times 487 = 1980\ \text{mm}^2 = 19.8\ \text{cm}^2$$

Choose four bars, 25 mm diameter, in two rows.

A_s provided $= 4 \times 4.9 = 19.6\ \text{cm}^2 \approx A_s$. Total depth is

$$h = d + 25\ \text{mm} + 60\ \text{mm}$$

$$= 487 + 25 + 60 = 572\ \text{mm}, \qquad \text{say, } 580\ \text{mm}$$

Check minimum width:

$$b_{\min} = 2D + 1S + 95\ \text{mm} = 3 \times 25 + 95 = 170\ \text{mm} < 250\ \text{mm}$$

Bars are placed in two rows.

Example 4.13

Calculate the required reinforcement for a beam that has a section of $b = 300$ mm and a total depth $h = 600$ mm to resist $M_u = 696$ kN·m. Given: $f'_c = 30\ \text{N/mm}^2$ and $f_y = 400$ N/mm.

Solution

1. Determine the ultimate moment capacity of the section using $\rho_{\max}$:

$$\rho_{\max} = 0.6375 \times \beta_1 \times \left(\frac{f'_c}{f_y}\right) \times \left(\frac{600}{600 + f_y}\right)$$

$$= 0.6375 \times 0.85 \times (30/400) \times (600/1000) = 0.02438$$

$$R_{u\max} = \phi\rho_{\max}f_y\left(1 - \frac{\rho_{\max}f_y}{1.7f'_c}\right)$$

$$= 0.9 \times 0.02438 \times 400\left(1 - \frac{0.02438 \times 400}{1.7 \times 30}\right)$$

$$= 7.1\ \text{N/mm}^2$$

$$d = h - 85 \text{ mm} \qquad \text{(assuming two rows of bars)}$$
$$= 600 - 85 = 515 \text{ mm}$$
$$\phi M_n = R_u bd^2 = 7.1 \times 300 \times (515)^2 \times 10^{-6} = 564.9 \text{ kN}\cdot\text{m}$$

This is less than the external moment; therefore, compression reinforcement is needed.

2. Calculate A_{s1}, M_{u1}, and M_{u2}:

$$A_{s1} = \rho_{\max} bd = 0.02438 \times 300 \times 515 = 3767 \text{ mm}^2$$
$$M_{u2} = M_u - M_{u1} = 696 - 564.9 = 131.1 \text{ kN}\cdot\text{m}$$

3. Calculate A_{s2} and A'_s due to M_{u2}. Assume $d' = 60$ mm:

$$M_{u2} = \phi A_{s2} f_y (d - d')$$
$$131.1 \times 10^6 = 0.9 A_{s2} \times 400(515 - 60), \quad A_{s2} = 800 \text{ mm}^2$$

Total tension steel is $3767 + 800 = 4567$ mm^2. The compression steel A'_s is 800 mm^2.

4. Compression steel yields if

$$(\rho - \rho') = \rho_1 \geq 0.85\beta_1 \times \frac{f'_c}{f_y} \times \frac{d'}{d} \times \frac{600}{600 - f_y} = \text{K}$$

$$\text{K} = (0.85)^2 \times \frac{30}{400} \times \frac{60}{515} \times \frac{600}{600 - 400} = 0.0189$$

Because $(\rho - \rho') = \rho_{\max} = 0.02438 > 0.0189$, therefore, compression steel yields. Therefore, $A'_s = A_{s2} = 800$ mm^2.

5. Choose steel bars as follows: For tension, choose seven bars 30 mm in diameter. The A_s provided (4940 mm) is greater than A_s, as required. For compression steel, choose two bars 25 mm in diameter:

$$A'_s = 982 \text{ mm}^2 > 800 \text{ mm}^2$$

SUMMARY

Flow charts are shown in Chapter 21.

Sections 4.1–4.3 Design of a Singly Reinforced Rectangular Section

Given: M_u(external ultimate moment), f'_c (compressive strength of concrete), and f_y (yield stress of steel).

Case 1: When b, d, and A_s or (ρ) are *not* given:

a. Assume $\rho_{\min} \leq \rho \leq \rho_{\max}$.
Choose $\rho_{\max}$ for a minimum concrete cross section (smallest) or choose ρ between $\rho_{\max}/2$ and $\rho_b/2$ for larger sections.
For example, if $f_y = 60$ Ksi, you may choose:

$$\rho = 1.2\% \qquad R_n = 618 \text{ psi} \qquad \text{for } f'_c = 3 \text{ Ksi}$$
$$\rho = 1.4\% \qquad R_n = 736 \text{ psi} \qquad \text{for } f'_c = 4 \text{ Ksi}$$
$$\rho = 1.4\% \qquad R_n = 757 \text{ psi} \qquad \text{for } f'_c = 5 \text{ Ksi}$$

For any other value of ρ, $R_n = \rho f_y\left(1 - \dfrac{\rho f_y}{1.7 f'_c}\right)$, $R_u = \phi R_n$.

b. Calculate $bd^2 = M_u/\phi R_n (\phi = 0.9)$.

c. Choose b and d. The ratio of d to b is approximately $1 \rightarrow 3$, or $d/b \approx 2.0$.

d. Calculate $A_s = \rho bd$; then choose bars to fit in b either in one row or two rows. (Check $b_{\min}$ from the tables.)

e. Calculate

$$h = d + 2.5 \text{ in.} \qquad \text{(for one row of bars)}$$

$$h = d + 3.5 \text{ in.} \qquad \text{(for two rows of bars)}$$

b and h must be to the nearest higher inch.

Note: If h is increased, calculate new $d = h - 2.5$ (or 3.5) and recalculate A_s to get a smaller value.

Case 2: When ρ is given, $d, b,$ and A_s are required. Repeat steps (a) through (e) from Case 1.

Case 3: When b and d (or h) are given, A_s is required.

a. Calculate $R_n = M_u/\phi bd^2 (\phi = 0.9)$.

b. Calculate

$$\rho = \left(\frac{0.85 f'_c}{f_y}\right)\left[1 - \sqrt{1 - \frac{2R_n}{0.85 f'_c}}\right]$$

(or get ρ from tables or equation (4.2)).

c. Calculate $A_s = \rho bd$, choose bars, and check $b_{\min}$.

d. Calculate h to the nearest higher inch (see note, Case 1(e)).

Case 4: When b and ρ are given, d and A_s are required.

a. Calculate

$$R_n = \rho f_y\left(1 - \frac{\rho f_y}{1.7 f'_c}\right), \quad R_u = \phi R_n (\phi = 0.9)$$

b. Calculate

$$d = \sqrt{\frac{M_u}{\phi R_n b}}$$

c. Calculate $A_s = \rho bd$, choose bars, and check $b_{\min}$.

d. Calculate h to the nearest higher inch (see note, Case 1(e)).

Note: Equations that may be used to check the moment capacity of the section after the final section is chosen are

$$M_u = \phi M_n = \phi A_s f_y\left(d - \frac{A_s f_y}{1.7 f'_c b}\right) = \phi A_s f_y d\left(1 - \frac{\rho f_y}{1.7 f'_c}\right)$$

$$= \phi \rho f_y (bd^2)\left(1 - \frac{\rho f_y}{1.7 f'_c}\right) = R_u bd^2$$

Section 4.4 Design of Rectangular Sections with Compression Steel

Given: $M_u, b, d, d', f'_c, f_y,$ and $\phi = 0.9$.
Required: A_s and A'_s.

1. *General*

a. Calculate $\rho_{\max}$ and $\rho_{\min}$ as singly reinforced from equations (or from tables).

b. Calculate $R_{n\,\max} = \rho_{\max} f_y\left(1 - \dfrac{\rho_{\max} f_y}{1.7 f'_c}\right)$ (or use tables).

c. Calculate the maximum capacity of the section as singly reinforced: $\phi M_n = \phi R_{n\,\max} bd^2$.

d. If $M_u > \phi M_n$, then compression steel is needed. If $M_u < \phi M_n$, it is a singly reinforced section.

2. If $M_u > \phi M_n$ and compression steel is needed:

a. Let $M_{u1} = \phi R_{n\,\max} bd^2$;

b. Calculate $A_{s1} = \rho_{\max} bd$ (basic section);

c. Calculate $M_{u2} = M_u - M_{u1}$ (for the steel section).

3. Calculate A_{s2} and A'_s as steel section.

a. $M_{u2} = \phi A_{s2} f_y (d - d')$,

b. Calculate total tension steel: $A_s = A_{s1} + A_{s2}$.

4. Calculate A'_s (compression steel area):

a. Calculate $a = \dfrac{A_{s1} f_y}{0.85 f'_c b}$ and $c = a/\beta_1$.

b. Calculate $f'_s = 87\left(\dfrac{c - d'}{c}\right) \le f_y$.

If $f'_s \ge f_y$, then $f'_s = f_y$ and $A'_s = A_{s2}$.

If $f'_s < f_y$, then $A'_s = A_{s2}\left(\dfrac{f_y}{f'_s}\right)$.

c. Check that total steel area is $A_s \ge \max A_s$, or

$$A_s \le \left[\rho_{\max}(bd) + A'_s\left(\frac{f'_s}{f_y}\right)\right]$$

Section 4.5 Design of T-Sections

Given: $M_u, f'_c, f_y, b, t,$ and b_w.

Required: A_s and d (if not given).

There are two cases:

Case 1: When d and A_s (or ρ) are *not* given:

a. Let $a \le t$ (as singly reinforced rectangular section).

If $a = t$ is assumed, then

$$M_u = (\text{total flange}) = \phi(0.85 f'_c) bt\left(d - \frac{t}{2}\right) = \phi A_s f_y\left(d - \frac{t}{2}\right)$$

Solve for d and then for A_s.

$$d = \frac{M_u}{\phi(0.85 f'_c) bt} + \frac{t}{2} \qquad A_s = \frac{M_u}{\phi f_y\left(d - \frac{t}{2}\right)}$$

b. If a is assumed less than t, then

$$d = \frac{M_u}{\phi(0.85 f'_c) ba} + \frac{a}{2} \quad \text{and} \quad A_s = \frac{M_u}{\phi f_y\left(d - \frac{a}{2}\right)}$$

Case 2: When d is given and A_s is required (one unknown):

a. Check if $a \le t$ by considering the moment capacity of the flange (bt).

$$(\text{Flange})\ \phi M_n = \phi(0.85 f'_c) bt\left(d - \frac{t}{2}\right)$$

If $\phi M_n > M_u$ (external), then $a < t$ (rectangular section).
If $\phi M_n < M_u$ (external), then $a > t$ (T-section).

b. If $a < t$, calculate $R_n = M_u/\phi bd^2$ and then calculate ρ (or determine ρ from tables or equation (4.2)):

$$\rho = \frac{0.85f'_c}{f_y}\left(1 - \sqrt{1 - \frac{2R_n}{0.85f'_c}}\right)$$

Then calculate $A_s = \rho bd$.

c. If $a > t$:

i. Calculate C_f and A_{sf}.

$$A_{sf} = 0.85f'_c t\frac{(b - b_w)}{f_y} = \frac{C_f}{f_y}\text{(flange)}$$

Then calculate M_{uf} (flange) $= \phi C_f(d - t/2)$.

ii. Calculate M_{uw} (web) $= M_u - M_{uf}$. Calculate R_n (web) $= M_{uw}/(\phi b_w d^2)$; then find ρ_w (use the equation or tables).
Calculate A_{sw} (web) $= \rho_w b_w d$.

iii. Total $A_s = A_{sf}$ (flange) $+ A_{sw}$ (web).
Total A_s must be less than or equal to A_s (max) and greater than or equal to A_s (min).

iv. $\rho_w = \left(\frac{0.85f'_c}{f_y}\right)\left(1 - \sqrt{1 - \frac{2R_n}{0.85f'_c}}\right)$

v. Check that $\rho_w = A_s/b_w d \geq \rho_{\min} \cdot (\rho_w = \text{steel ratio in web})$ or $A_s > A_{s\,\min}$, where $A_{s\,\min} = \rho_{\min}(b_w d)$.
Check that $A_s \leq \max A_s$:

$$\text{Max } A_s = 0.6375\left(\frac{f'_c}{f_y}\right)\left[t(b - b_w) + b_w\beta_1 d\left(\frac{87}{87 + f_y}\right)\right]$$

PROBLEMS

4.1 Based on the information given in the accompanying table and for each assigned problem, design a singly reinforced concrete section to resist the ultimate moment shown. Use $f'_c = 4$ Ksi, $f_y = 60$ Ksi, and draw a detailed, neat section.

No.	M_u (K · ft)	b (in.)	d (in.)	ρ%
a	272.7	12	21.5	—
b	969.2	18	32.0	—
c	816.0	16	—	1.70
d	657.0	16	—	1.50
e	559.4	14	—	1.75
f	254.5	10	21.5	—
g	451.4	14	—	2.09
h	832.0	18	28.0	—
i	345.0	15	—	1.77
j	510.0	0.5d	—	$\rho_{\max}$
k	720.0	—	2.5b	2.00
l	605.0	—	1.5b	1.80

For problems in SI units: 1 in. = 25.4 mm
1 Ksi = 6.9 MPa (N/mm^2)
1 M_u (K · ft) = 1.356 kN · m

4.2 Based on the information given in the following table and for each assigned problem, design a rectangular section with compression reinforcement to resist the ultimate moment shown. Use $f'_c = 4$ Ksi, $f_y = 60$ Ksi, and $d' = 2.5$ in. Draw detailed, neat sections. (*Note:* In the first seven problems, $f'_s = f_y$.)

No.	M_u (K · ft)	b (in.)	d (in.)
a	554	14	20.5
b	790	16	24.5
c	448	12	18.5
d	620	12	20.5
e	765	16	20.5
f	855	18	22.0
g	555	16	18.5
h	400	12	16.5
i	456	16	16.5
j	330	12	16.5
k	342	14	14.5
l	442	14	17.5

For problems in SI units: 1 in. = 25.4 mm
1 Ksi = 6.9 MPa (N/mm^2)
1 M_u (K · ft) = 1.356 kN · m

4.3 Based on the information given in the following table and for each assigned problem, calculate the tension steel and bars required to resist the ultimate moment shown. Use $f'_c = 3$ Ksi and $f_y = 60$ Ksi. Draw detailed, neat sections.

No.	M_u (K · ft)	b (in.)	b_w (in.)	t (in.)	d (in.)	Notes
a	394	48	14	3	18.5	
b	800	60	16	4	19.5	
c	250	44	15	3	15.0	
d	327	50	14	3	13.0	
e	577	54	16	4	18.5	
f	559	48	14	4	17.5	
g	388	44	12	3	16.0	
h	380	46	14	3	15.0	
i	537	60	16	3	16.5	
j	515	54	16	3	17.5	
k	361	44	15	3	15.0	
l	405	50	14	3	15.5	
m	378	44	16	3	—	Let $a = t$.
n	440	36	16	4	—	Let $a = t$.
o	567	48	12	3	—	Let $A_s = 6.0$ in^2.
p	507	46	14	3	—	Let $A_s = 7.0$ in^2.

For problems in SI units: 1 in. = 25.4 mm
1 Ksi = 6.9 MPa (N/mm^2)
1 M_u (K · ft) = 1.356 kN · m

4.4 Design a singly reinforced rectangular section to resist an ultimate moment of 232 K · ft (320 kN · m) if $f'_c = 3$ Ksi (20 MPa), $f_y = 50$ Ksi (350 MPa), and $b = 10$ in. (250 mm), using (a) ρ_{max}, (b) $\rho = 0.016$, and (c) $\rho = 0.012$.

4.5 Design a singly reinforced section to resist an ultimate moment of 186 K · ft (252 kN · m) if $b = 12$ in. (275 mm), $d = 20$ in. (500 mm), $f'_c = 3$ Ksi (20 MPa), and $f_y = 40$ Ksi (300 MPa).

4.6 Determine the reinforcement required for the section given in Problem 4.5 when $f'_c = 4$ Ksi (30 MPa), and $f_y = 60$ Ksi (400 MPa).

4.7 A simply supported beam has a 20-ft (6-m) span and carries a uniform dead load of 600 lb/ft (9 kN/m) and a concentrated live load at midspan of 9 kips (40 kN) (Figure 4.14). Design the beam if $b = 12$ in. (300 mm), $f'_c = 4$ Ksi (30 MPa), and $f_y = 60$ Ksi (400 MPa). (Beam self-weight is not included in the dead load.)

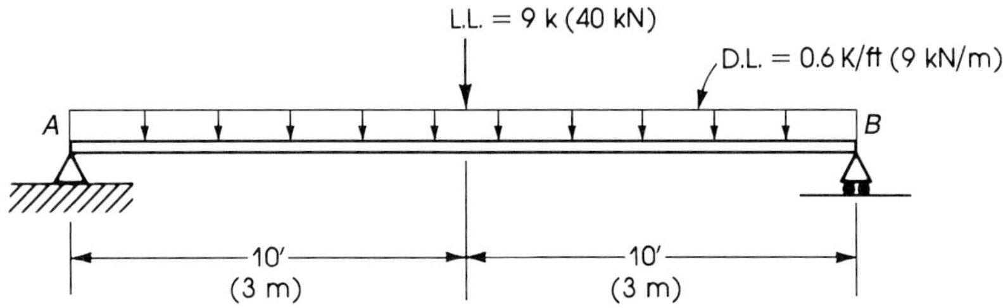

Figure 4.14 Problem 4.7.

4.8 A beam with a span of 24 ft (7.2 m) between supports has an overhanging extended part of 8 ft (2.4 m) on one side only. The beam carries a uniform dead load of 2 K/ft (30 kN/m) (including its own weight) and a uniform live load of 1.2 K/ft (18 kN/m) (Figure 4.15). Design the smallest singly reinforced rectangular section to be used for the entire beam. Select steel for positive and negative moments. Use $f'_c = 3$ Ksi (20 MPa), $f_y = 60$ Ksi (400 MPa), and $b = 12$ in. (300 mm). (Determine the maximum positive and maximum negative moments by placing the live load once on the span and once on the overhanging part.)

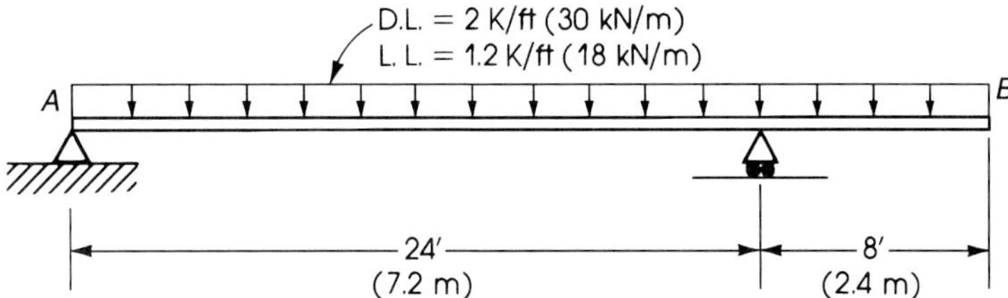

Figure 4.15 Problem 4.8.

4.9 Design a 15-ft (4.5-m) cantilever beam of uniform depth to carry a uniform dead load of 0.8 K/ft (12 kN/m) and a live load of 1 K/ft (15 kN · m). Assume a beam width $b = 14$ in. (350 mm), $f'_c = 4$ Ksi (30 MPa), and $f_y = 60$ Ksi (400 MPa).

4.10 A 10-ft (3-m) cantilever beam carries a uniform dead load of 1.35 K/ft (20 kN/m) (including its own weight) and a live load of 0.67 K/ft (10 kN/m) (Figure 4.16). Design the beam using a variable depth. Draw all details of the beam and reinforcement. Given: $f'_c = 3$ Ksi (20 MPa), and $f_y = 40$ Ksi (300 MPa), and $b = 12$ in. (300 mm). Assume h at the free end is 10 in. (250 mm).

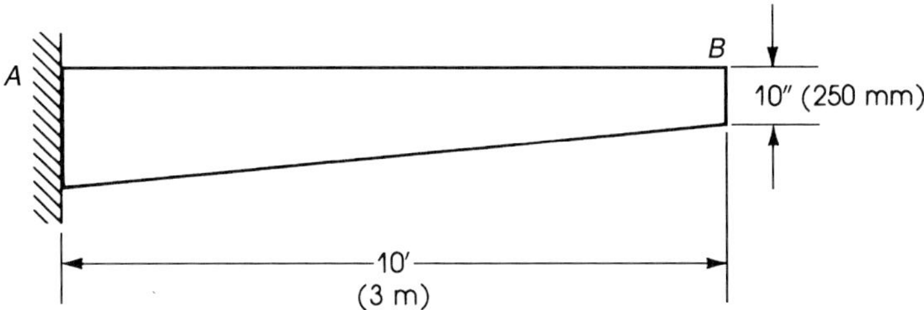

Figure 4.16 Problem 4.10.

4.11 Determine the necessary reinforcement for a concrete beam to resist an external ultimate moment of 290 K · ft (400 kN · m) if $b = 12$ in. (300 mm), $d = 19$ in. (475 mm), $d' = 2.5$ in. (65 mm), $f'_c = 3$ Ksi (20 MPa), and $f_y = 60$ Ksi (400 MPa).

4.12 Design a reinforced concrete section that can carry an ultimate moment of 260 K · ft (360 kN · m) as

a. Singly reinforced, $b = 10$ in. (250 mm);

b. Doubly reinforced, 25% of the moment to be resisted by compression steel, $b = 10$ in. (250 mm);

c. T-section, which has a flange thickness of 3 in. (75 mm), flange width of 20 in. (500 mm), and web width of 10 in. (250 mm); $f'_c = 3$ Ksi (20 MPa), and $f_y = 60$ Ksi (400 MPa).

Determine the quantities of concrete and steel designed per foot length (meter length) of beams and then determine the cost of each design if the price of the concrete equals \$50/yd^3 (\$67/m^3) and that of steel is \$0.30/lb (\$0.66/kg). Tabulate and compare results.

4.13 Determine the necessary reinforcement for a T-section that has a flange width $b = 40$ in. (1000 mm), flange thickness $t = 4$ in. (100 mm), and web width $b_w = 10$ in. (250 mm) to carry an ultimate moment of 545 K · ft (750 kN · m). Given: $f'_c = 3$ Ksi (20 MPa), and $f_y = 60$ Ksi (400 MPa).

4.14 The two-span continuous beam shown in Figure 4.17 is subjected to a uniform dead load of 2 K/ft (including its own weight) and a uniform live load of 3 K/ft. The reactions due to two different loadings are also shown. Calculate the maximum negative ultimate moment at the intermediate support B and the maximum positive ultimate moment within the span AB (at $0.42L$ from support A), design the critical section at B and D, and draw the reinforcement details for the entire beam ABC. Given: $L = 20$ ft, $b = 12$ in., $h = 21$ in., use $d = 18$ in. for one row of bars and $d = 17$ in. for two rows, $f'_c = 4$ Ksi, and $f_y = 60$ Ksi.

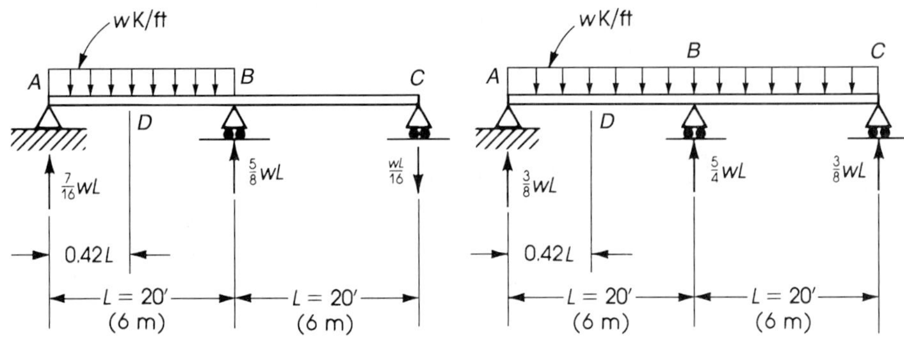

Figure 4.17 Problem 4.14.

4.15 The two-hinged frame shown in Figure 4.18 carries a uniform dead load of 3.2 K/ft and a uniform live load of 2.4 K/ft on BC. The reactions at A and D can be evaluated as follows: $HA = HD = wL/9$ and $RA = RD = wL/2$, where w = uniform load on BC. A typical cross section of the beam BC is also shown. Using the strength design method,

a. Draw the ultimate bending moment diagram for the frame $ABCD$;

b. Design the beam BC for the applied ultimate moments (positive and negative);

c. Draw the reinforcement details of BC.

Given: $f'_c = 4$ Ksi and $f_y = 60$ Ksi.

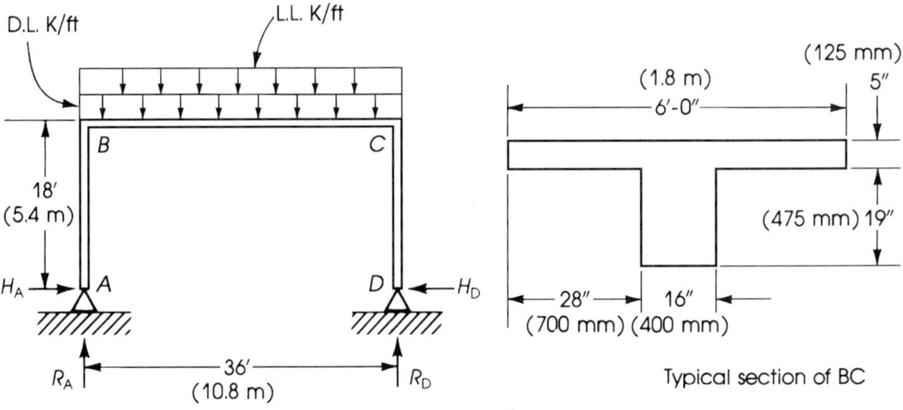

Figure 4.18 Problem 4.15.

5 ELASTIC CONCEPT: FLEXURAL ANALYSIS OF BEAMS

Office building, Minneapolis, MN

5.1 ASSUMPTIONS

Structural concrete members may be designed by a method that is based on the elastic concept. This method is called the *working stress, straight-line, service load,* or *alternate design method.* The ACI Code dealt with this concept in its Appendix A as the alternate design method. The elastic concept is based on the following assumptions [2]:

1. A cross section that is plane before loading remains plane after loading. This means that strains above and below the neutral axis are proportional to the distance from the neutral axis.
2. The stress-strain relation for concrete is a straight line under service loads within the allowable working stress limits. Stresses vary as the distance from the neutral axis (except for deep beams).
3. The concrete strength in tension is neglected in a cracked section, and the steel resists all tension due to flexure. Before cracking the entire concrete area is effective.
4. The tension reinforcement is replaced in design computations with an equivalent concrete tension area equal to n times that of the reinforcing steel, where n, the modular ratio, is equal to the ratio of the moduli of elasticity of steel and concrete $(n = E_s/E_c)$.
5. The modulus of elasticity of steel (E_s) is assumed to be equal to 29×10^6 psi (200,000 MPa).

The modulus of elasticity of concrete (E_c) can be assumed equal to $33W^{1.5}\sqrt{f'_c}$ psi $(E_c = 0.043W^{1.5}\sqrt{f'_c}$ MPa$)$, where W is the unit weight of concrete, which is between 90 and 155 pcf (1440 and 2420 kg/m^3). This chapter explains the elastic concept and gives the basic equations for calculating the moments of inertia of different sections that are needed to compute deflections, as is explained in the next chapter.

5.2 TRANSFORMED AREA CONCEPT

For short-term live loads, Hooke's law holds for both concrete and steel when stresses are below the allowable working stresses. In this case both materials are assumed to behave in an elastic manner, with unit stresses proportional to unit strains; that is, the stress is equal to the strain times the modulus of elasticity: stress $= \varepsilon E$.

The modular ratio $n = E_s/E_c$, which is used in the transformed area concept, may be used as the nearest whole number but may not be less than 6. For example, when

$$f'_c = 2500 \text{ psi } (17.5 \text{ N/mm}^2), \qquad n = 10$$

$$f'_c = 3000 \text{ psi } (20 \text{ N/mm}^2), \qquad n = 9 \quad (\text{or } 9.13)$$

$$f'_c = 4000 \text{ psi } (30 \text{ N/mm}^2), \qquad n = 8 \quad (\text{or } 7.9)$$

$$f'_c = 5000 \text{ psi } (35 \text{ N/mm}^2), \qquad n = 7$$

For normal-weight concrete, n can be taken as $500/\sqrt{f'_c}$. (psi units)

The basic concept of a transformed area is that the section of steel and concrete is transformed into a homogeneous section of concrete by replacing the actual steel area with an equivalent area of concrete. Two conditions must be satisfied for the transformation process.

1. *Equilibrium:* The force in the steel must be equal to the force in the transformed concrete, or

$$A_s \times f_s = A_{tc} \times f_{tc} \tag{5.1}$$

where

A_s = area of the steel in the section
f_s = the stress in the steel
A_{tc} = the transformed concrete area
f_{tc} = the stress in the transformed area

2. *Strain compatibility:* The unit strains must be the same,

$$\frac{f_s}{E_s} = \frac{f_{tc}}{E_c} \quad \text{or} \quad f_{tc} = f_s \frac{E_c}{E_s} = \frac{f_s}{n}$$

and

$$f_s = nf_{tc} \tag{5.2}$$

For the transformed area of concrete,

$$f_s = nf_c \tag{5.3}$$

and from equation (5.1), the transformed area is

$$A_{tc} = \frac{A_s f_s}{f_{tc}} = \frac{A_s n f_{tc}}{f_{tc}} = nA_s \tag{5.4}$$

For dead loads or long-term loads, the elastic stresses discussed here are complicated by the presence of shrinkage stresses and creep in the concrete, producing inelastic effects, mainly in the compression steel. The actual stress in the compression steel is much greater than nf'_s, where f'_s is the calculated stress in the compression steel. The ACI Code requires the use of $2n$ with compression steel in stress computations.

Note that as long as the maximum tensile stress in the concrete is smaller than the modulus of rupture, $(7.5\sqrt{f'_c})$, so that no tension cracks develop, the entire gross concrete area is used in computations. But when the maximum tensile stress exceeds the modulus of rupture, the cracked transformed area is used in computations. Figures 5.1 and 5.2 show the transformed areas for rectangular cracked sections with or without compression steel. For an uncracked section, the whole concrete area must be used.

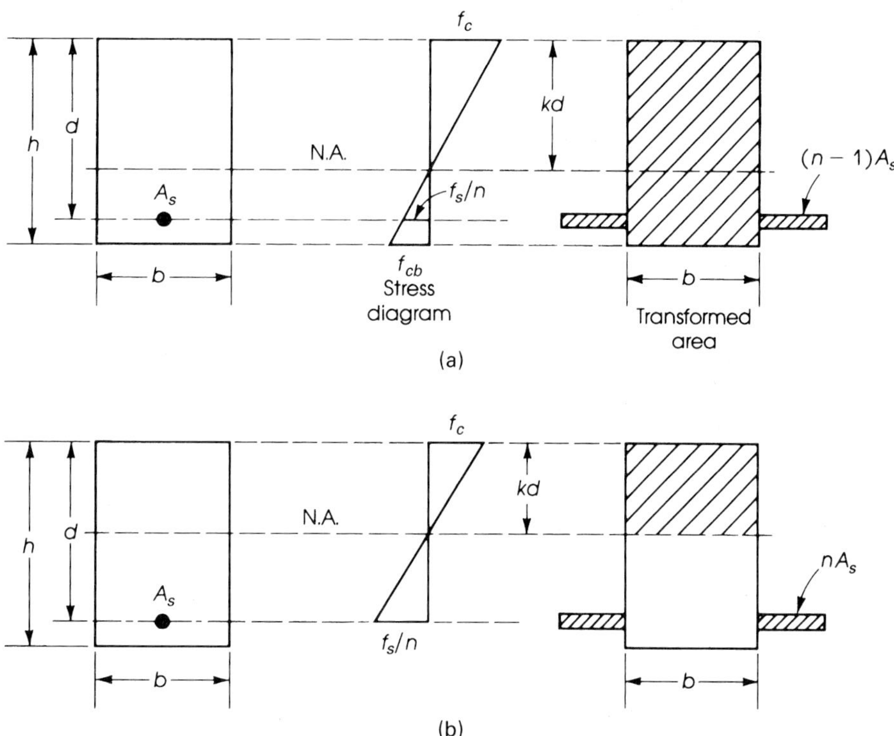

Figure 5.1 (a) Transformed area for uncracked singly reinforced section and (b) transformed area for a cracked singly reinforced section. (NA = Neutral axis.)

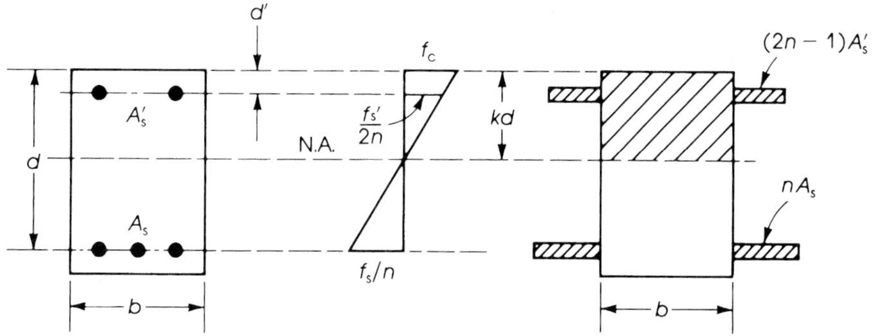

Figure 5.2 Transformed area for a cracked section with compression steel.

5.3 CRACKING MOMENT

The behavior of a simply supported structural concrete beam loaded to failure was explained in Section 3.3. At a low load, a small bending moment develops, and the stress at the extreme tensile fibers will be less than the modulus of rupture of concrete, $f_r = 7.5\sqrt{f'_c}$. The whole uncracked section will be effective in resisting the external moment. If the load is increased until the tensile stress reaches an average stress of the modulus of rupture f_r, cracks will develop. If the tensile stress is higher than f_r, the section will crack, and a cracked section case will develop. This means that there are three cases of analysis to be considered:

1. When the tensile stress, f_t, is less than f_r, the whole uncracked section resists the external moment; the tension steel will not be effective and can be neglected. In this case the stresses in the concrete section can be calculated by the flexural formula, $f = Mc/I$, using the gross moment of inertia, I_g (the whole concrete area, neglecting the tension steel). For example, for a rectangular section (bh), $f_t = f_c = Mc/I_g$, where $I_g = bh^3/12$ and $c = h/2$ and f_t and f_c are the tensile and compressive stresses in the concrete extreme fibers.
2. When the tensile stress, f_t, is equal to the modulus of rupture of concrete, $f_r = 7.5\sqrt{f'_c}$, a crack may start to develop, and the moment that causes this stress, f_r, is called the cracking moment. Using the flexural formula,

$$f_r = \frac{M_{cr} \cdot c}{I_g} \quad \text{or} \quad M_{cr} = f_r \cdot \frac{I_g}{c} \tag{5.5}$$

   where $f_r = 7.5\sqrt{f'_c}$, I_g = the gross moment of inertia, and c = the distance from the neutral axis to the extreme tension fibers. For example, for a rectangular section $I_g = bh^3/12$ and $c = h/2$.
3. When the applied external moment is greater than the cracking moment, M_{cr}, a cracked section case is developed, and the concrete in the tension zone is neglected. A transformed cracked section is used to calculate the stresses in concrete and steel.

Note that the stresses in the concrete and steel in a cracked section should not exceed the allowable stresses set by the ACI Code.

The ACI Code in Appendix A specifies the following allowable stresses for concrete and steel in bending.

For concrete, the allowable compressive stress is $f_{ca} = 0.45f'_c$, where f'_c is the minimum specified compressive strength of a standard concrete cylinder.

For steel, the allowable stress, f_{sa} is 20 Ksi (140 MPa) for grade 40 $(f_y = 40 \text{ Ksi} = 276 \text{ MPa})$ and grade 50 $(f_y = 50 \text{ Ksi} = 345 \text{ MPa})$ steels. For grade 60 $(f_y = 60 \text{ Ksi} = 414 \text{ MPa})$ and stronger steels, the allowable f_{sa} is 24 Ksi (170 MPa). For main reinforcement $\frac{3}{8}$ in. (10 mm) or less in diameter in one-way slabs of not more than a 12-ft (3.6-m) span, the allowable stress is $f_{sa} = 0.5f_y \leq 30$ Ksi (200 MPa). In SI units, allowable stresses of 140 MPa and 170 MPa are used for steel bars with yield strengths of 300 MPa and 400 MPa, respectively.

Example 5.1

A rectangular concrete section is reinforced with 3 no. 9 bars in one row and has a width $b = 12$ in., total depth $h = 25$ in., and $d = 22.5$ in. (Figure 5.3). Calculate the modulus of rupture f_r, the gross moment of inertia I_g, and the cracking moment M_{cr}. Use $f'_c = 4000$ psi and $f_y = 60$ Ksi.

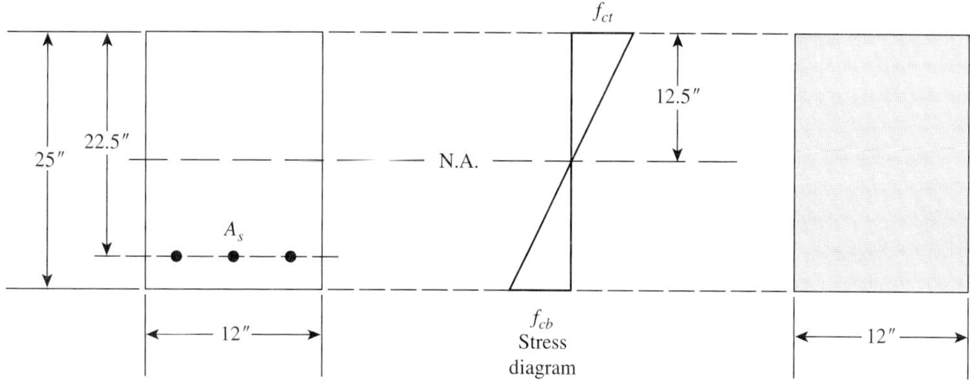

Figure 5.3 Example 5.1.

Solution

1. The modulus of rupture is $f_r = 7.5\sqrt{f'_c} = 7.5\sqrt{4000} = 474$ psi.
2. The gross moment of inertia for a rectangular section is

$$\frac{bh^3}{12} = \frac{12(25)^3}{12} = 15{,}625 \text{ in.}^4$$

3. The cracking moment is $M_{cr} = f_r \cdot I_g/c$.

$$f_r = 474 \text{ psi}, \qquad I_g = 15{,}625 \text{ in.}^4, \qquad c = \frac{h}{2} = 12.5 \text{ in.}$$

$$M_{cr} = \frac{474(15{,}625)}{12.5(1000)} = 592.5 \text{ K}\cdot\text{in.} = 49.38 \text{ K}\cdot\text{ft}$$

Note that the tensile and compressive stress in concrete is 474 psi. If the applied moment is greater than M_{cr}, then cracks will develop, and a cracked transform section must be used to analyze the section.

5.4 RECTANGULAR SECTIONS IN BENDING WITH TENSION REINFORCEMENT

In the analysis of structural concrete sections, the properties of the section and the allowable working stresses are given. It is usually required either to determine the stresses applied to the section and compare them with the allowable stresses or to determine the allowable bending moment that the section may carry. Two approaches may be used for analysis:

1. The internal couple method, in which the external bending moment is equated to the internal resisting couple. The internal couple consists of a compressive force C on one side of the neutral axis and an equal tensile force T on the other side. The arm is equal to jd, where j is less than 1.
2. The flexural formula approach, in which the stress at any point is calculated by the conventional flexural formula

$$f = \frac{Mc}{I}$$

where

f = bending stress
c = distance from the considered point to the neutral axis
M = the bending moment
I = the moment of inertia of the transformed area

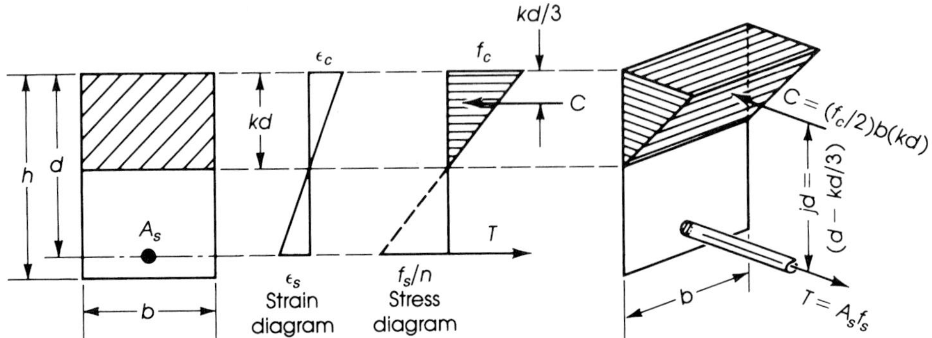

Figure 5.4 Internal forces in a singly reinforced rectangular section.

Figure 5.4 shows a rectangular section, the resisting internal forces, C and T, and the moment arm, jd. The force C is equal to $\frac{1}{2}bf_cKd$, which is the volume of half a prism. The distance Kd indicates the position of the neutral axis with respect to the extreme compressive fibers.

Two main equilibrium equations must be satisfied: First, the compressive and tensile forces are equal,

$$C = T \quad \text{or} \quad \tfrac{1}{2}f_cbKd = A_sf_s \tag{5.6}$$

and, second, the internal moment is equal to either the compressive force or the tensile force multiplied by its arm:

$$M = C(jd) = T(jd) \tag{5.7}$$

From these two equations, the following results can be deduced:

1. The distance of the neutral axis from the top compressive fibers (Kd) can be established for a rectangular section by setting the moments of the transformed area about the neutral axis equal to zero:

$$\frac{b(Kd)^2}{2} - nA_s(d - Kd) = 0 \tag{5.8}$$

   Solve for Kd.

2. The moment arm jd is equal to $(d - Kd/3)$, or

$$j = 1 - \frac{K}{3} \tag{5.9}$$

3. The internal moment M can be calculated using either the concrete compressive force C or the tension force T:

$$M_c = \tfrac{1}{2}f_cbKd(jd) = (\tfrac{1}{2}f_cKj)bd^2 = Rbd^2 \tag{5.10}$$

   where

$$R = \tfrac{1}{2}f_cKj \tag{5.11}$$

$$M_t = A_s \cdot f_s \cdot jd \tag{5.12}$$

4. The moment of inertia of a cracked singly reinforced rectangular section can be calculated as follows:

$$I_{cr} = \left(\frac{b}{3}\right)(Kd)^3 + nA_s(d - Kd)^2 \tag{5.13}$$

5. A balanced condition is achieved when the concrete extreme fibers reach the allowable stress, f_{ca}, at the same time as the tension steel reaches its allowable stress, f_{sa}. In this case,

$$K_b = \frac{f_{ca}}{f_{ca} + f_{sa}} \tag{5.14}$$

and $j_b = 1 - K_b/3$. It is a common practice to design a balanced selection in the alternate design method.

Example 5.2

Calculate I_{cr} and the stresses in the extreme concrete compression fibers f_c and in the steel reinforcement f_s for the rectangular beam section shown in Figure 5.5 when the external moment $M = 60\ K \cdot \text{ft}$, using the flexural formula $f = Mc/I$. Determine also the allowable moment capacity of the section. Given: $f'_c = 3$ Ksi and $f_y = 60$ Ksi $(f_{ca} = 1350\text{ psi}, f_{sa} = 24\text{ Ksi})\ (A_s = 2.37\text{ in.}^2, n = 9)$.

Solution

1. Determine the distance of the neutral axis (N.A.) from the top compressive fibers. The sum of the moments of the effective transformed areas about the neutral axis is equal to zero. Therefore,

$$b\frac{(Kd)^2}{2} - nA_s(d - Kd) = 0$$

$$\frac{10}{2}(Kd)^2 - 9 \times 2.37(19.5 - Kd) = 0 \tag{5.8}$$

$$(Kd)^2 + 4.26(Kd) = 83.19, \qquad Kd = 7.24\text{ in.}$$

2. Solution using the flexural formula: The moment of inertia of the cracked transformed section is

$$I_{cr} = b\frac{(Kd)^3}{3} + nA_s(d - Kd)^2$$

$$= (10)\frac{(7.24)^3}{3} + 9 \times (2.37)(19.5 - 7.24)^2 = 4471\text{ in.}^4$$

$$f_c = \frac{Mc}{I} = \frac{M(Kd)}{I} = \frac{60(12{,}000)(7.24)}{4471} = 1165\text{ psi}$$

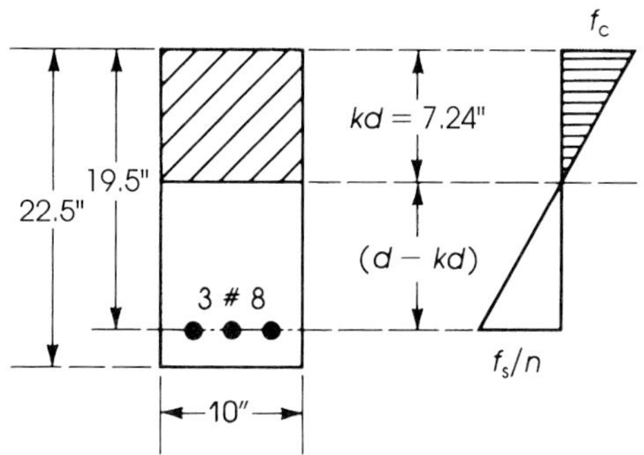

Figure 5.5 Example 5.2.

3. $\dfrac{f_s}{n} = \dfrac{M(d - Kd)}{I} = \dfrac{60(12{,}000)(19.5 - 7.24)}{4471} = 1975 \text{ psi}$

$f_s = 1.975(9) = 17.78$ Ksi

Both f_c and f_s are less than the allowable stresses of $f_{ca} = 1350$ psi and $f_{sa} = 24$ Ksi.

4. Calculate the allowable moment capacity of the section: allowable $f_{ca} = 1350$ psi, allowable $f_{sa} = 24$ Ksi.

$$C = \tfrac{1}{2} f_{ca} b(Kd) = \frac{1350}{2} \times \frac{10 \times 7.24}{1000} = 48.87 \text{ Kips}$$

$$T = A_s f_{sa} = 2.37 \times 24 = 56.88 \text{ Kips}$$

Because C is less than T, the allowable stress in the concrete will be reached before the steel bars reach their allowable stress; therefore, C controls.

Allowable moment is

$$M = C(jd) = C\left(d - \frac{Kd}{3}\right) = \frac{48.87}{12}\left(19.5 - \frac{7.24}{3}\right) = 69.6 \text{ K}\cdot\text{ft}$$

which is greater than the applied moment of 60 K · ft.

5.5 RECTANGULAR SECTIONS WITH COMPRESSION REINFORCEMENT

The elastic analysis of beams with compression reinforcement requires that compression steel, A'_s, has to be replaced by $(2n - 1)A'_s$ units of transformed area of concrete to account for the displaced concrete by A'_s and to account for shrinkage and creep stresses. The ACI Code (Appendix A) allows the use of $(2n)A'_s$ in stress computation without exceeding the allowable stress, f_{sa}.

The distance of the neutral axis from the extreme compressive fibers (kd) can be established by setting the moments of the transformed areas about the neutral axis equal to zero (neglecting the area of concrete in tension). Considering the tension steel transformed area, nA_s, and the net compression steel transformed area, $(2n - 1)A'_s$, then

$$b\frac{(kd)^2}{2} + (2n - 1)A'_s(kd - d') - nA_s(d - kd) = 0 \tag{5.15}$$

Solve this second-degree equation to get kd. The moment of inertia of a reinforced concrete rectangular section with compression reinforcement is calculated as follows:

$$I_{cr} = b\frac{(kd)^3}{3} + (2n - 1)A'_s(kd - d')^2 + nA_s(d - kd)^2 \tag{5.16}$$

The stresses can be calculated as follows. The stress in concrete is

$$f_c = M\frac{kd}{I} \le f_{ca}$$

The stress in tension steel is

$$f_s = (n)\frac{M(d - kd)}{I} \le f_{sa}$$

The stress in the compression steel is

$$f'_s = (2n)\frac{M(kd - d')}{I} \le f_{sa}$$

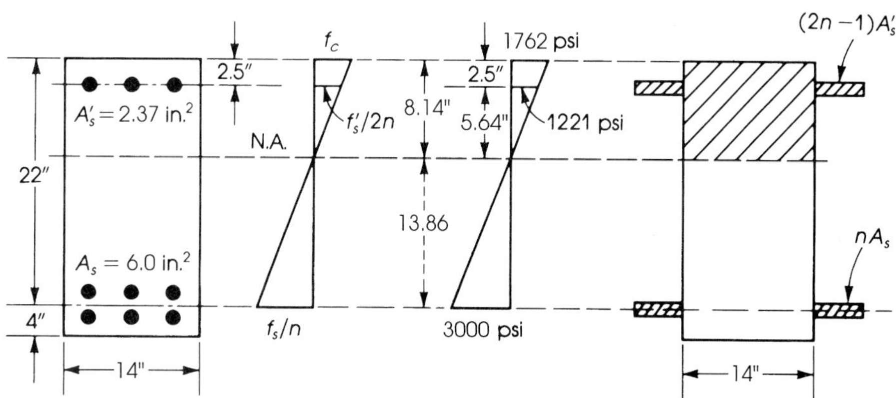

Figure 5.6 Example 5.3.

Example 5.3

Determine the allowable bending moment that the section shown in Figure 5.6 can carry. Given: $f'_c = 4$ Ksi, $f_y = 60$ Ksi $(f_{sa} = 24 \text{ Ksi})$, $A_s = 6.0$ in.^{2} (6 no. 9), $A'_s = 2.37$ in.^{2} (3 no. 8), $n = 8, b = 14$ in., and $d = 22$ in.

Solution

1. Determine the position of the neutral axis: $nA_s = 8(6) = 48$ in.^{2}, $(2n - 1)A'_s = (16 - 1)(2.37) = 35.55$ in.^{2} Take moments about the neutral axis:

$$\tfrac{14}{2}(kd)^2 + 35.55(kd - 2.5) - 48(22 - kd) = 0$$

Solve to get $kd = 8.14$ in.

2. Determine if steel or concrete controls. From the stress diagram, if the steel stress is 24 Ksi, then

$$\frac{f_{sa}}{n} = \frac{24}{8} = 3 \quad \text{and} \quad f_c\left(\frac{f_{sa}}{n}\right)\frac{(kd)}{(d - kd)} = \frac{3(8.14)}{(22 - 8.14)} = 1.762 \text{ Ksi} = 1762 \text{ psi}$$

which is less than $f_{ca} = 0.45f'_c = 1800$ psi. Therefore, steel controls.

3. From triangles,

$$\frac{f'_s}{2n} = f_c\left(\frac{5.64}{8.14}\right) = 1762\left(\frac{5.64}{8.14}\right) = 1221 \text{ psi}$$

$$f'_s = 2n(1.221) = 19.536 \text{ Ksi} < 24 \text{ Ksi}$$

So the value is acceptable.

4. Calculate the allowable moment:

$$C_c\,(\text{concrete}) = \tfrac{1}{2}f_c(kd)b = \tfrac{1}{2}(1.762)(8.14)(14) = 100.4 \text{ Kips}$$

$$C_s\,(\text{comp. steel}) = (2n - 1)A'_s\left(\frac{f'_s}{2n}\right) = 35.55(1.221) = 40.97 \text{ Kips}$$

Take moments about A_s:

$$\text{Allowable moment} = C_c\left(d - \frac{kd}{3}\right) + C_s(d - d')$$

$$= 100.4\left(22 - \frac{8.14}{3}\right) + 40.97(22 - 2.5)$$

$$= 2735.5 \text{ K}\cdot\text{in.} = 228 \text{ K}\cdot\text{ft}$$

5. The same results can be obtained using I_{cr} (equation (5.16)) and the flexural formula.

5.6 ANALYSIS OF T-SECTIONS

Concrete slabs and beams are usually cast together, forming T-shaped sections. The limitation of the effective flange width was discussed in Section 3.13. The effective flange width is the smallest of (1) one-fourth of the span, (2) $16t + b_w$, or (3) the distance between centerlines of adjacent slabs. The case when the neutral axis lies within the flange producing a rectangular section and the case when the neutral axis lies below the flange producing T-sections were discussed in Section 3.13.

Cracked section. The location of the neutral axis can be determined by setting the moments of the transformed areas about the neutral axis equal to zero, neglecting the area of concrete in tension (refer to Figure 5.7 and Example 5.4):

$$bt\left(kd - \frac{t}{2}\right) + b_w\frac{(kd - t)^2}{2} - nA_s(d - kd) = 0 \tag{5.17}$$

The moment of inertia of a cracked T-section can be calculated as follows:

$$I_{cr} = \left[\frac{bt^3}{12} + bt\left(kd - \frac{t}{2}\right)^2\right] + \left[b_w\frac{(kd - t)^3}{3}\right] + nA_s(d - kd)^2 \tag{5.18}$$

The stress in concrete is $f_c = M(kd)/I$, and the stress in the tension steel is $f_s = n \cdot M(d - kd)/I$.

Uncracked section. The gross moment of inertia I_g of the whole uncracked concrete T-section $(kd > t)$, neglecting A_s, can be determined as follows:

$$I_g = \left[\frac{bt^3}{12} + bt\left(kd - \frac{t}{2}\right)^2\right] + \left[b_w\frac{(kd - t)^3}{3}\right] + \left[b_w\frac{(h - kd)^3}{3}\right] \tag{5.19}$$

where kd can be calculated from the following equation:

$$bt\left(kd - \frac{t}{2}\right) + b_w\frac{(kd - t)^2}{2} - b_w\frac{(h - kd)^2}{2} = 0 \tag{5.20}$$

The following example explains the analysis of T-sections.

Example 5.4

Calculate f_c and f_s of the T-beam shown in Figure 5.7 for a moment of 100 K · ft. Given: $f'_c = 4$ Ksi, $f_{sa} = 20$ Ksi, $n = 8$, and $A_s = 3.8$ in.^{2}

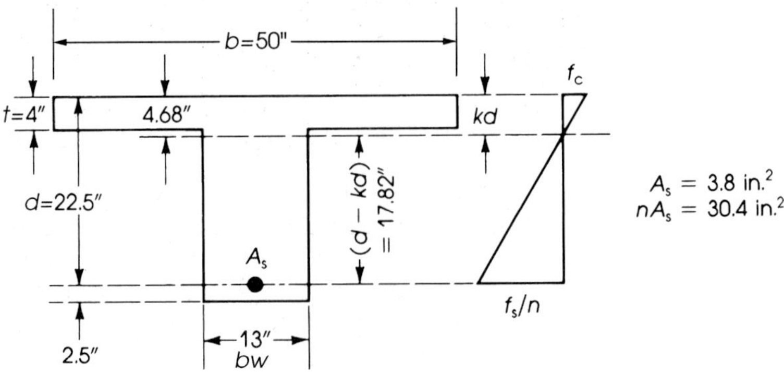

Figure 5.7 Example 5.4.

Solution

1. Determine whether the neutral axis lies within or below the flange by taking moments of areas about the bottom of the flange. For the concrete flange:

$$\text{Area} = 50(4) = 200 \text{ in.}^2$$

$$\text{Moment about bottom of flange} = 200(2) = 400 \text{ in.}^3$$

For the steel bars:

$$nA_s = (8 \times 3.8) = 30.4 \text{ in.}^2$$

$$\text{Moment about bottom of flange} = (8 \times 3.8)18.5 = 562 \text{ in.}^3$$

This indicates that the section behaves as a T-beam, because the neutral axis must lie below the flange so that the above two moments can be equal.

2. Determine the position of the neutral axis. The sum of the moments of the transformed areas about the neutral axis is equal to zero.

$$bt\left(Kd - \frac{t}{2}\right) + b_w \frac{(Kd - t)^2}{2} - nA_s(d - Kd) = 0$$

$$(50)(4)(kd - 2) + (13)\frac{(Kd - 4)^2}{2} - 30.4(22.5 - Kd) = 0$$

$$(Kd)^2 + 27.5(Kd) - 151 = 0, \qquad Kd = 4.68 \text{ in.}$$

3. Determine the moment of inertia of the cracked transformed section:

$$I_{cr} = 50 \times \frac{(4)^3}{12} + 50 \times 4 \times (2.68)^2 + \frac{13 \times (0.68)^3}{3} + 30.4 \times (17.82)^2 = 11{,}360 \text{ in.}^4$$

4. $f_c = \dfrac{M(Kd)}{I} = \dfrac{100 \times 12{,}000 \times 4.68}{11{,}360} = 494 \text{ psi}$

5. $\dfrac{f_s}{n} = \dfrac{M(d - Kd)}{I} = \dfrac{100 \times 12{,}000 \times 17.82}{11{,}360} = 1882 \text{ psi}$

   $f_s = 8(1882) = 15{,}060 \text{ psi}$

5.7 NONRECTANGULAR SECTIONS IN BENDING

A procedure similar to that explained here can be used for the analysis of nonrectangular beams. The location of the resultant compressive force C is the main part of analysis, because the tension portion of the concrete is always neglected. For nonsymmetrical sections loaded along their principal axis, lateral deflection is expected. But if these beams are cast monolithically with the slabs, such lack of symmetry may be ignored.

Example 5.5

The section shown in Figure 5.8 is that of a 22-ft simply supported beam whose floor system is composed of precast slabs. The dead load including its own weight is equal to 900 lb/ft. It is required to calculate the allowable live load on the beam, based on moment only. Given: $f'_c = 5$ Ksi, $f_{sa} = 24$ Ksi, $n = 7$, and $A_s = 6$ in.^{2}

Solution

1. To use the flexural formula $f = Mc/I$, first determine the position of the neutral axis:

$$nA_s = 7(6.0) = 42.0 \text{ in.}^2$$

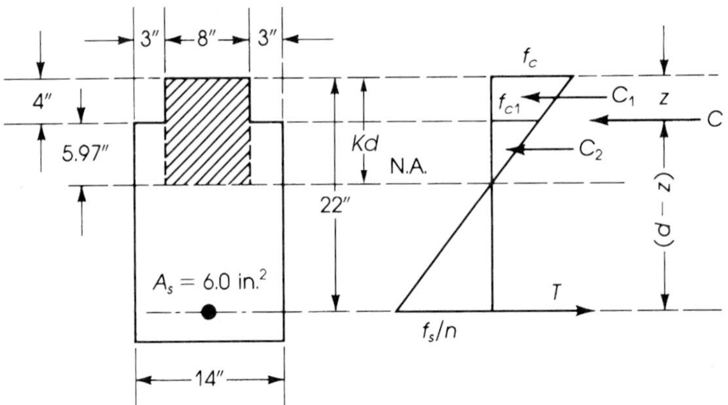

Figure 5.8 Example 5.5.

Subdividing the compression area into three rectangles and taking moments about the neutral axis,

$$(8)\left[\frac{(Kd)^2}{2}\right] + 2(3)\frac{(Kd - 4)^2}{2} - 42.0(22 - Kd) = 0$$

$$7(Kd)^2 + 18(Kd) = 876, \qquad Kd = 9.97 \text{ in.}$$

2. Calculate the moment of inertia (for the three rectangles plus A_s):

$$I_{cr} = (8)\frac{(9.97)^3}{3} + (2)(3)\left[\frac{(5.97)^3}{3}\right] + 42(22 - 9.97)^2 = 9177 \text{ in.}^4$$

$$\text{Allowable } f_c = 0.45f'_c = 0.45 \times 5000 = 2250 \text{ psi}$$

3. Calculate the allowable moment:

$$M = \frac{f_c I}{Kd} \quad \text{or} \quad M = \frac{f_s}{n} \times \frac{I}{(d - Kd)}$$

The internal allowable based on concrete stress f_c is

$$M_c = \frac{2250(9177)}{9.97} = 2.07 \times 10^6 \text{ lb}\cdot\text{in.} = 2071 \text{ K}\cdot\text{in.}$$

The internal moment based on steel stress f_{sa}(24 Ksi) is

$$M_s = \left(\frac{24}{7}\right)\frac{9177}{(22 - 9.97)} = 2616 \text{ K}\cdot\text{in.}$$

Because M_c is smaller than M_s, the internal moment based on concrete stress of 2250 psi controls. This means that the concrete will reach its allowable stress first and that the section is over-reinforced.

4. Determine the allowable live load on the beam. The external moment M is equal to the internal moment.

$$M = \frac{WL^2}{8}(12) = 2071 \text{ K}\cdot\text{in.}$$

$$W = \frac{2071 \times 8}{12 \times 22 \times 22} = 2.850 \text{ K/ft} = 2850 \text{ lb/ft}$$

Therefore, the allowable live load is $2850 - 900 = 1950$ lb/ft.

Example 5.6 SI Units

Calculate the stresses f_c and f_s for a rectangular section if $b = 250$ mm, $d = 475$ mm, and reinforced with 3 bars 25 mm in diameter due to an external moment of 81 kN·m. Use (1) the internal couple method and (2) the flexural formula. Then (3) determine the allowable moment. Given: $f'_c = 20$ MPa, $f_y = 400$ MPa, $f_{ca} = 9.0$ MPa, and $f_{sa} = 170$ MPa.

Solution

1. Solution using the internal couple method: Given that $n = 9$ and $A_s = 3(490.9) = 1473\ \text{mm}^2$,

$$b\frac{(Kd)^2}{2} - nA_s(d - Kd) = 0$$

$$250\frac{(Kd)^2}{2} - 9 \times 1473(475 - Kd) = 0 \quad Kd = 178\ \text{mm}$$

$$M = Cjd = \frac{1}{2}f_c bKd\left(d - \frac{Kd}{3}\right)$$

$$81 \times 10^6(\text{N}\cdot\text{mm}) = \tfrac{1}{2} \times f_c \times 250(178)\left(475 - \frac{178}{3}\right)$$

$$f_c = 8.8\ \text{MPa} < 9.0\ \text{MPa}$$

$$M = A_s f_s(jd) = A_s f_s\left(d - \frac{Kd}{3}\right)$$

$$81 \times 10^6 = 1473 f_s\left(475 - \frac{178}{3}\right)$$

$$f_s = 133\ \text{MPa} < 170\ \text{MPa}$$

2. Solution using the flexural formula:

$$I_{cr} = b\frac{(Kd)^3}{3} + nA_s(d - Kd)^2$$

$$= \frac{250}{3}(178)^3 + 9 \times 1473(475 - 178)^2 = 1639.15 \times 10^6\ \text{mm}^4$$

$$f_c = \frac{Mc}{I} = \frac{(81 \times 10^6) \times (178)}{1639.15 \times 10^6} = 8.8\ \text{MPa} < 9.0\ \text{MPa}$$

$$f_s = \frac{nM(d - Kd)}{I} = \frac{(9)(81 \times 10^6)}{1639.15 \times 10^6}(475 - 178) = 133\ \text{MPa} < 170\ \text{MPa}$$

3. Allowable moment: Given $f_{ca} = 9.0$ MPa and $f_{sa} = 170$ MPa, the compression force is

$$C = \frac{1}{2}f_{ca}bkd = \frac{9.0}{2} \times 250 \times 178 \times 10^{-3} = 200\ \text{kN}$$

$$T = A_s f_{sa} = 1473 \times 170 \times 10^{-3} = 250.4\ \text{kN}$$

C is less than T; thus the allowable stress in the concrete will be reached before the steel bars reach their allowable stress and C controls.

Allowable moment is

$$M = C(jd) = C\left(d - \frac{Kd}{3}\right) = 200\left(475 - \frac{178}{3}\right) \times 10^{-3} = 83.13\ \text{kN}\cdot\text{m}$$

which is greater than the applied moment of 81 kN·m.

SUMMARY

Sections 5.1–5.2

The modular ratio $n = E_s/E_c \approx 500/\sqrt{f'_c}$. A transformed steel area $= nA_s$.

Section 5.3

Cracking Moment

a. Neglect A_s (tension steel, also compression steel A'_s).

b. Cracking stress = modulus of rupture in concrete $= f_r$:

$$f_r = 7.5\sqrt{f'_c}, \qquad f_r = \frac{M_{\text{cr}} \cdot C_b}{I_g}, \qquad M_{\text{cr}} = \frac{f_r \cdot I_g}{C_b}$$

For a rectangular section, $C_b = h/2, I_g = bh^3/12$.

c. For T-sections: C_b = distance from centroid to extreme tension fibers.

Sections 5.4–5.6

1. *Uncracked Section* Check of stresses:
a. Neglect A_s (tension steel).
b. Use gross moment of inertia, I_g.
c. Calculate stresses in concrete (in tension and compression).
Tension:

$$f_t = \frac{M \cdot C_b}{I_g}, \qquad f_t = \frac{M(h/2)}{bh^3/12} = \frac{6M}{bh^2} \qquad \text{(for rectangular sections)}$$

Compression:

$$f_c = \frac{M \cdot C_t}{I_g}, \qquad f_c = \frac{6M}{bh^2} \qquad \text{(for rectangular sections)}$$

2. *Cracked Section* tensile stress $f_t > f_r(7.5\sqrt{fc'})$:
a. Determine $n = E_s/E_c$ = modular ratio.
b. Use a transformed section to change steel to concrete (nA_s).
c. Calculate the neutral axis distance x and then the cracked moment of inertia I_{cr}.
d. Calculate stresses in concrete and steel:

$$f_c = \frac{M \cdot x}{I_{\text{cr}}}, \qquad \frac{f_s}{n} = \frac{M(d - x)}{I_{\text{cr}}}$$

e. For the compression steel A'_s, use $(2n - 1)$ to calculate the transformed area. Calculate compressive stress:

$$\frac{f'_s}{2n} = \frac{M(x - d')}{I_{\text{cr}}}$$

3. Gross moment of inertia, I_g:

a. Neglect all the steel reinforcement in the section and consider only the whole concrete section.

b. Determine the centroid of the whole concrete section and then its moment of inertia about the centroidal axis.

c. For a rectangular section of width b and a total depth h, $I_g = bh^3/12$.

d. For a T-section, flange width b, web width b_w, and flange thickness t, calculate y, the distance to the centroidal axis from top of flange:

$$y = \frac{(bt^2/2) + b_w(h - t)[(h - t)/2 + t]}{bt + b_w(h - t)}$$

Then calculate I_g:

$$I_g = \left[\frac{bt^3}{12} + bt\left(y - \frac{t}{2}\right)^2\right] + \left[b_w\frac{(y - t)^3}{3}\right] + \left[b_w\frac{(h - y)^3}{3}\right]$$

4. Cracked moment of inertia, I_{cr}: Let x = the distance of the neutral axis from the extreme compression fibers (x = Kd).

a. Rectangular section with tension concrete, A_s, only:

i. Calculate x from the following equation:

$$\frac{bx^2}{2} - n \cdot A_s(d - x) = 0$$

ii. Calculate $I_{cr} = bx^3/3 + n \cdot A_s(d - x)^2$.

b. Rectangular section with tension steel A_s and compression steel A'_s:

i. Calculate x:

$$\frac{bx^2}{2} + (2n - 1)A'_s(x - d') - n \cdot A_s(d - x) = 0$$

ii. Calculate $I_{cr} = \dfrac{bx^3}{3} + (2n - 1)A'_s(x - d')^2 + n \cdot A_s(d - x)^2$.

c. T-sections with tension steel A_s:

i. Calculate x: $bt(x - t/2) + b_w(x - t)^2/2 - n \cdot A_s(d - x) = 0$.

ii. Calculate I_{cr}:

$$I_{cr} = \left[\frac{bt^3}{12} + bt\left(x - \frac{t}{2}\right)^2\right] + \left[b_w\frac{(x - t)^3}{3}\right] + n \cdot A_s(d - x)^2$$

REFERENCES

1. American Concrete Institute. *Reinforced Concrete Design Handbook (Working Stress Design)*. Publication SP-3, Detroit, 1965.
2. American Concrete Institute. *Building Code Requirements for Reinforced Concrete,* ACI 318-63. Detroit, 1963.
3. P.M. Ferguson. *Reinforced Concrete Fundamentals.* New York: John Wiley, 1979.
4. American Concrete Institute. *Building Code Requirements for Structural Concrete,* ACI 318-99. Detroit, 1999.

PROBLEMS

5.1 For each assigned problem of the rectangular sections and data given in the following table, calculate the gross moment of inertia, I_g, the cracked moment of inertia, I_{cr}, the cracking moment, the compressive stress in concrete, the stress in the tension steel, the compression steel (if applicable), and the allowable moment. Use $f'_c = 4$ Ksi, $f_y = 60$ Ksi, and $d' = 2.5$ in. $(n = 8, f_{ca} = 1800$ psi, $f_{sa} = 24$ Ksi$)$.

No.	Moment K·ft	b in.	d in.	h in.	A_s	A'_s
a	110	12	21.5	24	4 no. 8 (1)	—
b	220	18	32.0	35	6 no. 10 (1)	—
c	225	16	23.5	26	5 no. 10 (1)	—
d	105	10	20.5	23	3 no. 9 (1)	—
e	190	14	22.5	26	6 no. 9 (2)	—
f	140	14	20.5	23	4 no. 10 (1)	—
g	130	12	18.5	22	6 no. 9 (2)	2 no. 9
h	250	16	24.5	27	5 no. 10 (1)	2 no. 8
i	165	12	20.5	24	6 no. 9 (2)	2 no. 8
j	125	14	17.5	20	4 no. 10 (1)	2 no. 6

1. (1) or (2) indicates one or two rows of bars.
2. For SI problems, 1 in. = 25.4 mm, 1 in.^{2} = 645 mm^2; 1 Ksi = 6.9 MPa, 1 M_u (K·ft) = 1.356 kN·m.

5.2 For each assigned T-section problem and data given in the following table, calculate the gross moment of inertia, I_g, the moment inertia of a cracked section, I_{cr}, the compressive stress in concrete, the stress in the tension steel, and the allowable moment. Use $f'_c = 3$ Ksi and $f_y = 60$ Ksi $(n = 9, f_{ca} = 1350$ psi, and $f_{sa} = 24$ Ksi$)$.

No.	Moment K·ft	b in.	b_w in.	t in.	d in.	A_s
a	160	54	14	3	17.5	4 no. 10
b	125	48	14	4	16.5	4 no. 9
c	240	72	16	4	18.5	8 no. 10
d	95	32	16	3	15.5	3 no. 9
e	230	44	12	4	20.5	8 no. 8
f	155	50	14	3	16.5	5 no. 9
g	100	40	16	3	16.5	5 no. 8
h	110	42	12	3	17.5	4 no. 9

Note: $h = d + 2.5$ in. for all problems except c and e.
$h = d + 3.5$ in. for Problems c and e.

5.3 A rectangular section has width $b = 8$ in. (200 mm) and an effective depth $d = 20$ in. (500 mm) and is reinforced with 6 no. 6 bars (6 × 20 mm). Calculate the maximum stresses in the steel and concrete if a moment $M = 53$ K·ft (72 kN·m) is applied to the section. Determine the allowable moment on the section. Given $f'_c = 3$ Ksi (20 MPa), $f_{sa} = 20$ Ksi (140 MPa), and $n = 9$.

5.4 A 10-ft- (3-m-) span cantilever beam carries a uniform load of 4 K/ft (60 kN/m) and has the section shown in Figure 5.9. Calculate the bending stresses in the concrete and steel if $f'_c = 4$ Ksi (30 MPa), $f_{sa} = 20$ Ksi (140 MPa), and $n = 8$.

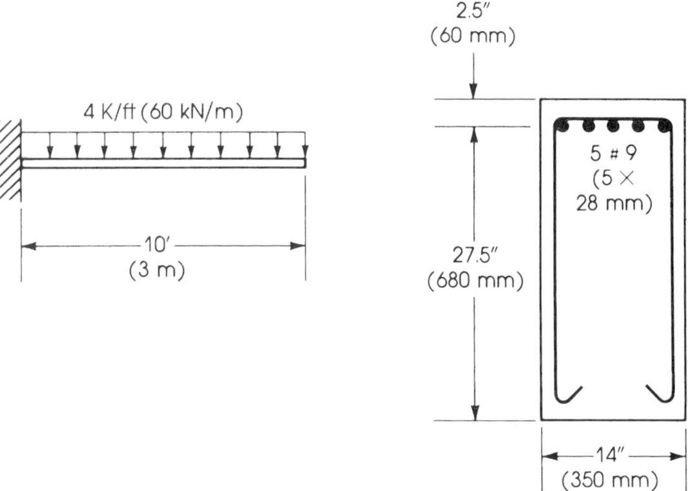

Figure 5.9 Problem 5.4.

5.5 A 15-ft- (4.5-m-) span simply supported beam carries a uniform load of 2 K/ft (30 kN/m) and a concentrated load of 6 K (27 kN) at midspan. The section of the beam is shown in Figure 5.10. Calculate the bending stresses in the concrete and steel, and compare with the allowable stresses. Given: $f'_c = 4$ Ksi (30 MPa), $f_{sa} = 24$ Ksi (170 MPa), and $n = 8$.

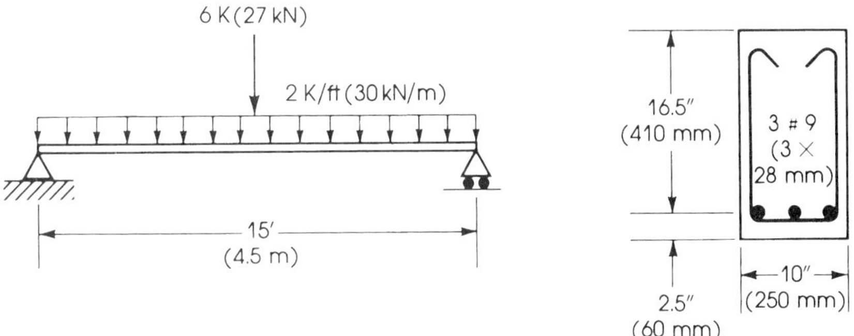

Figure 5.10 Problem 5.5.

5.6 Determine the allowable moment on the section of Problem 5.5 and calculate the additional uniform load that can be applied on the beam if the concentrated load P is removed.

5.7 If an 8-ft- (2.4-m-) span cantilever beam has the section given in Problem 5.4, determine the allowable uniform load that can be applied on the beam.

5.8 Determine the allowable bending moment on a rectangular section if $b = 12$ in. (300 mm), $d = 22$ in. (550 mm), $d' = 2.5$ in. (65 mm), $A_s = 8$ no. 7 bars (8 × 22 mm), $A'_s = 3$ no. 8 bars (3 × 25 mm), $f'_c = 3$ Ksi (20 MPa), and $f_{sa} = 20$ Ksi (140 MPa).

5.9 A T-section has a flange width $b = 40$ in. (1000 mm), a flange thickness $t = 4$ in. (100 mm), a web width $b_w = 10$ in. (250 mm), and $d = 19.5$ in. (480 mm) and is reinforced with 6 no. 8 bars (6 × 25 mm). For $M = 72.5$ K · ft (100 kN · m), calculate the maximum stresses in concrete and steel by the flexural formula. Also determine the allowable moment on the section. Given: $f'_c = 3$ Ksi (20 MPa), $f_{sa} = 24$ Ksi (170 MPa), and $n = 9$.

5.10 Calculate the resisting moment of the section shown in Figure 5.11. Use $f'_c = 4$ Ksi (30 MPa) and $f_{sa} = 24$ Ksi (170 MPa).

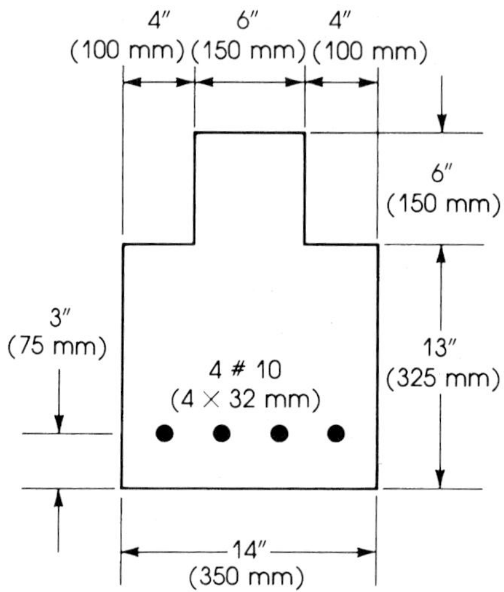

Figure 5.11 Problem 5.10.

5.11 Determine the resisting moment of the rectangular sections with compression reinforcement shown in Figure 5.12. Use $f'_c = 4$ Ksi (30 MPa) and $f_{sa} = 24$ Ksi (170 MPa).

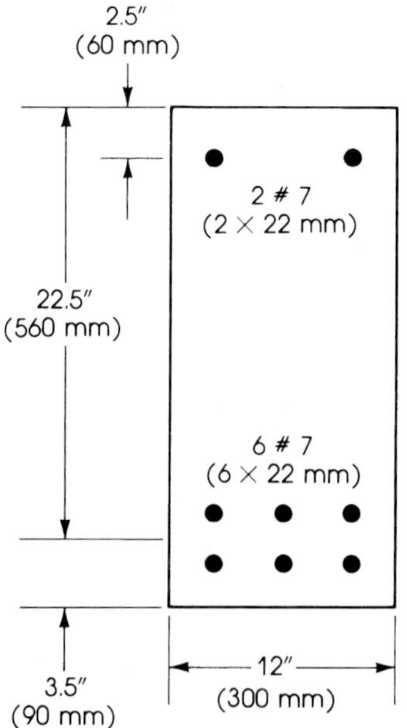

Figure 5.12 Problem 5.11.

5.12 Repeat Problem 5.11 for Figure 5.13.

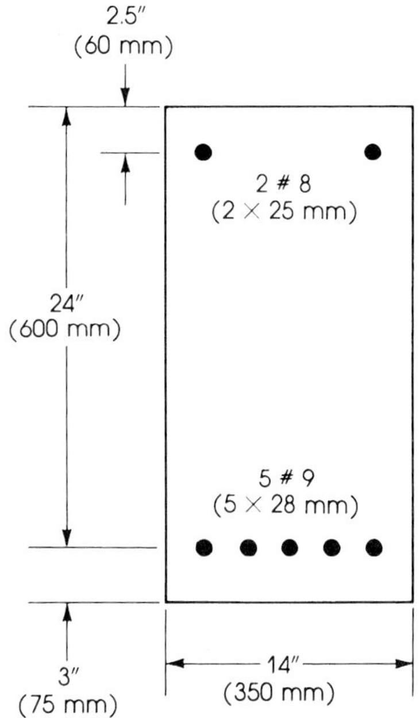

Figure 5.13 Problem 5.12.

5.13 Compute the resisting moment of the irregular section with compression reinforcement shown in Figure 5.14. Use $f'_c = 3$ Ksi (20 MPa) and $f_{sa} = 20$ Ksi (140 MPa).

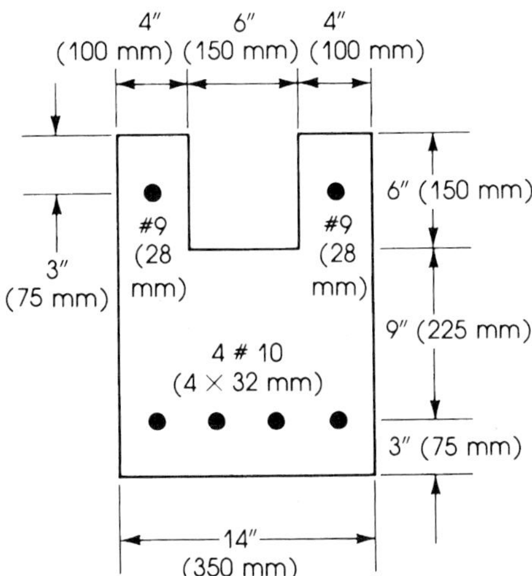

Figure 5.14 Problem 5.13.

6 DEFLECTION AND CONTROL OF CRACKING

High-rise building, Chicago

6.1 DEFLECTION OF STRUCTURAL CONCRETE MEMBERS

Flexural concrete members must be designed for safety and serviceability. The members will be safe if they are designed according to the ACI Code equations and limitations. Consequently, as explained in previous chapters, the size of each member is determined as well as the reinforcement required to maintain an internal moment capacity equal to or greater than that of the external moment. Once the final dimensions are determined, the beam must be checked for serviceability: cracks and deflection. Adequate stiffness of the member is necessary to prevent excessive cracks and deflection.

The use of the strength design method, taking into consideration the nonlinear relationship between stress and strain in concrete, has resulted in smaller sections than those designed by the elastic theory. The ACI Code, Section 9.4, recognizes the use of steel up to a yield strength of 80 Ksi (560 MPa) and the use of high-strength concrete. The use of high-strength steel and concrete results in smaller sections and a reduction in the stiffness of the flexural member and consequently increases its deflection.

The permissible deflection is governed by many factors, such as the type of the building, the appearance of the structure, the presence of plastered ceilings and partitions, the damage expected due to excessive deflection, and the type and magnitude of live load.

The ACI Code, Section 9.5, specifies minimum thickness for one-way flexural members and one-way slabs, as shown in Table 6.1. The values are for members not supporting or attached to partitions or other construction likely to be damaged by large deflections.

Table 6.1 Minimum Thickness of Beams and One-Way Slabs (L = span length)

Member	Yield Strength f_y (Ksi)	Simply Supported	One End Continuous	Both Ends Continuous	Cantilever
Solid one-way	40	$L/25$	$L/30$	$L/35$	$L/12.5$
slabs	50	$L/22$	$L/27$	$L/31$	$L/11$
	60	$L/20$	$L/24$	$L/28$	$L/10$
Beams or ribbed	40	$L/20$	$L/23$	$L/26$	$L/10$
one-way slabs	50	$L/18$	$L/20.5$	$L/23.5$	$L/9$
	60	$L/16$	$L/18.5$	$L/21$	$L/8$

The minimum thicknesses indicated in Table 6.1 are used for members made of normal-weight concrete, $W = 145$ pcf ($2320\ \mathrm{kg/m^3}$), and for steel reinforcement with yield strengths as mentioned in the table. The values are modified for cases of lightweight concrete or a steel yield strength different from 60 Ksi as follows:

- For lightweight concrete having unit weights in the range of 90–120 pcf, the values in the tables for $f_y = 60$ Ksi (420 MPa) shall be multiplied by the greater of $(1.65 - 0.005W_c)$ but not less than 1.09, where W_c is the unit weight of concrete in pounds per cubic foot.
- For yield strength of steel different from 60 Ksi (420 MPa), the values in the tables for 60 Ksi shall be multiplied by $(0.4 + f_y/100)$, where f_y is in Ksi.

6.1.1 Instantaneous Deflection

The deflection of structural members is due mainly to the dead load plus a fraction of or all the live load. The deflection that occurs immediately upon the application of the load is called the *immediate,* or *instantaneous, deflection.* Under sustained loads, the deflection increases appreciably with time. Various methods are available for computing deflections in statically determinate and indeterminate structures. The instantaneous deflection calculations are based on the elastic behavior of the flexural members. The elastic deflection Δ is a function of the load, W, span, L, moment of inertia, I, and the modulus of elasticity of the material, E:

$$\Delta = f\left(\frac{WL}{EI}\right) = \alpha\left(\frac{WL^3}{EI}\right) = K\left(\frac{ML^2}{EI}\right) \tag{6.1}$$

where W = total load on the span and α and K are coefficients that depend on the degree of fixity at the supports, the variation of moment of inertia along the span, and the distribution of load. For example, the maximum deflection on a uniformly loaded simply supported beam is

$$\Delta = \frac{5WL^3}{384EI} = \frac{5wL^4}{384EI} \tag{6.2}$$

while W = the total load on the span = wL (uniform load per unit length × span). Deflections of beams with different loadings and different end conditions as a function of the load, span, and EI are given in Appendix C and in books of structural analysis.

Because W and L are known, the problem is to calculate the modulus of elasticity, E, and the moment of inertia, I, of the concrete member or the flexural stiffness of the member EI.

6.1.2 Modulus of Elasticity

The ACI Code, Section 8.5, specifies that the modulus of elasticity of concrete, E_c, may be taken as

$$E_c = 33W_c^{1.5}\sqrt{f'_c}\ \text{psi} \tag{6.3}$$

for values of W_c between 90 and 155 pcf. For normal-weight concrete $(W_c = 145 \text{ pcf})$,

$$E_c = 57{,}400\sqrt{f'_c} \text{ psi} \qquad \left(\text{or } 57{,}000\sqrt{f'_c}\right) \tag{6.4}$$

The modulus of elasticity is usually determined by the short-term loading of a concrete cylinder. In actual members, creep due to sustained loading, at least for the dead load, affects the modulus on the compression side of the member. For the tension side, the modulus in tension is assumed to be the same as in compression when the stress magnitude is low. At high stresses the modulus decreases appreciably. Furthermore, the modulus varies along the span due to the variation of moments and shear forces.

6.1.3 Moment of Inertia

The moment of inertia, in addition to the modulus of elasticity, determines the stiffness of the flexural member. Under small loads, the produced maximum moment will be small, and the tension stresses at the extreme tension fibers will be less than the modulus of rupture of concrete; in this case the gross transformed cracked section will be effective in providing the rigidity. At working loads or higher, flexural tension cracks are formed. At the cracked section, the position of the neutral axis is high, whereas at sections midway between cracks along the beam, the position of the neutral axis is lower (nearer to the tension steel). In both locations only the transformed cracked sections are effective in determining the stiffness of the member; therefore, the effective moment of inertia varies considerably along the span. At maximum bending moment, the concrete is cracked, and its portion in the tension zone is neglected in the calculations of moment of inertia. Near the points of inflection the stresses are low, and the entire section may be uncracked. For this situation and in the case of beams with variable depth, exact solutions are complicated.

Figure 6.1(a) shows the load-deflection curve of a concrete beam tested to failure. The beam is a simply supported 17-ft span and loaded by two concentrated loads 5 ft apart, symmetrical about the centerline. The beam was subjected to two cycles of loading: In the first (curve cy 1), the load-deflection curve was a straight line up to a load $P = 1.7$ K when cracks started to occur in the beam. Line *a* represents the load-deflection relationship using a moment of inertia for the uncracked transformed section. It can be seen that the actual deflection of the beam under loads less than the cracking load, based on a homogeneous uncracked section, is very close to the calculated deflection (line *a*). Curve cy 1 represents the actual deflection curve when the load is increased to about half the ultimate load. The slope of the curve, at any level of load, is less than the slope of line *a* because cracks developed, and the cracked part of the concrete section reduces the stiffness of the beam. The load was then released, and a residual deflection was observed at midspan. Once cracks developed, the assumption of uncracked section behavior under small loads did not hold.

In the second cycle of loading, the deflection (curve *c*) increased at a rate greater than that of line *a* because the resistance of the concrete tension fibers was lost. When the load was increased, the load-deflection relationship was represented by curve cy 2. If the load in the first cycle is increased up to the ultimate load, curve cy 1 will take the path cy 2 at about 0.6 of the ultimate load. Curve *c* represents the actual behavior of the beam for any additional loading or unloading cycles.

Line *b* represents the load-deflection relationship based on a cracked transformed section; it can be seen that the deflection calculated on that basis differs from the actual deflection. Figure 6.1(c) shows the variation of the beam stiffness EI with an increase in moment. The ACI Code, Section 9.5, presents an equation to determine the effective moment of inertia used in calculating deflection in flexural members. The effective moment of inertia given by the ACI Code (equation (9.7)) is based on the expression proposed by Branson [3] and calculated as follows:

$$I_e = \left(\frac{M_{\text{cr}}}{M_a}\right)^3 I_g + \left[1 - \left(\frac{M_{\text{cr}}}{M_a}\right)^3\right] I_{\text{cr}} \le I_g \tag{6.5}$$

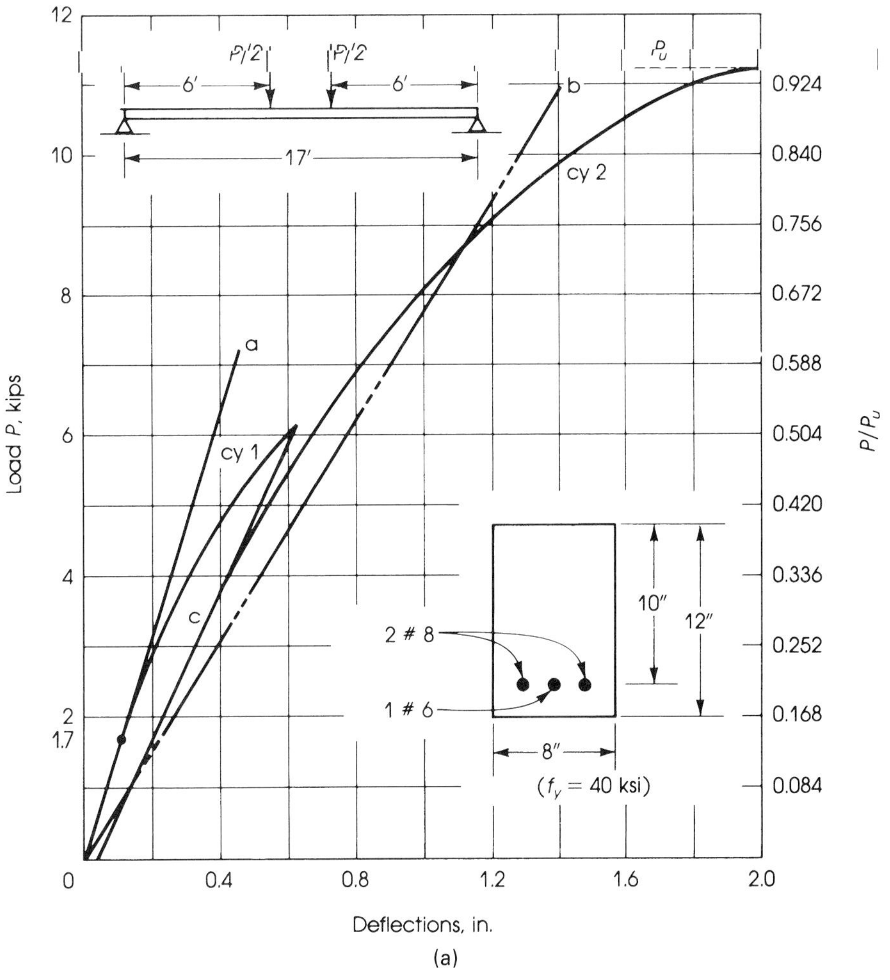

(a)

(b)

Figure 6.1 (a) Experimental and theoretical load-deflection curves for a beam of the section and load illustrated, (b) deflection of a reinforced concrete beam, and (c) variation of beam moment of inertia, I, with an increase in moment (E = constant). BC is a transition curve between I_g and I_{cr}.

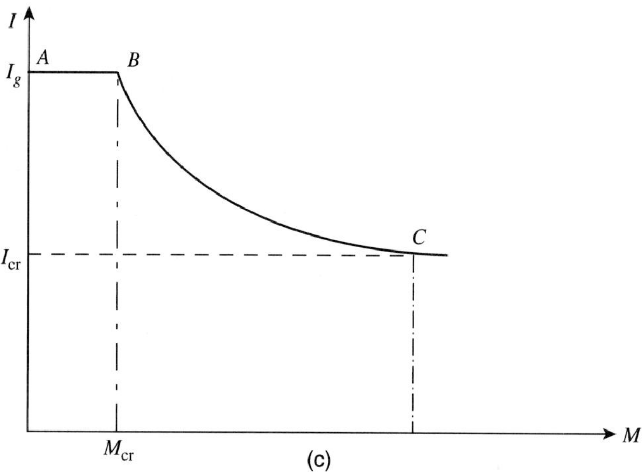

Figure 6.1 *(cont.)*

where

I_e = effective moment of inertia

M_{cr} = cracking moment, $\left(\dfrac{f_r I_g}{Y_t}\right)$ (6.6)

f_r = modulus of rupture of concrete

$= 7.5\sqrt{f'_c}$ psi, $(0.623\sqrt{f'_c}$ MPa$)$ (6.7)

M_a = maximum moment in member at stage for which deflection is being computed

I_g = moment of inertia of gross concrete section about the centroidal axis, neglecting the reinforcement

I_{cr} = moment of inertia of cracked transformed section

Y_t = distance from centroidal axis of cross section, neglecting steel, to extreme fiber in tension

The following limitations are specified by the code:

1. For continuous spans, the effective moment of inertia may be taken as the average of the moment of inertia of the critical positive and negative moment sections.
2. For lightweight concrete, the modulus of rupture f_r to be used in equation (6.6) is equal to

$$f_r = 7.5\left(\frac{f_{ct}}{6.7}\right) \quad \text{and} \quad \frac{f_{ct}}{6.7} \le \sqrt{f'_c} \tag{6.8a}$$

   where f_{ct} is the average splitting tensile strength, in psi. When f_{ct} is not known, f_r may be taken as follows: For all-lightweight concrete,

$$f_r = 5.6\sqrt{f'_c} \tag{6.8b}$$

   For sand-lightweight concrete,

$$f_r = 6.4\sqrt{f'_c} \tag{6.8c}$$

3. For prismatic members, I_e may be taken as the value obtained from equation (6.5) at midspan for simple and continuous spans and at the support section for cantilevers (ACI Code, Section 9.5.2).

   Note that I_e, as computed by equation (6.5), provides a transition between the upper and lower bounds of the gross moment of inertia, I_g, and the cracked moment of inertia, I_{cr}, as a function of the level of M_{cr}/M_a. Heavily reinforced concrete members may have an effective moment of inertia, I_e, very close to that of a cracked section, I_{cr}, whereas flanged members may have an effective moment of inertia close to the gross moment of inertia, I_g.

4. For continuous beams, an approximate value of the average I_e for prismatic or non-prismatic members for somewhat improved results is as follows: For beams with both ends continuous,

$$\text{Average } I_e = 0.70I_m + 0.15(I_{e1} + I_{e2}) \tag{6.9}$$

For beams with one end continuous,

$$\text{Average } I_e = 0.85I_m + 0.15(I_{\text{con}}) \tag{6.10}$$

where I_m = midspan $I_e, I_{e1}, I_{e2} = I_e$ at beam ends, and $I_{\text{con}} = I_e$ at the continuous end.

Also, I_e may be taken as the average value of the I_e's at the critical positive- and negative-moment sections. Moment envelopes should be used in computing both positive and negative values of I_e. In the case of a beam subjected to a single heavy concentrated load, only the midspan I_e should be used.

6.2 LONG-TIME DEFLECTION

Deflection of reinforced concrete members continues to increase under sustained load, although more slowly with time. Shrinkage and creep are the cause of this additional deflection, which is called long-time deflection [1]. It is influenced mainly by temperature, humidity, age at time of loading, curing, quantity of compression reinforcement, and magnitude of the sustained load. The ACI Code, Section 9.5, suggests that unless values are obtained by a more comprehensive analysis, the additional long-term deflection for both normal and lightweight concrete flexural members shall be obtained by multiplying the immediate deflection by the factor

$$\lambda = \frac{\zeta}{1 + 50\rho'} \tag{6.11}$$

where

$\rho' = A'_s/bd$ for the section at midspan of a simply supported or continuous beam or at the support of a cantilever beam

ζ = time-dependent factor for sustained loads that may be taken as shown in Table 6.2.

Table 6.2 Multipliers for Long-Time Deflections

Period (months)	1	3	6	12	24	36	48	60
ζ	0.5	1.0	1.2	1.4	1.7	1.8	1.9	2.0

The factor λ is used to compute deflection caused by the dead load and the portion of the live load that will be sustained for a sufficient period to cause significant time-dependent deflections. The factor λ is a function of the material property, represented by ζ, and the section property, represented by $(1 + 50\rho')$. In equation (6.11), the effect of compression reinforcement is related to the area of concrete rather than the ratio of compression to tension steel.

The ACI Code Commentary, Section 9.5, presents a curve to estimate ζ for periods less than 60 months. These values are estimated as shown in Table 6.2.

The total deflection is equal to the immediate deflection plus the additional long-time deflection. For instance, the total additional long-time deflection of a flexural beam with $\rho' = 0.01$ at a 5-year period is equal to λ times the immediate deflection, where $\lambda = 2/(1 + 50 \times 0.01) = 1.33$.

6.3 ALLOWABLE DEFLECTION

Deflection shall not exceed the following values according to the ACI Code, Section 9.5:

- $L/180$ for immediate deflection due to live load for flat roofs not supporting elements that are likely to be damaged
- $L/360$ for immediate deflection due to live load for floors not supporting elements likely to be damaged
- $L/480$ for the part of the total deflection that occurs after attachment of elements, that is, the sum of the long-time deflection due to all sustained loads and the immediate deflection due to any additional live load, for floors or roofs supporting elements likely to be damaged
- $L/240$ for the part of the total deflection occurring after elements are attached, for floors or roofs not supporting elements likely to be damaged

6.4 DEFLECTION DUE TO COMBINATIONS OF LOADS

If a beam is subjected to different types of loads (uniform, nonuniform, or concentrated loads) or subjected to end moments, the deflection may be calculated for each type of loading or force applied on the beam separately and the total deflection calculated by superposition. This means that all separate deflections are added up algebraically to get the total deflection. The deflections of beams under individual loads are shown in Table 6.3 and Appendix C.

Example 6.1

Calculate the instantaneous midspan deflection for the simply supported beam shown in Figure 6.2, which carries a uniform dead load of 0.4 K/ft and a live load of 0.6 K/ft in addition to a concentrated dead load of 5 kips at midspan. Given: $f'_c = 4$ Ksi, $f_y = 60$ Ksi, $b = 13$ in., $d = 21$ in., and total depth $= 25$ in. ($n = 8$).

Solution

1. Check minimum depth according to the ACI Code, Table 6.1.

$$\text{Minimum total depth} = \frac{L}{16} = \frac{40 \times 12}{16} = 30 \text{ in.}$$

The total thickness used is 25 in. $<$ 30 in.; therefore, deflection must be checked.

2. The deflection at midspan due to a distributed load is

$$\Delta_1 = \frac{5wL^4}{384E_cI_e}$$

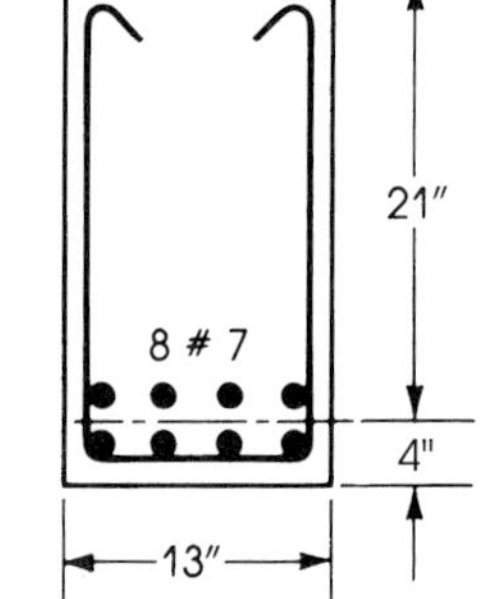

Figure 6.2 Example 6.1.

Table 6.3 Deflection of Beams

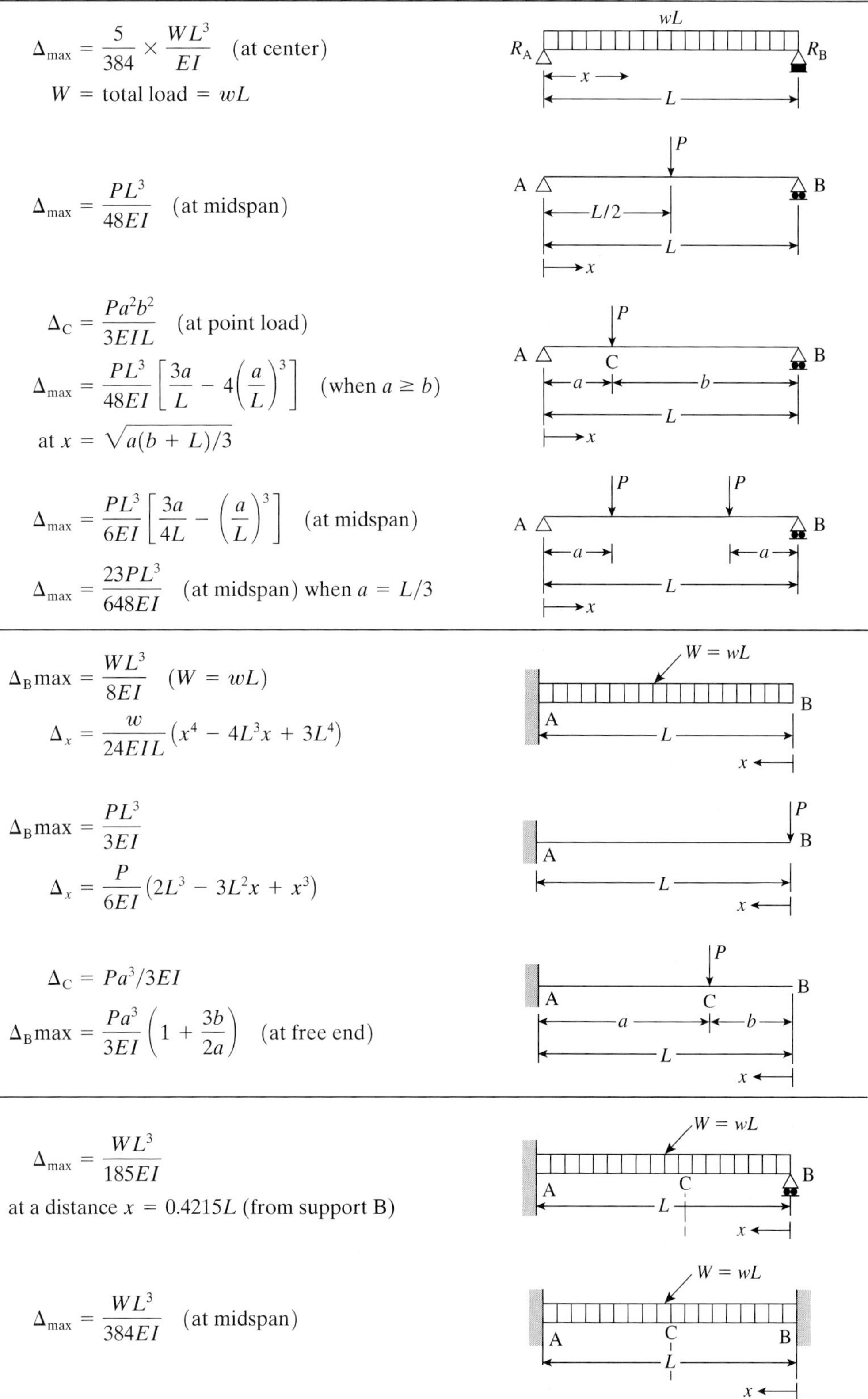

$$\Delta_{max} = \frac{5}{384} \times \frac{WL^3}{EI} \quad \text{(at center)}$$

$$W = \text{total load} = wL$$

$$\Delta_{max} = \frac{PL^3}{48EI} \quad \text{(at midspan)}$$

$$\Delta_C = \frac{Pa^2b^2}{3EIL} \quad \text{(at point load)}$$

$$\Delta_{max} = \frac{PL^3}{48EI}\left[\frac{3a}{L} - 4\left(\frac{a}{L}\right)^3\right] \quad \text{(when } a \geq b\text{)}$$

$$\text{at } x = \sqrt{a(b + L)/3}$$

$$\Delta_{max} = \frac{PL^3}{6EI}\left[\frac{3a}{4L} - \left(\frac{a}{L}\right)^3\right] \quad \text{(at midspan)}$$

$$\Delta_{max} = \frac{23PL^3}{648EI} \quad \text{(at midspan) when } a = L/3$$

$$\Delta_B\text{max} = \frac{WL^3}{8EI} \quad (W = wL)$$

$$\Delta_x = \frac{w}{24EIL}(x^4 - 4L^3x + 3L^4)$$

$$\Delta_B\text{max} = \frac{PL^3}{3EI}$$

$$\Delta_x = \frac{P}{6EI}(2L^3 - 3L^2x + x^3)$$

$$\Delta_C = Pa^3/3EI$$

$$\Delta_B\text{max} = \frac{Pa^3}{3EI}\left(1 + \frac{3b}{2a}\right) \quad \text{(at free end)}$$

$$\Delta_{max} = \frac{WL^3}{185EI}$$

at a distance $x = 0.4215L$ (from support B)

$$\Delta_{max} = \frac{WL^3}{384EI} \quad \text{(at midspan)}$$

The deflection at midspan due to a concentrated load is

$$\Delta_2 = \frac{PL^3}{48E_cI_e}$$

Because w, P, and L are known, we must determine the modulus of elasticity, E_c, and the effective moment of inertia, I_e.

3. The modulus of elasticity of concrete is

$$E_c = 57{,}400\sqrt{f'_c} = 57{,}400\sqrt{4000} = 3.63 \times 10^6 \text{ psi}$$

4. The effective moment of inertia is equal to

$$I_e = \left(\frac{M_{cr}}{M_a}\right)^3 I_g + \left[1 - \left(\frac{M_{cr}}{M_a}\right)^3\right] I_{cr} \leq I_g$$

Determine values of all terms on the right-hand side:

$$M_a = \frac{wL^2}{8} + \frac{PL}{4} = \frac{(0.6 + 0.4)}{8}(40)^2 \times 12 + \frac{5 \times 40}{4} \times 12 = 3000 \text{ K}\cdot\text{in.}$$

$$I_g = \frac{bh^3}{12} = \frac{13(25)^3}{12} = 16{,}927 \text{ in.}^4$$

$$M_{cr} = \frac{f_r I_g}{Y_t}, Y_t = \frac{h}{2} = 12.5 \text{ in.} \quad \text{and} \quad f_r = 7.5\sqrt{f'_c} = 474 \text{ psi}$$

$$M_{cr} = \frac{0.474 \times 16{,}927}{12.5} = 642 \text{ K}\cdot\text{in.}$$

The moment of inertia of the cracked transformed area, I_{cr}, is calculated as follows: Determine the position of the neutral axis for a cracked section by equating the moments of the transformed area about the neutral axis to zero, letting $x = Kd$ = distance to the neutral axis:

$$\frac{bx^2}{2} - nA_s(d - x) = 0, \qquad n = \frac{E_s}{E_c} = 8.0, \quad \text{and} \quad A_s = 4.8 \text{ in.}^2$$

$$\frac{13}{2}x^2 - (8)(4.8)(21 - x) = 0$$

$$x^2 + 5.9x - 124 = 0, \quad \text{and} \quad x = 8.8 \text{ in.}$$

$$I_{cr} = \frac{bx^3}{3} + nA_s(d - x)^2 = \frac{13(8.8)^3}{3} + 38.4(21 - 8.8)^2 = 8660 \text{ in.}^4$$

With all terms calculated,

$$I_e = \left(\frac{642}{3000}\right)^3 \times 16{,}927 + \left[1 - \left(\frac{642}{3000}\right)^3\right] \times 8660 = 8740 \text{ in.}^4$$

5. Calculate the deflections from the different loads:

$$\Delta_1 \text{ (due to distributed load)} = \frac{5wL^4}{384E_cI_e}$$

$$\Delta_1 = \left(\frac{5}{384}\right) \times \left(\frac{1000}{12}\right) \times \frac{(40 \times 12)^4}{3.63 \times 10^6 \times 8740} = 1.82 \text{ in.}$$

$$\Delta_2 \text{ (due to concentrated load)} = \frac{PL^3}{48E_cI_e}$$

$$\Delta_2 = \frac{5000 \times (40 \times 12)^3}{48 \times 3.63 \times 10^6 \times 8740} = 0.36 \text{ in.}$$

$$\text{Total immediate deflection} = \Delta_1 + \Delta_2 = 1.82 + 0.36 = 2.18 \text{ in.}$$

6. Compare the calculated values with the allowable deflection: The immediate deflection due to a uniform live load of 0.6 K/ft is equal to 0.6(1.82) = 1.09 in. If the member is part of a floor construction not supporting or attached to partitions or other elements likely to be damaged by large deflection, the allowable immediate deflection due to live load is equal to

$$\frac{L}{360} = \frac{40 \times 12}{360} = 1.33 \text{ in.} > 1.09 \text{ in.}$$

If the member is part of a flat roof and similar to the preceding, the allowable immediate deflection due to live load is $L/180 = 2.67$ in. Both allowable values are greater than the actual deflection of 1.09 in. due to the uniform applied live load.

Example 6.2

Determine the long-time deflection of the beam in Example 6.1 if the time-dependent factor equals 2.0.

Solution

1. The sustained load causing long-time deflection is that due to dead load, consisting of a distributed uniform dead load of 0.4 K/ft and a concentrated dead load of 5 K at midspan.

$$\text{Deflection due to uniform load} = 0.4 \times 1.82 = 0.728 \text{ in.}$$

Deflection is a linear function of load w, all other values (L, E_c, I_e) being the same.

$$\text{Deflection due to concentrated load} = 0.36 \text{ in.}$$

$$\text{Total immediate deflection due to sustained loads} = 0.728 + 0.36 = 1.088 \text{ in.}$$

2. For additional long-time deflection, the immediate deflection is multiplied by the factor λ:

$$\lambda = \frac{\zeta}{1 + 50\rho'} = \frac{2}{1 + 0}$$

In this problem, $A'_s = 0$; therefore, $\lambda = 2.0$.

$$\text{Additional long-time deflection} = 2 \times 1.088 = 2.176 \text{ in.}$$

3. Total long-time deflection is the immediate deflection plus additional long-time deflection: 2.18 + 2.176 = 4.356 in.
4. Deflection due to dead load plus additional long-time deflection due to shrinkage and creep is 1.088 + 2.176 = 3.264 in.

Example 6.3

Calculate the instantaneous and 1-year long-time deflection at the free end of the cantilever beam shown in Figure 6.3. The beam has a 20-ft span and carries a uniform dead load of 0.4 K/ft,

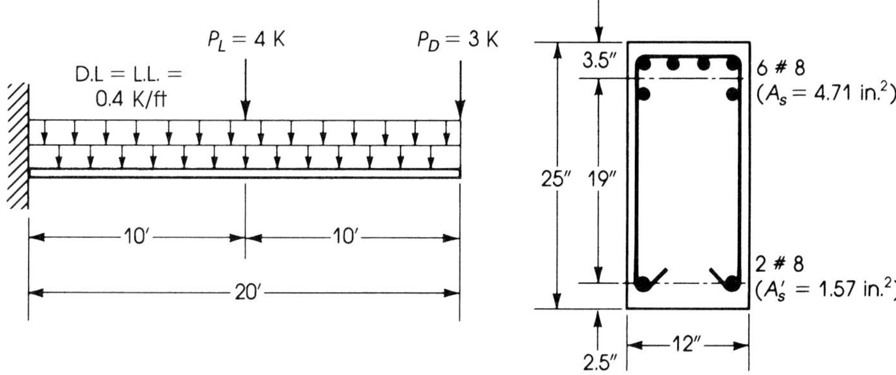

Figure 6.3 Example 6.3.

a uniform live load of 0.4 K/ft, a concentrated dead load, P_D, of 3 K at the free end, and a concentrated live load P_L of 4 K placed at 10 ft from the fixed end. Given: $f'_c = 4$ Ksi, $f_y = 60$ Ksi, $b = 12$ in., $d = 21.5$ in., and total depth of section $= 25$ in. (Tension steel is 6 no. 8 bars and compression steel is 2 no. 8 bars.)

Solution

1. Minimum depth $= L/8 = \frac{20}{8} = 2.5$ ft $= 30$ in., which is greater than the 25 in. used. Therefore, deflection must be checked. The maximum deflection of a cantilever beam is at the free end. The deflection at the free end is as follows.

Deflection due to distributed load:

$$\Delta_1 = \frac{wL^4}{8EI}$$

Deflection due to a concentrated dead load at the free end:

$$\Delta_2 = \frac{P_D L^3}{3EI}$$

Deflection due to concentrated live load at $a = 10$ ft from the fixed end is maximum at the free end:

$$\Delta_3 = \frac{P_L(a)^2}{6EI}(3L - a) \text{ or } = \frac{Pa^3}{3EI}\left(1 + \frac{3b}{2a}\right)$$

2. The modulus of elasticity of normal-weight concrete is

$$E_c = 57{,}400\sqrt{f'_c} = 57{,}400\sqrt{4000} = 3.63 \times 10^6 \text{ psi}$$

3. Maximum moment at the fixed end is

$$M_a = \frac{wL^2}{2} + P_D \times 20 + P_L \times 10$$

$$= \frac{(0.4 + 0.4)(400)}{2} + 3 \times 20 + 4 \times 10 = 260 \text{ K}\cdot\text{ft}$$

4. I_g = gross moment of inertia (concrete only)

$$= \frac{bh^3}{12} = \frac{12 \times (25)^3}{12} = 15{,}625 \text{ in.}^4$$

5. $$M_{cr} = \frac{f_r I_g}{Y_t} = \frac{(7.5\sqrt{4000}) \times 15{,}625}{\frac{25}{2}} = 592.9 \text{ K}\cdot\text{in.} = 49.40 \text{ K}\cdot\text{ft}$$

6. Determine the position of the neutral axis; then determine the moment of inertia of the cracked transformed section. Take moments of areas about the neutral axis and equate them to zero. Use $n = 8$ to calculate the transformed area of A_s and use $(n - 1) = 7$ to calculate the transformed area of A'_s (see Section 5.5). Let $Kd = x$.

$$b\frac{(x^2)}{2} + (n - 1)A'_s(x - d') - nA_s(d - x) = 0$$

For this section, $x = 8.44$ in.

$$I_{cr} = \frac{b}{3}x^3 + (n - 1)A'_s(x - d')^2 + nA_s(d - x)^2 = 9220 \text{ in.}^4$$

7. Effective moment of inertia is

$$I_e = \left(\frac{M_{cr}}{M_a}\right)^3 I_g + \left[1 - \left(\frac{M_{cr}}{M_a}\right)^3\right] I_{cr} \le I_g$$

$$= \left(\frac{49.40}{260}\right)^3 \times 15{,}625 + \left[1 - \left(\frac{49.40}{260}\right)^3\right] \times 9220 = 9264 \text{ in.}^4$$

8. Determine the components of the deflection:

$$\Delta_1 \text{ (due to uniform load of 0.8 K/ft)} = \frac{800}{12} \times \frac{(20 \times 12)^4}{8 \times 3.63 \times 10^6 \times 9264} = 0.82 \text{ in.}$$

$$\Delta_1 \text{ (due to dead load)} = 0.82 \times \frac{0.4}{0.8} = 0.41 \text{ in.}$$

$$\Delta_2 \text{ (due to concentrated dead load) at free end} = \frac{3000(20 \times 12)^3}{3 \times 3.63 \times 10^6 \times 9264} = 0.41 \text{ in.}$$

$$\Delta_3 \text{ (due to concentrated live load at 10 ft from fixed end)} = \frac{4000(10 \times 12)^2 \times (3 \times 20 \times 12 - 10 \times 12)}{6 \times 3.63 \times 10^6 \times 9264} = 0.17 \text{ in.}$$

The total immediate deflection is

$$\Delta = \Delta_1 + \Delta_2 + \Delta_3 = 0.82 + 0.41 + 0.17 = 1.40 \text{ in.}$$

9. For additional long-time deflection, the immediate deflection is multiplied by the factor λ. For a 1-year period, $\zeta = 1.4$.

$$\rho' = \frac{A'_s}{bd} = \frac{1.57}{12 \times 21.5} = 0.0061$$

$$\lambda = \frac{1.4}{1 + 50 \times 0.0061} = 1.073$$

Total immediate deflection Δ_s due to sustained load (here only the dead load of 0.4 K/ft and $P_D = 3$ K at free end): $\Delta_s = (0.41 + 0.41) = 0.82$ in. Additional long-time deflection $= 1.073 \times 0.82 = 0.88$ in.

10. Total long-time deflection is the immediate deflection plus long-time deflection due to shrinkage and creep.

$$\text{Total } \Delta = 1.40 + 0.88 = 2.28 \text{ in.}$$

Example 6.4

Calculate the instantaneous midspan deflection of beam AB in Figure 6.4, which has a span of 32 ft. The beam is continuous over several supports of different span lengths. The absolute bending moment diagram and cross sections of the beam at midspan and supports are also shown. The beam carries a uniform dead load of 4.2 K/ft and a live load of 3.6 K/ft. Given: $f'_c = 3$ Ksi, $f_y = 60$ Ksi, and $n = 9.2$.

Moment at midspan:	$M_D = 192$ K·ft	$M_{(D+L)} = 480$ K·ft
Moment at left support A:	$M_D = 179$ K·ft	$M_{(D+L)} = 420$ K·ft
Moment at right support B:	$M_D = 216$ K·ft	$M_{(D+L)} = 542$ K·ft

Solution

1. The beam AB is subjected to a positive moment that causes a deflection downward at midspan and negative moments at the two ends, causing a deflection upward at midspan. As was explained earlier, the deflection is a function of the effective moment of inertia, I_e. In a continuous beam, the value of I_e to be used is the average value for the positive and negative moment regions. Therefore, three sections will be considered, the section at midspan and the sections at the two supports.

2. Calculate I_e: For the gross area of all sections, $Kd = 13.5$ in. and $I_g = 114{,}300$ in.^{4} Also, $f_r = 7.5\sqrt{f'_c} = 410$ psi and $E_c = 57{,}400\sqrt{f'_c} = 3.15 \times 10^6$ for all sections. The values of Kd, I_{cr}, and M_{cr} for each cracked section, I_e for dead load only (using M_a of dead load), and I_e for dead and live loads (using M_a for dead and live loads) are calculated and tabulated as follows.

Section	*Kd* (in.)	I_{cr} (in.^{4})	M_{cr} (K · ft)	I_e (in.^{4}) (Dead load)	I_e (in.^{4}) ($D + L$)
Midspan	6.67	48,550	159.4	86,160	50,960
Support *A*	10.9	34,930	289.3	114,300	60,880
Support *B*	12.6	44,860	289.3	114,300	55,415

Note that when the beam is subjected to dead load only and the ratio M_{cr}/M_a is greater than 1.0, I_e is equal to I_g.

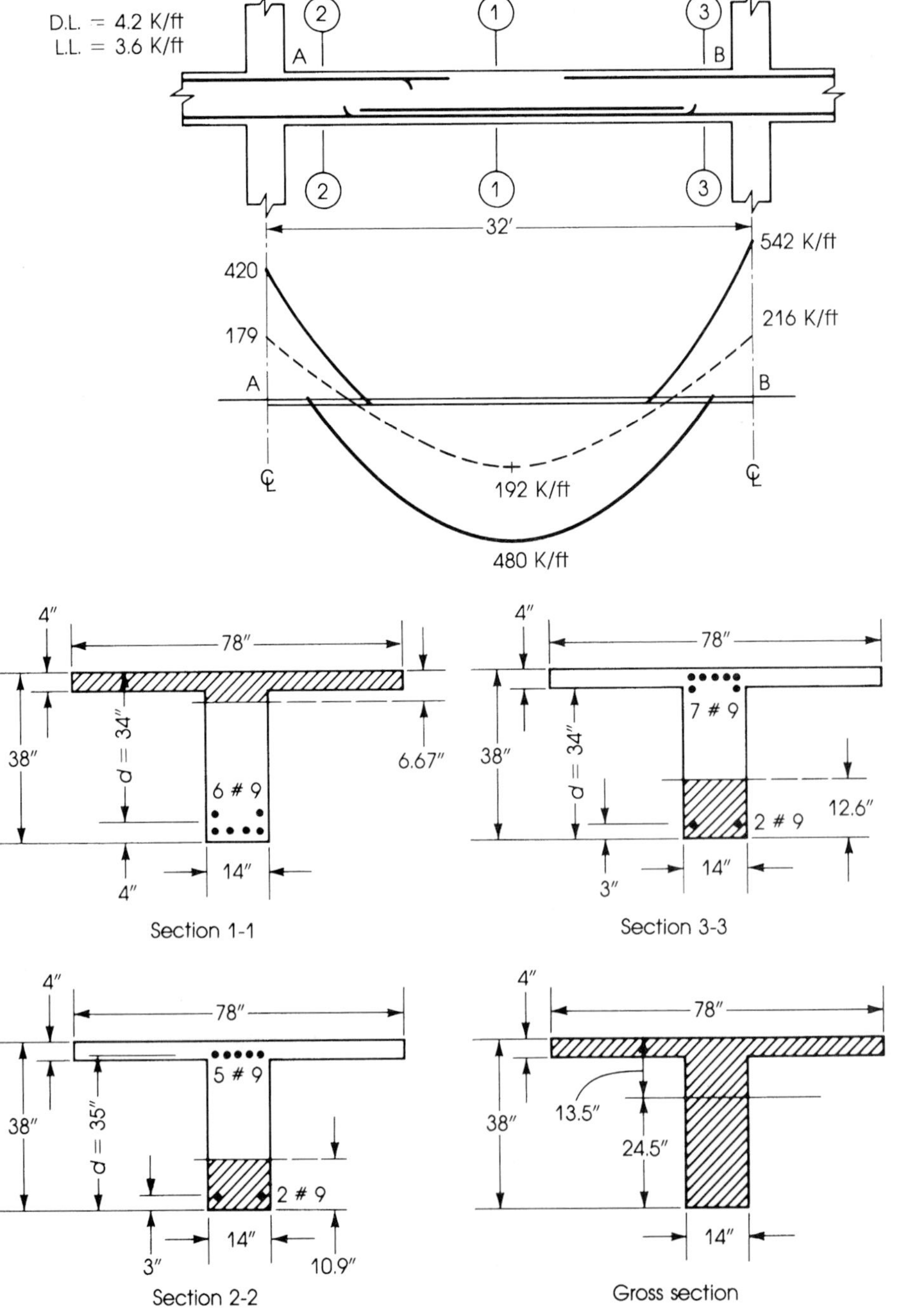

Figure 6.4 Example 6.4, deflection of a continuous beam.

3. Calculate average I_e from equation (6.9):

$$I_{e1}\,(\text{average}) = 0.7(50{,}960) + 0.15(60{,}880 + 55{,}415) = 53{,}116 \text{ in.}^4$$

For dead and live loads,

$$\text{Average } I_e \text{ for end sections} = \tfrac{1}{2}(60{,}880 + 55{,}415) = 58{,}150 \text{ in.}^4$$

$$I_{e2}\,(\text{average}) = \tfrac{1}{2}(50{,}960 + 58{,}150) = 54{,}550 \text{ in.}^4$$

For dead loads only,

$$\text{Average } I_e \text{ for end sections} = 114{,}300 \text{ in.}^4$$

$$I_{e3}\,(\text{average}) = \tfrac{1}{2}(86{,}160 + 114{,}300) = 100{,}230 \text{ in.}^4$$

4. Calculate immediate deflection at midspan:

$$\Delta_1 \text{ (due to uniform load)} = \frac{5wL^4}{384EI_e} \qquad \text{(downward)}$$

$$\Delta_2 \left(\text{due to a moment at } A,\, M_A\right) = -\frac{M_A L^2}{16EI_e} \qquad \text{(upward)}$$

$$\Delta_3 \left(\text{due to a moment at } B,\, M_B\right) = -\frac{M_B L^2}{16EI_e} \qquad \text{(upward)}$$

$$\text{Total deflection } \Delta = \Delta_1 - \Delta_2 - \Delta_3$$

The dead-load deflection for a uniform dead load of 4.2 K/ft, taking M_A(D.L.) = 179 K · ft, M_B(D.L.) = 216 K · ft, and I_{e3} = 100,230 in.^{4} and then substituting in the preceding equations, is

$$\Delta = 0.314 - 0.063 - 0.075 = 0.176 \text{ in.} \qquad \text{(downward)}$$

The deflection due to combined dead and live loads is found by taking dead plus live load = 7.8 K/ft, M_A = 420 K · ft, M_B = 542 K · ft, and I_{e2} = 54,550 in.^{4}:

$$\Delta = 1.071 - 0.270 - 0.349 = 0.452 \text{ in.} \qquad \text{(downward)}$$

The immediate deflection due to live load only is 0.542 − 0.176 = 0.276 in. (downward). If the limiting permissible deflection is $L/480 = (32 \times 12)/480 = 0.8$ in., then the section is adequate.

There are a few points to mention about the results.

- If I_e of the midspan section only is used $\left(I_e = 50{,}960 \text{ in.}^4\right)$, then the deflection due to dead plus live loads is calculated by multiplying the obtained value in Step 4 by the ratio of the two I_e:

$$\Delta\,(\text{dead} + \text{live}) = 0.452 \times \left(\frac{54{,}550}{50{,}960}\right) = 0.484 \text{ in.}$$

  The difference is small, about 7% on the conservative side.
- If I_{e1} (average) is used $\left(I_{e1} = 53{,}116 \text{ in.}^4\right)$, then Δ (dead + live) = 0.471 in. The difference is small, about 4% on the conservative side.
- It is believed that it is more convenient to use I_e at midspan section unless a more rigorous solution is required.

6.5 CRACKS IN FLEXURAL MEMBERS

The study of crack formation, behavior of cracks under increasing load, and control of cracking is necessary for proper design of reinforced concrete structures. In flexural members, cracks develop under working loads, and because concrete is weak in tension, reinforcement is placed in the cracked tension zone to resist the tension force produced by the external loads.

Flexural cracks develop when the stress at the extreme tension fibers exceeds the modulus of rupture of concrete. With the use of high-strength reinforcing bars, excessive cracking may develop in reinforced concrete members. The use of high-tensile steel has many advantages, yet the development of undesirable cracks seems to be inevitable. Wide cracks may allow corrosion of the reinforcement or leakage of water structures and may spoil the appearance of the structure.

A crack is formed in concrete when a narrow opening of indefinite dimension has developed in the concrete beam as the result of internal tensile stresses. These internal stresses may be due to one or more of the following:

- external forces such as direct axial tension, shear, flexure, or torsion
- shrinkage
- creep
- internal expansion resulting from a change of properties of the concrete constituents

In general, cracks may be divided into two main types, secondary cracks and main cracks.

6.5.1 Secondary Cracks

Secondary cracks, very small cracks that develop in the first stage of cracking, are produced by the internal expansion and contraction of the concrete constituents and by low flexural tension stresses due to the self-weight of the member and any other dead loads. There are three types of secondary cracks.

Shrinkage cracks *Shrinkage cracks* are important cracks, because they affect the pattern of cracking that is produced by loads in flexural members. When they develop, they form a weak path in the concrete. When load is applied, cracks start to appear at the weakest sections, such as along the reinforcing bars. The number of cracks formed is limited by the amount of shrinkage in concrete and the presence of restraints. Shrinkage cracks are difficult to control.

Secondary flexural cracks Usually *secondary flexural cracks* are widely spaced, and one crack does not influence the formation of others [8]. They are expected to occur under low loads, such as dead loads. When a load is applied gradually on a simple beam, tensile stress develops at the bottom fibers, and when it exceeds the flexural tensile stress of concrete, cracks start to develop. They widen gradually and extend toward the neutral axis. It is difficult to predict the sections at which secondary cracks start because concrete is not a homogeneous, isotropic material.

Salinger [9] and Billing [10] estimated the steel stress just before cracking to be from about 6000 to 7000 psi (42 to 49 MPa). An initial crack width of the order of 0.001 in. (0.025 mm) is expected at the extreme concrete tensile fibers. Once cracks are formed, the tensile stress of concrete at the cracked section decreases to zero, and the steel bars take all the tensile force. At this moment, some slip occurs between the steel bars and the concrete due to the differential elongation of concrete and steel and extends to a section where the concrete and steel strains are equal. Figure 6.5 shows the typical stress distribution between cracks in a member under axial tension.

Corrosion secondary cracks *Corrosion secondary cracks* form when moisture containing deleterious agents such as sodium chloride, carbon dioxide, and dissolved oxygen penetrates the concrete surface, corroding the steel reinforcement [11]. The oxide compounds formed by deterioration of steel bars occupy a larger volume than the steel and exert mechanical pressure that perpetuates extensive cracking [12], [13]. This type of cracking may

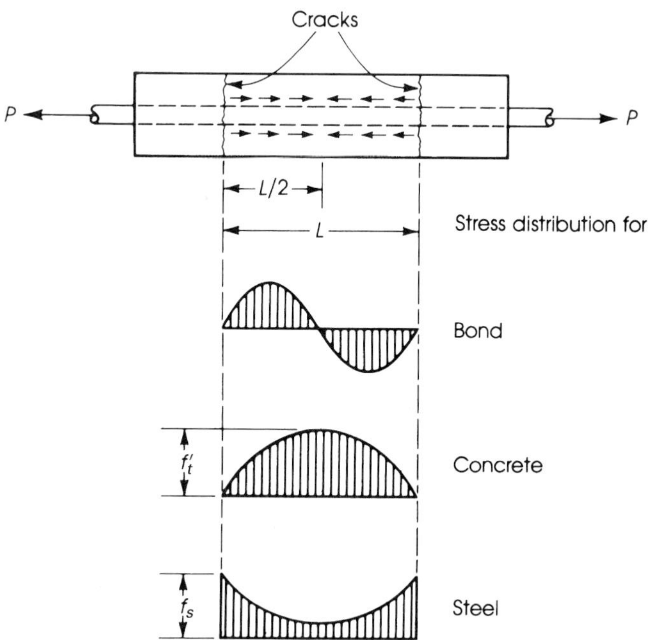

Figure 6.5 Typical stress distribution between cracks.

be severe enough to result in eventual failure of the structure. The failure of a roof in Muskegan, Michigan, in 1955 due to the corrosion of steel bars was reported by Shermer [13]. The extensive cracking and spalling of concrete in the San Mateo–Hayward Bridge in California within 7 years was reported by Stratfull [12]. Corrosion cracking may be forestalled by using proper construction methods and high-quality concrete. More details are discussed by Evans [14] and Mozer and others [15].

6.5.2 Main Cracks

Main cracks develop at a later stage than secondary cracks. They are caused by the difference in strains in steel and concrete at the section considered. The behavior of main cracks changes at two different stages. At low tensile stresses in steel bars, the number of cracks increases, whereas the widths of cracks remain small; as tensile stresses are increased, an equilibrium stage is reached. When stresses are further increased, the second stage of cracking develops, and crack widths increase without any significant increase in the number of cracks. Usually one or two cracks start to widen more than the others, forming critical cracks (Figure 6.6).

Main cracks in beams and axially tensioned members have been studied by many investigators; prediction of the width of cracks and crack control were among the problems studied. These are discussed here, along with the requirements of the ACI Code.

Crack width *Crack width* and *crack spacing,* according to existing crack theories, depend on many factors, which include steel percentage, its distribution in the concrete section, steel flexural stress at service load, concrete cover, and properties of the concrete constituents. Different equations for predicting the width and spacing of cracks in reinforced concrete members were presented at the Symposium on Bond and Crack Formation in Reinforced Concrete in Stockholm, Sweden, in 1957. Chi and Kirstein [16] presented equations for the crack width and spacing as a function of an effective area of concrete around the steel bar: A concrete circular area of diameter equal to four times the diameter of the bar was used to calculate crack width. Other equations were presented over the next decade [17]–[23].

(a)

(b)

Figure 6.6 (a) Main cracks in a reinforced concrete beam. (b) Spacing of cracks in a reinforced concrete beam.

An empirical formula (presented by Kaar and Mattock [20] as a result of extensive research) to estimate the maximum crack width W at the steel level is

$$W_{max} = 0.115 f_s \sqrt[4]{A} \times 10^{-6} \text{ (in.)} \tag{6.12}$$

where

A = area of concrete surrounding each bar, equal to the effective tension area of concrete having the same centroid as that of the reinforcing bars divided by the number of bars, in square inches

f_s = the stress in the steel bars, psi

At the tension concrete face, the following formula was suggested:

$$W_{max} = 0.115 \beta f_s \sqrt[4]{A} \times 10^{-6} \text{ (in.)} \tag{6.13}$$

where β is the ratio of distances from the neutral axis to the tension face and to the steel centroid and A and f_s are as defined previously.

Gergely and Lutz [23] presented the following formula for the limiting crack width:

$$W = 0.076\beta f_s \sqrt[3]{Ad_c} \times 10^{-6} \text{ (in.)} \tag{6.14}$$

where β, A, and f_s are as defined previously and d_c = thickness of concrete cover measured from the extreme tension fiber to the center of the closest bar. The value of β can be taken approximately equal to 1.2 for beams and 1.35 for slabs. Note that f_s is in psi and W is in inches.

The mean ratio of maximum crack width to average crack width was found to vary between 1.5 and 2.0, as reported by many investigators. An average value of 1.75 may be used.

In SI units (mm and MPa), equation (6.14) is

$$W = 11.0\beta f_s \sqrt[3]{Ad_c} \times 10^{-6} \tag{6.15}$$

Tolerable crack width The formation of cracks in reinforced concrete members is unavoidable. Hairline cracks occur even in carefully designed and constructed structures. Cracks are usually measured at the face of the concrete, but actually they are related to crack width at the steel level, where corrosion is expected. The permissible crack width is also influenced by aesthetic and appearance requirements. The naked eye can detect a crack about 0.006 in. (0.15 mm) wide, depending on the surface texture of concrete. Different values for permissible crack width at the steel level have been suggested by many investigators, ranging from 0.010 to 0.016 in. (0.25–0.40 mm) for interior members and from 0.006 to 0.010 in. (0.15–0.25 mm) for exterior exposed members. A limiting crack width of 0.016 in. (0.40 mm) for interior members and 0.013 in. (0.32 mm) for exterior members under dry conditions can be tolerated.

Crack control *Control* grows in importance with the use of high-strength steel in reinforced concrete members, as larger cracks develop under working loads because of the high allowable stresses. Control of cracking depends on the permissible crack width: It is always preferable to have a large number of fine cracks rather than a small number of large cracks. Secondary cracks are minimized by controlling the total amount of cement paste, water-cement ratio, permeability of aggregate and concrete, rate of curing, shrinkage, and end-restraint conditions.

The factors involved in controlling main cracks are the reinforcement stress, the bond characteristics of reinforcement, the distribution of reinforcement, the diameter of the steel bars used, the steel percentage, the concrete cover, and the properties of concrete constituents. Any improvement in these factors will help in reducing the width of cracks.

6.6 ACI CODE REQUIREMENTS

To control cracks in reinforced concrete members, the ACI Code, Chapter 10, specifies the following:

1. Only deformed bars are permitted as main reinforcement.
2. Tension reinforcement should be well distributed in the zones of maximum tension (Section 10.6.3).
3. When the flange of the section is under tension, part of the main reinforcement should be distributed over the effective flange width or one-tenth of the span, whichever is smaller. Some longitudinal reinforcement has to be provided in the outer portion of the flange (Section 10.6.6).
4. The design yield strength of reinforcement should not exceed 80 Ksi (560 MPa) (Section 9.4).

5. The maximum spacing **s** of reinforcement closest to a concrete surface in tension in reinforced concrete beams and one-way slabs is limited to

$$\mathbf{s}\text{ (in.)} = 540/f_s - 2.5C_c \tag{6.16}$$

but not greater than $12(36/f_s)$ where

f_s = calculated stress (Ksi) in reinforcement at service load computed as the unfactored moment divided by the product of steel area and the internal moment arm (refer to equation (5.12)), $f_s = M/(A_s \cdot jd)$ (Alternatively, $f_s = 0.6f_y$ may be used; An approximate value of $jd = 0.87d$ may be used)

C_c = clear cover from the nearest surface in tension to the surface of the flexural tension reinforcement (in.)

s = center to center spacing of flexural tension reinforcement nearest to the extreme concrete tension face (in.)

The preceding limitations are applicable to reinforced concrete beams and one-way slabs subject to normal environmental condition and do not apply to structures subjected to aggressive exposure. The spacing limitation just given is independent of the bar size, which may lead to the use of smaller bar sizes to satisfy the spacing criteria. For the case of concrete beams reinforced with grade 60 steel bars and $C_c = 2$ in., clear cover to the tension face, the maximum spacing is calculated as follows: Assume $f_s = 0.6f_y = 0.6 \times 60 = 36$ Ksi, $\mathbf{s} = (540/36) - 2.5 \times 2 = 10$ in. (controls), which is less than $12(36/36) = 12$ in.

6. In SI units, equation (6.16) becomes

$$\mathbf{s}\text{ (mm)} = 95{,}000/f_s - 2.5C_c \tag{6.17}$$

but not greater than $300(252/f_s)$, where f_s is in MPa and C_c is in mm. For example, if bars with $f_y = 420$ MPa (60 Ksi) and a clear cover equal to 50 mm are used, then the maximum spacing **s** is calculated as follows:

$$\mathbf{s} = (95{,}000/252) - 2.5 \times 50 = 300\text{ mm}$$

which is similar to $300(252/252) = 300$ mm in this example. This is assuming that $f_s = 0.6 \times 420 = 252$ MPa.

7. In the previous Codes, control of cracking was based on a factor Z defined as follows:

$$Z = f_s\sqrt[3]{Ad_c} \leq 175\text{ K/in.}\quad (31\text{ kN/mm})\qquad \text{for interior members}$$

$$Z \leq 140\text{ K/in.}\quad (26\text{ kN/mm})\qquad \text{for exterior members.} \tag{6.18}$$

where f_s = steel flexural stress at service load (Ksi) and may be taken as $0.6f_y$. A and d_c are the effective tension area of concrete and thickness of concrete cover, respectively. This expression is based on equation (6.14) assuming a limiting crack width of 0.016 in. for interior members and 0.013 in. for exterior members. It encouraged a decrease in the reinforcement cover to achieve a smaller Z, while unfortunately it penalized structures with concrete cover that exceeded 2 in.

8. *Skin reinforcement,* A_{sk}: For relatively deep girders, with an effective depth d equal to or greater than 36 in. (900 mm), light reinforcement should be added near the vertical faces in the tension zone to control cracking in the web above the main reinforcement. The ACI Code, Section 10.6.7, referred to this additional steel as skin reinforcement, A_{sk}. The skin reinforcement should be uniformly distributed along both side faces of the member for a distance $d/2$ nearest the flexural tension reinforcement. The minimum area of A_{sk} per foot of height on each side face of the member is evaluated as follows:

$$A_{sk} \geq 0.012(d - 30\text{ in.})\qquad \text{per foot, per side} \tag{6.19}$$

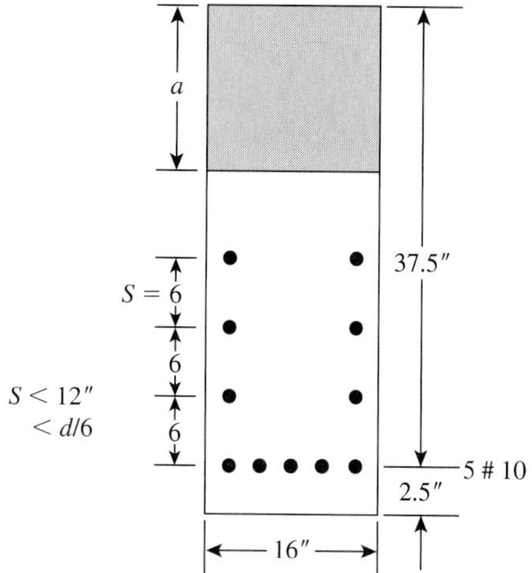

Figure 6.7 Skin reinforcement (6 no. 3 bars).

The maximum spacing of the skin-reinforcement bars shall not exceed the lesser of $d/6$ and 12 in. (Figure 6.7). The skin reinforcement may be included in the strength computations based on the strain and stress calculations. Moreover, the total skin reinforcement in both faces of the member should not exceed one-half of A_s. Referring to Figure 6.7, if $b = 16$ in. and $d = 37.5$ in., then the minimum $A_{sk} = 0.012(37.5 - 30) = 0.09$ in.^{2}/ft/side to be distributed at a height of $d/2 = 37.5/2 = 18.75$ in. $= 1.56$ ft. Maximum spacing is $d/6 = 37.5/6 = 6.25$ in., say, 6 in. which is less than 12 in. Choose 3 no. 3 bars (area $= 0.33$ in.^{2}) spaced at 6 in. on each side, as shown in Figure 6.7. Total A_{sk} on both sides used is $2(0.33) = 0.66$ in.^{2}, which is less than $0.5A_s = 0.5(6.33) = 3.16$ in.^{2}

Example 6.5

The sections of a simply supported beam are shown in Figure 6.8.

a. Check if the bar arrangement satisfies the ACI Code requirements.

b. Determine the expected crack width.

c. Check the Z-factor based on equation (6.18).
Given: $f'_c = 4$ Ksi, $f_y = 60$ Ksi, and no. 3 stirrups.

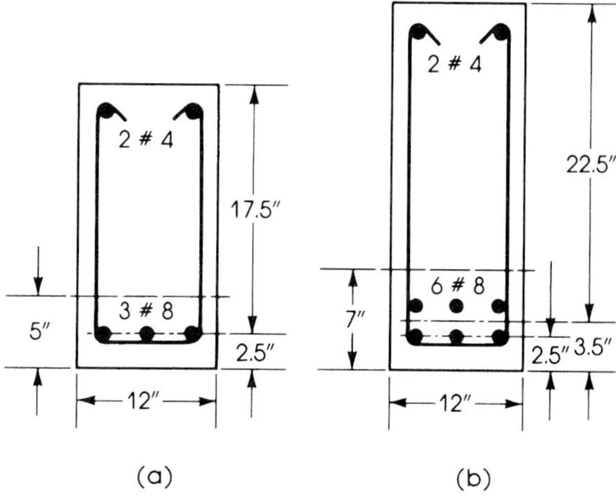

Figure 6.8 Two sections for Example 6.5.

Solution

1. Section 6.8a:

a. For 3 no. 8 bars, $A_s = 2.35$ in.^{2}, clear cover $C_c = 2.5 - 8/16 = 2.0$ in. Assume $f_s = 0.6f_y = 0.6 \times 60 = 36$ Ksi. Maximum spacing **s** $= 540/36 - 2.5 \times 2 = 10$ in., which is less than $12(36/36) = 12$ in. Spacing provided $= 0.5(12 - 2.5 - 2.5) = 3.5$ in., center to center of bars, which is less than 10 in.

b. For this section, $d_c = 2.5$ in. The effective tension area of concrete for one bar is

$$A = 12(2 \times 2.5)/3 = 20 \text{ in.}^2$$

Estimated crack width using equation (6.14) is

$$W = 0.076(12)(1.2)(36{,}000)\sqrt[3]{20 \times 2.5} \times 10^{-6} = 0.0121 \text{ in.}$$

This is assuming $\beta = 1.2$ for beams and $f_s = 36$ Ksi. The crack width above is less than 0.016 in. and 0.013 in. for interior and exterior members.

c. Check the Z-value, equation (6.18):

$$Z = f_s\sqrt[3]{A \cdot d_c} = 36\sqrt[3]{20 \times 2.5} = 132.6 \text{ K/in.}$$

which is less than 175 K/in. or 145 K/in. for interior and exterior members. Calculations (b) and (c) are not required based on the new ACI Code. See the discussion that follows.

2. Section 8.6b:

a. Calculations of spacing of bars are similar to those in Section 8.6a.

b. For this section, $d_c = 2.5$ in., and the steel bars are placed in two layers. The centroid of the steel bars is 3.5 in. from the bottom fibers. The effective tension concrete area is $A = 12(2 \times 3.5)/6 = 14$ in.^{2}

$$W = .076 \times 1.2 \times 36{,}000\sqrt[3]{14 \times 2.5} \times 10^{-6} = 0.0107 \text{ in.}$$

which is adequate.

c. $Z = 36\sqrt[3]{14 \times 2.5} = 117.7$ K/ft
which is adequate.

Discussion

It can be seen that the spacing **s** in equation (6.16) is a function of the stress in the tension bars or, indirectly, is a function of the strain in the tension steel, $f_s = E_s \times \varepsilon_s$ and E_s for steel is equal to 29,000 Ksi. The spacing also depends on the concrete cover C_c. An increase in the concrete cover will reduce the limited spacing **s**, which is independent of the bar size used in the section.

In this example, the expected crack width was calculated by equation (6.14) to give the student or the engineer a physical feeling for the crack width and crack control requirement. The crack width is usually measured in beams when tested in the laboratory or else in actual structures under loading when serious cracks develop in beams or slabs and testing is needed. If the crack width measured before and after loading is greater then the yield strain of steel, then the main reinforcement is in the plastic range and ineffective. Sheets with lines of different thickness representing crack widths are available in the market for easy comparisons with actual crack widths. The Z-value calculated in this example is a factor that relates to the crack width W. In addition to the steel stress and the concrete cover, both W and Z depend on a third factor A representing the tension area of concrete surrounding one bar in tension.

Example 6.6

Design a simply supported beam with a span of 24 ft to carry a uniform dead load of 1.3 K/ft and a live load of 1.1 K/ft. Choose adequate bars; then check their spacing arrangement to satisfy the ACI Code. Given: $b = 16$ in., $f'_c = 4$ Ksi, $f_y = 60$ Ksi, a steel percentage $= 0.8\%$, and a clear concrete cover of 2 in.

Solution

1. For a steel percentage of 0.8%, $R_u = 400$ psi $= 0.4$ Ksi. The external factored moment is $M_u = w_u \times L^2/8$, and $w_u = 1.4(1.3) + 1.7(1.1) = 3.69$ K/ft.

$$M_u = 3.69(24)^2/8 = 265.68 \text{ K}\cdot\text{ft} = 3188.2 \text{ K}\cdot\text{in.}$$

$$M_u = R_u \cdot bd^2, d = 22.32 \text{ in.}, A_s = .0008 \times 16 \times 22.32 = 2.86 \text{ in.}^2$$

Choose 3 no. 9 bars (area $= 3.0$ in.^{2}) in one row, and a total depth $h = 25.0$ in. Actual $d = 25 - 2 - 9/16 = 22.44$ in. (Figure 6.9(a)).

2. Check spacing of bars using equation (6.16). Calculate the service load and moment: $w = 1.3 + 1.1 = 2.4$ K/ft.

$$M = 2.4(24)^2/8 = 172.8 \text{ K}\cdot\text{ft} = 2073.6 \text{ K}\cdot\text{in.}$$

3. Calculate the neutral axis depth kd and the moment arm jd (refer to Chapter 5, Equations (5.8), (5.9), (5.12) and Example 5.2).

$$b(kd)^2/2 - nA_s(d - kd) = 0, \quad n = 8, A_s = 3.0, \text{ and } d = 22.44 \text{ in.}$$

$$kd = 6.85 \text{ in.}, jd = d - kd/3 = 20.16 \text{ in., and } j = 20.16/22.44 = 0.898$$

Note that an approximate value of $j = 0.87$ may be used if kd is not calculated.

4. Calculate the stress f_s from equation (5.12).

$$M = A_s \cdot f_s \cdot jd, \quad 2073.6 = 3(f_s)(20.16), \text{ and } f_s = 34.3 \text{ Ksi}$$

5. Calculate the spacing **s** by equation (6.16):

$$\mathbf{s} = 540/34.3 - 2.5 \times 2 = 10.74 \text{ in. (controls)}$$

which is less than $12(36/36) = 12.6$ in. Spacing provided $= 0.5(16 - 2.56 - 2.56) = 5.44$ in., which is less than 10.74 in.

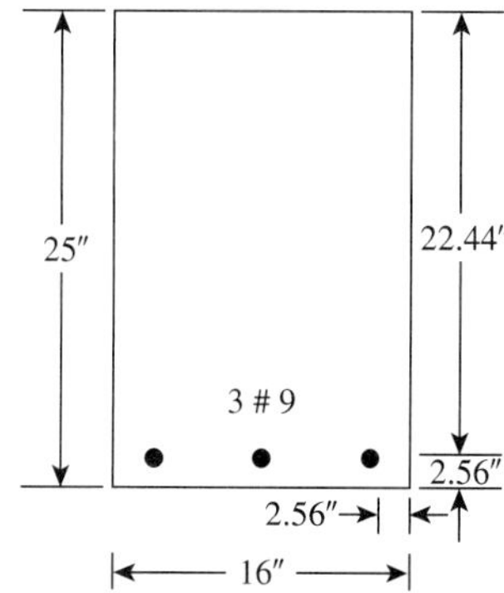

Figure 6.9(a) Example 6.6.

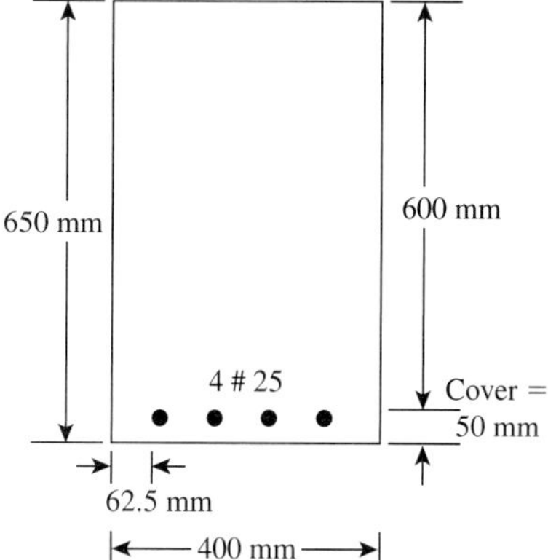

Figure 6.9(b) Example 6.7.

Example 6.7 SI Units

Design a simply supported beam of 7.2 m span to carry a uniform dead load of 19 kN/m and a live load of 16 kN/m. Choose adequate bars, then check their spacing arrangement to satisfy the ACI Code.

Given: $b = 400$ mm, $f'_c = 30$ MPa, $f_y = 400$ MPa, a steel percentage of 0.8%, and a clear concrete cover of 50 mm.

Solution

1. For a steel percentage of 0.008 and from equation (3.19), $R_u = 2.7$ MPa. Factored load $w_u = 1.4(19) + 1.7(16) = 53.8$ kN/m.

$$M_u = w_u \cdot L^2/8 = 53.8(7.2)^2/8 = 348.6 \text{ kN}\cdot\text{m}$$

$$M_u = R_u \cdot bd^2, \text{ or } 348.6 \times 10^6 = 2.7 \times 400d^2; \text{ then } d = 568 \text{ mm}$$

$$A_s = \rho bd = 0.008 \times 400 \times 568 = 1818 \text{ mm}^2$$

Choose 4 bars, 25 mm (no. 25 M), $A_s = 2040$ mm², in one row $(b_{\min} = 220 \text{ mm})$.

Let $h = 650$ mm, the actual $d = 650 - 50 - 25/2 = 587.5$ mm, say 585 mm.

Final section: $b = 400$ mm, $h = 650$ mm, with 4 no. 25 mm bars (Figure 6.9(b)).

2. Check spacing of bars using equation (6.17). Calculate the service load moment, $w = 19 + 16 = 35$ kN/m.

$$M = 35(7.2)^2/8 = 226.8 \text{ kN}\cdot\text{m}$$

Calculate kd and jd as in the previous example or Example 5.6. Alternatively, use a moment arm $jd = 0.87d = 0.87(585) = 509$ mm. From equation (5.12), $f_s = M/(A_s \cdot jd) = 226.8(10)^6/(2040 \times 585) = 190$ MPa. From equation (6.17), Maximum **s** $= (95{,}000/190) - 2.5(50) = 375$ mm (controls), which is less than $300((252/f_s) = 300(252)/190 = 398$ mm. Note that if $f_s = 0.6f_y = 0.6(400) = 240$ MPa is used, then maximum **s** $= 270$ mm. It is preferable to calculate f_s from the moment equation to reflect the actual stress in the bars. Spacing provided $= (1/3)(400 - 50 - 25) = 92$ mm, which is adequate.

SUMMARY

Section 6.1

1. Deflection $\Delta = \alpha(WL^3/EI) = 5WL^3/384EI = 5wL^4/384EI$ for a simply supported beam subjected to a uniform total load of $W = wL$.

$$E_c = 33w^{1.5}\sqrt{f'_c} = 57{,}400\sqrt{f'_c} \text{ psi}$$

for normal-weight concrete.

2. Effective moment of inertia is

$$I_e = \left(\frac{M_{cr}}{M_a}\right)^3 I_g + \left[1 - \left(\frac{M_{cr}}{M_a}\right)^3\right] I_{cr} \leq I_g \qquad (6.5)$$

$$M_{cr} = f_r \times \frac{I_g}{y_t} \quad \text{and} \quad f_r = 7.5\sqrt{f'_c}$$

Section 6.2

The deflection of reinforced concrete members continues to increase under sustained load.

Additional long-time deflection $= \lambda \times$ instantaneous deflection:

$$\lambda = \frac{\zeta}{1 + 50\rho'}$$

$\zeta = 1.0, 1.2, 1.4$, and 2.0 for periods of 3, 6, 12, and 60 months, respectively.

Sections 6.3 and 6.4

1. The allowable deflection varies between $L/180$ and $L/480$.
2. Deflections for different types of loads may be calculated for each type of loading separately and then added algebraically to obtain the total deflection.

Section 6.5

1. Cracks are classified as secondary cracks (shrinkage, corrosion, or secondary flexural cracks) and main cracks.
2. Maximum crack width is

$$W = 0.076\beta f_s \sqrt[3]{Ad_c} \times 10^{-6} \text{ (in.)} \tag{6.14}$$

   Approximate values for β, f_s, and d_c are $\beta = 1.2$ for beams and 1.35 for slabs, $d_c = 2.5$ in., and $f_s = 0.6f_y$.

3. The limiting crack width is 0.016 in. for interior members and 0.013 in. for exterior members.

Section 6.6

The maximum spacing **s** of bars closest to a concrete surface in tension is limited to

$$\mathbf{s} = 540/f_s - 2.5C_c \tag{6.16}$$

but not more than $12(36/f_s)$. Note that f_s may be taken as $0.6f_y$.

REFERENCES

1. Wei-Wen Yu and G. Winter. "Instantaneous and Longtime Deflections of Reinforced Concrete Beams under Working Loads." *ACI Journal* 57 (July 1960).
2. M. N. Hassoun. "Evaluation of Flexural Cracks in Reinforced Concrete Beams." *Journal of Engineering Sciences* 1, No. 1 (January 1975). College of Engineering, Riyadh, Saudi Arabia.
3. D. W. Branson. "Instantaneous and Time-Dependent Deflections of Simply and Continuous Reinforced Concrete Beams, Part 1," Alabama Highway Research Report No. 7, August 1963.
4. ACI Committee 435. "Allowable Deflections." *ACI Journal* 65 (June 1968).
5. ACI Committee 435. "Deflections of Continuous Beams." *ACI Journal* 70 (December 1973).
6. ACI Committee 435. "Variability of Deflections of Simply Supported Reinforced Concrete Beams." *ACI Journal* 69 (January 1972).
7. Dan E. Branson. *Deformation of Concrete Structures.* New York: McGraw-Hill, 1977.
8. American Concrete Institute. *Causes, Mechanism and Control of Cracking in Concrete.* ACI Publication SP-20, Detroit, 1968.
9. R. Salinger. "High Grade Steel in Reinforced Concrete." *Proceedings 2nd Congress of International Association for Bridge and Structural Engineering.* Berlin-Munich, 1936.
10. K. Billing. *Structural Concrete.* New York: St. Martins Press, 1960.
11. P. E. Halstead. "The Chemical and Physical Effects of Aggressive Substances on Concrete," *The Structural Engineer* 40 (1961).
12. R. E. Stratfull. "The Corrosion of Steel in a Reinforced Concrete Bridge." *Corrosion* 13, no. 3 (March 1957).

13. C. L. Shermer. "Corroded Reinforcement Destroys Concrete Beams," *Civil Engineering* 26 (December 1956).

14. V. R. Evans. *An Introduction to Metallic Corrosion.* London: Edward Arnold Publishers, 1948.

15. J. D. Mozer, A. Bianchini, and C. Kesler. "Corrosion of Steel Reinforcement in Concrete." University of Illinois Dept. of Theoretical and Applied Mechanics Report No. 259 (April 1964).

16. M. Chi and A. F. Kirstein. "Flexural Cracks in Reinforced Concrete Beams." *ACI Journal* 54 (April 1958).

17. R. C. Mathy and D. Watstein. "Effect of Tensile Properties of Reinforcement on the Flexural Characteristics of Beams." *ACI Journal* 56 (June 1960).

18. F. Levi. "Work of European Concrete Committee." *ACI Journal* 57 (March 1961).

19. E. Hognestad. "High-Strength Bars as Concrete Reinforcement." *Journal of the Portland Cement Association, Development Bulletin* 3, no. 3 (September 1961).

20. P. H. Kaar and A. H. Mattock. "High-Strength Bars as Concrete Reinforcement (Control of Cracking), Part 4." *PCA Journal* 5 (January 1963).

21. B. B. Broms. "Crack Width and Crack Spacing in Reinforced Concrete Members." *ACI Journal* 62 (October 1965).

22. G. D. Base, J. B. Reed, and H. P. Taylor. "Discussion on 'Crack and Crack Spacing in Reinforced Concrete Members.'" *ACI Journal* 63 (June 1966).

23. P. Gergely and L. A. Lutz. "Maximum Crack Width in Reinforced Concrete Flexural Members." In *Causes, Mechanism and Control of Cracking in Concrete.* ACI Publication SP-20, 1968.

24. ACI Committee 224. "Control of Cracking in Concrete Structures." *ACI Journal* 69 (December 1972).

25. ACI Committee 224. "Causes, Evaluation, and Repair of Cracks in Concrete Structures." SP-ACJ 224.1R-93, 1993.

26. M. N. Hassoun and K. Sahebjum. "Cracking of Partially Prestressed Concrete Beams." ACI Special Publications No. SP-113, 1989.

PROBLEMS

6.1 Determine the instantaneous and long-time deflection of a 20-ft-span simply supported beam for each of the following load conditions. Assume that 10% of the live loads are sustained and the dead loads include the self-weight of the beams. Use $f'_c = 4$ Ksi, $f_y = 60$ Ksi, $d' = 2.5$ in., and a time limit of 5 years. Refer to Figure 6.10.

No.	b (in.)	d (in.)	h (in.)	A_s (in.^{2})	A'_s (in.^{2})	W_D (K/ft)	W_L (K/ft)	P_D (K)	P_L (K)
a	14	17.5	20	5 no. 9	—	2.2	1.8	—	—
b	20	27.5	30	6 no. 10	—	7.0	3.6	—	—
c	12	19.5	23	6 no. 8	—	3.0	1.5	—	—
d	18	20.5	24	6 no. 10	2 no. 9	6.0	2.0	—	—
e	16	22.5	26	6 no. 11	2 no. 10	5.0	3.2	12	10
f	14	20.5	24	8 no. 9	2 no. 9	3.8	2.8	8	6

$h - d = 2.5$ in. indicates one row of bars, whereas $h - d = 3.5$ in. indicates two rows of bars. Concentrated loads are placed at midspan.

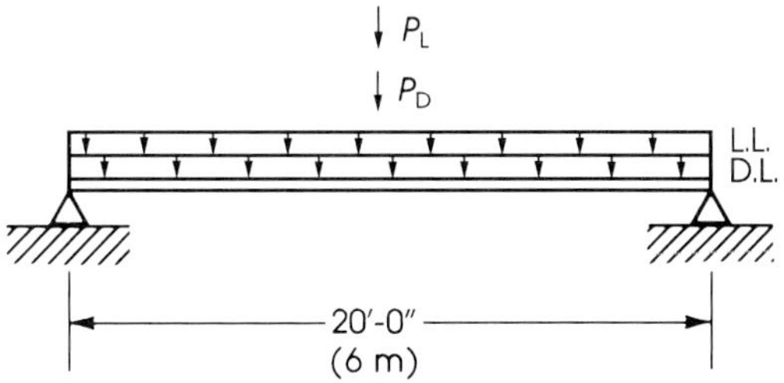

Figure 6.10 Problem 6.1.

6.2 Determine the instantaneous and long-term deflection of the free end of a 12-ft-span cantilever beam for each of the following load conditions. Assume that only dead loads are sustained, and the dead loads include the self-weight of the beams. Use $f'_c = 4$ Ksi, $f_y = 60$ Ksi, and a time limit of more than 5 years. Refer to Figure 6.11.

No.	b (in.)	d (in.)	h (in.)	A_s (in.^{2})	A'_s (in.^{2})	W_D (K/ft)	W_L (K/ft)	P_D (K)	P_L (K)
a	15	20.5	24	8 no. 9	2 no. 9	3.5	2.0	—	—
b	18	22.5	26	6 no. 10	—	2.0	1.5	7.4	5.0
c	12	19.5	23	8 no. 8	2 no. 8	2.4	1.6	—	—
d	14	20.5	24	8 no. 9	2 no. 9	3.0	1.1	5.5	4.0

$(h - d) = 2.5$ in. indicates one row of bars, whereas $(h - d) = 3.5$ in. indicates two rows of bars. Concentrated loads are placed as shown.

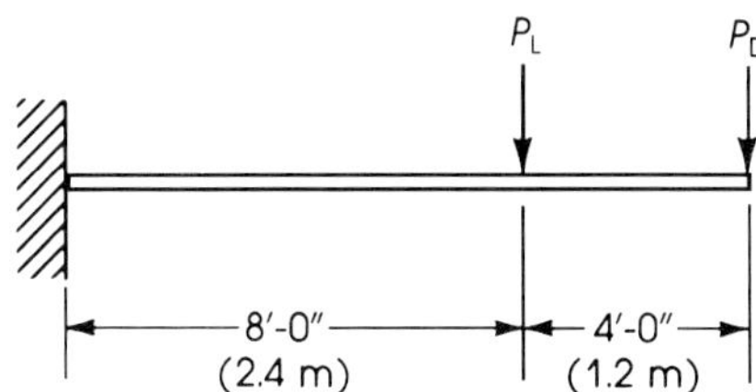

Figure 6.11 Problem 6.2.

6.3 A 28-ft simply supported beam carries a uniform dead load of 2 K/ft (including self-weight) and a live load of 1.4 K/ft. Design the critical section at midspan using the maximum steel ratio allowed by the ACI Code and then calculate the instantaneous deflection. Use $f'_c = 4$ Ksi, $f_y = 60$ Ksi, and $b = 12$ in. See Figure 6.12.

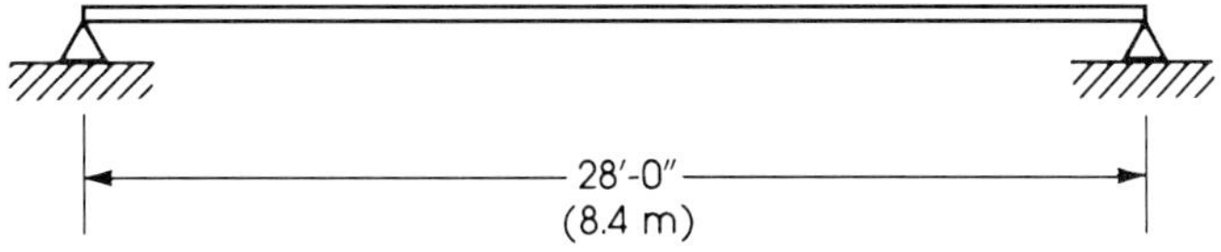

Figure 6.12 Problem 6.3: D.L. = 2 K/ft (30 kN/m) and L.L. = 1.4 K/ft (21 kN/m).

6.4 Design the beam in Problem 6.3 as doubly reinforced, considering that compression steel resists 20% of the maximum bending moment. Then calculate the maximum instantaneous deflection.

6.5 The four cross sections shown in Figure 6.13 belong to four different beams with $f'_c = 4$ Ksi and $f_y = 60$ Ksi. Check the spacing of the bars in each section according to the ACI Code requirement using $f_s = 0.6f_y$. Then calculate the tolerable crack width W.

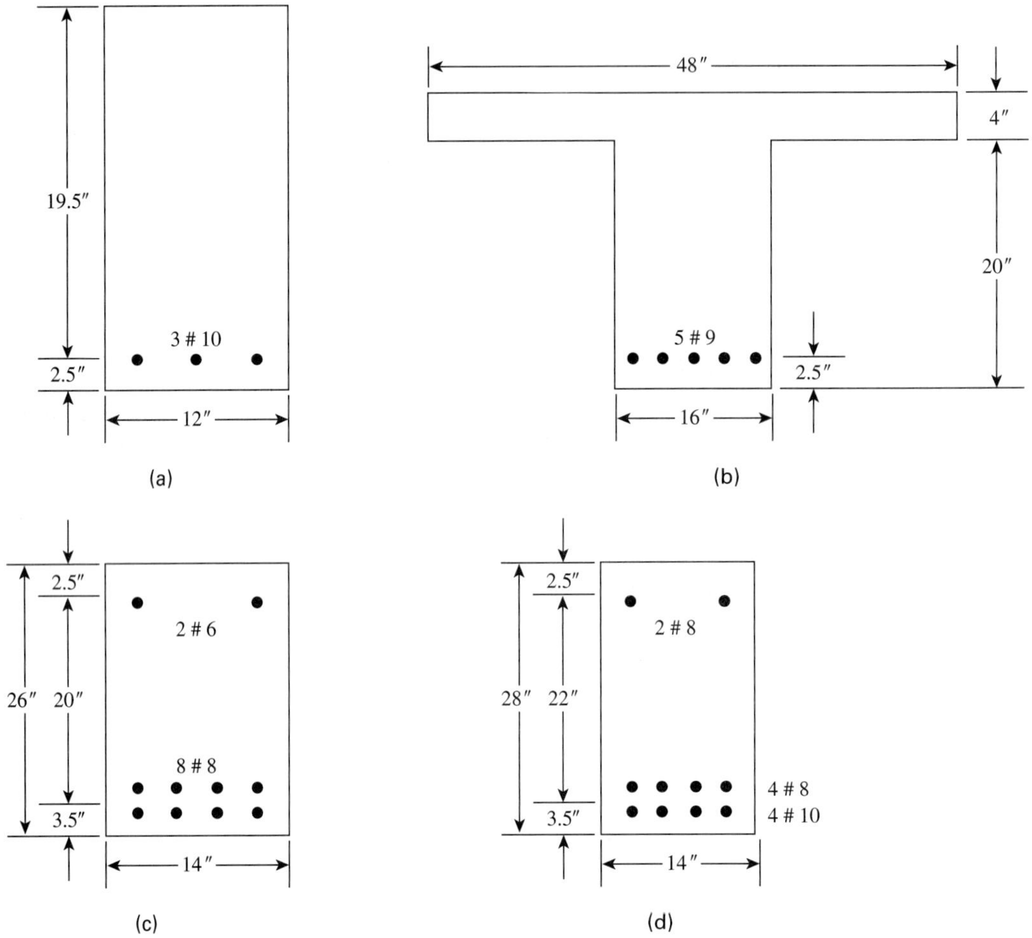

19.5″ 2.5″ 3 # 10 12″

(a)

48″ 4″ 20″ 5 # 9 2.5″ 16″

(b)

2.5″ 26″ 20″ 2 # 6 8 # 8 3.5″ 14″

(c)

2.5″ 28″ 22″ 2 # 8 4 # 8 4 # 10 3.5″ 14″

(d)

Figure 6.13 Problem 6.5.

6.6 Determine the necessary skin reinforcement for the beam section shown in Figure 6.14. Then choose adequate bars and spacings. Use $f'_c = 4$ Ksi and $f_y = 60$ Ksi.

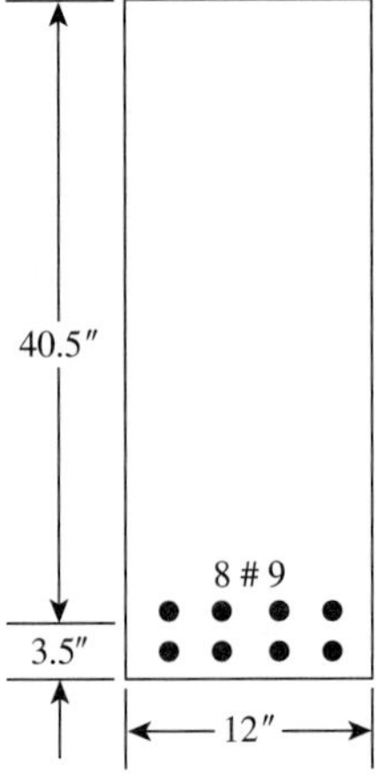

Figure 6.14 Problem 6.6 (skin reinforcement).

DEVELOPMENT LENGTH OF REINFORCING BARS

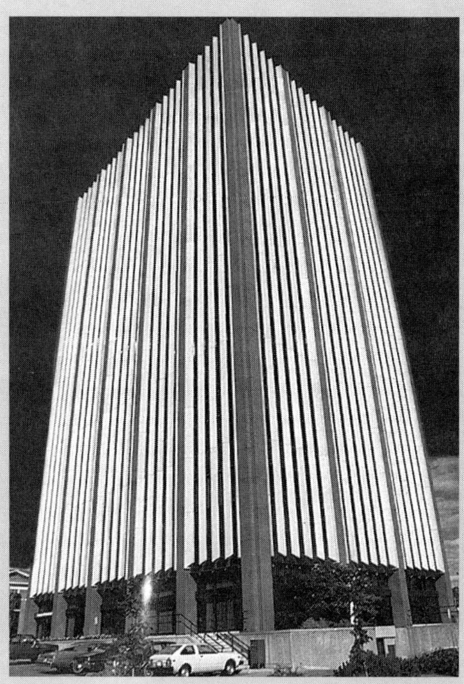

Reinforced concrete columns supporting an office building, Toronto, Canada.

7.1 INTRODUCTION

The joint behavior of steel and concrete in a reinforced concrete member is based on the fact that a bond is maintained between the two materials after the concrete hardens. If a straight bar of round section is embedded in concrete, a considerable force is required to pull the bar out of the concrete. If the embedded length of the bar is long enough, the steel bar may yield, leaving some length of the bar in the concrete. The bonding force depends on the friction between steel and concrete. It is influenced mainly by the roughness of the steel surface area, the concrete mix, shrinkage, and the cover of concrete. Deformed bars give a better bond than plain bars. Rich mixes have greater adhesion than weak mixes. An increase in the concrete cover will improve the ultimate bond stress of a steel bar [2].

In general, the bond strength is influenced by the following factors:

1. Yield strength of reinforcing bars, f_y. Longer development length is needed with higher f_y.
2. Quality of concrete and its compressive strength, f'_c. An increase in f'_c reduces the required development length of reinforcing bars.
3. Bar size, spacing, and location in the concrete section. Horizontal bars placed with more than 12 in. of concrete below them have lower bond strength due to the fact that concrete shrinks and settles during the hardening process. Also, wide spacings

of bars improve the bond strength, giving adequate effective concrete area around each bar.

4. Concrete cover to reinforcing bars. A small cover may cause the cracking and spalling of the concrete cover.
5. Confinement of bars by lateral ties. Adequate confinement by ties or stirrups prevents the spalling of concrete around bars.

7.2 DEVELOPMENT OF BOND STRESSES

7.2.1 Flexural Bond

Consider a length dx of a beam subjected to uniform loading. Let the moment produced on one side be M_1 and on the other side be M_2, with M_1 being greater than M_2. The moments will produce internal compression and tension forces, as shown in Figure 7.1. Because M_1 is greater than M_2, T_1 is greater than T_2; consequently, C_1 is greater than C_2.

At any section, $T = M/jd$, where jd is the moment arm:

$$T_1 - T_2 = dT = \frac{dM}{jd}, \qquad \text{but}$$

$$T_1 = T_2 + u\Sigma O\, dx$$

where u is the average bond stress and ΣO is the sum of perimeters of bars in the section at the tension side. Therefore,

$$T_1 - T_2 = u\Sigma O\, dx = \frac{dM}{jd}$$

$$u = \frac{dM}{dx} \times \frac{1}{jd\Sigma O}$$

The rate of change of the moment with respect to x is the shear, or $dM/dx = V$. Therefore,

$$u = \frac{V}{jd\Sigma O} \tag{7.1}$$

The value u is the average bond stress; for practical calculations, j can be taken approximately equal to 0.87:

$$u = \frac{V}{0.87d\Sigma O}$$

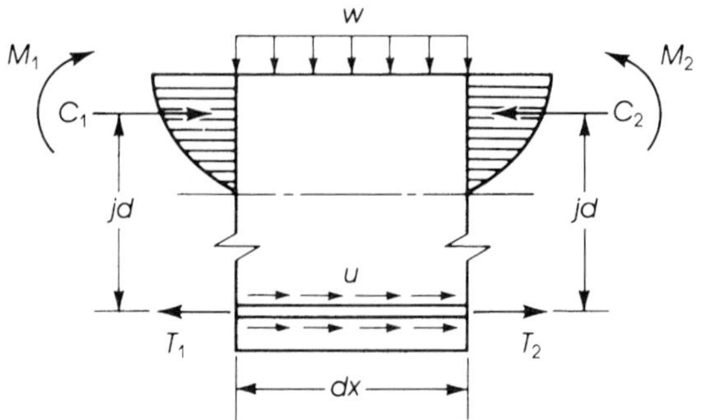

Figure 7.1 Flexural bond.

In the strength-design method, the nominal bond strength is reduced by the capacity reduction factor $\phi = 0.85$. Thus,

$$U_u = \frac{V_u}{\phi(0.87)d\Sigma O} \tag{7.2}$$

Based on the preceding analysis, the bond stress is developed along the surface of the reinforcing bar due to shear stresses and shear interlock.

7.2.2 Tests for Bond Efficiency

Tests to determine the bond stress capacity can be made using the pullout test (Figure 7.2). This test evaluates the bond capacity of various types of bar surfaces relative to a specific embedded length. The distribution of tensile stresses will be uniform around the reinforcing bar at a specific section and varies along the anchorage length of the bar and at a radial distance from the surface of the bar (Figure 7.2). However, this test does not represent the effective bond behavior in the surface of the bars in flexural members, because stresses vary along the depth of the concrete section. A second type of test can be performed on an embedded rod (Figure 7.3). In these tests, the tensile force P is increased gradually and the number of cracks and their spacings and widths are recorded. The bond stresses vary along the bar length between the cracks. The strain in the steel bar is maximum at the cracked section and decreases toward the middle section between cracks.

Tests on flexural members are also performed to study the bond effectiveness along the surface of the tension bars. The analysis of bond stresses in the bars of these members was explained earlier, and they are represented by equation (7.2).

Based on this discussion, it is important to choose an appropriate length in each reinforcing bar to develop its full yield strength without a failure in the bond strength. This length

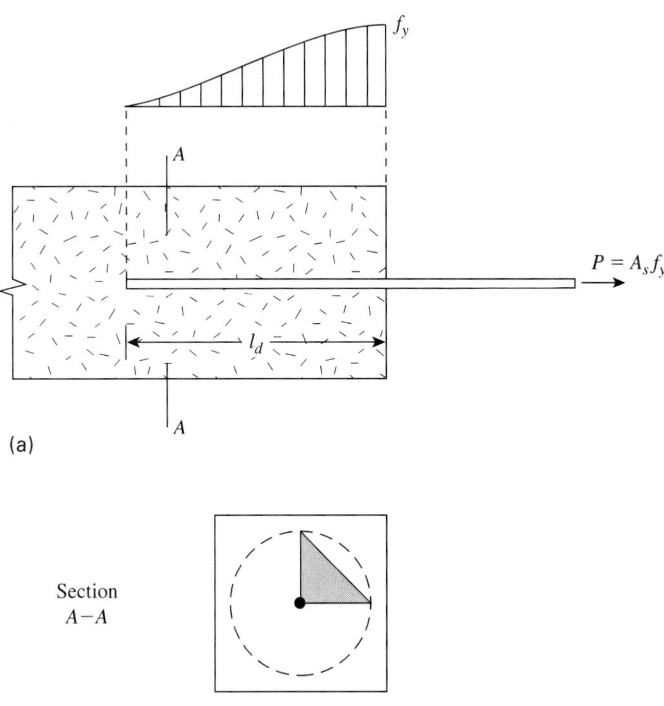

Figure 7.2 Bond stresses and development length. (a) Distribution of stress along l_d and (b) radial stress in concrete around the bar.

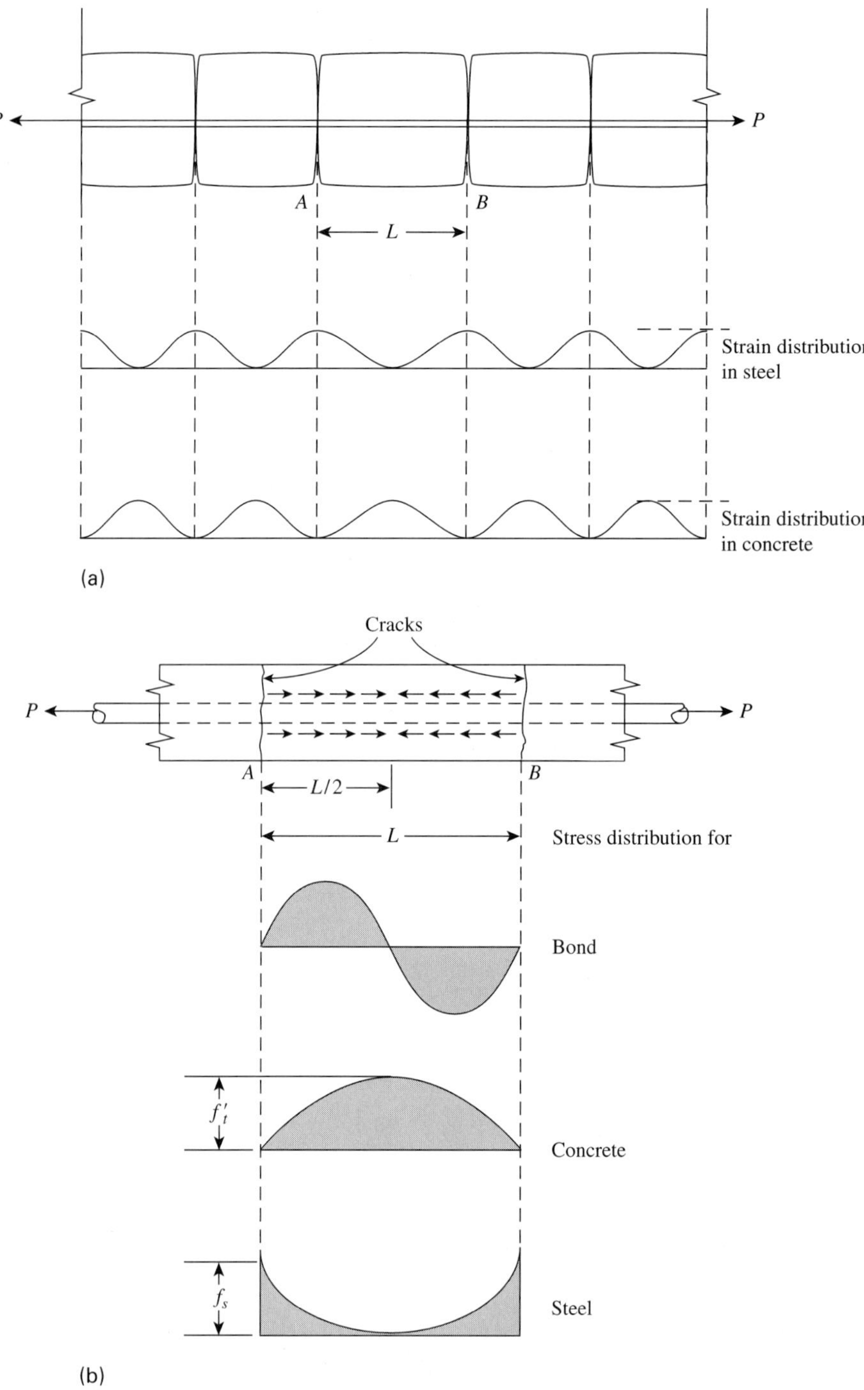

Figure 7.3 Bond mechanism in an embedded bar. Strain (a) and stress (b) distribution between cracks.

is called the *development length*, l_d. If this length is not provided, the bond stresses in the tension zone of a beam become high enough to cause cracking and splitting in the concrete cover around the tension bars (Figure 7.4). If the split continues to the end of the bar, the beam will eventually fail. Note that small spacings between tensile bars and a small concrete cover on the sides and bottom will reduce the bond capacity of the reinforcing bars (Figure 7.4).

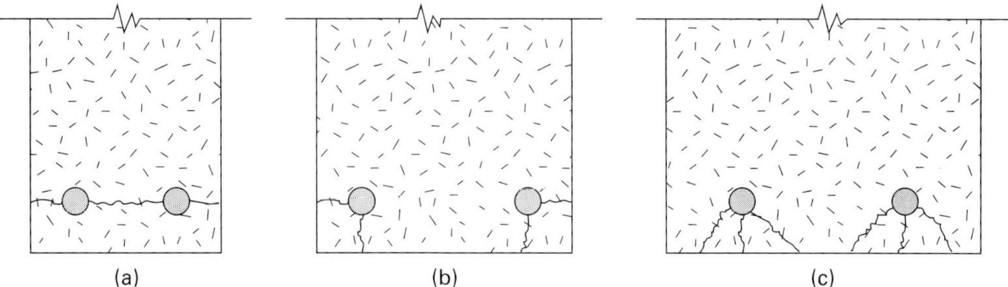

Figure 7.4 Examples of spalling of concrete cover. (a) High bottom cover, (b) wide spacing, and (c) small bottom cover.

7.3 DEVELOPMENT LENGTH IN TENSION

7.3.1 Basic Development Length, l_{db}

If a steel bar is embedded in concrete, as shown in Figure 7.2, and is subjected to a tension force T, then this force will be resisted by the bond stress between the steel bar and the concrete. The maximum tension force is equal to $A_s f_y$, where A_s is the area of the steel bar. This force is resisted by another internal force of magnitude $U_u O l_d$, where U_u is the ultimate average bond stress, l_d is the embedded length of the bar, and O is the perimeter of the bar (πD). The two forces must be equal for equilibrium:

$$A_s f_y = U_u O l_d \quad \text{and} \quad l_d = \frac{A_s f_y}{U_u O}$$

For a combination of bars,

$$l_d = \frac{A_s f_y}{U_u \Sigma O} \tag{7.3}$$

The length l_d is the minimum permissible anchorage length and is called the development length.

$$l_d = \frac{\pi d_b^2 f_y}{4U_u(\pi d_b)} = \frac{d_b f_y}{4U_u} \tag{7.4}$$

where d_b = diameter of reinforcing bars.

This means that the development length is a function of the size and yield strength of the reinforcing bars in addition to the ultimate bond stress, which in turn is a function of $\sqrt{f'_c}$. The bar length l_d given in equation (7.4) is called the *basic development length*, l_{db}. The final development length should also include the other factors mentioned in Section 7.1. Equation (7.4) may be written as follows:

$$\frac{l_{db}}{d_b} = K\left(\frac{f_y}{\sqrt{f'_c}}\right) \tag{7.5}$$

where K is a general factor that can be obtained from tests to include factors such as the bar characteristics (bar size, spacing, epoxy coated or uncoated, location in concrete section and bar splicing), amount of transverse reinforcement, and the provision of excess reinforcement compared to that required from design.

The ACI Code, Section 12.2.3, evaluated K as follows:

$$K = \left(\frac{3}{40}\right)\frac{\alpha\beta\gamma\lambda}{(c + K_{tr})/d_b} \tag{7.6}$$

and equation (7.5) becomes

$$\frac{l_d}{d_b} = \left(\frac{3}{40}\right) \frac{f_y}{\sqrt{f'_c}} \frac{\alpha\beta\gamma\lambda}{[(c + K_{tr})/d_b]} \tag{7.7}$$

where
α = bar location
β = coating factor
γ = bar-size factor
λ = lightweight aggregate concrete factor
c = spacing or cover dimension (in.), whichever is smaller
K_{tr} = transverse reinforcement index
= $(A_{tr} \cdot f_{yt})/(1500)$ sn (1500 carries units of psi)
n = number of bars or wires being developed along the plane of splitting
s = maximum spacing of transverse reinforcement within l_d, center to center (in.)
f_{yt} = yield strength of transverse reinforcement (psi)
A_{tr} = total sectional area of all transverse reinforcement within spacing s that crosses the potential plane of splitting through to the reinforcement being developed (in.^{2})

Notes:

1. $(c + K_{tr})/d_b$ shall not exceed 2.5 to safeguard against pullout type failures.
2. The value of $\sqrt{f'_c}$ shall not exceed 100 psi (ACI Code Section 12.2).

7.3.2 ACI Code Factors for Calculating l_d for Bars in Tension

1. α = bar-location factor

α_1 = 1.3 for top bars defined as horizontal reinforcement, placed so that more than 12 in. of fresh concrete is below the development length, or splice

α_2 = 1.0 for all other reinforcement

2. β = coating factor

β_1 = 1.5 for epoxy-coated bars or wires with cover less than $3d_b$ or clear spacing less than $6d_b$

β_2 = 1.2 for all other epoxy coated bars or wires

β_3 = 1.0 for uncoated reinforcement (However, the value of the product $\alpha\beta$ should not exceed 1.7.)

3. γ = bar-size factor

γ_1 = 0.8 for no. 6 bars or smaller bars and deformed wires

γ_2 = 1.0 for no. 7 bars and larger bars

4. λ = lightweight aggregate concrete factor

λ_1 = 1.3 for lightweight aggregate concrete (However, when f_{ct} is specified, use $\lambda = 6.7\sqrt{f'_c}/f_{ct}$, but $\lambda \geq 1$.)

λ_2 = 1.0 for normal-weight concrete

5. The ACI Code permits using $K_{tr} = 0$ even if transverse reinforcement is present. In this case

$$\frac{l_d}{d_b} = \left(\frac{3}{40}\right)\left(\frac{f_y}{\sqrt{f'_c}}\right)\frac{\alpha\beta\gamma\lambda}{(c/d_b)} \tag{7.8}$$

The value of $\sqrt{f'_c}$ should not exceed 100 psi.

6. R_s is the reduction factor due to excess reinforcement. The ACI Code, Section 12.2.5, permits the reduction of l_d by the factor R_s when the reinforcement in a flexural member exceeds that required by analysis, except where anchorage or development for f_y is specifically required or the reinforcement is designed considering seismic effects.

$$R_s = \frac{A_s \text{ (required)}}{A_s \text{ (provided)}}$$

7. The development length l_d in all cases shall not be less than 12 in.

7.3.3 Simplified Expressions for l_d

The ACI Code, Section 12.2.2, permits the use of simplified expressions to calculate the ratio l_d/d_b. This is based on the fact that current practical construction cases utilize spacing and cover values along with confining reinforcement, such as stirrups and ties, that produce a value of $(c + K_{tr})/d_b \geq 1.5$. Moreover, tests indicated that the development length l_d can be reduced by 20% for no. 6 and smaller bars. Based on these assumptions and assuming $(c + K_{tr})/d_b = 1.5$, equation (7.7) can be reduced to the following expressions:

1. For no. 7 and larger bars:

$$\frac{l_d}{d_b} = \left(\frac{f_y}{\sqrt{f'_c}}\right)\frac{\alpha\beta\lambda}{20} \tag{7.8}$$

For no. 6 and smaller bars and deformed wires:

$$\frac{l_d}{d_b} = \left(\frac{f_y}{\sqrt{f'_c}}\right)\frac{\alpha\beta\lambda}{25} \tag{7.9}$$

The ratio l_d/d_b in equation (7.9) represents 80% of that in equation (7.8). These equations are used when one of the following conditions is met:

a. Clear spacing of bars being developed or spliced not less than d_b, clear cover not less than d_b, and stirrups or ties throughout l_d not less than the code minimum.

b. Clear spacing of bars being developed or spliced not less than $2d_b$ and clear cover not less than d_b.

2. For all other cases, the value of l_d/d_b in equations (7.8) and (7.9) must be multiplied by 1.5 to restore them to equivalence with equation (7.7).

These equations are relatively simple to use for the general conditions involved in practical design and construction. For example, in all structures with normal-weight concrete ($\lambda = 1.0$), uncoated reinforcement ($\beta = 1.0$), no. 7 *or larger* bars ($\alpha = 1.0$), equation (7.8) becomes

$$\frac{l_d}{d_b} = \frac{f_y}{(20\sqrt{f'_c})} \tag{7.10}$$

This equation is used when conditions (a) and (b) are met, whereas for all other cases, l_d/d_b is multiplied by 1.5, or

$$\frac{l_d}{d_b} = \frac{3f_y}{(40\sqrt{f'_c})} \tag{7.11}$$

Similarly, for the same conditions and for no. 6 *or smaller* bars, equation (7.9) becomes

$$\frac{l_d}{d_b} = \frac{f_y}{(25\sqrt{f'_c})} \tag{7.12}$$

This is used when conditions (a) and (b) are met; for all other cases, l_d/d_b is multiplied by 1.5, or

$$\frac{l_d}{d_b} = \frac{3f_y}{(50\sqrt{f'_c})} \tag{7.13}$$

It is quite common to use $f'_c = 4$ Ksi and $f_y = 60$ Ksi in the design and construction of reinforced concrete buildings. If these values are substituted in the preceding equations, then

Equation (7.10) becomes $l_d = 47.5d_b$ ($\geq$ no. 7 bars) (7.10a)

Equation (7.11) becomes $l_d = 71.2d_b$ ($\geq$ no. 7 bars) (7.11a)

Equation (7.12) becomes $l_d = 38d_b$ ($\leq$ no. 6 bars) (7.12a)

Equation (7.13) becomes $l_d = 57d_b$ ($\leq$ no. 6 bars) (7.13a)

Other values of l_d/d_b ratios are shown in Table 7.1. Table 7.2 gives the development length, l_d, for different reinforcing bars (when $f_y = 60$ Ksi and $f'_c = 3$ Ksi and 4 Ksi) for both cases: when conditions (a) and (b) are met and for all other cases.

Table 7.1 Values of l_d/d_b for Various Values of f'_c and f_y (Tension Bars)

	$f_y = 40$ Ksi				**$f_y = 60$ Ksi**			
	≤No. 6 Bars		**≥No. 7 Bars**		**≤No. 6 Bars**		**≥No. 7 Bars**	
f'_c Ksi	***Conditions Met***	***Other Cases***	***Conditions Met***	***Other Cases***	***Conditions Met***	***Other Cases***	***Conditions Met***	***Other Cases***
3	29.3	43.9	36.6	54.8	43.9	65.8	54.8	82.2
4	25.3	38.0	31.7	47.5	38.0	57.0	47.5	71.2
5	22.7	34.0	28.3	42.5	34.0	51.0	42.5	63.7
6	20.7	31.0	25.9	38.8	31.0	46.5	38.8	58.1

Table 7.2 Development Length l_d (in.) for Tension Bars and $f_y = 60$ Ksi ($\alpha = \beta = \lambda = 1.0$)

		Development Length l_d (in.)—Tension Bars			
		$f'_c = 3$ Ksi		**$f'_c = 4$ Ksi**	
Bar Number	**Bar Diameter (in.)**	***Conditions Met***	***Other Cases***	***Conditions Met***	***Other Cases***
3	0.375	17	25	15	21
4	0.500	22	33	19	29
5	0.625	28	41	24	36
6	0.750	33	50	29	43
7	0.875	48	72	42	63
8	1.000	55	83	48	72
9	1.128	62	93	54	81
10	1.270	70	105	61	92
11	1.410	78	116	68	102

7.4 DEVELOPMENT LENGTH IN COMPRESSION

The development length of deformed bars in compression is generally smaller than that required for tension bars, due to the fact that compression bars do not have the cracks that develop in tension concrete members that cause a reduction in the bond between bars and the surrounding concrete. The ACI Code, Section 12.3.2, gives the basic development length in compression for all bars as follows:

$$l_{db} = \frac{0.02 d_b f_y}{\sqrt{f'_c}} \geq 0.0003 d_b f_y \tag{7.14}$$

which must not be less than 8 in. The development length, l_{db}, may be reduced by multiplying l_{db} by $R_s = (A_s\,(\text{required}))/(A_s\,(\text{provided}))$. For spirally reinforced concrete compression members with spirals of not less than $\frac{1}{4}$ in. diameter and a spacing of 4 in. or less, the value of l_{db} in equation (7.14) may be multiplied by $R_{sl} = 0.75$. In general, $l_d = l_{db} \times (R_s$ or R_{sl}, if applicable$) \geq 8$ in. Tables 7.3 and 7.4 give the values of l_d/d_b and the basic l_d when $f_y = 60$ Ksi.

Table 7.3 Values of l_d/d_b for Various Values of f'_c and f_y (Compression Bars), Minimum $l_d = 8$ in. $l_d/d_b = 0.02 f_y/\sqrt{f'_c} \geq 0.0003 f_y$

f'_c (Ksi)	3	4	5 or more
$f_y = 40$ Ksi	15	13	12
$f_y = 60$ Ksi	22	19	18

Table 7.4 Basic Development Length, l_d (in.), for Compression Bars ($f_y = 60$ Ksi)

		Development Length, l_d (in.) when f'_c =		
Bar Number	**Bar Diameter (in.)**	**3 (Ksi)**	**4 (Ksi)**	**5 (Ksi) or More**
3	0.375	9	8	8
4	0.500	11	10	9
5	0.625	14	12	12
6	0.750	17	15	14
7	0.875	20	17	16
8	1.000	22	19	18
9	1.128	25	22	21
10	1.270	28	25	23
11	1.410	31	27	26

7.5 SUMMARY FOR THE COMPUTATION OF l_d IN TENSION

Assuming normal construction practices, $(c + K_{tr})/d_b = 1.5$.

1. If one of the following two conditions is met,
   a. Clear spacing of bars $\geq d_b$, clear cover $\geq d_b$, and bars are confined with stirrups not less than the code minimum; or
   b. Clear spacing of bars $\geq 2d_b$ and clear cover $\geq d_b$; then

For no. 7 and larger bars:
$$\frac{l_d}{d_b} = \frac{\alpha\beta\lambda f_y}{(20\sqrt{f'_c})} \tag{7.8}$$

For no. 6 or smaller bars, $$\frac{l_d}{d_b} = \frac{\alpha\beta\lambda f_y}{(25\sqrt{f'_c})} \quad (7.9)$$

2. For all other cases, multiply these ratios by 1.5.
3. Note that $\sqrt{f'_c} \leq 100$ psi and $\alpha\beta \leq 1.7$; values of α, β, and λ are as explained earlier.
4. For bundled bars, either in tension or compression, l_d should be increased by 20% for three-bar bundles and by 33% for four-bar bundles. A unit of bundled bars is considered as a single bar of a diameter and area equivalent to the total area of all bars in the bundle. This equivalent diameter is used to check spacings and concrete cover.

Example 7.1

Figure 7.5 shows the cross section of a simply supported beam reinforced with 4 no. 8 bars that are confined with no. 3 stirrups spaced at 6 in. Determine the development length of the bars if the beam is made of normal-weight concrete, bars are not coated, $f'_c = 3$ Ksi, and $f_y = 60$ Ksi.

Solution

1. Check if conditions for spacing and concrete cover are met:

$$\text{For no. 8 bars, } d_b = 1.0 \text{ in.}$$

$$\text{Clear cover} = 2.5 - 0.5 = 2.0 \text{ in.} > d_b$$

$$\text{Clear spacing between bars} = \frac{12 - 5}{3} - 1.0 = 1.33 \text{ in.} > d_b$$

Bars are confined with no. 3 stirrup. The conditions are met. Then

$$\frac{l_d}{d_b} = \frac{\alpha\beta\lambda f_y}{(20\sqrt{f'_c})} \quad \text{(for bars > no. 7)} \quad (7.8)$$

2. Determine the multiplication factors: $\alpha = 1.0$ (bottom bars), $\beta = 1.0$ (no coating), and $\lambda = 1.0$ (normal-weight concrete). Also check that $\sqrt{f'_c} = 54.8$ psi < 100 psi.

$$\frac{l_d}{d_b} = \frac{60{,}000}{(20\sqrt{3000})} = 54.8$$

So, $l_d = 54.8(1.0) = 54.8$ in., say, 55 in. These values can be obtained directly from Tables 7.1 and 7.2. Note that if the general formula (7.7) for l_d/d_b is used, assuming $K_{tr} = 0$, then

$$\frac{l_d}{d_b} = \left(\frac{3}{40}\right)\left(\frac{f_y}{\sqrt{f'_c}}\right)\frac{\alpha\beta\gamma\lambda}{(c/d_b)} \quad (7.8)$$

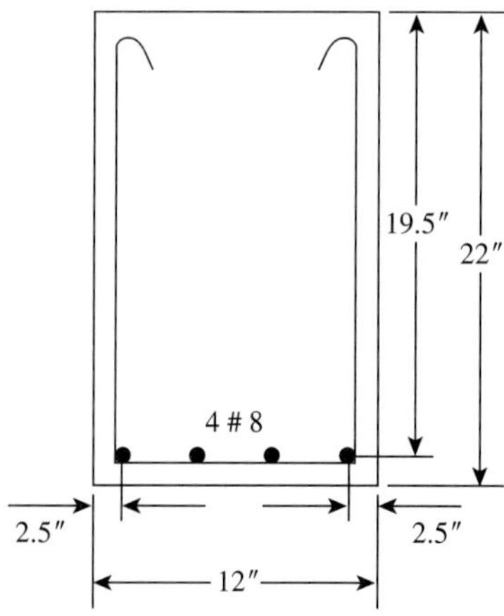

Figure 7.5 Example 7.1.

In this example, $\alpha = \beta = \gamma = \lambda = 1.0$.

Also,

c = smaller of distance from center of bar to the nearest concrete surface (c_1) or one-half the center-to-center of bars spacing (c_2)

$$c_1 = 2.5 \text{ in.}, \qquad c_2 = \frac{0.5(12 - 5)}{3} = 1.17 \text{ in. (controls)}$$

$(c + K_{tr})/d_b = 1.17/1.0 = 1.17 < 1.5$, so use $(c + K_{tr})/b = 1.5$. Consequently, $l_d/d_b = f_y/(20\sqrt{f'_c})$ as in Step 2, and $l_d = 55$ in.

Note: If the bars are not confined by stirrups, this value of l_d must be multiplied by 1.5 $(s = 1.33 \text{ in.} < 2d_b = 2.0 \text{ in.})$.

Example 7.2

Repeat Example 7.1 if the beam is made of lightweight aggregate concrete, the bars are epoxy coated, and A_s required from analysis is 2.79 in.^{2}

Solution

1. Determine the multiplication factors: $\alpha = 1.0$ (bottom bars), $\beta = 1.5$ (epoxy coated), $\lambda = 1.3$ (lightweight aggregate concrete), and $R_s = (A_s \text{ (required)})/(A_s \text{ (provided)}) = 2.79/3.14 = 0.89$. The value of β is 1.5, because the concrete cover is less than $3d_b = 3$ in. Check that $\alpha\beta = 1.0(1.5) = 1.5 < 1.7$.

2. $$l_d/d_b = \frac{R_s \alpha\beta\lambda f_y}{(20\sqrt{f'_c})}$$

$$= \frac{0.89(1.5)(1.3)(60{,}000)}{(20\sqrt{3000})} = 95.1 \text{ in.}, \qquad \text{say, 96 in.}$$

3. The basic l_d can be obtained from Table 7.2 ($l_d = 55$ in. for no. 8 bars) and then multiplied by the factors 1.5 and 1.3.

Example 7.3

A reinforced concrete column is reinforced with 8 no. 10 bars, which should extend to the footing. Determine the development length needed for the bars to extend down in the footing. Use $f'_c = 4$ Ksi and $f_y = 60$ Ksi.

Solution

The development length in compression is

$$l_{db} = \frac{0.02 d_b f_y}{\sqrt{f'_c}} \geq 0.0003 d_b f_y$$

$$l_{db} = \frac{0.02(1.27)(60{,}000)}{\sqrt{4000}} = 24.1 \text{ in.} \qquad \text{(controls)}$$

The minimum l_{db} is $0.0003(1.27)(60{,}000) = 22.86$ in., but it cannot be less than 8 in. Because there are no other multiplication factors, then $l_d = 24.1$ in., or 25 in. (The same value is shown in Table 7.4.)

7.6 CRITICAL SECTIONS IN FLEXURAL MEMBERS

The critical sections for development of reinforcement in flexural members are

- At points of maximum stress;
- At points where tension bars within the span are terminated or bent;
- At the face of the support;
- At points of inflection at which moment changes signs.

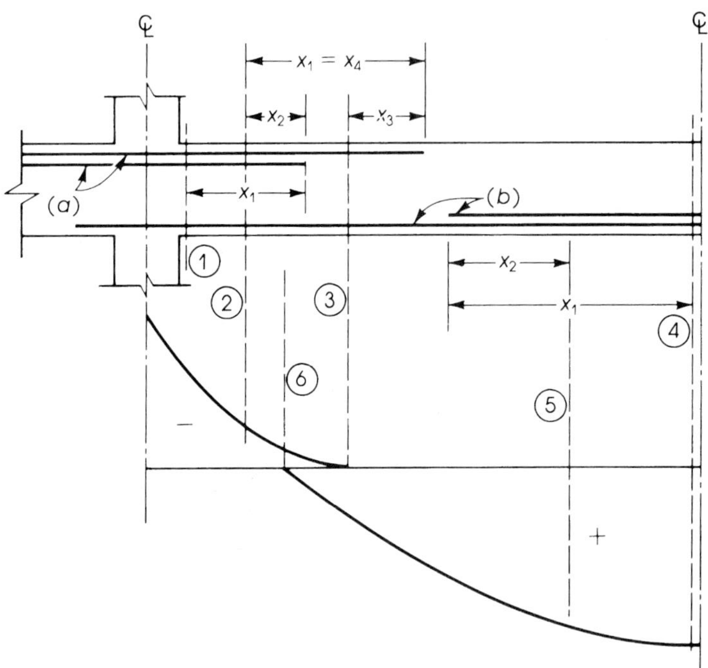

Figure 7.6 Critical sections (circled numbers) and development lengths ($x_1 - x_4$).

The critical sections for a typical uniformly loaded continuous beam are shown in Figure 7.6. The sections and the relative development lengths are explained as follows:

1. Three sections are critical for the negative moment reinforcement: Section 1 is at the face of the support, where the negative moment as well as stress are at maximum values. Two development lengths, x_1 and x_2, must be checked.

   Section 2 is the section where part of the negative reinforcement bars can be terminated. To develop full tensile force, the bars should extend a distance x_2 before they can be terminated. Once part of the bars are terminated, the remaining bars develop maximum stress.

   Section 3 is at the point of inflection. The bars shall extend a distance x_3 beyond section 3: x_3 must be equal to or greater than the effective depth d, 12 bar diameters, or $\frac{1}{16}$ clear span, whichever is greater. At least one-third of the total reinforcement provided for negative moment at the support shall be extended a distance x_3 beyond the point of inflection, according to the ACI Code, Section 12.12.3.
2. Three sections are critical for positive moment reinforcement: Section 4 is that of maximum positive moment and maximum stresses. Two development lengths, x_1 and x_2, have to be checked. The length x_1 is the development length l_d specified by the ACI Code, Section 12.11, as mentioned later. The length x_2 is equal to or greater than d or 12 bar diameters.

   Section 5 is where part of the positive reinforcement bars may be terminated. To develop full tensile force, the bars should extend a distance x_2. The remaining bars will have a maximum stress due to the termination of part of the bars. At the face of support, section 1, at least one-fourth of the positive moment reinforcement in continuous members shall extend along the same face of the member into the support, according to the ACI Code, Section 12.11.1. For simple members, at least one-third of the reinforcement shall extend into the support.

At points of inflection, section 6, limits are according to Section 12.11.3 of the ACI Code.

Example 7.4

A continuous beam has the bar details shown in Figure 7.7. The bending moments for maximum positive and negative moments are also shown. We must check the development lengths at all critical sections. Given: $f_c' = 3$ Ksi, $f_y = 40$ Ksi, $b = 12$ in., $d = 18$ in., and span $L = 24$ ft.

Solution

The critical sections are (a) at the face of the support for tension and compression reinforcement (section 1), (b) at points where tension bars are terminated within the span (sections 2 and 5), (c) at point of inflection (sections 3 and 6), and (d) at midspan (section 4).

1. Development lengths for negative-moment reinforcement, from Figure 7.7, are as follows: Three no. 9 bars are terminated at a distance $x_1 = 4.5$ ft from the face of the support, whereas the other three bars extend to a distance of 6 ft 0 in. (72 in.) from the face of the support.
   a. The development length of no. 9 tension bars is $36.3d_b$ (Table 7.1) if conditions of spacing and cover are met.

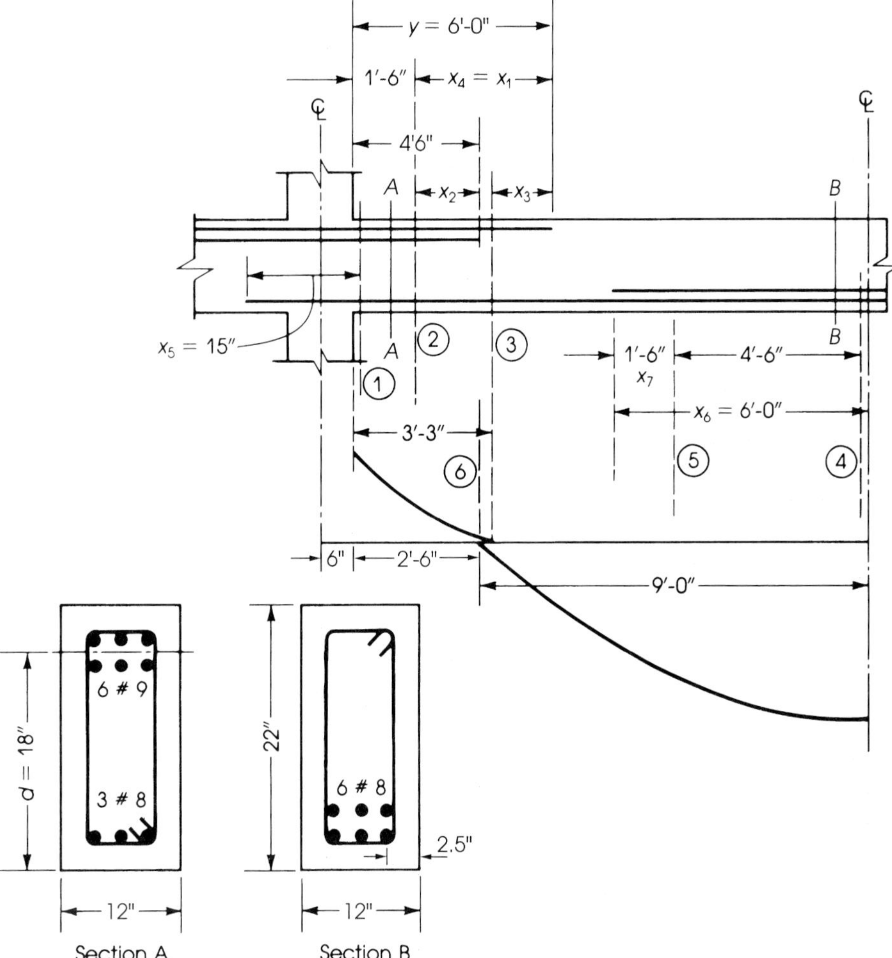

Figure 7.7 Development length of a continuous beam (Example 7.4).

For no. 9 bars, $d_b = 1.128$ in.

$$\text{Cover} = 2.5 - \frac{1.128}{2} = 1.94 \text{ in.} > d_b$$

$$\text{Clear spacing} = \frac{12 - 5}{2} - 1.128 = 2.37 \text{ in.} > 2d_b$$

Then conditions are met, and $l_d = 36.6(1.128) = 41.3$ in. For top bars, $x_1 = l_d = 1.3(41.3) = 54$ in. $= 4.50$ ft $= x_1 > 12$ in. (minimum).

b. The development length x_2 shall extend beyond the point where three no. 9 bars are not needed, either $d = 18$ in. or $12d_b = 13.6$ in., whichever is greater. Thus $x_2 = 18$ in. The required development length is $x_4 = 4.50$ ft, similar to x_1. Total length required is $y = x_1 + 1.5 \text{ ft} = 6.0$ ft.

c. Beyond the point of inflection (section 3), three no. 9 bars extend a length $x_3 = y - 39 = 72 - 39 = 33$ in. The ACI Code requires that at least one-third of the bars should extend beyond the inflection point. Three no. 9 bars are provided, which are adequate. The required development length of x_3 is the greatest of $d = 18$ in., $12d_b = 13.6$ in., or $L/16 = 24 \times \frac{12}{16}$ in. $= 18$ in., which is less than x_3 provided.

2. Compressive reinforcement at the face of the support (section 1) (no. 8 bars): The development length x_5 is equal to

$$l_d = \frac{0.02 d_b f_y}{\sqrt{f'_c}} = \frac{0.02 \times 1 \times 40{,}000}{\sqrt{3000}} = 14.6 \text{ in.}$$

So, we can use 15 in.

$$\text{Minimum } l_d = 0.0003 d_b f_y = 0.0003 \times 1 \times 40{,}000 = 12 \text{ in.}$$

but it cannot be less than 8 in. The length 15 in. controls. For no. 8 bar, $d_b = 1$ in.; l_d provided $= 15$ in., which is greater than that required.

3. Development length for positive moment reinforcement: Three no. 8 bars extend 6 ft. beyond the centerline, and the other bars extend to the support. The development length x_6 from the centerline is $l_d = 36.6d_b = 37$ in. (Table 7.1), but it cannot be less than 12 in. That is, x_6 provided is 6 ft $= 72$ in. > 37 in.

The length x_7 is equal to d or $12d_b$, that is, 18 in. or $12 \times 1 = 12$ in. The provided value is 18 in., which is adequate.

The actual position of the termination of bars within the span can be determined by the moment-resistance diagram, as will be explained later.

7.7 STANDARD HOOKS (ACI CODE, SECTIONS 12.5 AND 7.1)

A *hook* is used at the end of a bar when its straight embedment length is less than the necessary development length, l_d. Thus the full capacity of the bar can be maintained in the shortest distance of embedment. The minimum diameter of bend, measured on the inside of the main bar of a standard hook D_b, is as follows [9] (Figure 7.8):

- For no. 3 to no. 8 bars (10–25 mm), $D_b = 6d_b$.
- For no. 9 to no. 11 bars (28, 32, and 36 mm), $D_b = 8d_b$.
- For no. 14 and no. 18 bars (43 and 58 mm), $D_b = 10d_b$.

The ACI Code, Section 12.5.2, specifies a basic development length l_{hb} for a grade 60 hooked bar ($f_y = 60$ Ksi), as follows:

$$l_{hb} = \frac{1200 d_b}{\sqrt{f'_c}} \tag{7.15}$$

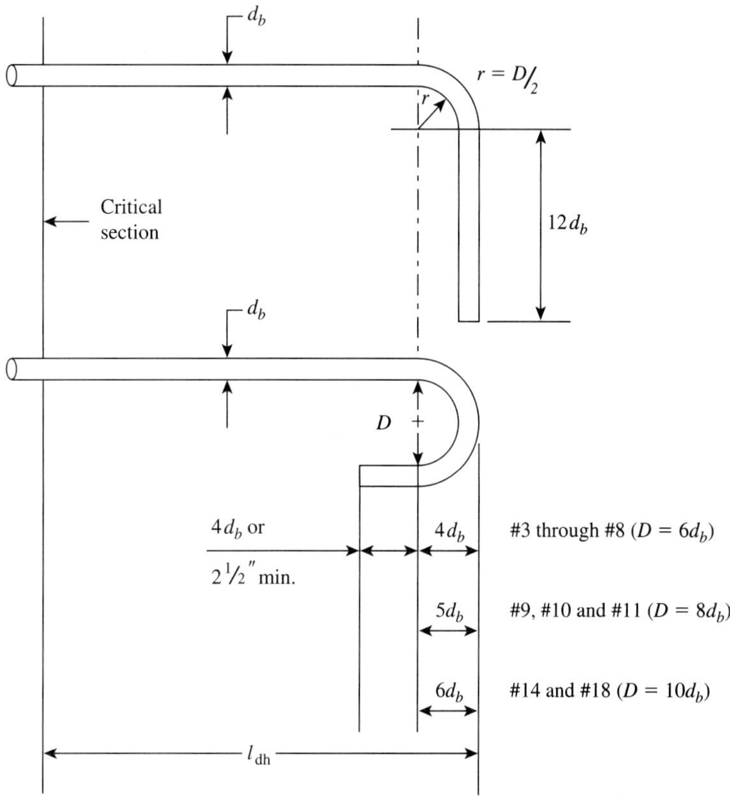

Figure 7.8 Hooked-bar details for the development of standard hooks [9].

Based on different conditions, the basic development length, l_{hb}, must be multiplied by one of the following factors:

1. When bars with yield strength other than 60 Ksi are used, the basic development length, l_{hb}, is multiplied by $f_y/60$, where f_y is in Ksi.
2. When no. 11 or smaller bars are used and the hook is enclosed vertically or horizontally within stirrups or ties spaced not greater than three times the diameter of the hooked bar, the basic development length is multiplied by 0.8.
3. When no. 11 or smaller bars are used and the side concrete cover, normal to the plane of the hook, is not less than 2.5 in., the basic development length is multiplied by 0.7. The same factor applies for a 90° hook when the concrete cover on bar extension beyond the hook is not less than 2 in.
4. When a bar anchorage is not required, the basic development length for the reinforcement in excess of that required is multiplied by the ratio

$$\frac{A_s \text{ (required)}}{A_s \text{ (provided)}}$$

5. For lightweight aggregate concrete, a modification factor of 1.3 must be used.
6. When standard hooks with less than a 2.5-in. concrete cover on the side and top or bottom are used at a discontinuous end of a member, the hooks shall be enclosed by ties or stirrups spaced at no greater than $3d_b$. Moreover, the factor 0.8 given in Item 2 shall not be used.

The development length, l_{dh}, of a standard hook for deformed bars in tension is equal to the basic development length, l_{hb}, multiplied by the applicable modified factor but must not be less than $8d_b$ or 6 in., whichever is greater. For example, if the side cover is large, so that splitting is effectively eliminated, and ties are provided, then $l_{dh} = l_{hb}(0.7)(0.8)$. Note that hooks are not effective for reinforcing bars in compression and may be *ignored* (ACI Code, Section 12.5).

Details of standard 90° and 180° hooks are shown in Figure 7.8 [9]. The dimensions given are needed to protect members against splitting and spalling of concrete cover. Figure 7.9 shows details of hooks at a discontinuous end with a concrete cover less than 2.5 in. that may produce concrete spalling [9]. The use of closed stirrups is necessary for proper design. Figure 7.10 shows the stress distribution along a 90° hooked bar under a tension force p.

The development length required for deformed welded wire fabric is covered in Section 12.7 in the ACI Code. The basic development length (measured from the critical section) with at least one cross wire within the development length and not less than 2 in. shall be the greater of $(f_y - 35{,}000)/f_y$ (units in psi) or $5d_b/S_w$ but should not be taken greater than 1.0, where S_w = spacing of wire to be developed or spliced (in.).

Example 7.5

Compute the development length required for the top no. 8 bars of the cantilever beam shown in Figure 7.11 that extend into the column support if the bars are

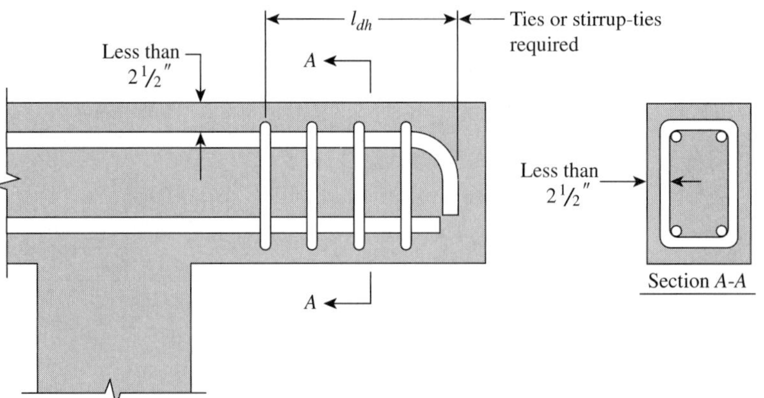

Figure 7.9 Concrete cover limitations [9].

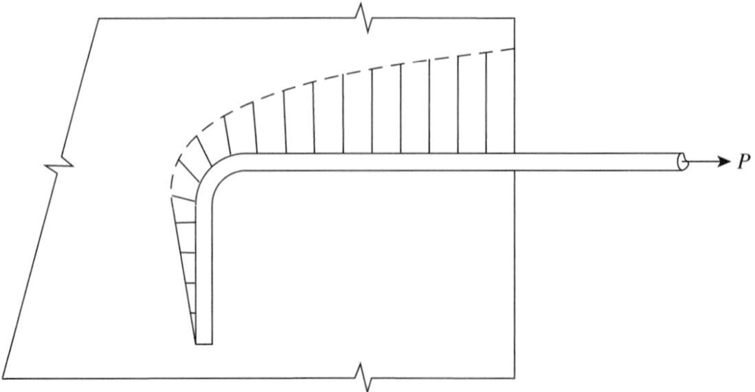

Figure 7.10 Stress distribution in 90° hooked bar.

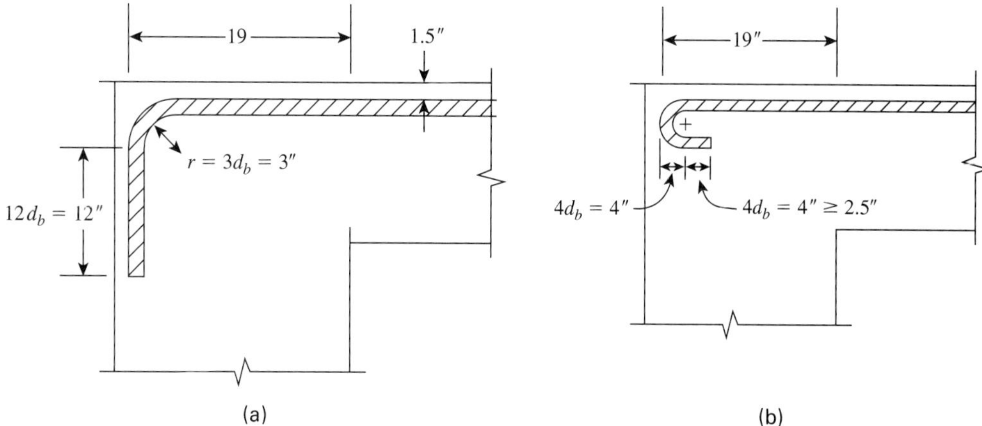

Figure 7.11 Example 7.5.

a. Straight;

b. Have a 90° hook at the end;

c. Have a 180° hook at the end.

The bars are confined by no. 3 stirrups spaced at 6 in. and have a clear cover = 1.5 in. and clear spacings = 2.0 in. Use $f'_c = 4$ Ksi and $f_y = 60$ Ksi.

Solution

a. Straight bars: For no. 8 bars, $d_b = 1.0$ in. Because clear spacing $= 2d_b$ and clear cover is greater than d_b with bars confined by stirrups, then conditions (a) and (b) are met. Equation (7.10) can be used to calculate the basic l_d or you can get it directly from Table 7.2: basic $l_d = 48$ in. For top bars $\alpha = 1.3$ and final $l_d = 1.3(48) = 63$ in.

b. Bars with 90° hook: For no. 8 bars, $d_b = 1.0$ in. Basic development length $l_{hb} = 1200d_b/\sqrt{f'_c} = 1200(1.0)/\sqrt{4000} = 19$ in. Because no other modifications apply, then $l_{dh} = 19$ in. $> 8d_b = 8$ in. or 6 in. Other details are shown in Figure 7.11. The factor $\alpha = 1.3$ for top bars does not apply to hooks.

c. Bars with 180° hook: Basic $l_{hb} = 19$ in., as calculated before. No other modifications apply; then $l_{dh} = 19$ in. $> 8d_b = 8$ in. Other details are shown in Figure 7.11.

7.8 SPLICES OF REINFORCEMENT

7.8.1 General

Steel bars that are used as reinforcement in structural members are fabricated in lengths of 20, 40, and 60 ft (6, 12, and 18 m), depending on the bar diameter, transportation facilities, and other reasons. Bars are usually tailored according to the reinforcement details of the structural members. When some bars are short, it is necessary to splice them by lapping the bars a sufficient distance to transfer stress through the bond from one bar to the other.

Splices may be made by lapping or welding or with mechanical devices that provide positive connection between bars. Lap splices should not be used for bars larger than no. 11 (36 mm). For noncontact lap splices in flexural members, bars should not be spaced transversely farther apart than one-fifth the required length or 6 in. (150 mm). An approved welded splice is one in which the bars are butted and welded to develop in tension at least 125% of the specified yield strength of the bar. The ACI Code, Section 12.14, also specifies that full positive mechanical connections must develop in tension or compression at least 125% of the specified yield strength of the bar.

Splices should not be made at or near sections of maximum moments or stresses. Also, it is recommended that all bars should not be spliced at the same location to avoid a

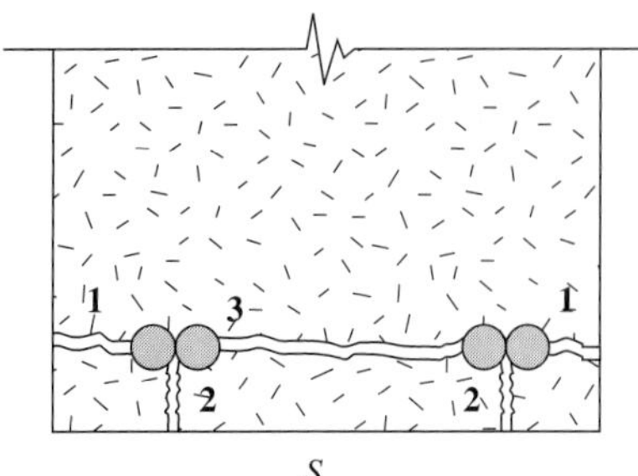

Figure 7.12 Lap splice failure due to the development of one or more cracks.

weakness in the concrete section and to avoid the congestion of bars at the same location, which may cause difficulty in placing the concrete around the bars.

The stresses developed at the end of a typical lap splice are equal to zero, whereas the lap length, l_d, embedded in concrete is needed to develop the full stress in the bar, f_y. Therefore, a minimum lap splice of l_d is needed to develop a continuity in the spliced tension or compression bars. If adequate splice length is not provided, splitting and spalling occurs in the concrete shell (Figure 7.12).

Splices in tension and compression are covered by Sections 12.15 and 12.16 of the ACI Code.

7.8.2 Lap Splices in Tension, l_{st}

Depending upon the percentage of bars spliced on the same location and the level of stress in the bars or deformed wires, the ACI Code introduces two classes of splices (with a minimum length of 12 in.):

1. Class A splices: These splices have a minimum length $l_{st} = l_d$ and are used when (a) one-half or less of the total reinforcement is spliced within the required lap length; and (b) the area of reinforcement provided is at least twice that required by analysis over the entire length of the splice. The length l_d is the development length of the bar, as calculated earlier.
2. Class B splices: These splices have a minimum length $l_{st} = 1.3l_d$ and are used for all other cases that are different from the aforementioned conditions. For example, class B splices are required when all bars or deformed wires are spliced at the same location with any ratio of $(A_s\text{ (provided)})/(A_s\text{ (required)})$. Splicing all the bars in one location should be avoided when possible.

7.8.3 Lap Splice in Compression, l_{sc}

The splice lap length of the reinforcing bars in compression, l_{sc}, should be equal to or greater than the development length of the bar in compression, l_d (including the modifiers), calculated earlier (equation (7.14)). Moreover, the lap length shall satisfy the following (ACI Code, Section 12.16.1):

$$l_{sc} \geq (0.0005 f_y d_b) \qquad (\text{for } f_y \leq 60{,}000 \text{ psi}) \tag{7.16}$$

$$l_{sc} = (0.0009 f_y - 24) d_b \quad (\text{for } f_y > 60{,}000 \text{ psi}) \tag{7.17}$$

For all cases, the lap length must not be less than 12 in. Table 7.5 gives the lap-splice length for various f_y values. If the concrete strength, f'_c, is less than 3000 psi, the lap length, l_{sc}, must be increased by one-third.

In spirally reinforced columns, lap-splice length within a spiral may be multiplied by 0.75 but may not be less than 12 in. In tied columns, with ties within the splice length having

Table 7.5 Lap-Splice Length in Compression, l_{sc} (in.), ($f'_c \geq 3$ Ksi and Minimum $l_{sc} = 12$ in.)

Bar Number	Bar Diameter (in.)	f_y (Ksi) 40	60	80
3	0.375	12	12	18
4	0.500	12	15	24
5	0.625	13	19	30
6	0.750	15	23	36
7	0.875	18	27	42
8	1.000	20	30	48
9	1.128	23	34	55
10	1.270	26	39	61
11	1.410	29	43	68

a minimum effective area of 0.0015 hs, lap splice may be multiplied by 0.83 but may not be less than 12 in., where h = overall thickness of column and s = spacing of ties (in.).

Example 7.6

Calculate the lap-splice length for six no. 8 tension bottom bars (in two rows) with clear spacing = 2.5 in. and clear cover = 1.5 in. for the following cases:

a. When three bars are spliced and $(A_s \text{ provided})/(A_s \text{ required}) > 2$

b. When four bars are spliced and $(A_s \text{ provided})/(A_s \text{ required}) < 2$

c. When all bars are spliced at the same location. Given: $f'_c = 5$ Ksi and $f_y = 60$ Ksi

Solution

a. For no. 8 bars, $d_b = 1.0$ in., and $\alpha = \beta = \gamma = \lambda = 1.0$ (equation (7.8)): $\sqrt{5000} = 70.7$ psi < 100 psi, and from Table 7.1, $l_d = 42.5d_b$, because conditions for clear spacings and cover are met.

$$l_d = 42.5(1.0) = 42.5 \text{ in.,} \quad \text{or} \quad 43 \text{ in.}$$

For $(A_s\,(\text{provided}))/(A_s\,(\text{required})) > 2$, class A splice applies, $l_{st} = 1.0l_d = 43$ in. > 12 in. (minimum). Bars spliced are less than half the total number.

b. $l_d = 43$ in., as calculated before. Because $(A_s\,(\text{provided}))/(A_s\,(\text{required}))$ is less than 2, class B splice applies, $l_{st} = 1.3l_d = 1.3(42.5) = 55.25$ in., say, 56 in., which is greater than 12 in.

c. Class B splice applies and $l_{st} = 56$ in. > 12 in.

Example 7.7

Calculate the lap-splice length for a no. 10 compression bar in a tied column when $f'_c = 5$ Ksi and when (a) $f_y = 60$ Ksi and (b) $f_y = 80$ Ksi.

Solution

a. For no. 10 bars, $d_b = 1.27$ in., and the basic development length from Table 7.4 or 7.3 is 23 in. Because no modifiers apply, $l_{sc} = 23$ in. > 12 in. Check that $l_s \geq 0.0005d_bf_y = 0.0005(1.27)(60{,}000) = 38.1$ in. Therefore, $l_{sc} = 39$ in. controls.

b. The basic l_d is 23 in., as calculated before. Check that $l_s \geq (0.0009f_y - 24)d_b = [0.0009(80{,}000) - 24](1.27) = 61$ in. Therefore, $l_{sc} = 61$ in. controls.

7.9 MOMENT-RESISTANCE DIAGRAM (BAR CUTOFF POINTS)

The moment capacity of a beam is a function of its effective depth, d, width, b, and the steel area for given strengths of concrete and steel. For a given beam, with constant width and depth, the amount of reinforcement can be varied according to the variation of the bending moment along the span. It is a common practice to cut off the steel bars where they are no longer needed to resist the flexural stresses. In some other cases, as in continuous beams, positive-moment steel bars may be bent up, usually at 45°, to provide tensile reinforcement for the negative moments over the supports.

The ultimate moment capacity of an under-reinforced concrete beam at any section is

$$M_u = \phi A_s f_y (d - a/2) \tag{7.18}$$

The lever arm $(d - a/2)$ varies for sections along the span as the amount of reinforcement varies; however, the variation in the lever arm along the beam length is small and is never less than the value obtained at the section of maximum bending moment. Thus, it may be assumed that the moment capacity of any section is proportional to the tensile force or the area of the steel reinforcement, assuming proper anchorage lengths are provided.

To determine the position of the cutoff or bent points, the moment diagram due to external loading is drawn first. A moment-resistance diagram is also drawn on the same graph, indicating points where some of the steel bars are no longer required. The ultimate moment resistance of one bar, M_{ub}, is

$$M_{ub} = \phi A_{sb} f_y \left(d - \frac{a}{2} \right) \tag{7.19}$$

where

$$a = \frac{A_s f_y}{0.85 f'_c b}$$

$$A_{sb} = \text{area of one bar}$$

The intersection of the moment-resistance lines with the external bending moment diagram indicates the theoretical points where each bar can be terminated. To illustrate this discussion, Figure 7.13 shows a uniformly loaded simple beam, its cross section, and the bending moment diagram. The bending moment curve is a parabola with a maximum moment at midspan of 2400 K · in. Because the beam is reinforced with four no. 8 bars, the ultimate moment resistance of one bar is

$$M_{ub} = \phi A_{sb} f_y \left(d - \frac{a}{2} \right)$$

$$a = \frac{A_s f_y}{0.85 f'_c b} = \frac{4 \times 0.79 \times 50}{0.85 \times 3 \times 12} = 5.2 \text{ in.}$$

$$M_{ub} = 0.9 \times 0.79 \times 50 \left(20 - \frac{5.2}{2} \right) = 620 \text{ K} \cdot \text{in.}$$

The ultimate moment resistance of four bars is thus 2480 K · in., which is greater than the external moment of 2400 K · in. If the moment diagram is drawn to scale on the base line A-A, it can be seen that one bar can be terminated at point a, a second bar at point b, the third bar at point c, and the fourth bar at the support end A. These points are the theoretical positions for the termination of the bars. However, it is necessary to develop part of the strength of the bar by bond, as explained earlier. The ACI Code specifies

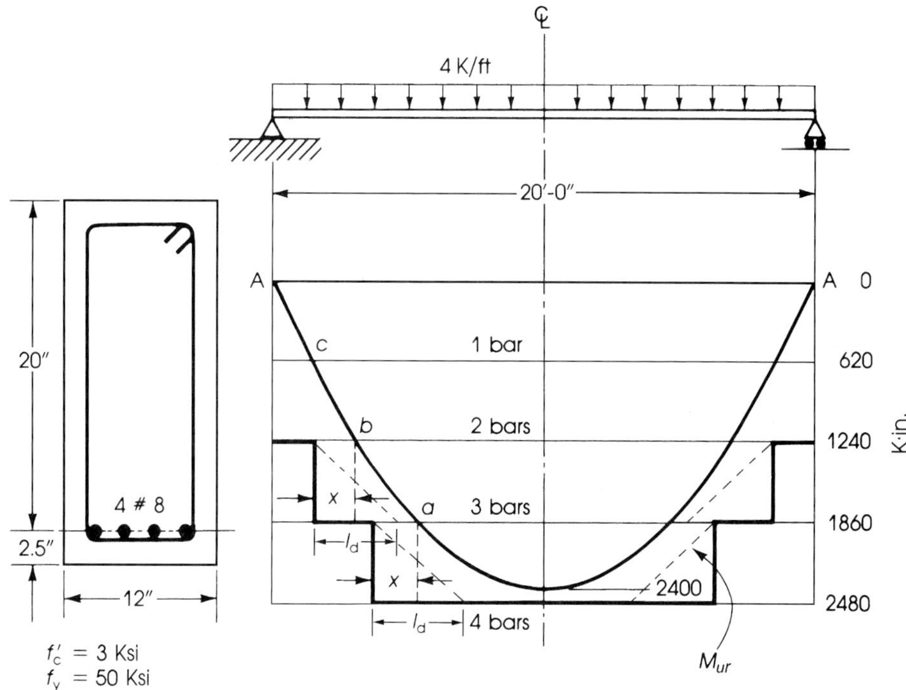

Figure 7.13 Moment-resistance diagram.

that every bar should be continued at least a distance equal to the effective depth, d, of the beam or 12 bar diameters, whichever is greater, beyond the theoretical points a, b, and c. The code also specifies (Section 12.11.1) that at least one-third of the positive moment reinforcement must be continued to the support for simple beams. Therefore, for the example discussed here, two bars must extend into the support, and the moment-resistance diagram, M_{ur}, shown in Figure 7.13, must enclose the external bending moment diagram at all points. Full load capacity of each bar is attained at a distance l_d from its end.

For continuous beams, the bars are bent at the required points and used to resist the negative moments at the supports. At least one-third of the total reinforcement provided for the negative moment at the support must be extended beyond the inflection points a distance not less than the effective depth, 12 bar diameters, or $\frac{1}{16}$ the clear span, whichever is greatest (ACI Code, Section 12.12.3).

Bent bars are also used to resist part of the shear stresses in beams. The moment-resistance diagram for a typical continuous beam is shown in Figure 7.14.

Example 7.8

For the simply supported beam shown in Figure 7.15, design the beam for the given ultimate loads and draw the moment-resistance diagram. Also, show where the reinforcing bars can be terminated. Use b = 10 in., a steel ratio of 0.018, f'_c = 3 Ksi, and f_y = 40 Ksi.

Solution

For $\rho = 0.018$, R_u = 556 psi and $M_u = R_u bd^2$. M_u = 132.5 K·ft. Now $132.5(12) = 0.556(10)d^2$, so d = 17 in.; let h = 20 in. $A_s = 0.018(10)(17) = 3.06$ in.^{2}; use four no. 8 bars (A_s = 3.14 in.^{2}). Actual $d = 20 - 2.5 = 17.5$ in.

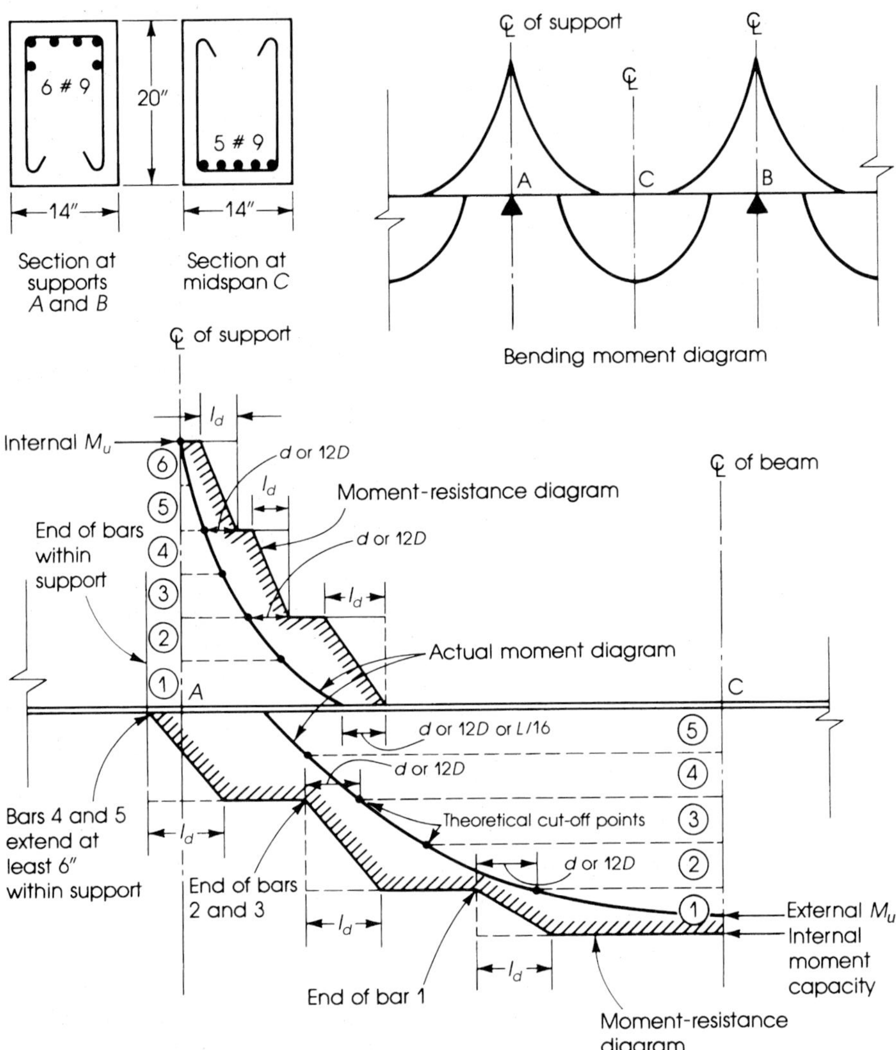

Figure 7.14 Sections and bending moment diagram (top) and moment-resistance diagram (bottom) of a continuous beam. Bar diameter is signified by *D*.

$$M_{ur} = \phi A_s f_y \left(d - \frac{a}{2} \right), \quad \text{and} \quad a = \frac{3.14(40)}{0.85(3)(10)} = 4.93 \text{ in.}$$

$$M_{ur} \text{ (for one bar)} = 0.9(0.79)(40)\left(17.5 - \frac{4.93}{2} \right)$$

$$= 427.7 \text{ K} \cdot \text{in.} = 35.64 \text{ K} \cdot \text{ft}$$

$$M_{ur} \text{ (for all 4 bars)} = 1710.8 \text{ K} \cdot \text{in.} = 142.6 \text{ K} \cdot \text{ft}$$

Details of the moment-resistance diagram are shown in Figure 7.15. Note that the bars can be bent or terminated at a distance of 17.5, say, 18 in. (or 12 bar diameters, whichever is greater), beyond the points, where (theoretically) the bars are not needed. The development length l_d for no. 8 bars is $36.6d_b = 37$ in. (Table 7.1). The cutoff points of the first and second bars are at points *A* and *B*, but the actual points are at *A*′ and *B*′, 18 in. beyond *A* and *B*. From *A*′, a

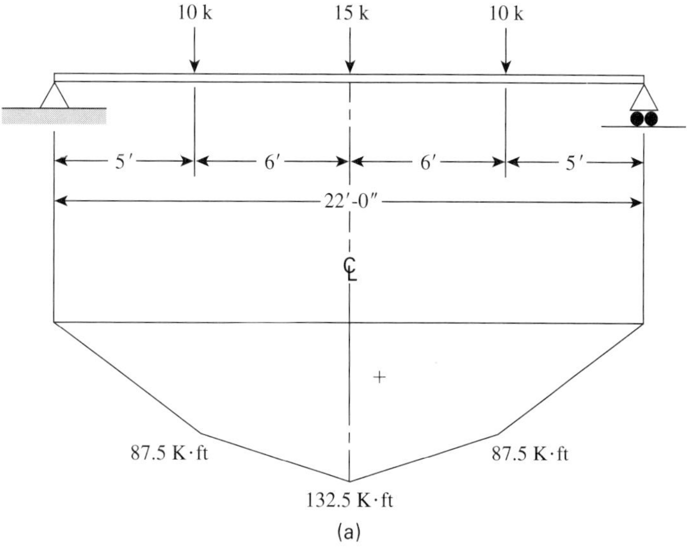

(a)

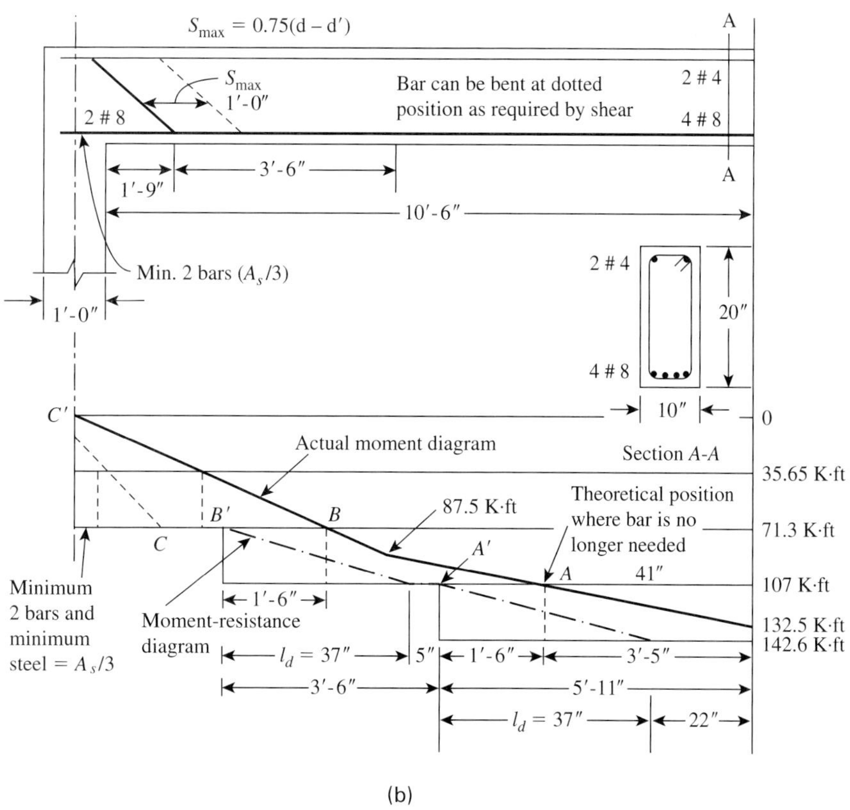

(b)

Figure 7.15 Details of reinforcing bars and the moment-resistance diagram, Example 7.8.

length l_d = 37 in. backward is shown to establish the moment-resistance diagram (the dashed line). The end of the last two bars extending to the support will depend on how far they extend inside the support, say, at C'. Normally, bars are terminated within the span at A' and B' as bent bars are not commonly used to resist shear.

SUMMARY

Sections 7.1 and 7.2

Bond is influenced mainly by the roughness of the steel surface area, the concrete mix, shrinkage, and the cover of concrete. In general,

$$l_d = \frac{A_s f_y}{U_u \Sigma O} \tag{7.3}$$

Sections 7.3 and 7.5

1. The general formula for the development length in tension for no. 11 bars or smaller is

$$\frac{l_d}{d_b} = \left(\frac{3}{40}\right)\left(\frac{f_y}{\sqrt{f'_c}}\right)\frac{\alpha\beta\gamma\lambda}{(c + K_{tr})/d_b} \tag{7.7}$$

   K_{tr} may be assumed to be 0. Other values of l_d/d_b are given in Tables 7.1 and 7.2. α, β, γ, and λ are multipliers defined in Section 7.3.

2. Simplified expressions are used when conditions for concrete cover and spacings are met. For no. 7 and larger bars,

$$\frac{l_d}{d_b} = \left(\frac{f_y}{\sqrt{f'_c}}\right)\left(\frac{\alpha\beta\lambda}{20}\right) = Q \tag{7.8}$$

   For no. 6 and smaller bars,

$$\frac{l_d}{d_b} = 0.8Q \tag{7.9}$$

3. For all other cases, multiply the previous Q by 1.5.
4. Minimum length is 12 in.

Section 7.4

Development length in compression for all bars is

$$l_d = \frac{0.02 d_b f_y}{\sqrt{f'_c}} \geq 0.0003 d_b f_y \geq 8 \text{ in.} \tag{7.14}$$

For specific values, refer to Tables 7.3 and 7.4.

Section 7.6

The critical sections for the development of reinforcement in flexural members are

- At points of maximum stress;
- At points where tension bars are terminated within the span;
- At the face of the support;
- At points of inflection.

Section 7.7

The minimum diameter of bends in standard hooks is

- For no. 3 to no. 8 bars, $6d_b$;
- For no. 9 to no. 11 bars, $8d_b$.

The basic development length l_{hb} of a standard hook is

$$l_{hb} = \frac{1200d_b}{\sqrt{f'_c}} \tag{7.15}$$

Section 7.8

1. For splices in tension, the minimum lap-splice length is 12 in. If (1) one-half or less of the total reinforcement is spliced within the required lap length and (2) the area of reinforcement provided is at least twice that required by analysis over the entire length of the splice, then $l_{st} = 1.0l_d$ = class A splice.
2. For all other cases, class B has to be used when $l_{st} = 1.3l_d$.
3. For splices in compression, the lap length should be equal to or greater than l_d in compression, but it also should satisfy the following: $l_{sc} \geq 0.0005f_y d_b$ (for $f_y \leq 60{,}000$ psi).

REFERENCES

1. L. A. Lutz and P. Gergely. "Mechanics of Bond and Slip of Deformed Bars in Concrete." *ACI Journal 68* (April 1967).
2. ACI Committee 408. "Bond Stress—The State of the Art." *ACI Journal 63* (November 1966).
3. ACI Committee 408. "Opportunities in Bond Research." *ACI Journal 67* (November 1970).
4. Y. Goto. "Cracks Formed in Concrete around Deformed Tensioned Bars." *ACI Journal 68* (April 1971).
5. T. D. Mylrea. "Bond and Anchorage." *ACI Journal 44* (March 1948).
6. E. S. Perry and J. N. Thompson. "Bond Stress Distribution on Reinforcing Steel in Beams and Pullout Specimens." *ACI Journal 63* (August 1966).
7. C. O. Orangum, J. O. Jirsa, and J. E. Breen. "A Reevaluation of Test Data on Development Length and Splices." *ACI Journal 74* (March 1977).
8. J. Minor and J. O. Jirsa. "Behavior of Bent Bar Anchorage." *ACI Journal 72* (April 1975).
9. American Concrete Institute. "Building Code Requirements for Structural Concrete." *ACI* (318–99). Detroit, Mich. (November 1999).

PROBLEMS

7.1 For each assigned problem, calculate the development length required for the following tension bars. All bars are bottom bars in normal-weight concrete unless specified otherwise in the notes.

No.	Bar No.	f'_c (Ksi)	f_y (Ksi)	Clear Cover (in.)	Clear Spacing (in.)	Notes
a	5	3	60	2.0	2.25	
b	6	4	60	2.0	2.50	Lightweight aggregate concrete
c	7	5	60	2.0	2.13	Epoxy coated
d	8	3	40	2.5	2.30	Top bars, lightweight aggregate concrete
e	9	4	60	1.5	1.5	
f	10	5	60	2.0	2.5	No. 3 stirrups at 6 in.
g	11	5	60	3.0	3.0	
h	9	3	40	2.0	1.5	Epoxy coated
i	8	4	60	2.0	1.75	$(A_s \text{ provided})/(A_s \text{ required}) = 1.5$
j	6	4	60	1.5	1.65	Top bars, expoxy coated and no. 3 stirrup at 4 in.

7.2 For each assigned problem, calculate the development length required for the following bars in compression.

No.	Bar No.	f'_c (Ksi)	f_y (Ksi)	Notes
a	8	3	60	
b	9	4	60	
c	10	4	40	
d	11	5	60	$(A_s\text{ (required)})/(A_s\text{ (provided)}) = 0.8$
e	7	6	60	$(A_s\text{ (required)})/(A_s\text{ (provided)}) = 0.9$
f	9	5	60	Column with spiral no. 3 at 2 in.

7.3 Compute the development length required for the top no. 9 bars of a cantilever beam that extend into the column support if the bars are

a. Straight;

b. Have a 90° hook at the end;

c. Have a 180° hook at the end.

The bars are confined with no. 3 stirrups spaced at 5 in. and have a clear cover of 2.0 in. Use $f'_c = 4$ Ksi and $f_y = 60$ Ksi. (Clear spacing = 2.5 in.)

7.4 Repeat Problem 7.3 when no. 7 bars are used.

7.5 Repeat Problem 7.3 when $f'_c = 3$ Ksi and $f_y = 40$ Ksi.

7.6 Repeat Problem 7.3 when no. 10 bars are used.

7.7 Calculate the lap splice length for no. 9 tension bottom bars with clear spacing of 2.0 in. and clear cover of 2.0 in. for the following cases:

a. When 50% of the reinforcement is spliced and $(A_s\text{ (provided)})/(A_s\text{ (required)}) = 2$

b. When 75% of the reinforcement is spliced and $(A_s\text{ (provided)})/(A_s\text{ (required)}) = 1.5$

c. When all bars are spliced at one location and $(A_s\text{ (provided)})/(A_s\text{ (required)}) = 2$

d. When all bars are spliced at one location and $(A_s\text{ (provided)})/(A_s\text{ (required)}) = 1.3$ Use $f'_c = 4$ Ksi and $f_y = 60$ Ksi.

7.8 Repeat Problem 7.7 using $f'_c = 3$ Ksi and $f_y = 60$ Ksi.

7.9 Calculate the lap splice length for no. 9 bars in compression when $f'_c = 5$ Ksi and $f_y = 60$ Ksi.

7.10 Repeat Problem 7.9 when no. 11 bars are used.

7.11 Repeat Problem 7.9 when $f_y = 80$ Ksi.

7.12 Repeat Problem 7.9 when $f'_c = 4$ Ksi and $f_y = 60$ Ksi.

7.13 A continuous beam has the typical steel reinforcement details shown in Figure 7.16. The sections at midspan and at the face of the support of a typical interior span are also shown. Check the development lengths of the reinforcing bars at all critical sections. Use $f'_c = 4$ Ksi and $f_y = 60$ Ksi.

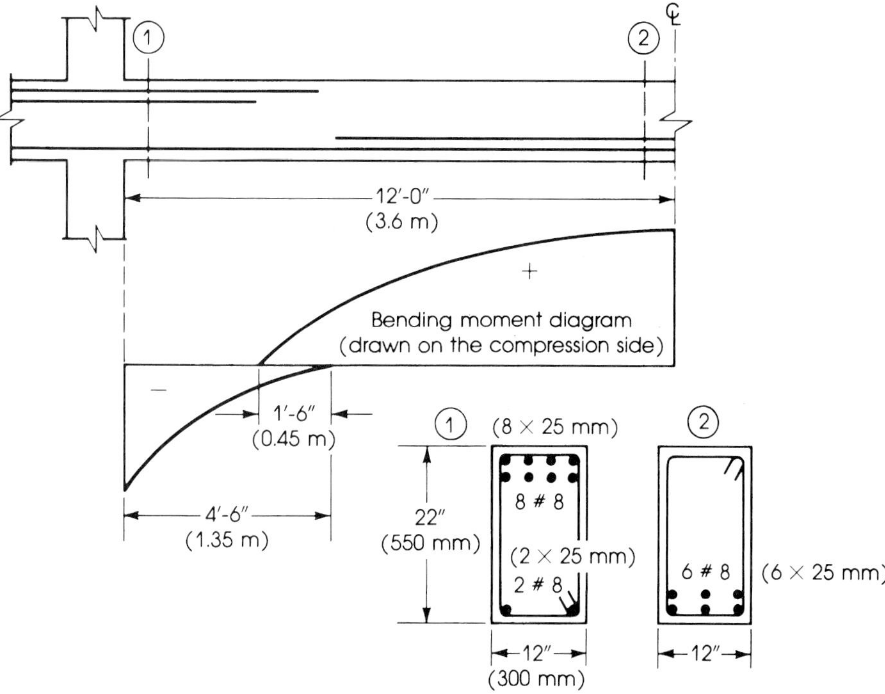

Figure 7.16 Problem 7.13.

7.14 Design the beam shown in Figure 7.17 using ρ_{max}. Draw the moment-resistance diagram and indicate where the reinforcing bars can be terminated. The beam carries a uniform dead load, including self-weight of 1.5 K/ft, and a live load of 2.2 K/ft. Use $f'_c = 4$ Ksi, $f_y = 60$ Ksi, and $b = 12$ in.

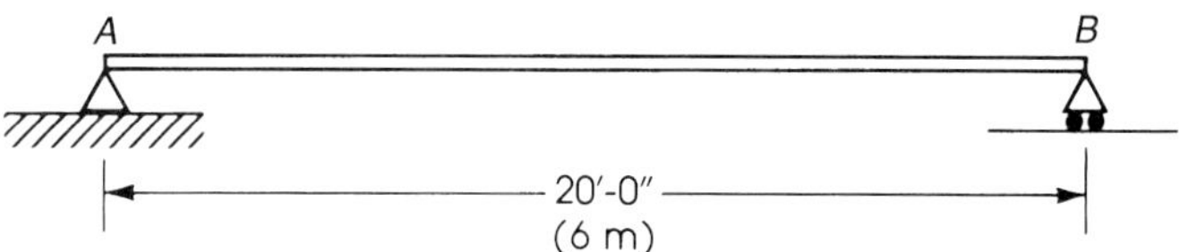

Figure 7.17 Problem 7.14: D.L. = 1.5 K/ft (22.5 kN/m), L.L. = 2.2 K/ft (33 kN/m).

7.15 Design the beam shown in Figure 7.18 using a steel ratio $\rho = 1/2\rho_b$. Draw the moment-resistance diagram and indicate the cutoff points. Use $f'_c = 3$ Ksi, $f_y = 60$ Ksi, and $b = 12$ in.

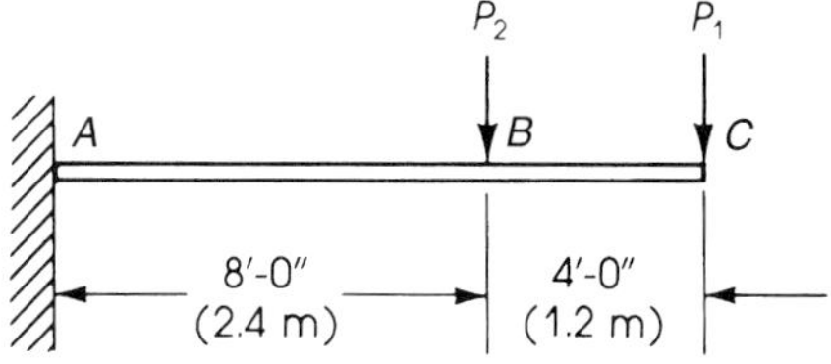

Figure 7.18 Problem 7.15: D.L. = 2 K/ft (30 kN/m), L.L. (concentrated loads only) is $P_1 = 10$ K (45 kN), $P_2 = 16$ K (72 kN).

7.16 Design the section at support B of the beam shown in Figure 7.19 using ρ_{max}. Adopting the same dimensions of the section at B for the entire beam ABC, determine the reinforcement required for part AB and draw the moment-resistance diagram for the beam ABC. Use $f'_c = 4$ Ksi, $f_y = 60$ Ksi, and $b = 12$ in.

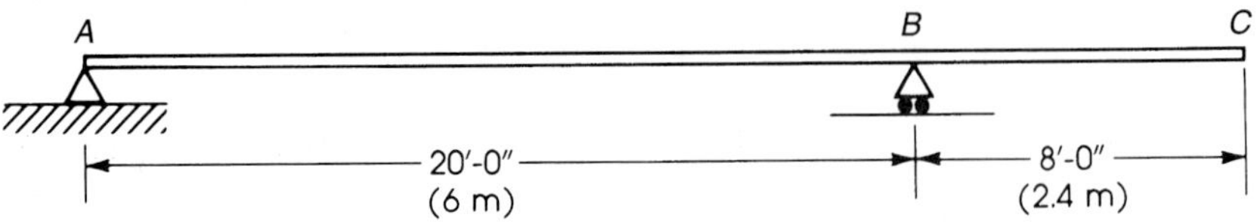

Figure 7.19 Problem 7.16: D.L. = 6 K/ft (90 kN/m), L.L. = 4 K/ft (60 kN/m).

Office building, Chicago.

SHEAR AND DIAGONAL TENSION

8.1 INTRODUCTION

When a simple beam is loaded as shown in Figure 8.1, bending moments and shear forces develop along the beam. To carry the loads safely, the beam must be designed for both types of forces. Flexural design is considered first to establish the dimensions of the beam section and the main reinforcement needed, as explained in the previous chapters.

The beam is then designed for shear. If shear reinforcement is not provided, shear failure may occur. *Shear failure* is characterized by small deflections and lack of ductility, giving little or no warning before failure. On the other hand, flexural failure is characterized by a gradual increase in deflection and cracking, thus giving warning before total failure. This is due to the ACI Code limitation on flexural reinforcement. The design for shear must ensure that shear failure does not occur before flexural failure.

8.2 SHEAR STRESSES IN CONCRETE BEAMS

The general formula for the shear stress in a homogeneous beam is

$$v = \frac{VQ}{Ib} \tag{8.1}$$

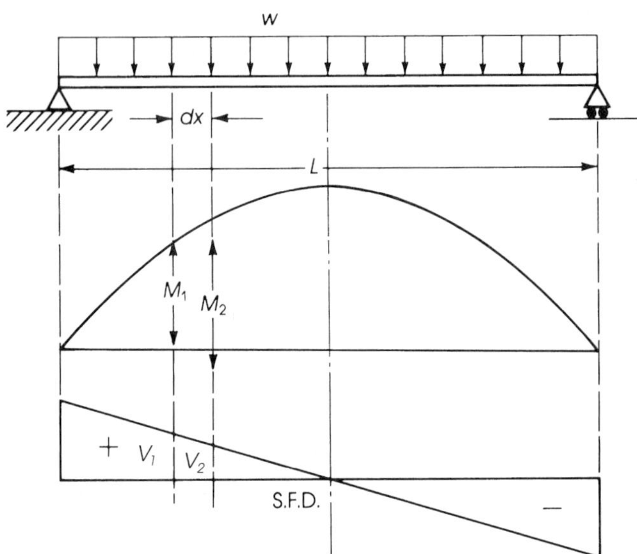

Figure 8.1 Bending moment and shearing force diagrams for a simple beam.

where

V = total shear at the section considered

Q = statical moment about the neutral axis of that portion of cross section lying between a line through the point in question parallel to the neutral axis and nearest face, upper or lower, of the beam

I = moment of inertia of cross section about the neutral axis

b = width of beam at the given point

The distribution of bending and shear stresses according to elastic theory for a homogeneous rectangular beam is as shown in Figure 8.2. The bending stresses are calculated from the flexural formula $f = Mc/I$, whereas the shear stress at any point is calculated by the shear formula of equation (8.1). The maximum shear stress is at the neutral axis and is equal to $1.5v_a$ (average shear), where $v_a = V/bh$. The shear stress curve is parabolic.

For a singly reinforced concrete beam, the distribution of shear stress above the neutral axis is a parabolic curve. Below the neutral axis, the maximum shear stress is maintained down to the level of the tension steel, because there is no change in the tensile force down to this point and the concrete in tension is neglected. The shear stress below the

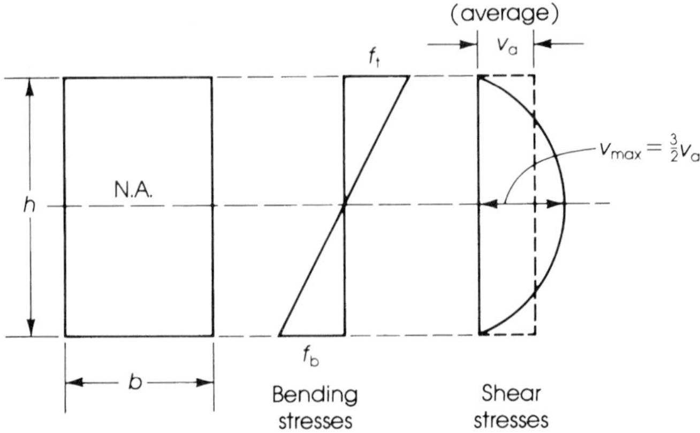

Figure 8.2 Bending and shear stresses in a homogeneous beam, according to elastic theory.

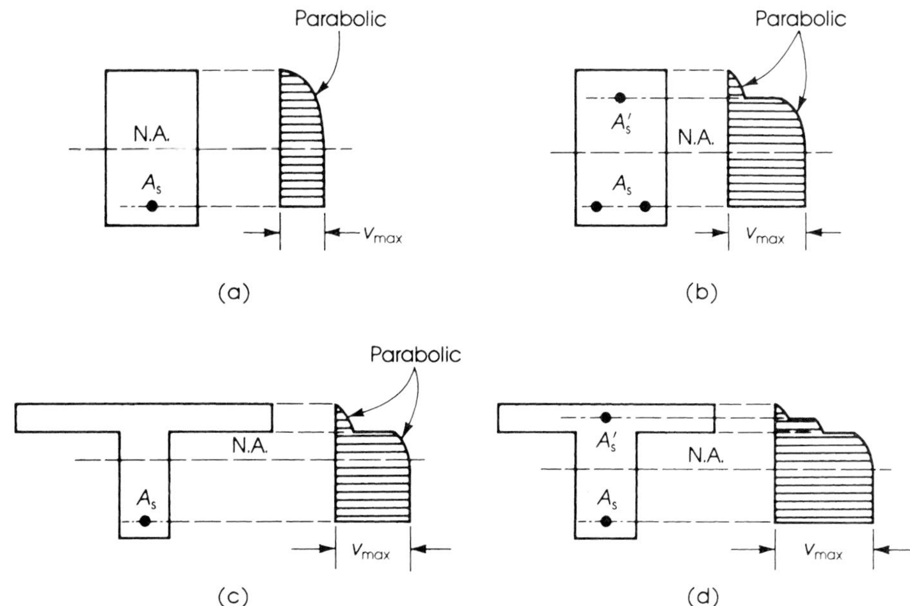

Figure 8.3 Distribution of shear stresses in reinforced concrete beams: (a) singly reinforced, (b) doubly reinforced, (c) T-section, (d) T-section with compression steel.

tension steel is zero (Figure 8.3). For doubly reinforced and T-sections, the distribution of shear stresses is as shown in Figure 8.3.

It can be observed that almost all the shear force is resisted by the web, whereas the flange resists a very small percentage; in most practical problems, the shear capacity of the flange is neglected.

Referring to Figure 8.1 and taking any portion of the beam dx, the bending moments at both ends of the element, M_1 and M_2, are not equal. Because $M_2 > M_1$ and to maintain the equilibrium of the beam portion dx, the compression force C_2 must be greater than C_1. Consequently, a shear stress v develops along any horizontal section *a-a* or *b-b* (Figure 8.4(a)). The normal and shear stresses on a small element at levels *a-a* and *b-b* are shown in Figure 8.4(b). Notice that the normal stress at the level of the neutral axis *b-b* is zero, whereas the shear stress is maximum. The horizontal shear stress is equal to the vertical shear stress, as shown in Figure 8.4(b). When the normal stress f is zero or low, a case of pure shear may occur. In this case, the maximum tensile stress, f_t, acts at 45° (Figure 8.4(c)).

The tensile stresses are equivalent to the principal stresses, as shown in Figure 8.4(d). Such principal stresses are traditionally called *diagonal tension stresses*. When the diagonal tension stresses reach the tensile strength of concrete, a diagonal crack develops. This brief analysis explains the concept of diagonal tension and diagonal cracking. The actual behavior is more complex, and it is affected by other factors, as explained later. For the combined action of shear and normal stresses at any point in a beam, the maximum and minimum diagonal tension (principal stresses) are given by the equation

$$f_p = \frac{f}{2} \pm \sqrt{\left(\frac{f}{2}\right)^2 + v^2} \tag{8.2}$$

where

f = intensity of normal stress due to bending
v = shear stress

The shear failure in a concrete beam is most likely to occur where shear forces are maximum, generally near the supports of the member. The first evidence of impending failure is the formation of diagonal cracks.

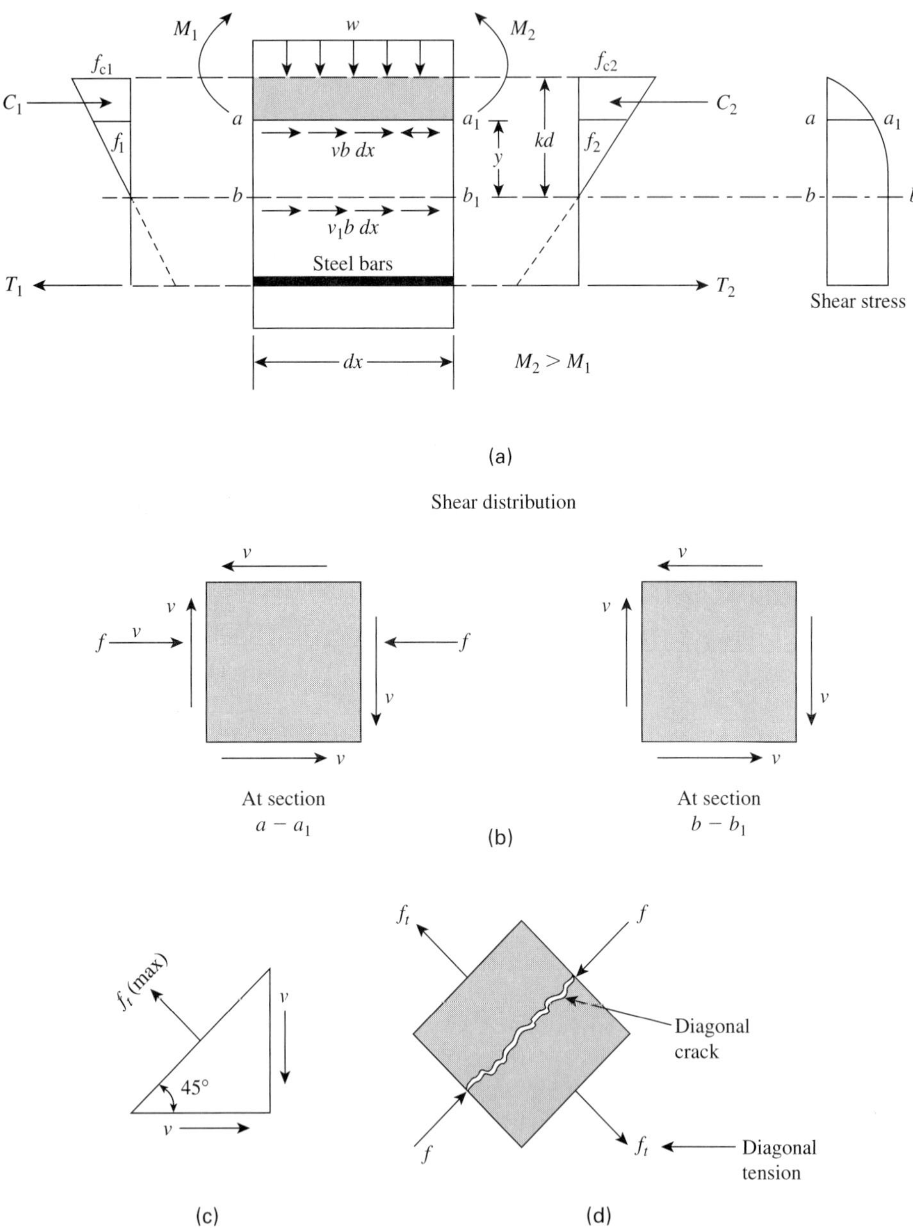

Figure 8.4 (a) Forces and stresses along the depth of the section, (b) normal and shear stresses, (c) pure shear, and (d) diagonal tension.

8.3 BEHAVIOR OF BEAMS WITHOUT SHEAR REINFORCEMENT

Concrete is weak in tension, and the beam will collapse if proper reinforcement is not provided. The tensile stresses develop in beams due to axial tension, bending, shear, torsion, or a combination of these forces. The location of cracks in the concrete beam depends on the direction of principal stresses. For the combined action of normal stresses and shear stresses, maximum diagonal tension may occur at about a distance d from the face of the support.

The behavior of reinforced concrete beams with and without shear reinforcement tested under increasing load was discussed in Section 3.3. In the tested beams, vertical flexural cracks developed at the section of maximum bending moment when the tensile stresses in concrete exceeded the modulus of rupture of concrete, or $f_r = 7.5\sqrt{f'_c}$. Inclined cracks in the web developed at a later stage at a location very close to the support.

An inclined crack occurring in a beam that was previously uncracked is generally referred to as a *web-shear crack*. If the inclined crack starts at the top of an existing flexural crack and propagates into the beam, the crack is referred to as *flexural-shear crack* (Figure 8.5). Web-shear cracks occur in beams with thin webs in regions with high shear and low moment. They are relatively uncommon cracks and may occur near the inflection points of continuous beams or adjacent to the supports of simple beams.

Flexural-shear cracks are the most common type found in reinforced concrete beams. A flexural crack extends vertically into the beam; then the inclined crack forms, starting from the top of the beam when shear stresses develop in that region. In regions of high shear stresses, beams must be reinforced by stirrups or by bent bars to produce ductile beams that do not rupture at a failure. To avoid a shear failure before a bending failure, a greater factor of safety must be provided against a shear failure. The ACI Code specifies a capacity reduction factor ϕ of 0.85 for shear and 0.9 for flexure.

Shear resistance in reinforced concrete members is developed by a combination of the following mechanisms [2] (Figure 8.5):

- Shear resistance of the uncracked concrete, V_z [3]
- Interface shear transfer, V_a, due to aggregate interlock tangentially along the rough surfaces of the crack [3]
- Arch action [4]
- Dowel action, V_d, due to the resistance of the longitudinal bars to the transverse shearing force [5]

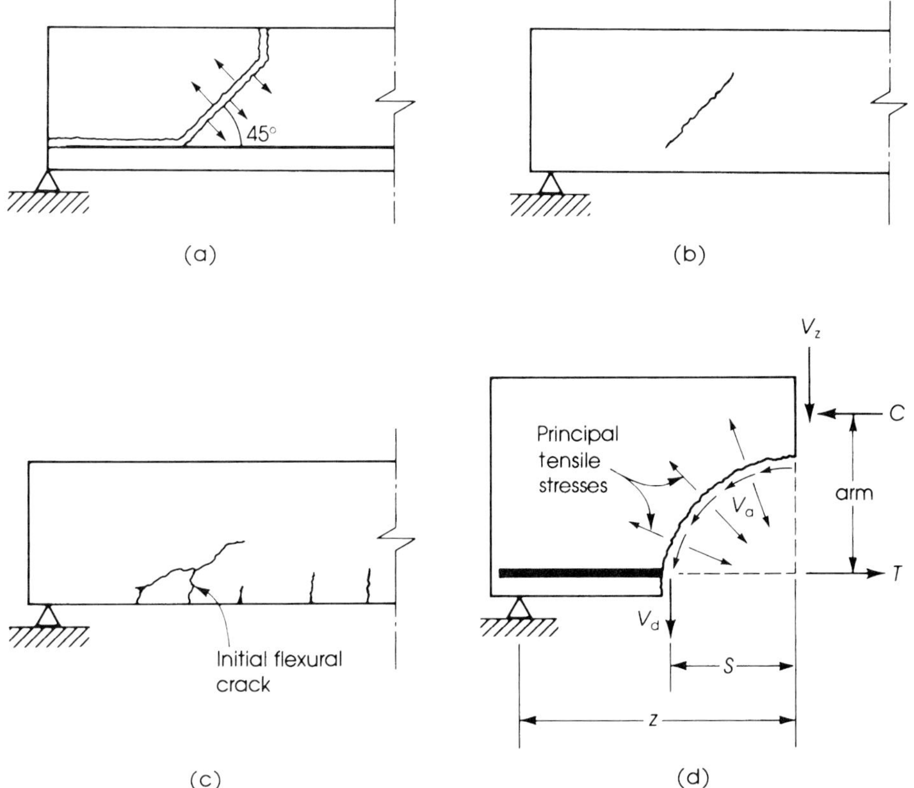

Figure 8.5 Shear failure: (a) general form, (b) web-shear crack, (c) flexural-shear crack, (d) analysis of forces involved in shear. V_a is interface shear, V_z is shear resistance, and V_d is dowel force.

In addition to these forces, shear reinforcement increases the shear resistance by V_s, which depends on the diameter and spacings of stirrups used in the concrete member. If shear reinforcement is not provided in a rectangular beam, the proportions of the shear resisted by the various mechanisms are 20% to 40% by V_z, 35% to 50% by V_a, and 15% to 25% by V_d [6].

8.4 MOMENT EFFECT ON SHEAR STRENGTH

In simply supported beams under uniformly distributed load, the midspan section is subjected to a large bending moment and zero or small shear, whereas sections near the ends are subjected to large shear and small bending moments (Figure 8.1). The shear and moment values are both high near the intermediate supports of a continuous beam. At a location of large shear force and small bending moment, there will be little flexural cracking, and an average stress v is equal to V/bd. The diagonal tensile stresses are inclined at about 45° (Figure 8.4(c)). Diagonal cracks can be expected when the diagonal tensile stress in the vicinity of the neutral axis reaches or exceeds the tensile strength of concrete. In general, the ultimate shear strength varies between $3.5\sqrt{f'_c}$ and $5\sqrt{f'_c}$. After completing a large number of beam tests on shear and diagonal tension [1], it was found that in regions with large shear and small moment, diagonal tension cracks were formed at an average shear force of

$$V_{cr} = 3.5\sqrt{f'_c}\, b_w d \tag{8.3}$$

where b_w is the width of the web in a T-section or the width of a rectangular section and d is the effective depth of the beam.

In locations where shear forces and bending moments are high, flexural cracks are formed first. At a later stage, some cracks bend in a diagonal direction when the diagonal tension stress at the upper end of such cracks exceeds the tensile strength of concrete. Given the presence of large moments on a beam, for which adequate reinforcement is provided, the nominal shear force at which diagonal tension cracks develop is given by

$$V_{cr} = 1.9\sqrt{f'_c}\, b_w d \tag{8.4}$$

This value is a little more than half of the value in equation (8.3) when bending moment is very small. This means that large bending moments reduce the value of shear stress for which cracking occurs. The following equation has been suggested to predict the nominal shear stress at which a diagonal crack is expected [1]:

$$v_{cr} = \frac{V}{b_w d} = \left(1.9\sqrt{f'_c} + 2500\rho\frac{Vd}{M}\right) \leq 3.5\sqrt{f'_c} \tag{8.5}$$

Shear failure near a middle support.

The ACI Code, Section 11.3.2, adopted this equation for the nominal ultimate shear force to be resisted by concrete:

$$V_c = \left[1.9\sqrt{f'_c} + 2500\rho_w \frac{V_u d}{M_u}\right] b_w d \leq (3.5\sqrt{f'_c}) b_w d \tag{8.6}$$

where $\rho_w = A_s/b_w d$ and b_w are the web width in a T-section or the width of a rectangular section and V_u and M_u are the ultimate shearing force and bending moment occurring simultaneously on the considered section.

The value of $V_u d/M_u$ must not exceed 1.0 in equation (8.6).

If M_u is large in equation (8.6), the second term becomes small and v_c approaches $1.9\sqrt{f'_c}$. If M_u is small, the second term becomes large and the upper limit of $3.5\sqrt{f'_c}$ controls. As an alternate to equation (8.6), the ACI Code, Section 11.3.1, permits evaluating the shear strength of concrete as follows:

$$V_c = (2\sqrt{f'_c}) b_w d \tag{8.7}$$

$$V_c = (0.17\sqrt{f'_c}) b_w d \qquad \text{(SI)}$$

In general, the value of $\sqrt{f'_c}$ shall not exceed 100 psi except in special conditions given in Section 11.1.2 in the code. For members subjected simultaneously to axial force and shear, the magnitude of the diagonal tension is modified. Axial compression will decrease the tensile principal stresses caused by shear, thus increasing the concrete shear capacity. Axial tension, on the other hand, will increase the principal stresses caused by shear, thus reducing the net concrete shear capacity. To compute the nominal shear strength of concrete V_c for a cross section subjected to axial force N_u and shear V_u simultaneously, the ACI Code, Section 11.3.2, presents the following equations:

1. In the case of an axial compression force N_u,

$$V_c = \left(1.9\sqrt{f'_c} + 2500\rho_w \frac{V_u d}{M_m}\right) b_w d \tag{8.8}$$

$$M_m = M_u - N_u\left(\frac{4h - d}{8}\right)$$

where $\rho_w = \dfrac{A_s}{b_w d}$

h = overall depth

$V_u d/M_u$ may be greater than 1.0, but V_c must not exceed

$$V_c = b_w d(3.5\sqrt{f'_c})\sqrt{1 + \frac{N_u}{500 A_g}} \tag{8.9}$$

where A_g is the gross area in in.^{2}

Alternatively, V_c may be computed by

$$V_c = b_w d\left(2 + 0.001\frac{N_u}{A_g}\right)\sqrt{f'_c} \tag{8.10}$$

2. In the case of an axial tensile force N_u,

$$V_c = b_w d\left(2 + 0.004\frac{N_u}{A_g}\right)\sqrt{f'_c} \tag{8.11}$$

where N_u is to be taken as negative for tension and N_u/A_g is in psi. If V_c is negative, V_c should be taken equal to zero.

8.5 BEAMS WITH SHEAR REINFORCEMENT

Different types of shear reinforcement may be used:

1. Stirrups, which can be placed either perpendicular to the longitudinal reinforcement or inclined, usually making a 45° angle and welded to the main longitudinal reinforcement. Vertical stirrups, using no. 3 or no. 4 U-shaped bars, are the most commonly used shear reinforcement in beams (Figure 8.6(a)).
2. Bent bars, which are part of the longitudinal reinforcement, bent up (where they are no longer needed) at an angle of 30° to 60°, usually at 45°.
3. Combinations of stirrups and bent bars.
4. Welded wire fabric with wires perpendicular to the axis of the member.
5. Spirals, circular ties, or hoops in circular sections, as columns.

The shear strength of a reinforced concrete beam is increased by the use of shear reinforcement. Prior to the formation of diagonal tension cracks, shear reinforcement contributes

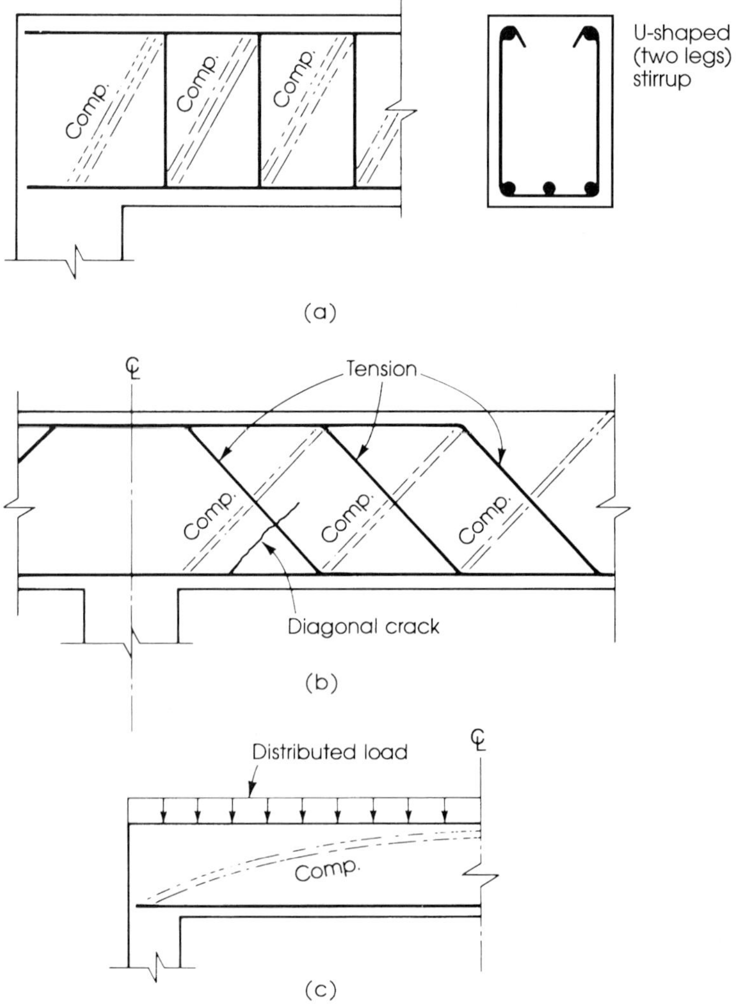

Figure 8.6 Truss action of web reinforcement with (a) stirrups, (b) bent bars, and (c) tension steel.

very little to the shear resistance. After diagonal cracks have developed, shear reinforcement augments the shear resistance of a beam, and a redistribution of internal forces occurs at the cracked section. When the amount of shear reinforcement provided is small, failure due to yielding of web steel may be expected, but if the amount of shear reinforcement is too high, a shear-compression failure may be expected, which should be avoided.

Concrete, stirrups, and bent bars act together to resist transverse shear. The concrete, by virtue of its high compressive strength, acts as the diagonal compression member of a lattice girder system, where the stirrups act as vertical tension members. The diagonal compression force is such that its vertical component is equal to the tension force in the stirrup. Bent-up reinforcement acts also as tension members in a truss, as shown in Figure 8.6.

In general, the contribution of shear reinforcement to the shear strength of a reinforced concrete beam can be described as follows [2]:

1. It resists part of the shear, V_s.
2. It increases the magnitude of the interface shear, V_a (Figure 8.5), by resisting the growth of the inclined crack.
3. It increases the dowel force, V_d (Figure 8.5), in the longitudinal bars.
4. The confining action of the stirrups on the compression concrete may increase its strength.
5. The confining action of stirrups on the concrete increases the rotation capacity of plastic hinges that develop in indeterminate structures at ultimate load and increases the length over which yielding takes place [7].

The total nominal shear strength of beams with shear reinforcement V_n is due partly to the shear strength attributed to the concrete V_c and partly to the shear strength contributed by the shear reinforcement V_s:

$$V_n = V_c + V_s \tag{8.12}$$

The shear force V_u produced by factored loads must be less than or equal to the total nominal shear strength V_n, or

$$V_u \leq \phi V_n = \phi(V_c + V_s) \tag{8.13}$$

where $V_u = 1.4V_D + 1.7V_L$ and $\phi = 0.85$.

An expression for V_s may be developed from the truss analogy (Figure 8.7). For a 45° crack and a series of inclined stirrups or bent bars, the vertical shear force V_s resisted by

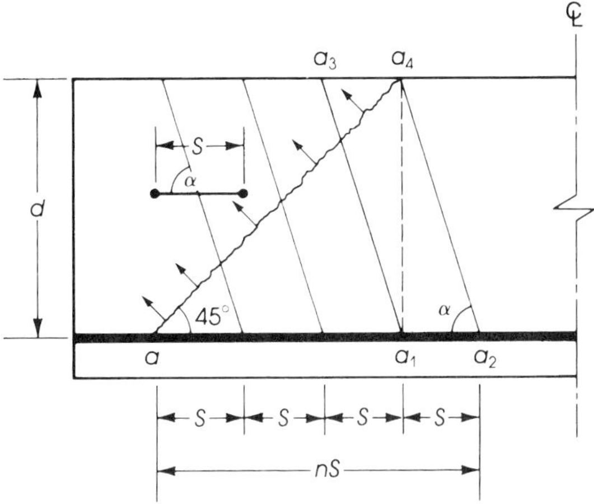

Figure 8.7 Factors in inclined shear reinforcement.

shear reinforcement is equal to the sum of the vertical components of the tensile forces developed in the inclined bars. Therefore,

$$V_s = nA_v f_y \sin\alpha \tag{8.14}$$

where A_v is the area of shear reinforcement with a spacing s and f_y is the yield strength of shear reinforcement. But ns is the distance aa_1a_2:

$$d = a_1a_4 = aa_1 \tan 45° \qquad \text{(from triangle } aa_1a_4\text{)}$$

$$d = a_1a_4 = a_1a_2 \tan\alpha \qquad \text{(from triangle } a_1a_2a_4\text{)}$$

$$ns = aa_1 + a_1a_2$$

$$= d(\cot 45° + \cot\alpha) = d(1 + \cot\alpha)$$

$$n = \frac{d}{s}(1 + \cot\alpha)$$

Substituting this value in equation (8.14) gives

$$V_s = \frac{A_v f_y d}{s} \sin\alpha(1 + \cot\alpha) = \frac{A_v f_y d}{s}(\sin\alpha + \cos\alpha) \tag{8.15}$$

For the case of vertical stirrups, $\alpha = 90°$ and

$$V_s = \frac{A_v f_y d}{s} \quad \text{or} \quad s = \frac{A_v f_y d}{V_s} \tag{8.16}$$

In the case of T-sections, b is replaced by the width of web b_w in all shear equations. When $\alpha = 45°$, equation (8.15) becomes

$$V_s = 1.4\left(\frac{A_v f_y d}{s}\right) \quad \text{or} \quad s = \frac{1.4 A_v f_y d}{V_s} \tag{8.17}$$

For a single bent bar or group of parallel bars in one position, the shearing force resisted by steel is

$$V_s = A_v f_y \sin\alpha \quad \text{or} \quad A_v = \frac{V_s}{f_y \sin\alpha} \tag{8.18}$$

For $\alpha = 45°$,

$$A_v = 1.4\left(\frac{V_s}{f_y}\right) \tag{8.19}$$

For circular sections, mainly in columns, V_s shall be computed from equation (8.16) using $d = 0.8 \times$ diameter and $A_v =$ two times the area of the bar in a circular tie, hoop, or spiral.

8.6 ACI CODE SHEAR DESIGN REQUIREMENTS

8.6.1 Critical Section for Nominal Shear Strength Calculation

The ACI Code, Section 11.1.3, permits taking the critical section for nominal shear strength calculation at a distance d from the face of the support. This recommendation is based on the fact that the first inclined crack is likely to form within the shear span of the beam at some distance d away from the support. The distance d is also based on experimental work and appeared in the testing of the beams discussed in Chapter 3. This critical section is permitted on the condition that the support reaction introduces compression into the end region, loads are applied at or near the top of the member, and no concentrated load occurs between the face of the support and the location of the critical section.

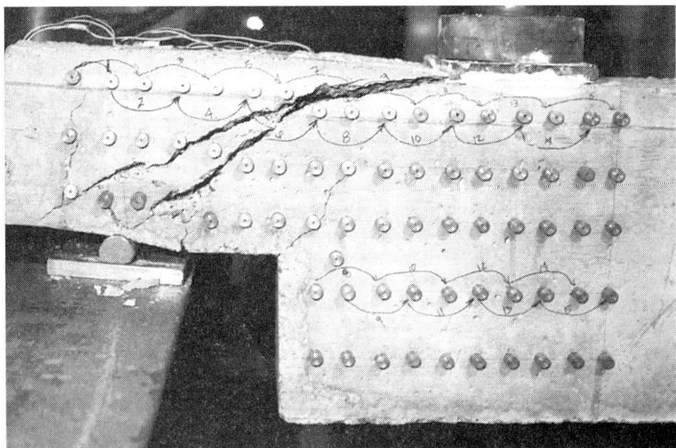

Shear failure in a dapped-end beam.

The Code also specifies that shear reinforcement must be provided between the face of the support and the distance d, using the same reinforcement adopted for the critical section.

8.6.2 Minimum Area of Shear Reinforcement

The presence of shear reinforcement in a concrete beam restrains the growth of inclined cracking. Moreover, ductility is increased, and a warning of failure is provided. If shear reinforcement is not provided, brittle failure will occur without warning. Accordingly, a minimum area of shear reinforcement is specified by the code. The ACI Code, Section 11.5.5, requires all stirrups to have a minimum shear reinforcement area, A_v, equal to

$$A_v = \frac{50 b_w s}{f_y} \tag{8.20}$$

where b_w is the width of the web and s is the spacing of the stirrups. The minimum amount of shear reinforcement is required whenever V_u exceeds $\phi V_c/2$, except in

- Slabs and footings;
- Concrete floor joist construction;
- Beams where the total depth does not exceed 10 in., 2.5 times the flange thickness for T-shaped flanged sections, or one-half the web width, whichever is greatest.

It is a common practice to increase the depth of a slab, footing, or shallow beam to increase its shear capacity. Stirrups may not be effective in shallow members, because their compression zones have relatively small depths and may not satisfy the anchorage requirements of stirrups. For beams that are not shallow, reinforcement is not required when V_u is less than $\phi V_c/2$.

The minimum shear reinforcement area can be achieved by using no. 3 stirrups placed at maximum spacing, $S_{\max}$. If $f_y = 60$ Ksi and U-shaped (two legs) no. 3 stirrups are used, then equation (8.20) becomes

$$S_{\max} = \frac{A_v f_y}{50 b_w} \tag{8.21}$$

or

$$S_{\max}(\text{in.}) = \frac{0.22(60{,}000)}{50 b_w} = \frac{264}{b_w} \tag{8.22}$$

Table 8.1 Values of $S_{max} = (A_v f_y/50b_w) \le 24$ in. when $f_y = 60$ Ksi

b_w (in.)	10	11	12	13	14	15	16	18	20	22	24	b_w
S_{max} (in.) no. 3 stirrups	24	24	22	20.3	18.9	17.6	16.5	14.7	13.2	12	11	$\frac{264}{b_w}$
S_{max} (in.) no. 4 stirrups	24	24	24	24	24	24	24	24	24	21.8	20	$\frac{480}{b_w}$

If U-shaped no. 4 stirrups are used, then

$$S_{max}(\text{in.}) = \frac{0.4(60{,}000)}{50b_w} = \frac{480}{b_w} \tag{8.23}$$

Note that S_{max} shall not exceed 24 in.

Table 8.1 gives S_{max} based on equations (8.22) and (8.23). Final spacings should be rounded to the lower inch. For example, $S = 20.3$ in. becomes 20 in.

8.6.3 Maximum Shear Carried by Web Reinforcement V_s

To prevent a shear-compression failure, where the concrete may crush due to high shear and compressive stresses in the critical region on top of a diagonal crack, the ACI Code, Section 11.5.6.8, requires that V_s shall not exceed $(8\sqrt{f'_c})b_w d$. If V_s exceeds this value, the section should be increased. Based on this limitation,

- If $f'_c = 3$ Ksi, then $V_s \le 0.438 b_w d$ (kips) or $V_s/b_w d \le 438$ psi;
- If $f'_c = 4$ Ksi, then $V_s \le 0.506 b_w d$ (kips) or $V_s/b_w d \le 506$ psi;
- If $f'_c = 5$ Ksi, then $V_s \le 0.565 b_w d$ (kips) or $V_s/b_w d \le 565$ psi.

8.6.4 Maximum Spacing of Stirrups

To ensure that a diagonal crack will always be intersected by at least one stirrup, the ACI Code, Section 11.5.4, requires that the spacings between stirrups shall not exceed $d/2$, provided that $V_s \le (4\sqrt{f'_c})b_w d$. This is based on the assumption that a diagonal crack develops at 45° and extends a horizontal distance of about d. In regions of high shear, where V_s exceeds $(4\sqrt{f'_c})b_w d$, the maximum spacing between stirrups must not exceed $d/4$. This limitation is necessary to ensure that the diagonal crack will be intersected by at least three stirrups. When V_s exceeds the maximum value of $8\sqrt{f'_c}\, b_w d$, this limitation of maximum stirrup spacing does not apply, and the dimensions of the concrete cross section should be increased.

A second limitation for the maximum spacing of stirrups may also be obtained from the condition of minimum area of shear reinforcement. A minimum A_v is obtained when the spacing s is maximum, or maximum $s = A_v f_y/50b_w$ (equation (8.21)).

A third limitation for maximum spacing is 24 in. when $V_s \le (4\sqrt{f'_c})b_w d$ and 12 in. when V_s is greater than $(4\sqrt{f'_c})b_w d$ but less than or equal to $(8\sqrt{f'_c})b_w d$. The least value of all maximum spacings must be adopted. The ACI Code maximum spacing requirement ensures closely spaced stirrups that hold the longitudinal tension steel in place within the beam, thereby increasing their dowel capacity, V_d (Figure 8.5).

8.6.5 Yield Strength of Shear Reinforcement

The ACI Code, Section 11.5.2, requires that the design yield strength of shear reinforcement shall not exceed 60 Ksi (420 MPa). The reason behind this decision is to limit the crack width caused by the diagonal tension and to ensure that the sides of the crack remain in close contact to improve the interface shear transfer V_a (Figure 8.5). For welded deformed wire fabric, the design yield strength shall not exceed 80 Ksi (560 MPa).

8.6.6 Anchorage of Stirrups

The ACI Code, Section 12.13, requires that shear reinforcement be carried as close as possible to the compression and tension extreme fibers, within the code requirements for concrete cover, because near ultimate load the flexural tension cracks penetrate deep in the beam. Also, for stirrups to achieve their full yield strength, they must be well anchored. Near ultimate load, the stress in a stirrup reaches its yield stress at the point where a diagonal crack intercepts that stirrup. The ACI Code requirements for stirrup anchorage, Section 12.13, are as follows:

1. Each bend in the continuous portion of a simple U-stirrup or multiple U-stirrup shall enclose a longitudinal bar (ACI Code, Section 12.13.3). See Figure 8.8(a).
2. The code allows the use of a standard hook of 90°, 135°, or 180° around longitudinal bars for no. 5 bars or D31 wire stirrups. If no. 6, 7, or 8 stirrups with $f_y > 40$ Ksi are used, the code (Section 12.13.2) requires a standard hook plus an embedment length of $0.014 d_b f_y / \sqrt{f'_c}$ between midheight of the member and the outside of the hook. If the bars are bent at 90°, extensions shall not be less than $12d_b$. For no. 5 bars or smaller stirrups, the extension is $6d_b$ (ACI Code, Section 7.1). See Figure 8.8.
3. If spliced double U-stirrups are used to form closed stirrups, the lap length shall not be less than $1.3l_d$ (ACI Code, Section 12.13.5). See Figure 8.8(c).
4. Welded wire fabric is used for shear reinforcement in the precast industry. Anchorage details are given in the ACI Code, Section 12.13.2.3, and in its commentary.
5. Closed stirrups are required for beams subjected to torsion or stress reversals (ACI Code, Section 7.11).
6. Beams at the perimeter of the structure should contain closed stirrups to maintain the structural integrity of the member (ACI Code, Section 7.13.2.2).

8.6.7 Stirrups Adjacent to the Support

The ACI Code, Section 11.1.3, specifies that shear reinforcement provided between the face of the support and the critical section at a distance d from it may be designed for the same shear V_u at the critical section. It is common practice to place the first stirrup at a distance $s/2$ from the face of the support, where s is the spacing calculated by equation (8.16) for V_u at the critical section.

8.6.8 Effective Length of Bent Bars

Only the center three-fourths of the inclined portion of any longitudinal bar shall be considered effective for shear reinforcement. This means that the maximum spacing of bent bars is $0.75(d - d')$. From Figure 8.9, the effective length of the bent bar is $0.75(d - d')/(\sin 45°) = 0.75(1.414)(d - d') = 1.06(d - d')$. The maximum spacing s is equal to the horizontal projection of the effective length of the bent bar. Thus $S_{max} = 1.06(d - d') \cos 45°$, or $S_{max} = 0.707[1.06(d - d')] = 0.75(d - d')$.

8.7 DESIGN OF VERTICAL STIRRUPS

Stirrups are needed when $V_u > \frac{1}{2}\phi V_c$. Minimum stirrups are used when V_u is greater than $\frac{1}{2}\phi V_c$ but less than ϕV_c. This is achieved by using no. 3 stirrups placed at maximum spacing. When V_u is greater than ϕV_c, then stirrups must be provided. The spacing of stirrups may be

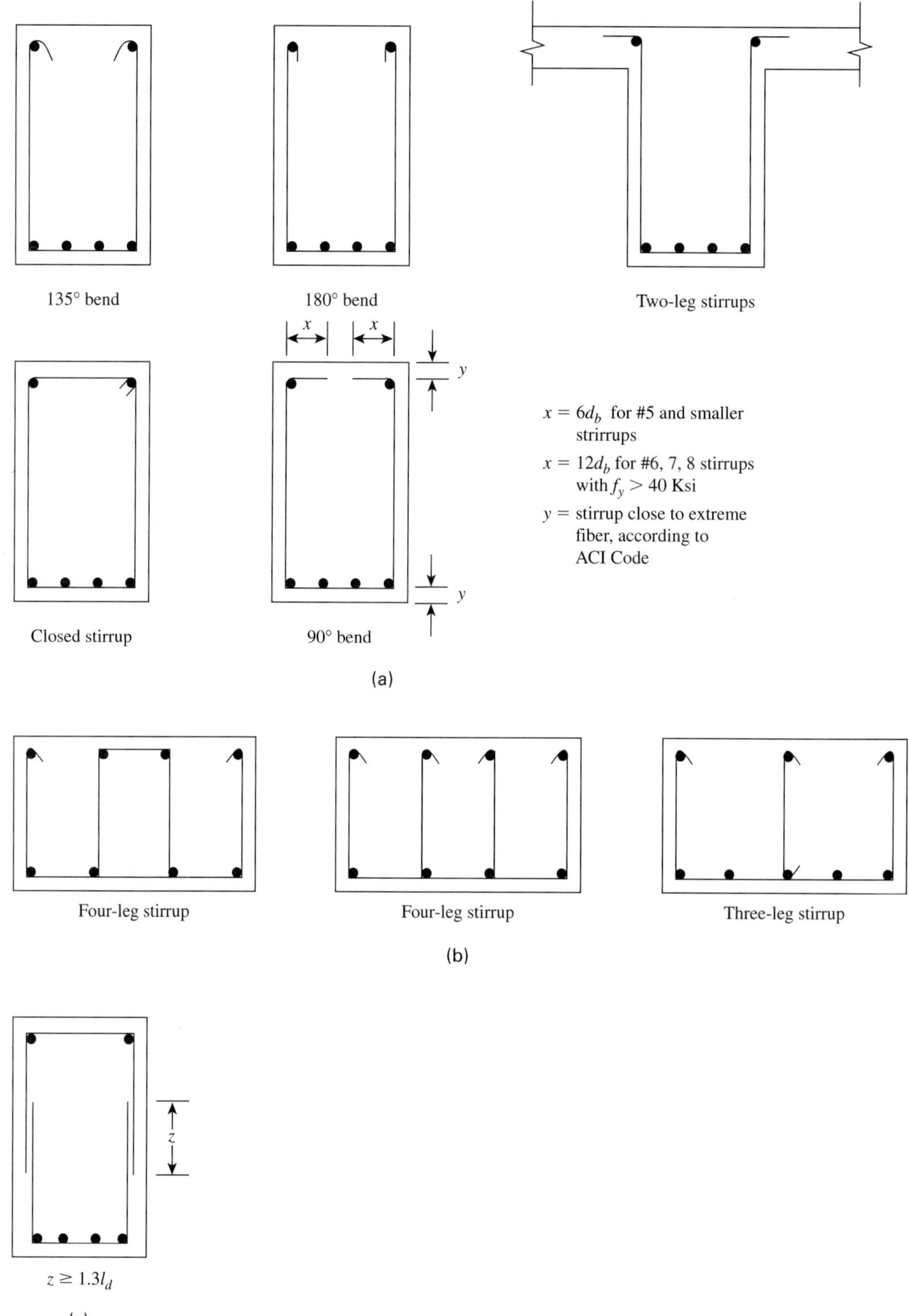

Figure 8.8 Stirrup types: (a) U-stirrups enclosing longitudinal bars, anchorage lengths, and closed stirrups, (b) multileg stirrups, and (c) spliced stirrups.

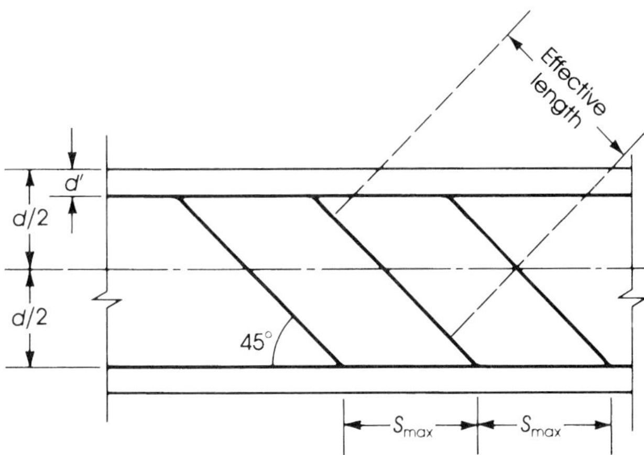

Figure 8.9 Effective length and spacing of bent bars.

less than the maximum spacing and can be calculated using equation (8.16): $S = A_v f_y d/V_s$. The stirrups that are commonly used in concrete sections are made of two-leg no. 3 or no. 4 U-stirrups with $f_y = 60$ Ksi. If no. 3 stirrups are used, then equation (8.16) becomes

$$\frac{S}{d} = \frac{A_v f_y}{V_s} = \frac{0.22(60)}{V_s} = \frac{13.2}{V_s} \tag{8.24}$$

If no. 4 stirrups are used, then

$$\frac{S}{d} = \frac{A_v f_y}{V_s} = \frac{0.4(60)}{V_s} = \frac{24}{V_s} \tag{8.25}$$

The ratio of stirrup spacings relative to the effective depth of the beam, d, depends on V_s. The values of S/d for different values of V_s when $f_y = 60$ Ksi are given in Tables 8.2 and 8.3 for no. 3 and no. 4 U-stirrups, respectively. The same values are plotted in Figures 8.10 and 8.11. The following observations can be made:

1. If no. 3 stirrups are used, $S = d/2$ when $V_s \leq 26.4$ K. When V_s increases, S/d decreases in a nonlinear curve to reach 0.132 at $V_s = 100$ K. If the minimum spacing is limited to 3 in., then d must be equal to or greater than 22.7 in. to maintain that 3-in. spacing. When V_s is equal to or greater than 52.8 K, then $S \leq d/4$.
2. If no. 4 U-stirrups are used, $S = d/2$ when $V_s \leq 48$ K. When V_s increases, S/d decreases to reach 0.16 at $V_s = 150$ K. If the minimum spacing is limited to 3 in., then $d \geq 18.75$ in. to maintain the 3-in. spacing. When V_s is equal to or greater than 96 K, then $S \leq d/4$.
3. If grade 40 U-stirrups are used ($f_y = 40$ Ksi), multiply the S/d values by $\frac{2}{3}$ or, in general, $f_y/60$.

Table 8.2 S/d Ratio for Different Values of V_s ($f_y = 60$ Ksi, $S/d = 13.2/V_s$), no. 3 Stirrups

V_s (K)	26.4	30	40	50	52.8	60	70	80	90	100	125
S/d	0.5	0.44	0.33	0.264	0.25	0.22	0.19	0.165	0.15	0.132	0.106

Table 8.3 S/d Ratio for Different Values of V_s ($f_y = 60$ Ksi, $S/d = 24/V_s$), no. 4 Stirrups

V_s (K)	48	50	60	70	80	90	96	100	110	120	150	175
S/d	0.50	0.48	0.40	0.34	0.3	0.27	0.25	0.24	0.22	0.20	0.16	0.137

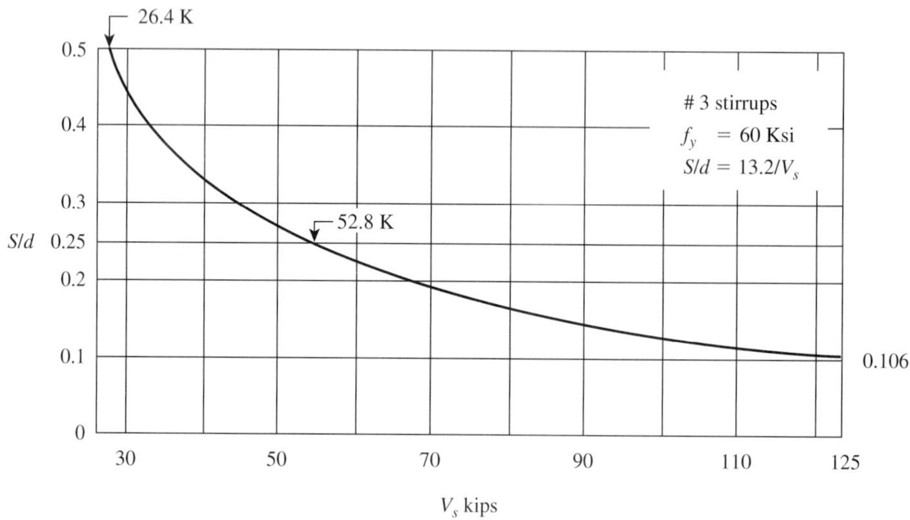

Figure 8.10 V_s versus S/d for no. 3 stirrups and $f_y = 60$ Ksi.

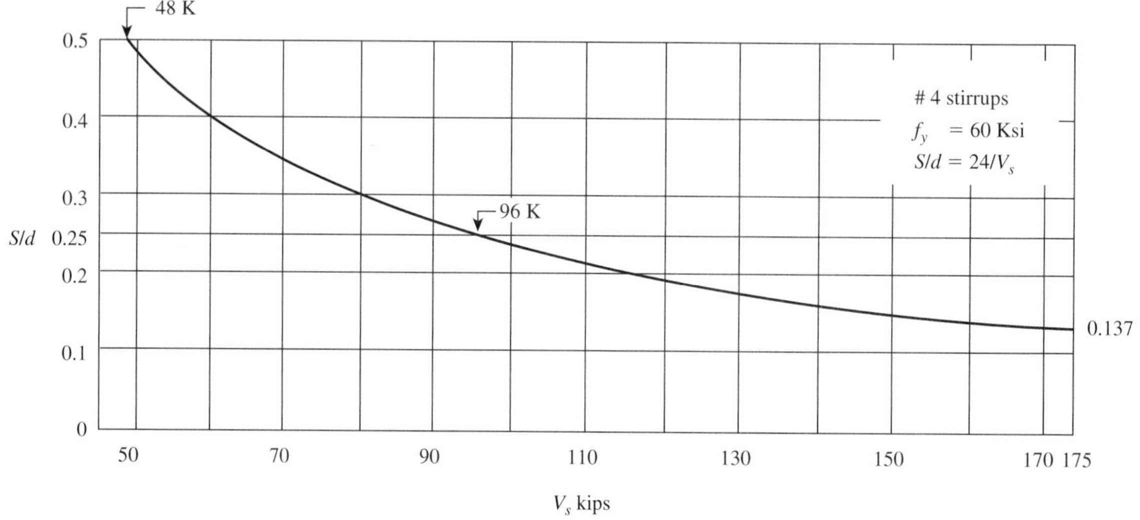

Figure 8.11 V_s versus S/d for no. 4 stirrups and $f_y = 60$ Ksi.

8.8 DESIGN SUMMARY

The design procedure for shear using vertical stirrups according to the ACI Code can be summarized as follows:

1. Calculate the ultimate shearing force V_u from the applied forces acting on the structural member. The critical design shear value is at a section located at a distance d from the face of the support.

2. Calculate $\phi V_c = (2\phi\sqrt{f'_c})b_w d$, or

$$\phi V_c = \left[1.9\sqrt{f'_c} + 2500\rho_w \frac{V_u d}{M_u}\right] b_w d \le (3.5\sqrt{f'_c})b_w d$$

Then calculate $\frac{1}{2}\phi V_c$.

3. a. If $V_u < \frac{1}{2}\phi V_c$, no shear reinforcement is needed.

b. If $\frac{1}{2}\phi V_c < V_u \le \phi V_c$, minimum shear reinforcement is required. Use no. 3 U-stirrups spaced at maximum spacings, as explained in Step 7.

c. If $V_u > \phi V_c$, shear reinforcement must be provided according to Steps 4 through 8.

4. If $V_u > \phi V_c$, calculate the shear to be carried by shear reinforcement:

$$V_u = \phi V_c + \phi V_s \quad \text{or} \quad V_s = \frac{V_u - \phi V_c}{\phi}$$

5. Calculate $V_{c1} = (4\sqrt{f'_c})b_w d$ and $V_{c2} = (8\sqrt{f'_c})b_w d = 2V_{c1}$. Compare the calculated V_s with the maximum permissible value of $V_{c2} = (8\sqrt{f'_c})b_w d$. If V_s is less than V_{c2}, proceed in the design; if not, increase the dimensions of the section.

6. Calculate the stirrup spacings based on the calculated V_s. $S_1 = A_v f_y d/V_s$, or use Figures 8.10 and 8.11 or Tables 8.2 and 8.3.

7. Determine the maximum spacing allowed by the ACI Code. The maximum spacing is the least of S_2 and S_3:

a. $S_2 = d/2 \le 24$ in. if $V_s \le V_{c1} = (4\sqrt{f'_c})b_w d$ or
$S_2 = d/4 \le 12$ in. if $V_{c1} < V_s \le V_{c2}$.

b. $S_3 = A_v f_y/50b_w$.

S_{max} is the smaller of S_2 and S_3. Values of S_3 are shown in Table 8.1.

8. If S_1 calculated in Step 6 is less than S_{max} (the smaller of S_2 and S_3), then use S_1 to the nearest smaller $\frac{1}{2}$ in. If $S_1 > S_{max}$, then use S_{max} as the adopted S.

9. The ACI Code did not specify a minimum spacing. Under normal conditions, a practical minimum S may be assumed equal to 3 in. for $d \le 20$ in. and 4 in. for deeper beams. If S is considered small, either increase the stirrup bar number or use multiple-leg stirrups (Figure 8.8).

10. For Circular Sections, the area used to compute V_c = diameter times the effective depth d, where $d = 0.8$ the diameter, ACI code, Section 11.3.3.

Example 8.1

A simply supported beam has a rectangular section $b = 12$ in., $d = 21.5$ in., and $h = 24$ in. and is reinforced with four no. 8 bars. Check if the section is adequate for each of the following ultimate shear forces. If it is not adequate, design the necessary shear reinforcement in the form of U-stirrups. Use $f'_c = 4$ Ksi and $f_y = 60$ Ksi.

(a) $V_u = 12$ K **(b)** $V_u = 25$ K **(c)** $V_u = 61$ K **(d)** $V_u = 87$ K **(e)** $V_u = 145$ K

Solution

In general, $b_w = b = 12$ in., $d = 21.5$ in., and

$$\phi V_c = \phi(2\sqrt{f'_c})bd = 0.85(2\sqrt{4000})(12)(21.5) = 27.74 \text{ K}$$

$$\tfrac{1}{2}\phi V_c = 13.87 \text{ K}$$

$$V_{c1} = (4\sqrt{f'_c})bd = (4\sqrt{4000})(12)(21.5)/1000 = 65.3 \text{ K}$$

$$V_{c2} = (8\sqrt{f'_c})bd = 130.6 \text{ K}$$

a. $V_u = 12 \text{ K} < \frac{1}{2}\phi V_c = 13.87$ K, section is adequate, shear reinforcement is not required.

b. $V_u = 25 \text{ K} > \frac{1}{2}\phi V_c$, but it is less than $\phi V_c = 27.74$ K. Therefore, $V_s = 0$ and minimum shear reinforcement is required. Choose no. 3 U-stirrup (two legs) at maximum spacing. $A_v = 2(0.11) = 0.22 \text{ in}^2$. Maximum spacing is the least of

1. $S_2 = d/2 = 21.5/2 = 10.75$ in., say, 10.5 in. (controls);

2. $S_3 = A_v f_y/50b_w = 0.22(60{,}000)/50(12) = 22$ in. (or use Table 8.1);

3. $S_4 = 24$ in. Use no. 3 U-stirrups spaced at 10.5 in.

c. $V_u = 61 \text{ K} > \phi V_c$. Shear reinforcement is needed. Calculation may be organized in steps:

1. Calculate $V_s = (V_u - \phi V_c)/\phi = (61 - 27.74)/0.85 = 39.1$ K.
2. Check if $V_s \le V_{c1} = (4\sqrt{f'_c})b_w d = 65.3$ K. Because $V_s < 65.3$ K, then $S_{max} \le d/2$, and the $d/4$ condition does not apply.
3. Choose no. 3 U-stirrups and calculate the required spacings based on V_s.

$$S_1 = \frac{A_v f_y d}{V_s} = \frac{0.22(60)(21.5)}{39.1} = 7.26 \text{ in.,} \qquad \text{say, 7 in.}$$

4. Calculate maximum spacings: $S_2 = 10.5$ in., $S_3 = 22$ in., and $S_4 = 24$ in. and maximum $S = 10.5$ in. (calculated in (b)).
5. Because $S = 7$ in. $< S_{max} = 10.5$ in., then use no. 3 U-stirrups spaced at 7 in.

d. $V_u = 87 \text{ K} > \phi V_c$, so stirrups must be provided.

1. Calculate $V_s = (V_u - \phi V_c)/\phi = (87 - 27.74)/0.85 = 69.72$ K.
2. Check if $V_s \le V_{c1} = (4\sqrt{f'_c})b_w d = 65.3$ K. Because $V_s > 65.3$ K, then $S_{max} \le d/4 \le 12$ in. must be used.
3. Check if $V_s \le V_{c2} = (8\sqrt{f_c})b_w d = 130.6$ K. Because $V_{c1} < V_s < V_{c2}$, then stirrups can be used without increasing the section.
4. Choose no. 3 U-stirrups and calculate S_1 based on V_s:

$$S_1 = \frac{A_v f_y d}{V_s} = \frac{0.22(60)(21.5)}{69.72} = 4.1 \text{ in.,} \qquad \text{say, 4 in.}$$

5. Calculate maximum spacings: $S_2 = d/4 = 21.5/4 = 5.3$ in., say, 5.0 in.; $S_3 = 22$ in.; and $S_4 = 12$ in. Hence $S_{max} = 5$-in. controls.
6. Because $S = 4$ in. $< S_{max} = 5$ in., then use no. 3 stirrups spaced at 4 in.

e. $V_u = 145 \text{ K} > \phi V_c$, so shear reinforcement is required.

1. Calculate $V_s = (V_u - \phi V_c)/\phi = (145 - 27.74)/0.85 = 138$ K.
2. Because $V_s > V_{c2} = 130.2$ K, the section is not adequate. Increase one or both dimensions of the beam section.

Notes: Table 8.2 and Figure 8.10 can be used to calculate the spacing S for (c) and (d).

1. For (c): $V_s = 39.1$ K, from Figure 8.10 (or Table 8.2 for no. 3 U-stirrups), $S/d = 0.34$ and $S_1 = 7.3$ in., which is less than $d/2 = 10.5$ in. Note that S_{max} based on V_s is $d/2$ and not $d/4$. Also, from Table 8.1, $S_3 = A_v f_y / 50 b_w = 22$ in.
2. For (d): $V_s = 69.72$ K, $S/d = 0.19$ and $S_1 = 4.1$ in. $V_s = 69.72$ is greater than 52.8 K, and $S_{max} = d/4$ is required.

Example 8.2

A 17-ft-span simply supported beam has a clear span of 16 ft and carries a uniformly distributed dead and live loads of 4.5 K/ft and 3 K/ft, respectively. The dimensions of the beam section and steel reinforcement are shown in Figure 8.12. Check the section for shear and design the necessary shear reinforcement. Given $f'_c = 3$ Ksi and $f_y = 60$ Ksi.

Solution

Given b_w (web) $= 14$ in., $d = 22.5$ in.

1. Calculate ultimate shear from external loading:

$$\text{Ultimate uniform load} = 1.4(4.5) + 1.7(3) = 11.4 \text{ K/ft}$$

$$V_u \text{ (at face of support)} = \frac{11.4(16)}{2} = 91.2 \text{ K}$$

Design V_u (at d distance from the face of support) $= 91.2 - 22.5(11.4)/12 = 69.83$ K.

2. Calculate ϕV_c:

$$\phi V_c = \phi(2\sqrt{f'_c})b_w d = \frac{0.85(2\sqrt{3000})(14)(22.5)}{1000} = 29.33 \text{ K}$$

$$\tfrac{1}{2}\phi V_c = 14.67 \text{ K}$$

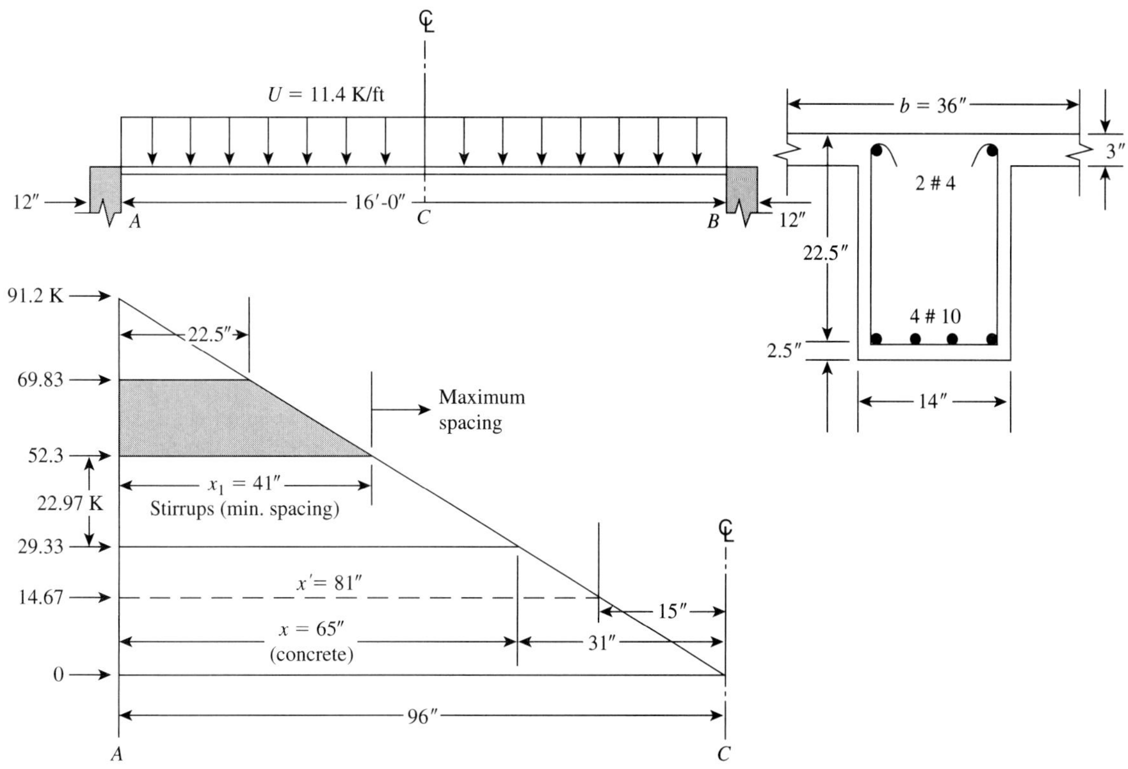

Figure 8.12 Example 8.2.

Calculate $V_{c1} = (4\sqrt{f'_c})b_w d = (4\sqrt{3000})(14)(22.5)/1000 = 69$ K. Calculate $V_{c2} = (8\sqrt{f'_c})b_w d = 138$ K.

3. Design $V_u = 69.83 \text{ K} > \phi V_c = 29.33$ K; therefore, shear reinforcement must be provided. The distance x' at which no shear reinforcement is needed $\left(\text{at } \frac{1}{2}\phi V_c\right)$ is

$$x' = \left(\frac{91.2 - 14.67}{91.2}\right)(8) = 6.71 \text{ ft} = 81 \text{ in.}$$

(from the triangles of shear diagram, Figure 8.12).

4. Calculate $V_s = (V_u - \phi V_c)/\phi = (69.83 - 29.33)/0.85 = 47.65$ K. Because V_s is less than $V_{c1} = (4\sqrt{f'_c})b_w d$, then $S_{max} \le d/2$ must be considered (or refer to Figure 8.10 or Table 8.2: $V_s < 52.8$ K).

5. Design of stirrups: Choose no. 3 U-stirrups, $A_v = 2(0.11) = 0.22 \text{ in.}^2$ Calculate S_1 based on $V_s = 47.65$ K, $S_1 = A_v f_y d/V_s = 13.2d/V_s = 6.23$ in., say, 6 in. (or get $s/d = 0.28$ from Table 8.2 or Figure 8.10).

6. Calculate maximum spacings: $S_2 = d/2 = 22.5/2 = 11.25$ in., say, 11.0 in.; $S_3 = A_v f_y/50b_w = 0.22(60{,}000)/50(14) = 18.9$ in. (or use Table 8.1); $S_4 = 24$ in.; $S_{max} = 11$-in. controls.

7. Because $S_1 = 6 \text{ in.} < S_{max} = 11$ in., use no. 3 U-stirrups spaced at 6 in.

8. Calculate V_s for maximum spacings of 11 in.:

$$V_s = \frac{A_v f_y d}{s} = \frac{0.22(60)(22.5)}{11} = 27 \text{ K}$$

$$\phi V_s = 22.95 \text{ K}$$

$$\phi V_c + \phi V_s = 29.33 + 22.95 = 52.3 \text{ K}$$

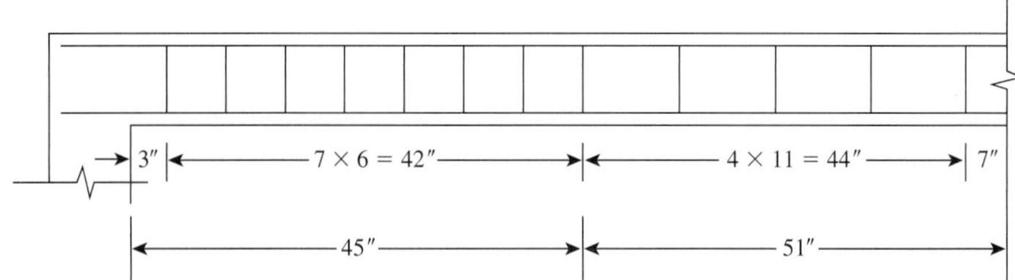

Figure 8.13 Distribution of stirrups (Example 8.2).

The distance x_1 at which $S = 11$ can be used is

$$\left(\frac{91.2 - 52.3}{91.2}\right)(96) = 41 \text{ in.}$$

Because x_1 is relatively small, use $S = 6$ in. for a distance greater than or equal to 41 and then use $S = 11$ for the rest of the beam. *Note:* If x_1 is long, then an intermediate spacing between 6 in. and 11 in. may be added.

9. Distribute stirrups as follows: Place the first stirrup at $S/2$ from the face of the support.

First stirrup at $S/2 = 6/2 =$	3 in.
7 stirrups at $S = 6$	= 42 in.
Total	= 45 in. > 41 in.
4 stirrups at $S = 11$	= 44 in.
Total	= 89 in. > 81 in.

The total number of stirrups for the beam is $2(1 + 7 + 4) = 24$. Distribution of stirrups is shown in Figure 8.13, whereas calculated shear forces are shown in Figure 8.12.

10. Place two no. 4 bars at the top of beam section to act as stirrup hangers.

8.9 SHEAR FORCE DUE TO LIVE LOADS

In Example 8.2, it was assumed that the dead and live loads are uniformly distributed along the full span, producing zero shear at midspan. Actually, the dead load does exist along the full span, but the live load may be applied to the full span or part of the span, as needed to develop the maximum shear at midspan or at any specific section. Figure 8.14(a) shows a simply supported beam with a uniform load acting on the full span. The shear force varies linearly along the beam, with maximum shear acting at support A. In the case of live load, $W_2 = 1.7W_L$, the maximum shear force acts at support A when W_2 is applied on the full span, Figure 8.14(a). The maximum shear at midspan develops if the live load is placed on half the beam, BC (Figure 8.14(b)), producing V_u at midspan equal to $W_2L/8$. Consequently, the design shear force is produced by adding the maximum shear force due to live load (placed at different lengths of the span) to the dead load shear force, Figure 8.14(c), to give the shear distribution shown in Figure 8.14(d). It is a common practice to consider the maximum shear at support A to be $W_uL/2 = (1.4W_D + 1.7W_L)L/2$, whereas V_u at midspan is $W_2L/8 = (1.7W_L)L/8$ with a straight-line variation along AC and CB, as shown in Figure 8.14(d). The design for shear in this case will follow the same procedure explained in Example 8.2. If the approach is applied to the beam in Example 8.2, then V_u (at A) = 91.2 K and V_u (at midspan) = $(1.7 \times 3)(16/8) = 10.2$ K.

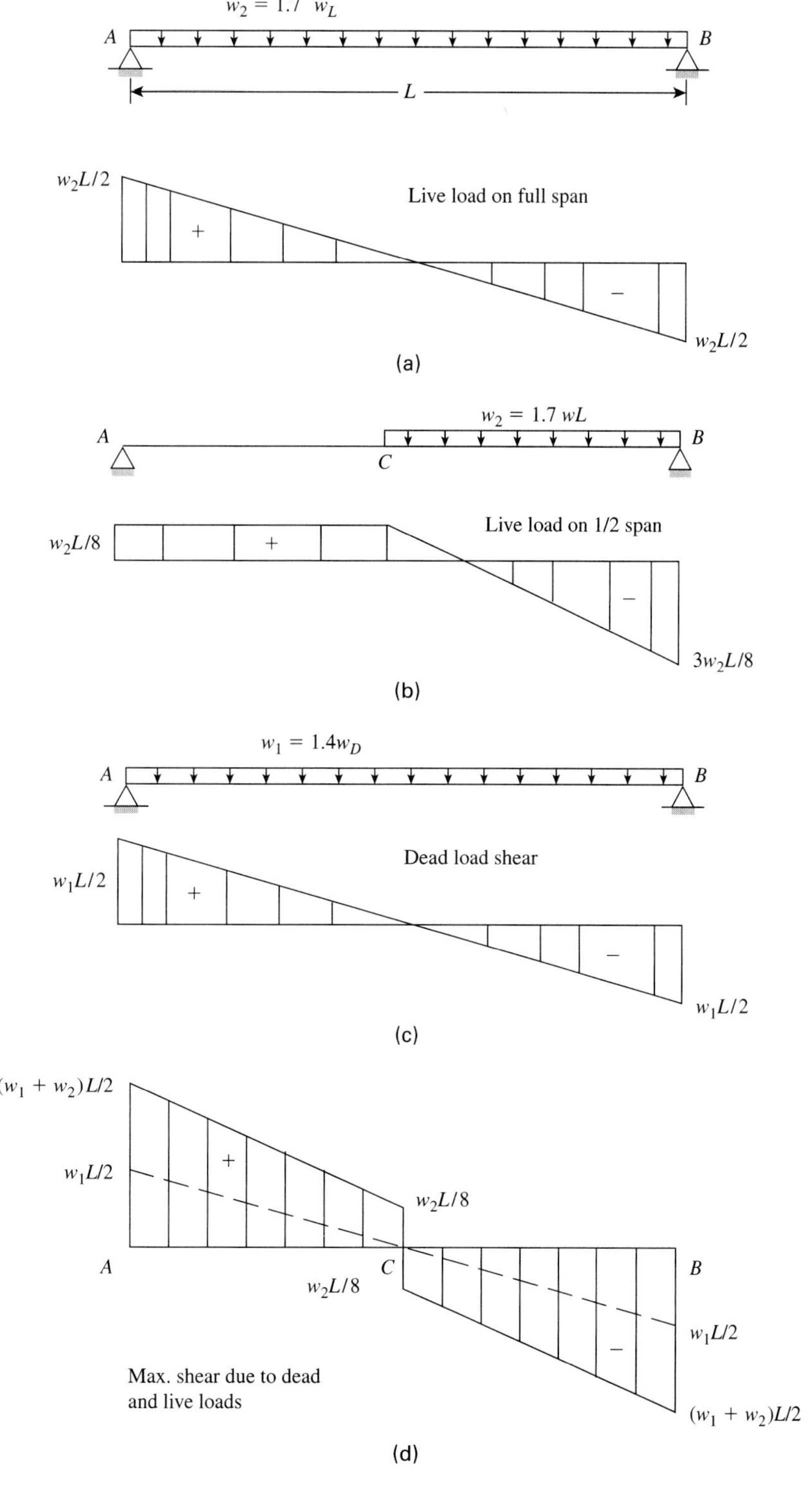

Figure 8.14 Effect of live load application on part of the span.

Example 8.3

A 10-ft-span cantilever beam has a rectangular section and carries uniform and concentrated ultimate loads (self-weight is included), as shown in Figure 8.15. Using $f'_c = 4$ Ksi and $f_y = 60$ Ksi, design the shear reinforcement required for the entire length of the beam according to the ACI Code.

Solution

1. Calculate the shear force along the beam due to external loads.

$$V_u \text{ (at support)} = 5.5(10) + 20 + 8 = 83 \text{ K}$$

$$V_{ud} \text{ (at } d \text{ distance)} = 83 - 5.5\left(\frac{18.5}{12}\right) = 74.52 \text{ K}$$

$$V_u \text{ (at 4 ft left)} = 83 - 4(5.5) = 61 \text{ K}$$

$$V_u \text{ (at 4 ft right)} = 61 - 20 = 41 \text{ K}$$

$$V_u \text{ (at free end)} = 8 \text{ K}$$

The shear diagram is shown in Figure 8.15.

2. Calculate ϕV_c:

$$\phi V_c = 2\sqrt{f'_c}\, bd = 2(0.85)\sqrt{4000}\,(12)(18.5) = 23.87 \text{ K}$$

$$\tfrac{1}{2}\phi V_c = 11.93 \text{ K}$$

Because $V_{ud} > \phi V_c$, shear reinforcement is required. Calculate

$$V_{c1} = 4\sqrt{f'_c}\, bd = 4\sqrt{4000}\,(12)(18.5) = 56.16 \text{ K}$$

$$V_{c2} = 8\sqrt{f'_c}\, bd = 2V_1 = 112.32 \text{ K}$$

The distance x at which no shear reinforcement is needed $\left(\text{at } \tfrac{1}{2}\phi V_c = 11.93 \text{ K}\right)$, measured from support A:

$$x = 4 + \left(\frac{41 - 11.93}{41 - 8}\right)6 = 9.285 \text{ ft}$$

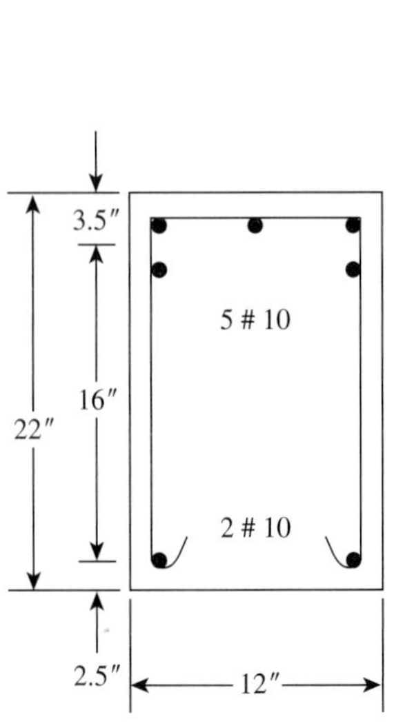

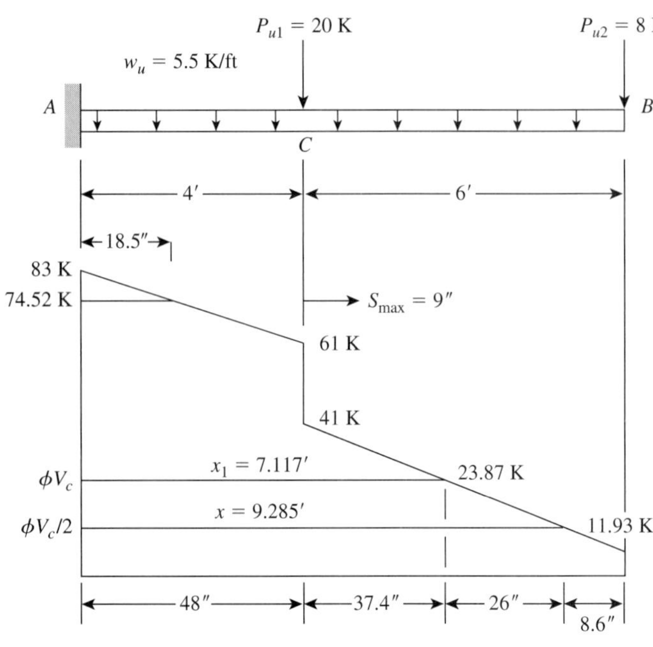

Figure 8.15 Example 8.3.

(8.57 in. from free end). Similarly, x_1 for ϕV_c is 7.117 ft from A (34.6 in. from the free end).

3. Part AC: Design shear $V_u = V_{ud} = 74.52$ K. Calculate $V_s = (V_u - \phi V_c)/\phi = (74.52 - 23.87)/0.85 = 59.6$ K. Because $V_{c1} < V_s < V_{c2}$, then $S_{max} \leq d/4$ must be considered (or check Figure 8.10).

4. Design stirrups: Choose no. 3 U-stirrups, $A_v = 0.22$ in.² Calculate S_1 (based on V_s):

$$S_1 = \frac{A_v f_y d}{V_s} = \frac{13.2d}{V_s} = \frac{13.2(18.5)}{59.6} = 4.1 \text{ in.}$$

Use 4.0 in. (or get $s/d = 0.22$ from Figure 8.10).

5. Calculate maximum spacings: $S_2 = d/4 = 18.5/4 = 4.625$ in., so use 4.5 in.

$$S_3 = \frac{A_v f_y}{50 b_w} = 22 \text{ in.} \qquad \text{(from Table 8.1 for } b = 12 \text{ in.)}$$

$$S_4 = 12 \text{ in.}$$

Then $S_{max} = 4.5$ in.

6. Because $S = 4$ in. $< S_{max} = 4.5$ in., use no. 3 stirrups spaced at 4 in.

7. At C, design shear $V_u = 61$ K $> \phi V_c$. Then $V_s = (61 - 23.87)/0.85 = 43.68$ K, $S_1 = A_v f_y d/V_s = 5.6$ in.

$$V_s = 43.68 \text{ K} < V_{c1} = 56.16 \text{ K}, \qquad S_2 = \frac{d}{2} = \frac{18.5}{2} = 9.25 \text{ in. (or 9 in.).}$$

$S_1 = 5.6$ in. $< S_2$; then $S_1 = 5.6$ or 5.5 in. controls.

8. Because spacing 5.5 in. and 4.0 in. are close, use no. 3 U-stirrups spaced at 4 in. for part AC.

9. Part BC:

a. $V_u = 41 \text{ K} > \phi V_c$

$V_s = (V_u - \phi V_c)/\phi = (41 - 23.87)/0.85 = 20.15 \text{ K} < V_{c1} = 56.16$ K

b. $S_1 = A_v f_y d/V_s = (13.2)(18.5)/20.15 = 12.12$ in.

c. $S_2 = d/2 = 18.5/2 = 9.25$ in. (or less than $S_3 = 22$ in. or $S_4 = 24$ in.). Let $S_{max} = 9$ in. Choose no. 3 stirrups spaced at 9 in. for part BC.

10. Distribution of stirrups measured from support A: Place the first stirrup at

$\frac{S}{2} = \frac{4}{2}$	=	2 in.
12 @ 4 in.	=	48 in.
		50 in.
7 @ 9 in.	=	63 in.
Total		113 in.

Distance left to the free end is 7 in., which is less than 8.57 in., where no stirrups are needed. Distribution of stirrups is shown in Figure 8.16. Total number of stirrups - 20.

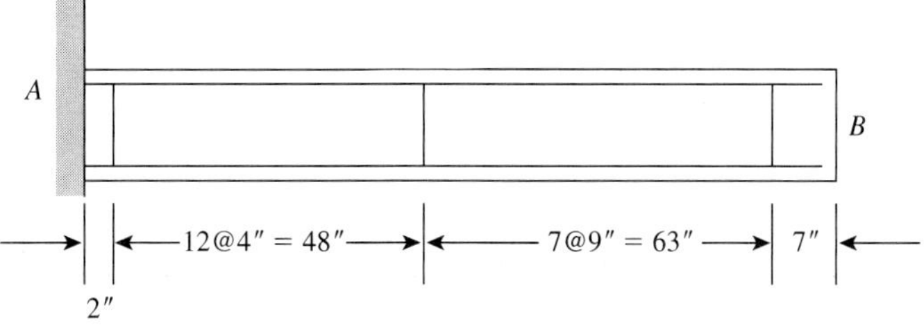

Figure 8.16 Distribution of stirrups, Example 8.3.

8.10 SHEAR STRESSES IN MEMBERS OF VARIABLE DEPTH

The shear stress, v, is a function of the effective depth, d; therefore, shear stresses vary along a reinforced concrete beam with variable depth [10]. In such a beam (Figure 8.17), consider a small element dx. The compression force C at any section is equal to the moment divided by its arm, or $C = M/y$. The first derivative of C is

$$dC = \frac{ydM - Mdy}{y^2}$$

If C_1 is greater than C_2, then $C_1 - C_2 = dC = vb\,dx$

$$vb\,dx = \frac{ydM - Mdy}{y^2} = \frac{dM}{y} - \frac{M}{y^2}dy$$

$$v = \frac{1}{yb}\left(\frac{dM}{dx}\right) - \frac{M}{by^2}\left(\frac{dy}{dx}\right)$$

Because $y = jd$, dM/dx is equal to the shearing force V and $d(jd)/dx$ is the slope,

$$v = \frac{V}{bjd} - \frac{M}{b(jd)^2}\left[\frac{d}{dx}(jd)\right] \quad \text{and} \quad v = \frac{V}{bjd} \pm \frac{M}{b(jd)^2}(\tan\alpha) \tag{8.26}$$

where V and M are the external shear and moment, respectively, and α is the slope angle of one face of the beam relative to the other face. The plus sign is used when the beam depth decreases as the moment increases, whereas the minus sign is used when the depth increases as the moment increases. This formula is used for small slopes, where the angle α is less than or equal to 30°.

A simple form of equation (8.26) can be formed by eliminating the j value:

$$v = \frac{V}{bd} \pm \frac{M}{bd^2}(\tan\alpha) \tag{8.27}$$

For the strength design method, the following equation may be used:

$$v_u = \frac{V_u}{\phi bd} \pm \frac{M_u}{\phi bd^2}(\tan\alpha) \tag{8.28}$$

For the shearing force,

$$\phi V_n = V_u \pm \frac{M_u}{d}(\tan\alpha) \tag{8.29}$$

Figure 8.18 shows a cantilever beam with a concentrated load P at the free end. The moment and the depth d increase toward the support. In this case a negative sign is used

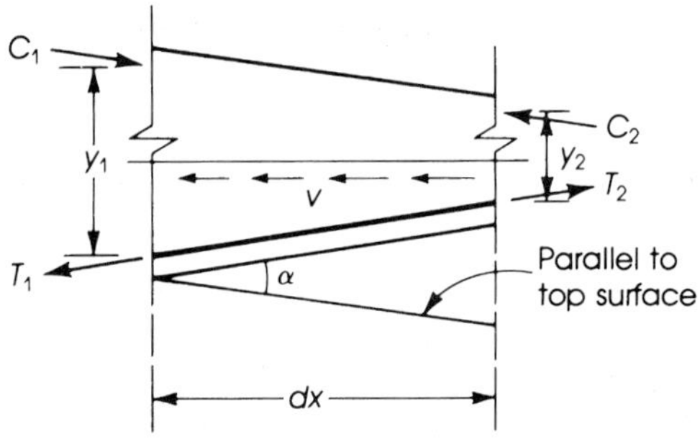

Figure 8.17 Shear stress in a beam with variable depth.

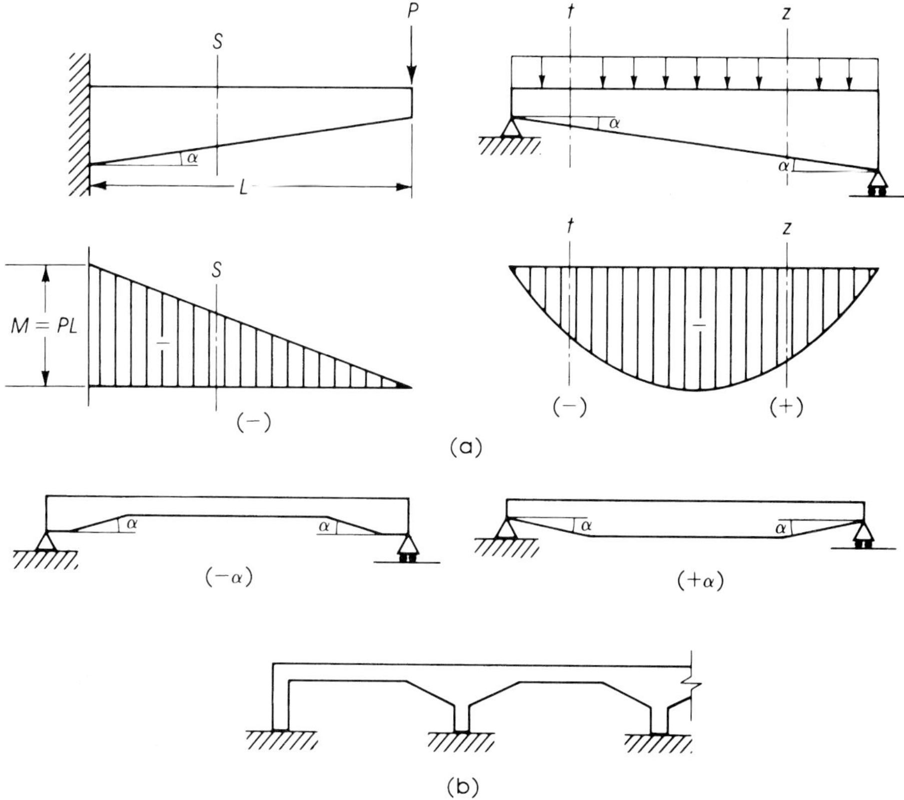

Figure 8.18 Beams with variable depth: (a) moment diagrams and (b) typical forms.

in equations (8.27), (8.28), and (8.29). Similarly, a negative sign is used for section t in the simply supported beam shown, and a positive sign is used for section Z, where moment increases as the depth decreases.

In many cases, the variation in the depth of beams occurs on parts of the beams near their supports (Figure 8.18). Tests [11] on beams with variable depth indicate that beams with greater depth at the support fail mainly by shear compression. Beams with smaller depth at the support fail generally by an instability type of failure, caused by the propagation of the major crack in the beam upward and then horizontally to the beam's top section. Tests also indicate that for beams with variable depth (Figure 8.18) with an inclination α of about 10° and subjected to shear and flexure, the concrete shear strength, V_{cv}, may be computed by

$$V_{cv} = V_c(1 + \tan\alpha) \tag{8.30}$$

where

V_{cv} = shear strength of beam with variable depth

V_c = ACI Code equation (11.6)

$$= \left(1.9\sqrt{f'_c} + 2500\rho_w \frac{V_u d_s}{M_u}\right) b_w d_s \leq 3.5\sqrt{f'_c}\, b_w d_s$$

α = inclination of beam at the support, considered positive for beams of small depth at the support and negative for beams with greater depth at the support (Figure 8.18)

d_s = effective depth of the beam at the support

The simplified ACI Code, equation (11.3), can also be used to compute V_c:

$$V_c = (2\sqrt{f'_c})b_w d_s \tag{8.31}$$

Example 8.4

Design the cantilever beam shown in Figure 8.19 under the ultimate loads applied if the total depth at the free end is 12 in., and it increases toward the support. Use a steel percentage $\rho = 1.5\%$, $f'_c = 3$ Ksi, $f_y = 60$ Ksi, and $b = 10$ in.

Solution

1. M_u (support) $= (2.5/2)(8)^2(12) + (14)(8)(12) = 2304$ K·in.
2. For $\rho = 1.5\%$, $R_u = 741$ psi (from tables).

$$d = \sqrt{\frac{M}{R_u b}} = \sqrt{\frac{2304}{0.741 \times 10}} = 17.6 \text{ in.}$$

$A_s = 0.015 \times 10 \times 17.6 = 2.64$ in.^{2} (use three no. 9 bars); let actual $d = 17.5$ in., $h = 20$ in.

3. Design for shear: Maximum shear at the support is $14 + 20 = 34$ K. Because the beam section is variable, moment effect shall be considered; because the beam depth increases as the moment increases, a minus sign is used in equation (8.28).

$$v_u = \frac{V_u}{\phi bd} - \frac{M_u}{\phi bd^2}(\tan\alpha)$$

To find $\tan\alpha$, let d at the free end be 9.5 in., and d at the support be 17.5 in.:

$$\tan\alpha = \frac{17.5 - 9.5}{8 \times 12} = 0.083$$

$$v_u \text{ (at the support)} = \frac{34{,}000}{(0.85 \times 10 \times 17.5)} - \frac{2304 \times 1000 \times 0.083}{[0.85 \times 10 \times (17.5)^2]}$$

$$= 155 \text{ psi}$$

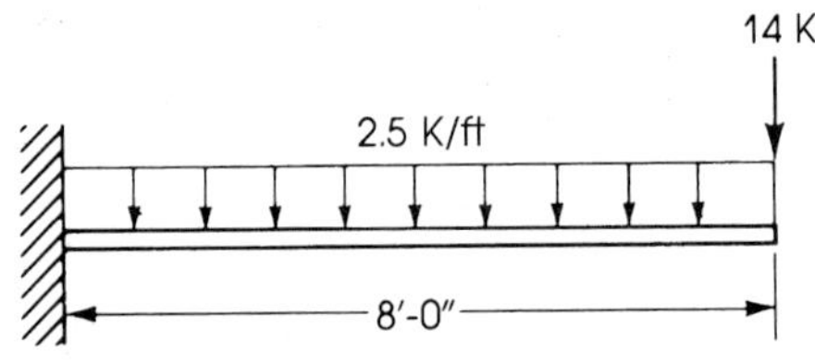

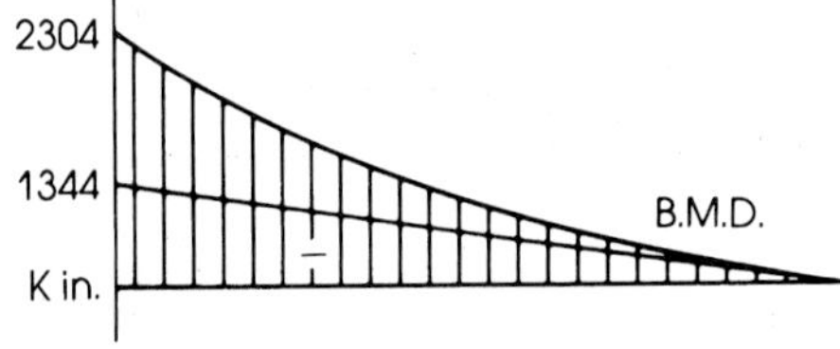

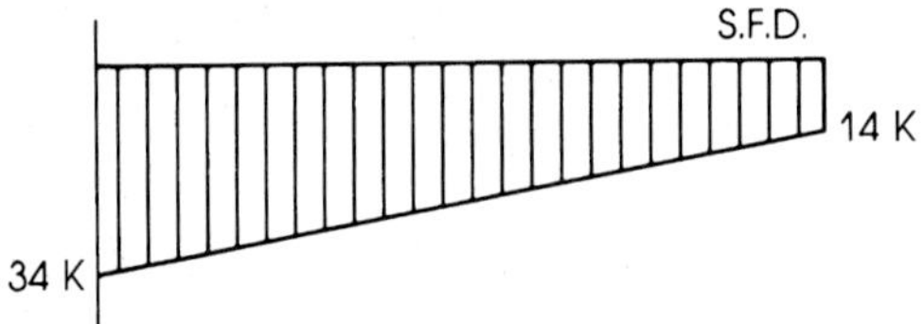

Figure 8.19 Example 8.4 with bending moment diagram (middle) and shear force diagram (bottom).

4. Shear stress at the free end is $V_u/\phi bd(M_u = 0)$.

$$v_u = \frac{14{,}000}{0.85 \times 10 \times 9.5} = 173 \text{ psi}$$

5. At a distance $d = 16.1$ in. from the face of the support, the effective depth is 16.1 in. (from geometry), so

$$V_u = 34 - 2.5 \times \frac{16.1}{12} = 30.6 \text{ K}$$

$$M_u \text{ (at 16.1 in. from support)} = 14 \times 79.9 + \frac{2.5}{12} \times \frac{(79.9)^2}{2}$$

$$= 1784 \text{ K} \cdot \text{in.}$$

$$v_u = \frac{30{,}600}{0.85 \times 10 \times 16.1} - \frac{1784 \times 1000 \times 0.083}{0.85 \times 10 \times (16.1)^2}$$

$$= 157 \text{ psi}$$

6. At midspan (48 in. from the support),

$$d = 13.5 \text{ in.}$$

$$V_u = 14 + 10 = 24 \text{ K}$$

$$M_u = 14 \times 48 + \frac{2.5}{12} \times \frac{(48)^2}{2} = 912 \text{ K} \cdot \text{in.}$$

$$v_u = \frac{24{,}000}{0.85 \times 10 \times 13.5} - \frac{912 \times 1000 \times 0.083}{0.85 \times 10 \times (13.5)^2} = 160 \text{ psi}$$

Similarly, at 6 ft from the support (2 ft from the free end),

$$d = 11.5 \text{ in.} \qquad V_u = 19 \text{ K} \qquad M_u = 396 \text{ K} \cdot \text{in.}$$

$$v_u = 165 \text{ psi}$$

At 1 ft from the free end,

$$d = 10.5 \text{ in.} \qquad V_u = 16.5 \text{ K} \qquad M_u = 183 \text{ K} \cdot \text{in.}$$

$$v_u = 169 \text{ psi}$$

These values are shown in Figure 8.20.

7. Shear stress resisted by concrete is

$$2\sqrt{f'_c} = 2\sqrt{3000} = 109 \text{ psi}$$

Minimum shear stress to be resisted by shear reinforcement

$$v_{us} = 173 - 109 = 64 \text{ psi}$$

(v_u and consequently v_{us} have already been increased by the ratio $1/\phi$ in equation (8.28)).

8. Choose no. 3 stirrups with two legs.

$$A_v = 2 \times 0.11 = 0.22 \text{ in.}^2$$

$$S \text{ (required)} = \frac{A_v f_y}{v_s b_w} = \frac{0.22 \times 60{,}000}{64 \times 10} = 20.6 \text{ in.}$$

$$S_{max}\left(\text{for } \frac{d}{2}\right) = 8.0 \text{ in. to } 4.5 \text{ in. at the free end}$$

$$S_{max} \text{ (for minimum } A_v) = \frac{A_v f_y}{50 b_w} = \frac{0.22 \times 60{,}000}{50 \times 10} = 26.4 \text{ in.}$$

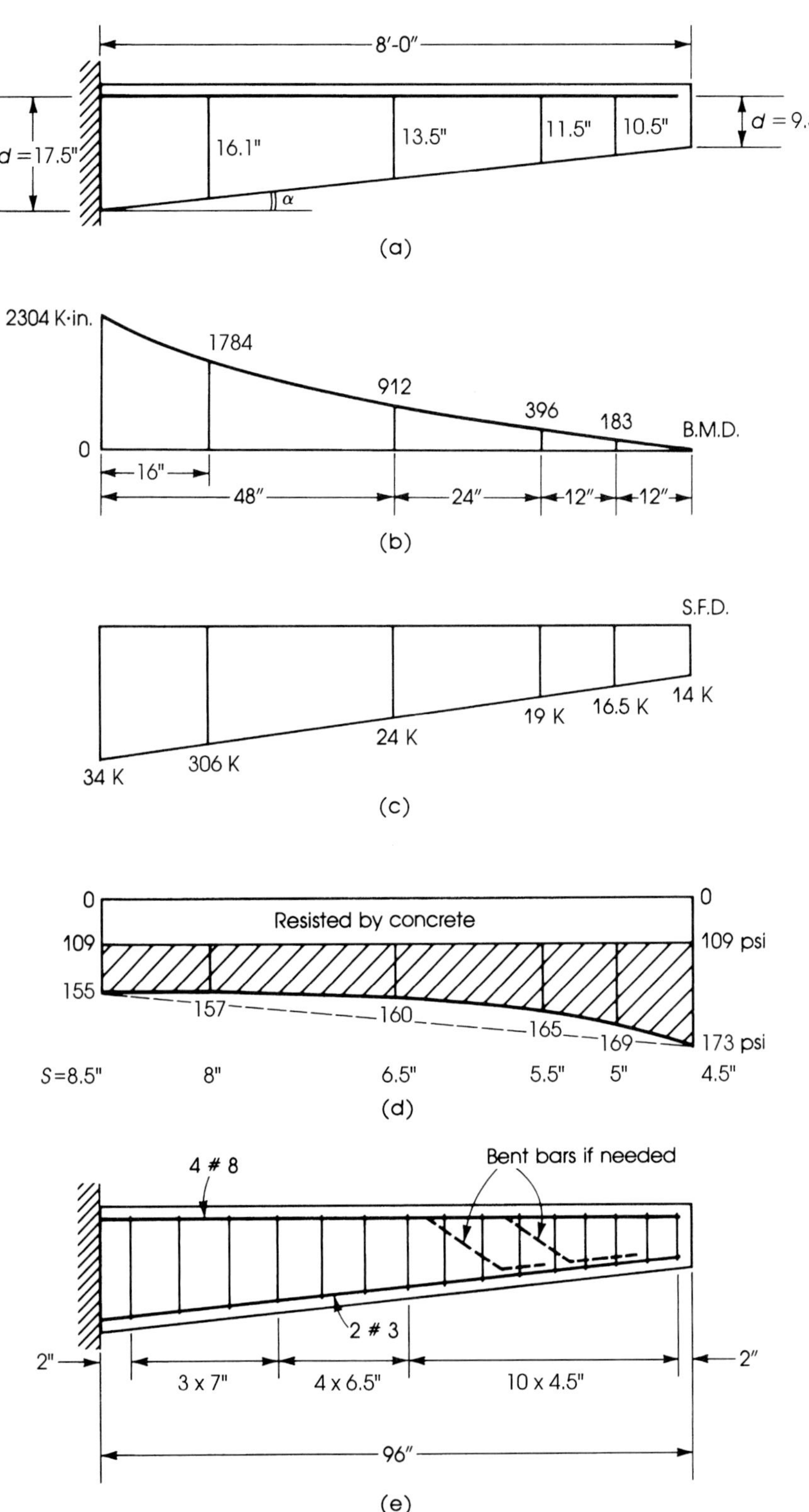

Figure 8.20 Web reinforcement for a beam of variable depth (Example 8.4).

9. Check for maximum spacing $(d/4)$: $v_{us} \leq 4\sqrt{f'_c}$.

$$4\sqrt{f'_c} = 4\sqrt{3000} = 218 > 64 \text{ in.}$$

10. Distribution of stirrups (distances from the free end):

1 stirrup at 2 in.	= 2 in.
10 stirrups at 4.5 in.	= 45 in.
4 stirrups at 6.5 in.	= 26 in.
3 stirrups at 7 in.	= 21 in.
Total	= 94 in.

There is 2 in. left to the face of the support.

8.11 DEEP FLEXURAL MEMBERS

Flexural members should be designed as deep beams if the ratio of the clear span, l_n (measured from face to face of the supports; Figure 8.21), to the effective depth, d, is less than 5 (ACI Code, Section 11.8). The members should be loaded on one face and supported on the opposite face so that compression struts can develop between the loads and supports (Figure 8.22). If the loads are applied through the bottom or sides of the deep beam, shear design equations for ordinary beams given earlier should be used. Examples of deep beams are short-span beams supporting heavy loads, vertical walls under gravity loads, shear walls, and floor slabs subjected to horizontal loads.

A second definition of deep flexural members is presented in the ACI Code, Section 10.7.1. It indicates that flexural members where the ratio of the clear span, l_n, to the overall depth, h (Figure 8.21), is less than 1.25 for simple beams or 2.5 for continuous beams are considered deep flexural members. Such beams should be designed taking into account nonlinear distribuion of stress and lateral buckling (Figure 8.22(a)).

Figure 8.22(a) shows the elastic stress distribution at the midspan section of a deep beam, and Figure 8.22(b) shows the principal trajectories in top-loaded deep beams. Solid lines indicate tensile stresses, whereas dashed lines indicate compressive stress distribution. Under heavy loads, inclined vertical cracks develop in the concrete in a direction perpendicular to the principal tensile stresses and almost parallel to the dashed trajectories (Figure 8.22(c)). Hence, both horizontal and vertical reinforcement is needed to resist principal stresses. Moreover, tensile flexural reinforcement is needed within about the bottom one-fifth of the beam along the tensile stress trajectories (Figure 8.22(b)). In general, the analysis of deep beams is complex and can be performed using truss models or more accurately using a finite-element approach or similar methods. The ACI Code, Section 11.8,

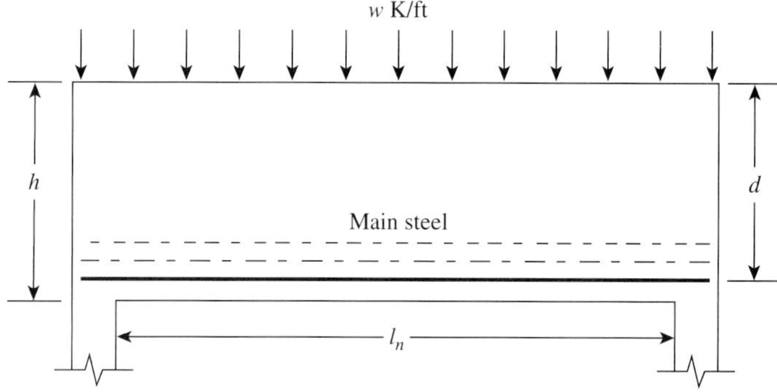

Figure 8.21 Single-span deep beam $(l_n/d < 5)$.

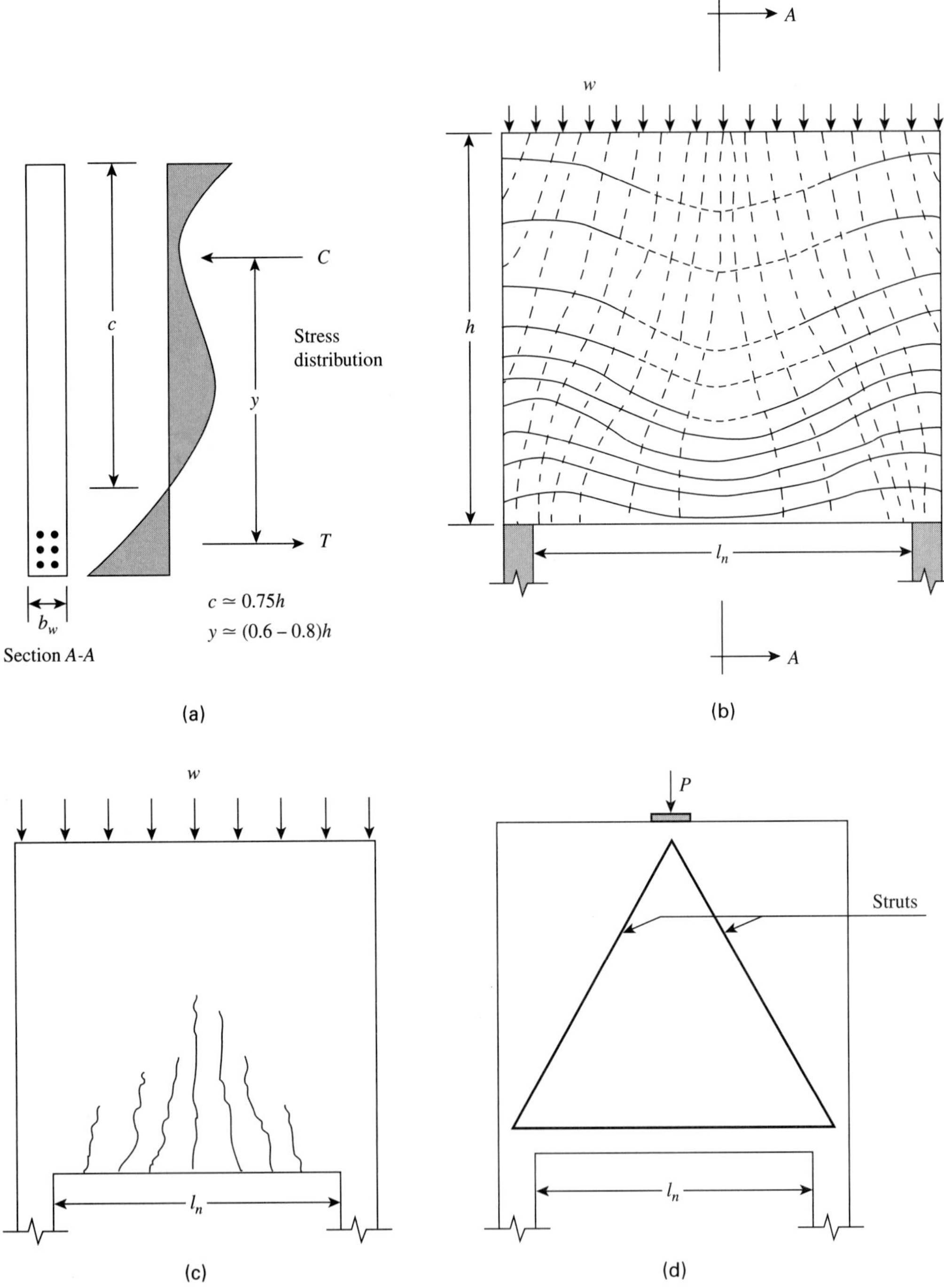

Figure 8.22 Stress distribution and cracking: (a) elastic stress distribution, (b) stress trajectories (tension, solid lines, and compression, dashed lines), (c) cracks pattern, and (d) truss model for a concentrated load applied at the wall upper surface.

introduced special simplified provisions for the shear design of deep beams. These provisions can be presented in steps as follows:

1. Critical section: If the critical section for shear design in deep beams supporting top vertical loads is located at a distance X from the face of the support, then the distance X can be determined as follows (Figure 8.23):

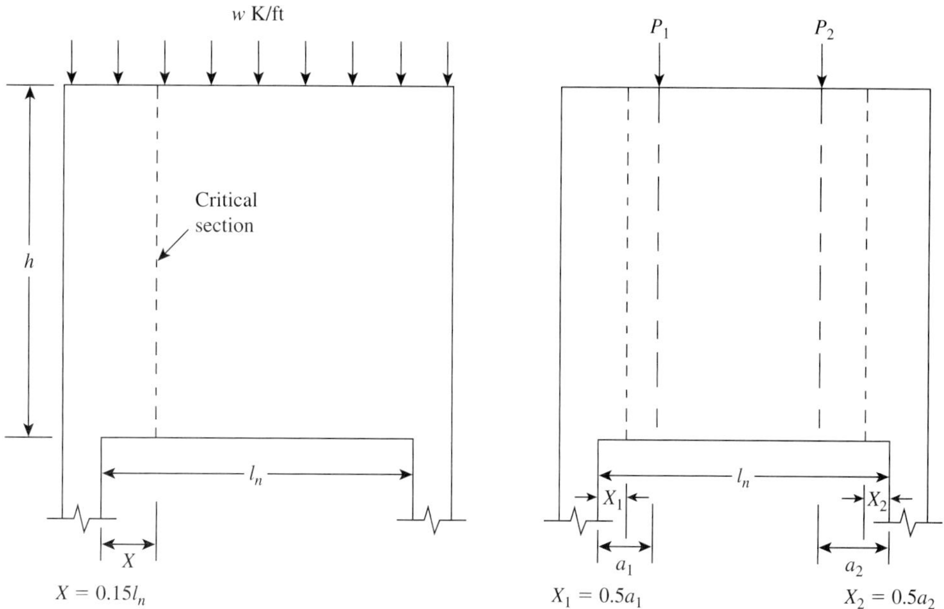

Figure 8.23 Critical sections for shear design.

a. For deep beams supporting uniformly distributed loads, $X = 0.15l_n$, where l_n = clear span.

b. For concentrated loads, $X_1 = 0.5a_1$ (left support) or $X_2 = 0.5a_2$ (right support) (Figure 8.23), where a_1 and a_2 equal the shear span near each support. The shear span is the distance between the concentrated load and the face of the support.

In all cases, the distances X, X_1, and X_2 must not exceed the effective depth, d.

2. Maximum shear strength ϕV_n: The maximum shear strength, ϕV_n, for deep flexural members shall not exceed the following values ($\phi = 0.85$): (ACI Code, Section 11.8.4)

$$\text{For } \frac{l_n}{d} < 2, \quad \phi V_n = \phi 8\sqrt{f'_c}\, b_w d \tag{8.32}$$

$$\text{For } 2 \leq \frac{l_n}{d} \leq 5, \quad \phi V_n = \phi \frac{2}{3}\left(10 + \frac{l_n}{d}\right)\sqrt{f'_c}\, b_w d \tag{8.33}$$

If V_u exceeds ϕV_n, then the section dimensions must be increased.

3. a. Concrete shear strength, V_c: The nominal shear strength, V_c, of concrete can be estimated as follows:

$$V_c = 2\sqrt{f'_c}\, b_w d \tag{8.34}$$

This V_c is similar to the concrete shear strength for regular beams, as in the previous sections of this chapter.

b. Alternatively, another expression may be used that takes into account the effect of the ultimate moment and shear at the critical section:

$$V_c = \left(3.5 - \frac{2.5M_u}{V_u d}\right)\left(1.9\sqrt{f'_c} + \frac{2500\rho_w V_u d}{M_u}\right) b_w d \tag{8.35}$$

But V_c should not exceed $6\sqrt{f'_c}\, b_w d$.

The value of $(3.5 - 2.5M_u/V_u d)$ may not be greater than 2.5 and must not be less than 1.0. The values of M_u and V_u are taken at the critical design section. The ACI

Code allows this higher shear capacity than that of equation (8.34), with the idea that minor unsightly cracking may occur in the deep beam and can be tolerated. Cracks may start to develop at about one-third the ultimate load.

4. Shear reinforcement: When the factored shear force, V_u, exceeds ϕV_c, shear reinforcement must be provided, considering that $V_u = \phi(V_c + V_s)$, or $V_s = (V_u - \phi V_c)/\phi$. The steps are as follows (ACI Code, Section 11.8):

a. Determine V_s: The force resisted by shear reinforcement V_s is determined from the following expression.

$$V_s = \left[\frac{A_v}{S_v}\left(\frac{1 + l_n/d}{12}\right) + \frac{A_{vh}}{S_h}\left(\frac{11 - l_n/d}{12}\right)\right] f_y d \tag{8.36}$$

where A_v = total area of vertical shear reinforcement spaced at S_v and perpendicular to the main flexural tensile reinforcement on both faces of the beam and A_{vh} = total area of horizontal shear reinforcement spaced at S_h parallel to the main flexural tensile reinforcement on both faces of the beam.

b. Spacing of shear reinforcement is

$$\text{Maximum vertical spacings} \qquad S_v \leq \frac{d}{5} \leq 18 \text{ in.}$$

$$\text{Maximum horizontal spacings} \qquad S_h \leq \frac{d}{3} \leq 18 \text{ in.}$$

c. Minimum shear reinforcement: The area of vertical shear reinforcement is $A_v \geq 0.0015 b_w S_v$. The area of horizontal shear reinforcement is $A_{vh} \geq 0.0025 b_w S_h$.

d. The shear reinforcement required at the critical section should be extended throughout the length and depth of the deep beam.

e. For continuous deep beams, the same shear reinforcement may be used in all spans if the spans are almost equal with similar loading.

5. Flexural reinforcement of deep beams: The flexural behavior of deep beams is complex and requires nonlinear analysis of stresses and strains along the depth of the beam. For a preliminary design, the following simplified approach may be used:

$$\phi M_n = \phi A_s f_y(y)$$

where y = moment arm = $(d - a/2)$. Because the value of $(d - a/2)$ is not easy to calculate, the moment arm y may be taken approximately equal to $0.6h$ for $l_n/h \leq 1.0$ and equal to $0.8h$ for $l_n/h = 2.0$. Linear interpolation may be used to estimate y when l_n/h varies between 1.0 and 2.0. Therefore,

$$A_s = \frac{M_u}{\phi y f_y} \tag{8.37}$$

The value of A_s may not be less than the minimum flexural reinforcement required for regular beams given next, assuming $d = 0.9h$:

$$\text{Minimum } A_s = \left(\frac{3\sqrt{f'_c}}{f_y}\right) b_w d \geq \left(\frac{200}{f_y}\right) b_w d \tag{8.38}$$

The second term controls when $f'_c < 4500$ psi. Note that f'_c and f_y are in psi.

The flexural tension reinforcement should be placed within $h/4$ to $h/5$ of the beam and should be adequately spaced along the bottom tension zone. Tension bars should be well anchored to the supports.

For more accurate analysis and design and for continuous deep beams, a rigorous nonlinear approach should be used to determine the proper amount and distribution of the tension reinforcement.

Example 8.5

A simply supported deep beam has a span = 14 ft, a clear span l_n = 12 ft, a total height h = 8 ft, and width b = 16 in. The deep beam supports a uniform service dead load of 36 K/ft (including self-weight) and a live load of 20 K/ft on top of the beam. Design the beam for moment and shear using f'_c = 4 Ksi and f_y = 60 Ksi. Refer to Figure 8.24.

Solution

1. Design for moment:

$$W_u = 1.4W_D + 1.7W_L = 1.4(36) + 1.7(20) = 84.4 \text{ K/ft}$$

$$M_u = \frac{W_u L^2}{8} = \frac{84.4(14)^2}{8} = 2067.8 \text{ K} \cdot \text{ft}$$

$$\frac{l_n}{h} = \frac{12}{8} = 1.5$$

Determine the moment arm, y. For $l_n/h = 1.0$, $y = 0.6d$, and for $l_n/h = 2.0$, $y = 0.8d$; hence for $l_n/h = 1.5$, $y = 0.7d$ (by interpolation) $= 0.7(8 \times 12) = 67.2$ in.

$$A_s = \frac{M_u}{\phi y f_y} = \frac{2067.8 \times 12}{0.9(67.2)(60)} = 6.84 \text{ in.}^2$$

Assume $d = 0.9h = 0.9(8 \times 12) = 86.4$ in. Because $f'_c < 4500$ psi,

$$A_s \text{ (minimum)} = \left(\frac{200}{f_y}\right) b_w d = \frac{200(16)(86.4)}{60{,}000} = 4.6 \text{ in.}^2$$

Therefore, $A_s = 6.84$ in.^{2} controls. Choose 10 no. 8 bars (7.85 in.$_2$), 5 on each face, distributed within $h/5 = 8(12)/5 = 19.2$ in. of the tension zone of the beam. Spacing of bars $= 19.2/5 = 3.84$ in., or 4 in. Bars should be well-anchored into the supports.

2. Design for shear:

a. Calculate V_u and M_u at the distance $x = 0.15l_n \leq d$ from the face of the support.

$$0.15l_n = 0.15(12 \times 12) = 21.6 \text{ in.} = 1.8 \text{ ft} < d = 86.4 \text{ in.}$$

Design $V_u = 84.4(12/2) - 84.4(1.8) = 354.5$ K.

$$M_u = 84.4(6)(1.8) - \frac{84.4(1.8)^2}{2} = 774.8 \text{ K} \cdot \text{ft}$$

$$\frac{M_u}{V_u d} = \frac{774.8(12)}{354.5(86.4)} = 0.304$$

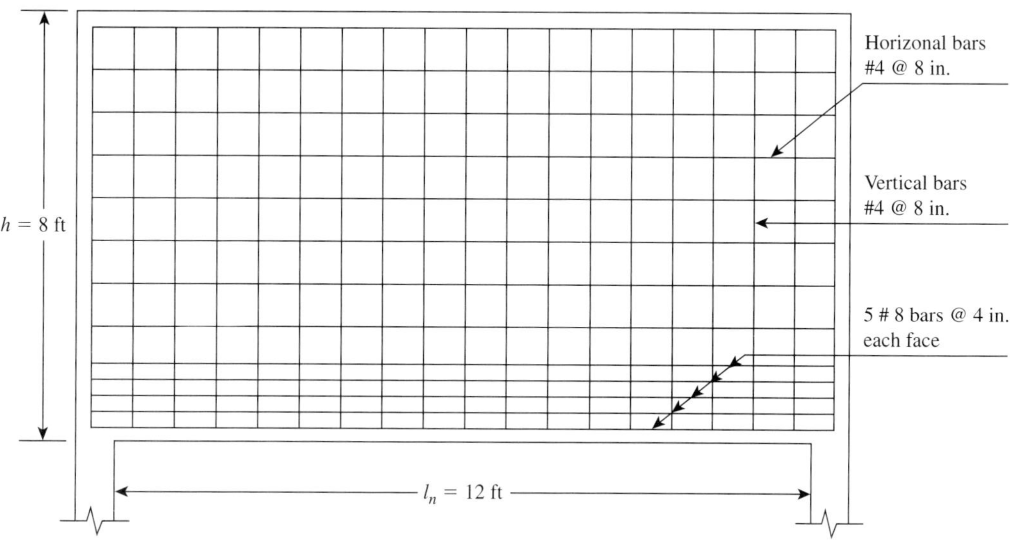

Figure 8.24 Example 8.5.

b. Calculate V_c:

$$3.5 - 2.5\frac{M_u}{V_u d} = 3.5 - 2.5(0.304) = 2.74 > 2.5$$

So, use 2.5. In this case determine $M_u/V_u d$ to be used to calculate V_c: $2.5 = 3.5 - 2.5M_u/(V_u d)$, and $M_u/(V_u d) = 0.4$.

$$\frac{V_u d}{M_u} = \frac{1}{0.4} = 2.5$$

$$\rho_w = \frac{A_s}{b_w d} = \frac{7.85}{16 \times 86.4} = 0.00496$$

$$V_c = 2.5\left(1.9\sqrt{f'_c} + \frac{2500\rho_w V_u d}{M_u}\right)b_w d$$

$$= 2.5\left[1.9\sqrt{4000} + (2500)(0.00496)(2.5)\right](16)(86.4)$$

$$= 522.4 \text{ K}$$

$$V_c \le 6\sqrt{f'_c}\, b_w d = 6\sqrt{4000}\,(16)(86.4) = 524.6 \text{ K}$$

Hence, $V_c = 522.4$ controls and $\phi V_c = 444.1$ K.

c. Calculate $V_s = (V_u - \phi V_c)/\phi$. Because $\phi V_c = 444.1 \text{ K} > V_u = 354.5$ K, then $V_s = 0$, and only minimum shear reinforcement is required.

d. Calculate shear reinforcement: Assume no. 4 bars placed on both faces in the horizontal and vertical directions; then $A_v = A_{vh} = 2(0.2) = 0.4 \text{ in.}^2$ Maximum allowable spacing of vertical bars is $S_v = d/5 \le 18$ in. $S_v = 86.4/5 = 17.3$ in. < 18; use $S_n = 16$ in. Maximum allowable spacing of horizontal bars is $S_h = d/3 \le 18$ in. $S_h = 86.4/3 = 28.8$ in. > 18 in.; use $S_h = 18$ in. Minimum A_v (vertical) $= 0.0015bS_v = 0.0015(16)(16) = 0.384 \text{ in.}^2 < 0.4 \text{ in.}^2$ Minimum A_{vh} (horizontal) $= 0.0025bS_h = 0.0025(16)(18) = 0.72 \text{ in.}^2 > 0.4 \text{ in.}^2$ Reduce spacing to $S_h = 0.4/(0.0025 \times 16) = 10$ in. Therefore, use no. 4 vertical bars spaced at 16 in., and use no. 4 horizontal bars spaced at 10 in.

3. If $V_c = 2\sqrt{f'_c}\, b_w d$ is used, then $V_c = 2\sqrt{4000}(16)(86.4) = 174.9$ K and $\phi V_c = 0.85V_c = 148.6$ K, which is less than $V_u = 354.5$ K. Hence shear reinforcement is required.

$$V_s = \frac{V_u - \phi V_c}{\phi} = \frac{354.1 - 148.6}{0.85} = 241.8 \text{ K}$$

Assuming no. 4 bars placed on both faces in the vertical and horizontal directions, then $A_v = A_{vh} = 2(0.2) = 0.4 \text{ in.}^2$ Assuming that the spacings of bars in both directions are equal, $S_v = S_h = S$ and $l_n/d = 12 \times 12/86.4 = 1.67$, then

$$V_s = \left[\frac{A_v}{S_v}\left(\frac{1 + l_n/d}{12}\right) + \frac{A_{vh}}{S_h}\left(\frac{11 - l_n/d}{12}\right)\right]f_y d \tag{8.36}$$

$$241.8 = \left[\frac{0.4}{S}\left(\frac{1 + 1.67}{12}\right) + \frac{0.4}{S}\left(\frac{11 - 1.67}{12}\right)\right](60)(86.4)$$

$S = 8.6$ in., which is less than the maximums $S_v = 17.3$ in. and $S_h = 18$ in. Use $S = 8$ in. for both vertical and horizontal spacing.

Minimum A_v (vertical) $= 0.0015(16)(8) = 0.192 \text{ in.}^2 < 0.4 \text{ in.}^2$

Minimum A_{vh} (horizontal) $= 0.0025(16)(8) = 0.32 \text{ in.}^2 < 0.4 \text{ in.}^2$

Then use no. 4 bars spaced at 8 in. on both faces in the horizontal and vertical directions. A welded wire fabric mesh may be adopted to replace the preceding bar arrangements. It can be seen that this solution is more conservative than that given in Step 2.

Reinforcement details are shown in Figure 8.24.

8.12 EXAMPLES USING SI UNITS

The general design requirements for shear reinforcement according to the ACI Code are summarized in Table 8.4, which gives the necessary design equations in both U.S. customary and SI units. The following example shows the design of shear reinforcement using SI units.

Table 8.4 Shear Reinforcement Formulas

U.S. Customary Units	SI Units
V_u = design shear	V_u = design shear
(Maximum design V_u is at a distance d from the face of the support)	
$V_c = (2.0\sqrt{f'_c})b_w d$	$V_c = (0.17\sqrt{f'_c})b_w d$
$V_c = \left[1.9\sqrt{f'_c} + \left(2500\rho_w \frac{V_u d}{M_u}\right)\right]b_w d$	$V_c = \left[0.16\sqrt{f'_c} + \left(17.2\rho_w \frac{V_u d}{M_u}\right)\right]b_w d$
$\rho_w = \frac{A_s}{b_w d} \qquad \frac{V_u d}{M_u} \le 1.0$	$\rho_w = \frac{A_s}{b_w d} \qquad \frac{V_u d}{M_u} \le 1.0$
$V_c \le (3.5\sqrt{f'_c})b_w d$	$V_c \le (0.29\sqrt{f'_c})b_w d$
$V_u = \phi V_c + \phi V_s$	$V_u = \phi V_c + \phi V_s$
Vertical stirrups	
$\phi V_s = V_u - \phi V_c$	$\phi V_s = V_u - \phi V_c$
$s = \frac{A_v f_y d}{V_s}$	$s = \frac{A_v f_y d}{V_s}$
Minimum $A_v = \frac{50 b_w s}{f_y}$	Minimum $A_v = \frac{0.35 b_w s}{f_y}$
Maximum $s = \frac{A_v f_y}{50 b_w}$	Maximum $s = \frac{A_v f_y}{0.35 b_w}$
For vertical web reinforcement	
Maximum $s = \frac{d}{2} \le 24$ in.	Maximum $s = \frac{d}{2} \le 600$ mm
if $V_s \le 4.0\sqrt{f'_c}(b_w d)$	If $V_s \le 0.33\sqrt{f'_c}(b_w d)$
Maximum $s = d/4 \le 12$ in.	Maximum $s = d/4 \le 300$ mm
if $V_s > 4.0\sqrt{f'_c}(b_w d)$	if $V_s > 0.33\sqrt{f'_c}(b_w d)$
$V_s \le 8\sqrt{f'_c}(b_w d)$	$V_s \le 0.67\sqrt{f'_c}(b_w d)$
Otherwise increase the dimensions of the section.	
Series of bent bars or inclined stirrups	
$A_v = \frac{V_s s}{f_y d(\sin\alpha + \cos\alpha)}$	$A_v = \frac{V_s s}{f_y d(\sin\alpha + \cos\alpha)}$
For $\alpha = 45°$, $s = \frac{1.4 A_v f_y d}{V_s}$	For $\alpha = 45°$, $s = \frac{1.4 A_v f_y d}{V_s}$
For a single bent bar or group of bars, parallel and bent in one position:	
$A_v = \frac{V_s}{f_y \sin\alpha}$	$A_v = \frac{V_s}{f_y \sin\alpha}$
For $\alpha = 45°$, $A_v = \frac{1.4 V_s}{f_y}$	For $\alpha = 45°$, $A_v = \frac{1.4 V_s}{f_y}$
$V_s \le (3\sqrt{f'_c})b_w d$	$V_s \le (0.25\sqrt{f'_c})b_w d$

Example 8.6

A 6-m clear span simply supported beam carries a uniform dead load of 45 kN/m and a live load of 20 kN/m (Figure 8.25). The dimensions of the beam section are $b = 350$ mm, $d = 550$ mm. The beam is reinforced with four bars of 25-mm diameter in one row. It is required to design the necessary shear reinforcement. Given: $f'_c = 21$ MPa and $f_y = 280$ MPa.

Solution

1. Ultimate load is

$$1.4D + 1.7L = 1.4 \times 45 + 1.7 \times 20 = 97 \text{ kN/m}$$

2. Ultimate shear force at the face of the support is

$$V_u = 97 \times \frac{6}{2} = 291 \text{ kN}$$

3. Maximum design shear at a distance d from the face of the support is

$$V_u \text{ (at distance } d) = 291 - 0.55 \times 97 = 237.65 \text{ kN}$$

4. The nominal shear strength provided by concrete is

$$V_c = (0.17\sqrt{f'_c})bd = (0.17\sqrt{21}) \times 350 \times 550 = 150 \text{ kN}$$

$$V_u = \phi V_c + \phi V_s$$

$$\phi V_c = 0.85 \times 150 = 127.5 \text{ kN}$$

$$\tfrac{1}{2}\phi V_c = 63.75 \text{ kN}$$

$$\phi V_s = 237.65 - 127.5 = 110.15 \text{ kN}$$

$$V_s = \frac{110.15}{0.85} = 129.6 \text{ kN}$$

5. Distance from the face of the support at which $\frac{1}{2}\phi V_c = 63.75$ kN is

$$x' = \frac{(291 - 63.75)}{291}(3) = 2.34 \text{ m} \qquad \text{(from triangles)}$$

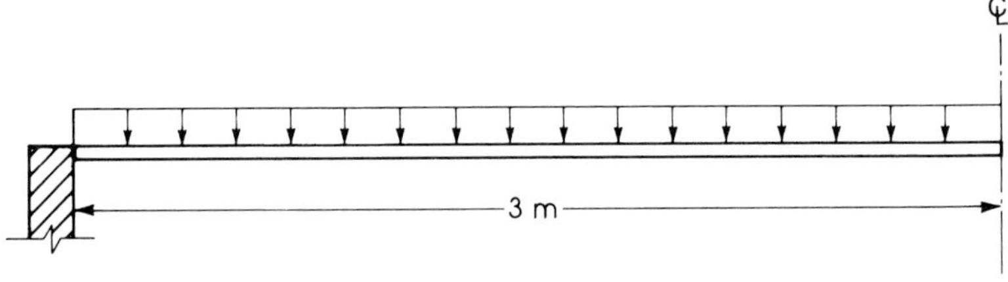

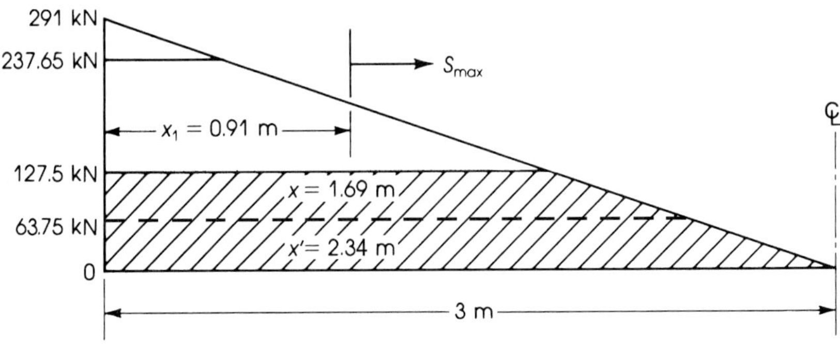

Figure 8.25 Example 8.6.

6. Design of stirrups:

a. Choose stirrups 10 mm in diameter with two branches.

$$A_v = 2 \times 78.5 = 157 \text{ mm}^2$$

$$\text{Spacing } s_1 = \frac{A_v f_y d}{V_s} = \frac{157 \times 280 \times 550}{129.6 \times 10^3} = 186.6 \text{ mm} < 600 \text{ mm}$$

So, use 180 mm. Check maximum spacing of stirrups:

$$\text{Maximum } s_2 = \frac{d}{2} = \frac{550}{2} = 275 \text{ mm}$$

$$S_3 = \frac{A_v f_y}{0.35b} = \frac{157 \times 280}{0.35 \times 350} = 359 \text{ mm}$$

$S = S_1 = 180$ mm controls.

b. Check for maximum spacing of $d/4$:

$$\text{If } V_s \leq (0.33\sqrt{f'_c})bd, \qquad S_{\text{max}} = \frac{d}{2}.$$

$$\text{If } V_s > (0.33\sqrt{f'_c})bd, \qquad S_{\text{max}} = \frac{d}{4}.$$

$$bd(0.33\sqrt{f'_c}) = 0.33\sqrt{21} \times 350 \times 550 = 291.1 \text{ kN}$$

Actual $V_s = 129.6$ kN < 291.1 kN. Therefore, S_{max} is limited to $d/2 = 275$ mm.

7. The shear reinforcement, stirrups 10 mm in diameter and spaced at 180 mm, will be needed only for a distance $d = 0.55$ m from the face of the support. Beyond that, the shear stress V_s decreases to zero at a distance $x = 1.69$ m when $\phi V_c = 127.5$ kN. It is not practical to provide stirrups at many different spacings. One simplification is to find out the distance from the face of support where maximum spacing can be used, and then only two different spacings may be adopted.

$$\text{Maximum spacing} = \frac{d}{2} = 275 \text{ mm}$$

$$V_s\,(\text{for } s_{\text{max}} = 275 \text{ mm}) = \frac{A_v f_y d}{s} = \frac{157 \times 0.280 \times 550}{275} = 87.9 \text{ kN}$$

$$\phi V_s = 87.9 \times 0.85 = 74.7 \text{ kN}$$

The distance from the face of the support where $s_{\text{max}} = 275$ mm can be used (from the triangles):

$$x_1 = \frac{291 - (127.5 + 74.7)}{291}(3) = 0.91 \text{ m}$$

Then, for 0.91 m from the face of support, use stirrups 10 mm diameter at 180 mm, and for the rest of the beam, minimum stirrups (with maximum spacings) can be used.

8. Distribution of stirrups:

$$\begin{aligned}
&1 \text{ stirrup at } \frac{s}{2} = \frac{180}{2} && = 90 \text{ mm} \\
&5 \text{ stirrups at } 180 \text{ mm} && = \underline{900 \text{ mm}} \\
&\text{Total} && = 990 \text{ mm} = 0.99 \text{ m} > 0.91 \text{ m} \\
&7 \text{ stirrups at } 270 \text{ mm} && = \underline{1890 \text{ mm}} \\
&\text{Total} && = 2880 \text{ mm} = 2.88 \text{ m} < 3 \text{ m}
\end{aligned}$$

The last stirrup is $(3 - 2.88) = 0.12$ m $= 120$ mm from the centerline of the beam, which is adequate. A similar stirrup distribution applies to the other half of the beam, giving a total number of stirrups of 26.

The other examples in this chapter can be worked out in a similar way using SI equations.

SUMMARY

Sections 8.1 and 8.2

The shear stress in a homogeneous beam is $v = VQ/Ib$. The distribution of the shear stress above the neutral axis in a singly reinforced concrete beam is parabolic. Below the neutral axis, the maximum shear stress is maintained down to the level of the steel bars.

Section 8.3

The development of shear resistance in reinforced concrete members occurs by

- Shear resistance of the uncracked concrete;
- Interface shear transfer;
- Arch action;
- Dowel action.

Section 8.4

The shear stress at which a diagonal crack is expected is

$$v_c = \frac{V}{bd} = \left(1.9\sqrt{f'_c} + 2500\rho_w \frac{V_u d}{M_u}\right) \le 3.5\sqrt{f'_c}$$

The nominal shear strength is

$$V_c = v_c b_w d = 2\sqrt{f'_c}\, b_w d$$

Sections 8.5 and 8.6

1. The common types of shear reinforcement are stirrups (perpendicular or inclined to the main bars), bent bars, or combinations of stirrups and bent bars.

$$V_u = \phi V_n = \phi V_c + \phi V_s \quad \text{and} \quad V_s = \frac{1}{\phi}(V_u - \phi V_c)$$

2. The ACI Code design requirements are summarized in Table 8.4.

Sections 8.7 and 8.8

Design of vertical stirrups and shear summary is given in these sections.

Sections 8.9 and 8.10

1. Variation of shear force along the span due to live load may be considered.
2. For members with variable depth,

$$\phi V_n = V_u \pm \frac{M_u(\tan\alpha)}{d} \tag{8.29}$$

Section 8.11

For deep beams, the shear capacity V_c may be determined from the following expressions:

$$V_c = 2\sqrt{f'_c}\,b_w d \tag{8.34}$$

or

$$V_c = \left(3.5 - \frac{2.5M_u}{V_u d}\right)\left(1.9\sqrt{f'_c} + \frac{2500\rho_w V_u d}{M_u}\right)b_w d \tag{8.35}$$

The critical section for shear design is at $X = 0.15l_n$ for uniform loads and $X = 0.5a$ for concentrated loads.

REFERENCES

1. "Report of the ACI-ASCE Committee 326." *ACI Journal* 59 (1962).
2. ACI-ASCE Committee 426. "The Shear Strength of Reinforced Concrete Members." *ASCE Journal, Structural Division* (June 1973).
3. R. C. Fenwick and T. Paulay. "Mechanisms of Shear Resistance of Concrete Beams." *ASCE Journal, Structural Division* (October 1968).
4. G. N. Kani. "Basic Facts Concerning Shear Failure." *ACI Journal* 63 (June 1966).
5. D. W. Johnson and P. Zia. "Analysis of Dowel Action." *ASCE Journal, Structural Division* 97 (May 1971).
6. H. P. Taylor. "The Fundamental Behavior of Reinforced Concrete Beams in Bending and Shear." *Shear in Reinforced Concrete,* vol. 1. American Concrete Institute Special Publication 42. Detroit, 1974.
7. A. L. L. Baker. *Limit-State Design of Reinforced Concrete.* Cement and Concrete Association. London, 1970.
8. P. E. Regan and M. H. Khan. "Bent-up Bars as Shear Reinforcement." *Shear in Reinforced Concrete,* vol. 1. American Concrete Institute Special Publication 42. Detroit, 1974.
9. J. G. MacGregor. "The Design of Reinforced Concrete Beams for Shear." *Shear in Reinforced Concrete,* vol. 2. American Concrete Institute Special Publication 42. Detroit, 1974.
10. German Code of Practice, DIN 1045.
11. S. Y. Debaiky and E. I. Elniema. "Behavior and Strength of Reinforced Concrete Beams in Shear." *ACI Journal* 79 (May–June 1982).
12. P. Marti. "Truss Models in Detailing." *Concrete International Design and Construction* 7, no. 12 (December 1985).
13. P. Marti. "Basic Tools of Beam Design." *ACI Journal* 82, no. 1 (January–February 1985).
14. CEB-FIP Model Code 1990. *Comité Euro-International du Béton.* London: Thomas Telford, 1993.
15. D. M. Rogowsky and J. MacGregor. "Design of Deep R.C. Continuous Beams." *Concrete International* 8, no. 8 (August 1986).
16. ACI Code. "Building Code Requirements for Structural Concrete." ACI (318-99) *American Concrete Institute.* Detroit, 1999.

PROBLEMS

8.1 Design the necessary shear reinforcement (if needed) in the form of U-stirrups (two legs) for the T-section shown in Figure 8.26. Use $f'_c = 4$ Ksi (28 MPa) and $f_y = 60$ Ksi (420 MPa).

a. $V_u = 24$ K (108 kN)

b. $V_u = 64$ K (285 kN)

c. $V_u = 79$ K (351 kN)

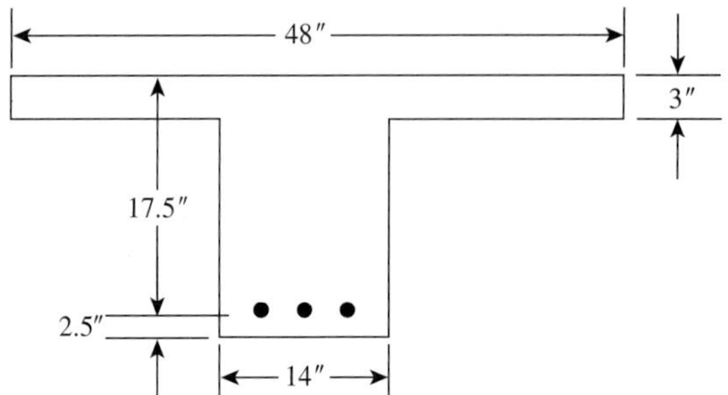

Figure 8.26 Problem 8.1.

8.2 Repeat Problem 8.1 for the section shown in Figure 8.27.

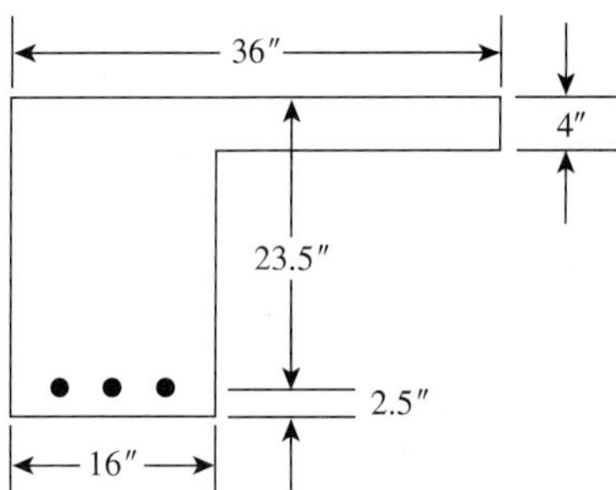

Figure 8.27 Problem 8.2.

8.3 Design the necessary shear reinforcement (if needed) in the form of U-stirrups (two legs) for the rectangular section shown in Figure 8.28 using $f'_c = 3$ Ksi (21 MPa) and $f_y = 60$ Ksi (420 MPa).

a. $V_u = 55$ K (245 kN)

b. $V_u = 110$ K (490 kN)

c. $V_u = 144$ K (640 kN)

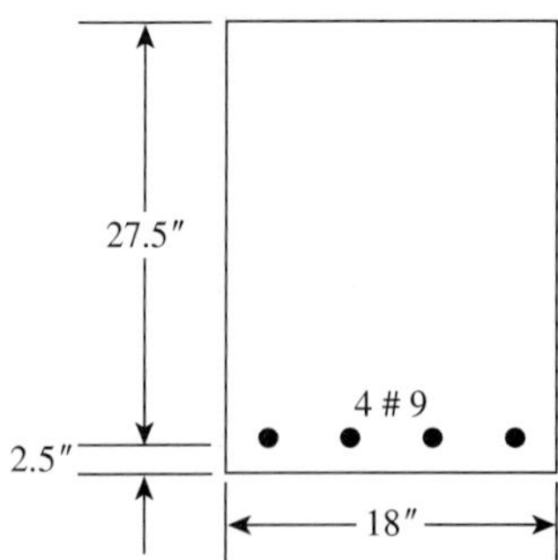

Figure 8.28 Problem 8.3.

8.4 A 16-ft- (4.8-m-) span simply supported beam has a clear span of 15 ft (4.5 m) and is supported by 12 × 12-in. (300 × 300-mm) columns. The beam carries an ultimate uniform load of 11.5 K/ft (172 kN/m). The dimensions of the beam section and the flexural steel reinforcement are shown in Figure 8.29. Design the necessary shear reinforcements using $f'_c = 3$ Ksi (21 MPa) and $f_y = 60$ Ksi (420 MPa). Show the distribution of stirrups along the beam.

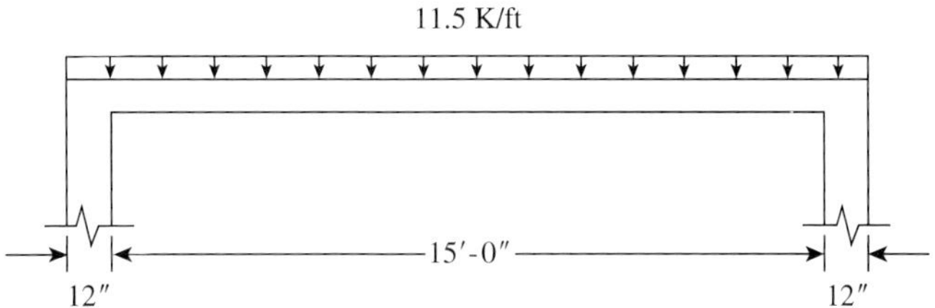

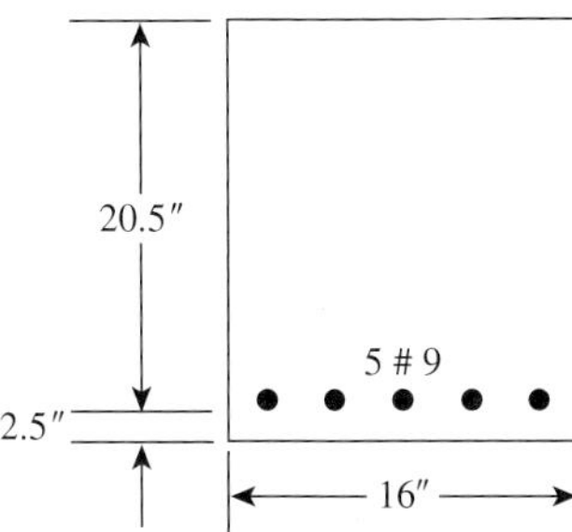

Figure 8.29 Problem 8.4.

8.5 An 18-ft- (5.4-m-) span simply supported beam carries a uniform dead load of 4 K/ft (60 kN/m) and a live load of 1.5 K/ft (22 kN/m). The beam has a width $b = 12$ in. (300 mm) and a depth $d = 24$ in. (600 mm) and is reinforced with six no. 9 bars (6 × 28 mm) in two rows. Check the beam for shear and design the necessary shear reinforcement. Given: $f'_c = 3$ Ksi (21 MPa) and $f_y = 40$ Ksi (280 MPa).

8.6 Design the necessary shear reinforcement for a 14-ft (4.2-m) simply supported beam that carries an ultimate uniform load of 10 K/ft (150 kN/m) (including self-weight) and an ultimate concentrated load at midspan $P_u = 24$ K (108 kN). The beam has a width $b = 14$ in. (350 mm) and a depth $d = 16.5$ (400 mm) and is reinforced with four no. 8 bars (4 × 25 mm). Given: $f'_c = 4$ Ksi (28 MPa) and $f_y = 60$ Ksi (420 MPa).

8.7 A cantilever beam with 7.4-ft (2.20-m) span carries a uniform dead load of 2 K/ft (29 kN/m) (including self-weight) and a concentrated live load of 18 K (80 kN) at a distance of 3 ft (0.9 m) from the face of the support. Design the beam for moment and shear using the strength design method. Given: $f'_c = 3$ Ksi (21 MPa), $f_y = 60$ Ksi (420 MPa), and $b = 12$ in. (200 mm), and use $\rho = 3/4\rho_{max}$.

8.8 Design the critical sections of an 11-ft (3.3-m-) span simply supported beam for bending moment and shearing forces using ρ_{max} and the strength design method. Given: $f'_c = 3$ Ksi (21 MPa), $f_y = 60$ Ksi (420 MPa), and $b = 10$ in. (250 mm). Dead load is 2.75 K/ft (40 kN/m) and live load is 1.375 K/ft (20 kN/m).

8.9 A rectangular beam is to be designed to carry an ultimate shearing force of 75 kips (335 kN). Determine the minimum beam section if controlled by shear $\left(V_c = 2\sqrt{f'_c}\,bd\right)$ using the minimum shear reinforcement as specified by the ACI Code and no. 3 stirrups. Given: $f'_c = 4$ Ksi (28 MPa), $f_y = 40$ Ksi (280 MPa), and $b = 16$ in. (400 mm).

8.10 Redesign Problem 8.5 using $f_y = 60$ Ksi.

8.11 Redesign the shear reinforcement of the beam in Problem 8.6 if the uniform ultimate load of 6 K/ft (90 kN/m) is due to dead load and the ultimate concentrated load $P_u = 24$ K (108 kN) is due to a moving live load. Change the position of the live load to cause maximum shear at the support and at midspan.

8.12 Design a cantilever beam that has a span of 9 ft (2.7 m) to carry an ultimate triangular load that varies from zero load at the free end to maximum load of 8 K/ft (120 kN/m) at the face of the support. The beam shall have a variable depth, with minimum depth at the free end of 10 in. (250 mm) and increasing linearly toward the support. Use maximum steel percentage ρ_{max} for flexural design. Given: $f'_c = 3$ Ksi (21 MPa), $f_y = 60$ Ksi (420 MPa) for flexural reinforcement, $f_y = 40$ Ksi (280 MPa) for stirrups, and $b = 11$ in. (275 mm).

9 ONE-WAY SLABS

The Westin Hotel, Toronto, Ontario, Canada.

9.1 TYPES OF SLABS

Structural concrete slabs are constructed to provide flat surfaces, usually horizontal, in building floors, roofs, bridges, and other types of structures. The slab may be supported by walls, by reinforced concrete beams usually cast monolithically with the slab, by structural steel beams, by columns, or by the ground. The depth of a slab is usually very small compared to its span. See Figure 9.1.

Structural concrete slabs in buildings may be classified as follows:

1. *One-way slabs:* If a slab is supported on two opposite sides only, it will bend or deflect in a direction perpendicular to the supported edges. The structural action is one way, and the loads are carried by the slab in the deflected short direction. This type of slab is called a *one-way slab* (Figure 9.1(a)). If the slab is supported on four sides and the ratio of the long side to the short side is equal to or greater than 2, most of the load (about 95% or more) is carried in the short direction, and one-way action is considered for all practical purposes (Figure 9.1(b)). If the slab is made of reinforced concrete with no voids, then it is called a *one-way solid slab.* Figure 9.1(c), (d), and (e) shows cross sections and bar distribution.
2. *One-way joist floor system:* This type of slab is also called a *ribbed slab.* It consists of a floor slab, usually 2 to 4 in. (50 to 100 mm) thick, supported by reinforced concrete

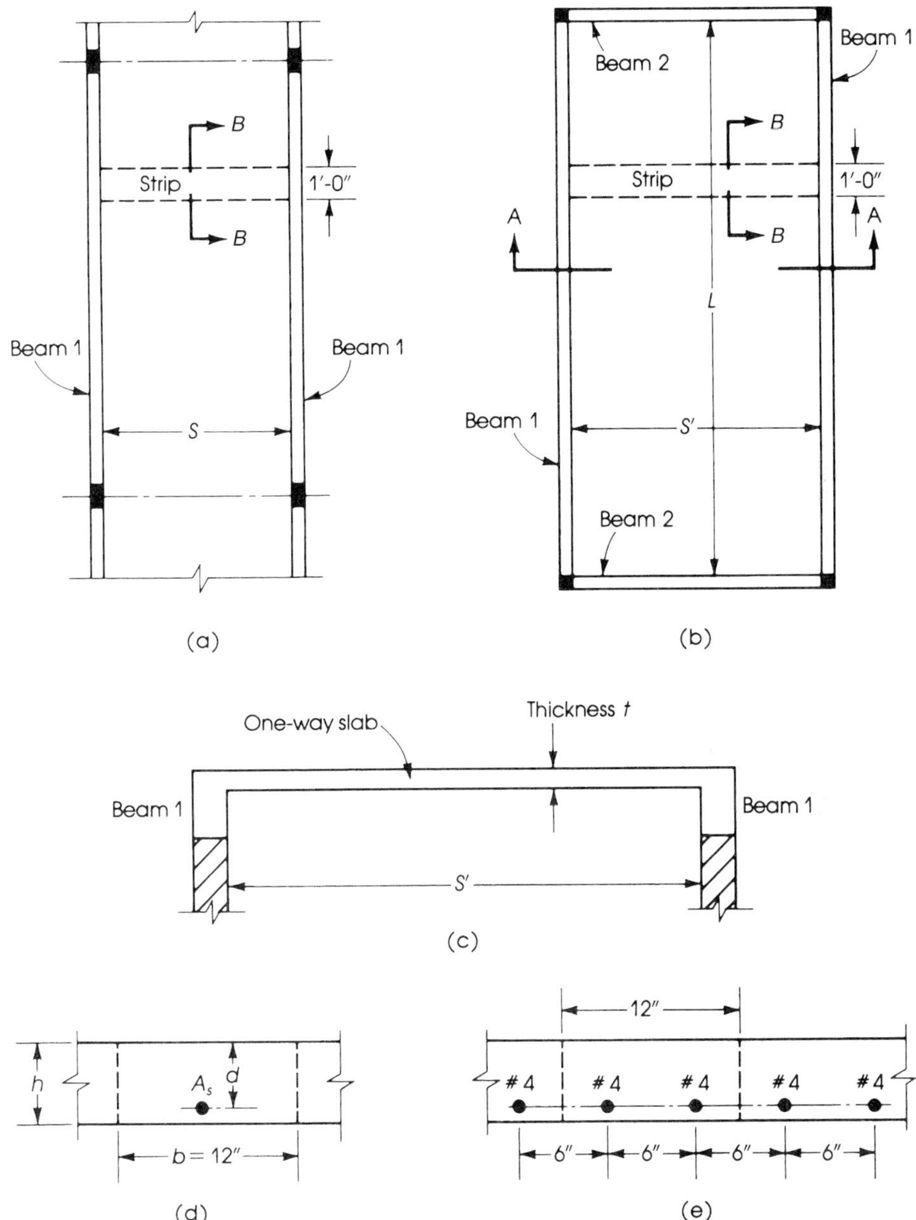

Figure 9.1 One-way slabs.

ribs (or joists). The ribs are usually tapered and are uniformly spaced at distances that do not exceed 30 in. (750 mm). The ribs are supported on girders that rest on columns. The spaces between the ribs may be formed using removable steel or fiberglass form fillers (pans), which may be used many times (Figure 9.2). In some ribbed slabs, the spaces between ribs may be filled with permanent fillers to provide a horizontal slab.

3. *Two-way floor systems:* When the slab is supported on four sides and the ratio of the long side to the short side is less than 2, the slab will deflect in double curvature in both directions. The floor load is carried in two directions to the four beams surrounding the slab (refer to Chapter 17). Other types of *two-way floor systems* are flat plate floors, flat slabs, and waffle slabs, all explained in Chapter 17. This chapter deals only with one-way floor systems.

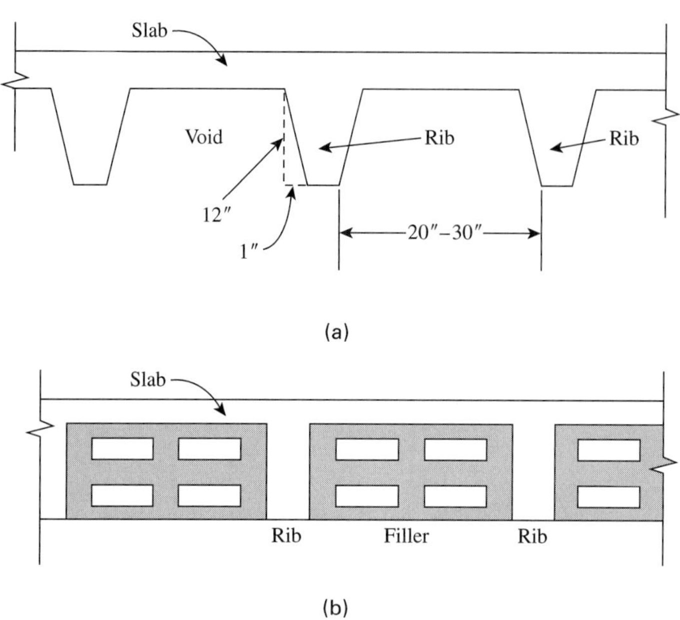

Figure 9.2 Cross sections of one-way ribbed slab: (a) without fillers and (b) with fillers.

9.2 DESIGN OF ONE-WAY SOLID SLABS

If the concrete slab is cast in one uniform thickness without any type of voids, it can be referred to as a *solid slab.* In a one-way slab, the ratio of the length of the slab to its width is greater than 2. Nearly all the loading is transferred in the short direction, and the slab may be treated as a beam. A unit strip of slab, usually 1 ft (or 1 m) at right angles to the supporting girders, is considered as a rectangular beam. The beam has a unit width with a depth equal to the thickness of the slab and a span length equal to the distance between the supports. A one-way slab thus consists of a series of rectangular beams placed side by side (Figure 9.1).

If the slab is one span only and rests freely on its supports, the maximum positive moment M for a uniformly distributed load of w psf is $M = (wL^2)/8$, where L is the span length between the supports. If the same slab is built monolithically with the supporting beams or is continuous over several supports, the positive and negative moments are calculated either by structural analysis or by moment coefficients as for continuous beams. The ACI Code, Section 8.3, permits the use of moment and shear coefficients in the case of two or more approximately equal spans (Figure 9.3). This condition is met when the larger of two adjacent spans does not exceed the shorter span by more than 20%. For uniformly distributed loads, the unit live load shall not exceed three times the unit dead load. When these conditions are not satisfied, structural analysis is required. In structural analysis, the negative bending moments at the centers of the supports are calculated. The value that may be considered in the design is the negative moment at the face of the support. To obtain this value, subtract from the maximum moment value at the center of the support a quantity equal to $Vb/3$, where V is the shearing force calculated from the analysis and b is the width of the support:

$$M_f \text{ (at face of the support)} = M_c \text{ (at centerline of support)} - \frac{Vb}{3} \qquad (9.1)$$

In addition to moment, diagonal tension and development length of bars should also be checked for proper design.

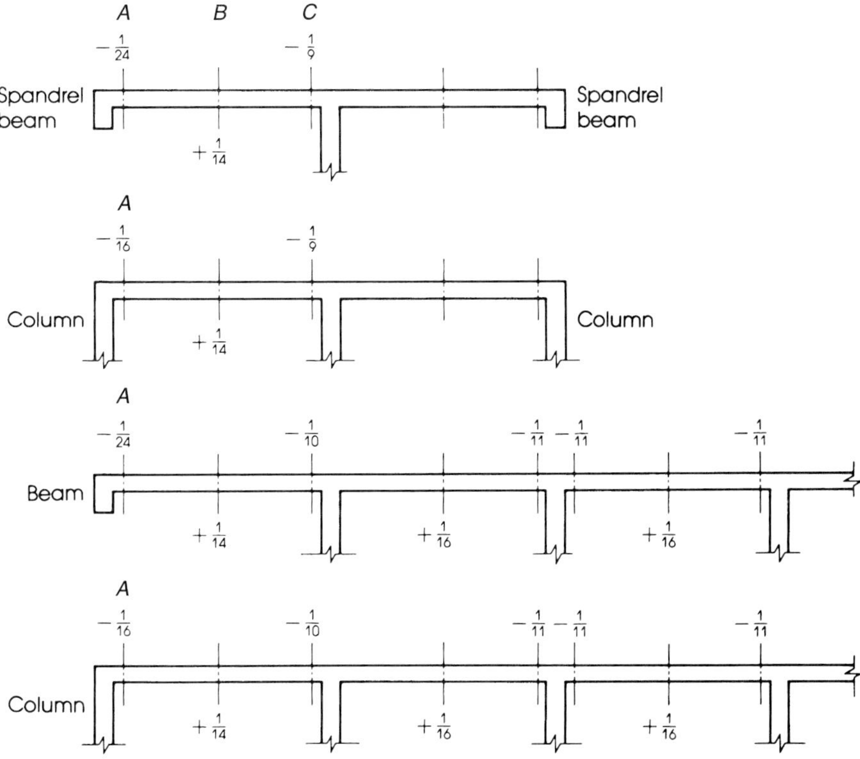

Figure 9.3 Moment coefficients for continuous beams and slabs (ACI Code, Section 8.3).

The conditions under which the moment coefficients for continuous beams and slabs given in Figure 9.3 should be used can be summarized as follows:

1. Spans are approximately equal: longer span $\leq$ 1.2 (shorter span).
2. Loads are uniformly distributed.
3. The ratio (live load/dead load) is less than or equal to 3.
4. For slabs with spans less than or equal to 10 ft, negative bending moment at face of all supports is $\left(\frac{1}{12}\right)w_u l_n^2$.
5. For an unrestrained discontinuous end at A, the coefficient is 0 at A and $+\frac{1}{11}$ at B.
6. Shearing force at C is $1.15w_u l_n/2$ and at the face of all other support is $\frac{1}{2}w_u l_n$.
7. M_u = (coefficient)$\left(w_u l_n^2\right)$ and l_n = clear span.

9.3 DESIGN LIMITATIONS ACCORDING TO THE ACI CODE

The following limitations are specified by the ACI Code.

1. A typical imaginary strip 1 ft (or 1 m) wide is assumed.
2. The minimum thickness of one-way slabs using grade 60 steel according to the ACI Code, Table 9.5(a), for solid slabs and for (beams or ribbed one-way slabs) should be equal to the following:
   - For simply supported spans: solid slabs, $h = L/20$ (ribbed slabs, $h = L/16$).
   - For one-end continuous spans: solid slabs, $h = L/24$ (ribbed slabs, $h = L/18.5$).

- For both-ends continuous spans: solid slabs, $h = L/28$ (ribbed slabs, $h = L/21$).
- For cantilever spans: solid slabs, $h = L/10$ (ribbed slabs, $h = L/8$).
- For f_y other than 60 Ksi, these values shall be multiplied by $0.4 + 0.01f_y$, where f_y is in Ksi. This minimum thickness should be used unless computation of deflection indicates a lesser thickness can be used without adverse effects.

3. Deflection is to be checked when the slab supports are attached to construction likely to be damaged by large deflections. Deflection limits are set by the ACI Code, Table 9.5(b).
4. It is preferable to choose slab depth to the nearest $\frac{1}{2}$ in. (or 10 mm).
5. Shear should be checked, although it does not usually control.
6. Concrete cover in slabs shall not be less than $\frac{3}{4}$ in. (20 mm) at surfaces not exposed to weather or ground. In this case, $d = h - (\frac{3}{4}\text{ in.}) - $ (half-bar diameter). Refer to Figure 9.1(d).
7. In structural slabs of uniform thickness, the minimum amount of reinforcement in the direction of the span shall not be less than that required for shrinkage and temperature reinforcement (ACI Code, Section 7.12).
8. The principal reinforcement shall be spaced not farther apart than three times the slab thickness nor more than 18 in. (ACI Code, Section 7.6.5).
9. Straight-bar systems may be used in both tops and bottoms of continuous slabs. An alternative bar system of straight and bent (trussed) bars placed alternately may also be used.
10. In addition to main reinforcement, steel bars at right angles to the main reinforcement must be provided. This additional steel is called *secondary, distribution, shrinkage,* or *temperature reinforcement.*

9.4 TEMPERATURE AND SHRINKAGE REINFORCEMENT

Concrete shrinks as the cement paste hardens, and a certain amount of shrinkage is usually anticipated. If a slab is left to move freely on its supports, it can contract to accommodate the shrinkage. However, slabs and other members are joined rigidly to other parts of the structure, causing a certain degree of restraint at the ends. This results in tension stresses known as *shrinkage stresses.* A decrease in temperature and shrinkage stresses is likely to cause hairline cracks. Reinforcement is placed in the slab to counteract contraction and distribute the cracks uniformly. As the concrete shrinks, the steel bars are subjected to compression.

Reinforcement for shrinkage and temperature stresses normal to the principal reinforcement should be provided in a structural slab in which the principal reinforcement extends in one direction only. The ACI Code, Section 7.12.2, specifies the following minimum steel ratios: For slabs in which grade 40 or 50 deformed bars are used, $\rho = 0.2\%$, and for slabs in which grade 60 deformed bars or welded bars or welded wire fabric are used, $\rho = 0.18\%$. In no case shall such reinforcement be placed farther apart than five times the slab thickness nor more than 18 in.

For temperature and shrinkage reinforcement, the whole concrete depth h exposed to shrinkage shall be used to calculate the steel area. For example, if a slab has a total depth $h = 6$ in. and $f_y = 60$ Ksi, then the area of steel required per 1-ft width of slab is $A_s = 6(12)(0.0018) = 0.129$ in.^{2}. The spacings of the bars S can be determined as follows:

$$S = \frac{12A_b}{A_s} \tag{9.2}$$

where A_b = area of the bar chosen and A_s = calculated area of steel.

For example, if no. 3 bars are used $(A_b = 0.11 \text{ in.}^2)$, then $S = 12(0.11)/0.129 = 10.33$ in., say, 10 in. If no. 4 bars are chosen $(A_b = 0.2 \text{ in.}^2)$, then $S = 12(0.2)/0.129 = 18.6$ in., say, 18 in. Maximum spacing is the smaller of five times slab thickness (30 in.) or 18 in. Then no. 4 bars spaced at 18 in. are adequate (or no. 3 bars at 10 in.). These bars act as secondary reinforcement and are placed normal to the main reinforcement calculated by flexural analysis. Note that areas of bars in slabs are given in Table A.14.

9.5 REINFORCEMENT DETAILS

In continuous one-way slabs, the steel area of the main reinforcement is calculated for all critical sections, at midspans, and at supports. The choice of bar diameter and detailing depends mainly on the steel areas, spacing requirements, and development length. Two bar systems may be adopted.

In the straight-bar system (Figure 9.4), straight bars are used for top and bottom reinforcement in all spans. The time and cost to produce straight bars is less than that required to produce bent bars; thus, the straight-bar system is widely used in construction.

In the bent-bar, or trussed, system, straight and bent bars are placed alternately in the floor slab. The location of bent points should be checked for flexural, shear, and development length requirements. For normal loading in buildings, the bar details at the end and interior spans of one-way solid slabs may be adopted as shown in Figure 9.4.

9.6 DISTRIBUTION OF LOADS FROM ONE-WAY SLABS TO SUPPORTING BEAMS

In one-way floor slab systems, the loads from slabs are transferred to the supporting beams along the long ends of the slabs. The beams transfer their loads in turn to the supporting columns.

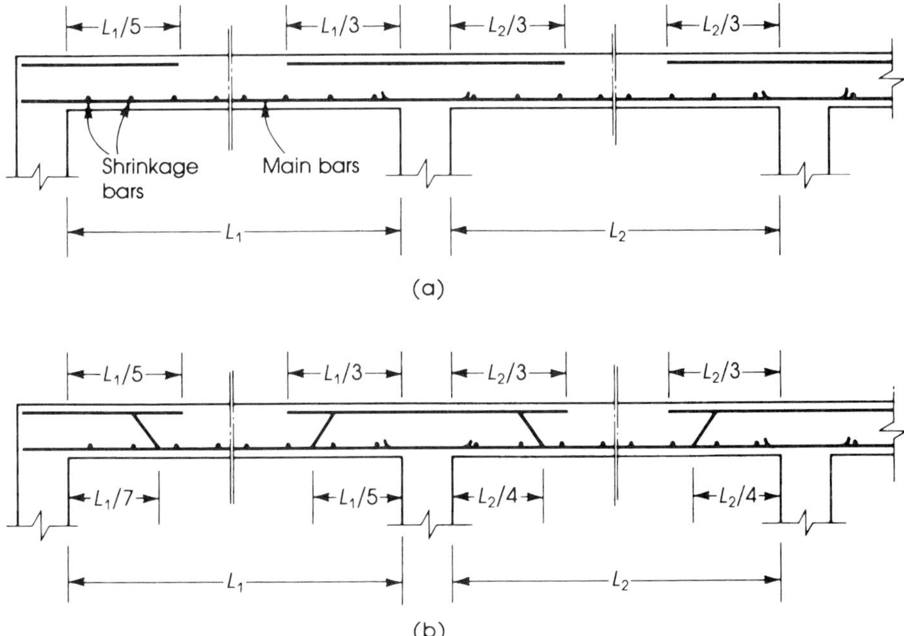

Figure 9.4 Reinforcement details in continuous one-way slabs: (a) straight bars and (b) bent bars.

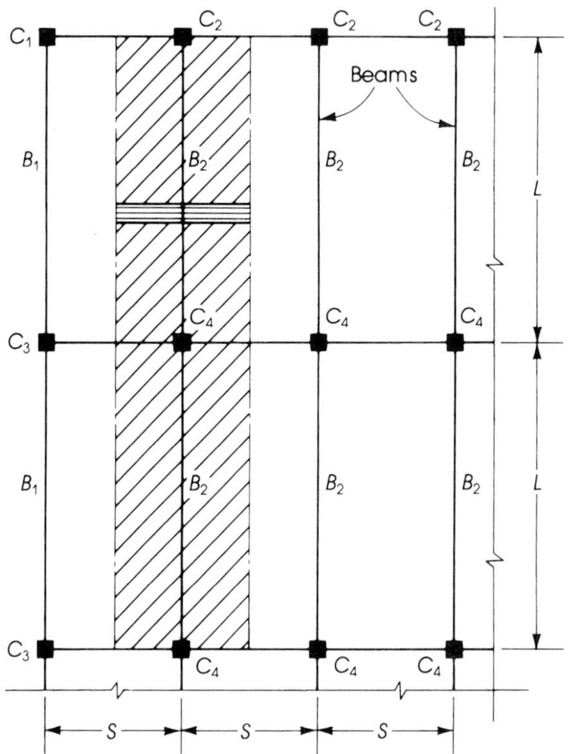

Figure 9.5 Distribution of loads on beams.

From Figure 9.5, it can be seen that beam B_2 carries loads from two adjacent slabs. Considering a 1-ft length of beam, the load transferred to the beam is equal to the area of a strip 1 ft wide and S feet in length multiplied by the intensity of load on the slab. This load produces a uniformly distributed load on the beam:

$$U_B = U_S \cdot S$$

The uniform load on the end beam, B_1, is half the load on B_2, because it supports a slab from one side only.

The load on column C_4 is equal to the reactions from two adjacent beams B_2:

$$\text{Load on column } C_4 = U_B L = U_S L S$$

The load on column C_3 is one-half the load on column C_4, because it supports loads from slabs on one side only. Similarly, the loads on columns C_2 and C_1 are

$$\text{Load on } C_2 = U_S \frac{L}{2} S = \text{load on } C_3$$

$$\text{Load on } C_1 = U_S \left(\frac{L}{2}\right)\left(\frac{S}{2}\right)$$

From this analysis, it can be seen that each column carries loads from slabs surrounding the column and up to the centerline of adjacent slabs: up to $L/2$ in the long direction and $S/2$ in the short direction.

Distribution of loads from two-way slabs to their supporting beams and columns is discussed in Chapter 17.

Example 9.1

Calculate the ultimate moment capacity of a one-way solid slab that has a total depth $h = 7$ in. and is reinforced with no. 6 bars spaced at $S = 7$ in. Use $f'_c = 3$ Ksi and $f_y = 60$ Ksi.

Solution

1. Determine the effective depth, d:

$$d = h - \tfrac{3}{4}\text{ in.(cover)} - \text{half-bar diameter} \qquad \text{(see Figure 9.1(d))}$$

$$d = 7 - \tfrac{3}{4} - \tfrac{6}{16} = 5.875 \text{ in.}$$

2. Determine the average A_s provided per 1-ft width (12 in.) of slab. The area of no. 6 bar is $A_b = 0.44$ in.^{2}.

$$A_s = \frac{12A_b}{S} = \frac{12(0.44)}{7} = 0.754 \text{ in.}^2/\text{ft}$$

Areas of bars in slabs are given in Table A.14 in Appendix A.

3. Compare the steel ratio used with $\rho_{\max}$ and $\rho_{\min}$. For $f'_c = 3$ Ksi and $f_y = 60$ Ksi, $\rho_{\max} = 0.0161$ and $\rho_{\min} = 0.00333$. ρ (used) $= 0.754/(12 \times 5.875) = 0.0107$, which is adequate.
4. Calculate $\phi M_n = \phi A_s f_y (d - a/2)$.

$$a = A_s f_y / (0.85 f'_c b) = 0.754(60)/(0.85 \times 3 \times 12) = 1.48 \text{ in.}$$

$$\phi M_n = 0.9(0.754)(60)(5.875 - 1.48/2) = 209 \text{ K}\cdot\text{in.} = 17.42 \text{ K}\cdot\text{ft}$$

Example 9.2

Determine the allowable uniform live load that can be applied on the slab of the previous example if the slab span is 16 ft between simple supports and carries a uniform dead load (excluding self-weight) of 100 psf.

Solution

1. The ultimate moment capacity of the slab is 17.42 K·ft per 1-ft width of slab.

$$M_u = \phi M_n = 17.42 = \frac{W_u L^2}{8} = \frac{W_u(16)^2}{8}$$

The ultimate uniform load is $W_u = 0.544 \text{ K/ft}^2 = 544$ psf.

2. $W_u = 1.4D + 1.7L$

$$D = 100 \text{ psf} + \text{self-weight} = 100 + \tfrac{7}{12}(150) = 187.5 \text{ psf}$$

$$544 = 1.4(187.5) + 1.7L, \qquad L = 166 \text{ psf}$$

Example 9.3

Design a 12-ft simply supported slab to carry a uniform dead load (excluding self-weight) of 120 psf and a uniform live load of 100 psf. Use $f'_c = 3$ Ksi, $f_y = 60$ Ksi, and the ACI Code limitations.

Solution

1. Assume a slab thickness. For $f_y = 60$ Ksi, the minimum depth to control deflection is $L/20 = 12(12)/20 = 7.2$ in. Assume a total depth $h = 7$ in. and assume $d = 6$ in. (to be checked later).
2. Calculate ultimate load: weight of slab $= \tfrac{7}{12}(150) = 87.5$ psf.

$$W_u = 1.4D + 1.7L = 1.4(87.5 + 120) + 1.7(100) = 432.5 \text{ psf}$$

For a 1-ft width of slab, $M_u = W_u L^2/8$.

$$M_u = \frac{0.4325(12)^2}{8} = 7.785 \text{ K}\cdot\text{ft}$$

3. Calculate A_s: For $M_u = 7.785$ K · ft, $b = 12$ in. and $d = 6$ in., then $R_u = M_u/bd^2 = 7.785(12{,}000)/(12)(6)^2 = 216$ psi. From tables in Appendix A, $\rho = 0.0042$.

$$A_s = \rho bd = 0.0042(12)(6) = 0.30 \text{ in.}^2$$

Choosing no. 4 bars ($A_b = 0.2$ in.^{2}), then $S = 12A_b/A_s = 12(0.2)/0.3 = 8$ in. Check actual $d = h - \frac{3}{4} - \frac{4}{16} = 6$ in. It is acceptable.

4. Check the moment capacity of the final section.

$$a = \frac{A_s f_y}{(0.85 f'_c b)} = \frac{0.3(60)}{0.85 \times 3 \times 12} = 0.59 \text{ in.}$$

$$\phi M_n = \phi A_s f_y\left(d - \frac{a}{2}\right) = 0.9(0.3)(60)(6 - 0.59/2) = 92.42 \text{ K·in.} = 7.7 \text{ K·ft}$$

Because ϕM_n is slightly less than $M_u = 7.785$ K·ft, use no. 4 bars spaced at $7\frac{1}{2}$ in. and check ϕM_n again.

$$\text{Actual } A_s = \frac{12A_b}{S} = \frac{12(0.2)}{7.5} = 0.32 \text{ in.}^2$$

$$a = 0.63 \text{ in.,} \qquad \phi M_n = 98.24 \text{ K·in.} = 8.16 \text{ K·ft} > M_u$$

$$\rho = \frac{0.32}{6 \times 12} = 0.00444$$

which is adequate.

5. Calculate the secondary (shrinkage) reinforcement normal to the main steel. For $f_y = 60$ Ksi,

$$\rho_{\min} = 0.0018$$

$$A_{sh} = \rho bh = 0.0018(12)(7) = 0.1512 \text{ in.}^2$$

Choose no. 4 bars, $A_b = 0.2$ in.^{2}, $S = 12A_b/A_s = 12(0.2)/0.1512 = 15.9$ in. Use no. 4 bars spaced at 15 in.

6. Check shear requirements: V_u at a distance d from the support is $0.4325(\frac{12}{2} - \frac{6}{12}) = 2.38$ K.

$$\phi V_c = \phi 2\sqrt{f'_c}\,bd = \frac{0.85(2)(\sqrt{3000})(12 \times 6)}{1000} = 6.7 \text{ K}$$

$\frac{1}{2}\phi V_c = 3.35 \text{ K} > V_u$, so the shear is adequate.

7. Final section: $h = 7$ in., main bars = no. 4 spaced at $7\frac{1}{2}$ in., and secondary bars = no. 4 spaced at 15 in.

Example 9.4

The cross section of a continuous one-way solid slab in a building is shown in Figure 9.6. The slabs are supported by beams that span 24 ft between simple supports. The dead load on the slabs is that due to self-weight plus 60 psf; the live load is 120 psf. Design the continuous slab and draw a detailed section. Given: $f'_c = 3$ Ksi, $f_y = 40$ Ksi.

Solution

1. The minimum thickness of the first slab is $L/30$, because one end is continuous and the second end is discontinuous. The distance between centers of beams may be considered the span L, here equal to 12 ft. For $f_y = 40$ Ksi,

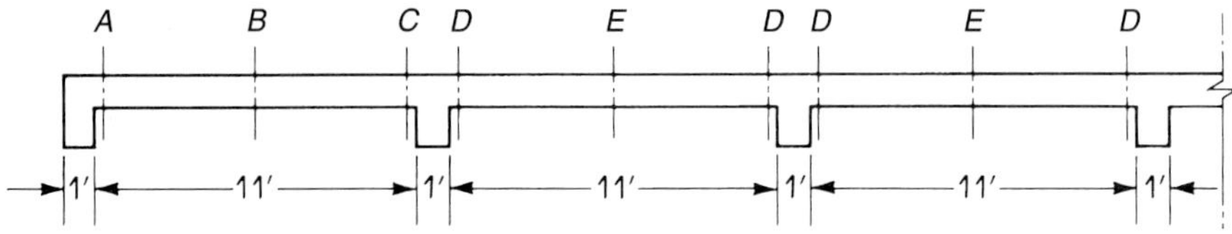

Figure 9.6 Example 9.4.

$$\text{Minimum total depth} = \frac{L}{30} = \frac{12 \times 12}{30} = 4.8 \text{ in.}$$

$$\text{Minimum total depth for interior span} = \frac{L}{35} = 4.1 \text{ in.}$$

Assume a uniform thickness of 5 in., which is greater than 4.8 in.; therefore, it is not necessary to check deflection.

2. Calculate loads and moments in a unit strip:

$$\text{Dead load} = \text{weight of slab} + 60 \text{ psf}$$

$$= \left(\frac{5}{12} \times 150\right) + 60 = 122.5 \text{ psf}$$

$$\text{Ultimate load } U = 1.4D + 1.7L = 1.4 \times 122.5 + 1.7 \times 120 = 375.5 \text{ psf}$$

The clear span is 11.0 ft. The ultimate moment in the first span is over the support and equals $UL^2/10$.

$$M_u = \frac{U(11)^2}{10} = (0.3755)\frac{121}{10} = 4.54 \text{ K}\cdot\text{ft} = 54.5 \text{ K}\cdot\text{in.}$$

3. Assume $\rho = 1.4\%$; then $R_u = 450$ psi $= 0.45$ Ksi. This value is less than $\rho_{\max}$ of 0.0278 and greater than $\rho_{\min}$ of 0.005.

$$d = \sqrt{\frac{M_u}{R_u b}} = \sqrt{\frac{54.5}{0.45 \times 12}} = 3.18 \text{ in.}$$

$$A_s = \rho bd = 0.014(12)(3.18) = 0.53 \text{ in.}^2$$

Choosing no. 5 bars,

$$\text{Total depth} = d + \tfrac{1}{2}\text{ bar diameter} + \text{cover} = 3.18 + \tfrac{5}{16} + \tfrac{3}{4} = 4.25 \text{ in.}$$

Use slab thickness of 5 in., as assumed earlier.

$$\text{Actual } d \text{ used} = 5 - \tfrac{3}{4} - \tfrac{5}{16} = 3.9 \text{ in.}$$

4. Moments and steel reinforcement required at other sections using $d = 3.9$ in. are as follows:

Location	Moment Coefficient	M_u (K · in.)	$R_u = M_u/bd^2$ (psi)	ρ (%)	A_s (in.^{2})	Bars and Spacings
A	$-\frac{1}{24}$	22.7	Small	0.50	0.23	No. 4 at 10 in.
B	$+\frac{1}{14}$	38.9	213	0.65	0.30	No. 5 at 12 in.
C	$-\frac{1}{10}$	54.5	300	0.90	0.44	No. 5 at 8 in.
D	$-\frac{1}{11}$	49.6	271	0.80	0.38	No. 5 at 8 in.
E	$+\frac{1}{16}$	34.1	187	0.55	0.26	No. 4 at 8 in.

The arrangement of bars is shown in Figure 9.7.

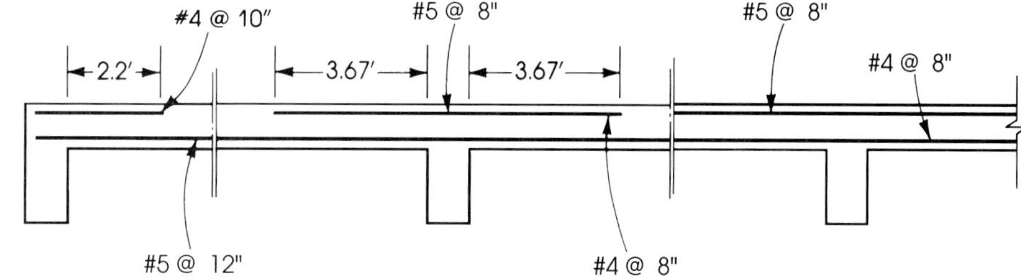

Figure 9.7 Reinforcement details of Example 9.4.

5. Maximum shear occurs at the exterior face of the second support, section C.

$$V_u \text{ (at C)} = 1.15UL_n/2 = \frac{1.15(0.3755)(11)}{2} = 2.375 \text{ K/ft of width}$$

$$\phi V_c = \phi 2\sqrt{f'_c}bd = \frac{0.85(2)(\sqrt{3000})(12)(3.9)}{1000} = 4.36 \text{ K}$$

This result is acceptable. Note that the provision of minimum area of shear reinforcement when V_u exceeds $\frac{1}{2}\phi V_c$ does not apply to slabs (ACI Code, Section 11.5.5).

Example 9.5

Determine the uniform ultimate load on an intermediate beam supporting the slabs of Example 9.4. Also calculate the axial load on an interior column; refer to the general plan of Figure 9.5.

Solution

1. The uniform ultimate load per foot length on an intermediate beam is equal to the ultimate uniform load on slab multiplied by S, the short dimension of the slab. Therefore,

$$U \text{ (beam)} = U \text{ (slab)} \times S = 0.3755 \times 12 = 4.5 \text{ K/ft}$$

The weight of the web of the beam shall be added to this value. Span of the beam is 24 ft.

$$\text{Estimated total depth} = \frac{L}{20} \times 0.8 = \left(\frac{24}{20} \times 0.8\right) \times 12 = 11.5 \text{ in.} \qquad \text{say, 12 in.}$$

Slab thickness is 5 in. and height of the web is $12 - 5 = 7$ in.

$$\text{Ultimate weight of beam web} = \left(\frac{7}{12} \times 150\right) \times 1.4 = 122.5 \text{ lb/ft}$$

$$\text{Total uniform load on beam} = 4.5 + 0.1225 = 4.62 \text{ K/ft}$$

2. Axial load on an interior column:

$$P_u = 4.62 \times 24 \text{ ft} = 110.9 \text{ K}$$

9.7 ONE-WAY JOIST FLOOR SYSTEM

A one-way joist floor system consists of hollow slabs with a total depth greater than that of solid slabs. The system is most economical for buildings where superimposed loads are small and spans are relatively large, such as schools, hospitals, and hotels. The concrete in the tension zone is ineffective; therefore, this area is left open between ribs or filled with lightweight material to reduce the self-weight of the slab.

The design procedure and requirements of ribbed slabs follow the same steps as those for rectangular and T-sections explained in Chapter 3. The following points apply to design of one-way ribbed slabs:

1. Ribs are usually tapered and uniformly spaced at about 16 to 30 in. (400 to 750 mm). Voids are usually formed by using pans (molds) 20 in. (500 mm) wide and 6 to 20 in. (150 to 500 mm) deep, depending on the design requirement. The standard increment in depth is 2 in. (50 mm).
2. The ribs shall not be less than 4 in. (100 mm) wide and must have a depth of not more than 3.5 times the width. Clear spacing between ribs shall not exceed 30 in. (750 mm) (ACI Code, Section 8.11).
3. Shear strength V_c provided by concrete for the ribs may be taken 10% greater than that for beams. This is mainly due to the interaction between the slab and the closely spaced ribs (ACI Code, Section 8.11.8).

One-way ribbed slab roof. The wide beams have the same total depth as the ribbed slab.

4. The thickness of the slab on top of the ribs is usually 2 to 4 in. (50 to 100 mm) and contains minimum reinforcement (shrinkage reinforcement). This thickness shall not be less than $\frac{1}{12}$ of the clear span between ribs or 2 in. (50 mm) (ACI Code, Section 8.11.6).
5. The ACI coefficients for calculating moments in continuous slabs can be used for continuous ribbed slab design.
6. There are additional practice limitations, which can be summarized as follows:
   - The minimum width of the rib is one-third of the total depth or 4 in. (100 mm), whichever is greater.
   - Secondary reinforcement in the slab in the transverse directions of ribs should not be less than the shrinkage reinforcement or one-fifth of the area of the main reinforcement in the ribs.
   - Secondary reinforcement parallel to the ribs shall be placed in the slab and spaced at distances not more than half of the spacings between ribs.
   - If the live load on the ribbed slab is less than 3 kN/m^2 (60 psf) and the span of ribs exceeds 5 m (17 ft), a secondary transverse rib should be provided at midspan (its direction is perpendicular to the direction of main ribs) and reinforced with the same amount of steel as the main ribs. Its top reinforcement shall not be less than half of the main reinforcement in the tension zone. These transverse ribs act as floor stiffeners.
   - If the live load exceeds 3 kN/m^2 (60 psf) and the span of ribs varies between 4 and 7 m (13 and 23 ft), one traverse rib must be provided, as indicated before. If the span exceeds 7 m (23 ft), at least two transverse ribs at one-third span must be provided with reinforcement, as explained before.

Example 9.6

Design an interior rib of a concrete joist floor system with the following description: span of rib = 20 ft (simply supported), dead load (excluding own weight) = 10 psf, live load = 80 psf, $f'_c = 3$ Ksi, and $f_y = 40$ Ksi.

Solution

1. Design of the slab: Assume a top slab thickness of 2 in. that is fixed to ribs that have a clear spacing of 20 in. No fillers are used. The self-weight of slab is $\frac{2}{12} \times 150 = 25$ psf.

$$\text{Total D.L.} = 25 + 10 = 35 \text{ psf}$$

$$U = 1.4D + 1.7L = 1.4 \times 35 + 1.7 \times 80 = 185 \text{ psf}$$

$$M_u = \frac{UL^2}{12} \qquad \text{(Slab is assumed fixed to ribs.)}$$

$$= \frac{0.185}{12}\left(\frac{20}{12}\right)^2 = 0.043 \text{ K}\cdot\text{ft} = 0.514 \text{ K}\cdot\text{in.}$$

Considering that the moment in slab will be carried by plain concrete only, the allowable flexural tensile strength is $f_t = 5\sqrt{f'_c}$, with a capacity-reduction factor $\phi = 0.65$, $f_t = 5\sqrt{3000} = 274$ psi.

$$\text{Flexural tensile strength} = \frac{Mc}{I} = \phi f_t, \quad I = \frac{bh^3}{12} = \frac{12(2)^3}{12} = 8 \text{ in.}^4, \quad c = \frac{h}{2} = \frac{2}{2} = 1 \text{ in.}$$

$$M = \phi f_t \frac{I}{c} = 0.65 \times 0.274 \times \frac{8}{1} = 1.42 \text{ K}\cdot\text{in.}$$

This value is greater than $M_u = 0.514$ K·in., and the slab is adequate. For shrinkage reinforcement, $A_s = 0.002 \times 12 \times 2 = 0.048$ in.^{2} Use no. 3 bars spaced at 12 in. laid transverse to the direction of the ribs. Welded wire fabric may be economically used for this low amount of steel reinforcement. Use similar shrinkage reinforcement no. 3 bars spaced at 12 in. laid parallel to the direction of ribs, one bar on top of each rib and one bar in the slab between ribs.

2. Calculate moment in a typical rib:

$$\text{Minimum depth} = \frac{L}{20} = \frac{20 \times 12}{20} = 12 \text{ in.}$$

The total depth of rib and slab is 10 + 2 = 12 in. Assume a rib width of 4 in. at the lower end that tapers to 6 in. at the level of the slab (Figure 9.8). The average width is 5 in. Note that the increase in the rib width using removable forms has a ratio of about 1 horizontal to 12 vertical.

$$\text{Weight of rib} = \frac{5}{12} \times \frac{10}{12} \times 150 = 52 \text{ lb/ft}$$

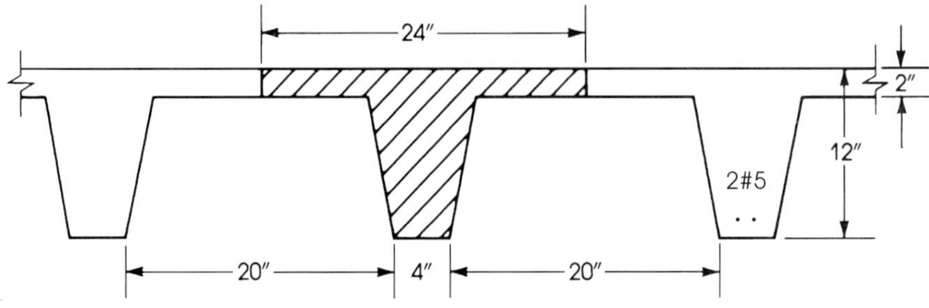

Figure 9.8 Example 9.6.

Rectangular steel pans used in one-way ribbed slab construction.

The rib carries a load from (20 + 4)-in.-wide slab plus its own weight:

$$U = \frac{24}{12} \times 185 + (1.4 \times 52) = 422 \text{ lb/ft}$$

$$M_u = \frac{UL^2}{8} = \frac{0.422}{8}(20)^2 \times 12 = 253 \text{ K} \cdot \text{in.}$$

3. Design of rib:

The total depth is 12 in. Assuming no. 5 bars and concrete cover of $\frac{3}{4}$ in., the effective depth d is $12 - \frac{3}{4} - \frac{5}{16} = 10.94$ in.

Check the moment capacity of the flange:

$$M_u \text{ (flange)} = \phi C\left(d - \frac{t}{2}\right), \quad \text{where} \quad C = 0.85 f'_c bt$$

$$M_u = 0.9(0.85 \times 3 \times 24 \times 2)\left(10.94 - \frac{2}{2}\right) = 1095 \text{ K} \cdot \text{in.}$$

The moment capacity of the flange is greater than the applied moment; thus, the rib acts as a rectangular section with $b = 24$ in., and the depth of the equivalent compressive block a is less than 2 in.

$$M_u = \phi A_s f_y\left(d - \frac{a}{2}\right) = \phi A_s f_y\left(d - \frac{A_s f_y}{1.7 f'_c b}\right)$$

$$253 = 0.9 A_s \times 40\left(10.94 - \frac{A_s \times 40}{1.7 \times 3 \times 24}\right) \quad \text{and} \quad A_s = 0.66 \text{ in.}^2$$

$$a = \frac{A_s f_y}{0.85 \times f'_c b} = 0.43 \text{ in.} < 2 \text{ in.}$$

Use two no. 5 bars per rib ($A_s = 0.65$ in.^{2}).

$$A_{s\,\min} = 0.005 b_w d = 0.005(5)(10.94) = 0.274 \text{ in.}^2 < 0.65 \text{ in.}^2$$

4. Calculate shear in the rib: The allowable shear strength of the rib web is

$$\phi V_c = \phi(1.1) \times 2\sqrt{f'_c}\, b_w d$$

$$= 0.85 \times 1.1 \times 2\sqrt{3000} \times 5 \times 10.94 = 5603 \text{ lb}$$

The ultimate shear at a distance d from the support is

$$V_u = 422\left(10 - \frac{10.94}{12}\right) = 3835 \text{ lb}$$

This is less than the shear capacity of the rib. Minimum stirrups may be used, and in this case an additional no. 4 bar will be placed within the slab above the rib to hold the stirrups in place.

It is advisable to add one transverse rib at midspan perpendicular to the direction of the ribs having the same reinforcement as that of the main ribs to act as a stiffener.

SUMMARY

Section 9.1

Slabs are of different types, one way (solid or joist floor systems) and two way (solid, ribbed, waffle, flat slabs, and flat plates).

Sections 9.2–9.3

1. The ACI Code moment and shear coefficients for continuous one-way slabs are given in Figure 9.3.
2. The minimum thickness of one-way slabs using grade 60 steel is $L/20, L/24, L/28, L/10$ for simply supported, one end continuous, both ends continuous, and cantilever slabs, respectively.

Section 9.4

The minimum shrinkage steel ratios $\rho_{\min}$ in slabs are 0.002 in. for slabs in which grade 40 or grade 50 bars are used and 0.0018 in. for slabs in which deformed bars of grade 60 are used.

$$\text{Maximum spacings between bars} \leq 5 \text{ times rib thickness} \leq 18 \text{ in.}$$

Sections 9.5–9.6

1. Reinforcement details are shown in Figure 9.4.
2. Distribution of loads from one-way slabs to the supporting beams is shown in Figure 9.5.

Section 9.7

The design procedure of ribbed slabs is similar to that of rectangular and T-sections. The width of ribs must be greater than or equal to 4 in., whereas the depth must be less than or equal to 3.5 times the width. The minimum thickness of the top slab is 2 in. or not less than one-twelfth of the clear span between ribs.

REFERENCES

1. Concrete Reinforcing Steel Institute. *CRSI Handbook.* Chicago, 1992.
2. Portland Cement Association. *Continuity in Concrete Building Frames.* Chicago, 1959.
3. American Concrete Institute. ACI Code 318-99, "Building Code Requirements for Structural Concrete." Detroit, Michigan, 1999.

PROBLEMS

9.1 For each problem, calculate the ultimate moment capacity of each concrete slab section using $f_y = 60$ Ksi.

Number	f'_c (Ksi)	h (in.)	Bars and Spacings (in.)	Answer ϕM_n (K · ft)
(a)	3	5	No. 4 at 6	6.35
(b)	3	6	No. 5 at 8	9.29
(c)	3	7	No. 6 at 9	14.06
(d)	3	8	No. 8 at 12	21.01
(e)	4	$5\frac{1}{2}$	No. 5 at 10	6.93
(f)	4	6	No. 7 at 12	11.80
(g)	4	$7\frac{1}{2}$	No. 6 at 6	22.68
(h)	4	8	No. 8 at 8	31.23
(i)	5	5	No. 5 at 10	6.19
(j)	5	6	No. 5 at 8	9.66

9.2 For each slab problem, determine the required steel reinforcement A_s and the total depth, if required; then choose adequate bars and their spacings. Use $f_y = 60$ Ksi for all problems, $b = 12$ in., and a steel ratio close to the steel ratio $\rho = A_s/bd$ given in some problems.

Number	f'_c (Ksi)	M_u (K · ft)	h (in.)	ρ (%)	One Answer	
					h (in.)	Bars
(a)	3	5.4	6	—	6	No. 4 at 9 in.
(b)	3	13.8	$7\frac{1}{2}$	—	$7\frac{1}{2}$	No. 6 at 10 in.
(c)	3	24.4	—	0.85	9	No. 8 at 12 in.
(d)	3	8.1	5	—	5	No. 5 at 7 in.
(e)	4	22.6	—	1.18	$7\frac{1}{2}$	No. 7 at 8 in.
(f)	4	13.9	$8\frac{1}{2}$	—	$8\frac{1}{2}$	No. 6 at 12 in.
(g)	4	13.0	—	1.10	6	No. 6 at 8 in.
(h)	4	11.2	—	0.51	$7\frac{1}{2}$	No. 5 at 9 in.
(i)	5	20.0	9	—	9	No. 7 at 12 in.
(j)	5	10.6	—	0.90	6	No. 6 at 10 in.

9.3 A 16-ft (4.8-m-) span simply supported slab carries a uniform dead load of 200 psf (10 kN/m^2) (excluding its own weight). The slab has a uniform thickness of 7 in. (175 mm) and is reinforced with no. 6 (20-mm) bars spaced at 5 in. (125 mm). Determine the allowable uniformly distributed load that can be applied on the slab if $f'_c = 3$ Ksi (21 MPa) and $f_y = 60$ Ksi (420 MPa).

9.4 Design a 10-ft (3-m) cantilever slab to carry a uniform total dead load of 120 psf (5.8 kN/m^2) and a concentrated live load at the free end of 2 K/ft (30 kN/m), when $f'_c = 4$ Ksi (28 MPa) and $f_y = 60$ Ksi (420 MPa).

9.5 A 6-in. (150-mm) solid one-way slab carries a uniform dead load of 160 psf (7.8 kN/m^2) (including its own weight) and a live load of 80 psf (3.9 kN/m^2). The slab spans 12 ft (3.6 m) between 10-in.- (250-mm-) wide simple supports. Determine the necessary slab reinforcement using $f'_c = 4$ Ksi (28 MPa) and $f_y = 50$ Ksi (350 MPa).

9.6 Repeat Problem 9.4 using a variable section with a minimum total depth at the free end of 4 in. (100 mm).

9.7 Design a continuous one-way solid slab supported on beams spaced at 14 ft (4.2 m) on centers. The width of the beams is 12 in. (300 mm), leaving clear slab spans of 13 ft (3.9 m). The slab carries a uniform dead load of 100 psf (4.8 kN/m^2) (including self-weight of slab) and a live load of 120 psf (5.8 kN/m^2). Use f'_c = 3 Ksi (21 MPa), f_y = 40 Ksi (280 MPa), and the ACI coefficients. Show bar arrangements using straight bars for all top and bottom reinforcement.

9.8 Repeat Problem 9.7 using equal clear spans of 10 ft (3 m), f'_c = 3 Ksi (21 MPa), and f_y = 60 Ksi (420 MPa).

9.9 Repeat Problem 9.7 using f'_c = 4 Ksi (28 MPa) and f_y = 60 Ksi (420 MPa).

9.10 Design an interior rib of a concrete joist floor system with the following description: span of ribbed slab is 18 ft (5.4 m) between simple supports; uniform dead load (excluding self-weight) is 15 psf (0.72 kN/m^2); live load is 100 psf (4.8 kN/m^2); support width is 14 in. (350 mm); f'_c = 3 Ksi (21 MPa) and f_y = 60 Ksi (420 MPa). Use 30-in.- (750-mm-) wide removable pans.

9.11 Repeat Problem 9.10 using 20-in.- (500-mm-) wide removable pans.

9.12 Use the information given in Problem 9.10 to design a continuous ribbed slab with three equal spans of 18 ft (5.4 m) each.

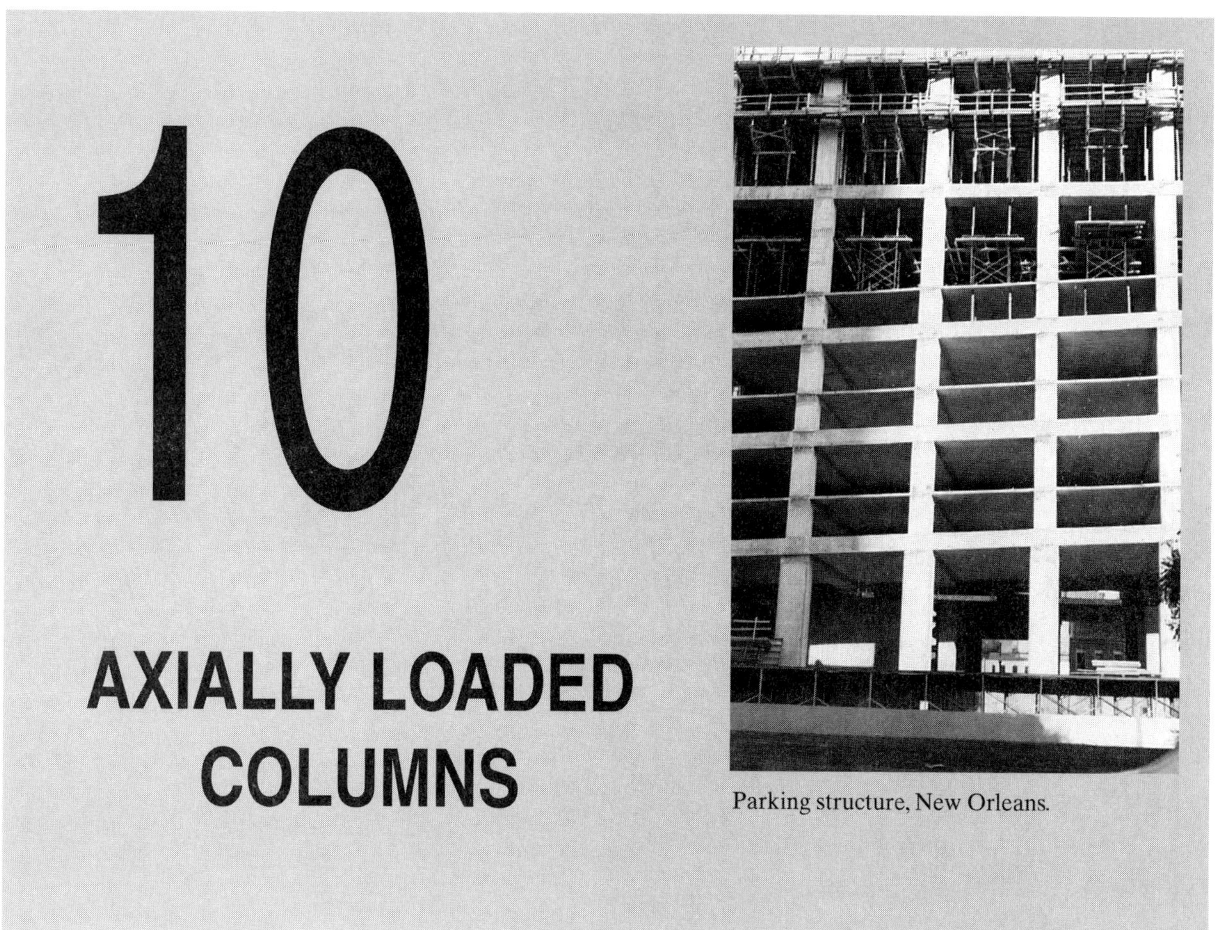

10 AXIALLY LOADED COLUMNS

Parking structure, New Orleans.

10.1 INTRODUCTION

Columns are members used primarily to support axial compressive loads and have a ratio of height to the least lateral dimension of 3 or greater. In reinforced concrete buildings, concrete beams, floors, and columns are cast monolithically, causing some moments in the columns due to end restraint. Moreover, perfect vertical alignment of columns in a multistory building is not possible, causing loads to be eccentric relative to the center of columns. The eccentric loads will cause moments in columns. Therefore, a column subjected to pure axial loads does not exist in concrete buildings. However, it can be assumed that axially loaded columns are those with relatively small eccentricity, e, of about $0.1h$ or less, where h is the total depth of the column and e is the eccentric distance from the center of the column. Because concrete has a high compressive strength and is an inexpensive material, it can be used in the design of compression members economically. This chapter deals only with short columns; slender columns are covered in detail in Chapter 12.

10.2 TYPES OF COLUMNS

Columns may be classified based on the following different categories (Figure 10.1):

1. Based on loading, columns may be classified as follows:
   a. Axially loaded columns, where loads are assumed acting at the center of the column section.

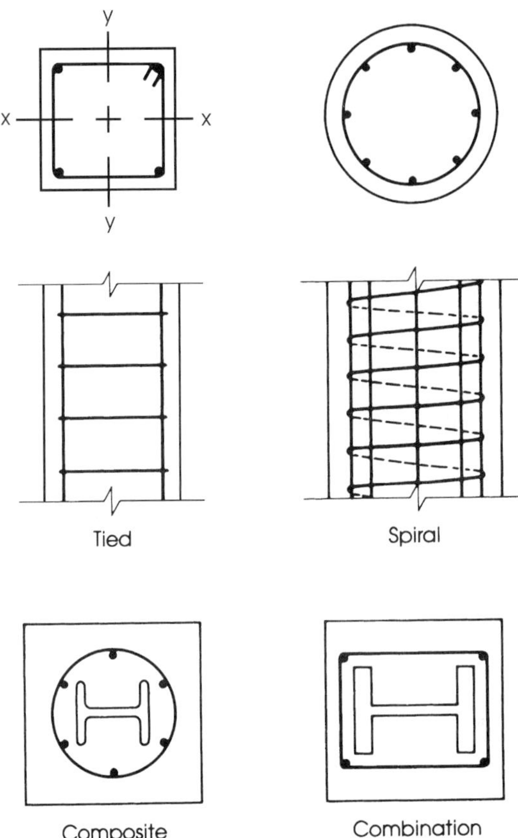

Figure 10.1 Types of columns.

   b. Eccentrically loaded columns, where loads are acting at a distance e from the center of the column section. The distance e could be along the x- or y-axis, causing moments either about the x- or y-axis.
   c. Biaxially loaded columns, where the load is applied at any point on the column section, causing moments about both the x- and y-axes simultaneously.
2. Based on length, columns may be classified as follows:
   a. Short columns, where the column's failure is due to the crushing of concrete or the yielding of the steel bars under the full load capacity of the column.
   b. Long columns, where buckling effect and slenderness ratio must be taken into consideration in the design, thus reducing the load capacity of the column relative to that of a short column.
3. Based on the shape of the cross section, column sections may be square, rectangular, round, L-shaped, octagonal, or any desired shape with an adequate side width or dimensions.
4. Based on column ties, columns may be classified as follows:
   a. Tied columns containing steel ties to confine the main longitudinal bars in the columns. Ties are normally spaced uniformly along the height of the column.
   b. Spiral columns containing spirals (spring type reinforcement) to hold the main longitudinal reinforcement and to help increase the column ductility before failure. In general, ties and spirals prevent the slender, highly stressed longitudinal bars from buckling and bursting the concrete cover.

5. Based on frame bracing, columns may be part of a frame that is braced against sidesway or unbraced against sidesway. Bracing may be achieved by using shear walls or bracings in the building frame. In braced frames, columns resist mainly gravity loads, and shear walls resist lateral loads and wind loads. In unbraced frames, columns resist both gravity and lateral loads, which reduce the load capacity of the columns.
6. Based on materials, columns may be reinforced, prestressed, composite (containing rolled steel sections such as I-sections), or a combination of rolled steel sections and reinforcing bars. Concrete columns reinforced with longitudinal reinforcing bars are the most common type used in concrete buildings.

10.3 BEHAVIOR OF AXIALLY LOADED COLUMNS

When an axial load is applied to a reinforced concrete short column, the concrete can be considered to behave elastically up to a low stress of about $(1/3)f'_c$. If the load on the column is increased to reach its ultimate strength, the concrete will reach the maximum strength and the steel will reach its yield strength, f_y. The ultimate nominal load capacity of the column can be written as follows:

$$P_o = 0.85f'_c A_n + A_{st} f_y \tag{10.1}$$

where A_n and A_{st} = the net concrete and total steel compressive areas, respectively.

$$A_n = A_g - A_{st}$$

$$A_g = \text{gross concrete area}$$

Two different types of failure occur in columns, depending on whether ties or spirals are used. For a tied column, the concrete fails by crushing and shearing outward, the longitudinal steel bars fail by buckling outward between ties, and the column failure occurs suddenly, much like the failure of a concrete cylinder.

A spiral column undergoes a marked yielding, followed by considerable deformation before complete failure. The concrete in the outer shell fails and spalls off. The concrete inside the spiral is confined and provides little strength before the initiation of column failure. A hoop tension develops in the spiral, and for a closely spaced spiral, the steel may yield. A sudden failure is not expected. Figure 10.2 shows typical load deformation curves for tied and spiral columns. Up to point a, both columns behave similarly. At point a, the

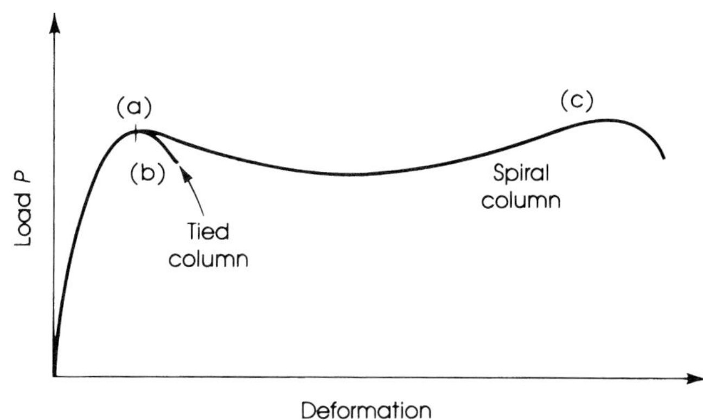

Figure 10.2 Behavior of tied and spiral columns.

longitudinal steel bars of the column yield, and the spiral column shell spalls off. After the ultimate load is reached, a tied column fails suddenly (curve b), whereas a spiral column deforms appreciably before failure (curve c).

10.4 ACI CODE LIMITATIONS

The ACI Code presents the following limitations for the design of compression members:

1. For axially as well as eccentrically loaded columns, the ACI Code sets the strength capacity-reduction factors at $\phi = 0.70$ for tied columns and $\phi = 0.75$ for spirally reinforced columns. The difference of 0.05 between the two values shows the additional ductility of spirally reinforced columns.

   The capacity-reduction factor for columns is much lower than those for flexure ($\phi = 0.9$) and shear ($\phi = 0.85$). This is because in axially loaded columns, the strength depends mainly on the concrete compression strength, whereas the strength of members in bending is less affected by the variation of concrete strength, especially in the case of an under-reinforced section. Furthermore, the concrete in columns is subjected to more segregation than in the case of beams. Columns are cast vertically in long, narrow forms, but the concrete in beams is cast in shallow, horizontal forms. Also, the failure of a column in a structure is more critical than that of a floor beam.
2. The minimum longitudinal steel percentage is 1%, and the maximum percentage is 8% of the gross area of the section (ACI Code, Section 10.9.1). Minimum reinforcement is necessary to provide resistance to bending, which may exist, and to reduce the effects of creep and shrinkage of the concrete under sustained compressive stresses. Practically, it is very difficult to fit more than 8% of steel reinforcement into a column and maintain sufficient space for concrete to flow between bars.
3. At least four bars are required for tied circular and rectangular members and six bars are needed for circular members enclosed by spirals (ACI Code, Section 10.9.2). For other shapes, one bar should be provided at each corner, and proper lateral reinforcement must be provided. For tied triangular columns, at least three bars are required. Bars shall not be located at a distance greater than 6 in. clear on either side from a laterally supported bar. Figure 10.3 shows the arrangement of longitudinal bars in tied columns and the distribution of ties. Ties shown in dotted lines are required when the clear distance on either side from laterally supported bars exceeds 6 in. The minimum concrete cover in columns is 1.5 in.
4. The minimum ratio of spiral reinforcement, ρ_s, according to the ACI Code, equation (10.6), and as explained in Section 10.9, is limited to

$$\rho_s = 0.45\left(\frac{A_g}{A_c} - 1\right)\frac{f'_c}{f_y} \tag{10.2}$$

where A_g = gross area of section

A_c = area of core of spirally reinforced column measured to the outside diameter of spiral

f_y = yield strength of spiral reinforcement $\leq$ 60 Ksi (ACI Code, Section 10.9.3)

5. The minimum diameter of spirals is $\frac{3}{8}$ in., and their clear spacing should not be more than 3 in. nor less than 1 in., according to the ACI Code, Section 7.10.4. Splices may be provided by welding or lapping the deformed uncoated spiral bars by 48 diameters or a minimum of 12 in. Lap splices for plain uncoated bar or wire $= 72d_p \leq 12$ in.

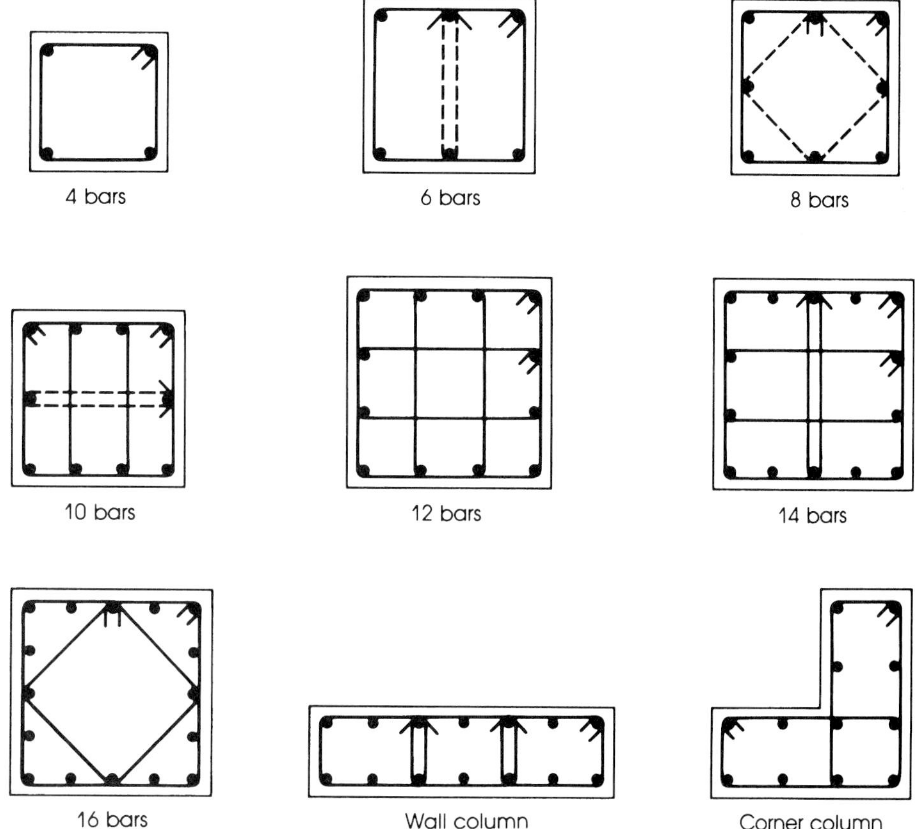

Figure 10.3 Arrangement of bars and ties in columns.

The same applies for epoxy-coated deformed bar or wire. The Code also allows full mechanical splices.

6. Ties for columns must have a minimum diameter of $\frac{3}{8}$ in. to enclose longitudinal bars of no. 10 size or smaller and a minimum diameter of $\frac{1}{2}$ in. for larger bar diameters (ACI Code, Section 7.10.5).

7. Spacing of ties shall not exceed the smallest of 48 times the tie diameter, 16 times the longitudinal bar diameter, or the least dimension of the column. Table 10.1 gives spacings for no. 3 and no. 4 ties. The Code does not give restrictions on the size of columns to allow wider utilization of reinforced concrete columns in smaller sizes.

Table 10.1 Maximum Spacings of Ties

Column Least Side or Diameter (in.)	Spacings of Ties (in.) for Bar					
	No. 6	**No. 7**	**No. 8**	**No. 9**	**No. 10**	**No. 11**
12	12	12	12	12	12	12
14	12	14	14	14	14	14
16	12	14	16	16	16	16
18	12	14	16	18	18	18
20	12	14	16	18	18	20
22–40	12	14	16	18	18	22
Ties	No. 3	No. 3	No. 3	No. 3	No. 3	No. 4

10.5 SPIRAL REINFORCEMENT

Spiral reinforcement in compression members prevents a sudden crushing of concrete and buckling of longitudinal steel bars. It has the advantage of producing a tough column that undergoes gradual and ductile failure. The minimum spiral ratio required by the ACI Code is meant to provide an additional compressive capacity to compensate for the spalling of the column shell. The strength contribution of the shell is

$$P_u\,(\text{shell}) = 0.85 f'_c (A_g - A_c) \tag{10.3}$$

where A_g is the gross concrete area and A_c is the core area (Figure 10.4).

In spirally reinforced columns, spiral steel is at least twice as effective as longitudinal bars; therefore, the strength contribution of spiral equals $2\rho_s A_c f_y$, where ρ_s is the ratio of volume of spiral reinforcement to total volume of core.

If the strength of the column shell is equated to the spiral strength contribution, then

$$0.85 f'_c (A_g - A_c) = 2\rho_s A_c f_y$$

$$\rho_s = 0.425\left(\frac{A_g}{A_c} - 1\right)\frac{f'_c}{f_y} \tag{10.4}$$

The ACI Code adopted a minimum ratio of ρ_s according to the following equation:

$$\text{Minimum } \rho_s = 0.45\left(\frac{A_g}{A_c} - 1\right)\frac{f'_c}{f_y} \tag{10.2}$$

The design relationship of spirals may be obtained as follows (Figure 10.4):

$$\rho_s = \frac{\text{volume of spiral in one loop}}{\text{volume of core for a spacing } S} = \frac{a_s \pi (D_c - d_s)}{\left(\frac{\pi}{4} D_c^2\right) S} = \frac{4a_s(D_c - d_s)}{S D_c^2} \tag{10.5}$$

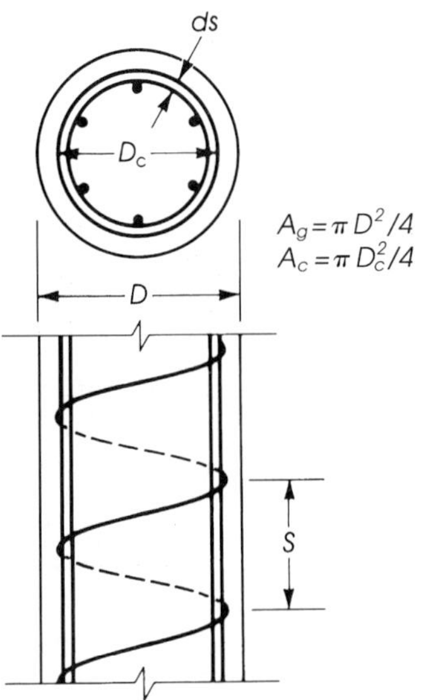

Figure 10.4 Dimensions of a column spiral.

where a_s = area of spiral reinforcement
D_c = diameter of the core measured to the outside diameter of spiral
D = diameter of the column
d_s = diameter of the spiral
S = spacing of the spiral

Table 10.2 gives spiral spacings for no. 3 and no. 4 spirals with $f_y = 60$ Ksi.

Table 10.2 Spirals For Circular Columns ($f_y = 60$ Ksi)

	$f'_c = 4$ Ksi No. 3 Spirals	$f'_c = 5$ Ksi No. 3 and No. 4 Spirals		$f'_c = 6$ Ksi No. 4 Spirals
Column Diameter (in.)	***Spacing (in.)***	***Spiral No.***	***Spacing (in.)***	***Spacing (in.)***
12	2.0	4	2.75	2.25
14	2.0	4	3.00	2.25
16	2.0	4	3.00	2.50
18	2.0	4	3.00	2.50
20	2.0	4	3.00	2.50
22	2.0	4	3.00	2.50
24	2.0	3	1.75	2.50
26 to 40	2.25	3	1.75	2.75

10.6 DESIGN EQUATIONS

The ultimate nominal load capacity of an axially loaded column was given in equation (10.1). Because a perfect axially loaded column does not exist, some eccentricity occurs on the column section, thus reducing its load capacity, P_o. To take that into consideration, the ACI Code specifies that the maximum nominal load, P_o, should be multiplied by a factor equal to 0.8 for tied columns and 0.85 for spirally reinforced columns. Introducing the capacity-reduction factor ϕ, the allowable ultimate loads on columns according to the ACI Code, Section 10.3.5, are as follows:

$$\phi P_n = \phi(0.80)\left[0.85f'_c(A_g - A_{st}) + A_{st}f_y\right] \tag{10.6}$$

for tied columns and

$$\phi P_n = \phi(0.85)\left[0.85f'_c(A_g - A_{st}) + A_{st}f_y\right] \tag{10.7}$$

for spiral columns, where

A_g = gross concrete area
A_{st} = total steel compressive area
ϕ = 0.7 for tied columns and 0.75 for spirally reinforced columns

Equations (10.6) and (10.7) may be written as follows:

$$\phi P_n = \phi K\left[0.85f'_c A_g + A_{st}(f_y - 0.85f'_c)\right] \tag{10.8}$$

where $\phi = 0.7$ and $K = 0.8$ for tied columns and $\phi = 0.75$ and $K = 0.85$ for spiral columns.

If the gross steel ratio is $\rho_g = A_{st}/A_g$, or $A_{st} = \rho_g A_g$, then equation (10.8) may be written as follows:

$$P_u = \phi P_n = \phi K A_g\left[0.85f'_c + \rho_g(f_y - 0.85f'_c)\right] \tag{10.9}$$

Equation (10.8) can be used to calculate the ultimate load capacity of the column, whereas equation (10.9) is used when the external ultimate load is given and it is required to calculate the size of the column section, A_g, based on an assumed steel ratio, ρ_g, between a minimum of 1% and a maximum of 8%.

It is a common practice to use grade 60 reinforcing steel bars in columns with a concrete compressive strength of 4 Ksi or greater to produce relatively small concrete column sections. For $f_y = 60$ Ksi, equations (10.8) and (10.9) can be simplified to the following equations:

$$P_u = \phi P_n = K_1 A_g + K_2 A_{st} \quad \text{and} \tag{10.10}$$

$$P_u = \phi P_n = A_g(K_1 + K_2 \rho_g) \tag{10.11}$$

where K_1 and K_2 are given in Table 10.3.

Table 10.3 Values of K_1 and K_2

	Tied Columns		Spiral Columns	
f'_c (Ksi)	K_1	K_2	K_1	K_2
4	1.90	31.70	2.17	36.00
5	2.38	31.22	2.71	35.54
6	2.85	30.74	3.25	35.00

10.7 AXIAL TENSION

Concrete will not crack as long as stresses are below its tensile strength; in this case both concrete and steel resist the tensile stresses. But when the tension force exceeds the tensile strength of concrete (about one-tenth of the compressive strength), cracks develop across the section, and the entire tension force is resisted by steel. The ultimate load that the member can carry is that due to tension steel only:

$$T_n = A_{st} f_y \quad \text{and} \tag{10.12}$$

$$T_u = \phi A_{st} f_y \tag{10.13}$$

where $\phi = 0.9$ for axial tension.

Tie rods in arches and similar structures are subjected to axial tension. Under working loads, the concrete cracks and the steel bars carry the whole tension force. The concrete acts as a fire and corrosion protector. Special provisions must be taken for water structures, as in the case of water tanks. In such designs, the concrete is not allowed to crack under the tension caused by the fluid pressure.

10.8 LONG COLUMNS

The equations developed in this chapter for the strength of axially loaded members are for short columns. In the case of long columns, the load capacity of the column is reduced by a reduction factor.

A long column is one with a high slenderness ratio, h'/r, where h' is the effective height of the column and r is the radius of gyration. The design of long columns is explained in detail in Chapter 12.

Example 10.1

Determine the allowable ultimate axial load on a 12-in. square, short tied column reinforced with four no. 9 bars. Ties are no. 3 spaced at 12 in. Use $f'_c = 4$ Ksi and $f_y = 60$ Ksi.

Solution

1. Using equation (10.8),

$$\phi P_n = \phi K[0.85f'_c A_g + A_{st}(f_y - 0.85f'_c)]$$

For a tied column, $\phi = 0.7$, $K = 0.8$, and $A_{st} = 4.0$ in.^{2}

$$\phi P_n = 0.7(0.8)[0.85(4)(12 \times 12) + 4(60 - 0.85 \times 4)] = 401 \text{ K}$$

2. Check steel percentage: $\rho_g = \frac{4}{144} = 0.02778 = 2.778\%$. This is less than 8% and greater than 1%.
3. Check tie spacings:

Minimum tie diameter is no. 3. Spacing is the smallest of the 48-tie diameter, 16-bar diameter, or least column side. $S_1 = 48(\frac{3}{8}) = 18$ in., $S_2 = 16(\frac{9}{8}) = 18$ in., $S_3 = 12.0$ in. Ties are adequate (Table 10.1).

Example 10.2

Design a square tied column to support an axial dead load of 400 K and a live load of 210 K using $f'_c = 5$ Ksi, $f_y = 60$ Ksi, and a steel ratio of about 5%. Design the necessary ties.

Solution

1. Calculate $P_u = 1.4P_D + 1.7P_L = 1.4(400) + 1.7(210) = 917$ K. Using equation (10.9), $P_u = 917 = 0.7(0.8)A_g[0.85 \times 5 + 0.05(60 - 0.85 \times 5)]$, $A_g = 232.7$ in.^{2}, and column side $= 15.25$ in., so use 16 in.
(Actual $A_g = 256$ in.^{2})
2. Because a larger section is adopted, the steel percentage may be reduced by using $A_g = 256$ in.^{2} in equation (10.8):

$$917 = 0.7(0.8)[0.85 \times 5 \times 256 + A_{st}(60 - 0.85 \times 5)]$$

$$A_{st} = 9.86 \text{ in.}^2$$

Use eight no. 10 bars ($A_{st} = 10.16$ in.^{2}). See Figure 10.5.
3. Design of ties (by calculation or from Table 10.1): Choose no. 3 ties with spacings equal to the least of $S_1 = 16(\frac{10}{8}) = 20$ in., $S_2 = 48(\frac{3}{8}) = 18$ in., or $S_3 =$ column side $= 16$ in. Use no. 3 ties spaced at 16 in. Clear distance between bars is 4.23 in., which is less than 6 in. Therefore, no additional ties are required.

Example 10.3

Repeat Example 10.2 using a rectangular section that has a width $b = 14$ in.

Solution

1. $P_u = 917$ K, calculated $A_g = 232.7$ in.^{2} For $b = 14$ in., $h = 232.7/14 = 16.62$ in. Choose a column 14 × 18 in., actual $A_g = 252$ in.^{2}.

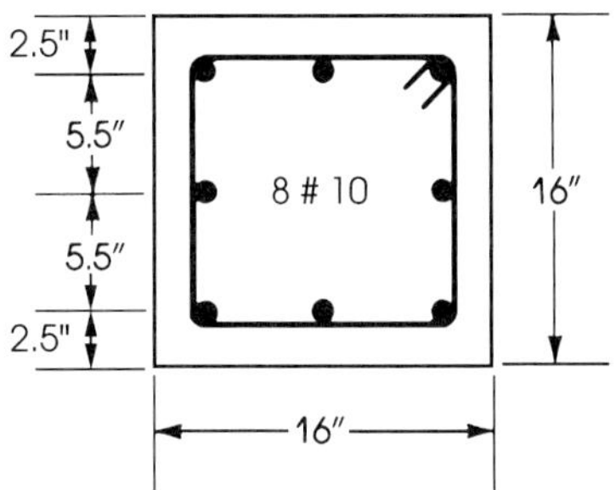

Figure 10.5 Example 10.2.

2. $P_u = 917 = 0.7(0.8)\left[0.85 \times 5 \times 252 + A_{st}(60 - 0.85 \times 5)\right]$

$$A_{st} = 10.16 \text{ in.}^2$$

Use eight no. 10 bars. $(A_{st}=10.16 \text{ in.}^2)$

3. Design of ties: Choose no. 3 ties, $S_1 = 20$ in., $S_2 = 18$ in., and $S_3 = 14$ in. (least side). Use no. 3 ties spaced at 14 in. Clear distance between bars in the long direction is $(18 - 5)/2$ − bar diameter of $1.27 = 5.23$ in. < 6 in. No additional ties are needed. Clear distance in the short direction is $(14 - 5)/2 - 1.27 = 3.23$ in. < 6 in.

Example 10.4

Design a circular spiral column to support an axial dead load of 500 K and a live load of 200 K using $f'_c = 4$ Ksi, $f_y = 60$ Ksi, and a steel ratio of about 3%. Also, design the necessary spirals.

Solution

1. Calculate $P_u = 1.4P_D + 1.7P_L = 1.4(500) + 1.7(200) = 1040$ K. Using equation (10.9), and spiral columns,

$$P_u = 1040 = 0.75(0.85)A_g\left[0.85 \times 4 + 0.03(60 - 0.85 \times 4)\right]$$

$A_g = 320 \text{ in.}^2$ and column diameter $= 20.2$ in., so use 20 in. Actual $A_g = 314.2 \text{ in.}^2$

2. Calculate A_{st} needed from equation (10.8):

$$P_u = 1040 = 0.75(0.85)\left[0.85 \times 4 \times 314.2 + A_{st}(60 - 0.85 \times 4)\right]$$

$$= 681.03 + 36.1A_{st}$$

$$A_{st} = 9.94 \text{ in.}^3$$

Use eight no. 10 bars $(A_{st} = 10.16 \text{ in.}^2)$.

3. Design of spirals: The diameter of core is $20 - 2(1.5) = 17$ in. The area of core is

$$A_c = \frac{\pi}{4}(17)^2, \qquad A_g = \frac{\pi}{4}(20)^2$$

$$\text{Minimum } \rho_s = 0.45\left(\frac{A_g}{A_c} - 1\right)\frac{f'_c}{f_y} = 0.45\left(\frac{20^2}{17^2} - 1\right)\left(\frac{4}{60}\right) = 0.01152$$

Assume no. 3 spiral, $a_s = 0.11 \text{ in.}^2$, and $d_s = 0.375$ in.

$$\rho_s = 0.01152 = \frac{4a_s(D_c - d_b)}{SD_c^2} = \frac{4(0.11)(17 - 0.375)}{S(17)^2}$$

Spacing s is equal to 2.2 in.; use no. 3 spiral at $s = 2$ in. (as shown in Table 10.2).

Example 10.5 (SI Units)

Design a rectangular tied short column to carry an ultimate axial load of 1900 kN. Use $f'_c = 30$ MPa, $f_y = 400$ MPa, column width $b = 300$ mm, and a steel ratio of about 2%.

Solution

1. Using equation (10.9),

$$P_u = 0.8\phi A_g\left[0.85f'_c + \rho_g(f_y - 0.85f'_c)\right]$$

Assuming a steel percentage of 2%,

$$1900 \times 10^3 = 0.8 \times 0.7A_g\left[0.85 \times 30 + 0.02(400 - 0.85 \times 30)\right]$$

$$A_g = 102{,}845 \text{ mm}^2$$

For $b = 300$ mm, the other side of the rectangular column is 343 mm. Therefore, use a section of 300 by 350 mm $(A_g = 105{,}000 \text{ mm}^2)$.

2. $A_s = 0.02 \times 102{,}845 = 2057 \text{ mm}^2$
   Choose six bars, 22 mm in diameter $(A_s = 2280 \text{ mm}^2)$.

3. Check the actual axial load capacity of the section using equation (10.6):

$$\begin{aligned} P_u &= 0.8\phi[0.85f'_c(A_g - A_{st}) + A_{st}f_y] \\ &= 0.8 \times 0.7[0.85 \times 30(105{,}000 - 2280) + 2280 \times 400] \times 10^{-3} \\ &= 1997 \text{ kN} \end{aligned}$$

   This meets the required P_u of 1900 kN.

4. Choose ties 10 mm in diameter. Spacing is the least of (1) $16 \times 22 = 352$ mm, (2) $48 \times 10 = 480$ mm, or (3) 300 mm. Choose 10-mm ties spaced at 300 mm.

SUMMARY

Sections 10.1–10.4

Columns may be tied or spirally reinforced.

$$\phi = 0.7 \quad \text{for tied columns}$$

$$\phi = 0.75 \quad \text{for spirally reinforced columns}$$

ρ_g must be $\leq 8\%$ and $\geq 1\%$.

Section 10.5

Minimum ratio of spirals is

$$\rho_s = 0.45\left(\frac{A_g}{A_c} - 1\right)\frac{f'_c}{f_y} \tag{10.2}$$

$$\rho_s = \frac{4a_s(D_c - d_s)}{D_c^2 S} \tag{10.5}$$

The minimum diameter of spirals is $\frac{3}{8}$ in., and their clear spacings should be not more than 3 in. nor less than 1 in.

Section 10.6

For tied columns,

$$\begin{aligned} P_u &= \phi P_n = 0.8\phi[0.85f'_c(A_g - A_{st}) + A_{st}f_y], \quad \text{or} \\ P_u &= \phi P_n = 0.8\phi A_g[0.85f'_c + \rho_g(f_y - 0.85f'_c)] \end{aligned} \tag{10.6}$$

For spiral columns,

$$\begin{aligned} P_u &= \phi P_n = 0.85\phi[0.85f'_c(A_g - A_{st}) + A_{st}f_y] \\ P_u &= \phi P_n = 0.85\phi A_g[0.85f'_c + \rho_g(f_y - 0.85f'_c)] \end{aligned} \tag{10.7}$$

where $\rho_g = A_{st}/A_g$.

Section 10.7

1. For axial tension,

$$T_u = \phi A_{st} f_y \qquad (\phi = 0.9) \tag{10.13}$$

2. Arrangements of vertical bars and ties in columns are shown in Figure 10.3.

REFERENCES

1. ACI Committee 315. *Manual of Standard Practice for Detailing Reinforced Concrete Structures*. Detroit, Mich.: American Concrete Institute, 1992.
2. Concrete Reinforcing Steel Institute. *CRSI Handbook*, 2d ed. Chicago, 1994.
3. B. Bresler and P. H. Gilbert. "Tie Requirements for Reinforced Concrete Columns." *ACI Journal* 58 (November 1961).
4. J. F. Pfister. "Influence of Ties on the Behavior of Reinforced Concrete Columns." *ACI Journal* 61 (May 1964).
5. N. G. Bunni. "Rectangular Ties in Reinforced Concrete Columns." *Publication No. SP-50*. Detroit, Mich.: American Concrete Institute, 1975.
6. Ti-Huang. "On the Formula for Spiral Reinforcement." *ACI Journal* 61 (March 1964).
7. American Concrete Institute. *Design Handbook, Vol. 2, Columns*. ACI Publication SP-17. Detroit, 1991.

PROBLEMS

10.1 For each problem, determine the allowable ultimate load-bearing capacity $(0.8\phi P_o)$ for each of the following short rectangular columns according to the ACI Code limitations. Assume $f_y = 60$ Ksi and properly tied columns (b = width of column, in., and h = total depth, in.).

Number	f'_c (Ksi)	b (in.)	h (in.)	Bars	Answer $(\phi k P_o)$ K
(a)	4	16	16	8 no. 9	741
(b)	4	20	20	16 no. 11	1553
(c)	4	12	12	8 no. 8	473
(d)	4	12	24	12 no. 10	1029
(e)	5	14	14	10 no. 9	778
(f)	5	16	16	4 no. 10	767
(g)	5	14	26	12 no. 10	1340
(h)	5	18	32	8 no. 11	1760
(i)	6	16	16	8 no. 10	1042
(j)	6	12	20	6 no. 10	918

10.2 For each problem, determine the allowable ultimate load-bearing capacity of each of the following short, spirally reinforced circular columns according to the ACI Code limitations. Assume $f_y = 60$ Ksi and the spirals are adequate (D = diameter of column, in.).

Number	f'_c (Ksi)	D (in.)	Bars	Answer (ϕkP_o) K
(a)	4	14	8 no. 9	622
(b)	4	16	6 no. 10	710
(c)	5	18	8 no. 10	1050
(d)	5	20	12 no. 10	1393
(e)	6	15	8 no. 9	854

10.3 For each problem, design a short square, rectangular, or circular column, as indicated, for each set of axial loads given, according to ACI limitations. Also, design the necessary ties or spirals and draw sketches of the column sections showing all bar arrangements. Use $f_y = 60$ Ksi and a steel ratio close to the ρ_g given $(P_D$ = dead load, P_L = live load, b = width of a rectangular column, and $= \rho_g = A_{st}/A_g)$.

Number	f'_c (Ksi)	P_D (K)	P_L (K)	ρ_g%	Section	One Solution
(a)	4	200	200	4	Square	14 × 14, 8 no. 9
(b)	4	750	400	3.5	Square	24 × 24, 16 no. 10
(c)	4	220	165	7	Square	12 × 12, 8 no. 10
(d)	5	330	230	3	Square	16 × 16, 8 no. 9
(e)	4	190	170	2	Rectangular, b = 12 in.	12 × 18, 6 no. 8
(f)	4	280	315	4.5	Rectangular, b = 14 in.	14 × 20, 10 no. 10
(g)	4	210	150	3	Rectangular, b = 12 in.	12 × 16, 6 no. 9
(h)	5	690	460	2	Rectangular, b = 18 in.	18 × 32, 8 no. 11
(i)	4	350	130	4	Circular—spiral	16, 8 no. 9
(j)	4	475	220	3.25	Circular—spiral	20, 8 no. 10
(k)	4	400	260	5	Circular—spiral	18, 10 no. 10
(l)	5	285	200	4.25	Circular—spiral	15, 6 no. 10

For SI units, use 1 psi = 0.0069 MPa, 1 K = 4.45 kN, and 1 in. = 25.4 mm.

MEMBERS IN COMPRESSION AND BENDING

Residential building, Minneapolis.

11.1 INTRODUCTION

Vertical members that are part of a building frame are subjected to combined axial loads and bending moments. These forces develop due to external loads, such as dead, live, and wind loads. The forces are determined by manual calculations or computer applications that are based on the principles of statics and structural analysis. For example, Figure 11.1 shows a two-hinged portal frame that carries a uniform ultimate load on BC. The bending moment is drawn on the tension side of the frame for clarification. Columns AB and CD are subjected to an axial compressive force and a bending moment. The ratio of the moment to the axial force is usually defined as the eccentricity, e, where $e = M_n/P_n$ (Figure 11.1). The eccentricity, e, represents the distance from the plastic centroid of the section to the point of application of the load. The plastic centroid is obtained by determining the location of the resultant force produced by the steel and the concrete, assuming that both are stressed in compression to f_y and $0.85f'_c$, respectively. For symmetrical sections, the plastic centroid coincides with the centroid of the section. For nonsymmetrical sections, the plastic centroid is determined by taking moments about an arbitrary axis, as explained in Example 11.1.

Example 11.1

Determine the plastic centroid of the section shown in Figure 11.2. Given: $f'_c = 4$ Ksi and $f_y = 60$ Ksi.

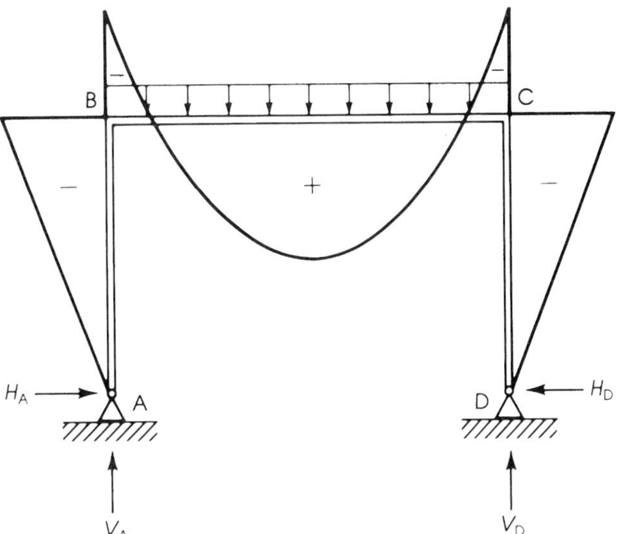

Figure 11.1 Two-hinged portal frame with bending moment diagram drawn on the tension side.

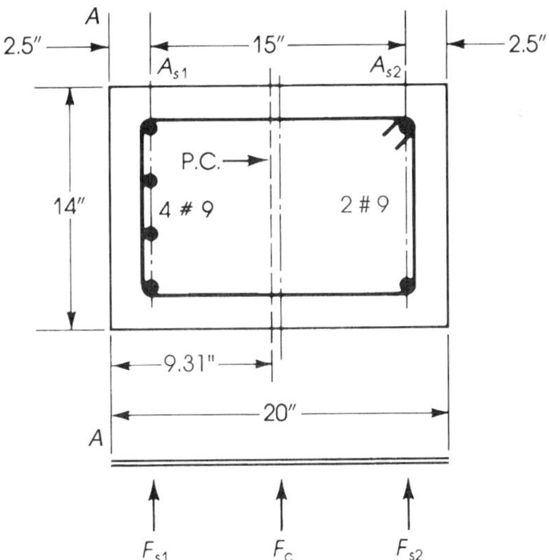

Figure 11.2 Plastic centroid (P.C.) of section in Example 11.1.

Solution

1. It is assumed that the concrete is stressed in compression to $0.85f'_c$:

$$F_c = \text{force in concrete} = (0.85f'_c)A_g$$
$$= (0.85 \times 4) \times 14 \times 20 = 952 \text{ K}$$

F_c is located at the centroid of the concrete section (at 10 in. from axis A-A).

2. Forces in steel bars:

$$F_{s1} = A_{s1}f_y = 4 \times 60 = 240 \text{ K}$$
$$F_{s2} = A_{s2}f_y = 2 \times 60 = 120 \text{ K}$$

3. Take moments about A-A:

$$x = \frac{(952 \times 10) + (240 \times 2.5) + (120 \times 17.5)}{952 + 240 + 120} = 9.31 \text{ in.}$$

Therefore, the plastic centroid lies at 9.31 in. from axis A-A.

4. If $A_{s1} = A_{s2}$ (symmetrical section), then $x = 10$ in. from axis A-A.

11.2 DESIGN ASSUMPTIONS FOR COLUMNS

The design limitations for columns, according to the ACI Code, Section 10.2, are as follows:

1. Strains in concrete and steel are proportional to the distance from the neutral axis.
2. Equilibrium of forces and strain compatibility must be satisfied.
3. The maximum usable compressive strain in concrete is 0.003.
4. Strength of concrete in tension can be neglected.
5. The stress in the steel is $f_s = \varepsilon E_s \leq f_y$.
6. The concrete stress block may be taken as a rectangular shape with concrete stress of $0.85f'_c$ that extends from the extreme compressive fibers a distance $a = \beta_1 c$, where c is the distance to the neutral axis and β_1 is 0.85 when $f'_c \leq 4000$ psi (30 MPa); β_1 decreases by 0.05 for each 1000 psi above 4000 psi (0.008 per 1 MPa above 30 MPa) but is not less than 0.65. (Refer to Figure 3.6, Chapter 3.)

11.3 LOAD-MOMENT INTERACTION DIAGRAM

When a normal force is applied on a short reinforced concrete column, the following cases may arise, according to the location of the normal force with respect to the plastic centroid. Refer to Figure 11.3(a) and 11.3(b):

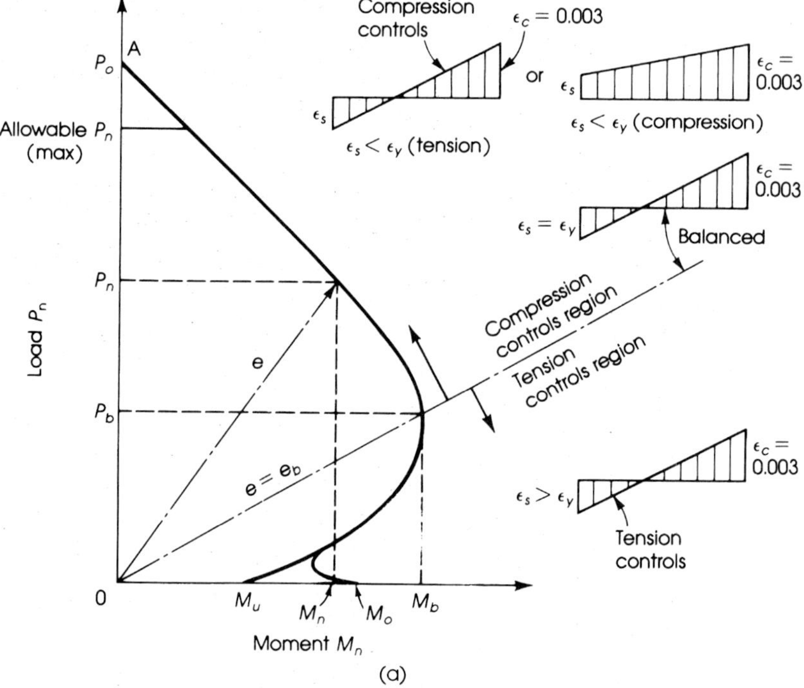

Figure 11.3 (a) Load-moment strength interaction diagram showing ranges of cases discussed in text, and (b) column sections showing the location of P_n for different load conditions.

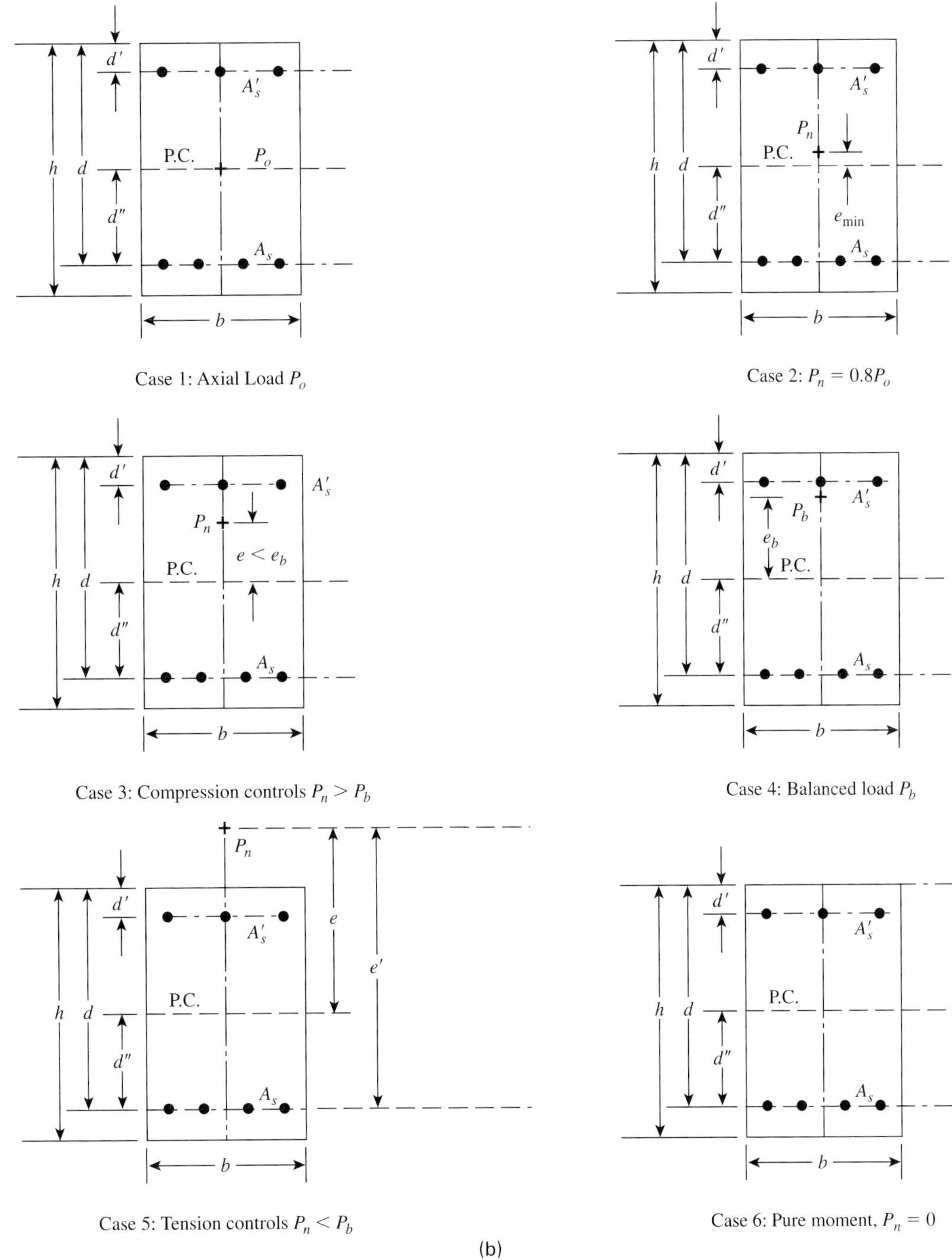

(b)

Figure 11.3 *(continued)*

1. *Axial compression* (P_o): This is a theoretical case assuming that a large axial load is acting at the plastic centroid; $e = 0$ and $M_n = 0$. Failure of the column occurs by crushing of the concrete and yielding of steel bars. This is represented by P_o on the curve of Figure 11.3(a).
2. *Maximum allowable axial load* $P_{n\,max}$: This is the case of a normal force acting on the section with minimum eccentricity. According to the ACI Code, $P_{n\,max} = 0.80P_o$ for tied columns and $0.85P_o$ for spirally reinforced columns, as explained in

Chapter 10. In this case, failure occurs by crushing of the concrete and the yielding of steel bars.

3. *Compression controls:* This is the case of a large axial load acting at a small eccentricity. The range of this case varies from a maximum value of $P_n = P_{n\,\max}$ to a minimum value of $P_n = P_b$ (balanced load). Failure occurs by crushing of the concrete on the compression side with a strain of 0.003, whereas the stress in the steel bars (on the tension side) is less than the yield strength, $f_y(f_s < f_y)$. In this case $P_n > P_b$ and $e < e_b$.
4. *Balanced condition* (P_b): A balanced condition is reached when the compression strain in the concrete reaches 0.003 and the strain in the tensile reinforcement reaches $\varepsilon_y = f_y/E_s$ simultaneously; failure of concrete occurs at the same time as the steel yields. The moment that accompanies this load is called the *balanced moment,* M_b, and the relevant balanced eccentricity is $e_b = M_b/P_b$.
5. *Tension controls:* This is the case of a small axial load with large eccentricity, that is, a large moment. Before failure, tension occurs in a large portion of the section, causing the tension steel bars to yield before actual crushing of the concrete. At failure, the strain in the tension steel is greater than the yield strain, ε_y, whereas the strain in the concrete reaches 0.003. The range of this case extends from the balanced to the case of pure flexure (Figure 11.3). When tension controls, $P_n < P_b$ and $e > e_b$.
6. *Pure flexure:* The section in this case is subjected to a bending moment M_n, whereas the axial load is $P_n = 0$. Failure occurs as in a beam subjected to bending moment only. The eccentricity is assumed to be at infinity. Note that radial lines from the origin represent constant ratios of $M_n/P_n = e =$ eccentricity of the load P_n from the plastic centroid.

Cases 1 and 2 were discussed in Chapter 10, and Case 6 was discussed in detail in Chapter 3. The other cases are discussed in this chapter.

11.4 SAFETY PROVISIONS

The safety provisions for load factors in the design of columns are as follows:

1. Load factors for gravity and wind loads are

$$U = 1.4D + 1.7L$$

$$U = 0.75(1.4D + 1.7L + 1.7W), \quad \text{or}$$

$$U = 0.9D + 1.3W$$

The most critical ultimate load should be used.

2. The strength-reduction factor ϕ to be used for columns may vary according to the following cases:
   a. When $P_u = \phi P_n \geq 0.1f'_c A_g$, then ϕ is 0.70 for tied columns and 0.75 for spirally reinforced columns. This case occurs generally when compression controls. A_g is the gross area of the concrete section.
   b. Between values of $0.1f'_c A_g$ and zero, P_u lies in the tension control zone and ϕ is larger than when compression controls. The ACI Code, Section 9.3.2, specifies that for members in which f_y does not exceed 60 Ksi, with symmetrical reinforcement and with the distance between compression and tension steel $(h - d' - d_s)$ or $(d - d')$ not less than $0.7h$ ($h =$ total depth of section) and $d = h - d_s$, the value of ϕ is

Columns supporting 52-story building, Minneapolis. (Columns are 96 × 64 in. with round ends.)

$$\phi = 0.9 - \frac{2.0P_u}{f'_c A_g} \geq 0.70 \quad \text{for tied columns} \tag{11.1}$$

$$\phi = 0.9 - \frac{1.5P_u}{f'_c A_g} \geq 0.75 \quad \text{for spirally reinforced columns} \tag{11.2}$$

(see Figure 11.4). For other reinforced members, ϕ may be increased linearly to 0.9 as ϕP_n decreases from $0.1f'_c A_g$ or ϕP_b (whichever is smaller) to zero.

c. When $P_u = 0$, then $\phi = 0.9$, because it is a case of pure flexure.

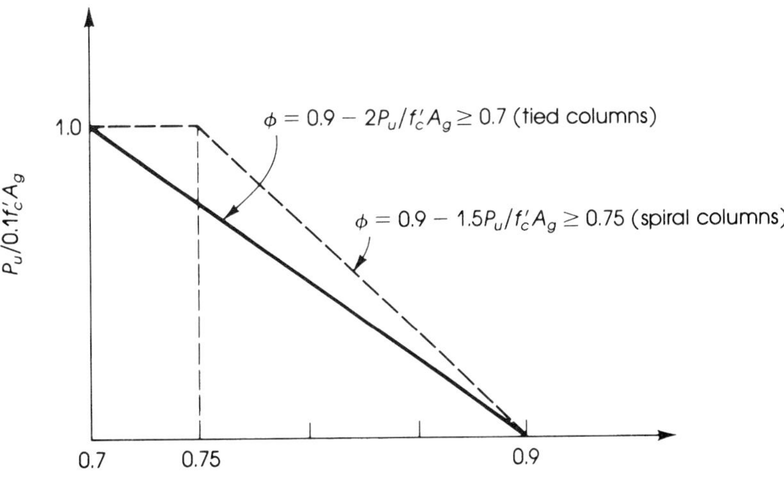

Figure 11.4 Variation in ϕ when tension controls.

11.5 BALANCED CONDITION—RECTANGULAR SECTIONS

A balanced condition occurs in a column section when a load is applied on the section and produces, at ultimate strength, a strain of 0.003 in the compressive fibers of concrete and a strain $\varepsilon_y = f_y/E_s$ in the tension steel bars simultaneously. This is a special case where the neutral axis can be determined from the strain diagram with known extreme values. When the ultimate applied eccentric load is greater than P_b, compression controls; if it is smaller than P_b, tension controls in the section.

The analysis of a balanced column section can be explained in steps (Figure 11.5):

1. Let c equal the distance from the extreme compressive fibers to the neutral axis. From the strain diagram,

$$\frac{c_b\,(\text{balanced})}{d} = \frac{0.003}{0.003 + f_y/E_s} \quad (\text{where } E_s = 29{,}000 \text{ Ksi}) \quad \text{and} \tag{11.3}$$

$$c_b = \frac{87d}{87 + f_y} \quad (\text{where } f_y \text{ is in Ksi})$$

The depth of the equivalent compressive block is

$$a_b = \beta_1 c_b = \left(\frac{87}{87 + f_y}\right)\beta_1 d \tag{11.4}$$

where $\beta_1 = 0.85$ for $f'_c \leq 4000$ psi and decreases by 0.05 for each 1000-psi increase in f'_c.

2. From equilibrium, the sum of the horizontal forces equals zero: $P_b - C_c - C_s + T = 0$, where

$$C_c = 0.85f'_c ab \quad \text{and} \quad T = A_s f_y \tag{11.5}$$

$$C_s = A'_s(f'_s - 0.85f'_c)$$

(Use $f'_s = f_y$ if compression steel yields.)

$$f'_s = 87\left(\frac{c - d'}{c}\right) \leq f_y$$

The expression of C_s takes the displaced concrete into account. Therefore, equation (11.5) becomes

$$P_b = 0.85f'_c ab + A'_s(f'_s - 0.85f'_c) - A_s f_y \tag{11.6}$$

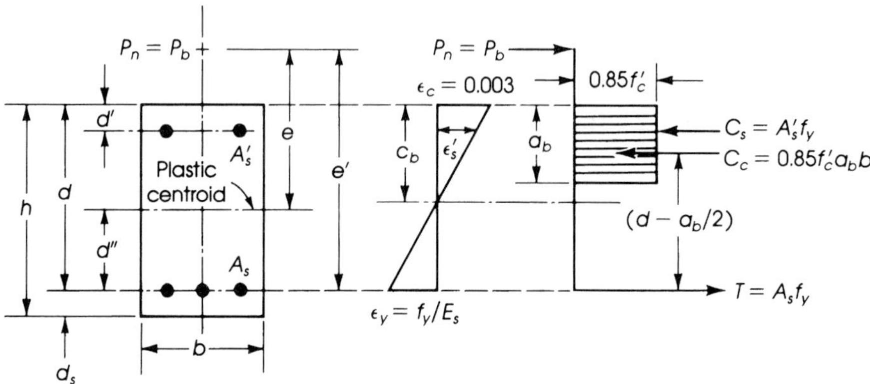

Figure 11.5 Balanced condition (rectangular section).

3. The eccentricity e_b is measured from the plastic centroid and e' is measured from the centroid of the tension steel: $e' = e + d''$ (in this case $e = e_b + d''$), where d'' is the distance from the plastic centroid to the centroid of the tension steel. The value of e_b can be determined by taking moments about the plastic centroid.

$$P_b e_b = C_c\left(d - \frac{a}{2} - d''\right) + C_s(d - d' - d'') + Td'' \tag{11.7}$$

or

$$P_b e_b = M_b = 0.85f'_c ab\left(d - \frac{a}{2} - d''\right) + A'_s(f_y - 0.85f'_c)(d - d' - d'') + A_s f_y d'' \tag{11.8}$$

The balanced eccentricity is

$$e_b = M_b/P_b \tag{11.9}$$

For nonrectangular sections, the same procedure applies, taking into consideration the actual area of concrete in compression.

Example 11.2

Determine the balanced compressive force P_b; then determine e_b and M_b for the section shown in Figure 11.6. Given: $f'_c = 4$ Ksi and $f_y = 60$ Ksi.

Solution

1. For a balanced condition, the strain in the concrete is 0.003 and the strain in the tension steel is

$$\varepsilon_y = \frac{f_y}{E_s} = \frac{60}{29{,}000} = 0.00207$$

2. Locate the neutral axis:

$$c_b = \frac{87}{87 + f_y} d = \frac{87}{87 + 60}(19.5) = 11.54 \text{ in.}$$

$$a_b = 0.85c_b = 0.85 \times 11.54 = 9.81 \text{ in.}$$

3. Check if compression steel yields. From the strain diagram,

$$\frac{\varepsilon'_s}{0.003} = \frac{c - d'}{c} = \frac{11.54 - 2.5}{11.54}, \quad \varepsilon'_s = 0.00235$$

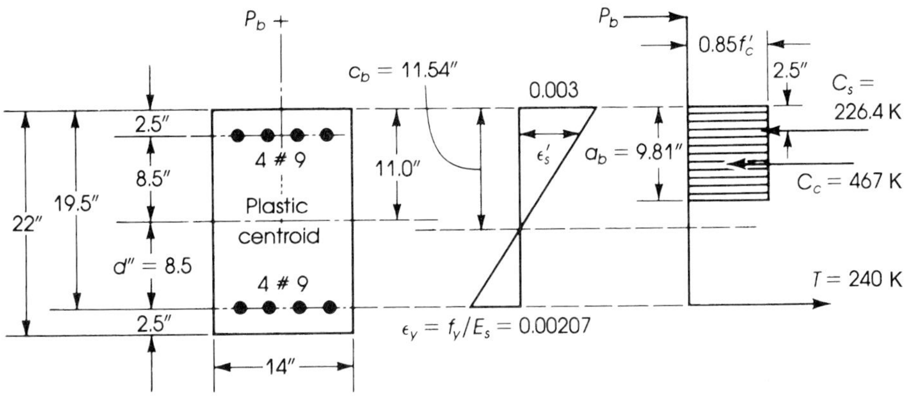

Figure 11.6 Example 11.2—balanced condition.

which exceeds ε_y of 0.00207; thus, compression steel yields. Or check that

$$f'_s = 87\left(\frac{c - d''}{c}\right) \leq f_y$$

$$f'_s = \frac{87(11.54 - 2.5)}{11.54} = 68 \text{ Ksi} > 60 \text{ Ksi}$$

Then $f'_s = f_y = 60$ Ksi.

4. Calculate the forces acting on the section:

$$C_c = 0.85f'_c ab = 0.85 \times 4 \times 9.81 \times 14 = 467 \text{ K}$$

$$T = A_s f_y = 4 \times 60 = 240 \text{ K}$$

$$C_s = A'_s(f_y - 0.85f'_c) = 4(60 - 3.4) = 226.4 \text{ K}$$

5. Calculate P_b and e_b:

$$P_b = C_c + C_s - T = 467 + 226.4 - 240 = 453.4 \text{ K}$$

From equation (11.7),

$$M_b = P_b e_b = C_c\left(d - \frac{a}{2} - d''\right) + C_s(d - d' - d'') + Td''$$

The plastic centroid is at the centroid of the section, and $d'' = 8.5$ in.

$$M_b = 453.4e_b = 467\left(19.5 - \frac{9.81}{2} - 8.5\right)$$

$$+ 226.4(19.5 - 2.5 - 8.5) + 240 \times 8.5$$

$$= 6810.8 \text{ K} \cdot \text{in.} = 567.6 \text{ K} \cdot \text{ft}$$

$$e_b = \frac{M_b}{P_b} = \frac{6810.8}{453.4} = 15.0 \text{ in.}$$

6. For a balanced condition $\phi = 0.7$, $\phi P_b = 317.4$ K and $\phi M_b = 397.3$ K·ft.

11.6 COLUMN SECTIONS UNDER ECCENTRIC LOADING

For the two cases when compression or tension controls, two basic equations of equilibrium can be used in the analysis of columns under eccentric loadings: (1) the sum of the horizontal or vertical forces = 0, and (2) the sum of moments about any axis = 0. Referring to Figure 11.7, the following equations may be established.

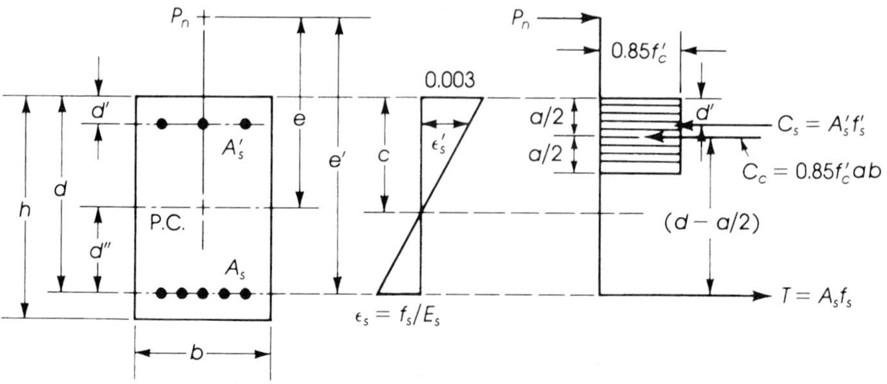

Figure 11.7 General case, rectangular section.

Reinforced concrete tied columns under construction. The two columns are separated by an expansion joint.

1. $P_n - C_c - C_s + T = 0$ (11.10)

where $C_c = 0.85 f'_c ab$

$C_s = A'_s(f'_s - 0.85 f'_c)$ (If compression steel yields, then $f'_s = f_y$.)

$T = A_s f_s$ (If tension steel yields, then $f_s = f_y$.)

2. Taking moments about A_s,

$$P_n e' - C_c\left(d - \frac{a}{2}\right) - C_s(d - d') = 0 \tag{11.11}$$

The quantity $e' = e + d''$, and $e' = (e + d - h/2)$ for symmetrical reinforcement (d'' is the distance from the plastic centroid to the centroid of the tension steel).

$$P_n = \frac{1}{e'}\left[C_c\left(d - \frac{a}{2}\right) + C_s(d - d')\right] \tag{11.12}$$

3. Taking moments about C_c,

$$P_n\left[e' - \left(d - \frac{a}{2}\right)\right] - T\left(d - \frac{a}{2}\right) - C_s\left(\frac{a}{2} - d'\right) = 0 \tag{11.13}$$

$$P_n = \frac{T\left(d - \frac{a}{2}\right) + C_s\left(\frac{a}{2} - d'\right)}{(e' + a/2 - d)} \tag{11.14}$$

If $A_s = A'_s$ and $f_s = f'_s = f_y$, then

$$P_n = \frac{A_s f_y(d - d')}{(e' + a/2 - d)} = \frac{A_s f_y(d - d')}{(e - h/2 + a/2)} \tag{11.15}$$

$$A_s = A'_s = \frac{P_n(e - h/2 + a/2)}{f_y(d - d')} \tag{11.16}$$

11.7 STRENGTH OF COLUMNS WHEN TENSION CONTROLS

When a column is subjected to an eccentric force with large eccentricity e, tension controls. The column section fails due to the yielding of steel and crushing of concrete when the strain in the steel exceeds $\varepsilon_y(\varepsilon_y = f_y/E_s)$. In this case the ultimate capacity, P_n, will be less than P_b, or the eccentricity, $e = M_n/P_n$, is greater than the balanced eccentricity, e_b. Because it is difficult in some cases to predict if tension or compression controls, it can be assumed (as a guide) that tension controls when $e > d$. This assumption should be checked later.

The general equations of equilibrium, (11.10) and (11.11), may be used to calculate the ultimate nominal strength of the column. This is illustrated in steps as follows:

1. When tension controls, the tension steel yields and its stress is $f_s = f_y$. Assume that stress in compression steel is $f'_s = f_y$.
2. Evaluate P_n from equilibrium conditions (equation (11.10)):

$$P_n = C_c + C_s - T$$

where $C_c = 0.85f'_c ab$, $C_s = A'_s(f_y - 0.85f'_c)$, and $T = A_s f_y$.

3. Calculate P_n by taking moments about A_s (equation (11.11)):

$$P_n \cdot e' = C_c\left(d - \frac{a}{2}\right) + C_s(d - d')$$

where $e' = e + d''$ and $e' = e + d - h/2$ when $A_s = A'_s$.

4. Equate P_n from Steps 2 and 3:

$$C_c + C_s - T = \frac{1}{e'}\left[C_c\left(d - \frac{a}{2}\right) + C_s(d - d')\right]$$

This is a second-degree equation in a. Substitute the values of C_c, C_s, and T and solve for a.

5. The second-degree equation, after the substitution of C_c, C_s, and T, is reduced to the following equation:

$$Aa^2 + Ba + C = 0$$

where

$$A = 0.425f'_c b$$
$$B = 0.85f'_c b(e' - d) = 2A(e' - d)$$
$$C = A'_s(f'_s - 0.85f'_c)(e' - d + d') - A_s f_y e'$$

Solve for a to get

$$a = \frac{-B \pm \sqrt{B^2 - 4AC}}{2A}$$

Note that the value of $f'_s - 0.85f'_c$ must be a positive value. If this value is negative, then let $f'_s - 0.85f'_c = 0$.

6. Substitute a in the equation of Step 2 to obtain P_n. The moment M_n can be calculated: $M_n = P_n \cdot e$.
7. Check if compression steel yields as assumed. If $\varepsilon'_s \geq \varepsilon_y$, then compression steel yields; otherwise, $f'_s = E_s\varepsilon'_s$. Repeat Steps 2 through 5. Note that $\varepsilon'_s = [(c - d')/c]0.003$, $\varepsilon_y = f_y/E_s$, and $c = a/\beta_1$.
8. Check that tension controls. Tension controls when $e > e_b$ or $P_n < P_b$. Example 11.3 illustrates this procedure.

Example 11.3

Determine the nominal compressive strength P_n for the section given in Example 11.2 if $e = 20$ in. (see Figure 11.8).

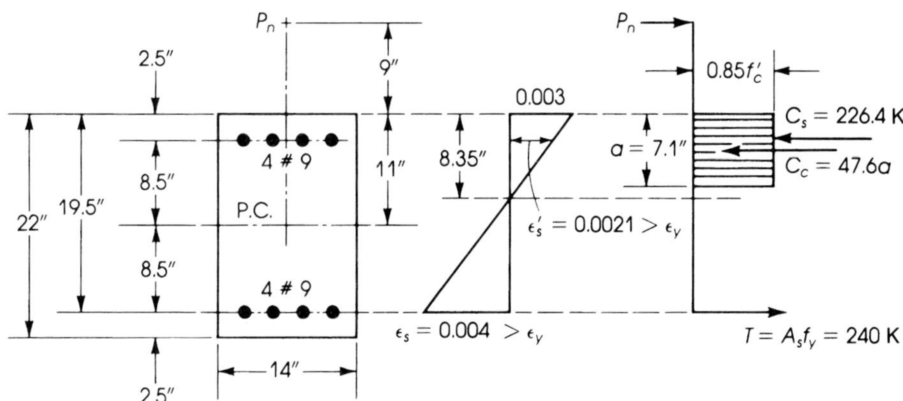

Figure 11.8 Example 11.3: tension controls.

Solution

1. Because $e = 20$ in. is greater than $d = 19.5$ in., assume that tension controls (to be checked later). The strain in the tension steel ε_s will be greater than ε_y and its stress is f_y. Assume that compression steel yields $f'_s = f_y$, which should be checked later.
2. From the equation of equilibrium, (11.10):

$$P_n = C_c + C_s - T$$

where

$$C_c = 0.85f'_c ab = 0.85 \times 4 \times 14a = 47.6a$$

$$C_s = A'_s(f_y - 0.85f'_c) = 4(60 - 0.85 \times 4) = 226.4 \text{ K}$$

$$T = A_s f_y = 4 \times 60 = 240 \text{ K}$$

$$P_n = 47.6a + 226.4 - 240 = (47.6a - 13.6) \quad \text{(I)}$$

3. Taking moments about A_s (equation (11.12)),

$$P_n = \frac{1}{e'}\left[C_c\left(d - \frac{a}{2}\right) + C_c(d - d')\right]$$

Note that for the plastic centroid at the center of the section, $d'' = 8.5$ in.

$$e' = e + d'' = 20 + 8.5 = 28.5 \text{ in.}$$

$$P_n = \frac{1}{28.5}\left[47.6a\left(19.5 - \frac{a}{2}\right) + 226.4 \times 17\right]$$

$$P_n = 32.56a - 0.835a^2 + 135.0 \quad \text{(II)}$$

4. Equating equations (I) and (II),

$$P_n = (47.6a - 13.6) = 32.56a - 0.835a^2 + 135.0, \quad \text{or}$$

$$a^2 + 18a - 178.0 = 0, \qquad a = 7.1 \text{ in.}$$

5. From equation (I) ($\phi = 0.7$),

$$P_n = 47.6 \times 7.1 - 13.6 = 324.4 \text{ K}, \qquad \phi P_n = 227.1 \text{ K}$$

$$M_n = P_n e = 324.4 \times \frac{20}{12} = 540.67 \text{ K}\cdot\text{ft}, \qquad \phi M_n = 378.5 \text{ K}\cdot\text{ft}$$

6. Check if compression steel has yielded:

$$c = \frac{a}{0.85} = \frac{7.1}{0.85} = 8.35 \text{ in.}, \qquad \varepsilon_y = \frac{60}{29{,}000} = 0.00207$$

$$\varepsilon_s' = \frac{(8.35 - 2.5)}{8.35}(0.003) = 0.0021 > \varepsilon_y$$

Compression steel yields. Check strain in tension steel:

$$\varepsilon_s = \left(\frac{19.5 - 8.35}{8.35}\right) \times 0.003 = 0.004 > \varepsilon_y$$

If compression steel does not yield, use f_s' as calculated from $f_s' = \varepsilon_s' E_s$ and revise the calculations.

7. Because $e = 20$ in. $> e_b = 15$ in. (Example 11.2), tension controls.

8. The same results can be obtained using the values of A, B, and C given earlier.

$$Aa^2 + Ba + C = 0$$

$$A = 0.425 f_c' b = 0.425(4)(14) = 23.8$$

$$B = 2A(e' - d) = 2(23.8)(28.5 - 19.5) = 428.4$$

$$C = 4(60 - 0.85 \times 4)(28.5 - 19.5 + 2.5) - 4(60)(28.5)$$

$$= -4236.4$$

Solve for a to get $a = 7.1$ in. and $P_n = 324.4$ K.

11.8 STRENGTH OF COLUMNS WHEN COMPRESSION CONTROLS

If the compressive applied force P_n exceeds the balanced force P_b or the eccentricity, $e = M_n/P_n$, is less than e_b, compression failure is expected. In this case compression controls, and the strain in the concrete will reach 0.003, whereas the strain in the steel is less than ε_y (Figure 11.9). A large part of the column will be in compression. The neutral axis moves toward the tension steel, increasing the compression area, and therefore the distance to the neutral axis c is greater than the balanced c_b (Figure 11.9).

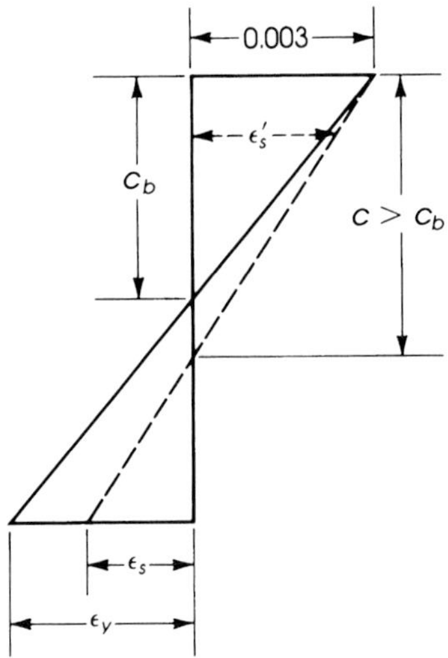

Figure 11.9 Strain diagram when compression controls. When $\varepsilon_s < \varepsilon_y$, then $c > c_b$ and $\varepsilon_s' \geq \varepsilon_y$.

Because it is difficult to predict if compression or tension controls whenever a section is given, it can be assumed that compression controls when $e < 2d/3$, which should be checked later. The ultimate load capacity P_n can be calculated using the principles of statics. The analysis of column sections when compression controls can be achieved using equations (11.10) and (11.11) given earlier and one of the following solutions.

11.8.1 Trial Solution

This solution can be summarized as follows:

1. Calculate the distance to the neutral axis for a balanced section c_b:

$$c_b = \left(\frac{87d}{87 + f_y}\right), \qquad \text{where } f_y \text{ is in Ksi} \tag{11.3}$$

2. Evaluate P_n using equilibrium conditions:

$$P_n = C_c + C_s - T \tag{11.10}$$

3. Evaluate P_n by taking moments about the tension steel A_s:

$$P_n \cdot e' = C_c\left(d - \frac{a}{2}\right) + C_s(d - d') \tag{11.11}$$

   where $e' = e + d - h/2$ when $A_s = A'_s$ or $e' = e + d''$ in general, $C_c = 0.85f'_c ab$, $C_s = A'_s(f'_s - 0.85f'_c)$, and $T = A_s f_s$.
4. Assume a value for c such that $c > c_b$ (calculated in Step 1). Calculate $a = \beta_1 c$. Assume $f'_s = f_y$.
5. Calculate f_s based on the assumed c:

$$f_s = \varepsilon_s E_s = 87\left(\frac{d - c}{c}\right) \text{Ksi} \le f_y$$

6. Substitute the preceding values in equation (11.10) to calculate P_{n1} and in equation (11.11) to calculate P_{n2}. If P_{n1} is close to P_{n2}, then choose the smaller or average of P_{n1} and P_{n2}. If P_{n1} is not close to P_{n2}, assume a new c or a and repeat the calculations starting from Step 4 until P_{n1} is close to P_{n2} (1% is quite reasonable).
7. Check that compression steel yields by calculating $\varepsilon'_s = 0.003[(c - d')/c]$ and comparing it with $\varepsilon_y = f_y/E_s$. When $\varepsilon'_s \ge \varepsilon_y$, then compression steel yields; otherwise, $f'_s = \varepsilon'_s \cdot E_s$ or, directly,

$$f'_s = 87\left(\frac{c - d'}{c}\right) \le f_y \text{ Ksi}$$

8. Check that $e < e_b$ or $P_n > P_b$ to prove that compression controls. Example 11.4 illustrates the procedure.

Example 11.4

Determine the nominal compressive strength, P_n, for the section given in Example 11.2 if $e = 10$ in. (see Figure 11.10).

Solution

1. Because $e = 10 \text{ in.} < (2/3)d = 13$ in., assume compression controls. This assumption will be checked later. Calculate the distance to the neutral axis for a balanced section, c_b:

$$c_b = \frac{87}{87 + f_y} d = \frac{87}{87 + 60}(19.5) = 11.54 \text{ in.}$$

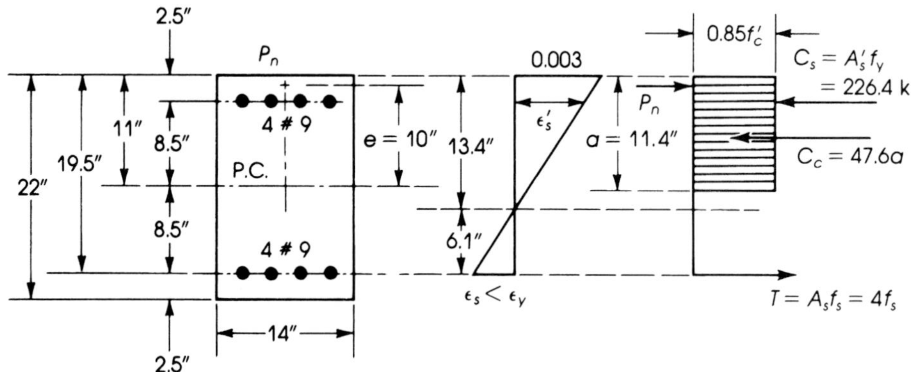

Figure 11.10 Example 11.4: compression controls.

2. From the equations of equilibrium,

$$P_n = C_c + C_s - T \tag{11.10}$$

where

$$C_c = 0.85f'_c ab = 0.85 \times 4 \times 14a = 47.6a$$

$$C_s = A'_s(f_y - 0.85f'_c) = 4(60 - 0.85 \times 4) = 226.4 \text{ K}$$

Assume compression steel yields (this assumption will be checked later).

$$T = A_s f_s = 4f_s \qquad (f_s < f_y)$$

$$P_n = 47.6a + 226.4 - 4f_s \tag{I}$$

3. Taking moments about A_s,

$$P_n = \frac{1}{e'}\left[C_c\left(d - \frac{a}{2}\right) + C_s(d - d')\right] \tag{11.11}$$

The plastic centroid is at the center of the section and $d'' = 8.5$ in.

$$e' = e + d'' = 10 + 8.5 = 18.5 \text{ in.}$$

$$P_n = \frac{1}{18.5}\left[47.6a\left(19.5 - \frac{a}{2}\right) + 226.4(19.5 - 2.5)\right]$$

$$P_n = 50.17a - 1.29a^2 + 208 \tag{II}$$

4. Assume $c = 13.45$ in., which exceeds c_b (11.54 in.).

$$a = 0.85 \times 13.45 = 11.43 \text{ in.}$$

Substitute $a = 11.43$ in equation (II):

$$P_{n1} = 50.17 \times 11.43 - 1.29(11.43)^2 + 208 = 612.9 \text{ K}$$

5. Calculate f_s from the strain diagram when $c = 13.45$ in.

$$f_s = \left(\frac{19.5 - 13.45}{13.45}\right)87 = 39.13 \text{ Ksi}$$

6. Substitute $a = 11.43$ in. and $f_s = 39.13$ Ksi in equation (I) to calculate P_{n2}:

$$P_{n2} = 47.6(11.43) + 226.4 - 4(39.13) = 613.9 \text{ K}$$

which is very close to the calculated P_{n1} of 612.9 K (less than 1% difference).

$$M_n = P_n \cdot e = 612.9\left(\frac{10}{12}\right) = 510.8 \text{ K}\cdot\text{ft}$$

7. Check if compression steel yields. From the strain diagram,

$$\varepsilon_s' = \frac{13.45 - 2.5}{13.45}(0.003) = 0.00244 > \varepsilon_y = 0.00207$$

Compression steel yields, as assumed.

8. $P_n = 612.9$ K is greater than $P_b = 453.4$ K, and $e = 10$ in. $< e_b = 15$ in., both calculated in the previous example, indicating that compression controls, as assumed. Note that it may take a few trials to get P_{n1} close to P_{n2}.

9. From equation (11.1), $\phi = 0.7$, $\phi P_n = 429$ K, and $\phi M_n = 357.5$ K·ft.

11.8.2 Numerical Analysis Solution

The analysis of columns when compression controls can also be performed by reducing the calculations into one cubic equation in the form

$$Aa^3 + Ba^2 + Ca + D = 0$$

and then solving for a by a numerical method, or a can be obtained directly by using one of many inexpensive scientific calculators with built-in programs that are available. From the equations of equilibrium,

$$\begin{aligned} P_n &= C_c + C_s - T \\ &= (0.85f_c'ab) + A_s'(f_y - 0.85f_c') - A_sf_s \end{aligned} \tag{11.10}$$

Taking moments about the tension steel A_s,

$$\begin{aligned} P_n &= \frac{1}{e'}\left[C_c\left(d - \frac{a}{2}\right) + C_s(d - d')\right] \\ &= \frac{1}{e'}\left[0.85f_c'ab\left(d - \frac{a}{2}\right) + A_s'(f_y - 0.85f_c')(d - d')\right] \end{aligned} \tag{11.11}$$

From the strain diagram,

$$\varepsilon_s = \left(\frac{d - c}{c}\right)(0.003) = \frac{(d - a/\beta_1)}{a/\beta_1}(0.003)$$

The stress in the tension steel is

$$f_s = \varepsilon_s E_s = 29{,}000\varepsilon_s = \frac{87}{a}(\beta_1 d - a)$$

Substituting this value of f_s in equation (11.10) and equating equations (11.10) and (11.11) and simplifying gives

$$\left(\frac{0.85f_c'b}{2}\right)a^3 + \left[0.85f_c'b(e' - d)\right]a^2 + \left[A_s'(f_y - 0.85f_c')(e' - d + d') + 87A_se'\right]a - 87A_se'\beta_1 d = 0$$

This is a cubic equation in terms of a:

$$Aa^3 + Ba^2 + Ca + D = 0$$

where

$$A = \frac{0.85f_c'b}{2}$$

$$B = 0.85f_c'b(e' - d)$$

$$C = A_s'(f_y - 0.85f_c')(e' - d + d') + 87A_se'$$

$$D = -87A_se'\beta_1 d$$

Once the values of $A, B, C,$ and D are calculated, a can be determined by trial or directly by a scientific calculator. Also, the solution of the cubic equation can be obtained by using the well-known Newton-Raphson method. This method is very powerful for finding a root of $f(x) = 0$. It involves a simple technique, and the solution converges rapidly by using the following steps:

1. Let $f(a) = Aa^3 + Ba^2 + Ca + D$, and calculate $A, B, C,$ and D.
2. Calculate the first derivative of $f(a)$:

$$f'(a) = 3Aa^2 + 2Ba + C$$

3. Assume any initial value of a, say, a_0, and compute the next value:

$$a_1 = a_0 - \frac{f(a_0)}{f'(a_0)}$$

4. Use the obtained value a_1 in the same way to get

$$a_2 = a_1 - \frac{f(a_1)}{f'(a_1)}$$

5. Repeat the same steps to get the answer up to the desired accuracy. In the case of the analysis of columns when compression controls, the value a is greater than the balanced $a(a_b)$. Therefore, start with $a_0 = a_b$ and repeat twice to get reasonable results.

Example 11.5

Repeat Example 11.4 using numerical solution.

Solution

1. Calculate $A, B, C,$ and D and determine $f(a)$.

$$A = 0.85 \times 4 \times \tfrac{14}{2} = 23.8$$

$$B = 0.85 \times 4 \times 14(18.5 - 19.5) = -47.6$$

$$C = 4(60 - 0.85 \times 4)(18.5 - 19.5 + 2.5) + 87 \times 4 \times 18.5$$

$$= 6777.6$$

$$D = -87 \times 4 \times 18.5 \times (0.85 \times 19.5) = -106{,}710$$

$$f(a) = 23.8a^3 - 47.6a^2 + 6777.6a - 106{,}710$$

2. Calculate the first derivative:

$$f'(a) = 71.4a^2 - 95.2a + 6777.6$$

3. Let $a_0 = a_b = 9.81$ in. For a balanced section, $c_b = 11.54$ in. and $a_b = 9.81$ in.

$$a_1 = 9.81 - \frac{f(9.81)}{f'(9.81)} = 9.81 - \frac{-22{,}334}{12{,}715} = 11.566 \text{ in.}$$

4. Calculate a_2:

$$a_2 = 11.566 - \frac{f(11.566)}{f'(11.566)} = 11.566 - \frac{2136}{15{,}228} = 11.43 \text{ in.}$$

This value of a is similar to that obtained earlier in Example 11.3. Substitute the value of a in equation (11.10) or (11.11) to get $P_n = 612.9$ K.

11.8.3 Approximate Solution

An approximate equation was suggested by Whitney to estimate the nominal compressive strength of short columns when compression controls, as follows [15]:

$$P_n = \frac{bhf'_c}{3he/d^2 + 1.18} + \frac{A'_s f_y}{e/(d - d') + 0.5} \tag{11.17}$$

This equation can be used only when the reinforcement is symmetrically placed in single layers parallel to the axis of bending.

A second approximate equation was suggested by Hsu [16]:

$$\frac{P_n - P_b}{P_o - P_b} + \left(\frac{M_n}{M_b}\right)^{1.5} = 1.0 \tag{11.18}$$

where

P_n = nominal axial strength of the column section
P_b, M_b = nominal load and moment of the balanced section
M_n = nominal bending moment = $P_n \cdot e$
P_o = nominal axial load at $e = 0$
$= 0.85f'_c(A_g - A_{st}) + A_{st}f_y$
A_g = gross area of the section = bh
A_{st} = total steel area

Example 11.6

Determine the nominal compressive strength P_n for the section given in Example 11.4 by equations (11.17) and (11.18) using the same eccentricity, $e = 10$ in., and compare results.

Solution

1. Solution by Whitney equation (11.17):
   a. Properties of the section shown in Figure 11.10 are $b = 14$ in., $h = 22$ in., $d = 19.5$ in., $d' = 2.5$ in., $A'_s = 4.0$ in.^{2}, and $(d - d') = 17$ in.
   b. Apply the Whitney equation:

$$P_n = \frac{14 \times 22 \times 4}{(3 \times 22 \times 10)/(19.5)^2 + 1.18} = \frac{4 \times 60}{(\frac{10}{17}) + 0.5} = 643 \text{ K}$$

$$\phi P_n = 0.7P_n = 450.1 \text{ K}$$

   c. P_n calculated by the Whitney equation is not a conservative value in this example, and the value of $P_n = 643$ K is greater than the more accurate value of 612.9 K calculated by statics in Example 11.4.
2. Solution by Hsu equation (11.18):
   a. For a balanced condition $P_b = 453.4$ K and $M_b = 6810.8$ K · in. (Example 11.2).
   b. $P_0 = 0.85f'_c(A_g - A_{st}) + A_{st}f_y$
      $= 0.85(4)(14 \times 22 - 8) + 8(60) = 1500$ K
   c. $\dfrac{P_n - 453.4}{1500 - 453.4} + \left(\dfrac{10P_n}{6810.8}\right)^{1.5} = 1$

      Multiply by 1000 and solve for P_n.

$$0.9555P_n + 0.05626P_n^{1.5} = 1433.2 \text{ K}$$

By trial, $P_n = 611$ K, which is very close to 612.9 K, as calculated by statics.

11.9 INTERACTION DIAGRAM EXAMPLE

In Example 11.2, the balanced loads P_b, M_b, and e_b were calculated for the section shown in Figure 11.6 $(e_b = 15 \text{ in.})$. Also, in Examples 11.3 and 11.4, the load capacity of the same section was calculated for the case when $e = 20$ in. (tension controls) and when $e = 10$ in. (compression controls). These values are shown in Table 11.1.

Table 11.1 Summary of the Load Capacity of the Column Section of the Previous Examples

e (in.)	*a* (in.)	ϕ	P_n (K)	ϕP_n (K)	ϕM_n (K·ft)	NOTES
0	—	0.7	1500	1050	0.0	ϕP_{n0}
2.25	19.39	0.7	1200	840	157.5	$0.8\phi P_{n0}$
4	16.82	0.7	1018	712.6	237.6	Compression
6	14.19	0.7	843.3	590.3	295.1	Compression
10*	11.43	0.7	612.9	429.0	357.5	Compression
12	10.63	0.7	538.0	376.6	376.6	Compression
15*	9.81	0.7	453.4	317.4	397.3	Balanced
20*	7.10	0.7	324.4	227.1	378.5	Tension
30	5.06	0.7	189.4	132.6	331.5	Tension
50	4.01	0.773	100.6	77.76	324.0	Tension
80	3.59	0.822	58.8	48.33	322.0	Tension
P.M.	3.08	0.9	0	0.0	352.0	Tension
P.M.	3.08	0.7	0	0.0	273.8	P.M. (X)

* = values calculated in Examples 11.2, 11.3 and 11.4 (P.M. = pure moment).
X = not applicable; needed only to plot the interaction diagram.

To plot the load-moment interaction diagram, different values of ϕP_n and ϕM_n were calculated for various e values that varied between $e = 0$ and $e =$ maximum for the case of pure moment when $P_n = 0$. These values are shown in Table 11.1. The interaction diagram is shown in Figure 11.11. The load $\phi P_{n0} = 1050$ K represents the theoretical axial load when $e = 0$, whereas $0.8\phi P_{n0} = 840$ K represents the maximum axial load allowed by the ACI Code based on minimum eccentricity. Note that when compression controls, $e < e_b$ and $P_n > P_b$, and when tension controls, $e > e_b$ and $P_n < P_b$. The last two cases in the table represent the pure moment (P.M.) or beam-action case for $\phi = 0.9$ and $\phi = 0.7$ $(M_n = 391 \text{ K}\cdot\text{ft})$. To be consistent with the design of beams due to bending moments, the ACI Code allows the use of $\phi = 0.9$ with pure moment, so $\phi M_n = 352$ K·ft instead of 273.8 K·ft. Also note that ϕ varies between 0.7 and 0.9 according to equation (11.1) for tied columns. Note that $M_n = 391.1$ K·ft.

11.10 RECTANGULAR COLUMNS WITH SIDE BARS

In some column sections, the steel reinforcement bars are distributed around the four sides of the column section. The side bars are those placed on the sides along the depth of the section in addition to the tension and compression steel A_s and A'_s and can be denoted by A_{ss} (Figure 11.12). In this case the same procedure explained earlier can be applied, taking into consideration the strain variation along the depth of the section and the relative force in each side bar either in the compression or tension zone of the section. These are added to those of C_c, C_s, and T to determine P_n. Equation (11.10) becomes

$$P_n = C_c + \Sigma C_s - \Sigma T \tag{11.10a}$$

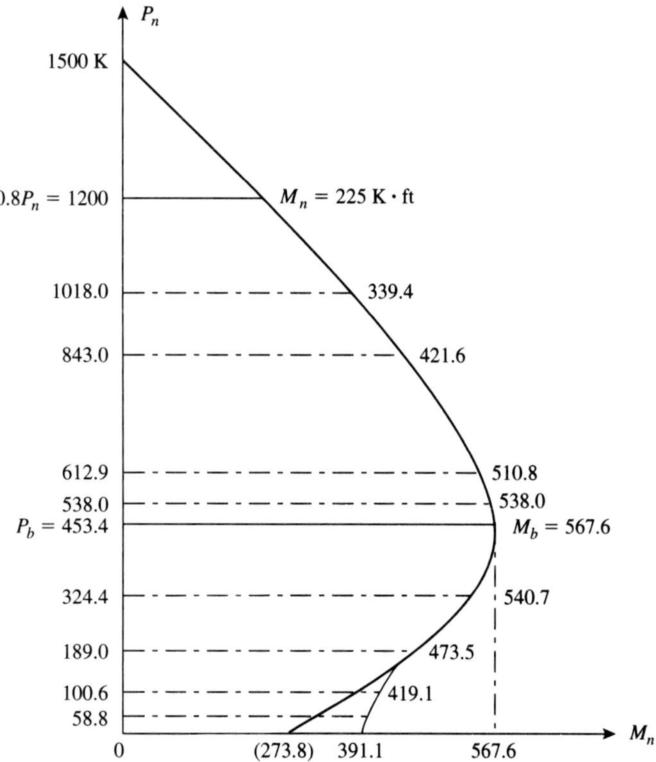

Figure 11.11 Interaction diagram of the column section shown in Figure 11.10.

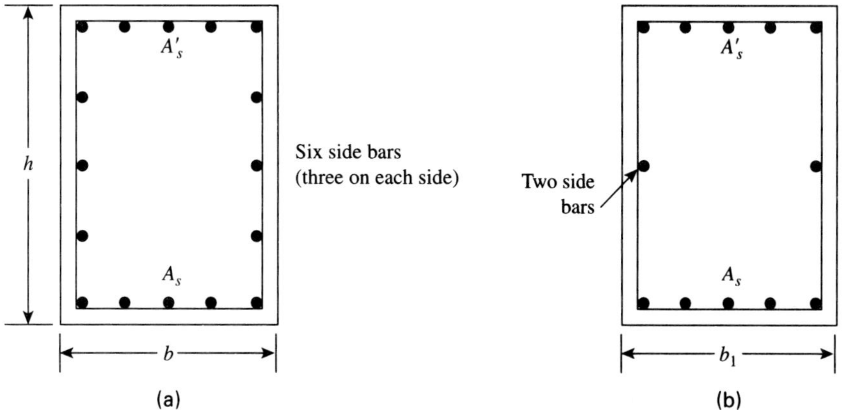

Figure 11.12 Side bars in rectangular sections: (a) six side bars and (b) two side bars (may be neglected).

Example 11.7 explains this analysis. Note that if the side bars are located near the neutral axis (Figure 11.12(b)) the strains—and, consequently, the forces—in these bars are very small and can be neglected. Those bars close to A_s and A'_s have appreciable force and increase the load capacity of the section.

Example 11.7

Determine the balanced load, P_b, moment, M_b, and e_b for the section shown in Figure 11.13. Use $f'_c = 4$ Ksi and $f_y = 60$ Ksi.

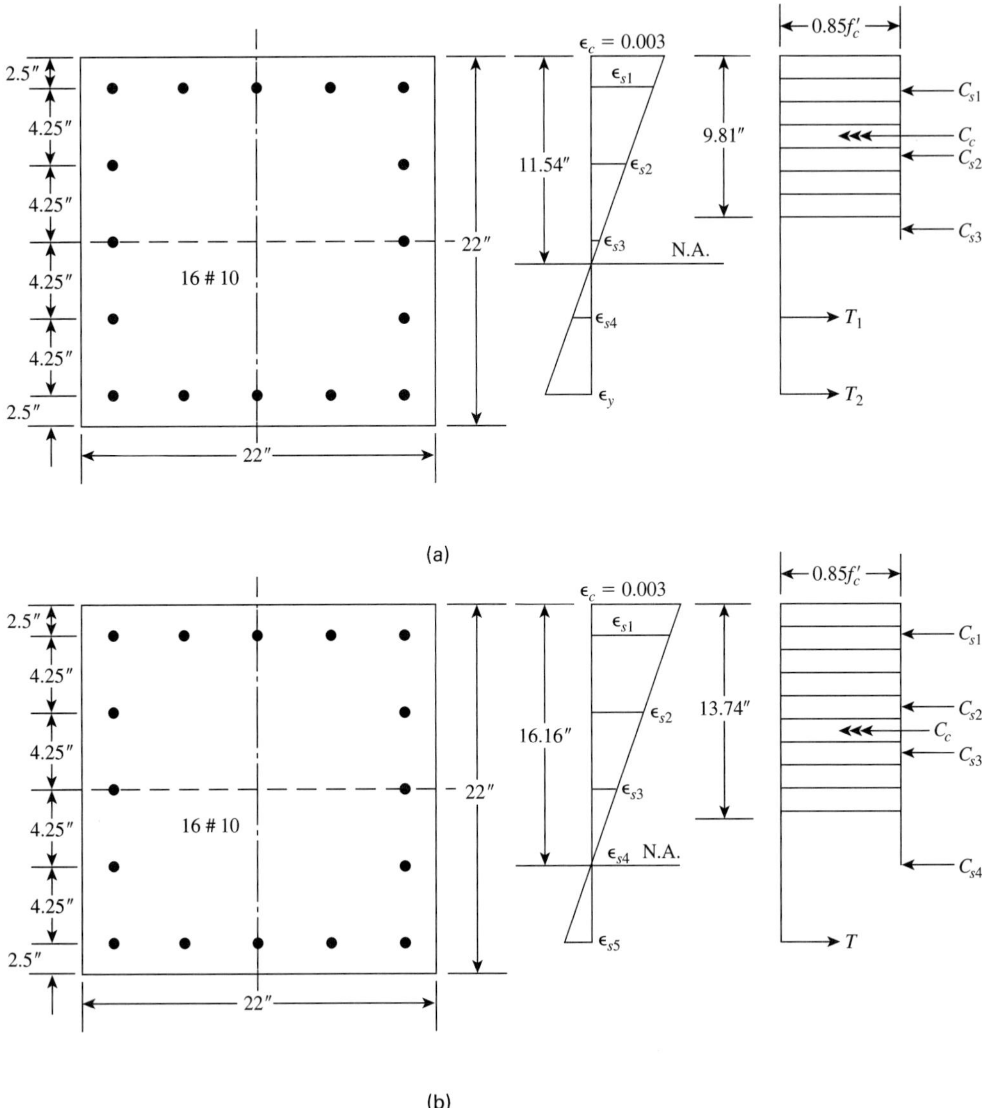

Figure 11.13 Example 11.7: (a) balanced section; Example 11.8: (b) when compression controls, $e = 6$ in.

Solution

The balanced section is similar to Example 11.2. Given: $b = h = 22$ in., $d = 19.5$ in., $d' = 2.5$ in., $A_s = A'_s = 6.35$ in.^{2} (five no. 10 bars), and six no. 10 side bars (three on each side).

1. Calculate the distance to the neutral axis:

$$c_b = \left(\frac{87}{87 + f_y}\right)d = \left(\frac{87}{87 + 60}\right)19.5 = 11.54 \text{ in.}$$

$$a_b = 0.85(11.54) = 9.81 \text{ in.}$$

2. Calculate the forces in concrete and steel bars; refer to Figure 11.13(a). In the compression zone, $C_c = 0.85f'_c ab = 0.85(4)(9.81)(22) = 733.8$ K.

$$f'_s = 87\left(\frac{c - d'}{c}\right) = 87\left(\frac{11.54 - 2.5}{11.54}\right) = 68.15 \text{ Ksi} > 60 \text{ Ksi}$$

Then $f'_s = 60$ Ksi.

$$C_{s1} = A'_s(f_y - 0.85f'_c) = 6.35(60 - 0.85 \times 4) = 359.4 \text{ K}$$

$$f_{s2} = 87\left(\frac{11.54 - 2.5 - 4.25}{11.54}\right) = 36.11 \text{ Ksi}$$

$$C_{s2} = 2(1.27)(36.11 - 0.85 \times 4) = 83.1 \text{ K}$$

Similarly, $f_{s3} = 4.07$ Ksi and $C_{s3} = 2(1.27)(4.07 - 0.85 \times 4) = 1.7$ K.
In the tension zone,

$$\varepsilon_{s4} = 964.50 \times 10^{-6}, \qquad f_{s4} = 28 \text{ Ksi}$$

$$T_1 = 2(1.27)(28) = 71 \text{ K}$$

$$T_2 = A_s f_y = 6.35(60) = 381 \text{ K}$$

3. Calculate $P_b = C_c + \Sigma C_s - \Sigma T$.

$$P_b = 733.8 + (359.4 + 83.1 + 1.7) - (71 + 381)$$

$$= 726 \text{ K}$$

4. Taking moments about the plastic centroid,

$$M_b = 733.8(6.095) + 359.4(8.5) + 83.1(4.25)$$

$$+ 71(4.25) + 381(8.5)$$

$$= 11{,}421 \text{ K}\cdot\text{in.} = 952 \text{ K}\cdot\text{ft}$$

$$e_b = \frac{M_b}{P_b} = 15.735 \text{ in.}$$

$$\phi P_b = 0.7_b = 508 \text{ K} \quad \text{and} \quad \phi M_b = 0.7M_b = 666.4 \text{ K}\cdot\text{ft}$$

Example 11.8

Repeat the previous example when $e = 6.0$ in.

Solution

1. Because $e = 6$ in. $< e_b = 15.735$ in., compression controls. Assume $c = 16.16$ in. (by trial) and $a = 0.85(16.16) = 13.74$ in. (Figure 11.13(b)).

2. Calculate the forces in concrete and steel bars:

$$C_c = 0.85(4)(13.74)(22) = 1027.75 \text{ K}$$

In a similar approach to the balanced case, $f_{s1} = 60$ Ksi and $C_{s1} = 359.41$.

$$f_{s2} = 50.66 \text{ Ksi}, \qquad C_{s2} = 120.0 \text{ K}$$

$$f_{s3} = 27.78 \text{ Ksi}, \qquad C_{s3} = 61.92 \text{ K}$$

$$f_{s4} = 4.9 \text{ Ksi}, \qquad C_{s4} = 3.81 \text{ K}$$

$$f_{s5} = 18 \text{ Ksi}, \qquad T = 6.35(18) = 114.2 \text{ K}$$

3. Calculate $P_n = C_c + \Sigma C_s - \Sigma T = 1458.7$ K.

$$M_n = P_n \cdot e = 729.35 \text{ K}\cdot\text{ft} \qquad (e = 6 \text{ in.})$$

$$\phi P_n = 0.7P_n = 1021 \text{ K}, \qquad \phi M_n = 510.5 \text{ K}\cdot\text{ft}$$

4. Check P_n by taking moments about A_s:

$$P_n = \frac{1}{e'}\left[C_c\left(d - \frac{a}{2}\right) + C_{s1}(d - d') + C_{s2}(d - d' - s) + C_{s3}(d - d' - 2s) + C_{s4}(d - d' - 3s)\right]$$

$$e' = e + d - \frac{h}{2} = 6 + 19.5 - \frac{22}{2} = 14.5 \text{ in.}$$

s = distance between side bars

= 4.25 in. (s = constant in this example.)

$$P_n = \frac{1}{14.5}\left[1027.75\left(19.5 - \frac{13.74}{2}\right) + 359.41(17) + 120(17 - 4.25) + 61.92(17 - 8.5) + 3.81(17 - 12.75)\right] = 1459 \text{ K}$$

Note: If side bars are neglected, then

$$P_b = 733.8 + 359.4 - 381 = 712.2 \text{ K}$$

$$P_n \text{ (at } e = 6 \text{ in.)} = 1027.75 + 359.4 - 114.2 = 1273 \text{ K}$$

If side bars are considered, the increase in P_b is about 2% and that in P_n is about 14.6%.

11.11 LOAD CAPACITY OF CIRCULAR COLUMNS

11.11.1 Balanced Condition

The values of the balanced load P_b and the balanced moment M_b for circular sections can be determined using the equations of equilibrium, as was done in the case of rectangular sections. The bars in a circular section are arranged in such a way that their distance from the axis of plastic centroid varies, depending on the number of bars in the section. The main problem is to find the depth of the compressive block a and the stresses in the reinforcing bars. The following example explains the analysis of circular sections under balanced conditions. A similar procedure can be adopted to analyze sections when tension or compression controls.

Example 11.9

Determine the balanced load P_b and the balanced moment M_b for the 16-in.-diameter circular spiral column reinforced with eight no. 9 bars, as shown in Figure 11.14. Given: $f'_c = 4$ Ksi and $f_y = 60$ Ksi.

Solution

1. Because the reinforcement bars are symmetrical about the axis A-A passing through the center of the circle, the plastic centroid lies on that axis.
2. Determine the location of the neutral axis:

$$d = 13.1 \text{ in.} \qquad \varepsilon_y = \frac{f_y}{E_s} \qquad (E_s = 29{,}000 \text{ Ksi})$$

$$\frac{c_b}{d} = \frac{0.003}{0.003 + \varepsilon_y} = \frac{0.003}{0.003 + f_y/E_s} = \frac{87}{87 + f_y}$$

$$c_b = \frac{87}{87 + 60}(13.1) = 7.75 \text{ in.}$$

$$a_b = 0.85 \times 7.75 = 6.59 \text{ in.}$$

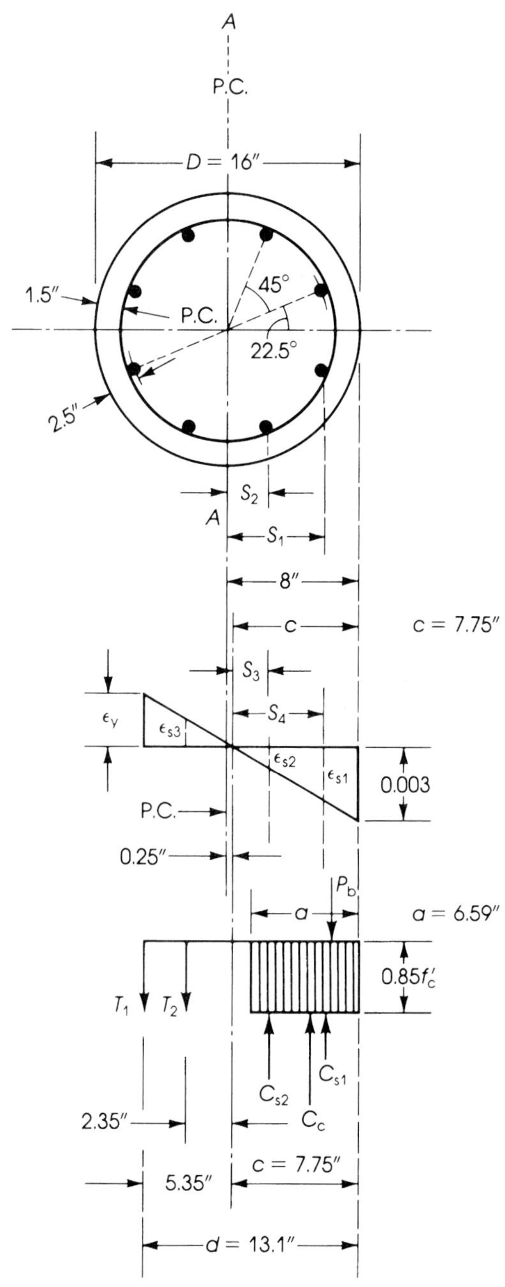

Figure 11.14 Example 11.9: eight no. 9 bars

$S = 8 - 2.5 = 5.5$ in.

$S_1 = S \cos 22.5° = 5.1$ in.

$S_2 = S \cos 67.5° = 2.1$ in.

$d = 8 + 5.1 = 13.1$ in.

$S_3 = 1.85$ in.

$S_4 = 4.85$ in.

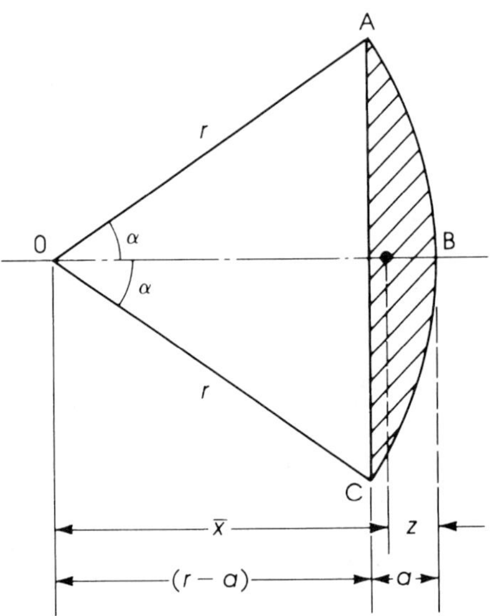

Figure 11.15 Properties of circular segments.

3. Calculate the properties of a circular segment (Figure 11.15):

$$\text{Area of segment} = r^2(\alpha - \sin\alpha\cos\alpha) \tag{11.19}$$

Location of centroid $\bar{x}$ (from the circle center 0):

$$\bar{x} = \frac{2}{3}\frac{(r\sin^3\alpha)}{(\alpha - \sin\alpha\cos\alpha)} \tag{11.20}$$

$$Z = r - \bar{x} \tag{11.21}$$

$$r\cos\alpha = (r - a) \quad \text{or} \quad \cos\alpha = \left(1 - \frac{a}{r}\right) \tag{11.22}$$

$$\cos\alpha = \left(1 - \frac{6.59}{8}\right) = 0.176$$

and $\alpha = 79.85°$, $\sin\alpha = 0.984$, and $\alpha = 1.394$ rad.

$$\text{Area for segment} = (8)^2(1.394 - 0.984 \times 0.176)$$
$$= 78.12 \text{ in.}^2$$

$$\bar{x} = \left(\frac{2}{3}\right)\frac{8(0.984)^3}{(1.394 - 0.984 \times 0.176)} = 4.16 \text{ in.}$$

$$Z = r - \bar{x} = 8 - 4.16 = 3.84 \text{ in.}$$

4. Calculate the compressive force C_c:

$$C_c = 0.85f'_c \times \text{area of segment}$$
$$= 0.85 \times 4 \times 78.12 = 265.6 \text{ K}$$

It acts at 4.16 in. from the center of the column.

5. Calculate the strains, stresses, and forces in the tension and the compression steel. Determine the strains from the strain diagram. For T_1,

$$\varepsilon = \varepsilon_y = 0.00207, \qquad f_s = f_y = 60 \text{ Ksi}$$

$$T_1 = 2 \times 60 = 120 \text{ K}$$

For T_2,

$$\varepsilon_{s3} = \frac{2.35}{5.35}\varepsilon_y = \frac{2.35}{5.35} \times 0.00207 = 0.00091$$

$$f_{s3} = 0.00091 \times 29{,}000 = 26.4 \text{ Ksi}$$

$$T_2 = 26.4 \times 2 = 52.8 \text{ K}$$

For C_{s1},

$$\varepsilon_{s1} = \frac{4.85}{7.75} \times 0.003 = 0.00188$$

$$f_{s1} = 0.00188 \times 29{,}000 = 54.5 \text{ Ksi} < 60 \text{ Ksi}$$

$$C_{s1} = 2(54.5 - 3.4) = 102.2 \text{ K}$$

For C_{s2},

$$\varepsilon_{s2} = \frac{1.85}{7.75} \times 0.003 = 0.000716$$

$$f_{s2} = 0.000716 \times 29{,}000 = 20.8 \text{ Ksi}$$

$$C_{s2} = 2(20.8 - 3.4) = 34.8 \text{ K}$$

The stresses in the compression steel have been reduced to take into account the concrete displaced by the steel bars.

6. The balanced force is $P_b = C_c + \Sigma C_s - \Sigma T$: ($\phi = 0.75$).

$$P_b = 265.6 + (102.2 + 34.8) - (120 + 52.8) = 230 \text{ K}$$

$$\phi P_b = 172.5 \text{ K}$$

7. Take moments about the plastic centroid (axis A-A through the center of the section) for all forces:

$$M_b = P_b e_b = [C_c \times 4.16 + C_{s1} \times 5.1 + C_{s2} \times 2.1 + T_1 \times 5.1 + T_2 \times 2.1]$$

$$= 2422.1 \text{ K}\cdot\text{in.} = 201.9 \text{ K}\cdot\text{ft},$$

$$\phi M_b = 151.4 \text{ K}\cdot\text{ft}$$

$$e_b = \frac{2422.1}{230} = 10.5 \text{ in.}$$

11.11.2 Strength of Circular Columns When Compression Controls:

A circular column section under eccentric load can be analyzed in similar steps as the balanced section. This is achieved by assuming a value for $c > c_b$ or $a > a_b$ and calculating the forces in concrete and steel at different locations to determine P_{n1}, $P_{n1} = C_c + \Sigma C_s - \Sigma T$. Also, M_n can be calculated by taking moments about the plastic centroid (center of the section) and determining $P_{n2} = M_n/e$. If they are not close enough, within about 1%, assume a new c or a and repeat the calculations (see also Section 11.8). Compression controls when $e < e_b$ or $P_n > P_b$.

For example, if it is required to determine the load capacity of the column section of Example 11.9 when $e = 6$ in., P_n can be determined in steps similar to those of Example 11.9:

1. Because $e = 6$ in. is less than $e_b = 10.5$ in., compression controls.
2. Assume $c = 9.0$ in. (by trial) $> c_b = 7.75$ in. and $a = 7.65$ in.
3. Calculate $\bar{x} = 3.585$ in., $Z = 4.415$ in., and the area of concrete segment $= 94.93$ in.^{2}

4–5. Calculate forces: $C_c = 322.7$ K, $C_{s1} = 110.7$ K, $C_{s2} = 53.1$ K, $T_1 = 21.6$ K, and $T_2 = 78.9$ K.

6. Calculate $P_{n1} = C_c + \Sigma C_s - \Sigma T = 386$ K.

7. Taking moments about the center of the column (plastic centroid): $M_n = 191 \text{ K}\cdot\text{ft}$, $P_{n2} = M_n/6 = 382$ K, which is close to P_{n1} (the difference is about 1%). Therefore, $P_n = 382$ K. Note that if the column is spirally reinforced, $\phi = 0.75$.

An approximate equation for estimating P_n in a circular section when compression controls was suggested by Whitney [15]:

$$P_n = \frac{A_g f'_c}{\left[\dfrac{9.6De}{(0.8D + 0.67D_s)^2} + 1.18\right]} + \frac{A_s f_y}{\left(\dfrac{3e}{D_s} + 1\right)} \tag{11.23}$$

where A_g = gross area of the section
D = external diameter
D_s = diameter measured through the centroid of the bar arrangement
A_{st} = total vertical steel area
e = eccentricity measured from the plastic centroid

Example 11.10

Calculate the nominal compressive strength P_n for the section of Example (11.9) using the Whitney equation if the eccentricity is $e = 6$ in.

Solution

1. $e = 6$ in. is less than $e_b = 10.5$ in. calculated earlier; thus compression controls.

2. Using the Whitney equation,

$$A_g = \frac{\pi}{4}D^2 = \frac{\pi}{4}(16)^2 = 20.1 \text{ in.}^2$$

$$D = 16 \text{ in.} \qquad D_s = 16 - 5 = 11.0 \text{ in.} \qquad A_{st} = 8 \times 1 = 8 \text{ in.}^2$$

$$P_n = \frac{(201.1 \times 4)}{\left[\dfrac{(9.6 \times 16 \times 6)}{(0.8 \times 16 + 0.67 \times 11)^2} + 1.18\right]} + \frac{8 \times 60}{\left(\dfrac{3 \times 6}{11}\right) + 1}$$

$$= 415.5 \text{ K}$$

3. $M_n = P_n e = 415.5 \times \frac{6}{12} = 207.8 \text{ K}\cdot\text{ft}$

The value of P_n here is greater than $P_n = 382$ K calculated earlier by statics.

11.11.3 Strength of Circular Columns When Tension Controls

Tension controls in circular columns when the load is applied at an eccentricity $e > e_b$, or $P_n < P_b$. In this case, the column section can be analyzed in steps similar to those of the balanced section and Example 11.8. This is achieved by assuming $c < c_b$ or $a < a_b$ and then following the steps explained in Section 11.11.1. Note that because the steel bars are uniformly distributed along the perimeter of the circular section, the tension steel A_s provided could be relatively low, and the load capacity becomes relatively small. Therefore, it is advisable to avoid the use of circular columns when tension controls.

11.12 ANALYSIS AND DESIGN OF COLUMNS USING CHARTS

The analysis of column sections explained earlier is based on the principles of statics. For preliminary analysis or design of columns, special charts or tables may be used either to determine ϕP_n and ϕM_n for a given section or determine the steel requirement for a given load P_u and moment M_u. These charts and tables are published by the American Concrete Institute (ACI), the Concrete Reinforcing Steel Institute (CRSI), and the Portland Cement

Association (PCA). Final design of columns must be based on statics by using manual calculations or computer programs. The use of the ACI charts is illustrated in the following examples. The charts are given in Figures 11.16 and 11.17. These data are limited to the column sections shown on the top right corner of the charts.

Example 11.11

Use charts to determine the ultimate strength ϕP_n of the short tied column shown in Figure 11.18 (a) acting at an eccentricity $e = 12$ in. Given: $f'_c = 5$ Ksi and $f_y = 60$ Ksi.

Solution

1. Properties of the section: $h = 24$ in., $\gamma h = 24 - 5 = 19$ in. = distance between A_s and A'_s.

$$\gamma = \frac{19}{24} = 0.79, \qquad \rho = \frac{A_{st}}{bh} = \frac{8 \times 1.27}{14 \times 24} = 0.03$$

$$\frac{e}{h} = \frac{12}{24} = 0.5, \qquad A_g = bh$$

2. From the charts (Figure 11.17),

$$\frac{P_u}{bh} = 1.5 \quad \text{for } \gamma = 0.75, \quad \text{and} \quad \frac{P_u}{bh} = 1.65 \quad \text{for } \gamma = 0.9$$

By interpolation for $\gamma = 0.79$,

$$\frac{P_u}{bh} = 1.56 \quad \text{and} \quad P_u = 1.56 \times 14 \times 24 = 524 \text{ K}$$

By analysis, $P_u = 526$ K ($a = 11.53$ in.).

Example 11.12

Determine the necessary reinforcement for a short tied column to carry an ultimate load of 483 K and an ultimate moment of 322 K·ft. The column section has a width of $b = 14$ in. and a total depth $h = 20$ in. Given: $f'_c = 4$ Ksi, $f_y = 60$ Ksi, and the charts of Figure 11.16.

Solution

1. The eccentricity is $e = M_u/P_u = (322 \times 12)/483 = 8$ in. Let $d = 20 - 2.5 = 17.5$ in. The distance between the tension and compression steel is $\gamma h = d - d'$.

$$\gamma h = 20 - (2.5 + 2.5) = 15.0 \text{ in.}$$

$$\gamma = \frac{15}{20} = 0.75 \qquad \frac{e}{h} = \frac{8}{20} = 0.40$$

$$\frac{P_u}{bh} = \frac{483}{14 \times 20} = 1.725$$

2. From graphs, for $P_u/bh = 1.725$, $e/h = 0.40$, and $\gamma = 0.75$,

$$\rho = 0.036$$

$$A_s = 0.036 \times 14 \times 20 = 10.08 \text{ in.}^2$$

Choose eight no. 10 bars ($A_s = 10.16$ in.^{2}), four bars on each short side, and use no. 3 ties spaced at 14 in, Figure 11.18 (b).

11.13 DESIGN OF COLUMNS UNDER ECCENTRIC LOADING

In the previous sections, the analysis, behavior, and the load-interaction diagram of columns subjected to an axial load and bending moment were discussed. The design of columns is more complicated, because the external load and moment, P_u and M_u, are given and it is required to determine many unknowns, such as b, h, A_s, and A'_s, all within the ACI Code

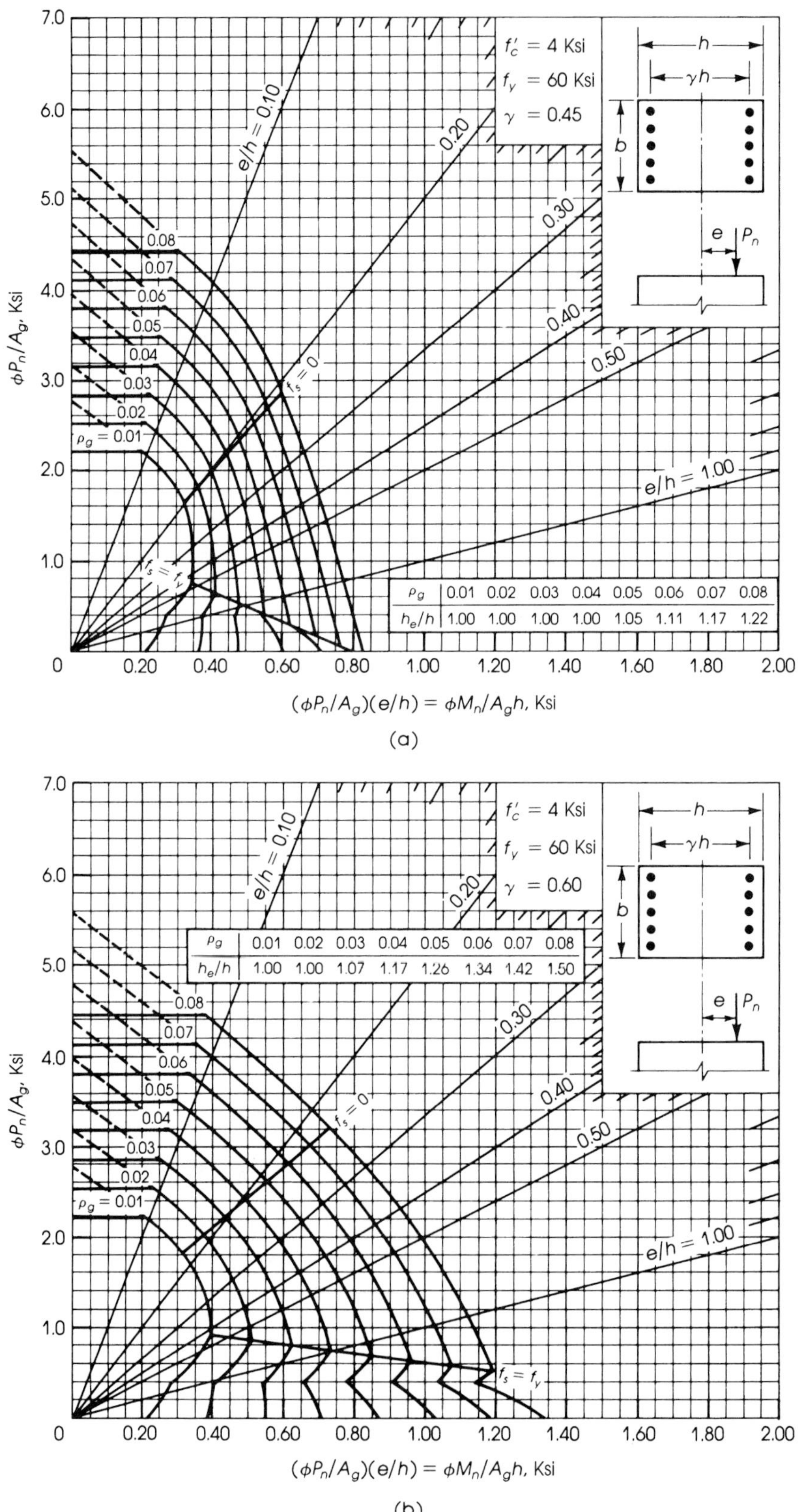

ρ_g	0.01	0.02	0.03	0.04	0.05	0.06	0.07	0.08
h_e/h	1.00	1.00	1.00	1.00	1.05	1.11	1.17	1.22

ρ_g	0.01	0.02	0.03	0.04	0.05	0.06	0.07	0.08
h_e/h	1.00	1.00	1.07	1.17	1.26	1.34	1.42	1.50

Figure 11.16 Load-moment strength interaction diagram for rectangular columns where f'_c = 4 Ksi, f_y = 60 Ksi, and (a) γ = 0.45, (b) γ = 0.60, (c) γ = 0.75, and (d) γ = 0.90. Courtesy of American Concrete Institute [7].

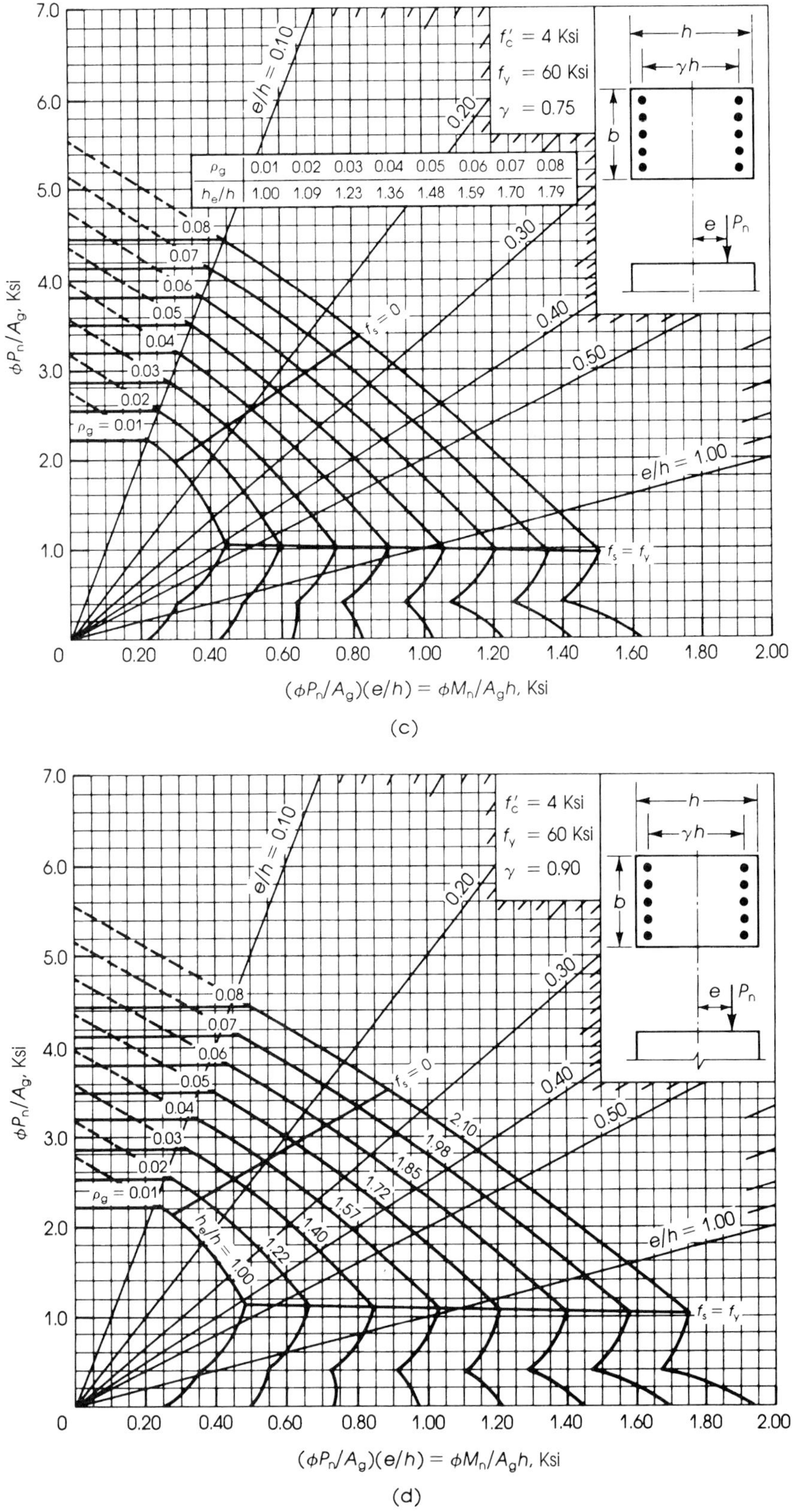

Figure 11.16 *(continued)*

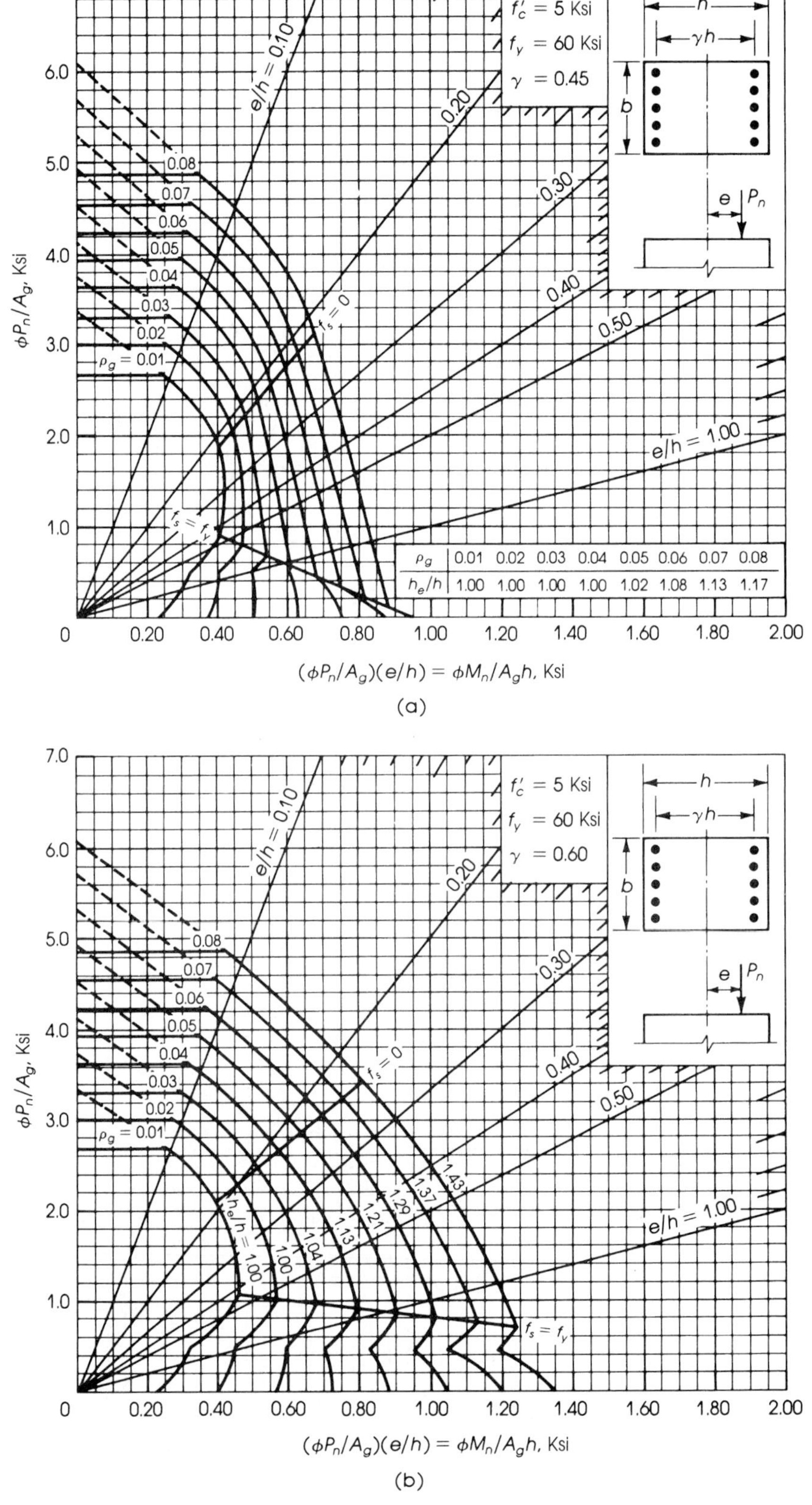

Figure 11.17 Load-moment strength-interaction diagram for rectangular columns where f'_c = 5 Ksi, f_y = 60 Ksi, and (a) γ = 0.45, (b) γ = 0.60, (c) γ = 0.75, and (d) γ = 0.90. Courtesy of American Concrete Institute [7].

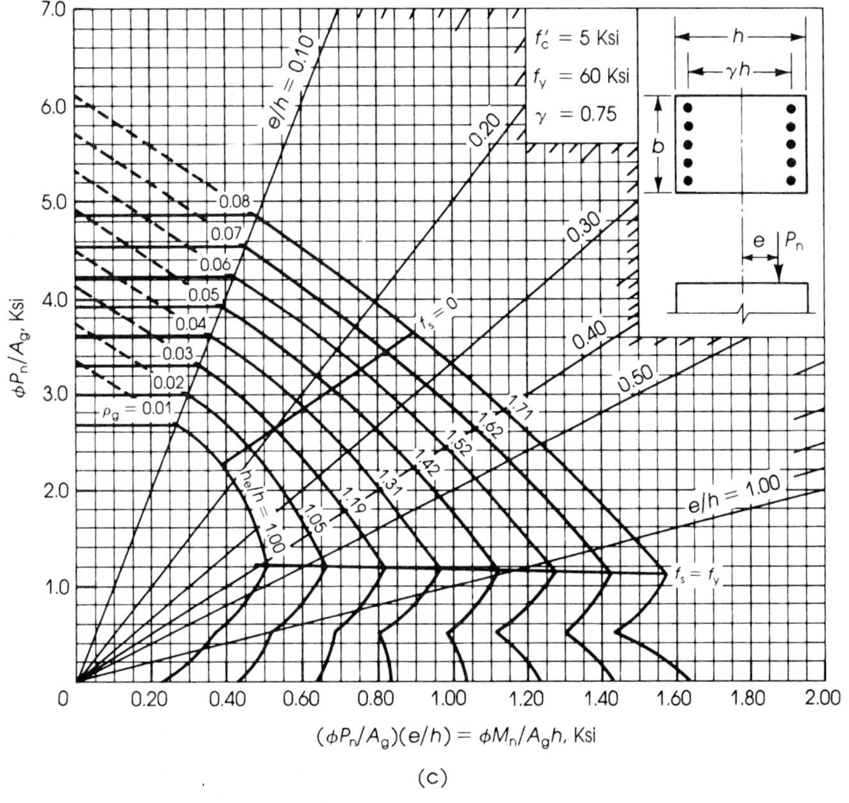

(c)

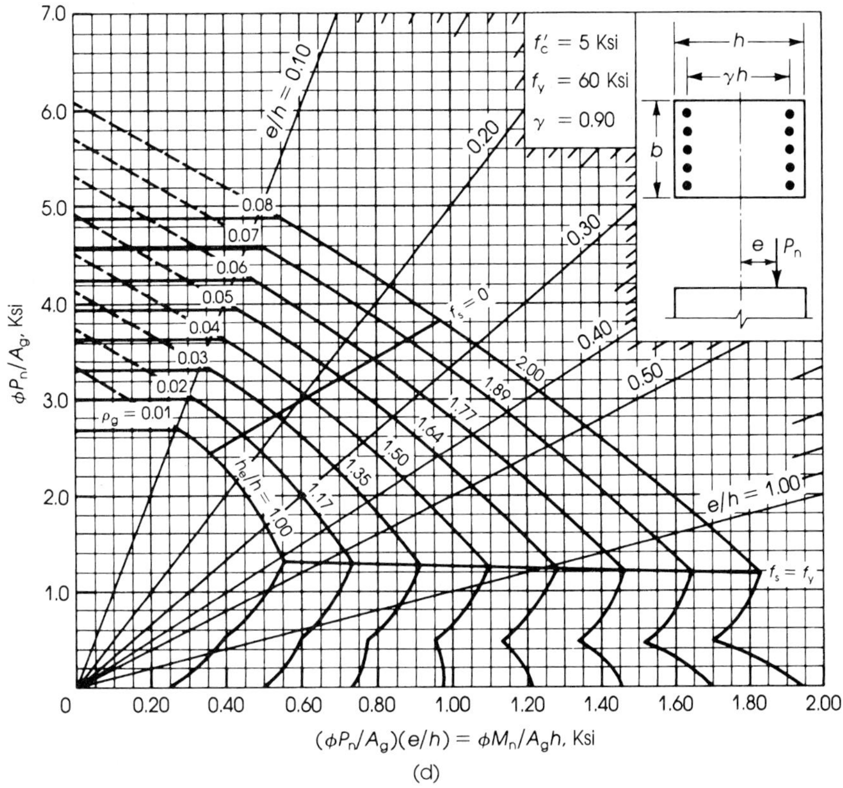

(d)

Figure 11.17 *(continued)*

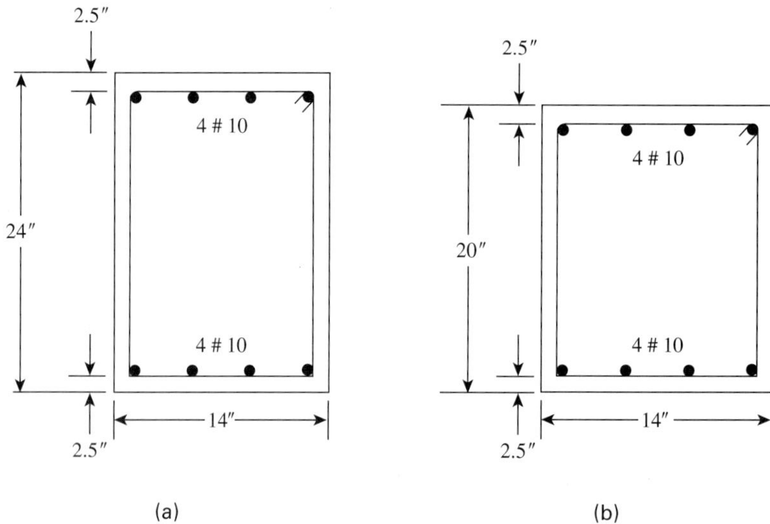

Figure 11.18 Column sections of (a) Example 11.11 and (b) Example 11.12.

limitations. It is a common practice to assume a column section first and then determine the amount of reinforcement needed. If the designer needs to change the steel reinforcement calculated, then the cross section may be adjusted accordingly. The following examples illustrate the design of columns.

11.13.1 Design of Columns When Compression Controls

When compression controls, it is preferable to use $A_s = A'_s$ for rectangular sections. The eccentricity, e, is equal to M_u/P_u. Based on the magnitude of e, two cases may develop.

1. When e is relatively very small (say, $e \leq 4$ in.), a minimum eccentricity case may develop that can be treated by using equation (10.8), as explained in the examples of Chapter 10. Alternatively, the designer may proceed as in Case 2. This loading case occurs in the design of the lower-floor columns in a multistory building, where the moment, M_u, develops from one floor system and the load, P_u, develops from all floor loads above the column section.
2. The compression control zone represents the range from the axial to the balanced load, as shown in Figures 11.3 and 11.11. In this case a cross section (bh) may be assumed and then the steel reinforcement is calculated for the given P_u and M_u. The steps can be summarized as follows:
   a. Assume a square or rectangular section (bh); then determine d, d', and $e = M_u/P_u$.
   b. Assuming $A_s = A'_s$, calculate A'_s from equation (11.17) using the dimensions of the assumed section, and $\phi = 0.7$ for tied columns. Let $A_s = A'_s$ and then choose adequate bars. Determine the actual areas used for A_s and A'_s. Alternatively, use the ACI charts.
   c. Check that $\rho_g = (A_s + A'_s)/bh$ is less than or equal to 8% and greater or equal to 1%. If ρ_g is small, reduce the assumed section, but increase the section if less steel is required.
   d. Check the adequacy of the final section by calculating ϕP_n from statics; as explained in the previous examples, ϕP_n should be greater than or equal to P_u.
   e. Determine the necessary ties.

A simple approximate formula for determining the initial size of the column bh or the total steel ratio ρ_g is

$$P_n = K_c bh^2, \quad \text{or} \quad P_u = \phi P_n = \phi K_c bh^2 \tag{11.24}$$

where K_c has the values shown in Table 11.2 and plotted in Figure 11.19 for $f_y = 60$ Ksi and $A_s = A'_s$. Units for K_c are in lb/in.^{3} and P_n units are in pounds.

Table 11.2 Values of K_c ($f_y = 60$ Ksi)

	K_c		
ρ_g **(%)**	$f'_c = 4$ **Ksi**	$f'_c = 5$ **Ksi**	$f'_c = 6$ **Ksi**
1%	0.090	0.110	0.130
4%	0.137	0.157	0.177
8%	0.200	0.220	0.240

The values of K_c shown in Table 11.2 are approximate and easy to use, because K_c increases by 0.02 for each increase of 1 Ksi in f'_c. For the same section, as the eccentricity, $e = M_u/P_u$, increases, P_n decreases, and, consequently, K_c decreases. Thus K_c values represent a load P_n on the interaction diagram between $0.8P_{n0}$ and P_b, as shown in Figure 11.3 or 11.11.

Linear interpolation can be used. For example, $K_c = 0.1685$ for $\rho_g = 6\%$ and $f'_c = 4$ Ksi. The steps in designing a column section can be summarized as follows:

1. Assume an initial size of the column section bh.
2. Calculate $K_c = P_u/(\phi bh^2)$.
3. Determine ρ_g from Table 11.2 for the given f'_c.
4. Determine $A_s = A'_s = \rho_g bh/2$ and choose bars and ties.
5. Determine ϕP_n of the final section by statics (accurate solution). The value of ϕP_n should be greater than or equal to P_u. If not, adjust bh or ρ_g.

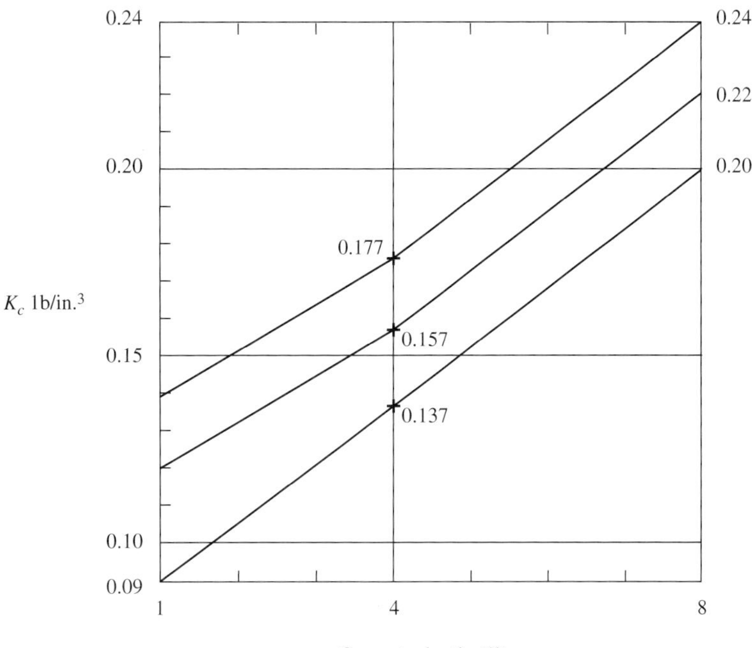

Figure 11.19 Values of K_c versus ρ_g (%).

Alternatively, if a specific steel ratio is desired, say $\rho_g = 6\%$, then proceed as follows:

1. Assume ρ_g as required and then calculate $e = M_u/P_u$.
2. Based on the given f'_c and ρ_g, determine K_c from Table 11.2.
3. Calculate $bh^2 = P_u/\phi K_c$; then choose b and h. Repeat Steps 4 and 5. It should be checked that ρ_g is less than or equal to 8% and greater than or equal to 1%. Also, check that c calculated by statics is greater than $c_b = 87d/(87 + f_y)$ for compression failure to control.

Example 11.13

Determine the tension and compression reinforcement for a 16- × 24-in. rectangular tied column to support $P_u = 840$ K and $M_u = 420$ K · ft. Use $f'_c = 4$ Ksi and $f_y = 60$ Ksi.

Solution

1. Calculate $e = M_u/P_u = 420(12)/840 = 6.0$ in. We have $h = 24$ in.; let $d = 21.5$ in. and $d' = 2.5$ in. Because e is less than $\frac{2}{3}d = 14.38$ in., assume compression controls.
2. Assume $A_s = A'_s$ and use equation (11.17) to determine the initial value of A'_s. $P_n = P_u/\phi = 840/0.7 = 1200$ K.

$$P_n = \frac{bhf'_c}{\left(\dfrac{3he}{d^2}\right) + 1.18} + \left[\frac{A'_s f_y}{\left(\dfrac{e}{d - d'}\right) + 0.5}\right] \tag{11.17}$$

For $P_n = 1200$ K, $e = 6$ in., $d = 21.5$ in., $d' = 2.5$ in., and $h = 24$ in., calculate $A'_s = 6.44$ in.$^2 = A_s$. Choose five no. 10 bars ($A_s = 6.35$ in.^{2}) for A_s and A'_s (Figure 11.20).

3. $\rho_g = 2(6.35)/(16 \times 24) = 0.033$, which is less than 0.08 and > 0.01.
4. Check the section by statics following the steps of Example 11.4 to get

$$a = 16.64 \text{ in.}, \qquad c = 19.58 \text{ in.}, \qquad C_c = 905.2 \text{ K}$$

$$C_s = 6.35(60 - 0.85 \times 4) = 359.4 \text{ K}$$

$$f_s = 87\left(\frac{d - c}{c}\right) = 8.55 \text{ Ksi}$$

$$T = A_s f_s = 6.35(8.55) = 54.3 \text{ K}$$

$$P_n = C_c + C_s - T = 1210.3 \text{ K} > 1200 \text{ K}$$

Note that if $\phi P_n < P_u$, increase A_s and A'_s, for example, to six no. 10 bars, and check the section again.

5. Check P_n based on moments about A_s (equation (11.12)) to get $P_n = 1210$ K.

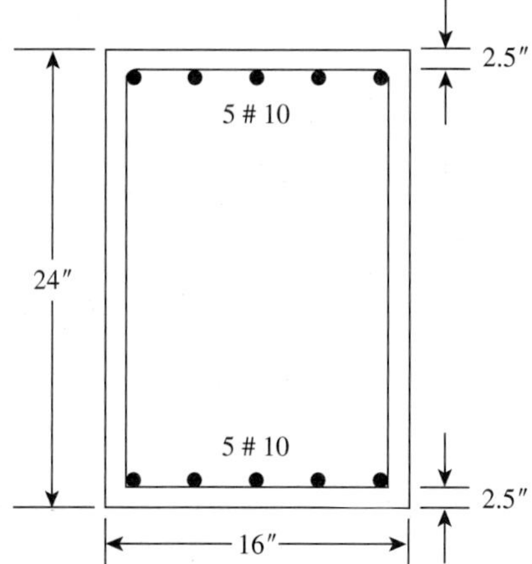

Figure 11.20 Example 11.13.

6. For a balanced section,

$$c_b = \left(\frac{87}{87 + f_y}\right)d = \left(\frac{87}{147}\right)21.5 = 12.7 \text{ in.}$$

Because $c = 19.58$ in. $> c_b = 12.7$ in., compression controls, as assumed.

7. Use no. 3 ties spaced at 16 in. (refer to Chapter 10).

Example 11.14

Repeat Example 11.13 using equation (11.24).

Solution

1. The column section is given: 16×24 in.
2. Determine K_c from equation (11.24):

$$K_c = \frac{P_u}{\phi bh^2} = \frac{840}{0.7 \times 16 \times 24^2} = 0.13 \text{ lb/in.}^3$$

3. From Table 11.2 or Figure 11.19, for $K_c = 0.13, f'_c = 4$ Ksi; by interpolation, get $\rho_g = 3.5\%$.
4. Calculate $A_s = A'_s = \rho bh/2 = 0.035(16)(24)/2 = 6.77$ in.^{2} Choose five no. 10 bars $(A_s = 6.35 \text{ in.}^2)$ for the first trial.
5. Determine ϕP_n using Steps 4–7 in Example 11.13. $\phi P_n = 1210.3 \text{ K} > P_u = 1200$ K, so the section is adequate.
6. If the section is not adequate, or $\phi P_n < P_u$, increase A_s and A'_s and check again to get closer values.

Example 11.15

Design a rectangular column section to support $P_u = 750$ K and $M_u = 500$ K·ft with a total steel ratio ρ_g of about 4%. Use $f'_c = 4$ Ksi, $f_y = 60$ Ksi, and $b = 18$ in.

Solution

1. Calculate $e = M_u/P_u = 500(12)/750 = 8$ in. Assume compression controls (to be checked later) and $A_s = A'_s$.
2. For $\rho_g = 4\%$ and $f'_c = 4$ Ksi, $K_c = 0.137$ (Table 11.2).
3. Calculate bh^2 from equation (11.24):

$$P_u = \phi K_c bh^2, \quad \text{or} \quad 750 = 0.7(0.137)(18)h^2, \quad \text{so} \quad h = 20.84 \text{ in.}$$

Let $h = 22$ in.

4. Calculate $A_s = A'_s = 0.04(18 \times 22)/2 = 7.92$ in.^{2} Choose five no. 11 bars $(A_s = 7.8 \text{ in.}^2)$ in one row for A_s and A'_s (Figure 11.21). Choose no. 4 ties spaced at 18 in.

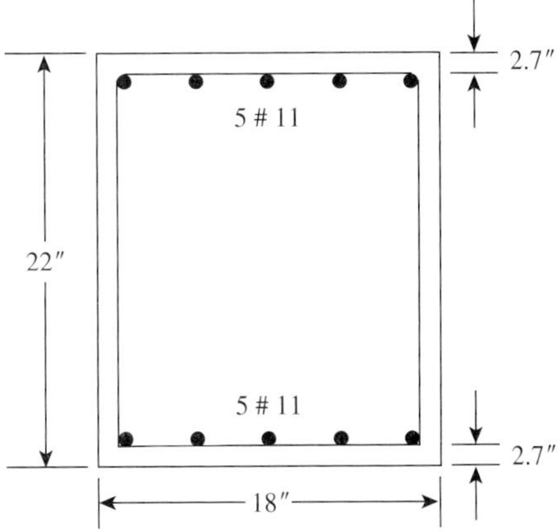

Figure 11.21 Example 11.15.

5. Check the final section by analysis, similar to Example 11.4, to get $a = 13.15$ in., $c = 15.47$ in., $C_c = 0.85f'_c ab = 804.8$ K, $f'_s = 60$ Ksi, $C_s = A'_s(f_y - 0.85f'_c) = 441.5$ K, $f_s = 87[(d - c)/c] = 21.24$ Ksi, and $T = A_s f_s = 168$ K. Also, $P_n = C_c + C_s - T = 1078.3$ K and $\phi P_n = 0.7P_n = 754.8 \text{ K} > 750$ K. The section is adequate.
6. For a balanced section,

$$c_b = \left(\frac{87}{87 + f_y}\right), \quad d = \left(\frac{87}{147}\right)19.3 = 11.42 \text{ in.} < c = 15.47$$

Then compression controls, as assumed.

11.13.2 Design of Columns When Tension Controls

Tension controls when $P_n < P_b$ or the eccentricity $e > e_b$, as explained in Section 11.7. In the design of columns, P_u and M_u are given, and it is required to determine the column size and its reinforcement. It may be assumed (as a guide) that tension controls when the ratio of M_u (K · ft) to P_u (kips) is greater than 1.75 for sections of $h < 24$ in. and 2.0 for $h \geq 24$ in. In this case, a section may be assumed, and then A_s and A'_s are determined. The ACI charts may be used to determine ρ_g for a given section with $A_s = A'_s$. Note that ϕ varies between 0.7 and 0.9, as explained in Section 11.4.

When tension controls, the tension steel yields, whereas the compression steel may or may not yield. Assuming initially $f'_s = f_y$ and $A_s = A'_s$, equation (11.16) (Section 11.6) may be used to determine the initial values of A_s and A'_s:

$$A_s = A'_s = \frac{P_n(e - h/2 + a/2)}{f_y(d - d')} \tag{11.16}$$

Because a is not known yet, assume $a = 0.4d$ and $P_u = \phi P_n$; then

$$A_s = A'_s = \frac{P_u(e - 0.5h + 0.2d)}{\phi f_y(d - d')} \tag{11.25}$$

The final column section should be checked by statics to prove that $\phi P_n \geq P_u$. Example 11.16 explains this approach.

When the load P_u is very small relative to M_u, the section dimensions may be determined due to M_u only, assuming $P_u = 0$. The final section should be checked by statics. This case occurs in single- or two-story building frames used mainly for exhibition halls or similar structures. In this case A'_s may be assumed less than A_s. A detailed design of a one-story, two-hinged frame exhibition hall is given in Chapter 16.

Example 11.16

Determine the necessary reinforcement for a 16- × 22-in. rectangular tied column to support an ultimate load $P_u = 200$ K and an ultimate moment $M_u = 500$ K · ft. Use $f'_c = 4$ Ksi and $f_y = 60$ Ksi.

Solution

1. Calculate $e = M_u/P_u = 500(12)/200 = 30$ in.; let $d = 22 - 2.5 = 19.5$ in. Because $M_u/P_u = 500/200 = 2.5 > 1.75$, or because $e > d$, assume tension controls (to be checked later).
2. Assume $A_s = A'_s$ and $f'_s = f_y$ and use equation (11.25) to determine A_s and A'_s. Let $P_u = 200$ K, $e = 30$ in., $\phi = 0.7$, $h = 22$ in., $d = 19.5$ in., and $d' = 2.5$ in.

$$A_s = A'_s = \frac{200(30 - 0.5 \times 22 + 0.2 \times 19.5)}{0.7(60)(17.0)} = 6.41 \text{ in.}^2$$

Choose five no. 10 bars (6.35 in.^{2}) in one row for each of A_s and A'_s (Figure 11.22).
3. Check $\rho_g = 2(6.35)(16 \times 22) = 0.036$, which is less than 0.08 and greater than 0.01.

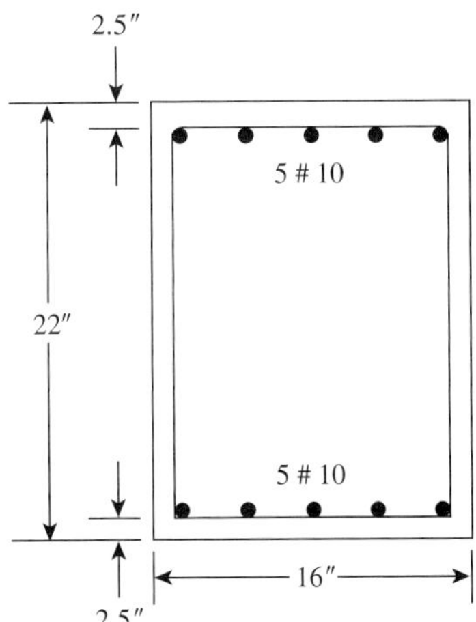

Figure 11.22 Example 11.16.

4. Check the chosen section by statics similar to Example 11.3.
   a. Determine the value of a using the general equation $Aa^2 + Ba + C = 0$ with $e' = e + d - h/2 = 38.5$ in., $A = 0.425f'_c b = 27.2$, $B = 2A(e' - d) = 1033.6$, $C = A'_s(f_y - 0.85f'_c)(e' - d + d') - A_s f_y e' = -6941.2$ Solve to get $a = 5.82$ in. and $c = a/0.85 = 6.85$.
   b. Check f'_s:

$$f'_s = 87\left(\frac{c - d'}{c}\right) = 87\left(\frac{6.85 - 2.5}{6.85}\right) = 55.26 \text{ Ksi}$$

   Let $f'_s = 57$ Ksi.
   c. Recalculate a:

$$C = A'_s(f'_s - 0.85f'_c)(e' - d + d') - A_s f_y e' = -7351$$

   Solve now for a to get $a = 6.13$ and $c = 7.21$ in.
   d. Check f'_s:

$$f'_s = 87\left(\frac{c - 2.5}{c}\right) = 56.83 \text{ Ksi}$$

   Calculate $C_c = 0.85(4)(6.13)(16) = 333.5$ K, $C_s = A'_s(f'_s - 0.85f'_c) = 6.35(57 - 0.85 \times 4) = 340.4$ K, $T = A_s f_y = 6.35(60) = 381$ K.
   e. $P_n = C_c + C_s - T = 292.9$ K.
5. Determine ϕ: $0.1f'_c A_g = 0.1(4)(16 \times 22) = 140.8$ K. Because $P_u = 200 \text{ K} > 140.8$ K, $\phi = 0.7$.
6. $\phi P_n = 0.7(292.9) = 205.0 \text{ K} > 200$ K, the section is adequate.

11.14 BIAXIAL BENDING

The analysis and design of columns under eccentric loading was discussed earlier in this chapter, considering a uniaxial case. This means that the load P_n was acting along the y-axis (Figure 11.23), causing a combination of axial load P_n and a moment about the x-axis equal to $M_{nx} = P_n e_y$ or acting along the x-axis (Figure 11.24) with an eccentricity e_x, causing a combination of an axial load P_n and a moment $M_{ny} = P_n e_x$.

If the load P_n is acting anywhere such that its distance from the x-axis is e_y and its distance from the y-axis is e_x, then the column section will be subjected to a combination of forces: an axial load P_n, a moment about the x-axis $= M_{nx} = P_n e_y$, and a moment about the

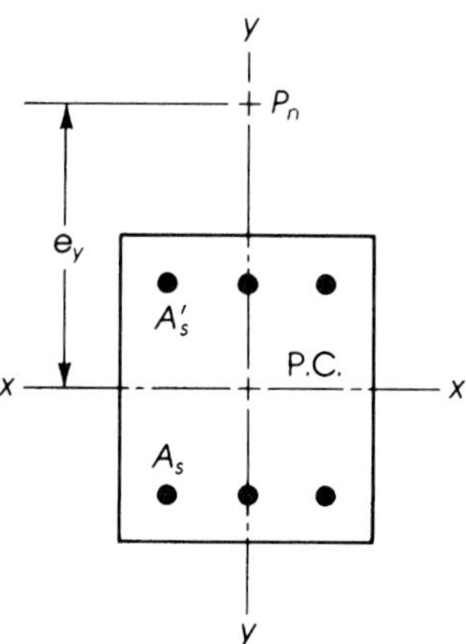

Figure 11.23 Uniaxial bending with load P_n along the y-axis with eccentricity e_y.

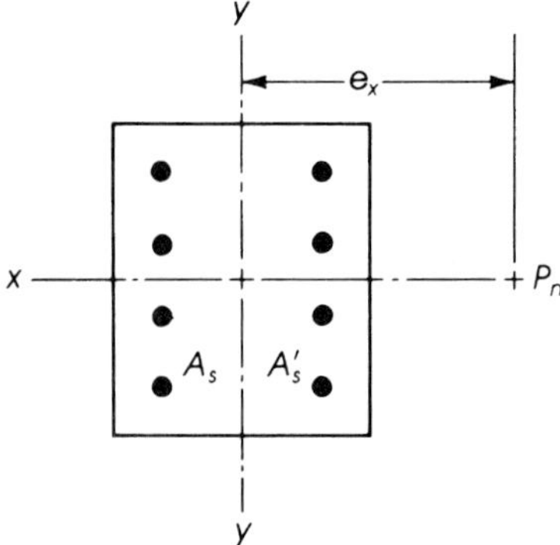

Figure 11.24 Uniaxial bending with load P_n along the x-axis, with eccentricity e_x.

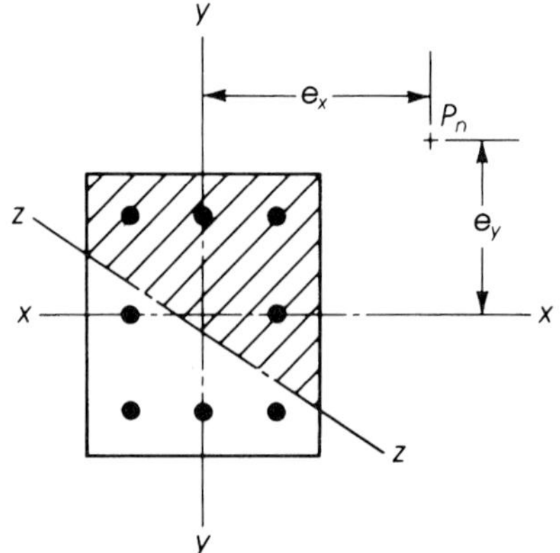

Figure 11.25 Biaxial bending.

y-axis $= M_{ny} = P_n e_x$ (Figure 11.25). The column section in this case is said to be subjected to *biaxial bending*. The analysis and design of columns under this combination of forces is not simple when the principles of statics are used. The neutral axis is at an angle with respect to both axes, and lengthy calculations are needed to determine the location of the

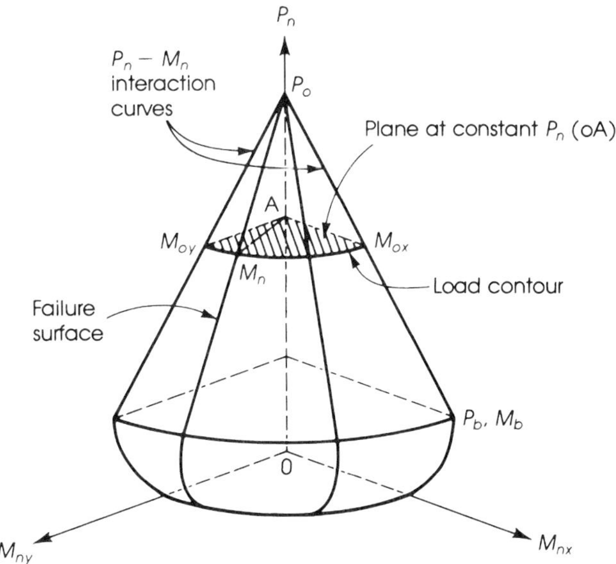

Figure 11.26 Biaxial interaction surface.

neutral axis, strains, concrete compression area, and internal forces and their point of application. Therefore, it was necessary to develop practical solutions to estimate the strength of columns under axial load and biaxial bending. The formulas developed relate the response of the column in biaxial bending to its uniaxial strength about each major axis.

The biaxial bending strength of an axially loaded column can be represented by a three-dimensional interaction curve, as shown in Figure 11.26. The surface is formed by a series of uniaxial interaction curves drawn radially from the P_n-axis. The curve M_{ox} represents the interaction curve in uniaxial bending about the x-axis, and the curve M_{oy} represents the curve in uniaxial bending about the y-axis. The plane at constant axial load P_n shown in Figure 11.26 represents the contour of the bending moment M_n about any axis.

Different shapes of columns may be used to resist axial loads and biaxial bending. Circular, square, or rectangular column cross sections may be used with equal or unequal bending capacities in the x- and y-directions.

11.15 CIRCULAR COLUMNS WITH UNIFORM REINFORCEMENT UNDER BIAXIAL BENDING

Circular columns with reinforcement distributed uniformly about the perimeter of the section have almost the same moment capacity in all directions. If a circular column is subjected to biaxial bending about the x- and y-axes, the equivalent uniaxial ultimate moment M_u can be calculated using the following equations:

$$M_u = \sqrt{(M_{ux})^2 + (M_{uy})^2} = P_u \cdot e \tag{11.26}$$

and

$$e = \sqrt{(e_x)^2 + (e_y)^2} = \frac{M_u}{P_u} \tag{11.27}$$

where $M_{ux} = P_u e_y$ = ultimate moment about the x-axis

$M_{uy} = P_u e_x$ = ultimate moment about the y-axis

$M_u = P_u e$ = equivalent uniaxial moment capacity of the section due to M_{ux} and M_{uy}

In circular columns, a minimum of six bars should be used, and these should be uniformly distributed in the section.

Example 11.17 Circular Column

Determine the load capacity P_n of a 20-in.-diameter column reinforced with 10 no. 10 bars when $e_x = 4$ in. and $e_y = 6$ in. Use $f'_c = 4$ Ksi and $f_y = 60$ Ksi.

Solution

1. Calculate the eccentricity that is equivalent to uniaxial loading by using equation (11.27).

$$e \text{ (for uniaxial loading)} = \sqrt{e_x^2 + e_y^2} = \sqrt{(4)^2 + (6)^2} = 7.211 \text{ in.}$$

2. Determine the load capacity of the column based on $e = 7.211$ in. Proceed as in Example 11.9:

$$d = 17.12 \text{ in.,} \qquad a = 9.81 \text{ in.} \qquad c = 11.54 \text{ in. (by trial)}$$

$$C_c = 521.2 \text{ K,} \qquad \Sigma C_s = 269.8 \text{ K,} \qquad \Sigma T = 132.1 \text{ K}$$

$$P_n = C_c + \Sigma C_s - \Sigma T = 650 \text{ K}$$

3. For a balanced condition,

$$c_b = \left(\frac{87}{87 + f_y}\right)d = \left(\frac{87}{147}\right)17.12 = 10.13 \text{ in.}$$

$c = 11.54$ in. $> c_b$, then compression controls

11.16 SQUARE AND RECTANGULAR COLUMNS UNDER BIAXIAL BENDING

11.16.1 Bresler Reciprocal Method

Square or rectangular columns with unequal bending moments about their major axes will require a different amount of reinforcement in each direction. An approximate method of analysis of such sections was developed by Boris Bresler and is called the Bresler reciprocal method [9], [12]. According to this method, the load capacity of the column under biaxial bending can be determined by using the following expression:

$$\frac{1}{P_u} = \frac{1}{P_{ux}} + \frac{1}{P_{uy}} - \frac{1}{P_{u0}} \tag{11.28}$$

or

$$\frac{1}{P_n} = \frac{1}{P_{nx}} + \frac{1}{P_{ny}} - \frac{1}{P_{n0}} \tag{11.29}$$

where P_u = load capacity under biaxial bending

P_{ux} = uniaxial load capacity when the load acts at an eccentricity e_y and $e_x = 0$

P_{uy} = uniaxial load capacity when the load acts at an eccentricity e_x and $e_y = 0$

P_{u0} = pure axial load when $e_x = e_y = 0$

$$P_n = \frac{P_u}{\phi}, \qquad P_{nx} = \frac{P_{ux}}{\phi}, \qquad P_{ny} = \frac{P_{uy}}{\phi}, \qquad P_{n0} = \frac{P_{u0}}{\phi}$$

The uniaxial load strengths P_{nx}, P_{ny}, and P_{n0} can be calculated according to the equations and method given earlier in this chapter. After that, they are substituted into equation (11.29) to calculate P_n.

The Bresler equation is valid for all cases when P_n is equal to or greater than $0.10P_{n0}$. When P_n is less than $0.10P_{n0}$, the axial force may be neglected and the section can be designed as a member subjected to pure biaxial bending according to the following equations:

$$\frac{M_{ux}}{M_x} + \frac{M_{uy}}{M_y} \le 1.0, \quad \text{or} \tag{11.30}$$

$$\frac{M_{nx}}{M_{ox}} + \frac{M_{ny}}{M_{oy}} \le 1.0 \tag{11.31}$$

where $M_{ux} = P_u e_y$ = design moment about the x-axis
$M_{uy} = P_u e_x$ = design moment about the y-axis
M_x and M_y = uniaxial moment capacities about the x- and y-axes

$$M_{nx} = \frac{M_{ux}}{\phi}, \qquad M_{ny} = \frac{M_{uy}}{\phi}, \qquad M_{ox} = \frac{M_x}{\phi}, \qquad M_{oy} = \frac{M_y}{\phi}$$

The Bresler equation is not recommended when the section is subjected to axial tension loads.

11.16.2 Bresler Load Contour Method

In this method, the failure surface shown in Figure 11.26 is cut at a constant value of P_n, giving the related values of M_{nx} and M_{ny}. The general nondimension expression for the load contour method is

$$\left(\frac{M_{nx}}{M_{ox}}\right)^{\alpha 1} + \left(\frac{M_{ny}}{M_{oy}}\right)^{\alpha 2} = 1.0 \tag{11.32}$$

Bresler indicated that the exponent α can have the same value in both terms of this expression $(\alpha_1 = \alpha_2)$. Furthermore, he indicated that the value of α varies between 1.15 and 1.55 and can be assumed to be 1.5 for rectangular sections. For square sections, α varies between 1.5 and 2.0, and an average value of $\alpha = 1.75$ may be used for practical designs. When the reinforcement is uniformly distributed around the four faces in square columns, α may be assumed to be 1.5.

$$\left(\frac{M_{nx}}{M_{ox}}\right)^{1.5} + \left(\frac{M_{ny}}{M_{oy}}\right)^{1.5} = 1.0 \tag{11.33}$$

The British Code CP110 assumes $\alpha = 1.0, 1.33, 1.67$, and 2.0 when the ratio $P_u//1.1P_{u0}$ is equal to 0.2, 0.4, 0.6, and ≥ 0.8, respectively.

11.17 PARME LOAD CONTOUR METHOD

The load contour approach, proposed by the Portland Cement Association (PCA), is an extension of the method developed by Bresler. In this approach, which is also called the *Parme method* [11], a point B on the load contour (of a horizontal plane at a constant P_n shown in Figure 11.26) is defined such that the biaxial moment capacities M_{nx} and M_{ny} are in the same ratio as the uniaxial moment capacities M_{ox} and M_{oy}; that is,

$$\frac{M_{nx}}{M_{ny}} = \frac{M_{ox}}{M_{oy}} \quad \text{or} \quad \frac{M_{nx}}{M_{ox}} = \frac{M_{ny}}{M_{oy}} = \beta$$

The ratio β is shown in Figure 11.27 and represents that constant portion of the uniaxial moment capacities that may be permitted to act simultaneously on the column section.

For practical design, the load contour shown in Figure 11.27 may be approximated by two straight lines, AB and BC. The slope of line AB is $(1 - \beta)/\beta$, and the slope of line BC is $\beta/(1 - \beta)$. Therefore, when

$$\frac{M_{ny}}{M_{oy}} > \frac{M_{nx}}{M_{ox}}, \quad \text{then} \quad \frac{M_{ny}}{M_{oy}} + \frac{M_{nx}}{M_{ox}}\left(\frac{1-\beta}{\beta}\right) = 1 \tag{11.34}$$

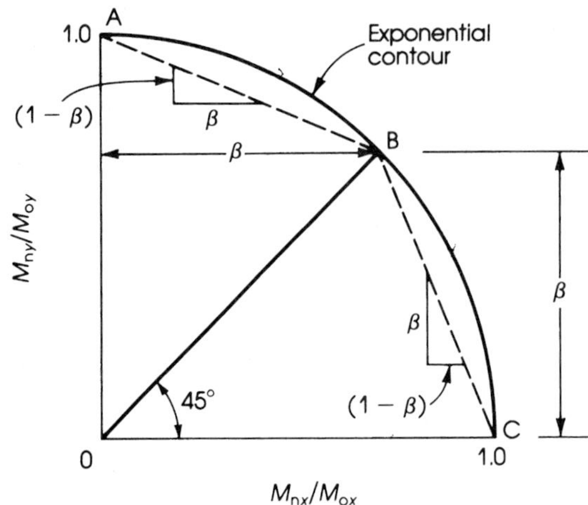

Figure 11.27 Nondimensional load contour at constant P_n (straight-line approximation).

and when

$$\frac{M_{ny}}{M_{oy}} < \frac{M_{nx}}{M_{ox}}, \quad \text{then} \quad \frac{M_{nx}}{M_{ox}} + \frac{M_{ny}}{M_{oy}}\left(\frac{1-\beta}{\beta}\right) = 1 \tag{11.35}$$

The actual value of β depends on the ratio P_n/P_0 as well as the material and properties of the cross section. For lightly loaded columns, β will vary from 0.55 to 0.7. An average value of $\beta = 0.65$ can be used for design purposes.

When uniformly distributed reinforcement is adopted along all faces of rectangular columns, the ratio M_{oy}/M_{ox} is approximately b/h, where b and h are the width and total depth of the rectangular section, respectively. Substituting this ratio in equations (11.34) and (11.35),

$$M_{ny} + M_{nx}\left(\frac{b}{h}\right)\left(\frac{1-\beta}{\beta}\right) \approx M_{oy} \tag{11.36}$$

and

$$M_{nx} + M_{ny}\left(\frac{h}{b}\right)\left(\frac{1-\beta}{\beta}\right) \approx M_{ox} \tag{11.37}$$

For $\beta = 0.65$ and $h/b = 1.5$, then

$$M_{oy} \approx M_{ny} + 0.36M_{nx} \tag{11.38}$$

and

$$M_{ox} \approx M_{nx} + 0.80M_{ny} \tag{11.39}$$

From this presentation, it can be seen that direct explicit equations for the design of columns under axial load and biaxial bending are not available. Therefore, the designer should have enough experience to make an initial estimate of the section using the values of P_n, M_{nx}, and M_{ny} and the uniaxial equations and then check the adequacy of the column section using the equations for biaxial bending or by computer.

Example 11.18

The section of a short tied column is 16 × 24 in. and is reinforced with eight no. 10 bars distributed as shown in Figure 11.28. Determine the allowable ultimate load on the section ϕP_n if it acts at $e_x = 8$ in. and $e_y = 12$ in. Use $f'_c = 5$ Ksi, $f_y = 60$ Ksi, and the Bresler reciprocal equation.

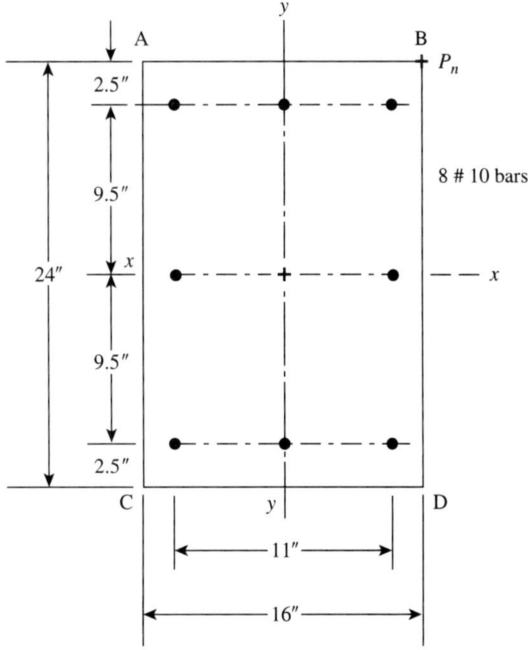

Figure 11.28 Biaxial load, Bresler method, Example 11.18: $P_n = 421.5$ K.

Solution

1. Determine the uniaxial load capacity P_{nx} about the x-axis when $e_y = 12$ in. In this case, $b = 16$ in., $h = 24$ in., $d = 21.5$ in., $d' = 2.5$ in., and $A_s = A'_s = 3.81$ in.^{2} The solution will be performed using statics following the steps of Examples 11.2 and 11.4 for balanced and compression-control conditions.

   a. For the balanced condition,

$$c_b = \left(\frac{87}{87 + f_y}\right), \qquad d = \left(\frac{87}{147}\right)21.5 = 12.72 \text{ in.}$$

$$a_b = 0.80(12.72) = 10.18 \text{ in.} \qquad (\beta_1 = 0.8 \text{ when } f'_c = 5 \text{ Ksi})$$

$$C_c = 0.85 f'_c ab = 692.3 \text{ K}, \qquad f'_s = 87\left(\frac{c - d'}{c}\right) = 69.9 \text{ Ksi}$$

   Then $f'_s = 60$ Ksi.

$$C_s = A'_s(f_y - 0.85 f'_c) = 212.4 \text{ K}, \qquad T = A_s f_y = 228.6 \text{ K}$$

$$P_{bx} = C_c + C_s - T = 676.1 \text{ K}, \qquad \phi P_{bx} = 0.7 P_{bx} = 473.3 \text{ K}$$

   b. For $e_y = 12$ in. $< d = 21.5$ in., assume compression controls and follow the steps of Example 11.4 to get $a = 10.65$ in. and $c = a/0.8 = 13.31$ in. $> C_b = 12.72$ in. Thus, compression controls. Check

$$f'_s = 87\left(\frac{c - d'}{c}\right) = 70 \text{ Ksi} > f_y; \qquad \text{then, } f'_s = 60 \text{ Ksi}$$

   Check

$$f_s = 87\left(\frac{d - c}{c}\right) = 53.53 \text{ Ksi} < 60 \text{ Ksi}$$

Calculate forces: $C_c = 0.85f'_c ab = 724.2$ K, $C_s = A'_s(f_y - 0.85f'_c) = 212.4$ K, $T = A_s f_s = 203.95$K, $P_{nx} = C_c + C_s - T = 732.6$ K, and $P_{ux} = 0.7P_{nx} = 512.8$ K. $P_{nx} > P_{bx}$, so compression controls as assumed.

c. Take moments about A_s using equation (11.11),

$$d'' = 9.5 \text{ in.}, \qquad e' = 21.5 \text{ in.}$$

$$P_{nx} = \frac{1}{e'}\left[C_c\left(d - \frac{a}{2}\right) + C_s(d - d')\right] = 732.5 \text{ K}$$

2. Determine the uniaxial load capacity P_{ny} about the y-axis when $e_x = 8$ in. In this case, $b = 24$ in., $h = 16$ in., $d = 13.5$ in., $d' = 2.5$ in., and $A_s = A'_s = 3.81$ in.^{2} The solution will be performed using statics, as explained in Step 1.

a. Balanced condition:

$$c_b = \left(\frac{87}{87 + f_y}\right)d = \left(\frac{87}{147}\right)13.5 = 7.99 \text{ in.} \qquad a_b = 0.8(7.99) = 6.39 \text{ in.}$$

$$C_c = 0.85f'_c ab = 651.8 \text{ K}, \qquad f'_s = 87\left(\frac{c - d'}{c}\right) = 59.8 \text{ Ksi}$$

$$C_s = A'_s(f'_s - 0.85f'_c) = 211.6 \text{ K}, \qquad T = A_s f_y = 228.6 \text{ K}$$

In a balanced load, $P_{by} = C_c + C_s - T = 634.8$ K, $\phi P_{by} = 0.7P_{by} = 444.4$ K.

b. For $e_x = 8$ in., assume compression controls and follow the steps of Example 11.4 to get $a = 6.65$ in. and $c = a/0.8 = 8.31$ in. $> c_b$ (compression controls). Check

$$f'_s = 87\left(\frac{c - d'}{c}\right) = 60.8 \text{ Ksi} \quad \text{then} \quad f'_s = 60 \text{ Ksi}$$

Check

$$f_s = 87\left(\frac{d - c}{c}\right) = 54.3 \text{ Ksi}$$

Calculate forces: $C_c = 0.85f'_c ab = 678.3$ K, $C_s = A'_s(60 - 0.85f'_c) = 212.4$ K, $T = A_s f_s = 206.9$ K, $P_{ny} = C_c + C_s - T = 683.8$ K, and $\phi P_{ny} = P_{uy} = 0.7P_{ny} = 478.7$ K. Because $P_{ny} > P_{by}$, compression controls, as assumed.

c. Take moments about A_s using equation (11.11):

$$d'' = 5.5 \text{ in.}, \qquad e' = 13.5 \text{ in.}$$

$$P_{ny} = \frac{1}{e'}\left[C_c\left(d - \frac{a}{2}\right) + C_s(d - d')\right] = 684 \text{ K}$$

3. Determine the theoretical axial load P_{n0}:

$$P_{n0} = 0.85f'_c A_g + A_{st}(f_y - 0.85f'_c)$$
$$= 0.85(5)(16 \times 24) + 10.16(60 - 0.85 \times 5) = 2198.4 \text{ K}$$

4. Using Bresler equation (11.28), multiply by 100:

$$\frac{100}{P_u} = \frac{100}{512.8} + \frac{100}{487.7} - \frac{100}{1538.9} = 0.3389$$

$$P_u = 295 \text{ K} \quad \text{and} \quad P_n = P_u/0.7 = 421.5 \text{ K}$$

Notes:

1. Approximate equations or the ACI charts may be used to calculate P_{nx} and P_{ny}. However, since the Bresler equation is an approximate solution, it is preferable to use accurate procedures, as was done in this example, to calculate P_{nx} and P_{ny}. Many approximations in the solution will produce inaccurate results. Computer programs based on statics are available and may be used with proper checking of the output.

2. In Example 11.18, the areas of the corner bars were used twice, once to calculate P_{nx} and once to calculate P_{ny}. The results obtained are consistent with similar solutions. A conservative solution is to use half of the corner bars in each direction, giving $A_s = A'_s = 2(1.27) = 2.54$ in.^{2}, which will reduce the values of P_{nx} and P_{ny}.

Example 11.19

Determine the nominal ultimate load P_n for the column section of the previous example using the Parme load contour method; see Figure 11.29.

Solution

1. Assume $\beta = 0.65$. The uniaxial load capacities in the direction of x- and y-axes were calculated in Example 11.18:

$$P_{ux} = 512.8 \text{ K}, \qquad P_{uy} = 478.7 \text{ K}, \qquad P_{nx} = 732.6 \text{ K}, \qquad P_{ny} = 683.8 \text{ K}$$

2. The moment capacity of the section about the x-axis is

$$M_{ox} = P_{nx} \cdot e_y = 732.6 \times 12$$

The moment capacity of the section about the y-axis is

$$M_{oy} = P_{ny} e_x = 683.8 \times 8 \text{ K} \cdot \text{in.}$$

3. Let the nominal load capacity be P_n. The nominal design moment on the section about the x-axis is

$$M_{nx} = P_n e_y = P_n \times 12 \text{ K} \cdot \text{in.}$$

and that about the y-axis is

$$M_{ny} = P_n e_x = 8P_n$$

4. Check if $M_{ny}/M_{oy} > M_{nx}/M_{ox}$:

$$\frac{8P_n}{683.8 \times 8} > \frac{12P_n}{732.6 \times 12} \quad \text{or} \quad 1.463 \times 10^{-3} P_n > 1.365 \times 10^{-3} P_n$$

Then $M_{ny}/M_{oy} > M_{nx}/M_{ox}$. Therefore, use equation (11.34).

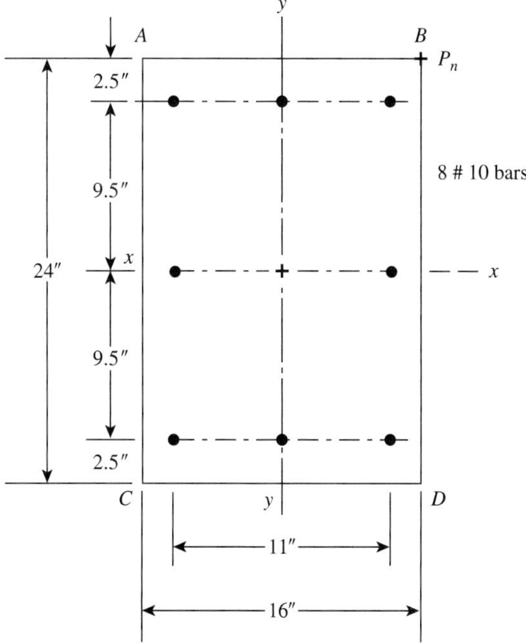

Figure 11.29 Biaxial load, PCA method: $P_n = 455$ K (Example 11.19).

5. $\dfrac{8P_n}{683.8 \times 8} + \dfrac{12P_n}{732.6 \times 12}\left(\dfrac{1 - 0.65}{0.65}\right) = 1$

Multiply by 1000 to simplify calculations.

$$1.463P_n + 0.735P_n = 1000, \qquad P_n = 455 \text{ K}$$

$$P_u = \phi P_n = 318.5 \text{ K} \qquad (\phi = 0.7)$$

Note that P_u is greater than that value of 295 K obtained by the Bresler reciprocal method (equation (11.28)) in the previous example by about 8%.

11.18 EQUATION OF FAILURE SURFACE

A general equation for the analysis and design of reinforced concrete short and tied rectangular columns was suggested by Hsu [16]. The equation is supposed to represent the failure surface and interaction diagrams of columns subjected to combined biaxial bending and axial load, as shown in Figure 11.26. The axial load can be compressive or a tensile force. The equation is presented as follows:

$$\left(\frac{P_n - P_b}{P_0 - P_b}\right) + \left(\frac{M_{nx}}{M_{bx}}\right)^{1.5} + \left(\frac{M_{ny}}{M_{by}}\right)^{1.5} = 1.0 \tag{11.40}$$

where

P_n = nominal axial strength (positive if compression and negative if tension) for a given eccentricity

P_0 = nominal axial load (positive if compression and negative if tension) at zero eccentricity

P_b = nominal axial compressive load at balanced strain condition

M_{nx}, M_{ny} = nominal bending moments about the x- and y-axes, respectively

M_{bx}, M_{by} = nominal balanced bending moments about the x- and y-axes, respectively, at balanced strain conditions

To use equation (11.40), all terms must have a positive sign. The value of P_0 was given earlier (equation (10.1)):

$$P_0 = 0.85f'_c(A_g - A_{st}) + A_{st} \cdot f_y \tag{11.41}$$

The nominal balanced load, P_b, and the nominal balanced moment, $M_b = P_b e_b$, were given in equations (11.6) and (11.7), respectively, for sections with tension and compression reinforcement only. For other sections, these values can be obtained by using the principles of statics.

Note that the equation of failure surface can also be used for uniaxial bending representing the interaction diagram. In this case, the third term will be omitted when $e_x = 0$, and the second term will be omitted when $e_y = 0$.

When $e_x = 0$ (moment about the x-axis only),

$$\left(\frac{P_n - P_b}{P_0 - P_b}\right) + \left(\frac{M_{nx}}{M_{bx}}\right)^{1.5} = 1.0 \tag{11.42}$$

(This is equation (11.18), given earlier.) When $e_y = 0$ (moment about the y-axis only),

$$\left(\frac{P_n - P_b}{P_0 - P_b}\right) + \left(\frac{M_{ny}}{M_{by}}\right)^{1.5} = 1.0 \tag{11.43}$$

Applying equation (11.42) to Examples 11.2 and 11.4: $P_b = 453.4$ K, $M_{bx} = 6810.8$ K·in., $e_y = 10$ in., and $P_0 = 0.85(4)(14 \times 22 - 8) + 8(60) = 1500$ K.

$$\frac{P_n - 453.4}{1500 - 453.4} + \left(\frac{10P_n}{6810.8}\right)^{1.5} = 1.0$$

Multiply by 1000 and solve for P_n:

$$(0.9555P_n - 433.2) + 0.05626P_n^{1.5} = 1000$$

$$0.9555P_n + 0.05626P_n^{1.5} = 1433.2$$

$P_n = 611$ K, which is close to that obtained by analysis.

Example 11.20

Determine the nominal ultimate load P_n for the column section of Example 11.18 using the equation of failure surface.

Solution

1. Compute

$$\begin{aligned} P_0 &= 0.85f'_c(A_g - A_{st}) + A_{st}f_y \\ &= 0.85(5)(16 \times 24 - 10.16) + (10.16 \times 60) \\ &= 2198.4 \text{ K} \end{aligned}$$

2. Compute P_b and M_b using equations (11.8) and (11.9) about the x- and y-axes, respectively.

a. About the x-axis:

$$c_{bx} = \frac{87d}{87 + f_y} = \frac{87(21.5)}{87 + 60} = 12.72 \text{ in.}$$

$$a_{bx} = 0.8(12.72) = 10.18 \text{ in.}$$

$$f'_s = 87\left(\frac{c - d'}{c}\right) = 69.9 \text{ Ksi}, \qquad f'_s = 60 \text{ Ksi}$$

$$d''_x = 9.5 \text{ in.}, \qquad A_s = A'_s = 3.81 \text{ in.}^2$$

$$\begin{aligned} P_{bx} &= 0.85f'_c a_x b + A'_s(f_y - 0.85f'_c) - A_s f_y \\ &= 0.85(5)(10.18)(16) + 3.81(60 - 0.85 \times 5) - 3.81(60) \\ &= 676.1 \text{ K} \end{aligned}$$

$$\begin{aligned} M_{bx} &= 0.85(5)(10.18)(16)\left(21.5 - \frac{10.18}{2} - 9.5\right) \\ &\quad + 3.81(60 - 0.85 \times 5) \times (21.5 - 2.5 - 9.5) + 3.81(60)(9.5) \\ &= 8973 \text{ K} \cdot \text{in.} = 747.8 \text{ K} \cdot \text{ft} \end{aligned}$$

b. About the y-axis: $d = 13.5$ in., $d''_y = 5.5$ in., $A_s = A'_s = 3.81$ in.^{2}

$$c_{by} = \frac{87(13.5)}{87 + 60} = 7.99 \text{ in.}$$

$$a_{by} = 0.8(7.99) = 6.39 \text{ in.}, \qquad f'_s = 87\left(\frac{c - d'}{c}\right) = 59.8 \text{ Ksi}$$

$$\begin{aligned} P_{by} &= 0.85(5)(6.39)(24) + 3.81(59.8 - 0.85 \times 5) - 3.81(60) \\ &= 634.8 \text{ K} \end{aligned}$$

$$\begin{aligned} M_{by} &= 0.85(5)(6.39)(24)\left(13.5 - \frac{6.39}{2} 5.5\right) \\ &\quad + 3.81(59.8 - 0.85 \times 5)(13.5 - 2.5 - 5.5) + 3.81(60)(5.5) \\ &= 5557.3 \text{ K} \cdot \text{in.} = 463 \text{ K} \cdot \text{ft} \end{aligned}$$

3. Compute the nominal balanced load for biaxial bending, P_{bb}:

$$\tan\alpha = \frac{M_{ny}}{M_{nx}} = \frac{P_n \cdot e_x}{P_n \cdot e_y} = \frac{e_x}{e_y} = \frac{8}{12}, \qquad \alpha = 33.7°$$

$$\frac{P_{bx} - P_{by}}{90°} = \frac{\Delta P_b}{90° - \alpha°} \quad \text{or} \quad \frac{676.1 - 634.8}{90} = \frac{\Delta P_b}{90 - 33.7}$$

$$\Delta P_b = 25.8 \text{ K}$$

$$P_{bb} = P_{by} + \Delta P_b = 634.8 + 25.8 = 660.6 \text{ K}$$

4. Compute P_n from the equation of failure surface:

$$\frac{P_n - 660.6}{2198.4 - 660.6} + \left(\frac{P_n \times 12}{8973}\right)^{1.5} + \left(\frac{P_n \times 8}{5557.3}\right)^{1.5} = 1.0$$

Multiply by 1000 and solve for P_n:

$$(0.65P_n - 429.85) + 0.0489P_n^{1.5} + 0.0546P_n^{1.5} = 1000$$

$$0.65P_n + 0.1035P_n^{1.5} = 1429.85$$

By trial, $P_n = 487$ K. Because $P_n < P_{bb}$, tension controls for biaxial bending, and so $P_0 = -2198.4$ K (to keep the first term positive).

$$1000\left(\frac{P_n - 660.9}{-2198.4 - 660.9}\right) + 0.0489P_n^{1.5} + 0.0546P_n^{1.5} = 1000$$

$$0.35P_n + 0.1035P_n^{1.5} = 769.1$$

$$P_n = 429 \text{ K} \quad \text{and} \quad P_u = 0.7P_n = 300.3 \text{ K}$$

(Because $P_u > 0.1f'_c A_g = 192$ K, then $\phi = 0.7$.)

Note: The strength capacity ϕP_n of the same rectangular section was calculated using the Bresler reciprocal equation (Example 11.18), Parme method (Example 11.19), and Hsu method (Example 11.21) to get $\phi P_n = 295$ K, 318.5 K, and 300.3 K, respectively. The Parme method gave the highest value for this example.

11.19 SI EXAMPLE

Example 11.21

Determine the balanced compressive forces P_b, e_b, and M_b for the section shown in Figure 11.30. Use $f'_c = 30$ MPa and $f_y = 400$ MPa ($b = 350$ mm, $d = 490$ mm).

Solution

1. For a balanced condition, the strain in the concrete is 0.003 and the strain in the tension steel is $\varepsilon_y = f_y/E_s = 400/200{,}000 = 0.002$, where $E_s = 200{,}000$ MPa.

$$A_s = A'_s = 4(700) = 2800 \text{ mm}^2$$

2. Locate the neutral axis depth c_b:

$$c_b = \left(\frac{600}{600 + f_y}\right)d \qquad \text{(where } f_y \text{ is in MPa)}$$

$$= \left(\frac{600}{600 + 420}\right)(490) = 288 \text{ mm}$$

$$a_b = 0.85c_b = 0.85 \times 288 = 245 \text{ mm}$$

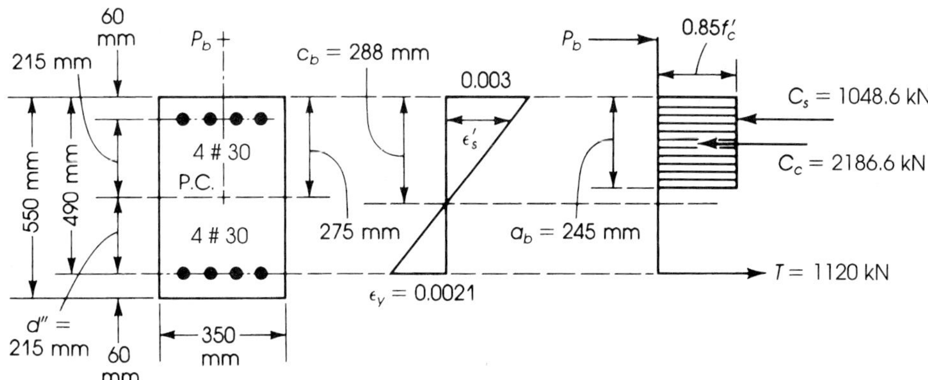

Figure 11.30 Example 11.21.

3. Check if compression steel yields. From the strain diagram,

$$\frac{\varepsilon'_s}{0.003} = \frac{c - d'}{c} = \frac{288 - 60}{288}$$

$$\varepsilon'_s = 0.00238 > \varepsilon_y$$

Therefore, compression steel yields.

4. Calculate the forces acting on the section:

$$C_c = 0.85f'_c ab = \frac{0.85}{1000} \times 30 \times 245 \times 350 = 2186.6 \text{ kN}$$

$$T = A_s f_y = 2800 \times 0.400 = 1120 \text{ kN}$$

$$C_s = A'_s(f_y - 0.85f'_c) = \frac{2800 \text{ mm}^2}{1000}(400 - 0.85 \times 30) = 1048.6 \text{ kN}$$

5. Calculate P_b and M_b:

$$P_b = C_c + C_s - T = 2115.2 \text{ kN}$$

From equation (11.7):

$$M_b = P_b e_b = C_c\left(d - \frac{a}{2} - d''\right) + C_s(d - d' - d'') + Td''$$

The plastic centroid is at the centroid of the section and $d'' = 215$ mm.

$$M_b = 2186.6\left(490 - \frac{245}{2} - 215\right) + 1048.6(490 - 60 - 215)$$

$$+ 1120 \times 215 = 799.7 \text{ kN}\cdot\text{m}$$

$$e_b = \frac{M_b}{P_b} = \frac{799.7}{2115.2} = 0.378 \text{ m} = 378 \text{ mm}$$

SUMMARY

Sections 11.1–11.3

1. The plastic centroid can be obtained by determining the location of the resultant force produced by the steel and the concrete, assuming both are stressed in compression to f_y and $0.85f'_c$, respectively.

2. On a load-moment interaction diagram the following cases of analysis are developed:
   a. Axial compression P_0
   b. Maximum allowable axial load P_n (max) $= 0.8P_0$ (for tied columns) and P_n (max) $= 0.85P_0$ (for spiral columns)
   c. Compression controls when $P_n > P_b$ or $e < e_b$
   d. Balanced condition, P_b and M_b
   e. Tension controls when $P_n < P_b$ or $e > e_b$
   f. Pure flexure

Section 11.4

1. When $P_u = \phi P_n \geq 0.1 f'_c A_g$, $\phi = 0.7$ for tied columns and 0.75 for spiral columns.
2. When $0.1 f'_c A_g > P_u > 0$,

$$\phi = 0.9 - \frac{2P_u}{f'_c A_g} \geq 0.7 \qquad \text{(for tied columns)}$$

$$\phi = 0.9 - \frac{1.5P_u}{f'_c A_g} \geq 0.75 \qquad \text{(for spiral columns)}$$

Section 11.5

For a balanced section,

$$c_b = \frac{87d}{87 + f_y} \quad \text{and} \quad a_b = \beta_1 c_b$$

$$\beta_1 = 0.85 \text{ for } f'_c \leq 4 \text{ Ksi}$$

$$P_b = C_c + C_s - T = 0.85 f'_c ab + A'_s(f_y - 0.85 f'_c) - A_s f_y$$

$$M_b = P_b e_b = C_c\left(d - \frac{a}{2} - d''\right) + Td''$$

$$e_b = \frac{M_b}{P_b}$$

Section 11.6

The equations for the general analysis of rectangular sections under eccentric forces are summarized.

Sections 11.7–11.8

Examples for the cases when tension and compression controls are given.

Sections 11.9–11.10

Examples are given for the interaction diagram and for the case when side bars are used.

Section 11.11

This section gives the load capacity of circular columns. The cases of a balanced section when compression controls and when compression controls are explained by examples.

Section 11.12

This section gives examples of the analysis and design of columns using charts.

Section 11.13

This section gives examples of the design of column sections.

Sections 11.14–11.18

Biaxial bending:

1. For circular columns with uniform reinforcement,

$$M_u = \sqrt{(M_{ux})^2 + (M_{uy})^2}, \qquad e = \sqrt{(e_x)^2 + (e_y)^2}$$

2. For square and rectangular sections,

$$\frac{1}{P_n} = \frac{1}{P_{nx}} + \frac{1}{P_{ny}} - \frac{1}{P_{n0}}$$

$$\frac{M_{nx}}{M_{ox}} + \frac{M_{ny}}{M_{oy}} \leq 1.0$$

3. In the Bresler load contour method,

$$\left(\frac{M_{nx}}{M_{ox}}\right)^{1.5} + \left(\frac{M_{ny}}{M_{oy}}\right)^{1.5} = 1.0$$

4. In the PCA load contour method,

$$M_{ny} + M_{nx}\left(\frac{b}{h}\right)\left(\frac{1-\beta}{\beta}\right) = M_{oy}$$

$$M_{nx} + M_{ny}\left(\frac{h}{b}\right)\left(\frac{1-\beta}{\beta}\right) = M_{ox}$$

5. Equations of failure surface method are given with applications.

REFERENCES

1. B. Brester. *Reinforced Concrete Engineering,* Vol. 1. New York: John Wiley, 1974.
2. E. O. Pfrang, C. P. Siess, and M. A. Sozen. "Load-Moment-Curvature Characteristics of Reinforced Concrete Cross-Sections." *ACI Journal* 61 (July 1964).
3. F. E. Richart, J. O. Draffin, T. A. Olson, and R. H. Heitman. "The Effect of Eccentric Loading, Protective Shells, Slenderness Ratio, and Other Variables in Reinforced Concrete Columns." *Bulletin no. 368.* Engineering Experiment Station, University of Illinois, Urbana, 1947.
4. N. G. Bunni. "Rectangular Ties in Reinforced Concrete Columns." In *Reinforced Concrete Columns,* Publication no. SP-50. American Concrete Institute, 1975.
5. C. S. Whitney. "Plastic Theory of Reinforced Concrete." *Transactions ASCE* 107 (1942).
6. Concrete Reinforcing Steel Institute. *CRSI Handbook.* Chicago, 1992.
7. American Concrete Institute. *Design Handbook,* Vol. 2, *Columns.* Publication SP-17a. Detroit, 1990.
8. Portland Cement Association. "Ultimate Load Tables for Circular Columns." Chicago, 1968.
9. B. Bresler. "Design Criteria for Reinforced Concrete Columns." *ACI Journal* 57 (November 1960).
10. R. Furlong. "Ultimate Strength of Square Columns under Biaxially Eccentric Loads." *ACI Journal* 57 (March 1961).

11. A. L. Parme, J. M. Nieves, and A. Gouwens. "Capacity of Reinforced Rectangular Columns Subjected to Biaxial Bending." *ACI Journal* 63 (September 1966).

12. J. F. Fleming and S. D. Werner. "Design of Columns Subjected to Biaxial Bending." *ACI Journal* 62 (March 1965).

13. M. N. Hassoun. "Ultimate-Load Design of Reinforced Concrete." *View Point Publication.* Cement and Concrete Association. London, 1981.

14. M. N. Hassoun. *Design Tables of Reinforced Concrete Members.* Cement and Concrete Association. London, 1978.

15. American Concrete Institute. *Building Code Requirements for Reinforced Concrete (ACI 318-63).* Detroit, 1963.

16. C. T. Hsu. "Analysis and Design of Square and Rectangular Columns by Equation of Failure Surface." *ACI Structural Journal* (March–April 1988): 167–179.

17. American Concrete Institute. *Building Code Requirements for Structure Concrete (ACI 318-99).* Detroit, 1999.

PROBLEMS

Note: For all problems, use $f_y = 60$ Ksi, $d' = 2.5$ in., and $A_s = A'_s$ where applicable. Slight variation in answers are expected.

11.1 (Rectangular Sections: Balanced Condition) For the rectangular column sections given in Table 11.3, determine the balanced compressive load, ϕP_b, the balanced moment, ϕM_b, and the balanced eccentricity, e_b, for each assigned problem. (Answers are given in Table 11.3.)

Table 11.3

					Answers to Problems			
					11.1		11.2	11.3
Number	f'_c (Ksi)	b (in.)	h (in.)	$A_s = A'_s$	ϕP_b	e_b	ϕP_n ($e = 6$ in.)	ϕP_n ($e = 24$ in.)
(a)	4	20	20	6 no. 10	400	17.4	834	275
(b)	4	14	14	4 no. 8	174	10.9	285	68
(c)	4	24	24	8 no. 10	593	20.1	1301	487
(d)	4	18	26	6 no. 10	488	20.6	1066	414
(e)	4	12	18	4 no. 9	212	15.2	412	124
(f)	4	14	18	4 no. 10	247	16.2	499	154
(g)	5	16	16	5 no. 10	284	15.3	565	159
(h)	5	18	18	5 no. 9	378	12.5	651	162
(i)	5	14	20	4 no. 9	334	13.4	593	154
(j)	5	16	22	4 no. 10	424	14.8	797	229
(k)	6	16	24	5 no. 10	522	16.8	1072	323
(l)	6	14	20	4 no. 9	372	12.8	661	160

11.2 (Rectangular Sections: Compression Controls) For the rectangular column sections given in Table 11.3, determine the load capacity ϕP_n for each assigned problem when the eccentricity is $e = 6$ in. (Answers are given in Table 11.3.)

11.3 (Rectangular Sections: Tension Controls) For the rectangular column sections given in Table 11.3, determine the load capacity, ϕP_n, for each assigned problem when the eccentricity is $e = 24$ in. (Answers are given in Table 11.3.)

11.4 (Rectangular Sections with Side Bars) Determine the load capacity, ϕP_n, for the column section shown in Figure 11.31 considering all side bars when the eccentricity is $e_y = 8$ in. Use $f'_c = 4$ Ksi and $f_y = 60$ Ksi. (Answer: 709 K)

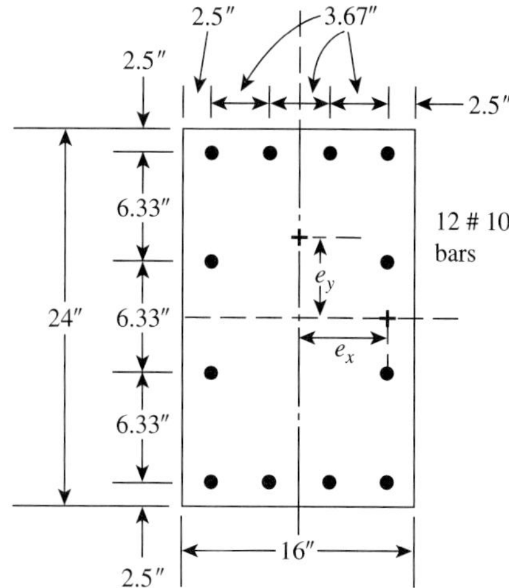

Figure 11.31 Problem 11.4.

11.5 Repeat Problem 11.4 with Figure 11.32. (Answer: 708 K)

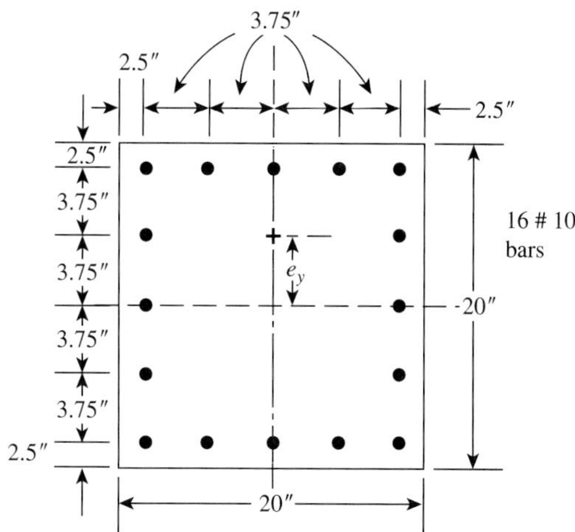

Figure 11.32 Problem 11.5.

11.6 Repeat Problem 11.4 with Figure 11.33. (Answer: 396 K)

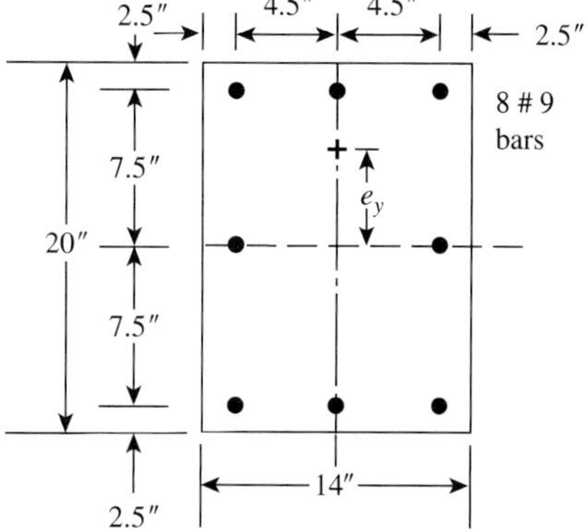

Figure 11.33 Problem 11.6.

11.7 Repeat Problem 11.4 with Figure 11.34. (Answer: 822 K)

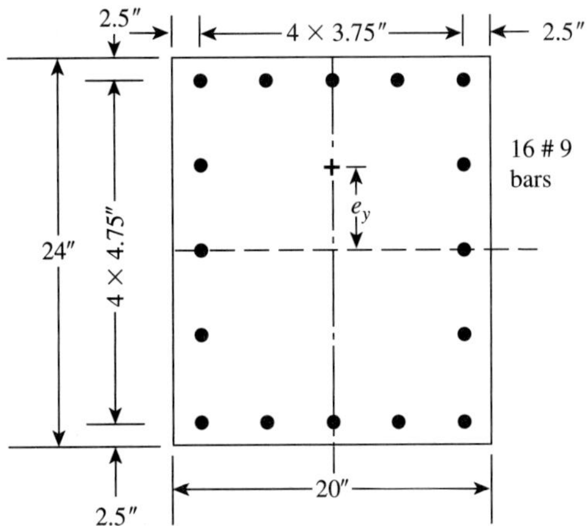

Figure 11.34 Problem 11.7.

11.8 (Design of Rectangular Column Sections) For each assigned problem in Table 11.4, design a rectangular column section to support the ultimate load and moment shown. Determine A_s, A'_s, and h if not given; then choose adequate bars considering that $A_s = A'_s$. The final total steel ratio, ρ_g, should be close to the given values where applicable. Check the load capacity ϕP_n of the final section using statics and equilibrium equations. One solution for each problem is given in Table 11.4.

Table 11.4

							One Solution	
Number	f'_c (Ksi)	P_u (K)	M_u (K · ft)	b (in.)	h (in.)	ρ_g%	h (in.)	$A_s = A'_s$
(a)	4	559	373	16	—	4.0	20	5 no. 10
(b)	4	440	220	14	18	—	18	5 no. 8
(c)	4	515	687	18	—	3.5	24	6 no. 10
(d)	4	470	470	20	20	—	20	6 no. 9
(e)	4	1215	405	20	24	—	24	6 no. 10
(f)	4	765	510	18	—	3.0	24	5 no. 10
(g)	5	310	310	14	—	2.0	20	3 no. 9
(h)	5	1100	735	20	26	—	26	6 no. 10
(i)	6	621	207	14	—	2.0	18	2 no. 10
(j)	6	714	357	16	20	—	20	4 no. 9

11.9 (ACI Charts) Repeat Problems 11.2(b), 11.2(d), 11.2(f), 11.8(a), 11.8(c), and 11.8(e) using the ACI charts.

11.10 (Circular Columns: Balanced Condition) Determine the balanced load capacity, ϕP_b, the balanced moment, ϕM_b, and the balanced eccentricity, e_b, for the circular tied sections shown in Figure 11.35. Use $f'_c = 4$ Ksi and $f_y = 60$ Ksi.

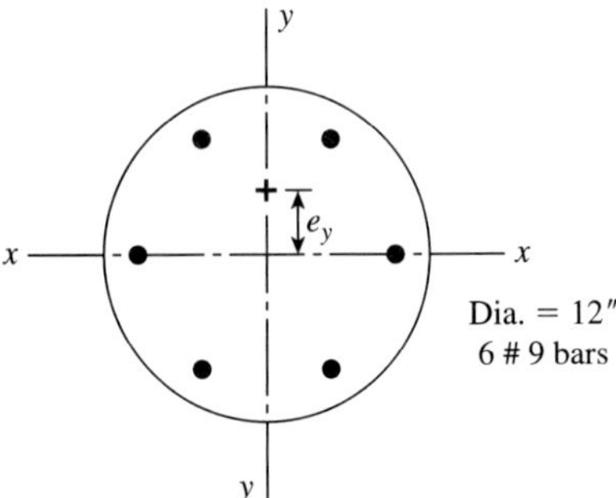

Figure 11.35 Problem 11.10.

11.11 Repeat Problem 11.10 for Figure 11.36.

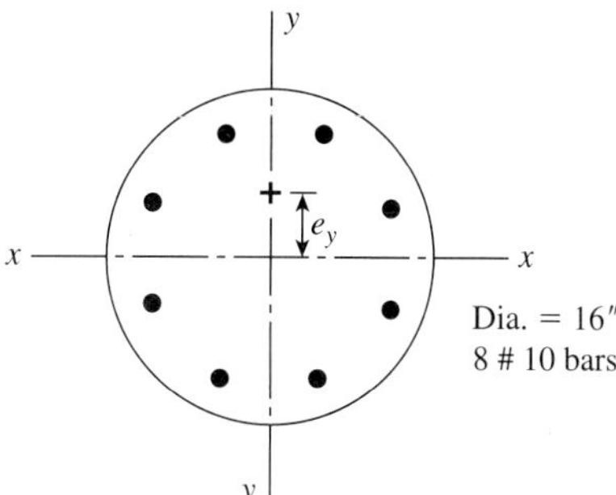

Figure 11.36 Problem 11.11.

11.12 Repeat Problem 11.10 for Figure 11.37.

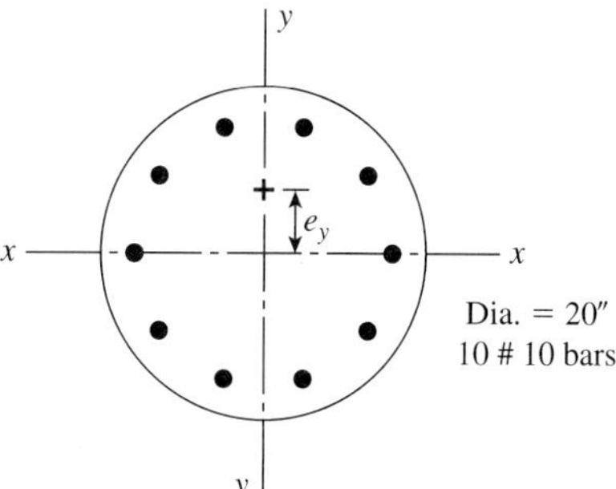

Figure 11.37 Problem 11.12.

11.13 Repeat Problem 11.11 for Figure 11.38.

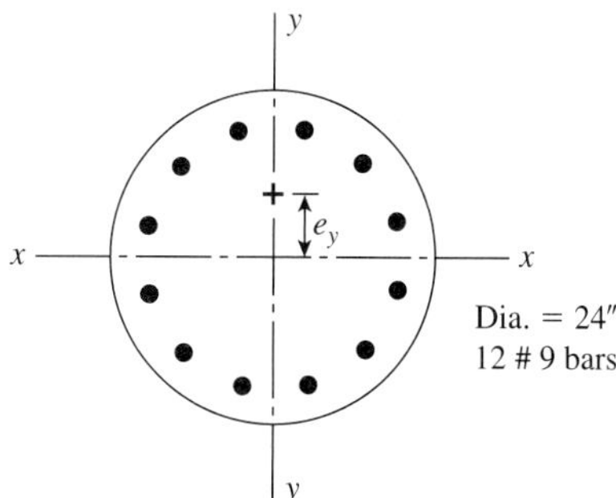

Figure 11.38 Problem 11.13.

11.14 (Circular Columns) Determine the load capacity ϕP_n for the circular tied column sections shown in Figures 11.35 through 11.38 when the eccentricity is $e_y = 6$ in. Use $f'_c = 4$ Ksi and $f_y = 60$ Ksi.

11.15 (Biaxial Bending) Determine the load capacity, ϕP_n, for the column sections shown in Figures 11.31 through 11.34 if $e_y = 8$ in. and $e_x = 6$ in. using the Bresler reciprocal method. Use $f'_c = 4$ Ksi and $f_y = 60$ Ksi. For each problem the values of P_{nx}, P_{ny}, P_{n0}, (P_{bx}, M_{bx}), and (P_{by}, M_{by}) are as follows:

a. Figure 11.31: 952 K, 835 K, 2168 K, (571 K, 792 K · ft), (536 K, 483 K · ft)

b. Figure 11.32: 930 K, 1108 K, 2505 K, (577 K, 742 K · ft), (577 K, 742 K · ft)

c. Figure 11.33: 558 K, 495 K, 1408 K, (408 K, 414 K · ft), (368 K, 260 K · ft)

d. Figure 11.34: 1093 K, 1145 K, 2538 K, (718 K, 865 K · ft), (701 K, 699 K · ft)

11.16 Repeat Problem 11.15 using the Parme method.

11.17 Repeat Problem 11.15 using the Hsu method.

11.18 For the column sections shown in Figure 11.31, determine

a. The uniaxial load capacities about the x- and y-axes, P_{nx}, and P_{ny} using $e_y = 6$ in. and $e_x = 6$ in.

b. The uniaxial balanced load and moment capacities about the x- and y-axes, P_{bx}, P_{by}, M_{bx}, and M_{by}.

c. The axial load P_{n0}.

d. The biaxial load capacity ϕP_n when $e_y = e_x = 6$ in., using the Bresler reciprocal method, the Hsu method, or both.

11.19 Repeat Problem 11.18 for Figure 11.32.

11.20 Repeat Problem 11.18 for Figure 11.33.

11.21 Repeat Problem 11.18 for Figure 11.34.

12 SLENDER COLUMNS

Columns in a high-rise building, Toronto, Ontario, Canada.

12.1 INTRODUCTION

In the analysis and design of short columns discussed in the previous two chapters, it was assumed that buckling, elastic shortening, and secondary moment due to lateral deflection had minimal effect on the ultimate strength of the column; thus, these factors were not included in the design procedure. However, when the column is long, these factors must be considered: The extra length will cause a reduction in the column strength that varies with the column effective height, width of the section, the slenderness ratio, and the column end conditions.

A column with a high slenderness ratio will have a considerable reduction in strength, whereas a low slenderness ratio means that the column is relatively short and the reduction in strength may not be significant. The slenderness ratio is the ratio of the column height, l, to the radius of gyration, r, where $r = \sqrt{I/A}$, I the moment of inertia of the section and A the sectional area.

For a rectangular section of width b and a depth h (Figure 12.1), $I_x = bh^3/12$ and $A = bh$. Therefore, $r_x = \sqrt{I/A} = 0.288h$ (or, approximately, $r_x = 0.3h$). Similarly, $I_y = hb^3/12$ and $r_y = 0.288b$ (or, approximately, $0.3b$). For a circular column with diameter D, $I_x = Iy = \pi D^2/64$ and $A = \pi D^2/4$; therefore, $r_x = r_y = 0.25D$.

In general, columns may be considered as follows:

1. Long with a relatively high slenderness ratio, where lateral bracing or shear walls are required.

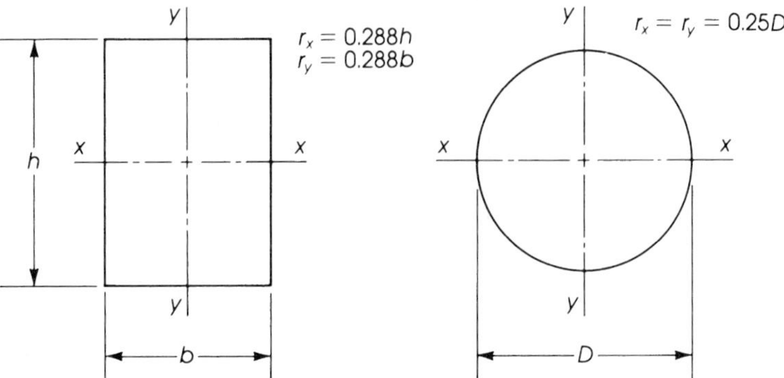

Figure 12.1 Rectangular and circular sections of columns, with radii of gyration r.

2. Long with a medium slenderness ratio that causes a reduction in the column strength. Lateral bracing may not be required, but strength reduction must be considered.
3. Short where the slenderness ratio is relatively small, causing a slight reduction in strength. This reduction may be neglected, as discussed in previous chapters.

12.2 EFFECTIVE COLUMN LENGTH (KL_U)

The slenderness ratio l/r can be calculated accurately when the effective length of the column (Kl_u) is used. This effective length is a function of two main factors.

1. The unsupported length, l_u, represents the unsupported height of the column between two floors. It is measured as the clear distance between slabs, beams, or any structural member providing lateral support to the column. In a flat slab system with column capitals, the unsupported height of the column is measured from the top of the lower floor slab to the bottom of the column capital. If the column is supported with a deeper beam in one direction than in the other direction, l_u should be calculated in both directions (about the x- and y-axes) of the column section. The critical (greater) value must be considered in the design.
2. The effective length factor, K, represents the ratio of the distance between points of zero moment in the column and the unsupported height of the column in one direction. For example, if the unsupported length of a column hinged at both ends, on which sidesway is prevented, is l_u, the points of zero moment will be at the top and bottom of the column—that is, at the two hinged ends. Therefore, the factor $K = l_u/l_u$ is 1.0. If a column is fixed at both ends and sidesway is prevented, the points of inflection (points of zero moment) are at $l_u/4$ from each end. Therefore, $K = 0.5l_u/l_u = 0.5$ (Figure 12.2). To evaluate the proper value of K, two main cases are considered.

   When structural frames are braced, the frame, which consists of beams and columns, is braced against sidesway by shear walls, rigid bracing, or lateral support from an adjoining structure. The ends of the columns will stay in position, and lateral translation of joints is prevented. The range of K in braced frames is always equal to or less than 1.0. The ACI Code, Section 10.12, recommends the use of $K = 1.0$ for braced frames.

   When the structural frames are unbraced, the frame is not supported against sidesway, and it depends on the stiffness of the beams and columns to prevent lateral

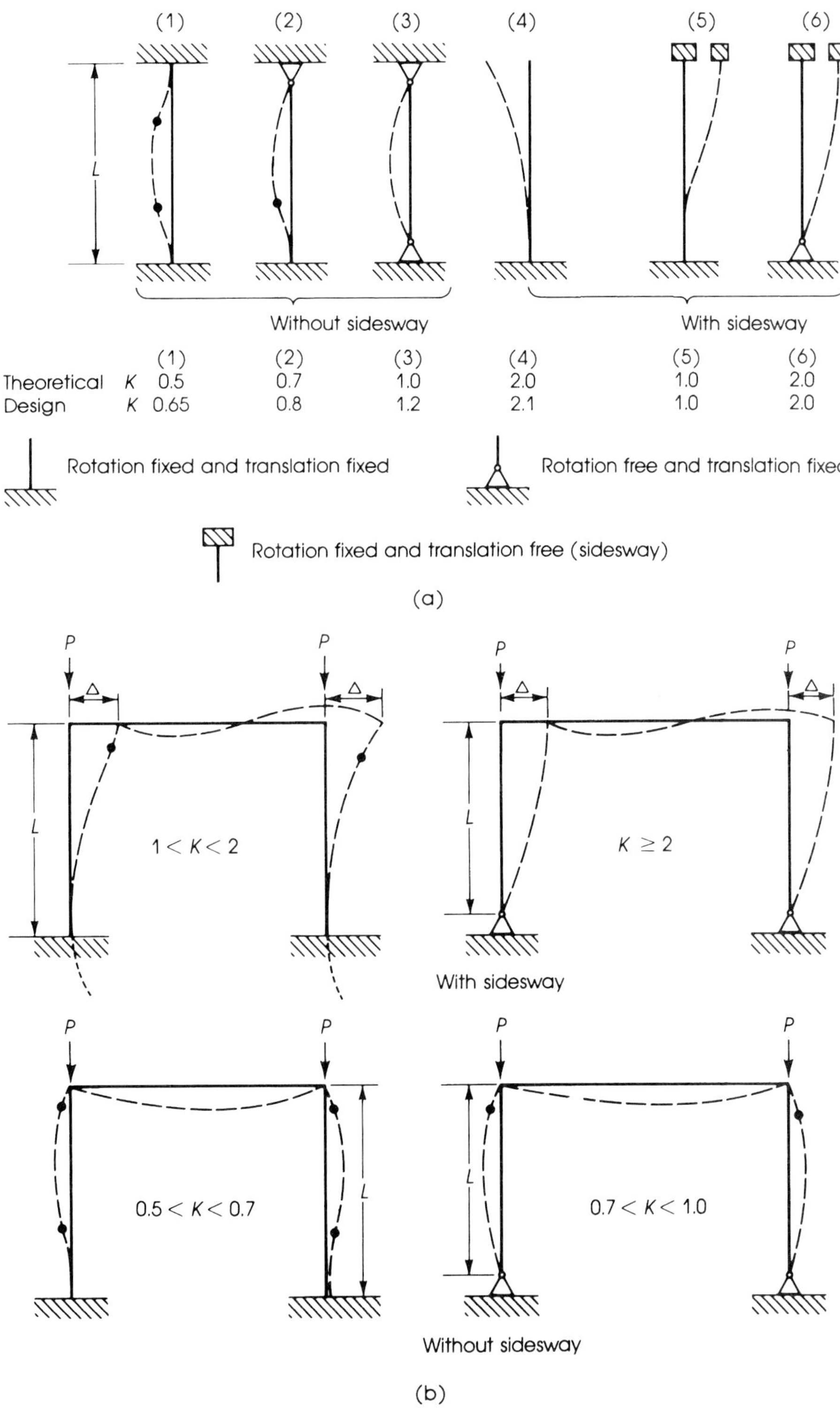

Figure 12.2 (a) Effective lengths of columns and length factor K and (b) effective lengths and K for portal columns.

deflection. Joint translations are not prevented, and the frame sways in the direction of lateral loads. The range of K for different columns and frames is given in Figure 12.2, considering the two cases when sidesway is prevented or not prevented.

12.3 EFFECTIVE LENGTH FACTOR (*K*)

The effective length of columns can be estimated by using the alignment chart shown in Figure 12.3 [10]. To find the effective length factor K, it is necessary first to calculate the end restraint factors ψ_A and ψ_B at the top and bottom of the column, respectively, where

$$\psi = \frac{\Sigma EI/l_u \text{ of columns}}{\Sigma EI/l_u \text{ of beams}} \tag{12.1}$$

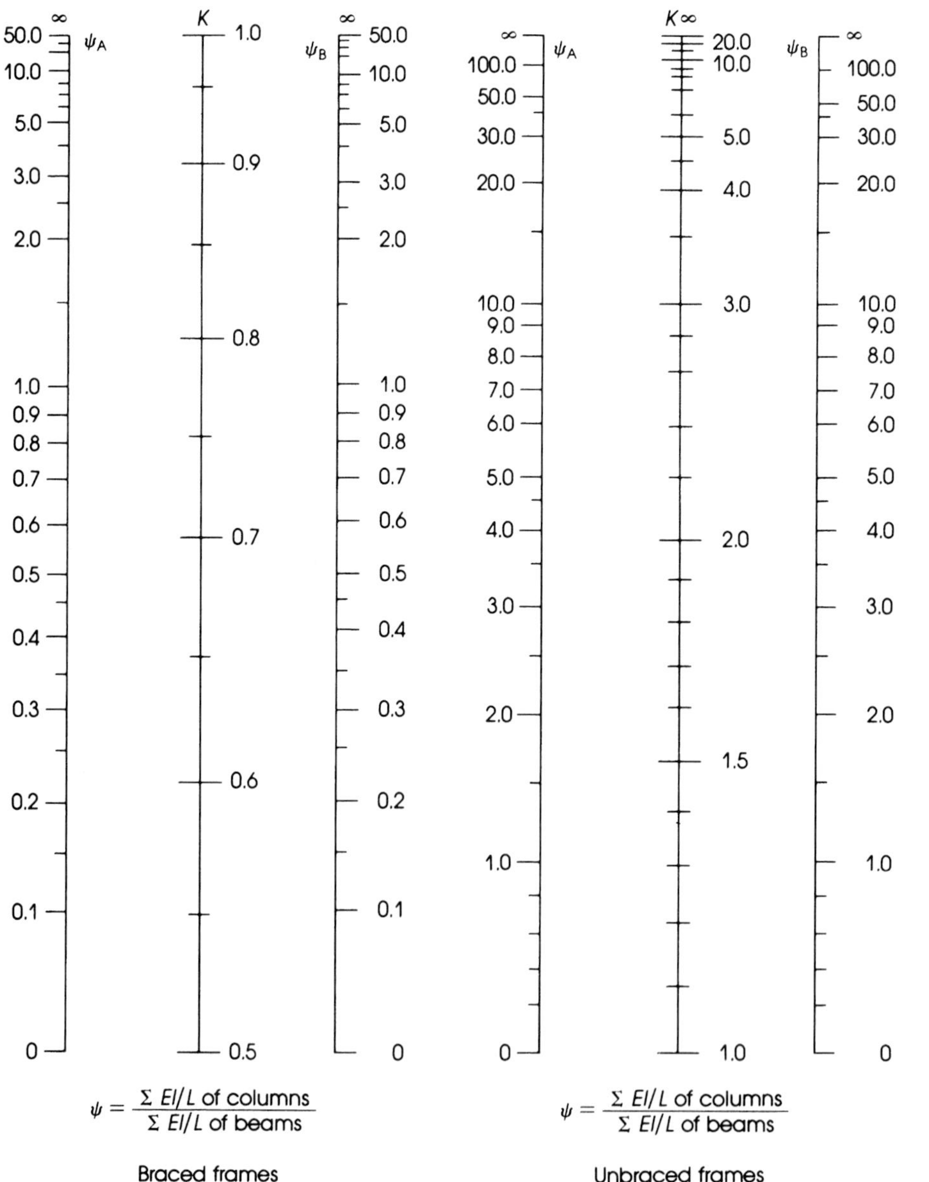

Figure 12.3 Alignment chart.

(both in the plane of bending). The ψ factor at one end shall include all columns and beams meeting at the joint. For a hinged end, ψ is infinite and may be assumed to be 10.0. For a fixed end, ψ is zero and may be assumed to be 1.0. Those assumed values may be used because neither a perfect frictionless hinge nor perfectly fixed ends can exist in reinforced concrete frames.

The procedure for estimating K is to calculate ψ_A for the top end of the column and ψ_B for the bottom end of the column. Plot ψ_A and ψ_B on the alignment chart of Figure 12.3 and connect the two points to intersect the middle line, which indicates the K-value. Two nomograms are shown, one for braced frames where sidesway is prevented, and the second for unbraced frames, where sidesway is not prevented. The development of the charts is based on the assumptions that (1) the structure consists of symmetrical rectangular frames, (2) the girder moment at a joint is distributed to columns according to their relative stiffnesses, and (3) all columns reach their critical loads at the same time.

In lieu of the alignment charts shown in Figure 12.3, the ACI Code Commentary (R10.12) proposed the following simplified expressions for evaluating the effective length factor, K:

1. For braced compression members, an upper bound of K may be taken as the smaller of the following two expressions:

$$k = 0.7 + 0.05(\psi_A + \psi_B) \leq 1.0 \tag{12.2}$$

$$k = 0.85 + 0.05\psi_{\min} \leq 1.0 \tag{12.3}$$

   where ψ_A and ψ_B are the values of ψ at the two ends of the column and $\psi_{\min}$ is the smaller of the two values.

2. For unbraced compression members restrained at both ends, the effective length factor, K, may be assumed as follows:

$$\text{For } \psi_m < 2: \quad k = \frac{20 - \psi_m}{20}\sqrt{1 + \psi_m} \tag{12.4}$$

$$\text{For } \psi_m \geq 2: \quad k = 0.9\sqrt{1 + \psi_m} \tag{12.5}$$

   where ψ_m is the average of the ψ values at the two ends of the compression member.

3. For unbraced compression members hinged at one end, K may be assumed as follows:

$$K = 2.0 + 0.3\psi \tag{12.6}$$

   where ψ is the value at the restrained end.

Example 12.1

Using the preceding equations, determine the effective length factor K for the compression member in a frame with the following conditions:

1. The frame is braced against sidesway and $\psi_A = 2.0$ and $\psi_B = 3.0$ at the upper and lower ends of members.
2. The frame is unbraced against sidesway and $\psi_A = 2.0$ and $\psi_B = 3.0$. (The member is fixed on both ends.)
3. The frame is unbraced against sidesway and $\psi_A = 0.0$ (hinged) and $\psi_B = 3.0$.

Solution

1. From equations (12.2) and (12.3),

$$K_1 = 0.7 + 0.05(2 + 3) = 0.95 < 1.0$$

$$K_2 = 0.85 + 0.05(2) = 0.95 < 1.0$$

   Use the smaller value of K_1 and K_2. In this case $K = 0.95$.

Long columns in an office building.

2. The average $\psi_m(2 + 3)/2 = 2.5$. Because $\psi_m > 2$, use equation (12.5):

$$K = 0.9\sqrt{1 + 2.5} = 1.684$$

3. From equation (12.6), $K = 2 + 0.3(3) = 2.9$.

12.4 MEMBER STIFFNESS (*EI*)

The stiffness of a structural member is equal to the modulus of elasticity E times the moment of inertia I of the section. The values of E and I for reinforced concrete members can be estimated as follows:

1. The modulus of elasticity of concrete was discussed in Chapter 2; the ACI Code gives the following expression:

$$E_c = 33w^{1.5}\sqrt{f'_c} \quad \text{or} \quad E_c = 57{,}000\sqrt{f'_c} \quad \text{(psi)}$$

for normal-weight concrete. The modulus of elasticity of steel is $E_s = 29 \times 10^6$ psi.

2. For reinforced concrete members, the moment of inertia I varies along the member, depending on the degree of cracking and the percentage of reinforcement in the section considered.

To evaluate the factor ψ, EI must be calculated for beams and columns. For this purpose, EI can be estimated as follows (ACI Code, Section 10.11.1):

For beams:	$I = 0.35I_g$
For columns:	$I = 0.70$
For walls — uncracked:	$I = 0.70I_g$
For walls — cracked:	$I = 0.35I_g$
For flat plates and flat slabs:	$I = 0.25I_g$

where I_g is the moment of inertia of the gross concrete section about centroidal axis, neglecting reinforcement.

3. Area, $A = 1.0A_g$ (gross-sectional area).
4. The moments of inertia shall be divided by $(1 + \beta_d)$ when sustained lateral loads act on the structure or for stability check, where

$$\beta_d = \text{(maximum factored sustained load)/(total factored axial load)}$$

12.5 LIMITATION OF THE SLENDERNESS RATIO (KL_u/r)

12.5.1 Nonsway Frames

The ACI Code, Section 10.12, recommends the following limitations between short and long columns in braced (nonsway) frames:

1. The effect of slenderness may be neglected and the column may be designed as a short column when

$$\frac{Kl_u}{r} \le 34 - \frac{12M_1}{M_2} \tag{12.7}$$

where M_1 and M_2 are the factored end moments of the column and M_2 is greater than M_1.

2. The ratio M_1/M_2 is considered positive if the member is bent in single curvature and negative for double curvature (Figure 12.4).

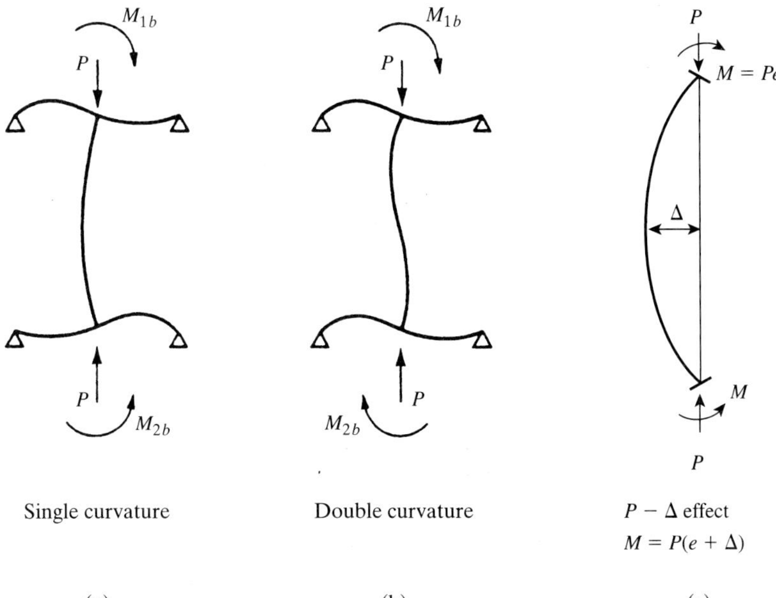

Figure 12.4 Single and double curvatures.

3. The term $(34 - 12M_1/M_2)$ shall not be taken greater than 40.
4. If the factored column moments are zero or $e = M_u/P_u < e_{min}$, the value of M_2 should be calculated using the minimum eccentricity:

$$e_{min} = (0.6 + 0.03h) \quad \text{(inch)} \tag{12.8}$$

$$M_2 = P_u(0.6 + 0.03h) \tag{12.9}$$

where M_2 is the minimum moment. The moment M_2 shall be considered about each axis of the column separately. The value of K may be assumed to be equal to 1.0 for a braced frame unless it is calculated on the basis of EI analysis.

12.5.2 Sway Frames

In compression members not braced against sidesway, the effect of the slenderness ratio may be neglected when

$$\frac{Kl_u}{r} < 22 \qquad \text{(Code Section 10.13)} \tag{12.10}$$

12.5.3 High Slenderness Ratio

When an individual compression member in a frame has a slenderness ratio $Kl_u/r > 100$, the moment magnifier method of the ACI Code cannot be used, and a rigorous second-order analysis is needed. However, the section may be increased to reduce the Kl_u/r ratio. This limit of 100 represents the upper range of actual tests. (ACI Code 10.11.5)

12.6 MOMENT-MAGNIFIER DESIGN METHOD

12.6.1 Introduction

The first step in determining the design moments in a long column is to determine if the frame is braced or unbraced against sidesway. If lateral bracing elements, such as shear walls and shear trusses, are provided or the columns have substantial lateral stiffness, then the lateral deflections produced are relatively small and their effect on the column strength is substantially low. It can be assumed a story within a structure is braced if

$$Q = \frac{\Sigma P_u \Delta_0}{V_u l_c} \leq 0.05 \tag{12.11}$$

where ΣP_u and V_u are the story total vertical load and story shear, respectively, and Δ_0 is the first-order relative deflection between the top and bottom of the story due to V_u. The length l_c is that of the compression member in a frame, measured from center to center of the joints in the frame.

In general, compression members may be subjected to lateral deflections that cause secondary moments. If the secondary moment, M', is added to the applied moment on the column, M_a, the final moment is $M = M_a + M'$. An approximate method for estimating the final moment M is to multiply the applied moment M_a by a factor called the *magnifying moment factor* δ, which must be equal to or greater than 1.0, or $M_{max} = \delta M_a$ and $\delta \geq 1.0$. The moment M_a is obtained from the elastic structural analysis using factored loads, and it is the maximum moment that acts on the column at either end or within the column if transverse loadings are present.

If the $P - \Delta$ effect is taken into consideration, it becomes necessary to use a second-order analysis to account for the nonlinear relationship between the load, lateral displacement, and the moment. This is normally performed using computer programs. The ACI Code permits the use of first-order or second-order analysis of columns. Second-order analysis is required if $Kl_u/r > 100$. The ACI Code *moment-magnifier design method* is a simplified approach for calculating the moment-magnifier factor in both braced and unbraced frames.

12.6.2 Magnified Moments in Nonsway Frames

The effect of slenderness ratio Kl_u/r in a compression member of a braced frame may be ignored if $Kl_u/r \leq 34 - 12M_1/M_2$, as given in Section 12.5.1. If Kl_u/r is greater than $(34 - 12M_1/M_2)$, then slenderness effect must be considered. The procedure for determining the magnification factor δ_{ns} in nonsway frames can be summarized as follows (ACI Code, Section 10.12):

1. Determine if the frame is braced against sidesway and find the unsupported length, l_u, and the effective length factor, K (K may be assumed to be 1.0).

2. Calculate the member stiffness, EI, using the equation

$$EI = \frac{0.2E_cI_g + E_sI_{se}}{1 + \beta_d} \tag{12.12}$$

or the more simplified equations

$$EI = \frac{0.4E_cI_g}{1 + \beta_d} \tag{12.13}$$

$$EI = 0.25E_cI_g \tag{12.14}$$

where

$E_c = 57{,}000\sqrt{f'_c}$

$E_s = 24 \times 10^6$ psi

I_g = gross moment of inertia of the section about the axis considered, neglecting A_s

I_s = moment of inertia of the reinforcing steel

$$\beta_d = \frac{\text{maximum factored axial sustained load}}{\text{maximum factored axial load}} = \frac{1.4D \quad \text{(sustained)}}{1.4D + 1.7L}$$

Note: The above β_d applies to nonsway frames and for stability checks of sway frames. For sway frames, β_d is the ratio of the maximum sustained shear within a story; to the maximum factored shear in that story.

Equations (12.13) and (12.14) are less accurate than equation (12.12). Moreover, equation (12.14) given in the ACI Commentary is obtained by assuming $\beta_d = 0.6$ in equation (12.13).

3. Determine the Euler buckling load, P_c:

$$P_c = \frac{\pi^2 EI}{(Kl_u)^2} \tag{12.15}$$

Use the values of EI, K, and l_u as calculated from Steps 1 and 2.

4. Calculate the value of the factor C_m to be used in the equation of the moment-magnifier factor. For braced members without transverse loads,

$$C_m = 0.6 + \frac{0.4M_1}{M_2} \geq 0.4 \tag{12.16}$$

where M_1/M_2 is positive if the column is bent in single curvature. For members with transverse loads between supports, C_m shall be taken as 1.0.

5. Calculate the moment magnifier factor δ_{ns}:

$$\delta_{ns} = \frac{C_m}{1 - (P_u/0.75P_c)} \geq 1.0 \tag{12.17}$$

where P_u is the applied factored load and P_c and C_m are as calculated previously.

6. Design the compression member using the axial factored load, P_u, from the conventional frame analysis and a magnified moment, M_c, computed as follows:

$$M_c = \delta_{ns} M_2 \tag{12.18}$$

where M_2 is the larger factored end moment due to loads that result in no sidesway. For frames braced against sidesway, the sway factor is $\delta_s = 0$. In nonsway frames, the lateral deflection is expected to be less than or equal to $H/1500$, where H is the total height of the frame.

12.6.3 Magnified Moments in Sway Frames

The effect of slenderness may be ignored in sway (unbraced) frames when $Kl_u/r < 22$. The procedure for determining the magnification factor, δ_s, in sway (unbraced) frames may be summarized as follows (ACI Code, Section 10.13):

1. Determine if the frame is unbraced against sidesway and find the unsupported length l_u and K, which can be obtained from equations (12.4), (12.5), or (12.6) or the alignment charts (Figure 12.3).

2–4. Calculate EI, P_c, and C_m as given by equations (12.12) through (12.16). Note that β_d (to calculate I) is the ratio of maximum factored sustained shear within a story to the total factored shear in that story.

5. Calculate the moment-magnifier factor, δ_s:

$$\delta_s = \frac{1.0}{1 - (\Sigma P_u/0.75\Sigma P_c)} \geq 1.0 \tag{12.19}$$

where $\delta_s \leq 2.5$ and ΣP_u is the summation for all vertical loads in a story and ΣP_c is the summation for all sway resisting columns in a story. Also,

$$\delta_s M_s = \frac{M_s}{1 - (\Sigma P_u/0.75\Sigma P_c)} \geq M_s \tag{12.20}$$

where M_s is the moment due to loads causing appreciable sway.

6. Calculate the magnified end moments M_1 and M_2 at the ends of an individual compression member, as follows:

$$M_1 = M_{1ns} + \delta_s M_{1s} \tag{12.21}$$

$$M_2 = M_{2ns} + \delta_s M_{2s} \tag{12.22}$$

where M_{1ns} and M_{2ns} are the moments obtained from the no-sway condition, whereas M_{1s} and M_{2s} are the moments obtained from the sway condition.

If M_2 is greater than M_1 from structural analysis, then the design magnified moment is

$$M_c = M_{2ns} + \delta_s M_{2s} \tag{12.23}$$

The end moments, M_1 and M_2, in equations (12.21), (12.22), and (12.23) contain unmagnified nonsway moments plus magnified sway moments, with the condition that

$$\frac{l_u}{r} < \frac{35}{\sqrt{P_u/f'_c A_g}} \tag{12.24}$$

Columns, University of Wisconsin, Madison.

In this case, the compression member should be designed for the axial factored load, P_u, and M_c. However, in the case when

$$\frac{l_u}{r} > \frac{35}{\sqrt{(P_u/f'_c A_g)}} \tag{12.25}$$

the compression member should be designed for P_u and a magnified nonsway moment $(\delta_{ns} M_2)$ plus a magnified sway moment, $\delta_s M_{2s}$, with a design moment $M_c = \delta_{ns} M_{2ns} + \delta_s M_{2s}$. This case may develop in slender columns with high axial loads when maximum moment occurs between the ends of the columns and not at the ends.

The ACI Code, Section 10.13.4, allows an alternative method for calculating $\delta_s M_s$ of equation (12.20) by using the stability index Q given in equation (12.11), provided that $\delta_s \leq 1.5$:

$$\delta_s M_s = \frac{M_s}{1 - Q} \geq M_s \tag{12.26}$$

Example 12.2

The column section shown in Figure 12.5 carries an axial load $P_D = 125$ K and a moment $M_D = 105$ K·ft due to dead load and an axial load $P_L = 112$ K and a moment $M_L = 96$ K·ft due to live load. The column is part of a frame that is braced against sidesway and bent in single curvature about its major axis. The unsupported length of the column is $l_u = 19$ ft, and the moments at both ends of the column are equal. Check the adequacy of the column using $f'_c = 4$ Ksi and $f_y = 60$ Ksi.

Solution

1. Calculate ultimate loads:

$$P_u = 1.4P_D + 1.7P_L = 1.4 \times 125 + 1.7 \times 112 = 365.4 \text{ K}$$

$$M_u = 1.4M_D + 1.7M_L = 1.4 \times 105 + 1.7 \times 96 = 310.2 \text{ K}\cdot\text{ft}$$

$$e = \frac{M_u}{P_u} = \frac{310.2 \times 12}{365.4} = 10.2 \text{ in.}$$

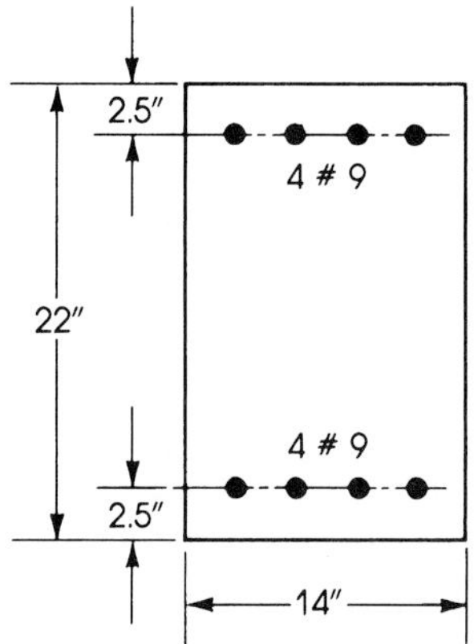

Figure 12.5 Example 12.2.

2. Check if the column is long: Because the frame is braced against sidesway, assume $K = 1.0, r = 0.3h = 0.3 \times 22 = 6.6$ in., and $l_u = 19$ ft.

$$\frac{Kl_u}{r} = \frac{1 \times 19 \times 12}{6.6} = 34.5$$

For braced columns, if $Kl_u/r \leq 34 - 12M_1/M_2$, slenderness effect may be neglected. Given end moments $M_1 = M_2$ and M_1/M_2 positive for single curvature,

$$\text{Right-hand side} = 34 - 12\frac{M_1}{M_2} = 34 - 12 \times 1 = 22$$

Because $Kl_u/r = 34.5 > 22$, slenderness effect must be considered.

3. Calculate EI from equation (12.12):

a. Calculate E_c:

$$E_c = 57{,}000\sqrt{f'_c} = 57{,}000\sqrt{4000} = 3605 \text{ Ksi}$$

$$E_s = 29{,}000 \text{ Ksi}$$

b. The moment of inertia is

$$I_g = \frac{14(22)^3}{12} = 12{,}422 \text{ in.}^4, \qquad A_s = A'_s = 4.0\ in.^2$$

$$I_{se} = 2 \times 4.0\left(\frac{22-5}{2}\right)^2 = 578 \text{ in.}^4$$

The dead-load moment ratio is

$$\beta_d = \frac{1.4 \times 125}{365.4} = 0.478$$

c. The stiffness is

$$EI = \frac{0.2E_cI_g + E_sI_{se}}{1 + \beta_d}$$

$$= \frac{(0.2 \times 3605 \times 12{,}422) + (29{,}000 \times 578)}{1 + 0.478}$$

$$= 17.45 \times 10^6 \text{ K} \cdot \text{in.}^2$$

4. Calculate P_c:

$$P_c = \frac{\pi^2 EI}{(Kl_u)^2} = \frac{\pi^2(17.45 \times 10^6)}{(12 \times 19)^2} = 3313 \text{ K}$$

5. Calculate C_m from equation (12.16):

$$C_m = 0.6 + \frac{0.4M_1}{M_2} \geq 0.4$$

$$= 0.6 + 0.4(1) = 1.0$$

6. Calculate the moment-magnifier factor from equation (12.17):

$$\delta_{ns} = \frac{C_m}{1 - (P_u/0.75P_c)} = \frac{1}{1 - (365.4/0.75 \times 3313)} = 1.17$$

7. Calculate the design moment and load:

$$P_n = \frac{365.4}{0.7} = 522 \text{ K} \qquad (\phi = 0.7)$$

$$M_n = \frac{310.2}{0.7} = 443.1 \text{ K}\cdot\text{ft}$$

Design $M_c = 443.1(1.17) = 518.5$ K·ft.
Design eccentricity $= 518.5/522 = 0.933$ ft $= 11.92$ in., or 12 in.

8. Determine the nominal load capacity of the section using $e = 12$ in. according to Example 11.4:

$$P_n = 47.6a + 226.4 - 4f_s \qquad \text{(I)}$$

$$e' = e + d - \frac{h}{2} = 12 + 19.5 - \frac{22}{2} = 20.50 \text{ in.}$$

$$P_n = \frac{1}{20.50}\left[47.6a\left(19.5 - \frac{a}{2}\right) + 226.4(19.5 - 2.5)\right]$$

$$= 45a - 1.15a^2 + 186.6 \qquad \text{(II)}$$

Solving for a from equations (I) and (II), $a = 10.6$ in. and $P_n = 535$ K. The load capacity, P_n, is greater than the required load of 522 K; therefore, the section is adequate. If the section is not adequate, increase steel reinforcement.

Example 12.3

Check the adequacy of the column in Example 12.2 if the unsupported length is $l_u = 10$ ft. Determine the maximum allowable nominal load on the column.

Solution

1. Applied loads are $P_n = 522$ K and $M_n = 443.1$ K.
2. Check if the column is long: $l_u = 10$ ft, $r = 0.3h = 0.3 \times 22 = 6.6$ in., and $K = 1.0$ (frame is braced against sidesway).

$$\frac{Kl_u}{r} = \frac{1 \times (10 \times 12)}{6.6} = 18.2$$

Check if $Kl_u/r \leq 34 - 12M_{1b}/M_{2b}$:

$$\text{Right-hand side} = 34 - 12 \times 1 = 22$$

$$\frac{Kl_u}{r} = 18.2 < 22$$

Therefore, the slenderness effect can be neglected.

3. Determine the nominal load capacity of the short column, as explained in Example 11.4. From Example 11.4, the nominal compressive strength is $P_n = 612.1$ K (for $e = 10$ in.), which is greater than the required load of 522 K, because the column is short with $e = 10.2$ in. (Example 12.2).

Example 12.4

Check the adequacy of the column in Example 12.2 if the frame is unbraced against sidesway, the end-restraint factors are $\psi_A = 0.8$ and $\psi_B = 2.0$, and the unsupported length is $l_u = 16$ ft.

Solution

1. Determine the value of K from the alignment chart (Figure 12.3) for unbraced frames. Connect the values of $\psi_A = 0.8$ and $\psi_B = 2.0$ to intersect the K-line at $K = 1.4$.

$$\frac{Kl_u}{r} = \frac{1.4 \times (16 \times 12)}{6.6} = 40.7$$

2. For unbraced frames, if $Kl_u/r \leq 22$, the column can be designed as a short column. Because actual $Kl_u r = 40.7 > 22$, the slenderness effect must be considered.
3. Calculate the moment magnifier δ_{ns}, given $C_m = 1.0, K = 1.4, EI = 17.45 \times 10^6\ \text{K} \cdot \text{in.}^2$ (from Example 12.2), and

$$P_c = \frac{\pi^2 EI}{(Kl_u)^2} = \frac{\pi^2 \times 17.45 \times 10^6}{(1.4 \times 16 \times 12)^2} = 2384\ \text{K}$$

$$\delta_{ns} = \frac{C_m}{1 - \left(\dfrac{P_u}{0.75 \times P_c}\right)} = \frac{1.0}{1 - \left(\dfrac{365.4}{0.75 \times 2384}\right)} = 1.26$$

4. From Example 12.2, the applied loads are $P_u = 365.4$ K and $M_u = 310.2\ \text{K} \cdot \text{ft}$, or

$$P_n = 522\ \text{K} \quad \text{and} \quad M_n = 443.1\ \text{K} \cdot \text{ft}$$

The design moment $M_c = 1.26(443.1) = 558.3\ \text{K} \cdot \text{ft}$

$$e = \frac{\delta_{ns} M_n}{P_n} = 558.3 \times \frac{12}{522} = 12.84\ \text{in.}, \qquad \text{say, 13 in.}$$

5. The requirement now is to check the adequacy of a short column for $P_n = 522$ K, $M_c = 558.3\ \text{K} \cdot \text{ft}$, and $e = 13$ in. The procedure is explained in Example 11.4.
6. From Example 11.4:

$$P_n = 47.6a + 226.4 - 4f_s$$

$$e' = e + d - \frac{h}{2} = 13 + 19.5 - \frac{22}{2} = 21.5\ \text{in.}$$

$$P_n = \frac{1}{21.5}\left[47.6a\left(19.5 - \frac{a}{2}\right) + 226.4(19.5 - 2.5)\right]$$

$$= 43.16a - 1.1a^2 + 179, \qquad a = 10.4\ \text{in.}$$

Thus, $c = 12.24$ in. and $P_n = 508$ K. This load capacity of the column is less than the required P_n of 522 K. Therefore, the section is not adequate.

7. Increase steel reinforcement to four no. 10 bars on each side and repeat the calculations to get $P_n = 568$ K.

Example 12.5

Design an interior square column for the first story of an 8-story office building. The clear height of the first floor is 16 ft, and the height of all other floors is 11 ft. The building layout is in 24 bays (Figure 12.6), and the columns are not braced against sidesway. The loads acting on a first floor interior column due to gravity and wind are as follows:

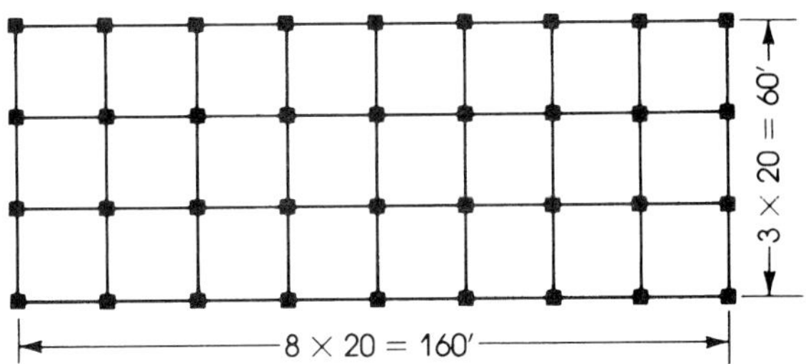

Figure 12.6 Example 12.5.

$$\text{Axial dead load} = 380 \text{ K}$$
$$\text{Axial live load} = 140 \text{ K}$$
$$\text{Axial wind load} = 0 \text{ K}$$
$$\text{Dead load moments} = 32 \text{ K}\cdot\text{ft (top) and } 54 \text{ K}\cdot\text{ft (bottom)}$$
$$\text{Live load moments} = 20 \text{ K}\cdot\text{ft (top) and } 36 \text{ K}\cdot\text{ft (bottom)}$$
$$\text{Wind load moments} = 60 \text{ K}\cdot\text{ft (top) and } 60 \text{ K}\cdot\text{ft (bottom)}$$
$$EI/l \text{ for beams} = 360 \times 10^3 \text{ K}\cdot\text{in.}$$

Use $f'_c = 5$ Ksi, $f_y = 60$ Ksi, and the ACI Code requirements. Assume an exterior column load of two-thirds the interior column load, a corner column load of one-third the interior column load, and $\beta_d = 0.55$.

Solution

1. Calculate ultimate forces using load combinations. For gravity loads,

$$P_u = 1.4D + 1.7L = 1.4(380) + 1.7(140) = 770 \text{ K}$$
$$M_u = M_{2ns} = 1.4M_D + 1.7M_L = 1.4(54) + 1.7(36) = 136.8 \text{ K}\cdot\text{ft}$$

For gravity plus wind load,

$$P_u = 0.75(1.4D + 1.7L + 1.7W)$$
$$0.75[1.4(380) + 1.7(140 + 0)] = 577.5 \text{ K}$$
$$M_{uns} = M_{2ns} = 0.75(1.4 \times 54 + 1.7 \times 36) = 102.6 \text{ K}\cdot\text{ft}$$
$$M_{us} = M_{2s} = 0.75(1.7M_w) = 0.75(1.7 \times 60) = 76.5 \text{ K}\cdot\text{ft}$$

Other combinations are not critical:

$$P_u = 0.9D + 1.3W = 0.9(380) + 1.3(0) = 342 \text{ K}$$
$$M_2 = 0.9M_D = 0.9(54) = 48.6 \text{ K}\cdot\text{ft}$$
$$M_{2s} = 1.3M_w = 1.3(60) = 78 \text{ K}\cdot\text{ft}$$
$$e = \frac{M_u}{P_u} = \frac{M_{2ns}}{P_u} = 136.8 \times \frac{12}{770} = 2.13 \text{ in.}$$
$$\min e = 0.6 + 0.03(18) = 1.14 \text{ in.} < 2.13$$

2. Select a preliminary section of column based on gravity load combination using tables or charts. Select a section 18 by 18 in. reinforced by four no. 10 bars (Figure 12.7). From the charts of Figure 11.17 and for $P_u = 770$ K, the allowable ultimate moment is $M_u = 206$ K · ft, which is greater than 136.8 K · ft.

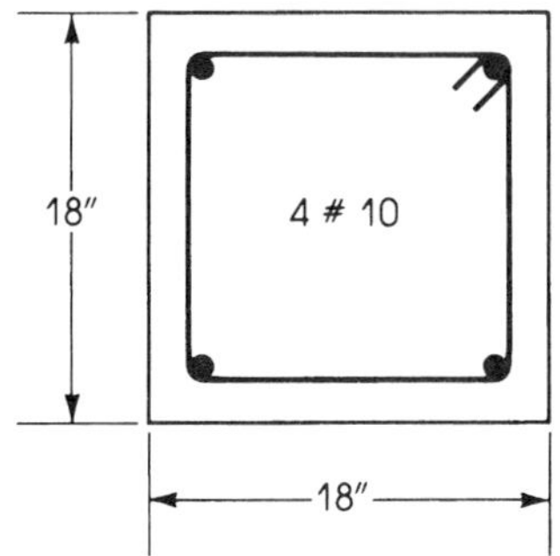

Figure 12.7 Column cross section, Example 12.5.

3. Check Kl_u/r:

$$I_g = \frac{(18)^4}{12} = 8748 \text{ in.}^4, \qquad E_c = 4.03 \times 10^6 \text{ psi}$$

for columns, $I = 0.7\,I_g$.

For a 16-ft column,

$$\frac{EI}{l_c} = \frac{(0.7)(8748)(4.03 \times 10^6)}{16 \times 12} = 128.5 \times 10^6$$

For an 11-ft column,

$$\frac{EI}{l_c} = \frac{(0.7)(8748)(4.03 \times 10^6)}{11 \times 12} = 187 \times 10^6$$

For beams, $EI_g/l_b = 360 \times 10^6$, $I = 0.35\,I_g$ and $EI/l_b = 0.35EI_g/l_b = 126 \times 10^6$

$$\psi(\text{top}) = \psi(\text{bottom}) = \frac{\Sigma(EI/l_c)}{\Sigma(EI/l_b)} = \frac{(128.5 + 187)}{2(126)} = 1.25$$

From the chart (Figure 12.3), K is 1.37 for an unbraced frame and 0.8 for a braced frame.

$$\frac{Kl_u}{r} = \frac{1.37(16 \times 12)}{0.3 \times 18} = 48.7$$

which is more than 22 and less than 100. Therefore, the slenderness ratio must be considered.

4. Check $l_u/r = 16 \times 12/0.3(18) = 35.6$.

$$\frac{35}{\sqrt{P_u/f'_c A_g}} = \frac{35}{\sqrt{770/5(18 \times 18)}} = 50.8 \qquad (12.24)$$

Because $l_u/r < 50.8$, nonsway moment M_{2ns} need not be magnified.

5. Compute P_c:

$$E_c = 4.03 \times 10^3 \text{ Ksi} \qquad E_s = 29 \times 10^3 \text{ Ksi}$$

$$I_g = 8748 \text{ in.}^4, \qquad I_{se} = 5.06\left(\frac{13}{2}\right)^2 = 214 \text{ in.}^4$$

$$\beta_d = 0.55$$

$$EI = \frac{0.2E_c I_g + E_s I_{se}}{1 + \beta_d}$$

$$EI = \frac{0.2(4.03 \times 10^3 \times 8748) + 29 \times 10^3(214)}{1 + 0.55} = 8.55 \times 10^6 \text{ K}\cdot\text{in.}^2$$

For calculation of $\delta_s, \beta_d = 0$ and $EI = 8.55 \times 10^6(1.55) = 13.25 \times 10^6 \text{ K} \cdot \text{in.}^2$

$$P_c = \frac{\pi^2 EI}{(Kl_u)^2} = \frac{\pi^2(8.55 \times 10^6)}{(0.8 \times 16 \times 12)^2} = 3577 \text{ K} \qquad \text{(braced)}$$

$$P_c = \frac{\pi^2(13.25 \times 10^6)}{(1.37 \times 16 \times 12)^2} = 1890 \text{ K} \qquad \text{(unbraced)}$$

For one floor in the building, there are 14 interior columns, 18 exterior columns, and 4 corner columns.

$$\Sigma P_u = 14(557.5) + 18\left(\frac{2}{3} \times 577.5\right) + 4\left(\frac{1}{3} \times 577.5\right) = 15{,}785 \text{ K}$$

$$\Sigma P_c = 14(1890) + 22\left(\frac{2}{3} \times 1890\right) = 54{,}180 \text{ K}$$

$$\delta_s = \frac{1.0}{1 - \left(\dfrac{15{,}785}{0.75 \times 54{,}180}\right)} = 1.64 \text{ which is greater than 1.0 and less than 2.5 (equation 12.19)}$$

$$M_c = M_{2ns} + \delta_s M_{2s} = (102.6) + 1.64(76.5) = 228 \text{ K} \cdot \text{ft}$$

6. Design loads are $P_u = 577.5$ K and $M_c = 228$ K · ft.

$$e = \frac{228(12)}{577.5} = 4.74 \text{ in.}$$

$$e_{\min} = 0.6 + 0.03(18) = 1.14 \text{ in.} < e$$

By analysis, for $e = 4.74$ in. and $A_s = A'_s = 2.53$ in.^{2}, the load capacity of the 18- × 18-in. column is $\phi P_n = 623$ K and $\phi M_n = 246$ K · ft, so the section is adequate. (Solution steps are similar to Example 11.4. Values are $a = 10.37$ in., $c = 13$ in., $f_s = 17$ Ksi, $f'_s = 60$ Ksi, $\phi P_b = 385$ K, and $e_b = 8.9$ in.)

SUMMARY

Sections 12.1–12.3

1. The radius of gyration is $r = \sqrt{I/A}$, where $r = 0.3h$ for rectangular sections and $0.25D$ for circular sections.
2. The effective column length is Kl_u. For braced frames, $K = 1.0$; for unbraced frames, K varies as shown in Figure 12.2.
3. K can be determined from the alignment chart (Figure 12.3) or equations (12.2) through (12.6).

Section 12.4

Member stiffness is EI:

$$E_c = 33w^{1.5}\sqrt{f'_c}$$

The moment of inertia, I, may be taken as $I = 0.35I_g$ for beams, $0.70I_g$ for columns, 0.70 for uncracked walls, $0.35I_g$ for cracked walls, and $0.25I_g$ for plates and flat slabs.

Section 12.5

The effect of slenderness may be neglected when

$$\frac{Kl_u}{r} \leq 22 \qquad \text{(for unbraced frames)} \tag{12.10}$$

$$\frac{Kl_u}{r} \leq 34 - \frac{12M_1}{M_2} \qquad \text{(for braced columns)} \tag{12.7}$$

where M_1 and M_2 are the end moments and $M_2 > M_1$.

Section 12.6

1. For nonsway frames:

$$EI = \frac{(0.2E_cI_g + E_sI_s)}{1 + \beta_d} \quad \text{or} \tag{12.12}$$

$$EI = \frac{0.4E_cI_g}{1 + \beta_d} \tag{12.13}$$

More simply,

$$EI = 0.25E_cI_g \tag{12.14}$$

$$\beta_d = \frac{1.4D}{1.4D + 1.7L} \tag{12.14}$$

The Euler buckling load is

$$P_c = \frac{\pi^2 EI}{(Kl_u)^2} \tag{12.15}$$

$$C_m = 0.6 + \frac{0.4M_1}{M_2} \geq 0.4 \tag{12.16}$$

The moment-magnifier factor (nonsway frames) is

$$\delta_{ns} = \frac{C_m}{1 - (P_u/0.75P_c)} \tag{12.17}$$

The design moment is

$$M_c = \delta_{ns} M_2 \tag{12.18}$$

2. For sway (unbraced) frames, the moment-magnifier factor is

$$\delta_s = \frac{1.0}{1 - (\Sigma P_u/0.75\Sigma P_c)} \tag{12.19}$$

and

$$\delta_s M_2 = \frac{M_s}{1 - (\Sigma P_u/0.75\Sigma P_c)} \geq M_s \tag{12.20}$$

In the case when $l_u/r \leq 35/\sqrt{(P_u/f'_c A_g)}$,

$$M_c = M_{2ns} + \delta_s M_{2s} \tag{12.23}$$

where M_{2ns} is the unmagnified moment due to gravity loads (nonsway moment) and $\delta_s M_{2s}$ is the magnified moment due to sway frame loads.

In cases when $l_u/r > 35/\sqrt{(P_u/f'_c A_g)}$, then both M_{2ns} and M_{2s} due to nonsway loads and sway loads should be magnified by δ_{ns} and δ_s, respectively. The design moment is

$$M_c = \delta_{ns} M_{2ns} + \delta_s M_{2s}$$

REFERENCES

1. B. B. Broms. "Design of Long Reinforced Concrete Columns." *Journal of Structural Division ASCE* 84 (July 1958).
2. J. G. MacGregor, J. Breen, and E. O. Pfrang. "Design of Slender Concrete Columns." *ACI Journal* 67 (January 1970).
3. American Concrete Institute. *Design Handbook, Vol. 2, Columns.* ACI Publication SP-17a. Detroit, 1978.
4. W. F. Chang and P. M. Ferguson. "Long Hinged Reinforced Concrete Columns." *ACI Journal* 60 (January 1963).
5. R. W. Furlong and P. M. Ferguson. "Tests of Frames with Columns in Single Curvature." *Symposium on Reinforced Columns.* American Concrete Institute SP-13. 1966.
6. J. G. MacGregor and S. L. Barter. "Long Eccentrically Loaded Concrete Columns Bent in Double Curvature." *Symposium on Reinforced Concrete Columns.* American Concrete Institute SP-13. 1966.
7. R. Green and J. E. Breen. "Eccentrically Loaded Concrete Columns Under Sustained Load." *ACI Journal* 66 (November 1969).
8. B. G. Johnston. *Guide to Stability Design for Metal Structures,* 3d ed. New York: John Wiley, 1976.
9. T. C. Kavanaugh. "Effective Length of Frames Columns." *Transactions ASCE* 127, Part II (1962).
10. American Concrete Institute. "Commentary on Building Code Requirements for Reinforced Concrete." ACI 318-95. 1995.
11. R. G. Drysdale and M. W. Huggins. "Sustained Biaxial Load on Slender Concrete Columns." *Journal of Structural Division ASCE* 97 (May 1971).
12. American Concrete Institute. "Building Code Requirements for Structural Concrete." ACI 318-99. Detroit, Michigan, 1999.
13. J. G. MacGregor. "Design of Slender Concrete Columns." *ACI Structural Journal* 90, no. 3 (May—June 1993).
14. J. S. Grossman. "Reinforced Concrete Design." Chapter 22, *Building Structural Design Handbook.* New York: John Wiley, 1987.

PROBLEMS

12.1 The column section in Figure 12.8 carries an axial load $P_D = 120$ K and a moment $M_D = 110$ K · ft due to dead load and an axial load $P_L = 95$ K and a moment $M_1 = 100$ K · ft due to live load. The column is part of a frame, braced against sidesway, and bent in single curvature about its major axis. The unsupported length of the column is $l_u = 18$ ft, and the moments at both ends are equal. Check the adequacy of the section using $f'_c = 4$ Ksi and $f_y = 60$ Ksi.

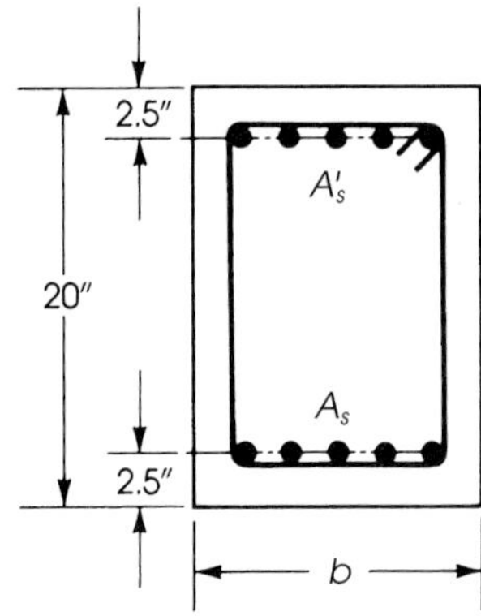

Figure 12.8 Problem 12.1 $(A_s = A_s') = 5$ # 9 bars and $b = 14$ in.

12.2 Repeat Problem 12.1 if $l_u = 12$ ft.

12.3 Repeat Problem 12.1 if the frame is unbraced against sidesway and the end-restraint factors are ψ (top) $= 0.7$ and ψ (bottom) $= 1.8$ and the unsupported height is $l_u = 14$ ft.

12.4 The column section shown in Figure 12.9 is part of a frame unbraced against sidesway and supports an axial load $P_D = 155$ K and a moment $M_D = 100$ K·ft due to dead load and $P_L = 115$ K and $M_L = 80$ K·ft due to live load. The column is bent in single curvature and has an unsupported length $l_u = 16$ ft. The moment at the top of the column is $M_2 = 1.5M_1$, the moment at the bottom of the column. Check if the section is adequate using $f_c' = 5$ Ksi, $f_y = 60$ Ksi, ψ (top) $= 2.0$ and ψ (bottom) $= 1.0$.

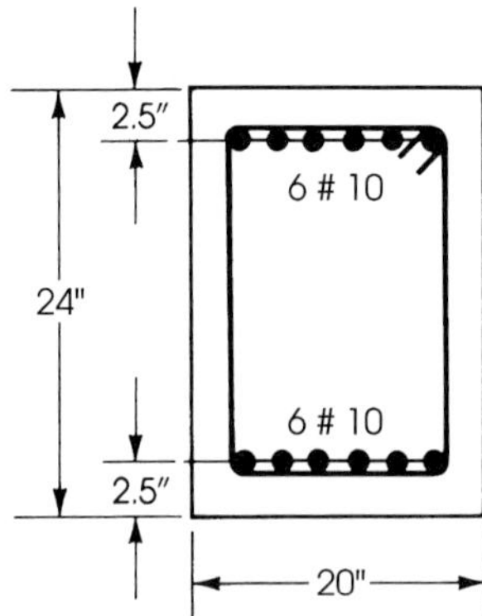

Figure 12.9 Problem 12.4.

12.5 Repeat Problem 12.4 if the column length is $l_u = 14$ ft.

12.6 Repeat Problem 12.4 if the frame is braced against sidesway and $M_1 = M_2$.

12.7 Repeat Problem 12.4 using $f_c' = 4$ Ksi and $f_y = 60$ Ksi.

12.8 Design a 20-ft-long rectangular tied column for an axial load $P_D = 200$ K and a moment $M_D = 60$ K·ft due to dead load and an axial load $P_L = 120$ K and a moment $M_L = 40$ K·ft due to live load. The column is bent in single curvature about its major axis, braced against sidesway, and the end moments are equal. The end-restraint factors are ψ (top) $= 2.5$ and ψ (bottom) $= 1.4$. Use $f_c' = 5$ Ksi, $f_y = 60$ Ksi, and $b = 15$ in.

12.9 Design the column in Problem 12.8 if the column length is 10 ft.

12.10 Repeat Problem 12.8 if the column is unbraced against sidesway.

FOOTINGS

Office building under construction, New Orleans.

13.1 INTRODUCTION

Reinforced concrete footings are structural members used to support columns and walls and to transmit and distribute their loads to the soil. The design is based on the assumption that the footing is rigid, so that the variation of the soil pressure under the footing is linear. Uniform soil pressure is achieved when the column load coincides with the centroid of the footing. Although this assumption is acceptable for rigid footings, such an assumption becomes less accurate as the footing becomes relatively more flexible. The proper design of footings requires that

1. The load capacity of the soil is not exceeded;
2. Excessive settlement, differential settlement, or rotations are avoided;
3. Adequate safety against sliding and/or overturning is maintained.

The most common types of footings used in buildings are the single footings and wall footings (Figures 13.1 and 13.2). When a column load is transmitted to the soil by the footing, the soil becomes compressed. The amount of settlement depends on many factors, such as the type of soil, the load intensity, the depth below ground level, and the type of footing. If different footings of the same structure have different settlements, new stresses develop in the structure. Excessive differential settlement may lead to the damage of nonstructural members in the buildings or even failure of the affected parts.

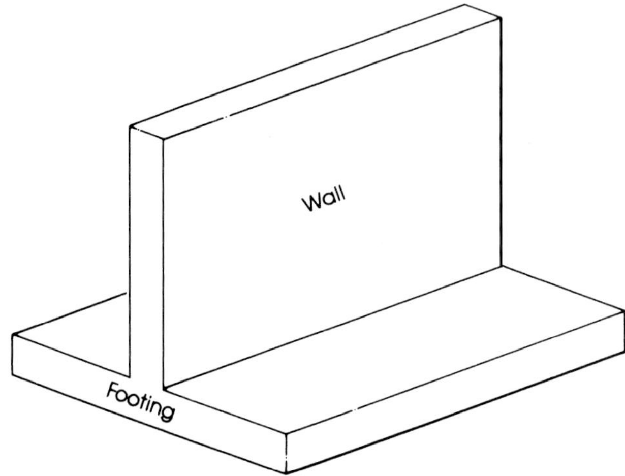

Figure 13.1 Wall footing.

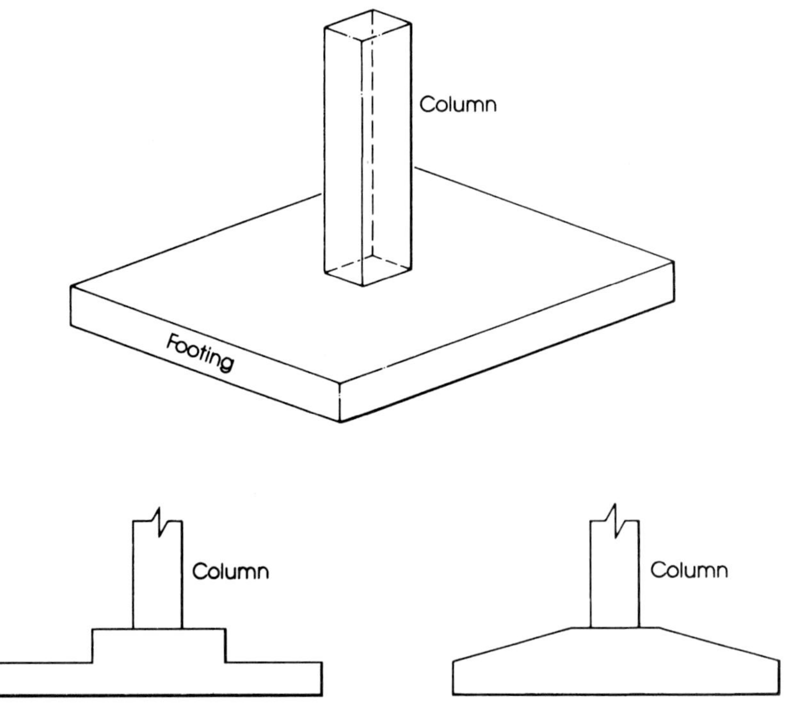

Figure 13.2 Single footing.

Vertical loads are usually applied at the centroid of the footing. If the resultant of the applied loads does not coincide with the centroid of the bearing area, a bending moment develops. In this case, the pressure on one side of the footing will be greater than the pressure on the other side.

If the bearing soil capacity is different under different footings—for example, if the footings of a building are partly on soil and partly on rock—a differential settlement will occur. It is usual in such cases to provide a joint between the two parts to separate them, allowing for independent settlement.

The depth of the footing below the ground level is an important factor in the design of footings. This depth should be determined from soil tests, which should provide reliable information on safe bearing capacity at different layers below ground level. Soil test reports specify the allowable bearing capacity to be used in the design. In cold areas where freezing occurs, frost action may cause heaving or subsidence. It is necessary to place footings below freezing depth to avoid movements.

13.2 TYPES OF FOOTINGS

Different types of footings may be used to support building columns or walls. The most common types are as follows:

1. *Wall footings* are used to support structural walls that carry loads from other floors or to support nonstructural walls. They have a limited width and a continuous length under the wall (Figure 13.1). Wall footings may have one thickness, be stepped, or have a sloped top.
2. *Isolated,* or *single, footings* are used to support single columns (Figure 13.2). They may be square, rectangular, or circular. Again, the footing may be of uniform thickness, stepped, or have a sloped top. This is one of the most economical types of footings, and it is used when columns are spaced at relatively long distances. The most commonly used are square or rectangular footings with uniform thickness.
3. *Combined footings* (Figure 13.3) usually support two columns or three columns not in a row. The shape of the footing in plan may be rectangular or trapezoidal, depending on column loads. Combined footings are used when two columns are so close that single footings cannot be used or when one column is located at or near a property line.
4. *Cantilever,* or *strap, footings* (Figure 13.4) consist of two single footings connected with a beam or a strap and support two single columns. They are used when one footing supports an eccentric column and the nearest adjacent footing lies at quite a distance from it. This type replaces a combined footing and is sometimes more economical.

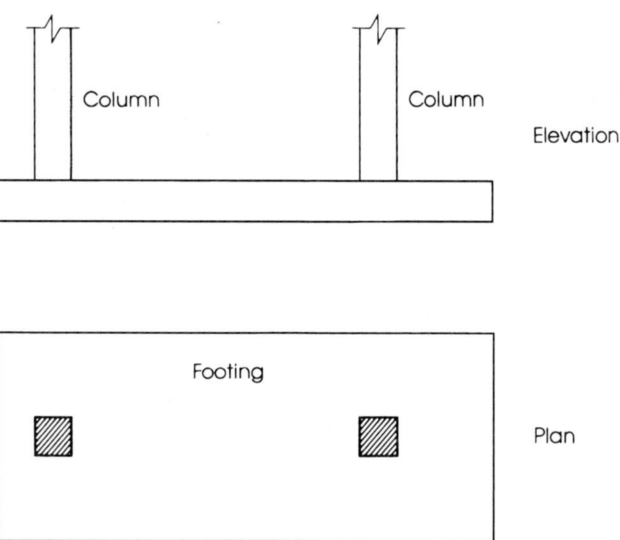

Figure 13.3 Combined footing.

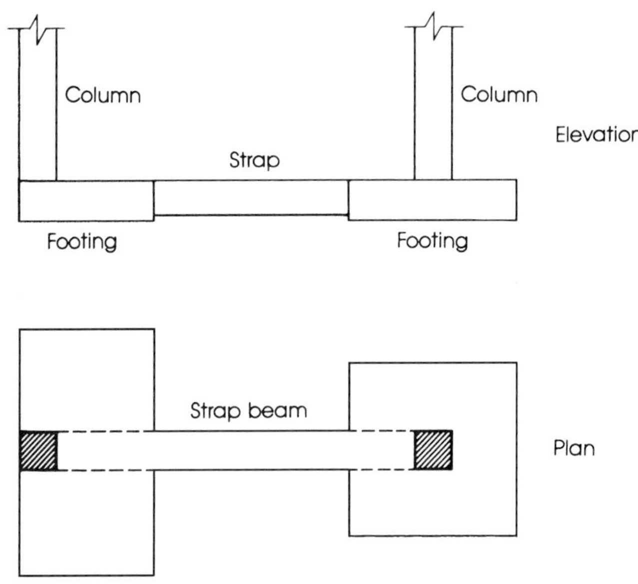

Figure 13.4 Strap footing.

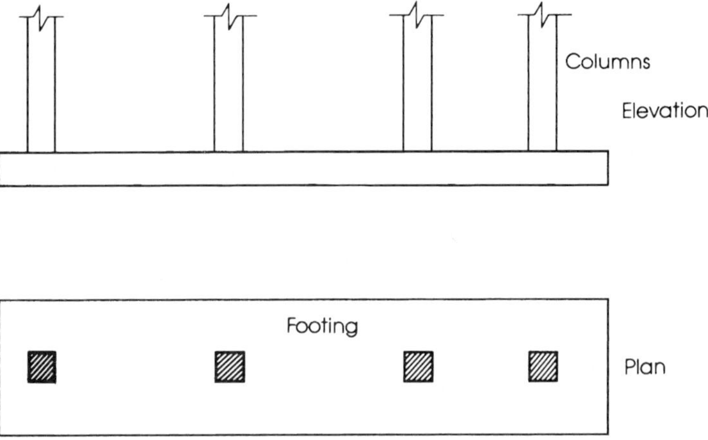

Figure 13.5 Continuous footing.

5. *Continuous footings* (Figure 13.5) support a row of three or more columns. They have limited width and continue under all columns.
6. *Raft,* or *mat, foundations* (Figure 13.6) consist of one footing, usually placed under the entire building area, and support the columns of the building. They are used when
   - **a.** The soil-bearing capacity is low;
   - **b.** Column loads are heavy;
   - **c.** Single footings cannot be used;
   - **d.** Piles are not used;
   - **e.** Differential settlement must be reduced through the entire footing system.
7. *Pile caps* (Figure 13.7) are thick slabs used to tie a group of piles together and to support and transmit column loads to the piles.

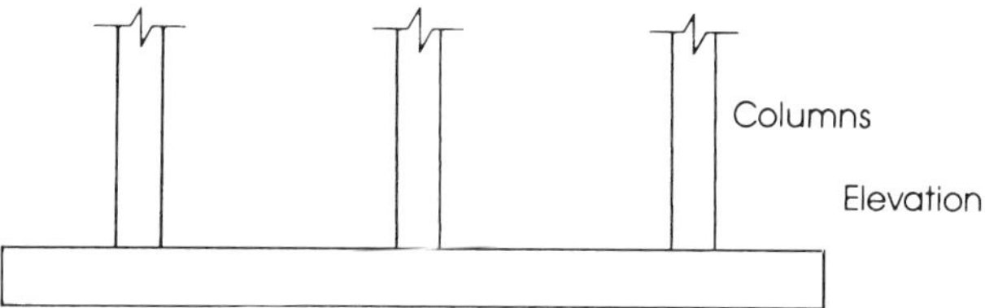

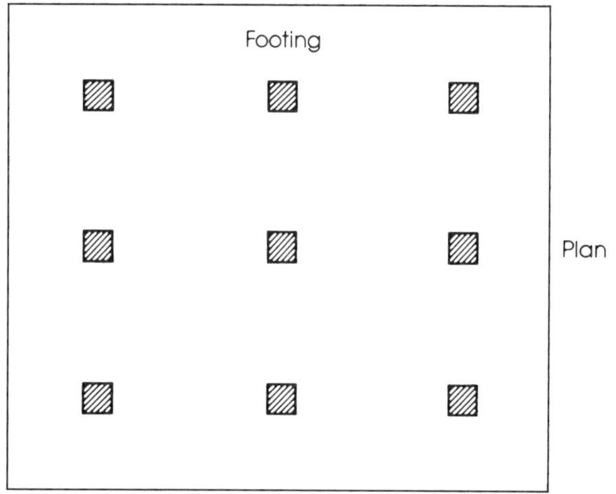

Figure 13.6 Raft, or mat, foundation.

Reinforcing rebars placed in two layers in a raft foundation.

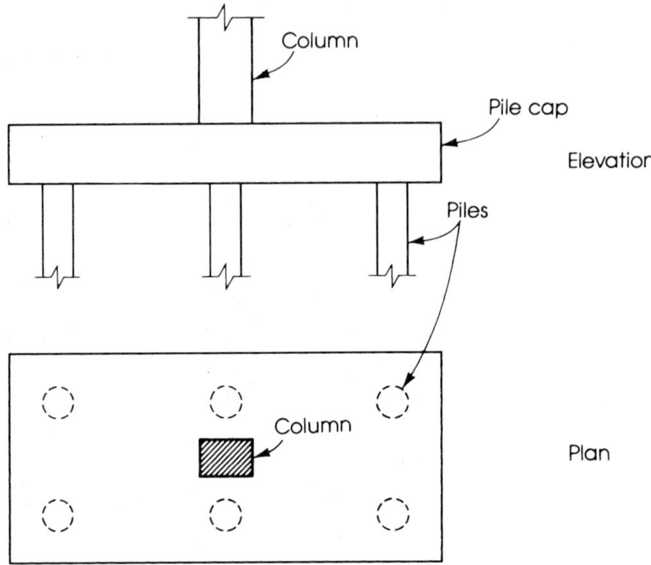

Figure 13.7 Pile cap footing.

13.3 DISTRIBUTION OF SOIL PRESSURE

Figure 13.8 shows a footing supporting a single column. When the column load, P, is applied on the centroid of the footing, a uniform pressure is assumed to develop on the soil surface below the footing area. However, the actual distribution of soil pressure is not uniform but depends on many factors, especially the composition of the soil and the degree of flexibility of the footing.

For example, the distribution of pressure on cohesionless soil (sand) under a rigid footing is shown in Figure 13.9. The pressure is maximum under the center of the footing

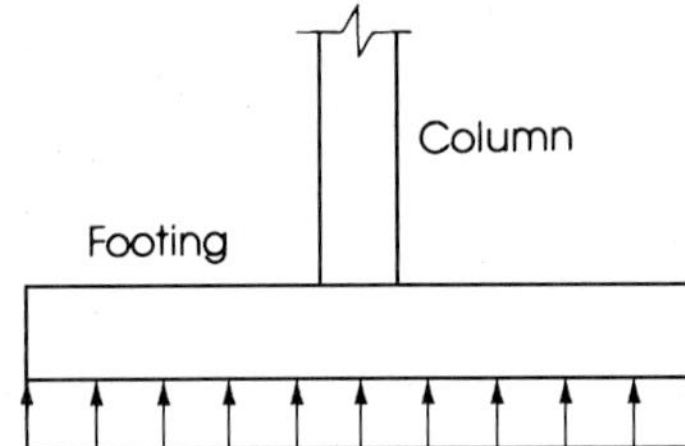

Figure 13.8 Distribution of soil pressure assuming uniform pressure.

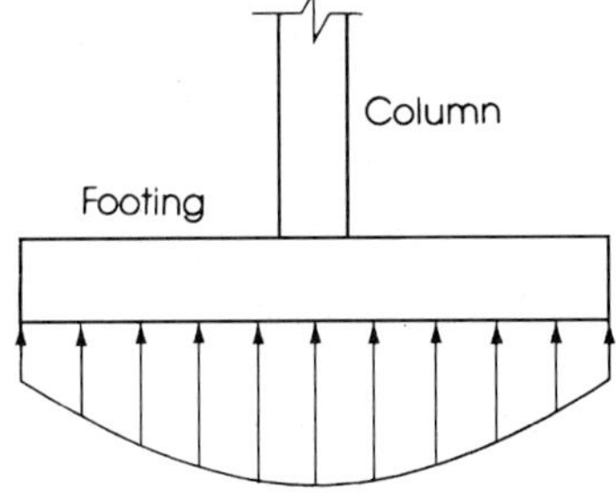

Figure 13.9 Soil pressure distribution in cohesionless soil (sand).

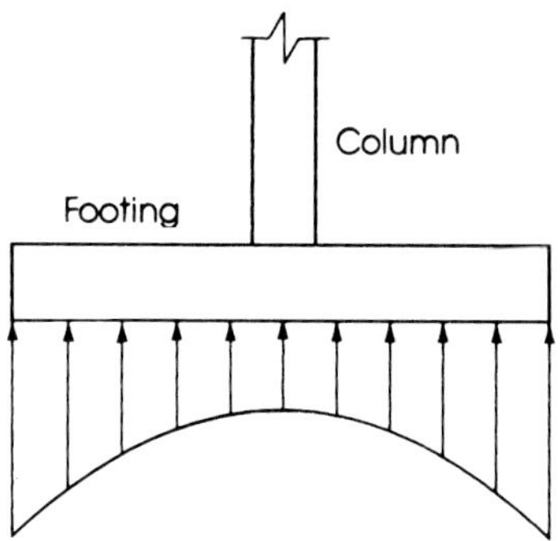

Figure 13.10 Soil pressure distribution in cohesive soil (clay).

and decreases toward the ends of the footing. The cohesionless soil tends to move from the edges of the footing, causing a reduction in pressure, whereas the pressure increases around the center to satisfy equilibrium conditions. If the footing is resting on a cohesive soil such as clay, the pressure under the edges is greater than at the center of the footing (Figure 13.10). The clay near the edges has a strong cohesion with the adjacent clay surrounding the footing, causing the nonuniform pressure distribution.

The allowable bearing soil pressure, q_a, is usually determined from soil tests. The allowable values vary with the type of soil, from extremely high in rocky beds to low in silty soils. For example, q_a for sedimentary rock is 30 Ksf, for compacted gravel is 8 Ksf, for well-graded compacted sand is 6 Ksf, and for silty-gravel soils is 3 Ksf.

Referring to Figure 13.8, when the load P is applied, the part of the footing below the column tends to settle downward. The footing will tend to take a uniform curved shape, causing an upward pressure on the projected parts of the footing. Each part acts as a cantilever and must be designed for both bending moments and shearing forces. The design of footings is explained in detail later.

13.4 DESIGN CONSIDERATIONS

Footings must be designed to carry the column loads and transmit them to the soil safely. The design procedure must take the following strength requirements into consideration:

1. The area of the footing based on the allowable bearing soil capacity
2. One-way shear
3. Two-way shear, or punching shear
4. Bending moment and steel reinforcement required
5. Bearing capacity of columns at their base and dowel requirements
6. Development length of bars
7. Differential settlement

These strength requirements are explained in the following sections.

13.4.1 Size of Footings

The area of the footings can be determined from the actual external loads (unfactored forces and moments) such that the allowable soil pressure is not exceeded. In general, for vertical loads

Wall and column footings, partly covered.

$$\text{Area of footing} = \frac{\text{total service load (including self-weight)}}{\text{allowable soil pressure, } q_a}, \text{ or}$$

$$\text{Area} = \frac{P\text{ (total)}}{q_a} \qquad (13.1)$$

where the total service load is the unfactored design load. Once the area is determined, an ultimate soil pressure is obtained by dividing the factored load, $P_u = 1.4D + 1.7L$, by the area of the footing. This is required to design the footing by the strength design method.

$$q_u = \frac{P_u}{\text{area of footing}} \qquad (13.2)$$

The allowable soil pressure, q_a, is obtained from soil test and it is based on service load conditions.

13.4.2 One-Way Shear (Beam Shear) (V_{u1})

For footings with bending action in one direction, the critical section is located at a distance d from the face of the column. The diagonal tension at section m-m in Figure 13.11 can be checked as was done before in beams. The allowable shear in this case is equal to

$$\phi V_c = 2\phi\sqrt{f'_c}\,bd \qquad (\phi = 0.85) \qquad (13.3)$$

where b = width of section m-m. The ultimate shearing force at section m-m can be calculated as follows:

$$V_{u1} = q_u b\left(\frac{L}{2} - \frac{C}{2} - d\right) \qquad (13.4)$$

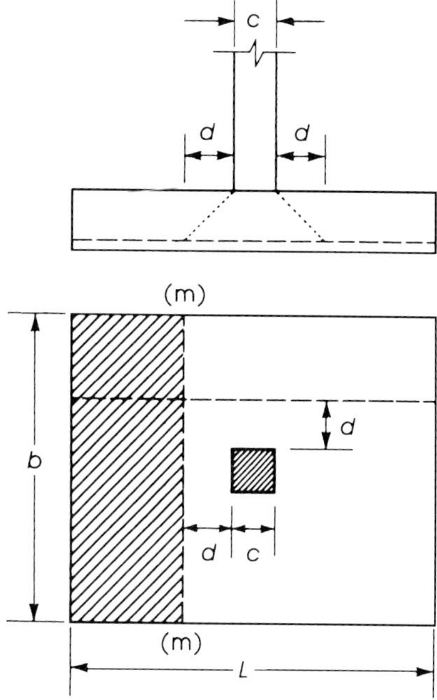

Figure 13.11 One-way shear.

If no shear reinforcement is to be used, then d can be determined, assuming $V_u = \phi V_c$:

$$d = \frac{V_{u1}}{2\phi\sqrt{f'_c}b} \tag{13.5}$$

13.4.3 Two-Way Shear (Punching Shear) (V_{u2})

Two-way shear is a measure of the diagonal tension caused by the effect of the column load on the footing. Inclined cracks may occur in the footing at a distance $d/2$ from the face of the column on all sides. The footing will fail as the column tries to punch out part of the footing (Figure 13.12).

The ACI Code, Section 11.12.2, allows a shear strength V_c in footings without shear reinforcement for two-way shear action, the smallest of

$$V_{c1} = 4\sqrt{f'_c}b_0 d \tag{13.6}$$

$$V_{c2} = \left(2 + \frac{4}{\beta_c}\right)\sqrt{f'_c}b_0 d \tag{13.7}$$

$$V_{c3} = \left(\frac{\alpha_s d}{b_0} + 2\right)\sqrt{f'_c}b_0 d \tag{13.8}$$

where

β_c = ratio of long side to short side of rectangular area

b_0 = perimeter of the critical section taken at $d/2$ from the loaded area (column section) (see Figure 13.12)

d = effective depth of footing

For the values of V_{c1} and V_{c2}, it can be observed that V_{c1} controls (less than V_{c2}) whenever $\beta_c \le 2$, whereas V_{c2} controls (less than V_{c1}) whenever $\beta_c > 2$. This indicates

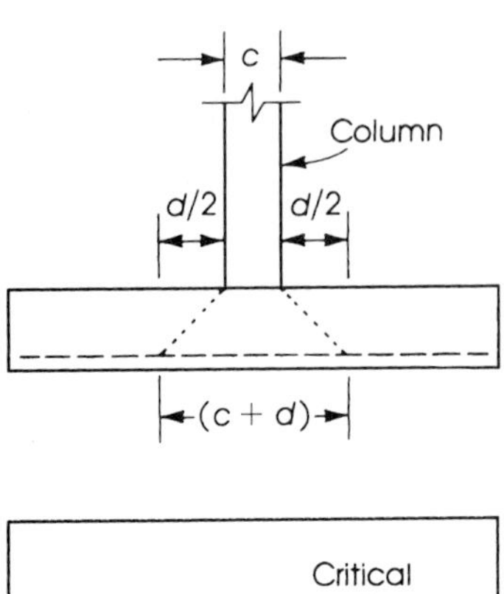

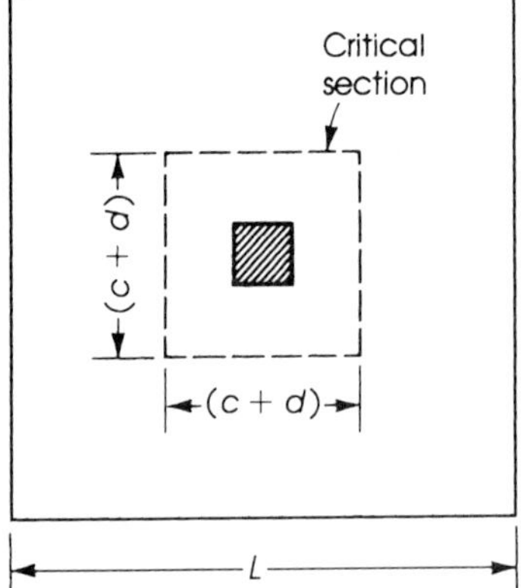

Figure 13.12 Punching shear (two-way).

Reinforced concrete single footings.

that the allowable shear V_c is reduced for relatively long footings. The actual soil pressure variation along the long side increases with an increase in β_c. For shapes other than rectangular, β_c is taken to be the ratio of the longest dimension of the effective loaded area in the long direction to the largest width in the short direction (perpendicular to the long direction).

For equation (13.8), α_s is assumed to be 40 for interior columns, 30 for edge columns, and 20 for corner columns. The concrete shear strength, V_{c3}, represents the effect of an increase in b_0 relative to d. For a high ratio of b_0/d, V_{c3} may control.

Based on the preceding three values of V_c, the effective depth, d, required for two-way shear is the largest obtained from the following formulas ($\phi = 0.85$):

$$d_1 = \frac{V_{u2}}{\phi 4\sqrt{f'_c}\, b_0} \qquad (\text{when } \beta_c \leq 2) \tag{13.9}$$

or

$$d_1 = \frac{V_{u2}}{\phi\left(2 + \dfrac{4}{\beta_c}\right)\sqrt{f'_c}\, b_0} \qquad (\text{when } \beta_c > 2) \tag{13.10}$$

$$d_2 = \frac{V_{u2}}{\phi\left(\dfrac{\alpha_s d}{b_0} + 2\right)\sqrt{f'_c}\, b_0} \tag{13.11}$$

The two-way shearing force, V_{u2}, and the effective depth, d, required (if shear reinforcement is not provided) can be calculated as follows (refer to Figure 13.12):

1. Assume d.
2. Determine b_0: $b_0 = 4(c + d)$ for square columns, where one side $= c$; $b_0 = 2(c_1 + d) + 2(c_2 + d)$ for rectangular columns of sides c_1 and c_2.
3. The shearing force V_{u2} acts at a section that has a length $b_0 = 4(c + d)$ or $[2(c_1 + d) + 2(c_2 + d)]$ and a depth d; the section is subjected to a vertical downward load P_u and a vertical upward pressure q_u (equation (13.2)). Therefore,

$$V_{u2} = P_u - q_u(c + d)^2 \qquad \text{for square columns} \tag{13.12a}$$

$$V_{u2} = P_u - q_u(c_1 + d)(c_2 + d) \qquad \text{for rectangular columns} \tag{13.12b}$$

4. Determine the largest d (of d_1 and d_2). If d is not close to the assumed d, revise your assumption and repeat.

13.4.4 Flexural Strength and Footing Reinforcement

The critical sections for moment occur at the face of the column (section n-n, Figure 13.13). The bending moment in each direction of the footing must be checked and the appropriate reinforcement must be provided. In square footings and square columns, the bending moments in both directions are equal. To determine the reinforcement required, the depth of the footing in each direction may be used. Because the bars in one direction rest on top of the bars in the other direction, the effective depth, d, varies with the diameter of the bars used. An average value of d may be adopted. A practical value of d may be assumed to be $(h - 4.5)$ in.

The depth of the footing is often controlled by shear, which requires a depth greater than that required by the bending moment. The steel reinforcement in each direction can be calculated in the case of flexural members as follows:

$$M_u = \phi A_s f_y\left(d - \frac{A_s f_y}{1.7 f'_c b}\right) \tag{13.13}$$

Also, the steel ratio, ρ, can be determined as follows: (equation 4.2)

$$\rho = \frac{0.85 f'_c}{f_y}\left[1 - \sqrt{1 - \frac{2R_u}{\phi(0.85 f'_c)}}\right] \tag{13.14}$$

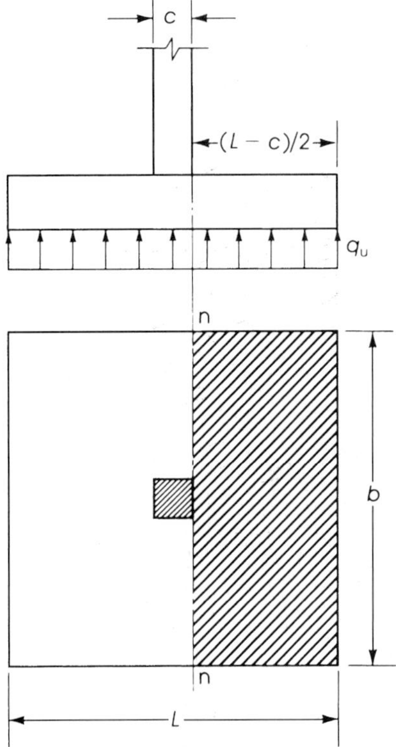

Figure 13.13 Critical section of bending moment.

where $R_u = M_u/bd^2$. When R_u is determined, ρ can also be obtained from the tables in Appendix A.

The minimum steel ratio requirement in flexural members is equal to $200/f_y$ when $f'_c < 4500$ psi and equal to $3\sqrt{f'_c}/f_y$ when $f'_c \geq 4500$ psi. However, the ACI Code, Section 10.5, indicates that for structural slabs of uniform thickness, the minimum area and maximum spacing of steel bars in the direction of bending shall be as required for shrinkage and temperature reinforcement. This last minimum steel requirement is very small, and a higher minimum reinforcement ratio is recommended, but it should not be greater than $200/f_y$.

The reinforcement in one-way footings and two-way footings must be distributed across the entire width of the footing. In the case of two-way rectangular footings, the ACI Code, Section 15.4.4, specifies that in the long direction, reinforcement must be distributed uniformly along the width of the footing. In the short direction, a certain ratio of the total reinforcement in this direction must be placed uniformly within a bandwidth equal to the length of the short side of the footing according to

$$\frac{\text{Reinforcement in bandwidth}}{\text{Total reinforcement in the short direction}} = \frac{2}{\beta + 1} \tag{13.15}$$

where

$$\beta = \frac{\text{long side of footing}}{\text{short side of footing}} \tag{13.16}$$

The bandwidth must be centered on the centerline of the column (Figure 13.14). The remaining reinforcement in the short direction must be uniformly distributed outside the bandwidth. This remaining reinforcement percentage shall not be less than that required for shrinkage and temperature.

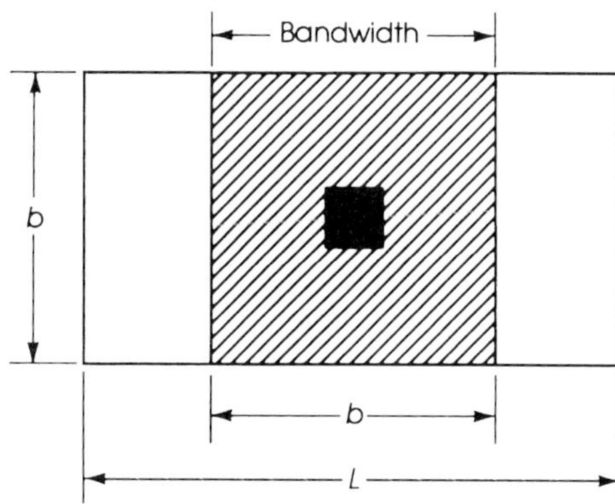

Figure 13.14 Bandwidth for reinforcement distribution.

When structural steel columns or masonry walls are used, then the critical sections for moments in footings are taken at halfway between the middle and the edge of masonry walls and halfway between the face of the column and the edge of the steel base place (ACI Code, Section 15.4.2).

13.4.5 Bearing Capacity of Column at Base

The loads from the column act on the footing at the base of the column, on an area equal to the area of the column cross section. Compressive forces are transferred to the footing directly by bearing on the concrete.

Forces acting on the concrete at the base of the column must not exceed the bearing strength of concrete as specified by the ACI Code, Section 10.17:

$$\text{Bearing strength } N_1 = \phi(0.85 f'_c A_1) \tag{13.17}$$

where $\phi = 0.7$ and $A_1 =$ the bearing area of the column. The value of the bearing strength given in equation (13.17) may be multiplied by a factor $\sqrt{A_2/A_1} \leq 2.0$ for bearing on footings when the supporting surface is wider on all sides than the loaded area. Here A_2 is the area of the part of the supporting footing that is geometrically similar to and concentric with the loaded area (Figure 13.15). Because $A_2 > A_1$, the factor $\sqrt{A_2/A_1}$ is greater than unity, indicating that the allowable bearing strength is increased because of the lateral support from the footing area surrounding the column base. The modified bearing strength is

$$N_2 = \phi(0.85 f'_c A_1)\sqrt{\frac{A_2}{A_1}} \leq 2\phi(0.85 f'_c A_1) \tag{13.18}$$

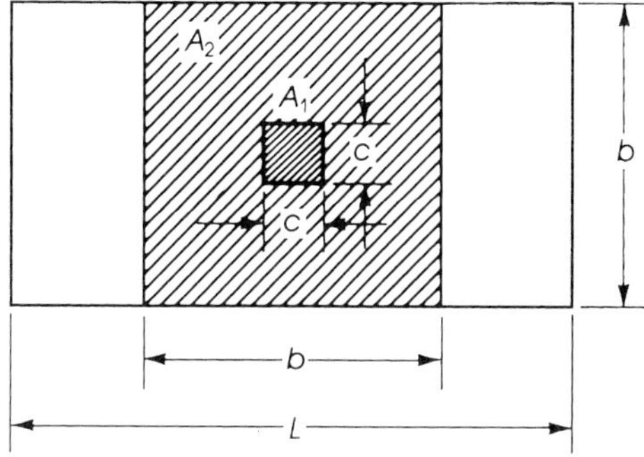

Figure 13.15 Bearing areas on footings. $A_1 = c^2$, $A_2 = b^2$.

If the ultimate force, P_u, is greater than either N_1 or N_2, reinforcement must be provided to transfer the excess force. This is achieved by providing dowels or extending the column bars into the footing. The excess force is $P_{ex} = P_u - N_1$, and the area of the dowel bars is $A_{sd} = (P_{ex}/f_y) \geq 0.005A_1$, where A_1 is the area of the column section. At least four bars should be used at the four corners of the column. If the ultimate force is less than either N_1 or N_2, then minimum reinforcement must be provided. The ACI Code, Section 15.8.2, indicates that the minimum area of the dowel reinforcement is at least 0.005 A_g (and not less than four bars), where A_g is the gross area of the column section. The minimum reinforcement requirements apply also to the case when the ultimate forces are greater than N_1 and N_2. The dowel bars may be placed at the four corners of the column and extended in both the column and footing. The dowel diameter shall not exceed the diameter of the longitudinal bars in the columns by more than 0.15 in. This requirement is necessary to ensure proper action between the column and footing. The development length of the dowels must be checked to determine proper transfer of the compression force into the footing.

13.4.6 Development Length of the Reinforcing Bars

The critical sections for checking the development length of the reinforcing bars are the same as those for bending moments. The development length for compression bars was given in Chapter 7:

$$l_d = \frac{0.02 f_y d_b}{\sqrt{f'_c}}$$

but this value cannot be less than $0.0003 f_y d_b \geq 8$ in. For other values, refer to Chapter 7.

13.4.7 Differential Settlement (Balanced Footing Design)

Footings usually support the following loads:

- Dead loads from the substructure and superstructure
- Live load resulting from occupancy
- Weight of materials used in backfilling
- Wind loads

Each footing in a building is designed to support the maximum load that may occur on any column due to the critical combination of loadings, using the allowable soil pressure.

The dead load, and maybe a small portion of the live load (called the *usual* live load), may act continuously on the structure. The rest of the live load may occur at intervals and on some parts of the structure only, causing different loadings on columns. Consequently, the pressure on the soil under different footings will vary according to the loads on the different columns, and differential settlement will occur under the various footings of one structure. Because partial settlement is inevitable, the problem turns out to be the amount of differential settlement that the structure can tolerate. The amount of differential settlement depends on the variation in the compressibility of the soils, the thickness of the compressible material below foundation level, and the stiffness of the combined footing and superstructure. Excessive differential settlement results in cracking of concrete and damage to claddings, partitions, ceilings, and finishes.

Differential settlement may be expressed in terms of angular distortion of the structure. Bjerrum [5] indicated the danger limits of distortion for some conditions vary between $\frac{1}{600}$ to $\frac{1}{150}$, depending on the damage that will develop in the building.

For practical purposes it can be assumed that the soil pressure under the effect of sustained loadings is the same for all footings, thus causing equal settlements. The sustained load (or the usual load) can be assumed equal to the dead load plus a percentage of the live load, which occurs very frequently on the structure. Footings then are proportioned for these sustained loads to produce the same soil pressure under all footings. In no case is the allowable soil bearing capacity to be exceeded under the dead load plus the maximum live load for each footing. Example 13.4 explains the procedure for calculating the areas of footings, taking into consideration the effect of differential settlement.

13.5 PLAIN CONCRETE FOOTINGS

Plain concrete footings may be used to support masonry walls or other light loads and transfer them to the supporting soil. The ACI Code allows the use of plain concrete pedestals and footings on soil, provided that the design stresses shall not exceed the following:

1. Maximum flexural stress in tension is less than or equal to $5\phi\sqrt{f'_c}$ (where $\phi = 0.65$).
2. Maximum stress in one-way shear (beam action) is less than or equal to $\frac{4}{3}\phi\sqrt{f'_c}$ (where $\phi = 0.65$).
3. Maximum shear stress in two-way action is

$$\left(\frac{4}{3} + \frac{8}{3\beta_c}\right)\phi\sqrt{f'_c} \leq 2.66\phi\sqrt{f'_c} \qquad (\text{where } \phi = 0.65)$$

4. Maximum compressive strength shall not exceed the concrete bearing strengths specified; f'_c of plain concrete should not be less than 2500 psi.
5. The minimum thickness of plain concrete footings shall not be less than 8 in.
6. The critical sections for bending moments are at the face of the column or wall.
7. The critical sections for one-way shear and two-way shear action are at distances d and $d/2$ from the face of the column or wall, respectively. Although plain concrete footings do not require steel reinforcement, it will be advantageous to provide shrinkage reinforcement in the two directions of the footing.
8. Stresses due to factored loads are computed assuming a linear distribution in concrete.
9. The effective depth, d, must be taken equal to the overall thickness minus 3 in.
10. For flexure and one-way shear, use a gross section bh, whereas for two-way shear, use b_0h to calculate ϕV_c.

Example 13.1

Design a reinforced concrete footing to support a 20-in.-wide concrete wall carrying a dead load of 26 K/ft, including the weight of the wall, and a live load of 20 K/ft. The bottom of the footing is 6 ft below final grade. Use $f'_c = 4$ Ksi, $f_y = 60$ Ksi, and an allowable soil pressure of 5 Ksf.

Solution

1. Calculate the effective soil pressure. Assume a total depth of footing of 20 in. Weight of footing is $(\frac{20}{12})(150) = 250$ psf. Weight of the soil fill on top of the footing, assuming that soil weighs 100 lb/ft^3, is $(6 - \frac{20}{12}) \times 100 = 433$ psf. Effective soil pressure at the bottom of the footing is $5000 - 250 - 433 = 4317$ psf $= 4.32$ Ksf.
2. Calculate the width of the footing for a 1-ft length of the wall:

$$\text{Width of footing} = \frac{\text{total load}}{\text{effective soil pressure}}$$

$$= \frac{26 + 20}{4.32} = 10.7 \text{ ft}$$

Use 11 ft.

3. Net upward pressure = (ultimate load)/(footing width)(per 1 ft):

$$P_u = 1.4D + 1.7L = 1.4 \times 26 + 1.7 \times 20 = 70.4 \text{ K}$$

$$\text{Net pressure} = q_u = \frac{70.4}{11} = 6.4 \text{ Ksf}$$

4. Check the assumed depth for shear requirements. The concrete cover in footings is 3 in., and assume no. 8 bars; then $d = 20 - 3.5 = 16.5$. The critical section for one-way shear is at a distance d from the face of the wall:

$$V_u = q_u\left(\frac{B}{2} - d - \frac{c}{2}\right) = 6.4\left(\frac{11}{2} - \frac{16.5}{12} - \frac{20}{2 \times 12}\right) = 21.06 \text{ K}$$

$$\text{Allowable one-way shear} = 2\sqrt{f'_c} = 2\sqrt{4000} = 126.5 \text{ psi}$$

$$\text{Required } d = \frac{V_u}{\phi(2\sqrt{f'_c})b} = \frac{21.06 \times 1000}{0.85(126.5)(12)} = 16.32 \text{ in.}$$

$$b = \text{1-ft length of footing} = 12 \text{ in.}$$

Total depth is $16.32 + 3.5 = 19.82$ in., or 20 in. Actual d is $20 - 3.5 = 16.5$ in. (as assumed). Note that few trials are needed to get the assumed and calculated d quite close.

5. Calculate the bending moment and steel reinforcement. The critical section is at the face of the wall:

$$M_u = \frac{1}{2}q_u\left(\frac{B}{2} - \frac{c}{2}\right)^2 = \frac{6.4}{2}\left(\frac{11}{2} - \frac{20}{24}\right)^2 = 69.7 \text{ K}\cdot\text{ft}$$

$$R_u = \frac{M_u}{bd^2} = \frac{69.7 \times 12{,}000}{12(16.5)^2} = 256 \text{ psi}$$

From Table A.1 in Appendix A, for $R_u = 256$ psi, $f'_c = 4$ Ksi, and $f_y = 60$ Ksi, the steel percentage is $\rho = 0.005$ (or from equation (13.14)). Minimum steel percentage for flexural members is

$$\rho_{\min} = \frac{200}{f_y} = \frac{200}{60{,}000} = 0.0033$$

Percentage of shrinkage reinforcement is 0.18% (for $f_y = 60$ Ksi). Therefore, use $\rho = 0.005$ as calculated.

$$A_s = 0.005 \times 12 \times 16.5 = 1.0 \text{ in.}^2$$

Use no. 8 bars spaced at 9 in. ($A_s = 1.05$ in.^{2}). (table A14)

6. Check the development length for no. 8 bars:

$$l_d = 48d_b = 48(1) = 48 \text{ in.} \qquad \text{(refer to Chapter 7)}$$

Provided

$$l_d = \frac{B}{2} - \frac{c}{2} - 3 \text{ in.} = \frac{11(12)}{2} - \frac{20}{2} - 3 = 53 \text{ in.}$$

7. Calculate secondary reinforcement in the longitudinal direction: $A_s = 0.0018(12)(20) = 0.43$ in.^{2}/ft. Choose no. 5 bars spaced at 8 in. ($A_s = 0.46$ in.^{2}). Details are shown in Figure 13.16.

Example 13.2

Design a square single footing to support an 18-in.-square tied interior column reinforced with eight no. 9 bars. The column carries an unfactored axial dead load of 245 K and an axial live load of 200 K. The base of the footing is 4 ft below final grade and the allowable soil pressure is 5 Ksf. Use $f'_c = 4$ Ksi and $f_y = 60$ Ksi.

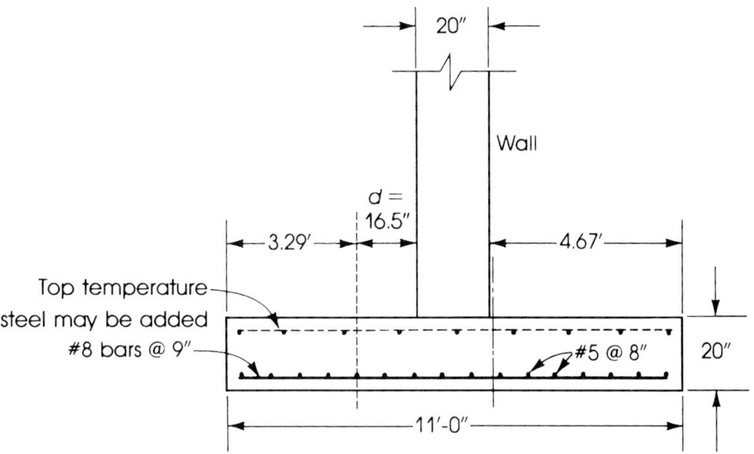

Figure 13.16 Wall footing, Example 13.1.

Solution

1. Calculate the effective soil pressure. Assume a total depth of footing of 2 ft. The weight of the footing is $2 \times 150 = 300$ psf. The weight of the soil on top of the footing (assuming the weight of soil $= 100$ pcf is $(2) \times 100 = 200$ psf.

$$\text{Effective soil pressure} = 5000 - 300 - 200 = 4500 \text{ psf}$$

2. Calculate the area of the footing:

$$\text{Actual loads} = D + L = 245 + 200 = 445 \text{ K}$$

$$\text{Area of footing} = \frac{445}{4.5} = 98.9 \text{ ft.}^2$$

$$\text{Side of footing} = 9.94 \text{ ft}$$

So, use 10 ft (Figure 13.17).

3. Net upward pressure equals (ultimate load)/(area of footing).

$$P_u = 1.4D + 1.7L$$

$$= 1.4 \times 245 + 1.7 \times 200 = 683 \text{ K}$$

$$\text{Net upward pressure } q_u = \frac{683}{10 \times 10} = 6.83 \text{ Ksf}$$

4. Check depth due to two-way shear. If no shear reinforcement is used, two-way shear determines the critical footing depth required. For an assumed total depth of 24 in., calculate d to the centroid of the top layer of the steel bars to be placed in the two directions within the footing. Let the bars to be used be no. 8 bars for calculating d.

$$d = 24 - 3 \text{ (cover)} - 1.5 \text{ (bar diameters)} = 19.5 \text{ in.}$$

It is quite practical to assume $d = h - 4.5$ in.

$$b_0 = 4(c + d) = 4(18 + 19.5) = 150 \text{ in.}$$

$$c + d = 18 + 19.5 = 37.5 \text{ in.} = 3.125 \text{ ft}$$

$$V_{u2} = P_u - q_u(c + d)^2 = 683 - 6.83(3.125)^2 = 616.3 \text{ K}$$

$$\text{Required } d_1 = \frac{V_{u2}}{4\phi\sqrt{f'_c}b_0}$$

$$= \frac{616.3(1000)}{(4)0.85\sqrt{4000}(150)} = 19.1 \text{ in.} \qquad (\beta_c = 1), \text{ equation (13.9)}$$

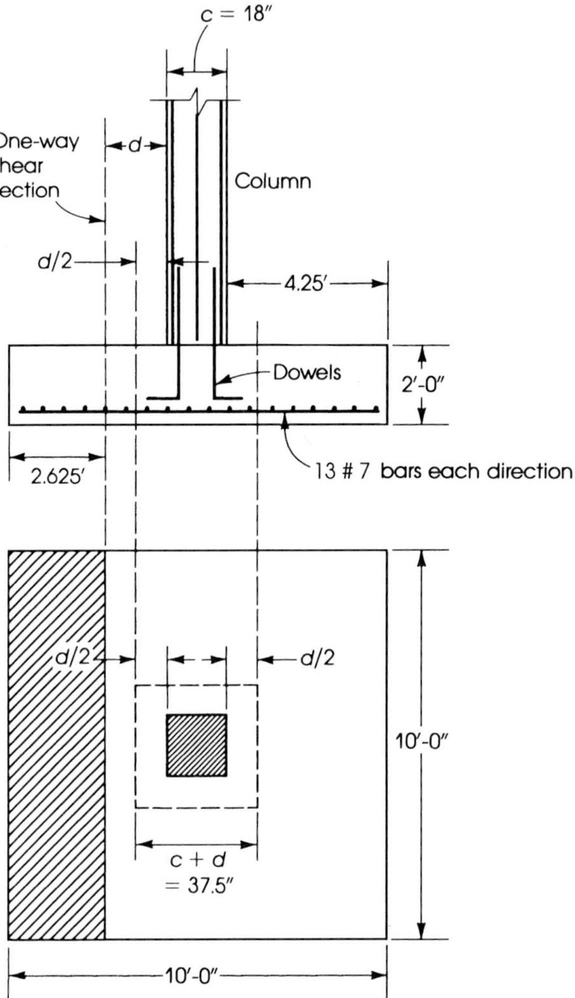

Figure 13.17 Square footing, Example 13.2.

$$\text{Required } d_2 = \frac{616.3(1000)}{0.85\left(\dfrac{40 \times 19.5}{150} + 2\right)(\sqrt{4000})(150)}$$

$$= 10.6 \text{ in.} \qquad \text{(not critical)}$$

($\alpha_s = 40$ for interior columns.) Thus the assumed depth is adequate. Two or more trials may be needed to reach an acceptable d that is close to the assumed one.

5. Check depth due to one-way shear action: The critical section is at a distance d from the face of the column.

$$\text{Distance from edge of footing} = \left(\frac{L}{2} - \frac{c}{2} - d\right) = 2.625 \text{ ft}$$

$$V_{u2} = 6.83 \times (2.625)(10) = 179.3 \text{ K}$$

The depth required for one-way shear is

$$d = \frac{V_{u1}}{(0.85)(2)\sqrt{f'_c}b}$$

$$= \frac{179.3(1000)}{(0.85)(2)(\sqrt{4000})(10 \times 12)} = 13.9 \text{ in.} < 19.5 \text{ in.}$$

6. Calculate the bending moment and steel reinforcement. The critical section is at the face of the column. The distance from edge of footing is

$$\left(\frac{L}{2} - \frac{c}{2}\right) = 5 - \frac{1.5}{2} = 4.25 \text{ ft}$$

$$M_u = \frac{1}{2} q_u \left(\frac{L}{2} - \frac{c}{2}\right)^2 b = \frac{1}{2}(6.83)(4.25)^2(10) = 617 \text{ K} \cdot \text{ft}$$

$$R_u = \frac{M_u}{bd^2} = \frac{617(12{,}000)}{(10 \times 12)(19.5)^2} = 162.3 \text{ psi}$$

Applying equation (13.14), $\rho = 0.0031$.

$$A_s = \rho bd = 0.0031(10 \times 12)(19.5) = 7.25 \text{ in.}^2$$

$$\text{Minimum } A_s \text{ (shrinkage steel)} = 0.0018(10 \times 12)(24) = 5.18 \text{ in.}^2 < 7.25 \text{ in.}^2$$

$$\text{Minimum } A_s \text{ (flexure)} = 0.0033(10 \times 12)(19.5) = 7.72 \text{ in.}^2$$

Therefore, $A_s = 7.25$ in.^{2} can be adopted. Use 13 no. 7 bars ($A_s = 7.82$ in.^{2}), spaced at $s = (120 - 6)/12 = 9.5$ in. in both directions.

7. Check bearing stress:

a. Bearing strength, N_1, at the base of the column ($A_1 = 18 \times 18$ in.) is

$$N_1 = \phi(0.85 f'_c A_1) = 0.7(0.85 \times 4)(18 \times 18) = 771 \text{ K}$$

b. Bearing strength, N_2, at the top of footing ($A_2 = 10 \times 10$ ft) is

$$N_2 = N_1 \sqrt{\frac{A_2}{A_1}} \le 2N_1$$

$$A_2 = 10 \times 10 = 100 \text{ ft}^2, \qquad A_1 = \frac{18 \times 18}{144} = 2.25 \text{ ft}^2$$

$$\sqrt{\frac{A_2}{A_1}} = 6.27 > 2$$

Therefore, $N_2 = 2N_1 = 1542$ K. Because $P_u = 683 \text{ K} < N_1$, bearing stress is adequate. The minimum area of dowels required is $0.005A_1 = 0.005(18 \times 18) = 1.62$ in.^{2} The minimum number of bars is 4, so use 4 no. 8 bars placed at the four corners of the column.

c. Development length of dowels in compression:

$$l_d = \frac{0.02 d_b f_y}{\sqrt{f'_c}} = \frac{0.02(1)(60{,}000)}{\sqrt{4000}} = 19 \text{ in.}$$

(controls). Minimum l_d is $0.0003 d_b f_y = 0.0003(1)(60{,}000) = 18$ in. ≥ 8 in. Therefore, use four no. 8 dowels extending 19 in. into column and footing. Note that l_d is less than d of 19.5 in., which is adequate.

8. The development length of main bars in footing for no. 7 bars is $l_d = 48d_b = 42$ in. (refer to Chapter 7); provided $l_d = L/2 - c/2 - 3$ in. $= 48$ in. Details of the footing are shown in Figure 13.17.

Example 13.3

Design a rectangular footing for the column of Example 13.2 if one side of the footing is limited to 8.5 ft.

Solution

1. The design procedure for rectangular footings is similar to that of square footings, taking into consideration the forces acting on the footing in each direction separately.

2. From the previous example, the area of the footing required is 98.9 ft^2:

$$\text{Length of footing} = \frac{98.9}{8.5} = 11.63 \text{ ft}$$

so use 12 ft (Figure 13.18). Footing dimensions are 8.5×12 ft.

3. $P_u = 683$ K. Thus, net upward pressure is

$$q_u = \frac{683}{8.5 \times 12} = 6.7 \text{ Ksf}$$

4. Check the depth due to one-way shear. The critical section is at a distance d from the face of the column. In the longitudinal direction,

$$V_{u1} = \left(\frac{L}{2} - \frac{c}{2} - d\right) \times q_u b$$

$$= \left(\frac{12}{2} - \frac{1.5}{2} - \frac{19.5}{12}\right) \times 6.7 \times 8.5 = 206.4 \text{ K}$$

This shear controls. In the short direction, $V_u = 150.7$ K (not critical).

$$\text{Required } d = \frac{V_{u1}}{2\phi\sqrt{f'_c}b} = \frac{206.4 \times 1000}{2 \times 0.85\sqrt{4000} \times (8.5 \times 12)} = 18.8 \text{ in.}$$

$$d \text{ provided} = 19.5 \text{ in.} > 18.8 \text{ in.}$$

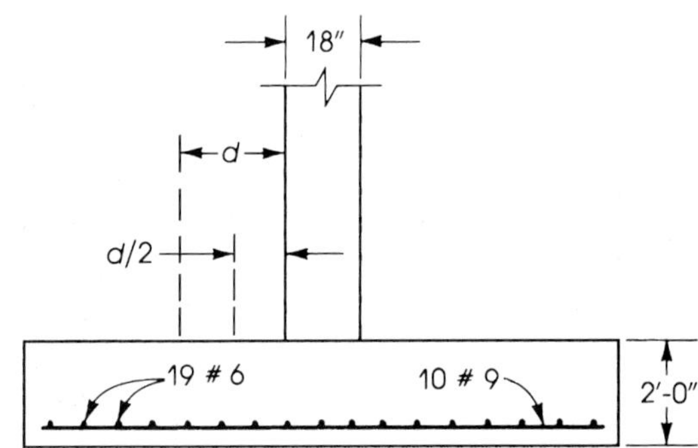

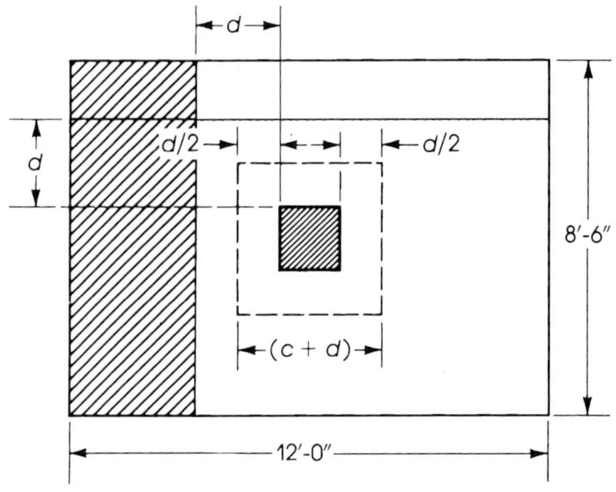

Figure 13.18 Rectangular footing, Example 13.3.

5. Check the depth for two-way shear action (punching shear). The critical section is at a distance $d/2$ from the face of the column on four sides.

$$b_0 = 4(18 + 19.5) = 150 \text{ in.}$$

$$(c + d) = 18 + 19.5 = 37.5 \text{ in.} = 3.125 \text{ ft}$$

$$\beta_c = \frac{12}{8.5} = 1.41 < 2 \qquad (\text{Use } V_c = 4\phi\sqrt{f'_c}\,b_0 d.)$$

$$V_{u2} = P_u - q_u(c + d)^2 = 683 - 6.84(3.125)^2 = 616.2 \text{ K}$$

$$d_1 = \frac{V_{u2}}{4\phi\sqrt{f'_c}\,b_0} = \frac{616.2 \times 1000}{4(0.85)\sqrt{4000} \times 150} = 19.1 \text{ in.}$$

$$d_2 = 10.6 \text{ in.} \qquad (\text{Does not control.})$$

6. Design steel reinforcement in the longitudinal direction. The critical section is at the face of the support. The distance from the edge of the footing is

$$\frac{L}{2} - \frac{c}{2} = \frac{12}{2} - \frac{1.5}{2} = 5.25 \text{ ft}$$

$$M_u = \tfrac{1}{2}(6.7)(5.25)^2(8.5) = 784.8 \text{ K}\cdot\text{ft}$$

$$R_u = \frac{M_u}{bd^2} = \frac{784.8(12{,}000)}{(8.5 \times 12)(19.5)^2} = 243 \text{ psi}$$

Applying equation (13.14), $\rho = 0.0047$:

$$A_s = 0.0047(8.5 \times 12)(19.5) = 9.34 \text{ in.}^2$$

$$\text{Min } A_s \text{ (shrinkage)} = 0.0018(8.5 \times 12)(24) = 4.4 \text{ in.}^2$$

$$\text{Min } A_s \text{ (flexure)} = 0.0033(8.5 \times 12)(19.5) = 6.56 \text{ in.}^2$$

Use $A_s = 9.34$ in. and 10 no. 9 bars $(A_s = 10 \text{ in.}^2)$ spaced at $S = (102 - 6)/9 = 10.7$ in.

7. Design steel reinforcement in the short direction. The distance from the face of the column to the edge of the footing is

$$\frac{8.5}{2} - \frac{1.5}{2} = 3.5 \text{ ft}$$

$$M_u = \tfrac{1}{2}(6.7)(3.5)^2(12) = 492.5 \text{ K}\cdot\text{ft}$$

$$R_u = \frac{M_u}{bd^2} = \frac{492.5(12{,}000)}{(12 \times 12)(19.5)^2} = 108 \text{ psi}$$

Applying equation (13.4), $\rho = 0.00204$:

$$A_s = 0.00204(12 \times 12)(19.5) = 5.72 \text{ in.}^2$$

$$\text{Min } A_s \text{ (shrinkage)} = 0.0018(12 \times 12)(24) = 6.22 \text{ in.}^2$$

$$\text{Min } A_s \text{ (flexure)} = 0.0033(12 \times 12)(19.5) = 9.26 \text{ in.}^2$$

The value of A_s to be used must be greater than or equal to 6.22 in.² Use 18 no. 6 bars $(A_s = 7.92 \text{ in.}^2)$.

$$\left(\frac{\text{Reinforcement in band width}}{\text{Total reinforcement}}\right) = \frac{2}{\beta + 1} = \frac{2}{(12/8.5) + 1} = 0.83$$

The number of bars in an 8.5-ft band is $18(0.83) = 15$ bars. The number of bars left on each side is $(18 - 15)/2 = 2$ bars. Therefore, place 15 no. 6 bars within the 8.5-ft band; then place 2 no. 6 bars $(A_s = 0.88 \text{ in.}^2)$ within $(12 - 8.5)/2 = 1.625$ ft on each side of the band.

The total number of bars is 19 no. 6 bars $(A_s = 8.36 \text{ in.}^2)$. In this example, the bars may be distributed at equal spacings all over the 12-ft length; $S = (144 - 6)/18 = 7.67$ in. Details of reinforcement are shown in Figure 13.18.

8. Check the bearing stress at the base of the column, as explained in the previous example. Use four no. 8 dowel bars.

9. Development length of the main reinforcement: $l_d = 29$ in. for no. 6 bars and 54 in. for no. 9 bars.

$$\text{Provided } l_d \text{ (long direction)} = \left(\frac{l}{2} - \frac{c}{2} - 3 \text{ in.}\right) = 60 \text{ in.}$$

$$\text{Provided } l_d \text{ (short direction)} = 39 \text{ in.} > 29 \text{ in.}$$

Example 13.4

Determine the footing areas required for equal settlement (balanced footing design) if the usual live load is 20% for all footings. The footings are subjected to dead loads and live loads as indicated in the following table. The allowable net soil pressure is 6 Ksf.

	Footing Number				
	1	**2**	**3**	**4**	**5**
Dead load	120 K	180 K	140 K	190 K	210 K
Live load	150 K	220 K	200 K	170 K	240 K

Solution

1. Determine the footing that has the largest ratio of live load to dead load. In this example, the footing 3 ratio of 1.43 is higher than the other ratios.

2. Calculate the usual load for all footings. The usual load is the dead load the portion of live load that most commonly occurs on the structure. In this example,

$$\text{Usual load} = \text{D.L.} + 0.2 \text{ (L.L)}$$

The values of the usual loads are shown in the following table.

3. Determine the area of the footing that has the highest ratio of L.L./D.L.:

$$\text{Area of footing 3} = \frac{\text{D.L.} + \text{L.L.}}{\text{allowable soil pressure}} = \frac{140 + 200}{6} = 56.7 \text{ ft}^2$$

The usual soil pressure under footing 3 is

$$\frac{\text{Usual load}}{\text{Area of footing}} = \frac{180}{56.7} = 3.18 \text{ Ksf}$$

4. Calculate the area required for each footing by dividing its usual load by the soil pressure of footing 3. The areas are tabulated in the following table. For footing 1, for example, the required area is $150/3.18 = 47.2 \text{ ft}^2$.

5. Calculate the maximum soil pressure under each footing:

$$q_{\max} = \frac{D + L}{\text{area}} \leq 6 \text{ Ksf} \qquad \text{(allowable soil pressure)}$$

Description	Footing Number 1	2	3	4	5
$\frac{\text{Live load}}{\text{Dead load}}$	1.25	1.22	1.43	0.90	1.14
Usual load = D.L. + 0.2(L.L.) (Kips)	150	224	180	224	258
Area required = $\frac{\text{usual load}}{3.18\text{ Ksf}}$ (ft^2)	47.2	70.4	56.7	70.4	81.1
Max. soil pressure = $\frac{\text{D + L}}{\text{area}}$ (Ksf)	5.72	5.68	6.00	5.11	5.55

Example 13.5

Design a plain concrete footing to support a 16-in.-thick concrete wall. The loads on the wall consist of a 16-K/ft dead load (including the self-weight of wall) and a 10-K/ft live load. The base of the footing is 4 ft below final grade. Use $f'_c = 3$ Ksi and an allowable soil pressure of 5 Ksf.

Solution

1. Calculate the effective soil pressure. Assume a total depth of footing of 28 in.

$$\text{Weight of footing} = \frac{28}{12} \times 145 = 338 \text{ psf}$$

The weight of the soil, assuming that soil weighs 100 pcf, is $(4 - 2.33) \times 100 = 167$ psf. Effective soil pressure is $500 - 338 - 167 = 4495$ psf.

2. Calculate the width of the footing for a 1-ft length of the wall ($b = 1$ ft):

$$\text{Width of footing} = \frac{\text{total load}}{\text{effective soil pressure}}$$

$$= \frac{16 + 10}{4.495} = 5.79 \text{ ft}$$

Use 6.0 ft (Figure 13.19).

3. $U = 1.4D + 1.7L = 1.4 \times 16 + 1.7 \times 10 = 39.4$ K/ft.
The net upward pressure is $q_u = 39.4/6 = 6.57$ Ksf.

4. Check bending stresses. The critical section is at the face of the wall. For a 1-ft length of wall and footing,

$$M_u = \frac{1}{2} q_u \left(\frac{L}{2} - \frac{c}{2}\right)^2 = \frac{1}{2}(6.57)\left(\frac{6}{2} - \frac{16}{2 \times 12}\right)^2 = 17.88 \text{ K} \cdot \text{ft}$$

Let the effective depth, d, be $28 - 3 = 25$ in., assuming that the bottom 3 in. is not effective.

$$I_g = \frac{bd^3}{12} = \frac{12}{12}(25)^3 = 15{,}625 \text{ in.}^4$$

The flexural tensile stress is

$$f_t = \frac{M_u c}{I} = \frac{(17.88 \times 12{,}000)}{15{,}625}\left(\frac{25}{2}\right) = 172 \text{ psi}$$

The allowable flexural tensile stress is $5\phi\sqrt{f'_c} = 5 \times 0.65\sqrt{3000} = 178 \text{ psi} > 172$ psi.

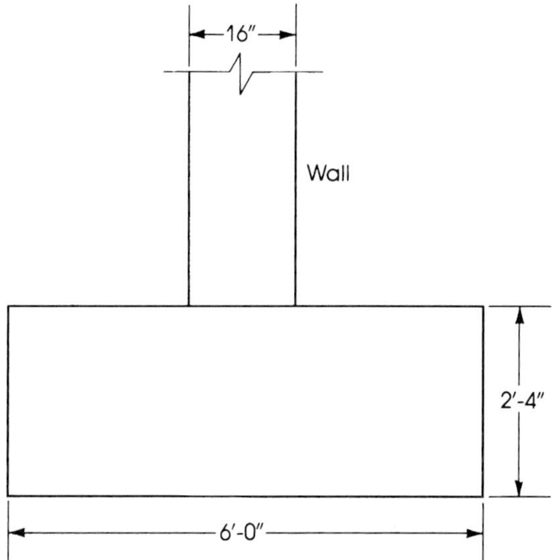

Figure 13.19 Plain concrete wall footing, Example 13.5.

5. Check shear stress: The critical section is at a distance $d = 25$ in. from the face of the wall.

$$V_u = q_u\left(\frac{L}{2} - \frac{c}{2} - d\right) = 6.57\left(\frac{6}{2} - \frac{16}{2 \times 12} - \frac{25}{12}\right) = 1.65 \text{ K}$$

$$\phi V_c = \phi\left(\frac{4}{3}\right)\sqrt{f'_c}\,bd = \frac{(0.65)(\frac{4}{3})\sqrt{3000}(12)(25)}{1000} = 14.24 \text{ K}$$

Therefore, the section is adequate. It is advisable to use minimum reinforcement in both directions.

13.6 COMBINED FOOTINGS

When a column is located near a property line, part of the single footing might extend into the neighboring property. To avoid this situation, the column may be placed on one side or edge of the footing, causing eccentric loading. This may not be possible under certain conditions, and sometimes it is not an economical solution. A better design can be achieved by combining the footing with the nearest internal column footing, forming a combined footing. The center of gravity of the combined footing coincides with the resultant of the loads on the two columns.

Another case where combined footings become necessary is when the soil is poor and the footing of one column overlaps the adjacent footing. The shape of the combined footing may be rectangular or trapezoidal (Figure 13.20). When the load of the external column near the property line is greater than the load of the interior column, a trapezoidal footing may be used to keep the centroid of footing in line with the resultant of the two column loads. In most other cases a rectangular footing is preferable.

The length and width of the combined footing are chosen to the nearest 3 in., which may cause a small variation in the uniform pressure under the footing, but it can be tolerated. For a uniform upward pressure, the footing will deflect, as shown in Figure 13.21. The ACI Code, Section 15.10, does not provide a detailed approach for the design of combined footings. The design, in general, is based on structural analysis.

A simple method of analysis is to treat the footing as a beam in the longitudinal direction, loaded with uniform upward pressure q_u. For the transverse direction, it is assumed that the column load is spread over a width under the column equal to the column width

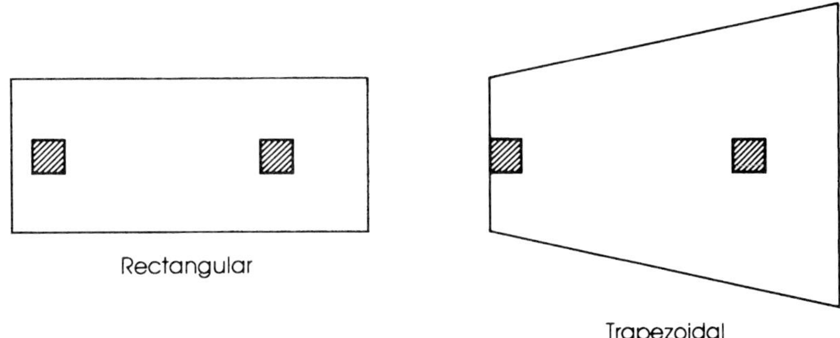

Figure 13.20 Combined footings.

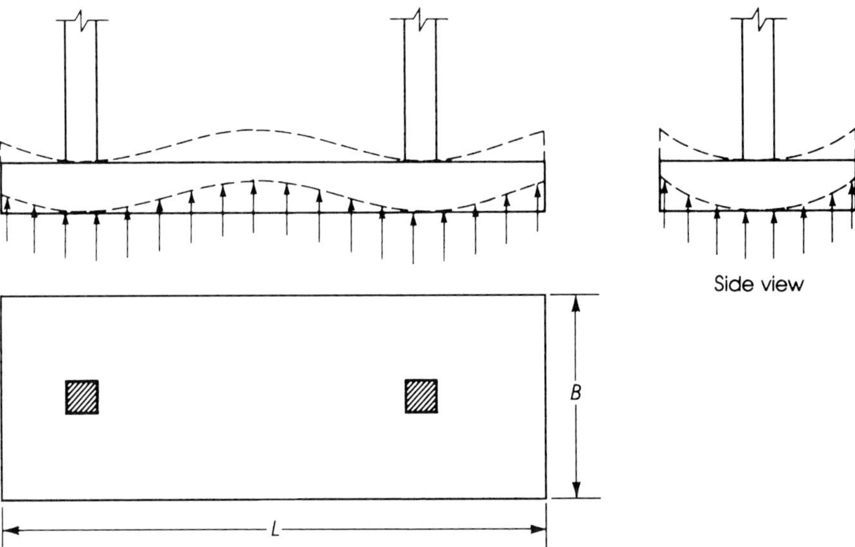

Figure 13.21 Upward deflection of a combined footing in two directions.

plus d on each side, whenever that is available. In other words, the column load acts on a beam under the column within the footing, which has a maximum width of $(c + 2d)$ and a length equal to the short side of the footing (Figure 13.22). A smaller width, down to $(c + d)$, may be used. The next example explains the design method in detail.

Example 13.6

Design a rectangular combined footing to support two columns, as shown in Figure 13.23. The edge column, I, has a section 16 by 16 in. and carries a D.L. of 180 K and an L.L. of 120 K. The interior column, II, has a section 20 by 20 in. and carries a D.L. of 250 K and an L.L. of 140 K. The allowable soil pressure is 5 Ksf and the bottom of the footing is 5 ft below final grade. Design the footing using $f'_c = 4$ Ksi, $f_y = 60$ Ksi, and the ACI strength design method.

Solution

1. Determine the location of the resultant of the column loads. Take moments about the center of the exterior column I:

$$x = \frac{(250 + 140) \times 16}{(250 + 140) + (180 + 120)} = 9 \text{ ft from column I}$$

The distance of the resultant from the property line is $9 + 2 = 11.0$ ft. The length of the footing is $2 \times 11 = 22.0$ ft. In this case the resultant of column loads will coincide with the resultant of the upward pressure on the footing.

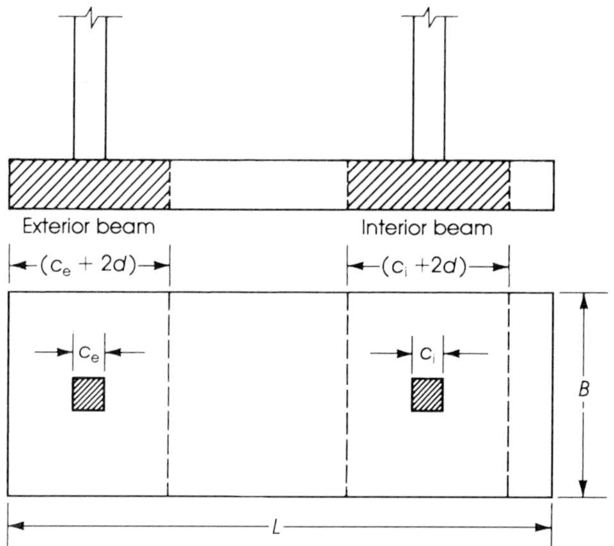

Figure 13.22 Analysis of combined footing in the transverse direction.

2. Determine the area of the footing. Assume the footing total depth is 36 in. ($d = 36 - 4.5 = 31.5$ in.).

   Total actual (working) loads = 300 + 390 = 690 K

$$\text{Net upward pressure} = 5000 - \left(\frac{36}{12} \times 150\right) - (2 \times 100) = 4500 \text{ psf}$$

   (Assumed weight of soil is 100 psf.)

$$\text{Required area} = \frac{690}{4.5} = 153.3 \text{ ft}^2$$

$$\text{Width of footing} = \frac{153.3}{22} = 6.97 \text{ ft}$$

   Use 7 ft. Choose a footing 22 by 7 ft (area = 154 ft²).

3. Determine the ultimate upward pressure using ultimate loads:

$$P_{u1} \text{ (column I)} = 1.4 \times 180 + 1.7 \times 120 = 456 \text{ K}$$

$$P_{u2} \text{ (column II)} = 1.4 \times 250 + 1.7 \times 140 = 588 \text{ K}$$

   The net ultimate soil pressure is $q_u = (456 + 588)/154 = 6.78$ Ksf.

4. Draw the factored shearing force diagram as for a beam of $L = 22$ ft supported on two columns and subjected to an upward pressure of 6.78 Ksf × 7 (width of footing) = 47.5 K/ft (per foot length of footing).

$$V_u \text{ (at outer face column I)} = 47.5\left(2 - \tfrac{8}{12}\right) = 63.3 \text{ K}$$

$$V_u \text{ (at interior face column I)} = 456 - 47.5\left(2 + \tfrac{8}{12}\right) = 329.3 \text{ K}$$

$$V_u \text{ (at outer face column II)} = 47.5\left(4 - \tfrac{10}{12}\right) = 150.4 \text{ K}$$

$$V_u \text{ (at interior face column I)} = 588 - \left(4 + \tfrac{10}{12}\right) \times 47.5 = 358.4 \text{ K}$$

   Find the point of zero shear, x: distance between interior faces of columns I and II is

$$16 - \frac{8}{12} - \frac{10}{12} = 14.5 \text{ ft}$$

$$x = \frac{329.3}{(329.3 + 358.4)}(14.5) = 6.9 \text{ ft}$$

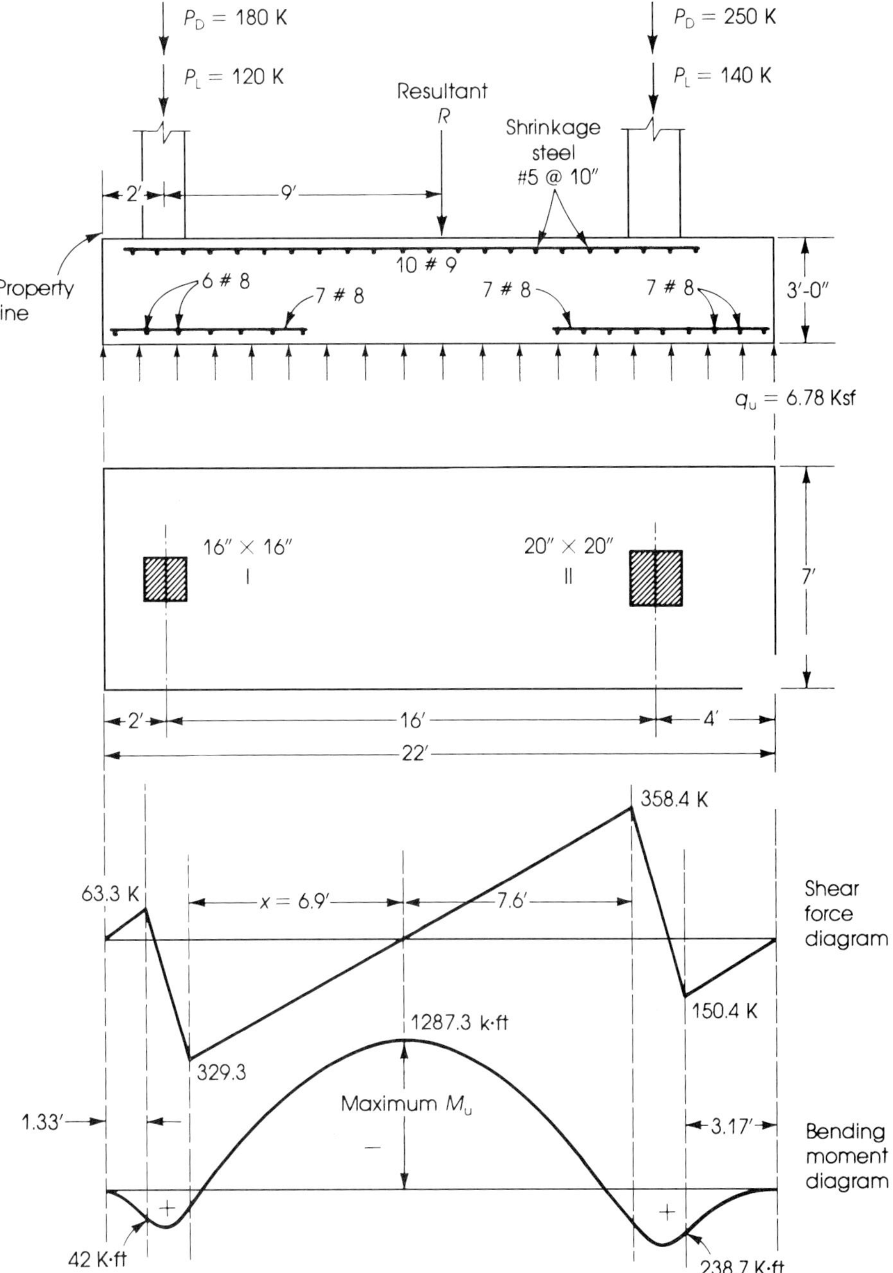

Figure 13.23 Design of a combined footing, Example 13.6.

5. Draw the factored moment diagram considering the footing as a beam of $L = 22$ ft supported by the two columns. The uniform upward pressure is 47.5 K/ft.

$$M_{u1} \text{ (at outer face column I)} = 47.5\frac{(1.33)^2}{2} = 42 \text{ K}\cdot\text{ft}$$

$$M_{u2} \text{ (at outer face column II)} = 47.5\frac{(3.17)^2}{2} = 238.7 \text{ K}\cdot\text{ft}$$

The maximum moment occurs at zero shear:

$$\text{Maximum } M_u \text{ (calculated from column I side)} = 456\left(6.9 + \frac{8}{12}\right) - \frac{47.5}{2}\left(6.9 + \frac{8}{12} + 2\right)^2$$

$$= 1276.8 \text{ K}\cdot\text{ft}$$

$$\text{Maximum } M_u \text{ (from column II side)} = 588\left(7.6 + \frac{10}{12}\right) - \frac{47.5}{2}\left(7.6 + \frac{10}{12} + 4\right)^2$$

$$= 1287.3 \text{ K}\cdot\text{ft}$$

The moments calculated from both sides of the footings are close enough, and M_u (max) $= 1287.3$ K·ft may be adopted. This variation occurred mainly because of the adjustment of the length and width of the footing.

6. Check the depth for one-way shear. Maximum shear occurs at a distance $d = 31.5$ in. from the interior face of column II (Figure 13.23).

$$V_{u1} = 358.4 - \frac{31.5}{12}(47.5) = 233.7 \text{ K}$$

$$d = \frac{V_{u1}}{\phi(2\sqrt{f'_c})b} = \frac{233.7 \times 100}{0.85(2\sqrt{4000})(7 \times 12)} = 25.85 \text{ in.}$$

The effective depth provided is 31.5 in. > 25.85 in.; thus, the footing is adequate.

7. Check depth for two-way shear (punching shear). For the interior column,

$$b_0 = 4(c + d) = \left(\frac{4}{12}\right)(20 + 31.5) = 17.17 \text{ ft}$$

$$(c + d) = \frac{20 + 31.5}{12} = 4.29 \text{ ft}$$

The shear V_{u2} at a section $d/2$ from all sides of the column is equal to

$$V_{u2} = P_{u2} - q_u(c + d)^2 = 588 - 6.78(4.29)^2 = 463 \text{ K}$$

$$d = \frac{V_{u2}}{\phi(4\sqrt{f'_c})b_0} = \frac{463(1000)}{0.85(4\sqrt{4000})(17.7 \times 12)} = 10.5 \text{ in.} < 31.5 \text{ in.}$$

The exterior column is checked and proved not to be critical.

8. Check the depth for moment and determine the required reinforcement in the long direction.

$$\text{Maximum bending moment} = 1287.3 \text{ K}\cdot\text{ft}$$

$$R_u = \frac{M_u}{bd^2} = \frac{1287.3(12{,}000)}{(7 \times 12)(31.5)^2} = 185 \text{ psi}$$

Applying equation (13.14), the steel percentage is $\rho = 0.0036 > 0.0033(\rho_{\min})$.

$$A_s = 0.0036(84 \times 31.5) = 9.52 \text{ in.}^2$$

$$\text{Min. } A_s \text{ (shrinkage)} = 0.0018(84)(36) = 5.44 \text{ in.}^2$$

$A_s = 9.52$ in.^{2} controls. Use 10 no. 9 bars ($A_s = 10$ in.^{2}).

$$\text{Spacing of bars} = \frac{84 - 6 \text{ (concrete cover)}}{9 \text{ (no. of spacings)}} = 8.67 \text{ in.}$$

The bars are extended between the columns at the top of the footing with a concrete cover of 3 in. Place minimum reinforcement at the bottom of the projecting ends of the footing beyond the columns to take care of the positive moments. Extend the bars a development length l_d beyond the side of the column.

The minimum shrinkage reinforcement is 5.44 in.^{2} Use seven no. 8 bars ($A_s = 5.5$ in.^{2}).

The development length required for the main top bars is $1.3l_d = 1.3(54) = 70$ in. beyond the point of maximum moment. Development lengths provided to both columns are adequate.

9. For reinforcement in the short direction, calculate the bending moment in the short (transverse) direction, as in the case of single footings. The reinforcement under each column is to be placed within a maximum bandwidth equal to the column width twice the effective depth d of the footing (Figure 13.24).

a. Reinforcement under the exterior column I:

$$\begin{aligned}\text{Bandwidth} &= 16 \text{ in. (column width)}\\ &\quad + 16 \text{ in. (on exterior side of column)}\\ &\quad + 31.5 \text{ in. } (d)\\ &= 63.5 \text{ in.} = 5.3 \text{ ft}\end{aligned}$$

Use 5.5 ft. Net upward pressure in the short direction under column I is

$$\frac{P_{u1}}{\text{width of footing}} = \frac{456}{7} = 65.1 \text{ K/ft}$$

Distance from the free end to the face of the column is $\frac{7}{2} - \frac{8}{12} = 2.83$ ft.

$$M_u \text{ (at face of column I)} = \frac{65.1}{2}(2.83)^2 = 260.7 \text{ K}\cdot\text{ft}$$

$$R_u = \frac{M_u}{bd^2} = \frac{260.7 \times 12{,}000}{(5.5 \times 12)(31.5)^2} = 48 \text{ psi}$$

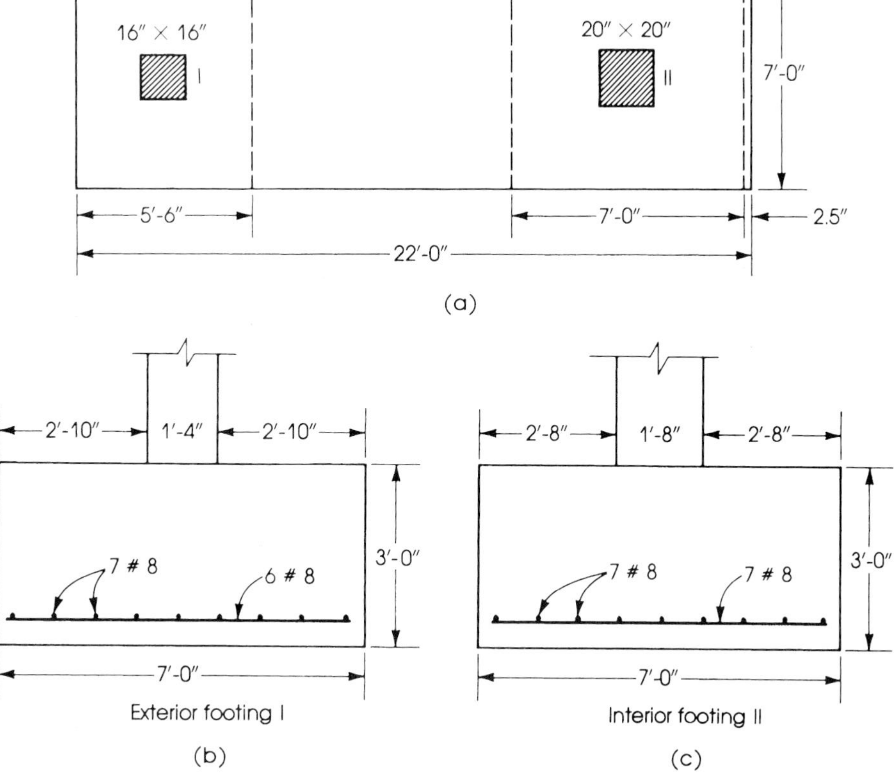

Figure 13.24 Design of combined footing, transverse direction: (a) plan, (b) exterior footing, and (c) interior footing.

The steel percentage, ρ, is less than minimum ρ for shrinkage reinforcement ratio of 0.0018.

$$A_s\,(\text{min.}) = (0.0018)(5.5 \times 12)(36) = 4.3\ \text{in.}^2$$

Use six no. 8 bars $\left(A_s = 4.71\ \text{in.}^2\right)$ placed within the bandwidth of 66 in.

b. Reinforcement under the interior column II:

$$\text{Bandwidth} = 20 + 31.5 + 31.5 = 83\ \text{in.} = 6.91\ \text{ft}$$

Use 7 ft (84 in.).

$$\text{Net upward pressure} = \frac{P_{u2}}{\text{width of footing}} = \frac{588}{7} = 84\ \text{K/ft}$$

$$\text{Distance to face of column} = \frac{7}{2} - \frac{10}{12} = 2.67\ \text{ft}$$

$$M_u\ (\text{at face of column II in short direction}) = \tfrac{1}{2}(84)(2.67)^2 = 299.4\ \text{K}\cdot\text{ft}$$

$$R_u = \frac{(299.4)(12{,}000)}{(84)(31.5)^2} = 43\ \text{psi}$$

which is very small. Use a minimum shrinkage reinforcement ratio of 0.0018.

$$A_s = (0.0018)(84)(36) = 5.44\ \text{in.}^2$$

Use seven no. 8 bars placed within the bandwidth of 84 in. under column II, as shown in Figures 13.23 and 13.24. The development length l_d of no. 8 bars in the short direction is 48 in.

13.7 FOOTINGS UNDER ECCENTRIC COLUMN LOADS

When a column transmits axial loads only, the footing can be designed such that the load acts at the centroid of the footing, producing uniform pressure under the footing. However, in some cases, the column transmits an axial load and a bending moment, as in the case of the footings of fixed-end frames. The pressure q that develops on the soil will not be uniform and can be evaluated from the following equation:

$$q = \frac{P}{A} \pm \frac{Mc}{I} \geq 0 \tag{13.19}$$

where A and I are the area and moment of inertia of the footing, respectively.

Different soil conditions exist, depending on the magnitudes of P, M, and allowable soil pressure. The different design conditions are shown in Figure 13.25 and are summarized as follows:

1. When $e = M/P < L/6$, the soil pressure is trapezoidal.

$$q_{\max} = \frac{P}{A} + \frac{Mc}{I} = \frac{P}{LB} + \frac{6M}{BL^2} \tag{13.20}$$

$$q_{\min} = \frac{P}{A} - \frac{Mc}{I} = \frac{P}{LB} - \frac{6M}{BL^2} \tag{13.21}$$

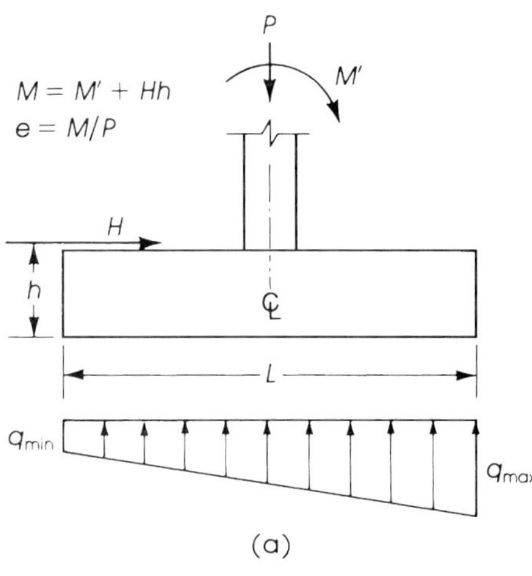

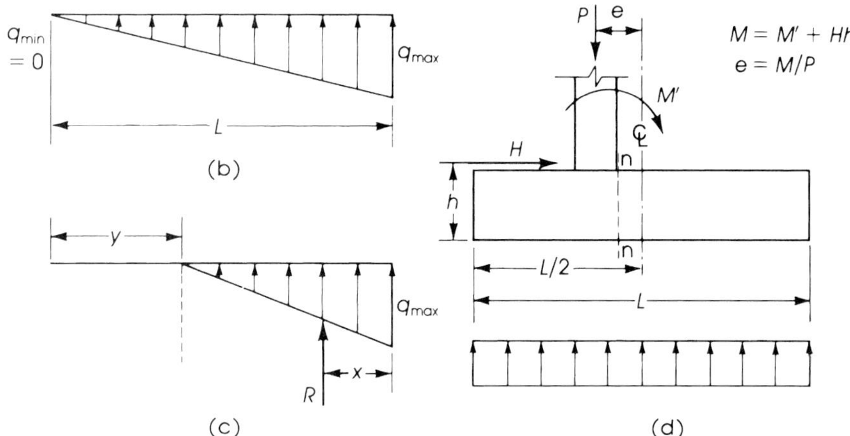

Figure 13.25 Single footing subjected to eccentric loading: L = length of footing, B = width, and h = height.

2. When $e = M/P = L/6$, the soil pressure is triangular.

$$q_{\max} = \frac{P}{LB} + \frac{6M}{BL^2} = \frac{2P}{LB} \tag{13.22}$$

$$q_{\min} = 0 = \frac{P}{LB} - \frac{6M}{BL^2} \quad \text{or} \quad \frac{P}{LB} = \frac{6M}{BL^2} \tag{13.23}$$

3. When $e > L/6$, the soil pressure is triangular.

$$x = \frac{L - y}{3} = \frac{L}{2} - e$$

$$P = q_{\max}\left(\frac{3x}{2}\right)B \tag{13.24}$$

$$q_{\max} = \frac{2P}{3xB} = \frac{4P}{3B(L - 2e)}$$

4. When the footing is moved a distance e from the axis of the column to produce uniform soil pressure under the footing. Maximum moment occurs at section n-n.

$$M = M' - Hh, \quad \text{and} \quad e = \frac{M}{P}$$

13.8 FOOTINGS UNDER BIAXIAL MOMENT

In some cases, a footing may be subjected to an axial force and biaxial moments about its x- and y-axes; such a footing may be needed for a factory crane that rotates 360°. The footing then must be designed for the critical loading.

Referring to Figure 13.26, if the axial load P acts at a distance e_x from the y-axis and e_y from the x-axis, then

$$M_x = Pe_y \quad \text{and} \quad M_y = Pe_x$$

The soil pressure at corner 1 is

$$q_{\text{max}} = \frac{P}{A} + \frac{M_x c_y}{I_x} + \frac{M_y c_x}{I_y}$$

At corner 2,

$$q_2 = \frac{P}{A} - \frac{M_x c_y}{I_x} + \frac{M_y c_x}{I_y}$$

At corner 3,

$$q_3 = \frac{P}{A} - \frac{M_x c_y}{I_x} - \frac{M_y c_x}{I_y}$$

At corner 4,

$$q_4 = \frac{P}{A} + \frac{M_x c_y}{I_x} - \frac{M_y c_x}{I_y}$$

Again, note that the allowable soil pressure must not be exceeded and the soil cannot take any tension; that is, $q \geq 0$.

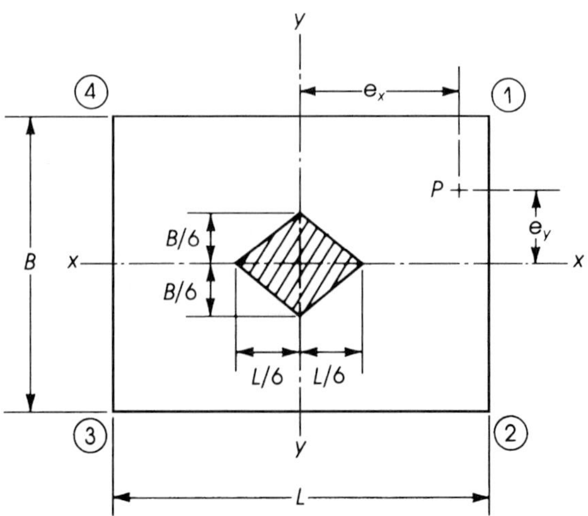

Figure 13.26 Footing subjected to P and biaxial moment. If $e_x < L/6$ and $e_y < B/6$, footing will be subjected to upward soil pressure on all bottom surface (nonuniform pressure).

Example 13.7

A 12- by 24-in. column of an unsymmetrical shed shown in Figure 13.27(a) is subjected to an axial load $P_D = 220$ K and a moment $M_d = 180$ K·ft due to dead load and an axial load $P_L = 165$ K and a moment $M_L = 140$ K·ft due to live load. The base of the footing is 5 ft below final grade, and the allowable soil bearing pressure is 5 Ksf. Design the footing using $f'_c = 4$ Ksi and $f_y = 60$ Ksi.

Solution

The footing is subjected to an axial load and a moment

$$P = 220 + 165 = 385 \text{ K}$$

$$M = 180 + 140 = 320 \text{ K}\cdot\text{ft}$$

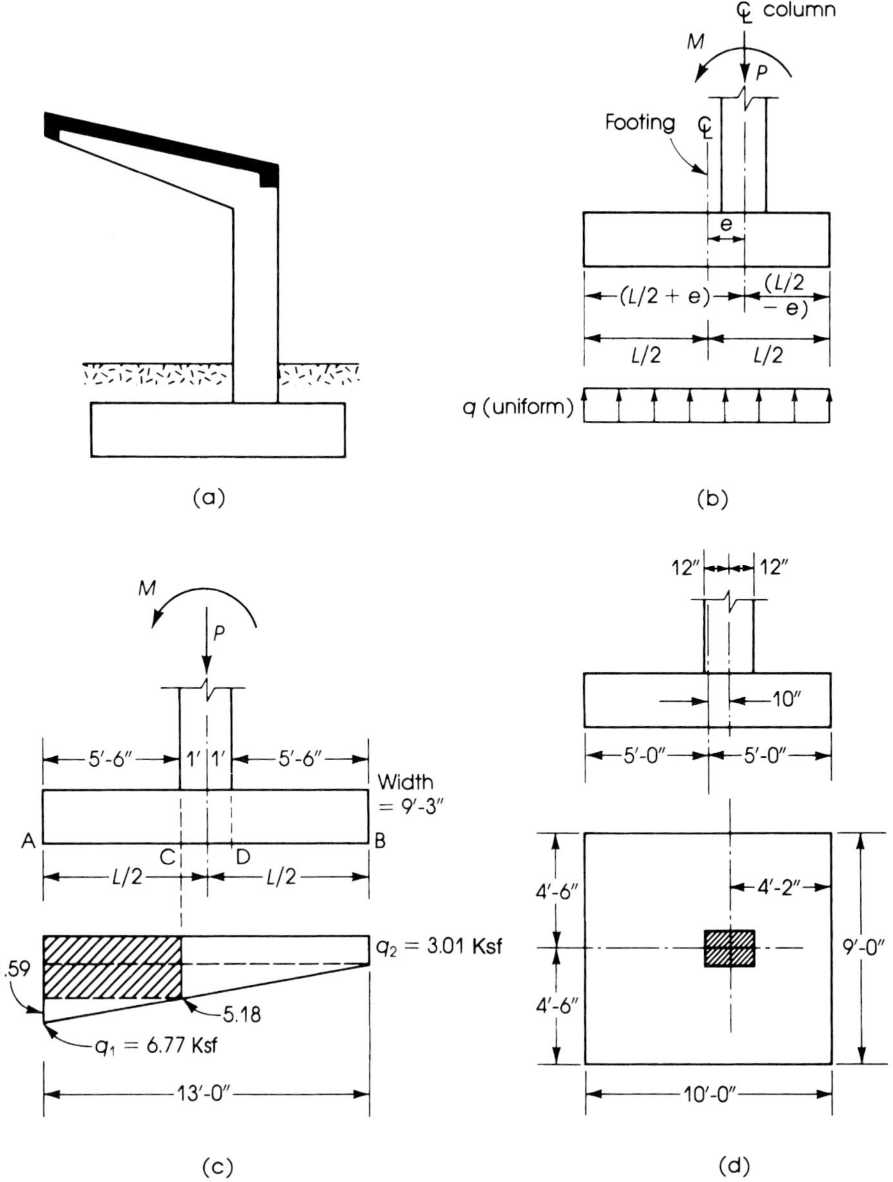

Figure 13.27 Example 13.7.

The eccentricity is

$$e = \frac{M}{P} = \frac{320 \times 12}{385} = 9.97 \text{ in.}, \qquad \text{say, 10 in.}$$

The footing may be designed by two methods.

Method 1: Move the center of the footing a distance $e = 10$ in. from the center of the column. In this case the soil pressure will be considered uniformly distributed under the footing (Figure 13.27(b)).

Method 2: The footing is placed concentric with the center of the column. In this case, the soil pressure will be trapezoidal or triangular (Figure 13.27(c)), and the maximum and minimum values can be calculated as shown in Figure 13.27.

The application of the two methods to Example 13.7 can be explained briefly as follows:

1. For the first method, assume a footing depth of 20 in. ($d = 16.5$ in.) and assume the weight of soil is 100 pcf. Net upward pressure is $5000 - \frac{20}{12} \times 150$ (footing) $- \left(5 - \frac{20}{12}\right) \times 100 = 4417$ psf.

$$\text{Area of footing} = \frac{385}{4.42} = 87.1 \text{ ft}^2$$

Assume a footing width of 9 ft; then the footing length is $87.1/9 = 9.7$ ft, say, 10 ft. Choose a footing 9 by 10 ft and place the column eccentrically, as shown in Figure 13.27(d). The center of the footing is 10 in. away from the center of the column.

2. The design procedure now is similar to that for a single footing. Check the depth for two-way and one-way shear action. Determine the bending moment at the face of the column for the longitudinal and transverse directions. Due to the eccentricity of the footing, the critical section will be on the left face of the column in Figure 13.27(d). The distance to the end of footing is $(5 \times 12) - 2 = 58$ in. $= 4.833$ ft.

$$P_u = 1.4D + 1.7L = 1.4 \times 200 + 1.7 \times 165 = 588.5 \text{ K}$$

$$q_u = \frac{588.5}{9 \times 10} = 6.54 \text{ Ksf}$$

$$\text{Maximum } M_u = (6.54 \times 9) \times \frac{(4.833)^2}{2} = 687.4 \text{ K}\cdot\text{ft} \qquad \text{(in 9-ft width)}$$

In the transverse direction,

$$M_u = (6.54 \times 10) \times \frac{(4)^2}{2} = 523.2 \text{ K}\cdot\text{ft}$$

Revise the assumed depth if needed and choose the required reinforcement in both directions of the footing, as was explained in the single-footing example.

3. For the second method, calculate the area of the footing in the same way as explained in the first method; then calculate the maximum soil pressure and compare it with that allowable using actual loads.

$$\text{Total load } P = 385 \text{ K}$$

$$\text{Size of footing} = 10 \times 9 \text{ ft}$$

Because the eccentricity is $e = 10$ in. $< L/6 = 10 \times \frac{12}{6} = 20$ in., the shape of the upward soil pressure is trapezoidal. Calculate the maximum and minimum soil pressure:

$$q_{max} = \frac{P}{LB} + \frac{6M}{BL^2} = \frac{385}{10 \times 9} + \frac{6 \times 320}{9(10)^2} = 6.42 \text{ Ksf} > 4.42 \text{ Ksf}$$

The footing is not safe. Try a footing 9.25 × 13 ft (area = 120.25 ft²).

$$q_{max} = \frac{385}{120.25} + \frac{6 \times 320}{9.25(13)^2} = 4.22 \text{ Ksf} < 4.42 \text{ Ksf}$$

$$q_{min} = 3.2 - 1.22 = 1.98 \text{ Ksf}$$

4. Calculate the ultimate upward pressure using factored loads; then calculate moments and shears, as explained in previous examples.

13.9 SLABS ON GROUND

A concrete slab laid directly on ground may be subjected to

1. Uniform load over its surface, producing small internal forces; or
2. Nonuniform or concentrated loads, producing some moments and shearing forces. Tensile stresses develop, and cracks will occur in some parts of the slab.

Tensile stresses are generally induced by a combination of

1. Contraction due to temperature and shrinkage, restricted by the friction between the slab and the subgrade, causing tensile stresses;
2. Warping of the slab;
3. Loading conditions;
4. Settlement.

Contraction joints may be formed to reduce the tensile stresses in the slab. Expansion joints may be provided in thin slabs up to a thickness of 10 in.

Basement floors in residential structures may be made of 4- to 6-in. concrete slabs reinforced in both directions with a wire fabric reinforcement. In warehouses, slabs may be 6 to 12 in. thick, depending on the loading on the slab. Reinforcement in both directions must be provided, usually in the form of wire fabric reinforcement. Basement floors are designed to resist upward earth pressure and any water pressure. If the slab rests on very stable or incompressible soils, then differential settlement is negligible. In this case the slab thickness will be a minimum if no water table exists. Columns in the basement will have independent footings. If there is any appreciable differential settlement, the floor slab must be designed as a stiff raft foundation.

13.10 FOOTINGS ON PILES

When the ground consists of soft material for a great depth, and its bearing capacity is very low, it is not advisable to place the footings directly on the soil. It may be better to transmit the loads through piles to a deep stratum that is strong enough to bear the loads or to develop sufficient friction around the surface of the piles.

Many different kinds of piles are used for foundations. The choice depends on ground conditions, presence of ground water, function of the pile, and cost. Piles may be made of concrete, steel, or timber.

In general, a pile cap (or footing) is necessary to distribute the load from a column to the heads of a number of piles. The cap should be of sufficient size to accommodate deviation in the position of the pile heads. The caps are designed as beams spanning between the pile heads and carrying concentrated loads from columns. When the column is supported by two piles, the cap may be designed as a reinforced concrete truss of a triangular shape.

The ACI Code, Section 15.2, indicates that computations for moments and shears for footings on piles may be based on the assumption that the reaction from any pile is concentrated at the pile center. The base area of the footing or number of piles shall be determined from the unfactored forces and moments.

The minimum concrete thickness above the reinforcement in a pile footing is limited to 12 in. (ACI Code, Section 15.7). For more design details of piles and pile footings, refer to books on foundation engineering.

13.11 SI EQUATIONS

1. One-way shear:

$$\phi V_c = \frac{2\phi\sqrt{f'_c}bd}{12} \tag{13.3}$$

2. Two-way shear:

$$V_{c1} = \frac{4\sqrt{f'_c}b_0 d}{12} \tag{13.6}$$

$$V_{c2} = \frac{(2 + 4/\beta_c)\sqrt{f'_c}b_0 d}{12} \tag{13.7}$$

$$V_{c3} = \frac{(2 + \alpha_s/b_0)\sqrt{f'_c}b_0 d}{12} \tag{13.8}$$

Other equations remain the same.

SUMMARY

Sections 13.1–13.4

1. *General:*

H = distance of the bottom of footing from final grade (ft)

h = total depth of footing (in.)

c = wall thickness (in.)

q_a = allowable soil pressure (Ksf)

q_e = effective soil pressure

W_s = weight of soil (pcf) (Assume 100 pcf if not given.)

2. *Design of wall footings:* The design steps can be summarized as follows.

a. Assume a total depth of footing h (in.). Consider 1-ft length of footing.

b. Calculate $q_e = q_a - (h/12)(150) - W_s(H - h/12)(q_a \text{in psf})$.

c. Calculate width of footing: $B = (\text{total service load})/q_e = (P_D + P_n)/q_e$. (Round to the nearest higher half foot.) The footing size is $(B \times 1)$ ft.

d. Calculate the ultimate upward pressure, $q_u = P_u/B$

$$P_u = 1.4P_D + 1.7P_L$$

e. Check the assumed depth for one-way shear requirements considering $d_a = (h - 3.5)$ in. (two-way shear does not apply).

$$V_u = q_u\left(\frac{B}{2} - d - \frac{c}{2}\right) \qquad \text{(Use kips.)}$$

$$\text{Required } d = \frac{V_u(1000)}{\phi(2\sqrt{f'_c})(12)} \geq d_a$$

f. Calculate the bending moment and main steel. The critical section is at the face of the wall.

1. $M_u = 0.5q_u(L/2 - c/2)^2$; get $R_u = M_u/bd^2$.

2. Determine ρ from tables in Appendix A or from equation (13.14).

3. $A_s = \rho bd = 12\rho d$ in.^{2}/ft; $A_s \geq A_s$ (min).

4. Minimum steel for shrinkage is

$$A_{sh} = 0.0018\text{ (bh) for } f_y = 60 \text{ Ksi} \quad \text{and}$$

$$A_{sh} = 0.0020\text{ (bh) for } f_y = 40, \quad \text{or} \quad 50 \text{ Ksi}$$

Minimum steel for flexure is

$$A_{sf} = \left(\frac{200}{f_y}\right)bd = \left(\frac{200}{f_y}\right)(12d) \qquad \text{when } f'_c < 4500 \text{ psi}$$

$$A_{sf} = \frac{(3\sqrt{f'_c})(12d)}{f_y} \qquad \text{when } f'_c > 4500 \text{ psi}$$

A_s calculated must be greater than A_{sh} (shrinkage). However, if $A_s < A_{sf}$, it is recommended to use $A_s = A_{sf}$ and then choose bars and spacings.

g. Check development length: Refer to Tables 7.1, 7.2, 7.3, and 7.4.

h. Calculate secondary reinforcement in the direction of the wall. $A_s = A_{sh}$ as calculated in Step 6(d) using $b = 12$ in. Choose bars and spacings.

3. *Design of square/rectangular footings:* The design steps are as follows.

a. Assume a total depth h (in.); let d_a (assumed) $= (h - 4.5)$ in. Calculate $q_e = q_a - (h/12)(150) - W_s(H - h/12)$ (use psf).

b. Calculate the area of the footing, $AF = (P_D + P_L)/q_e$. Choose either a square footing, side $= \sqrt{AF}$, or a rectangular footing of length L and width B (short length); then round dimensions to the higher half ft.

c. Calculate $q_u = P_u/(LB)$, where $P_u = 1.4P_D + P_L$.

d. Check footing depth due to two-way shear first. Maximum V_{u2} occurs at a section located at a distance equal to $d/2$ around the column.

1. Calculate $b_0 = 4(c + d)$ for square columns and $b_0 = 2(c_1 + d) + 2(c_2 + d)$ for rectangular columns.

$$V_{u2} = P_u - q_u(c + d)^2 \qquad \text{for square columns}$$

$$V_{u2} = P_u - q_u(c_1 + d)(c_2 + d) \qquad \text{for rectangular columns}$$

2. Calculate $d_1 = V_{u2}/4\phi\sqrt{f'_c}\,b_0$ when $\beta_c = L/B \le 2$.

$$d_1 = \frac{V_{u2}}{\phi(2 + 4/\beta_c)\sqrt{f'_c}\,b_0} \qquad \text{when } \beta_c > 2$$

3. Calculate

$$d_2 = \frac{V_{u2}}{\phi(\alpha_s d/b_0 + 2)\sqrt{f'_c}\,b_0}$$

Let d = the larger of d_1 and d_2. If d is less than d_a (assumed), increase d_a (or h) and repeat. The required d should be close to the assumed d_a (within 5% or 1 in. higher).

e. Check one-way shear (normally does not control in single footings):

1. $V_{u11} = q_u B(L/2 - c/2 - d)$ in the long direction (or for square footings).

$$d_{11} = \frac{V_{u1}}{2\phi\sqrt{f'_c}\,B}$$

2. $V_{u12} = q_u L(B/2 - c/2 - d)$ in the short direction.

$$d_{12} = \frac{V_{u12}}{2\phi\sqrt{f'_c}\,L} \qquad \text{(for rectangular footings)}$$

3. Let d be the larger of d_{11} and d_{12}; then use the larger d from Steps 4 and 5.
4. Determine $h = (d + 4.5)$ in.; round to the nearest higher inch.
5. Calculate the final $d = (h - 4.5)$ in.

f. Calculate the bending moment and the main steel in one direction only for square footings and two directions for rectangular footings.

1. In the long direction (or for square footings)

$$M_{uL} = 0.5q_u\left(\frac{L}{2} - \frac{c}{2}\right)^2, \qquad R_u = \frac{M_{uL}}{Bd^2}$$

2. In the short direction (for rectangular footings):

$$M_{us} = 0.5q_u\left(\frac{B}{2} - \frac{c}{2}\right)^2, \qquad R_{us} = \frac{M_{us}}{Ld^2}$$

3. Calculate the reinforcement in the long direction A_{sL} and in the short direction A_{ss} using equation (13.14).
4. Check that A_{sL} and A_{ss} are greater than the minimum steel reinforcement. Choose bars and spacings. For square footings, the same bars are used in both directions. Distribute bars in the bandwidth of rectangular columns according to equation (13.15).

g. Check bearing stress:

1. Calculate N_1 and N_2: $N_1 = \phi(0.85f'_c A_s)$, where $\phi = 0.7$ and A_1 = area of column section; $N_2 = N_1\sqrt{A_2/A_1} \le 2N_1$, where A_2 = square area of footing under column $(A_2 = B^2)$.
2. If $P_u \le N_1$, bearing stress is adequate. Minimum area of dowels is $0.005A_1$. Choose four bars to be placed at the four corners of column section.

3. If $P_u > N_1$, determine the excess load, $P_{ex} = (P_u - N_1)$, and then calculate A_{sd} (dowels) $= P_{ex}/f_y$. A_{sd} must be equal to or greater than $0.005A_1$. Choose at least four dowel bars.

4. Determine the development length in compression for dowels in the column and in the footing.

h. Calculate the development lengths, l_d, of the main bars in the footings. The calculated l_d must be greater than or equal to l_d provided in the footing. Provided l_d is $(L/2 - c/2 - 3)$ in. in the long direction and $l_d = (B/2 - c/2 - 3)$ in the short direction. Examples 13.2 and 13.3 explain these steps.

Section 13.5

Plain concrete may be used to support walls. The maximum flexural stress in tension should be calculated and must be less than the allowable stress of $5\phi\sqrt{f'_c}$.

Section 13.6

A combined footing is used when a column is located near a property line. Design of such footings is explained in Example 13.6.

Sections 13.7 to 13.9

Footings under eccentric column loads are explained in Example 13.7.

REFERENCES

1. R. W. Furlong. "Design Aids for Square Footings." *ACI Journal* 62 (March 1965).
2. R. Peck, W. Hanson, and T. Thornburn. *Foundation Engineering,* 2d ed. New York: John Wiley, 1974.
3. ACI Committee 436. "Suggested Design Procedures for Combined Footings and Mats." *ACI Journal* 63 (October 1966).
4. F. Kramrisch and P. Rogers. "Simplified Design of Combined Footings." *Journal of Soil Mechanics and Foundation Division, ASCE* 85 (October 1961).
5. L. S. Blake (ed). *Civil Engineer's Reference Book,* 3d ed. London: Butterworths, 1975.
6. American Concrete Institute. "Building Code Requirements for Structural Concrete." *ACI Code* 318–99. Detroit, Michigan, 1999.

PROBLEMS

For all problems in this chapter, use the following:

H_a = distance from bottom of footing to the final grade

h = depth of concrete footing

q_a = allowable soil pressure in Ksf

Assume the weight of the soil is 100 pcf and $f_y = 60$ Ksi.

13.1 For each problem in Table 13.1, design a wall footing to support the given reinforced concrete wall loads. Design for shear and moment; check the development-length requirements. Also, determine the footing bars and their distribution. (Assume $d = h - 3.5$ in.)

Table 13.1

Number	Wall Thickness (in.)	Dead Load (K/ft)	Live Load (K/ft)	f'_c (Ksi)	q_a (Ksf)	H (ft)	Part Answers L (ft)	h (in.)
(a)	12	22	12	3	4	5	10	19
(b)	12	18	14	3	5	4	7.5	17
(c)	14	28	16	3	6	6	8.5	20
(d)	14	26	24	3	4	5	14.5	27
(e)	16	32	16	3	5	5	11	23
(f)	16	24	20	4	6	8	9	19
(g)	14	20	18	4	4	6	11.5	19
(h)	14	28	20	4	5	4	10.5	21
(i)	12	18	14	4	6	5	6	14
(j)	14	16	20	4	6	5	7	16

13.2 For each problem in Table 13.2, design a square single footing to support the given square and round column loads. Design for moments, shear, load transfer, dowel length, and development lengths for footing main bars. Choose adequate bars and spacings. (Assume $d = h - 4.5$ in. for all problems.)

Table 13.2

Number	Column (in.)	Column bars	Dead Load (K)	Live Load (K)	f'_c (Ksi)	q_a (Ksf)	H (ft)	Part Answers L (ft)	h (in.)
(a)	16 × 16	8 no. 8	150	115	3	5	6	8	20
(b)	18 × 18	8 no. 9	160	100	3	6	5	7.5	19
(c)	20 × 20	12 no. 9	245	159	3	6	7	9	23
(d)	12 × 12	8 no. 8	180	140	3	5	8	9	24
(e)	14 × 14	8 no. 9	140	160	4	5	6	8.5	21
(f)	16 × 16	8 no. 9	190	140	4	4	5	10	21
(g)	18 × 18	12 no. 8	200	120	4	6	7	8	20
(h)	20 × 20	12 no. 9	195	195	4	5	8	10	22
(i)	Dia. 20	8 no. 9	120	85	4	5	5	7	16
(j)	Dia. 16	8 no. 8	110	90	3	4	6	8	18

13.3 Repeat Problems 13.2(a)–(h) using rectangular footings with widths of 6, 6, 8, 8, 7, 8, 6, and 9 ft, respectively.

13.4 Repeat Problems 13.2(a)–(d) using rectangular columns of 14 × 20 in., 16 × 20 in., 16 × 24 in., and 12 × 18 in., respectively, and rectangular footings with the length equal to about 1.5 times the width.

13.5 Repeat Problems 13.1(a)–(d) using plain concrete wall footings and one-half the applied dead and live loads.

13.6 Design a rectangular combined footing to support the two columns shown in Figure 13.28. The center of the exterior column is 1 ft away from the property line and 14 ft from the center of the interior column. The exterior column is square with 18-in. sides, is reinforced with no. 8 bars, and carries an axial dead load of 160 K and a live load of 140 K. The interior column is square with 20-in. sides, is reinforced with no. 9 bars, and carries an axial dead load of 240 K and a live load of 150 K. The bottom of the footing is 5 ft below final grade. Use $f'_c = 4$ Ksi, $f_y = 50$ Ksi, and an allowable soil pressure of 5 Ksf.

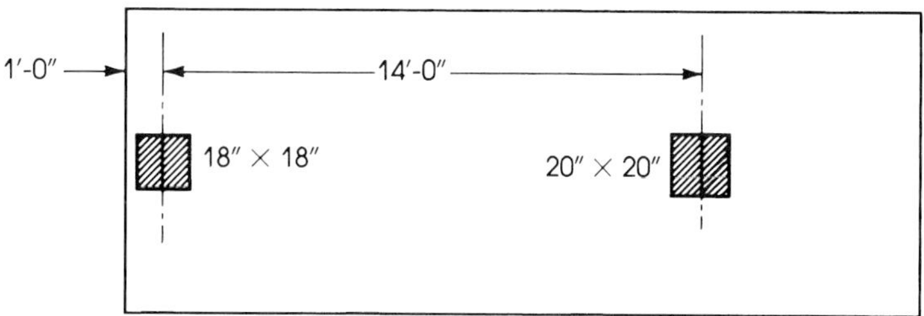

Figure 13.28 Problem 13.6.

13.7 Determine the footing areas required for a balanced footing design (equal settlement approach) if the usual load is 25% for all footings. The allowable soil pressure is 5 Ksi and the dead and live loads are given in Table 13.3.

Table 13.3

	Footing no.					
	1	**2**	**3**	**4**	**5**	**6**
Dead load	130 K	220 K	150 K	180 K	200 K	240 K
Live load	160 K	220 K	210 K	180 K	220 K	200 K

13.8 The 12- by 20-in. (300- by 500-mm) column of the frame shown in Figure 13.29 is subjected to an axial load $P_D = 200$ K and a moment $M_D = 120$ K·ft due to dead load and an axial load $P_L = 160$ K and a moment $M_L = 110$ K·ft due to live load. The base of the footing is 4 ft below final grade. Design the footing using $f'_c = 4$ Ksi, $f_y = 40$ Ksi, and an allowable soil pressure of 4 Ksi. Use a uniform pressure and eccentric footing approach.

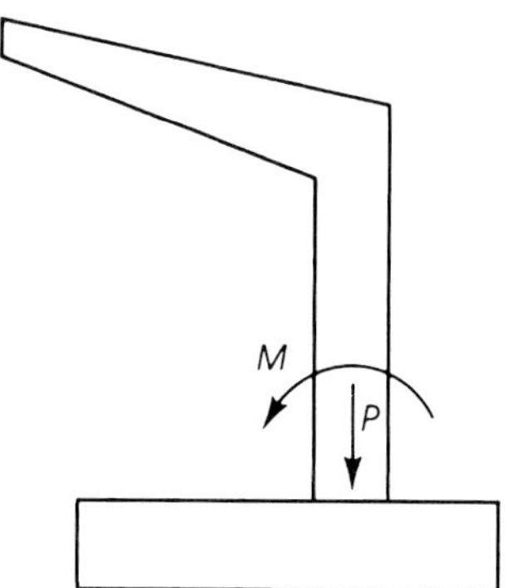

Figure 13.29 Problem 13.8.

13.9 Repeat Problem 13.8 if both the column and the footing have the same centerline (concentric case).

13.10. Determine the size of a square or round footing for the case of Problem 13.9, assuming that the loads and moments on the footing are for a rotating crane fixed at its base.

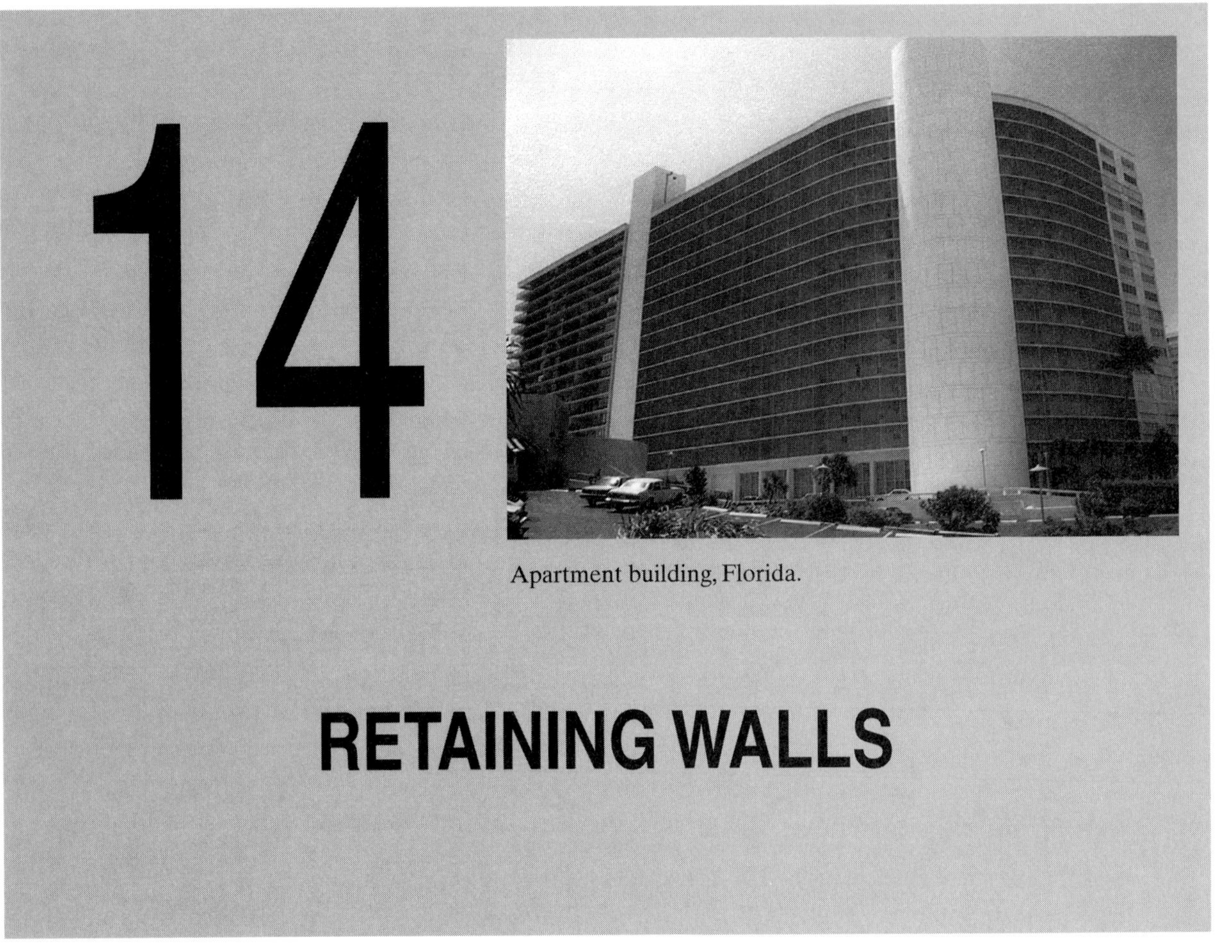

Apartment building, Florida.

14 RETAINING WALLS

14.1 INTRODUCTION

Retaining walls are structural members used to provide stability for soil or other materials and to prevent them from assuming their natural slope. In this sense, the retaining wall maintains unequal levels of earth on its two faces. The retained material on the higher level exerts a force on the retaining wall that may cause its overturning or failure. Retaining walls are used in bridges as abutments, in buildings as basement walls, and in embankments. They are also used to retain liquids, as in water tanks and sewage-treatment tanks.

14.2 TYPES OF RETAINING WALLS

Retaining walls may be classified as follows (refer to Figure 14.1):

1. *Gravity walls* usually consist of plain concrete or masonry and depend entirely on their own weight to provide stability against the thrust of the retained material. These walls are proportioned so that tensile stresses do not develop in the concrete or masonry due to the exerted forces on the wall. The practical height of a gravity wall does not exceed 10 ft.

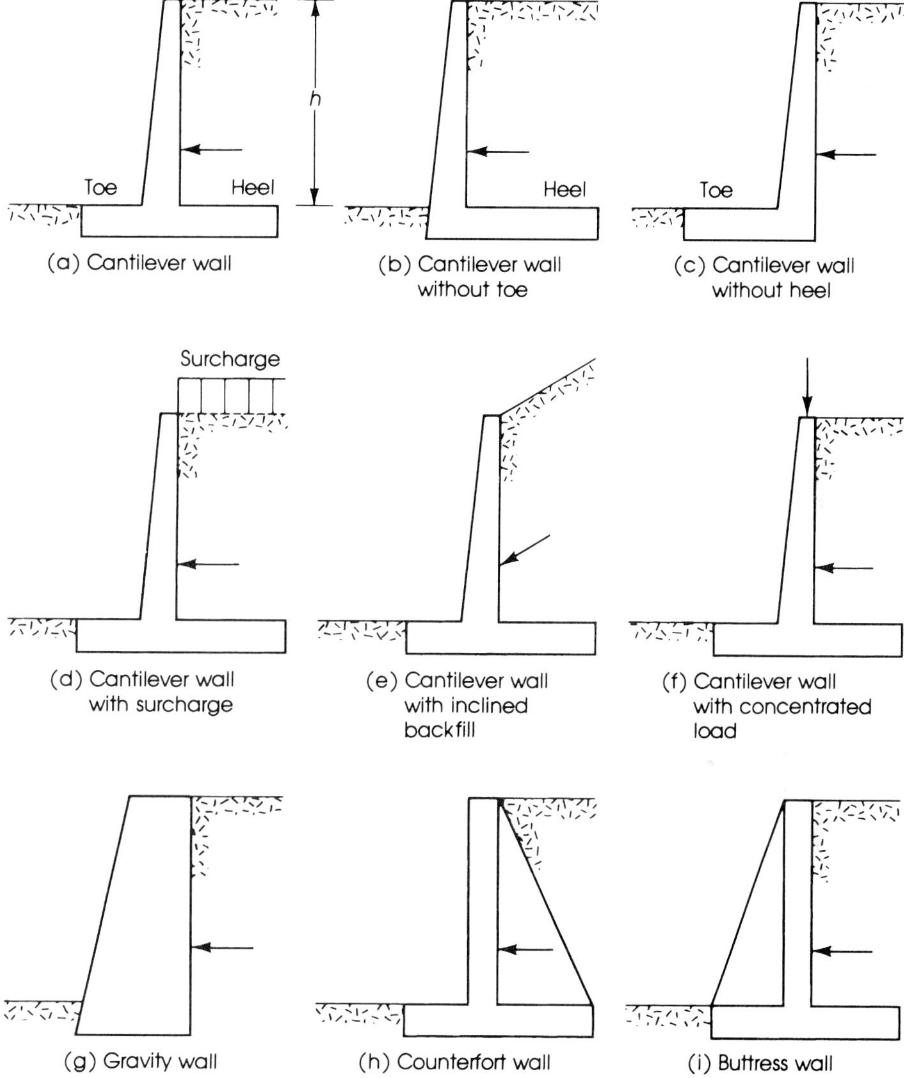

Figure 14.1 Types of retaining walls.

2. *Semigravity walls* are gravity walls that have a wider base to improve the stability of the wall and to prevent the development of tensile stresses in the base. Light reinforcement is sometimes used in the base or stem to reduce the large section of the wall.
3. *The cantilever retaining wall* is a reinforced concrete wall that is generally used for heights from 8 to 20 ft. It is the most common type of retaining structure because of economy and simplicity of construction. Various types of cantilever retaining walls are shown in Figure 14.1.
4. *Counterfort retaining walls* higher than 20 ft develop a relatively large bending moment at the base of the stem, which makes the design of such walls uneconomical. One solution in this case is to introduce transverse walls (or counterforts) that tie the stem and the base together at intervals. The counterforts act as tension ties supporting the vertical walls. Economy is achieved because the stem is designed as a continuous slab spanning horizontally between counterforts, whereas the heel is designed as a slab supported on three sides (Figure 14.1(h)).

5. *The buttressed retaining wall* is similar to the counterfort wall, but in this case the transverse walls are located on the opposite, visible side of the stem and act in compression (Figure 14.1(i)). The design of such walls becomes economical for heights greater than 20 ft. They are not popular because of the exposed buttresses.
6. *Bridge abutments* are retaining walls that are supported at the top by the bridge deck. The wall may be assumed fixed at the base and simply supported at the top.
7. *Basement walls* resist earth pressure from one side of the wall and span vertically from the basement-floor slab to the first-floor slab. The wall may be assumed fixed at the base and simply supported or partially restrained at the top.

14.3 FORCES ON RETAINING WALLS

Retaining walls are generally subjected to gravity loads and to earth pressure due to the retained material on the wall. Gravity loads due to the weights of the materials are well defined and can be calculated easily and directly. The magnitude and direction of the earth pressure on a retaining wall depends on the type and condition of soil retained and on other factors and cannot be determined as accurately as gravity loads. Several references on soil mechanics [1], [2] explain the theories and procedure for determining the soil pressure on retaining walls. The stability of retaining walls and the effect of dynamic reaction on walls are discussed in [3] and [4].

Granular materials, such as sand, behave differently from cohesive materials, such as clay, or from any combination of both types of soils. Although the pressure intensity of soil on a retaining wall is complex, it is common to assume a linear pressure distribution on the wall. The pressure intensity increases with depth linearly, and its value is a function of the height of the wall and the weight and type of soil. The pressure intensity, p, at a depth, h, below the earth's surface may be calculated as follows:

$$p = Cwh \tag{14.1}$$

where w is the unit weight of soil and C is a coefficient that depends on the physical properties of soil. The value of the coefficient C varies from 0.3 for loose granular soil, such as sand, to about 1.0 for cohesive soil, such as wet clay. If the retaining wall is assumed absolutely rigid, a case of earth pressure at rest develops. Under soil pressure, the wall may deflect or move a small amount from the earth, and active soil pressure develops, as shown in Figure 14.2. If the wall moves toward the soil, a passive soil pressure develops. Both the active and passive soil pressures are assumed to vary linearly with the depth of wall (Figure 14.2). For dry, granular, noncohesive materials, the assumed linear pressure diagram is fairly satisfactory; cohesive soils or saturated sands behave in a different, nonlinear manner. Therefore, it is very common to use granular materials as backfill to provide

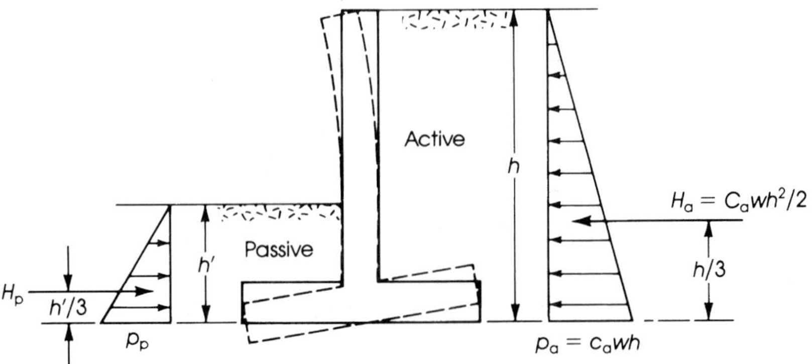

Figure 14.2 Active and passive earth pressure.

an approximately linear pressure diagram and also to provide for the release or drainage of water from behind the wall.

For a linear pressure, the active and passive pressure intensities are determined as follows:

$$P_a = C_a wh \tag{14.2}$$

$$P_p = C_p wh \tag{14.3}$$

where C_a and C_p are the approximate coefficients of the active and passive pressures, respectively.

14.4 ACTIVE AND PASSIVE SOIL PRESSURES

The two theories most commonly used in the calculation of earth pressure are those of Rankine and Coulomb [1], [6].

1. In Rankine's approach, the retaining wall is assumed to yield a sufficient amount to develop a state of plastic equilibrium in the soil mass at the wall surface. The rest of the soil remains in the state of elastic equilibrium. The theory applies mainly to a homogeneous, incompressible, cohesionless soil and neglects the friction between soil and wall. The active soil pressure at a depth h on a retaining wall with a horizontal backfill based on Rankine's theory is determined as follows:

$$P_a = C_a wh = wh\left(\frac{1 - \sin\phi}{1 + \sin\phi}\right) \tag{14.4}$$

where

$$C_a = \left(\frac{1 - \sin\phi}{1 + \sin\phi}\right)$$

$$\phi = \text{angle of internal friction of the soil (Table 14.3)}$$

$$\text{Total active pressure } H_a = \frac{wh^2}{2}\left(\frac{1 - \sin\phi}{1 + \sin\phi}\right) \tag{14.5}$$

The resultant, H_a, acts at $h/3$ from the base (Figure 14.2). When the earth is surcharged at an angle δ to the horizontal, then

$$C_a = \cos\delta\left(\frac{\cos\delta - \sqrt{\cos^2\delta - \cos^2\phi}}{\cos\delta + \sqrt{\cos^2\delta - \cos^2\phi}}\right) \tag{14.6}$$

$$P_a = C_a wh \quad \text{and} \quad H_a = C_a\frac{wh^2}{2}$$

The resultant, H_a, acts at $h/3$ and is inclined at an angle δ to the horizontal (Figure 14.3). The values of C_a expressed by equation (14.6) for different values of δ and ϕ are shown in Table 14.1.

Table 14.1 Values of C_a

	ϕ (Angle of Internal Friction)						
δ	28°	30°	32°	34°	36°	38°	40°
0°	0.361	0.333	0.307	0.283	0.260	0.238	0.217
10°	0.380	0.350	0.321	0.294	0.270	0.246	0.225
20°	0.461	0.414	0.374	0.338	0.306	0.277	0.250
25°	0.573	0.494	0.434	0.385	0.343	0.307	0.275
30°	0	0.866	0.574	0.478	0.411	0.358	0.315

Reinforced concrete retaining wall.

Retaining wall in a parking area.

Passive soil pressure develops when the retaining wall moves against and compresses the soil. The passive soil pressure at a depth h on a retaining wall with horizontal backfill is determined as follows:

$$P_p = C_p wh = wh\left(\frac{1 + \sin\phi}{1 - \sin\phi}\right) \tag{14.7}$$

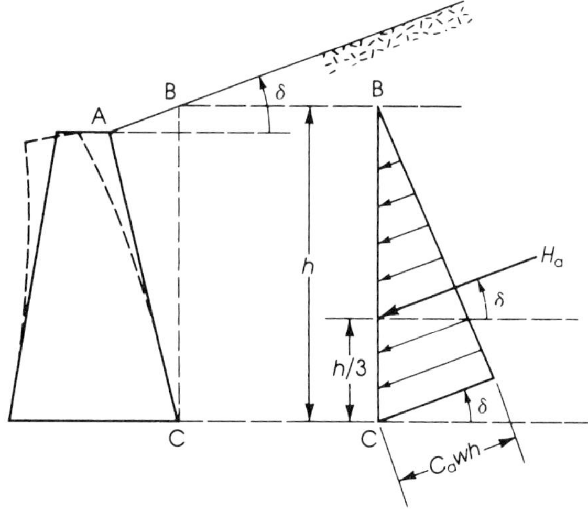

Figure 14.3 Active soil pressure with surcharge.

where

$$C_p = \left(\frac{1 + \sin\phi}{1 - \sin\phi}\right) = \frac{1}{C_a}$$

Total passive pressure is

$$H_p = \frac{wh^2}{2}\left(\frac{1 + \sin\phi}{1 - \sin\phi}\right) \tag{14.8}$$

The resultant, H_p, acts at $h'/3$ from the base (Figure 14.2). When the earth is surcharged at an angle δ to the horizontal, then

$$C_p = \cos\delta\left(\frac{\cos\delta + \sqrt{\cos^2\delta - \cos^2\phi}}{\cos\delta - \sqrt{\cos^2\delta - \cos^2\phi}}\right) \tag{14.9}$$

$$P_p = C_p wh \quad \text{and} \quad H_p = C_p \frac{wh^2}{2}$$

H_p acts at $h'/3$ and is inclined at an angle δ to the horizontal (Figure 14.4). The values of C_p expressed by equation (14.9) for different values of δ and ϕ are shown in Table 14.2.

Table 14.2 Values of C_p

	ϕ (Angle of Internal Friction)						
δ	28°	30°	32°	34°	36°	38°	40°
0°	2.77	3.00	3.25	3.54	3.85	4.20	4.60
10°	2.55	2.78	3.02	3.30	3.60	3.94	4.32
20°	1.92	2.13	2.36	2.61	2.89	3.19	3.53
25°	1.43	1.66	1.90	2.14	2.40	2.68	3.00
30°	0	0.87	1.31	1.57	1.83	2.10	2.38

The values of ϕ and w vary with the type of backfill used. As a guide, common values of ϕ and w are given in Table 14.3.

2. In Coulomb's theory, the active soil pressure is assumed to be the result of the tendency of a wedge of soil to slide against the surface of a retaining wall. Hence

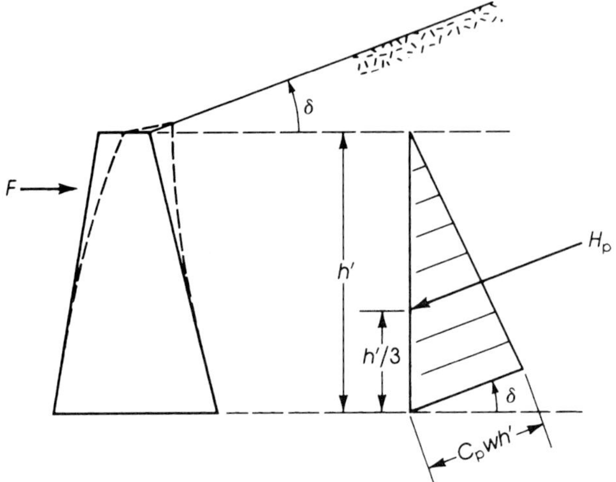

Figure 14.4 Passive soil pressure with surcharge.

Table 14.3 Values of w and ϕ

Type of Backfill	Unit Weight, W pcf	Unit Weight, W kg/m³	Angle of Internal Friction ϕ
Soft clay	90–120	1440–1920	0°–15°
Medium clay	100–120	1600–1920	15°–30°
Dry loose silt	100–120	1600–1920	27°–30°
Dry dense silt	110–120	1760–1920	30°–35°
Loose sand and gravel	100–130	1600–2100	30°–40°
Dense sand and gravel	120–130	1920–2100	25°–35°
Dry loose sand, graded	115–130	1840–2100	33°–35°
Dry dense sand, graded	120–130	1920–2100	42°–46°

Coulomb's theory is referred to as the wedge theory. While it takes into consideration the friction of the soil on the retaining wall, it assumes that the surface of sliding is a plane, whereas in reality it is slightly curved. The error in this assumption is negligible in calculating the active soil pressure. Coulomb's equations to calculate the active and passive soil pressure are as follows:

The active soil pressure is

$$P_a = C_a wh$$

where

$$C_a = \frac{\cos^2(\phi - \theta)}{\cos^2\theta\cos(\theta + \beta)\left[1 + \sqrt{\dfrac{\sin(\phi + \beta)\sin(\phi - \delta)}{\cos(\theta + \beta)\cos(\theta - \delta)}}\right]^2} \tag{14.10a}$$

where

ϕ = angle of internal friction of soil

θ = angle of the soil pressure surface from the vertical

β = angle of friction along the wall surface (angle between soil and concrete)

δ = angle of surcharge to the horizontal

The total active soil pressure

$$H_a = C_a \frac{wh^2}{2} = p_a \frac{h}{2}$$

When the wall surface is vertical, $\theta = 0°$, and if $\beta = \delta$, then C_a in equation (14.10a) reduces to equation (14.6) of Rankine.

Passive soil pressure is

$$P_p = C_p wh' \quad \text{and} \quad H_p = \left(\frac{wh'^2}{2}\right) C_p = P_p \frac{h'}{2}$$

where

$$C_p = \frac{\cos^2(\phi + \theta)}{\cos^2\theta \cos(\theta - \beta)\left[1 - \sqrt{\dfrac{\sin(\phi + \beta)\sin(\phi + \delta)}{\cos(\theta - \beta)\cos(\phi - \delta)}}\right]^2} \tag{14.10b}$$

The values of ϕ and w vary with the type of backfill used. As a guide, common values of ϕ and w are given in Table 14.3.

3. When the soil is saturated, the pores of the permeable soil are filled with water, which exerts hydrostatic pressure. In this case the buoyed unit weight of soil must be used. The buoyed unit weight (or submerged unit weight) is a reduced unit weight of soil and equals w minus the weight of water displaced by the soil. The effect of the hydrostatic water pressure must be included in the design of retaining walls subjected to a high water table and submerged soil. The value of the angle of internal friction may be used, as shown in Table 14.3.

14.5 EFFECT OF SURCHARGE

Different types of loads are often imposed on the surface of the backfill behind a retaining wall. If the load is uniform, an equivalent height of soil, h_s, may be assumed acting on the wall to account for the increased pressure. For the wall shown in Figure 14.5, the horizontal pressure due to the surcharge is constant throughout the depth of the retaining wall.

$$h_s = \frac{w_s}{w} \tag{14.11}$$

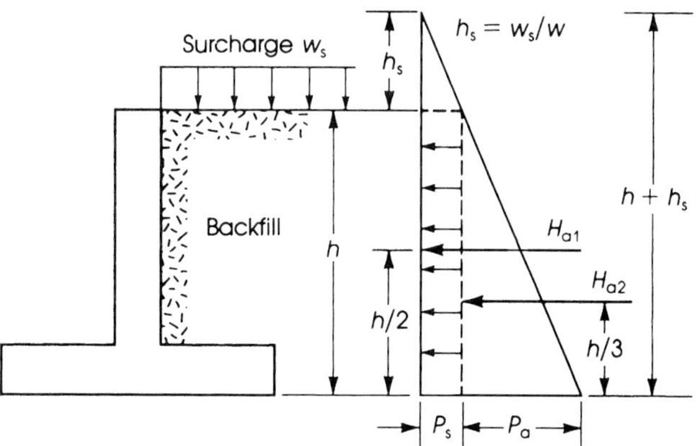

Figure 14.5 Surcharge effect under a uniform load.

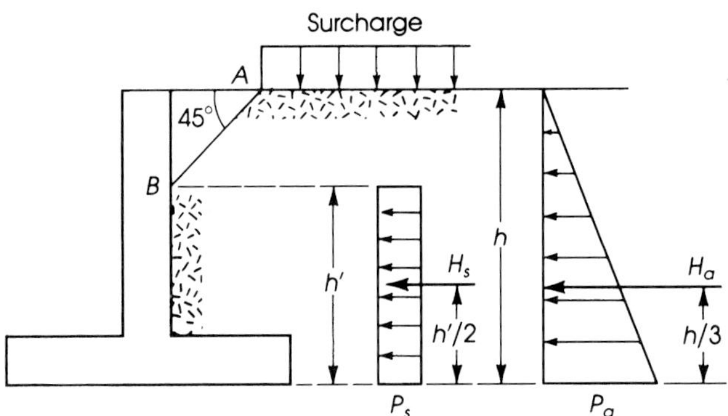

Figure 14.6 Surcharge effect under a partial uniform load at a distance from the wall.

where h_s = equivalent height of soil (ft)
w_s = pressure of the surcharge (psf)
w = unit weight of soil (pcf)

The total pressure is

$$H_a = H_{a1} + H_{a2} = C_a w\left(\frac{h^2}{2} + hh_s\right) \tag{14.12}$$

In the case of a partial uniform load acting at a distance from the wall, only a portion of the total surcharge pressure affects the wall (Figure 14.6).

It is a common practice to assume that the effective height of pressure due to partial surcharge is h', measured from point B to the base of the retaining wall [1]. The line AB forms an angle of 45° with the horizontal.

In the case of a wheel load acting at a distance from the wall, the load is to be distributed over a specific area, which is usually defined by known specifications such as AASHTO and AREA [4] specifications.

14.6 FRICTION ON THE RETAINING WALL BASE

The horizontal component of all forces acting on a retaining wall tends to push the wall in a horizontal direction. The retaining wall base must be wide enough to resist the sliding of the wall. The coefficient of friction to be used is that of soil on concrete for coarse granular soils and the shear strength of cohesive soils [4]. The coefficients of friction μ that may be adopted for different types of soil are as follows:

- Coarse-grained soils without silt, $\mu = 0.55$
- Coarse-grained soils with silt, $\mu = 0.45$
- Silt, $\mu = 0.35$
- Sound rock, $\mu = 0.60$

The total frictional force F on the base to resist the sliding effect is

$$F = \mu R + H_p \tag{14.13}$$

where μ = the coefficient of friction
R = the vertical force acting on the base
H_p = passive resisting force

The factor of safety against sliding is

$$\text{Factor of safety} = \frac{F}{H_{ah}} = \frac{\mu R + H_p}{H_{ah}} \geq 1.5 \tag{14.14}$$

where H_{ah} is the horizontal component of the active pressure H_a. The factor of safety against sliding should not be less than 1.5 if the passive resistance H_p is neglected and should not be less than 2.0 if H_p is taken into consideration.

14.7 STABILITY AGAINST OVERTURNING

The horizontal component of the active pressure H_a tends to overturn the retaining wall about the point O on the toe (Figure 14.7). The overturning moment is equal to $M_o = H_a h/3$. The weight of the concrete and soil tends to develop a balancing moment, or rightening moment, to resist the overturning moment. The balancing moment for the case of the wall shown in Figure 14.7 is equal to

$$M_b = w_1 x_1 + w_2 x_2 + w_3 x_3 = \Sigma wx$$

The factor of safety against overturning is

$$\text{Factor of safety} = \frac{M_b}{M_o} = \frac{\Sigma wx}{Hh/3} \geq 2.0 \tag{14.15}$$

This factor of safety should not be less than 2.0.

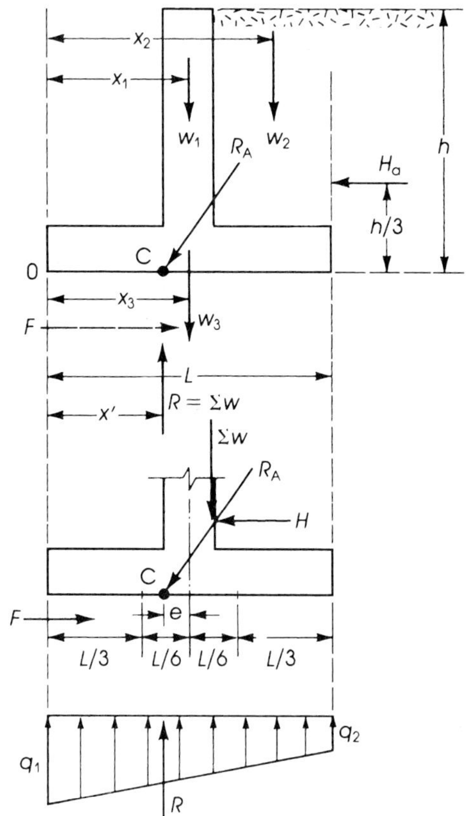

Figure 14.7 Overturning of a cantilever retaining wall.

The resultant of all forces acting on the retaining wall, R_A, intersects the base at point C (Figure 14.7). In general, point C does not coincide with the center of the base, L, thus causing eccentric loading on the footing. It is desirable to keep point C within the middle third to get the whole footing under soil pressure. (The case of a footing under eccentric load was discussed in Chapter 13.)

14.8 PROPORTIONS OF RETAINING WALLS

The design of a retaining wall begins with a trial section and approximate dimensions. The assumed section is then checked for stability and structural adequacy. The following rules may be used to determine the approximate sizes of the different parts of a cantilever retaining wall.

1. *Height of the wall:* The overall height of the wall is equal to the difference in elevation required plus 3 to 4 ft, which is the estimated frost penetration depth in northern states.
2. *Thickness of the stem:* The intensity of the pressure increases with the depth of the stem and reaches its maximum value at the base level. Consequently the maximum bending moment and shear in the stem occur at its base. The stem base thickness may be estimated as $\frac{1}{12}$ to $\frac{1}{10}$ of the height h. The thickness at the top of the stem may be assumed to be 8 to 12 in. Because retaining walls are designed for active earth pressure, causing a small deflection of the wall, it is advisable to provide the face of the wall with a batter (taper) of $\frac{1}{4}$ in. per foot of height h to compensate for the forward deflection. For short walls up to 10 ft high, a constant thickness may be adopted.
3. *Length of the base:* An initial estimate for the length of the base of $\frac{2}{5}$ to $\frac{2}{3}$ of the wall height, h, may be adopted.
4. *Thickness of the base:* The base thickness below the stem is estimated as the same thickness of the stem at its base, that is, $\frac{1}{12}$ to $\frac{1}{10}$ of the wall height. A minimum thickness of about 12 in. is recommended. The wall base may be of uniform thickness or tapered to the ends of the toe and heel, where the bending moment is zero.

The approximate initial proportions of a cantilever retaining wall are shown in Figure 14.8.

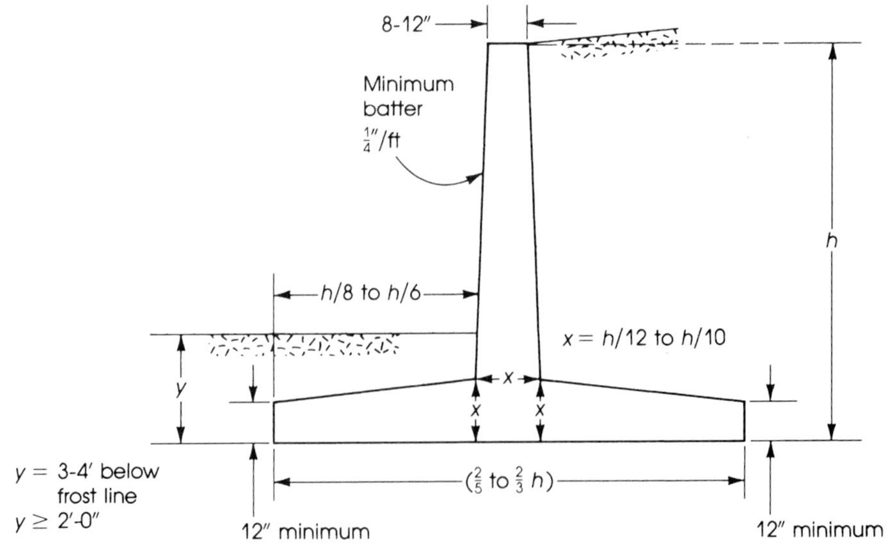

Figure 14.8 Trial proportions of a cantilever retaining wall.

14.9 DESIGN REQUIREMENTS

The ACI Code, Chapter 14, provides methods for bearing wall design. The main requirements are as follows:

1. The minimum thickness of bearing walls is $\frac{1}{25}$ the supported height or length, whichever is shorter, but not less than 4 in.
2. The minimum area of the horizontal reinforcement in the wall is $0.0025bh$, where bh is the gross concrete wall area. This value may be reduced to $0.0020bh$ if no. 5 or smaller deformed bars with $f_y \geq 60$ Ksi are used. For welded wire fabric (plain or deformed), the minimum steel area is $0.0020bh$.
3. The minimum area of the vertical reinforcement is $0.0015bh$, but it may be reduced to $0.0012bh$ if no. 5 or smaller deformed bars with $f_y \geq 60$ Ksi are used. For welded wire fabric (plain or deformed), the minimum steel area is $0.0012bh$.
4. The maximum spacing of the vertical or the horizontal reinforcing bars is the smaller of 18 in. or three times the wall thickness.
5. If the wall thickness exceeds 10 in., the vertical and horizontal reinforcement should be placed in two layers parallel to the exterior and interior wall surfaces, as follows:
   a. For exterior wall surfaces, at least $\frac{1}{2}$ of the reinforcement A_s (but not more than $\frac{2}{3}A_s$) should have a minimum concrete cover of 2 in. but not more than $\frac{1}{3}$ of the wall thickness. This is because the exterior surface of the wall is normally exposed to different weather conditions and temperature changes.
   b. For interior wall surfaces, the balance of the required reinforcement in each direction should have a minimum concrete cover of $\frac{3}{4}$ in. but not more than $\frac{1}{3}$ of the wall thickness.
   c. The minimum steel area in the wall footing (heel or toe), according to ACI Code, Section 10.5.3, is that required for shrinkage and temperature reinforcement, which is $0.0018bh$ when $f_y = 60$ Ksi and $0.0020bh$ when $f_y = 40$ Ksi or 50 Ksi. Because this minimum steel area is relatively small, it is a common practice to increase it to that minimum A_s required for flexure:

$$A_{s\,\min} = \left(\frac{3\sqrt{f'_c}}{f_y}\right)bd \geq \left(\frac{200}{f_y}\right)bd \qquad (14.16)$$

14.10 DRAINAGE

The earth pressure discussed in the previous sections does not include any hydrostatic pressure. If water accumulates behind the retaining wall, the water pressure must be included in the design. Surface or underground water may seep into the backfill and develop the case of submerged soil. To avoid hydrostatic pressure, drainage should be provided behind the wall. If well-drained cohesionless soil is used as a backfill, the wall can be designed for earth pressure only. The drainage system may consist of one or a combination of the following:

1. Weep holes in the retaining wall of 4 in. or more in diameter and spaced about 5 ft on centers horizontally and vertically (Figure 14.9(a)).
2. Perforated pipe 8 in. in diameter laid along the base of the wall and surrounded by gravel (Figure 14.9(b)).
3. Blanketing or paving the surface of the backfill with asphalt to prevent seepage of water from the surface.
4. Any other method to drain surface water.

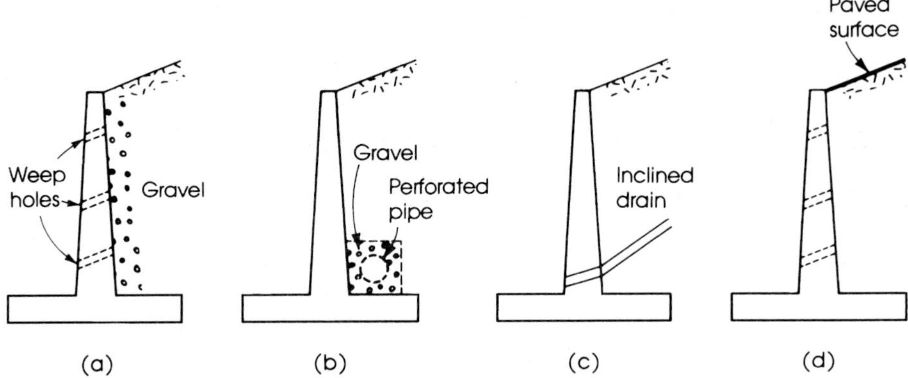

Figure 14.9 Drainage systems.

Example 14.1

The trial section of a semigravity plain concrete retaining wall is shown in Figure 14.10. It is required to check the safety of the wall against overturning, sliding, and bearing pressure under the footing. Given: Weight of backfill is 110 pcf, angle of internal friction is $\phi = 35°$, coefficient of friction between concrete and soil is $\mu = 0.5$, allowable soil pressure is 2.5 Ksf, and $f'_c = 3$ Ksi.

Solution

1. Using the Rankine equation,

$$C_a = \frac{1 - \sin\phi}{1 + \sin\phi} = \frac{1 - 0.574}{1 + 0.574} = 0.271$$

The passive pressure on the toe is that for a height of 1 ft, which is small and can be neglected.

$$H_a = \frac{C_a wh^2}{2} = \frac{0.271}{2}(110)(11)^2 = 1804 \text{ lb}$$

H_a acts at a distance $h/3 = \frac{11}{3} = 3.67$ ft from the bottom of the base.

2. The overturning moment is $M_o = 1.804 \times 3.67 = 6.62$ K · ft.
3. Calculate the balancing moment, M_b, taken about the toe end O (Figure 14.10):

Weight (lb)	Arm (ft)	Moment (K · ft)
$w_1 = 1 \times 10 \times 145 = 1450$	1.25	1.81
$w_2 = \frac{1}{2} \times 2.5 \times 10 \times 145 = 1812$	2.60	4.71
$w_3 = 5.25 \times 1 \times 145 = 725$	2.625	2.00
$w_4 = \frac{1}{2} \times 2.5 \times 10 \times 110 = 1375$	3.42	4.70
$w_5 = \frac{12}{12} \times 10 \times 110 = 1100$	4.75	5.22

$$\Sigma w = R = 6.50 \text{ K} \qquad M_b = \Sigma M = 18.44 \text{ K}\cdot\text{ft}$$

4. The factor of safety against overturning is $18.44/6.50 = 2.82 > 2.0$.
5. The force resisting sliding, $F = \mu R$, is $F = 0.5(6.46) = 3.25$ K. The factor of safety against sliding is $F/H_a = 3.25/1.804 = 1.8 > 1.5$.
6. Calculate the soil pressure under the base:
   a. The distance of the resultant from toe end O is

$$x = \frac{M_b - M_o}{R} = \frac{18.44 - 6.62}{6.50} = 1.82 \text{ ft}$$

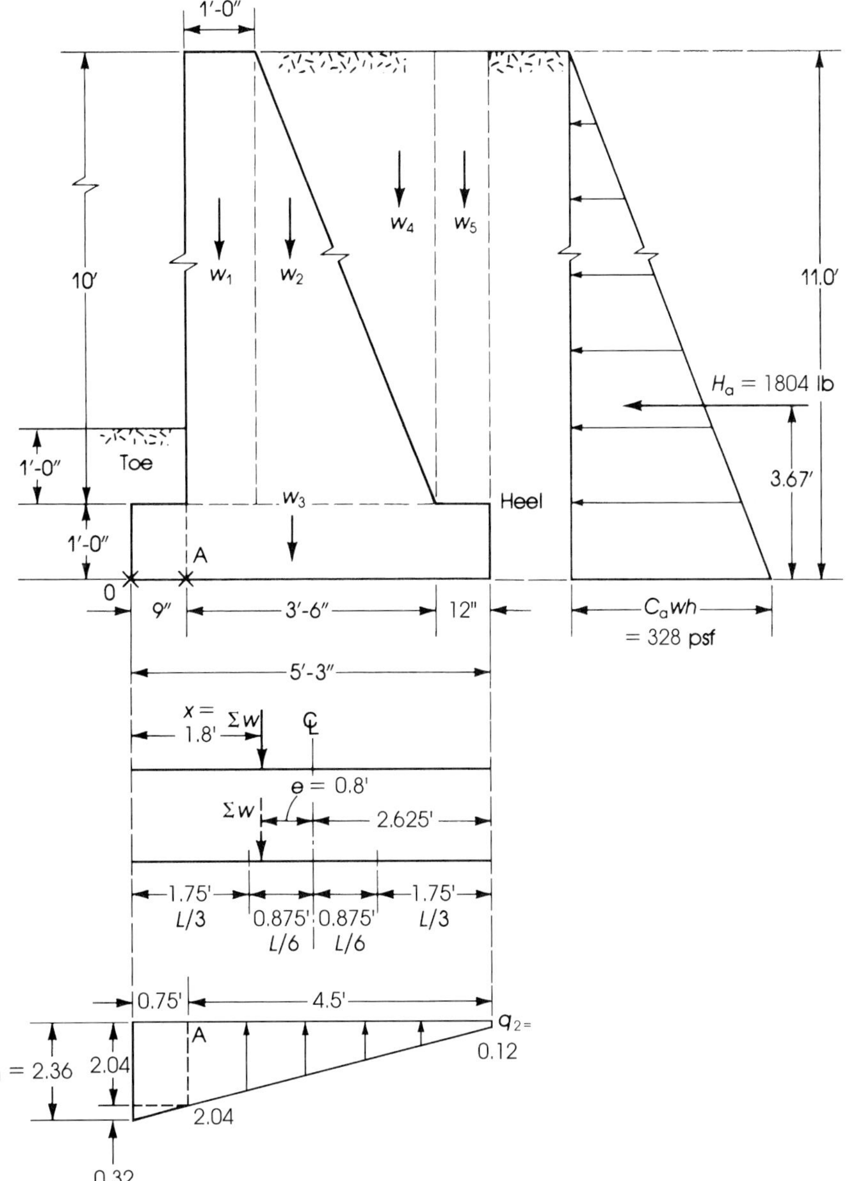

Figure 14.10 Example 14.1.

The eccentricity is $e = 2.62 - 1.82 = 0.80$ ft. The resultant R acts just inside the middle third of the base and has an eccentricity $e = 0.8$ ft from the center of the base (Figure 14.10). For a 1-ft length of the footing, the effective length of footing is 5.25 ft.

b. The moment of inertia is $I = 1.0(5.25)^3/12 = 12.1 \text{ ft}^4$. Area $= 5.25 \text{ ft}^2$.

c. The soil pressures at the two extreme ends of the footing are $q_1, q_2 = R/A \pm Mc/I$. The moment M is $Re = 6.50(0.8) = 5.2 \text{ K}\cdot\text{ft}$; $c = 2.62$ ft.

$$q_1 = \frac{6.50}{5.25} + \frac{5.2(2.62)}{12.1} = 1.24 + 1.12 = 2.36 \text{ Ksf}$$

$$q_2 = 1.24 - 1.12 = 0.12 \text{ Ksf}$$

7. Check the bending stress in concrete at point A of the toe.

a. Soil pressure at A (from geometry) is $q_A = 0.12 + \left(\frac{4.5}{5.25}\right)(2.36 - 0.12) = 2.04$ Ksf.

b. M_A is calculated at A due to a rectangular stress and a triangular stress.

$$M_A = \frac{2.04(0.75)^2}{2} + (0.32 \times 0.75 \times 0.5)(0.75 \times \tfrac{2}{3})$$
$$= 0.63 \text{ K}\cdot\text{ft}$$

c. The flexural stress in concrete is

$$Mc/I = 0.63(12{,}000)(6)/1728 = 26 \text{ psi}$$

where $c = h/2 = 12/2 = 6$ in. and $I = 12(12)^3/12 = 1728$ in.^{4}

d. The modulus of rupture of concrete is $7.5\sqrt{f'_c} = 410$ psi > 26 psi. The factor of safety against cracking is $410/26 = 16$. Therefore, the section is adequate. No other sections need to be checked.

Example 14.2

Design a cantilever retaining wall to support a bank of earth 16.5 ft high. The top of the earth is to be level with a surcharge of 330 psf. Given: The weight of the backfill is 110 pcf, the angle of internal friction is $\phi = 35°$, the coefficient of friction between concrete and soil is $\mu = 0.5$, the coefficient of friction between soil layers is $\mu = 0.7$, allowable soil bearing capacity is 4 Ksf, $f'_c = 3$ Ksi, and $f_y = 60$ Ksi.

Solution

1. Determine the dimensions of the retaining wall using the approximate relationships shown in Figure 14.8.

a. Height of wall: Allowing 3 ft for frost penetration, the height of the wall becomes $h = 16.5 + 3 = 19.5$ ft.

b. Base thickness: Assume base thickness is $0.08h = 0.08 \times 19.5 = 1.56$ ft, or 1.5 ft. The height of the stem is $19.5 - 1.5 = 18$ ft.

c. Base length: The base length varies between $0.4h$ and $0.67h$. Assuming an average value of $0.53h$, then the base length equals $0.53 \times 19.5 = 10.3$ ft, say, 10.5 ft. The projection of the base in front of the stem varies between $0.17h$ and $0.125h$. Assume a projection of $0.17h = 0.17 \times 19.5 = 3.3$ ft, say, 3.5 ft.

d. Stem thickness: The maximum stem thickness is at the bottom of the wall and varies between $0.08h$ and $0.1h$. Choose a maximum stem thickness equal to that of the base, or 1.5 ft. Select a practical minimum thickness of the stem at the top of the wall of 1.0 ft. The minimum batter of the face of the wall is $\frac{1}{4}$ in./ft. For an 18-ft-high wall, the minimum batter is $\frac{3}{4} \times 18 = 4.5$ in., which is less than the $1.5 - 1.0 = 0.5$ ft (6 in.) provided. The trial dimensions of the wall are shown in Figure 14.11.

2. Using the Rankine equation:

$$C_a = \frac{1 - \sin\phi}{1 + \sin\phi} = \frac{1 - 0.574}{1 + 0.574} = 0.271$$

3. The factor of safety against overturning can be determined as follows:

a. Calculate the actual unfactored forces acting on the retaining wall. First, find those acting to overturn the wall:

$$h_s \text{ (due to surcharge)} = \frac{w_s}{w} = \frac{330}{110} = 3 \text{ ft}$$

$$p_1 = C_a w h_s = 0.271 \times (110 \times 3) = 90 \text{ psf}$$

$$p_2 = C_a w h = 0.271 \times (110 \times 19.5) = 581 \text{ psf}$$

$$H_{a1} = 90 \times 19.5 = 1755 \text{ lb}, \qquad \text{arm} = \frac{19.5}{2} = 9.75 \text{ ft}$$

$$H_{a2} = \frac{1}{2} \times 581 \times 19.5 = 5665 \text{ lb}, \qquad \text{arm} = \frac{19.5}{3} = 6.5 \text{ ft}$$

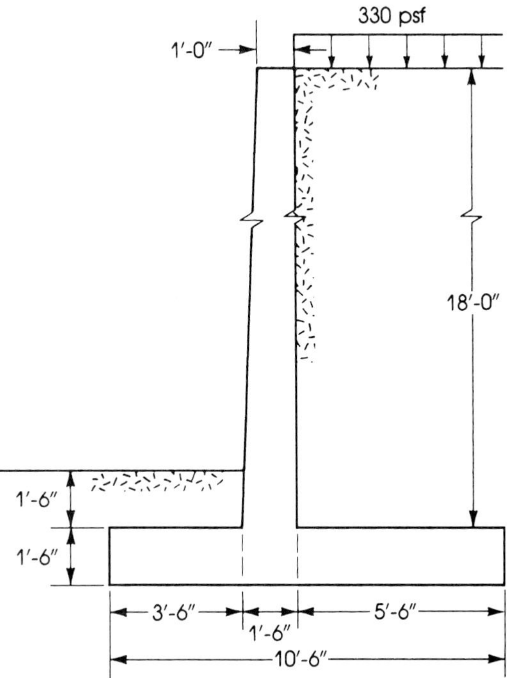

Figure 14.11 Example 14.2: trial configuration of retaining wall.

b. The overturning moment is $1.755 \times 9.75 + 5.665 \times 6.5 = 53.93$ K·ft.

c. Calculate the balancing moment against overturning (see Figure 14.12):

Force (lb)	Arm (ft)	Moment (K·ft)
$w_1 = 1 \times 18 \times 150 = 2{,}700$	4.50	12.15
$w_2 = \frac{1}{2} \times 18 \times \frac{1}{2} \times 150 = 675$	3.83	2.59
$w_3 = 10.5 \times 1.5 \times 150 = 2{,}363$	5.25	12.41
$w_4 = 5.5 \times 21 \times 110 = 12{,}705$	7.75	98.46

$$\Sigma w = R = 18.44 \text{ K} \qquad \Sigma M = 125.61 \text{ K}\cdot\text{ft}$$

$$\text{Factor of safety against overturning} = \frac{125.61}{53.93} = 2.33 > 2.0$$

4. Calculate the base soil pressure. Take moments about the toe end O (Figure 14.12) to determine the location of the resultant R of the vertical forces.

$$x = \frac{\Sigma M - \Sigma H y}{R} = \frac{\text{balancing } M - \text{overturning } M}{R}$$

$$= \frac{125.61 - 53.93}{18.44} = 3.89 \text{ ft} > \frac{10.5}{3}, \quad \text{or} \quad 3.5 \text{ ft}$$

The eccentricity is $e = 10.5/2 - 3.89 = 1.36$ ft. The resultant R acts within the middle third of the base and has an eccentricity $e = 1.36$ ft from the center of the base. For a 1-ft length of the footing, area $= 10.5 \times 1 = 10.5 \text{ ft}^2$.

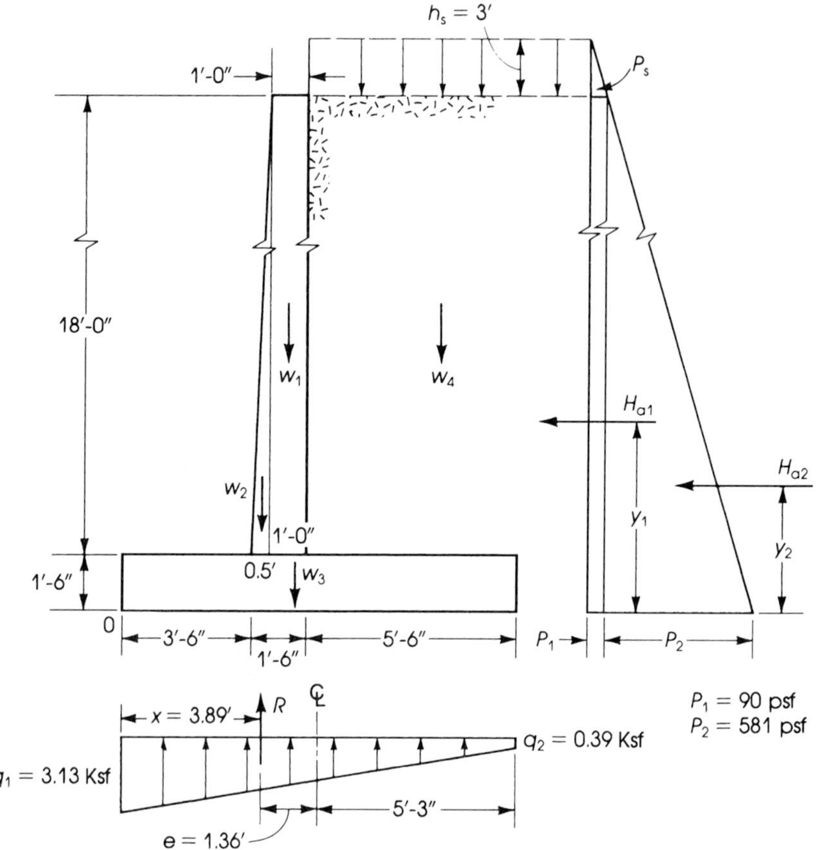

Figure 14.12 Example 14.2: forces acting on retaining wall.

$$I = 1 \times \frac{(10.5)^3}{12} = 96.47 \text{ ft}^4$$

$$q_1 = \frac{R}{A} + \frac{(Re)C}{I} = \frac{18.44}{10.5} + \frac{(18.44 \times 1.36) \times 5.25}{96.47}$$

$$= 1.76 + 1.37 = 3.13 \text{ Ksf} < 4 \text{ Ksf}$$

$$q_2 = 1.76 - 1.37 = 0.39 \text{ Ksf}$$

Soil pressure is adequate.

5. Calculate the factor of safety against sliding. A minimum factor of safety of 1.5 must be maintained.

$$\text{Force causing sliding} = H_{a1} + H_{a2} = 1.76 + 5.67 = 7.43 \text{ K}$$

$$\text{Resisting force} = \mu R = 0.5 \times 18.44 = 9.22 \text{ K}$$

$$\text{Factor of safety against sliding} = \frac{9.22}{7.43} = 1.24 < 1.5$$

The resistance provided does not give an adequate safety against sliding. In this case, a key should be provided to develop a passive pressure large enough to resist the excess force that causes sliding. Another function of the key is to provide sufficient development length for the dowels of the stem. The key is therefore placed such that its face is about 6 in. from the back face of the stem (Figure 14.13). In the calculation of the passive pressure, the top foot of the earth at the toe side is usually neglected, leaving a height of 2 ft in this example. Assume a key depth $t = 1.5$ ft and a width $b = 1.5$ ft.

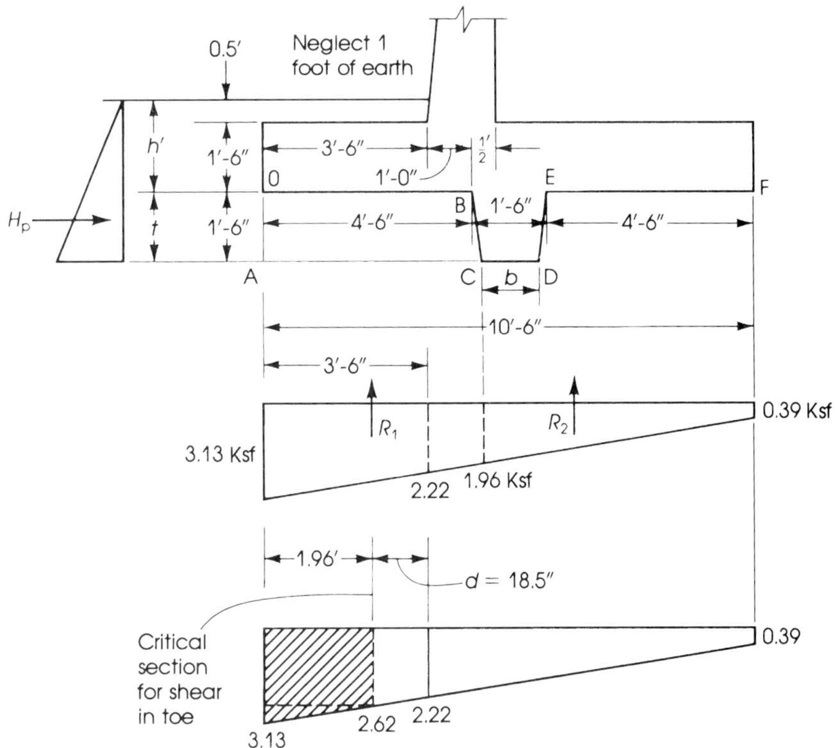

Figure 14.13 Footing details, Example 14.2.

$$C_p = \frac{1 + \sin\phi}{1 - \sin\phi} = \frac{1}{C_a} = \frac{1}{0.271} = 3.69$$

$$H_p = \frac{1}{2}C_p w(h' + t)^2 = \frac{1}{2} \times 3.69 \times 110(2 + 1.5)^2 = 2486 \text{ lb}$$

The sliding may occur now on the surfaces AC, CD, and EF (Figure 14.13). The sliding surface AC lies within the soil layers with a coefficient of internal friction $= \tan\phi = \tan 35° = 0.7$, whereas the surfaces CD and EF are those between concrete and soil with a coefficient of internal friction of 0.5, as given in this example. The frictional resistance is $F = \mu_1 R_1 + \mu_2 R_2$.

$$R_1 = \text{reaction on } AC = \left(\frac{3.13 + 1.96}{2}\right) \times 4.5 = 11.44 \text{ K}$$

$$R_2 = R - R_1 = 18.44 - 11.44 = 7.0 \text{ K}$$

$$R_2 = \text{reaction on } CDF = \left(\frac{1.96 + 0.39}{2}\right) \times 6 = 7.05 \text{ K}$$

$$F = 0.7(11.44) + 0.5(7.00) = 11.50 \text{ K}$$

The total resisting force is

$$F + H_p = 11.50 + 2.49 = 13.99 \text{ K}$$

The factor of safety against sliding is

$$\frac{13.99}{7.43} = 1.9 \quad \text{or} \quad \frac{11.5}{7.43} = 1.55 > 1.5$$

The factor is greater than 1.5, which is recommended when passive resistance against sliding is not included.

6. Design the wall (stem). The design of the different reinforced concrete structural elements can be performed now using the ACI Code strength-design method.

a. Main reinforcement: The lateral forces applied to the wall are calculated using a load factor of 1.7. The critical section for bending moment is at the bottom of the wall, height = 18 ft. Calculate the applied ultimate forces:

$$P_1 = 1.7(C_a w h_s) = 1.7(0.271 \times 110 \times 3) = 152 \text{ lb}$$

$$P_2 = 1.7(C_a w h) = 1.7(0.271 \times 110 \times 18) = 912 \text{ lb}$$

$$H_{a1} = 0.152 \times 18 = 2.74 \text{ K}, \qquad \text{arm} = \tfrac{18}{2} = 9 \text{ ft}$$

$$H_{a2} = \tfrac{1}{2} \times 0.912 \times 18 = 8.21 \text{ K}, \qquad \text{arm} = \tfrac{18}{3} = 6 \text{ ft}$$

$$M_u \text{ (at bottom of wall)} = 2.74 \times 9 + 8.21 \times 6 = 73.92 \text{ K} \cdot \text{ft}$$

The total depth used is 18 in., $b = 12$ in., and $d = 18 - 2$ (concrete cover) $- 0.5$ (half the bar diameter) $= 15.5$ in.

$$R_u = \frac{M_u}{bd^2} = \frac{73.92 \times 12{,}000}{12(15.5)^2} = 308 \text{ psi}$$

The steel ratio, ρ, can be obtained from Table A.1 in Appendix A or from

$$\rho = \frac{0.85 f'_c}{f_y}\left[1 - \sqrt{1 - \frac{2R_u}{\phi 0.85 f'_c}}\right] = 0.0062$$

$$A_s = 0.0062(12)(15.5) = 1.153 \text{ in.}^2$$

Use no. 8 bars spaced at 8 in. (1.18 in.^{2}). The minimum vertical A_s according to the ACI Code, Section 14.3 is

$$A_{s\,\min} = 0.0015(12)(18) = 0.32 \text{ in.}^2 < 1.18 \text{ in.}^2$$

Because the moment decreases along the height of the wall, A_s may be reduced according to the moment requirements. It is practical to use one A_s, or spacing, for the lower half and a second A_s, or spacing, for the upper half of the wall. To calculate the moment at midheight of the wall, 9 ft from the top:

$$P_1 = 1.7(0.271 \times 110 \times 3) = 152 \text{ lb}$$

$$P_2 = 1.7(0.271 \times 110 \times 9) = 456 \text{ lb}$$

$$H_{a1} = 0.152 \times 9 = 1.37 \text{ K}, \qquad \text{arm} = \tfrac{9}{2} = 4.5 \text{ ft}$$

$$H_{a2} = \tfrac{1}{2} \times 0.456 \times 9 = 2.1 \text{ K}, \qquad \text{arm} = \tfrac{9}{3} = 3 \text{ ft}$$

$$M_u = 1.37 \times 4.5 + 2.1 \times 3 = 12.47 \text{ K} \cdot \text{ft}$$

The total depth at midheight of wall is

$$\frac{12 + 18}{2} = 15 \text{ in.}$$

$$d = 15 - 2 - 0.5 = 12.5 \text{ in.}$$

$$R_u = \frac{M_u}{bd^2} = \frac{12.47 \times 12{,}000}{12 \times (12.5)^2} = 80 \text{ psi}$$

$$\rho = 0.00151 \text{ and } A_s = 0.00151(12)(12.5) = 0.23 \text{ in.}^2$$

$$A_{s\,\min} = 0.0015 \times 12 \times 15 = 0.27 \text{ in.}^2 > 0.23 \text{ in.}^2$$

Use no. 4 vertical bars spaced at 8 in. (0.29 in.^{2}) with similar spacing to the lower vertical steel bars in the wall.

b. Temperature and shrinkage reinforcement: The minimum horizontal reinforcement at the base of the wall according to ACI Code, Section 14.3 is

$$A_{s\,\min} = 0.0020 \times 12 \times 18 = 0.432 \text{ in.}^2$$

(for the bottom third), assuming no. 5 bars or smaller.

$$A_{s\,\min} = 0.0020 \times 12 \times 15 = 0.36 \text{ in.}^2$$

(for the upper two-thirds). Because the front face of the wall is mostly exposed to temperature changes, use half to two-thirds of the horizontal bars at the external face of the wall and place the balance at the internal face.

$$0.5A_s = 0.5 \times 0.432 = 0.22 \text{ in.}^2$$

Use no. 4 horizontal bars spaced at 8 in. $(A_s = 0.29 \text{ in.}^2)$ at both the internal and external surfaces of the wall. Use no. 4 vertical bars spaced at 12 in. at the front face of the wall to support the horizontal temperature and shrinkage reinforcement.

c. Dowels for the wall vertical bars: The anchorage length of no. 8 bars into the footing must be at least 22 in. Use an embedment length of 2 ft into the footing and the key below the stem.

d. Design for shear: The critical section for shear is at a distance $d = 15.5$ in. from the bottom of the stem. At this section, the distance from the top equals $18 - 15.5/12 = 16.7$ ft.

$$P_1 = 152 \text{ lb}$$

$$P_2 = 1.7(0.271 \times 110 \times 16.7) = 846 \text{ lb}$$

$$H_{a1} = 0.152 \times 16.7 = 2.54 \text{ K}$$

$$H_{a2} = \tfrac{1}{2} \times 0.846 \times 16.7 = 7.07 \text{ K}$$

$$\text{Total } H = 2.54 + 7.07 = 9.61 \text{ K}$$

$$\phi V_c = \phi(2\sqrt{f'_c})bd = \frac{0.85 \times 2}{100} \times \sqrt{3000} \times 12 \times 15.5$$

$$= 17.32 \text{ K} > 9.61 \text{ K}$$

7. Design of the heel: A load factor of 1.4 is used to calculate the factored bending moment and shearing force due to the backfill and concrete, whereas a load factor of 1.7 is used for the surcharge. The upward soil pressure is neglected, because it will reduce the effect of the backfill and concrete on the heel. Referring to Figure 14.12, the total load on the heel is

$$V_u = 1.4[(18 \times 5.5 \times 110) + (1.5 \times 5.5 \times 150)]$$
$$+ 1.7(3 \times 5.5 \times 110) = 20 \text{ K}$$

$$M_u \text{ (at face of wall)} = V_u \times \frac{5.5}{2} = 55 \text{ K}\cdot\text{ft}$$

The critical section for shear is usually at a distance d from the face of the wall when the reaction introduces compression into the end region of the member. In this case, the critical section will be considered at the face of the wall, because tension and not compression develops in the concrete.

$$V_u = 20.0 \text{ K}$$

$$\phi V_c = \phi(2\sqrt{f'_c})bd = \frac{0.85 \times 2}{1000} \times \sqrt{3000} \times 12 \times 14.5$$

$$= 16.2 \text{ K}$$

ϕV_c is less than V_u of 20.0 K, and the section must be increased by the ratio 20.0/16.2 or shear reinforcement must be provided.

$$\text{Required } d = \frac{20.0}{16.2} \times 14.5 = 17.9 \text{ in.}$$

$$\text{Total thickness required} = 17.9 + 3.5 = 21.4 \text{ in.}$$

Use a base thickness of 22 in. and $d = 18.5$ in.

$$R_u = \frac{M_u}{bd^2} = \frac{55 \times 12{,}000}{12 \times (18.5)^2} = 161 \text{ psi}, \qquad \rho = 0.0031$$

$$A_s = \rho bd = 0.69 \text{ in.}^2$$

$$\text{Min. shrinkage } A_s = 0.0018(12)(22) = 0.475 \text{ in.}^2$$

$$\text{Min. flexural } A_s = 0.0033(12)(18.5) = 0.733 \text{ in.}^2$$

Use no. 6 bars spaced at 7 in. $(A_s = 0.76 \text{ in.}^2)$. The development length for the no. 6 top bars equals $1.4l_d = 35$ in. Therefore, the bars must be extended 3 ft into the toe of the base.

Temperature and shrinkage reinforcement in the longitudinal direction is not needed in the heel or toe, but it may be preferable to use minimal amounts of reinforcement in that direction, say, no. 4 bars spaced at 12 in.

8. Design of the toe: The toe of the base acts as a cantilever beam subjected to upward pressures, as calculated in Step 4. The factored soil pressure is obtained by multiplying the service load soil pressure by a load factor of 1.7, because it is primarily caused by the lateral forces. The critical section for the bending moment is at the front face of the stem. The critical section for shear is at a distance d from the front face of the stem, because the reaction in the direction of shear introduces compression into the toe.

Referring to Figure 14.13, the toe is subjected to an upward pressure from the soil and downward pressure due to self-weight of the toe slab.

$$V_u = 1.7\left(\frac{3.13 + 2.62}{2}\right) \times 1.96 - 1.4\left(\frac{22}{12} \times 0.150\right) \times 1.96$$

$$= 8.83 \text{ K}$$

This is less than ϕV_c of 16.2 K calculated for the heel in Step 7.

$$M_u = 1.7\left[\frac{2.22}{2} \times (3.5)^2 + (3.13 - 2.22) \times 3.5 \times 0.5\left(\frac{2}{3} \times 3.5\right)\right]$$

$$- 1.4\left[\left(\frac{22}{12} \times 0.150\right) \times \frac{(3.5)^2}{2}\right] = 27.10 \text{ K} \cdot \text{ft}$$

$$R_u = \frac{M_u}{bd^2} = \frac{27.1 \times 12{,}000}{12 \times (18.5)^2} = 79 \text{ psi}, \qquad \rho = 0.0015$$

$$A_s = 0.0015(12)(18.5) = 0.333 \text{ in.}^2$$

$$\text{Min. shrinkage } A_s = 0.0018(12)(22) = 0.475 \text{ in.}^2$$

$$\text{Min. flexural } A_s = 0.0033(12)(18.5) = 0.733 \text{ in.}^2$$

Use no. 6 bars spaced at 7 in., similar to the heel reinforcement. Development length of no. 6 bars equals 25 in. Extend the bars into the heel 25 in. The final reinforcement details are shown in Figure 14.14.

9. Shear keyway between wall and footing: In the construction of retaining walls, the footing is cast first and then the wall is cast on top of the footing at a later date. A construction joint is used at the base of the wall. The joint surface takes the form of a keyway, as shown in Figure 14.15, or is left in a very rough condition (Figure 14.14). The joint must be capable of transmitting the stem shear into the footing.

10. Proper drainage of the backfill is essential in this design, because the earth pressure used is for drained backfill. Weep holes should be provided in the wall, 4 in. in diameter and spaced at 5 ft in the horizontal and vertical directions.

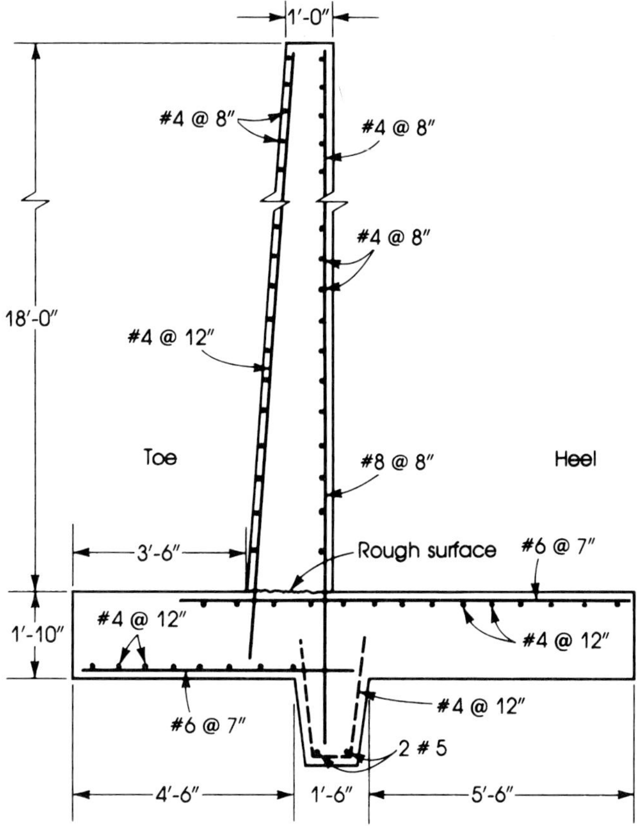

Figure 14.14 Reinforcement details, Example 14.2.

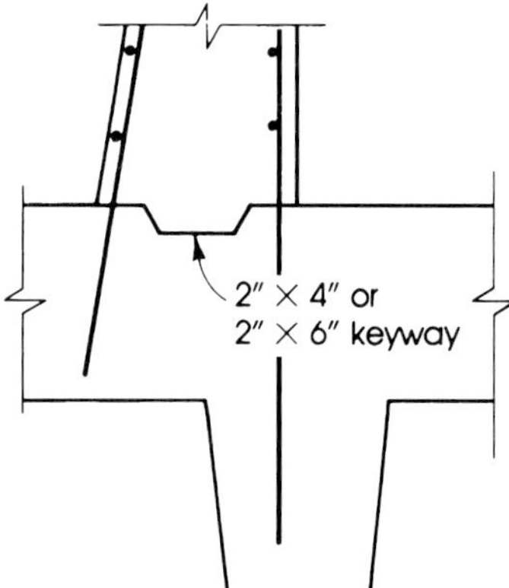

Figure 14.15 Keyway details.

14.11 BASEMENT WALLS

It is a common practice to assume that basement walls span vertically between the basement floor slab and the first floor slab. Two possible cases of design should be investigated for a basement wall.

First, when the wall only has been built on top of the basement floor slab, the wall will be subjected to lateral earth pressure with no vertical loads except its own weight. The wall in this case acts as a cantilever, and adequate reinforcement should be provided for a cantilever wall design. This case can be avoided by installing the basement and the first-floor slabs before backfilling against the wall.

Second, when the first-floor and the other floor slabs have been constructed and the building is fully loaded, the wall in this case will be designed as a propped cantilever wall subjected to earth pressure and to vertical load.

For an angle of internal friction of 35°, the coefficient of active pressure is $C_a = 0.271$. The horizontal earth pressure at the base is $\rho_a = C_a wh$. For $w = 110$ pcf and an average height of a basement of $h = 10$ ft, then

$$P_a = 0.271 \times 0.110 \times 10 = 0.3 \text{ Ksf}$$

$$H_a = 0.271 \times 0.110 \times \frac{100}{2} = 1.49 \text{ K/ft of wall}$$

H_a acts at $h/3 = 10/3 = 3.33$ ft from the base. An additional pressure must be added to allow for a surcharge of about 200 psf on the ground behind the wall. The equivalent height to the surcharge is

$$h_s = \frac{200}{110} = 1.82 \text{ ft}$$

$$\rho_s = C_a wh_s = 0.271 \times 0.110 \times 1.82 = 0.054 \text{ Ksf}$$

$$H_s = (C_a wh_s) \times h = 0.054 \times 10 = 0.54 \text{ K/ft of wall}$$

H_s of the surcharge acts at $h/2 = 5$ ft from the base.

In the preceding calculations, it is assumed that the backfill is dry, but it is necessary to investigate the presence of water pressure behind the wall. The maximum water pressure

Basement wall

occurs when the whole height of the basement wall is subjected to water pressure, and $p_w = wh = 62.5 \times 10 = 625$ psf.

$$H_w = \frac{wh^2}{2} = 0.625 \times 5 = 3.125 \text{ K/ft of wall}$$

The maximum pressure may not be present continuously behind the wall. Therefore, if the ground is intermittently wet, a percentage of the preceding pressure may be adopted, say, 50%:

$$\frac{P_w}{2} = \frac{0.625}{2} = 0.31 \text{ Ksf}$$

$$H'_w = \frac{H_w}{2} = (0.5wh)\frac{h}{2} = \frac{3.125}{2} = 1.56 \text{ K/ft of wall}$$

H'_w acts at $h/3 = \frac{10}{3} = 3.33$ ft from the base. Water may be prevented from collecting against the wall by providing drains at the lower end of the wall.

In addition to drainage, a waterproofing or damp-proofing membrane must be laid or applied to the external face of the wall. The ACI Code, Section 14.5, specifies that the minimum thickness of an exterior basement wall and its foundation is 7.5 in. In general, the minimum thickness of bearing walls is $\frac{1}{25}$ of the supported height or length, whichever is shorter, or 4 in.

Example 14.3

Determine the thickness and necessary reinforcement for the basement retaining wall shown in Figure 14.16. Given: Weight of backfill = 110 pcf, angle of internal friction = 35°, $f'_c = 3$ Ksi, and $f_y = 60$ Ksi.

Solution

1. The wall spans vertically and will be considered as fixed at the bottom end and propped at the top. Consider a span $L = 9.75$ ft, as shown in Figure 14.16. For these data, the different lateral pressures on a 1-ft length of the wall are as follows: Due to active soil pressure, $p_a = 0.3$ Ksf and $H_a = 1.49$ K. Due to water pressure, $p_w = 0.31$ Ksf and $H_w = 1.56$ K. Due to surcharge, $p_s = 0.054$ Ksf and $H_s = 0.54$ K. H_a and H_w are due to triangular loadings, whereas H_s is due to uniform loading. Referring to Figure 14.16 and using moment coefficients of a propped beam subjected to triangular and uniform loads,

$$M_u = 1.7(H_a + H_w)\frac{L}{7.5} + 1.7H_s\frac{L}{8}$$

$$= 1.7\left(\frac{3.05}{7.5} \times 9.75 + 0.54 \times \frac{9.75}{8}\right) = 7.87 \text{ K}\cdot\text{ft}$$

$$R_B = 1.7\left(\frac{3.05}{3} + \frac{0.54}{2}\right) - \frac{7.87}{9.75} = 1.38 \text{ K}$$

$$R_A = 4.73 \text{ K}$$

Maximum positive bending moment within the span occurs at the section of zero shear.

$$V_u = 1.38 - 1.7(0.054x) - 1.7\left(0.063\frac{x^2}{2}\right) = 0$$

$$x = 4.3 \text{ ft}$$

$$M_c = 1.38 \times 4.3 - 1.7\left[\frac{0.054}{2}(4.3)^2 + \frac{0.27}{2}\frac{(4.3)^2}{3}\right]$$

$$= +3.66 \text{ K}\cdot\text{ft}$$

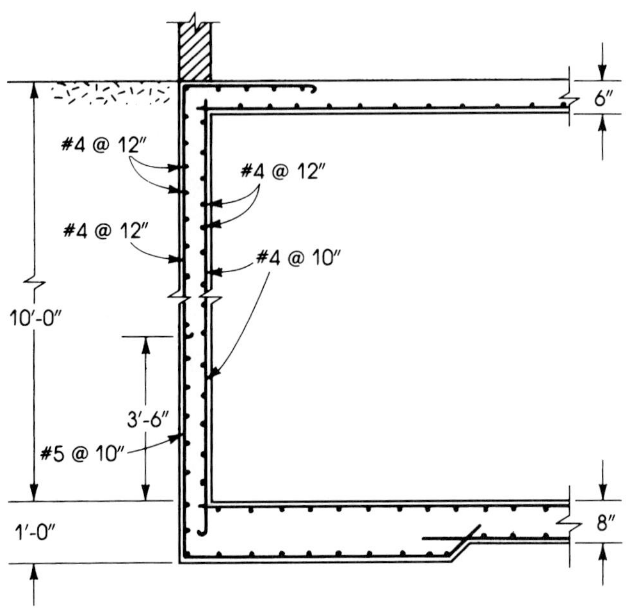

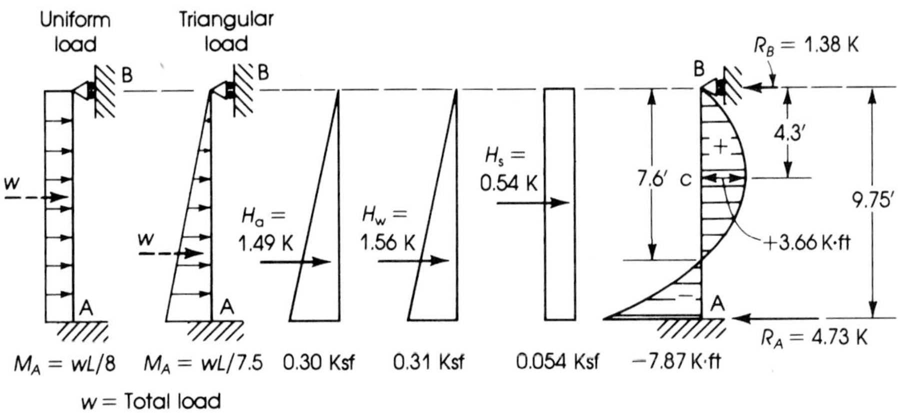

Figure 14.16 Basement wall, Example 14.3.

2. Assuming 0.01 steel ratio, R_u = 332 psi,

$$d = \sqrt{\frac{M_u}{R_u b}} = \sqrt{\frac{7.87 \times 12}{0.332 \times 12}} = 4.87 \text{ in.}$$

Total depth = 4.87 + 1.5 (concrete cover) + 0.25 = 6.62 in. Use a $7\frac{1}{2}$-in. slab, d = 5.75 in.

$$R_u = \frac{M_u}{bd^2} = \frac{7.87 \times 12{,}000}{12 \times (5.75)^2} = 240 \text{ psi}$$

The steel ratio is $\rho = 0.0047$ and $A_s = 0.0047 \times 12 \times 5.75 = 0.325$ in.^{2}

$$\text{Minimum } A_s = 0.0015\, bh = 0.0015(12)(7.5) = 0.135 \text{ in.}^2 \quad \text{(vertical bars)}$$

$$\text{Minimum } A_s \text{ (flexsure)} = 0.0033(12)(5.75) = 0.23 \text{ in.}^2$$

Use no. 5 bars spaced at 10 in. (A_s = 0.37 in.^{2}).

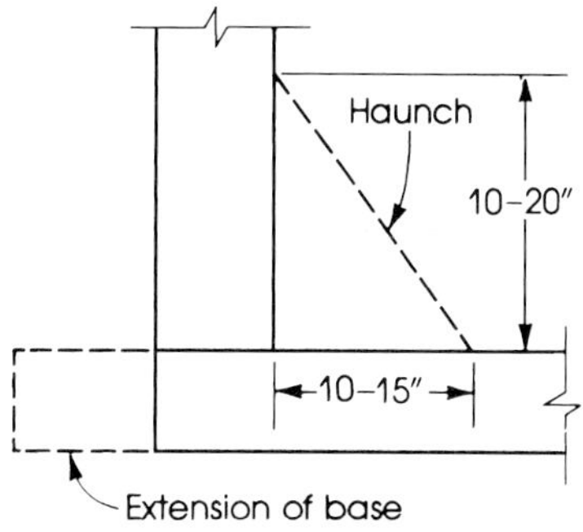

Figure 14.17 Adjustment of wall base, Example 14.3.

3. For the positive moment, $M_c = 3.66\ \text{K}\cdot\text{ft}$:

$$R_u = \frac{3.66 \times 12{,}000}{12 \times (5.75)^2} = 111 \text{ psi}, \qquad \rho = 0.0021$$

$$A_s = 0.0021 \times 12 \times 5.75 = 0.15 \text{ in.}^2 < 0.33 \text{ in.}^2$$

Use no. 4 bars spaced at 10 in. $(A_s = 0.24 \text{ in.}^2)$.

4. Zero moment occurs at a distance of 7.6 ft from the top and 2.15 ft from the base. The development length of no. 5 bars is 14 in. Therefore, extend the main no. 5 bars to a distance of $2.15 + 1.2 = 3.35$ ft, or 3.5 ft, above the base; then use no. 4 bars spaced at 12 in. at the exterior face. For the interior face, use no. 4 bars spaced at 10 in. throughout.
5. Longitudinal reinforcement: Use a minimum steel ratio of 0.0020 (ACI Code, Section 14.3), or $A_s = 0.0020 \times 7 \times 12 = 0.17 \text{ in.}^2$ Use no. 4 bars spaced at 12 in. on each side of the wall.
6. If the bending moment at the base of the wall is quite high, it may require a thick wall slab, for example, 12 in. or more. In this case a haunch may be adopted, as shown in Figure 14.17. This solution will reduce the thickness of the wall, because it will be designed for the moment at the section exactly above the haunch.
7. The basement slab may have a thickness greater than the wall thickness and may be extended outside the wall by about 10 in. or more, as required.

SUMMARY

Sections 14.1–14.3

1. A retaining wall maintains unequal levels of earth on its two faces. The most common types of retaining walls are gravity, semigravity, cantilever, counterfort, buttressed, and basement walls.
2. For a linear pressure, the active and passive pressure intensities are

$$P_a = C_a wh \quad \text{and} \quad P_p = C_p wh$$

According to Rankine's theory,

$$C_a = \left(\frac{1 - \sin\phi}{1 + \sin\phi}\right) \quad \text{and} \quad C_p = \left(\frac{1 + \sin\phi}{1 - \sin\phi}\right)$$

Values of C_a and C_p for different values of ϕ and δ are given in Tables 14.1 and 14.2.

Sections 14.4–14.5

1. When soil is saturated, the submerged unit weight must be used to calculate earth pressure. The hydrostatic water pressure must also be considered.
2. A uniform surcharge on a retaining wall causes an additional pressure height, $h_s = w_s/w$.

Sections 14.6–14.8

1. A total frictional force F to resist sliding effect is

$$F = \mu R + H_p \tag{14.13}$$

$$\text{Factor of safety against sliding} = \frac{F}{H_{ah}} \geq 1.5 \tag{14.14}$$

2. Factor of safety against overturning $= \dfrac{M_b}{M_o} = \dfrac{\Sigma wx}{H_h/3} \geq 2.0$ (14.15)
3. Approximate dimensions of a cantilever retaining wall are shown in Figure 14.8.

Sections 14.9–14.10

1. Minimum reinforcement is needed in retaining walls.
2. To avoid hydrostatic pressure on a retaining wall, a drainage system should be used that consists of weep holes, perforated pipe, or any other adequate device.
3. Basement walls in buildings may be designed as propped cantilever walls subjected to earth pressure and vertical loads. This case occurs only if the first floor slab has been constructed. A surcharge of 200 psf may be adopted.

REFERENCES

1. K. Terzaghi and R. B. Peck. *Soil Mechanics in Engineering Practice.* New York: John Wiley, 1968.
2. W. C. Huntington. *Earth Pressures and Retaining Walls.* New York: John Wiley, 1957.
3. G. P. Fisher and R. M. Mains. "Sliding Stability of Retaining Walls." *Civil Engineering* (July 1952).
4. "Retaining Walls and Abutments." AREA *Manual,* vol. 1. Chicago: American Railway and Engineering Association, 1958.
5. M. S. Ketchum. *The Design of Walls, Bins and Grain Elevators,* 3d ed. New York: McGraw-Hill, 1949.
6. J. E. Bowles. *Foundation Analysis and Design.* New York: McGraw-Hill, 1980.
7. American Concrete Institute. "Building Code Requirements for Structural Concrete." ACI Code (318–99). Detroit, Michigan, 1999.

PROBLEMS

14.1 Check the adequacy of the retaining wall shown in Figure 14.18 with regard to overturning, sliding, and the allowable soil pressure. Given: Weight of backfill = 110 pcf, the angle of internal friction is $\phi = 30°$, the coefficient of friction between concrete and soil is $\mu = 0.5$, allowable soil pressure = 3.5 Ksf, and $f'_c = 3$ Ksi.

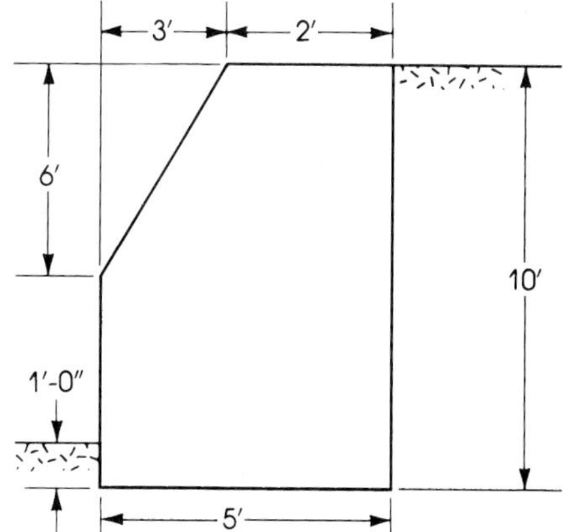

Figure 14.18 Problem 14.1: gravity wall.

14.2 Repeat Problem 14.1 with Figure 14.19.

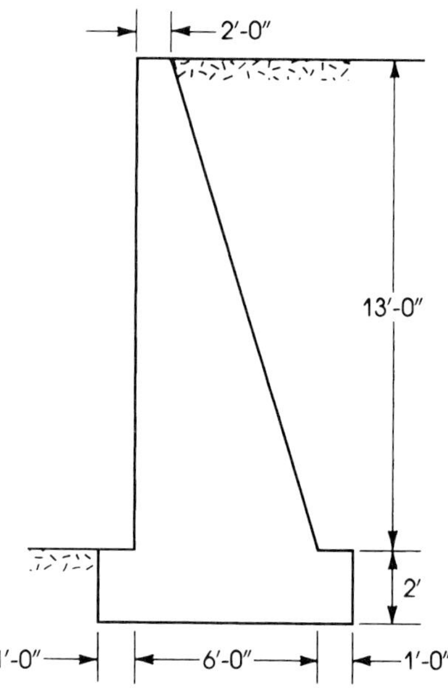

Figure 14.19 Problem 14.2: semigravity wall.

14.3 For each problem in Table 14.4, determine the factor of safety against overturning and sliding. Also, determine the soil pressure under the wall footing and check if all calculated values are adequate (equal or below the allowable values). Given: Weight of soil = 110 pcf, weight of concrete = 150 pcf, and coefficient of friction between concrete and soil is 0.5 and between soil layers is 0.6. Consider that the allowable soil pressure of 4 Ksf and the top of the backfill is level without surcharge. Neglect the passive soil resistance. See Figure 14.20. (ϕ = 35°.)

Table 14.4

Problem No.	*H*	h_f	*A*	*B*	*C*	*L*
(a)	12	1.00	2.0	1.0	4.0	7
(b)	14	1.50	2.0	1.5	4.5	8
(c)	15	1.50	2.0	1.5	4.5	8
(d)	16	1.50	3.0	1.5	4.5	9
(e)	17	1.50	3.0	1.5	4.5	9
(f)	18	1.75	3.0	1.75	5.25	10
(g)	19	1.75	3.0	1.75	5.25	10
(h)	20	2.00	3.0	2.0	6.0	11
(i)	21	2.00	3.5	2.0	6.5	12
(j)	22	2.00	3.5	2.0	6.5	12

Refer to Figure 14.20. All dimensions are in feet.

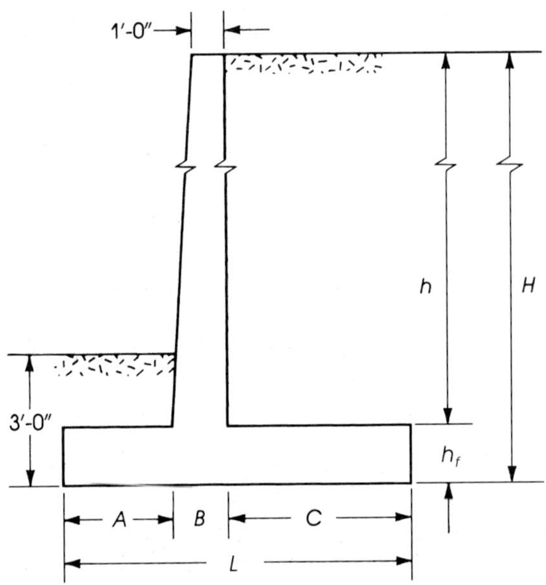

Figure 14.20 Problem 14.3.

14.4 Repeat Problems 14.3(e)–(h), assuming a surcharge of 300 psf.

14.5 Repeat Problems 14.3(e)–(h), assuming that the backfill slopes at 10° to the horizontal. (Add key if needed.)

14.6 For Problems 14.3(e)–(h), determine the reinforcement required for the stem, heel, and toe, and choose adequate bars and distribution. Use $f'_c = 3$ Ksi and $f_y = 60$ Ksi.

14.7 Determine the dimensions of a cantilever retaining wall to support a bank of earth 16 ft high. Assume that frost penetration depth is 4 ft. Check the safety of the retaining wall against overturning and sliding only. Given: Weight of backfill = 120 pcf, angle of internal friction = 33°, coefficient of friction between concrete and soil = 0.45, coefficient of friction between soil layers = 0.65, and allowable soil pressure = 4 Ksf. Use a 1.5- × 1.5-ft key if needed.

14.8 A complete design is required for the retaining wall shown in Figure 14.21. The top of the backfill is to be level without surcharge. Given: Weight of backfill soil = 110 pcf, angle of internal friction = 35°, the coefficient of friction between concrete and soil is 0.55, and that between soil layers is 0.6. Use $f'_c = 3$ Ksi, and $f_y = 60$ Ksi, and an allowable soil pressure of 4 Ksf.

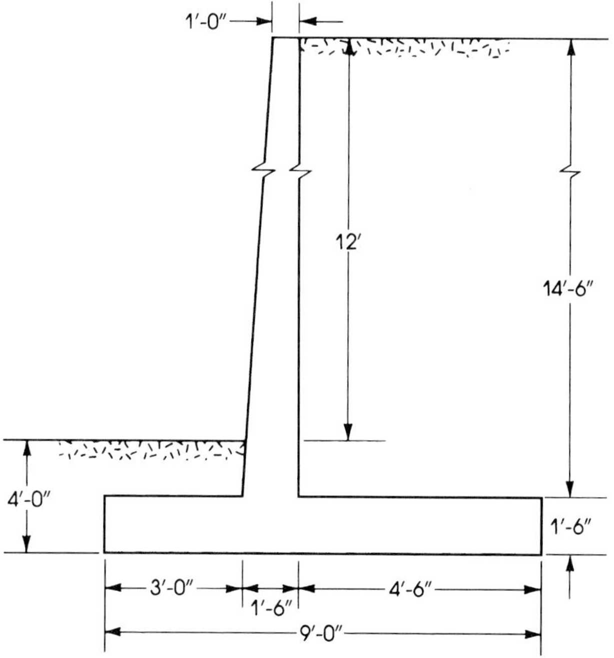

Figure 14.21 Problem 14.8: cantilever retaining wall.

14.9 Check the adequacy of the cantilever retaining wall shown in Figure 14.22 for both sliding and overturning conditions. Use a key of 1.5 × 1.5 ft if needed. Then determine reinforcement needed for the stem, heel, and toe, and choose adequate bars and distribution. Given: Weight of soil = 120 pcf, the angle of internal friction is $\phi = 35°$, the coefficient of friction between concrete and soil is 0.52 and that between soil layers is 0.70. Use $f'_c = 3$ Ksi, $f_y = 60$ Ksi, an allowable soil pressure of 4 Ksf, and a surcharge of 300 psf.

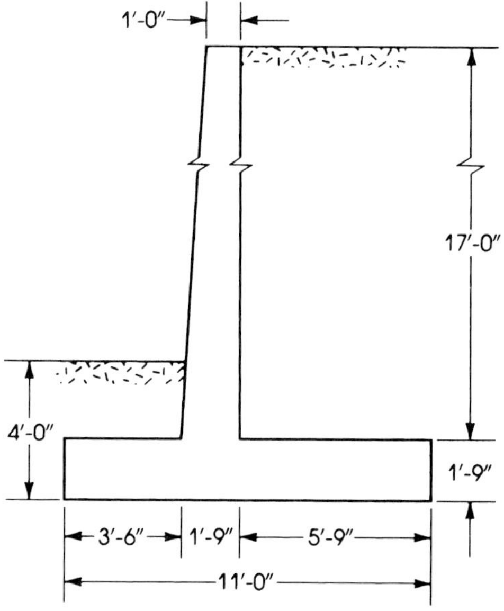

Figure 14.22 Problem 14.9: cantilever retaining wall.

14.10 Repeat Problem 14.9, assuming the backfill slopes at 30° to the horizontal.

14.11 Determine the thickness and necessary reinforcement for the basement wall shown in Figure 14.23. The weight of backfill is 120 pcf and the angle of internal friction is $\phi = 30°$. Assume a surcharge of 400 psf and use $f'_c = 3$ Ksi and $f_y = 60$ Ksi.

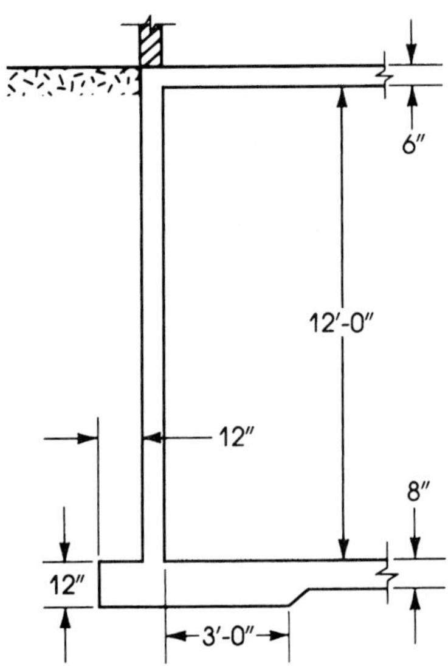

Figure 14.23 Problem 14.11: basement wall.

14.12 Repeat Problem 14.11 using a basement clear height of 14 ft.

Apartment building, Habitat 67, Montreal, Canada.

15 DESIGN FOR TORSION

15.1 INTRODUCTION

Torsional stresses develop in a beam section when a moment acts on that section parallel to its surface. Such moments, called torsional moments, cause a rotation in the structural member and cracking on its surface, usually in the shape of a spiral. To illustrate torsional stresses, let a torque T be applied on a circular cantilever beam made of elastic homogeneous material, as shown in Figure 15.1. The torque will cause a rotation of the beam. Point B moves to point B' at one end of the beam, whereas the other end is fixed. The angle θ is called the angle of twist. The plane $AO'OB$ will be distorted to the shape $AO'OB'$. Assuming that all longitudinal elements have the same length, the shear strain is

$$\gamma = \frac{(BB')}{L} = \frac{r\theta}{L}$$

where L is the length of the beam and r is the radius of the circular section.

In reinforced concrete structures, members may be subjected to torsional moments when they are curved in plan, support cantilever slabs, act as spandrel beams (end beams), or are part of a spiral stairway.

Structural members may be subjected to pure torsion only or, as in most cases, subjected simultaneously to shearing forces and bending moments. Example 15.1 illustrates the different forces that may act at different sections of a cantilever beam.

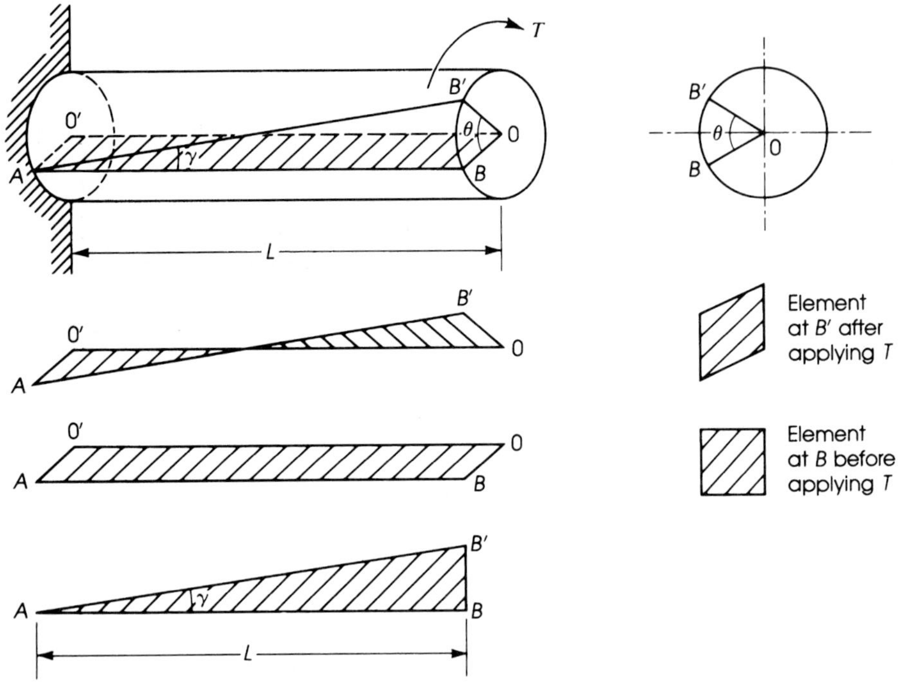

Figure 15.1 Torque applied to a cantilever beam.

Example 15.1

Calculate the forces acting at sections 1, 2, and 3 of the cantilever beam shown in Figure 15.2. The beam is subjected to a vertical force $P_1 = 15$ K, a horizontal force $P_2 = 12$ K acting at C, and a horizontal force $P_3 = 20$ K acting at B and perpendicular to the direction of the force P_2.

Solution

Let N = normal force, V = shearing force, M = bending moment, and T = torsional moment. The forces are as follows.

Section	N (K)	M_x (K · ft)	M_y (K · ft)	V_x (K)	V_y (K)	T (K · ft)
1	0	−135 (15 × 9)	+108 (12 × 9)	+12	+15	0
2	−12 Compression	0	+108	+20	+15	135 (15 × 9)
3	−12 Compression	−180	+348	+20	+15	135 (15 × 9)

If P_1, P_2, and P_3 are ultimate loads $(P_u = 1.4P_D + 1.7P_L)$, then the values in the table will be the ultimate design forces.

15.2 TORSIONAL MOMENTS IN BEAMS

It was shown in Example 15.1 that forces can act on building frames, causing torsional moments. If a concentrated load P is acting at point C in the frame ABC shown in Figure 15.3(a), it develops a torsional moment in beam AB of $T = PZ$ acting at D. When D is at midspan of AB, then the torsional design moment in AD equals that in DB, or $\frac{1}{2}T$. If a cantilever slab is supported by the beam AB in Figure 15.3(b), the slab causes a uniform

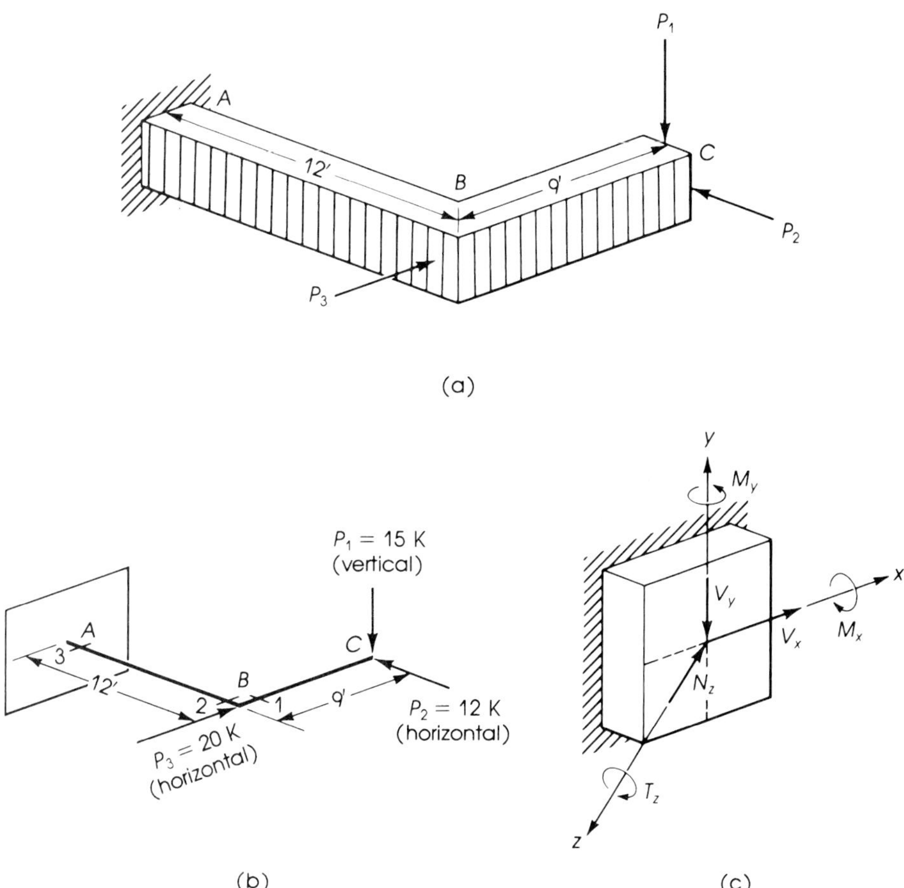

Figure 15.2 Example 15.1.

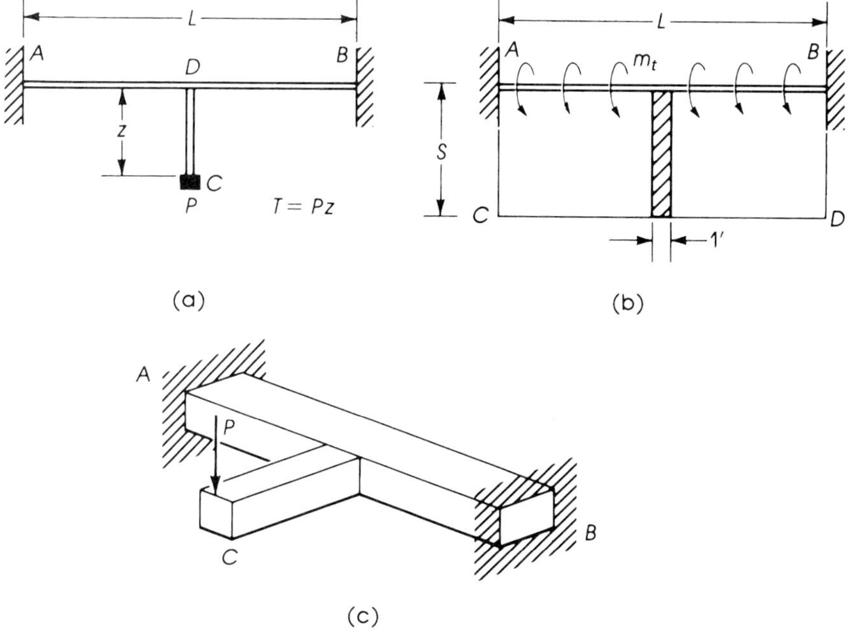

Figure 15.3 Torsional moments on *AB*.

torsional moment m_t along AB. This uniform torsional moment is due to the load on a unit width strip of the slab. If S is the width of the cantilever slab and w is load on the slab (psf), then $m_t = wS^2/2$ K·ft/ft of beam AB. The maximum torsion design moment in beam AB is $T = \frac{1}{2}m_t L$ acting at A and B. Other cases of loading are explained in Table 15.1. In general, the distribution of torsional moments in beams has the same shape and numerically has the same values as the shear diagrams for beams subjected to a load m_t or T.

Table 15.1 Torsion diagrams

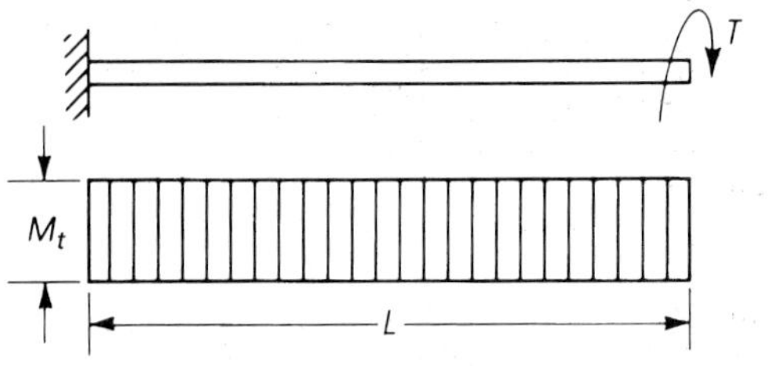

Diagram		
	At support $M_t = T$	For a circular section $J = \dfrac{\pi R^4}{2}$
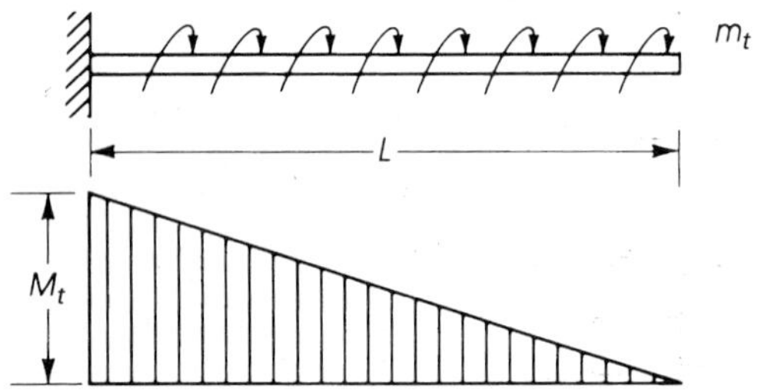	At support $M_t = m_t L$ $m_t =$ uniform torque	
	$M_{t1} = M_{t2} = \dfrac{T}{2}$	
	$M_{t1} = \dfrac{Tb}{L}$ $M_{t2} = \dfrac{Ta}{L}$	
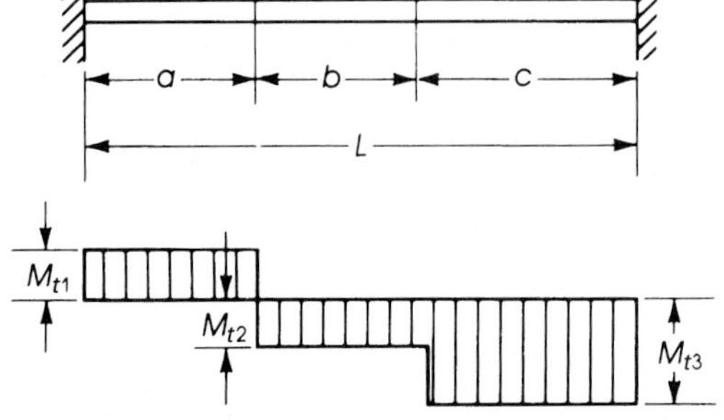	$M_{t1} = \dfrac{T_1(b - c) + T_2 c}{L}$ $M_{t2} = \dfrac{T_2 c - T_1 a}{L}$ $M_{t3} = \dfrac{T_1 a - T_2(a + b)}{L}$ Note: When $a = b = c = L/3$ and $T_1 = T_2 = M_{t1} = -M_{t3} = T$. $M_{t2} = 0$	

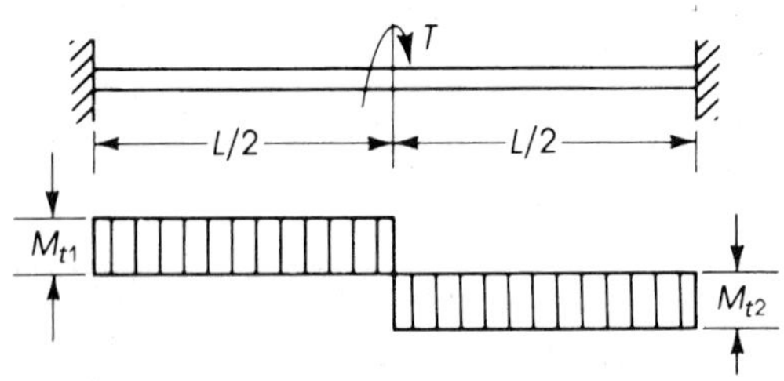

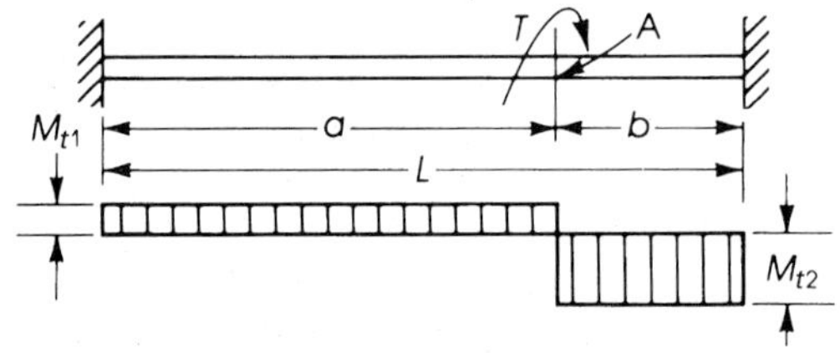

Table 15.1 *(Continued)*

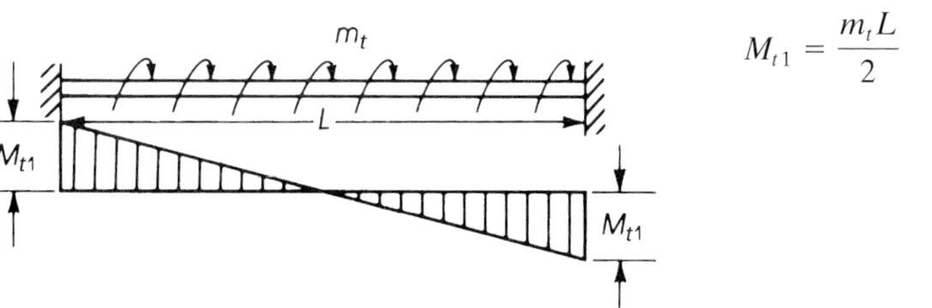

$$M_{t1} = \frac{m_t L}{2}$$

15.3 TORSIONAL STRESSES

Considering the cantilever beam with circular section of Figure 15.1, the torsional moment T will cause a shearing force dV perpendicular to the radius of the section. From the conditions of equilibrium, the external torsional moment is resisted by an internal torque equal to and opposite to T. If dV is the shearing force acting on the area dA (Figure 15.4), then the magnitude of the torque is $T = \int r\, dV$. Let the shearing stress be v; then

$$dV = v\, dA \quad \text{and} \quad T = \int rv\, dA$$

The maximum elastic shear occurs at the external surface of the circular section at radius r with thickness dr; then the torque T can be evaluated by taking moments about the center O for the ring area:

$$dT = (2\pi r\, dr)vr$$

where $(2\pi r\, dr)$ is the area of the ring and v is the shear stress in the ring. Thus

$$T = \int_0^R (2\pi r\, dr)vr = \int_0^R 2\pi r^2 v\, dr \tag{15.1}$$

For a hollow section with internal radius R_1,

$$T = \int_{R_1}^R 2\pi r^2 v\, dr \tag{15.2}$$

For a solid section, using equation (15.1) and using $v = v_{\max} r/R$,

$$\begin{aligned} T &= \int_0^R 2\pi r^2 \left(\frac{v_{\max}\, r}{R}\right) dr = \left(\frac{2\pi}{R}\right) v_{\max} \int_0^R r^3\, dr \\ &= \left(\frac{2\pi}{R}\right) v_{\max} \times \frac{R^4}{4} = \left(\frac{\pi}{2}\right) v_{\max} R^3 \end{aligned} \tag{15.3}$$

$$v_{\max} = \frac{2T}{\pi R^3}$$

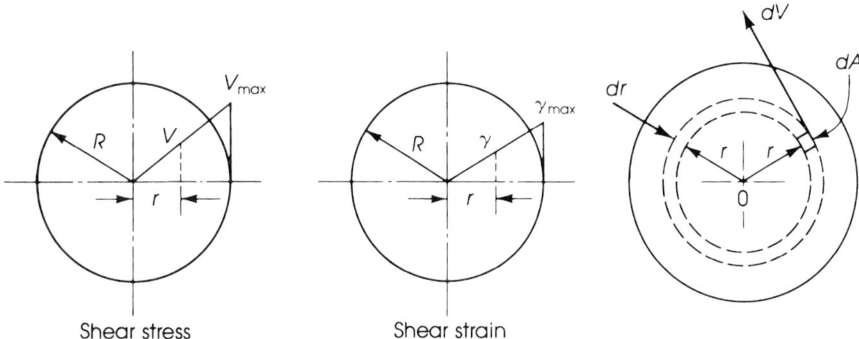

Figure 15.4 Torque in circular sections.

The polar moment of inertia of a circular section is $J = \pi R^4/2$. Therefore, the shear stress can be written as a function of the polar moment of inertia J as follows:

$$v_{\text{max}} = \frac{TR}{J} \tag{15.4}$$

15.4 TORSIONAL MOMENT IN RECTANGULAR SECTIONS

The determination of the stress in noncircular members subjected to torsional loading is not as simple as that for circular sections. However, results obtained from the theory of elasticity indicate that the maximum shearing stress for rectangular sections can be calculated as follows:

$$v_{\text{max}} = \frac{T}{\alpha x^2 y} \tag{15.5}$$

where

T = the applied torque
x = the short side of the rectangular section
y = the long side of the rectangular section
α = coefficient that depends on the ratio of y/x; its value is given in the following table.

y/x	**1.0**	**1.2**	**1.5**	**2.0**	**4**	**10**
α	0.208	0.219	0.231	0.246	0.282	0.312

The maximum shearing stress occurs along the centerline of the longer side y (Figure 15.5).

For members composed of rectangles, such as T-, L-, or I-sections, the value of α can be assumed equal to be $\frac{1}{3}$, and the section may be divided into several rectangular components having a long side y_i and a short side x_i. The maximum shearing stress can be calculated from

$$v_{\text{max}} = \frac{3T}{\Sigma x_i^2 y_i} \tag{15.6}$$

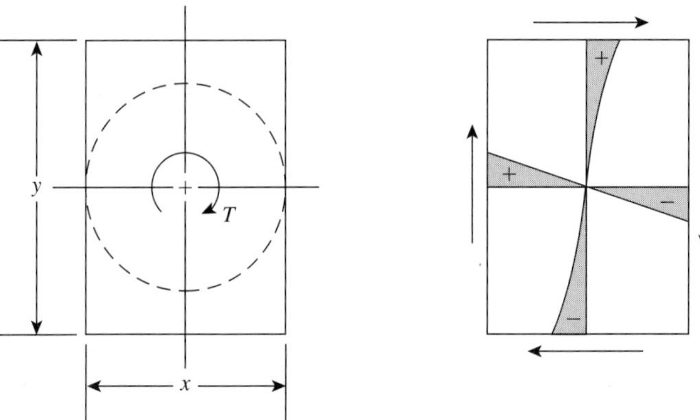

Figure 15.5 Stress distribution in rectangular sections due to pure torsion.

where $\Sigma x_i^2 y_i$ is the value obtained from the rectangular components of the section. When $y/x \leq 10$, a better expression may be used:

$$v_{max} = \frac{3T}{\Sigma x^2 y(1 - 0.63x/y)} \tag{15.7}$$

15.5 COMBINED SHEAR AND TORSION

In most practical cases, a structural member may be subjected simultaneously to both shear and torsional forces. Shear stresses will be developed in the section, as was explained in Chapter 8, with an average shear stress $= v_1$ in the direction of the shear force V, Figure 15.6(a). The torque T produces torsional stresses along all sides of the rectangular section $ABCD$, Figure 15.6(a), with $v_3 > v_2$. The final stress distribution is obtained by adding the effect of both shear and torsion stresses to produce maximum value of $(v_1 + v_3)$ on side CD, whereas side AB will have a final stress of $(v_1 - v_3)$. Both sides AD and BC will be subjected to torsional stress v_2 only. The section must be designed for the maximum $v = (v_1 + v_3)$.

15.6 TORSION THEORIES FOR CONCRETE MEMBERS

Various methods are available for the analysis of reinforced concrete members subjected to torsion or simultaneous torsion, bending, and shear. The design methods rely generally on two basic theories: the skew bending theory and the space truss analogy.

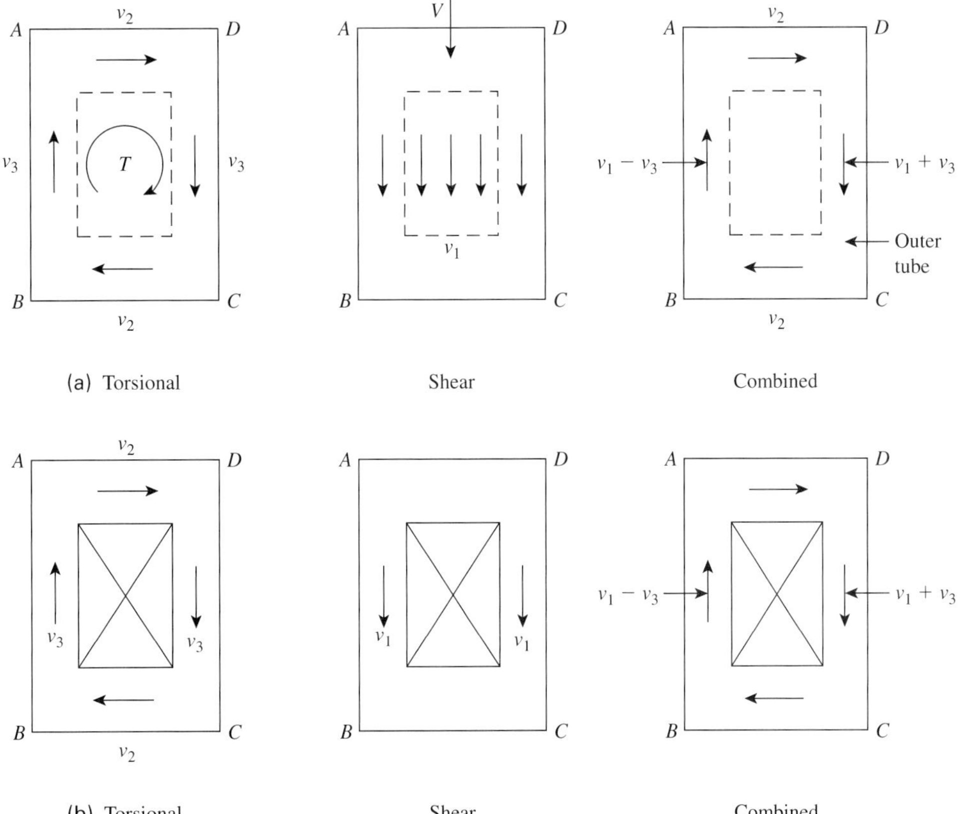

Figure 15.6 Combined shear and torsional stresses: (a) solid sections and (b) hollow sections.

15.6.1 Skew Bending Theory

The skew bending concept was first presented by Lessig in 1959 [2] and was further developed by Goode and Helmy [3], Collins et al. in 1968 [4], and Below, Rangan, and Hall in 1975 [5]. The concept was applied to reinforced concrete beams subjected to torsion and bending. Expressions for evaluating the torsional capacity of rectangular sections were presented by Hsu in 1968 [6], [7] and were adopted by ACI Code of 1971. Torsion theories for concrete members were discussed by Zia [8]. Empirical design formulas were also presented by Victor et al. in 1976 [9].

The basic approach of the skew bending theory, as presented by Hsu, is that failure of a rectangular section in torsion occurs by bending about an axis parallel to the wider face of the section y and inclined at about 45° to the longitudinal axis of the beam (Figure 15.7). Based on this approach, the minimum torsional moment, T_n, can be evaluated as follows:

$$T_n = \left(\frac{x^2 y}{3}\right) f_r \tag{15.8}$$

where f_r is the modulus of rupture of concrete; f_r is assumed to be $5\sqrt{f'_c}$ in this case, as compared to $7.5\sqrt{f'_c}$ adopted by the ACI Code for the computation of deflection in beams.

The torque resisted by concrete is expressed as follows:

$$T_c = \left(\frac{2.4}{\sqrt{x}}\right) x^2 y \sqrt{f'_c} \tag{15.9}$$

and the torque resisted by torsional reinforcement is

$$T_s = \frac{\alpha_1 (x_1 y_1 A_t f_y)}{s} \tag{15.10}$$

Thus, $T_n = T_c + T_s$, where T_n is the nominal torsional moment capacity of the section.

15.6.2 Space Truss Analogy

The space truss analogy was first presented by Rausch in 1929 and was further developed by Lampert [10], [11], who supported his theoretical approach with extensive experimental work. The Canadian Code provisions for the design of reinforced concrete beams in

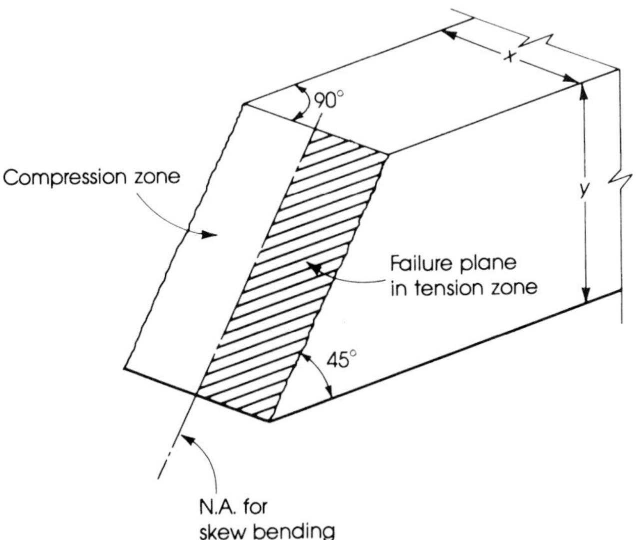

Figure 15.7 Failure surface due to skew bending.

torsion and bending are based on the space truss analogy. Mitchell and Collins [12] presented a theoretical model for structural concrete in pure torsion. McMullen and Rangan [13] discussed the design concepts of rectangular sections subjected to pure torsion. In 1983, Solanki [14] presented a simplified design approach based on the theory presented by Mitchell and Collins.

The concept of the space truss analogy is based on the assumption that the torsional capacity of a reinforced concrete rectangular section is derived from the reinforcement and the concrete surrounding the steel only. In this case, a thin-walled section is assumed to act as a space truss (Figure 15.8). The inclined spiral concrete strips between cracks resist the compressive forces, whereas the longitudinal bars at the corners and stirrups resist the tensile forces produced by the torsional moment.

The behavior of a reinforced concrete beam subjected to pure torsion can be represented by an idealized graph relating the torque to the angle of twist, as shown in Figure 15.9.

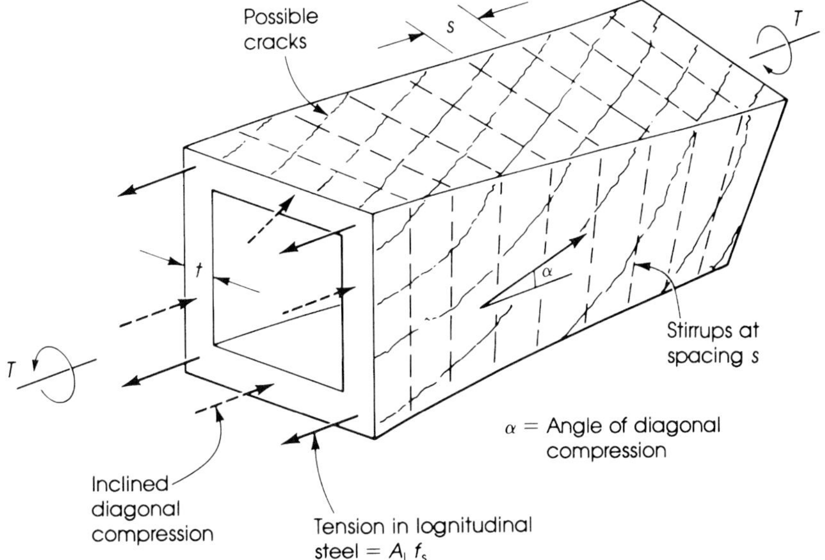

Figure 15.8 Forces on section in torsion (space truss analogy).

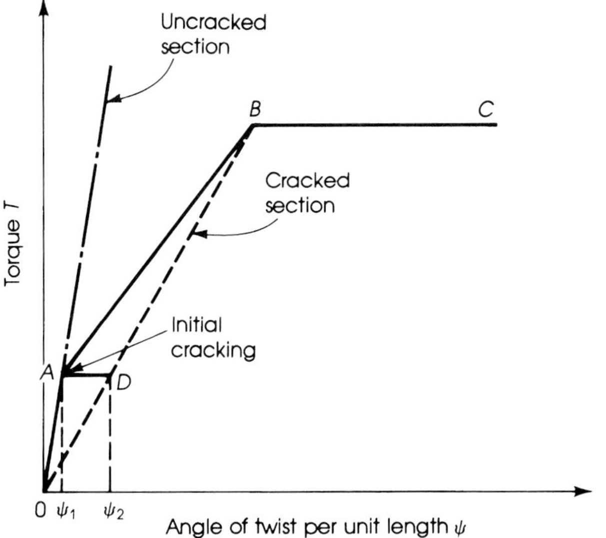

Figure 15.9 Idealized torque versus twist relationship.

It can be seen that prior to cracking, the concrete resists the torsional stresses and the steel is virtually unstressed. After cracking, the elastic behavior of the beam is not applicable, and hence a sudden change in the angle of twist occurs, which continues to increase until the maximum torsional capacity is reached. An approximate evaluation of the torsional capacity of a cracked section may be expressed as follows:

$$T_n = 2\left(\frac{A_t f_y}{s}\right) x_1 y_1 \tag{15.11}$$

where A_t = area of one leg of stirrups
s = spacing of stirrups
x_1 and y_1 = short and long distances, center to center of closed rectangular stirrups or corner bars

The preceding expression neglects the torsional capacity due to concrete. Mitchell and Collins [12] presented the following expression to evaluate the angle of twist per unit length ψ:

$$\psi = \left(\frac{P_0}{2A_0}\right)\left[\left(\frac{\varepsilon_1}{\tan\alpha}\right) + \left(\frac{P_h(\varepsilon_h \tan\alpha)}{P_0}\right) + \frac{2\varepsilon_d}{\sin\alpha}\right] \tag{15.12}$$

where

ε_1 = strain in the longitudinal reinforcing steel
ε_h = strain in the hoop steel (stirrups)
ε_d = concrete diagonal strain at the position of the resultant shear flow
P_h = hoop centerline perimeter
α = angle of diagonal compression $= (\varepsilon_d + \varepsilon_1)/\left[\varepsilon_d + \varepsilon_h\left(\frac{P_h}{P_0}\right)\right]$
A_0 = area enclosed by shear, or
= torque/$2q$ (where q = shear flow)
P_0 = perimeter of the shear flow path (perimeter of A_0)

The preceding twist expression is analogous to the curvature expression in flexure (Figure 15.10):

$$\phi = \text{curvature} = \frac{\varepsilon_c + \varepsilon_s}{d} \tag{15.13}$$

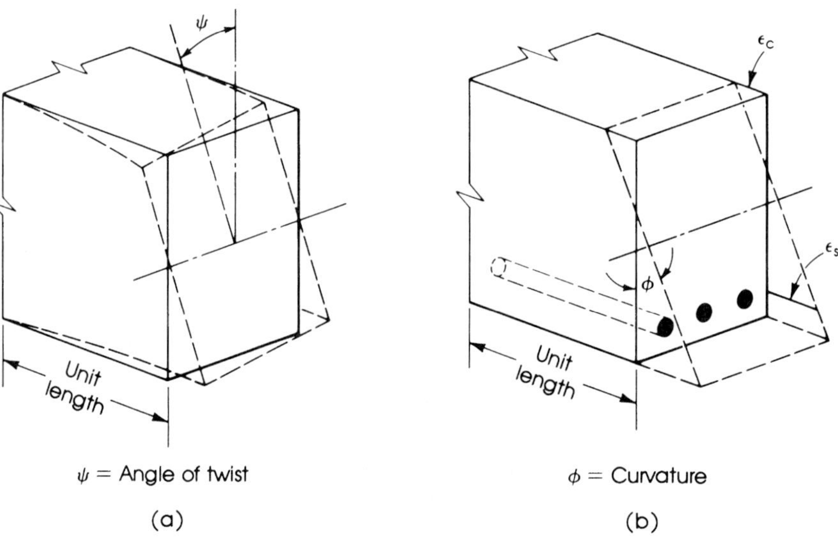

Figure 15.10 (a) Torsion and (b) flexure.

where ε_c and ε_s are the strains in concrete and steel, respectively. A simple equation is presented by Solanki [14] to determine the torsional capacity of a reinforced concrete beam in pure torsion, based on the space truss analogy, as follows:

$$T_u = (2A_0)\left[\left(\frac{\Sigma A_s f_{sy}}{P_0}\right) \times \left(\frac{A_h f_{hy}}{s}\right)\right]^{1/2} \tag{15.14}$$

where A_0, P_0, and s are as explained before and

$$\Sigma A_s f_{sy} = \text{yield force of all the longitudinal steel bars}$$

$$A_h f_{hy} = \text{yield force of the stirrups}$$

The ACI Code adapted this theory to design concrete structural members subjected to torsion or shear and torsion in a simplified approach.

15.7 TORSIONAL STRENGTH OF PLAIN CONCRETE MEMBERS

Concrete structural members subjected to torsion will normally be reinforced with special torsional reinforcement. In case that the torsional stresses are relatively low and need to be calculated for plain concrete members, the shear stress, v_{tc}, can be estimated using equation (15.6):

$$v_{tc} = \frac{3T}{\phi \Sigma x^2 y} \leq 6\sqrt{f'_c}$$

and the angle of twist is $\theta = 3TL/x^3 yG$, where T is the torque applied on the section (less than the cracking torsional moment) and G is the shear modulus and can be assumed equal to 0.45 times the modulus of elastic of concrete E_c; that is, $G = 25{,}700\sqrt{f'_c}$. The torsional cracking shear v_c in plain concrete may be assumed equal to $6\sqrt{f'_c}$. Therefore, for plain concrete rectangular sections,

$$T_c = 2\phi x^2 y \sqrt{f'_c} \tag{15.15}$$

and for compound rectangular sections,

$$T_c = 2\phi \sqrt{f'_c} \Sigma x^2 y \tag{15.16}$$

15.8 TORSION IN REINFORCED CONCRETE MEMBERS (ACI CODE PROCEDURE)

15.8.1 General

The design procedure for torsion is similar to that for flexural shear. When the ultimate torsional moment applied on a section exceeds that which the concrete can resist, torsional cracks develop, and consequently torsional reinforcement in the form of closed stirrups or hoop reinforcement must be provided. In addition to the closed stirrups, longitudinal steel bars are provided in the corners of the stirrups and are well distributed around the section. Both types of reinforcement, closed stirrups and longitudinal bars, are essential to resist the diagonal tension forces caused by torsion; one type will not be effective without the other. The stirrups must be closed, because torsional stresses occur on all faces of the section.

The reinforcement required for torsion must be added to that required for shear, bending moment, and axial forces. The reinforcement required for torsion must be provided such that the torsional moment strength of the section ϕT_n is equal to or exceeds the applied ultimate torsional moment T_u computed from factored loads.

$$\phi T_n \geq T_u \tag{15.17}$$

When torsional reinforcement is required, the torsional moment strength ϕT_n must be calculated assuming that all the applied torque, T_u, is to be resisted by stirrups and longitudinal bars with concrete torsional strength, $T_c = 0$. At the same time, the shear resisted by concrete, v_c, is assumed to remain unchanged by the presence of torsion.

15.8.2 Torsional Geometric Parameters

In the ACI Code, the design for torsion is based on the space truss analogy, as shown in Figure 15.8. After torsional cracking occurs, the torque is resisted by closed stirrups, longitudinal bars, and concrete compression diagonals. The concrete shell outside the stirrups becomes relatively ineffective and is normally neglected in design. The area enclosed by the centerline of the outermost closed stirrups is denoted by A_{oh}, the shaded area in Figure 15.11. Because other terms are used in the design equations, they are introduced here first to make the equation easier to comprehend. Referring to Figure 15.11, the given terms are defined as follows:

A_{cp} = area enclosed by outside perimeter of concrete section, in.^{2}

P_{cp} = outside perimeter of concrete gross area, A_{cp}, in.

A_{oh} = area enclosed by centerline of the outermost closed transverse torsional reinforcement, in.^{2} (shaded area in Figure 15.11)

A_0 = gross area enclosed by shear flow path and may be taken equal to $0.85A_{oh}$ (A_0 may also be determined from analysis [18], [19].)

P_h = perimeter of concrete of outermost closed transverse torsional reinforcement

θ = angle of compression diagonals between 30° and 60° (may be taken equal to 45° for reinforced concrete members)

In T- and L-sections, the effective overhang width of the flange on one side is limited to the projection of the beam above or below the slab, whichever is greater, but not greater than four times the slab thickness (ACI Code, Sections 11.6.1 and 13.2.4).

15.8.3 Cracking torsional moment T_{cr}

The cracking moment under pure torsion, T_{cr}, may be derived by replacing the actual section, prior to cracking, with an equivalent thin-walled tube, $t = 0.75A_{cp}/P_{cp}$, and an area enclosed by the wall centerline, $A_0 = 2A_{cp}/3$. When the maximum tensile stress (principal stress) reaches $4\sqrt{f'_c}$, cracks start to occur and the torque T in general is equal to

$$T = 2A_0\tau t \tag{15.18}$$

where τ = the torsional shear stress = $4\sqrt{f'_c}$ for torsional cracking.

Replacing τ by $4\sqrt{f'_c}$,

$$T_{cr} = 4\sqrt{f'_c}\left(\frac{A_{cp}^2}{P_{cp}}\right) = T_n \quad \text{and} \quad T_u = \phi T_{cr} \tag{15.19}$$

Assuming that a torque less than or equal to $T_{cr}/4$ will not cause a significant reduction in the flexural or shear strength in a structural member, the ACI Code, Section 11.6.1, permits neglect of torsion effects in reinforced concrete members when the factored torsional moment $T_u \leq \phi T_{cr}/4$, or

$$T_u \leq \phi\sqrt{f'_c}\left(\frac{A_{cp}^2}{P_{cp}}\right) = T_a \tag{15.20}$$

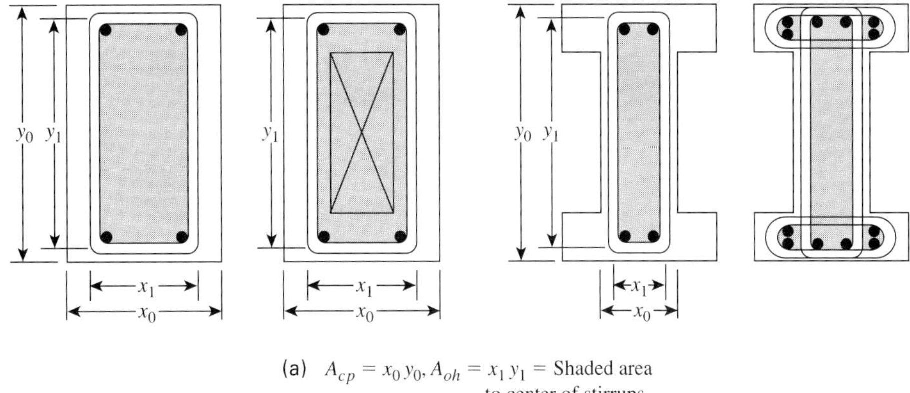

(a) $A_{cp} = x_0 y_0, A_{oh} = x_1 y_1 =$ Shaded area to center of stirrups

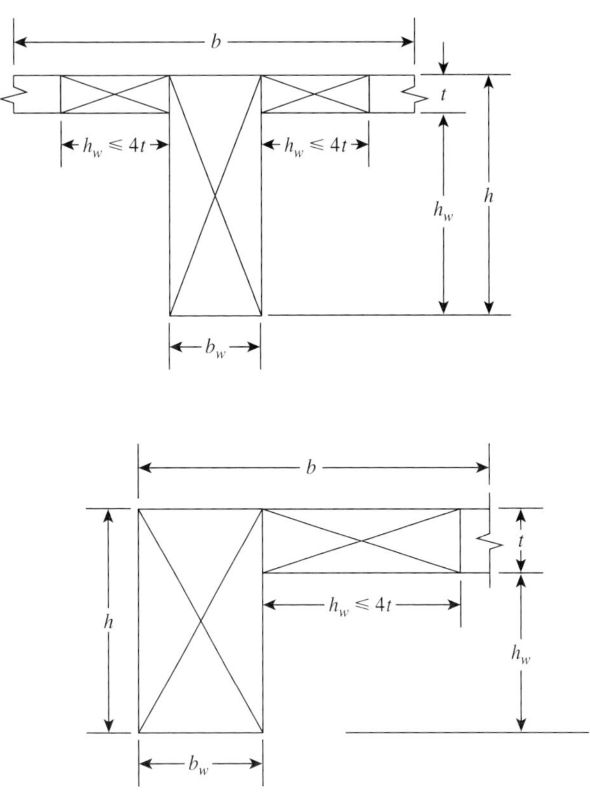

(b)

Figure 15.11 (a) Torsional geometric parameters; (b) effective flange width for T- and L-sections and component rectangles.

When T_u exceeds the value in equation (15.20), all T_u must be resisted by closed-stirrup and longitudinal bars. The torque, T_u, is calculated at a section located at distance d from the face of the support and $T_u = \phi T_n$, where $\phi = 0.85$.

Example 15.2

For the three sections shown in Figure 15.12, and based on the ACI Code limitations, it is required to compute the following:

a. The cracking moment ϕT_{cr}

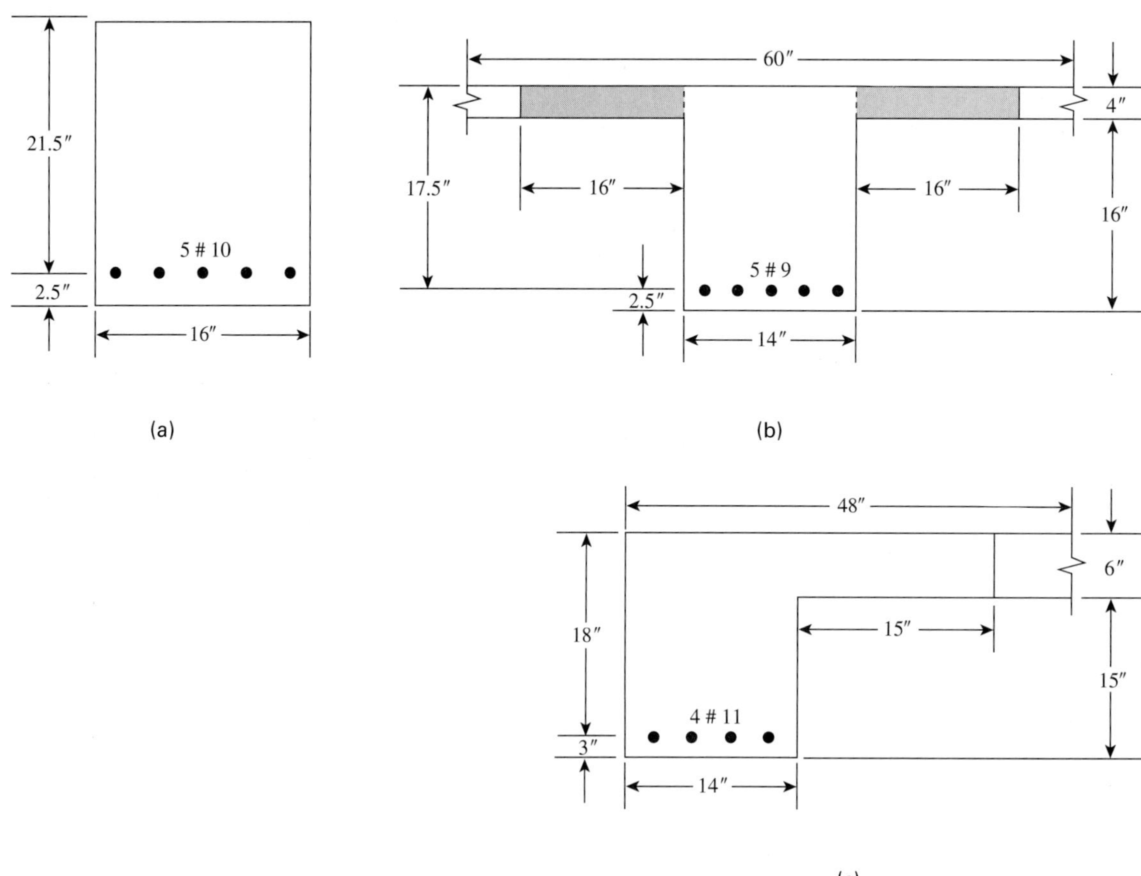

Figure 15.12 Example 15.2.

b. The maximum ultimate torque ϕT_n that can be applied to each section without using torsional web reinforcement

Assume $f'_c = 4$ Ksi, $f_y = 60$ Ksi, a 1.5-in. concrete cover, and no. 4 stirrups.

Solution

1. Section 1

a. Cracking moment, ϕT_{cr}, can be calculated from equation (15.19).

$$\phi T_{cr} = \phi 4\sqrt{f'_c}\left(\frac{A_{cp}^2}{P_{cp}}\right)$$

For this section, $A_{cp} = x_0 y_0$, the gross area of the section, where $x_0 = 16$ in. and $y_0 = 24$ in.

$$A_{cp} = 16(24) = 384 \text{ in.}^2$$

$$P_{cp} = \text{perimeter of the gross section}$$

$$= 2(x_0 + y_0) = 2(16 + 24) = 80 \text{ in.}$$

$$\phi T_{cr} = \frac{0.85(4)\sqrt{4000}\,(384)^2}{80} = 396.4 \text{ K}\cdot\text{in.}$$

b. The allowable ϕT_n that can be applied without using torsional reinforcement is computed from equation (15.20):

$$T_a = \frac{\phi T_r}{4} = \frac{396.4}{4} = 99.1 \text{ K}\cdot\text{in.}$$

2. Section 2

a. First calculate A_{cp} and P_{cp} for this section and apply equation (15.19) to calculate ϕT_r. Assuming flanges are confined with closed stirrups, the effective flange part to be used on each side of the web is equal to four times the flange thickness, or $4(4) = 16$ in. $= h_w = 16$ in.

$$A_{cp} = \text{web area } (b_w h) + \text{area of effective flanges}$$

$$= (14 \times 20) + 2(16 \times 4) = 408 \text{ in.}^2$$

$$P_{cp} = 2(b + h) = 2(14 + 2 \times 16 + 20) = 132 \text{ in.}^2$$

$$\phi T_{cr} = \frac{0.85(4)\sqrt{(4000)}\,(408)^2}{132} = 271.2 \text{ K}\cdot\text{in.}$$

Note: If the flanges are neglected and the torsional reinforcement is confined in the web only, then

$$A_{cp} = 14(20) = 280 \text{ in.}^2, \qquad P_{cp} = 2(14 + 20) = 68 \text{ in.}, \quad \text{and} \quad \phi T_{cr} = 248 \text{ K}\cdot\text{in.}$$

b. The allowable ϕT_n that can be applied without using torsional reinforcement is $\phi T_{cr}/4 = 271.2/4 = 67.8$ K·in.

3. Section 3

a. Assuming flange is confined with closed stirrups, effective flange width is equal to $b_w = 15$ in. $< 4 \times 6 = 24$ in.

$$A_{cp} = (14 \times 21) + (15 \times 6) = 384 \text{ in.}^2$$

$$P_{cp} = 2(b + h) = 2(14 + 15 + 21) = 100 \text{ in.}$$

$$\phi T_{cr} = \frac{0.85(4)\sqrt{(4000)}\,(384)^2}{100} = 317.1 \text{ K}\cdot\text{in.}$$

Note: If the flanges are neglected, then $A_{cp} = 294$ in.^{2}, $P_{cp} = 70$ in., and $\phi T_{cr} = 265.5$ K·in.

b. The allowable $\phi T_n = \phi T_{cr}/4 = 317.1/4 = 79.3$ K·in.

15.8.4 Equilibrium Torsion and Compatibility Torsion

Structural analysis of concrete members gives the different forces acting on the member, such as normal forces, bending moments, shear forces, and torsional moments, as explained in the simple problem of Example 15.1. The design of a concrete member is based on failure of the member under ultimate loads. In statically indeterminate members, a redistribution of moments occurs before failure; consequently, design moments may be reduced, whereas in statically determinate members, such as a simple beam or a cantilever beam, no moment redistribution occurs.

In the design of structural members subjected to torsional moments two possible cases may apply after cracking.

1. The *equilibrium torsion case* occurs when the torsional moment is required for the structure to be in equilibrium and T_u cannot be reduced by redistribution of moments, as in the case of simple beams. In this case torsion reinforcement must be provided to resist all of T_u. Figure 15.13 shows an edge beam supporting a cantilever slab where no redistribution of moments will occur [18] [19].
2. The *compatibility torsion case* occurs when the torsional moment, T_u, can be reduced by the redistribution of internal forces after cracking while compatibility of deformation is maintained in the structural member. Figure 15.14 shows an example of this case, where two transverse beams are acting on an edge beam producing twisting moments. At torsional cracking, a large twist occurs, resulting in a large distribution of forces in the structure [18] [19]. The cracking torque, T_{cr}, under combined

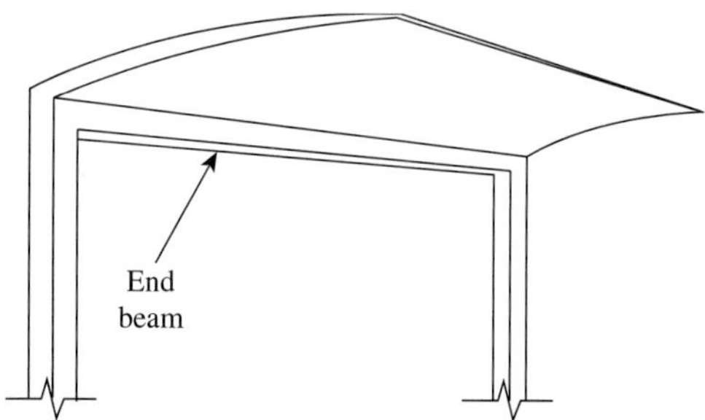

Figure 15.13 Design torque may not be reduced. Moment redistribution is not possible [19].

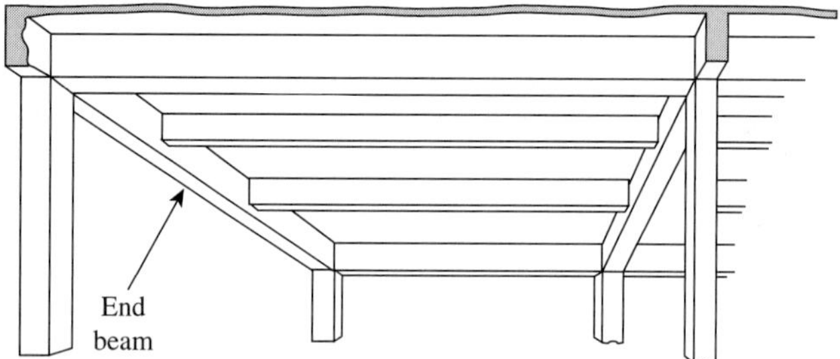

Figure 15.14 Design torque may be reduced in a spandrel beam. Moment redistribution is possible [19].

flexure, shear, and torsion is reached when the principal stress in concrete is about $4\sqrt{f'_c}$. When $T_u > T_{cr}$, a torque equal to T_{cr}, equation (15.19), may be assumed to occur at the critical sections near the faces of the supports.

The ACI Code limits the design torque to the smaller of T_u from factored loads or ϕT_{cr} from equation (15.19).

15.8.5 Limitation of Torsional Moment Strength

The ACI Code, Section 11.6.3, limits the size of the cross-sectional dimension by the following two equations:

1. For solid sections:

$$\sqrt{\left(\frac{V_u}{b_w d}\right)^2 + \left(\frac{T_u P_h}{1.7A_{oh}^2}\right)^2} \leq \phi\left[\left(\frac{V_c}{b_w d}\right) + 8\sqrt{f'_c}\right] \tag{15.21}$$

2. For hollow sections:

$$\left(\frac{V_u}{b_w d}\right) + \left(\frac{T_u P_h}{1.7A_{oh}^2}\right) \leq \phi\left[\left(\frac{V_c}{b_w d}\right) + 8\sqrt{f'_c}\right] \tag{15.22}$$

where $V_c = 2\sqrt{f'_c}b_w d$ = shear strength for normal weight concrete. All other terms were defined in Section 15.8.2.

This limitation is based on the fact that the sum of the stresses due to shear and torsion (on the left-hand side) may not exceed the cracking stress plus $8\sqrt{f'_c}$. The same condition was applied to the design of shear without torsion in Chapter 8. The limitation is needed to reduce cracking and to prevent crushing of the concrete surface due to inclined shear and torsion stresses.

15.8.6 Hollow Sections

Combined shear and torsional stresses in a hollow section is shown in Figure 15.6, where the wall thickness is assumed constant. In some hollow sections, the wall thickness may vary around the perimeter. In this case equation (15.22) should be evaluated at the location where the left-hand side is maximum. Note that at the top and bottom flanges, the shear stresses are usually negligible. In general, if the wall thickness of a hollow section t is less than A_{oh}/P_h, then equation (15.22) becomes

$$\frac{V_u}{b_w d} + \frac{T_u}{1.7A_{oh}t} \leq \phi\left[\left(\frac{V_c}{b_w d}\right) + 8\sqrt{f'_c}\right] \tag{15.23}$$

(ACI Code, Section 11.6.3).

15.8.7 Web Reinforcement

As was explained earlier, the ACI Code approach for the design of the members due to torsion is based on the space truss analogy in Figure 15.8. After torsional cracking, two types of reinforcement are required to resist the applied torque T_u: transverse reinforcement, A_t, in the form of closed stirrups, and longitudinal reinforcement, A_l, in the form of longitudinal bars. The ACI Code, Section 11.6.3, presented the following expression to compute A_t and A_l:

1. Closed stirrups A_t can be calculated as follows.

$$T_n = \frac{2A_0 A_t f_{yv} \cot\theta}{s} \tag{15.24}$$

where $T_n = \dfrac{T_u}{\phi}$ and $\phi = 0.85$

A_t = area of one leg of the transverse closed stirrups

f_{yv} = yield strength of $A_t \leq 60$ Ksi

s = spacing of stirrups

A_0 and θ were defined in Section 15.8.2. Equation (15.24) can be written as follows:

$$\frac{A_t}{s} = \frac{T_n}{2A_0 f_{yv} \cot\theta} \tag{15.25}$$

If $\theta = 45°$, then $\cot\theta = 1.0$, and if $f_{yv} = 60$ Ksi, then equation (15.25) becomes

$$\frac{A_t}{s} = \frac{T_n}{120A_0} \tag{15.26}$$

where T_n is in kips · in. Spacing of stirrups, s, should not exceed the smaller of $P_h/8$ or 12 in. For hollow sections in torsion, the distance measured from the centerline of stirrups to the inside face of the wall shall not be less than $0.5A_{oh}/P_h$.

2. The additional longitudinal reinforcement, A_l, required for torsion should not be less the following:

$$A_l = \left(\frac{A_t}{s}\right)P_h\left(\frac{f_{yv}}{f_{yl}}\right)\cot^2\theta \tag{15.27}$$

If $\theta = 45°$ and $f_{yv} = f_{yl} = 60$ Ksi for both stirrups and longitudinal bars, then equation (15.27) becomes

$$A_l = \left(\frac{A_t}{s}\right)P_h = 2\left(\frac{A_t}{s}\right)(x_1 + y_1) \tag{15.28}$$

P_h was defined in Section 15.8.2. Note that reinforcement required for torsion should be added to that required for the shear, moment, and axial force that act in combination with torsion. Other limitations for the longitudinal reinforcement, A_l, are as follows:

a. The smallest bar diameter of a longitudinal bar is that of no. 3 or stirrup spacing $s/24$, whichever is greater.
b. The longitudinal bars should be distributed around the perimeter of the closed stirrups with a maximum spacing of 12 in.
c. The longitudinal bars must be inside the stirrups with at least one bar in each corner of the stirrups. Corner bars are found to be effective in developing torsional strength and in controlling cracking.
d. Torsional reinforcement should be provided for a distance $(b_t + d)$ beyond the point theoretically required, where b_t is the width of that part of the cross section containing the stirrups resisting torsion.

15.8.8 Minimum Torsional Reinforcement

Where torsional reinforcement is required, the minimum torsional reinforcement may be computed as follows (ACI Code, Section 11.6.5):

1. Minimum transverse closed stirrups for combined shear and torsion:

$$A_v + 2A_t \geq \frac{50b_w s}{f_{yv}} \tag{15.29}$$

where
A_v = area of two legs of a closed stirrup determined from shear
A_t = area of one leg of closed stirrup determined from torsion
s = spacing of stirrups
f_{yv} = yield strength of closed stirrups $\leq$ 60 Ksi

Spacing of stirrups s should not exceed $P_h/8$ or 12 in., whichever is smaller. This spacing is needed to control cracking width.

2. Minimum total area of longitudinal torsional reinforcement:

$$A_{l\,\min} = \left(\frac{5\sqrt{f'_c}\,A_{cp}}{f_{yl}}\right) - \left(\frac{A_t}{s}\right)P_h\left(\frac{f_{yv}}{f_{yl}}\right) \tag{15.30}$$

where A_t/s shall not be taken less than $25b_w/f_{yv}$.

The minimum A_l in equation (15.30) was determined to provide a minimum ratio of the volume of torsional reinforcement to the volume of concrete of about 1% for reinforced concrete subjected to pure torsion.

15.9 SUMMARY OF ACI CODE PROCEDURE

The design procedure for combined shear and torsion can be summarized as follows:

1. Calculate the ultimate shearing force, V_u, and the ultimate torsional moment, T_u, from the applied forces on the structural member. Critical values for shear and torsion are at a section distance d from the face of the support.
2. **a.** Shear reinforcement is needed when $V_u > \phi V_c/2$, where $V_c = 2\sqrt{f'_c}\,b_w d$.
   b. Torsional reinforcement is needed when

$$T_u > \phi\sqrt{f'_c}\left(\frac{A_{cp}^2}{P_{cp}}\right) \tag{15.20}$$

If web reinforcement is needed, proceed as follows.

3. Design for shear:

a. Calculate the nominal shearing strength provided by the concrete, V_c. Determine the shear to be carried by web reinforcement,

$$V_u = \phi V_c + V_s \quad \text{or} \quad V_s - \frac{V_u - \phi V_c}{\phi}$$

b. Compare the calculated V_s with maximum permited value of $(8\sqrt{f'_c}\,b_w d)$ according to the ACI Code. If calculated V_s is less, proceed with the design; if not, increase the dimensions of the concrete section.

c. The shear web reinforcement is calculated as follows:

$$A_v = \frac{V_s s}{f_y d}$$

where

A_v = area of two legs of the stirrup and s = spacing of stirrups.

The shear reinforcement per unit length of beam is

$$\frac{A_v}{s} = \frac{V_s}{f_y d}$$

d. Check A_v/s calculated with the minimum A_v/s:

$$(\text{min})\,\frac{A_v}{s} = \frac{50 b_w}{f_y}$$

The minimum $A_v = 50 b_w s/f_y$, specified by the code under the combined action of shear and torsion, is given in Step 5.

4. Design for torsion:

a. Check if the factored torsional moment, T_u, causes equilibrium or computability torsion. For equilibrium torsion, use T_u. For compatibility torsion, the design torsional moment is the smaller of T_u from factored load and

$$T_{u2} = \phi 4\sqrt{f'_c}\left(\frac{A_{cp}^2}{P_{cp}}\right) \tag{15.19}$$

b. Check that the size of the section is adequate. This is achieved by checking either equation (15.21) for solid sections or equation (15.22) for hollow sections. If the left-hand-side value is greater than $\phi(V_c/b_w d + 8\sqrt{f'_c})$, then increase the cross section. If it is less than that value, proceed. For hollow sections, check if the wall thickness t is less than A_{oh}/P_h. If it is less, use equation (15.23) instead of equation (15.22); otherwise use equation (15.22).

c. Determine the closed stirrups required from equation (15.25):

$$\frac{A_t}{s} = \frac{T_n}{2A_0 f_{fyv}\cot\theta} \tag{15.25}$$

A_t/s should not be less than $25 b_w/f_{yv}$. Also, the angle θ may be assumed to be 45°, $T_n = T_u/\phi$, and $\phi = 0.85$.

Assume $A_0 = 0.85 A_{oh} = 0.85(x_1 y_1)$, where x_1 and y_1 are the width and depth of the section to the centerline of stirrups; see Figure 15.11. Values of A_0 and θ may be obtained from analysis [18]. For $\theta = 45°$ and $f_{yv} = 60$ Ksi,

$$\frac{A_t}{s} = \frac{T_n}{120 A_0} \tag{15.26}$$

The maximum allowable spacing s is the smaller of 12 in. or $P_h/8$.

d. Determine the additional longitudinal reinforcement from equation (15.27):

$$A_l = \left(\frac{A_t}{s}\right)P_h\left(\frac{f_{yv}}{f_{yt}}\right)\cot^2\theta \qquad (15.27a)$$

but not less than

$$A_{l\,min} = \left(\frac{5\sqrt{f'_c}A_{cp}}{f_{yl}}\right) - \left(\frac{A_t}{s}\right)P_h\left(\frac{f_{yv}}{f_{yl}}\right) \qquad (15.27b)$$

For $\theta = 45°$ and $f_{yv} = 60$ Ksi, then $A_l = (A_t/s)P_h$. (15.28)

Bars should have a diameter of at least stirrup spacing, $s/24$, but not less than no. 3 bars. The longitudinal bars should be placed inside the closed stirrups with maximum spacing of 12 in. At least one bar should be placed at each corner of stirrups. Normally, one-third of A_l is added to the tension reinforcement, one-third at mid-height of the section, and one-third at the compression side.

5. Determine the total area of closed stirrups due to V_u and T_u.

$$A_{vt} = (A_v + 2A_t) \geq 50b_w s/f_{yv} \qquad (15.29)$$

Choose proper closed stirrups with a spacing s as the smaller of 12 in. or $P_h/8$.

The stirrups should be extended a distance $(b_t + d)$ beyond the point theoretically no longer required, where b_t = width of cross section resisting torsion.

Example 15.3 (Equilibrium Torsion)

Determine the necessary web reinforcement for the rectangular section shown in Figure 15.15. The section is subjected to an ultimate shear $V_u = 54$ K and an equilibrium torsion $T_u = 400$ K·in at a section located at a distance d from the face of the support. Use $f'_c = 4$ Ksi and $f_y = 60$ Ksi.

Solution

The following steps explain the design procedure:

1. Design forces are $V_u = 54$ K and an equilibrium torsion $T_u = 400$ K·in.
2. **a.** Shear reinforcement is needed when $V_u > \phi V_c/2$.

$$\phi V_c = \phi 2\sqrt{f'_c}bd = 0.85(2)\sqrt{4000}\,(16)(20.5) = 35.26 \text{ K}$$

$$V_u = 54 \text{ K} > \frac{\phi V_c}{2} = 17.63 \text{ K}$$

Shear reinforcement is required.

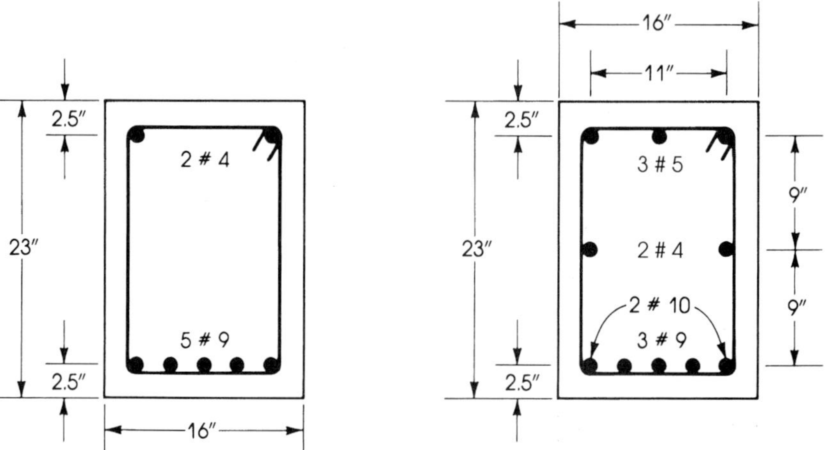

Figure 15.15 Example 15.3.

b. Torsional reinforcement is needed when

$$T_u > \phi\sqrt{f'_c}\left(\frac{A_{cp}^2}{P_{cp}}\right) = T_a$$

$$A_{cp} = x_0 y_0 = 16(23) = 368 \text{ in.}^2$$

$$P_{cp} = 2(x_0 + y_0) = 2(16 + 23) = 78 \text{ in.} \tag{15.20}$$

$$T_a = \frac{0.85\sqrt{4000}\,(368)^2}{78} = 93.33 \text{ K}\cdot\text{in.}$$

$$T_u = 400 \text{ K}\cdot\text{in.} > 93.33 \text{ K}\cdot\text{in.}$$

Torsional reinforcement is needed. Note that if T_u is less than 93.33 K · in., torsional reinforcement is not required, but shear reinforcement may be required.

3. Design for shear:

a. $V_u = \phi V_c + \phi V_s$, $\phi V_c = 35.26$ K
$54 = 35.26 + 0.85V_s$, $V_s = 22.0$ K

b. Maximum $V_s = 8\sqrt{f'_c}bd = 8\sqrt{4000}\,(16)(20.5) = 166 \text{ K} > V_s$

c. $A_v/s = V_s/f_y d = 22/(60 \times 20.5) = 0.018 \text{ in.}^2/\text{in.}$ (2 legs)
$A_v/2s = 0.018/2 = 0.009 \text{ in.}^2/\text{in.}$ (one leg)

4. Design for torsion:

a. Design $T_u = 400$ K · in. Determine sectional properties, assuming 1.5-in. concrete cover and no. 4 stirrups:

x_1 = width to center of stirrups $= 16 - 2(1.5 + 0.25) = 12.5$ in.

y_1 = depth to center of stirrups $= 23 - 2(1.5 + 0.25) = 19.5$ in.

Practically, x_1 can be assumed to be $b - 3.5$ in. and $y_1 = h - 3.5$ in.

$$A_{oh} = x_1 y_1 = (12.5 \times 19.5) = 244 \text{ in.}^2$$

$$A_0 = 0.85A_{oh} = 207.2 \text{ in.}^2$$

$$P_h = 2(x_1 + y_1) = 2(12.5 + 19.5) = 64 \text{ in.}$$

For $\theta = 45°$, $\cot\theta = 1.0$.

b. Check the adequacy of the size of the section using equation (15.21):

$$\sqrt{\left(\frac{V_u}{b_w d}\right)^2 + \left(\frac{T_u P_h}{1.7A_{oh}^2}\right)^2} \le \phi\left[\left(\frac{V_c}{b_w d}\right) + 8\sqrt{f'_c}\right]$$

$$\phi V_c = 35.26 \text{ K} \quad \text{and} \quad V_c = \frac{35.26}{0.85} = 41.5 \text{ K}$$

$$\text{Left-hand side} = \sqrt{\left(\frac{54{,}000}{16 \times 20.5}\right)^2 + \left(\frac{400{,}000 \times 64}{1.7(244)^2}\right)^2} = 302 \text{ psi} \tag{15.21}$$

$$\text{Right-hand side} = 0.85\left(\frac{41{,}500}{16 \times 20.5} + 8\sqrt{4000}\right) = 538 \text{ psi} > 302 \text{ psi}$$

The section is adequate.

c. Determine the required closed stirrups due to torsion from equation (15.25):

$$\frac{A_t}{s} = \frac{T_n}{2A_0 f_{yv}\cot\phi}$$

$$T_n = \frac{T_u}{\phi} = \frac{400}{0.85} = 470.6 \text{ K}\cdot\text{in.}, \quad \cot\theta = 1.0, \quad \text{and} \quad A_0 = 207.2 \text{ in.}^2$$

$$\frac{A_t}{s} = \frac{470.6}{2 \times 207.2 \times 60} = 0.019 \text{ in.}^2/\text{in.} \quad \text{(per one leg)}$$

d. Determine the additional longitudinal reinforcement from equation (15.27):

$$A_l = \left(\frac{A_t}{s}\right)P_h\left(\frac{f_{yv}}{f_{yl}}\right)\cot^2\theta$$

$$\frac{A_t}{s} = 0.019, \quad P_h = 64 \text{ in.}, \quad f_{yv} = f_{yl} = 60 \text{ Ksi}, \quad \text{and} \quad \cot\theta = 1.0$$

$$A_l = 0.019(64) = 1.21 \text{ in.}^2$$

$$\text{Min. } A_l = 5\sqrt{f'_c}A_{cp}/f_{yl} - \left(\frac{A_t}{s}\right)P_h\left(\frac{f_{yv}}{f_{yl}}\right)$$

$$A_{cp} = 368 \text{ in.}^2, \quad \frac{A_t}{s} = 0.019$$

$$f_{yv} = f_{yl} = 60 \text{ Ksi}$$

$$\text{Min. } A_l = \left[\frac{5\sqrt{4000}\,(368)}{60{,}000}\right] - (0.019 \times 64 \times 1.0) = 0.72 \text{ in.}^2$$

$$A_l = 1.21 \text{ in.}^2 \text{ controls}$$

5. Determine total area of closed stirrups:

a. For one leg of stirrups, $A_{vt}/s = A_t/s + A_v/2s$.

$$\text{Required } A_{vt} = \frac{0.018}{2} + 0.019 = 0.028 \text{ in.}^2/\text{in.} \qquad \text{(per one leg)}$$

Using no. 4 stirrups, area of one leg is 0.2 in.^{2}

$$\text{Spacing of stirrups} = \frac{0.2}{0.028} = 7.14 \text{ in.} \quad \text{or} \quad 7.0 \text{ in.}$$

b. Maximum $s = P_h/8 = \frac{64}{8} = 8$ in. or 12 in., whichever is smaller. The value of s used is 7.0 in. < 8 in.

c. Minimum $A_{vt}/s = 50b_w/f_{yv} = 50(16)/60{,}000 = 0.0133$ in.^{2}/in. This is less than 0.028 in.^{2}/in. provided.

6. To find the distribution of longitudinal bars, note that total $A_l = 1.21$ in.^{2} Use one-third at the top, or $1.21/3 = 0.4$ in.^{2}, to be added to the compression steel A'_s. Use one-third, or 0.4 in.^{2}, at the bottom, to be added to the tension steel, and one-third, or 0.4 in.^{2}, at mid-depth.

a. The total area of top bars is 0.4 (2 no. 4) + 0.4 = 0.8 in.^{2}; use three no. 5 bars $(A_s = 0.91 \text{ in.}^2)$.

b. The total area of bottom bars is 5 (5 no. 9) + 0.4 = 5.4 in.^{2}; use three no. 9 and two no. 10 bars at the corners (total $A_s = 5.53$ in.^{2}).

c. At middepth, use two no. 4 bars $(A_s = 0.4 \text{ in.}^2)$.

Reinforcement details are shown in Figure 15.15. Spacing of longitudinal bars is equal to 9 in., which is less than the maximum required of 12 in. The diameter of no. 4 bars used is greater than the minimum of no. 3 or stirrup spacing, or $s/24 = 0.21$ in.

Example 15.4 (Compatibility Torsion)

Repeat Example 15.3 if the factored torsional torque is a compatibility torsion.

Solution

Referring to the solution of Example 15.3:

1. Design forces are $V_u = 54$ K and compatibility torsion is 400 K·in.
2. Steps (a) and (b) are the same as in Example 15.3. Web reinforcement is required.
3. Step (c) is the same.
4. Design for torsion:

a. Because this is a compatibility torsion of 400 K · in., the design T_u is the smaller of 400 K · in. or ϕT_{cr} given in equation (15.19).

$$\phi T_{cr} = \phi 4\sqrt{f'_c}\left(\frac{A_{cp}^2}{P_{cp}}\right) = \frac{0.85(4)\sqrt{4000}\,(368)^2}{78} = 373.3\ \text{K}\cdot\text{in.} \tag{15.19}$$

Because $\phi T_{cr} < 400$ K · in., use $T_u = 373.3$ K · in. Repeat all the steps of Example 15.3 using $T_u = 373.3$ K · in. to determine that the section is adequate.

$$\frac{A_t}{s} = 0.018\ \text{in.}^2/\text{in.} \qquad \text{(one leg)}$$

$$A_l = 0.018(64) = 1.152\ \text{in.}^2$$

Use 1.2 in.^{2} > min. A_l.

5. Required $A_{vt} = 0.018/2 + 0.018 = 0.027$ in.^{2}/in. (one leg).

$$s = \frac{0.2}{0.027} = 7.4\ \text{in.}$$

Use 7 in. Choose bars, stirrups, and spacing similar to Example 15.3.

Example 15.5 (L-section with Equilibrium Torsion)

The edge beam of a building floor system is shown in Figure 15.16. The section at a distance d from the force of the support is subjected to $V_u = 60$ K and an equilibrium torque $T_u = 270$ K · in. Design the necessary web reinforcement using $f'_c = 4$ Ksi and $f_y = 60$ Ksi for all steel bars and stirrups.

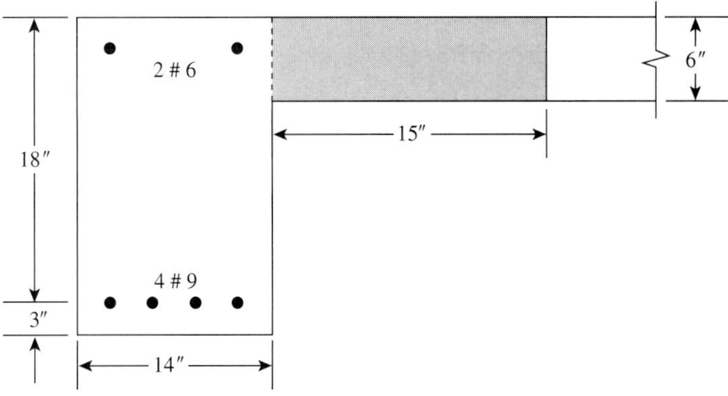

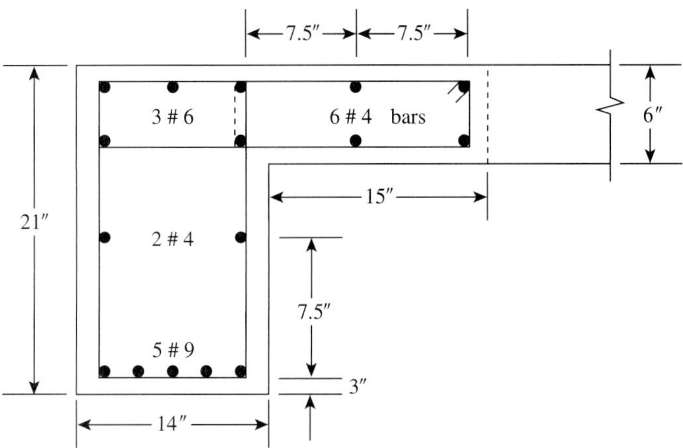

Figure 15.16 Example 15.4.

Solution

1. Design forces are $V_u = 60$ K and $T_u = 270$ K·in. = 22.5 K·ft.

2. a. Shear reinforcement is needed when $V_u > \phi V_c/2$.

$$\phi V_c = \phi 2\sqrt{f'_c}\, b_w d = 0.85(2)\sqrt{4000}\,(14)(18) = 27.1 \text{ K}$$

$$V_u > \frac{\phi V_c}{2} = 13.55 \text{ K}$$

Shear reinforcement is required.

b. Check if torsional reinforcement is needed. Assuming that flange is contributing to resist torsion, the effective flange length is $h_w = 15$ in. $< 4 \times 6 = 24$ in.

$$x_0 = 14 \text{ in.,} \quad \text{and} \quad y_0 = 21 \text{ in.}$$

$$A_{cp} = (14 \times 21)\ (\text{web}) + (15 \times 6)\ (\text{flange}) = 384 \text{ in.}^2$$

$$P_{cp} = 2(21 + 29) = 100 \text{ in.}$$

$$T_a\ (\text{equation (15.20)}) = \frac{0.85\sqrt{4000}\,(384)^2}{100} = 79.3 \text{ K·in.}$$

$$T_u > T_a$$

Torsional reinforcement is required.

3. Design for shear:

a. $V_u = \phi V_c + \phi V_s, \qquad 60 = 27.1 + 0.85 V_s, \qquad V_s = 38.7$ K

b. Maximum $V_s = 8\sqrt{f'_c}\, b_w d = 127.5 \text{ K} > V_s$

c. $$A_v/s = \frac{V_s}{f_y d} = \frac{38.7}{60 \times 18} = 0.036 \text{ in.}^2/\text{in.} \qquad (2 \text{ legs})$$

$$A_v/2s = \frac{0.036}{2} = 0.018 \text{ in.}^2/\text{in.}$$

4. Design for torsion: $T_u = 270$ K·in.

a. Determine section properties assuming a concrete cover of 1.5 in. and no. 4 stirrups.

Web: $x_1 = b - 3.5$ in. $= 14 - 3.5 = 10.5$ in., $\qquad y_1 = h - 3.5 = 21 - 3.5 = 17.5$ in.

Flange: $x_1 = 15$ in. (stirrups extend to the web), $\qquad y_1 = 6 - 3.5 = 2.5$ in.

$$A_{oh} = (15 \times 2.5) + (10.5 \times 17.5) = 221 \text{ in.}^2, \quad A_0 = 0.85 A_{oh} = 188 \text{ in.}^2$$

$$P_h = 2(15 + 2.5) + 2(10.5 + 17.5) = 91 \text{ in.,} \quad \theta = 45°, \quad \cot\theta = 1.0$$

b. Check the adequacy of the section using equation (15.21): $V_u = 60$ K, $\phi V_c = 27.1$ K, $V_c = 31.9$ K, $T_u = 270$ K·in.

$$\text{left-hand side} = \sqrt{\left(\frac{60{,}000}{14 \times 18}\right)^2 + \left[\frac{270{,}000 \times 91}{1.7(184)^2}\right]^2} = 380 \text{ psi}$$

$$\text{right-hand side} = 0.85\left[\frac{31{,}900}{14 \times 18} + 8\sqrt{4000}\right] = 538 \text{ psi}$$

The section is adequate.

c. Determine the torsional closed stirrups, A_t/s, from equation (15.25):

$$\frac{A_t}{s} = \frac{T_n}{2A_0 f_{yv}} = \frac{270}{0.85 \times 2 \times 188 \times 60} = 0.014 \text{ in.}^2/\text{in.} \qquad (\text{for one leg})$$

d. Calculate the additional longitudinal reinforcement from equation (15.28) (for $f_{yl} = 60$ Ksi and $\cot\theta = 1.0$):

$$A_l = \left(\frac{A_t}{s}\right)P_h = 0.014(91) = 1.28 \text{ in.}^2$$

$A_{l\,\min}$ (from equation (15.30)) is

$$A_l = \left[\frac{5\sqrt{4000}\,(384)}{60{,}000}\right] - (0.014 \times 91) = 0.75 \text{ in.}^2$$

The contribution of the flange may be neglected with slight difference in results, and less labor cost.

5. Determine the total area of the closed stirrups.

a. For one leg, $A_{vt} = A_t/s + A_v/2s$.

$$\text{Required}A_{vt} = 0.014 + 0.018 = 0.032 \text{ in.}^2/\text{in.} \qquad \text{(per leg)}$$

Choose no. 4 closed stirrups, area $= 0.2$ in.^{2}

$$\text{Spacing of stirrups} = \frac{0.2}{0.032} = 6.25 \text{ in.}$$

Use 6 in.

b. Max. $s = P_h/8 = 91/8 = 11.4$ in. Use $s = 6$ in., as calculated.

c. $A_{vt}/s = 50b_w/f_{yv} = 50(14)/60{,}000 = 0.017$ in.^{2}/in., which is less than the 0.032 in.^{2}/in. used. Use no. 4 closed stirrups spaced at 6 in.

6. Find the distribution of longitudinal bars. Total A_l is 1.28 in.^{2} Use one-third, or 0.43 in.^{2}, at the top, at the bottom, and at middepth.

a. Total top bars $= 0.88 + 0.43 = 1.31$ in.^{2}; use three no. 6 bars (1.32 in.^{2}).

b. Total bottom bars $= 4.0 + 0.43 = 4.43$ in.^{2}; use five no. 9 bars (5.0 in.^{2}).
Total A_l used $= (1.32 - 0.88) + (5 - 4) = 1.44$ in.^{2}

c. Use two no. 4 bars at middepth (0.40 in.^{2}).
Reinforcement details are shown in Figure 15.16. Spacing of longitudinal bars is at 7.5 in. < 12 in. The diameter of no. 4 bars used is 0.5 in., which is greater than no. 3 or stirrup spacing, $S/24 = \frac{6}{24} = 0.25$ in. Add no. 4 longitudinal bars on all corners of closed stirrups in beam web and flange.

SUMMARY

Sections 15.1–15.7

1. Torsional stresses develop in a beam when a moment acts on the beam section parallel to its surface.
2. In most practical cases, a structural member may be subjected to combined shear and torsional moments.
3. The design methods for torsion rely generally on two basic theories: the skew bending theory and the space truss theory. The ACI Code adopted the space truss theory.

Sections 15.8–15.9

A summary of the relative equations in U.S. customary units and SI units is given here.

Note that $(1.0\sqrt{f'_c})$ in psi is equivalent to $(0.08\sqrt{f'_c})$ in Mpa (N/mm^2), 1 in. $\approx$ 25 mm, and $f_{yv} \le 400$ MPa.

Equation	U.S. customary units	SI units
15.16	$T_c = 2\phi\sqrt{f'_c}\Sigma x^2 y$	$T_c = 0.17\phi\sqrt{f'_c}\Sigma x^2 y$
15.17	$\phi T_n \geq T_u$	Same
15.19	$T_{cr} = 4\sqrt{f'_c}\left(\frac{A_{cp}^2}{P_{cp}}\right)$	$T_{cr} = (\sqrt{f'_c}/3)(A_{cp}^2/P_{cp})$
15.20	$T_u \leq \phi\sqrt{f'_c}\left(\frac{A_{cp}^2}{P_{cp}}\right)$	$T_u \leq \phi(\sqrt{f'_c}/12)(A_{cp}^2/P_{cp})$
15.21	$\sqrt{\left(\frac{V_u}{b_w d}\right)^2 + \left(\frac{T_u P_h}{1.7A_{oh}^2}\right)^2} \leq \phi\left[\left(\frac{V_c}{b_w d}\right) + 8\sqrt{f'_c}\right]$ (U.S.)	
	$\sqrt{\left(\frac{V_u}{b_w d}\right)^2 + \left(\frac{T_u P_h}{1.7A_{oh}^2}\right)^2} \leq \phi\left[\left(\frac{V_c}{b_w d}\right) + (2\sqrt{f'_c}/3)\right]$ (S.I.)	
15.24	$T_n = \frac{2A_0 A_t f_{yv} \cot\theta}{s}$	Same
(Note that f_{yv} is in MPa, S is in mm, A_0 and A_t are in mm^2, and T_n is in kN · m.)		
15.25	$\frac{A_t}{s} = \frac{T_n}{2A_0 f_{yv}\cot\theta}$	Same
15.27	$A_l = \frac{A_t P_h(f_{yv}/f_{yl})\cot^2\theta}{s}$	Same
15.29	$A_v + 2A_t \geq \frac{50b_w s}{f_{yv}}$	$(A_v + 2A_t) \geq 0.35b_w s/f_{yv}$
15.30	$A_{l\,\min} = \left[\frac{5\sqrt{f'_c}A_{cp}}{f_{yl}}\right] - (A_t/s)P_h(f_{yv}/f_{yl})$	$A_{l\,\min} = [(5\sqrt{f'_c}A_{cp})/12f_{yl}] - \left(\frac{A_t}{s}\right)P_h\left(\frac{f_{yv}}{f_{yl}}\right)$

REFERENCES

1. S. Timoshenko and J. N. Goodier. *Theory of Elasticity.* New York: McGraw-Hill, 1951.
2. N. N. Lessig. "The Determination of the Load-bearing Capacity of Reinforced Concrete Elements with Rectangular Sections Subjected to Flexure and Torsion." Concrete and Reinforced Concrete Institute. Moscow, Work No. 5, 1959. (Translation: Foreign Literature Study No. 371, PCA Research and Development Laboratories, Skokie, III.)
3. C. D. Goode and M. A. Helmy. "Ultimate Strength of Reinforced Concrete Beams Subjected to Combined Torsion and Bending." In *Torsion of Structural Concrete,* Special Publication 18. American Concrete Institute. Detroit, 1968.
4. M. P. Collins, P. F. Walsh, F. E. Archer, and A. S. Hall. "Ultimate Strength of Reinforced Concrete Beams Subjected to Combined Torsion and Bending." In *Torsion of Structural Concrete,* Special Publication 18. American Concrete Institute. Detroit, 1968.
5. K. D. Below, B. V. Rangan, and A. S. Hall. "Theory for Concrete Beams in Torsion and Bending." *Journal of Structural Division,* ASCE, no. 98 (August 1975).
6. T. C. Hsu. "Ultimate Torque of Reinforced Concrete Beams." *Journal of Structural Division,* ASCE, no. 94 (February 1968).
7. T. C. Hsu. "Torsion of Structural Concrete—A Summary of Pure Torsion." In *Torsion of Structural Concrete,* Special Publication 18. American Concrete Institute. Detroit, 1968.

8. P. Zia. "Torsion Theories for Concrete Members." In *Torsion of Structural Concrete,* Special Publication 18. American Concrete Institute. Detroit, 1968.

9. J. Victor, N. Lakshmanan, and N. Rajagopalan. "Ultimate Torque of Reinforced Concrete Beams." *Journal of Structural Division,* ASCE, no. 102 (July 1976).

10. P. Lampert. "Postcracking Stiffness of Reinforced Concrete Beams in Torsion and Bending." Special Publication 35. American Concrete Institute. Detroit, 1973.

11. P. Lampert and B. Thurlimann. *Ultimate Strength and Design of Reinforced Concrete Beams in Torsion and Bending.* International Association for Bridge and Structural Engineering. Publication No. 31-I, 1971.

12. D. Mitchell and M. P. Collins. "The Behavior of Structural Concrete Beams in Pure Torsion." Department of Civil Engineering, University of Toronto. Publication No. 74-06, 1974.

13. A. E. McMullen and V. S. Rangan. "Pure Torsion on Rectangular Sections, a Reexamination." *ACI Journal* 75 (October 1978).

14. H. T. Solanki. "Behavior of Reinforced Concrete Beams in Torsion." Proceedings of the Institution of Civil Engineers 75 (March 1983).

15. Alan H. Mattock. "How to Design for Torsion." In *Torsion of Structural Concrete,* Special Publication 18. American Concrete Institute. Detroit, 1968.

16. T. C. Hsu and E. L. Kemp. "Background and Practical Application of Tentative Design Criteria for Torsion." *ACI Journal* 66 (January 1969).

17. D. Mitchell and M. P. Collins. "Detailing for Torsion." *ACI Journal* 73 (September 1976).

18. T. C. Hsu. *Unified Theory of Reinforced Concrete.* Boca Raton, Fla.: CRC Publication, 1993.

19. American Concrete Institute. "Building Code Requirements for Structural Concrete." ACI 318-99. Detroit, 1999.

PROBLEMS

For each problem, compute the cracking moment ϕT_{cr} and the maximum ultimate torque ϕT_n that can be applied without using torsional web reinforcement. Use $f'_c = 4$ Ksi and $f_y = 60$ Ksi.

15.1 A rectangular section with $b = 16$ in. and $h = 24$ in.

15.2 A rectangular section with $b = 12$ in. and $h = 20$ in.

15.3 A T-section with $b = 48$ in., $b_w = 12$ in., $t = 4$ in., and $h = 25$ in. Assume flanges are confined with closed stirrups.

15.4 A T-section with $b = 60$ in., $b_w = 16$ in., $t = 4$ in., and $h = 30$ in. Assume flanges are confined with closed stirrups.

15.5 An inverted L-section with $b = 32$ in., $b_w = 14$ in., $t = 6$ in., and $h = 24$ in. The flange does not have closed stirrups.

15.6 An inverted L-section with $b = 40$ in., $b_w = 12$ in., $t = 6$ in., and $h = 30$ in. The flange contains confined closed stirrups.

15.7 Determine the necessary web reinforcement for a simple beam subjected to an equilibrium torque $T_u = 220$ K·in. and $V_u = 36$ K. The beam section has $b = 14$ in., $h = 22$ in., $d = 19.5$ in., and is reinforced on the tension side by four no. 9 bars. Use $f'_c = 3$ Ksi and $f_y = 60$ Ksi.

15.8 Repeat Problem 15.7 using $f'_c = 4$ Ksi and $f_y = 60$ Ksi.

15.9 The section of an edge (spandrel) beam is shown in Figure 15.17. The critical section of the beam is subjected to an equilibrium torque $T_u = 300$ K·in and a shear $V_u = 60$ K. Determine the necessary web reinforcement using $f'_c = 4$ Ksi and $f_y = 60$ Ksi. Consider that the flange is not reinforced with closed stirrups.

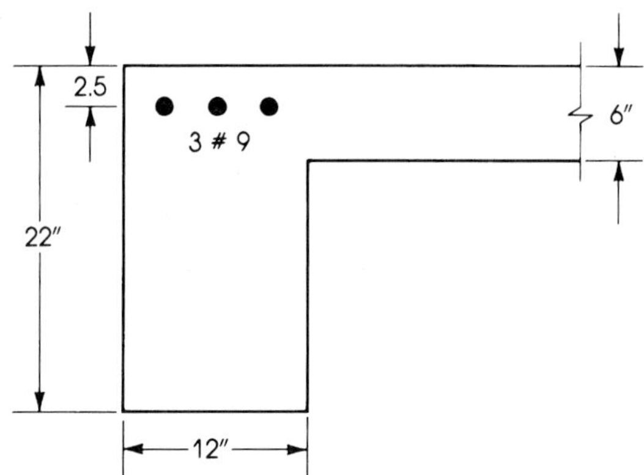

Figure 15.17 Problem 15.9.

15.10 Repeat Problem 15.9. Considering that the flange is effective and contains closed stirrups.

15.11 The T-section shown in Figure 15.18 is subjected to an ultimate shear $V_u = 28$ K and an ultimate equilibrium torque $T_u = 300$ K·in. and $M_u = 250$ K·ft. Design the necessary flexural and web reinforcement. Use $f'_c = 4$ Ksi and $f_y = 60$ Ksi.

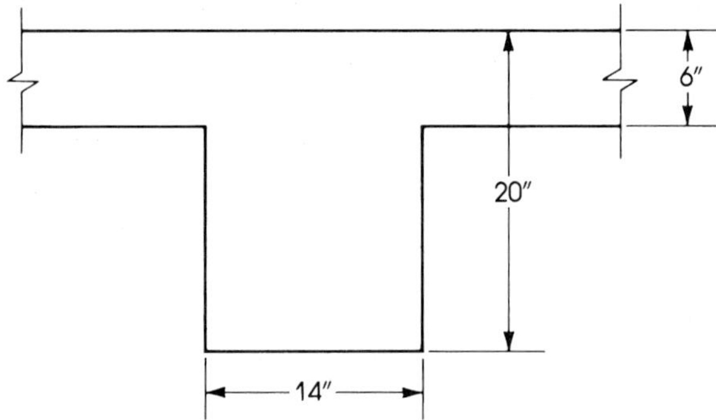

Figure 15.18 Problem 15.11.

15.12 Repeat Problem 15.11 if $V_u = 36$ K, $T_u = 360$ K·in., $M_u = 400$ K·ft., and $h = 24$ in.

15.13 Repeat Problem 15.11 using $f'_c = 3$ Ksi and $f_y = 60$ Ksi.

15.14 Repeat Problem 15.11 if T_u is a compatibility torsion.

15.15 Repeat Problem 15.13 if T_u is a compatibility torsion.

15.16 Repeat Problem 15.7 if T_u is a compatibility torsion.

15.17 The cantilever beam shown in Figure 15.19 is subjected to the ultimate forces shown.

a. Draw the axial and shearing forces and the bending and torsional moment diagrams.

b. Design the beam section at A using a steel percentage less than or equal to ρ_{max} for bending moment. Use $b = 16$ in. (300 mm), $f'_c = 4$ Ksi, and $f_y = 60$ Ksi.

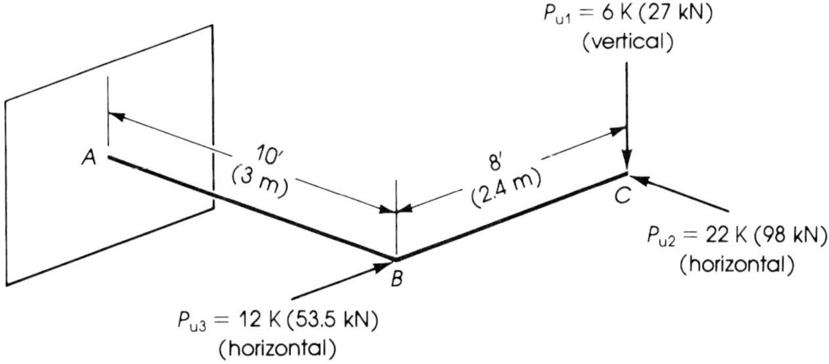

Figure 15.19 Problem 15.17.

15.18 The size of the slab shown in Figure 15.20 is 16 by 8 ft; it is supported by the beam AB, which is fixed at both ends. The uniform dead load on the slab (including its own weight) equals 100 psf, and the uniform live load equals 80 psf. Design the section at support A of beam AB using $f'_c = 4$ Ksi, $f_y = 60$ Ksi, $b_w = 14$ in., $h = 20$ in., a slab thickness of 5 in., and the ACI Code requirements.

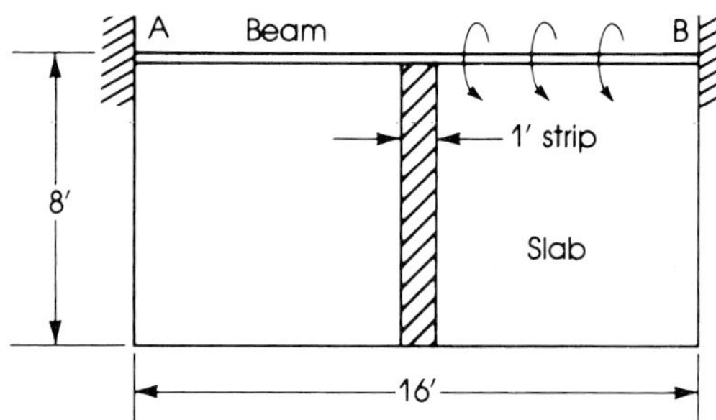

Figure 15.20 Problem 15.18.

Reinforced concrete parking structure, Minneapolis.

16 CONTINUOUS BEAMS AND FRAMES

16.1 INTRODUCTION

Reinforced concrete buildings consist of different types of structural members, such as slabs, beams, columns, and footings. These structural members may be cast in separate units as precast concrete slabs, beams, and columns or with the steel bars extending from one member to the other, forming a monolithic structure. Precast units are designed as structural members on simple supports unless some type of continuity is provided at their ends. In monolithic members, continuity in the different elements is provided, and the structural members are analyzed as statically indeterminate structures.

The analysis and design of continuous one-way slabs were discussed in Chapter 9, and the design coefficients and reinforcement details were shown in Figures 9.8 and 9.9. In one-way floor systems, the loads from slabs are transferred to the supporting beams, as shown in Figure 16.1(a). If the factored load on the slab is w_u psf, the uniform load on beams AB and BC per unit length is $w_u s$ plus the self-weight of the beam. The uniform load on beams DE and EF is $w_u s/2$ plus the self-weight of the beam. The load on column B equals $W_u LS$, whereas the loads on columns E, A, and D are $W_u LS/2$, $W_u SL/2$, and $W_u LS/4$, respectively.

In two-way rectangular slabs supported by adequate beams on four sides, the floor loads are transferred to the beam from tributary areas bounded by 45° lines, as shown in Figure 16.1(b). Part of the floor loads are transferred to the long beams AB, BC, DE, and EF from trapezoidal areas, whereas the rest of the floor loads are transferred to the short

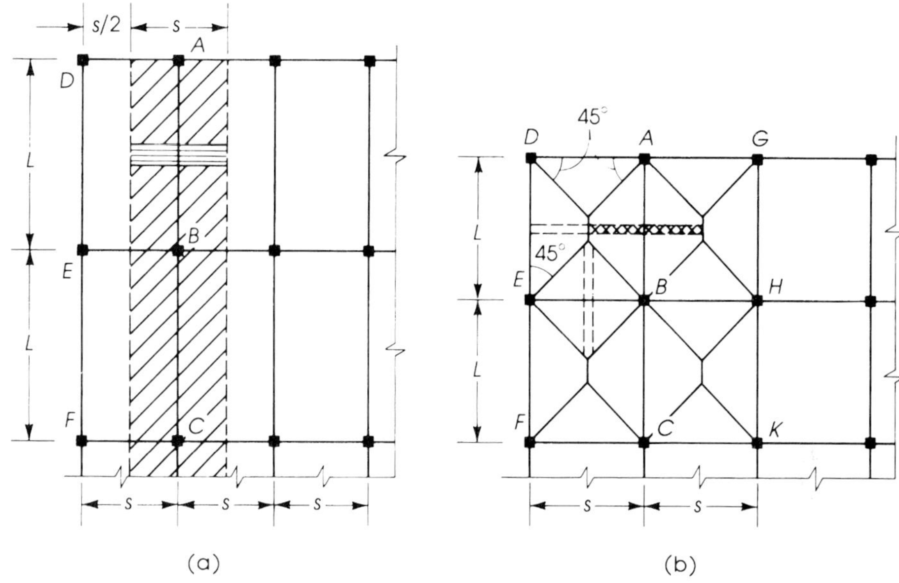

Figure 16.1 Slab loads on supporting beams: (a) one-way direction, $L/s > 2$; and (b) two-way direction, $L/s \leq 2$.

beams AD, BE, CF from triangular areas. In square slabs, loads are transferred to all surrounding beams from triangular floor areas. Interior beams carry loads from both sides, whereas end beams carry loads from one side only. Beams in both directions are usually cast monolithically with the slabs; therefore they should be analyzed as statically indeterminate continuous beams. The beams transfer their loads in turn to the supporting columns. The load on column B equals $W_u LS$, while the loads on columns E, A, and D are $W_u LS/2$, $W_u SL/2$, and $W_u LS/4$, respectively. The tributary area for each column extends from the centerlines of adjacent spans in each direction.

16.2 MAXIMUM MOMENTS IN CONTINUOUS BEAMS

16.2.1 Basic Analysis

The computation of bending moments and shear forces in reinforced concrete continuous beams is generally based on the elastic theory. When reinforced concrete sections are designed using the strength design method, the results are not entirely consistent with the elastic analysis. However, the ACI Code does not include provisions for a plastic design or limit design of reinforced concrete continuous structures except in allowing moment redistribution, as is explained later in this chapter.

16.2.2 Loading Application

The bending moment at any point in a continuous beam depends not only on the position of loads on the same span, but also on the loads on the other spans. In the case of dead loads, all spans must be loaded simultaneously, because the dead load is fixed in position and magnitude. In the case of moving loads or occasional live loads, the pattern of loading must be considered to determine the maximum moments at the critical sections. Influence lines may be used to determine the position of the live load to calculate the maximum and minimum moments. However, in this chapter, simple rules based on load-deflection curves are used to determine the loading pattern that produces maximum moments.

16.2.3 Maximum and Minimum Positive Moments Within a Span

The maximum positive bending moment in a simply supported beam subjected to a uniform load w K/ft is at midspan, and $M = wl_2/8$. If one or both ends are continuous, the restraint at the continuous end will produce a negative moment at the support and slightly shift the location of the maximum positive moment from midspan. The deflected shape of the continuous beam for a single-span loading is shown in Figure 16.2(a); downward deflection indicates a positive moment and upward deflection indicates a negative moment. If all spans deflected downward are loaded, each load will increase the positive moment at the considered span AB (Figure 16.2(d)). Therefore, to calculate the maximum positive moment within a span, the live load is placed on that span and on every alternate span on both sides. The factored live load moment, calculated as explained before, must be added to the factored dead-load moment at the same section to obtain the maximum positive moment.

The bending moment diagram due to a uniform load on AB is shown in Figure 16.2(b). The deflections and the bending moments decrease rapidly with the distance from the loaded span AB. Therefore, to simplify the analysis of continuous beams, the moments in any span can be computed by considering the loaded span and two spans on either side of the considered span AB, assuming fixed supports at the far ends (Figure 16.2(c)).

If the spans adjacent to span AB are loaded, the deflection curve will be as shown in Figure 16.2(e). The deflection within span AB will be upward, and a negative moment will

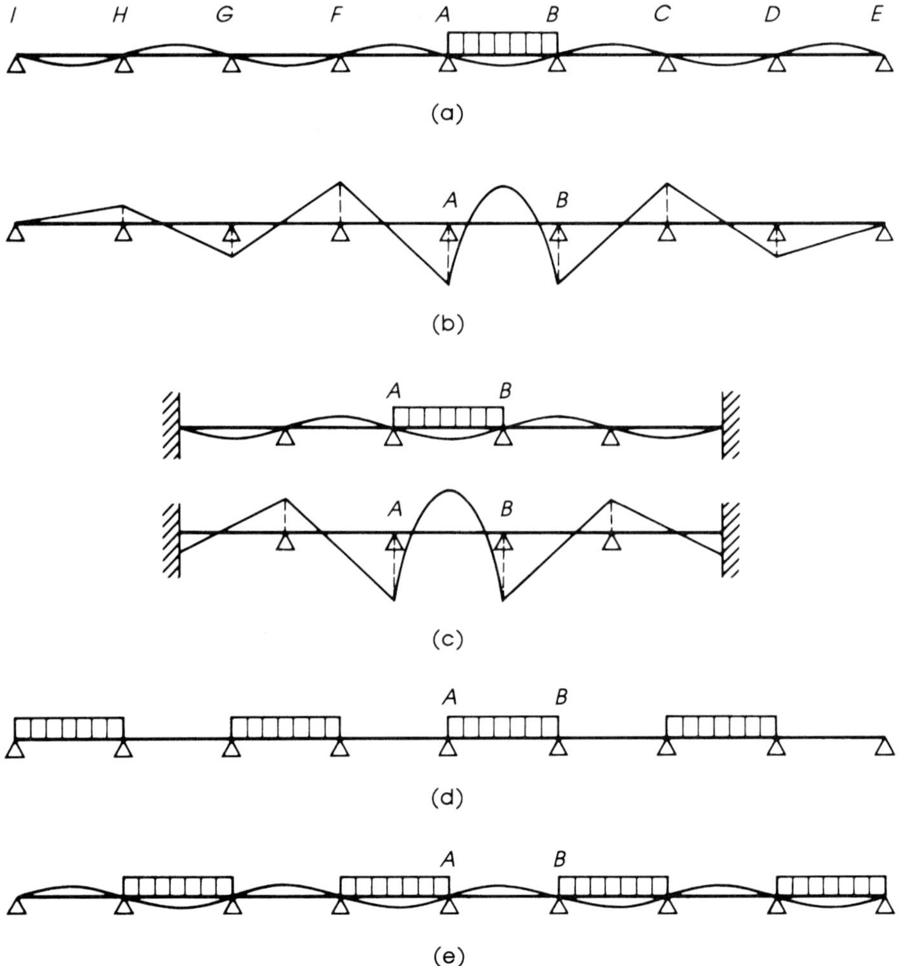

Figure 16.2 Loadings for maximum and minimum moment within span AB.

be produced in span AB. This negative moment must be added to the positive moment due to dead load to obtain the final bending moment. Therefore, to calculate the minimum positive moment (or maximum negative moment) within a span AB, the live load is placed on the adjacent spans and on every alternate span on both sides of AB (Figure 16.2(e)).

16.2.4 Maximum Negative Moments at Supports

In this case, it is required to determine the maximum negative moment at any support, say, support A (Figure 16.3). When span AB is loaded, a negative moment is produced at support A. Similarly, the loading of span AF will produce a negative moment at A. Therefore, to calculate the maximum negative moment at any support, the live load is placed on the two adjacent spans and on every alternate span on both sides (Figure 16.3).

In the structural analysis of continuous beams, the span length is taken from center to center of the supports, which are treated as knife-edge supports. In practice, the supports are always made wide enough to take the loads transmitted by the beam, usually the moments acting at the face of supports. To calculate the design moment at the face of the support, it is quite reasonable to deduct a moment equal to $V_u c/3$ from the factored moment at the centerline of the support, where V_u is the factored shear and c is the column width.

16.2.5 Moments in Continuous Beams

Continuous beams and frames can be analyzed using approximate methods or computer programs, which are available commercially. Other methods, such as the displacement and force methods of analysis based on the calculation of the stiffness and flexibility matrices, may also be adopted. Slope deflection and moment-distribution methods may also be used. These methods are explained in books dealing with the structural analysis of beams and frames. However, the ACI Code, Section 8.3, gives approximate coefficients for calculating the bending moments and shear forces in continuous beams and slabs. These coefficients were given in Chapter 9. The moments obtained using the ACI coefficients will be somewhat larger than those arrived at by exact analysis. The limitations stated in the use of these coefficients must be met.

Example 16.1

The slab-beam floor system shown in Figure 16.4 carries a uniform live load of 120 psf and a dead load that consists of the slab's own weight plus 60 psf. Using the ACI moment coefficients, design a typical interior continuous beam and draw detailed sections. Use $f'_c = 3$ Ksi, $f_y = 60$ Ksi, beam width $b = 12$ in., columns 12 by 12 in., and a slab thickness of 5.0 in.

Solution

1. Design of slabs: The floor slabs act as one-way slabs, because the ratio of the long to the short side is greater than 2. The design of a typical continuous slab was discussed in Example 9.4.
2. Loads on slabs:

$$\text{Dead load} = \frac{5}{12} \times 150 + 60 = 122.5 \text{ psf}$$

$$\text{Live load} = 120 \text{ psf}$$

$$\text{Ultimate load } w_u = 1.4(122.5) + 1.7(120) = 375.5 \text{ psf}$$

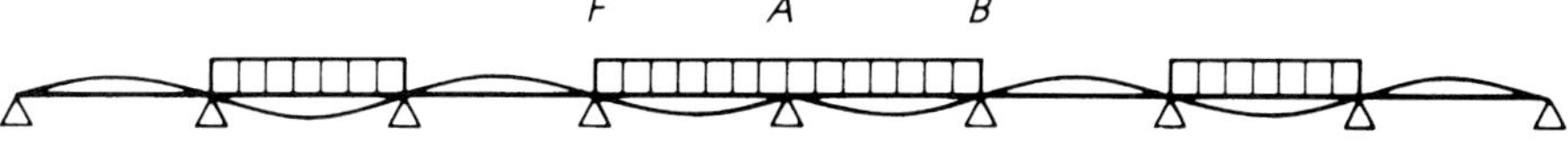

Figure 16.3 Loading for maximum negative moment at support A.

Test on a continuous reinforced concrete beam. Plastic hinges developed in the positive and negative maximum moment regions.

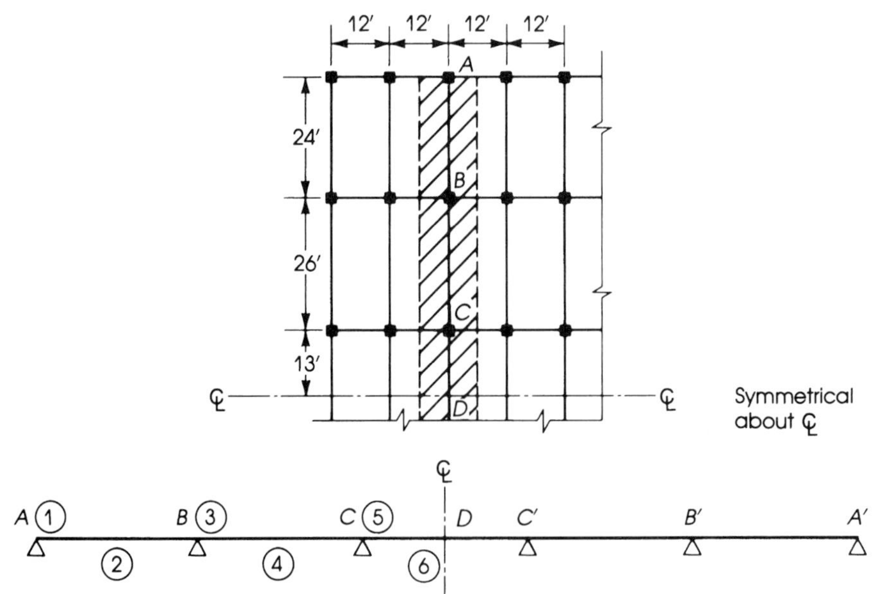

Figure 16.4 Example 16.1.

Loads on beams: A typical interior beam *ABC* carries slab loads from both sides of the beam, with a total slab width of 12 ft.

$$\text{Ultimate load on beam} = 12 \times 375.5 + 1.4 \times (\text{self-weight of beam web})$$

The depth of the beam can be estimated using the coefficients of minimum thickness of beams shown in Table A.6. For $f_y = 60$ Ksi, the minimum thickness of the first beam *AB* is $L/18.5 = (24 \times 12)/18.5 = 15.6$ in. Assume a total depth of 22 in. and a web depth of $22 - 5 = 17$ in. Therefore, the ultimate load on beam $ABCD$ is

$$w_u = 12(375.5) + 1.4\left(\frac{17 \times 12}{144} \times 150\right) = 4804 \text{ lb/ft}$$

Use 4.8 K/ft.

3. Moments in beam ABC: Moment coefficients are shown in Figure 9.8. The beam is continuous on five spans and symmetrical about the centerline at D. Therefore, it is sufficient to design half of the beam $ABCD$, because the other half will have similar dimensions and reinforcement. Because the spans AB and BC are not equal and the ratio $\frac{26}{24}$ is less than 1.2, the ACI moment coefficients can be applied to this beam. Moreover, the average of the adjacent clear span is used to calculate the negative moments at the supports.

Moments at critical sections are calculated as follows (Figure 16.4):

$$M_u = \text{coefficient} \times w_u l_n^2$$

Location	1	2	3	4	5	6
Moment coefficient	$-\frac{1}{16}$	$+\frac{1}{14}$	$-\frac{1}{10}$	$+\frac{1}{16}$	$-\frac{1}{11}$	$+\frac{1}{16}$
M_u (K · ft)	−158.7	181.4	−276.5	187.5	−272.7	187.5

4. Determine beam dimensions and reinforcement.

a. Maximum negative moment is −276.5 K · ft. Using $\rho_{\max} = 0.0161$,

$$R_{u\,\max} = 720 \text{ psi}$$

$$d = \sqrt{\frac{M_u}{R_u b}} = \sqrt{\frac{276.5 \times 12}{0.702 \times 12}} = 19.8 \text{ in.}$$

For one row of reinforcement, total depth is $19.8 + 2.5 = 22.3$ in., say, 23 in., and actual d is 20.5 in. $A_s = 0.0161 \times 12 \times 19.8 = 3.8$ in.^{2}; use four no. 9 bars in one row. Note that total depth used here is 23 in., which is more than the 22 in. assumed to calculate the weight of the beam. The additional load is negligible, and there is no need to revise the calculations.

b. The sections at the supports act as rectangular sections with tension reinforcement placed within the flange. The reinforcements required at the supports are as follows:

Location	1	3	5
M_u (K · ft)	−158.7	−276.5	−272.7
$R_u = \frac{M_u}{bd^2}$ (psi)	378	658	649
ρ (%)	0.77	1.48	1.45
A_s (in.^{2})	1.9	3.7	3.6
No. 9 bars	2	4	4

c. For the midspan T-sections, $M_u = +187.5$ K · ft. For $a = 1.0$ in. and flange width = 72 in.,

$$A_s = \frac{M_u}{\phi f_y\left(d - \frac{a}{2}\right)} = \frac{187.5 \times 12}{0.9 \times 60(20.5 - 1/2)} = 2.1 \text{ in.}^2$$

$$\text{Check } a\text{:} \quad a = \frac{A_s f_y}{0.85 f'_c b} = \frac{2.1 \times 60}{0.85 \times 3 \times 72} = 0.7 \text{ in.}$$

Revised a gives $A_s = 2.07$ in.^{2} Therefore, use three no. 8 bars $\left(A_s = 2.35 \text{ in.}^2\right)$ for all midspan sections. Reinforcement details are shown in Figure 16.5.

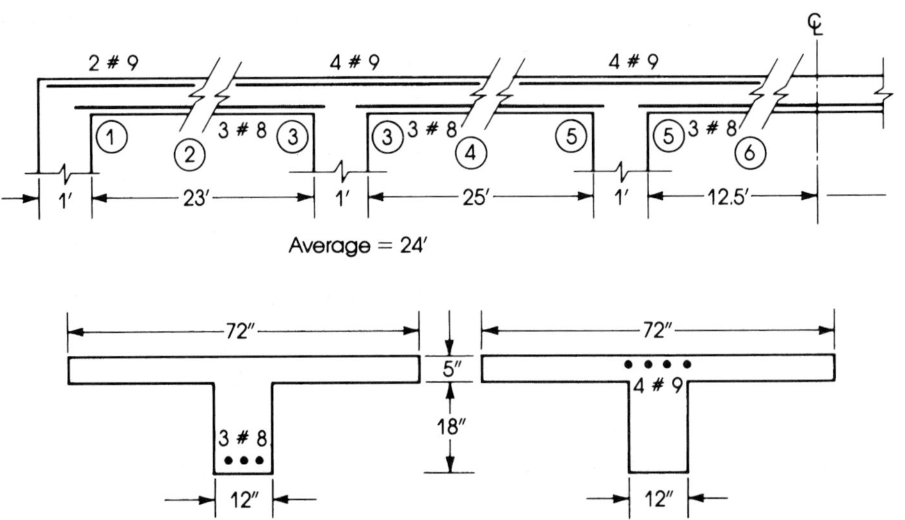

Figure 16.5 Reinforcement details, Example 16.1.

5. Design the beam for shear, as explained in Chapter 8.
6. Check deflection and cracking, as explained in Chapter 6.

16.3 BUILDING FRAMES

A building frame is a three-dimensional structural system consisting of straight members that are built monolithically and have rigid joints. The frame may be one bay long and one story high, such as the portal frames and gable frames shown in Figure 16.6(a), or it may consist of multiple bays and stories, as shown in Figure 16.6(b). All members of the frame are considered continuous in the three directions, and the columns participate with the beams in resisting external loads. Besides reducing moments due to continuity, a building frame tends to distribute the loads more uniformly on the frame. The effects of lateral loads, such as wind and earthquakes, are also spread over the whole frame, increasing its safety. For design purposes, approximate methods may be used by assuming a two-dimensional frame system.

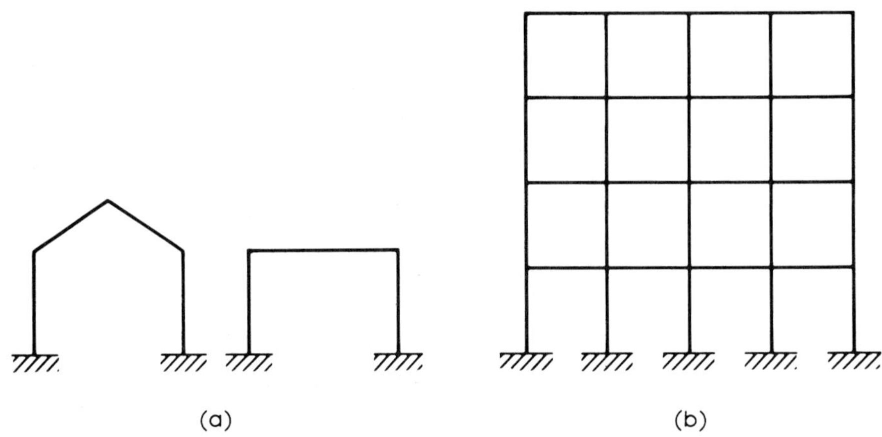

Figure 16.6 (a) Gable and portal frames (schematic) and (b) multibay, multistory frame.

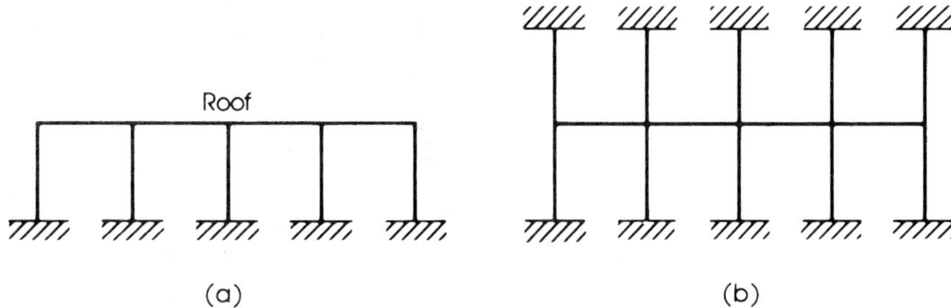

Figure 16.7 Assumption of fixed column ends for frame analysis.

A frame subjected to a system of loads may be analyzed by the equivalent frame method. In this method, the analysis of the floor under consideration is made assuming that the far ends of the columns above and below the slab level are fixed (Figure 16.7). Usually, the analysis is performed using the moment-distribution method.

In practice, the size of panels, distance between columns, number of stories, and the height of each story are known because they are based upon architectural design and utility considerations. The sizes of beams and columns are estimated first, and their relative stiffnesses based on the gross concrete sections are used. Once the moments are calculated, the sections assumed previously are checked and adjusted as necessary. More accurate analysis can be performed using computers, which is recommended in the structural analysis of statically indeterminate structures with several redundants. Methods of analysis are described in many books on structural analysis.

16.4 PORTAL FRAMES

A portal frame consists of a reinforced concrete stiff girder poured monolithically with its supporting columns. The joints between the girder and the columns are considered rigidly fixed, with the sum of moments at the joint equal to zero. Portal frames are used in building large-span halls, sheds, bridges, and viaducts. The top member of the frame may be horizontal (portal frame) or inclined (gable frame) (Figure 16.8). The frames may be fixed or hinged at the base.

A statically indeterminate portal frame may be analyzed by the moment-distribution method or any other method used to analyze statically indeterminate structures. The frame members are designed for moments, shear, and axial forces, whereas the footings are designed to carry the forces acting at the column base.

Girders and columns of frames may be of uniform or variable depths, as shown in Figure 16.8. The forces in a single-bay portal frame of uniform sections may be calculated as follows.

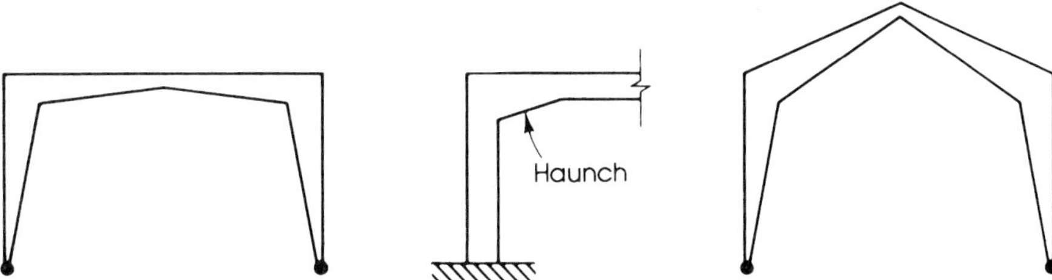

Figure 16.8 Portal and gable frames.

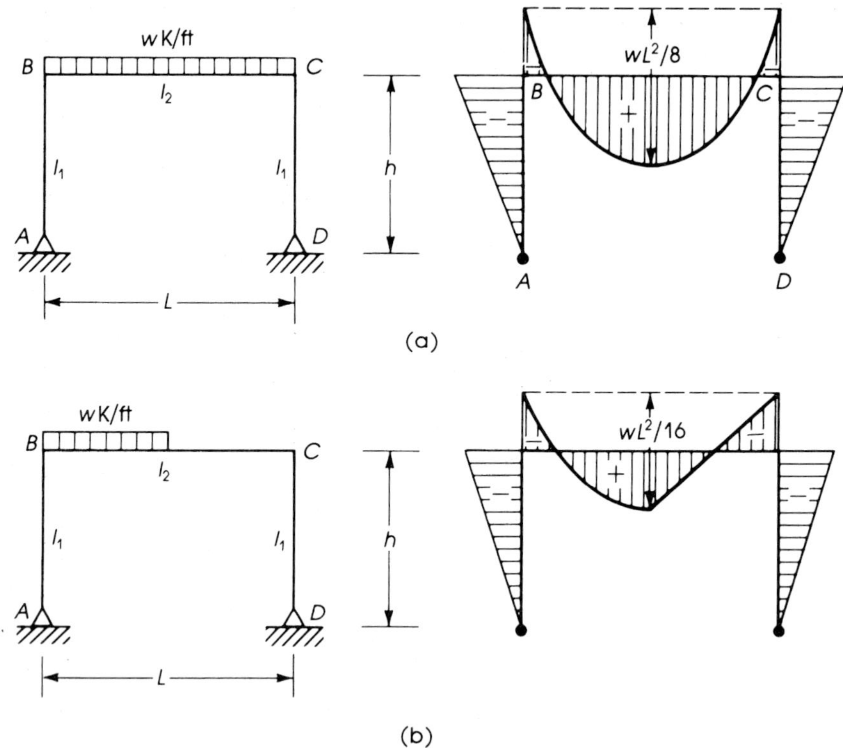

Figure 16.9 Portal frame with two hinged ends. Bending moments are drawn on the tension side.

16.4.1 Two Hinged Ends

The forces in the members of a portal frame with two hinged ends [2] can be calculated using the following expressions (Figure 16.9).

For the case of a uniform load on top member BC, let

$$K = 3 + 2\left(\frac{I_2}{I_1} \times \frac{h}{L}\right)$$

where

I_1 and I_2 = column and beam moments of inertia
h and L = height and span of frame

The bending moments at joints B and C are

$$M_B = M_C = -\frac{wL^2}{4K}$$

$$\text{Maximum positive moment at midspan } BC = \frac{wL^2}{8} + M_B$$

The horizontal reaction at A is $H_A = M_B/h = H_D$. The vertical reaction at A is $V_A = WL/2 = V_D$. For a uniform load on half the beam BC, Figure 16.9(b): $M_B = M_C = -WL_2/8K, H_A = H_D = M_B/h, V_A = 3WL/8$, and $V_D = WL/8$.

16.4.2 Two Fixed Ends

The forces in the members of a portal frame with two fixed ends [2] can be calculated as follows (Figure 16.10).

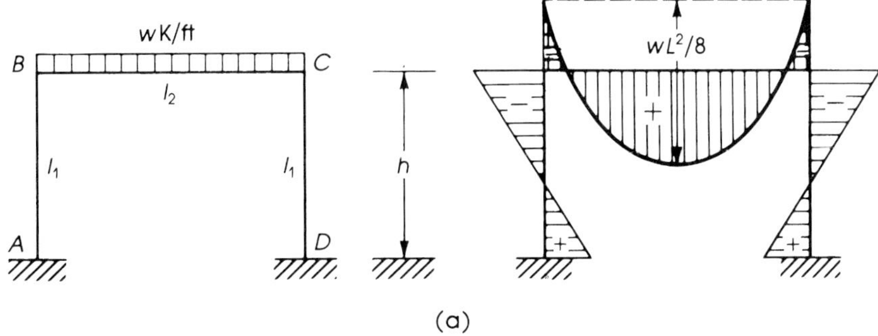

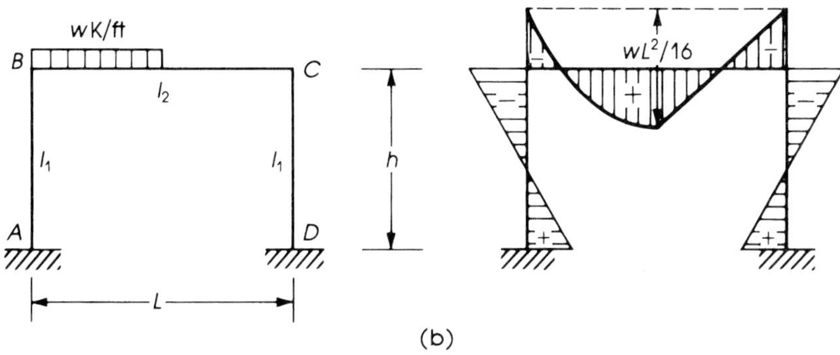

Figure 16.10 Portal frame with fixed ends. Bending moments are drawn on the tension side.

For a uniform load on top member BC, let

$$K_1 = 2 + \left(\frac{I_2}{I_1} \times \frac{h}{L}\right)$$

$$M_B = M_C = -\frac{wL^2}{6K_1}$$

$$M_A = M_D = \frac{M_B}{2}, \qquad M\ (\text{midspan}) = \frac{wL^2}{8} + M_B$$

$$H_A = H_D = \frac{3M_A}{h} \quad \text{and} \quad V_A = V_D = \frac{wL}{2}$$

For a uniform load on half the top member BC, let

$$K_2 = 1 + 6\left(\frac{I_2}{I_1} \times \frac{h}{L}\right)$$

Then

$$M_A = \frac{wL^2}{8}\left(\frac{1}{3K_1} - \frac{1}{8K_2}\right), \qquad M_B = \frac{wl^2}{8}\left(\frac{2}{3K_1} + \frac{1}{8K_2}\right)$$

$$M_C = \frac{wL^2}{8}\left(\frac{2}{3K_1} - \frac{1}{8K_2}\right), \qquad M_D = \frac{wL^2}{8}\left(\frac{1}{3K_1} - \frac{1}{8K_2}\right)$$

$$H_A = H_D = \frac{wl^2}{8} \times \frac{1}{K_1 h}$$

$$V_A = \frac{wL}{2} - V_D \quad \text{and} \quad V_D = \frac{wL}{8}\left(1 - \frac{1}{4K_2}\right)$$

16.5 GENERAL FRAMES

The main feature of a frame is its rigid joints, which connect the horizontal or inclined girders of the roof to the supporting structural members. The continuity between the members tends to distribute the bending moments inherent in any loading system to the different structural elements according to their relative stiffnesses. Frames may be classified as

1. Statically determinate frames (Figure 16.11(a));
2. Statically indeterminate frames (Figure 16.12);
3. Statically indeterminate frames with ties (Figure 16.13).

Different methods for the analysis of frames and other statically indeterminate structures are described in books dealing with structural analysis. Once the bending moments, shear, and axial forces are determined, the sections can be designed as the examples in this book are. Analysis may also be performed using computer programs.

16.6 DESIGN OF FRAME HINGES

The main types of hinges used in concrete structures are Mesnager hinges, Considère hinges, and lead hinges [19]. The description of each type is given next.

16.6.1 Mesnager Hinge

The forces that usually act on a hinge are a horizontal force, H, and a vertical force, P. The resultant of the two forces, R, is transferred to the footing through the crossing bars A and B shown in Figure 16.14. The inclination of bars A and B to the horizontal varies between 30° and 60°, with a minimum distance a, measured from the lower end of the frame column, equal to $8D$, where D is the diameter of the inclined bars. The gap between the frame colunm and the top of the footing y varies between 1 in. and $1.3h'$, where h' is the width of the concrete section at the hinge level. A practical gap height ranges between 2 and 4 in. The rotation of the frame ends is taken by the hinges, and the gap is usually filled with bituminous cork or similar flexible material. The bitumen protects the cork in contact with the soil from deterioration. The crossing bars A and B are subjected to compressive stresses that must not exceed one-third the yield strength of the steel bars f_y under service loads or $0.55f_y$ under factored loads. The low stress is assumed because any rotation at the hinge tends to bend the bars and induces secondary flexural stresses. It is generally satisfactory to keep the compression stresses low rather than to compute secondary stresses. The areas of bars A and B are calculated as follows:

$$\text{Area of bars } A\text{:} \quad A_{s1} = \frac{R_1}{0.55f_y} \tag{16.1}$$

$$\text{Area of bars } B\text{:} \quad A_{s2} = \frac{R_2}{0.55f_y} \tag{16.2}$$

where R_1 and R_2 are the components of the resultant R in the direction of the inclined bars A and B using factored loads. The components R_1 and R_2 are usually obtained by statics as follows:

$$H + R_2 \sin\theta = R_1 \sin\theta \quad \text{and} \quad R_2 = R_1 - \frac{H}{\sin\theta} \tag{16.3a}$$

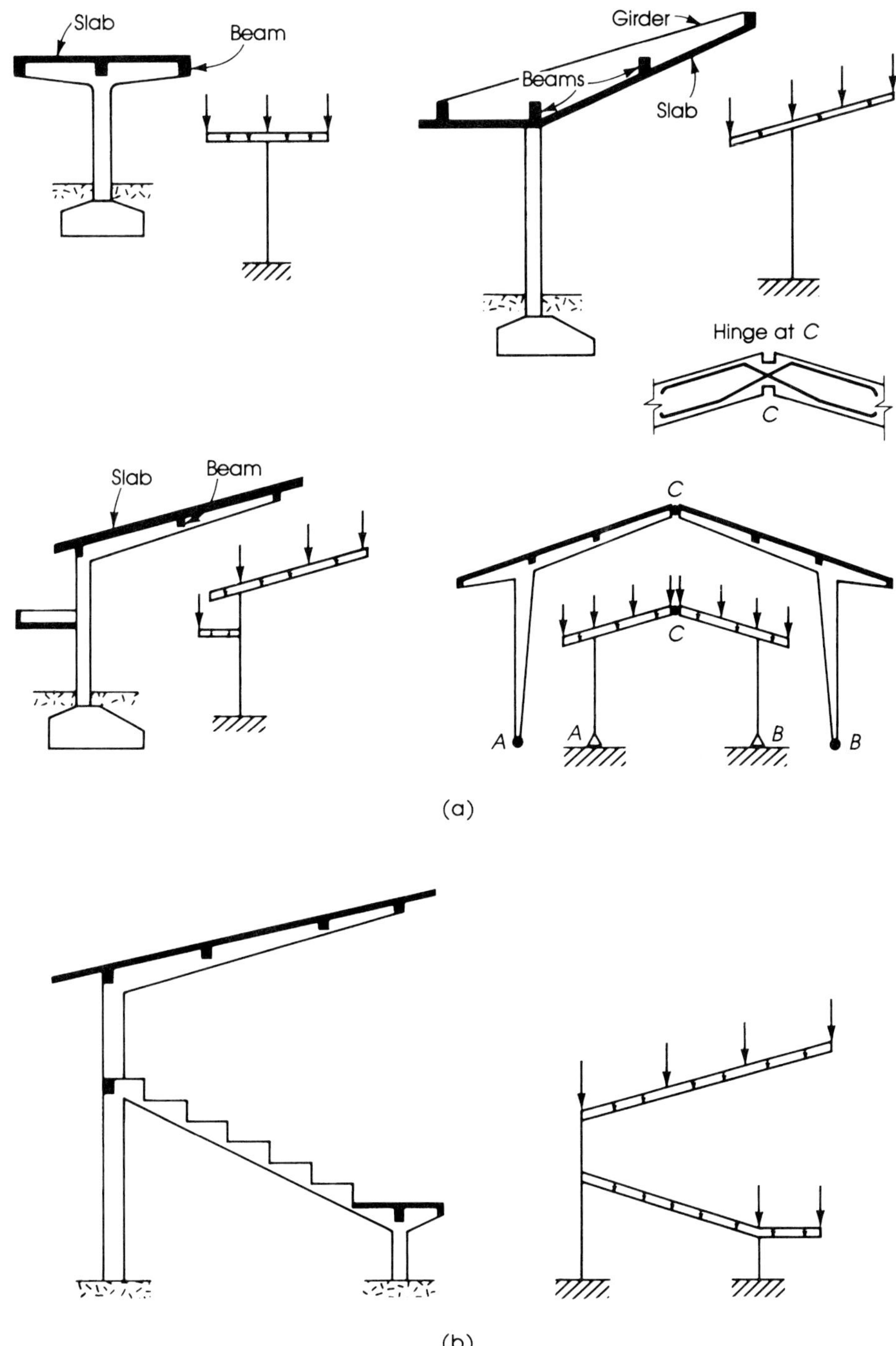

Figure 16.11 (a) Statically determinate frames and (b) reinforced concrete stadium.

Also, $(R_1 + R_2)\cos\theta = P_u$, so

$$R_1 = \frac{P_u}{\cos\theta} - R_2 = \frac{P_u}{\cos\theta} - \left[R_1 - \frac{H}{\sin\theta}\right] \tag{16.3b}$$

$$R_1 = \frac{1}{2}\left[\frac{P_u}{\cos\theta} + \frac{H}{\sin\theta}\right]$$

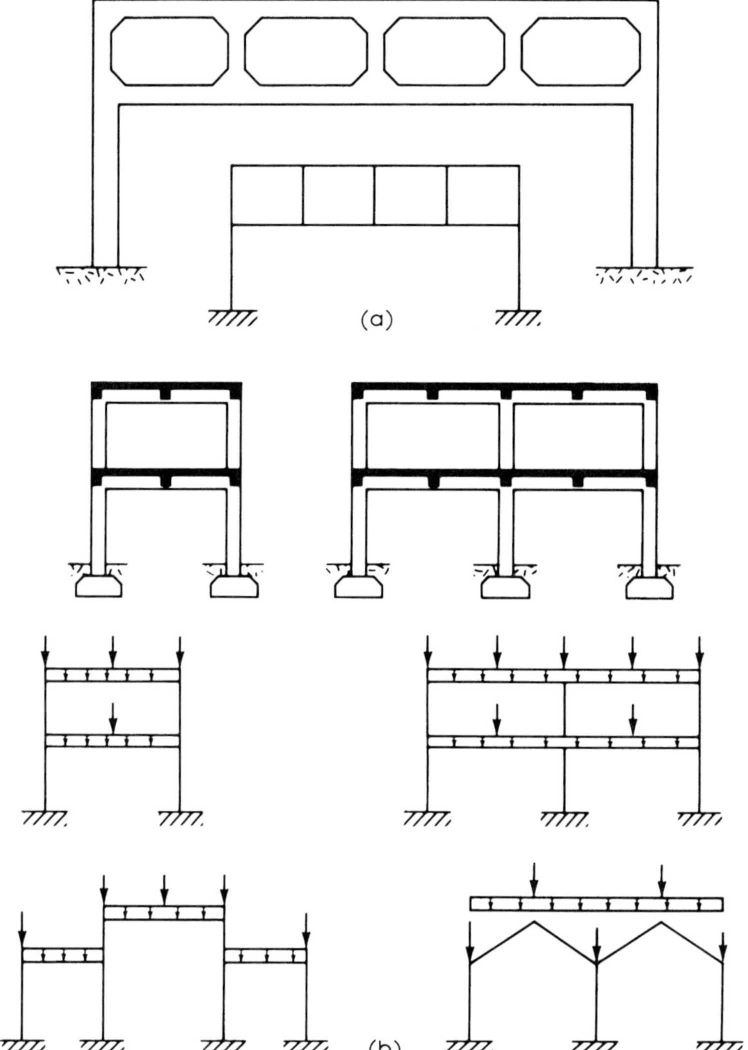

Figure 16.12 (a) Vierendeel girder and (b) statically indeterminate frames.

The inclined hinge bars transmit their force through the bond along the embedded lengths in the frame columns and footings. Consequently, the bars exert a bursting force, which must be resisted by ties. The ties should extend a distance $a = 8D$ (the larger bar diameter of bars A and B) in both columns and footings. The bursting force F can be estimated as

$$F = \frac{P_u}{2}\tan\theta + \frac{Ha}{0.85d} \tag{16.4}$$

If the contribution of concrete is neglected, then the area of tie reinforcement, A_{st}, required to resist F is

$$A_{st} = \frac{F}{\phi f_y} = \frac{F}{0.85 f_y} \tag{16.5}$$

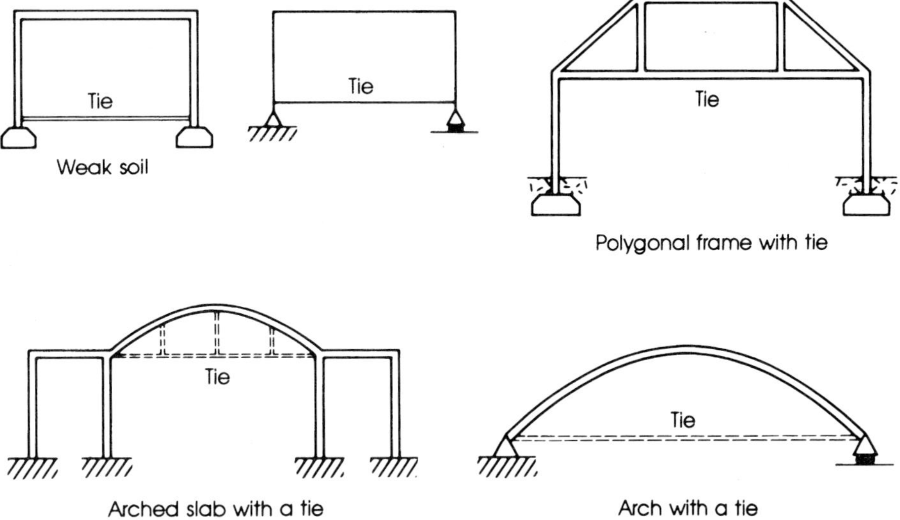

Figure 16.13 Structures with ties.

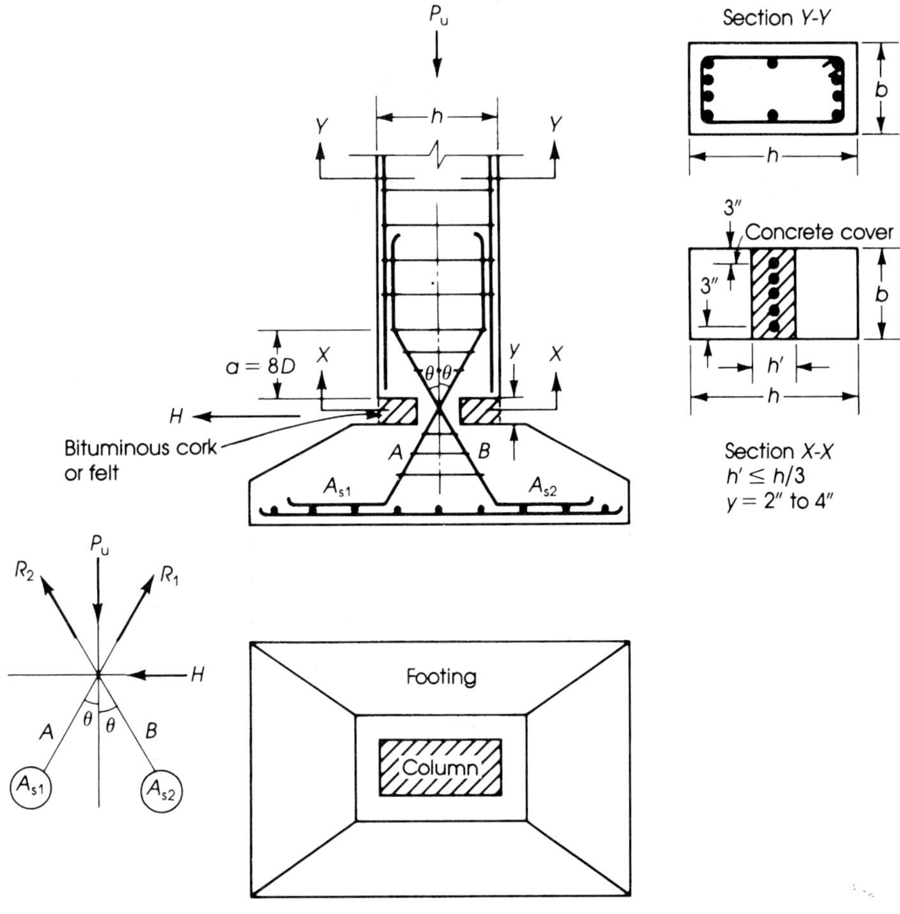

Figure 16.14 Hinge details.

The stress in the ties can also be computed as follows:

$$f_s\,(\text{ties}) = \frac{\dfrac{P_u}{2}\tan\theta + \dfrac{Ha}{0.85d}}{0.005ab + A_{st}\,(\text{ties})} \le 0.85f_y \tag{16.6}$$

where A_{st} = area of ties within a distance $a = 8D$

d = effective depth of column section

b = width of column section

This type of hinge is used for moderate forces and limited by the maximum number of inclined bars that can be placed within the column width.

16.6.2 Considère Hinge

The difference between the *Considère hinge* and the Mesnager one is that the normal force P_u is assumed to be transmitted to the footing by one or more short, spirally reinforced columns extending deep into the footing, whereas the horizontal force H is assumed to be resisted by the inclined bars A and B (Figure 16.15). The load capacity of the spirally reinforced short column may be calculated using equation (10.7), neglecting the factor 0.85 for minimum eccentricity.

$$P_u = \phi P_n = 0.75\left[0.85f'_c(A_g - A_{st}) + A_{st}f_y\right] \tag{16.7}$$

where A_g is the area of concrete hinge section, or bh', and A_{st} is the area of longitudinal bars within the spirals. Ties should be provided in the column up to a distance equal to the long side of the column section h.

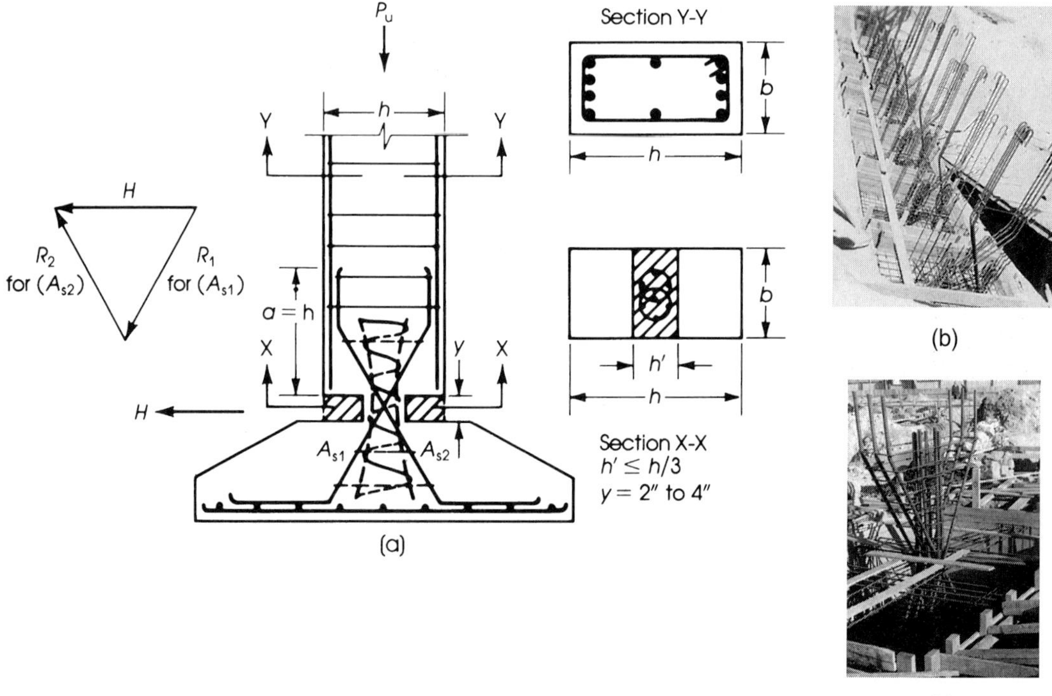

Figure 16.15 (a) Considère hinge, (b) Mesnager hinges for a series of portal frames, and (c) Considère hinge.

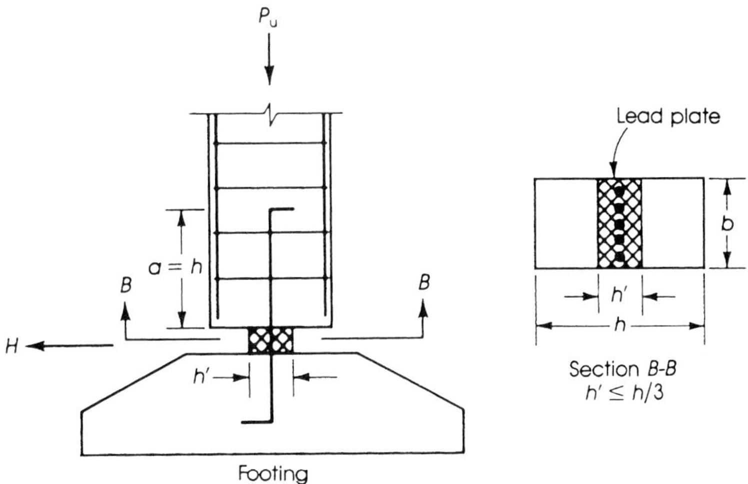

Figure 16.16 Lead hinge.

16.6.3 Lead Hinges

Lead hinges are sometimes used in reinforced concrete frames. In this type of hinge a lead plate, usually 0.75 to 1.0 in. thick, is used to transmit the normal force, P_u, to the footing. The horizontal force H is resisted by vertical bars placed at the center of the column and extended to the footing (Figure 16.16). At the base of the column, the axial load P_u should not exceed the bearing strength specified by the ACI Code, Section 10.15, of $\phi(0.85f'_c A_1)$, where $\phi = 0.7$ and $A_1 = bh'$. The area of the vertical bars is $A_s = H/0.6f_y$, where H = factored horizontal force.

Example 16.2

An 84- by 40-ft hall is to be covered by reinforced concrete slabs supported on hinged-end portal frames spaced at 12 ft on centers (Figure 16.17). The frame height is 15 ft, and no columns

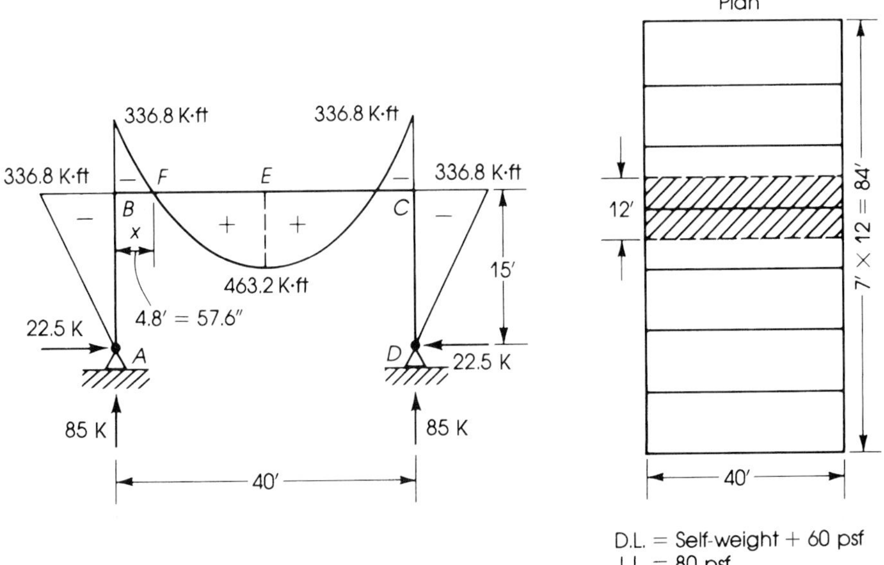

Figure 16.17 Design of portal frame, Example 16.2.

are allowed within the hall area. The dead load on the slabs is that due to self-weight plus 60 psf from roof finish. The live load on the slab is 80 psf. Design a typical interior frame using $f'_c = 4$ Ksi and $f_y = 60$ Ksi for the frame and a column width of $b = 16$ in.

Solution

The main structural design of the building will consist of the following:

- Design of one-way slabs
- Analysis of the portal frame
- Design of the frame girder due to moment
- Design of the frame girder due to shear
- Design of columns
- Design of hinges
- Design of footings

1. One-way roof slab: The minimum thickness of the first slab is $L/30$, because one end is continuous and the other end is discontinuous (Table A.6 in Appendix A).

$$\text{Minimum depth} = \frac{12 \times 12}{30} = 4.8 \text{ in.}$$

Assume a slab thickness of 5.0 in. and design the slab following the steps of Example 9.5.

2. Analysis of an interior portal frame:

a. The loads on slabs are

$$\text{Dead load on slabs} = 60 + \left(\frac{5}{12} \times 150\right) = 122.5 \text{ psf}$$

$$\text{Ultimate load on slabs} = 1.4 \times 122.5 + 1.7 \times 80 = 308 \text{ psf}$$

b. Determine loads on frames: The interior frame carries a load from a 12-ft slab in addition to its own weight. Assume that the depth of the beam is $L/24 = (40 \times 12)/24 = 20$ in. Use a projection below the slab of 16 in., giving a total beam depth of 21 in.

$$\text{Dead load from self-weight of beam} = \left(\frac{16}{12}\right)^2 \times 150 = 267 \text{ lb/ft}$$

$$\text{Total ultimate load on frame} = 308 \times 12 + 1.4 \times 267$$

$$= 4063 \text{ lb/ft}$$

$$w_u = 4.0 \text{ K/ft} \qquad (\text{or } 4.1 \text{ K/ft})$$

c. Determine the moment of inertia of the beam and columns sections. The beam acts as a T-section. The effective width of slab acting with the beam is the smallest of span/4 $= 40 \times 12/4 = 120$ in., $16h_s + b_w = 16 \times 5 + 16 = 96$ in., or 12 ft $\times$ 12 $= 144$ in. Use $b = 96$ in. The centroid of the section from the top fibers is

$$y = \frac{96 \times 5 \times 2.5 + 16 \times 16 \times 13}{96 \times 5 + 16 \times 16} = 6.2 \text{ in.}$$

$$I_b \text{ (beam)} = \left[\frac{96}{12}(5)^3 + 96 \times 5(3.7)^2\right] + \left[\frac{16}{12}(16)^3 + 16 \times 16(6.8)^2\right]$$

$$= 24{,}870 \text{ in.}^4$$

It is a common practice to consider an approximate moment of inertia of a T-beam as equal to twice the moment of inertia of a rectangular section having the total depth of the web and slab:

$$I_b \text{ (beam)} = 2 \times \frac{16}{12}(21)^3 = 24{,}696 \text{ in.}^4$$

(For an edge beam, approximate $I = 1.5 \times bh_3/12$.) Assume a column section 16 by 20 in. (having the same width as the beam).

$$I_c \text{ (column)} = \frac{16}{12}(20)^3 = 10{,}667 \text{ in.}^4$$

d. Let the factor

$$K = 3 + 2\left(\frac{I_b}{L} \times \frac{h}{I_c}\right) = 3 + 2\left(\frac{24{,}870}{40} \times \frac{15}{10{,}667}\right) = 4.75$$

Referring to Figure 16.17 and for a uniform load $w_u = 4.0$ K/ft on BC,

$$M_B = M_C = -\frac{w_u L^2}{4K} = -\frac{4.0(40)^2}{4 \times 4.75} = -336.8 \text{ K}\cdot\text{ft}$$

The maximum positive bending moment at midspan of BC equals

$$w_u \frac{L^2}{8} + M_B = \frac{4.0(40)^2}{8} - 336.8 = 463.2 \text{ K}\cdot\text{ft}$$

The horizontal reaction at A is

$$H_A = H_D = \frac{M_B}{h} = \frac{336.8}{15} = 22.5 \text{ K}$$

The vertical reaction at A is

$$V_A = V_D = \frac{w_u L}{2} + \text{weight of column}$$

$$V_A = 4.0 \times \frac{40}{2} + \frac{20}{12} \times \frac{16}{12} \times 0.150 \times 15 \text{ ft} = 85.0 \text{ K}$$

The bending moment diagram is shown in Figure 16.17.

e. To consider the sidesway effect on the frame, the live load is placed on half the beam BC, and the moments are calculated at the critical sections. This case is not critical in this example.

f. The maximum shear at the two ends of beam BC occurs when the beam is loaded with the factored load w_u, but the maximum shear at midspan occurs when the beam is loaded with half the live load and with the full dead load:

$$V_u \text{ at support} = 4.0 \times \frac{40}{2} = 80.0 \text{ K}$$

$$V_u \text{ at midspan} = W_1 \frac{L}{8} = (1.7 \times 80 \times 12) \times \frac{40}{8}$$

$$= 8160 \text{ lb} = 8.16 \text{ K}$$

g. The axial force in each column is $V_A = V_D = 85.0$ K.

h. Let the point of zero moment in BC be at a distance x from B; then

$$M_B = w_u L \frac{x}{2} - w_u \frac{x^2}{2}$$

$$336.8 = 4.0\left(\frac{40x}{2} - \frac{x^2}{2}\right) \quad \text{or} \quad x^2 - 40x + 168.4 = 0$$

$$x = 4.8 \text{ ft} = 57.6 \text{ in.} \qquad \text{from } B$$

3. Design of girder BC:

a. Design the critical section at midspan. $M_u = 463.2$ K·ft, web width is $b_w = 16$ in., flange width is $b = 96$ in., and $d = 21 - 3.5 = 17.5$ in. (assuming two rows of steel bars). Check if the section acts as a rectangular section with effective $b = 96$ in. Assume $a = 1.0$ in.; then

$$A_s = \frac{M_u}{\phi f_y(d - a/2)} = \frac{463.2 \times 12}{0.9 \times 60(17.5 - 1.0/2)} = 6.05 \text{ in.}^2$$

$$a = \frac{A_s f_y}{0.85 f'_c b} = \frac{6.05 \times 60}{0.85 \times 4 \times 96} = 1.1 \text{ in.} < 5.0 \text{ in.}$$

The assumed a equals approximately the calculated a. The section acts as a rectangular section; therefore, use six no. 9 bars. Check b_{min} (to place bars in one row):

$$b_{min} = 11(\tfrac{9}{8}) + 2(\tfrac{3}{8}) + 3 = 16.13 \text{ in.} > 16 \text{ in.}$$

Place bars in two rows, as shown in Figure 16.18.

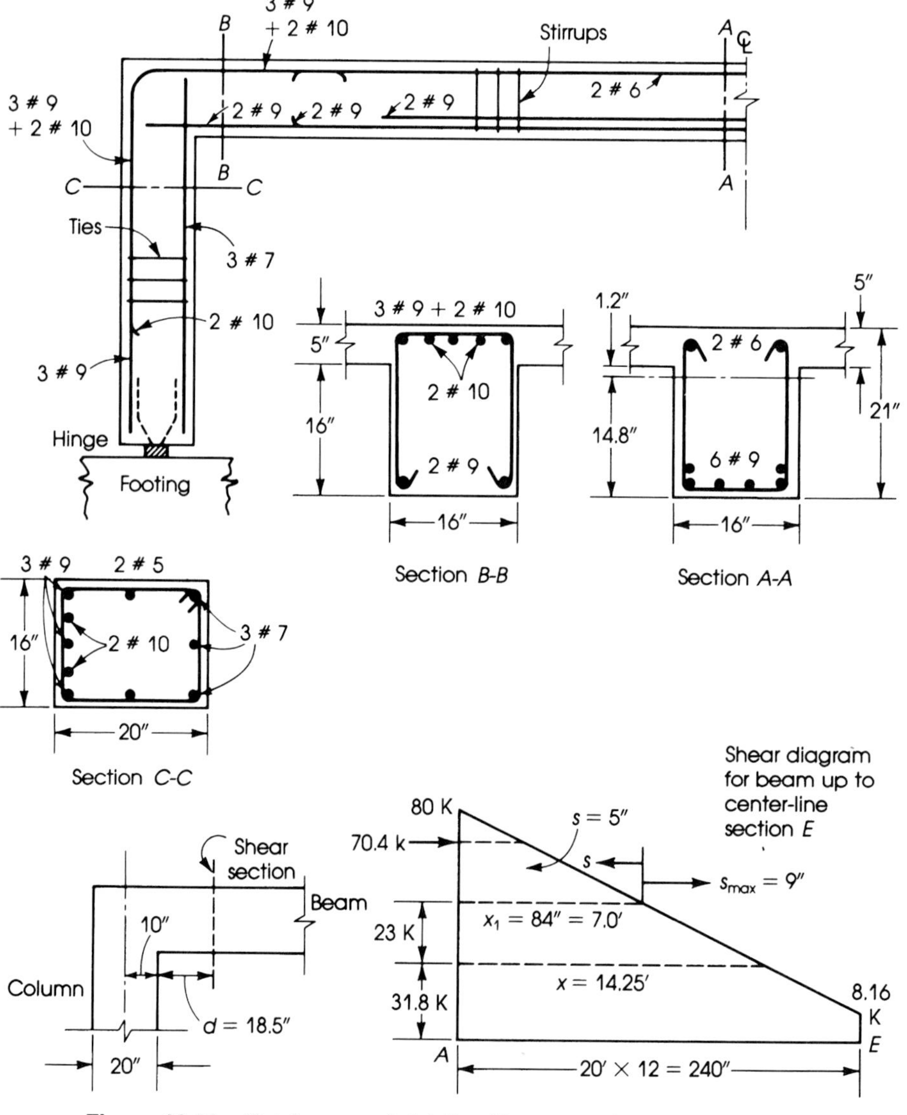

Figure 16.18 Reinforcement details of frame sections, Example 16.2.

b. Design the critical section at joint B: $M_u = 336.8$ K·ft, $b = 16$ in., and $d = 21 - 2.5 = 18.5$ in. (for one row of steel bars). The slab is under tension, and reinforcement bars are placed on top of the section.

$$R_u = \frac{M_u}{bd^2} = \frac{336.8 \times 12{,}000}{16(18.5)^2} = 738 \text{ psi}$$

From tables in Appendix A, $\rho = 0.016 < \rho_{max} = 0.0214$.

$$A_s = 0.016 \times 16 \times 18.5 = 4.73 \text{ in.}^2$$

Use five no. 9 bars in one row.

4. Design the girder BC due to shear:

a. The critical section is at a distance d from the face of the column with a distance from the column centerline of $10 + 18.5 = 28.5$ in. $= 2.4$ ft. Thus,

$$V_u \text{ (at distance } d) = 80 - 4 \times 2.4 = 70.4 \text{ K}$$

b. The shear strength provided by concrete is

$$\phi V_c = \phi(2\sqrt{f'_c})b_w d$$

$$\phi V_c = \frac{0.85 \times 2}{1000} \times \sqrt{4000} \times 16 \times 18.5 = 31.8 \text{ K}$$

The shear force to be provided by web reinforcement is

$$\phi V_s = V_u - \phi V_c = 70.4 - 31.8 = 38.6 \text{ K}$$

$$V_s = \frac{38.6}{0.85} = 45.4 \text{ K}$$

c. Choose no. 3 stirrups and $A_v = 2 \times 0.11 = 0.22$ in.^{2} Thus

$$s = \frac{A_v f_y d}{V_s} = \frac{0.22 \times 60 \times 18.5}{45.4} = 5.4 \text{ in.}$$

Use 5 in., or use no. 4 stirrups spaced at

$$\left(\frac{0.4 \times 60 \times 18.5}{45.4}\right) = 9.8 \text{ in.,} \qquad \text{say, 9.5 in.}$$

d. Maximum spacing of no. 3 stirrups is

$$s_{max} = \frac{d}{2} = \frac{18.5}{2} = 9.25 \text{ in.,} \qquad \text{say, 9 in.}$$

or

$$s_{max} = \frac{A_v f_y}{50 b_w} = \frac{0.22 \times 60{,}000}{50 \times 16} = 16.5 \text{ in.}$$

Check for maximum spacing of $d/2$: $V_s \leq 4\sqrt{f'_c}\, b_w d$ or

$$V_s \leq 4\sqrt{4000} \times \frac{16 \times 18.5}{1000} = 74.9 \text{ K}$$

The value V_s of 45.4 is less than 74.9 K, so use $s_{max} = 9$ in.

$$V'_s(\text{for } s_{max} = 9 \text{ in.}) = \frac{A_v f_y d}{s} = \frac{0.22 \times 60 \times 18.5}{9} = 27.1 \text{ K}$$

$$\phi V_s = 27.1 \times 0.85 = 23 \text{ K}$$

The distance from the centerline of the column where $s_{max} = 9$ in. can be used is equal to 84 in. $= 7.0$ ft (from the triangle of shear forces).

e. Distribution of stirrups:

First stirrups at $s/2 = 2.5$ in.

17 stirrups at 5 in. $= 85.0$ in.

15 stirrups at 9 in. $= 135.0$ in.

The distance from the face of the column to the centerline of the beam is $240 - 10 = 230$ in.
Use the same distribution for the second half of the beam, and place one stirrup at midspan.

5. Design the column section at joint B: $M_u = 336.8$ K·ft, $P_u = 80$ K, $b = 16$ in., and $h = 20$ in.

a. Assuming that the frame under the given loads will not be subjected to sidesway, then the effect of slenderness may be neglected, and the column can be designed as a short column when

$$\frac{Kl_u}{4} < 34 - \frac{12M_1}{M_2} \qquad \text{(see Section 12.5)}$$

$$M_1 = 0 \qquad \text{and} \qquad M_2 = 336.8 \text{ K·ft}$$

Let $K = 0.8$ (Figure 12.2), $L_u = 15 - 21/(2 \times 12) = 14.125$ ft, and $r = 0.3h = 0.3 \times 20 = 6$ in.; then

$$\frac{KL_u}{r} = 0.8 \times \frac{14.125 \times 12}{6} = 22.6 < 34$$

If K is assumed equal to 1.0, then

$$\frac{KL_u}{r} = 28.25 < 34$$

Therefore, design the member as a short column.

b. The design procedure is similar to Examples 11.16 and 11.3.

$$\text{Eccentricity } e = \frac{M_u}{P_u} = \frac{336.8 \times 12}{80} = 50.5 \text{ in.}$$

This is a large eccentricity, and it will be assumed that tension controls.

$$\phi = 0.9 - \frac{2P_u}{f'_c A_g} \geq 0.7$$

$$= 0.9 - \frac{2 \times 80}{4 \times (16 \times 20)} = 0.77$$

$$M_n = \frac{336.8}{0.77} = 437.4 \text{ K·ft} \quad \text{and} \quad P_n = \frac{80}{0.77} = 103.9 \text{ K}$$

$$d = 20 - 2.5 = 17.5 \text{ in.}$$

c. Because $e = 50.5$ in. is much greater than d, determine approximate A_s and A'_s from the M_u only and then check the final section by statics, as was explained in Example 11.3. For $M_u = 336.8$ K·ft, $b = 16$ in., $h = 20$ in., and $d = 17.5$ in., $R_u = M_u/bd^2 = 336.8(12{,}000)/16(17.5)^2 = 825$ psi.

$$\rho = 0.0183 \quad \text{and} \quad A_s = \rho bd = 0.0183(16)(17.5) = 5.12 \text{ in.}^2$$

Choose three no. 9 and two no. 10 bars and $A_s = 5.53$ in.^{2} Choose $A'_s = A_s/3 = 5.13/3 = 1.7$ in.^{2} and three no. 7 bars $(A'_s = 1.8 \text{ in.}^2)$ (Figure 16.18). When the eccentricity, e, is quite large, it is a common practice to use $A'_s = A_s/3$ or $A_s/2$ instead of $A_s = A'_s$.

d. Check the load capacity of the final section using $A_s = 5.53$ in.^{2} and $A'_s = 1.8$ in.^{2}, similar to Example 11.3, according to the following steps:

i. $P_n = C_c + C_s - T$

$$C_c = 0.85f'_c ab = 0.85(4)(16)a = 54.4a$$

$$C_s = A'_s(f'_s - 0.85f'_c) = 1.8(60 - 0.85 \times 4) = 101.8 \text{ K}$$

$$T = A_s f_y = 5.53(60) = 331.8 \text{ K}$$

$$P_n = 54.4a + 101.8 - 331.8 = (54.4a - 230) \quad \text{(I)}$$

ii. Take moments about A_s:

$$P_n = \frac{1}{e'}\left[C_c\left(d - \frac{a}{2}\right) + C_s(d - d')\right]$$

$e' = e + d''$, where d'' is the distance from A_s to the plastic centroid of the section. The plastic centroid occurs at 11.1 in. from the extreme compression fibers and $d'' = d - x = 6.4$ in. (refer to Example 11.1).

$$e' = 50.5 + 6.4 = 56.9 \text{ in.}$$

$$P_n = \frac{1}{56.9}\left[54.4a\left(17.5 - \frac{a}{2}\right) + 101.8(15)\right]$$

$$= 16.73a - 0.478a^2 + 26.86 \quad \text{(II)}$$

iii. Equate equations (I) and (II) and solve to get $a = 6.313$ in. and $P_n = 113.5 \text{ K} > 103.9$ K. Therefore, the section is adequate. Check $f'_s = 87(c - d')/c \le f_y$: $c = a/0.85 = 7.43$ in. and $f'_s = 87(7.43 - 2.5)/7.43 = 58$ Ksi, which is close to the 60 Ksi assumed in the calculations. Choose no. 3 ties spaced at 16 in.

6. Check the adequacy of the column section at midheight, 7.5 ft from A: $M_u = 336.8/2 = 168.4$ K · ft.

$$P_u = 80 + 2.5 \text{ (half the column weight)} = 82.5 \text{ K}$$

Use A_s = three no. 9 bars and A'_s = three no. 7 bars. In an approach similar to Step 5, $\phi P_n = 122 \text{ K} > 82.5$ K (no. 10 bars can be terminated, and they have to be extended a development length below the midheight of the column).

7. Design the hinge at A: $M_u = 0$, $H = 22.5$ K, $P_u = 85$ K.

a. Choose a Mesnager hinge. Using equations (16.3a) and (16.3b), $R_1 = 72$ K and $R_2 = 27$ K. (Refer to Figure 16.19 with $\theta = 30°$.)

$$A_{s1} = \frac{R_1}{0.55f_y} = \frac{72}{0.55 \times 60} = 2.2 \text{ in.}^2$$

Choose three no. 8 bars ($A_s = 2.35$ in.^{2}).

$$A_{s2} = \frac{R_2}{0.55 \times f_y} = \frac{27}{0.55 \times 60} = 0.82 \text{ in.}^2$$

Choose two no. 7 bars ($A_s = 1.2$ in.^{2}). Arrange the crossing bars by placing one no. 8 bar and then one no. 7 bar, as shown in Figure 16.19 (or use 5 no. 8 bars.)

b. Lateral ties should be placed along a distance a = 8-bar diameter = 8.0 in. within the column and footing. The bursting force is

$$F = \frac{P_u}{2}\tan\theta + \frac{Ha}{0.85d}$$

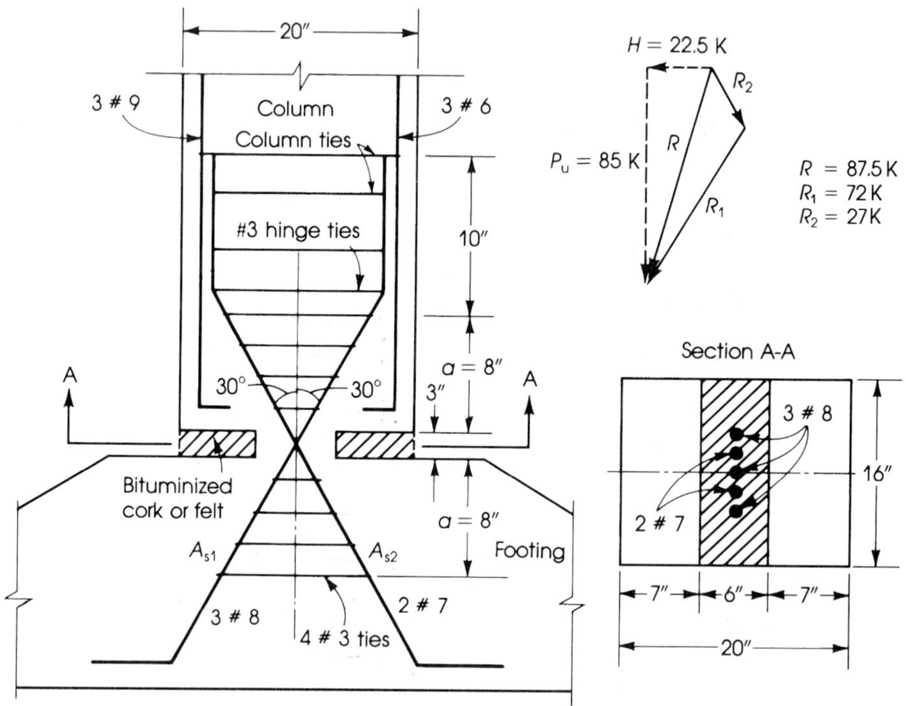

Figure 16.19 Hinge details, Example 16.2.

For $\theta = 30°$, $d = 17.5$ in., and $a = 8.0$ in.,

$$F = \frac{85}{2}\tan 30° + \frac{22.5 \times 8}{0.85 \times 17.5} = 36.6 \text{ K}$$

$$\text{Area of ties} = \frac{36.6}{0.85 \times 60} = 0.72 \text{ in.}^2$$

If no. 3 closed ties (two branches) are chosen, then the area of one tie is $2 \times 0.11 = 0.22$ in.^{2} The number of ties is $0.72/0.22 = 3.27$, say, 4 ties spaced at $\frac{8}{3} = 2.7$ in., as shown in Figure 16.19.

8. Design the footing: If the height of the footing is assumed to be h', then the forces acting on the footing are the axial load P and a moment $M = H/h'$. The soil pressure is calculated from equation (13.14) of Chapter 13:

$$q = +\frac{P}{A} \pm \frac{Mc}{I} \leq \text{allowable soil pressure}$$

The design procedure of the footing is similar to that of Example 13.7.

16.7 INTRODUCTION TO LIMIT DESIGN

16.7.1 General

Limit state design of a structure falls into three distinct steps:

1. Determination of the ultimate design load or factored load, obtained by multiplying the dead and live loads by load factors. The ACI Code adopted the load factors given in Chapter 3.

2. Analysis of the structure under ultimate or factored loads to determine the ultimate moments and forces at failure or collapse of the structure. This method of analysis has proved satisfactory for steel design; in reinforced concrete design, this type of analysis has not been fully adopted by the ACI Code because of the lack of ductility of reinforced concrete members. The code allows only partial redistribution of moments in the structure based on an empirical percentage, as will be explained later in this chapter.
3. Design of each member of the structure to fail at the ultimate moments and forces determined from structural analysis. This method is fully established now for reinforced concrete design and the ACI Code permits the use of the strength design method, as was explained in earlier chapters.

16.7.2 Limit Design Concept

Limit design in reinforced concrete refers to the redistribution of moments that occurs throughout a structure as the steel reinforcement at a critical section reaches its yield strength. The ultimate strength of the structure can be increased as more sections reach their ultimate capacity. Although the yielding of the reinforcement introduces large deflections, which should be avoided under service loads, a statically indeterminate structure does not collapse when the reinforcement of the first section yields. Furthermore, a large reserve of strength is present between the initial yielding and the collapse of the structure.

In steel design, the term *plastic design* is used to indicate the change in the distribution of moments in the structure as the steel fibers, at a critical section, are stressed to their yield strength. The development of stresses along the depth of a steel section under increasing load is shown in Figure 16.20. Limit analysis of reinforced concrete developed as a result of earlier research on steel structures and was based mainly on the investigations of Prager [4], Beedle [5], and J. B. Baker [6]. A. L. L. Baker [7] worked on the principles of limit design, whereas Cranston [8] tested portal frames to investigate the rotation capacity of reinforced concrete plastic hinges. However, more research work is needed before limit design can be adopted by the ACI Code.

16.7.3 Plastic Hinge Concept

The curvature ϕ of a member increases with the applied bending moment M. For an under-reinforced concrete beam, the typical moment-curvature and the load-deflection curves are shown in Figure 16.21. A balanced or an over-reinforced concrete beam is not permitted by the ACI Code, because it fails by the crushing of concrete and shows a small curvature range at ultimate moment (Figure 16.22).

The significant part of the moment-curvature curve in Figure 16.21 is that between B and C, in which M_u remains substantially constant for a wide range of values of ϕ. In limit analysis, the moment-curvature curve can be assumed to be of the idealized form shown in Figure 16.23, where the curvature ϕ between B and C is assumed to be constant,

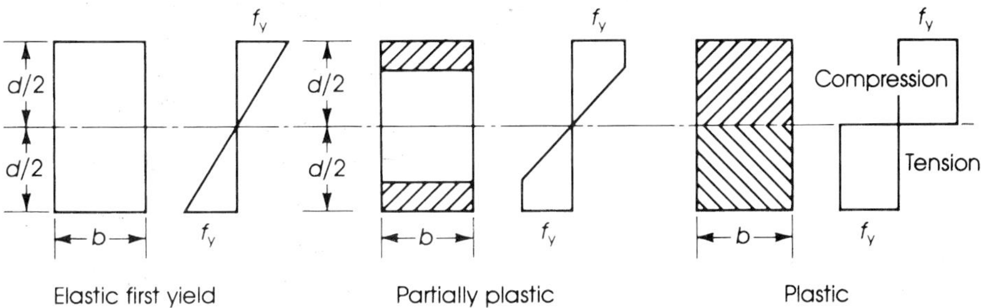

Figure 16.20 Distribution of yield stresses in a yielding steel rectangular section.

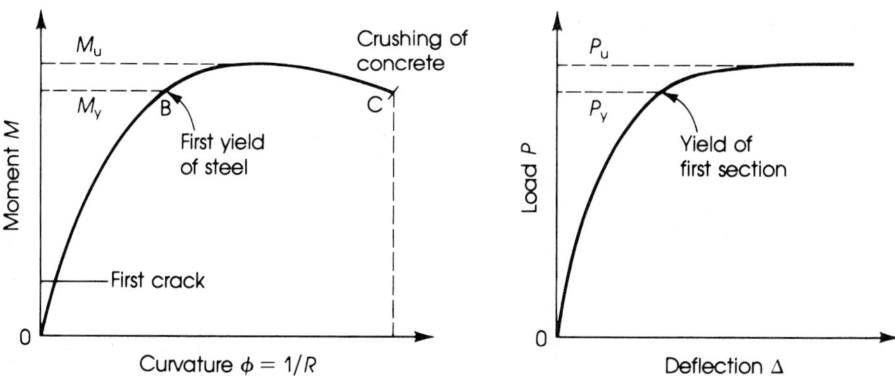

Figure 16.21 Yielding behavior of an under-reinforced concrete beam.

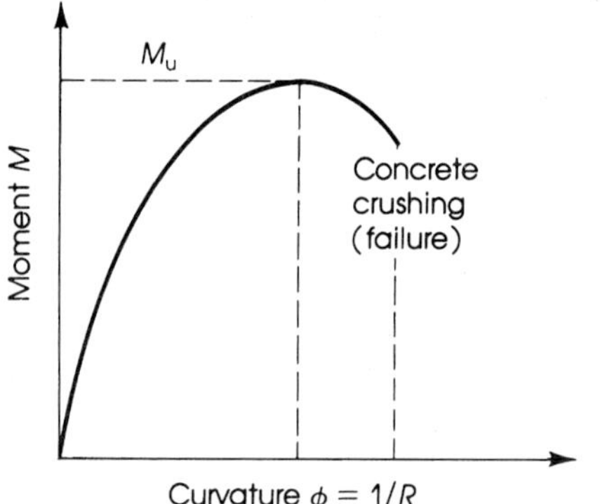

Figure 16.22 Yielding behavior of an over-reinforced concrete beam.

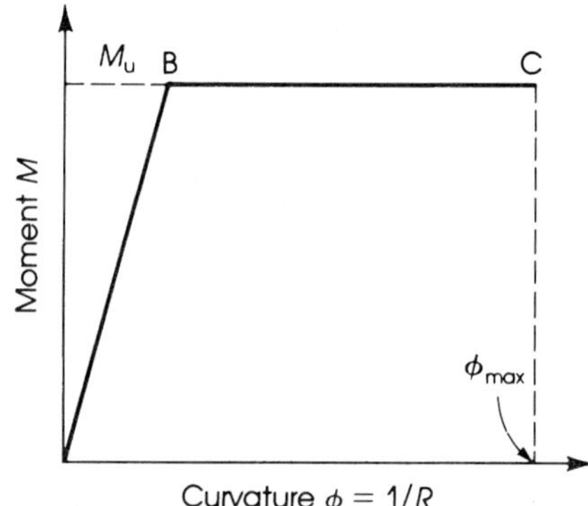

Figure 16.23 Idealized moment-curvature behavior of reinforced concrete beams.

forming a plastic hinge. Because concrete is a brittle material, there is usually considered to be a limit at which the member fails completely at maximum curvature at C.

Cranston [8] reported that in normally designed reinforced concrete frames, ample rotation capacity is available, and the maximum curvature at point C will not be reached until the failure or collapse of the frame. Therefore, when the member carries a moment equal to its ultimate moment, M_u, the curvature continues to increase between B and C without a change in the moment, producing a plastic hinge. The increase in curvature allows other parts of the statically indeterminate structure to carry additional loading.

16.8 THE COLLAPSE MECHANISM

In limit design, the ultimate strength of a reinforced concrete member is reached when it is on the verge of collapse. The member collapses when there are sufficient numbers of plastic hinges to transform it into a mechanism. The required number of plastic hinges, n, depends upon the degree of redundancy, r, of the structure. The relation between n and r to develop a mechanism is

$$n = 1 + r \tag{16.8}$$

For example, in a simply supported beam no redundants exist, and $r = 0$. Therefore, the beam becomes unstable and collapses when one plastic hinge develops at the section of

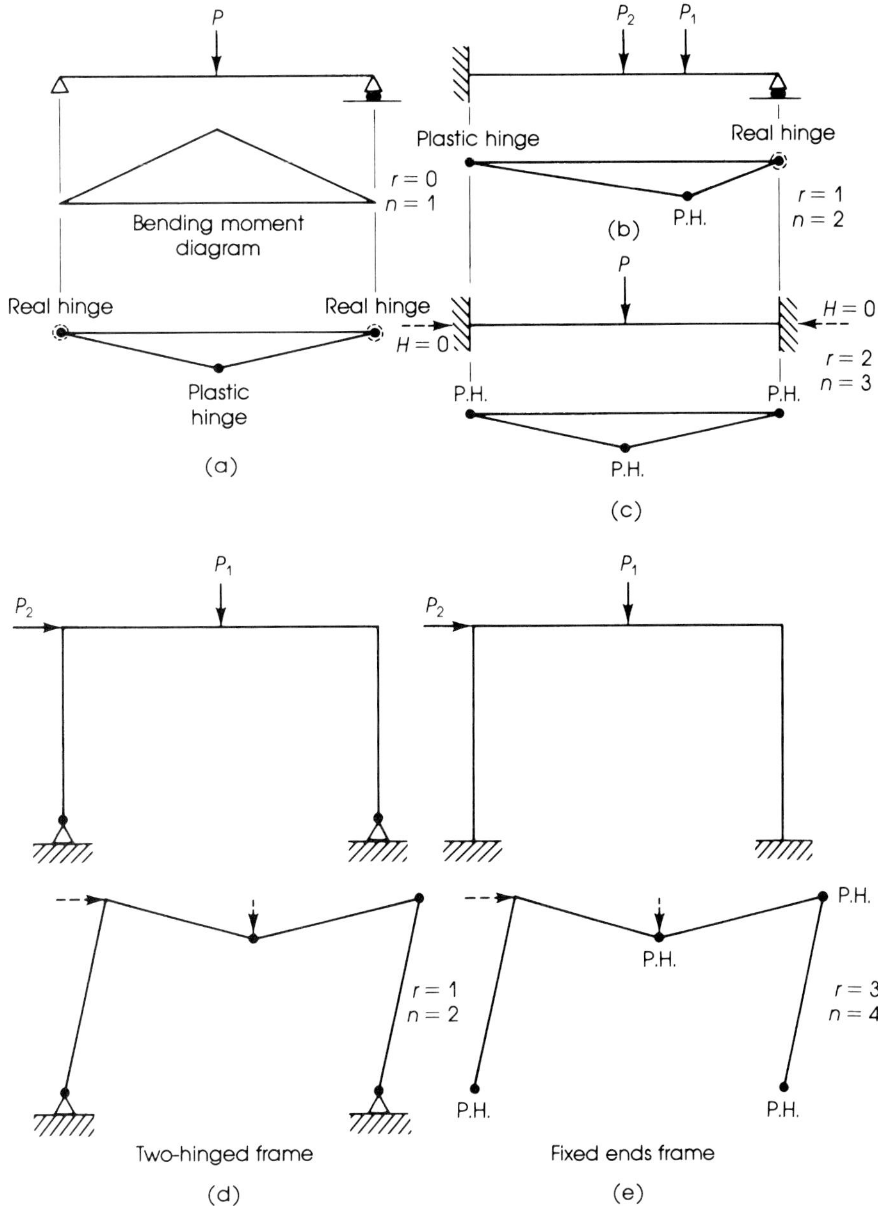

Figure 16.24 Development of plastic hinges (P.H.).

maximum moment, as shown in Figure 16.24(a). Applications to beams and frames are also shown in Figure 16.24.

16.9 PRINCIPLES OF LIMIT DESIGN

Under working loads, the distribution of moments in a statically indeterminate structure is based on elastic theory, and the whole structure remains in the elastic range. In limit design, where factored loads are used, the distribution of moments at failure, when a mechanism is reached, is different from that distribution based on elastic theory. This change reflects moment redistribution.

For limit design to be valid, four conditions must be satisfied.

1. *Mechanism condition:* Sufficient plastic hinges must be formed to transform the whole or part of the structure into a mechanism.
2. *Equilibrium condition:* The bending moment distribution must be in equilibrium with the applied loads.
3. *Yield condition:* The ultimate moment must not be exceeded at any point in the structure.
4. *Rotation condition:* Plastic hinges must have enough rotation capacity to permit the development of a mechanism.

Only the first three conditions apply to plastic design, because sufficient rotation capacity exists in ductile materials as steel. The fourth condition puts more limitations on the limit design of reinforced concrete members as compared to plastic design.

16.10 UPPER AND LOWER BOUNDS OF LOAD FACTORS

A structure on the verge of collapse must have developed the required number of plastic hinges to transform it into a mechanism. For arbitrary locations of the plastic hinges on the structure, the collapse loads can be calculated, which may be equal to or greater than the actual loads. Because the calculated loads cannot exceed the true collapse loads for the structure, then this approach indicates an upper or kinematic bound of the true collapse loads [10]. Therefore, if all possible mechanisms are investigated, the lowest M_u will be caused by the actual loads. Horne [11] explained the upper bound by assuming a mechanism and then calculating the external work, W_e, done by the applied loads and the internal work, W_i, done at the plastic hinges. If $W_e = W_i$, then the applied loads are either equal to or greater than the collapse loads.

If any arbitrary moment diagram is developed to satisfy the static equilibrium under the applied loads at failure, then the applied loads are either equal to or less than the true collapse loads. For different moment diagrams, different ultimate loads can be obtained. Higher values of the lower, or static, bound are obtained when the moments at several sections for the assumed moment diagram reach the collapse moment. Horne [11] explained the lower bound by assuming different moment distributions to obtain the one that is in equilibrium with the applied loads and satisfies the yield condition all over the structure. In this case, the applied loads are either equal to or less than the collapse loads.

16.11 LIMIT ANALYSIS

For the analysis of structures by the limit design procedure, two methods can be used, the virtual work method and the equilibrium method.

In the virtual work method, the work done by the ultimate load, P_u (or w_u), to produce a given virtual deflection, Δ, is equated to the work absorbed at the plastic hinges. The external work done by loads is $W_e = \Sigma(w_u \Delta)$ or $\Sigma(P_u \Delta)$. The work absorbed by the plastic hinges is internal work $= W_i = \Sigma(M_u \theta)$.

Example 16.3

The beam shown in Figure 16.25 carries a concentrated load at midspan. Calculate the collapse moment at the critical sections.

Solution

1. The beam is once statically indeterminate ($r = 1$), and the number of plastic hinges needed to transform the beam into a mechanism is $n = 1 + 1 = 2$ plastic hinges, at A

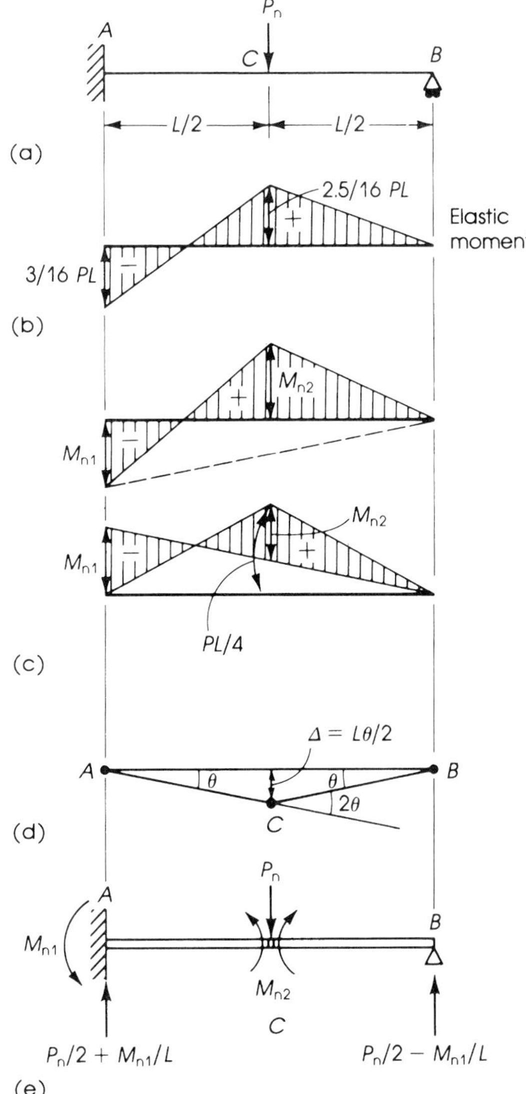

Figure 16.25 Example 16.3: $P_u = \phi P_n$ and $M_u = \phi M_n$.

and C. The first plastic hinge develops at A, and the beam acts as a simply supported member until a mechanism is reached.

2. If a rotation θ occurs at the plastic hinge at the fixed end, A, the rotation at the sagging hinge is $C = 2\theta$. The deflection of C under the load is $(L/2)\theta$ (Figure 16.25).

$$W_e = \text{external work} = \Sigma P_u \Delta = P_u\left(\frac{L\theta}{2}\right)$$

$$W_i = \text{internal work} = \Sigma M_u \theta = M_{u1}(\theta) + M_{u2}(2\theta)$$

If the two sections at A and C have the same dimensions and reinforcement, then $M_{u1} = M_{u2} = M_u$, and $W_i = 3M_u\theta$. Equating W_e and W_i,

$$M_{u1} + 2M_{u2} = P_u\frac{L}{2} = 3M_u \quad \text{and} \quad M_u = \frac{P_u L}{6}$$

Example 16.4

Calculate the collapse moments at the critical sections for the beam shown in Figure 16.26 due to a uniform load w_u.

Solution

1. The number of plastic hinges is 2.
2. For a deflection at $C = 1.0$, the rotation at A, θ_A, is $1/a; \theta_B = 1/b$, and

$$\theta_c = \theta_A + \theta_B = \frac{1}{a} + \frac{1}{b} = \frac{a+b}{ab} = \frac{L}{ab}$$

3. External work is

$$W_e = \Sigma w_u \Delta = w_u\left(\frac{1 \times L}{2}\right) = \frac{w_u L}{2}$$

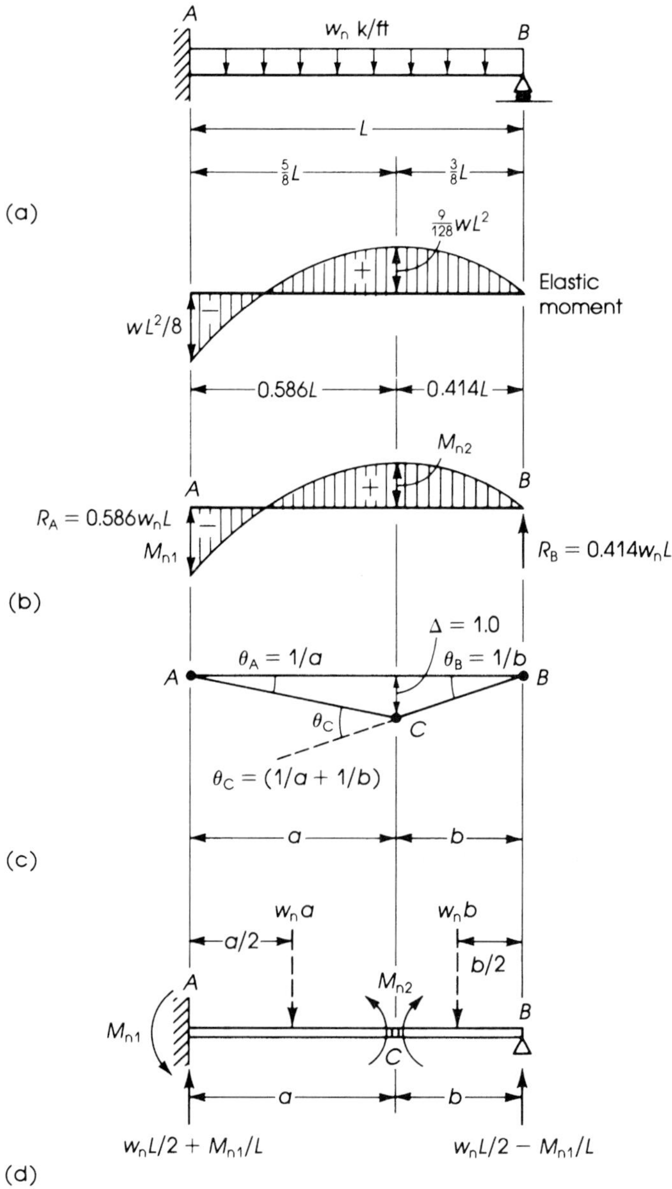

Figure 16.26 Example 16.4: $M_u = \phi M_n$ and $w_u = \phi w_n$.

Internal work is

$$W_i = \Sigma M_u \theta = M_{u1}\theta_A + M_{u2}\theta_c$$

$$= M_{u1}\left(\frac{1}{a}\right) + M_{u2}\left(\frac{1}{a} + \frac{1}{b}\right)$$

Equating W_e and W_i,

$$w_u = \frac{2}{L}\left(\frac{M_{u1}}{a} + \frac{M_{u2}}{a} + \frac{M_{u2}}{L - a}\right) \tag{16.9}$$

If both moments are equal, then

$$w_u = \frac{2M_u}{L}\left[\frac{2}{a} + \frac{1}{(L - a)}\right] = \frac{2M_u}{L}\left[\frac{(2L - a)}{a(L - a)}\right] \tag{16.10}$$

4. To determine the position of the plastic hinge at C that produces the minimum value of the collapse load w_u, differentiate equation (16.9) with respect to a and equate to zero:

$$\frac{\delta w_u}{\delta a} = 0; \qquad -\left(\frac{M_{u1}}{a^2} + \frac{M_{u2}}{a^2} - \frac{M_{u2}}{(L - a)^2}\right) = 0$$

If $M_{u1} = M_{u2} = M_u$, then

$$\frac{2}{a^2} = \frac{1}{(L - a)^2} \quad \text{or} \quad a = L(2 - \sqrt{2}) = 0.586L$$

From equation (16.10), the collapse load is $w_u = 11.66(M_u/L^2)$, and the collapse moment is $M_u = 0.0858 w_u L^2$. The reaction at A is $0.586 w_u L$, and the reaction at B is $0.414 w_u L$.

In the equilibrium method, the equilibrium of the beam or of separate segments of the beam is studied under the forces present at collapse. To illustrate analysis by this method, the two previous examples are repeated here.

Example 16.5

For the beam shown in Figure 16.25, calculate the collapse moments using the equilibrium method.

Solution

Two plastic hinges will develop at A and C. Referring to Figure 16.25(e), the reaction at A is $(P_u/2) + (M_{u1}L)$ and the reaction at B is $(P_u/2) - (M_{u1}/L)$. Considering the equilibrium of beam BC and taking moments about C,

$$\left(\frac{P_u}{2} - \frac{M_{u1}}{L}\right)\left(\frac{L}{2}\right) = M_{u2}$$

$$M_{u1} + 2M_{u2} = P_u\frac{L}{2}$$

which is the same equation obtained in Example 16.3. when $M_{u1} = M_{u2} = M_u$, then

$$3M_u = P_u\frac{L}{2} \quad \text{or} \quad M_u = P_u\frac{L}{6}$$

Example 16.6

Calculate the collapse moments for the beam shown in Figure 16.26 by the equilibrium method.

Solution

1. Two plastic hinges will develop in this beam at A and C. Referring to Figure 16.26(d), the reaction at $A = w_u(L/2) + (M_{u1}/L)$ and the reaction at $B = w_u(L/2) - (M_{u1}/L)$. The load on BC is $w_u b$ acting at $b/2$ from B, and $b = (L - a)$. Considering the equilibrium of segment BC and taking moments about C,

$$\left(w_u\frac{L}{2} - \frac{M_{u1}}{L}\right)b - (w_u b)\frac{b}{2} = M_{u2}$$

If $M_{u1} = M_{u2} = M_u$, then

$$w_u \frac{b}{2}(L - b) = M_u\left(1 + \frac{b}{L}\right) = \frac{M_u}{L}(2L - a)$$

$$w_u = \frac{2M_u}{L} \times \frac{(2L - a)}{a(L - a)}$$

which is similar to the results obtained in Example 16.4.

$$M_u = \frac{w_u L}{2} \times \frac{a(L - a)}{(2L - a)}$$

2. The position of a can be determined as before, where $a = 0.586L, M_u = 0.0858 w_u L_2$, and $w_u = 11.66(M_u/L^2)$.

16.12 ROTATION OF PLASTIC HINGES

16.12.1 Plastic Hinge Length

The assumption that the inelastic rotation of concrete occurs at the point of maximum moment while other portions of the member act elastically is a theoretical one; in fact, the plastic rotation occurs on both sides of the maximum moment section over a finite length. This length is called the plastic hinge length, l_p. The hinge length l_p is a function of the effective depth d, and the distance from the section of highest moment to the point of contraflexure (zero moment).

Referring to Figure 16.27(a), the length $l_p/2$ represents the plastic hinge length on one side of the center of support. M_u and ϕ_u indicate the ultimate moment and ultimate curvature at the critical section, whereas M_y and ϕ_y indicate the moment and curvature at first yield. The plastic curvature at the critical section ϕ_p is equal to $(\phi_u - \phi_y)$ and the rotation capacity is equal to $(\phi_p l_p)$.

The estimated length of the plastic hinge was reported by many investigators. A. L. L. Baker [7] assumed that the length of the plastic hinge is approximately equal to the effective depth d. Corley [12] proposed the following expression for the equivalent length of the plastic hinge:

$$l_p = 0.5d + 0.2\sqrt{d}\left(\frac{z}{d}\right) \quad (16.11)$$

where z = distance of the critical section to the point of contraflexure and d = effective depth of the section. Mattock [13] suggested a simpler form:

$$l_p = 0.5d + 0.05z \quad (16.12)$$

Tests [14] on reinforced concrete continuous beams showed that l_p can be assumed equal to $1.06d$. They also showed that the length of the plastic hinge, in reinforced concrete continuous beams containing hooked-end steel fibers, increases with the increase in the amount of the steel fibers and the main reinforcing steel according to the following expression:

$$l_p = (1.06 + 0.13\rho\rho_s)d \quad (16.13)$$

where ρ = percentage of main steel in the section and ρ_s = percentage of steel fibers by volume, $0 \le \rho_s \le 1.2$. For example, if $\rho = 1.0\%$ and $\rho_s = 0.8\%$, then $l_p = 1.164d$.

16.12.2 Curvature Distribution Factor

Another important factor involving the calculation of plastic rotations is the curvature distribution factor, β. The curvature along the plastic hinge varies significantly, and in most rotation estimations this factor is ignored, which leads to an overestimation of the plastic

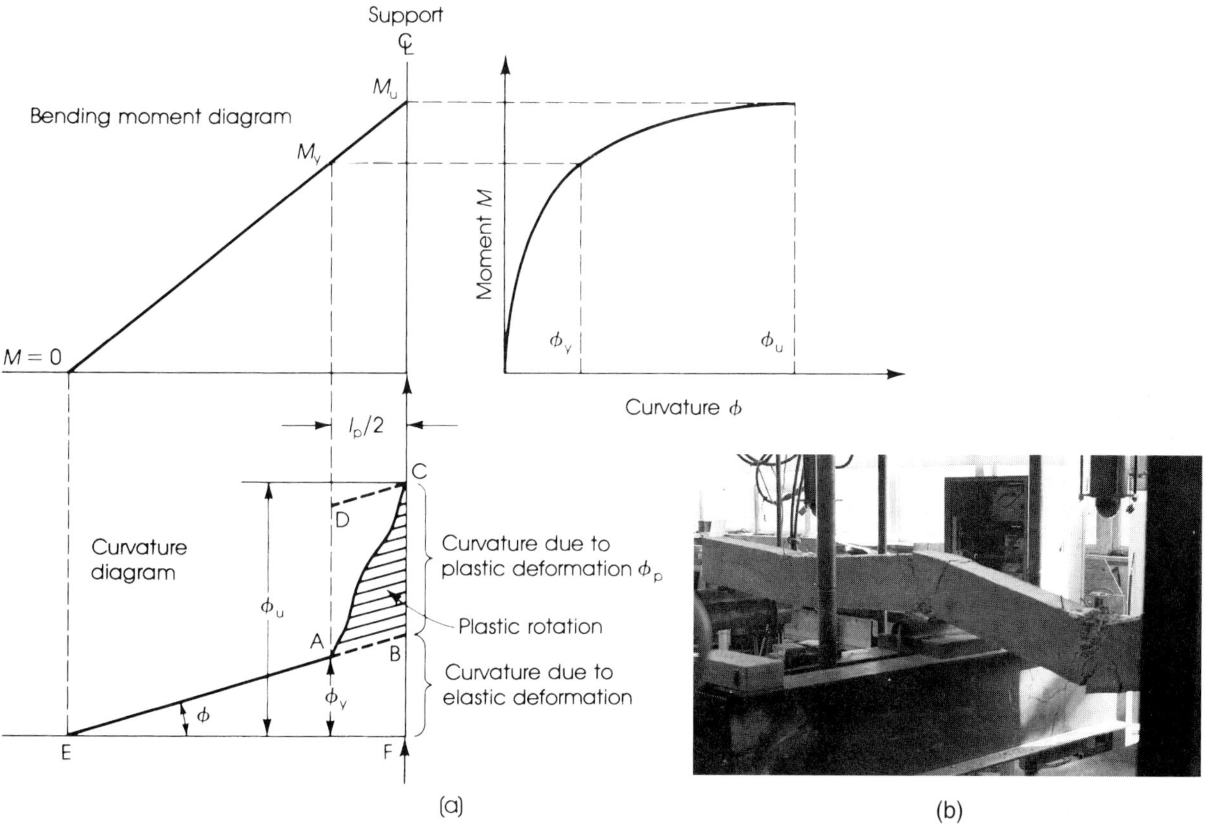

(a) (b)

Figure 16.27 (a) Plastic rotation from moment-curvature and moment gradient and (b) development of plastic hinges in a reinforced concrete continuous beam.

rotations. Referring to Figure 16.27, the shaded area, ABC, represents the inelastic rotation that can occur at the plastic hinge, whereas the unshaded area, EBF, represents the elastic contribution to the rotation over the length of the member. The shaded area ABC can be assumed equal to β times the total area $ABCD$ within the plastic hinge length, $l_p/2$, on one side of the critical section. The curvature distribution factor, β, represents the ratio of the actual plastic rotation, θ_{pc}, to ϕl_p, where ϕ is the curvature and l_p is the length of the plastic hinge. The value of β was reported to vary between 0.5 and 0.6. Tests [14] have showed that β can be assumed equal to 0.56. When hooked-end steel fibers were used in concrete beams, the value of β decreased according to the following expression:

$$\beta = 0.56 - 0.16\rho_s \tag{16.14}$$

where ρ_s is the percentage of steel fibers, $0 \leq \rho_s \leq 1.2\%$. The reduction of the curvature distribution factor of fibrous concrete does not imply that the rotation capacity is reduced: The plastic curvature of fibrous concrete is substantially higher than that of concrete without fibers. Figure 16.28 shows the distribution of the curvature along the plastic hinge length. The area ABC_1 represents the plastic rotation for concrete that does not contain steel fibers, $\beta = 0.56$, whereas the areas ABC_2 and ABC_3 represent the plastic rotation for concretes containing 0.8% and 1.2% steel fibers, respectively.

16.12.3 Ductility Index

The ratio of ultimate to first-yield curvature is called the *ductility index*, $\mu = \phi_u/\phi_y$. The ductility index of reinforced concrete beams was reported [15] to vary between 4 and 6. If steel

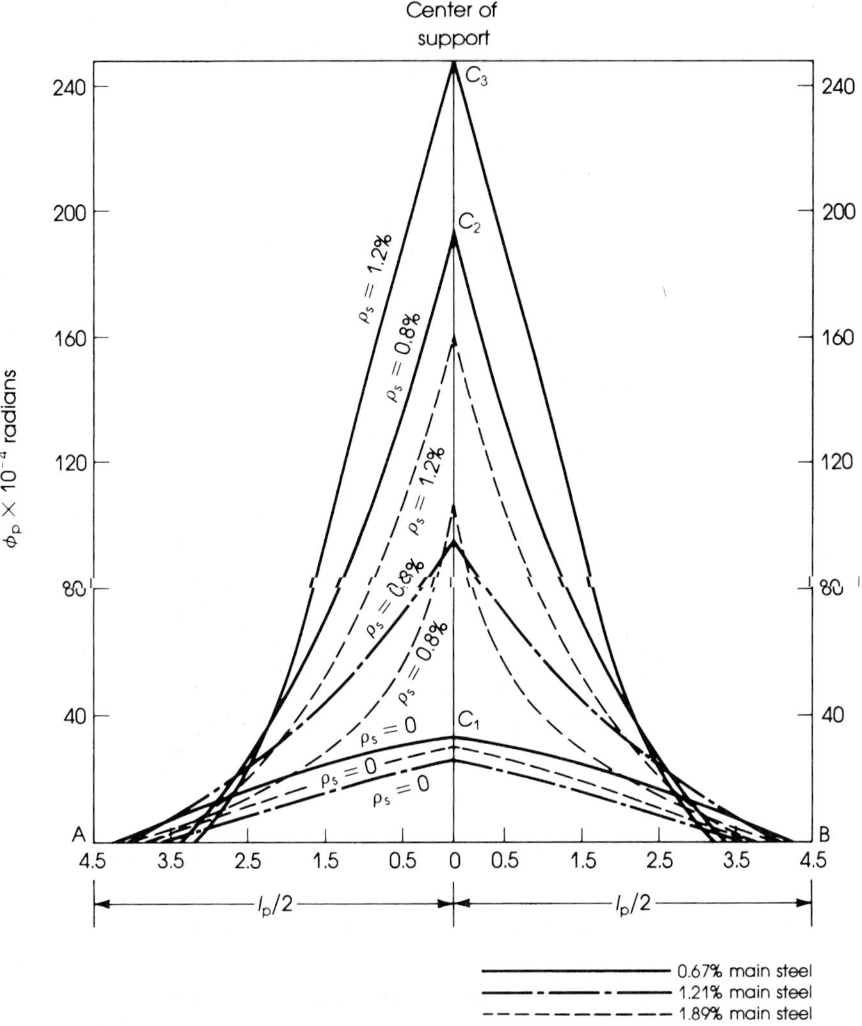

Figure 16.28 Curvature distribution along the plastic hinge.

fibers are used in concrete beams, the ductility index increases according to the following expression [14]:

$$\mu' = (1.0 + 3.8\rho_s)\mu \tag{16.15}$$

where

μ = the ratio of ultimate to first-yield curvature
μ' = ductility index of the fibrous concrete
ρ_s = percentage of steel fibers by volume, $0 \leq \rho_s \leq 1.2\%$

16.12.4 Required Rotation

The rotation of a plastic hinge in a reinforced concrete indeterminate structure is required to allow other plastic hinges to develop, and the structure to reach a mechanism can be determined by slope deflection from the following expression [7], [20]. For a segment AB between two plastic hinges, the rotation at A is

$$\theta_A = \frac{L}{6E_cI}\left[2(M_A - M_{FA}) + (M_B - M_{FB})\right] \tag{16.16}$$

where M_A and M_B = ultimate moments at A and B, respectively
M_{FA} and M_{FB} = elastic fixed-end moments at A and B
E_c = modulus of elasticity of concrete $= 33w^{1.5}\sqrt{f'_c}$
I = moment of inertia of a cracked section (Chapter 5)

16.12.5 Rotation Capacity Provided

Typical tensile plastic hinges at the support and midspan sections of a frame are shown in Figure 16.29. The rotation capacity depends mainly on the following:

1. The ultimate strain capacity of concrete, ε'_c, which may be assumed to be 0.003 or 0.0035, as used by Baker [7].
2. The length, l_p, over which yielding occurs at the plastic hinge, which can be assumed to be approximately equal to the effective depth of the section where the plastic hinge developed $(l_p = d)$.
3. The depth of the compressive block c in concrete at failure at the section of the plastic hinge. Baker [7] estimated the angle of rotation θ of a tensile plastic hinge as follows:

$$\theta = \frac{\varepsilon_p l_p}{c} \tag{16.17}$$

where ε_p is the increase in the strain in the concrete measured from the initial yielding of steel reinforcement in the section (see Figure 16.29(c)):

$$\varepsilon_p = \varepsilon'_c - \varepsilon_{c1} = 0.0035 - \varepsilon_{c1}$$

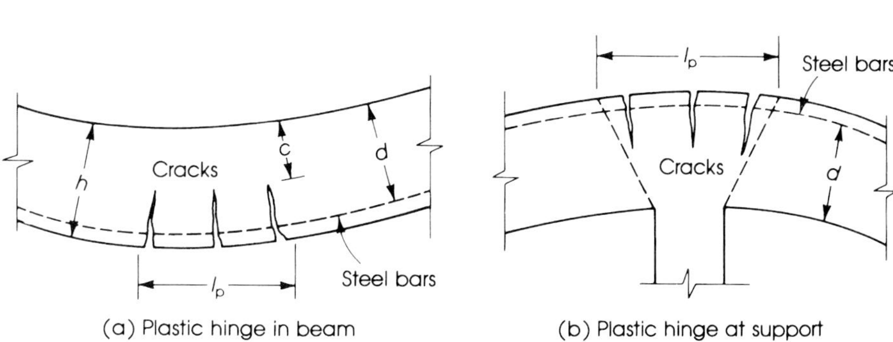

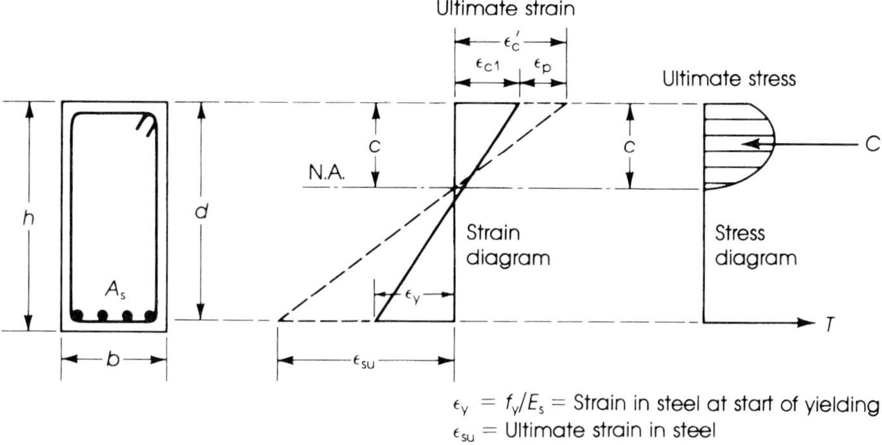

Figure 16.29 Plastic hinge and typical stress and strain distribution [2].

Plastic hinge in the maximum negative moment region.

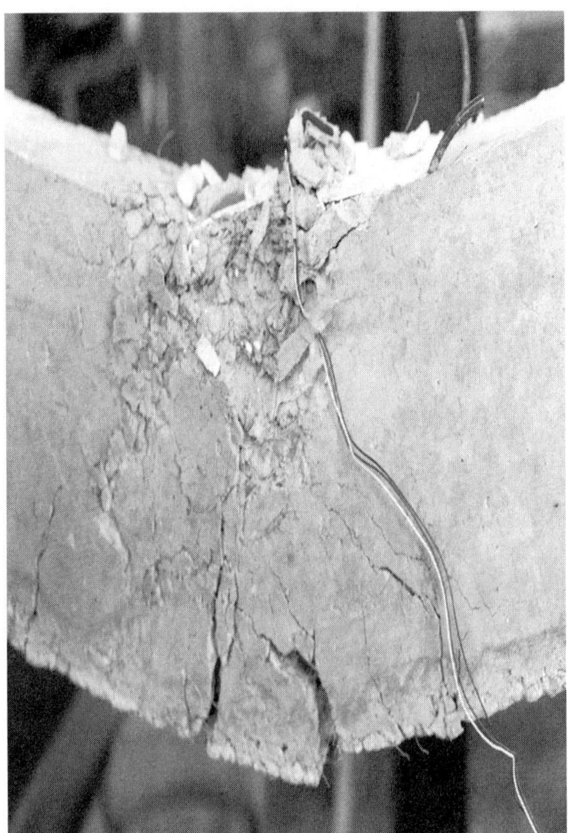

Plastic hinge in the maximum positive moment region.

If $l_p = d$ and the ratio c/d equals $\lambda \leq 0.5$,

$$\theta = \frac{(0.0035 - \varepsilon_{c1})d}{\lambda d} = \frac{0.0035 - \varepsilon_{c1}}{\lambda}$$

From strain triangles (Figure 16.29),

$$\varepsilon_{c1} = \varepsilon_y\left(\frac{c}{d - c}\right) = \frac{f_y}{E_s}\left(\frac{\lambda d}{d - \lambda d}\right) = \frac{f_y}{E_s}\left(\frac{\lambda}{1 - \lambda}\right)$$

where f_y = yield strength of steel bars and E_s = modulus of elasticity of steel = 29×10^6 psi. Therefore,

$$\theta = \frac{0.0035}{\lambda} - \frac{\varepsilon_{c1}}{\lambda} = \frac{0.0035}{\lambda} - \frac{f_y}{E_s(1 - \lambda)} \tag{16.18}$$

For grade 40 steel, $f_y = 40$ Ksi, and using a maximum value of λ of 0.50, then

$$\theta_{\min} = \frac{0.0035}{0.50} - \frac{40}{29{,}000 \times (1 - 0.50)} = 0.00424 \text{ rad}$$

For grade 60 steel, $f_y = 60$ Ksi and $\lambda_{\max} = 0.44$;

$$\theta_{\min} = \frac{0.0035}{0.44} - \frac{60}{29{,}000(1 - 0.44)} = 0.00426 \text{ rad}$$

The θ_{min} calculated here is from one side only, and the total permissible rotation at the plastic hinge equals 2θ or $2\theta_{min}$. The actual λ can be calculated as follows, given $a = \beta_1 c$ and $\beta_1 = 0.85$ for $f'_c \leq 4$ Ksi,

$$c = \frac{a}{0.85} = \frac{A_s f_y}{(0.85)^2 f'_c b} \tag{16.19}$$

$$\lambda = \frac{c}{d} = \frac{A_s f_y}{0.72 f'_c bd} = \frac{\rho f_y}{0.72 f'_c} \leq 0.5$$

where $\rho = A_s/bd$. (λ_{max} is obtained when ρ_{max} is used.)

If the rotation provided is not adequate, one can increase the section dimensions or reduce the percentage of steel reinforcement to obtain a smaller c, a smaller λ, and greater θ. A. L. L. Baker [3] indicated that if special binding or spirals are used, the ultimate crushing strain in bound concrete may be as high as 0.012. For a compression plastic hinge (as in columns),

$$\theta = \frac{\varepsilon_p l_p}{h} \tag{16.20}$$

where h = overall depth of the section and l_p = length over which yielding occurs. In compression hinges, l_p varies between $0.5h$ and h.

At a concrete ultimate stress of f'_c, $\varepsilon_c = 0.002$; thus $\varepsilon_p = \varepsilon'_c - 0.002 = 0.0035 - 0.002 = 0.0015$ is the minimum angle of rotation on one side. Therefore,

$$\theta_{min} = \frac{0.0015 \times 0.5h}{h} = 0.00075 \text{ rad}$$

With special binding or spirals, θ may be increased to

$$\theta_{max} = (0.012 - 0.002) \times \frac{0.5h}{h} = 0.005 \text{ rad}$$

The extreme value of $\varepsilon'_c = 0.012$ is quite high, and a smaller value may be used with proper spirals; otherwise a different section must be adopted.

In reinforced concrete continuous beams containing steel fibers, the plastic rotation may be estimated as follows [14]:

$$\theta_p = \lambda\beta\left(\frac{0.0035}{\lambda} - \frac{f_y}{E_s(1 - \lambda)}\right) \tag{16.21}$$

where

$$\lambda = \left(4.3 + 2.24\rho_s - 0.043 f_y + 4.17\rho\rho_s\right) \tag{16.22}$$

$$\beta = 0.56 - 0.16\rho_s \tag{16.14}$$

f_y = yield strength of steel, Ksi
E_s = modulus of elasticity of main steel
ρ = percentage of main steel
ρ_s = percentage of steel fibers

From equation (16.21), it is obvious that the plastic rotation of fibrous reinforced concrete is dependent upon the percentage of steel fibers and percentage of the main steel and its yield strength. Raising the yield strength of the main steel reduces the plastic rotation. Equation (16.21) also includes the effect of the plastic hinge length on rotation. A simplified form can be presented [14]:

$$\theta_p = \lambda\beta\left(\frac{0.003}{\lambda}\right) \tag{16.23}$$

For example, if $\rho_s = 0$ and $f_y = 60$ Ksi, then $\theta_{p1} = 0.00289/\lambda$, and for $\rho_s = 1.0\%, \rho = 1.5\%$, and $f_y = 60$ Ksi, then $\theta_{p2} = 0.01222/\lambda$. This means that the rotation capacity of a concrete beam may be increased by about four times if 1% of steel fibers is used.

16.13 SUMMARY OF LIMIT DESIGN PROCEDURE

1. Compute the factored loads using the load factors given in Chapter 3:

$$w_u = 1.4D + 1.7L$$

2. Determine the mechanism, plastic hinges, and ultimate moments M_u.
3. Design the critical sections using the strength design method.
4. Determine the required rotation of plastic hinges.
5. Calculate the rotation capacity provided at the sections of plastic hinges. The rotation capacity must exceed that required.
6. Check the factor against yielding of steel and excessive cracking, that is, $\phi M_u/$ elastic moment at service load.
7. Check deflection and cracking under service loads.
8. Check that adequate shear reinforcement is provided at all sections.

For more details, see [21].

Example 16.7

The beam shown in Figure 16.30 is fixed at both ends and carries a uniform dead load of 1.5 K/ft, a uniform live load of 2.0 K/ft, a concentrated dead load of 10 K, and a concentrated live load of 20 K. Design the beam using the limit design procedure. Use $b = 14$ in., $f'_c = 3$ Ksi, and $f_y = 40$ Ksi.

Solution

1. Ultimate uniform load $w_u = 1.4D + 1.7L = 1.4 \times 1.5 + 1.7 \times 2.0 = 5.5$ K/ft. Ultimate concentrated load $P_u = 1.4 \times 10 + 1.7 \times 20 = 48$ K.
2. The plastic hinges will develop at A, B, and C, causing the mechanism shown in Figure 16.30. Using the virtual work method of analysis and assuming a unit deflection at C, then the external work is equal to

$$W_e = 48 \times 1 + 5.5\left(24 \times \frac{1}{2}\right) = 114 \text{ K} \cdot \text{ft}$$

The internal work absorbed by the plastic hinges is

$$W_i = M_u\theta \text{ (at } A) + M_u\theta \text{ (at } B) + M_u(2\theta) \text{ at } C$$

$$= 4M_u\theta = 4M_u\left(\frac{1}{12}\right) = \frac{M_u}{3}$$

Equating W_e and W_i gives $M_u = 342$ K · ft. The general analysis gives directly

$$M_u = \frac{w_u L^2}{16} + P_u\frac{L}{8} = \frac{5.5}{16}(24)^2 + 48\frac{24}{8} = 342 \text{ K} \cdot \text{ft}$$

3. Design the critical sections at A, B, and C for $M_u = 342$ K · ft. From tables in Appendix A and for $f'_c = 3$ Ksi, $f_y = 40$ Ksi, and a steel percentage $\rho = 0.013$, $R_u = 420$ psi.

$$M_u = R_u bd^2$$

$$342 \times 12 = 0.42 \times 14(d)^2$$

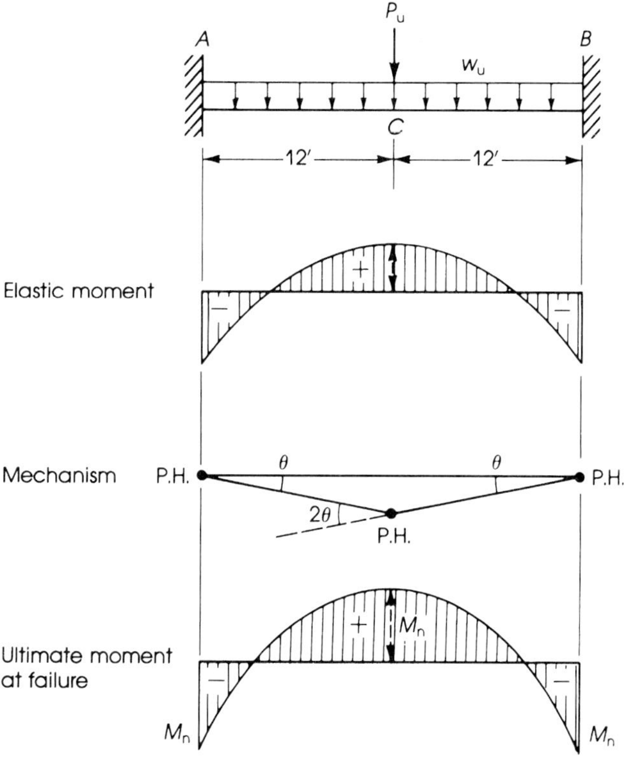

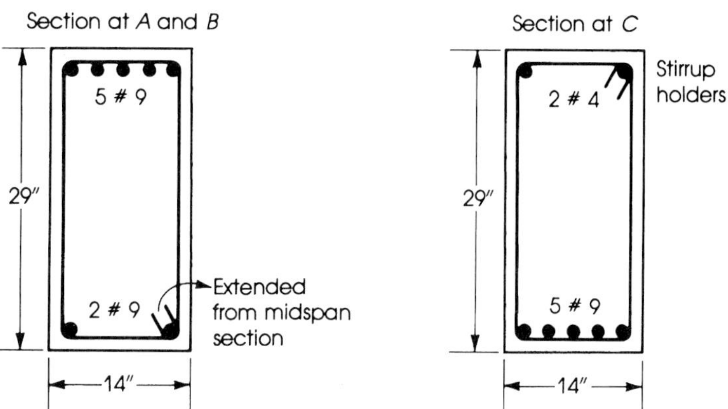

Figure 16.30 Example 16.7.

$d = 26.4$ in. and the total depth is $h = 26.4 + 2.5 = 28.9$ in., say, 29 in.

$$A_s = \rho bd = 0.013 \times 14 \times 26.4 = 4.8 \text{ in.}^2$$

Use five no. 9 bars in one row; A_s provided $= 5.0$ in.^{2}, $b_{\min} = 13.875$ in. < 14 in.

$$a = \frac{A_s f_y}{0.85 f'_c b} = \frac{5.0 \times 40}{0.85 \times 3 \times 14} = 5.6 \text{ in.}$$

$$c = \frac{a}{0.85} = 6.6 \text{ in.,} \qquad \lambda = \frac{c}{d} = \frac{6.6}{26.4} = 0.25$$

4. The required rotation of plastic hinges is as follows:

a $\theta_A = \dfrac{L}{6E_cI}[2(M_A - M_{FA}) + (M_B - M_{FB})]$

$$E_c = 57{,}400\sqrt{f'_c} = 3.144 \times 10^6 \text{ psi}$$

$$E_s = 29 \times 10^6 \text{ psi} \quad \text{and} \quad n = \frac{E_s}{E_c} = 9.2$$

b. Determine the fixed end moments at A and B using factored loads:

$$M_{FA} = M_{FB} = \frac{w_u L^2}{12}(\text{uniform load}) + \frac{P_u L}{8}(\text{concentrated load})$$

$$= 5.5\frac{(24)^2}{12} + 48 \times \frac{24}{8} = 408 \text{ K}\cdot\text{ft}$$

Plastic M_A = plastic $M_B = 342$ K·ft

c. The cracked moment of inertia can be calculated from

$$I_{cr} = b\frac{x^3}{3} + nA_s(d - x)^2$$

where x is the distance from compression fibers to the neutral axis (kd). To determine x (see Chapter 5), $x = 10.3$ in. and $I_{cr} = 17{,}172$ in.^{4}

d. Required minimum rotation: Considering all moments at supports A and B are negative, then

$$\theta_A = \frac{24 \times 12}{6 \times 3.144 \times 10^6 \times 17{,}172}[2(-342 + 408) + (-342 + 408)](12{,}000) = 0.00211 \text{ rad}$$

5. The rotation capacity provided is

$$\theta_A = \frac{0.0035}{\lambda} - \frac{f_y}{E_s(1 - \lambda)} = \frac{0.0035}{0.25} - \frac{40}{29{,}000(1 - 0.25)}$$

$$= 0.0122 \text{ rad} > 0.00211 \quad \text{required}$$

The rotation capacity provided is about 5.5 times that required, indicating that the section is adequate.

6. Check the ratio of ultimate to elastic moment at service load:

$$M_A = M_B = \frac{wL^2}{12} + \frac{PL}{8}$$

$$= 3.5\frac{(24)^2}{12} + \frac{30 \times 24}{8} = 258 \text{ K}\cdot\text{ft}$$

Actual $\phi M_n = \phi A_s f_y[d - (a/2)] = 0.9 \times 5 \times 40[26.5 - (5.6/2)]/12 = 356$K·ft. The ratio is $356/258 = 1.38$, which represents the factor of safety against the yielding of steel bars at the support.

7. Check maximum deflection due to service load (at midspan): For a uniform load w,

$$\Delta_1 = \frac{wL^4}{384EI}$$

For a concentrated load at midspan,

$$\Delta_2 = \frac{PL^3}{192EI}$$

and total deflection is

$$\Delta = \frac{(3500/12)(24 \times 12)^4}{384(17{,}172)(3.144 \times 10^6)} + \frac{30{,}000(24 \times 12)^3}{192(17{,}172)(3.144 \times 10^6)} = 0.166 \text{ in.}$$

$$\frac{\Delta}{L} = \frac{0.166}{24 \times 12} = \frac{1}{1735}$$

which is a very small ratio.

8. Adequate shear reinforcement must be provided to avoid any possible shear failure.

16.14 MOMENT REDISTRIBUTION

The ACI Code does not cover the limit design procedure, as explained here; however, it specifies another conservative approach to take into account the redistribution of moments in indeterminate structures at failure. Section 8.4.1 of the ACI Code indicates that the maximum negative moments calculated by elastic theory at the supports of continuous flexural members may be increased or decreased by not more than

$$q = 20\left(1 - \frac{\rho - \rho'}{\rho_b}\right)$$

$$\rho = \frac{A_s}{bd}, \quad \text{and} \quad \rho' = \frac{A'_s}{bd}$$

$$\rho_b = 0.85\beta_1 \frac{f'_c}{f_y}\left(\frac{87}{87 + f_y}\right) \tag{16.24}$$

Equation (16.24) must not be applied to the moments calculated from approximate structural analysis or to the alternative design method of reinforced concrete based on elastic theory. The ACI Code specifies a different factor for the redistribution of negative moments in continuous prestressed concrete members.

The code limits the steel ratio ρ (or $\rho - \rho'$) at the section where the moment is reduced to a maximum ratio of $0.5\rho_b$. The minimum steel ratio allowed in the section for flexural design is $\rho_{min} = 200/f_y$ for $fc' < 4500$ psi and $3\sqrt{f'_c}/f_y$ when $f'_c \geq 4500$ psi. If the extreme values, $0.5\rho_b$ and ρ_{min}, are substituted in equation (16.24), the maximum and minimum redistribution percentages will be as shown in Table 16.1.

Table 16.1 Maximum and Minimum q

f'_c (Ksi)	f_y (Ksi)	ρ_b	ρ_{min}	q_{max} (Percent) (for ρ_{min})	q_{min} (Percent) (for $0.5\rho_b$)
3	40	0.0371	0.0050	17.3	10
	60	0.0215	0.0033	16.9	10
4	40	0.0496	0.0050	18.0	10
	60	0.0285	0.0033	17.7	10
5	40	0.0581	0.0053	18.2	10
	60	0.0339	0.0035	18.0	10

In practice, a minimum steel percentage, ρ_{min}, is not adopted for a good design, and therefore a maximum redistribution percentage will not practically be used. A steel percentage of $0.5\rho_b$ or $0.5\rho_{max}$ is very common in the design of sections; thus a redistribution percentage of 10% to 15% may be expected.

Whatever percentage of moment redistribution is used, it is essential to ensure that no section is likely to suffer local damage or excessive cracking at service loads and that adequate rotation capacity is maintained at every critical section in the structure. The redistribution of moments in a statically indeterminate structure will result in a reduction in the negative moments at the supports and in the positive moments within the spans. This reduction will not imply that the safety of the structure has been reduced or jeopardized as compared with determinate structures. In fact, continuity in structures provides additional strength, stability, and economy in the design.

Example 16.8

Determine the maximum elastic moments at the supports and midspans of the continuous beam of four equal spans shown in Figure 16.31(a). The beam has a uniform section and carries a uniform dead load of 8 K/ft and a live load of 6 K/ft. Assume 10% maximum redistribution of moments and consider the following two cases: (1) When the live load is placed on alternate spans, calculate the maximum positive moments within the spans, and (2) when the live load is placed on adjacent spans, calculate the maximum negative moments at the supports.

Solution

1. The beam has a uniform moment of inertia I and has the same E; thus EI is constant. The three-moment equation to analyze the beam and for a constant EI is

$$M_A L_1 + 2M_B(L_1 + L_2) + M_c L_2 = -\frac{w_1 L_1^3}{4} - \frac{w_2 L_2^3}{4}$$

Because the spans are equal,

$$M_A + 4M_B + M_C = -\frac{L^2}{4}(w_1 + w_2) \tag{16.25}$$

In this example $M_A = M_E = 0$. Six different cases of loading will be considered, as shown in Figure 16.31:

Case 1. Dead load is placed on the whole beam $ABCDE$ (Figure 16.31(b)).

Case 2. Live load is placed on AB and CD for maximum positive moments within AB and CD (Figure 16.31(c)).

Case 3. Similar to Case 2 for beams BC and DE (Figure 16.31(d)).

Case 4. Live load is placed on AB, BC, and DE for a maximum negative moment at B (Figure 16.31(e)).

Case 5. Live load is placed on spans CD and DE (Figure 16.31(f)).

Case 6. Live load is placed on BC and CD for a maximum negative moment at C (Figure 16.31(g)).

2. Case 1. Apply equation (16.25) to the beam segments ABC, BCD, and CDE, respectively:

$$4M_B + M_C = -\frac{(20)^2}{4}(8 + 8) = -1600 \text{ K}\cdot\text{ft}$$

$$M_B + 4M_C + M_D = -1600 \text{ K}\cdot\text{ft}$$

$$M_C + 4M_D = -1600 \text{ K}\cdot\text{ft}$$

Solve the three equations to get

$$M_B = M_D = -342.8 \text{ K}\cdot\text{ft} \quad \text{and} \quad M_C = -228.6 \text{ K}\cdot\text{ft}$$

For a 10% reduction in moments,

$$M'_B = M'_D = 0.9(-342.8) = -308.5 \text{ K}\cdot\text{ft}$$

$$M'_C = 0.9(-228.6) = -205.7 \text{ K}\cdot\text{ft}$$

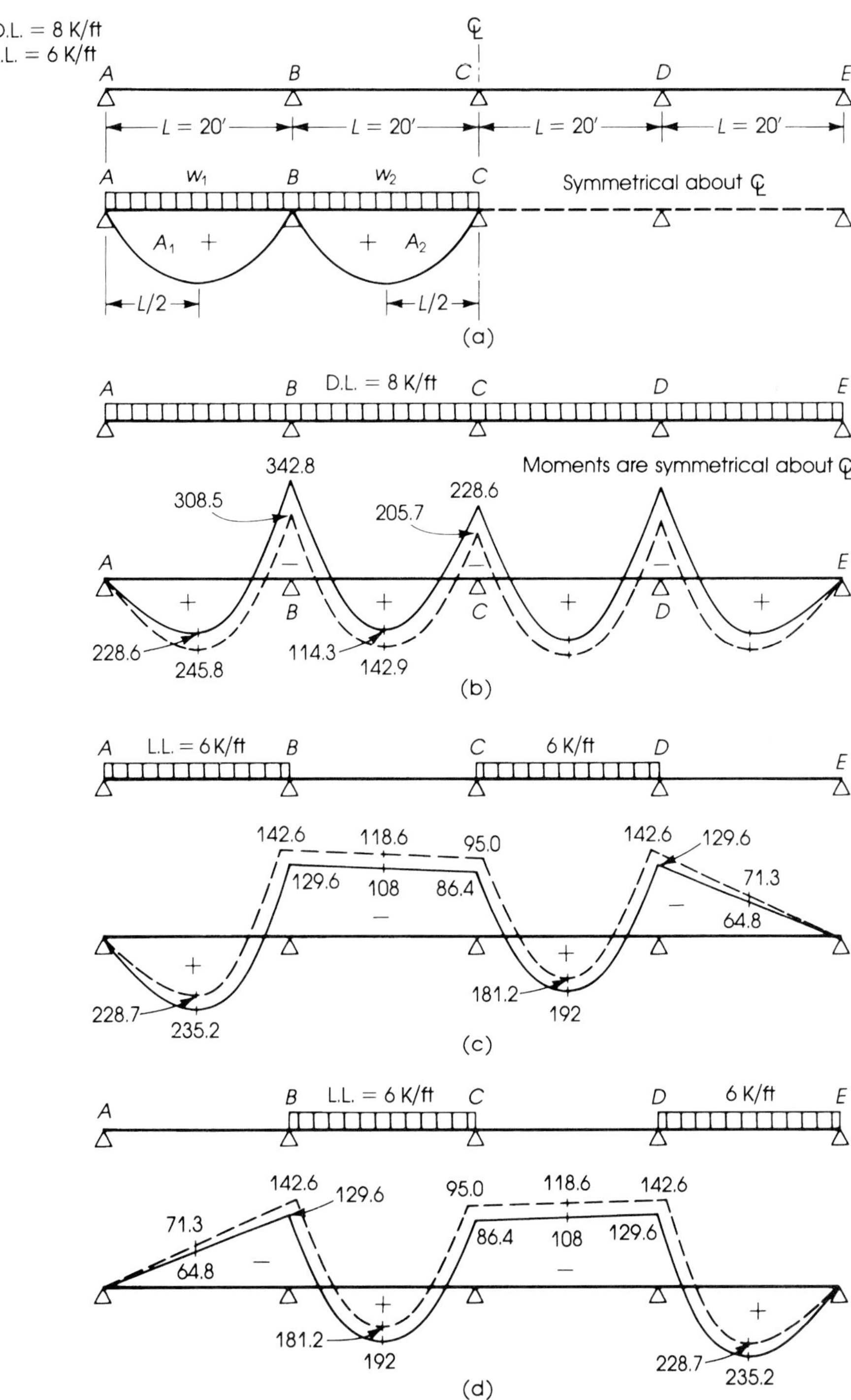

Figure 16.31 Example 16.8: Bending moments are drawn on the tension side.

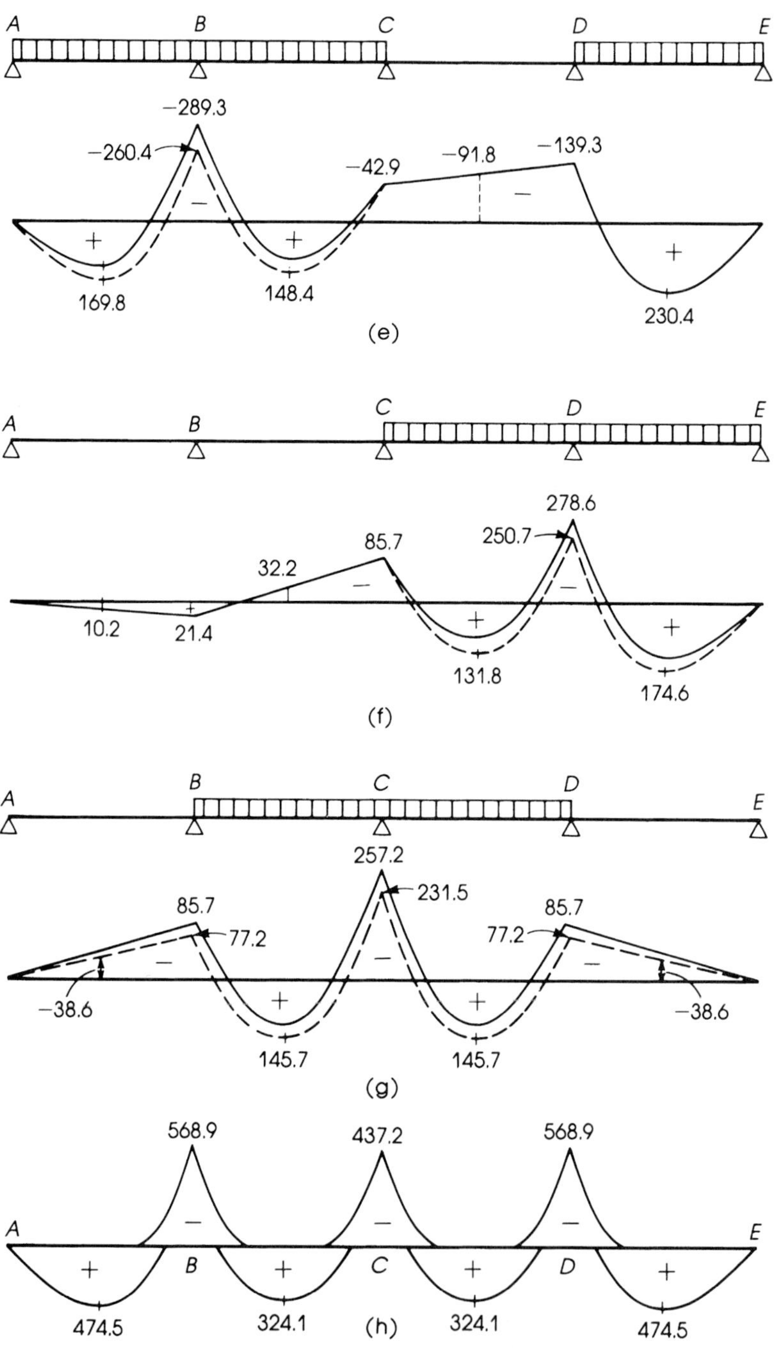

Figure 16.31e-h *(continued)*

The corresponding midspan moments are

$$\text{Span } AB = DE = \frac{w_D L^2}{8} + \frac{1}{2} M_B = \frac{8(20)^2}{8} - \frac{1}{2} \times 308.5 = 245.8 \text{ K} \cdot \text{ft}$$

$$\text{Span } BC = CD = \frac{w_D L^2}{8} - \frac{1}{2}(308.5 + 205.7) = \frac{8(20)^2}{8} - 257.1 = 142.9 \text{ K} \cdot \text{ft}$$

3. Case 2. Apply equation (16.25) to ABC, BCD, and CDE, respectively:

$$4M_B + M_C = -\frac{(20)^2}{4}(6) = -600 \text{ K}\cdot\text{ft}$$

$$M_B + 4M_C + M_D = -600 \text{ K}\cdot\text{ft}$$

$$M_C + 4M_D = -600 \text{ K}\cdot\text{ft}$$

Solve the three equations to get

$$M_B = M_D = -129.6 \text{ K}\cdot\text{ft} \qquad M_C = -86.4 \text{ K}\cdot\text{ft}$$

The corresponding elastic midspan moments are

$$\text{Beam } AB = \frac{w_L L^2}{8} + \frac{M_B}{2} = \frac{6(20)^2}{8} - \frac{129.6}{2} = +235.2 \text{ K}\cdot\text{ft}$$

$$BC = 0 - \frac{1}{2}(129.6 + 86.4) = -108 \text{ K}\cdot\text{ft}$$

$$CD = \frac{w_L L^2}{8} - \frac{1}{2}(129.6 + 86.4) = \frac{6(20)^2}{8} - 108 = +192 \text{ K}\cdot\text{ft}$$

$$DE = 0 - \frac{1}{2} \times 129.6 = -64.8 \text{ K}\cdot\text{ft}$$

To reduce the positive span moment, increase the support moments by 10% and calculate the corresponding positive span moments. The resulting positive moment must be at least 90% of the first calculated moments given previously.

$$M'_B = M'_D = 1.1(-129.6) = -142.6 \text{ K}\cdot\text{ft}$$

$$M'_C = 1.1(-86.4) = -95.0 \text{ K}\cdot\text{ft}$$

The corresponding midspan moments are

$$\text{Beam } AB = \frac{w_L L^2}{8} + \frac{M'_B}{2} = \frac{6(20)^2}{8} - \frac{142.6}{2} = +228.7 \text{ K}\cdot\text{ft}$$

$$BC = -\frac{1}{2}(142.6 + 95) = -118.8 \text{ K}\cdot\text{ft}$$

$$CD = \frac{w_L L^2}{8} + \frac{1}{2}\left(M'_C + M'_D\right) = \frac{6(20)^2}{8} - \frac{1}{2}(95 + 142.6) = 181.2 \text{ K}\cdot\text{ft}$$

$$DE = -\frac{1}{2} \times 142.6 = -71.3 \text{ K}\cdot\text{ft}$$

4. Case 3. This case is similar to Case 2, and the moments are shown in Figure 16.31(d).

5. Case 4. Consider the spans AB, BC, and DE loaded with live load to determine the maximum negative moment at support B:

$$4M_B + M_C = -\frac{w_L L^2}{2} = -\frac{6(20)^2}{2} = -1200 \text{ K}\cdot\text{ft}$$

$$M_B + 4M_C + M_D = -\frac{w_L L^2}{4} = -\frac{6(20)^2}{4} = -600 \text{ K}\cdot\text{ft}$$

$$M_C + 4M_D = -\frac{6(20)^2}{4} = -600 \text{ K}\cdot\text{ft}$$

Solve the three equations to get

$$M_C = -42.9 \text{ K·ft}$$

$$M_B = -289.3 \text{ K·ft}$$

$$M_D = -139.3 \text{ K·ft}$$

For 10% reduction in moment at support B,

$$M'_B = 0.9 \times (-289.3) = -260.4 \text{ K·ft}$$

The corresponding midspan moments are

$$\text{Beam } AB = \frac{w_L L^2}{8} + \frac{M_B}{2} = \frac{6(20)^2}{8} - \frac{260.4}{2} = 169.8 \text{ K·ft}$$

$$BC = \frac{w_L L^2}{8} - \frac{1}{2}(260.4 + 42.9) = 148.4 \text{ K·ft}$$

$$CD = -\frac{1}{2}(42.9 + 139.3) = -91.1 \text{ K·ft}$$

$$DE = 300 - \frac{1}{2} \times 139.3 = +230.4 \text{ K·ft}$$

6. Case 5. This is similar to Case 4, except that one end span is not loaded to produce maximum positive moment at support B (or support D for similar loading). The bending moment diagrams are shown in Figure 16.31(f).
7. Case 6. Consider the spans BC and CD loaded with live load to determine the maximum negative moment at support C:

$$4M_B + M_C = \frac{w_L L^2}{4} = -600 \text{ K·ft}$$

$$M_B + 4M_C + M_D = -\frac{w_L L^2}{2} = -1200 \text{ K·ft}$$

$$M_C + 4M_D = -\frac{w_L L^2}{4} = -600 \text{ K·ft}$$

Solve the three equations to get

$$M_C = -257.2 \text{ K·ft}$$

$$M_B = M_D = -85.7 \text{ K·ft}$$

For 10% reduction in support moments,

$$M'_C = 0.9 \times (-257.2) = -231.5 \text{ K·ft}$$

$$M'_B = M'_D = 0.9 \times (-85.7) = -77.2 \text{ K·ft}$$

The corresponding midspan moments are

$$\text{Beam } AB = DE = -\frac{77.2}{2} = -38.6 \text{ K·ft}$$

$$BC = CD = \frac{w_L L^2}{8} - \frac{1}{2}(231.5 + 77.2) = \frac{6(20)^2}{8} - 154.3 = 145.7 \text{ K·ft}$$

8. The final maximum and minimum moments after moment redistribution are shown in Table 16.2. The moment envelope is shown in Figure 16.31(h).

Table 16.2 Final Moments of Example 16.8 After Moment Redistribution

Case	1	2	3	4	5
Section Location	**D.L. Moments**	**L.L. Maximum Negative**	**L.L. Maximum Positive**	**D.L. + L.L. (1) + (2) Maximum Negative**	**D.L. + L.L. (1) + (3) Maximum Positive**
Support					
A	0	0	0	0	0
B	−308.5	−260.4	+21.4	−568.9*	−287.1
C	−205.7	−231.5	—	−437.2*	−205.7
D	−308.5	−260.4	+21.4	−568.9*	−287.1
E	0	0	0	0	0
Midspan					
AB	245.8	−71.3	228.7	+174.5	+474.5*
BC	142.9	−118.6	181.2	+24.3	+324.1*
CD	142.9	−118.6	181.2	+24.3	+324.1*
DE	245.8	−71.3	228.7	+174.5	+474.5*

* Final maximum and minimum design moments.

9. In this example, the midspan sections are used for simplicity: The midspan moments are not necessarily the maximum positive moments. In the case of the end spans AB and DE, the maximum moment after 10% moment redistribution is equal to $(w_D L_2)/12.2$ and occurs at $0.4L$ from A and D.

SUMMARY

Sections 16.1–16.3

In continuous beams, the maximum and minimum moments are obtained by considering the dead load acting on all spans, whereas pattern loading is considered for live or moving loads, as shown in Figures 16.2 and 16.3. The ACI moment coefficients given in Chapter 9 may be used to compute approximate values for the maximum and minimum moments and shears.

Sections 16.4–16.5

A frame subjected to a system of loads may be analyzed by the equivalent frame method. Frames may be statically determinate or indeterminate.

Section 16.6

There are several types of frame hinges: Mesnager, Considère, lead, and concrete hinges. The steel for a Mesnager hinge is calculated as follows:

$$A_{s1} = \frac{R_1}{0.55f_y} \quad \text{and} \quad A_{s2} = \frac{R_2}{0.55f_y} \tag{16.2}$$

$$\text{Burst force:} \quad F = \frac{P_u}{2}\tan\theta + \frac{Ha}{0.85d} \tag{16.4}$$

$$\text{Stress in ties:} \quad f_s = \frac{F}{0.005ab + A_{st}\,(\text{ties})} \le 0.85f_y \tag{16.6}$$

Sections 16.7–16.8

Limit design in reinforced concrete refers to redistribution of moments, which occurs throughout the structure as steel reinforcement reaches its yield strength. Ultimate strength is reached when the structure is on the verge of collapse. This case occurs when a number of plastic hinges, n, develop in a structure with redundants r, such that $n = 1 + r$.

Sections 16.9–16.11

For limit design to be valid, four conditions must be satisfied: mechanism, equilibrium, yield, and rotation. Two methods of analysis may be used: the virtual work method and the equilibrium method, which are both explained in Examples 16.3 through 16.6.

Sections 16.12–16.13

The plastic hinge length, l_p, can be considered equal to the effective depth d. In fibrous concrete,

$$l_p = (1.06 + 0.13\rho\rho_s)d \tag{16.13}$$

$$\text{Ductility index:} \quad \mu = \frac{\phi_u}{\phi_y}$$

For fibrous concrete,

$$\mu' = (1.0 + 3.8\rho_s)\mu \tag{16.15}$$

$$\text{Angle of rotation:} \quad \theta = \frac{0.0035}{\lambda} - \frac{f_y}{E_s(1 - \lambda)} \tag{16.18}$$

$$\lambda = \frac{\rho f_y}{0.72 f'_c} \leq 0.5 \tag{16.19}$$

A summary of the limit design procedure is given in Section 16.14.

Section 16.14

Moment redistribution may be taken into account in the analysis of statically indeterminate structures. In this case, the maximum negative moments calculated by the elastic theory may be increased or decreased by not more than the ratio q, where

$$q = 20\left(1 - \frac{\rho - \rho'}{\rho_b}\right) \tag{16.24}$$

Table 16.1 gives the different values of q. Moment redistribution is explained in detail in Example 16.8.

REFERENCES

1. J. C. McCormac. *Structural Analysis.* New York: Intext, 1985.
2. M. N. Hassoun. *Ultimate Load Design of Reinforced Concrete.* Cement and Concrete Association. London, 1981.
3. A. L. L. Baker. *The Ultimate Load Theory Applied to the Design of Reinforced and Prestressed Concrete Frames.* London: Concrete Publications, 1956.
4. W. Prager and P. G. Hodge. *Theory of Perfectly Plastic Solids.* New York: John Wiley, 1951.
5. L. S. Beedle, B. Thürlimann, and R. L. Ketter. *Plastic Design in Structural Steel.* New York: American Institute of Steel Construction, 1955.
6. J. F. Baker, M. R. Horne, and J. Heyman. *The Steel Skeleton,* Vol. 2. London: Cambridge University Press, 1956.

7. A. L. L. Baker. *Limit-State Design of Reinforced Concrete.* London: Cement and Concrete Association, 1970.
8. W. B. Cranston. "Tests on Reinforced Concrete Portal Frames." Cement and Concrete Association Technical Report TRA-392. London, 1965.
9. B. G. Neal. *The Plastic Methods of Structural Analysis.* London: Chapman and Hall, 1956.
10. L. S. Bedle. *Plastic Design of Steel Frames.* New York: John Wiley, 1958.
11. M. R. Horne. *Plastic Theory of Structures.* Cambridge, Mass.: M. I. T. Press, 1971, pp. 16–17.
12. W. G. Corley. "Rotational Capacity of Reinforced Concrete Beams." *ASCE Journal, Structural Division* 92 (October 1966).
13. A. H. Mattock. "Discussion" [12]. *ASCE Journal, Structural Division* 93 (April 1967).
14. K. Sahebjam. "The Effect of Steel Fibers on the Plastic Rotation of Reinforced Concrete Continuous Beams." M. Sc. Thesis, South Dakota State University, Brookings, So. Dakota, 1984.
15. W. Chan. "The Ultimate Strength and Deformation of Plastic Hinges in Reinforced Concrete Frameworks." *Magazine of Concrete Research* (November 1955).
16. S. K. Kaushik, L. M. Ramamurthy, and C. B. Kukreja. "Plasticity in Reinforced Concrete Continuous Beams, with Parabolic Soffits." *ACI Journal* (September–October 1980).
17. R. W. Furlong. "Design of Concrete Frames by Assigned Moment Limits." *ACI Journal* 67 (April 1970).
18. P. G. Hodge. *Plastic Analysis of Structures.* New York: McGraw–Hill, 1959.
19. M. Hilal. *Design of Reinforced Concrete Halls.* Cairo: J. Marcou and Company, 1971.
20. V. Ramakrishnan and P. D. Arthur. *Ultimate Strength Design for Structural Concrete.* India: Wheeler Publishing, 1977.
21. *Structural Design of Tall Concrete and Masonry Buildings.* New York: American Society of Civil Engineers, 1978.
22. ACI Code (318-99). "Building Code Requirements for Structural Concrete." American Concrete Institute. Detroit, November 1999.

PROBLEMS

16.1 The slab-beam floor system shown in Figure 16.32 carries a uniformly distributed dead load (excluding weight of slab and beam) of 40 psf and a live load of 100 psf. Using the ACI Code coefficients, design the interior continuous beam *ABCD* and draw detailed sections. Given: $f'_c = 3$ Ksi, $f_y = 40$ Ksi, width of beam web = 12 in., slab thickness = 4.0 in., and column dimensions = 14 by 14 in.

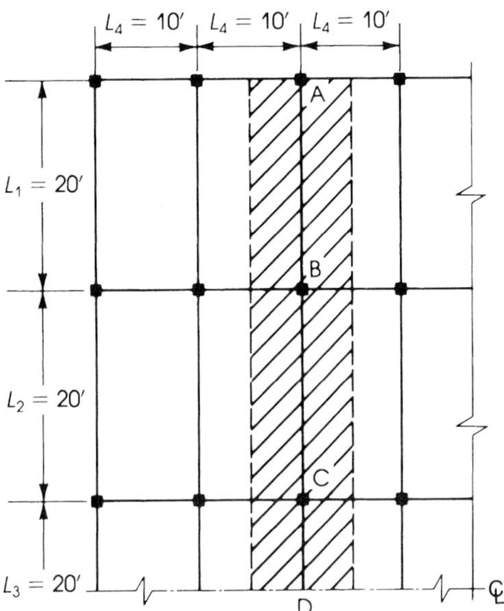

Figure 16.32 Problem 16.1.

16.2 Repeat Problem 16.1 using span lengths of the beams shown in Figure 16.32 as follows:

$$L_1 = 20 \text{ ft}, \qquad L_2 = 24 \text{ ft},$$

$$L_3 = 20 \text{ ft}, \quad \text{and} \quad L_4 = 10 \text{ ft}$$

16.3 For the beam shown in Figure 16.33, compute the reactions at A, B, and C using constant EI. Draw the shear and bending moment diagrams and design all critical sections, using $b = 14$ in., $h = 25$ in., $f'_c = 4$ Ksi, $f_y = 60$ Ksi, and a load factor $= 1.6$.

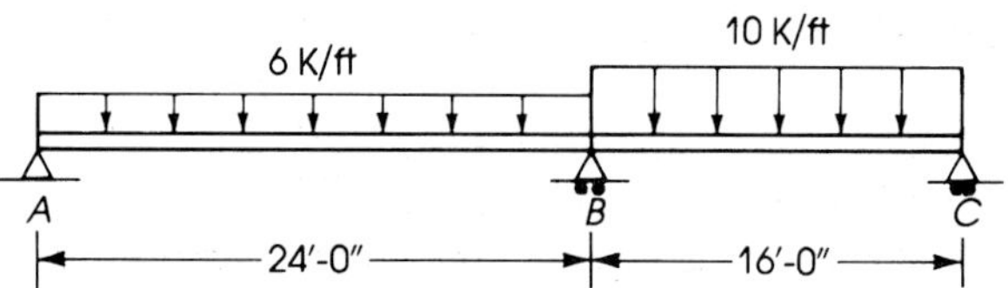

Figure 16.33 Problem 16.3.

16.4 Repeat Problem 16.3 using span lengths of beams as follows: span $AB = 20$ ft and span $BC = 16$ ft.

16.5 The two-hinged portal frame $ABCD$ shown in Figure 16.34 carries a uniform dead load (excluding self-weight) $= 2.0$ K/ft and a uniform live load of 1.8 K/ft. Design the frame $ABCD$, the hinges, and footings using $f'_c = 4$ Ksi, $f_y = 60$ Ksi, and a beam width $b = 16$ in. The footing is placed 5 ft below ground level and the allowable bearing soil pressure is 5 Ksf. Use a slab thickness of 6 in.

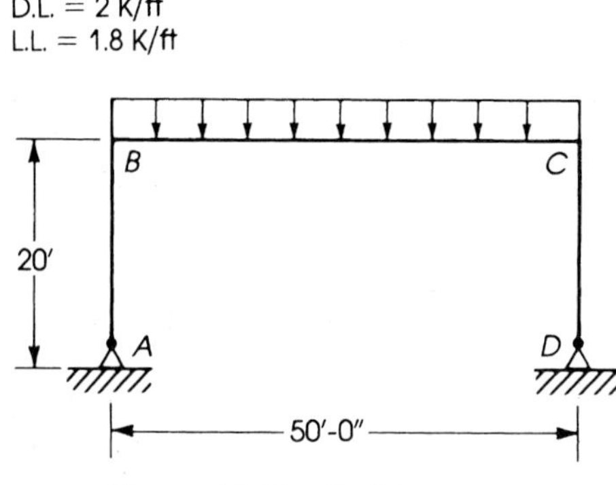

Figure 16.34 Problem 16.5.

16.6 Design the portal frame $ABCD$ of Problem 16.5 if the frame ends at A and D are fixed.

16.7 Calculate the collapse moments at the critical sections of the beams shown in Figure 16.35.

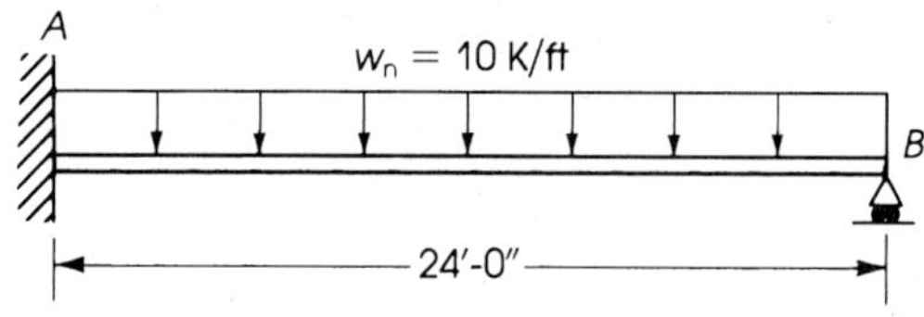

Figure 16.35 Problem 16.7.

16.8 Repeat Problem 16.7 for Figure 16.36.

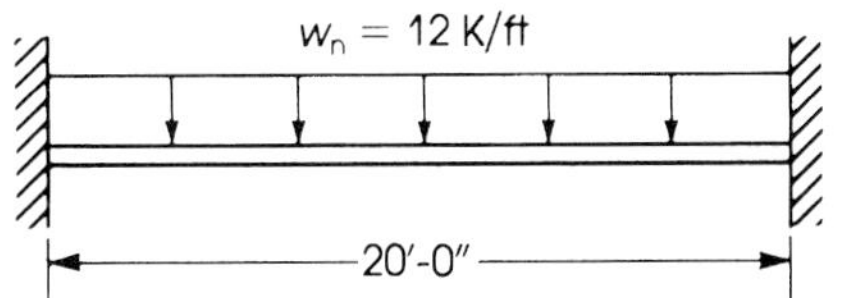

Figure 16.36 Problem 16.8.

16.9 If the beam shown in Figure 16.36 carries a uniform dead load of 2 K/ft and a live load of 2.4 K/ft, design the beam using the limit design procedure. Use $f'_c = 4$ Ksi, $f_y = 60$ Ksi, and a beam width $b = 14$ in.

16.10 Determine the maximum and minimum elastic moments at the supports and midspans of the three-span continuous beam shown in Figure 16.37. The beam has a uniform rectangular section and carries a uniform dead load of 6 K/ft and a live load of 5 K/ft. Assuming 10% maximum redistribution of moments, recalculate the maximum and minimum moments at the supports and midspans of the beam *ABC*. *Note:* Place the live load on alternate spans to calculate maximum positive moments and on adjacent spans to calculate the maximum negative (minimum) moments (Example 16.8).

16.11 Repeat Problem 16.10 if the beam consists of four equal spans, each 24 ft in length.

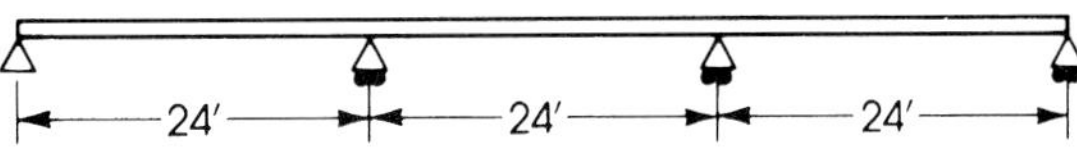

Figure 16.37 Problem 16.10.

The Bonaventure complex and the Bonaventure Hilton Hotel, Montreal, Canada.

17 DESIGN OF TWO-WAY SLABS

17.1 INTRODUCTION

Slabs can be considered as structural members whose depth, h, is small as compared to their length, L, and width, S. The simplest form of a slab is one supported on two opposite sides, which primarily deflects in one direction and is referred to as a *one-way slab*. The design of one-way slabs was discussed in Chapter 9.

When the slab is supported on all four sides and the length, L, is less than twice the width, S, the slab will deflect in two directions, and the loads on the slab are transferred to all four supports. This slab is referred to as a *two-way slab*. The bending moments and deflections in such slabs are less than those in one-way slabs; thus, the same slab can carry more load when supported on four sides. The load in this case is carried in two directions, and the bending moment in each direction is much less than the bending moment in the slab if the load were carried in one direction only. Typical slab-beam-girder arrangements of one-way and two-way slabs are shown in Figure 17.1.

17.2 TYPES OF TWO-WAY SLABS

Structural two-way concrete slabs may be classified as follows:

1. *Two-way slabs on beams:* This case occurs when the two-way slab is supported by beams on all four sides (Figure 17.1). The loads from the slab are transferred to all four supporting beams, which, in turn, transfer the loads to the columns.

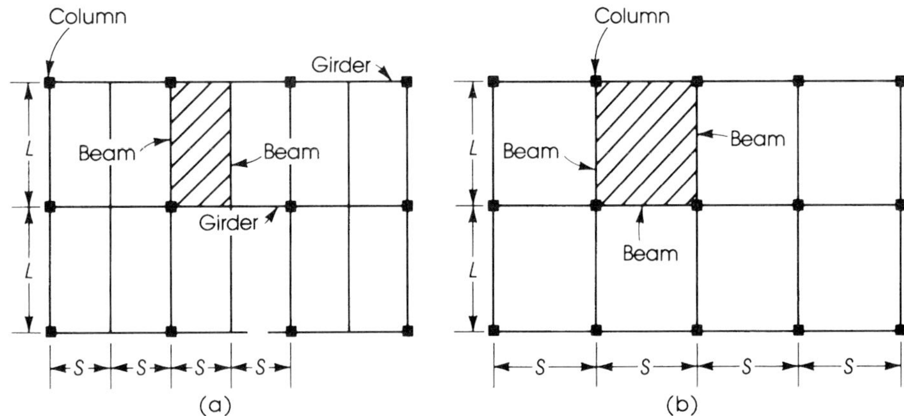

Figure 17.1 (a) One-way slab, $L/S > 2$, and (b) two-way slab, $L/S \leq 2$.

2. *Flat slabs:* A flat slab is a two-way slab reinforced in two directions that usually does not have beams or girders, and the loads are transferred directly to the supporting columns. The column tends to punch through the slab, which can be treated by three methods (refer to Figures 17.2 and 17.3):
   a. Using a drop panel and a column capital.
   b. Using a drop panel without a column capital. The concrete panel around the column capital should be thick enough to withstand the diagonal tensile stresses arising from the punching shear.
   c. Using a column capital without drop panel, which is not common.
3. *Flat-plate floors:* A flat-plate floor is a two-way slab system consisting of a uniform slab that rests directly on columns and does not have beams or column capitals (Figure 17.2(a)). In this case the column tends to punch through the slab, producing diagonal tensile stresses. Therefore, a general increase in the slab thickness is required or special reinforcement is used.
4. *Two-way ribbed slabs and the waffle slab system:* This type of slab consists of a floor slab with a length-to-width ratio less than 2. The thickness of the slab is usually 2 to 4 in. and is supported by ribs (or joists) in two directions. The ribs are arranged in each direction at spacings of about 20 to 30 in., producing square or rectangular shapes (Figure 17.2(c)). The ribs can also be arranged at 45° or 60° from the centerline of slabs, producing architectural shapes at the soffit of the slab. In two-way ribbed slabs, different systems can be adopted:
   a. A two-way rib system with voids between the ribs, obtained by using special removable and usable forms (pans) that are normally square in shape. The ribs are supported on four sides by girders that rest on columns. This type is called a *two-way ribbed (joist) slab system.*
   b. A two-way rib system with permanent fillers between ribs that produce horizontal slab soffits. The fillers may be of hollow, lightweight or normal-weight concrete or any other lightweight material. The ribs are supported by girders on four sides, which in turn are supported by columns. This type is also called a *two-way ribbed (joist) slab system* or a *hollow-block two-way ribbed system.*
   c. A two-way rib system with voids between the ribs with the ribs continuing in both directions without supporting beams and resting directly on columns through solid panels above the columns. This type is called a *waffle slab system.*

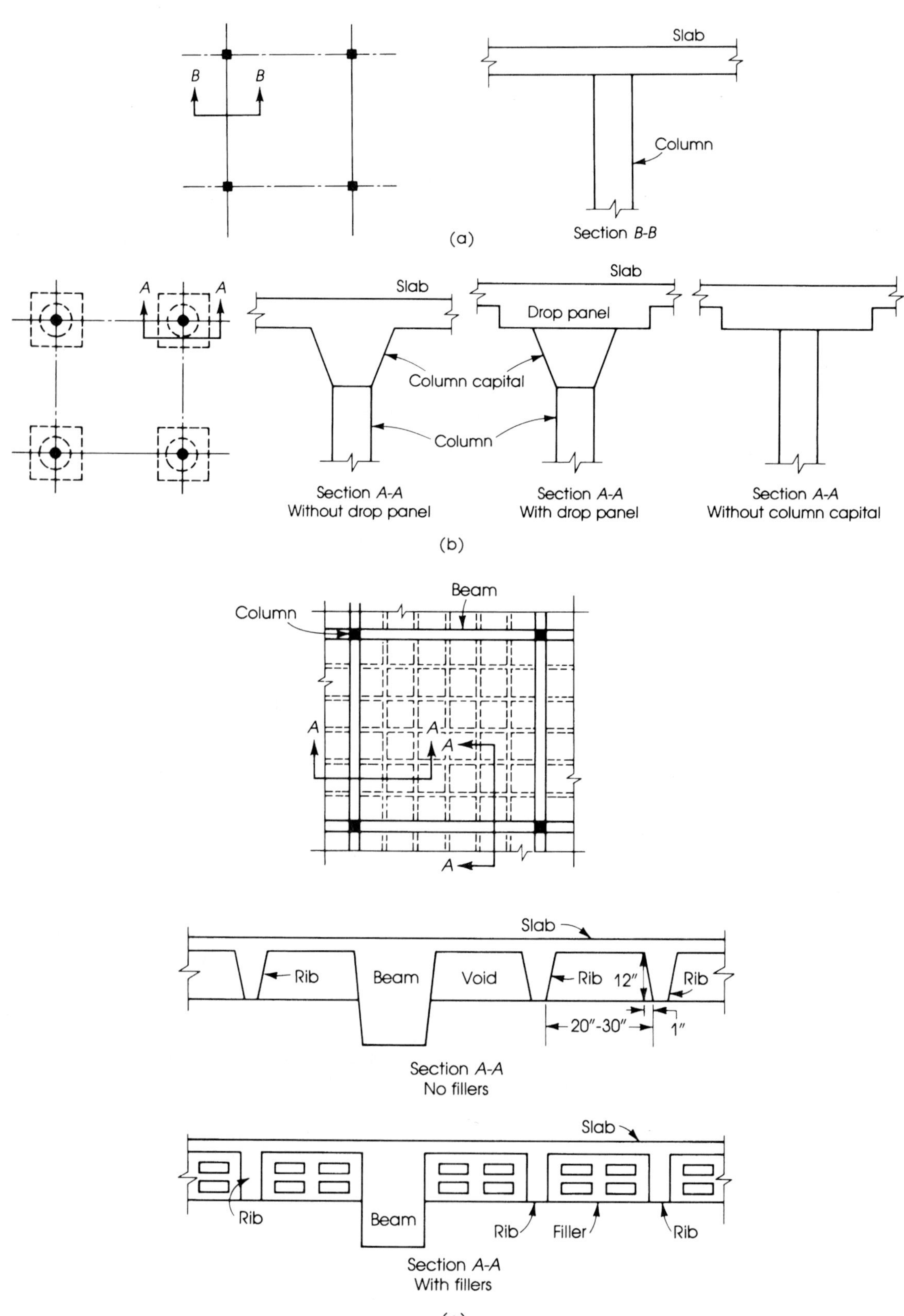

Figure 17.2 Two-way slabs without beams: (a) flat plate floor and section; (b) flat slab floor and sections; (c) ribbed slab and sections.

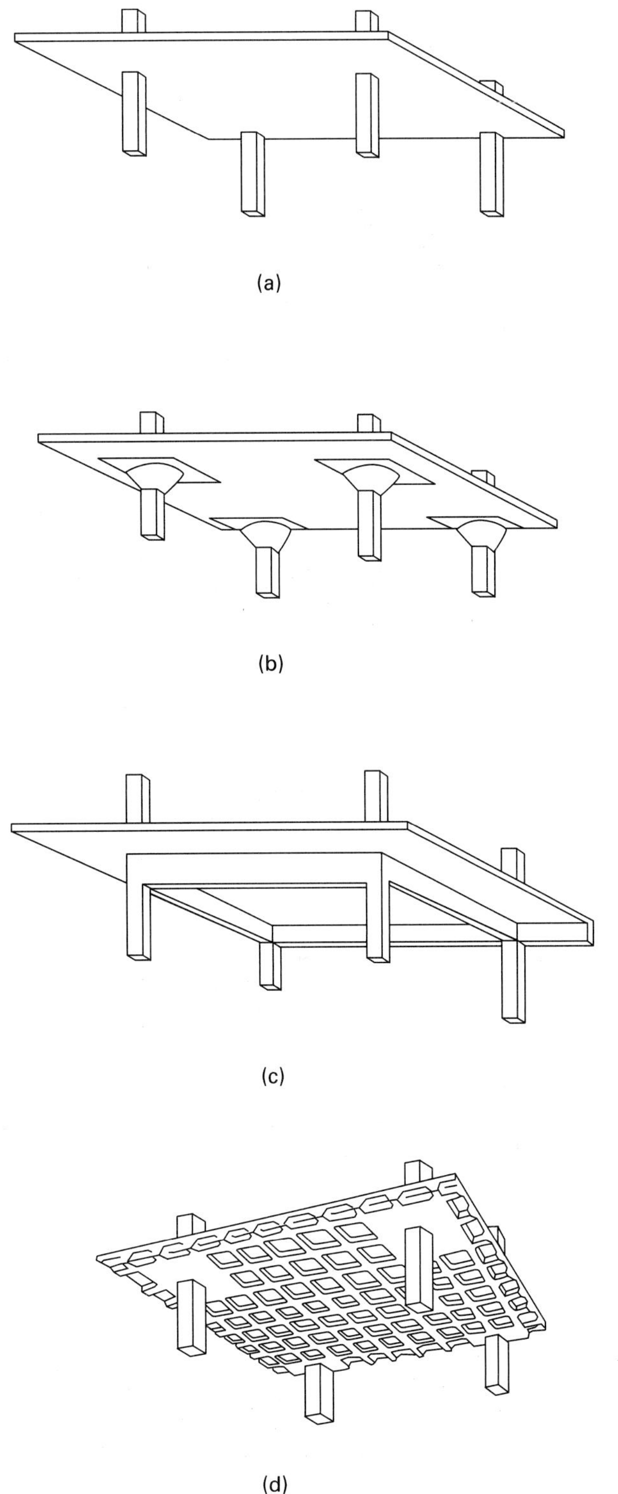

Figure 17.3 Types of two-way slab systems: (a) flat plate, (b) flat slab, (c) slab on beams, and (d) waffle slab.

Flat-plate floor system.

Slabs on beams.

17.3 ECONOMICAL CHOICE OF CONCRETE FLOOR SYSTEMS

Various types of floor systems can be used for general buildings, such as residential, office, and institutional buildings. The choice of an adequate and economic floor system depends on the type of building, architectural layout, aesthetic features, and the span length between columns. In general, the superimposed live load on buildings varies between 80 and 150 psf. A general guide for the economical use of floor systems can be summarized as follows:

1. *Flat plates:* Flat plates are most suitable for spans of 20 to 25 ft and live loads between 60 and 100 psf. The advantages of adopting flat plates include low-cost formwork, exposed flat ceilings, and fast construction. Flat plates have low shear capacity and relatively low stiffness, which may cause noticeable deflection. Flat plates are widely used in buildings either as reinforced or prestressed concrete slabs.
2. *Flat slabs:* Flat slabs are most suitable for spans of 20 to 30 ft and for live loads of 80 to 150 psf. They need more formwork than flat plates, especially for column capitals. In most cases, only drop panels without column capitals are used.
3. *Waffle slabs:* Waffle slabs are suitable for spans of 30 to 48 ft and live loads of 80 to 150 psf. They carry heavier loads than flat plates and have attractive exposed ceilings. Formwork, including the use of pans, is quite expensive.
4. *Slabs on beams:* Slabs on beams are suitable for spans between 20 and 30 ft and live loads of 60 to 120 psf. The beams increase the stiffness of the slabs, producing relatively low deflection. Additional formwork for the beams is needed.
5. *One-way slabs on beams:* One-way slabs on beams are most suitable for spans of 10 to 20 ft and a live load of 60 to 100 psf. They can be used for larger spans with relatively higher cost and higher slab deflection. Additional formwork for the beams is needed.
6. *One-way joist floor system:* A one-way joist floor system is most suitable for spans of 20 to 30 ft and live loads of 80 to 120 psf. Because of the deep ribs, the concrete and steel quantities are relatively low, but expensive formwork is expected. The exposed ceiling of the slabs may look attractive.

17.4 DESIGN CONCEPTS

An exact analysis of forces and displacements in a two-way slab is complex, due to its highly indeterminate nature; this is true even when the effects of creep and nonlinear behavior of the concrete are neglected. Numerical methods such as finite elements can be used, but simplified methods such as those presented by the ACI Code are more suitable for practical design. The ACI Code, Chapter 13, assumes that the slabs behave as wide, shallow beams that form, with the columns above and below them, a rigid frame. The validity of this assumption of dividing the structure into equivalent frames has been verified by analytical [1], [2] and experimental [3], [4] research. It is also established [3], [5] that ultimate load capacity of two-way slabs with restrained boundaries is about twice that calculated by theoretical analysis, because a great deal of moment redistribution occurs in the slab before failure. At high loads, large deformations and deflections are expected; thus a minimum slab thickness is required to maintain adequate deflection and cracking conditions under service loads.

The ACI Code specifies two methods for the design of two-way slabs:

1. The direct design method, DDM (ACI Code, Section 13.6), is an approximate procedure for the analysis and design of two-way slabs. It is limited to slab systems subjected to uniformly distributed loads and supported on equally or

Flat-slab floor system with drop panels (no column capitals).

nearly equally spaced columns. The method uses a set of coefficients to determine the design moments at critical sections. Two-way slab systems that do not meet the limitations of the ACI Code, Section 13.6.1, must be analyzed by more accurate procedures.

2. The equivalent frame method, EFM (ACI Code, Section 13.7), is one in which a three-dimensional building is divided into a series of two-dimensional equivalent frames by cutting the building along lines midway between columns. The resulting frames are considered separately in the longitudinal and transverse directions of the building and treated floor by floor, as shown in Figure 17.4.

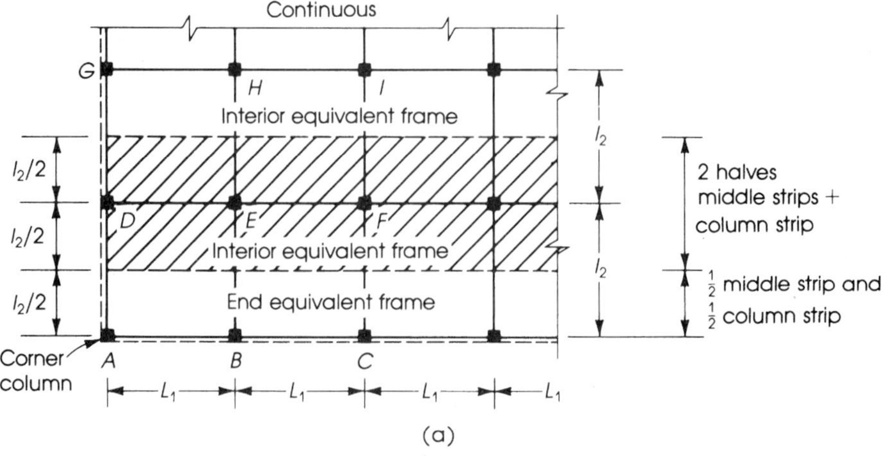

Figure 17.4 (a) Longitudinal and (b) transverse equivalent frames in plan view and (c) in elevation and perspective views.

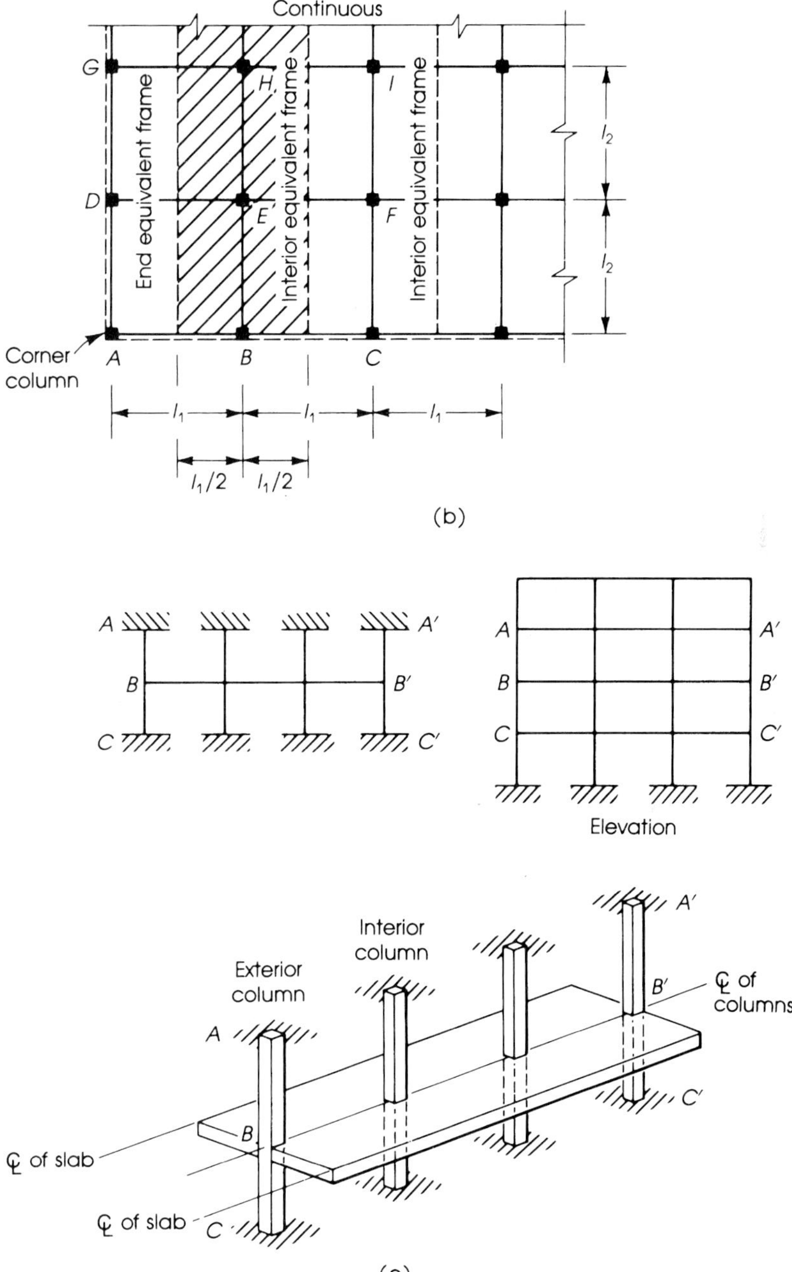

Figure 17.4 *(continued)*

The two ACI Code procedures are based on the results of elastic analysis of the structure as a whole using factored loads. A modified approach to the direct design method was presented in the commentary of the 1989 code as the modified stiffness method, or MSM. It is based on specific distribution factors introduced as a function of the stiffness ratio, α_{ec}, for proportioning the total static moment in an end span. This method is explained later.

In addition to the ACI Code procedures, a number of other alternatives are available for the analysis and design of slabs. The resulting slabs may have a greater or lesser amount of reinforcement. The analytical methods may be classified in terms of the basic relationship between load and deformation as elastic, plastic, and nonlinear.

Waffle slab with light fixtures at the centers of the squares.

1. In *elastic analysis,* a concrete slab may be treated as an elastic plate. The flexure, shear, and deflection may be calculated by the fourth differential equation relating load to deflection for thin plates with small displacements, as presented by Timoshenko [6]. Finite-difference as well as finite-element solutions have been proposed to analyze slabs and plates [7], [8]. In the finite-element method, the slab is divided into a mesh of triangles or quadrilaterals. The displacement functions of the nodes (intersecting mesh points) are usually established, and the stiffness matrices are developed for computer analysis.
2. For *plastic analysis,* three methods are available. The *yield line* method was developed by Johansen [9] to determine the limit state of the slab by considering the yield lines that occur in the slab as a collapse mechanism. The *strip* method was developed by Hillerborg [10]. The slab is divided into strips, and the load on the slab is distributed in two orthogonal directions. The strips are analyzed as simple beams. The third method is *optimal analysis.* There has been considerable research into optimal solutions. Rozvany and others [11] presented methods for minimizing reinforcement based on plastic analysis. Optimal solutions are complex in analysis and produce complex patterns of reinforcement.
3. *Nonlinear analysis* simulates the true load-deformation characteristics of a reinforced concrete slab when the finite-element method takes into consideration the nonlinearity of the stress-strain relationship of the individual elements [11], [12]. In this case, the solution becomes complex unless simplified empirical relationships are assumed.

The preceding methods are presented very briefly to introduce the reader to the different methods of analysis of slabs. Experimental work on slabs has not been extensive in recent years, but more research is probably needed to simplify current design procedures with adequate safety, serviceability, and economy [11].

17.5 COLUMN AND MIDDLE STRIPS

Figure 17.5 shows an interior panel of a two-way slab supported on columns A, B, C, and D. If the panel is loaded uniformly, the slab will deflect in both directions, with maximum deflection at the center, O. The highest points will be at the columns A, B, C, and D; thus, the part of the slab around the columns will have a convex shape. A gradual change in the

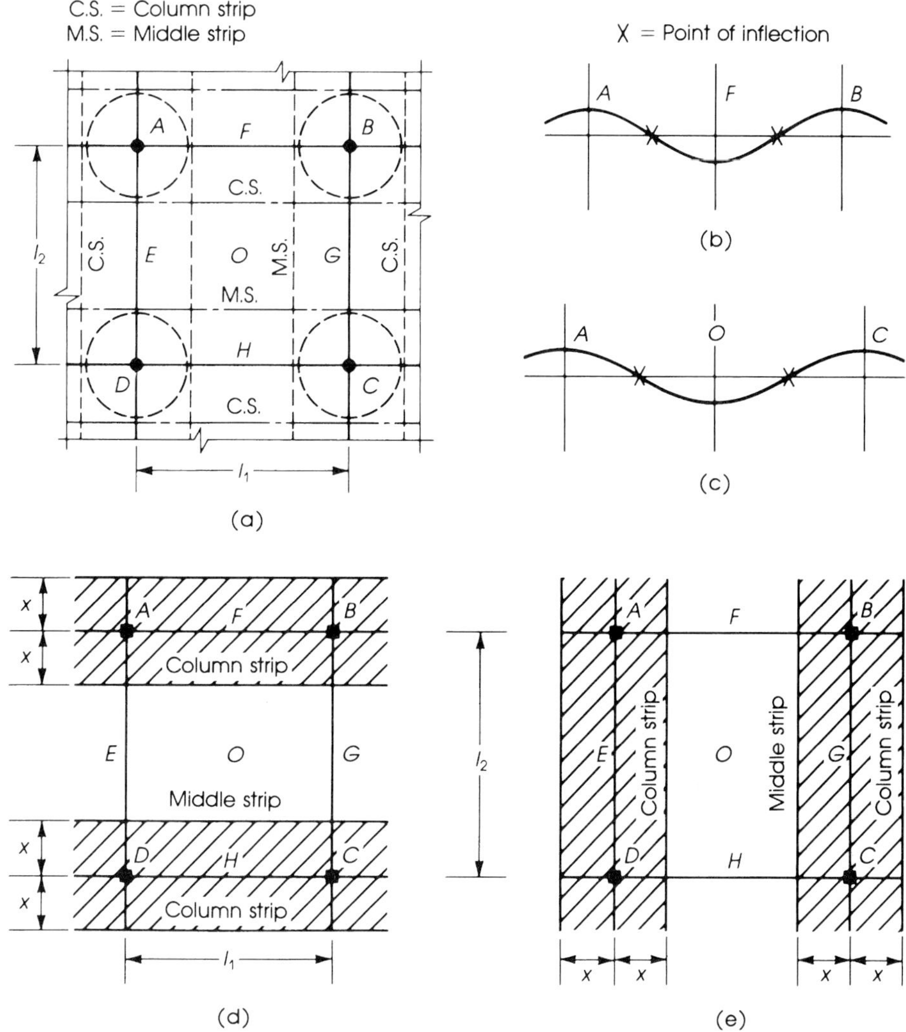

Figure 17.5 Column and middle strips; $x = 0.25l_1$ or $0.25l_2$, whichever is smaller.

shape of the slab occurs, from convexity at the columns to concavity at the center of the panel O, each radial line crossing a point of inflection. Sections at O, E, F, G, and H will have positive bending moments, whereas the periphery of the columns will have maximum negative bending moments. Considering a strip along AFB, the strip bends like a continuous beam (Figure 17.5(b)), having negative moments at A and B and positive bending moment at F. This strip extends between the two columns A and B and continues on both sides of the panel, forming a column strip.

Similarly, a strip along EOG will have negative bending moments at E and G and a positive moment at O, forming a middle strip. A third strip along DHC will behave similarly to strip AFB. Therefore, the panel can be divided into three strips, one in the middle along EOG, referred to as the *middle strip,* and one on each side, along AFB and DHC, referred to as *column strips* (Figure 17.5(a)). Each of the three strips behaves as a continuous beam. In a similar way, the panel is divided into three strips in the other direction, one middle strip along FOH and two column strips along AED and BGC, respectively (Figure 17.5(e)).

Referring to Figure 17.5(a), it can be seen that the middle strips are supported on the column strips, which in turn transfer the loads onto the columns, A, B, C, and D in this panel. Therefore, the column strips carry more load than the middle strips. Consequently, the positive bending moment in each column strip (at E, F, G, and H) is greater than the positive bending moment at O in the middle strip. Also, the negative moments at the columns A, B, C, and D in the column strips are greater than the negative moments at E, F, G, and H in the middle strips. The portions of the design moments assigned to each critical section of the column and middle strips are discussed in Section 17.8.

The extent of each of the column and middle strips in a panel is defined by the ACI Code, Section 13.2. The column strip is defined by a slab width on each side of the column centerline, x in Figure 17.5, equal to one-fourth the smaller of the panel dimensions l_1 and l_2, including beams if they are present, where

l_1 = span length, center to center of supports, in the direction moments are being determined

l_2 = span length, center to center of supports, in the direction perpendicular to l_1

The portion of the panel between two column strips defines the middle strip.

17.6 MINIMUM SLAB THICKNESS TO CONTROL DEFLECTION

The ACI Code, Section 9.5.3, specifies a minimum slab thickness in two-way slabs to control deflection. The magnitude of a slab's deflection depends on many variables, including the flexural stiffness of the slab, which in turn is a function of the slab thickness, h. By increasing the slab thickness, the flexural stiffness of the slab is increased, and consequently the slab deflection is reduced [13]. Because the calculation of deflections in two-way slabs is complicated and to avoid excessive deflections, the ACI Code limits the thickness of these slabs by adopting the following three empirical limiations, which are based on experimental research. If these limitations are not met, it will be necessary to compute deflections.

a. For $0.2 \le \alpha_m \le 2$:

$$h = \frac{l_n\left(0.8 + \dfrac{f_y}{200{,}000}\right)}{36 + 5\beta(\alpha_m - 0.2)} \quad (f_y \text{ in psi}) \tag{17.1}$$

but not less than 5 in.

b. For $\alpha_m > 2$:

$$h = \frac{l_n\left(0.8 + \dfrac{f_y}{200{,}000}\right)}{36 + 9\beta} \tag{17.2}$$

but not less than 3.5 in.

c. For $\alpha_m < 0.2$:

$$h = \text{minimum slab thickness without interior beams (Table 17.1)} \tag{17.3}$$

where l_n = clear span in the long direction measured face to face of columns (or face to face of beams for slabs with beams)

β = the ratio of the long to the short clear spans

α_m = the average value of α for all beams on the sides of a panel

α = the ratio of flexural stiffness of a beam section $E_{cb}I_b$ to the flexural stiffness of the slab $E_{cs}I_s$, bounded laterally by the centerlines of the panels on each side of the beam

$$\alpha = \frac{E_{cb}I_b}{E_{cs}I_s} \tag{17.4}$$

Table 17.1 Minimum Thickness of Slabs Without Interior Beams

	Without Drop Panels[b]			With Drop Panels[b]		
	Exterior Panels			Exterior Panels		
Yield Stress f_y psi (1)[a]	*Without Edge Beams*	*With Edge Beams*	*Interior Panels*	*Without Edge Beams*	*With Edge Beams[c]*	*Interior Panels*
40,000	$\frac{l_n}{33}$	$\frac{l_n}{36}$	$\frac{l_n}{36}$	$\frac{l_n}{36}$	$\frac{l_n}{40}$	$\frac{l_n}{40}$
60,000	$\frac{l_n}{30}$	$\frac{l_n}{33}$	$\frac{l_n}{33}$	$\frac{l_n}{33}$	$\frac{l_n}{36}$	$\frac{l_n}{36}$

[a] For values of reinforcement, yield stress between 40,000 and 60,000 psi minimum thickness shall be obtained by linear interpolation.

[b] Drop panel is defined in ACI Sections 13.3.7.1 and 13.3.7.2.

[c] Slabs with beams between columns along exterior edges. The value of α for the edge beam shall be not less than 0.8.

where E_{cb} and E_{cs} are the moduli of elasticity of concrete in the beam and the slab, respectively, and

I_b = the gross moment of inertia of the beam section about the centroidal axis (the beam section includes a slab length on each side of the beam equal to the projection of the beam above or below the slab, whichever is greater, but not more than four times the slab thickness)

I_s = the moment of inertia of the gross section of the slab

However, the thickness of any slab shall not be less than the following:

1. For slabs with $\alpha_m < 2.0$: 5 in.
2. For slabs with $\alpha_m \geq 2.0$: $3\frac{1}{2}$ in.

If no beams are used, as in the case of flat plates, then $\alpha = 0$ and $\alpha_m = 0$. The ACI Code equations for calculating slab thickness, h, take into account the effect of the span length, the panel shape, the steel reinforcement yield stress, f_y, and the flexural stiffness of beams. When very stiff beams are used, equation (17.1) may give a small slab thickness, and equation (17.2) may control. For flat plates and flat slabs, when no interior beams are used, the minimum slab thickness may be determined directly from Table 9.5(c) of the ACI Code, which is shown here as Table 17.1.

Other ACI Code limitations are summarized as follows:

1. For panels with discontinuous edges, end beams with a minimum α equal to 0.8 must be used; otherwise the minimum slab thickness calculated by equations (17.1) and (17.2) must be increased by at least 10% (ACI Code, Section 9.5.3).
2. When drop panels are used without beams, the minimum slab thickness may be reduced by 10%. The drop panels should extend in each direction from the centerline of support a distance not less than one-sixth of the span length in that direction between center to center of supports and also project below the slab at least $h/4$. This reduction is included in Table 17.1.
3. Regardless of the values obtained by equations (17.1) and (17.2), the thickness of two-way slabs shall not be less than the following: (1) for slabs without beams or drop panels, 5 in.; (2) for slabs without beams but with drop panels, 4 in.; (3) for slabs with beams on all four sides with $\alpha_m \geq 2.0$, $3\frac{1}{2}$ in., and for $\alpha_m < 2.0$, 5 in. (ACI Code, Section 9.5.3.).

The following steps summarize these calculations:

1. For slabs without interior beams (flat plates and flat slabs),
   a. Calculate the minimum slab thickness directly from Table 17.1. However, equations (17.1) and (17.2) may be used, and equation (17.1) normally controls. Minimum slab thickness shall be greater than or equal to 5 in. for slabs without drop panels and greater than or equal to 4 in. for slabs with drop panels.
   b. At discontinuous edges, an edge beam with $\alpha \geq 0.8$ should be used. Otherwise, the minimum slab thickness calculated by equations (17.1) and (17.2) should be increased by 10%. This increase of 10% has already been included in the second columns of Table 17.1.
   c. If drop panels are used in flat slabs, the minimum slab thickness may be reduced by 10% on the condition that the drop panel extends in each direction from the centerline of the support a distance not less than one-sixth of the span and projects below the slab at least $h/4$. This reduction is included in the factors of Table 17.1.
2. For slabs with beams on all sides $(\alpha_m > 0)$,
   a. Calculate α_m and then calculate the minimum slab thickness from equations (17.1) and (17.2). In most cases, equation (17.2) controls.
   b. The slab thickness should be greater than or equal to 5 in. for slabs with $\alpha_m < 2.0$ and should be greater than or equal to 3.5 in. for slabs with $\alpha_m \geq 2.0$.
3. For all slabs: A slab thickness less than the minimum thickness given in Steps 1 and 2 may be used if shown by computation that deflection will not exceed the ACI Code, Table 9.5(b) limitations explained earlier in Chapter 6.

Example 17.1

A flat-plate floor system with panels 24 by 20 ft is supported on 20-in. square columns. Using the ACI Code equations, determine the minimum slab thickness required for the interior and corner panels shown in Figure 17.6. Edge beams are not used. Use $f'_c = 4$ Ksi and $f_y = 60$ Ksi.

Solution

1. For corner panel no. 1, the minimum thickness is $l_n/30$ ($f_y = 60$ Ksi, and no edge beams are used; see Table 17.1).

$$l_{n1} = 24 - \tfrac{20}{12} = 22.33 \text{ ft} \qquad \text{(long direction)}$$

$$h_{\min} = \frac{22.33 \times 12}{30} = 8.93 \text{ in., say, } 9.0 \text{ in.}$$

Alternatively, equations (17.1) and (17.2) can be used to calculate the minimum thickness with $\alpha = \alpha_m = 0$.

2. For the interior panel no. 3 and $f_y = 60$ Ksi, the minimum slab thickness is $l_n/33 = (22.33 \times 12)/33 = 8.12$ in., say, 8.5 in. Alternatively, equations (17.1) and (17.2) can be used. If a uniform slab thickness is used for all panels, then $h = 9.0$ in. will be adopted.

Example 17.2

The floor system shown in Figure 17.7 consists of solid slabs and beams in two directions supported on 20-in. square columns. Using the ACI Code equations, determine the minimum slab thickness required for an interior panel. Use $f'_c = 3$ Ksi and $f_y = 60$ Ksi.

Solution

1. To use equation (17.1), α_m should be calculated first. Therefore, it is required to determine I_b, I_s, and α for the beams and slabs in the long and short directions.
2. The gross moment of inertia of the beam I_b is calculated for the section shown in Figure 17.7(b), which is made up of the beam and the extension of the slab on each side

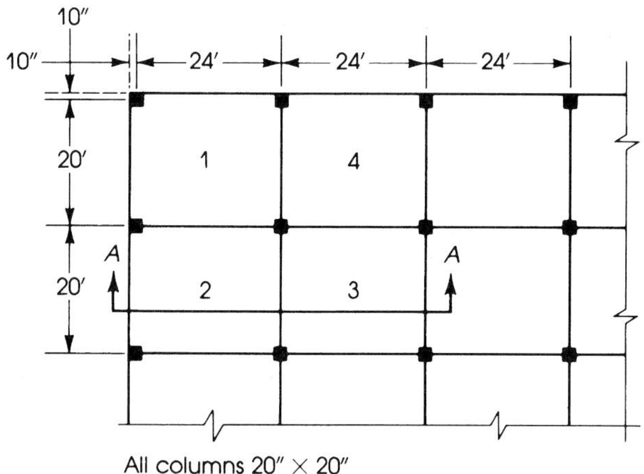

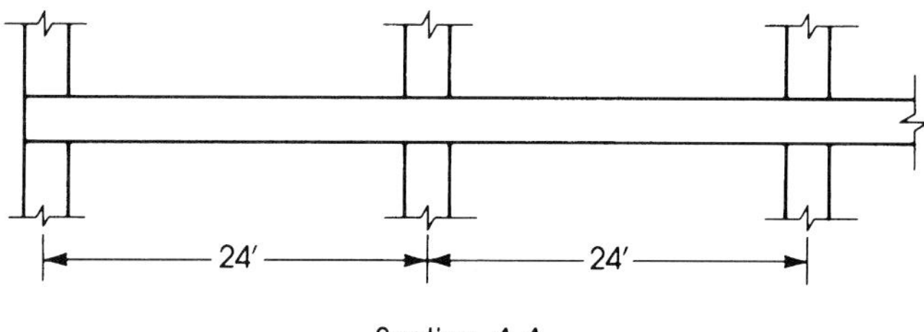

Section *A-A*

Figure 17.6 Example 17.1.

of the beam $x = y$ but not more than four times the slab thickness. Assume $h = 7$ in., to be checked later; then $x = y = 22 - 7 = 15$ in. $< 4 \times 7 = 28$ in. Therefore, $b_e = 16 + 2 \times 15 = 46$ in., and the T-section is shown in Figure 17.7(c). Determine the centroid of the section by taking moments about the top of the flange:

$$\text{Area of flange} = 7 \times 46 = 322 \text{ in.}^2$$

$$\text{Area of web} = 16 \times 15 = 240 \text{ in.}^2$$

$$\text{Total area} = 562 \text{ in.}^2$$

$$(322 \times 3.5) + 240 \times (7 + 7.5) = 562y$$

$$y = 8.20 \text{ in.}$$

$$I_b = \left[\frac{46}{12}(7)^3 + 322 \times (4.7)^2\right] + \left[\frac{16(15)^3}{12} + 240(7.5 - 1.2)^2\right] = 22{,}453 \text{ in.}^4$$

3. The moment of inertia of the slab in the long direction is $I_s = (bh^3)/12$, where $b = 20$ ft and $h = 7$ in.

$$I_l = \frac{(20 \times 12)(7)^3}{12} = 6860 \text{ in.}^4$$

$$\alpha_1 \text{ (in the long direction)} = \frac{EI_b}{EI_s} = \frac{22{,}453}{6860} = 3.27$$

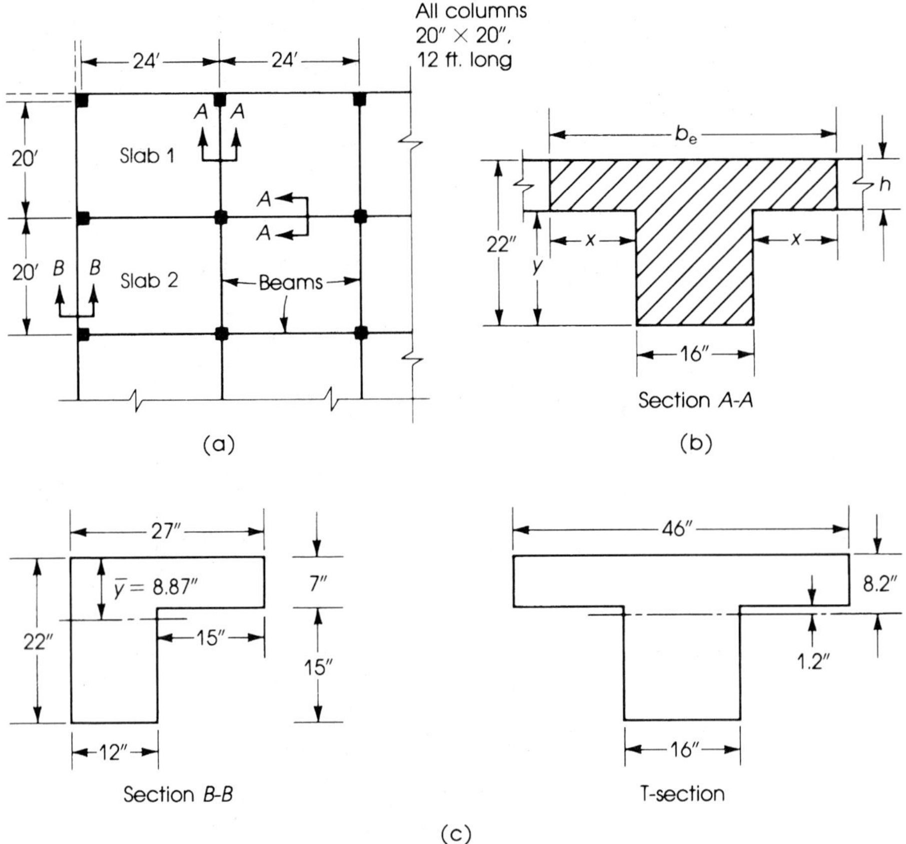

Figure 17.7 Example 17.2.

4. The moment of inertia of the slab in the short direction is $I_s = (bh^3)/12$ where $b = 24$ ft and $h = 7$ in.

$$I_s = \frac{(24 \times 12)(7)^3}{12} = 8232 \text{ in.}^4$$

$$\alpha_s = \frac{EI_b}{EI_s} = \frac{22{,}453}{8232} = 2.27$$

5. α_m is the average pf α_1 and α_s:

$$\alpha_m = \frac{3.27 + 2.72}{2} = 3.0$$

6. $\beta = \dfrac{\left(24 - \frac{20}{12}\right)}{\left(20 - \frac{20}{12}\right)} = \dfrac{22.33}{18.33} = 1.22$

7. Determine $h_{\min}$ using equation (17.1) $(l_n = 22.33 \text{ ft})$:

$$h_{\min} = \frac{(22.33 \times 12)(0.8 + 0.005 \times 60)}{36 + (5 \times 1.22)[3.0 - 0.2]} = 5.57 \text{ in.}$$

However, this value must not be less than h given by equation (17.2) $(\alpha_m > 2.0)$:

$$h = \frac{294.8}{36 + 9(1.22)} = 6.27 \text{ in.}$$

Also, $h_{\min} = 3.5$ in. Therefore, $h = 6.27$ in. controls. A slab thickness of 6.5 in. or 7.0 in. may be adopted. Note that in most practical cases, equation (17.2) controls.

17.7 SHEAR STRENGTH OF SLABS

In a two-way floor system, the slab must have adequate thickness to resist both bending moments and shear forces at the critical sections. To investigate the shear capacity of two-way slabs, the following cases should be considered.

17.7.1 Two-Way Slabs Supported on Beams

In two-way slabs supported on beams, the critical sections are at a distance d from the face of the supporting beams, and the shear capacity of each section is $\phi V_c = \phi(2\sqrt{f'_c}bd)$. When the supporting beams are stiff and are capable of transmitting floor loads to the columns, they are assumed to carry loads acting on floor areas bounded by 45° lines drawn from the corners, as shown in Figure 17.8. The loads on the trapezoidal areas will be carried by the long beams AB and CD, whereas the loads on the triangular areas will be carried by the short beams AC and BD. The shear per unit width of slab is highest between E and F in both directions, and $V_u = w_u(l_2/2)$, where w_u is the uniform factored load per unit area.

If no shear reinforcement is provided, the shearing force at a distance d from the face of the beam, V_{ud}, must be equal to

$$V_{ud} \leq \phi V_c \leq \phi(2\sqrt{f'_c}bd), \qquad \text{where } V_{ud} = w_u\left(\frac{l_2}{2} - d\right)$$

17.7.2 Two-Way Slabs without Beams

In flat plates and flat slabs, beams are not provided, and the slabs are directly supported by columns. In such slabs, two types of shear stresses must be investigated; the first is one-way shear, or beam shear. The critical sections are taken at a distance d from the face of the column, and the slab is considered as a wide beam spanning between supports, as in the case of one-way beams. The shear capacity of the concrete section is $\phi V_c = \phi(2\sqrt{f'_c}bd)$. The

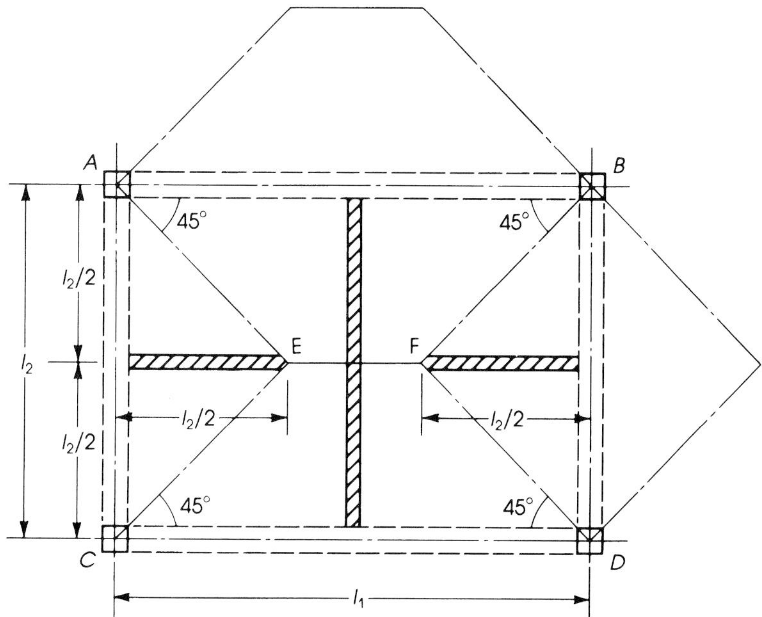

Figure 17.8 Areas supported by beams in two-way slab floor system.

second type of shear to be studied is two-way, or punching, shear, as was previously discussed in the design of footings. Shear failure occurs along a truncated cone or pyramid around the column. The critical section is located at a distance $d/2$ from the face of the column, column capital, or drop panel (Figure 17.9(a)). If shear reinforcement is not provided, the shear strength of concrete is the smaller of (17.5) and (17.6):

$$\phi V_c = \left(2 + \frac{4}{\beta_c}\right)\sqrt{f'_c}\,b_o d \leq 4\phi\sqrt{f'_c}\,b_o d \tag{17.5}$$

where

b_o = perimeter of the critical section

β_c = ratio of the long side of column (or loaded area) to the short side

$$\phi V_c = \phi\left(\frac{\alpha_s d}{b_o} + 2\right)\sqrt{f'_c}\,b_o d \tag{17.6}$$

where α_s is 40 for interior columns, 30 for edge columns, and 20 for corner columns.

When shear reinforcement is provided, the shear strength should not exceed

$$\phi V_n \leq \phi(6\sqrt{f'_c}\,b_o d) \tag{17.7}$$

17.7.3 Shear Reinforcement in Two-Way Slabs without Beams

In flat-slab and flat-plate floor systems, the thickness of the slab selected may not be adequate to resist the applied shear stresses. In this case, either the slab thickness must be increased or shear reinforcement must be provided. The ACI Code allows the use of shear reinforcement by shearheads and anchored bars or wires.

Shearheads consist of steel I-shapes or channel shapes welded into four cross arms and placed in the slabs above the column (Figure 17.9(c), (d)). Shearhead designs do not apply to exterior columns, where large torsional and bending moments must be transferred between slab and column. The ACI Code, Section 11.12.4, indicates that on the critical section the nominal shear strength V_n should not exceed $4\sqrt{f'_c}\,b_o d$, but if shearhead reinforcement is provided, V_n should not exceed $7\sqrt{f'_c}\,b_o d$. To determine the size of the shearhead, the ACI Code, Section 11.12.4, gives the following limitations:

1. The ratio α_v between the stiffness of shearhead arm, $E_s I$, and that of the surrounding composite cracked section of width, $c_2 + d$, must not be less than 0.15.
2. The compression flange of the steel shape must be located within $0.3d$ of the compression surface of the slab.
3. The depth of the steel shape must not exceed 70 times the web thickness.
4. The plastic moment capacity M_p of each arm of the shearhead is computed by

$$\phi M_p = \frac{V_u}{2n}\left[h_v + \alpha_v\left(l_v - \frac{c_1}{2}\right)\right] \quad \text{(ACI Code, equation (11.40))} \tag{17.8}$$

where $\phi = 0.9$

V_u = factored shear force around the periphery of the column face

n = number of arms

h_v = depth of the shearhead

l_v = length of the shearhead measured from the centerline of the column.

5. The critical slab section for shear must cross each shearhead arm at a distance equal to $(3/4)(l_v - c_1/2)$ from the column face to the end of the shearhead arm, as shown

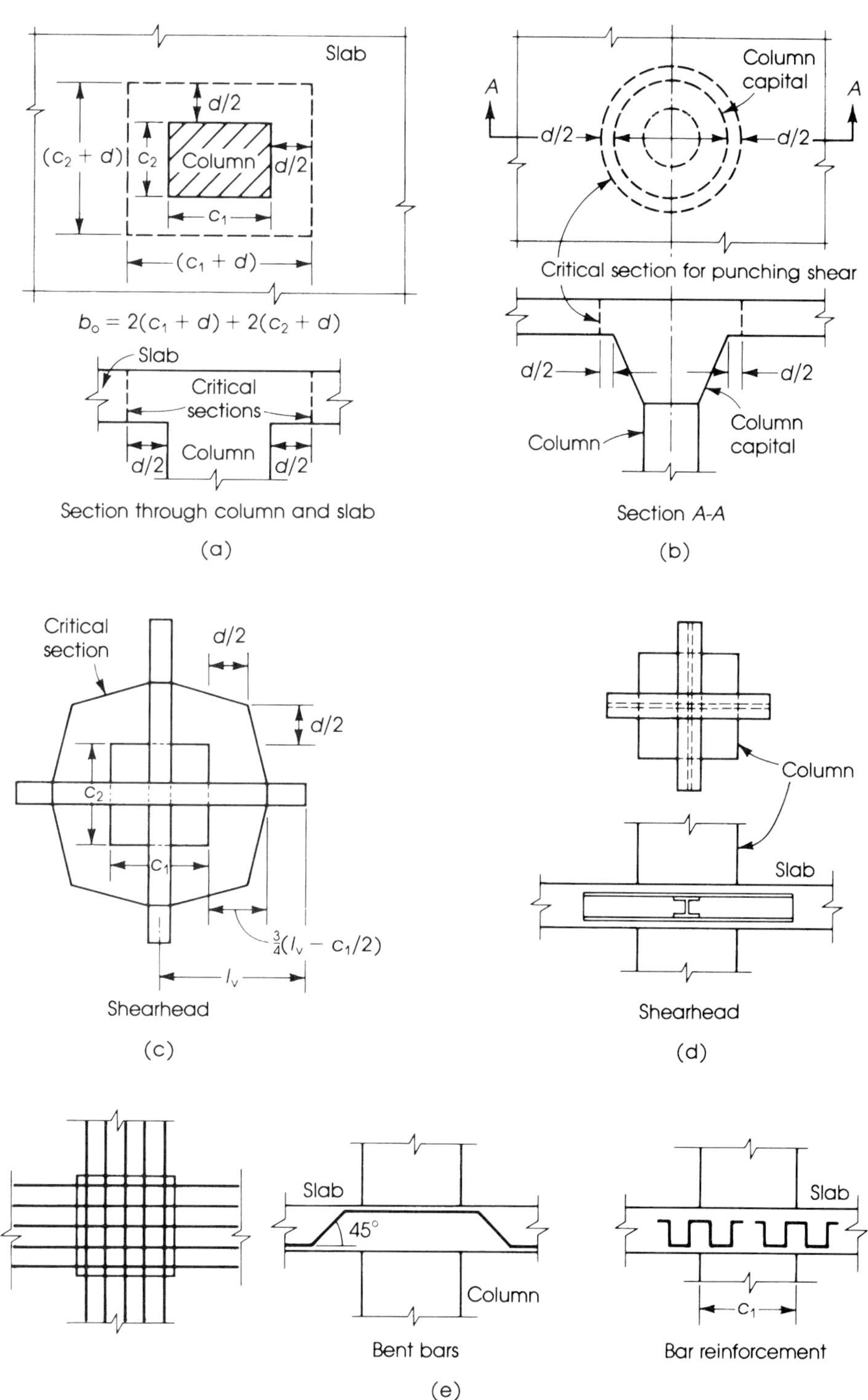

Figure 17.9 Critical section for punching shear in (a) flat plates and (b) flat slabs, reinforcement by (c, d) shearheads and (e) anchored bars, (f) conventional stirrup cages, and (g) studded steel strips used as shear reinforcement.

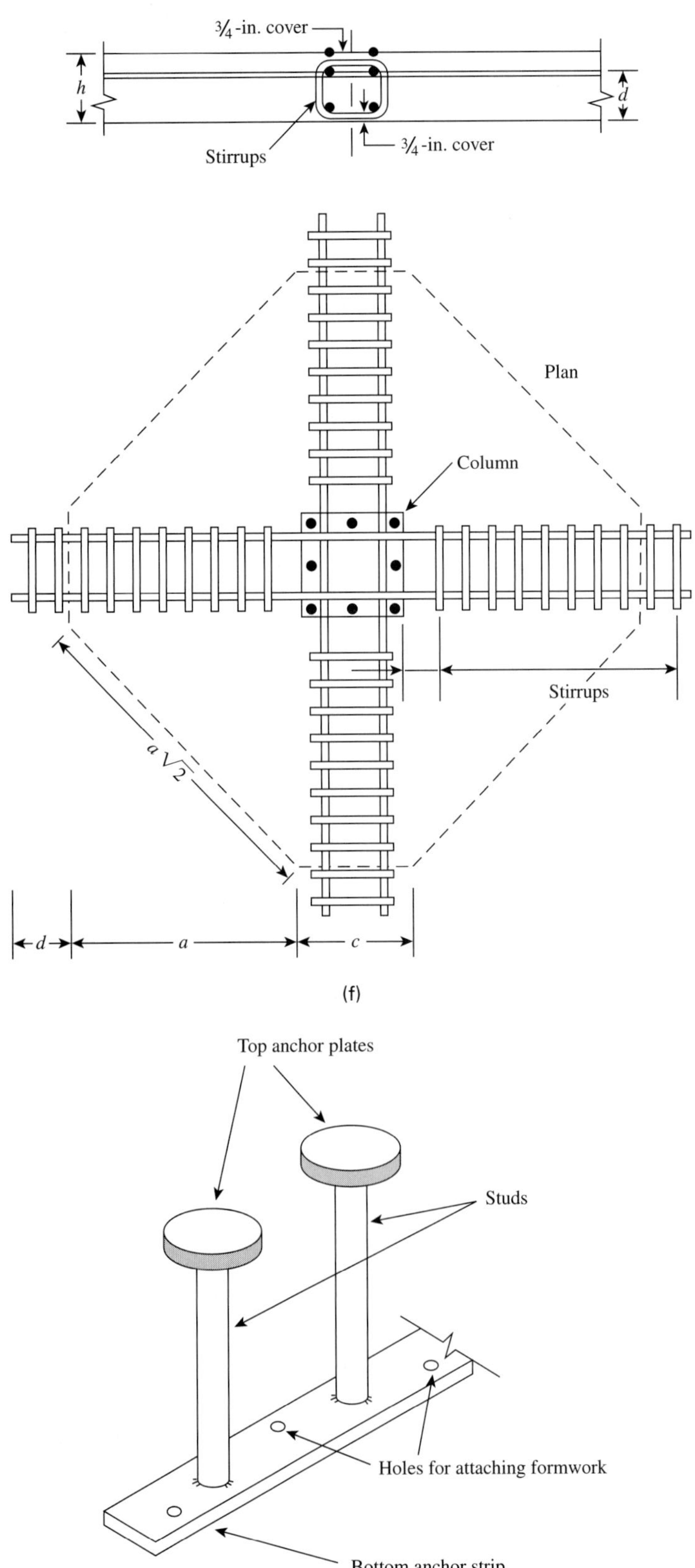

Figure 17.9 *(continued)*

in Figure 17.9(c). The critical section must have a minimum perimeter b_o, but it should not be closer than $d/2$ from the face of the column.

6. The shearhead is considered to contribute a moment resistance M_v to each slab column strip as follows:

$$M_v = \frac{\phi}{2n}\alpha_v V_u\left(l_v - \frac{c_1}{2}\right) \quad \text{(ACI Code, equation (11.40))} \qquad (17.9)$$

but it should not be more than the smallest of 30% of the factored moment required in the column strip or the change in the column strip moment over the length l_v or M_p given in equation (17.8).

The use of anchored bent bars or wires is permitted by the ACI Code, Section 11.12.3. The bars are placed on top of the column, and the possible arrangements are shown in Figure 17.9(e). When bars or wires are used as shear reinforcement, the nominal shear strength is

$$V_n = V_c + V_s = (2\sqrt{f'_c})b_o d + \frac{A_v f_y d}{s} \qquad (17.10)$$

where A_v is the total stirrup bar area and b_o is the length of the critical section of two-way shear at a distance $d/2$ from the face of the column. The nominal shear strength V_n should not exceed $6\sqrt{f'_c}\,b_o d$.

The use of shear reinforcement in flat plates reduces the slab thickness and still maintains the flat ceiling to reduce the cost of formwork. Typical stirrup cages for shear reinforcement are shown in Figure 17.9(f). Another type of shear reinforcement consists of studded steel strips (Figure 17.9(g)). The steel strip is positioned with bar chairs and fastened to the formwork, replacing the stirrup gages. The yield strength of the stud material is specified between 40 and 60 Ksi to achieve complete anchorage at ultimate load.

17.8 ANALYSIS OF TWO-WAY SLABS BY THE DIRECT DESIGN METHOD

The direct design method is an approximate method established by the ACI Code to determine the design moments in uniformly loaded two-way slabs. To use this method, some limitations must be met, as indicated by the ACI Code, Section 13.6.1.

17.8.1 Limitations

1. There must be a minimum of three continuous spans in each direction.
2. The panels must be square or rectangular; the ratio of the longer to the shorter span within a panel must not exceed 2.0.
3. Adjacent spans in each direction must not differ by more than one-third of the longer span.
4. Columns must not be offset by a maximum of 10% of the span length, in the direction of offset, from either axis between centerlines of successive columns.
5. All loads must be uniform, and the ratio of the live to dead load must not exceed 2.0.
6. If beams are present along all sides, the ratio of the relative stiffness of beams in two perpendicular directions, $\alpha_1 l_2^2/\alpha_2 l_1^2$, must not be less than 0.2 nor greater than 5.0.

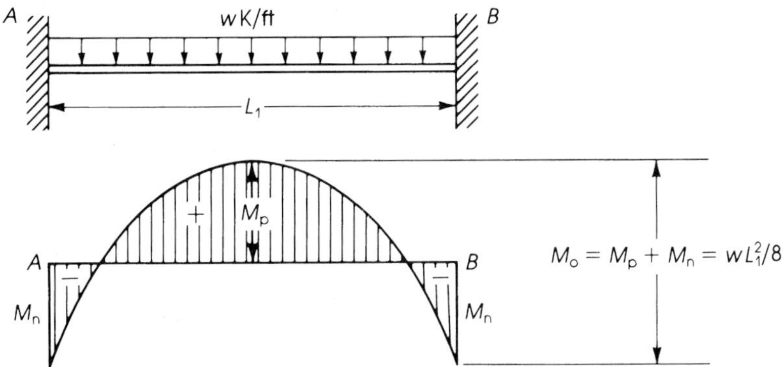

Figure 17.10 Bending moment in a fixed-end beam.

17.8.2 Total Factored Static Moment

If a simply supported beam carries a uniformly distributed load w K/ft, then the maximum positive bending moment occurs at midspan and equals $M_o = wl_1^2/8$, where l_1 is the span length. If the beam is fixed at both ends or continuous with equal negative moments at both ends, then the total moment $M_o = M_p$ (positive moment at midspan) + M_n (negative moment at support) $= wl_1^2/8$ (Figure 17.10). Now, if the beam AB carries the load W from a slab that has a width l_2 perpendicular to l_1, then $W = w_u l_2$, and the total moment is

$$M_o = \frac{(w_u l_2) l_1^2}{8}$$

where w_u = load intensity in K/ft². In this expression, the actual moment occurs when l_1 equals the clear span between supports A and B. If the clear span is denoted by l_n, then

$$M_o = (w_u l_2)\frac{l_n^2}{8} \qquad \text{(ACI Code, equation (13.3))} \qquad (17.11)$$

The clear span l_n is measured face to face of supports in the direction in which moments are considered, but not less than 0.65 times the span length from center to center of supports. The face of the support where the negative moments should be calculated is illustrated in Figure 17.11. The length l_2 is measured in a direction perpendicular to l_n and equals the direction between center to center of supports (width of slab). The total moment M_o calculated in the long direction will be referred to here as M_{ol} and that in the short direction, as M_{os}.

Once the total moment, M_o, is calculated in one direction, it is divided into a positive moment M_p and a negative moment M_n such that $M_o = M_p + M_n$ (Figure 17.10). Then each moment, M_p and M_n, is distributed across the width of the slab between the column and middle strips, as is explained shortly.

17.8.3 Longitudinal Distribution of Moments in Slabs

In a typical *interior panel,* the total static moment M_o is divided into two moments, the positive moment, M_p, at midspan, equal to $0.35M_o$, and the negative moment, M_n, at each support, equal to $0.65M_o$, as shown in Figure 17.12. These values of moment are based on the assumption that the interior panel is continuous in both directions, with approximately equal spans and loads, so that the interior joints have no significant rotation. Moreover, the moment values are approximately the same as those in a fixed-end beam subjected to uniform loading, where the negative moment at the support is twice the positive moment

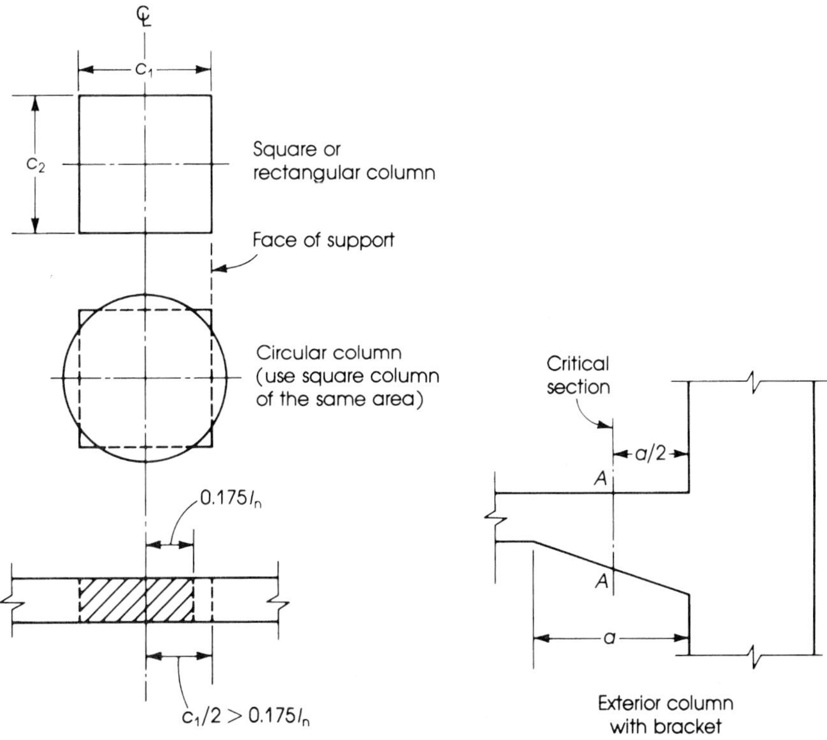

Figure 17.11 Critical sections for negative design moments. *A-A*, section for negative moment at exterior support with bracket.

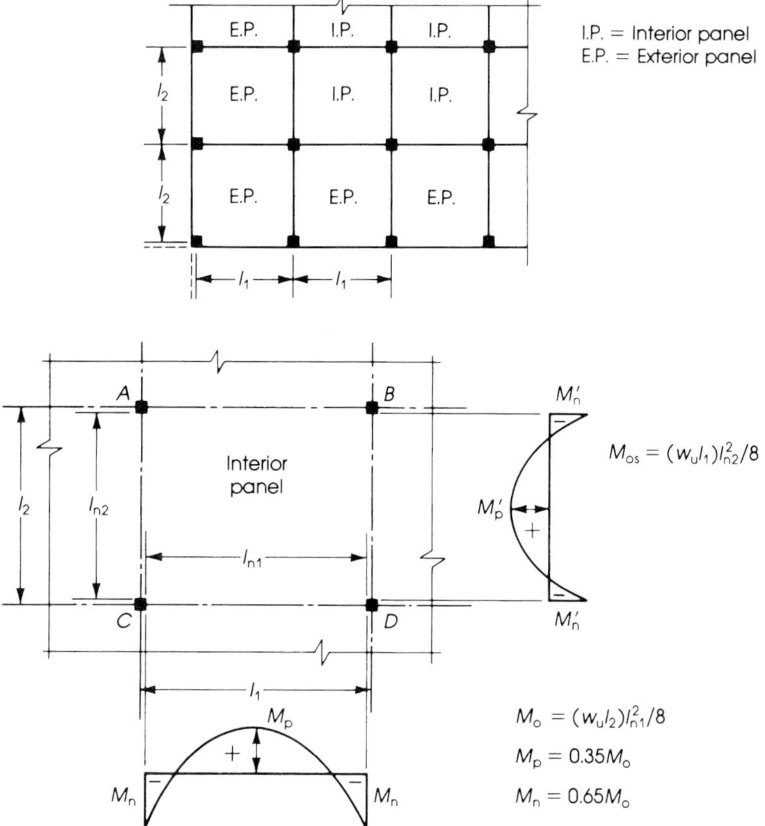

Figure 17.12 Distribution of moments in an interior panel.

at midspan. In Figure 17.12, if $l_1 > l_2$, then the distribution of moments in the long and short directions is as follows:

$$M_{ol} = (w_u l_2)\frac{l_{n1}^2}{8}, \qquad M_{pl} = 0.35M_{ol} \quad \text{and} \quad M_{n1} = 0.65M_{ol}$$

$$M_{os} = (w_u l_1)\frac{l_{n2}^2}{8}, \qquad M_{ps} = 0.35M_{os} \quad \text{and} \quad M_{ns} = 0.65M_{os}$$

If the magnitudes of the negative moments on opposite sides of an interior support are different because of unequal span lengths, the ACI Code specifies that the larger moment should be considered to calculate the required reinforcement.

In an *exterior panel,* the slab load is applied to the exterior column from one side only, causing an unbalanced moment and a rotation at the exterior joint. Consequently, there will be an increase in the positive moment at midspan and in the negative moment at the first interior support. The magnitude of the rotation of the exterior joint determines the increase in the moments at midspan and at the interior support. For example, if the exterior edge is a simple support, as in the case of a slab resting on a wall (Figure 17.13), the slab moment at the face of the wall there is zero, the positive moment at midspan can be taken as $M_p = 0.63M_o$, and the negative moment at the interior support is $M_n = 0.75M_o$. These values satisfy the static equilibrium equation

$$M_o = M_p + \tfrac{1}{2}M_n = 0.63M_o + \tfrac{1}{2}(0.75M_o)$$

In a slab-column floor system, there is some restraint at the exterior joint provided by the flexural stiffness of the slab and by the flexural stiffness of the exterior columns.

According to Section 13.6.3 of the ACI Code, the total static moment M_o in an end span is distributed in different ratios according to Table 17.2 and Figure 17.14. The moment

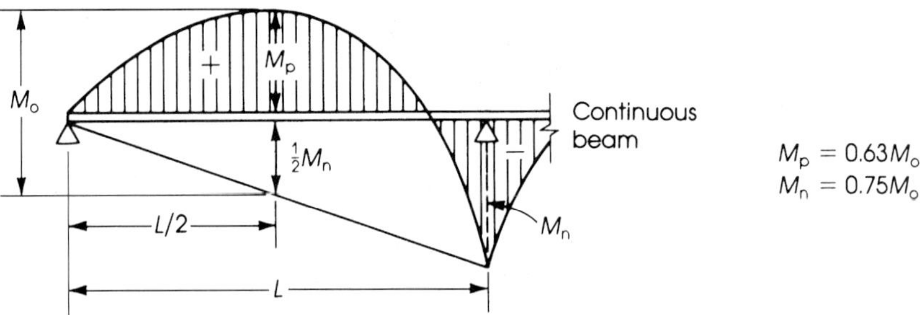

Figure 17.13 Exterior panel.

Table 17.2 Distribution of Moments in an end Panel

	Exterior Edge		Slab with Beams Between All Supports	Slab without Beams Between Interior Supports	
	Unrestrained (1)	***Fully Restrained (2)***	***(3)***	***With Edge Beam (4)***	***Without Edge Beam (5)***
Exterior negative factored moment	0	0.65	0.16	0.30	0.26
Positive factored moment	0.63	0.35	0.57	0.50	0.52
Interior negative factored moment	0.75	0.65	0.70	0.70	0.70

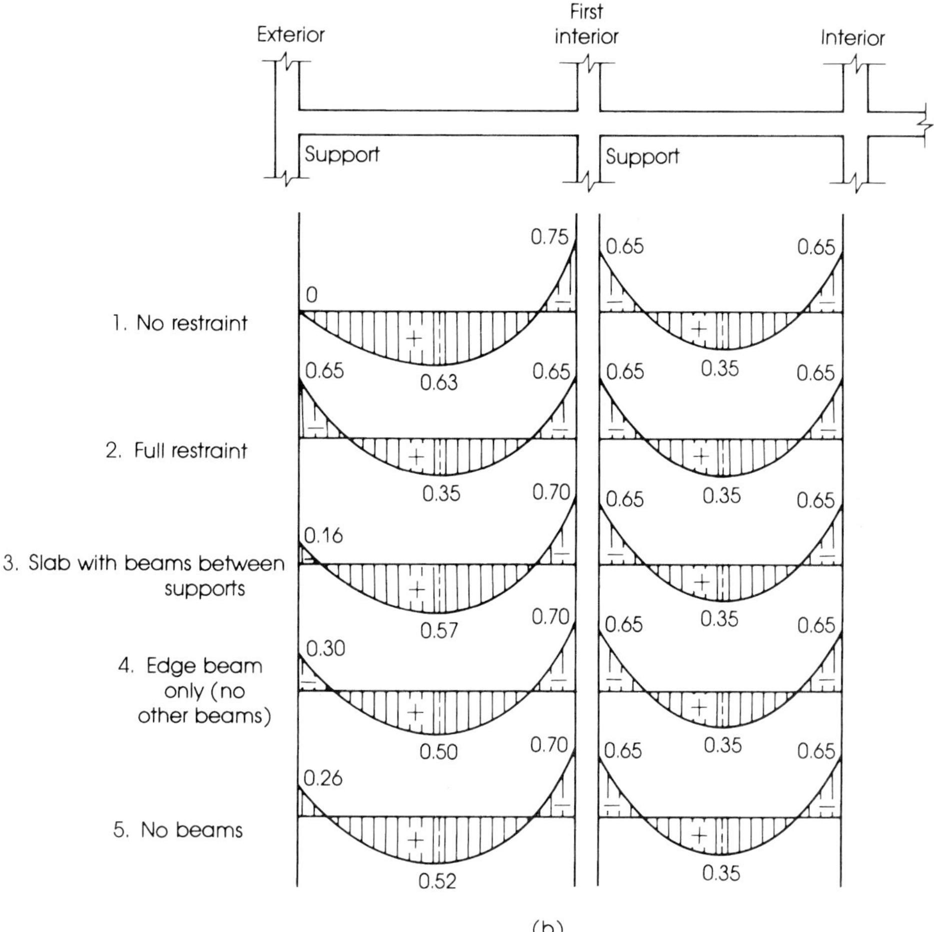

Figure 17.14 Distribution of total static moment into negative and positive span moments.

coefficients in column 1 for an unrestrained edge are based on the assumption that the ratio of the flexural stiffness of columns to the combined flexural stiffness of slabs and beams at a joint, α_{ec}, is equal to zero. The coefficients of column 2 are based on the assumption that the ratio α_{ec} is equal to infinity. The moment coefficients in columns 3, 4, and 5 have been established by analyzing the slab systems with different geometries and support conditions.

17.8.4 Transverse Distribution of Moments

The longitudinal moment values mentioned in the previous section are for the entire width of the equivalent building frame. This frame width is the sum of the widths of two half-column strips and two half-middle strips of two adjacent panels, as shown in Figure 17.15. The transverse distribution of the longitudinal moments to the middle and column strips is a function of the ratios l_2/l_1,

$$\alpha_1 = \frac{E_{cb}I_b}{E_{cs}I_s} = \frac{\text{beam stiffness}}{\text{slab stiffness}} \tag{17.12}$$

$$\beta_t = \frac{E_{cb}C}{2E_{cs}I_s} = \frac{\text{torsional rigidity of edge beam section}}{\text{flexural rigidity of a slab of width equal to beam span length}} \tag{17.13}$$

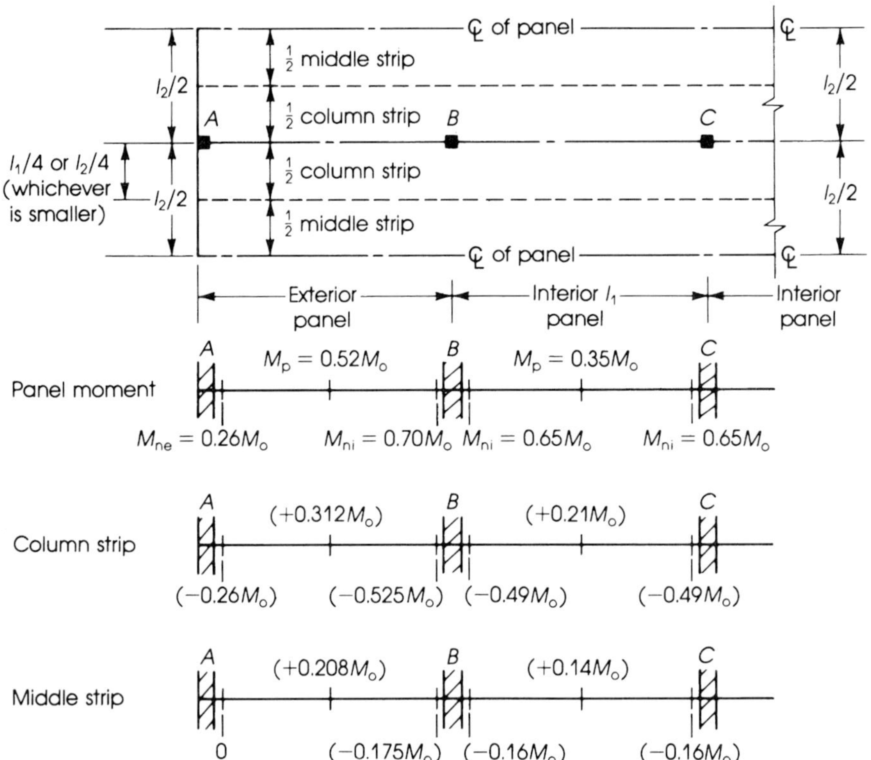

Figure 17.15 Width of the equivalent rigid frame (equal spans in this figure) and distribution of moments in flat plates, flat slabs, and waffle slabs with no beams.

where

$$C = \text{torsional constant} = \Sigma\left(1 - \frac{0.63x}{y}\right)\left(\frac{x^3y}{3}\right) \tag{17.14}$$

where x and y are the shorter and longer dimension of each rectangular component of the section. The percentages of each design moment to be distributed to column and middle strips for interior and exterior panels are given in Tables 17.3 through 17.6.

In a typical *interior* panel, the portion of the design moment that is not assigned to the column strip (Table 17.3) must be resisted by the corresponding half-middle strips. Linear interpolation of values of l_2/l_1 between 0.5 and 2.0 and of $\alpha_1 l_2/l_1$ between 0 and 1 is permitted by the ACI Code. From Table 17.3 it can be seen that when no beams are used, as in the case of flat plates or flat slabs, $\alpha_1 = 0$. The final percentage of moments in the column and middle strips as a function of M_o are given in Table 17.4.

Table 17.3 Percentage of Longitudinal Moment in Column Strips, Interior Panels (ACI Code, Section 13.6.4)

		Aspect Ratio, l_2/l_1		
	$\alpha_1 l_2/l_1$	**0.5**	**1.0**	**2.0**
Negative moment at interior support	0	75	75	75
	≥1.0	90	75	45
Positive moment near midspan	0	60	60	60
	≥1.0	90	75	45

Table 17.4 Percentage of Moments in Two-Way Interior Slabs without Beams ($\alpha_1 = 0$)

Total design moment $= M_o = (w_u l_2)\left(\frac{l_{nl}^2}{8}\right)$

	Negative Moment	**Positive Moment**
Longitudinal moments in one panel	$-0.65M_o$	$+0.35M_o$
Column strip	$0.75(-0.65M_o) = -0.49M_o$	$0.60(0.35M_o) = 0.21M_o$
Middle strip	$0.25(-0.65M_o) = 0.16M_o$	$0.40(0.35M_o) = 0.14M_o$

Table 17.5 Percentage of Longitudinal Moment in Column Strips, Exterior Panels (ACI Code, Section 13.6.4)

			Aspect Ratio l_2/l_1		
	$\alpha_1 l_2/l_1$	β_t	***0.5***	***1.0***	***2.0***
Negative moment at exterior support	0	0	100	100	100
		≥ 2.5	75	75	75
	≥ 1.0	0	100	100	100
		≥ 2.5	90	75	45
Positive moment near midspan	0		60	60	60
	≥ 1.0		90	75	45
Negative moment at interior support	0		75	75	75
	≥ 1.0		90	75	45

Table 17.6 Percentage of Longitudinal Moment in Column and Middle Strips, Exterior Panels (for all ratios of l_2/l_1), Given $\alpha_1 = \beta_t = 0$

	%	**Column Strip**	**Middle Strip**	**Final Moment as a Function of M_o and α_{ec} (Column Strip)**
Negative moment at exterior support	100	$0.26M_o$	0	$\left[\frac{0.65}{(1 + 1/\alpha_{ec})}\right](M_o)$
Positive moment $(0.6 \times 0.52M_o)$	60	$0.312M_o$	$0.208M_o$	$\left[0.63 - \frac{0.28}{(1 + 1/\alpha_{ec})}\right](M_o)$
Negative moment at interior support $(0.75 \times 0.70M_o)$	75	$0.525M_o$	$0.175M_o$	$\left[0.75 - \frac{0.10}{(1 + 1/\alpha_{ec})}\right](M_o)$

For *exterior* panels, the portion of the design moment that is not assigned to the column strip (Table 17.5) must be resisted by the corresponding half-middle strips. Again, linear interpolation between values shown in Table 17.5 is permitted by the ACI Code, Section 13.6.4.2. When no beams are used in an exterior panel, as in the case of flat slabs or flat plates with no edge (spandrel) beam, $\alpha_1 = 0$, $C = 0$, and $\beta_t = 0$. This means that the end column provides the restraint to the exterior end of the slab. The applicable values of Table 17.5 for this special case are shown in Table 17.6 and Figure 17.15.

From Table 17.6 it can be seen that when no edge beam is used at the exterior end of the slab, $\beta_t = 0$ and 100% of the design moment is resisted by the column strip. The middle strip will not resist any moment; therefore, minimum steel reinforcement must be provided.

The ACI Code, Section 13.6.4.3, specifies that when the exterior support is a column or wall extending for a distance equal to or greater than three-fourths the transverse span length, l_2, used to compute M_o, the exterior negative moment is to be uniformly distributed across l_2. When beams are provided along the centerlines of columns, the ACI Code, Section 13.6.5, requires that beams must be proportioned to resist 85% of the moment in the column strip if $\alpha_1(l_2/l_1) \geq 1.0$. For values of $\alpha_1(l_2/l_1)$ between 1.0 and 0, the moment assigned to the beam is determined by linear interpolation. Beams must also be proportioned to resist additional moments caused by all loads applied directly to the beams, including the weight of the projecting beam stem. The portion of the moment that is not assigned to the beam must be resisted by the slab in the column strip.

17.8.5 ACI Provisions for Effects of Pattern Loadings

In continuous structures, the maximum and minimum bending moments at the critical sections are obtained by placing the live load in specific patterns to produce the extreme values. Placing the live load on all spans will not produce either the maximum positive or negative bending moments. The maximum and minimum moments depend mainly on the following:

1. The ratio of live to dead load. A high ratio will increase the effect of pattern loadings.
2. The ratio of column to beam stiffnesses. A low ratio will increase the effect of pattern loadings.
3. Pattern loadings. Maximum positive moments within the spans are less affected by pattern loadings.

To determine the design-factored moments in continuous structures, the ACI Code, Section 13.7.6, specifies the following:

1. When the loading pattern is known, the equivalent frame shall be analyzed for that load.
2. When the live load is variable but does not exceed $\frac{3}{4}$ of the dead load, $w_L \leq 0.75w_D$, or when all the panels is almost loaded simultaneously with the live load, it is permitted to analyze the frame with full factored live load on the entire slab system.
3. For other loading conditions, it is permitted to assume that the maximum positive factored moment near a midspan occurs with 0.75 of the full factored live load on the panel and alternate panels. For the maximum negative factored moment in the slab at a support, it is permitted to assume that 0.75 of the full factored live load is applied on adjacent panels only.
4. Factored moments shall not be taken less than the moments occurring with full factored live load on all continuous panels.

17.8.6 Reinforcement Details

After all the percentages of the static moments in the column and middle strips are determined, the steel reinforcement can be calculated for the negative and positive moments in each strip, as was done for beam sections in Chapter 4:

$$M_u = \phi A_s f_y \left(d - \frac{a}{2} \right) = R_u b d^2 \tag{17.15}$$

Calculate R_u and determine the steel ratio ρ using the tables in Appendix A or use the following equation:

$$R_u = \phi \rho f_y \left(1 - \frac{\rho f_y}{1.7 f'_c} \right) \tag{17.16}$$

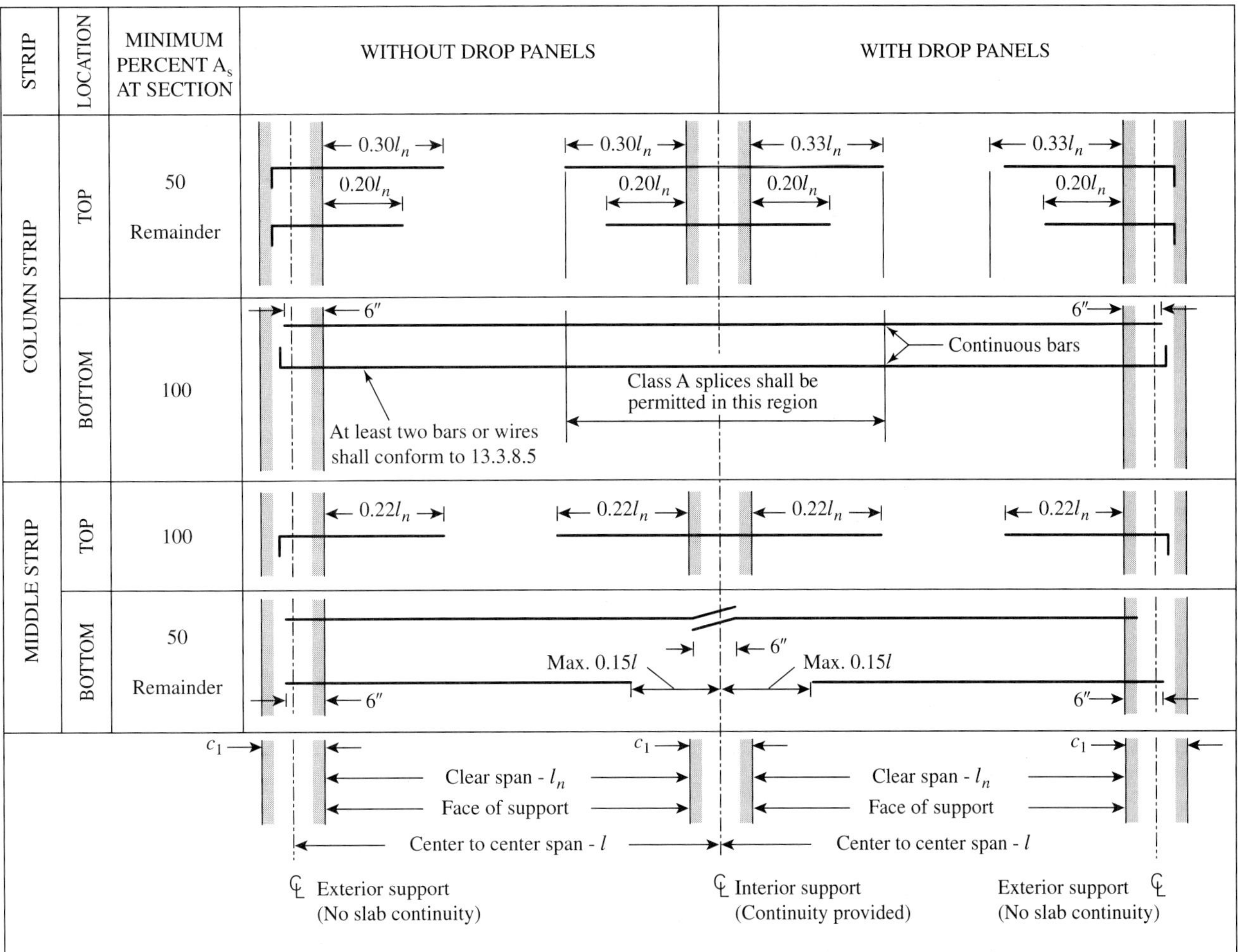

Figure 17.16 Minimum extensions for reinforcement in slabs without beams (ACI Code, Figure 13.3.8). Courtesy American Concrete Institute [18].

where $\phi = 0.9$. The steel area is $A_s = \rho bd$. When the slab thickness limitations, as discussed in Section 17.4, are met, no compression reinforcement will be required. Figure 13.3.8 of the ACI Code indicates the minimum length of reinforcing bars and reinforcement details for slabs without beams; it is reproduced here as Figure 17.16. The spacing of bars in the slabs must not exceed the ACI limits of maximum spacing: 18 in. (450 mm) or twice the slab thickness, whichever is smaller.

17.8.7 Modified Stiffness Method for End Spans

In this method, the stiffnesses of the slab end beam and of the exterior column are replaced by the stiffness of an equivalent column K_{ec}. The flexural stiffness of the equivalent column, K_{ec}, can be calculated from the following expression:

$$\frac{1}{K_{ec}} = \frac{1}{\Sigma K_c} + \frac{1}{K_t} \quad \text{or} \quad K_{ec} = \frac{\Sigma K_c}{1 + \Sigma K_c/K_t} \tag{17.17}$$

where K_{ec} = flexural stiffness of the equivalent column

K_c = flexural stiffness of the actual column

K_t = torsional stiffness of the edge beam

The sum of the flexural stiffness of the columns above and below the floor slab can be taken as follows:

$$\Sigma K_c = 4E\left(\frac{I_{c1}}{L_{c1}} + \frac{I_{c2}}{L_{c2}}\right) \tag{17.18}$$

where I_{c1} and L_{c1} = the moment of inertia and length of column above slab level

I_{c2} and L_{c2} = the moment of inertia and length of column below slab level

The torsional stiffness of the end beam, K_t, may be calculated as follows:

$$K_t = \Sigma \frac{9E_{cs}C}{l_2(1 - c_2/l_2)^3} \tag{17.19}$$

where c_2 = size of the rectangular or equivalent rectangular column, capital, or bracket measured on transverse spans on each side of the column

E_{cs} = modulus of elasticity of the slab concrete

C = torsion constant determined from the following expression:

$$C = \Sigma\left(1 - 0.63\frac{x}{y}\right)\left(\frac{x^3y}{3}\right) \tag{17.20}$$

where x is the shorter dimension of each component rectangle and y is the longer dimension of each component rectangle. In calculating C, the component rectangles of the cross section must be taken in such a way as to produce the largest value of C.

The preceding expressions are introduced here and will also be used in Section 17.12, "Equivalent Frame Method."

If a panel contains a beam parallel to the direction in which moments are being determined, the torsional stiffness, K_t, given in equation (17.19) must be replaced by a greater value, K_{ta}, computed as follows:

$$K_{ta} = K_t \times \frac{I_{sb}}{I_s}$$

where $I_s = \dfrac{l_2h^3}{12}$ = moment of inertia of a slab that has a width equal to the full width between panel centerlines (excluding that portion of the beam stem extending above or below the slab)

I_{sb} = I_s, including the portion of the beam stem extending above or below the slab

Cross sections of some attached torsional members are shown in Figure 17.17. Once K_{ec} is calculated, the stiffness ratio, α_{ec}, is obtained as follows:

$$\alpha_{ec} = \frac{K_{ec}}{\Sigma(K_s + K_b)} \tag{17.21}$$

where

$K_s = \dfrac{4E_{cs}I_s}{l_1}$ = flexural stiffness of the slab

$K_b = \dfrac{4E_{cb}I_b}{l_1}$ = flexural stiffness of the beam

I_b = gross moment of inertia of the longitudinal beam section

The distribution of the total static moment M_o in an exterior panel is given as a function of α_{ec} as follows:

$$\text{Interior negative factored moment} = \left[0.75 - \frac{0.1}{(1 + 1/\alpha_{ec})}\right] M_o$$

$$\text{Positive factored moment} = \left[0.63 - \frac{0.28}{(1 + 1/\alpha_{ec})}\right] M_o$$

$$\text{Exterior negative factored moment} = \left[\frac{0.65}{(1 + 1/\alpha_{ec})}\right] M_o$$

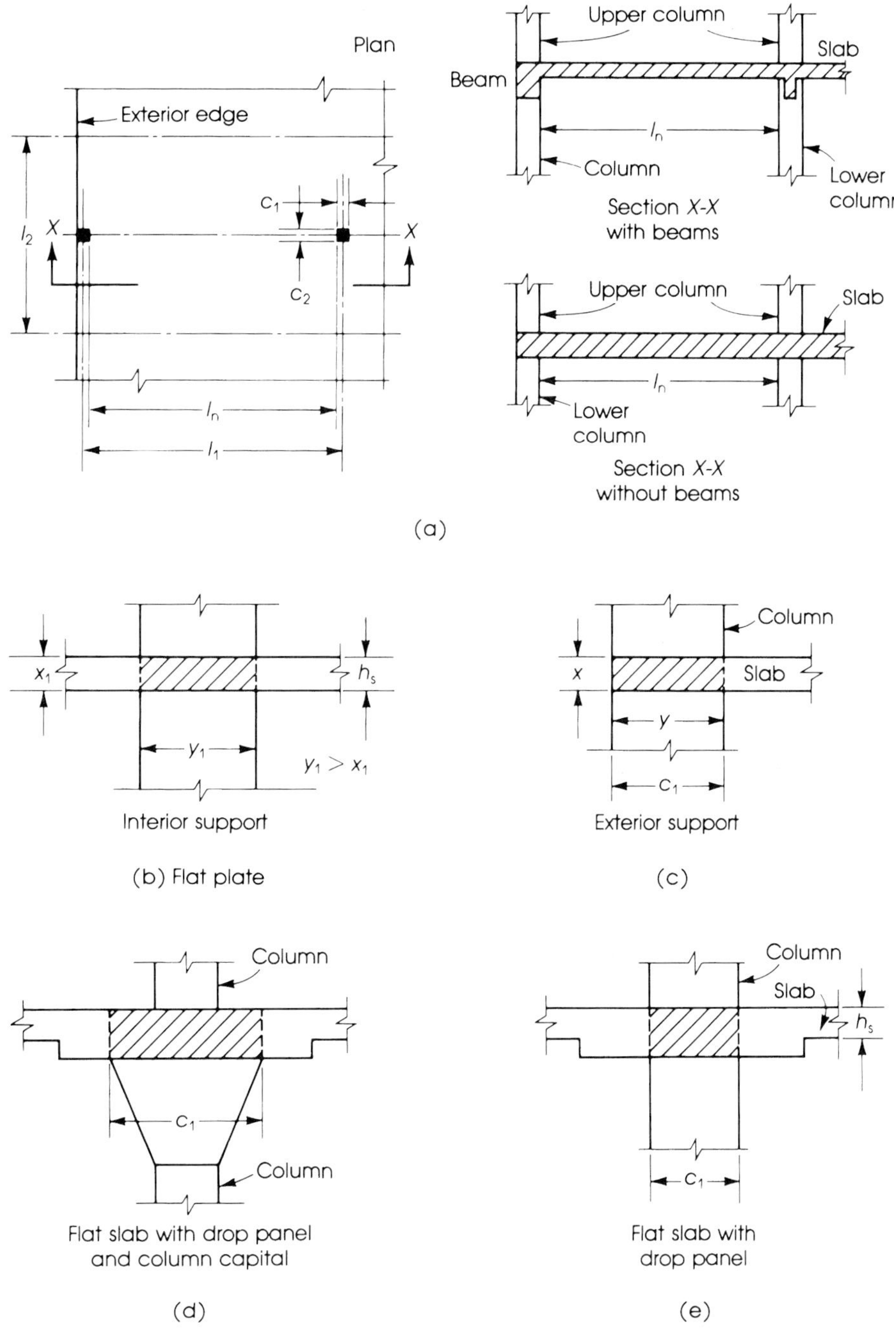

Figure 17.17 Cross sections of some attached torsional members.

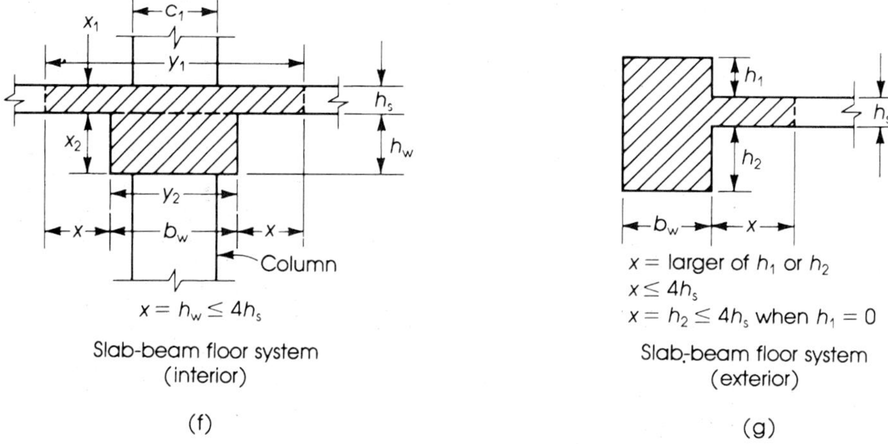

Figure 17.17 *(continued)*

These values are shown for a typical exterior panel in Figure 17.18. These factors take into consideration the effect of the stiffness of the exterior column as well as the slab end beam giving adequate distribution of moments.

17.8.8 Summary of the Direct Design Method (DDM)

Case 1: Slabs without beams (flat slabs and flat plates)

1. Check the limitation requirements explained in Section 17.8.1. If limitations are not met, DDM cannot be used.
2. Determine the minimum slab thickness $(h_{\min})$ to control deflection using values in Table 17.1. Exterior panels without edge beams give the highest $h_{\min}(l_n/30$ for

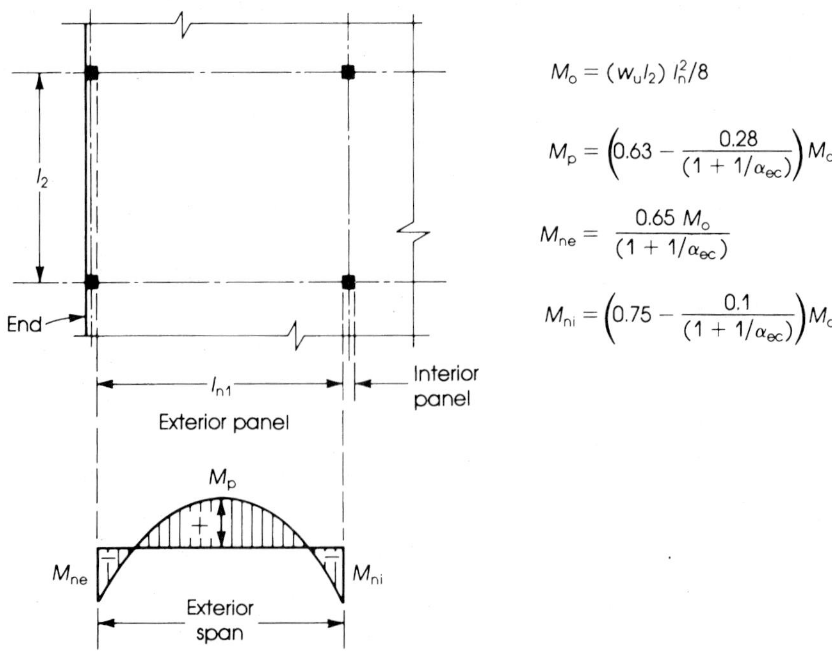

Figure 17.18 Distribution of moments in an exterior panel.

$f_y = 60$ Ksi). It is a common practice to use the same slab depth for all exterior and interior panels.

3. Calculate the factored loads, $W_u = 1.4W_D + 1.7W_L$.
4. Check the slab thickness, h, as required by one-way and two-way shear. If the slab thickness, h, is not adequate, either increase h or provide shear reinforcement.
5. Calculate the total static moment M_o in both directions (equation (17.11)).
6. Determine the distribution factors for the positive and negative moments in the longitudinal and transverse directions for each column and middle strip in both interior and exterior panels as follows:
   a. For interior panels, use the moment factors given in Table 17.4 or Figure 17.15.
   b. For exterior panels without edge beams, the panel moment factors are given in Table 17.2 or Figure 17.14 (Case 5). For the distribution of moments in the transverse direction, use Table 17.6 or Figure 17.15 for column-strip ratios. The middle strip will resist the portion of the moment that is not assigned to the column strip.
   c. For exterior panels with edge beams, the panel moment factors are given in Table 17.2 or Figure 17.14 (Case 4). For the distribution of moments in the transverse direction, use Table 17.5 for the column strip. The middle strip will resist the balance of the panel moment.
7. Determine the steel reinforcement for all critical sections of the column and middle strips and extend the bars throughout the slab according to Figure 17.16.
8. Compute the unbalanced moment and check if transfer of unbalanced moment by flexure is adequate. If not, determine the additional reinforcement required in the critical width. (Refer to Section 17.10.)
9. Check if transfer of the unbalanced moment by shear is adequate. If not, increase h or provide shear reinforcement. (Refer to Section 17.10.)

Case 2: Slabs with interior and exterior beams

1. Check the limitation requirements as explained in Section 17.8.1.
2. Determine the minimum slab thickness $(h_{\min})$ to control deflection using equations (17.1) through (17.3). In most cases, equation (17.2) controls. For equation (17.1), α_m should be calculated first, as shown in Example 17.1.
3. Calculate the factored load, W_u.
4. Check the slab thickness, h, according to one-way and two-way shear requirements. In general, shear is not critical for slabs supported on beams.
5. Calculate the total static moment M_o in both directions (equation (17.11)).
6. Determine the distribution factors for the positive and negative moments in the longitudinal and transverse directions for each column and middle strips in both interior and exterior panels as follows:
   a. For interior panels, use moment factors in Figure 17.14 (Case 3) or Figure 17.12. For the distribution of moments in the transverse direction, use Table 17.3 for column strips. The middle strips will resist the portion of the moments not assigned to the column strips. Calculate α_1 from equation (17.12).
   b. For exterior panels, use moment factors in Table 17.2 or Figure 17.14 (Case 3). For the distribution of moments in the transverse direction, use Table 17.5 for the column strip. The middle strip will resist the balance of the panel moment.
   c. In both cases (a) and (b), the beams must resist 85% of the moment in the column strip when $\alpha_1(l_2/l_1) \geq 1.0$, whereas the ratio varies between 85% and 0% when $\alpha_1(l_2/l_1)$ varies between 1.0 and 0.

7. Determine the steel reinforcement for all critical sections in the column strip, beam, and middle strip; then extend the bars throughout the slab according to Figure 17.16.

8. Compute the unbalanced moment and then check the transfer of moment by flexure and shear. (Refer to Section 17.10.)

Example 17.3

Using the direct design method, design the typical *interior flat-plate* panel shown in Figures 17.6 and 17.19. The floor system consists of four panels in each direction with a panel size of 24 by 20 ft. All panels are supported by 20- by 20-in. columns, 12 ft long. The slab carries a uniform service live load of 80 psf and a service dead load that consists of 24 psf of floor finish in addition to the slab self-weight. Use $f'_c = 4$ Ksi and $f_y = 60$ Ksi.

Solution

1. Determine the minimum slab thickness using Table 17.1 for flat plates. From Example 17.1, a 9-in. slab thickness is adopted.

2. Calculate the factored loads:

$$w_D = 24 + \text{weight of slab} = 24 + \frac{9.0}{12} \times 150 = 136.5 \text{ psf}$$

$$w_u = 1.4 \times (136.5) + 1.7 \times (80) = 327, \quad \text{say, } 330 \text{ psf}$$

3. Check one- and two-way shears:

a. Check punching shear at a distance $d/2$ from the face of the column (two-way action): Assuming $\frac{3}{4}$ in. concrete cover and no. 5 bars, then the average d is $9.0 - 0.75 - \frac{5}{8} = 7.6$ in. and $b_o = 4(20 + 7.6) = 110$ in. (see Figure 17.19(c)).

$$V_u = \left[l_1 l_2 - \left(\frac{27.6}{12} \times \frac{27.6}{12}\right)\right] \times w_u = (24 \times 20 - 5.3) \times 0.330 = 156.7 \text{ K}$$

$$\phi V_c = \phi(4\sqrt{f'_c})b_o d = \frac{0.85 \times 4}{1000} \times \sqrt{4000} \times 110 \times 7.6 = 179.8 \text{ K}$$

which is greater than V_u.

b. Check beam shear at a distance d from the face of the column; average d is 7.6 in. Consider a 1-ft strip (Figure 17.19(d)), with the length of the strip being

$$x = 12 - \frac{10}{12} - \frac{7.6}{12} = 10.5 \text{ ft}$$

$$V_u = w_u(1 \times 10.5) = 0.330 \times 1 \times 10.5 = 3.47 \text{ K}$$

$$\phi V_c = \phi(2\sqrt{f'_c})bd = \frac{0.85 \times 2}{1000} \times \sqrt{4000} \times (12 \times 7.6) = 9.8 \text{ K}$$

which is greater than $V_u = 3.47$ K. In normal loadings, one-way shear does not control.

4. Calculate the total static moments in the long and short directions. In the long direction,

$$M_{ol} = \frac{w_u l_2 l_{n1}^2}{8} = \frac{0.33}{8} \times 20(22.33)^2 = 411.4 \text{ K}\cdot\text{ft}$$

In the short direction,

$$M_{os} = \frac{w_u l_1 l_{n2}^2}{8} = \frac{0.33}{8} \times 24 \times (18.33)^2 = 333 \text{ K}\cdot\text{ft}$$

Because $l_2 < l_1$, the width of half a column strip in the long direction is $0.25 \times 20 = 5$ ft, and the width of the middle strip is $20 - 2 \times 5 = 10$ ft. The width

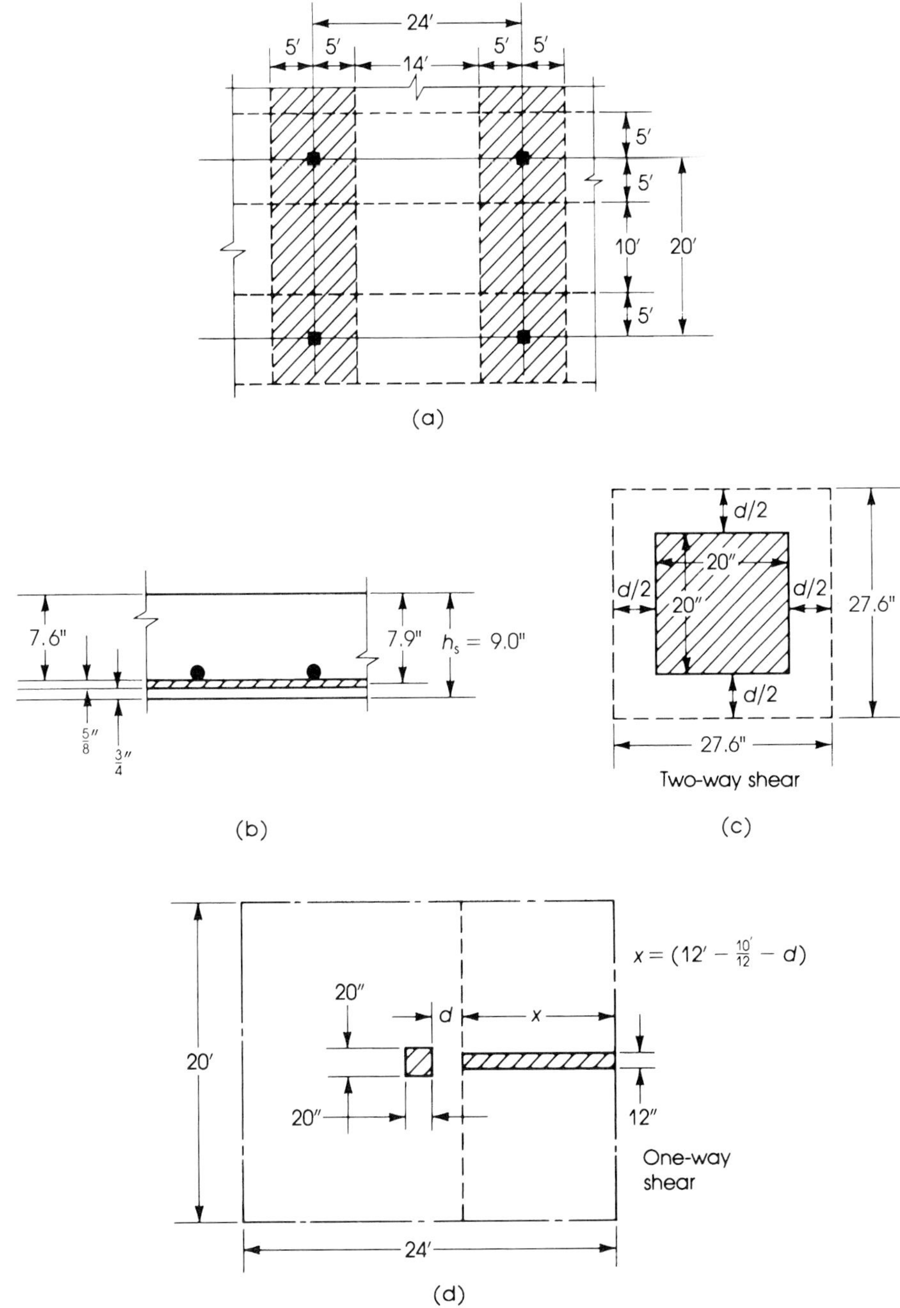

Figure 17.19 Example 17.3: Interior flat plate.

of half the column strip in the short direction is 5 ft, and the width of the middle strip is $24 - 2 \times 5 = 14$ ft. To calculate the effective depth, d, in each direction, assume that steel bars in the short direction are placed on top of the bars in the long direction. Therefore, d (long direction) $= 9.0 - 0.75 - \frac{5}{16} = 7.9$ in. and d (short direction) $= 9.0 - 0.75 - \frac{5}{8} - \frac{5}{16} = 7.3$ in. For practical applications, an average $d = 9 - 1.5 = 7.5$ in. can be used for both directions.

The design procedure can be conveniently arranged in a table form, as in Tables 17.7 and 17.8. The details for the bars selected for this interior slab are shown in Figure 17.20

Table 17.7 Design of an Interior Flat-Plate Panel

	$M_o = 411.4$ K·ft $M_n = 0.65M_o = -267.4$ K·ft $M_p = +0.35M_o = +144$ K·ft			
	Column Strip		***Middle Strip***	
Long Direction	***Negative***	***Positive***	***Negative***	***Positive***
Moment distribution (%)	75	60	25	40
M_u (K·ft)	$0.75M_n = -201.6$	$0.6M_p = +86.4$	$0.25M_n = -66.8$	$0.4M_p = +57.6$
Width of strip b (in.)	120	120	120	120
Effective depth d (in.)	7.9	7.9	7.9	7.9
$R_u = \frac{M_u}{bd^2}$ (psi)	323	128	107	93
Steel ratio ρ (%)	0.633	0.262	0.2	0.175
$A_s = \rho bd$ (in.^{2})	6.00	2.48	1.92	1.66
Min. $A_s = 0.0018bh_s$ (in.^{2})	1.94	1.94	1.94	1.94
Bars selected (Straight)	20 no. 5	10 no. 5	10 no. 4	10 no. 4
Spacing $\leq 2h_s = 18$ in.	6 in.	12	12	12

Table 17.8

	$M_o = 333$ K·ft $M_n = 0.65M_o = -216.5$ K·ft $M_p = +0.35M_o = +116.5$ K·ft			
	Column Strip		***Middle Strip***	
Short Direction	***Negative***	***Positive***	***Negative***	***Positive***
Moment distribution (%)	75	60	25	40
M_u (K·ft)	$0.75M_n = -162.4$	$0.6M_p = +69.9$	$0.25M_n = -54.1$	$0.4M_p = +46.6$
Width of strip b (in.)	120	120	168	168
Effective depth d (in.)	7.3	7.3	7.3	7.3
$R_u = \frac{M_u}{bd^2}$ (psi)	305	131	73	62
Steel ratio ρ (%)	0.60	0.25	0.14	0.12
$A_s = \rho bd$ (in.^{2})	5.23	2.18	1.72	1.46
Min. $A_s = 0.0018bh_s$ (in.^{2})	1.94	1.94	2.72	2.72
Bars selected (Straight)	18 no. 5	10 no. 5	14 no. 4	14 no. 4
Spacing $\leq 2h_s = 18$ in.	6.7	12	12	12

using the staight-bar system. Minimum lengths of the bars must meet those shown in Figure 17.16.

Straight bars and $f_y = 60$ Ksi steel bars are more often preferred by contractors.

$$\text{Maximum spacing} = \frac{\text{width of panel}}{\text{no. of bars}} = \frac{168}{12} = 14 \text{ in.}$$

occurs at the middle strip in the short direction; this spacing of 12 in. is adequate, because it is less than $2h_s = 18$ in. and less than 18 in. specified by the ACI Code.

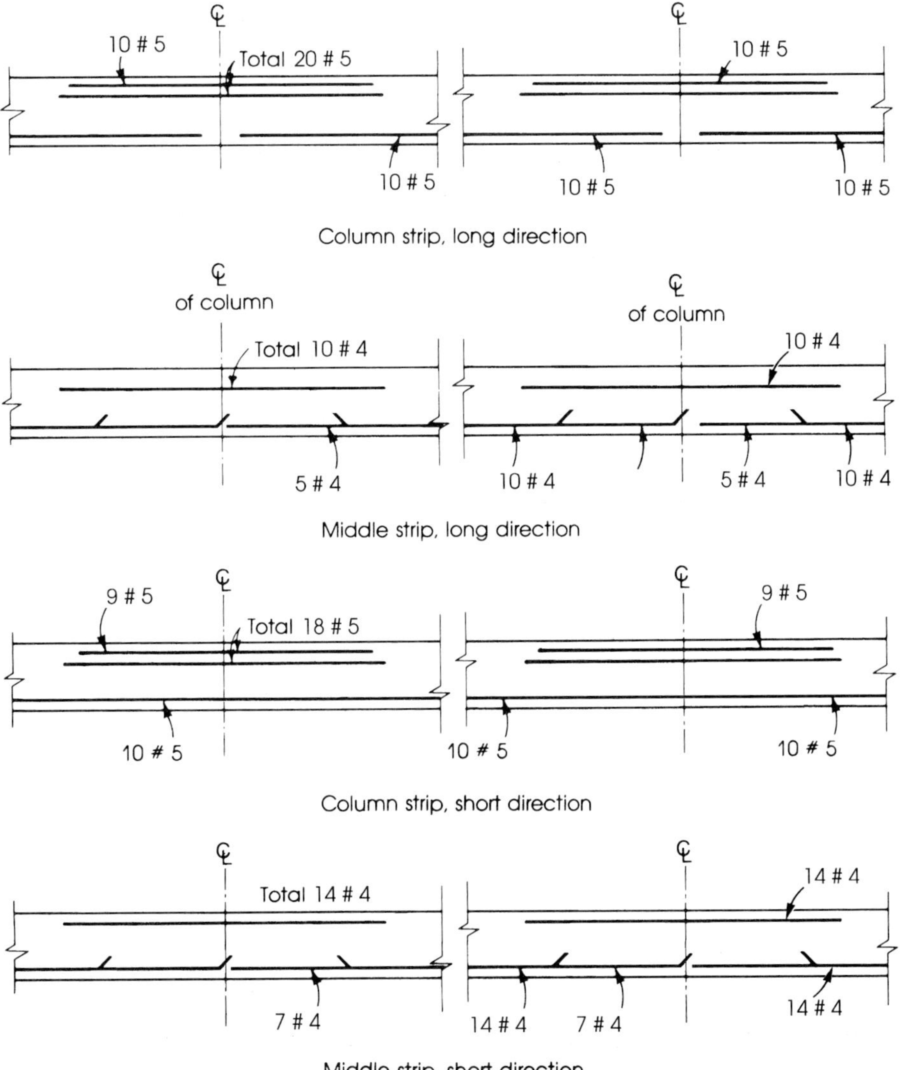

Figure 17.20 Example 17.3: Reinforcement details. For bar length, refer to Figure 17.16.

Example 17.4

Using the direct design method, design an *exterior flat-plate* panel that has the same dimensions, loads, and concrete and steel strengths given in Example 17.3. No beams are used along the edges (Figure 17.21).

Solution

1. Determine the minimum slab thickness using Table 17.1 for flat plates. From Example 17.1, a 9.0-in. slab thickness is adopted.
2. Calculate factored loads: $W_u = 330$ psf. (See Example 17.3.)
3. Check one- and two-way shear (refer to Example 17.3 and Figure 17.19).
   a. Check punching shear at an interior column, $V_u = 156.7 < \phi V_c = 179.8$ K.
   b. Check one-way shear: $V_u = 3.47 \text{ K} < \phi V_c = 9.8$ K.

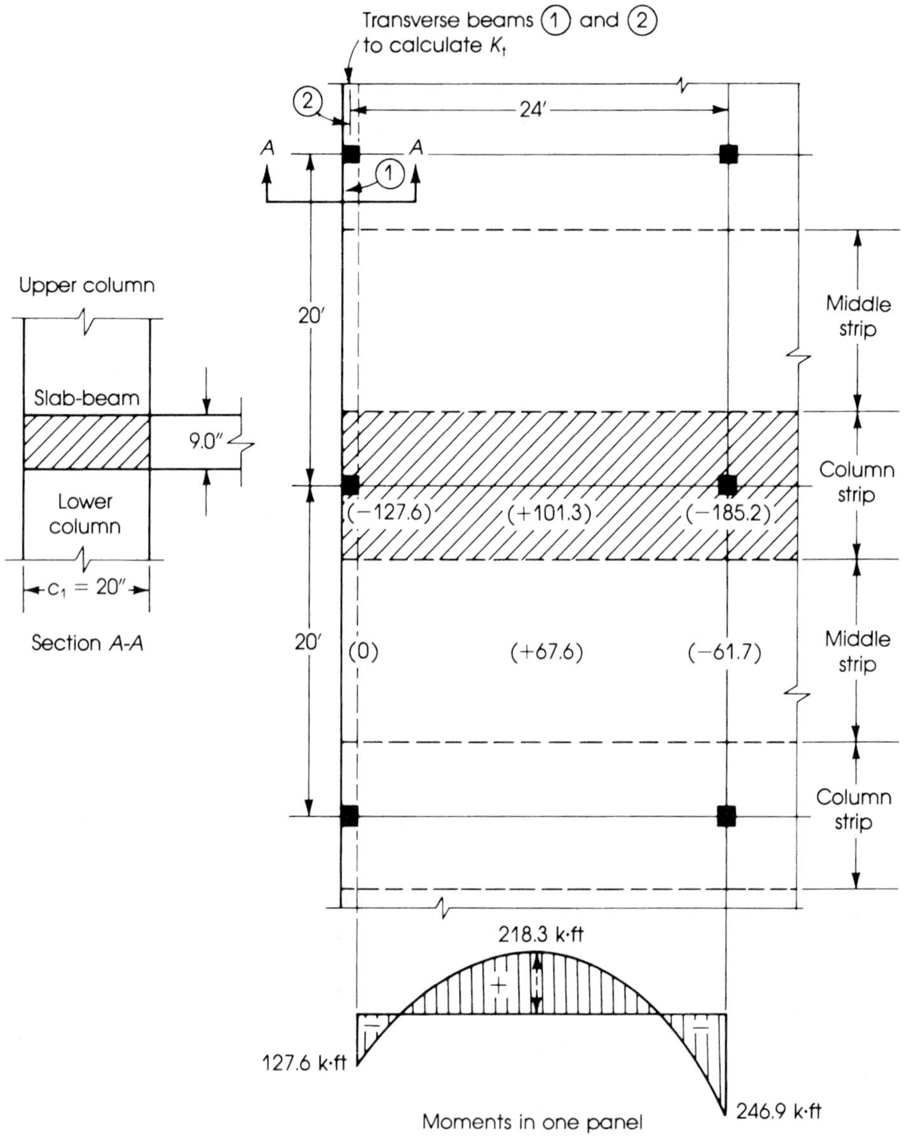

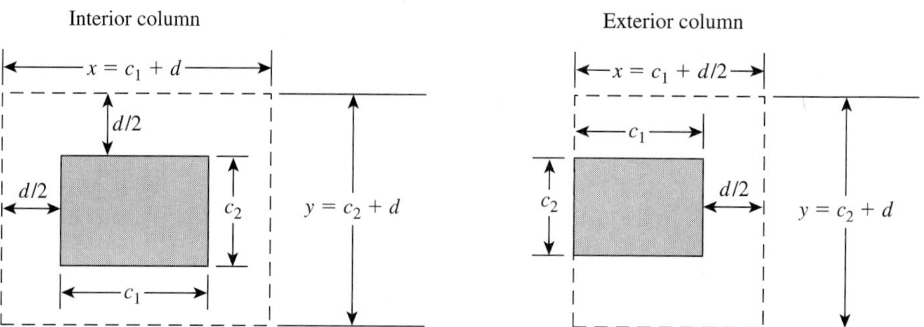

Figure 17.21 Distribution of bending moments, Example 17.4.

c. Check punching shear at the exterior column: $d = 7.6$ in.

$$x = 20 + \frac{d}{2} = 20 + \frac{7.6}{2} = 23.8 \text{ in.} = 1.98 \text{ ft}$$

$$y = 20 + d = 20 + 7.6 = 27.6 \text{ in.} = 2.30 \text{ ft}$$

$$b_o = 2x + y = 75.2 \text{ in.}$$

$$V_u = \left[20\left(12 + \frac{10}{24}\right) - 1.98(2.30)\right]0.33 = 80.45 \text{ K}$$

$$\phi V_c = \phi 4\sqrt{f'_c}\,b_o d = 122.9 \text{ K} > V_u$$

d. Check punching shear at a corner column: $d = 7.6$ in.

$$x = y = 20 + \frac{d}{2} = 23.8 \text{ in.} = 1.98 \text{ ft}$$

$$b_o = x + y = 47.6 \text{ in.}$$

$$V_u = \left[\left(10 + \frac{10}{24}\right)\left(12 + \frac{10}{24}\right) - (1.98)(1.98)\right]0.33 = 41.4 \text{ K}$$

$$\phi V_c = \phi 4\sqrt{f'_c}\,b_o d = 77.8 \text{ K} > V_u$$

4. Calculate the total static moments. From Example 17.3:

$$M_{ol} \text{ (long direction)} = 411.4 \text{ K}\cdot\text{ft}, \qquad d = 7.9 \text{ in.}$$

$$M_{os} \text{ (short direction)} = 333 \text{ K}\cdot\text{ft}, \qquad d = 7.3 \text{ in.}$$

The width of the column strip is 120 in., and the width of the middle strip is 168 in.

5. Calculate the design moments in the long direction: $l_1 = 24$ ft (refer to Table 17.5 or Figure 17.15). The distribution of the total moment, M_{ol}, in the column and middle strips is computed as follows:

a. Column strip:

$$\text{Interior negative moment} = -0.525M_o = -0.525(411.4) = -216 \text{ K}\cdot\text{ft}$$

$$\text{Positive moment within span} = 0.312M_o = 0.312(411.4) = +128.4 \text{ K}\cdot\text{ft}$$

$$\text{Exterior negative moment} = -0.26M_o = -0.26(411.4) = -107 \text{ K}\cdot\text{ft}$$

b. Middle strip:

$$\text{Interior negative moment} = -0.175M_o = -0.175 \times 411.4 = -72 \text{ K}\cdot\text{ft}$$

$$\text{Positive moment within span} = 0.208M_o = 0.208 \times 411.4 = +85.6 \text{ K}\cdot\text{ft}$$

$$\text{Exterior negative moment} = 0$$

6. Calculate the design moments in the short direction: $l_s = 20$ ft. It will be treated as an interior panel because it is continuous on both sides. Referring to Table 17.4 or Figure 17.15, the distribution of the total moment, M_{os}, in the column and middle strips is computed as follows:

a. Column strip:

$$\text{Negative moment} = 0.49M_o = -0.49(333) = -163.2 \text{ K}\cdot\text{ft}$$

$$\text{Positive moment} = +0.21M_o = +0.21(333) = +70.0 \text{ K}\cdot\text{ft}$$

b. Middle strip:

$$\text{Negative moment} = -0.16M_o = -0.16(333) = -53.3 \text{ K}\cdot\text{ft}$$

$$\text{Positive moment} = +0.14M_o = +0.14(333) = +46.6 \text{ K}\cdot\text{ft}$$

The design procedure can be conveniently arranged in Table 17.9. The details for bars selected are shown in Figure 17.22 using the straight-bar system in the long direction. Details of reinforcement in the short direction will be similar to Figure 17.20 using the bars chosen in Table 17.9.

Table 17.9 Design of Exterior Flat-Plate Panel for Example 17.4 ($d = 7.9$ in.)

	Column Strip			Middle Strip		
Long Direction	***Exterior***	***Positive***	***Interior***	***Exterior***	***Positive***	***Interior***
M_u (K·ft)	−107.06	+128.4	−216.0	0	+85.6	−72.0
b (in.)	120	120	120	120	120	120
$R_u = \frac{M_u}{bd^2}$ (psi)	172	206	346	0	138	116
Steel ratio ρ (%)	0.33	0.4	0.682	0	0.262	0.22
$A_s = \rho bd$	3.11	3.75	6.47	0	2.48	2.10
Min. $A_s = 0.0018bh_s$	1.94	1.94	1.94	1.94	1.94	1.94
Bars selected (Straight)	12 no. 5	12 no. 5	22 no. 5	10 no. 4	14 no. 4	14 no. 4
Spacing ≤ 18 in.	10	10	5.5	12	8.5	8.5

Short Direction	**Column Strip**		**Middle Strip**	
M_u (K·ft)	−163.2	+70.0	−53.3	+46.6
Width of strip b (in.)	120	120	168	168
d (in.)	7.3	7.3	7.3	7.3
$R_u = \frac{M_u}{bd^2}$ (psi)	306	131	71	63
Steel ratio ρ (%)	0.6	0.25	0.133	0.12
$A_s = \rho bd$ (in.^{2})	5.26	2.20	1.63	1.47
Min. $A_s = 0.0018bh_s$	1.94	1.94	2.72	2.72
Bars selected (Straight)	18 no. 5	8 no. 5	14 no. 4	14 no. 4
Spacing ≤ 18 in.	6.67	15	12	12

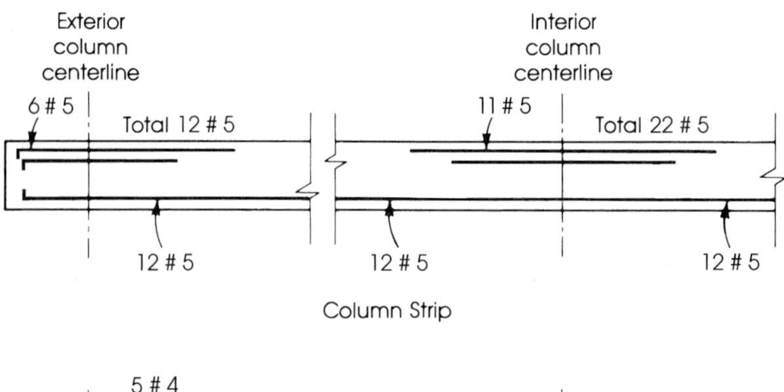

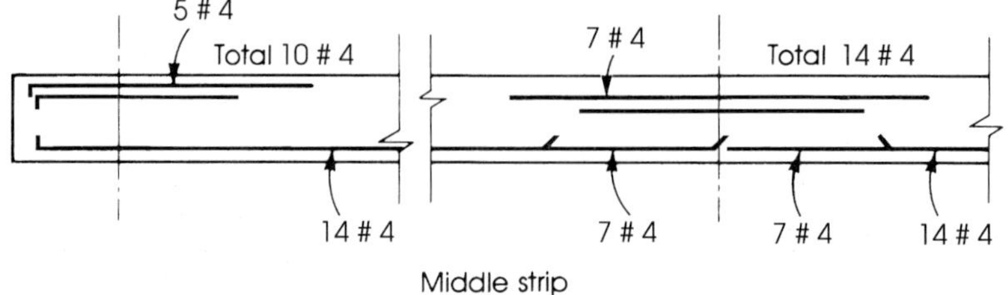

Figure 17.22 Reinforcement details (longitudinal direction), Example 17.4. For bar lengths, refer to Figure 17.16.

Example 17.5

Repeat Example 17.4 using the modified stiffness method. (Similar calculations are needed for the equivalent frame method, Section 17.12.)

Solution

1. Steps 1 through 4 will be the same as in Example 17.4.
2. Calculate the equivalent column stiffness K_{ec}:

$$\frac{1}{K_{ec}} = \frac{1}{\Sigma K_c} + \frac{1}{K_t}$$

It can be assumed that the part of the slab strip between exterior columns acts as a beam resisting torsion. The section of the slab-beam is 20 in. (width of the column) × 9.0 in. (thickness of the slab), as shown in Figure 17.21.

a. Determine the torsional stiffness K_t from equation (17.20):

$$C = \left(1 - 0.63\frac{x}{y}\right)\frac{x^3 y}{3}, \qquad x = 9 \text{ in.}, \quad \text{and} \quad y = 20 \text{ in.}$$

$$C = \left(1 - 0.63 \times \frac{9}{20}\right)\frac{9^3 \times 20}{3} = 3482 \text{ in.}^4$$

$$K_t = \frac{9E_c C}{l_2\left(1 - \frac{c_2}{l_2}\right)^3} = \frac{9E_c \times 3482}{(20 \times 12)\left(1 - \frac{20}{20 \times 12}\right)^3} = 170E_c$$

For the two adjacent slabs (on both sides of the column) acting as transverse beams,

$$K_t = 2 \times 170E_c = 340E_c$$

b. Calculate the column stiffness K_c; the column height $L_c = 12$ ft:

$$K_c = \frac{4E_c I_c}{L_c} = \frac{4E_c}{(12 \times 12)} \times \frac{(20)^4}{12} = 370.4E_c$$

For two columns above and below the floor slab,

$$K_c = 2 \times 370.4E_c = 740.8E_c$$

c. Calculate K_{ec}:

$$\frac{1}{K_{ec}} = \frac{1}{740.8E_c} + \frac{1}{340E_c}$$

To simplify the calculations, multiply by $1000E_c$:

$$\frac{1000E_c}{K_{ec}} = \frac{1000}{740.8} + \frac{1000}{340} = 4.29, \qquad K_{ec} = 233E_c$$

3. Calculate slab stiffness and the ratio α_{ec}:

$$K_s = \frac{4E_c I_s}{l_1}, \qquad h_s = 9 \text{ in.}, \qquad l_2 = 20 \text{ ft}, \qquad I_s = \frac{l_2 h_s^3}{12}$$

$$K_s = \frac{4E_c}{(24 \times 12)} \times \frac{(20 \times 12)(9.0)^3}{12} = 202.5E_c$$

$$\alpha_{ec} = \frac{K_{ec}}{\Sigma(K_s + K_b)} \tag{17.21}$$

$K_b = 0$ (no beams are provided); thus

$$\alpha_{ec} = \frac{233E_c}{202.5_c} = 1.15$$

Let

$$Q = \left(1 + \frac{1}{\alpha_{ec}}\right) = 1 + \frac{1}{1.15} = 1.87$$

4. Calculate the design moments in the long direction: $l_1 = 24$ ft. The distribution of moments in one panel is shown in Figure 17.18. The interior negative moment is

$$M_{ni} = \left[0.75 - \frac{0.10}{Q}\right]M_{ol} = \left(0.75 - \frac{0.10}{1.87}\right)(411.4) = -286.6\text{ K}\cdot\text{ft}$$

The positive moment is

$$M_p = \left[0.63 - \frac{0.28}{Q}\right]M_{ol}$$

$$= \left(0.63 - \frac{0.28}{1.87}\right)(411.4) = 197.6\text{ K}\cdot\text{ft}$$

The exterior negative moment is

$$M_{ne} = \frac{0.65}{Q}(M_{ol}) = \frac{0.65}{1.87}(411.4) = -143.0\text{ K}\cdot\text{ft}$$

5. Calculate the distribution of panel moments in the transverse direction to column and middle strips. The moments M_{ni}, M_p, and M_{ne} are distributed as follows (refer to Table 17.6):

a. The interior moment $M_{ni} = -286.6$ K · ft is distributed 75% for the column strip and 25% for the middle strip.

$$\text{Column strip} = 0.75(-286.6) = -215\text{ K}\cdot\text{ft}$$

$$\text{Middle strip} = 0.25(-286.6) = -71.6\text{ K}\cdot\text{ft}$$

b. The positive moment $M_p = 197.6$ K · ft is distributed 60% for the column strip and 40% for the middle strip.

$$\text{Column strip} = 0.6(197.6) = 118.5\text{ K}\cdot\text{ft}$$

$$\text{Middle strip} = 0.4(197.6) = 79.1\text{ K}\cdot\text{ft}$$

c. The exterior negative moment $M_{ne} = -143$ K · ft is distributed according to Table 17.5:

$$\beta_t = \frac{E_c C}{2E_c I_s} = \frac{C}{2I_s}$$

The concrete of slab and column are the same.

$$I_s = (20 \times 12)\frac{(9.0)^3}{12} = 14{,}580\text{ in.}^4$$

$$\beta_t = \frac{3482}{2 \times 14{,}580} = 0.119$$

$$\alpha_1 = \frac{E_{cb}I_b}{E_{cs}I_s} = 0, \qquad \alpha_1\frac{l_2}{l_1} = 0, \qquad \frac{l_2}{l_1} = 0.83$$

From Table 17.5 and by interpolation between $\beta_t = 0$ (percentage = 100%) and $\beta_t = 2.5$ (percentage = 75%), for $\beta_t = 0.119$, the percentage is 99%. The exterior negative moment in the column strip is $0.99(-143.0) = -142$ K · ft and in the middle strip, it is -1.0 K · ft. It is practical to consider that the column strip carries in this case 100% of $M_{ne} = -143$ K · ft.

6. Determine the reinforcement required in the long direction in a table form similar to Example 17.4. Results will vary slightly from those of Table 17.9.
7. Comparison of results between Examples 17.4 and 17.5 shows that the exterior moment in the column strip (-143 K · ft) is greater than that calculated in Example 17.4 (-107 K · ft) by about 34%, whereas the positive moment (+118.5) is reduced by about 8% (relative to +128.4). Other values are almost compatible.

Example 17.6

Design an interior panel of the two-way slab floor system shown in Figure 17.7. The floor consists of six panels in each direction, with a panel size of 24 by 20 ft. All panels are supported on 20- by 20-in. columns, 12 ft long. The slabs are supported by beams along the column lines with the cross sections shown in the figure. The service live load is to be taken as 80 psf, and the service dead load consists of 24 psf of floor finish in addition to the slab weight. Use $f'_c = 3$ Ksi, $f_y = 60$ Ksi, and the direct design method.

Solution

1. The limitations required by the ACI Code are met. Determine the minimum slab thickness using equations (17.1) and (17.2). The slab thickness has been already calculated in Example 17.2, and a 7.0-in. slab can be adopted. Generally, the slab thickness on a floor system is controlled by a corner panel, as the calculations of $h_{\min}$ for an exterior panel give greater slab thickness than for an interior panel.
2. Calculate factored loads:

$$w_D = 24 + \text{weight of slab} = 24 + \tfrac{7}{12} \times 150 = 111.5 \text{ psf}$$

$$w_u = 1.4(111.5) + 1.7(80) = 292 \text{ psf}$$

3. The shear stresses in the slab are not critical. The critical section is at a distance d from the face of the beam. For a 1-ft width:

$$V_u = w_u\left(10 - \frac{1}{2}\text{beam width} - d\right)$$

$$= 0.292\left(10 - \frac{16}{2 \times 12} - \frac{6}{12}\right) = 2.58 \text{ K}$$

$$\phi V_c = \phi(2\sqrt{f'_c})bd = \frac{0.85 \times 2 \times \sqrt{3000} \times 12 \times 6}{1000} = 7.1 \text{ K} > V_u$$

4. Calculate the total static moments in the long and short directions:

$$M_{ol} = \frac{w_u}{8} l_2(l_{n1})^2 = \frac{0.292}{8}(20)(22.33)^2 = 364.0 \text{ K} \cdot \text{ft}$$

$$M_{os} = \frac{w_u}{8} l_1(l_{n2})^2 = \frac{0.292}{8}(24)(18.33)^2 = 294.3 \text{ K} \cdot \text{ft}$$

5. Calculate the design moments in the long direction: $l_1 = 24$ ft.
   a. Distribution of moments in one panel:

$$\text{Negative moment } M_n = 0.65 M_{ol} = 0.65 \times 364 = -236.6 \text{ K} \cdot \text{ft}$$

$$\text{Positive moment } M_p = 0.35 M_{ol} = 0.35 \times 364 = 127.4 \text{ K} \cdot \text{ft}$$

b. Distributions of panel moments in the transverse direction to the beam, column, and middle strips are as follows:

$$\frac{l_2}{l_1} = \frac{20}{24} = 0.83, \qquad \alpha_1 = \alpha_s = \frac{EI_b}{EI_s} = 3.27 \qquad \text{(from Example 17.2)}$$

$$\alpha_1 \frac{l_2}{l_1} = 3.27 \times 0.83 = 2.71 > 1.0$$

c. Distribute the negative moment, M_n. The portion of the interior negative moment to be resisted by the column strip is obtained from Table 17.3 by interpolation and is equal to 80% (for $l_2/l_1 = 0.83$ and $\alpha_1(l_2/l_1) > 1.0$).

$$\text{Column strip} = 0.8M_n = 0.8 \times 236.6 = -189.3 \text{ K}\cdot\text{ft}$$

$$\text{Middle strip} = 0.2M_n = 0.2 \times 236.6 = -47.3 \text{ K}\cdot\text{ft}$$

Because $\alpha_1(l_2/l_1) > 1.0$, the ACI Code, Section 13.6.5, indicates that 85% of the moment in the column strip is assigned to the beam and the balance of 15% is assigned to the slab in the column strip.

$$\text{Beam} = 0.85 \times 189.3 = -160.9 \text{ K}\cdot\text{ft}$$

$$\text{Column strip} = 0.15 \times 189.3 = -28.4 \text{ K}\cdot\text{ft}$$

$$\text{Middle strip} = -47.3 \text{ K}\cdot\text{ft}$$

d. Distribute the positive moment M_p. The portion of the interior positive moment to be resisted by the column strip is obtained from Table 17.3 by interpolation and is equal to 80% (for $l_2/l_1 = 0.83$ and $\alpha_1(l_1/l_2) > 1.0$).

$$\text{Column strip} = 0.8M_p = 0.8 \times 127.4 = +101.9 \text{ K}\cdot\text{ft}$$

$$\text{Middle strip} = 0.2M_p = 0.2 \times 127.4 = +25.5 \text{ K}\cdot\text{ft}$$

Since $\alpha_1(l_2/l_1) > 1.0$, 85% of the moment in the column strip is assigned to the beam and the balance of 15% is assigned to the slab in the column strip:

$$\text{Beam} = 0.85 \times 101.9 = +86.6 \text{ K}\cdot\text{ft}$$

$$\text{Column strip} = 0.15 \times 101.9 = +15.3 \text{ K}\cdot\text{ft}$$

$$\text{Middle strip} = +25.5 \text{ K}\cdot\text{ft}$$

Moment details are shown in Figure 17.23.

6. Calculate the design moment in the short direction: span = 20 ft. The procedure is similar to Step 5.

$$\text{Negative moment } M_n = 0.65M_{os} = 0.65 \times 294.3 = -191.3 \text{ K}\cdot\text{ft}$$

$$\text{Positive moment } M_p = 0.35M_{os} = 0.35 \times 294.3 = +103.0 \text{ K}\cdot\text{ft}$$

Distribution of M_n and M_p to beam, column, and middle strips:

$$\frac{l_2}{l_1} = \frac{24}{20} = 1.2, \qquad \alpha_1 = \alpha_s = \frac{EI_b}{EI_s} = 2.72 \qquad \text{(from Example 17.2)}$$

$$\alpha_1 \frac{l_2}{l_1} = 2.72 \times 1.2 = 3.26 > 1.0$$

The percentages of the column strip negative and positive moments are obtained from Table 17.3 by interpolation. (For $l_2/l_1 = 1.2$ and $\alpha_1(l_2/l_1) > 1.0$, the percentage is 69%.)

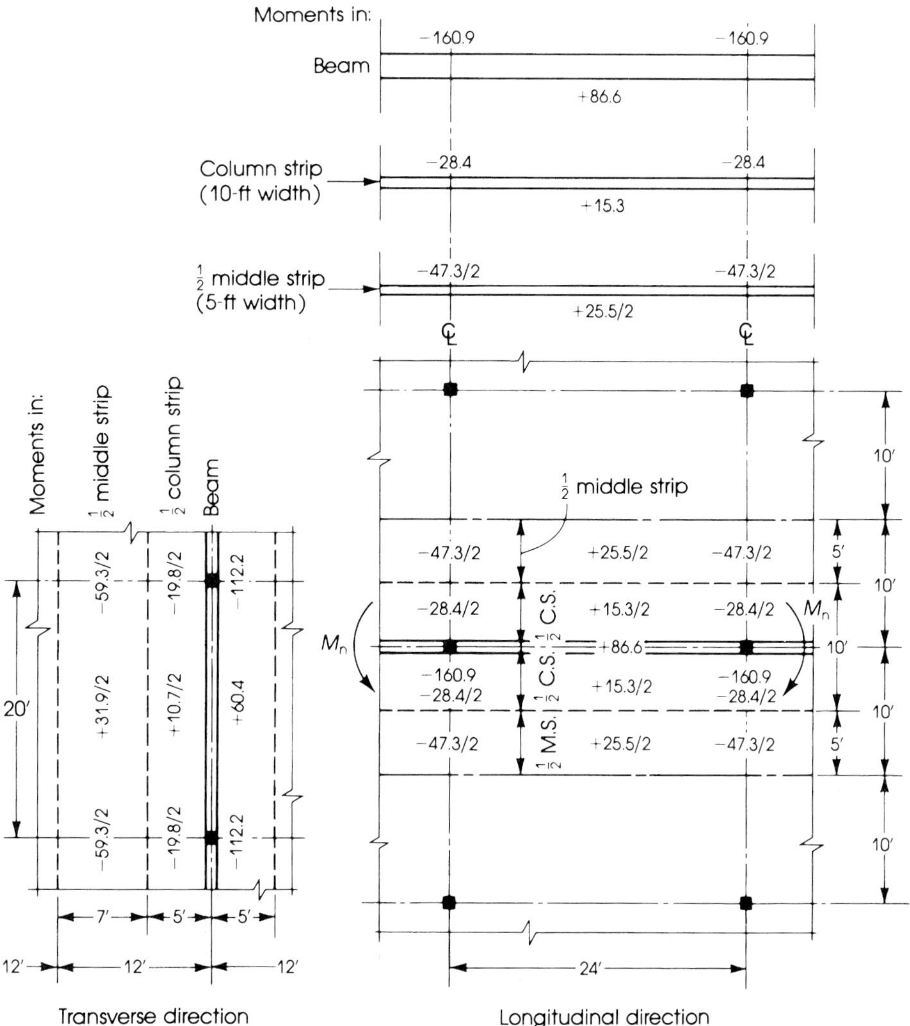

Figure 17.23 Example 17.6: Interior slab with beams. All moments are in K · ft.

$$\text{Column strip negative moment} = 0.69M_n = 0.69 \times 191.3 = -132 \text{ K} \cdot \text{ft}$$

$$\text{Middle strip negative moment} = 0.31M_n = 0.31 \times 191.3 = -59.3 \text{ K} \cdot \text{ft}$$

Since $\alpha_1(l_2/l_1) > 1.0$, 85% of −132 K · ft is assigned to the beam. Therefore,

$$\text{Beam negative moment} = 0.85 \times 132 = -112.2 \text{ K} \cdot \text{ft}$$

$$\text{Column strip negative moment} = 0.15 \times 132 = -19.8 \text{ K} \cdot \text{ft}$$

$$\text{Beam positive moment} = (0.85)(0.69 \times 103.0) = +60.4 \text{ K} \cdot \text{ft}$$

$$\text{Column strip positive moment} = (0.15)(0.69 \times 103.0) = +10.7 \text{ K} \cdot \text{ft}$$

$$\text{Middle strip positive moment} = (1 - 0.69)(103.0) = +31.9 \text{ K} \cdot \text{ft}$$

7. The steel reinforcement required and number of bars are shown in Table 17.10.

Table 17.10 Design of an Interior Two-Way Slab With Beams

	Long Direction			
	Column Strip		*Middle Strip*	
M_u (K·ft)	−28.4	+15.3	−47.3	+25.5
Width of strip (in.)	120	120	120	120
Effective depth (in.)	6.0	6.0	6.0	6.0
$R_u = \frac{M_u}{bd^2}$ (psi)	79	43	132	71
Steel ratio ρ	0.0016	Low	0.0026	0.0015
$A_s = \rho bd$ (in.^{2})	1.15	Low	1.87	1.08
Min. $A_s = 0.0018bh_s$ (in.^{2})	1.52	1.52	1.52	1.52
Selected bars	8 no. 4	8 no. 4	10 no. 4	8 no. 4
	Short Direction			
	Column Strip		*Middle Strip*	
M_u (K·ft)	−19.8	+10.7	−59.3	+31.9
Width of strip (in.)	120	120	168	168
Effective depth (in.)	5.5	5.5	5.5	5.5
$R_u = \frac{M_u}{bd^2}$ (psi)	65	35	196	105
Steel ratio ρ	Low	Low	0.0039	0.002
$A_s = \rho bd$ (in.^{2})	Low	Low	3.60	1.85
Min. $A_s = 0.0018bh_s$ (in.^{2})	1.52	1.52	2.10	2.10
Selected bars	8 no. 4	8 no. 4	18 no. 4	10 no. 4

Example 17.7

Using the direct design method, determine the negative and positive moments required for the design of the exterior panel (no. 2) of the two-way slab system with beams shown in Figure 17.7. Use the loads and the data given in Example 17.6.

Solution

1. Limitations required by the ACI Code are satisfied in this problem. Determine the minimum slab thickness h_s using equations (17.1) and (17.2) and the following steps:

Assume $h_s = 7.0$ in. The sections of the interior and exterior beams are shown in Figure 17.7. Note that the extension of the slab on each side of the beam $x = y = 15$ in.

2. a. The moments of inertia for the *interior* beams and slabs were calculated earlier in Example 17.2:

$$I_b \text{ (in both directions)} = 22{,}453 \text{ in.}^4$$
$$I_s \text{ (in the long direction)} = 6860 \text{ in.}^4$$
$$I_s \text{ (in the short direction)} = 8232 \text{ in.}^4$$

b. Calculate I_b and I_s for the *edge* beam and end slab.

$$I_b \text{ (edge beam)} = \left[\tfrac{27}{12}(7)^3 + (27 \times 7)(5.37)^2\right] + \left[\tfrac{12}{12}(15)^3 + (12 \times 15)(5.63)^2\right] = 15{,}302 \text{ in}^4$$

Calculate I_s for the end strip parallel to the edge beam, which has a width $= \frac{24}{2}$ ft $+ \frac{1}{2}$ column width $= 12 + \frac{10}{12} = 12.83$ ft.

$$I_s \text{ (end slab)} = \frac{(12.83 \times 12)}{12}(7)^3 = 4401 \text{ in.}^4$$

3. a. Calculate α $(\alpha = EI_b/EI_s)$:

$$\alpha_l \text{ (long direction)} = \frac{22{,}453}{6860} = 3.27$$

$$\alpha_s \text{ (short direction)} = \frac{22{,}453}{8232} = 2.72$$

$$\alpha \text{ (edge beam)} = \frac{15{,}302}{4401} = 3.48$$

$$\text{Average } \alpha = \alpha_m = \frac{3.27 + 2.72 \times 2 + 3.48}{4} = 3.05$$

b. β = ratio of long to short clear span

$$= \frac{22.33}{18.33} = 1.22$$

c. Calculate h_s:

$$\text{Min. } h_s = \frac{(22.33 \times 12)(0.8 + 0.005 \times 60)}{36 + (5 \times 1.22)[3.05 - 0.2]} = 5.52 \text{ in.}$$

But this value must not be less than

$$\text{Min. } h_s = \frac{294.756}{36 + 9(1.22)} = 6.30 \text{ in.} \qquad \text{(controls)}$$

Use $h_s = 7$ in. > 3.5 in. (minimum code limitations).

4. Calculate ultimate loads:

$$w_u = 292 \text{ psf} \qquad \text{(from Example 17.6)}$$

5. Calculate total static moments:

$$M_{ol} = 364.0 \text{ K}\cdot\text{ft}, \qquad M_{os} = 294.3 \text{ K}\cdot\text{ft} \qquad \text{(from previous example)}$$

6. Calculate the design moments in the short direction (span = 20 ft): Because the slab is continuous in this direction, the moments are the same as those calculated in Example 17.6 and shown in Figure 17.23 for an interior panel.

7. Calculate the moments in one panel using the coefficients given in Table 17.2 or Figure 17.14 (case 3):

$$\text{Interior negative moment } M_{ni} = 0.7M_o = 0.7 \times 364 = -254.8 \text{ K}\cdot\text{ft}$$

$$\text{Positive moment within span } M_p = 0.57M_o = 0.57 \times 364 = +207.5 \text{ K}\cdot\text{ft}$$

$$\text{Exterior negative moment } M_{ne} = 0.16M_o = 0.16 \times 364 = -58.2 \text{ K}\cdot\text{ft}$$

Note: If the modified stiffness method is used, then $C = 9528$, $K_t = 1520E_c$, $K_c = 370E_c$, $K_b = 312E_c$, $K_s = 95E_c$, $K_{ec} = 498E_c$, and $\alpha_{ec} = 1.22$. The interior negative moment becomes −253.3 K · ft (same as before). The positive moment becomes +173.9 K · ft (16% decrease) and the exterior moment becomes −128.6 K · ft (220% increase).

8. Distribute the panel moments to beam, column, and middle strips:

$$\frac{l_2}{l_1} = \frac{20}{24} = 0.83, \qquad \alpha_1 = a_s = 3.27$$

$$\alpha_1 \frac{l_2}{l_1} = 3.27 \times 0.83 = 2.71 > 1.0$$

$$\text{Calculate } C\text{:} \quad C = \Sigma\left(1 - 0.63\frac{x}{y}\right)\frac{x^3y}{3}$$

Divide the section of the edge beam into two rectangles in such a way as to obtain maximum C. Use for a beam section 12 by 22 in., $x_1 = 12$ in., $y_1 = 22$ in., and a slab section 7 by 15 in., $x_2 = 7$ in., and $y_2 = 15$ in.

$$C = \left(1 - 0.63 \times \frac{12}{22}\right)\left(\frac{12^3 \times 22}{3}\right) + \left(1 - 0.63 \times \frac{7}{15}\right)\left(\frac{7^3 \times 15}{3}\right)$$

$$= 9528 \text{ in.}^4$$

$$\beta_t = \frac{E_{cb}C}{2E_{cs}I_s} = \frac{9528}{2 \times 6860} = 0.69$$

a. Distribute the interior negative moment M_{ni}: Referring to Table 17.3 and by interpolation, the percentage of moment assigned to the column strip (for $l_2/l_1 = 0.83$ and $\alpha_1 l_2/l_1 > 1.0$) is 80%.

$$\text{Column strip} = 0.8 \times 254.8 = -203.8 \text{ K}\cdot\text{ft}$$

$$\text{Middle strip} = 0.2 \times 254.8 = -51.0 \text{ K}\cdot\text{ft}$$

Because $\alpha_1 l_2/l_1 > 1.0$, 85% of the moment in the column strip is assigned to the beam. Therefore,

$$\text{Beam} = 0.85 \times 203.8 = -173.3 \text{ K}\cdot\text{ft}$$

$$\text{Column strip} = 0.15 \times 203.8 = -30.6 \text{ K}\cdot\text{ft}$$

$$\text{Middle strip} = -51.0 \text{ K}\cdot\text{ft}$$

b. Distribute the positive moment M_p: Referring to Table 17.3 and by interpolation, the percentage of moment assigned to the column strip is 80% (85% of this value is assigned to the beam). Therefore,

$$\text{Beam} = (0.85)(0.8 \times 207.5) = +141.1 \text{ K}\cdot\text{ft}$$

$$\text{Column strip} = (0.15)(0.8 \times 207.5) = 24.9 \text{ K}\cdot\text{ft}$$

$$\text{Middle strip} = 0.2 \times 207.5 = +41.5 \text{ K}\cdot\text{ft}$$

c. Distribute the exterior negative moment M_{ne}: Referring to Table 17.5 and by interpolation, the percentage of moment assigned to the column strip (for $l_2/l_1 = 0.83, \alpha_1 l_2/l_1 > 1.0$, and $\beta_t = 0.69$) is 94%, and 85% of the moment is assigned to the beam. Therefore,

$$\text{Beam} = (0.85)(0.94 \times 58.2) = -46.5 \text{ K}\cdot\text{ft}$$

$$\text{Column strip} = (0.15)(0.94 \times 58.2) = -8.2 \text{ K}\cdot\text{ft}$$

$$\text{Middle strip} = 0.06 \times 58.2 = -3.5 \text{ K}\cdot\text{ft}$$

17.9 DESIGN MOMENTS IN COLUMNS

When the analysis of the equivalent frames is carried out by the direct design method, the moments in columns due to the unbalanced loads on adjacent panels are obtained from the following equation, which is specified by the ACI Code, Section 13.6.9:

$$M_u = 0.07\left[(w_d + 0.5w_l)l_2 l_n^2 - w_d' l_2' (l_n')^2\right] \tag{17.22a}$$

If the modified stiffness method using K_{ec} and α_{ec} is used, then the moment M_u is computed as follows:

$$M_u = \frac{0.08\left[(w_d + 0.5w_l)l_2 l_n^2 - w_d' l_2' (l_n')^2\right]}{(1 + 1/\alpha_{ec})} \tag{17.22b}$$

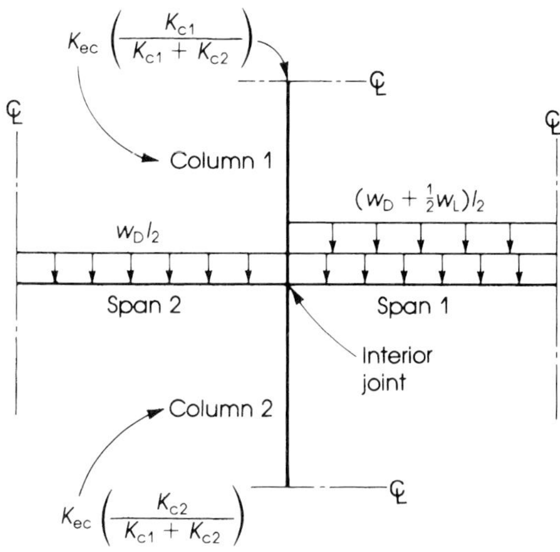

Figure 17.24 Interior column loading.

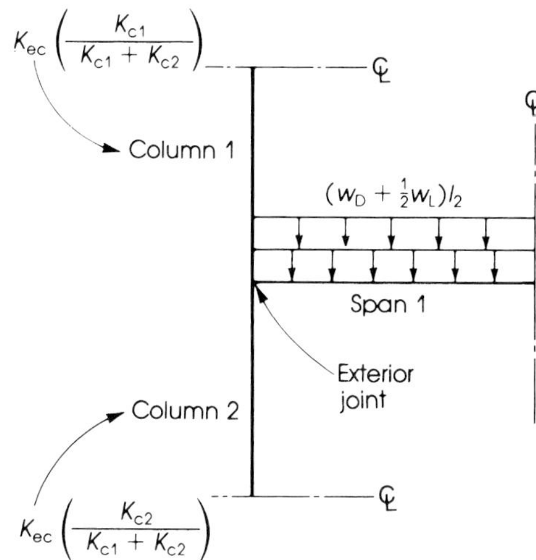

Figure 17.25 Exterior column loading.

where w_d and w_l = factored dead and live loads on the longer span
w'_d = factored dead load on the shorter span
l_n and l'_n = length of the longer and shorter spans, respectively

$$\alpha_{ec} = \frac{K_{ec}}{\Sigma(K_s + K_b)} \tag{17.21}$$

The moment in equations (17.22) should be distributed between the columns above and below the slab at the joint in proportion to their flexural stiffnesses (Figure 17.24). For equal spans $l_2 = l'_2$ and $l_n = l'_n$

$$M_u = 0.07(0.5w_l l_2 l_n^2) \tag{17.23a}$$

$$M_u = \frac{0.08(0.5w_l l_2 l_n^2)}{(1 + 1/\alpha_{ec})} \tag{17.23b}$$

The development of these equations is based on the assumption that half the live load acts on the longer span, whereas the dead load acts on both spans. Equation (17.22) can also be applied to an exterior column by assuming the shorter span length is zero (Figure 17.25).

17.10 TRANSFER OF UNBALANCED MOMENTS TO COLUMNS

17.10.1 Transfer of Moments

In the analysis of an equivalent frame in a building, moments develop at the slab-column joints due to lateral loads, such as wind, earthquakes, or unbalanced gravity loads, causing unequal moments in the slab on opposite sides of columns. A fraction of the unbalanced moment in the slabs must be transferred to the columns by flexure, and the balance must be transferred by vertical shear acting on the critical sections for punching shear. Approximately 60% of the moment transferred to both ends of the column at a joint is transferred by flexure, and the remaining 40% is transferred by eccentric shear (or torque) at the section located at $d/2$ from the face of the column [14], [15]. The ACI Code,

Section 13.5.3, states that the fraction of the unbalanced moment transferred by flexure M_f at a slab-column connection is determined as follows (ACI equation (13.1)):

$$M_f = \gamma_f M_u \tag{17.24}$$

$$\gamma_f = \frac{1}{1 + \left(\frac{2}{3}\sqrt{\frac{c_1 + d}{c_2 + d}}\right)} = \frac{1}{1 + (2/3)\sqrt{b_1/b_2}} \tag{17.25}$$

and the moment transferred by shear is

$$M_v = (1 - \gamma_f)M_u = M_u - M_f \tag{17.26}$$

where c_1 and c_2 are the lengths of the two sides of a rectangular or equivalent rectangular column, $b_1 = (c_1 + d)$, and $b_2 = (c_2 + d)$. When $c_1 = c_2$, $M_f = 0.6M_u$, and $M_v = 0.4M_u$.

17.10.2 Concentration of Reinforcement Over the Column

For a direct transfer of moment to the column, it is necessary to concentrate part of the steel reinforcement in the column strip within a specified width over the column. The part of the moment transferred by flexure, M_f, is considered acting through a slab width equal to the transverse column width c_2 plus $1.5h_s$ on each side of the column or to the width $(c_2 + 3h_s)$ (ACI Code, Section 13.5.3). Reinforcement can be concentrated over the column by closer spacing of bars or the use of additional reinforcement.

17.10.3 Shear Stresses Due to M_v

The shear stresses produced by the portion of the unbalanced moment, M_v, must be combined with the shear stresses produced by the shearing force, V_u, due to vertical loads. Both shear stresses are assumed acting around a periphery plane located at a distance $d/2$ from the face of the column [16], as shown in Figure 17.26. The equation for computing the shear stresses is

$$v_{1,2} = \frac{V_u}{A_c} \pm \frac{M_v C}{J_c} \tag{17.27}$$

where

A_c = area of critical section around the column

J_c = polar moment of inertia of the areas parallel to the applied moment in addition to that of the end area about the centroidal axis of the critical section

For an interior column,

$$A_c = 2d(x + y) \tag{17.28}$$

and

$$J_c = \frac{d}{2}\left(\frac{x^3}{3} + x^2 y\right) + \frac{xd^3}{6} \tag{17.29}$$

For an exterior column,

$$A_c = d(2x + y) \tag{17.30}$$

and

$$J_c = \frac{2dx^3}{3} - (2x + y)\,dx_1^2 + \frac{xd^3}{6} \tag{17.31}$$

where x, x_1, and y are as shown in Figure 17.26. The maximum shear stress, $v_1 = V_u/A_c + M_v C/J_c$, must be less than $\phi(4\sqrt{f'_c})$; otherwise, shear reinforcement should be provided.

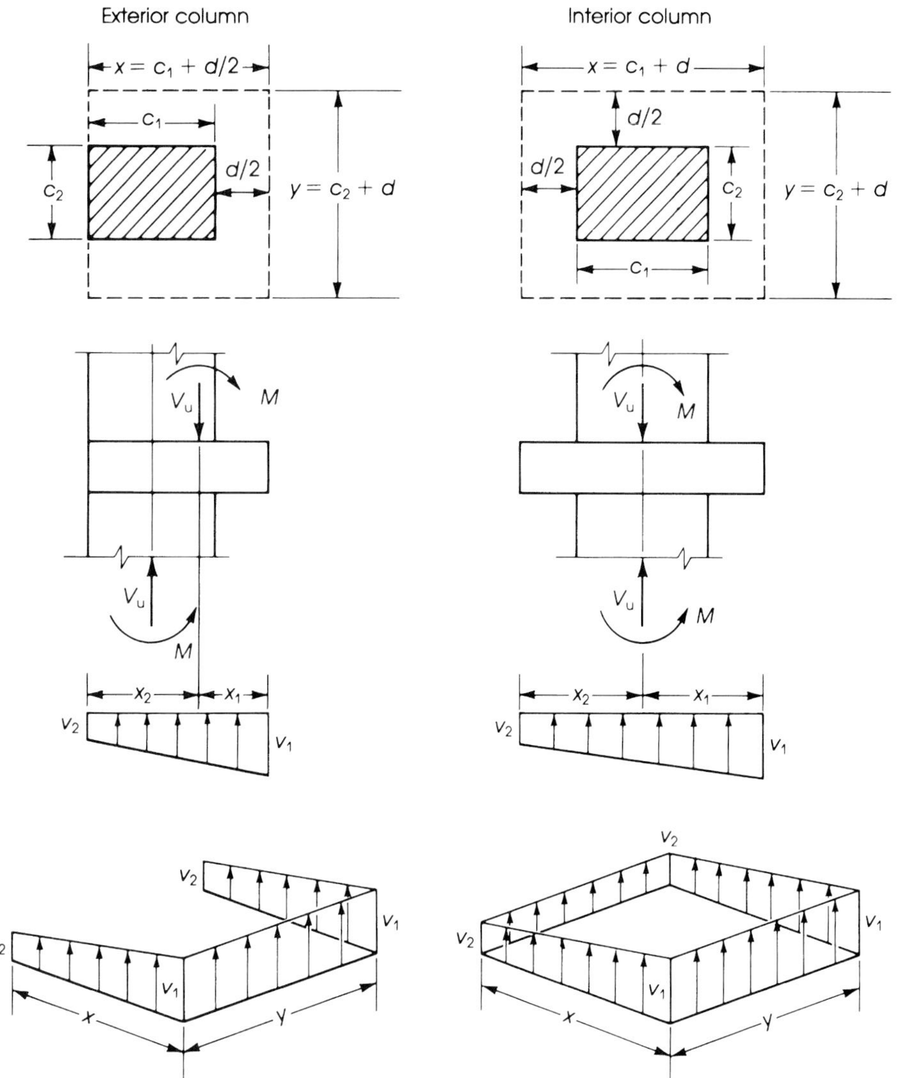

Figure 17.26 Shear stresses due to V_u and M.

Example 17.8

Determine the moments at the exterior and interior columns in the long direction of the flat plate in Example 17.4.

Solution

1. Find the exterior column moment. From Examples 17.4 and 17.5,

$$w_d = (136.5)(1.4) = 0.19 \text{ Ksf}$$

$$0.5w_l = 0.5 \times (1.7 \times 80) = 68 \text{ psf}$$

$$l_2 = l_2' = 20 \text{ ft}, \qquad l_n = l_n' = 22.33 \text{ ft}, \qquad \left(1 + \frac{1}{\alpha_{ec}}\right) = 1.87$$

The unbalanced moment to be transferred to the exterior column using Equation (17.22b) is

$$M_u = \frac{0.08}{1.87}\left[(0.19 + 0.068)(20)(22.33)^3 - 0\right] = 110 \text{ K} \cdot \text{ft}$$

If equation (17.22a) is used, $M_u = 180 \text{ K} \cdot \text{ft}$, which is a conservative value.

2. At an interior support, the slab stiffness on both sides of the column must be used to compute α_{ec}

$$\alpha_{ec} = \frac{K_{ec}}{\Sigma(K_s + K_b)} \tag{17.21}$$

From Example 17.5, $K_{ec} = 233E_c$, $K_s = 202.5E_c$, and $K_b = 0$. Therefore,

$$\alpha_{ec} = \frac{233E_c}{(2)202.5E_c} = 0.58$$

$$\left(1 + \frac{1}{\alpha_{ec}}\right) = 1 + \frac{1}{0.58} = 2.72$$

From equation (17.22b), the unbalanced moment at an interior support is

$$M_u = \frac{0.08}{2.72}\left[(0.19 + 0.068)(20)(22.33)^2 - 0.19(20)(22.33)^2\right] = 20 \text{ K} \cdot \text{ft}$$

If equation (17.22a) is used, $M_u = 47.5$ K · ft, which is a conservative value.

Example 17.9

For the flat plate in Example 17.4, calculate the shear stresses in the slab at the critical sections due to unbalanced moments and shearing forces at an interior and exterior column. Check the concentration of reinforcement and torsional requirements at the exterior column. Use $f'_c = 4$ Ksi and $f_y = 60$ Ksi.

Solution

1. The unbalanced moment at the interior support is $M_u = 20$ K · ft (Example 17.8), where $\gamma_f = 0.6$ (because $c_1 = c_2 = 20$ in.). The moment to be transferred by flexure is

$$M_f = \gamma_f M_u = 0.6 \times 20 = 12 \text{ K} \cdot \text{ft}$$

The moment to be transferred by shear is

$$M_v = 20 - 12 = 8 \text{ K} \cdot \text{ft}$$

Alternatively, moments calculated from equation (17.22a) may be used producing higher shear stresses.

Using $d = 7.9$ in. (Example 17.4),

$$V_u = 0.33\left[20 \times 24 - \left(\frac{27.9}{12}\right)^2\right] = 156.6 \text{ K}$$

From Figure 17.27,

$$A_c = 4(27.9)(7.9) = 882 \text{ in.}$$

$$J_c = \frac{d}{2}\left(\frac{x^3}{3} + x^2y\right) + \frac{xd^3}{6}$$

$$= \frac{7.9}{2}\left[\frac{(27.9)^3}{3} + (27.9)^2(27.9)\right] + \frac{27.9}{6}(7.9)^3 = 114{,}670 \text{ in.}^4$$

$$v_{max} = \frac{156{,}600}{882} + \frac{8(12{,}000)(27.9/2)}{114{,}670}$$

$$= 178 + 12 = 190 \text{ psi}$$

$$v_{min} = 178 - 12 = 166 \text{ psi}$$

Allowable v_c is $\phi 4\sqrt{f'_c} = 0.85 \times 4\sqrt{4000} = 215 \text{ psi} > 190$ psi.

2. For the exterior column, the unbalanced moment to be transferred by flexure M_f at a slab-column joint is equal to $\gamma_f M_u$, where $M_u = 110$ K · ft. Note that $c_1 = c_2 = 20$ in., $d = 7.9$ in. in the longitudinal direction, and $\gamma_f = 0.6$ for square columns.

$$M_f = 0.6(110) = 66 \text{ K} \cdot \text{ft}$$

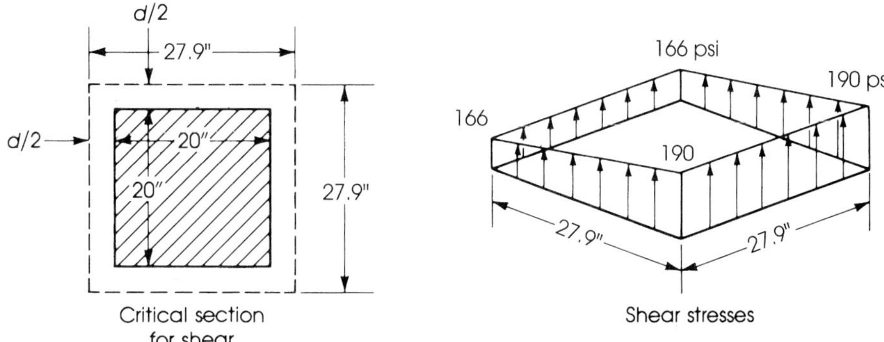

Figure 17.27 Example 17.9 shear stresses at interior column due to unbalanced moment.

The moment to be transferred by shear is

$$M_v = M_u - M_f = 110 - 66 = 44 \text{ K} \cdot \text{ft}$$

3. For transfer by shear at exterior column, the critical section is located at a distance $d/2$ from the face of the column (Figure 17.28).

$$w_u = 330 \text{ psf}$$

$$V_u = 0.33\left(20 \times 12.83 - \frac{23.95}{12} \times \frac{27.9}{12}\right) = 83.1 \text{ K}$$

Locate the centroid of the critical section by taking moments about AB:

$$2\left(23.95 \times \frac{23.95}{2}\right) = (2 \times 23.95 + 27.9)x_l$$

Therefore, $x_l = 7.6$ in. The area of the critical section A_c is $2(23.95 \times 7.9) + (27.9 \times 7.9) = 599 \text{ in.}^2$ Calculate $J_c = I_x + I_y$ for the two equal areas (7.9×23.95) with sides parallel to the direction of moment and the area (7.9×27.9) perpendicular to the direction of moment, all about the axis through CD.

$$J_c = I_y + I_x = \Sigma\left(\frac{bh^3}{12} + Ax^2\right)$$

$$= 2\left[7.9\frac{(23.95)^3}{12} + (7.9 \times 23.95)\left(\frac{23.95}{2} - 7.6\right)^2\right]$$

$$+ 2\left[\frac{23.95}{12}(7.9)^3\right] + \left[(27.9 \times 7.9)(7.6)^2\right] = 52{,}760 \text{ in.}^4$$

or by using equation (17.31) for an exterior column. Calculate the maximum and minimum nominal shear stresses using equation (17.27):

$$v_{\max} = \frac{V_u}{A_c} + \frac{M_v c}{J_c} = \frac{83{,}100}{599} + \frac{44(12{,}000)(7.6)}{52{,}760} = 214.7 \text{ psi}$$

$$v_{\min} = 62.7 \text{ psi}$$

Allowable $v_c = \phi 4\sqrt{f'_c} = 0.85 \times 4\sqrt{4000} = 215$ psi.

If shear stress is greater than the allowable v_c, increase the slab thickness or use shear reinforcement.

4. Check the concentration of reinforcement at the exterior column; that is, check that the flexural capacity of the section is adequate to transfer the negative moment into the exterior column. The critical area of the slab extends $1.5h_s$ on either side of the column, giving an area $(20 + 3 \times 9) = 47$ in. wide and 9 in. deep. The total moment in

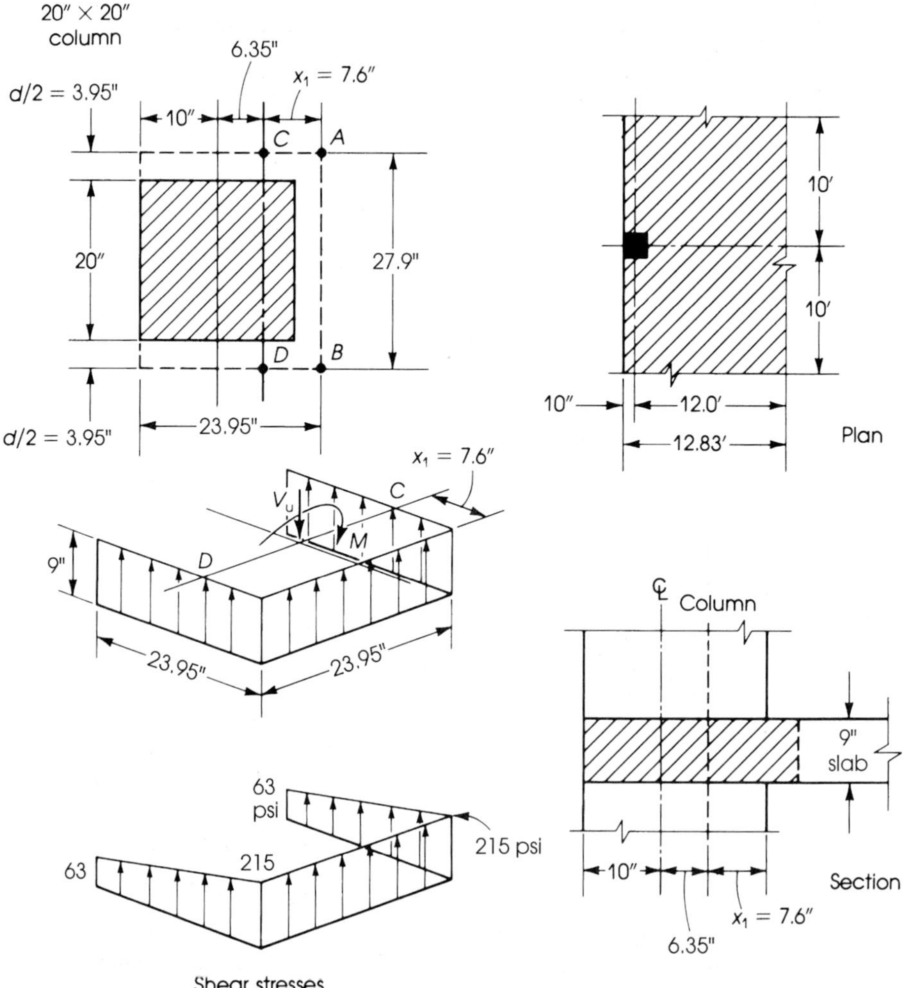

Figure 17.28 Example 17.9: Shear stresses at exterior column due to unbalanced moment.

the 120-in.-wide column strip is 107 K · ft, as calculated in Example 17.4 (Step 5). The moment in a width, $c_2 + 3h_s = 47$ in., is equal to $107(\frac{47}{120}) = 41.9$ K · ft.

If equal spacing in the column strip is used, then the additional reinforcement within the 47-in. width will be needed for a moment equal to $M_f - 41.9 = 66 - 41.9 = 24.1$ K · ft. The required $A_s = 0.73$ in.^{2} and four no. 4 bars $(A_s = 0.8 \text{ in.}^2)$ may be used. An alternative solution is to arrange the reinforcement within the column strip to increase the reinforcement within a width of 47 in. The amount of steel needed within this width should be enough to resist a moment of 0.6 times the negative moment in the column strip, or $0.6 \times 107 = 64.2$ K · ft.

$$A_s = \frac{M_u}{\phi f_y(d - a/2)}, \quad \text{assume } a = 1.0 \text{ in.}$$

$$A_s = \frac{64.2(12)}{0.9 \times 60(7.9 - 0.5)} = 1.93 \text{ in.}^2$$

$$\text{Check:} \quad a = \frac{A_s f_y}{0.85 f'_c b} = \frac{1.93 \times 60}{0.85 \times 4 \times 47} = 0.73 \text{ in.}$$

Use 10 no. 4 bars within a width 47 in. divided equally at both sides from the center of the column (Figure 17.29). Additional reinforcement of 4 no. 4 bars, as indicated before, provides a better solution.

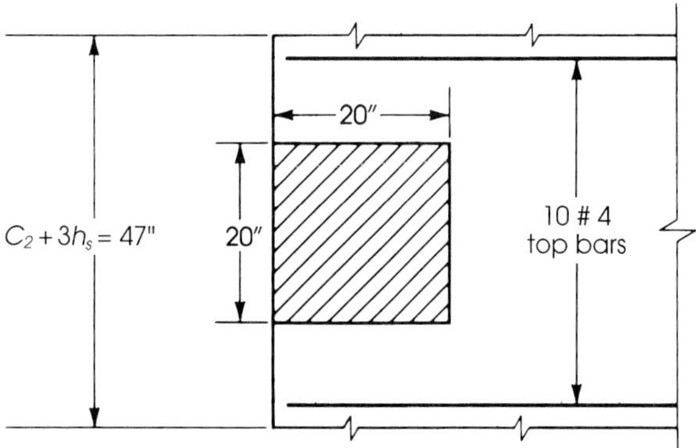

Figure 17.29 Concentration of reinforcement within exterior column strip, Example 17.9.

5. Torque on slab: The torque from both sides of the exterior column is equal to 40% of the column strip moment.

$$T_u = 0.4(107) = 42.8 \text{ K}\cdot\text{ft}$$

$$\text{Torque on each side: } \frac{42.8}{2} = 21.4 \text{ K}\cdot\text{ft} = 257 \text{ K}\cdot\text{in.}$$

A slab section of width equal to the column width will be assumed to resist the torsional stresses:

$$T_u = \tfrac{1}{3} v_{tu} \Sigma x^3 y$$

where $x = 9$ in. and $y = 20$ in. The critical section is at a distance d from the face of the column (Figure 17.30). Assuming that the torque varies in a parabolic curve to the center of the slab, then the torque at a distance d is

$$T_u = 257\left(\frac{140 - 7.9}{140}\right)^2 = 229 \text{ K}\cdot\text{in.}$$

For torsional strength of concrete,

$$T_c = 0.8\sqrt{f'_c}(x^2 y) = 0.8\sqrt{4000}\,(9)^2(20) = 82 \text{ K}\cdot\text{in.}$$

$$T_a = 30 \text{ K}\cdot\text{in.} < T_u$$

Torsional reinforcement is needed. The required closed stirrups and the additional longitudinal bars are determined as explained in Chapter 15. The final section is shown in Figure 17.30. It is advisable to provide an edge beam between the exterior columns to increase the torsional stiffness of the slab.

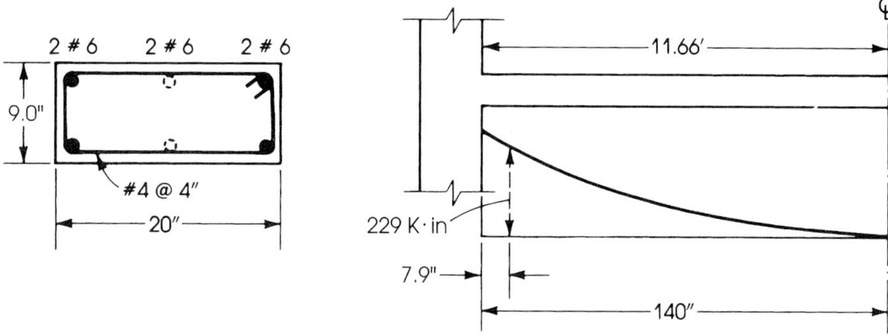

Figure 17.30 Reinforcement in edge of slab to resist torque, Example 17.9.

Example 17.10

Determine the shear reinforcement required for an interior flat plate panel considering the following: punching shear is $V_u = 195$ K, slab thickness = 9 in., $d = 7.5$ in., $f'_c = 3$ Ksi, $f_y = 60$ Ksi, and column size is 20×20 in.

Solution

1. Determine $\phi V_c = \phi 4\sqrt{f'_c}\, b_o d$ for two-way shear.

$$b_o = 4(20 + d) = 4(20 + 7.5) = 110 \text{ in.}$$

$$\phi V_c = 0.85(4)\sqrt{3000}\,(110)(7.5) = 153.6 \text{ K}$$

Because $V_u = 195 \text{ K} > \phi V_c$, shear reinforcement is required.

2. Maximum allowable ϕV_n using shear reinforcement is equal to $\phi 6\sqrt{f'_c}\, b_o d = 1.5(\phi V_c) = 1.5(153.6) = 230.5$ K. Because $\phi V_n > V_u$, shear reinforcement can be used.
3. Shear reinforcement may consist of reinforcing bars, structural steel sections such as I-beams, or special large-head studs welded to a steel strip. In this example, an inexpensive solution using normal shear reinforcement will be adopted. See Figure 17.9(f). Shear reinforcement must be provided on the four sides of the interior column (or three sides of an exterior column) for a distance of $d + a$. See Figure 17.31. The distance a is determined by equating $\phi V_c = V_u$ at section b_o, indicated by the dashed line, and assuming $\phi V_c = \phi 2\sqrt{f'_c}\, b_o d$.

$$b_o = 4(c + \sqrt{2}a) = 4(20 + \sqrt{2}a)$$

$$0.85(2)\sqrt{3000}\,(4)(20 + \sqrt{2}a)(7.5) = 195{,}000 \text{ lb}$$

Here, $a = 35.2$ in., and $(a + d) = 35.2 + 7.5 = 42.7$ in., so use 43 in.

4. Calculate shear reinforcement:

$$\phi V_s = (V_u - \phi V_c) = 195 - 153.6 = 41.4 \text{ K}, \qquad V_s = 48.7 \text{ K}$$

$$V_s \text{ (for one face of critical section)} = \frac{V_s}{4} = \frac{48.7}{4} = 12.2 \text{ K}$$

Use no. 3 U-stirrups, $A_v = 0.22$ in.^{2} (for 2 legs). The spacing is $S = A_v f_y d / V_s = 0.22(60)(7.5)/12.2 = 8.1$ in. Maximum spacing is $d/2 = 7.5/2 = 3.75$ in.; let $s = 3.5$ in.

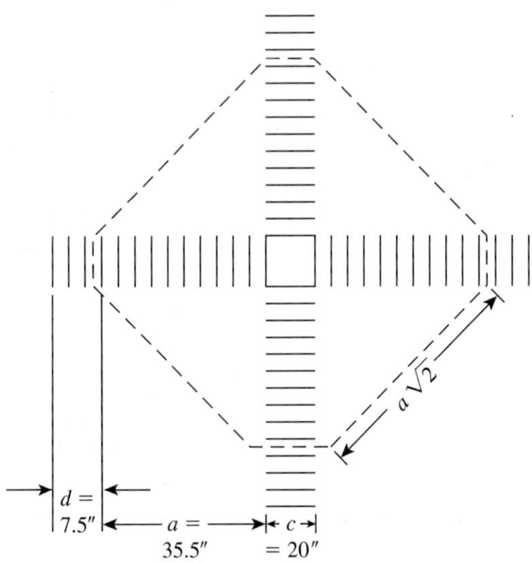

Figure 17.31 Shear reinforcement no. 3 at 3.5 in.

5. Distribution of stirrups: The number of stirrups per one side of column is $43/3.5 = 12.3$, or 13 stirrups. Total distance is $13(3.5) = 45.5$ in. (Figure 17.31).

Example 17.11 (Flat-slab floor system)

Using the direct design method, design a typical 24- × 20-ft interior flat-slab panel with drop panels only (Figure 17.32). All panels are supported by 20- × 20-in. columns, 12 ft long. The slab carries a uniform-service live load of 80 psf and a service dead load of 24 psf, excluding self-weight. Use $f'_c = 4$ Ksi, and $f_y = 60$ Ksi. (The solution is similar to Example 17.3.)

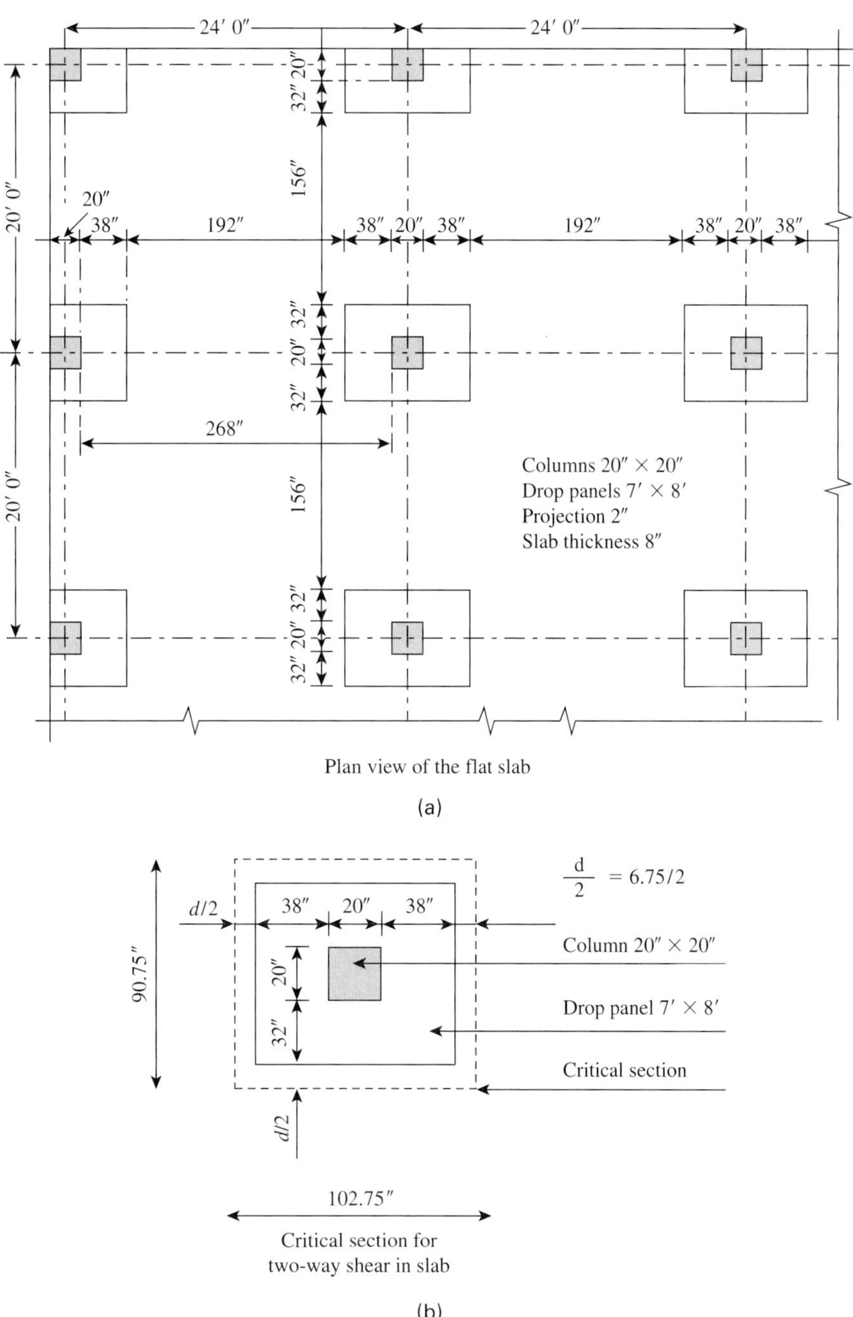

Figure 17.32 Flat slab with drop panel, Example 17.11.

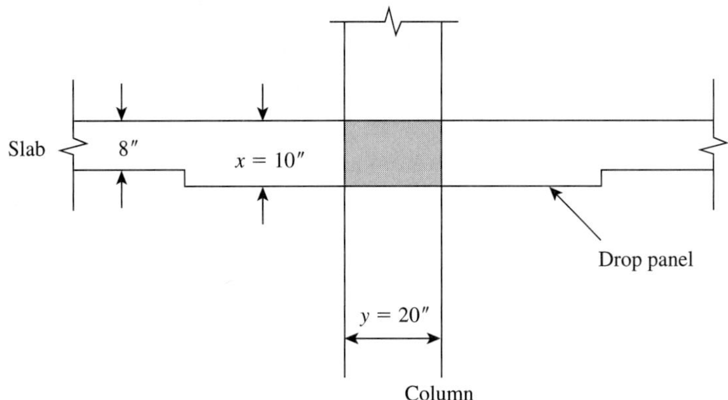

Torsional part
of the drop panel

(c)

Figure 17.32 *(continued)*

Solution

1. Determine slab and drop panel thicknesses using Table 17.1.
   a. The clear span is $24 - \frac{20}{12} = 22.33$ ft. For an exterior panel, minimum $h = l_n/33 = 8.12$ in., whereas for an interior panel, minimum $h = l_n/36 = 7.44$ in. Use a slab thickness of 8 in. The projection below the slab is $h/4 = \frac{8}{4} = 2.0$ in.; thus, the drop panel thickness is 10 in.
   b. Extend the drop panels $L/6 = \frac{24}{6} = 4$ ft in each direction from the centerline of support in the long direction and $\frac{20}{6} = 3.33$ ft, or 3.5 ft, in the short direction. Thus, the total size of one drop panel is 8 × 7 ft (Figure 17.32).
2. Calculate ultimate loads:

$$\text{Slab load} = 24 + \frac{8(150)}{12} = 124 \text{ psf}$$

$$W_u = 1.4(124) + 1.7(80) = 310 \text{ psf}$$

$$\text{Drop panel load} = 24 + \frac{10(150)}{12} = 149 \text{ psf}$$

$$W_u = 1.4(149) + 1.7(80) = 345 \text{ psf}$$

   Because the drop panel length is $L/3$ in each direction, then the average W_u is $(\frac{2}{3})(310) + (\frac{1}{3})(345) = 322$ psf.
3. Check two-way shear (at distance $d/2$ from the face of column):
   a. In the drop panel: $d = 10 - 0.75 - 0.5 = 8.75$ in.

$$b_o = 4(20 + 8.75) = 115 \text{ in.}$$

$$V_u = 0.322\left[24 \times 20 - \left(\frac{28.75}{12}\right)^2\right] = 152.7 \text{ K}$$

$$\phi V_c = \phi 4\sqrt{f'_c}\,b_o d = 0.85(4)\sqrt{4000}\,(115)(8.75) = 214.4 \text{ K} > V_u$$

   b. In the slab: $d = 8 - 0.75 - 0.5 = 6.75$ in. and b_o is measured at 6.75/2 in. (in slab) beyond the drop panel.

$$b_o = 2(8 \times 12 + 6.75) + 2(7 \times 12 + 6.75) = 387 \text{ in.}$$

$$V_u = 0.322[24 \times 20 - (102.75)(90.75)/144] = 133.7 \text{ K}$$

$$\phi V_c = 0.85(4)\sqrt{4000}\,(387)(6.75) = 561.7 \text{ K} > V_u$$

c. One-way shear is not critical.

4. Calculate the total static moments in the long and short directions:

$$M_{ol} = \frac{0.322(20)(22.33)^2}{8} = 401.4 \text{ K}\cdot\text{ft}$$

$$M_{os} = \frac{0.322(24)(18.33)^2}{8} = 324.6 \text{ K}\cdot\text{ft}$$

The width of column strip in each direction is $\frac{20}{2} = 10$ ft, whereas the width of the middle strip is 10 ft in the long direction and 14 ft in the short direction.

5. Calculations of moments and steel reinforcement are shown in Table 17.11. Use an average $d = 10 - 1.5 = 8.5$ in. in the column strip and $d = 8 - 1.5 = 6.5$ in. in the middle strip.

Bars are chosen for adequate distribution in both the column and middle strip. Reinforcement details are similar to those in flat-plate examples.

Table 17.11 Design of an Interior Flat-Slab Floor System

	Long Direction			
$M_{ol} = 401.4$ K·ft	*Column Strip*		*Middle Strip*	
M factor	$-0.49M_o$	$0.21M_o$	$-0.16M_o$	$0.14M_o$
M_u (K·ft)	−196.7	+84.3	−64.2	+56.2
Width of strip (in.), b	120	120	120	120
Effective depth (in.), d	8.5	6.5	6.5	6.5
$R_u = \frac{M_u}{bd^2}$ (psi)	272	199	152	133
Steel ratio ρ (%)	0.53	0.38	0.29	0.25
$A_s = \rho bd$ (in.^{2})	5.40	2.95	2.26	1.95
Min. $A_s = 0.0018bh_s$ (in.^{2})	2.16	2.16	1.73	1.73
Selected bars	24 no. 5	15 no. 4	12 no. 4	12 no. 4
	Short Direction			
$M_{os} = 324.6$ K·ft	*Column Strip*		*Middle Strip*	
M factor	$-0.49M_o$	$0.21M_o$	$-0.16M_o$	$0.14M_o$
M_u (K·ft)	−159	+68.2	−52.0	+45.4
Width of strip (in.), b	120	120	168	168
Effective depth (in.), d	8.5	6.5	6.5	6.5
$R_u = \frac{M_u}{bd^2}$ (psi)	220	161	88	77
Steel ratio ρ (%)	0.43	0.30	0.17	0.15
$A_s = \rho bd$ (in.^{2})	4.39	2.34	1.86	1.64
Min. $A_s = 0.0018bh_s$ (in.^{2})	2.16	2.16	2.42	2.42
Selected bars	16 no. 5	12 no. 4	12 no. 4	12 no. 4

17.11 WAFFLE SLABS

A two-way waffle slab system consists of concrete ribs that normally intersect at right angles. These slabs might be constructed without beams, in which case a solid column head is made over the column to prevent any punching due to shear. Wide beams can also be used on the column centerlines for uniform depth construction. Square metal or fiberglass pans are commonly used to form these joists. A thin slab of 3 to 5 in. is cast with these joists to form the waffle slab.

Each panel is divided into a column and a middle strip. The column strip includes all joists that frame into the solid head; the middle strip is located between consecutive column strips. Straight or bent bars could be used as a reinforcement in a waffle slab. The design of a two-way waffle slab is similar to that of flat slabs by considering the solid head as a drop panel. To prevent any excess in the diagonal tension in the head, a sufficient size of column must be used or a shear cap must be provided.

In the design of a waffle slab, the top slabs with each rib form a T-section, with considerable depth relative to flat plates. Consequently, long spans carrying heavy loads may be designed with great savings in concrete. Waffle slabs also provide an attractive ceiling, which is achieved by leaving the rib pattern or by integrating lighting fixtures.

The standard pans that are commonly used in waffle slabs can be one of the following two types:

1. 30- × 30-in. square pans with a 3-in. top slab, from which 6-in.-wide ribs at 36 in. (3 ft) on centers are formed. These are available in standard depths of 8 to 20 in. in 2-in. increments. Refer to Example 17.12 and Figure 17.33.
2. 19- × 19-in. square pans with a 3-in. top slab, from which 5-in.-wide ribs at 24 in. (2 ft) on centers are formed. These are available in standard depths of 4, 6, 8, 10, and 12 in. Other information about pans is shown in Table 17.12 [21]. Other types, ranging from 19- × 19-in. pans to 40- × 40-in. pans, are available in the construction industry.

Table 17.12 Gross Section Properties [21]

Top Slab (in.)	*Rib Depth (in.)*	*Volume (cf/pan)*	*Gross Area, (in.²)*	Y_{cg} *(in.)*	I_g *(in.⁴)*
For the Joists (30-in. × 30-in. pans)					
3	8	3.85	161.3	3.28	1,393
3	10	4.78	176.3	3.95	2,307
3	12	5.53	192	4.66	3,541
3	14	6.54	208.3	5.42	5,135
3	16	7.44	223.3	6.20	7,127
3	20	9.16	261.3	7.83	12,469
4.5	8	3.85	215.3	3.77	2,058
4.5	10	4.78	230.3	4.35	3,227
4.5	12	5.53	246.0	4.97	4,783
4.5	14	6.54	262.3	5.66	6,773
4.5	16	7.44	279.3	6.36	9,238
4.5	20	9.16	315.3	7.86	15,768
For the Joists (19-in. × 19-in. pans)					
3	6	1.09	105	2.886	598
3	8	1.41	117.4	3.564	1,098
3	10	1.9	130.4	4.303	1,824
3	12	2.14	144	5.083	2,807
4.5	6	1.09	141	3.457	957
4.5	8	1.41	153	4.051	1,618
4.5	10	1.9	166.4	4.709	2,550
4.5	12	2.14	180	5.417	3,794

Example 17.12 (Waffle Slab)

Design a waffle floor system that consists of square panels without beams considering the following data (Figure 17.33):

Span, center to center of columns = 33 ft

Width of rib = 6 in., spaced at 36 in. on centers

Depth of rib = 14 in. and slab thickness = 3 in.

Column size = 20 × 20 in.

Dead load (excluding self-weight) = 20 psf

Live load = 100 psf, $f'_c = 4$ Ksi, and $f_y = 60$ Ksi

Solution

1. Determine minimum slab thickness using Table 17.1: Minimum $h = l_n/30$, $l_n = 33 - \frac{20}{12} = 31.33$ ft, $h = 31.33(12)/30 = 12.5$ in. for exterior panels, and $h = l_n/33 = 11.4$ in. for interior panels. Equations (17.1) and (17.2) may be used. Assume the total depth is 17 in. consisting of 3-in. slab thickness and 14-in. rib depth.
2. Calculate loads on the waffle slab:
   - **a.** Ultimate load of solid head part = 1.4(150)(17/12) = 298 psf.
   - **b.** Voided volume of 14-in. rib = 6.54 ft^3 on (3 × 3)-ft^2 area. Total weight of 9-ft^2 area is $1.4(150)(9 \times \frac{17}{12} - 6.54) = 1304$ lb. Weight per square foot is $\frac{1304}{9} = 145$ psf.

Waffle slab (looking upward).

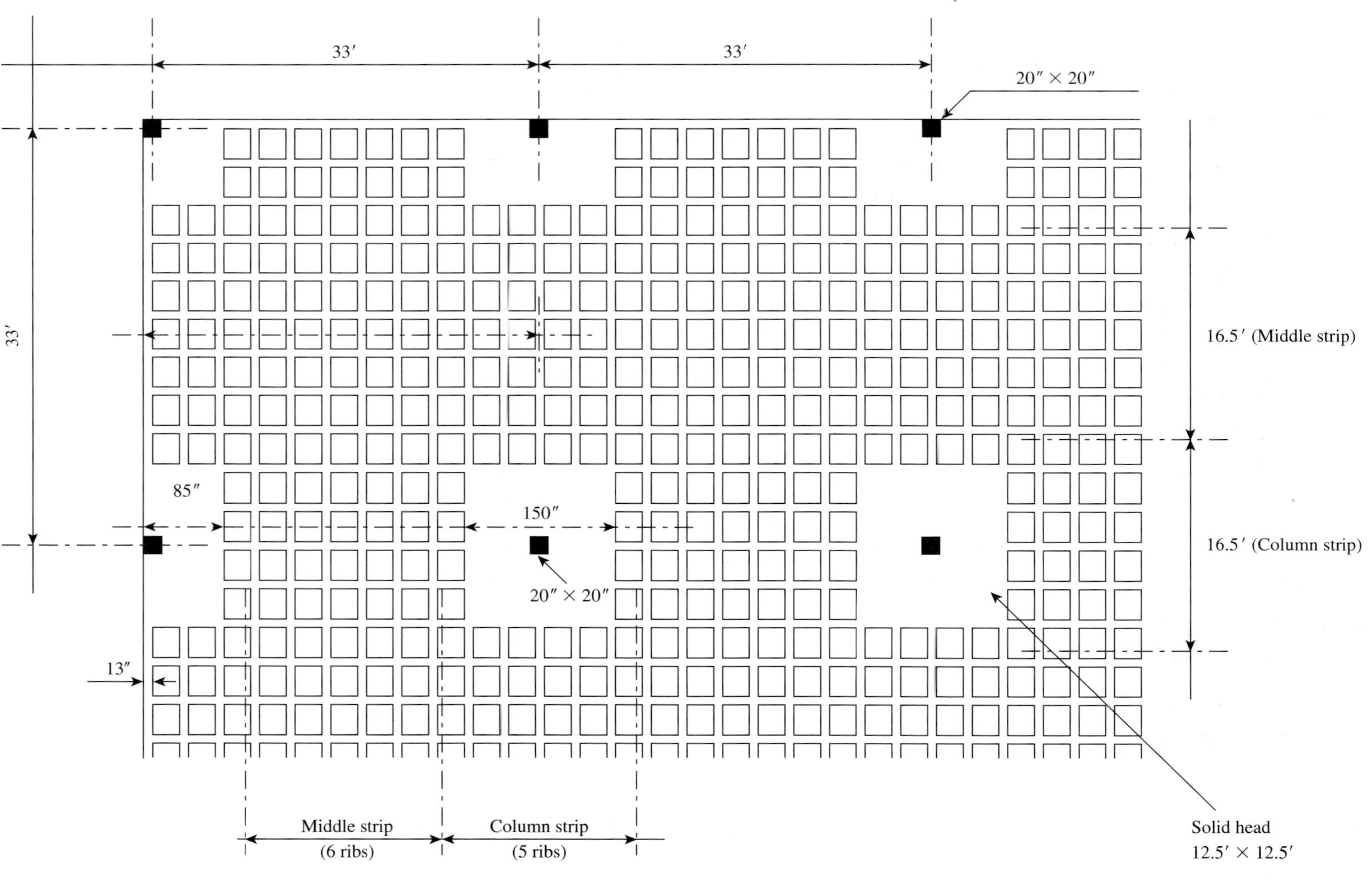

(a)

Figure 17.33 (a) Plan of the waffle slab, (b) cross section, (c) pan and rib dimensions, and (d) spacing and dimensions of solid heads (Example 17.12).

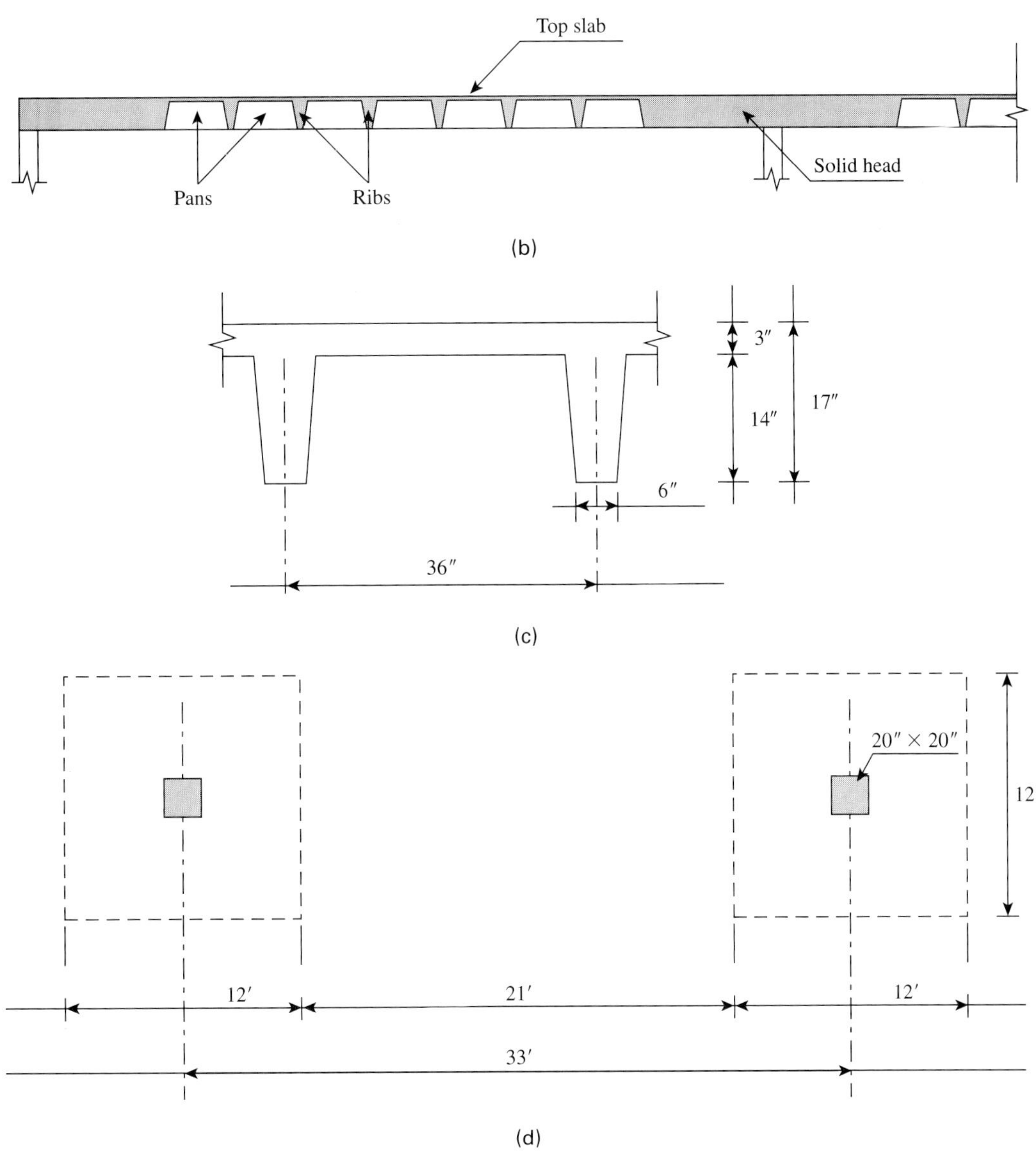

Figure 17.33 *(continued)*

c. Ultimate additional dead plus live load is 1.4(20) + 1.7(100) = 198 psf. Uniform w_u (at solid head) = 298 + 198 ≈ 500 psf. Uniform w_u (at ribbed area) = 145 + 198 ≈ 345 psf.

d. Loads on one panel (refer to Figure 17.34): At the solid head, $W = 0.5(12) + 0.345(21) = 13.22$ K/ft. At the ribbed area, $W = 0.345(33) = 11.39$ K/ft.

3. Calculate shear and total static moment:

$$V_u \text{ (at the face of column)} = 13.22(5.17) + \frac{(11.39)(21)}{2} = 188 \text{ K}$$

$$M_o \text{ (at midspan)} = 188(15.67) - 13.22(5.17)(13.09) - \frac{11.39(10.5)^2}{2} = 1424 \text{ K}\cdot\text{ft}$$

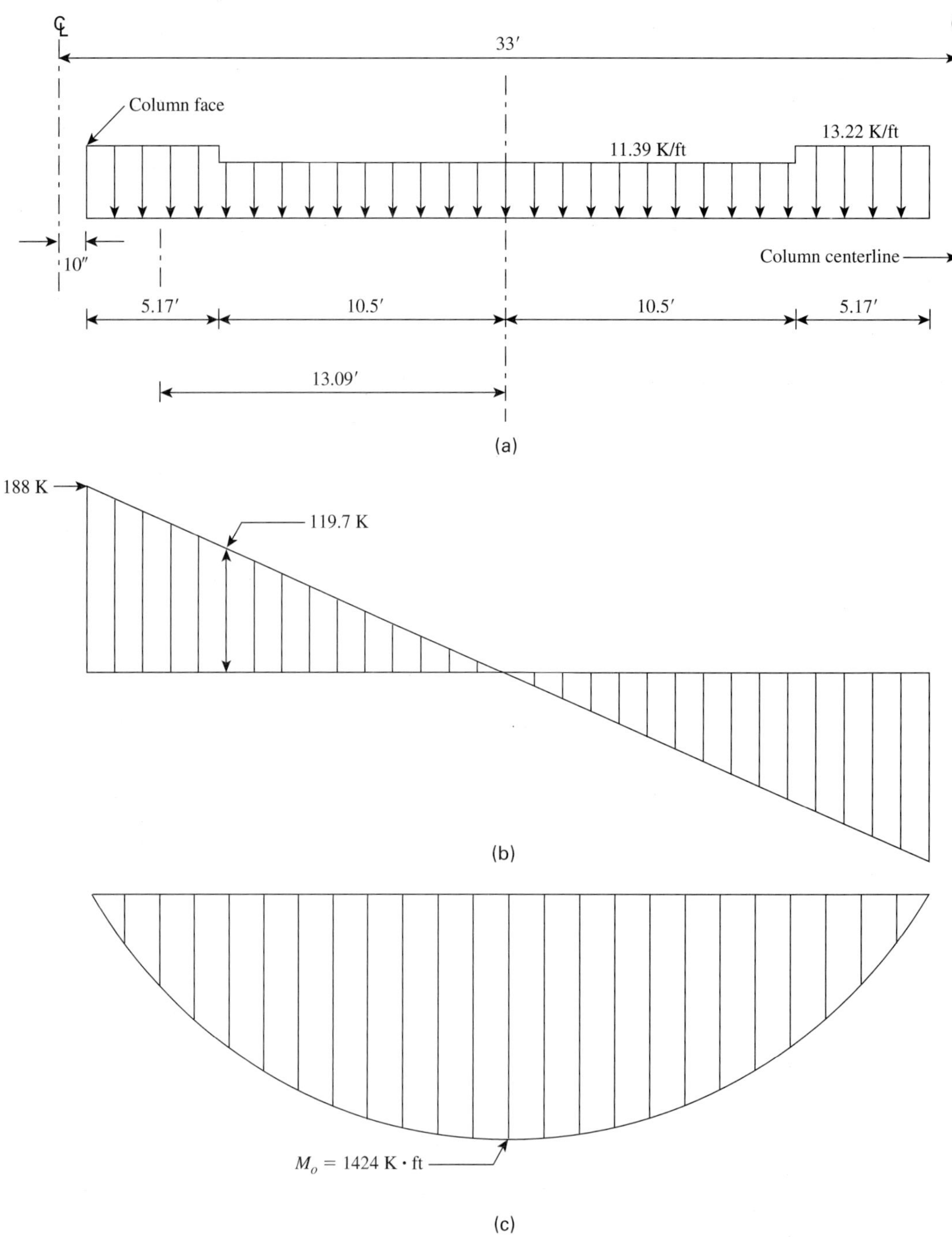

Figure 17.34 Load, shear, and moment diagrams: (a) load distribution on the span, (b) shear force diagram, and (c) bending moment diagram.

4. Check punching shear (refer to Figure 17.35):

a. In solid head at $d/2$ from column face: $h = 17$ in., $d = 17 - 1.25 = 15.75$ in., c (column) $= 20$ in., $b_o = 4(20 + 15.75) = 143$ in., $V_u = 11.39(21\text{ ft}) + 13.22(12\text{ ft}) - 0.5(37.75/12)^2 = 393.4$ K, and $\phi V_c = \phi 4\sqrt{f'_c}\, b_o d = 0.85(4)(\sqrt{4000})(143)(15.75) = 484.3\text{ K} > V_u$.

b. In the slab at distance $d/2$ from the edge of the solid head, slab thickness is 3 in.; let $d = 2.5$ in. Then

$$b_o = 4(150 + 2.5) = 610 \text{ in.}$$

$$V_u = 11.39(21) + 13.22(12) - 0.5\left(\frac{152.5}{12}\right)^2 = 317.4 \text{ K}$$

$$\phi V_c = 0.85(4)(\sqrt{4000})(610)(2.5) = 328 \text{ K} > V_u$$

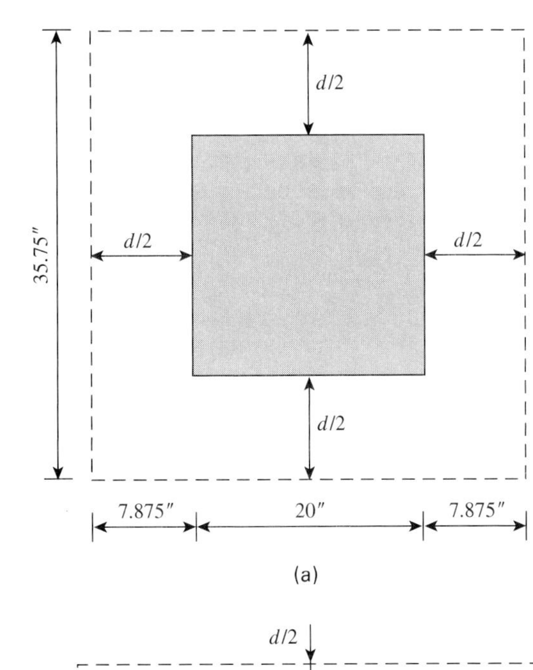

(a)

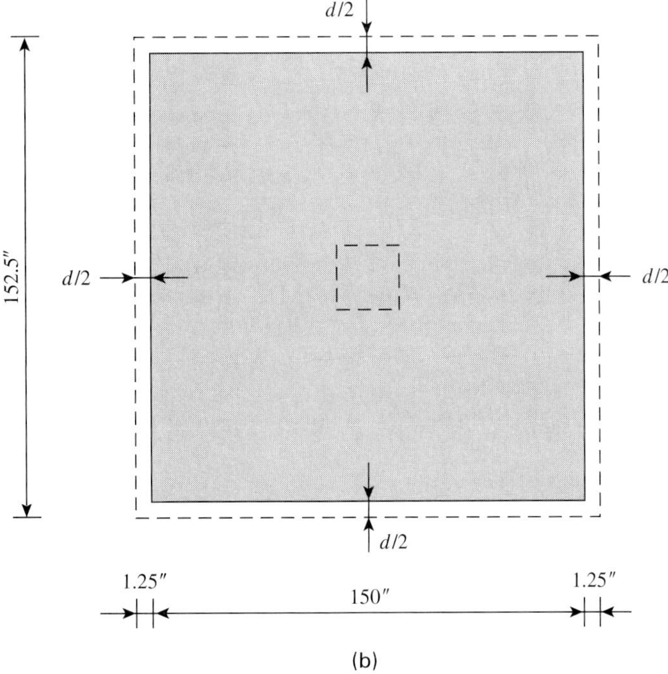

(b)

Figure 17.35 Punching shear locations: (a) punching shear in column head and (b) punching shear in slab.

5. Design moments and reinforcement

a. Exterior panel: $M_o = 1424\ \text{K}\cdot\text{ft}$

$$\text{Exterior negative moment} = 0.26M_o = -370\ \text{K}\cdot\text{ft}$$

$$\text{Positive moment} = 0.52M_o = +740\ \text{K}\cdot\text{ft}$$

$$\text{Interior negative moment} = 0.7M_o = -997\ \text{K}\cdot\text{ft}$$

b. Interior panel: $M_o = 1424\ \text{K}\cdot\text{ft}$

$$\text{Negative moment} = 0.65(1424) = -925.6\ \text{K}\cdot\text{ft}$$

$$\text{Positive moment} = 0.35(1424) = 498.4\ \text{K}\cdot\text{ft}$$

Design details are shown in Table 17.13 and Figure 17.36.

6. Calculate the unbalanced moments in columns and check shear for V_u and M_v, as in Examples 17.8 and 17.9.

Table 17.13 Design of an Exterior and an Interior Waffle Slab (5 ribs in column strip and 6 ribs in middle strips)

	Column Strip			Middle Strip	
Exterior Panel	***Exterior* (−*M*)**	**+*M***	***Interior* −*M***	**−*M***	**+*M***
Moment factor, (%)	100	60	75	25	40
M_u (K·ft)	370	444	748	249	296
Strip width, b (in.)	150	198	150	36 (6 ribs)	198
d (in.)	15.75	15.75	15.75	15.75	15.75
$R_u = \dfrac{M_u}{bd^2}$ (psi)	120	108	241	334	72
Steel ratio, ρ (%)	0.226	0.204	0.465	0.657	0.135
$A_s = \rho bd$ (in.^{2})	5.33	6.36	11.0	3.73	4.2
Min. $A_s = 0.0018bh$	2.6	1.22	4.6	1.1	1.47
Bars selected	14 no. 6	2 no. 8/rib	26 no. 6	10 no. 6	2 no. 7/rib

	Column Strip			Middle Strip	
Interior Panel	***Exterior* (−*M*)**	**+*M***	***Interior* −*M***	**−*M***	**+*M***
Moment factor, (%)	—	60	75	25	40
M_u (K·ft)	—	299	694.2	231.4	200
Strip width, b (in.)	—	198	150	36 (6 ribs)	198
d (in.)	—	15.75	15.75	15.75	15.75
$R_u = \dfrac{M_u}{bd^2}$ (psi)	—	73	224	311	49
Steel ratio, ρ (%)	—	0.137	0.431	0.61	0.091
$A_s = \rho bd$ (in.^{2})	—	4.27	10.18	3.45	2.84
Min. $A_s = 0.0018bh$	—	1.22/rib	4.6	1.1	1.47
Bars selected	—	2 no. 7/rib	24 no. 6	10 no. 6	2 no. 6/rib

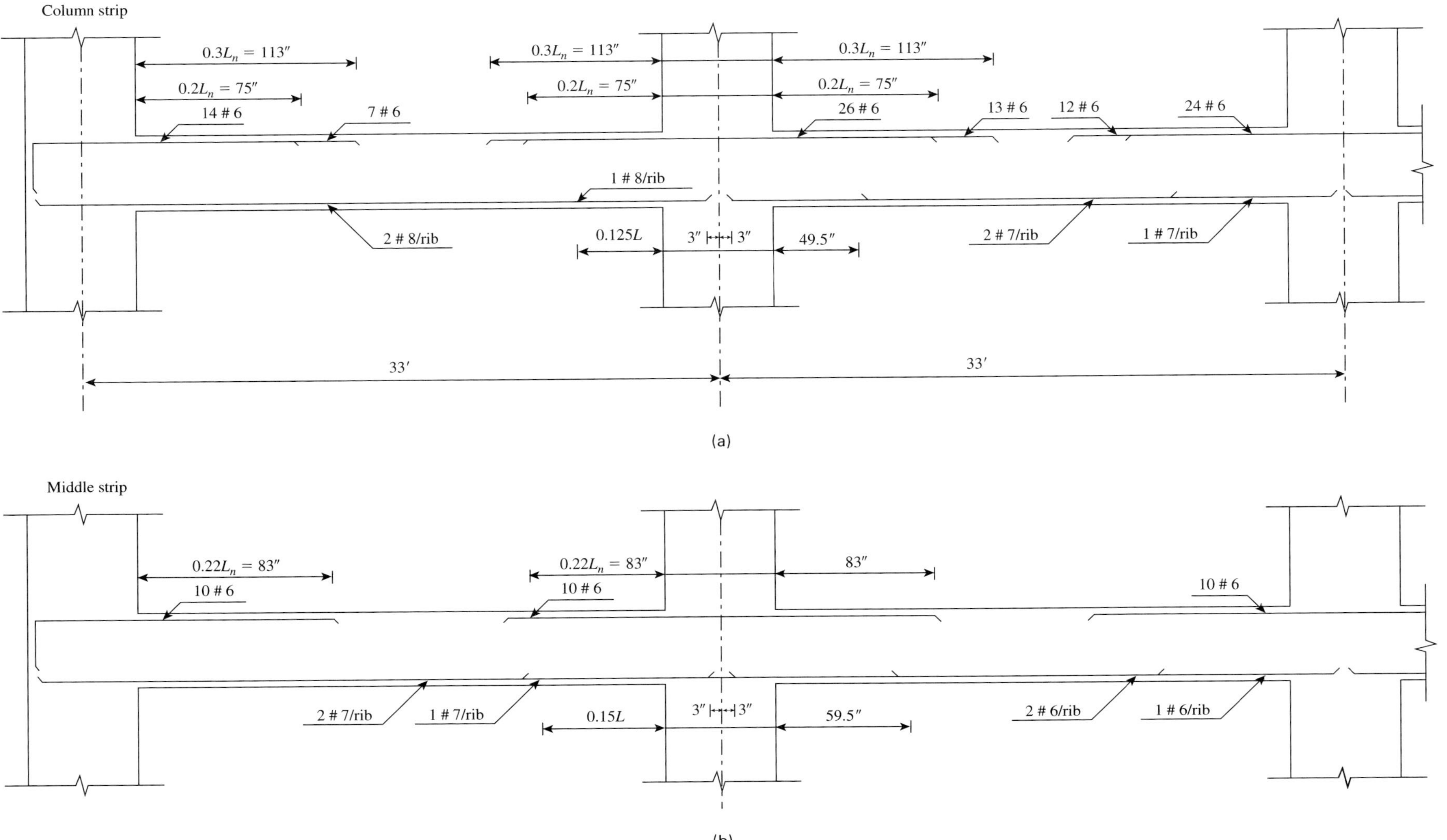

Figure 17.36 Reinforcement details of the waffle slab (Example 17.12).

17.12 EQUIVALENT FRAME METHOD

When two-way floor systems do not satisfy the limitations of the direct design method, the design moments must be computed by the equivalent frame method. In the latter method, the building is divided into equivalent frames in two directions and then analyzed elastically for all conditions of loadings. The difference between the direct design and equivalent frame methods lies in the way by which the longitudinal moments along the spans of the equivalent rigid frame are determined. The design requirements can be explained as follows.

1. Description of the equivalent frame: An equivalent frame is a two-dimensional building frame obtained by cutting the three-dimensional building along lines midway between columns (Figure 17.4). The resulting equivalent frames are considered separately in the longitudinal and transverse directions of the building. For vertical loads, each floor is analyzed separately, with the far ends of the upper and lower columns assumed to be fixed. The slab-beam may be assumed to be fixed at any support two panels away from the support considered, because the vertical loads contribute very little to the moment at that support. For lateral loads, the equivalent frame consists of all the floors and extends for the full height of the building, because the forces at each floor are a function of the lateral forces on all floors above the considered level. Analysis of frames can also be made using computer programs.
2. Load assumptions: When the ratio of the service live load to the service dead load is less than or equal to 0.75, the structural analysis of the frame can be made with the factored dead and live loads acting on all spans instead of a pattern loading. When the ratio of the service live load to the service dead load is greater than 0.75, pattern loading must be used, considering the following conditions:
   a. Only 75% of the full-factored live load may be used for the pattern-loading analysis.
   b. The maximum negative bending moment in the slab at the support is obtained by loading only the two adjacent spans.
   c. The maximum positive moment near a midspan is obtained by loading only alternate spans.
   d. The design moments must not be less than those occurring with a full-factored live load on all panels (ACI Code, Section 13.7.6).
   e. The critical negative moments are considered to be acting at the face of a rectangular column or at the face of the equivalent square column having the same area for nonrectangular sections.
3. Slab-beam moment of inertia: The ACI Code specifies that the variation in moment of inertia along the longitudinal axes of the columns and slab-beams must be taken into account in the analysis of frames. The critical region is located between the centerline of the column and the face of the column, bracket, or capital. This region may be considered as a thickened section of the floor slab. To account for the large depth of the column and its reduced effective width in contact with the slab-beam, the ACI Code, Section 13.7.3.3, specifies that the moment of inertia of the slab-beam between the center of the column and the face of the support is to be assumed equal to that of the slab-beam at the face of the column divided by the quantity $(1 - c_2/l_2)^2$, where c_2 is the column width in the transverse direction and l_2 is the width of the slab-beam. The area of the gross section can be used to calculate the moment of inertia of the slab-beam.
4. Column moment of inertia: The ACI Code, Section 13.7.4, states that the moment of inertia of the column is to be assumed infinite from the top of the slab to the bottom of the column capital or slab-beams (Figure 17.37).

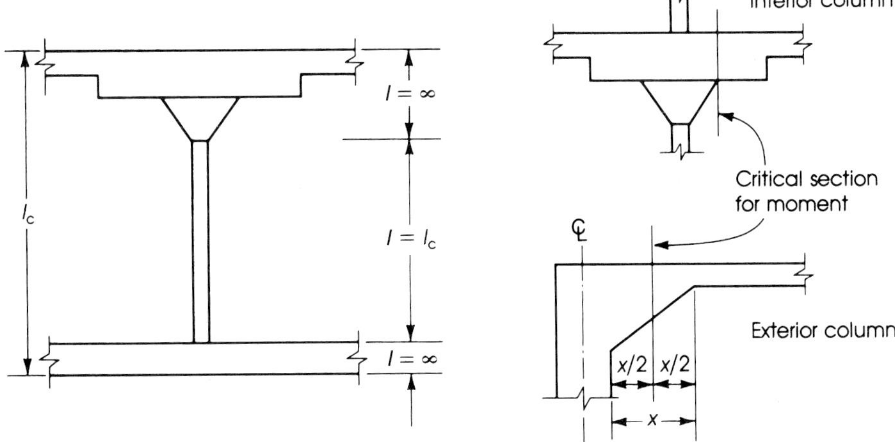

Figure 17.37 Critical sections for column moment, equivalent frame method.

5. Column stiffness K_{ec} is defined by

$$\frac{1}{K_{ec}} = \frac{1}{\Sigma K_c} + \frac{1}{K_t} \tag{17.19}$$

where ΣK_c is the sum of the stiffness of the upper and lower columns at their ends,

$$K_t = \Sigma \frac{9E_{es}C}{l_2(1 - c_2/l_2)^2} \tag{17.17}$$

$$C = \Sigma\left(1 - 0.63\frac{x}{y}\right)\left(\frac{x^3y}{3}\right) \tag{17.20}$$

6. Column moments: In frame analysis, moments determined for the equivalent columns at the upper end of the column below the slab and at the lower end of the column above the slab must be used in the design of a column.

7. Negative moments at the supports: The ACI Code, Section 13.7.7, states that for an interior column, the factored negative moment is to be taken at the face of the column or capital but at a distance not greater than $0.175l_1$ from the center of the column. For an exterior column, the factored negative moment is to be taken at a section located at half the distance between the face of the column and the edge of the support. Circular section columns must be treated as square columns with the same area.

8. Sum of moments: A two-way slab floor system that satisfied the limitations of the direct design method can also be analyzed by the equivalent frame method. To ensure that both methods will produce similar results, the ACI Code, Section 13.7.7, states that the computed moments determined by the equivalent frame method may be reduced in such proportion that the numerical sum of the positive and average negative moments used in the design must not exceed the total statical moment, M_o.

Example 17.13

By the equivalent frame method, analyze a typical interior frame of the flat-plate floor system given in Example 17.3 in the longitudinal direction only. The floor system consists of four panels in each direction with a panel size of 25 by 20 ft. All panels are supported by a 20- by 20-in. columns, 12 ft long. The service live load is 60 psf and the service dead load is 124 psf (including the weight of the slab). Use f'_c = 3 Ksi and f_y = 60 Ksi. Edge beams are not used. Refer to Figure 17.38.

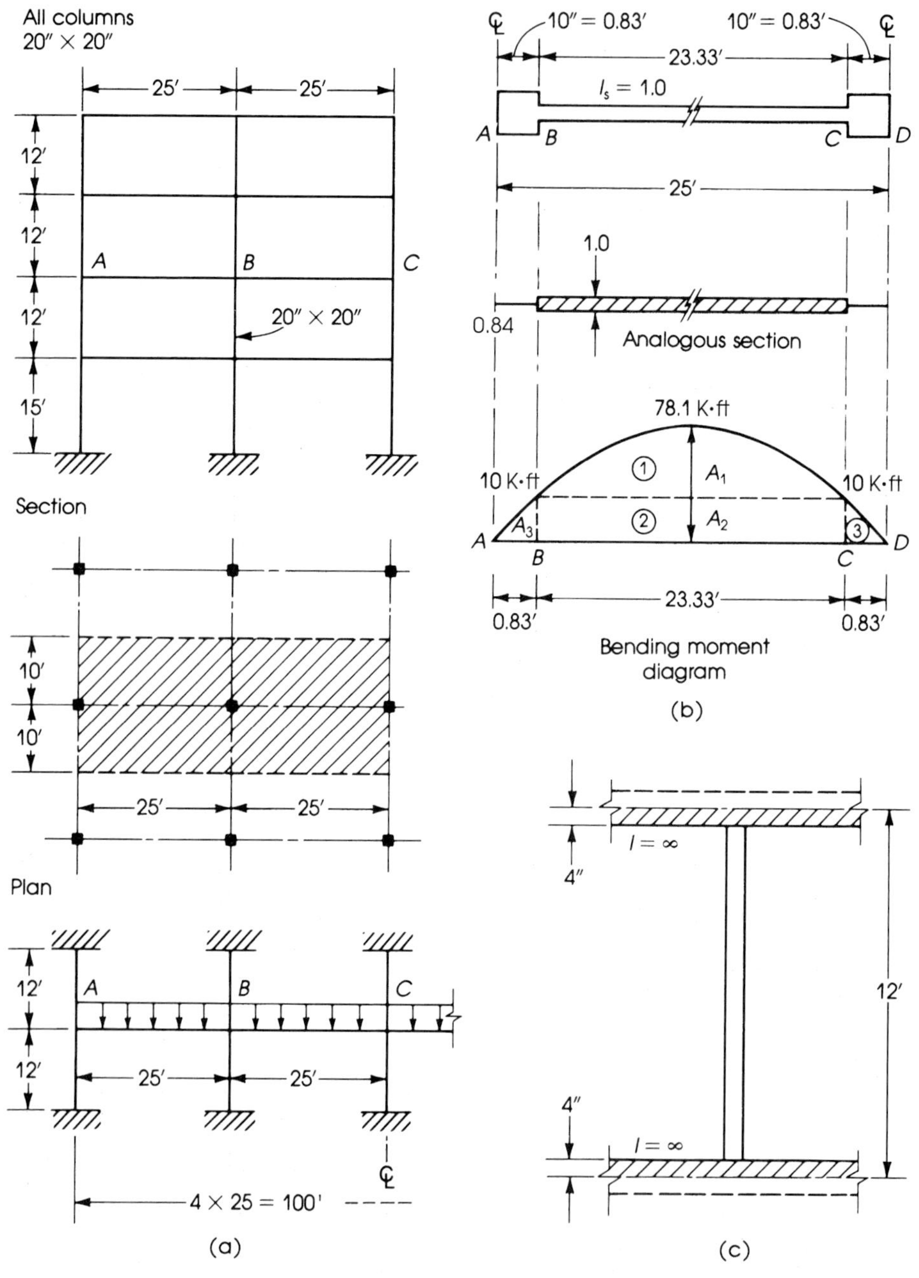

Figure 17.38 Example 17.13.

Solution

1. A slab thickness of 8.0 in. is chosen, as explained in Example 17.3.
2. Ultimate load is $w_u = 1.4 \times 124 + 1.7 \times 60 = 276$ psf. The ratio of service live load to service dead load is $60/124 = 0.48 < 0.75$; therefore, the frame can be analyzed with the full factored load w_u acting on all spans instead of pattern loading.
3. Determine the slab stiffness, K_s:

$$K_s = k\frac{EI_s}{l_s}$$

where k is the stiffness factor and

$$I_s = \frac{l_2 h_s^3}{12} = \frac{(20 \times 12)}{12}(8)^3 = 10{,}240 \text{ in.}^4$$

The stiffness factor can be determined by the column analogy method described in books on structural analysis. Considering the moment of inertia for the slab I_s to be 1.0 as a reference, the moment of inertia between the column centerline and the face of the column is

$$\frac{1.0}{(1 - c_2/l_2)^2} = \frac{1.0}{(1 - 20/(20 \times 12))^2} = 1.19$$

The width of the analogous column varies with $1/I$, as shown in Figure 17.38(b): $(1/1.19) = 0.84$.

$$\text{Slab stiffness factor } k = l_1\left(\frac{1}{A_a} + \frac{Mc}{I_a}\right) \tag{17.32}$$

where A_a = area of the analogous column section
I_a = moment of inertia of analogous column
M = moment due to a unit load at the extreme fiber of the analogous column located at the center of the slab

$$M = 1.0 \times \frac{l_1}{2}$$

$$A_a = 23.33 + 2 \times (0.83 \text{ ft})(0.84) = 23.33 + 1.40 = 24.72$$

$I_a = I$ (for slab portion of 23.33) $+ I$ (of end portion) about the centerline

$$I_a = \frac{(23.33)^3}{12} + 1.4\left(12.5 - \frac{0.83}{2}\right)^2 = 1263$$

neglecting the moment of inertia of the short end segments about their own centroid.

$$\text{Stiffness factor } k = 25\left[\frac{1}{24.72} + \frac{1.0 \times 12.5(12.5)}{1263}\right]$$

$$= 1.01 + 3.09 = 4.1$$

$$\text{Carryover factor} = \frac{3.09 - 1.01}{4.1} = 0.509$$

Therefore, slab stiffness is

$$K_s = \frac{4.1E \times 10{,}240}{(25 \times 12)} = 140E$$

4. Determine the column stiffness, K_c:

$$K_c = k'\left(\frac{EI_c}{l_c}\right) \times 2$$

for columns above and below the slab.

$$k' = \text{column stiffness factor}$$

$$l_c = 12 \text{ ft}, \qquad I_c = \frac{(20)^4}{12} = 13{,}333 \text{ in.}^4$$

The stiffness factor, k', can be determined as follows:

$$k' = l_c\left(\frac{1}{A_a} + \frac{Mc}{I_a}\right)$$

For the column, $c = l_c/2$ and $M = 1.0(l_c/2) = l_c/2$.

$$A_a = l_c - h_s = 12 - \frac{8}{12} = 11.33$$

$$I_a = \frac{(l_c - h_s)^3}{12} = \frac{(11.33)^3}{12} = 121.2$$

$$k' = 12\left[\frac{1}{11.33} + \frac{(1 \times \frac{12}{2})(\frac{12}{2})}{121.2}\right] = 4.62$$

$$K_c = 4.62E \times \frac{13{,}333}{12 \times 12} \times 2 = 856E$$

In a flat-plate floor system, the column stiffness, K_c, can be calculated directly as follows:

$$\frac{K_c}{E_c} = \frac{I_c}{(l_c - h_s)} + \frac{3I_c l_c^2}{(l_c - h_s)^3} \tag{17.33}$$

5. Calculate the torsional stiffness, K_t, of the slab at the side of the column:

$$K_t = \frac{\Sigma 9R_{cs}C}{l_2\left(1 - \frac{c_2}{l_2}\right)} \quad \text{and} \quad C = \Sigma\left(1 - 0.63\frac{x}{y}\right)\frac{x^3 y}{3}$$

In this example, $x = 8.0$ in. (slab thickness) and $y = 20$ in. (column width). See Figure 17.17.

$$c = \left(1 - 0.63 \times \frac{8}{20}\right)\left(\frac{(8)^3 \times 20}{3}\right) = 2553 \text{ in.}^4$$

$$K_t = \frac{9E_{cs} \times 2553}{(20 \times 12)(1 - 20/(20 \times 12))^3} = 124E_c$$

For two adjacent slabs, $K_t = 2 \times 124E_c = 248E_c$.

6. Calculate the equivalent column stiffness, K_{ec}:

$$\frac{1}{K_{ec}} = \frac{1}{\Sigma K_c} + \frac{1}{K_t} = \frac{1}{856E_c} + \frac{1}{248E_c}$$

or $K_{ec} = 192E_c$.

7. Moment distribution factors (D.F.): For the exterior joint,

$$\text{D.F. (slab)} = \frac{K_s}{K_s + K_{ec}} = \frac{140}{140 + 192} = 0.42$$

$$\text{D.F. (columns)} = \frac{K_{ec}}{\Sigma K} = 0.58$$

The columns above and below the slab have the same stiffness; therefore, the distribution factor of 0.58 is divided equally between both columns, and each takes a D.F. of $0.58/2 = 0.29$. For the interior joint,

$$\text{D.F. (slab)} = \frac{K_s}{2K_s + K_{ec}} = \frac{140}{2 \times 140 + 192} = 0.295$$

$$\text{D.F. (columns)} = \frac{K_{ec}}{\Sigma K} = \frac{192}{2 \times 140 + 192} = 0.41$$

Each column will have a D.F. of $0.41/2 = 0.205$.

8. Fixed-end moments: Because the actual L.L./D.L. is less than 0.75, the full-factored load is assumed to act on all spans.

$$\text{Fixed-end moment} = k'' w_u l_2 (L_1)^2$$

The factor k'' can be determined by the column analogy method: For a unit load $w = 1.0$ K/ft over the longitudinal span of 25 ft, the simple moment diagram is shown in Figure 17.38(b). The area of the bending moment diagram, considering the variation of the moment of inertia along the span, is

$$\text{Total area } A_m = A_1 + A_2 + 2A_3$$
$$= \frac{2}{3} \times 23.33(78.1 - 10) + 23.33 \times 10$$
$$+ 2\left(\frac{1}{2} \times 0.83 \times 10\right)(0.84) = 1300$$

$$\text{Fixed-end moment coefficient} = \frac{A_m}{A_a l_1^2}$$

where A_a for the slab is 24.72, as calculated in Step 3.

$$k'' = \frac{1300}{27.32(25)^2} = 0.084$$

It can be seen that the fixed-end moment coefficient, $k'' = 0.084$, is very close to the coefficient $\frac{1}{12} = 0.0833$ usually used to calculate the fixed-end moments in beams. This is expected, because the part of the span that has a variable moment of inertia is very small in flat plates where no column capital or drop panels are used. In this example only parts AB and CD, each equal to 0.83 ft, have a higher moment of inertia than I_s. In flat plates where the ratio of the span to column width is high, say, at least 20, the coefficient 0.0833 may be used to calculate approximately the fixed-end moments.

$$\text{Fixed-end moment } (\text{due to } w_u = 276 \text{ psf}) = 0.084(0.276)(20)(25)^2 = 289 \text{ K} \cdot \text{ft}$$

The factors K, K', and K'' can be obtained from tables prepared by Simmonds and Misic [17] to meet the ACI requirements for the equivalent frame method.

9. Moment distribution can be performed on half the frame due to symmetry. Once the end negative moments are computed, the positive moments at the center of any span can be obtained by subtracting the average value of the negative end moments from the simple beam positive moment. The moment distribution is shown in Figure 17.39. The final bending moments and shear forces are shown in Figure 17.40.

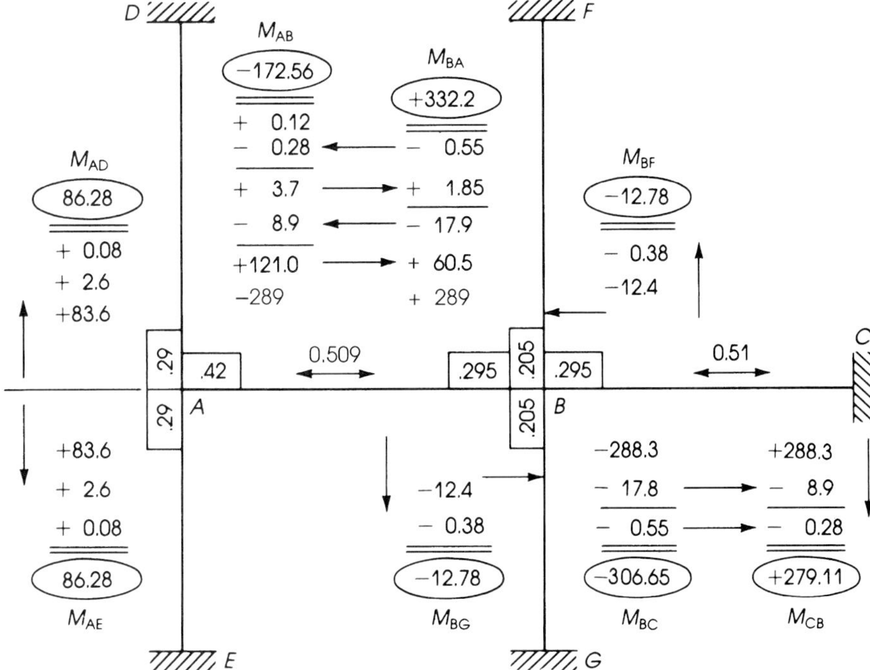

Figure 17.39 Analysis by moment distribution, Example 17.13. All moments are in K · ft.

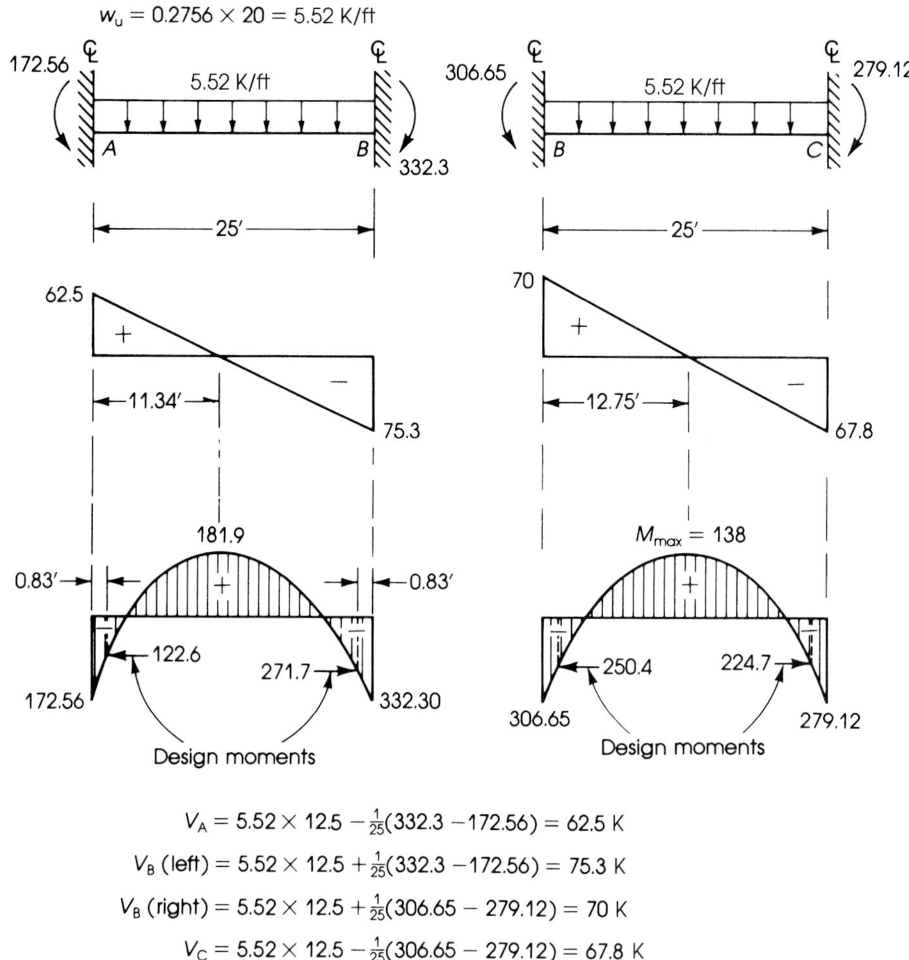

Figure 17.40 Example 17.13: Equivalent frame method—final bending moments and shear forces. (Slabs can be designed for the negative moments at the face of the columns as shown.)

10. Slabs can be designed for the negative moments at the face of the columns as shown in Figure 17.40.

Example 17.14 (SI Units)

Use the direct design method to design a typical interior flat slab with drop panels to carry a dead load of 7.86 kN/m^2 and a live load of 10 kN/m^2. The floor system consists of six panels in each direction, with a panel size of 6.0 by 5.4 m. All panels are supported by 0.4-m-diameter columns with 1.0-m-diameter column capitals. The story height is 3.0 m. Use $f'_c = 21$ MPa and $f_y = 400$ MPa.

Solution

1. All the ACI limitations to using the direct design method are met. Determine the minimum slab thickness h_s using equations (17.1) and (17.2). The diameter of the column capital equals 1.0 m. The equivalent square column section of the same area will have a side of $\sqrt{\pi r^2} = \sqrt{\pi(500)^2} = 885$ mm or 900 mm.

$$\text{Clear span (long direction)} = 6.0 - 0.9 = 5.1 \text{ m}$$

$$\text{Clear span (short direction)} = 5.4 - 0.9 = 4.5 \text{ m}$$

Because no beams are used $\alpha_m = 0$, $\beta_s = 1.0$, and $\beta = 6.0\text{ m}/5.4\text{ m} = 1.11$. From Table 17.1, minimum $h_s = l_n/33 = 5100/33 = 155$ mm. But because a drop panel is used, h_s may be reduced by 10% if drop panels extend a distance of at least $l/6$ in each direction from the centerline of support and project below the slab a distance of at least $h_s/4$. Therefore, use a slab thickness $h_s = 0.9 \times 155 = 140$ mm and a drop panel length and width as follows:

$$\text{Long direction:}\quad \frac{l_1}{3} = \frac{6.0}{3} = 2.0\text{ m}$$

$$\text{Short direction:}\quad \frac{l_2}{3} = \frac{5.4}{3} = 1.8\text{ m}$$

The thickness of the drop panel is $1.25h_s = 1.25 \times 140 = 175$ mm. Increase drop panel thickness to 220 mm to provide adequate thickness for punching shear and to avoid the use of a high percentage of steel reinforcement. All dimensions are shown in Figure 17.41.

2. Calculate ultimate loads:

$$w_u = 1.4 \times 7.86 + 1.7 \times 10 = 28\text{ kN/m}^2$$

3. Check two-way shear, first in the drop panel: The critical section is at a distance $d/2$ around the column capital. Let $d = 220 - 30\text{ mm} = 190$ mm.

$$\text{Diameter of shear section} = 1.0\text{ m} + d = 1.19\text{ m}$$

$$V_u = 28\left[6.0 \times 5.4 - \frac{\pi}{4}(1.19)^2\right] = 876\text{ kN}$$

$$b_o = 2\pi\left(\frac{1.19}{2}\right) = 3.74\text{ m}$$

$$\phi V_c = \phi 0.33 \times \sqrt{f'_c}\, b_o d$$

$$= \frac{0{,}85 \times 0.33}{1000}\sqrt{21} \times 3740 \times 190 = 913\text{ kN}$$

which is greater than V_u of 876 kN. Then check the two-way shear in the slab; the critical section is at a distance $d/2$ outside the drop panel.

$$d\text{ (slab)} = 140 - 30 = 110\text{ mm}$$

$$\text{Critical area} = (2.0 + 0.11)(1.8 + 0.11) = 4.03\text{ m}^2$$

$$b_o = 2(2.11 + 1.91) = 8.04\text{ m}$$

$$V_u = 28(6 \times 5.4 - 4.03) = 794\text{ kN}$$

$$\phi V_c = \frac{0.85 \times 0.33}{1000}\sqrt{21} \times 8040 \times 110 = 1137\text{ kN} > V_u$$

One-way shear is not critical.

4. Calculate the total static moments in the long and short directions:

$$M_{ol} = \frac{w_u}{8} l_2 l_{n1}^2 = \frac{28}{8}(5.4)(5.1)^2 = 491.6\text{ kN}\cdot\text{m}$$

$$M_{os} = \frac{w_u}{8} l_1 l_{n2}^2 = \frac{28}{8}(6)(4.5)^2 = 425.2\text{ kN}\cdot\text{m}$$

Because $l_2 < l_1$, the width of the column strip in the long direction is $2(0.25 \times 5.4) = 2.7$ m. The width of the column strip in the short direction is 2.7 m. Assuming that the steel bars are 12 mm in diameter and those in the short direction are placed on top of the bars in the long direction, then the effective depth in the short direction will be about 10 mm less than

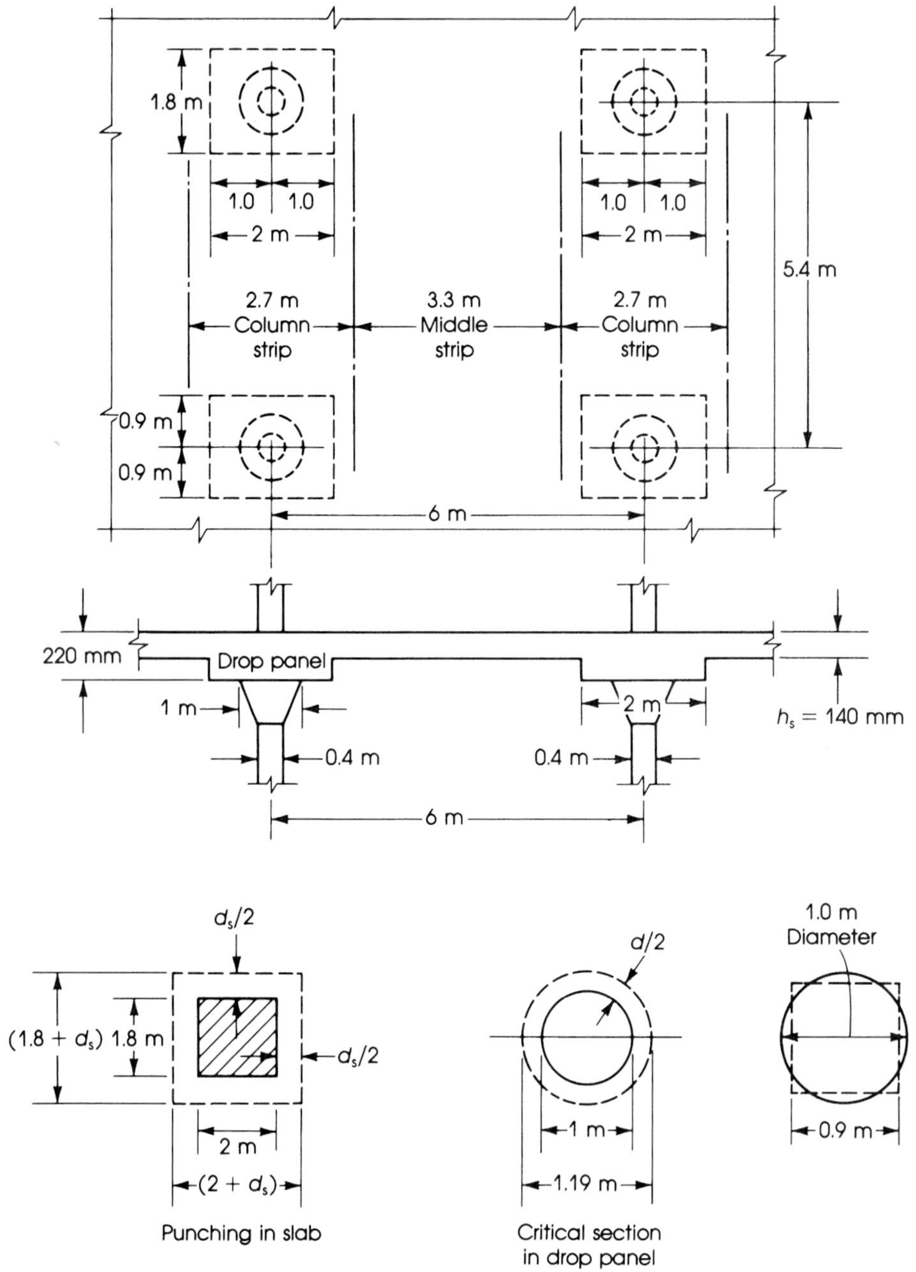

Figure 17.41 Example 17.14: Interior flat slab with drop panel.

the effective depth in the long direction. The d values and the design procedure are shown in Table 17.14. Minimum lengths of the selected reinforcement bars should meet the ACI Code length requirements shown in Figure 17.16.

5. The column stiffness is

$$\text{Ratio: } \frac{\text{D.L.}}{L.L.} = \frac{7.86}{10} = 0.786 \quad \text{and} \quad \frac{l_2}{l_1} = 1.11$$

Determine α_{min} from Table 17.7, taking into account that the relative beam stiffness is zero because no beams are used. By interpolation, $\alpha_{min} = 1.15$. An approximate method is used here to determine the stiffness of the column with its capital.

Table 17.14 Design of an Interior Flab Slab with Drop Panels

	$M_o = 491.6$ kN·m $M_n = 0.65M_o = -319.5$ kN·m $M_p = +0.35M_o = +172.1$ kN·m			
Long Direction	***Column Strip***		***Middle Strip***	
Moment factor	$0.75M_o$	$0.60M_p$	$0.25M_n$	$0.40M_p$
M_u (kN·m)	−239.6	+103.3	−79.9	+68.8
d (mm)	190	110	110	110
Strip width b (m)	2.7	2.7	2.7	2.7
$R_u = \frac{M_u}{bd^2}(MP_a)$	2.46	3.16	2.44	2.10
Steel ratio ρ (%)	0.71	0.93	0.7	0.6
$A_s = \rho bd$ (mm^2)	3642	2762	2079	1782
Min. $A_s = 0.0018bh$ (mm^2)	1070	680	680	680
Bars selected (straight bars)	18 × 16 mm	14 × 16 mm	20 × 12 mm	16 × 12 mm
Spacing (mm)	150	193	135	170
	$M_o = 425.2$ kN·m $M_n = -0.65M_o = -276.4$ kN·m $M_p = +0.35M_o = +148.8$ kN·m			
Short Direction	***Column Strip***		***Middle Strip***	
Moment factor	$0.75M_n$	$0.60M_p$	$0.25M_n$	$0.40M_p$
M_u (kN·m)	−207.3	+89.3	−69.1	+59.5
d (mm)	180	100	100	100
Strip width b (m)	2.7	2.7	3.3	3.3
$R_u = \frac{M_u}{bd^2}(MP_a)$	2.37	3.30	2.10	1.80
Steel ratio ρ (%)	0.69	1.00	0.6	0.5
$A_s = \rho bd$ (mm^2)	3353	2700	1980	1650
Min. $A_s = 0.0018bh$ (mm^2)	1070	680	832	832
Bars selected (straight bars)	18 × 16 mm	14 × 16 mm	18 × 12 mm	16 × 12 mm
Spacing (mm)	150	195	185	205

I_s (moment of inertia of slab, short direction)

$$= 6000\frac{(140)^3}{12} = 1372 \times 10^6 \text{ mm}^4$$

$$K_s = \frac{4E_cI_s}{l_2} = \frac{4E_c \times 1372 \times 10^6}{5400} = 1016 \times 10^3 E_c$$

$$I_c \text{ (for circular column, diameter 400 mm)} = \frac{\pi D^4}{64} = \frac{\pi}{64}(400)^4 = 1257 \times 10^6 \text{ mm}^4$$

$$K_c = \frac{4E_cI_c}{l_c} = \frac{4E_c \times 1257 \times 10^6}{3000 \text{ mm}} = 1676 \times 10^3 E_c$$

$$\text{Ratio of column stiffness/slab stiffness} = \frac{K_c}{K_s} = \frac{1676 \times 10^3}{1016 \times 10^3} = 1.65$$

which is greater than $\alpha_{\min}$ of 1.15. If I_s in the long direction is used, the calculated ratio of column to slab stiffness will be greater than 1.65. Therefore, the column is adequate.

6. Determine the balanced moment in the column and check the shear stresses in the slab, as explained in Examples 17.8 and 17.9.

SUMMARY

Sections 17.1–17.5

1. A two-way slab is one that has a ratio of length to width less than 2. Two-way slabs may be classified as flat slabs, flat plates, waffle slabs, or slabs on beams.
2. The ACI Code specifies two methods for the design of two-way slabs: the direct design method and the equivalent frame method. In the direct design method, the slab panel is divided (in each direction) into three strips, one in the middle (referred to as the *middle strip*) and one on each side (referred to as *column strips*).

Section 17.6

To control deflection, the minimum slab thickness, h, is limited to the values computed by Table 17.1 or equations (17.1) and (17.2) and as explained in Examples 17.1 and 17.2.

Section 17.7

For two-way slabs without beams, the shear capacity of the concrete section in one-way shear is

$$V_c = 2\sqrt{f'_c}\,bd$$

The shear capacity of the concrete section in two-way shear is

$$V_c = \left(2 + \frac{4}{\beta_c}\right)\sqrt{f'_c}\,b_o d \le 4\sqrt{f'_c}\,b_o d$$

When shear reinforcement is provided, $V_n \le 6\sqrt{f'_c}\,b_o d$.

Section 17.8

In the direct design method, approximate coefficients are used to compute the moments in the column and middle strips of two-way slabs. The total factored moment is

$$M_o = (w_u l_2)\frac{l_1^2}{8} \tag{17.11}$$

The distribution of M_o into negative and positive span moments is given in Figure 17.14. A summary of the direct design method is given in Section 17.8.8. The modified stiffness method is explained in Section 17.8.7.

Sections 17.9–17.11

1. Unbalanced loads on adjacent panels cause a moment in columns that can be computed by equation (17.22).
2. Approximately 60% of the moment transferred to both ends of a column at a joint is transferred by flexure, M_f, and 40% is transferred by eccentric shear, M_v. The fraction of the unbalanced moment transferred by flexure, M_f is $\gamma_f M_u$, where γ_f is computed from equation (17.25). The shear stresses produced by M_v must be combined with shear stresses produced by the shearing force V_u.
3. Waffle slabs are covered in Section 17.11.

Section 17.12

1. In the equivalent frame method, the building is divided into equivalent frames in two directions and then analyzed for all conditions of loadings. Example 17.13 explains this procedure.
2. Example 17.14 is an example of a two-way flat slab with drop panel (SI units).

REFERENCES

1. W. G. Corley and J. O. Jirsa. "Equivalent Frame Analysis for Slab Design." *ACI Journal* 67 (November 1970).
2. W. L. Gamble. "Moments in Beam Supported Slabs." *ACI Journal* 69 (March 1972).
3. M. A. Sozen and C. P. Siess. "Investigation of Multipanel Reinforced Concrete Floor Slabs." *ACI Journal* 60 (August 1963).
4. W. L. Gamble, M. A. Sozen, and C. P. Siess. "Tests of a Two-way Reinforced Concrete Slab." *Journal of Structural Division,* ASCE 95 (June 1969).
5. R. Park and W. Gamble. *Reinforced Concrete Slabs.* New York: John Wiley, 1980.
6. S. P. Timoshenko and S. W. Krieger. *Theory of Plates and Shells,* 2d ed. New York: McGraw-Hill, 1959.
7. R. H. Wood. *Plastic and Elastic Design of Slabs and Plates.* London: Thames and Hudson, 1961.
8. O. C. Zienkiewicz. *The Finite Element Method in Engineering Science.* New York: McGraw-Hill, 1971.
9. K. W. Johansen. *Yield-Line Formulae for Slabs.* London: Cement and Concrete Association, 1972.
10. A. Hillerborg. *Strip Method of Design.* London: Cement and Concrete Association, 1975.
11. R. K. Dhir and J. G. Munday. *Advances in Concrete Slab Technology.* New York: Pergamon, 1980.
12. W. C. Schnorbrich. "Finite Element Determination of Non-linear Behavior of Reinforced Concrete Plates and Shells." In *Proceedings of Symposium on Structural Analysis.* TRRL, 164 VC. Crowthorne, 1975.
13. M. Fintel, ed. *Handbook of Concrete Engineering.* New York: Van Nostrand Reinhold, 1974.
14. American Concrete Institute. *Commentary of Building Code Requirements of Structural Concrete.* ACI 318-95. Detroit, 1995.
15. N. W. Hanson and J. M. Hanson. "Shear and Moment Transfer Between Concrete Slabs and Columns." *PCA Journal* 10 (January 1968).
16. J. Moe. "Shear Strength of Reinforced Concrete Slabs and Footings under Concentrated Loads." *PCA Bulletin* D47 (April 1961).
17. S. H. Simmonds and J. Misic. "Design Factors for the Equivalent Frame Method." *ACI Journal* 68 (November 1971).
18. American Concrete Institute. *Building Code Requirements for Structural Concrete.* ACI 318-99. Detroit, 1999.
19. A. W. Hago and P. Bhatt. "Tests on Reinforced Concrete Slabs Designed by the Direct Design Method Procedure." *ACI Journal* 83 (November–December 1986).
20. D. F. Fraser. "Simplified Frame Analysis for Flat Plate Construction." *Concrete International* 6 (September 1984).
21. Brochure, Molded Fiber Glass Concrete Forms Co. (MFG), Union City, PA, 1990.

PROBLEMS

17.1 (Flat plates) Determine the minimum slab thickness according to the ACI Code for the flat-plate panels shown in Figure 17.42 and Table 17.15. The floor panels are supported by 24- × 24-in. columns, 12 ft long, with no edge beams at the end of the slab. Use $f'_c = 4$ Ksi, $f_y = 60$ Ksi, dead load (excluding self-weight) = 25 psf, and live load = 120 psf.

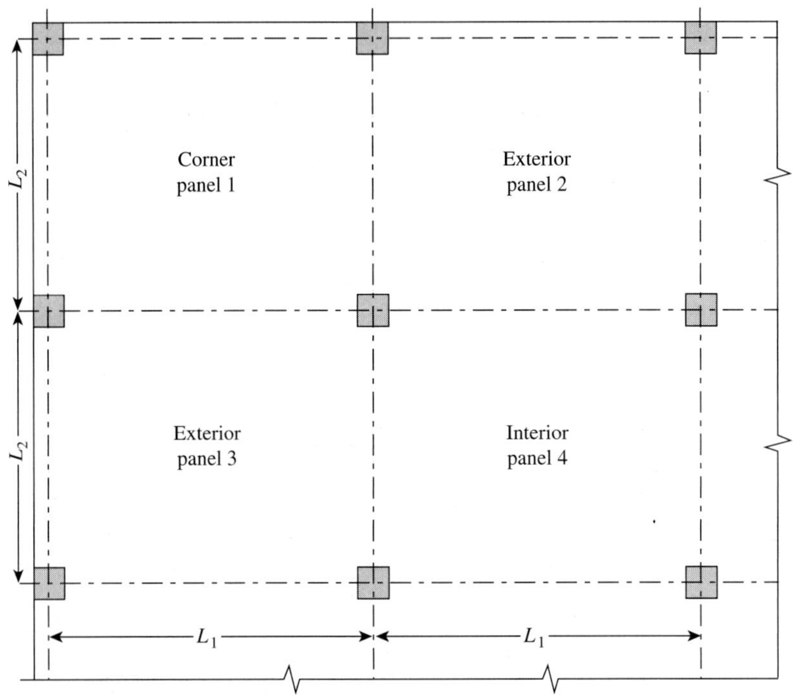

Figure 17.42

Table 17.15 Problem 17.1

Number (Flat Plate)	Panel Dimensions (ft) L_1	L_2	Panel Numbers
(a)	20	20	1 and 4
(b)	24	24	2 and 4
(c)	26	26	3 and 4
(d)	20	16	1 and 2
(e)	24	20	3 and 4
(f)	26	22	1 and 4
(g)	30	24	1 and 2
(h)	30	30	1 and 4

17.2 (Flat plates) Use the direct design method to design the interior flat-plate panel (no. 4) of Problems 17.1(a), (b), (c), and (e), using the data given earlier. Check the shear and moment transfer at an interior column. Draw sketches showing the reinforcement distribution and the shear stresses.

17.3 (Flat plates) Repeat Problem 17.2 for the exterior panel, no. 3. Check the shear and moment transfer at the exterior column. If shear stresses are not adequate, use shear reinforcement involving stirrups.

17.4 (Flat slabs with drop panels) Determine the minimum slab and drop panel thicknesses according to the ACI Code for the slabs shown in Figure 17.42 and Table 17.15. The floor panels are supported by 24- × 24-in. columns with no edge beams. Use $f'_c = 4$ Ksi, $f_y = 60$ Ksi, additional dead load (excluding self-weight) = 30 psf, and live load = 120 psf.

17.5 (Flat slabs) Use the direct design method to design the interior flat slab panel, no. 4, of Problems 17.4(a), (b), (c), and (e), using the data given in Problem 17.4. Check the shear and moment transfer at an interior column. Draw sketches showing the reinforcement distribution and the shear stresses. Use a 4-ft-column capital diameter for part (c) only.

17.6 (Flat slabs) Repeat Problem 17.5 for the exterior panel no. 3.

17.7 (Slabs on beams) Redesign the slabs in Problem 17.2, using the same data when the slabs are supported by beams on all four sides. Each beam has a width $b_w = 14$ in. and a projection below the bottom of the slab of 18 in.

17.8 (Slabs on beams) Redesign the slabs in Problem 17.7 as exterior panels.

17.9 (Waffle slabs) Repeat Example 17.12 when the spans are (a) 36 ft and (b) 42 ft. Use the same data and 24- × 24-in. columns.

17.10 (Waffle slabs) Redesign the waffle slabs in Problem 17.9 as exterior panels.

17.11 (Equivalent frame method) Redesign the flat-plate floor system of Problems 17.2(a) and (b) using the equivalent frame method.

17.12 (Equivalent frame method) Redesign the waffle slabs of Problem 17.9 using the equivalent frame method.

18 STAIRS

Office building under construction, Chicago.

18.1 INTRODUCTION

Stairs must be provided in almost all buildings, either low-rise or high-rise, even if adequate numbers of elevators are provided. Stairs consist of rises, runs (or treads), and landings. The total steps and landings are called a *staircase.* The *rise* is defined as the vertical distance between two steps, and the *run* is the depth of the step. The *landing* is the horizontal part of the staircase without rises (Figure 18.1).

The normal dimensions of the rises and runs in a building are related by some empirical rules.

$$\text{Rise} + \text{run} = 17 \text{ in.} \quad (430 \text{ mm})$$

$$2 \times \text{rise} + \text{run} = 25 \text{ in.} \quad (635 \text{ mm})$$

$$\text{rise} \times \text{run} = 75 \text{ in.}^2 \quad (0.05 \text{ m}^2)$$

The rise depends on the use of the building. For example, in public buildings the rise is about 6 in., whereas in residential buildings it varies between 6 and 7.5 in. The run is about 1 ft in public buildings and varies between 9 in. and 12 in. in residential buildings. In general, a rise should not exceed 8 in. or be less than 4 in., and the number of rises is obtained by dividing the structural floor-to-floor dimension by the assumed rise.

The finishing on the stairs varies from troweling Alundum grits to adding asphalt tiles, terrazzo tiles, marble, or carpets. In addition to dead loads, stairs must be designed for a minimum live load of 100 psf.

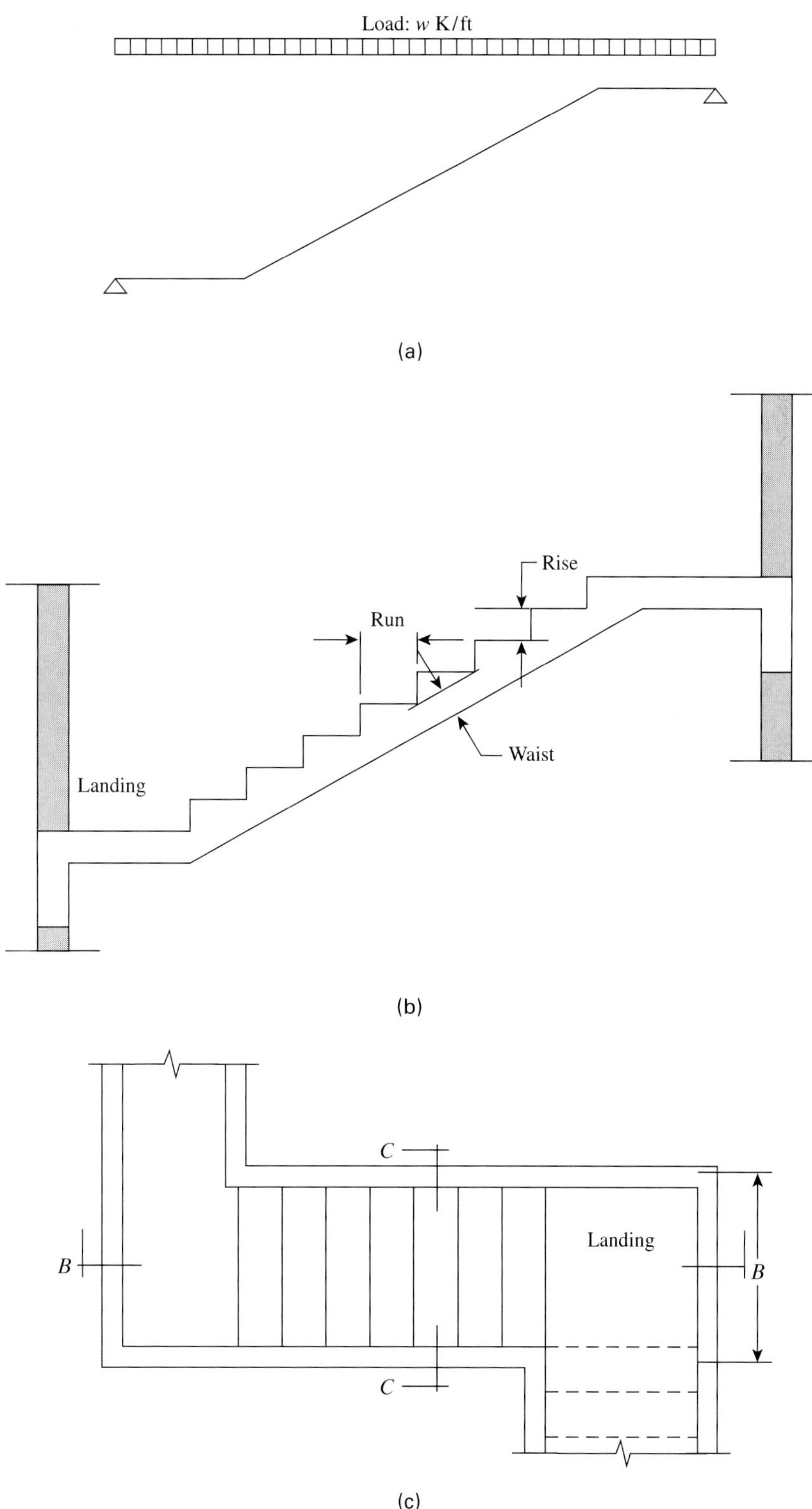

Figure 18.1 Plan of a single-flight staircase: (a) loads, (b) section *B-B*, and (c) plan.

18.2 TYPES OF STAIRS

There are different types of stairs, which depend mainly on the type and function of the building and on the architectural requirements. The most common types are as follows.

1. *Single-flight stairs:* The structural behavior of a flight of stairs is similar to that of a one-way slab supported at both ends. The thickness of the slab is referred to as the waist (Figure 18.1). When the flight of stairs contains landings, it may be more economical to provide beams at B and C between landings (Figure 18.2). If such supports are not provided, which is quite common, the span of the staircase will increase by the width of two landings and will extend between A and D. In residential buildings, the landing width is in the range of 4 to 6 ft, and the total distance between A and D is about 20 ft.

   An alternative method of supporting a single flight of stairs is to use stringers, or edge beams, at the two sides of the stairs; the steps are then supported between the beams (Figure 18.3).
2. *Double-flight stairs:* It is more convenient in most buildings to build the staircase in double flights between floors. The types commonly used are quarter-turn (Figure 18.4) and closed- or open-well stairs, as shown in Figure 18.5. For the structural analysis of the stairs, each flight is treated as a single flight and is considered supported on two or more beams, as shown in Figure 18.2. The landing extends in the transverse direction between two supports and is designed as a one-way slab. In the case of open-well stairs, the middle part of the landing carries a full load, whereas the two end parts carry half-loading only, as shown in Figure 18.5(d). The other half-loading is carried in the longitudinal direction by the stair flights, sections A-A and B-B.
3. *Three or more flights of stairs:* In some cases, where the overall dimensions of the staircase are limited, three or four flights may be adopted (Figure 18.6). Each flight will be treated separately, as in the case of double-flight staircases.
4. *Cantilever stairs:* Cantilever stairs are used mostly in fire-escape stairs, and they are supported by concrete walls or beams. The stairsteps may be of the full-flight type, projecting from one side of the wall, the half-flight type, projecting from both sides of the supporting wall, or of the semispiral type, as shown in Figure 18.7. In this type of stairs, each step acts as a cantilever, and the main reinforcement is placed in the tension side of the run and the bars are anchored within the concrete wall. Shrinkage and temperature reinforcement is provided in the transverse direction.

   Another form of a cantilever stair is that using open-riser steps supported by a central beam, as shown in Figure 18.8. The beam has a slope similar to the flight of stairs and receives the steps on its horizontally prepared portions. In most cases, precast concrete steps are used, with special provisions for anchor bolts that fix the steps into the beam.
5. *Precast flights of stairs:* The speed of construction in some projects requires the use of precast flights of stairs (Figure 18.8). The flights may be cast separately and then fixed to cast-in-place landings. In other cases, the flights, including the landings, are cast and then placed in position on their supporting walls or beams. They are designed as simply supported one-way slabs with the main reinforcement at the bottom of the stair waist. Adequate reinforcement must be provided at the joints, as shown in Figure 18.9.

   Provisions must be made for lifting and handling the precast stair units by providing lifting holes or inserting special lifting hooks into the concrete. Special reinforcement must be provided at critical locations to account for tensile stresses that will occur in the stairs from the lifting and handling process.

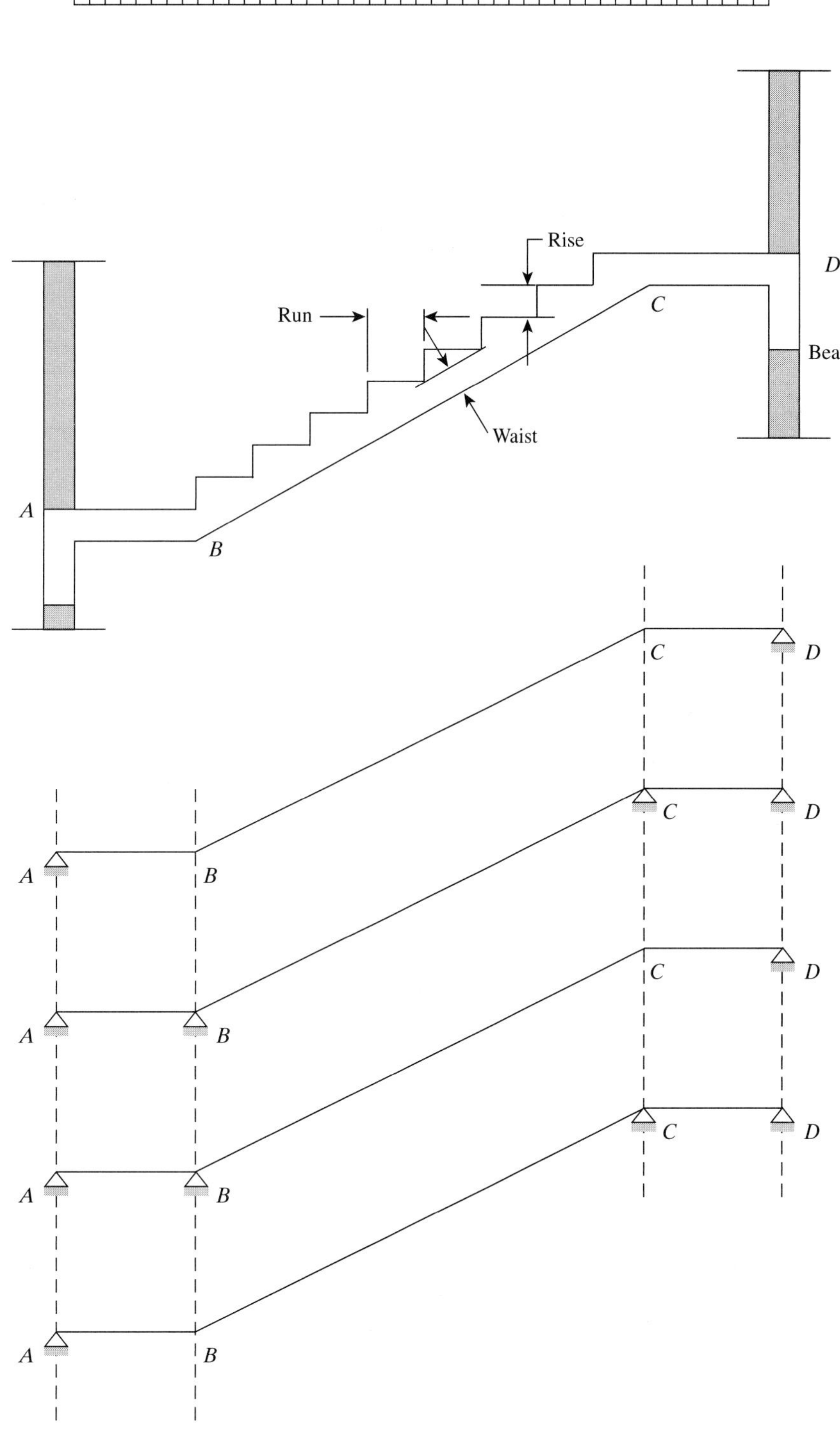

Figure 18.2 Supporting systems of one flight.

6. *Free-standing staircase:* In this type of stairs, the landing projects in the air without any support at its end (Figure 18.10). The stairs behave in a springboard manner, causing torsional stresses in the slab.

Free-standing staircase

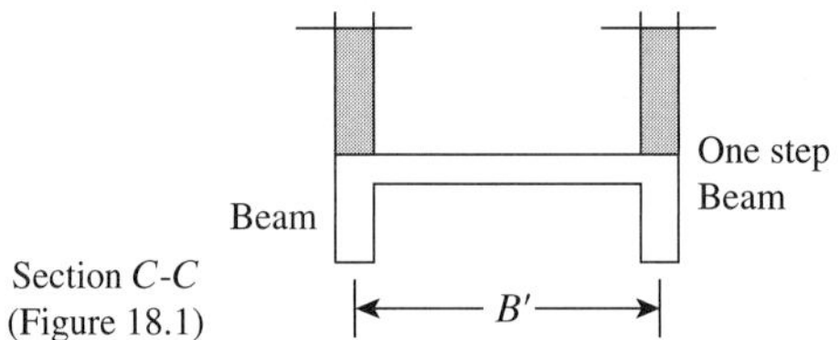

Figure 18.3 Steps supported by stringer beams.

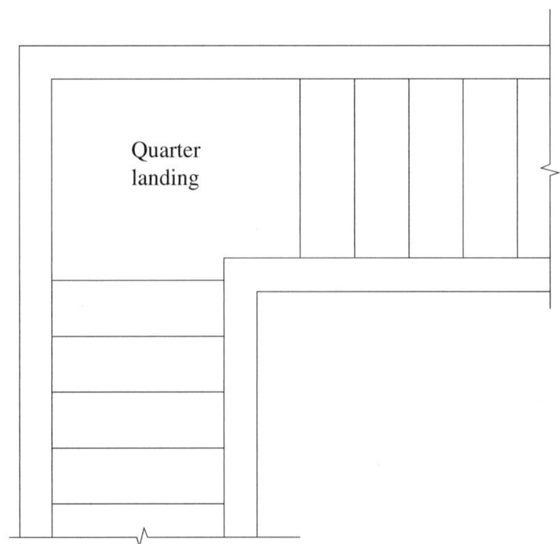

Figure 18.4 Quarter-turn staircase.

Three systems of loading must be considered in the design of this type of stairs, taking into consideration that torsional moments will develop in the slab in all cases:

a. When the live load acts on the upper flight and half the landing only (Figure 18.11), the upper flight slab will be subjected to tensile forces in addition to bending

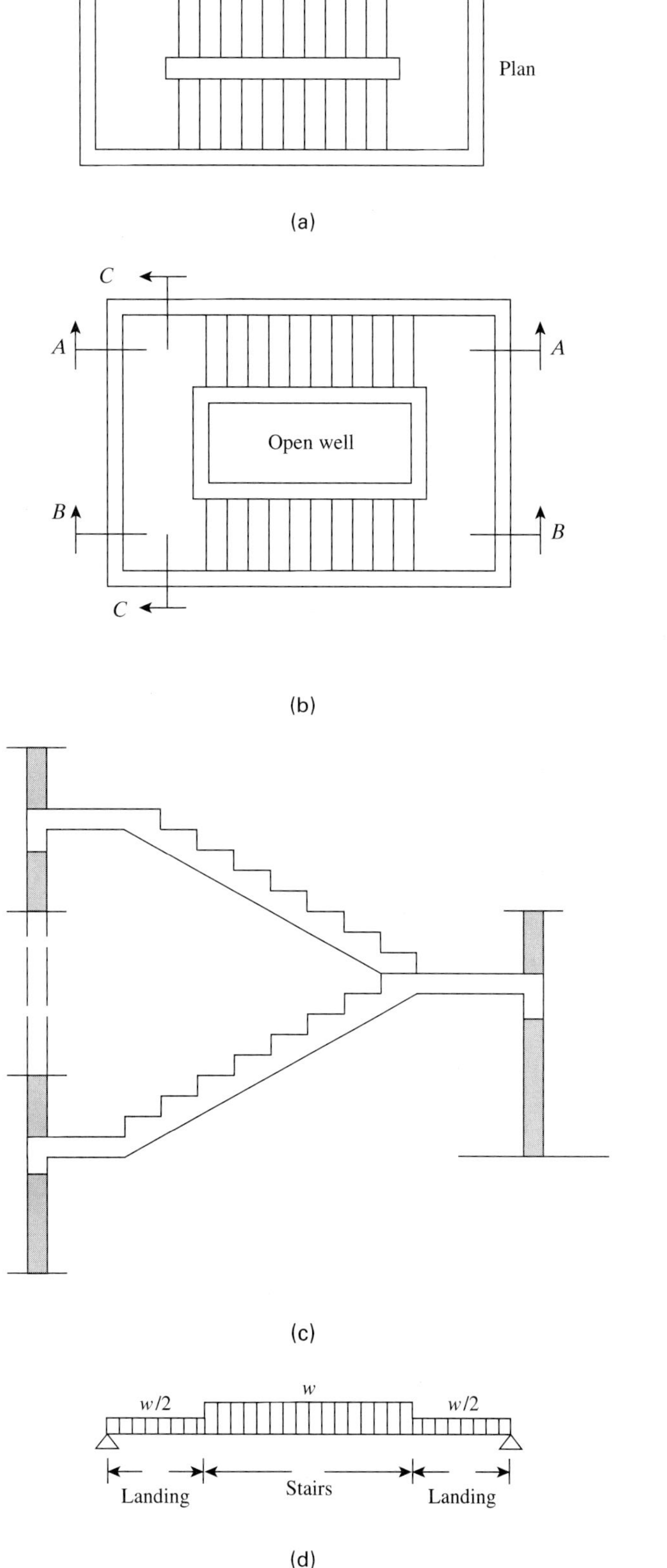

Figure 18.5 Double-flight stairs: (a) closed-well staircase, (b) open-well staircase, (c) section *B-B*, and (d) section *C-C*.

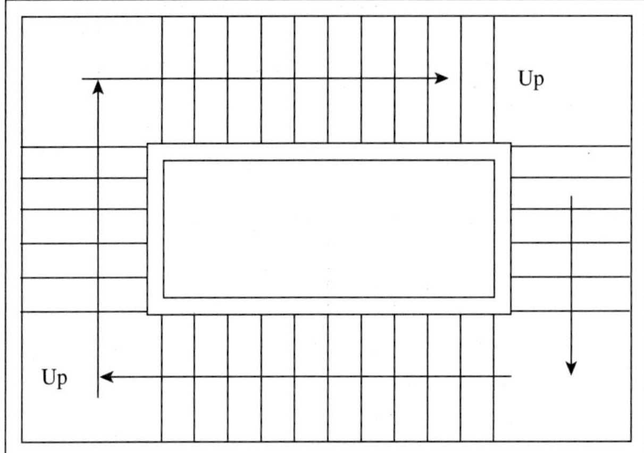

Four-stair flight

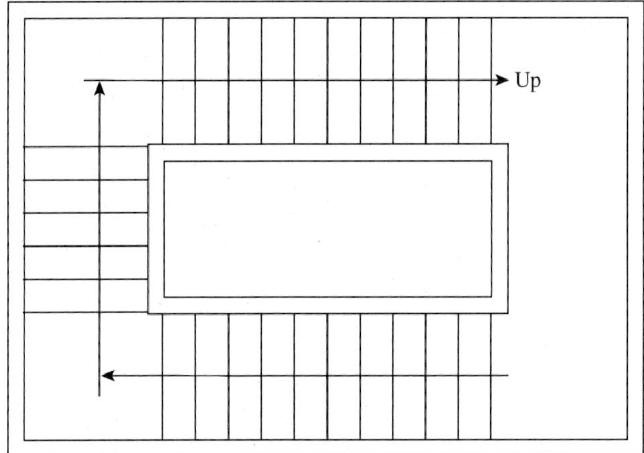

Three-stair flight

Figure 18.6 Three- and four-stair flights.

moments, whereas the lower flight will be subjected to compression forces, which may cause buckling of the slab.

b. When the live load acts on the lower flight and half the landing only (Figure 18.12), the upper flight slab will be subjected to tensile forces, whereas the lower flight will be subjected to bending moment and compression forces.

c. When the live load acts on both upper and lower flights, the loading of one flight will cause the twisting of the other. The torsional stresses developed in the stairs require adequate reinforcement in both faces of the stair slabs and the landing. Transverse reinforcement in the slab and the landing must be provided in both faces of the concrete in the shape of closed U-bars lapping at midwidth of the stairs. Typical reinforcement details are shown in Figure 18.13.

This type of stairs is favored by architects and sometimes called a pliers-shaped staircase or jackknife staircase.

7. *Run-riser stairs:* Run-riser stairs are stepped underside stairs that consist of a number of runs and risers rigidly connected without the provision of the normal waist slab (Figure 18.14(a)). This type of stairs has an elegant appearance and is sometimes favored by architects. The structural analysis of run-riser stairs can be simplified by assuming that the effect of axial forces is negligible and that the load on each run is

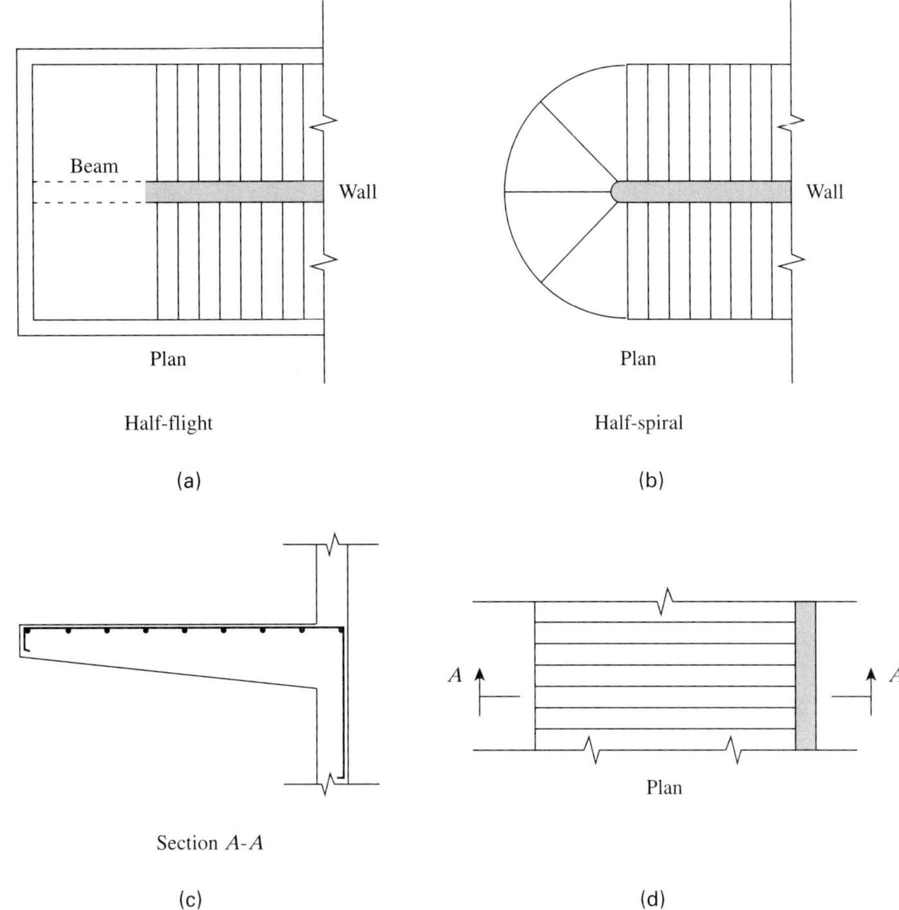

Figure 18.7 Steps projecting from one or two sides of the supporting wall.

concentrated at the end of the run (Figure 18.14(b)). For the analysis of a simply supported flight of stairs, consider a simple flight of two runs, ABC, subjected to a concentrated load P at B' (Figure 18.14(b)). Because joints B and B' are rigid, the moment at joint B is equal to the moment at B', or

$$M_B = M'_B = \frac{PS}{2}$$

where S is the width of the run. The moment in rise, BB', is constant and is equal to $PS/2$.

When the rise is absent, the stairs, ABC, act as a simply supported beam, and the maximum bending moment occurs at midspan with value

$$M_B = \frac{PL}{4} = \frac{PS}{2}$$

For a flight of stairs that consists of a number of runs and risers, the same approach can be used; the bending moment diagram is shown in Figure 18.15(a). The moment in BB' is constant and is equal to the moment at joint B, or $2PS$. Similarly, $M_C = M'_C = 3PS$, $M_D = M'_D = 3PS$, and $M_E = M'_E = 2PS$.

If a landing is present at one or both ends, the load on the landing practically may be represented by concentrated loads similar to the runs. The structural analysis may also be performed by considering a load uniformly distributed on the flight of

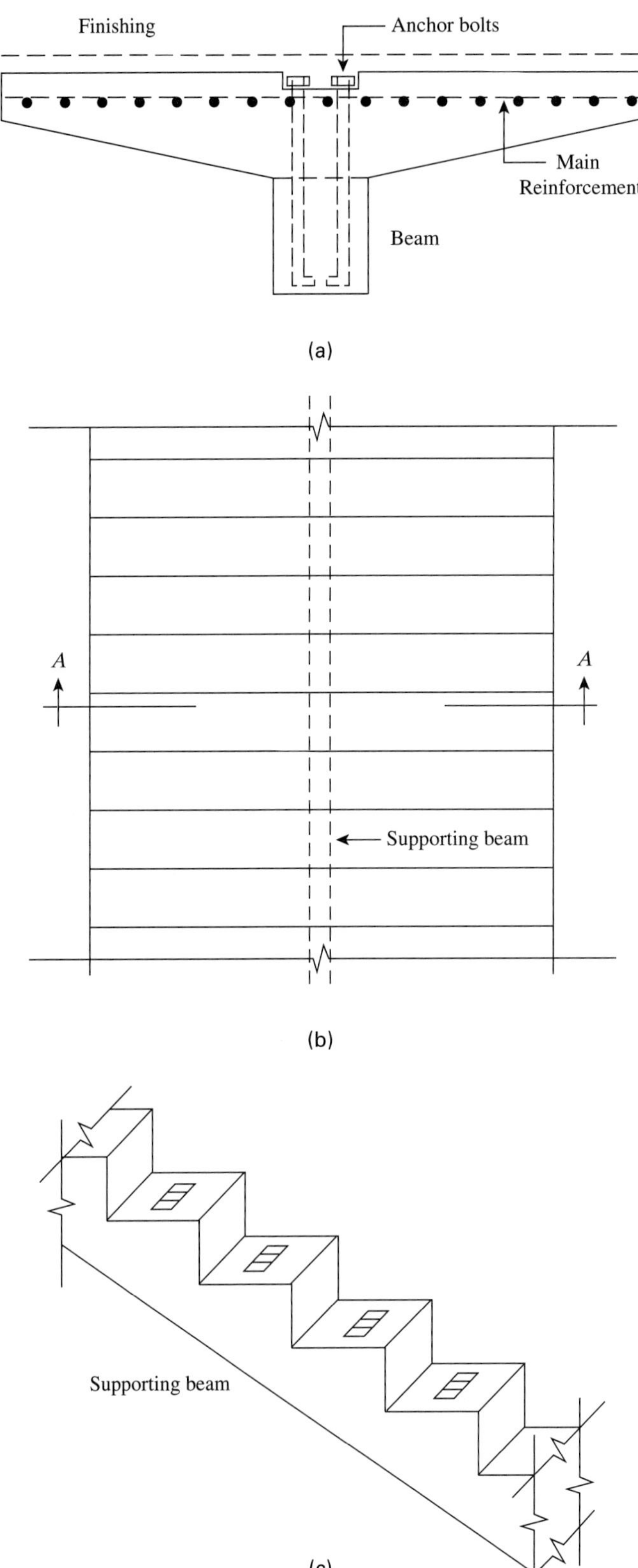

Figure 18.8 Precast cantilever stair supported by central beam: (a) section *A-A*, (b) part plan, and (c) supporting beam.

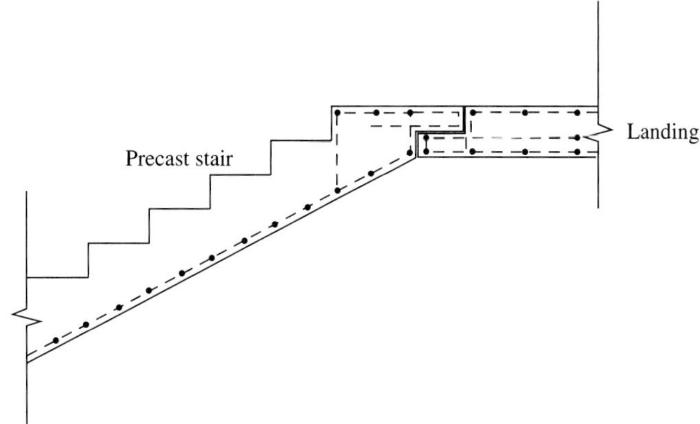

Figure 18.9 Joint of a precast concrete flight of stairs.

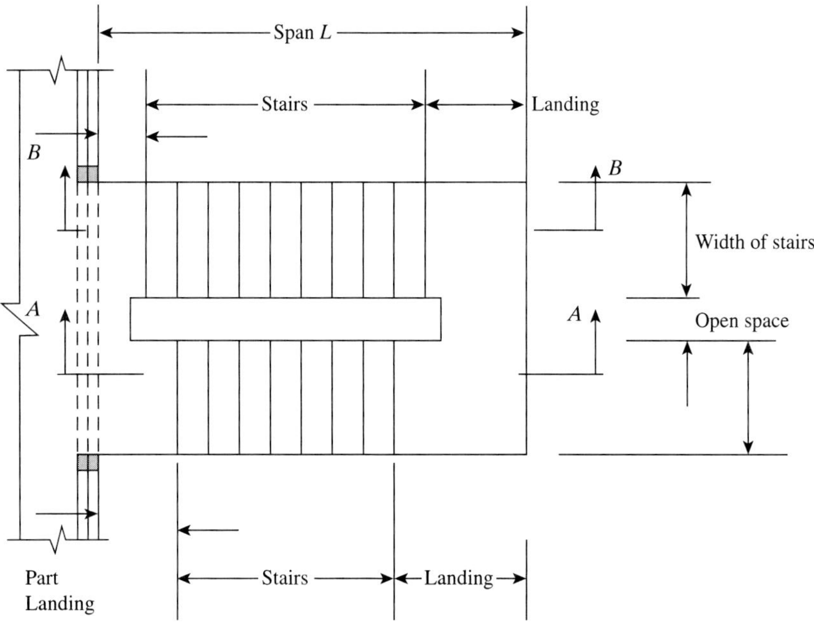

Figure 18.10 Plan of a free-standing staircase.

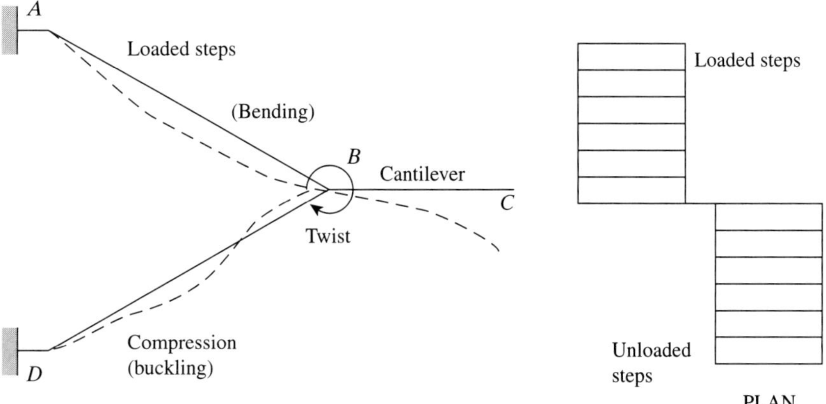

Figure 18.11 Case 1, *ABC* loaded.

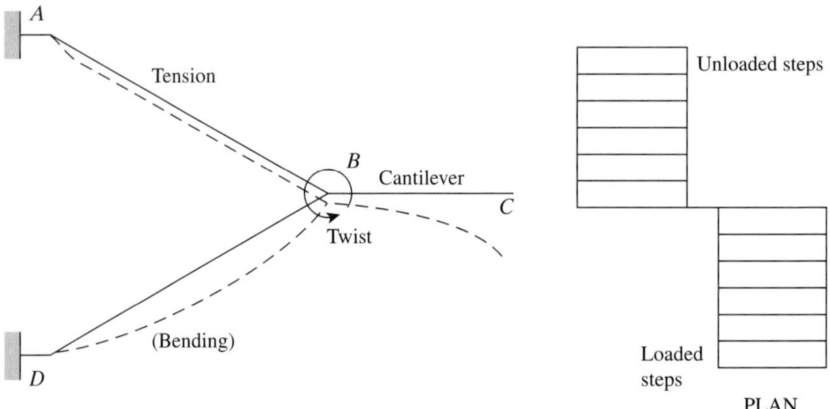

Figure 18.12 Case 2, *DBC* loaded.

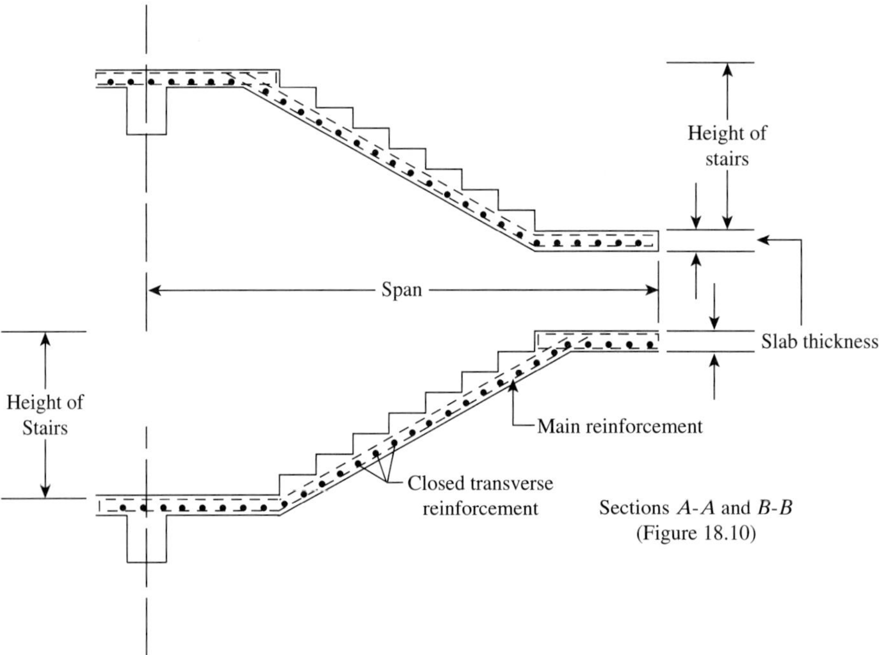

Figure 18.13 Section of a free-standing staircase.

stairs. The moment in every riser is constant and is obtained from the bending moment diagram of a simply supported beam subjected to a uniform load (Figure 18.15(b)). Example 18.3 illustrates the design of a staircase using the two assumptions of concentrated loads and uniform loads.

If the stair flight is fixed or continuous at one or both ends, the moments can be obtained using any method of structural analysis. To explain this case, consider a flight of stairs that consists of two runs and is fixed at both ends (Figure 18.16(a)). The moments at the fixed ends, A and B, due to a concentrated load at B are equal to $PL/8 = PS/4$. This result is obtained by assuming that the rise does not exist and the stairs, ABC, act as a fixed-end beam subjected to a concentrated load at midspan (Figure 18.16(b)). The moment at midspan, section B, is equal to

$$\frac{PL}{4} - M_A = \frac{PS}{2} - \frac{PS}{4} = \frac{PS}{4}$$

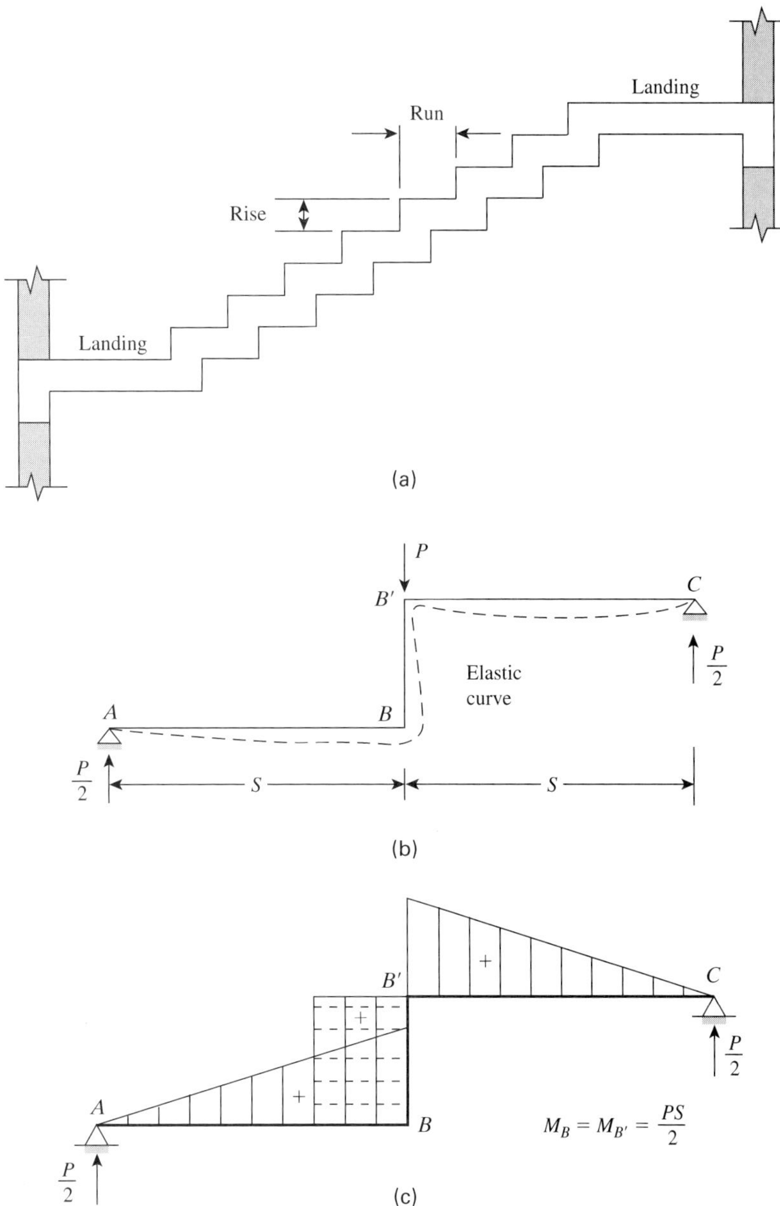

Figure 18.14 Run-riser staircase: (a) cross section, (b) elastic curve, and (c) bending moment diagram.

The bending moment of a flight of stairs with one riser is shown in Figure 18.16(a). Note that the moment in the riser BB' is constant, and $M_B = M'_B = PS/4$.

For a symmetrical stair flight, fixed at both ends and subjected to a number of concentrated loads at the node of each run, the moment at the fixed end can be calculated as follows:

$$M \text{ (fixed end)} = \frac{PS}{12}(n^2 - 1)$$

where P = concentrated load at the node of the run
S = width of run
n = number of runs

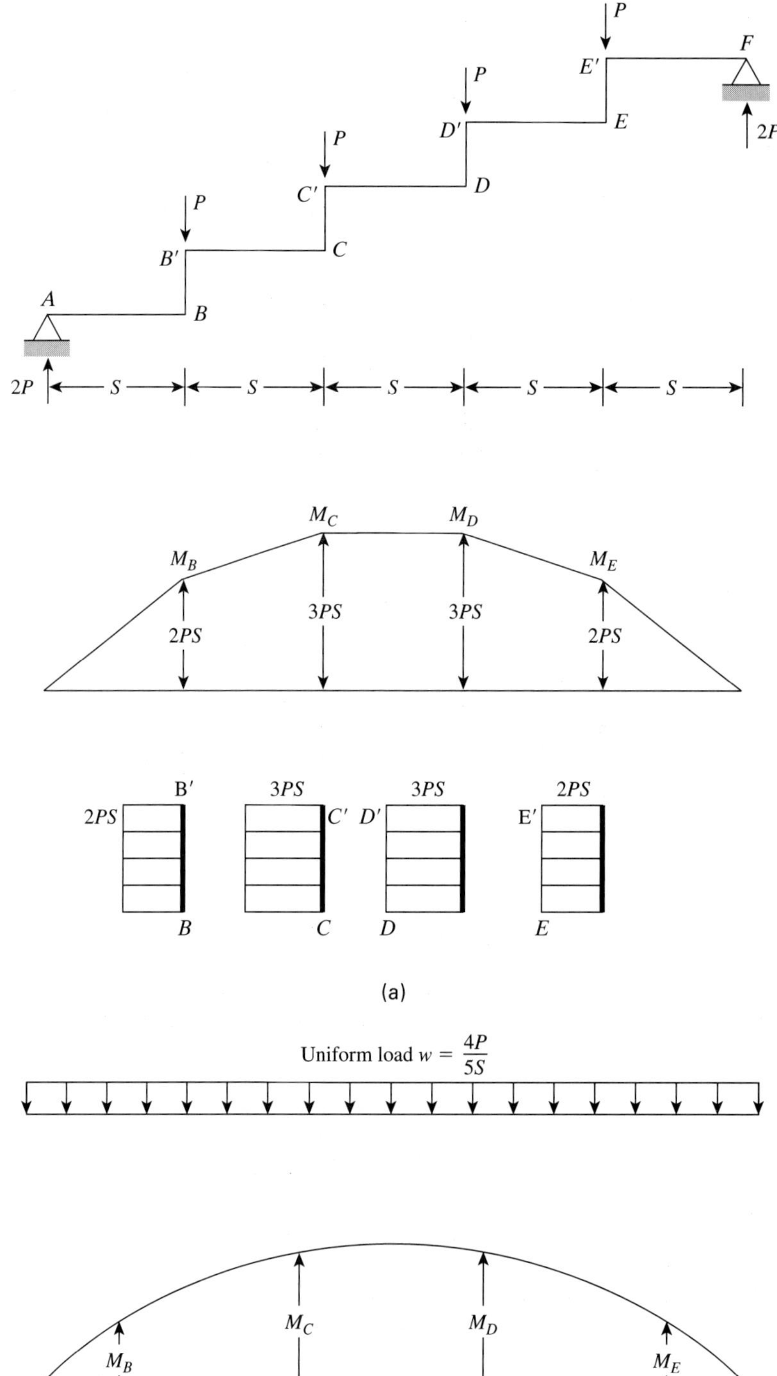

Figure 18.15 Distribution of moments: (a) bending moment due to concentrated loads and (b) bending moment due to uniform load.

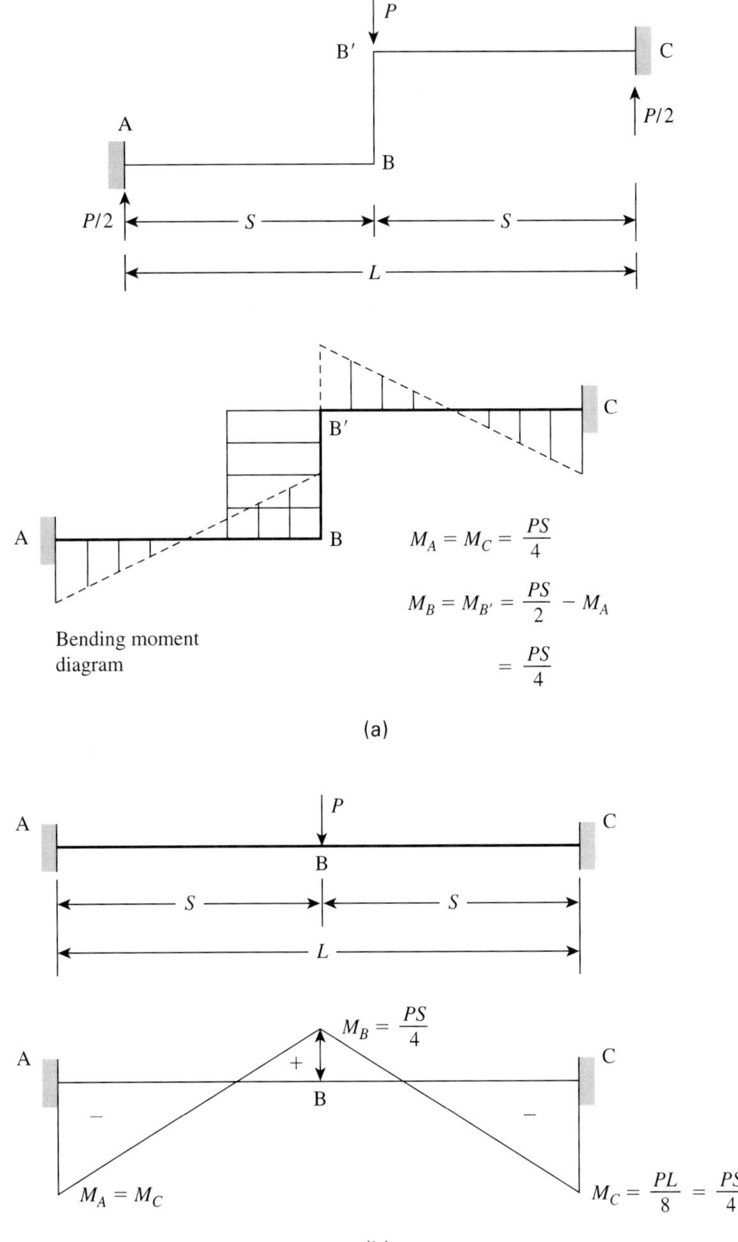

Figure 18.16 Fixed-end staircase: (a) loaded steps and (b) loaded beam.

When $n = 2$, then

$$M \text{ (fixed end)} = \frac{PS}{12}(4 - 1) = \frac{PS}{4}$$

which is the same result obtained earlier.

If a landing is present at one or both ends, the load on the landing may be represented by concentrated loads at spacing S.

Reinforced concrete helical staircase.

8. *Helical stairs (open-spiral stairs):* A helical staircase is a three-dimensional structure, which usually has a circular shape in plan (Figure 18.17). It is a distinctive type of stairs used mainly in entrance halls, theater foyers, and special low-rise office buildings. The cost of a helical stair is much higher than that of a normal staircase.

The stairs may be supported at some edges within adjacent walls or may be designed as a free-standing helical staircase, which is most popular. The structural analysis of helical staircases is complicated and was discussed by Morgan [1] and Scordelis [2] using the principles of strain energy. Design charts for helical stairs are also prepared by Cusens [3]. Under load, the flight slab will be subjected to torsional stresses throughout. The upper landing will be subjected to tensile stresses, whereas compressive stresses occur at the bottom of the flight. The forces acting at any section may consist of vertical moment, lateral moment, torsional moment, axial force, shearing force across the waist of the stairs, and radial horizontal shearing force. The main longitudinal reinforcement consists of helical bars placed in the concrete waist of the stairs and runs from the top landing to the bottom support. The transverse reinforcement must be in a closed stirrup form to resist torsional stresses or in a U-shape lapped at about the midwidth of the stairs.

An alternative method of providing a helical stair is to use a central helical girder located at the midwidth of the stairs and have the steps project equally on both sides of the girder. Each step is analyzed as a cantilever, and the reinforcement bars extend all along the top of the run. Precast concrete steps may be used and can be fixed to specially prepared horizontal faces at the top surfaces of the girder.

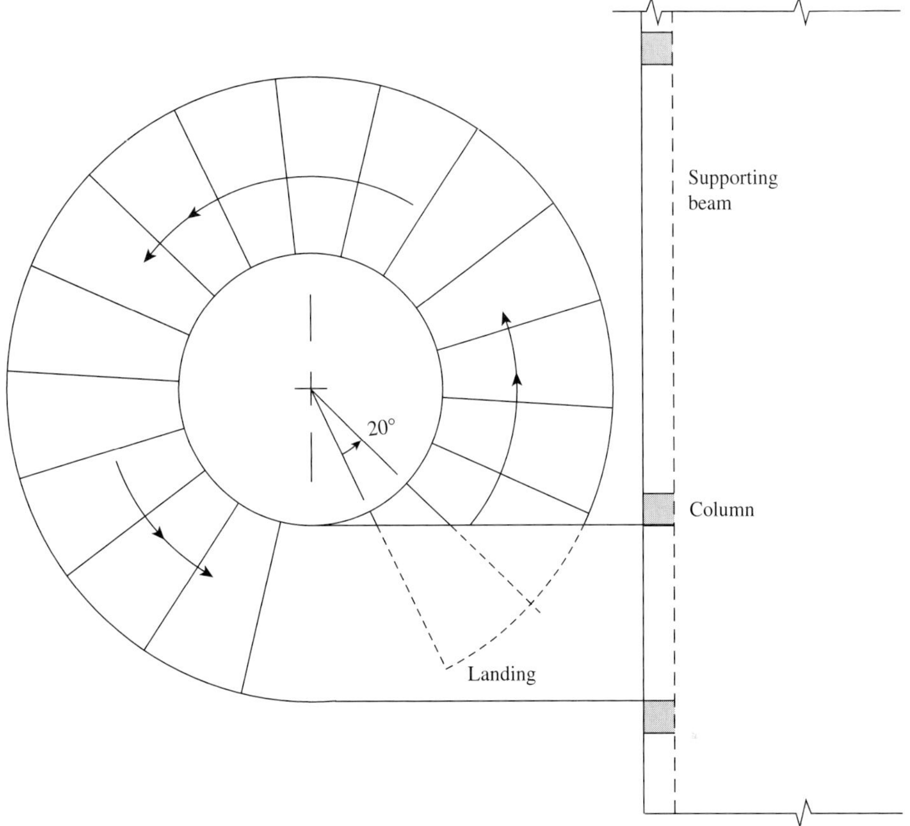

Figure 18.17 Plan of a helical staircase (16 equal runs at 20° pitch).

18.3 EXAMPLES

Example 18.1

Design the cantilever stairs shown in Figure 18.18 to carry a uniform live load of 100 psf. Assume the rise of the steps equals 6.0 in. and the run equals 12 in. Use $f'_c = 3$ Ksi and $f_y = 60$ Ksi.

Solution

1. Loads: Assume the thickness of the slab (waist) is 4.0 in. Weight of the assumed slab (areas A_1 and A_2) is

$$\text{trapezoidal area } mnn'm' = \left(\frac{4.9 + 10.9}{2 \times 12}\right)(1)(150) = 98.8 \text{ lb/ft}$$

Refer to Figure 18.18(b). Assume the weight of the step cover is 5 lb/ft. Total D.L. = 103.8 lb/ft.

$$W_u = 1.4D + 1.7L = 1.4 \times 103.8 + 1.7 \times 100 = 315 \text{ lb/ft}$$

2. Maximum bending moment per step is $W_u l^2/2$.

$$M_u = \frac{0.315}{2}(6)^2 = 5.67 \text{ K·ft}$$

Average thickness of a step is $(10.9 + 4.9)/2 = 7.9$ in. Let $d = 7.9 - 0.75$ (concrete cover) $- 0.25$ ($\frac{1}{2}$ bar diameter) $= 6.9$ in.

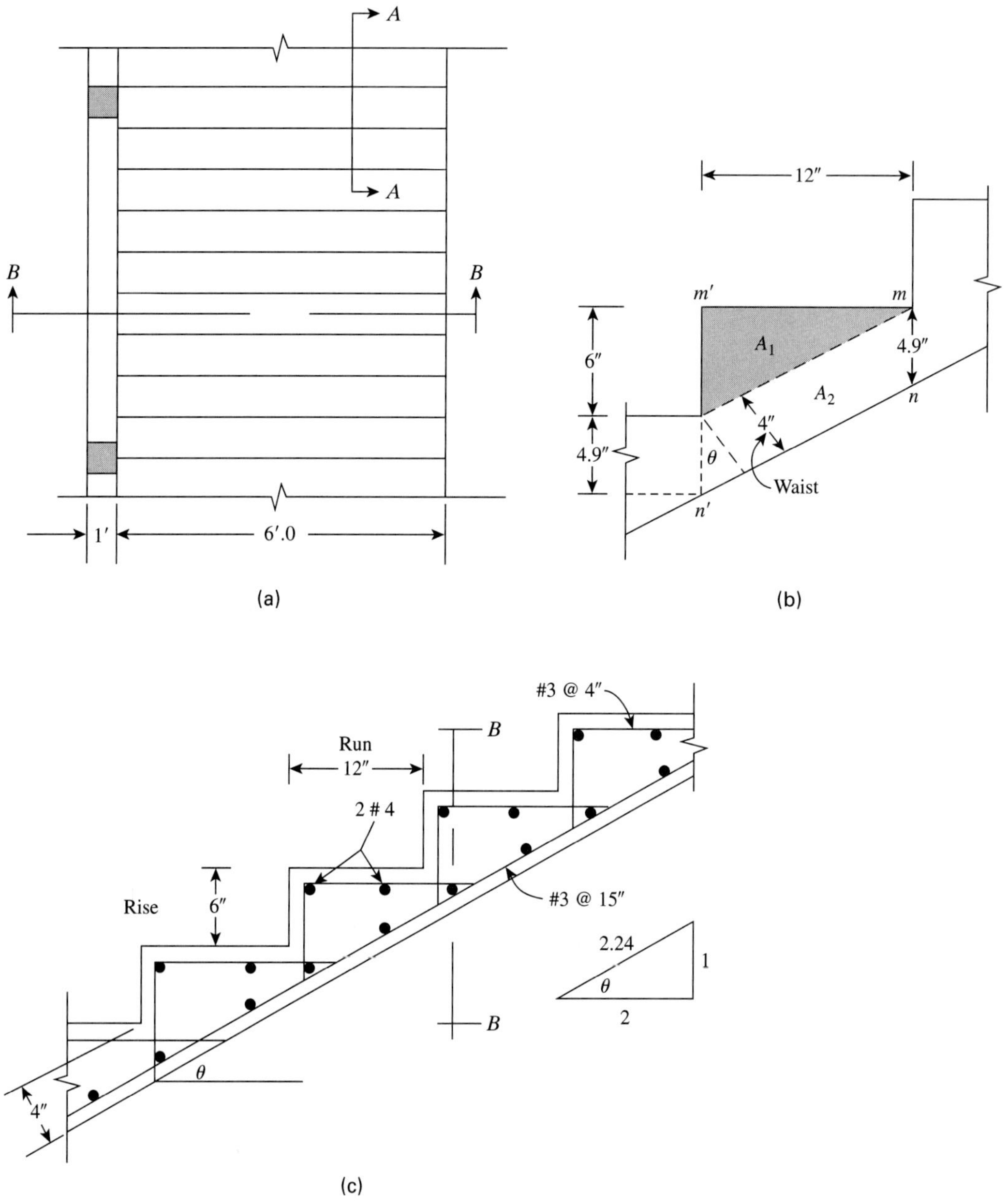

(a)

(b)

(c)

Figure 18.18 Cantilever stairs, Example 18.1: (a) plan, (b) section in one step, (c) section *A-A*, and (d) section *B-B*.

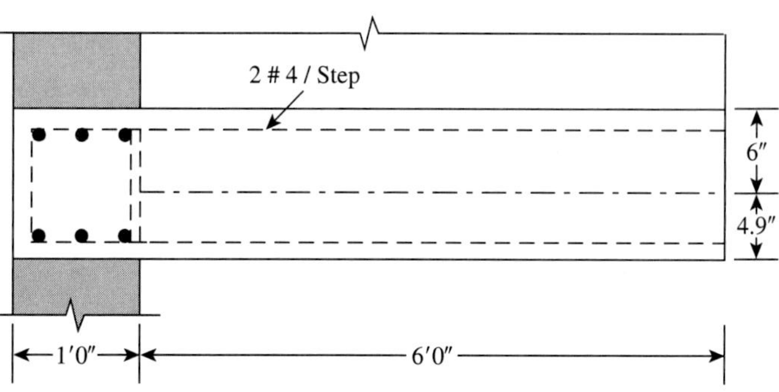

(d)

$$M_u = \phi A_s f_y\left(d - \frac{a}{2}\right), \qquad \text{Assume } a = 0.5 \text{ in.}$$

$$A_s = \frac{5.67 \times 12}{0.9 \times 60(6.9 - 0.25)} = 0.19 \text{ in.}^2$$

$$\text{Check:} \quad a = \frac{A_s f_y}{0.85 f'_c b} = \frac{0.19 \times 60}{0.85 \times 3 \times 12} = 0.38 \text{ in.} \quad \text{(close to 0.5 in.)}$$

$$\text{Minimum } A_s = 0.00333(12)(6.9) = 0.28 \text{ in.}^2$$

Use two no. 4 bars per step. A smaller depth may be adopted, but to avoid excessive deflection and vibration of stairs, a reasonable depth must be chosen.

3. Check flexural shear at a distance d from the face of the support.

$$V_u = 0.315\left(6 - \frac{6.9}{12}\right) = 1.7 \text{ K}$$

$$\phi V_c = 0.85(2\sqrt{f'_c}bd) = \frac{0.85}{1000} \times 2 \times \sqrt{3000}\,(12 \times 6.9) = 7.7 \text{ K}$$

Because $V_u < \phi V_c/2$, no shear reinforcement is required. But it is recommended to use no. 3 stirrups spaced at 4 in. to hold the main reinforcement.

4. The stairs must remain in equilibrium either by the weight of the wall or by a reinforced concrete beam within the wall. In this case, the beam will be subjected to torsional moment of 5.7 K · ft/ft.
5. Reinforcement details are shown in Figure 18.18.

Example 18.2

Design the staircase shown in Figure 18.19, which carries a uniform live load of 100 psf. Assume a rise of 7.0 in. and a run of 10.75 in. Use $f'_c = 3$ Ksi and $f_y = 60$ Ksi.

Solution

1. Structural system: If no stringer beam is used, one of the four possible solutions shown in Figure 18.2 may be adopted. When no intermediate supports are used, the flight of stairs will be supported at the ends of the upper and lower landings. This structural system will be adopted in this example.

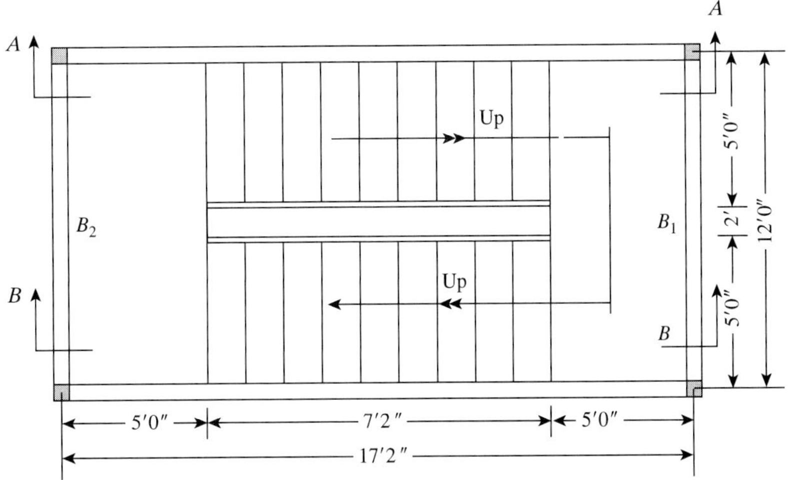

Figure 18.19 Example 18.2.

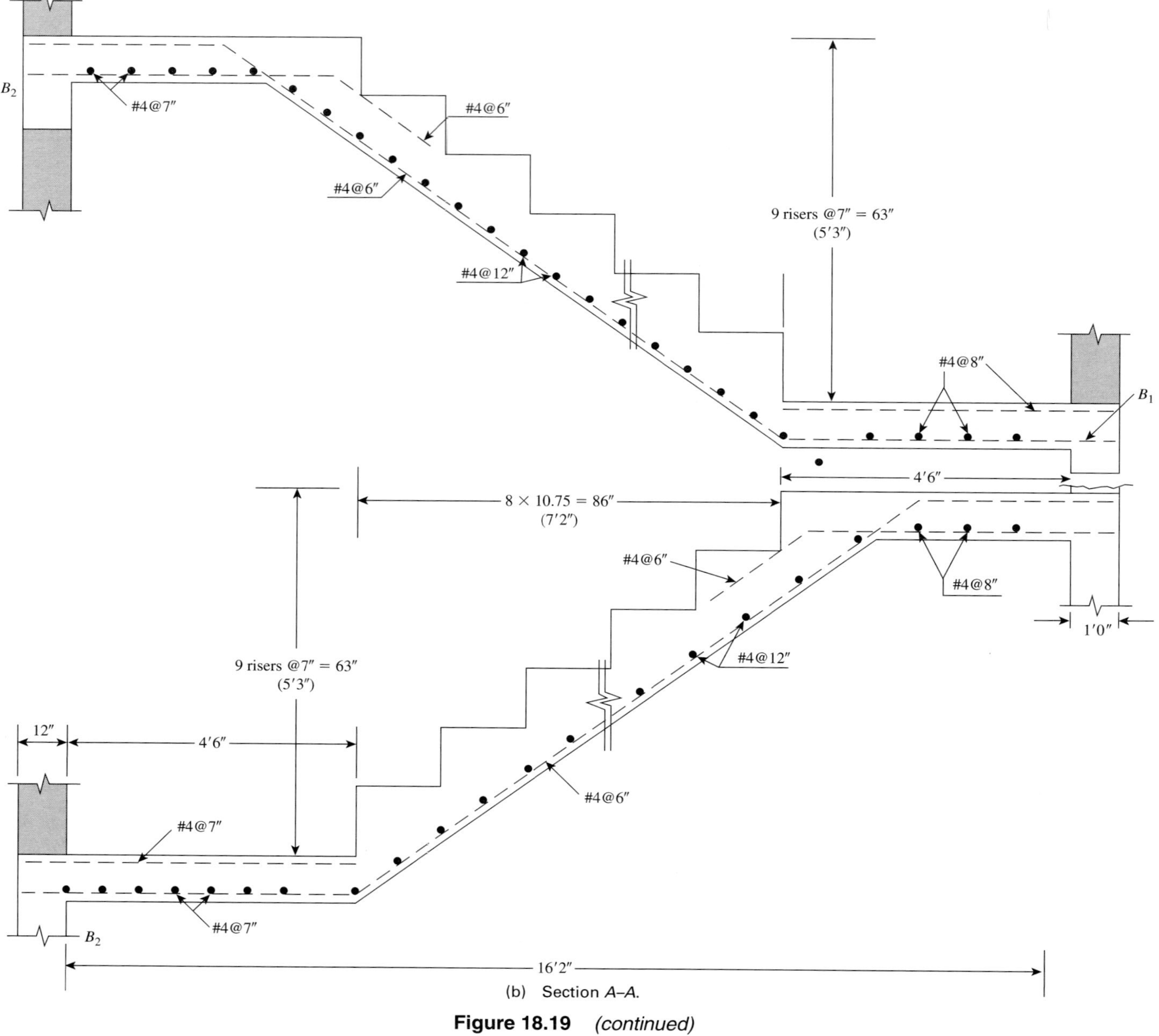

(b) Section *A–A*.

Figure 18.19 *(continued)*

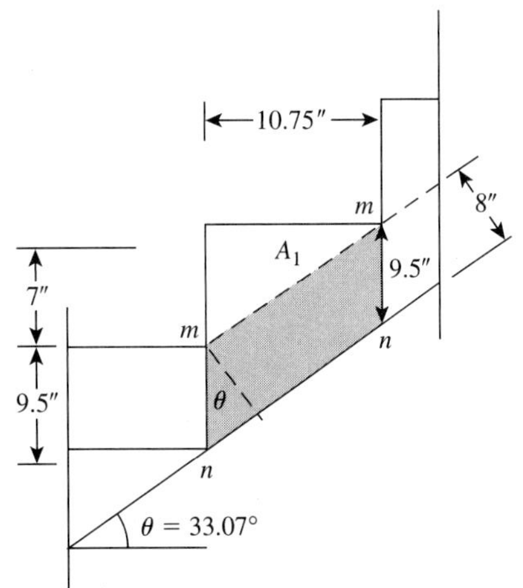

(c) Section in steps.

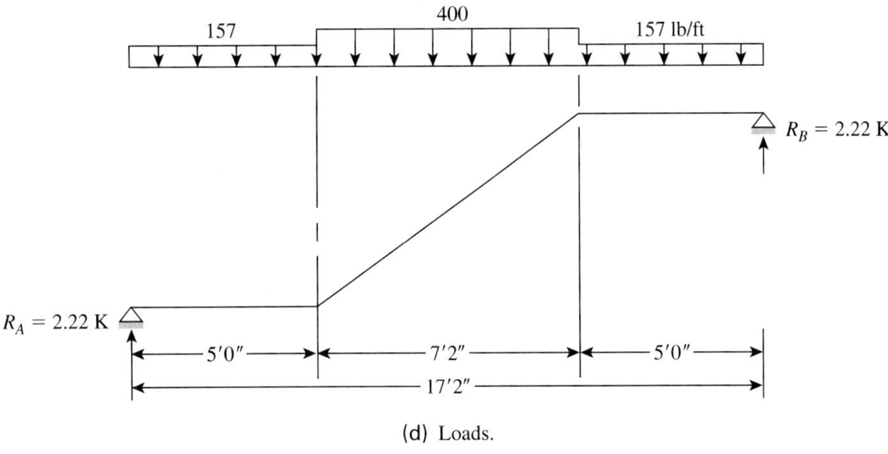

(d) Loads.

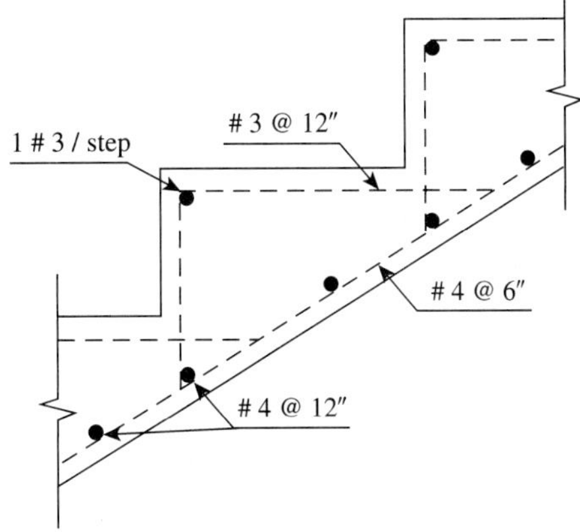

(e) Reinforcement details.

Figure 18.19 *(continued)*

2. Loads: Assume the thickness of the slab (waist) is 8.0 in.

$$\text{Weight of one step} = \text{trapezoidal area} \times 150 \text{ pcf}$$

$$= \left(\frac{9.5 + 16.5}{2 \times 12}\right)\left(\frac{10.75}{12}\right)(150) = 145.6 \text{ lb per step}$$

$$\text{Average weight per foot length} = 145.6\left(\frac{12}{10.75}\right) = 162.5 \text{ lb/ft}$$

$$\text{Weight of 8-in. landing} = \frac{8}{12} \times 150 = 100 \text{ lb/ft}$$

Assume weight of step cover is 2 lb/ft. The total D.L. on stairs is $162.5 + 2 = 164.5$ lb/ft, say, 165. The total D.L. on landing is $100 + 2 = 102$ lb/ft.

$$W_u \text{ (on stairs)} = 1.4 \times 165 + 1.7 \times 100 = 400 \text{ lb/ft}$$

$$W_u \text{ (on landing)} = 1.4 \times 102 + 1.7 \times 100 = 314 \text{ lb/ft}$$

Because the load on the landing is carried into two directions, only half the load will be considered in each direction.

3. Calculate the maximum bending moment and steel reinforcement (Figure 18.19(d)):

a. The moment at midspan is

$$M_u = 2.22\left(\frac{17.2}{2}\right) - (0.157 \times 5)(6.1) - (0.400)\frac{(3.6)^2}{2} = 11.71 \text{ K}\cdot\text{ft}$$

Let $d = 8.0 - 0.75$ (concrete cover) $- 0.25$ ($\frac{1}{2}$ bar diameter) $= 7.0$ in.

b. $M_u = \phi A_s f_y(d - a/2)$; assume $a = 0.8$ in.

$$A_s = \frac{11.71 \times 12}{0.9 \times 60(7 - 0.4)} = 0.4 \text{ in.}^2$$

$$\text{Check:} \quad a = \frac{A_s f_y}{0.85 f'_c b} = \frac{0.4 \times 60}{0.85 \times 3 \times 12} = 0.78 \text{ in.}$$

$$\text{Minimum } A_s = 0.0033 \times 12 \times 8 = 0.32 \text{ in.}^2 < 0.4 \text{ in.}^2$$

Use no. 4 bars spaced at 6 in. (A_s provided $= 0.4$ in.^{2}). For 5-ft-wide stairs, use 10 no. 4 bars.

c. Transverse reinforcement must be provided to account for shrinkage.

$$A_s = 0.0018 \times 12 \times 8 = 0.18 \text{ in.}^2/\text{ft}$$

Use no. 4 bars spaced at 12 in. ($A_s = 0.2$ in.^{2}).

d. If the slab will be cast monolithically with its supporting beams, additional reinforcement must be provided at the top of the upper and lower landings. Details of stair reinforcement are shown in Figure 18.19.

4. Minimum slab thickness for deflection is

$$\frac{L}{25} = \frac{17.2 \times 12}{25} = 8.26 \text{ in.}$$

(for a simply supported slab). In the case presented here, where the slab ends are cast with the supporting beams and additional negative reinforcement is provided, minimum thickness can be assumed to be

$$\frac{L}{28} = 7.4 \text{ in.} < 8 \text{ in. used.}$$

5. Design of landings: Considering a 1-ft length of the landing, the load on the landing is as shown in Figure 18.20. The middle 2 ft will carry a full load, whereas the two 5-ft lengths on each side will carry half the ultimate load.

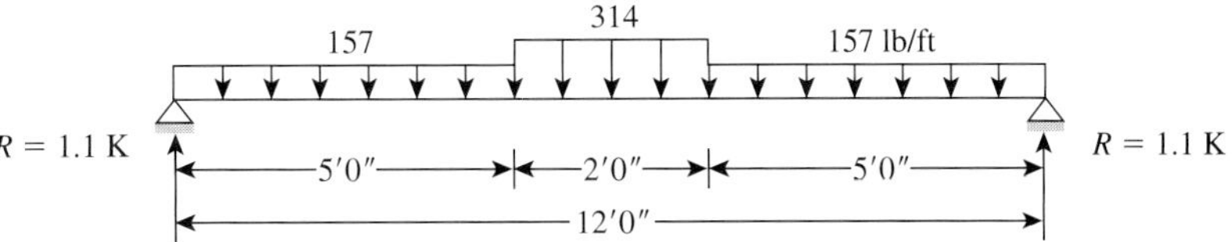

Figure 18.20 Loads on landing, Example 18.2.

$$\text{Maximum bending moment} = (1.1 \times 6) - (0.157 \times 5)(3.5) - (0.314)\frac{(1)^2}{2} = 3.7 \text{ K}\cdot\text{ft}$$

Because the bars in the landing will be placed on top of the main stair reinforcement,

$$d = 8.0 - 0.75 - \frac{4}{8} - 0.25 = 6.375 \text{ in.,} \qquad \text{say } 6.3 \text{ in.}$$

Assume $a = 0.4$ in.

$$A_s = \frac{3.7 \times 12}{0.9 \times 60(6.3 - 0.2)} = 0.14 \text{ in.}^2 < A_s\,(\text{min})\ of\ 0.32 \text{ in.}^2$$

Use $A_s = 0.32$ in.^{2} Use no. 4 bars spaced at 7 in. $(A_s = 0.34 \text{ in.}^2)$.

6. The transverse beams at the landing levels must be designed to carry loads from stairs (2.3 K/ft) in addition to their own weight and the weight of the wall above.
7. Check shear as usual.

Example 18.3

Design the simply supported run-riser stairs shown in Figure 18.21 for a uniform live load of 100 psf. Use $f'_c = 3$ Ksi and $f_y = 60$ Ksi.

Solution

1. Loads: Assume the thickness of runs and risers is 6 in. The concentrated load at each riser is calculated as follows (refer to Figure 18.21(b)). Due to dead load per foot depth of run,

$$P_D = \left(\frac{16}{12} \times \frac{6}{12} + \frac{1}{12} \times \frac{6}{12}\right)150 = 106 \text{ lb}$$

Note that the node dead load on the landing is less than 106 lb but can be assumed to be equal to P_D to simplify calculations. Due to live load per foot depth of run, $P_L = \frac{10}{12} \times (100) = 83.33$ lb.

$$\text{Ultimate load } P_u = 1.4P_D + 1.7P_L$$

$$= 1.4 \times 106 + 1.7 \times 83.33 = 290 \text{ lb}$$

2. Calculate the bending moments at midspan: Loads in this example are symmetrical about midspan section B. Reaction at A, R_A, is $\frac{1}{2}(15)(290) = 2175$ lb $= \left(7\frac{1}{2}P \times 290\right)$

$$\text{Moment at B} = R_A(8S) - 7P_u(4S)$$

$$= 2.175(8 \times 10) - 7(0.29)(4 \times 10) = 92.8 \text{ K}\cdot\text{in.}$$

3. Calculate the reinforcement required at midspan section: For $h = 6$ in., $d = 6 - 1.0 = 5.0$ in.

$$R_u = \frac{M_u}{bd^2} = \frac{92.8 \times 1000}{12(5.0)^2} = 309 \text{ psi}$$

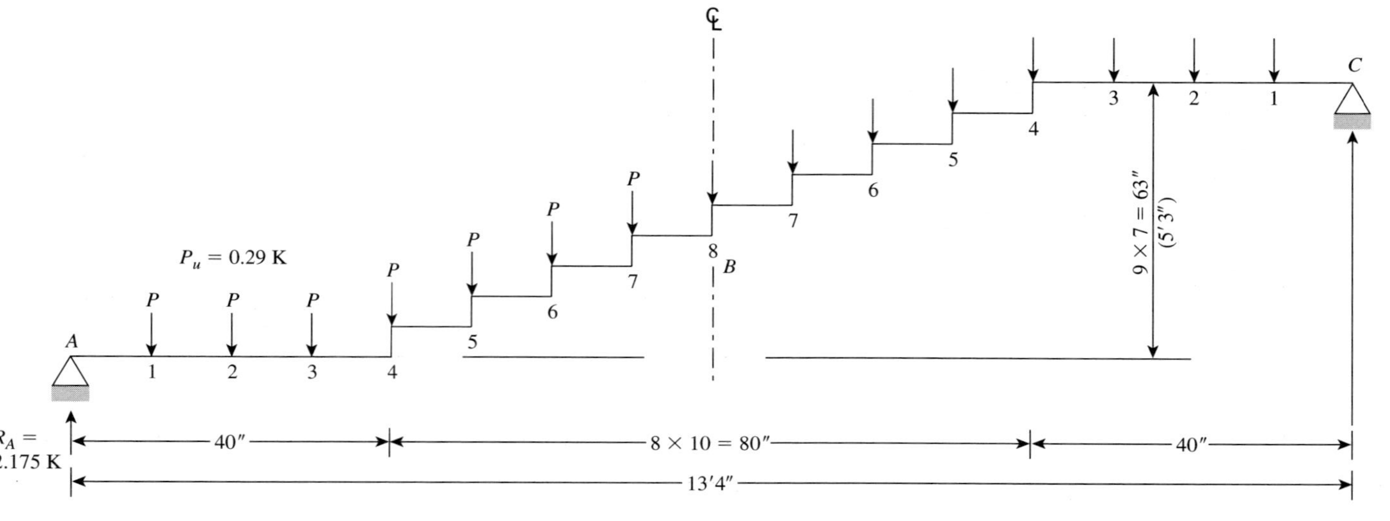

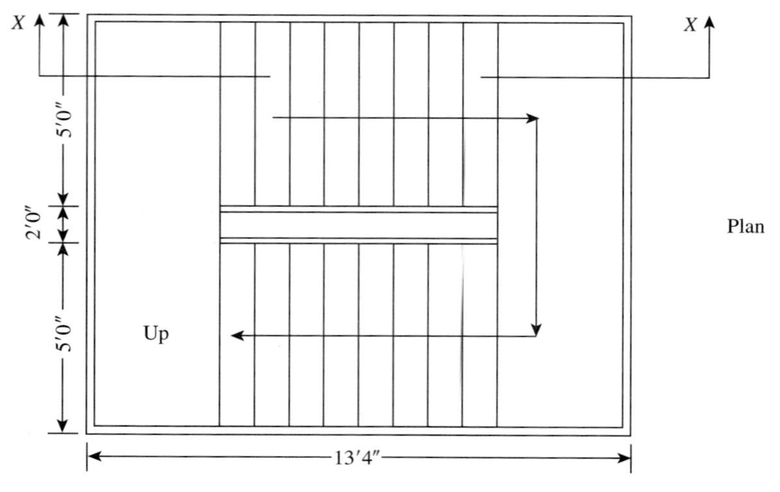

(a) Plan and section X–X

Figure 18.21 Example 18.3.

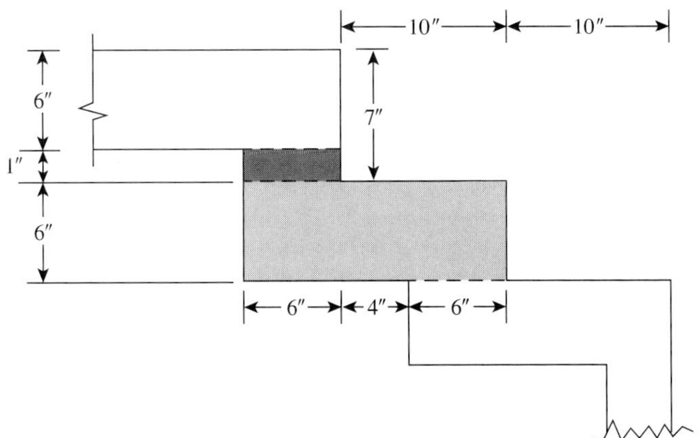

(b) Section in steps.

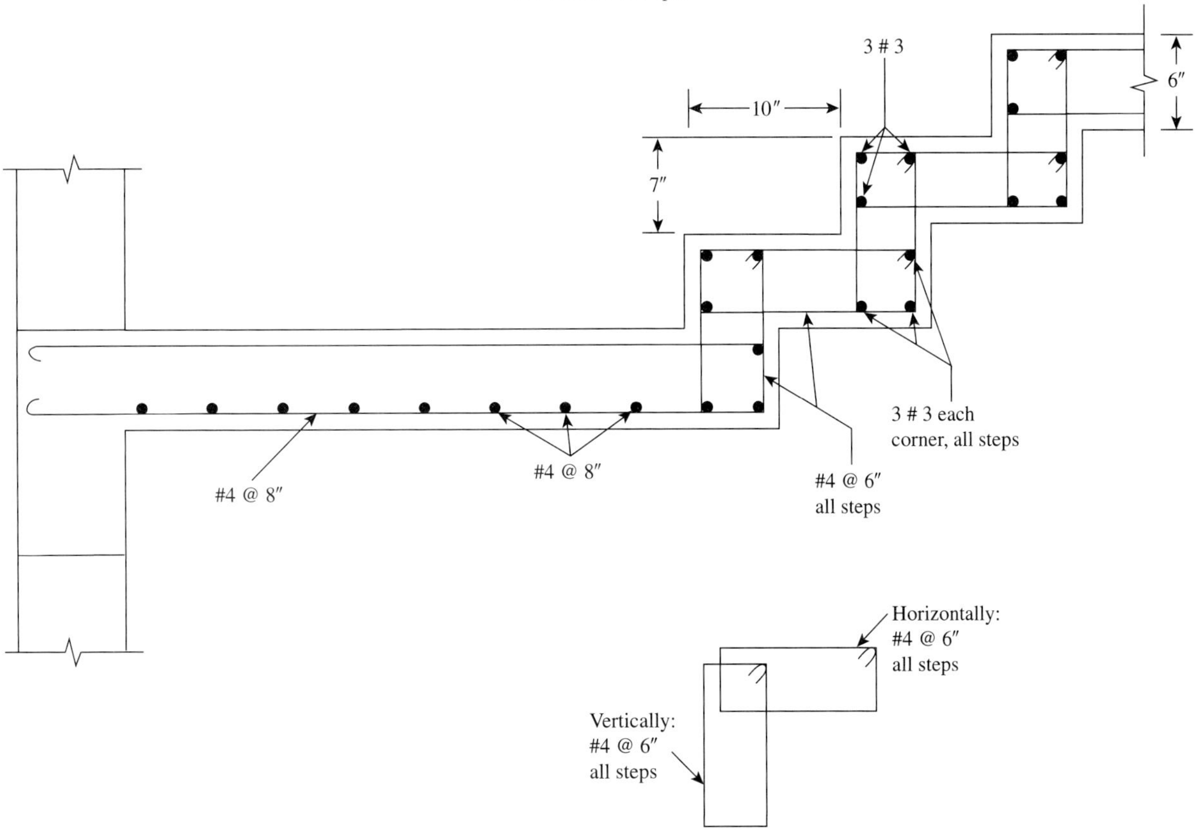

3 # 3
10″
7″
6″
#4 @ 8″
#4 @ 8″
#4 @ 6″
all steps
3 # 3 each
corner, all steps
Horizontally:
#4 @ 6″
all steps
Vertically:
#4 @ 6″
all steps

(c) Reinforcement details.

Figure 18.21 *(continued)*

For $f'_c = 3$ Ksi, $f_y = 60$ Ksi, and $R_u = 309$ psi, the steel ratio is $\rho = 0.0061$.

$$A_s = 0.0061 \times 12 \times 5.0 = 0.366 \text{ in.}^2$$

Use no. 4 bars spaced at 6 in. $(A_s = 0.39 \text{ in.}^2)$ horizontally and vertically in closed stirrup form. For distribution bars, use minimum ρ of 0.0018.

$$A_s = 0.0018 \times 16 \times 6 = 0.18 \text{ in.}^2$$

Use no. 3 bars spaced at 6 in. ($A_s = 0.22$ in.^{2}). For each step corner, use three no. 3 bars ($A_s = 0.33$ in.^{2}), as shown in Figure 18.21(c).

4. The moments and reinforcement required for other sections can be prepared in table form, as follows:

Location		A	1	2	3	4	5	6	7	8
B.M.	K·in.	0	22	41	57	70	80	87	91	92.8
R_u	psi	0	73	137	190	233	267	290	303	309
ρ	%	0	0.18	0.26	0.38	0.46	0.52	0.58	0.60	0.61
A_s	in.^{2}	0	0.11	0.16	0.23	0.28	0.31	0.35	0.36	0.37

Use no. 4 bars at 8 in. for the landing and no. 4 bars at 6 in. for the steps. For distribution bars, use minimum ρ of 0.0018. For $A_s = 0.18$ in.^{2}, use no. 4 bars spaced at 8 in. in the landing. Details of reinforcement are shown in Figure 18.21(c).

5. Check reinforcement required in the transverse direction of landing: Load per square foot on the landing is $\frac{290}{10} \times 12 = 348$ psf.

$$M_u = \frac{0.348}{8}(12)^2 \times 12 = 75 \text{ K}\cdot\text{in.}$$

$$R_u = \frac{75 \times 1000}{12(5.0)^2} = 250 \text{ psi}, \qquad \rho = 0.0049, \qquad A_s = 0.29 \text{ in.}^2$$

Use no. 4 bars spaced at 8 in. ($A_s = 0.29$ in.^{2}).

6. If a uniform load is assumed to be acting on the flight of stairs, similar results will be obtained. For example, ultimate node load was calculated to be 290 lb acting over a 10-in. run width. Load per foot is $\frac{290}{10} \times 12 = 348$ lb/ft. Maximum moment is at midspan, section B:

$$M_u = \frac{0.348}{8}(13.33)^2 = 92.8 \text{ K}\cdot\text{in.}$$

Moments at other sections can be easily calculated, and the design can be arranged in a table form, as explained in Step 4.

SUMMARY

Sections 18.1–18.2

The different types of stairs are: single and multiple flights, cantilever and precast concrete flights, free-standing and helical staircases, and run-riser stairs.

Section 18.3

Design examples are presented in this section.

REFERENCES

1. V. A. Morgan, "Comparison of Analyses of Helical Stairs." *Concrete and Constructional Engineering*, London (March 1960).
2. A. C. Scordelis, "Internal Forces in Uniformly Loaded Helicoidal Girders." *Journal of the American Concrete Institute* (April 1960).
3. A. R. Cusens and Jing-gwo Kuang. "A Simplified Method of Analyzing Free Standing Stairs." *Concrete & Constructional Engineering* (May 1965): 167–72.
4. A. R. Cusens and Jing-gwo Kuang. "Experimental Study of a Free Standing Staircase." *Journal of the American Concrete Institute* (May 1966): 587–603.

5. Franz Sauter. "Free Standing Stairs." *Journal of the American Concrete Institute* (July 1961): 847–69.

6. Cusens & Santathadapon. "Design Charts for Helical Stairs with Fixed Supports." Concrete Publications Ltd. 60 Buckingham Gate, London, S.W.I. England.

7. A. H. Mattock. "Design and Construction of a Helical Staircase." *Concrete & Constructional Engineering* (March 1957): 99–105.

PROBLEMS

18.1 Design a typical flight of the staircase shown in Figure 18.22, which is a part of a multistory building. The height between the concrete floors is 10 ft (3.0 m). The stairs are supported at the ends of the landings and carry a live load equal to 120 psf (5.75 kN/m^2); f'_c = 3 Ksi (20 MPa) and f_y = 60 Ksi (400 MPa).

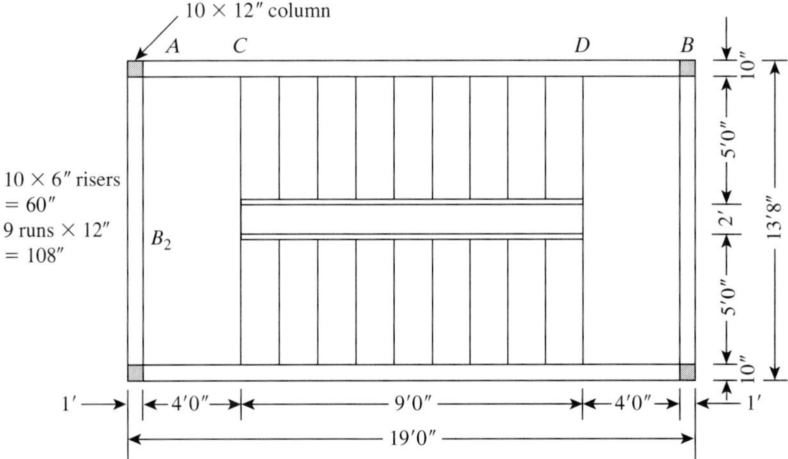

Figure 18.22 Problem 18.1.

18.2 Repeat Problem 18.1 if the stairs are supported by four transverse beams at *A*, *B*, *C*, and *D* and the live load is increased to 150 psf (7.2 kN/m^2).

18.3 The stairs shown in Figure 18.23 are to be designed for a live load equal to 100 psf (4.8 kN/m). The stairs are supported by beams, as shown. Design the stairs and the supporting beams for f'_c = 3 Ksi (20 MPa) and f_y = 60 Ksi (400 MPa).

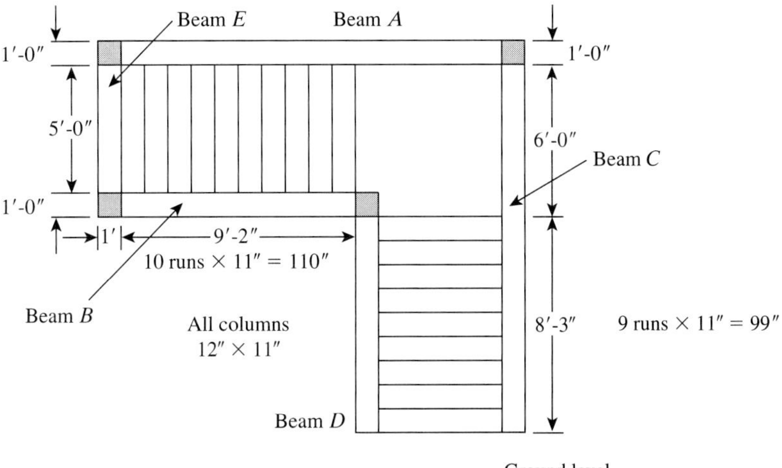

Figure 18.23 Problem 18.3.

18.4 Design a typical flight of stairs in a public building for the staircase arrangement shown in Figure 18.24. The stairs are supported by central beams, *A* and *B*. Design only one flight and the supporting beams *A* and *B*. The runs are 1.0 ft (300 m) deep and the rises are 6.5 in. high. Use $f'_c = 3$ Ksi (20 MPa), $f_y = 60$ Ksi (400 MPa), and a live load equal to 80 psf (3.85 kN/m^2).

Note: Design the beams for bending moments and shear, and neglect torsional moments caused by loading one-half of the steps.

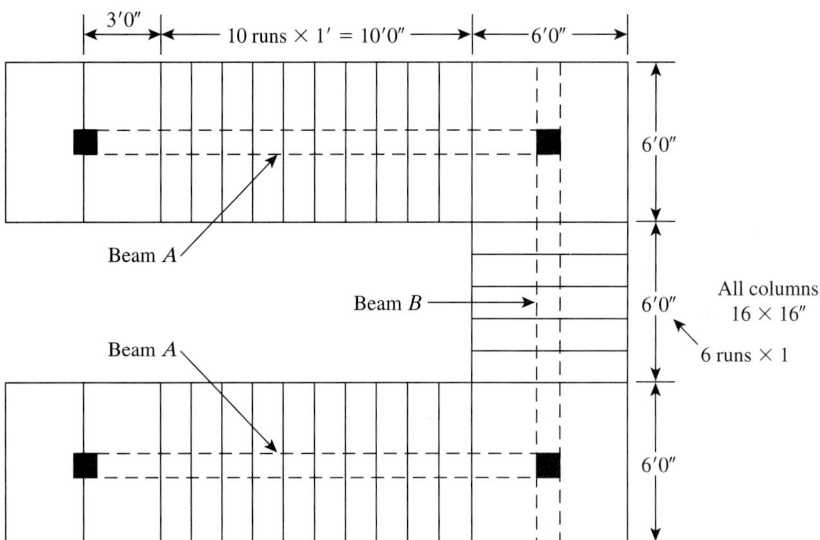

Figure 18.24 Problem 18.4.

18.5 Repeat Example 18.3 if the run is 12 in. (300 mm) and the rise is 6 in. (150 mm).

18.6 Repeat Example 18.3 if the landing is 5 ft (6 × 10″), runs are 8.33 ft (10 × 10″), risers at 5.5 ft (11 × 6″), and the live load is 120 psf.

19 BEAMS CURVED IN PLAN

Curved beams in an office building.

19.1 INTRODUCTION

Beams curved in plan are used to support curved floors in buildings, balconies, curved ramps and halls, circular reservoirs, and similar structures. In a curved beam, the center of gravity of the loads acting normal to the plane of curvature lies outside the line joining its supports. This situation develops torsional moments in the beam, in addition to bending moments and shearing forces. To maintain the stability of the beam against overturning, the supports must be fixed or continuous. In this chapter, the design of curved beams subjected to loads normal to the plane of curvature is presented. Analysis of curved beams subjected to loads in the plane of curvature is usually discussed in books dealing with mechanics of solids.

Analysis of beams curved in plan was discussed by Wilson [1]. He introduced formulas and coefficients to compute stresses in curved flexural members. Timoshinko [2], [3] also introduced several expressions for calculating bending stresses in square and rectangular sections. Tables and formulas for the calculation of bending and torsional moments, shear, and deflections for different cases of loadings on curved beams and rings are presented by Roark and Young [5].

19.2 UNIFORMLY LOADED CIRCULAR BEAMS

The first case to be considered here is that of a circular beam supported on columns placed at equal distances along the circumference of the beam and subjected to normal loads. Due to symmetry, the column reactions will be equal, and each reaction will be equal to the

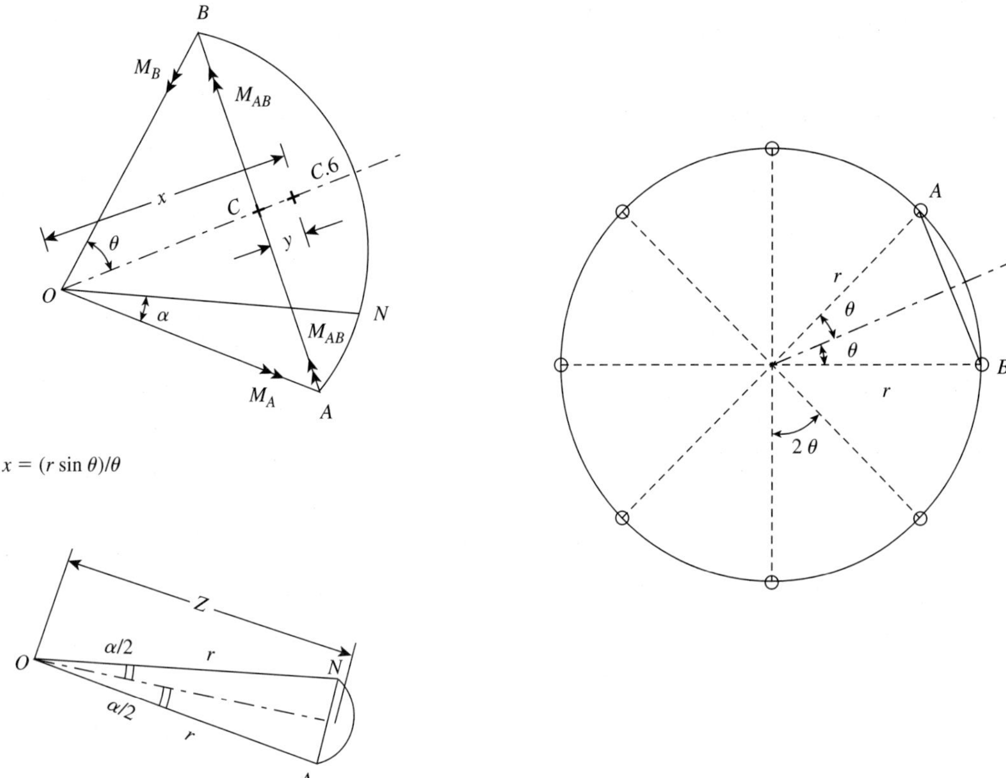

Figure 19.1 Circular beam.

total load on the beam divided by the number of columns. Referring to Figure 19.1, consider the part AB between two consecutive columns of the ring beam. The length of the curve AB is $r(2\theta)$, and the total load on each column is $P_u = w_u r(2\theta)$, where r is the radius of the ring beam and w_u is the factored load on the beam per unit length. The center of gravity of the load on AB lies at a distance

$$x = \left(\frac{r\sin\theta}{\theta}\right)$$

from the center O. The moment of the load P_u about AB is

$$M_{AB} = P_u \times y = P_u(x - r\cos\theta) = w_u r(2\theta)\left(\frac{r\sin\theta}{\theta} - r\cos\theta\right)$$

Consequently, the two reaction moments, M_A and M_B, are developed at supports A and B, respectively. The component of the moment at support A about AB is $M_A \sin\theta = M_B \sin\theta$. Equating the applied moment, M_{AB}, to the reaction moments components at A and B,

$$2M_A \sin\theta = M_{AB} = w_u r(2\theta)\left(\frac{r\sin\theta}{\theta} - r\cos\theta\right)$$

$$M_A = M_B = w_u r^2(1 - \theta\cot\theta) \tag{19.1}$$

The shearing force at support A is

$$V_A = \frac{P_u}{2} = w_u r\theta \tag{19.2}$$

The shearing force at any point N, V_N, is $V_A - w_u(r\alpha)$, or

$$V_N = w_u r(\theta - \alpha) \tag{19.3}$$

The load on AN is $w_u(r\alpha)$ and acts at a distance equal to

$$Z = \frac{r \sin \alpha/2}{\alpha/2}$$

from the center 0. The bending moment at point N on curve AB is equal to the moment of all forces on one side of N about the radial axis ON.

$$M_N = V_A(r \sin \alpha) - M_A \cos \alpha - (\text{load on the curve } AN)\left(Z \sin \frac{\alpha}{2}\right)$$

$$M_N = w_u r\theta(r \sin \alpha) - w_u r^2(1 - \theta \cot \theta) \cos \alpha$$

$$- (w_u r\alpha)\left(\frac{r \sin \alpha/2}{\alpha/2} \times \sin \frac{\alpha}{2}\right)$$

$$= w_u r^2\left[\theta \sin \alpha - \cos \alpha + (\theta \cot \theta \cos \alpha) - 2 \sin^2 \frac{\alpha}{2}\right]$$

$$M_N = w_u r^2[\theta \sin \alpha + (\theta \cot \theta \sin \alpha) - 1] \tag{19.4}$$

(Note that $\cos \alpha = 1 - 2 \sin^2 \alpha/2$.) The torsional moment at any point N on curve AB is equal to the moment of all forces on one side of N about the tangential axis at N.

$$T_N = M_A \sin \alpha - V_A \times r(1 - \cos \alpha) + w_r r\alpha\left(r - \frac{r \sin \alpha/2}{\alpha/2} \times \cos \alpha/2\right)$$

$$= w_u r^2(1 - \theta \cot \theta) \sin \alpha - w_u r^2\theta(1 - \cos \alpha) + w_u r^2(\alpha - \sin \alpha)$$

$$T_n = w_u r^2(\alpha - \theta + \theta \cos \alpha - \theta \cot \theta \sin \alpha) \tag{19.5}$$

To obtain the maximum value of the torsional moment T_N, differentiate equation (19.5) with respect to α and equate it to 0. This step will give the value of α for maximum T_N.

$$\sin \alpha = \frac{1}{\theta}\left[\sin^2 \theta \pm \cos \theta \sqrt{\theta^2 - \sin^2 \theta}\right] \tag{19.6}$$

The values of the support moment, midspan moment, the torsional moment, and its angle α from the support can be calculated from equations (19.1) through (19.6). Once the number of supports n is chosen, the angle θ is known,

$$2\theta = \frac{2\pi}{n} \quad \text{and} \quad \theta = \frac{\pi}{n}$$

and the moment coefficients can be calculated as shown in Table 19.1. Note that the angle α is half the central angle between two consecutive columns.

Table 19.1 Force Coefficients of Circular Beams

Number of Supports, n	$\theta = \frac{\pi}{n}$	K_1	K_2	K_3	$\alpha°$ for T_u (max)
4	90	0.215	0.110	0.0330	19.25
5	72	0.136	0.068	0.0176	15.25
6	60	0.093	0.047	0.0094	12.75
8	45	0.052	0.026	0.0040	9.50
9	40	0.042	0.021	0.0029	8.50
10	36	0.034	0.017	0.0019	7.50
12	30	0.024	0.012	0.0012	6.25

$$\text{Load on each column is } P_u = w_u r(2\theta) = w_u r\left(\frac{2\pi}{n}\right)$$

$$\text{Maximum shearing force is } V_u = \frac{P_u}{2}$$

$$\text{Negative moment at any support} = K_1 w_u r^2 \tag{19.7}$$

$$\text{Positive moment at midspan} = K_2 w_u r^2 \tag{19.8}$$

$$\text{Maximum torsional moment} = K_3 w_u r^2 \tag{19.9}$$

The variation of the shearing force and bending and torsional moments along a typical curved beam AB are shown in Figure 19.2.

Example 19.1

Design a circular beam supported on eight equally spaced columns. The centerline of the columns lies on a 40-ft-diameter circle. The beam carries a uniform dead load of 6 K/ft and a live load of 4 K/ft. Use $f'_c = 4$ Ksi, $f_y = 60$ Ksi, and $b = 14$ in.

Solution

1. Assume a beam size of 14 × 24 in. The weight of the beam is

$$\frac{14 \times 24}{12 \times 12}(0.150) = 0.35 \text{ K/ft}$$

The ultimate uniform load is $w_u = 1.4(6 + 0.35) + 1.7(4) = 15.7$ K/ft.

2. Because the beam is symmetrically supported on eight columns, the moments can be calculated by using equations (19.7) through (19.9) and Table 19.1. Negative moment at

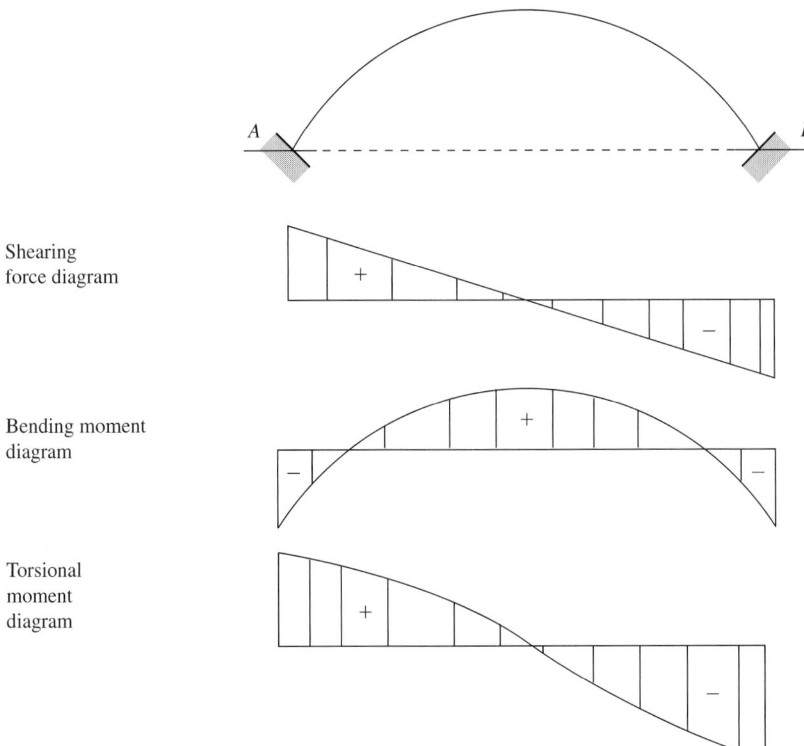

Figure 19.2 Forces in a circular beam.

any support is $K_1 w_u r^2 = 0.052(15.7)(20)^2 = 326.6$ K·ft. The positive moment at midspan is $K_2 w_u r^2 = 0.026(15.7)(20)^2 = 163.3$ K·ft. The maximum torsional moment is $K_3 w_u r^2 = 0.004(15.7)(20)^2 = 25.12$ K·ft. Maximum shear is

$$V_u = \frac{P_u}{2} = \frac{w_u r}{2}\left(\frac{2\pi}{n}\right) = (15.7)(20)\left(\frac{\pi}{8}\right) = 123.3 \text{ K}$$

3. For the section at support, $M_u = 326.6$ K·ft. Let $d = 21.5$ in.; then

$$R_u = \frac{M_u}{bd^2} = \frac{326.6 \times 12{,}000}{14(21.5)^2} = 605 \text{ psi}$$

For $f'_c = 4$ Ksi and $f_y = 60$ Ksi, $\rho = 0.0126 < \rho_{\max} = 0.0214$.

$$A_s = 0.0126 \times 14 \times 21.5 = 3.8 \text{ in.}^2$$

4. For the section at midspan, $M_u = 163.3$ K·ft.

$$R_u = \frac{163.3 \times 12{,}000}{14(21.5)^2} = 303 \text{ psi}$$

$$\rho = 0.006 \qquad \text{and} \qquad A_s = 0.006 \times 14 \times 21.5 = 1.81 \text{ in.}^2$$

Use two no. 9 bars.

5. Maximum torsional moment is $T_u = 25.12$ K·ft, and it occurs at an angle $\alpha = 9.5°$ from the support (Table 19.1). Shear at the point of maximum torsional moment is equal to the shear at the support minus $w_u r\alpha$.

$$V_u = 123.3 - 15.7(20)\left(\frac{9.5}{180} \times \pi\right) = 71.24 \text{ K}$$

The procedure for calculation of the shear and torsional reinforcement for $T_u = 25.12$ K·ft and $V_u = 71.24$ K is similar to Example 15.2.

a. Shear reinforcement is needed when $V_u > \phi V_c/2$.

$$\phi V_c = 2\phi\sqrt{f_c}\,bd = 2(0.85)\sqrt{4000}\,(14 \times 21.5) = 32.4 \text{ K}$$

$$\phi V_c/2 = 16.2 \text{ K} < V_u = 71.24 \text{ K}$$

Shear reinforcement is needed.

b. Torsional reinforcement is not needed when

$$T_u \le T_a = \phi\sqrt{f'_c}\left(\frac{A^2_{cP}}{p_{cp}}\right)$$

$$A_{cp} = x_0 y_0 = 14 \times 24 = 336 \text{ in.}^2$$

$$P_{cp} = 2(x_0 + y_0) = 2(14 + 24) = 76 \text{ in.}$$

$$T_a = 0.85\sqrt{4000}\left(\frac{336^2}{76}\right) = 79.86 \text{ K·in.}$$

$$T_u = 25.12 \text{ K·ft} = 301.4 \text{ K·in.} > T_a$$

Torsional reinforcement is needed.

c. Design for shear:

i. $V_u = \phi V_c + \phi V_s$ and $\phi V_c = 32.4$ K. $71.24 = 32.4 + 0.85V_s$, so $V_s = 45.7$ K

ii. Maximum $V_s = 8\sqrt{f'_c}\,bd = 8\sqrt{4000}\,(14 \times 21.5) = 152.3 \text{ K} > V_u$

iii. $A_v/S = V_s/f_y d = 45.7/(60 \times 21.5) = 0.0354$ in.^{2}/in. (2 legs)

$A_v/2S = 0.0177$ in.^{2}/in. (one leg)

d. Design for torsion:

i. Choose no. 4 stirrups and a 1.5-in. concrete cover:

$$x_1 = 14 - 3.5 = 10.5 \text{ in.}, \qquad y_1 = 24 - 3.5 = 20.5 \text{ in.}$$

$$A_{oh} = x_1 y_1 = 10.5(20.5) = 215.25 \text{ in.}^2$$

$$A_o = 0.85 A_{oh} = 183 \text{ in.}^2$$

$$p_h = 2(x_1 + y_1) = 2(10.5 + 20.5) = 62 \text{ in.}$$

For $\theta = 45°, \cot\theta = 1.0$.

ii. Check the adequacy of the size of the section using equation (15.21):

$$\sqrt{\left(\frac{V_u}{b_w d}\right)^2 + \left(\frac{T_u p_h}{1.7A_{oh}^2}\right)^2} \le \phi\left(\frac{V_c}{b_w d} + 8\sqrt{f'_c}\right)$$

$$\phi V_c = 32.4 \text{ K}, \qquad V_c = 38.12 \text{ K}$$

$$\text{Left-hand side} = \sqrt{\left(\frac{71{,}240}{14 \times 21.5}\right)^2 + \left[\frac{301{,}400 \times 62}{1.7(215.25)^2}\right]^2} = 335 \text{ psi}$$

$$\text{Right-hand side} = 0.85\left(\frac{38{,}120}{14 \times 21.5} + 8\sqrt{4000}\right) = 633 \text{ psi} > 335 \text{ psi}$$

The section is adequate.

iii. Determine the required closed stirrups due to T_u:

$$\frac{A_t}{S} = \frac{T_n \cot\theta}{2A_o f_y}$$

$$= \frac{301.4}{0.85 \times 2 \times 183 \times 60} = 0.0162 \text{ in.}^2/\text{in.} \qquad \text{(one leg)}$$

iv. The total area of one leg stirrup is $0.0177 + 0.0162 = 0.0339$ in.^{2}/in. For no. 4 stirrups, area of one leg $= 0.2$ in.^{2} Spacing of closed stirrups is $0.2/0.0339 = 5.9$ in., say, 5.5 in.

$$\text{Minimum } S = \frac{p_h}{8} = \frac{62}{8} = 7.75 \text{ in.} > 5.5 \text{ in.}$$

$$\text{Minimum } \frac{A_{vt}}{S} = \frac{50b_w}{f_y} = \frac{50(14)}{60{,}000} = 0.0117 \text{ in.}^2/\text{in.}$$

This is less than the A_t/s provided. Use no. 4 closed stirrups spaced at 5.5 in.

e. Longitudinal bars A_l equal $(A_t/s)p_h(f_{yv}/f_{yl})\cot^2\theta$ (equation (15.27)).

$$A_l = 0.0162(62)\left(\frac{60}{60}\right) = 1.0 \text{ in.}^2$$

$$\text{Min. } A_l = \frac{5\sqrt{f'_c}A_{cp}}{f_{yl}} - \left(\frac{A_t}{S}\right)p_h\left(\frac{f_{yv}}{f_{yl}}\right)$$

$$= \frac{(5\sqrt{4000})(336)}{60{,}000} - 0.0162(62)\left(\frac{60}{60}\right) = 0.77 \text{ in.}^2 < 1.0$$

Use $A_l = 1.0$ in.^{2}, with one-third at the top, one-third at middepth, and one-third at the bottom, or 0.33 in.^{2} in each location. For the section at the support, $A_s = 3.8$ in.$^2 + 0.33 = 4.13$ in.^{2} Choose two no. 10 and two no. 9 bars ($A_s = 4.53$ in.^{2}) as top bars. At middepth, use two no. 4 bars ($A_s = 0.4$ in.^{2}). Extend two no. 9 bars of the midspan section to the support. At middepth use two no. 4 bars ($A_s = 0.4$ in.^{2}). Details of the section are shown in Figure 19.3.

Circular beams in an office building.

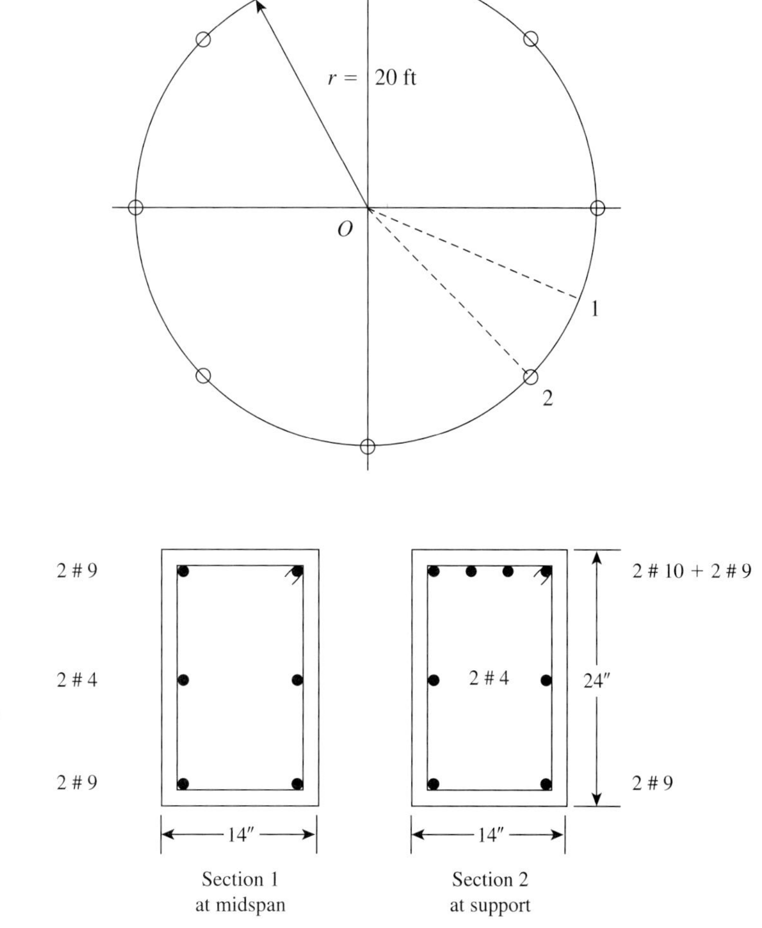

Figure 19.3 Example 19.1.

19.3 SEMICIRCULAR BEAM FIXED AT END SUPPORTS

If a semicircular beam supports a concrete slab, as shown in Figure 19.4, the ratio of the length to the width of the slab is $2r/r = 2$, and the slab is considered a one-way slab. The beam will be subjected to a distributed load, which causes torsional moments in addition to the bending moments and shearing forces. The structural analysis of the curved beam can be performed in steps as follows.

1. Load on beam: The load on the curved beam will be proportional to its distance from the support AB. If the uniform load on the slab equals w psf, the load on the curved beam at any section N is equal to half the load on the area $NCDE$ (Figure 19.4). The lengths are $CN = r\sin\theta, OC = r\cos\theta$, and $CD = (d/d\theta)(r\cos\theta) = (r\sin\theta\, d\theta)$, and the arc NE is $r\,d\theta$.

   The load on the curved beam per unit length is equal to

$$w' = \frac{w(r\sin\theta)r\sin\theta\, d\theta}{2(r\,d\theta)} = \frac{wr\sin^2\theta}{2} \tag{19.10}$$

2. Shearing force at A: For a uniform symmetrical load on the slab, the shearing force at A is equal to

$$V_A = V_B = \int_0^{\pi/2}\left(\frac{wr}{2}\sin^2\theta\right)(r\,d\theta) = \frac{wr^2}{2}\left[\frac{\theta}{2} - \frac{1}{4}\sin 2\theta\right]$$
$$= \left(\frac{\pi}{8}\right)wr^2 = 0.39wr^2 \tag{19.11}$$

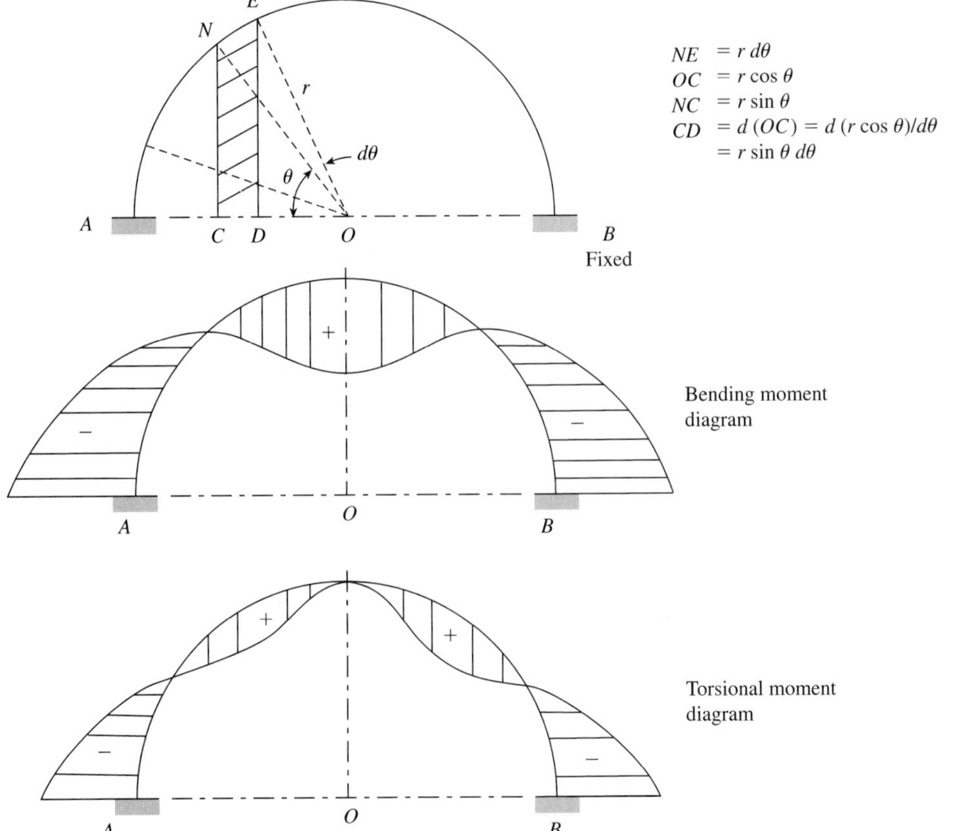

Figure 19.4 Semicircular beam fixed at the supports.

3. Bending moment at A: Taking moments about the line AB, the bending moment at A is equal to

$$
\begin{aligned}
M_A = M_B &= \int_0^{\pi/2} w'(r\,d\theta) \times (r \sin\theta) \\
&= \int_0^{\pi/2} \left(\frac{wr}{2}\sin^2\theta\right)(r\sin\theta)(r\,d\theta) = -\frac{wr^3}{3}
\end{aligned}
\tag{19.12}
$$

4. Torsional moment at support A: T_A can be obtained by differentiating the strain energy of the beam with respect to T_A and equating it to 0. Considering that T_A is acting clockwise at A, then the bending moment at any section N is calculated as follows:

$$M_N = V_A(r\sin\theta) - M_A\cos\theta + T_A\sin\theta - \int_0^{\theta}\left(\frac{wr}{2}\sin^2\alpha\right)(r\,d\alpha) \times r\sin(\theta - \alpha)$$

$$M_N = wr^3\left[\frac{\pi}{8}\sin\theta - \left(\frac{1}{6}\right)(1 + \cos^2\theta)\right] + T_A\sin\theta \tag{19.13}$$

The torsional moment at any station N on the curved beam is equal to

$$
\begin{aligned}
T_n &= -V_A r(1 - \cos\theta) + M_A\sin\theta \\
&\quad + T_A\cos\theta + \int_0^{\theta}\left(\frac{wr}{2}\sin^2\alpha\right)(r\,d\alpha) \times r[1 - \cos(\theta - \alpha)]
\end{aligned}
$$

$$T_N = wr^3\left[\frac{\pi}{8}(\cos\theta - 1) + \frac{\theta}{4} + \frac{1}{24}\sin 2\theta\right] + T_A\cos\theta \tag{19.14}$$

The strain energy is

$$U = \int \frac{M_N^2\,ds}{2EI} + \int \frac{T_N^2\,ds}{2GJ} \tag{19.15}$$

where $ds = r\,d\theta$
G = the modulus of rigidity
E = the modulus of elasticity
I = the moment of inertia of the section
J = the rotational constant of the section
= polar moment of inertia

To obtain T_A, differentiate U with respect to T_A:

$$\frac{\delta_U}{\delta T_A} = \int \frac{M_N}{EI} \times \frac{dM_N}{dT_A}(r\,d\theta) + \int \frac{T_N}{GJ} \times \frac{dT_N}{dT_A} \times (r\,d\theta) = 0$$

$$\frac{dM_N}{dT_A} = \sin\theta \quad \text{and} \quad \frac{dT_N}{dT_A} = \cos\theta$$

Therefore,

$$
\begin{aligned}
\frac{\delta_U}{\delta T_A} &= \frac{r}{EI}\int_0^{\pi/2} \sin\theta\left\{wr^2\left[\frac{\pi}{8}\sin\theta - \frac{1}{6}(1 + \cos^2\theta)\right] + T_A\sin\theta\right\} d\theta \\
&\quad + \frac{r}{GJ}\int_0^{\pi/2}\left\{wr^3\left[\frac{\pi}{8}(\cos\theta - 1) + \frac{\theta}{4} + \frac{1}{24}\sin 2\theta\right] + T_A\cos\theta\right\}\cos\theta \times d\theta = 0
\end{aligned}
$$

$$\frac{r}{EI}\left[wr^3\left(\frac{\pi^2}{32} - \frac{2}{9}\right) + T_A\left(\frac{\pi}{4}\right)\right] + \frac{r}{GJ}\left[wr^3\left(\frac{\pi^2}{32} - \frac{2}{9}\right) + T_A\left(\frac{\pi}{4}\right)\right] = 0$$

Let $EI/GJ = \lambda$; then

$$T_A\left(\frac{\pi}{4}\right)(1 + \lambda) = wr^3\left[\left(\frac{2}{9} - \frac{\pi^2}{32}\right) + \lambda\left(\frac{2}{9} - \frac{\pi^2}{32}\right)\right]$$

$$= wr^3\left(\frac{2}{9} - \frac{\pi^2}{32}\right)(1 + \lambda) = -0.0862wr^3(1 + \lambda)$$

Therefore,

$$T_A = -0.11wr^3 \tag{9.16}$$

Substituting the value of T_A in equation (19.13), the bending moment at any point N is equal to

$$M_N = wr^3\left[\frac{\pi}{8}\sin\theta - \frac{1}{6}(1 + \cos^2\theta) - 0.11\sin\theta\right] \tag{19.17}$$

Substituting the value of T_A in equation (19.14),

$$T_N = wr^3\left[\frac{\pi}{8}(\cos\theta - 1) + \frac{\theta}{4} + \frac{1}{24}\sin 2\theta - 0.11\cos\theta\right] \tag{19.18}$$

5. The value of G/E for concrete may be assumed to be equal to 0.43. The value of J for a circular section is $(\pi/2)r^4$, whereas J for a square section of side x is equal to $0.141x^4$. For a rectangular section with short and long sides x and y, respectively, J can be calculated as follows:

$$J = K' \times y^3 \tag{19.19}$$

The values of K' are calculated as follows:

$$K' = \frac{1}{16}\left[\frac{16}{3} - 3.36\frac{x}{y}\left(1 - \frac{x^4}{12y^4}\right)\right] \tag{19.20}$$

whereas

$$\lambda = \frac{EI}{GJ} = \left(\frac{1}{0.43}\right)\left(\frac{xy^3}{12}\right)\left(\frac{1}{K'yx^3}\right) = \frac{1}{5.16K'}\left(\frac{y}{x}\right)^2$$

Values of K' and λ are both shown in Table 19.2.

Table 19.2 Values of K' and λ for Different Values of y/x

y/x	0.5	1.0	1.1	1.2	1.25	1.3	1.4	1.5	1.6
K'	0.473	0.141	0.154	0.166	0.172	0.177	0.187	0.196	0.204
λ	0.102	1.37	1.52	1.68	1.76	1.85	2.03	2.22	2.43
y/x	1.7	1.75	2.0	2.5	3.0	4.0	5.0	6.0	10
K'	0.211	0.214	0.229	0.249	0.263	0.281	0.291	0.300	0.312
λ	2.65	2.77	3.39	4.86	6.63	11.03	16.5	23.3	62.1

Example 19.2

Determine the ultimate bending and torsional moments in sections C and D of the 10-ft-radius semicircular beam $ADCB$ shown in Figure 19.5. The beam is part of a floor slab that carries a uniform dead load of 120 psf (including its own weight) and a live load of 80 psf.

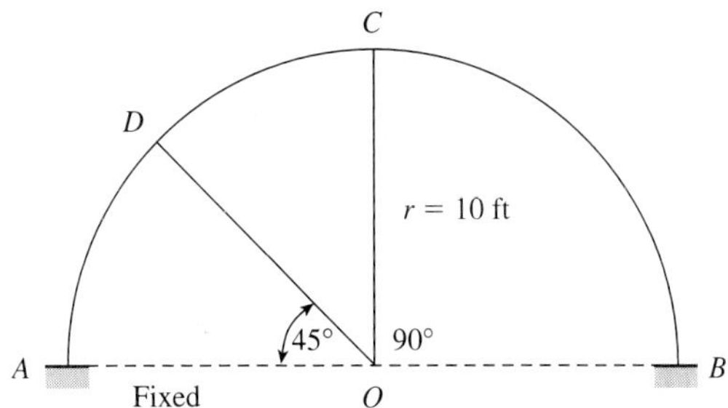

Figure 19.5 Example 19.2.

Solution

1. Ultimate load w_u is $1.4D + 1.7L = 304$ psf.
2. For the section at $C, \theta = \pi/2$ and $w_u r^3 = 0.304(10)^3 = 304$. From equation (19.17),

$$M_c = 304\left[\frac{\pi}{8}\sin\frac{\pi}{2} - \frac{1}{6}\left(1 + \cos^2\frac{\pi}{2}\right) - 0.11\sin\frac{\pi}{2}\right] = 35.3\ \text{K}\cdot\text{ft}$$

From equation (19.18),

$$T_c = 304\left[\frac{\pi}{8}\left(\cos\frac{\pi}{2} - 1\right) + \frac{\pi}{8} + \frac{1}{24}\sin\pi - 0.11\cos\frac{\pi}{2}\right] = 0$$

3. For the section at $D, \theta = \pi/4$.

$$M_D = 304\left[\frac{\pi}{8}\sin\frac{\pi}{4} - \frac{1}{6}\left(1 + \cos^2\frac{\pi}{4}\right) - 0.11\sin\frac{\pi}{4}\right] = -15.2\ \text{K}\cdot\text{ft}$$

$$T_D = 304\left[\frac{\pi}{8}\left(\cos\frac{\pi}{4} - 1\right) + \frac{\pi}{16} + \frac{1}{24}\sin\frac{\pi}{2} - 0.11\cos\frac{\pi}{4}\right] = 13.7\ \text{K}\cdot\text{ft}$$

4. Maximum shearing force occurs at the supports.

$$V_A = 0.39w_u r^2 = 0.39(0.304)(100) = 11.9\ \text{K}$$

Maximum positive moment occurs at C, whereas the maximum negative moment occurs at the supports.

$$M_A = -\frac{w_u r^3}{3} = -\frac{304}{3} = 101.3\ \text{K}\cdot\text{ft}$$

5. Design the critical sections for shear, bending, and torsional moments, as explained in Example 19.1.

19.4 FIXED-END SEMICIRCULAR BEAM UNDER UNIFORM LOADING

The previous section dealt with a semicircular beam fixed at both ends and subjected to a variable distributed load. If the load is uniform, then the beam will be subjected to a uniformly distributed load w K/ft, as shown in Figure 19.6. The forces in the curved beam can be determined as follows:

1. Shearing force at A:

$$V_A = V_B = \int_0^{\pi/2} wr\,d\theta = wr\frac{\pi}{2} = 1.57wr \qquad (19.21)$$

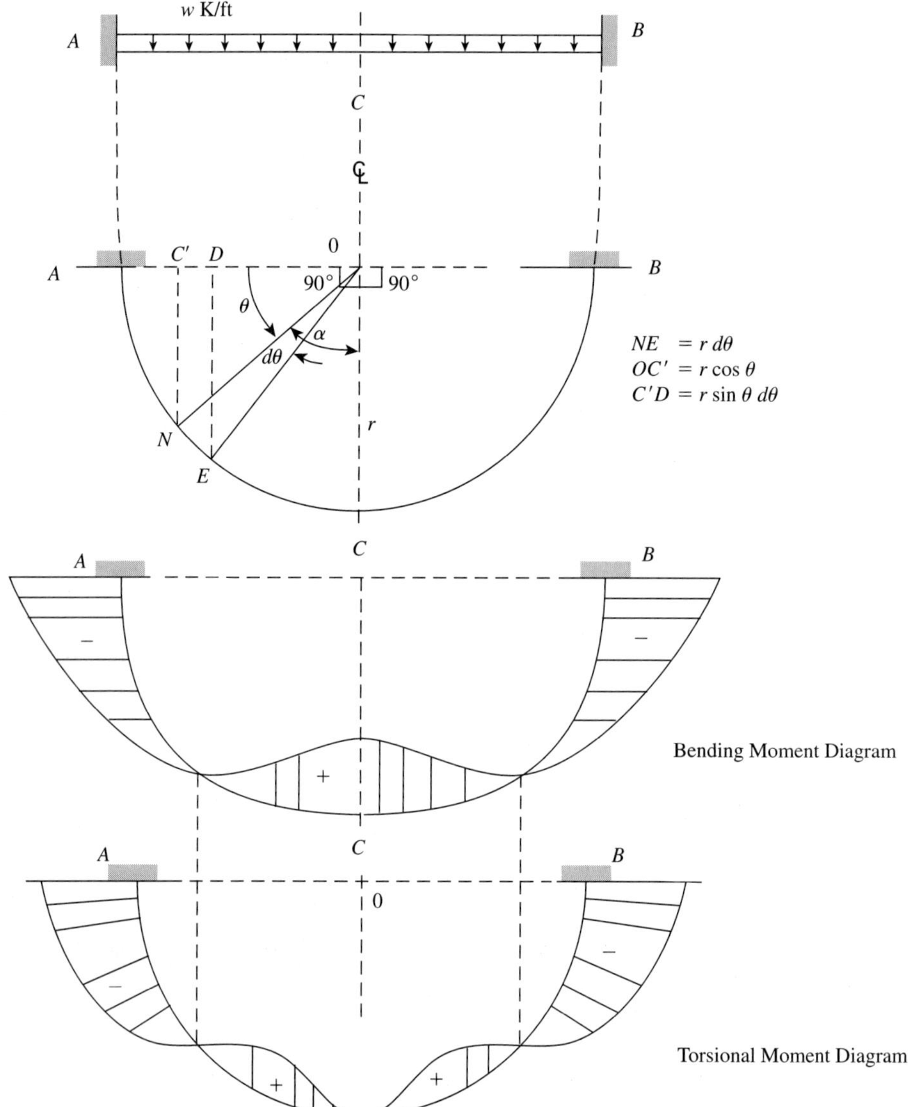

Figure 19.6 Semicircular beam under uniform load.

2. Bending moment at A:

$$M_A = M_B = \int_0^{\pi/2} w(r\,d\theta) \times (r \sin\theta) = wr^2 \tag{19.22}$$

3. Bending moment at any section N on the curved beam when the torsional moment at $A\,(T_A)$ acts clockwise:

$$M_N = V_A(r\sin\theta) - M_A\cos\theta + T_A\sin\theta - \int_0^{\theta}(wr\,d\alpha)\left[r\sin(\theta-\alpha)\right]$$

$$= \frac{\pi}{2}wr^2\sin\theta - wr^2\cos\theta + T_A\sin\theta - \left[wr^2 - wr^2\cos\theta\right]$$

$$M_N = wr^2\left[\frac{\pi}{2}\sin\theta - 1\right] + T_A\sin\theta \tag{19.23}$$

4. Torsional moment at any section N:

$$T_N = -V_A r(1 - \cos\theta) + M_A \sin\theta + T_A \cos\theta + \int_0^\theta (wr\, d\alpha) r[1 - \cos(\theta - \alpha)]$$

$$= -\frac{\pi}{2} wr^2 + \frac{\pi}{2} wr^2 \cos\theta + T_A \cos\theta + M_A \sin\theta$$

$$+ wr^2\theta - wr^2 \sin\theta$$

Substitute $M_A = wr^2$:

$$T_N = wr^2\left[\frac{\pi}{2}\cos\theta - \frac{\pi}{2} + \theta\right] + T_A \cos\theta \tag{19.24}$$

5. The strain energy expression was given in the previous section:

$$U = \int \frac{M_N^2\, ds}{2EI} + \int \frac{T_N^2\, ds}{2GJ} \tag{19.25}$$

To obtain T_A, differentiate U with respect to T_A.

$$\frac{\delta_U}{\delta T_A} = \int \frac{M_N}{EI} \times \frac{dM_N}{dT_A}(r\, d\theta) + \int \frac{T_N}{GJ} \times \frac{dT_N}{dT_A} \times (r\, d\theta) = 0$$

$$\frac{dM_N}{dT_A} = \sin\theta \quad \text{and} \quad \frac{dT_N}{dT_A} = \cos\theta \qquad \text{(from the preceding equations)}$$

$$\frac{\delta_U}{\delta T_A} = \frac{r}{EI}\int_0^{\pi/2}\left[wr^2\left(\frac{\pi}{2}\sin\theta - 1\right) + T_A \sin\theta\right]\sin\theta\, d\theta$$

$$+ \frac{r}{GJ}\int_0^{\pi/2}\left[wr^2\left(\frac{\pi}{2}\cos\theta - \frac{\pi}{2} + \theta\right) + T_A \cos\theta\right]\cos\theta\, d\theta = 0$$

The integration of the preceding equation produces the following:

$$\frac{\delta U}{\delta T_A} = \frac{r}{EI}\left[wr^2\left(\frac{\pi^2}{8} - 1\right) + \frac{\pi}{4}T_A\right] + \frac{r}{GJ}\left[wr^2\left(\frac{\pi^2}{8} - 1\right) + \frac{\pi}{4}T_A\right] = 0$$

$$r\left[wr^2\left(\frac{\pi^2}{8} - 1\right) + \frac{\pi}{4}T_A\right]\left(\frac{EI}{GJ} + 1\right) = 0$$

Because EI/GJ is not equal to zero,

$$wr^2\left(\frac{\pi^2}{8} - 1\right) + \frac{\pi}{4}T_A = 0$$

$$T_A = -wr^2\left(\frac{4}{\pi}\right)\left(\frac{\pi^2}{8} - 1\right) = -0.3wr^2 \tag{19.26}$$

6. Substitute T_A in equation (19.23):

$$M_N = wr^2\left[\left(\frac{\pi}{2}\sin\theta - 1\right) - \left(\frac{\pi}{2} - \frac{4}{\pi}\right)\sin\theta\right]$$

$$= wr^2\left(\frac{4}{\pi}\sin\theta - 1\right) \tag{19.27}$$

$$T_N = wr^2\left[\left(\frac{\pi}{2}\cos\theta + \theta - \frac{\pi}{2}\right) - \left(\frac{\pi}{2} - \frac{4}{\pi}\right)\cos\theta\right]$$

$$= wr^2\left(\theta - \frac{\pi}{2} + \frac{4}{\pi}\cos\theta\right) \tag{19.28}$$

The values of the bending and torsional moments at any section N are independent of λ ($\lambda = EI/GJ$).

7. Bending and torsional moments at midspan, section C, can be found by substituting $\theta = \pi/2$ in equations (19.27) and (19.28):

$$M_c = wr^2\left(\frac{4}{\pi} - 1\right) = 0.273wr^2 \tag{19.29}$$

$$T_c = wr^2\left(\frac{\pi}{2} - \frac{\pi}{2} + 0\right) = 0 \tag{19.30}$$

19.5 CIRCULAR BEAM SUBJECTED TO UNIFORM LOADING

The previous section dealt with a semicircular beam subjected to a uniformly distributed load. The forces acting on the beam at any section vary with the intensity of load, the span (or the radius of the circular beam), and the angle α measured from the centerline axis of the beam.

Considering the general case of a circular beam fixed at both ends and subjected to a uniform load w (K/ft), as shown in Figure 19.7, the bending and torsional moments can be calculated from the following expressions:

1. The moment at the centerline of the beam, M_c, can be derived using the strain energy expression, equation (19.25), and can be expressed as follows:

$$M_c = \frac{wr^2}{K_4}\left[\lambda(K_1 + K_2 - K_3) + (K_1 - K_2)\right] \tag{19.31}$$

Curved-beam bridge, Washington, D.C.

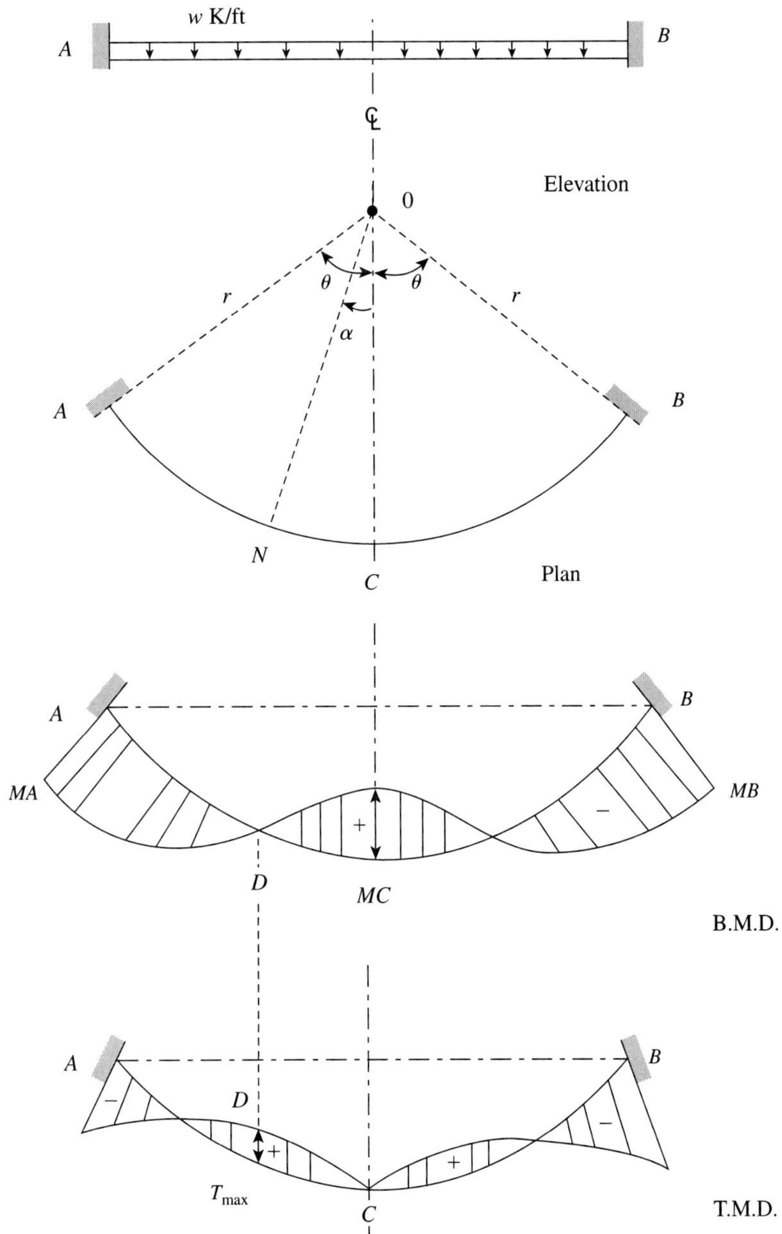

Figure 19.7 Circular beam subjected to uniform load, showing the bending moment diagram (BMD) and the torsional moment diagram (TMD).

where $\lambda = \dfrac{EI}{GJ}$

$$K_1 = 2(2\sin\theta - \theta)$$
$$K_2 = 2\sin\theta\cos\theta = \sin 2\theta$$
$$K_3 = 4\theta\cos\theta$$
$$K_4 = 2\theta(\lambda + 1) - (\lambda - 1)\sin 2\theta$$

2θ = the total central angle of the ends of the beam, angle AOB (Figure 19.7)

The torsional moment at the centerline section, T_c, is 0.

2. The moment at any section N on the curved beam where ON makes an angle α with the centerline axis (Figure 19.7) is

$$M_N = M_c \cos\alpha - wr^2(1 - \cos\alpha) \tag{19.32}$$

3. The torsional moment at any section N on the curved beam as a function of the angle α was derived earlier:

$$T_N = M_c \sin\alpha - wr^2(\alpha - \sin\alpha) \tag{19.33}$$

4. To compute the bending moment and torsional moment at the supports, substitute θ for α in the preceding equations:

$$M_A = M_c \cos\theta - wr^2(1 - \cos\theta) \tag{19.34}$$

$$T_A = M_c \sin\theta - wr^2(\theta - \sin\theta) \tag{19.35}$$

Example 19.3

A curved beam has a quarter-circle shape in plan with a 10-ft radius. The beam has a rectangular section with the ratio of the long to the short side of 2.0 and is subjected to a factored load of 8 K/ft. Determine the bending and torsional moments at the centerline of the beam, supports, and maximum values.

Solution

1. For a rectangular section with $y/x = 2$, $\lambda = EI/GJ = 3.39$ (Table 19.2).
2. The bending and torsional moments can be calculated using equations (19.31) through (19.35) for $\theta = \pi/4$. From equation (19.31),

$$K_1 = 2\left(2\sin\frac{\pi}{4} - \frac{\pi}{4}\right) = 1.2576, \qquad K_2 = \sin\frac{\pi}{2} = 1.0$$

$$K_3 = 4\left(\frac{\pi}{4}\right)\cos\frac{\pi}{4} = 2.2214$$

$$K_4 = 2\left(\frac{\pi}{4}\right)(3.39 + 1) - (3.39 - 1)\sin\frac{\pi}{2} = 4.506$$

$$M_c = \frac{wr^2}{4.506}\left[3.39(1.2576 + 1.0 - 2.2214) + (1.2576 - 1.0)\right]$$

$$= 0.0844wr^2$$

For $w = 8$ K/ft and $r = 10$ ft, $M_c = 64$ K·ft; $T_c = 0$.

3. $M_N = M_c\cos\alpha - wr^2(1 - \cos\alpha) = wr^2(1.08\cos\alpha - 1)$

$T_N = M_c\sin\alpha - wr^2(\alpha - \sin\alpha) = wr^2(1.08\sin\alpha - \alpha)$

For the moments at the supports, $\alpha = \theta = \pi/4$.

$$M_A = wr^2\left(1.08\cos\frac{\pi}{4} - 1\right) = -0.236wr^2$$

$$= -0.236 \times 8 \times (10)^2 = -189 \text{ K}\cdot\text{ft}$$

$$T_A = wr^2\left(1.08\sin\frac{\pi}{4} - \frac{\pi}{4}\right) = 0.022wr^2 = -17.4 \text{ K}\cdot\text{ft}$$

For $M_N = 0$, $1.08\cos\alpha - 1 = 0$, or $\cos\alpha = 0.926$ and $\alpha = 22.2° = 0.387$ rad. To calculate $T_{N\,max}$, let $dT_N/d\alpha = 0$, or $(1.08\cos\alpha - 1) = 0$. Then $\cos\alpha = 0.926$ and $\alpha = 22.2°$.

$$T_N\,(\text{max}) = wr^2(1.08\sin 22.2 - 0.387) = 0.0211wr^2$$

$$T_{N\,max} = 0.0211 \times 800 = 16.85 \text{ K}\cdot\text{ft}$$

19.6 CIRCULAR BEAM SUBJECTED TO A CONCENTRATED LOAD AT MIDSPAN

If a concentrated load is applied at the midspan of a circular beam, the resulting moments vary with the magnitude of the load, the span, and the coefficient $\lambda = EI/GJ$.

Considering the general case of a circular beam fixed at both ends and subjected to a concentrated load P at midspan (Figure 19.8), the bending and torsional moments can be calculated from the following expressions:

1. The moment at the centerline of the beam, section C, can be expressed as follows:

$$M_c = \frac{\lambda(2 - 2\cos\theta - \sin^2\theta) + \sin^2\theta}{2\theta(\lambda + 1) - (\lambda - 1)\sin 2\theta}(Pr)$$

$$M_c = \frac{Pr}{K_3}(\lambda K_1 + K_2) \tag{19.36}$$

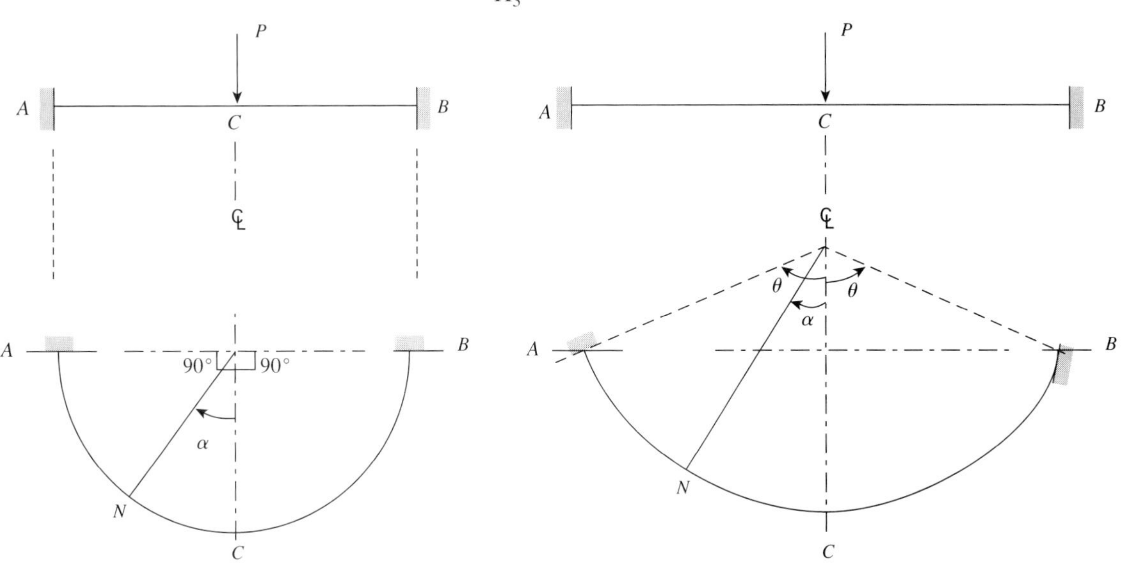

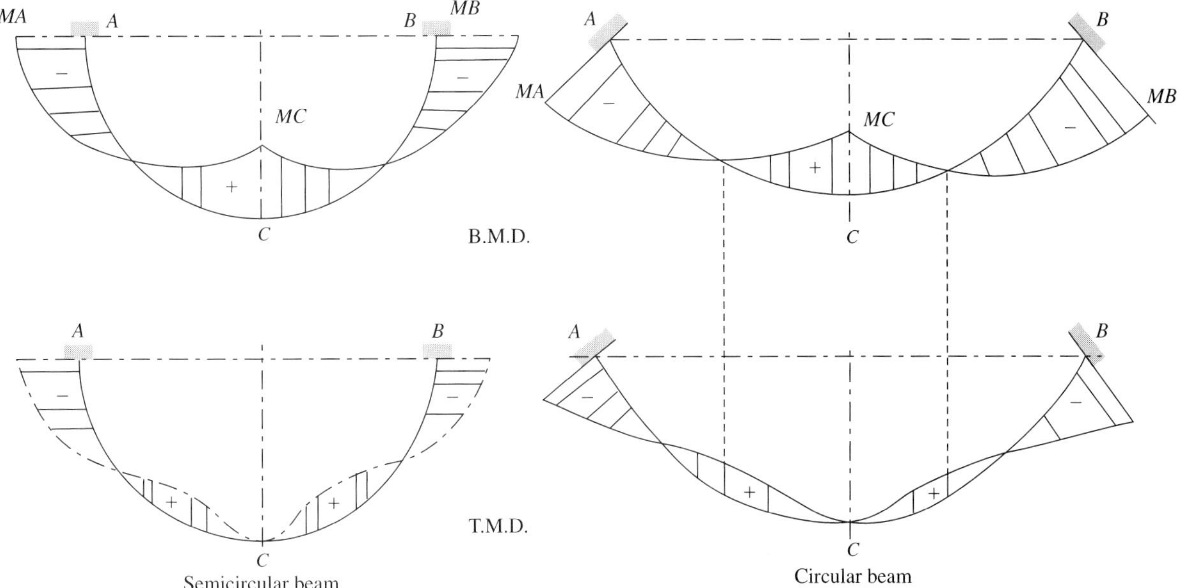

Figure 19.8 Circular beam subjected to a concentrated load at midspan, showing the bending moment diagram (BMD) and the torsional moment diagram (TMD).

where $$\lambda = \frac{EI}{GJ}$$
$$K_1 = (2 - 2\cos\theta - \sin^2\theta)$$
$$K_2 = \sin^2\theta$$
$$K_3 = 2\theta(\lambda + 1) - (\lambda - 1)\sin^2\theta$$

The torsional moment at the centerline is $T_c = 0$.

2. The bending and torsional moments at any section N on the curved beam where ON makes an angle α with the centerline axis are calculated as follows:

$$M_N = M_c \cos\alpha - \left(\frac{P}{2}r\right)\sin\alpha \tag{19.37}$$

$$T_N = M_c \sin\alpha - \left(\frac{P}{2}r\right)(1 - \cos\alpha) \tag{19.38}$$

3. To compute the bending and torsional moments at the supports, substitute θ for α:

$$M_A = M_c \cos\theta - \left(\frac{P}{2}r\right)\sin\theta \tag{19.39}$$

$$T_A = M_c \sin\theta - \left(\frac{P}{2}r\right)(1 - \cos\theta) \tag{19.40}$$

Example 19.4

Determine the bending and torsional moments of the quarter-circle beam of Example 19.3 if $\lambda = 1.0$ and the beam is subjected to a concentrated load at midspan of $P = 20$ K.

Solution

1. Given: $\lambda = 1.0$ and $\theta = \pi/4$. Therefore,

$$M_c = \left(\frac{Pr}{2}\right)\left(\frac{1 - \cos\theta}{\theta}\right)$$

(equation (19.36)) and $T_c = 0$. For $\theta = \pi/4$,

$$M_c = 0.187Pr = 0.187(20 \times 10) = 37.4 \text{ K}\cdot\text{ft}$$

2. From equations (19.39) and (19.40),

$$M_A = 0.187Pr\cos\frac{\pi}{4} - \frac{Pr}{2}\sin\frac{\pi}{4} = -0.22Pr$$
$$= -0.22 \times (200) = -44 \text{ K}\cdot\text{ft}$$

$$T_A = 0.187Pr\sin\frac{\pi}{4} - 0.5Pr\left(1 - \cos\frac{\pi}{4}\right) = -0.0142Pr$$
$$= -0.0142 \times 200 = -2.84 \text{ K}\cdot\text{ft}$$

3. $M_N = 0$ when

$$M_c\cos\alpha - \frac{Pr}{2}\sin\alpha = 0 \qquad \text{(equation (19.37))}$$

$$0.187Pr\cos\alpha - 0.5Pr\sin\alpha = 0$$

$$\tan\alpha = 0.374 \quad \text{and} \quad \alpha = 20.5°$$

$T_n = 0$ when $M_c\sin\alpha - (P/2)r(1 - \cos\alpha) = 0$ (equation (19.38)), from which $\alpha = 37.7°$.

4. To compute $T_{\max}$, let $dT_N/d\alpha = 0$ (equation (19.38)).

$$0.187Pr\cos\alpha - 0.5Pr\sin\alpha = 0, \qquad \tan\alpha = 0.374$$

and $\alpha = 20.5°$. Substitute $\alpha = 20.5°$ in equation (19.38) to get $T_{\max} = 0.035Pr = 7\ \text{K}\cdot\text{ft}$.

19.7 V-SHAPED BEAMS SUBJECTED TO UNIFORM LOADING

Beams that have a V-shape in plan and are subjected to loads normal to the plane of the beam may be analyzed using the *strain-energy principles.* Figure 19.9 shows typical bending moment diagram for a V-shaped beam subjected to a uniform load w.

Considering the general case of a V-shaped beam fixed at both ends and subjected to a uniform load w (K/ft), the bending and torsional moments can be calculated from the following expressions:

1. The moment at the centerline of the beam, section C, is calculated as follows:

$$M_c = (wa^2)\left[\frac{\sin^2\theta}{6(\sin^2\theta + \lambda\cos^2\theta)}\right] \tag{19.41}$$

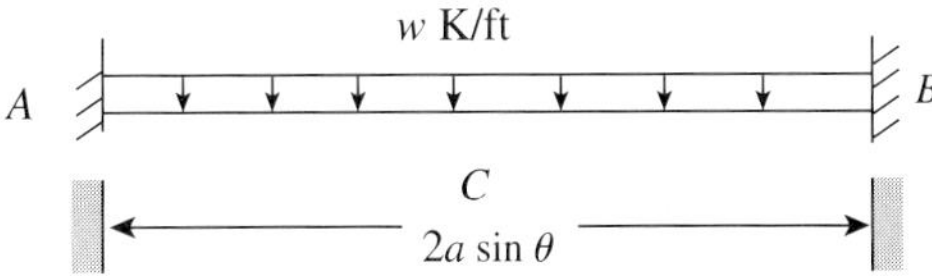

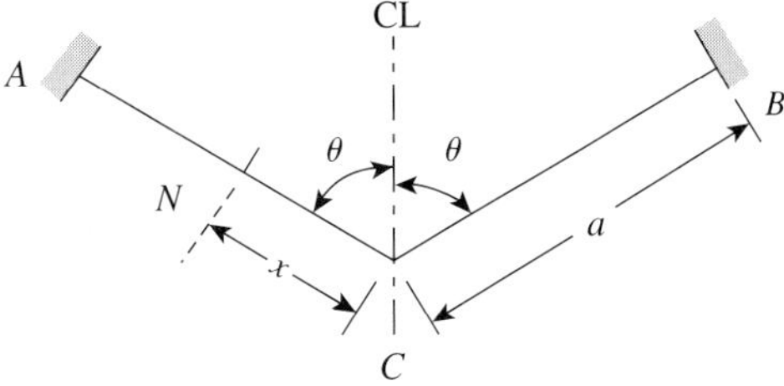

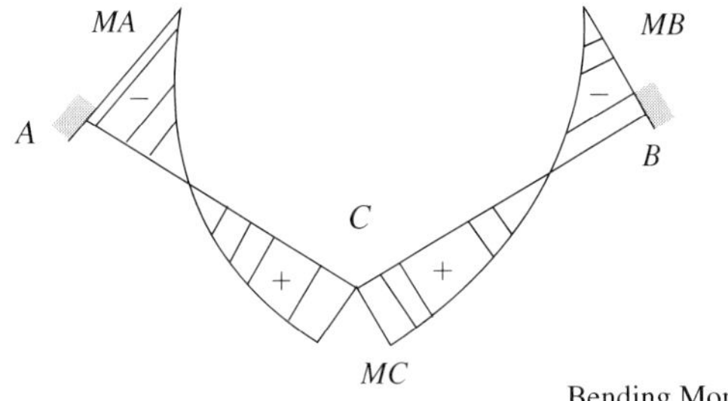

Figure 19.9 V-shaped beam under uniform load.

90° V-shaped beams, London, Ontario, Canada.

where

$$\lambda = \frac{EI}{GJ}$$

a = half the total length of the beam (length AC)

θ = half the angle between the two sides of the V-shaped beam.

The torsional moment at the centerline section is

$$T_c = \frac{M_c}{\sin\theta} \times \cos\theta = M_c \cot\theta \tag{19.42}$$

2. The bending and torsional moments at any section N along half the beam AC or BC at a distance x measured from section C are calculated as follows:

$$M_N = M_c - w\frac{x^2}{2} \tag{19.43}$$

$$T_N = T_c = \frac{M_c}{\sin\theta} \times \cos\theta = M_c \cot\theta \tag{19.44}$$

To compute the moments at the supports, let $x = a$. Then

$$M_A = M_c - w\frac{a^2}{2}$$

$$T_A = T_c = M_c \cot\theta$$

Example 19.5

Determine the bending and torsional moments in a V-shaped beam subjected to a uniform load of 6 K/ft. The length of half the beam is $a = 10$ ft and the angle between the V-shaped members is $2\theta = \pi/2$. The beam section is rectangular with a ratio of long side to short side of 2.

Apartment building.

Solution

1. For a rectangular section with the sides ratio, $y/x = 2, \lambda = 3.39$. For this beam $\theta = \pi/4$.

2. $$M_c = \frac{wa^2}{6}\left[\frac{\sin^2\theta}{(\sin^2\theta + \lambda\cos^2\theta)}\right] \qquad \text{(equation (19.41))}$$

$$M_c = \frac{wa^2}{6}\left(\frac{0.5}{(0.5 + 3.39 \times 0.5)}\right) = 0.038wa^2$$

$$= 0.038 \times 6(10)^2 = 22.8\ \text{K}\cdot\text{ft}$$

$$M_A = M_c - w\frac{a^2}{2} = 0.038wa^2 - 0.5wa^2 = -0.462wa^2$$

$$= -277.2\ \text{K}\cdot\text{ft}$$

$$M_N = 0 \quad \text{when} \quad M_c - w\frac{x^2}{2} = 0$$

or $0.038wa^2 - 0.5wx^2 = 0$, so $x = 0.276a = 2.76$ ft measured from c.

3. $$T_A = T_c = M_c \cot\theta$$

$$= 0.038wa^2 = 0.038 \times 600 = 22.8\ \text{K}\cdot\text{ft}$$

19.8 V-SHAPED BEAMS SUBJECTED TO A CONCENTRATED LOAD AT THE CENTERLINE OF THE BEAM

The general equations for computing the bending and torsional moments in a V-shaped beam fixed at both ends and subjected to a concentrated load P at the centerline of the beam (Figure 19.10) are as follows:

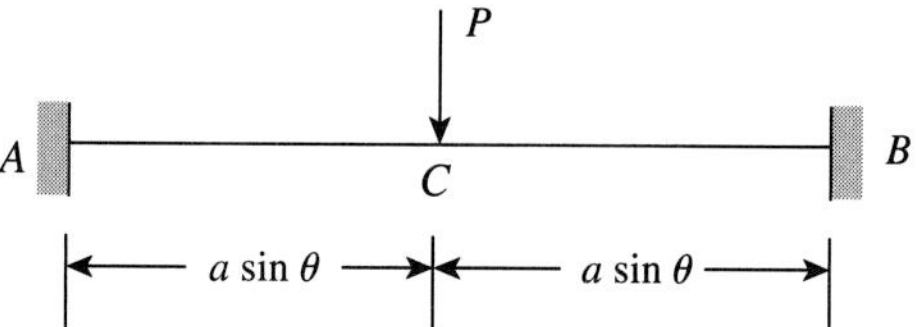

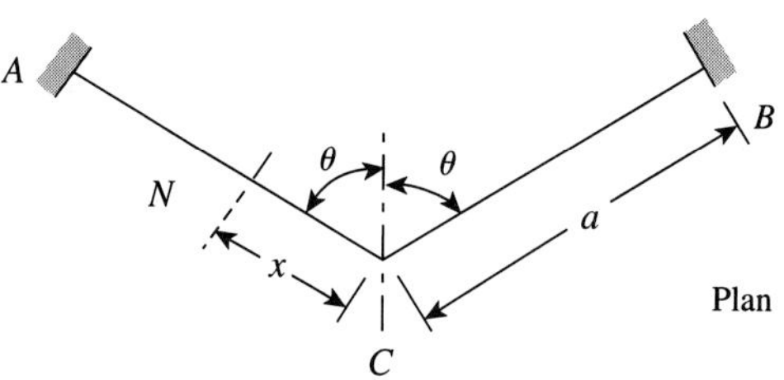

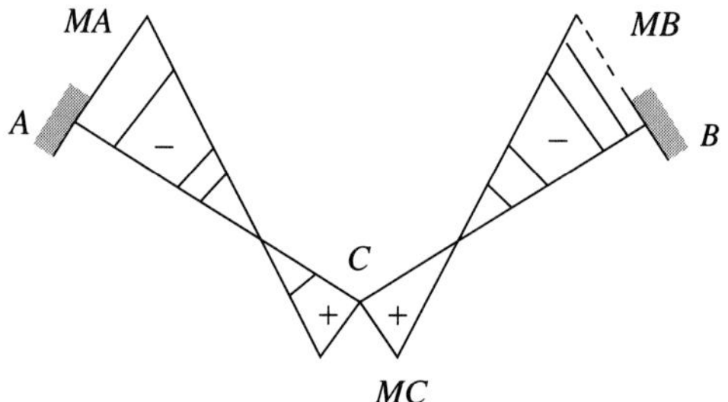

Figure 19.10 V-shaped beam under concentrated load.

1. The moment at the centerline of the beam, section C, for any value of λ, is

$$M_c = \left(\frac{Pa}{4}\right)\left(\frac{\sin^2\theta}{(\sin^2\theta + \lambda\cos^2\theta)}\right] \tag{19.45}$$

where

$$\lambda = \frac{EI}{GJ}$$

a = half the total length of the beam (part AB or BC)
θ = half the angle between the two sides of the V-shaped beam

The torsional moment at the centerline section is

$$T_c = \frac{M_c}{\sin\theta}\cos\theta = M_c \cot\theta \tag{19.46}$$

2. The bending and torsional moments at any section N along half the beam AC or BC at a distance x measured from C are calculated as follows:

$$M_N = M_c - \frac{Px}{2} \tag{19.47}$$

$$T_N = T_c = M_c \cot\theta \tag{19.48}$$

The moments at the supports are determined by assuming $x - a$:

$$M_A = M_c - \frac{Pa}{2} \tag{19.49}$$

$$T_A = T_c = M_c \cot\theta \tag{19.50}$$

Example 19.6

Determine the bending and torsional moments in a V-shaped beam subjected to a concentrated load $P = 30$ K acting at the centerline of the beam. Given: $\theta = \pi/4$, $y/x = 2.0$, and $a = 12$ ft.

Solution

1. For a rectangular section with $y/x = 2.0$, $\lambda = 3.39$.

2. $M_c = \dfrac{Pa}{4}\left(\dfrac{\sin^2 \pi/4}{\sin^2 \pi/4 + 3.39\cos^2 \pi/4}\right) = 0.057(Pa)$

$= 0.057 \times 30 \times 12 = 20.5$ K·ft

$M_A = M_c - \dfrac{Pa}{2} = (0.057 - 0.5)Pa = -0.443(Pa)$

$= -0.0443 \times 360 = -159.5$ K·ft

$M_N = 0 \quad \text{when} \quad M_c - \dfrac{Px}{2} = 0$

Hence $0.057Pa - 0.5Px = 0$ and $x = 0.114a = 0.114 \times 12 = 1.37$ ft measured from c.

3. $T_A = T_c = T_N = M_c \cot\dfrac{\pi}{4} = 0.057(Pa) = 20.5$ K·ft

---

Example 19.7

Determine the bending and torsional moments in the beam of Example 19.6 if the angle θ is $\pi/2$ (a straight beam fixed at both ends).

Solution

Given $\theta = \pi/2$ and the span $L = 2a =$ the distance between the two supports. The bending moment at the centerline is

$$M_c = \frac{Pa}{4}\left(\frac{1}{1}\right) = \frac{Pa}{4} = \frac{PL}{8} = +90 \text{ K·ft}$$

$$M_A = M_c - \frac{Pa}{2} = \frac{PL}{8} - \frac{P}{2}\left(\frac{L}{2}\right) = -\frac{PL}{8} = -90 \text{ K·ft}$$

$$T_A = T_c = 0$$

These values are similar to those obtained from the structural analysis of the fixed-end beam subjected to a concentrated load at midspan.

Example 19.8

The beam shown in Figure 19.11 has a V-shape in plan and carries a uniform dead load of 4 K/ft and a live load of 2 K/ft. The inclined length of half the beam is $a = 10$ ft and $\theta = 60°$. Design the beam for shear, bending, and torsional moments using $f'_c = 3$ Ksi and $f_y = 60$ Ksi.

Solution

1. $w_u = 1.4D + 1.7L = 1.4 \times 4 + 1.7 \times 2 = 9.0$ K/ft.

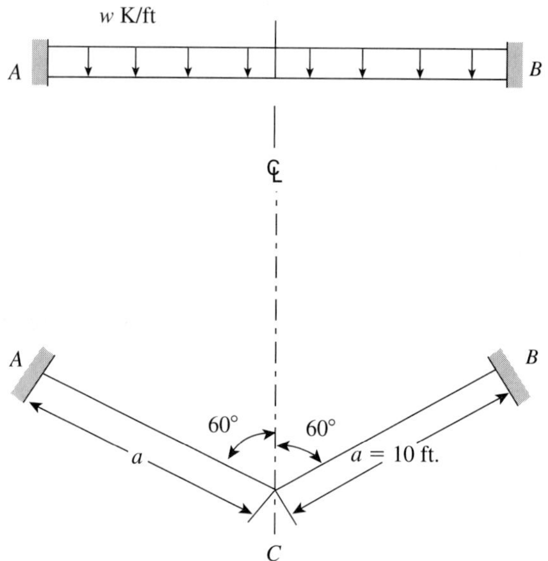

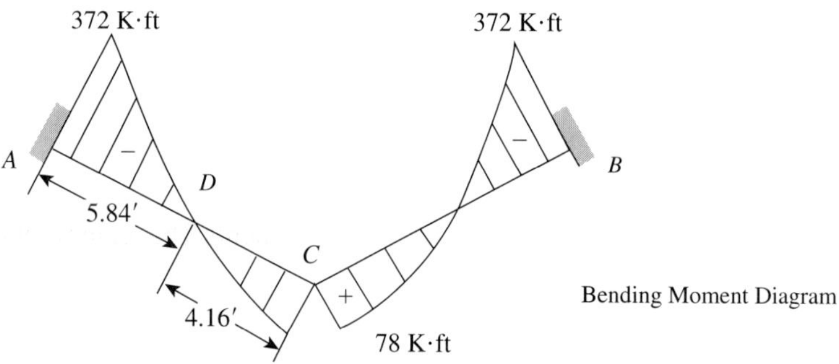

Figure 19.11 Example 19.8.

2. Assuming a rectangular section with a ratio of long to short side of $y/x = 1.75$, the value of λ is 2.77 (from Table 19.2). For $\theta = 60° = \pi/3$,

$$M_c = \frac{w_u a^2 \sin^2\theta}{6(\sin^2\theta + \lambda\cos^2\theta)} = \frac{9(100)(0.75)}{6(0.75 + 2.77 \times 0.25)} = +78 \text{ K}\cdot\text{ft}$$

$$M_A = M_c - w_u \frac{a^2}{2} = 78 - 9\left(\frac{100}{2}\right) = -372 \text{ K}\cdot\text{ft}$$

$$T_A = M_c \cot\theta = 78 \times 0.577 = 45 \text{ K}\cdot\text{ft} = 540 \text{ K}\cdot\text{in.}$$

$$T_c \text{ (at } x = 0) = M_c \cot\theta = 45 \text{ K}\cdot\text{ft} = 540 \text{ K}\cdot\text{in.}$$

$$V_A = 9 \times 10 = 90 \text{ K}$$

The bending moment is zero at $M_N = 0 = M_c - w_u x^2/2$. Hence, $78 - \frac{9}{2}x^2 = 0$ and $x = 4.16$ ft measured from c. The bending moment diagram is shown in Figure 19.11.

3. Design for a bending moment, M_u, equal to -372 K · ft.

a. For $f'_c = 3$ Ksi, $f_y = 60$ Ksi, $\rho_{max} = 0.0161$, and R_u (max) $= 705$ psi (Appendix A)

$$bd^2 = \frac{M_u}{R_u} = \frac{372 \times 12}{0.705} = 6332 \text{ in.}^3$$

For a ratio,

$$\frac{y}{x} = \frac{(d + 3)}{b} = 1.75$$

as assumed, then $d = 21.4$ in. and $b = 13.8$ in. Use a section 14×24 in.

$$A_s = \rho_{max}bd = 0.0161(14 \times 21.4) = 4.75 \text{ in.}^2$$

b. For the section at midspan, $M_u = 78$ K·ft and actual $d = 21.5$ in.

$$R_u = \frac{M_u}{bd^2} = \frac{78{,}000 \times 12}{14 \times (21.5)} = 145 \text{ psi}$$

$$\rho < \rho_{min} = 0.0033$$

Use $A_s = 0.0033 \times 14 \times 21.5 = 1.0$ in.^{2}

c. Design for torsional moment and shear: $T_u = 45$ K·ft for all sections.

$$V_u \text{ (at distance } d) = 90 - \frac{21.5}{12} \times 9 = 74.0 \text{ K}$$

The design procedure will be similar to that of Example 19.1. Details of the final section are shown in Figure 19.12.

SUMMARY

Sections 19.1–19.5

In a curved beam in plan, the center of gravity of normal loads lies outside the line joining the supports developing torsional moments. The analysis of uniformly loaded circular beams is presented in these sections.

Section 19.6

The analysis of circular beams subjected to concentrated loads is presented in this section.

Section 19.7

V-shape beams subjected to gravity loads may be analyzed using the strain energy principles. Equations to calculate the torsional moments of these types of beams are presented.

REFERENCES

1. B. J. Wilson and J. F. Quereau. "A Simple Method of Determining Stress in Curved Flexural Members." University of Illinois Engineering Exp. Station, Circular 16. 1925.
2. S. Timoshenko. "Bending Stresses in Curved Tubes of Rectangular Cross Sections." *Trans. ASME* 45 (1923).
3. S. Timoshenko. *Strength of Materials.* New York: D. Van Nostrand, 1930.
4. R. Levy. "Displacement of Circular Rings with Normal Loads." *Proc. ASCE Structural Journal* 88, no. 1 (February 1962).
5. R. J. Roark and W. C. Young. *Formulas for Stress and Strain,* 5th ed. New York: McGraw-Hill, 1975, pp. 209–285.

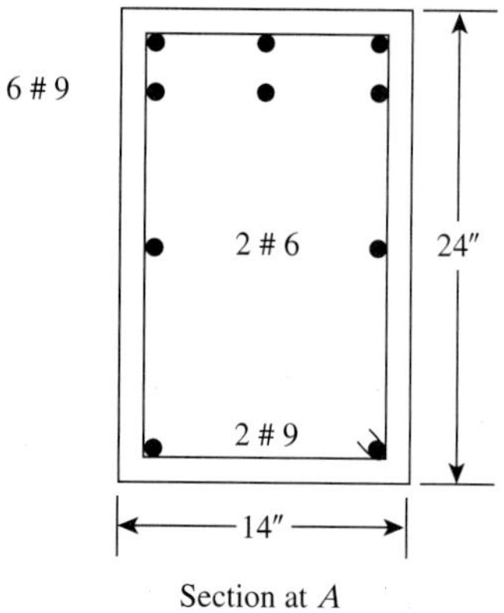

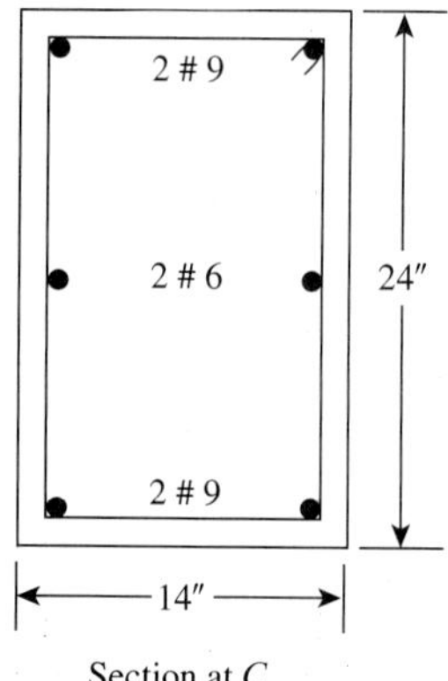

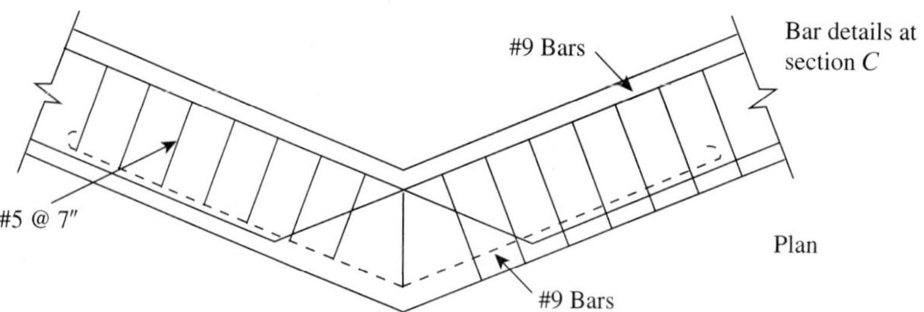

Figure 19.12 Example 19.8.

6. G. Glushkov. *Formulas for Designing Frames.* Moscow: MIR Publications, 1975.

7. P. Charon. *Calcul Pratique des Poutres, Portiques et Caders.* Paris: Eyrolles, 1974.

8. American Concrete Institute. "Building Code Requirements for Structural Concrete." ACI (318-99). Detroit, 1999.

PROBLEMS

19.1 A circular beam is supported on six equally spaced columns, and its centerline lies on a circle 20 ft in diameter. The beam carries a uniform dead load of 8 K/ft and a live load of 5 K/ft. Design the beam using $f'_c = 3$ Ksi, $f_y = 60$ Ksi, and $b = 14$ in.

19.2 Design a semicircular beam fixed on both ends. The center of columns lies on a circle 12 ft in diameter. The beam carries uniform dead and live loads of 4 K/ft and 3 K/ft, respectively. Use $f'_c = 3$ Ksi, $f_y = 60$ Ksi, and $b = 20$ in.

19.3 Determine the ultimate bending and torsional moment at sections C and D of the fixed-end beam shown in Figure 19.5 if the diameter of the circle is 30 ft. The beam is part of a floor slab that carries a uniform dead load (including its own weight) of 100 psf and a live load of 120 psf.

19.4 A quarter-circle cantilever beam has a radius of 8 ft and carries a uniform dead load of 5.5 K/ft and a concentrated live load of 4 K at its free end. Design the beam using $f'_c = 4$ Ksi, $f_y = 60$ Ksi, and $b = 14$ in.

19.5 Design the beam shown in Figure 19.11 if the inclined length of half the beam is $a = 8$ ft. The beam has a 60° V-shape in plan and carries uniform dead and live loads of 3 K/ft and 4 K/ft. Assume the ratio of the long to the short side of the rectangular section is 2. Use $f'_c = 4$ Ksi and $f_y = 60$ Ksi.

Briggs Library building, South Dakota State University, Brookings, S.D.

20 INTRODUCTION TO PRESTRESSED CONCRETE

20.1 PRESTRESSED CONCRETE

20.1.1 Principles of Prestressing

To prestress a structural member is to induce internal, permanent stresses that counteract the tensile stresses in the concrete resulting from external loads; this extends the range of stress that the member can safely withstand. Prestressing force may be applied either before or at the same time as the application of the external loads. Stresses in the structural member must remain, everywhere and for all states of loading, within the limits of stress that the material can sustain indefinitely. The induced stresses, primarily compressive, are usually created by means of high-tensile steel tendons, which are tensioned and anchored to the concrete member. Stresses are transferred to the concrete either by the bond along the surface of the tendon or by anchorages at the ends of the tendon.

To explain this discussion, consider a beam made of plain concrete, which has to resist the external gravity load shown in Figure 20.1(a). The beam section is chosen with the tensile flexural stress as the critical criterion for design; therefore, an uneconomical section results. This is because concrete is considerably stronger in compression than in tension. The maximum flexural tensile strength of concrete, the modulus of rupture f_r, is equal to $7.5\sqrt{f'_c}$ (Figure 20.1(a)).

In normal reinforced concrete design, the tensile strength of concrete is ignored and steel bars are placed in the tension zone of the beam to resist the tensile stresses, whereas the concrete resists the compressive stresses (Figure 20.1(b)).

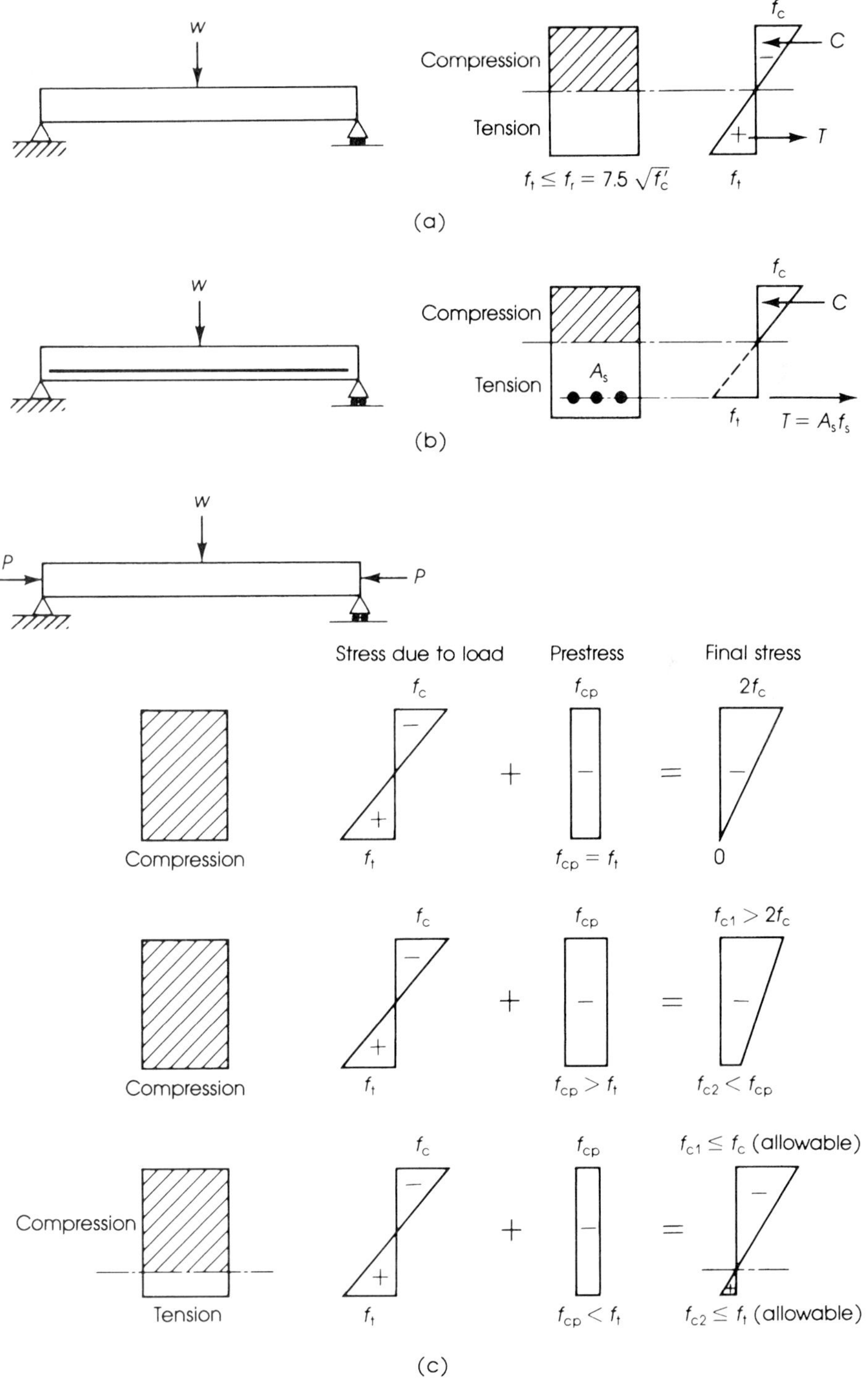

Figure 20.1 Effect of prestressing: (a) plain concrete, (b) reinforced concrete, and (c) prestressed concrete.

In prestressed concrete design, an initial compressive stress is introduced to the beam to offset or counteract the tensile stresses produced by the external loads (Figure 20.1(c)). If the induced compressive stress is equal to the tensile stress at the bottom fibers, then both stresses cancel themselves, whereas the compressive stress in the top fibers is doubled; in this case the whole section is in compression. If the induced compressive stress is less than the tensile stress at the bottom fibers, these fibers will be in tension, whereas the top fibers are in compression.

In practice, a concrete member may be prestressed in one of the following methods.

1. *Posttensioning:* In posttensioning, the steel tendons are tensioned after the concrete has been cast and hardened. Posttensioning is performed by two main operations: tensioning the steel wires or strands by hydraulic jacks that stretch the strands while bearing against the ends of the member and then replacing the jacks by permanent anchorages that bear on the member and maintain the steel strands in tension. A tendon is generally made of wires, strands, or bars. Wires and strands can be tensioned in groups, whereas bars are tensioned one at a time.

   In the posttensioning process, the steel tendons are placed in the formwork before the concrete is cast and the tendons are prevented from bonding to the concrete by waterproof paper wrapping or a metal duct (sheath).

   Tendons bonded to the concrete are called bonded tendons. Unbonded tendons, left without grout or coated with grease, have no bond throughout the length of the tendon.

2. *Pretensioning:* In pretensioning, the steel tendons are tensioned before the concrete is cast. The tendons are temporarily anchored against some abutments and then cut or released after the concrete has been placed and hardened. The prestressing force is transferred to the concrete by the bond along the length of the tendon. Pretensioning is generally done in precasting plants in permanent beds, which are used to produce pretensioned precast concrete elements for the building industry.

3. *External prestressing:* In external prestressing, the prestressing force is applied by flat jacks placed between the concrete member ends and permanent rigid abutments. The member does not contain prestressing tendons, as in the previous two methods (also called internal prestressing). External prestressing is not easy in practice because shrinkage and creep in concrete tend to reduce the induced compressive stresses unless the prestressing force can be adjusted.

The profile of the tendons may be straight, curved (bent), or circular, depending on the design of the structural member. Straight tendons are generally used in solid and hollow-cored slabs, whereas bent tendons are used in beams and most structural members. Circular tendons are used in circular structures such as tanks, silos, and pipes. The prestressing force may be applied in one or more stages, either to avoid overstressing concrete or in cases when the loads are applied in stages. In this case, part of the tendons are fully prestressed at each stage.

A considerable number of prestressing systems have been devised, among them Freyssinet, Magnel Blaton, B.B.R.V., Dywidag, CCL, Morandi, VSL, Western Concrete, Prescon, and INRYCO. The choice of the prestressing system for a particular job can sometimes be a problem. The engineer should consider three main factors that govern the choice of the system:

1. The magnitude of the prestressing force required
2. The geometry of the section and the space available for the tendons
3. Cost of the prestressing system (materials and labor)

The following example illustrates some of the features of prestressed concrete.

Example 20.1

For the simply supported beam shown in Figure 20.2, determine the maximum stresses at midspan section due to its own weight and the following cases of loading and prestressing:

1. A uniform live load of 900 lb/ft
2. A uniform live load of 900 lb/ft and an axial centroidal longitudinal compressive force of $P = 259.2$ K
3. A uniform live load of 2100 lb/ft and an eccentric longitudinal compressive force $P = 259.2$ K acting at an eccentricity $e = 4$ in.
4. A uniform live load of 2733 lb/ft and an eccentric longitudinal compressive force $P = 259.2$ K acting at the maximum practical eccentricity for this section ($e = 6$ in.)
5. The maximum live load when $P = 259.2$ K acting at $e = 6$ in.

Use $b = 12$ in., $h = 24$ in., $f'_c = 4500$ psi, and an allowable $f'_c = 2050$ psi.

Solution

1. Stresses due to dead and live loads only are

$$\text{Self-weight of beam} = (1 \times 2) \times 150 = 300 \text{ lb/ft}$$

$$\text{Dead-load moment } M_{\text{D.L.}} = \frac{wL^2}{8} = \frac{0.300(24)^2}{8} = 21.6 \text{ K}\cdot\text{ft}$$

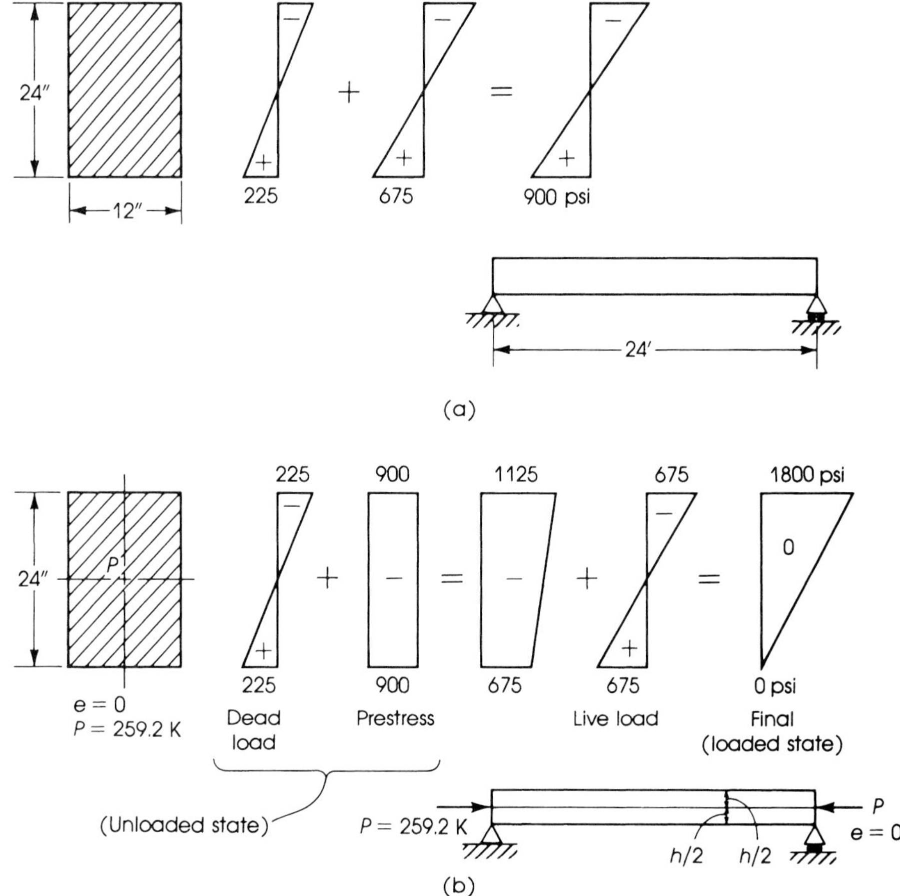

Figure 20.2 Example 20.1.

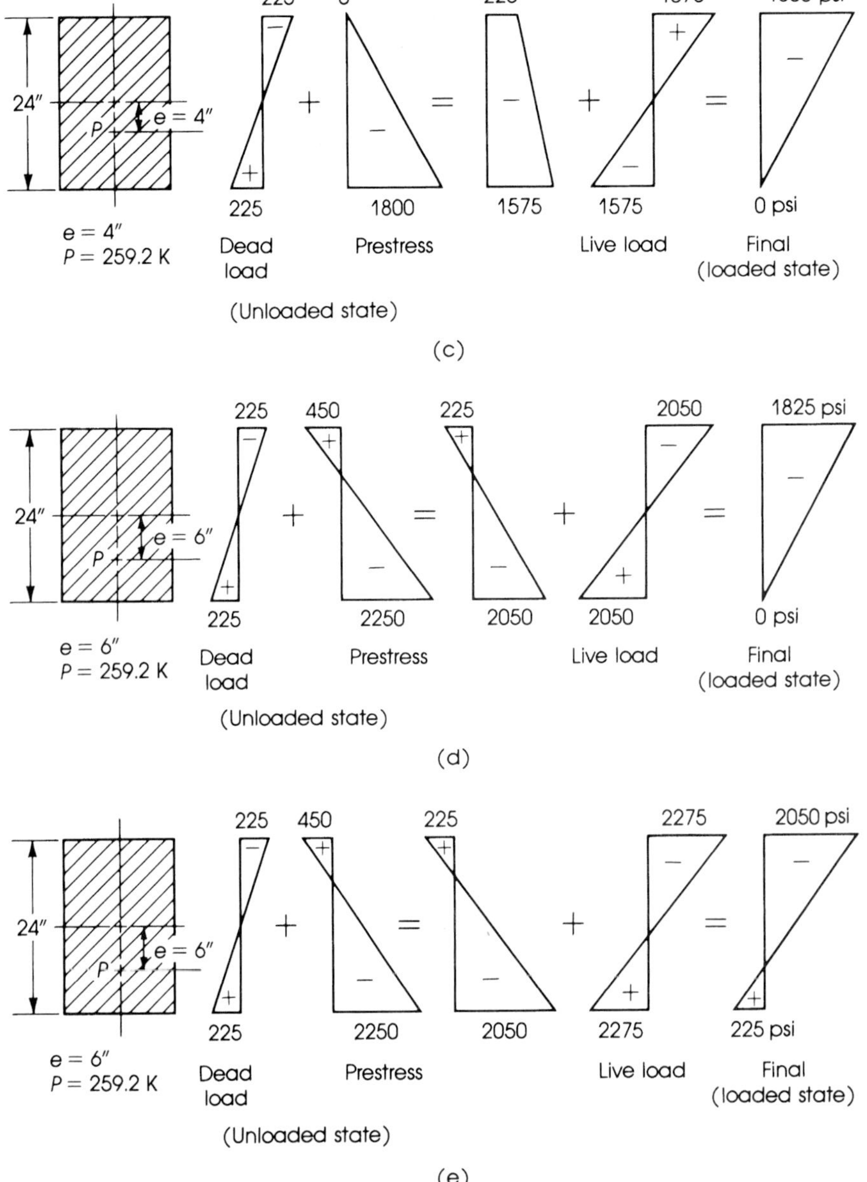

Figure 20.2 *(continued)*

Stresses at the extreme fibers are

$$\sigma = \frac{Mc}{I} = \frac{M(h/2)}{bh^3/12} = \frac{6M}{bh^2}$$

$$\sigma_D = \frac{6 \times 21.6 \times 12{,}000}{12(24)^2} = \pm 225 \text{ psi}$$

Stresses due to the live load $L_1 = 900$ lb/ft are

$$M_{\text{L.L.}} = \frac{0.9(24)^2}{8} = 64.8 \text{ K} \cdot \text{ft}$$

$$\sigma_{L1} = \frac{6M}{bh^2} = \frac{6 \times 64.8 \times 12{,}000}{12(24)^2} = \pm 675 \text{ psi}$$

Adding stresses due to the dead and live loads (Figure 20.2(a)) gives

$$\text{Top stress} = -225 - 675 = -900 \text{ psi} \qquad \text{(compression)}$$

$$\text{Bottom stress} = +225 + 675 = +900 \text{ psi} \qquad \text{(tension)}$$

The tensile stress is higher than the modulus of rupture of concrete, $f_r = 7.5\sqrt{f'_c} =$ 503 psi; hence, the beam will collapse.

2. In the case of stresses due to uniform prestress, if a compressive force $P = 259.2$ K is applied at the centroid of the section, then a uniform stress is induced on any section along the beam.

$$\sigma_p = \frac{P}{\text{area}} = \frac{259.2 \times 1000}{12 \times 24} = -900 \text{ psi} \qquad \text{(compression)}$$

Final stresses due to live and dead loads plus prestress load at the top and bottom fibers are 1800 psi and 0, respectively (Figure 20.2(b)). In this case, the prestressing force has doubled the compressive stress at the top fibers and reduced the tensile stress at the bottom fibers to 0. The maximum compressive stress of 1800 psi is less than the allowable stress of 2050 psi.

3. For stresses due to an eccentric prestress ($e = 4$ in.), if the prestressing force $P = 259.2$ K is placed at an eccentricity $e = 4$ in. below the centroid of the section, the stresses at the top and bottom fibers are calculated as follows.

Moment due to eccentric prestress is Pe:

$$\sigma_p = -\frac{P}{A} \pm \frac{(Pe)c}{I} = -\frac{P}{A} \pm \frac{6(Pe)}{bh^2}$$

$$= -\frac{259.2 \times 1000}{12 \times 24} \pm \frac{6(259.2 \times 1000 \times 4)}{12(24)^2}$$

$$= -900 \pm 900$$

$$= -1800 \text{ psi}$$

at the bottom fibers and $\sigma_p = 0$ at the top fibers. Consider now an increase in the live load of $L_2' = 2100$ lb/ft:

$$M_{\text{L.L.}} = \frac{2.1 \times (24)^2}{8} = 151.2 \text{ K}\cdot\text{ft}$$

$$\sigma_{L2} = \frac{6(151.2 \times 12{,}000)}{12(24)^2} = \pm 1575 \text{ psi}$$

Final stresses due to the dead, live, and prestressing loads at the top and bottom fibers are 1800 psi and 0, respectively (Figure 20.2(c)). Note that the final stresses are exactly the same as those of the previous case when the live load was 900 lb/ft; by applying the same prestressing force but at an eccentricity of 4 in., the same beam can now support a greater live load (by 1200 lb/ft).

4. For stresses due to eccentric prestress with maximum eccentricity, assume that the maximum practical eccentricity for this section is at $e = 6$ in., leaving a 2-in. concrete cover; then the bending moment induced is

$$Pe = 259.2 \times 6 = 1555.2 \text{ K}\cdot\text{in.} = 129.6 \text{ K}\cdot\text{ft}$$

Stresses due to the prestressing force are

$$\sigma_p = -\frac{259.2 \times 1000}{12 \times 24} \pm \frac{6 \times (129.6 \times 12{,}000)}{12(24)^2}$$

$$= -900 \pm 1350 \text{ psi}$$

$$= -2250 \text{ psi} \quad \text{and} \quad +450 \text{ psi}$$

Increase the live load now to $L_3 = 2733$ lb/ft. The stresses due to the live load, L_3, are

$$M_{\text{L.L.}} = \frac{2.733 \times (24)^2}{8} = 196.8\ \text{K}\cdot\text{ft}$$

$$\sigma_{L3} = \frac{6(196.8 \times 12{,}000)}{12(24)^2} = \pm 2050\ \text{psi}$$

The final stresses at the top and bottom fibers due to the dead load, live load (L_3), and the prestressing force are 1825 psi and 0, respectively (Figure 20.2(d)). Note that the final stresses are about the same as those in the previous cases, yet the live load has been increased to 2733 lb/ft. A tensile stress of 225 psi is developed when the prestressing force is applied on the beam. This stress is less than the modulus of rupture of concrete, $f_r = 503$ psi; hence, cracks will not develop in the beam.

5. The maximum live load when the eccentric force P acts at $e = 6$ in. is determined as follows. In the previous case, the final compressive stress is equal to 1825 psi, which is less than the allowable stress of 2050 psi. Therefore, the live load may be increased to $L_4 = 3033$ lb/ft.

$$M_{\text{L.L.}} = \frac{3.033 \times (24)^2}{8}$$

$$\sigma_{L4} = \frac{6(218.4 \times 12{,}000)}{12(24)^2} = \pm 2275\ \text{psi}$$

Partially prestressed concrete beams.

Final stresses due to the dead load, live load (L_4), and the prestressing force are −2050 psi and +225 psi (Figure 20.2(e)). The compressive stress is equal to the allowable stress of 2050 psi, and the tensile stress is less than the modulus of rupture of concrete of 503 psi. In this case, the uniform live load of 3033 lb/ft has been calculated as follows: Add the maximum allowable compressive stress of 2050 psi to the initial tensile stress at the top

Prestressing jack with a load cell.

Prestressing bed for T-beam sections.

fibers of 225 psi to get 2275 psi. The moment that will produce a stress at the top fibers of 2275 psi is equal to

$$M = \sigma\left(\frac{bh^2}{6}\right)$$

$$= \frac{2.275}{6}(12)(24)^2 = 2620.8 \text{ K}\cdot\text{in.} = 218.4 \text{ K}\cdot\text{ft}$$

$$M = \frac{W_L L^2}{8} \quad \text{and} \quad W_L = \frac{8 \times 218.4}{(24)^2} = 3.033 \text{ K/ft}$$

Notes:

1. The entire concrete section is active in resisting the external loads.
2. The final tensile stress in the section is less than the modulus of rupture of concrete, which indicates that a crackless concrete section can be achieved under full load.
3. The allowable load on the beam has been increased appreciably due to the application of the prestressing force.
4. An increase in the eccentricity of the prestressing force will increase the allowable applied load, provided that the allowable stresses on the section are not exceeded.

20.1.2 Partial Prestressing

A partially prestressed concrete member can be defined as one in which (1) there have been introduced internal stresses to counteract part of the stresses resulting from external loadings, (2) tensile stresses are developed in the concrete under working loads, and (3) nonprestressed reinforcement may be added to increase the moment capacity of the member. That definition implies that there are two cases that could be considered as partially prestressed concrete:

1. A combination of prestressed and nonprestressed steel is used in the same section. The prestressed cables induce internal stresses designed to take only part of the ultimate capacity of the concrete section. The rest of the capacity is taken by nonprestressed steel placed along the same direction as the prestressed cables. The steel used as nonprestressed steel could be any common grade of carbon steel or high-tensile-strength steel of the same kind as the prestressing cables with ultimate strength of 250 Ksi (1725 N/mm^2). The choice depends on two main factors: allowable deflection and allowable crack width. As for deflection, the ACI Code specifies a maximum ratio of span to depth of reinforced concrete members. With the smaller depth expected in partially prestressed concrete, and because a smaller steel percentage is used, excessive deflection under working loads must not be allowed. Cracks develop on the tension side of the concrete section or at the steel level because tensile stresses are allowed to occur under working loads. The maximum crack width that may be allowed is 0.016 in. (0.41 mm) for interior members and 0.013 in. (0.33 mm) for exterior members.
2. Internal stresses act on the member from prestressed steel only, but tensioned to a lower limit. In this case cracking develops earlier than in a fully prestressed member under similar loadings.

Partially prestressed concrete can be considered an intermediate form between reinforced and fully prestressed concrete. In reinforced concrete members, cracks develop under working loads; therefore reinforcement is placed in the tension zone. In prestressed concrete members, cracks do not usually develop under working loads. The compressive stresses due to prestressing may equal or exceed the tensile stresses due to external loadings. Therefore, a partially prestressed concrete member may be considered as a reinforced concrete member in which internal stresses are introduced to counteract part of the stress from external loadings so that tensile stresses in the concrete do not exceed a limited value under working load. It reduces to reinforced concrete when no internal stresses act on the member. Full prestressing is an upper extreme of partial prestressing in which nonprestressed reinforcing steel reduces to zero.

Between a reinforced cracked member and a fully prestressed uncracked member, there exists a wide range of design in partial prestressing (Figure 20.3). A proper choice of the degree of prestressing will produce a safe and economical structure.

Figure 20.3 shows the load deflection curves of concrete beams containing different amounts and types of reinforcement. Curve a represents a reinforced concrete beam, which normally cracks at a small load W_{cr}. The cracking moment M_{cr} can be determined as follows:

$$M_{cr} = \frac{f_r I}{c}$$

where

f_r = the modulus of rupture of concrete $= 7.5\sqrt{f'_c}$
I = moment of inertia of the gross concrete section
c = distance from the neutral axis to the tensile extreme fibers

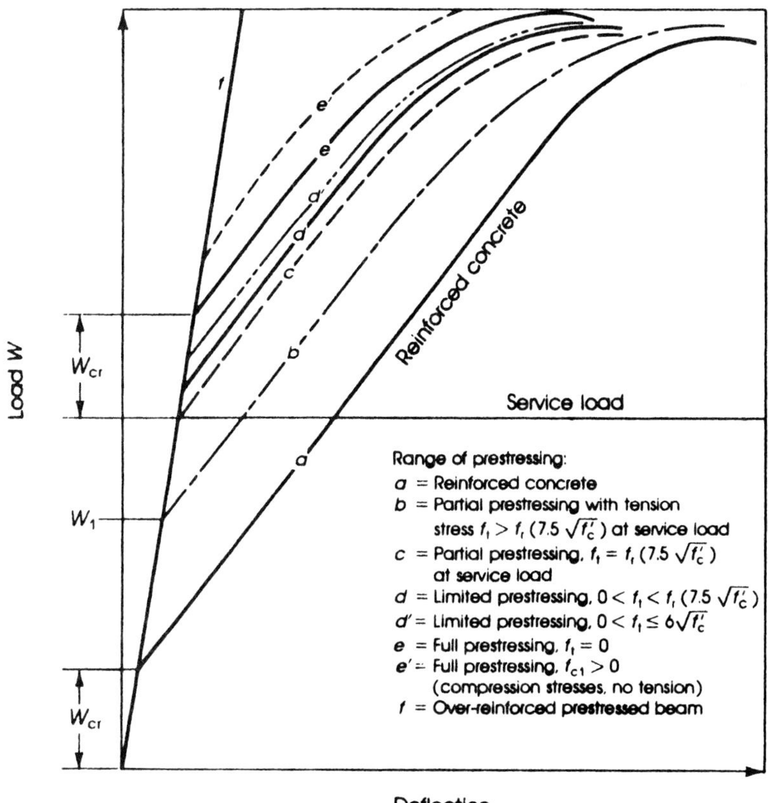

Figure 20.3 Load-deflection curves of concrete beams with different prestressing. The cracking load is W_{cr}.

The cracking load can be determined from the cracking moment when the span and the type of loading are specified. For a simply supported beam subjected to a concentrated load at midspan, $W_{cr} = (4M_{cr})/L$.

Curves e and f represent under-reinforced and over-reinforced fully prestressed concrete beams, respectively. The over-reinforced concrete beam fails by crushing of the concrete before the steel reaches its yield strength or proof stress. The beam has small deflection and undergoes brittle failure. The under-reinforced beam fails by the steel reaching its yield or ultimate strength. It shows appreciable deflection and cracking due to elongation of the steel before the gradual crushing of the concrete and the collapse of the beam.

Between curves a and e is a wide range of concrete beams with varying amounts of reinforcement and subjected to varying amounts of prestress. The beam with little prestressing is closer to curve a, while the beam with a large prestress is closer to curve e. Depending upon the allowable concrete stress, deflection, and maximum crack width, a suitable combination of prestressed and nonprestressed reinforcement may be chosen for the required design.

Curve b represents a beam which will crack under full working load. If only part of the live load L_1 occurs frequently on the structure, then W_1 represents the total dead load and that part of the live load L_1.

Curve c represents a beam that starts cracking at working load. The maximum tensile stress in the concrete $= f_r = 7.5\sqrt{f'_c}$.

Curve d represents a beam with limited prestress. The critical section of the beam will not crack under full working load, but it will have a maximum tensile stress $0 < f_t < 7.5\sqrt{f'_c}$. The maximum tensile stress in concrete allowed by the current ACI Code is $6\sqrt{f'_c}$.

Curves e and e' represent fully prestressed concrete beams with no tensile stress under working loads (see Figure 20.4).

The most important advantage of partial prestressing is the possibility of controlling camber. By reducing the prestressing force, the camber will be reduced and a saving in the amount of the prestressing steel, the amount of work in tensioning, and the number of end anchorages is realized.

Depending on the magnitude of the prestressing force, earlier cracking may occur in partially prestressed rather than in fully prestressed concrete members under service loads.

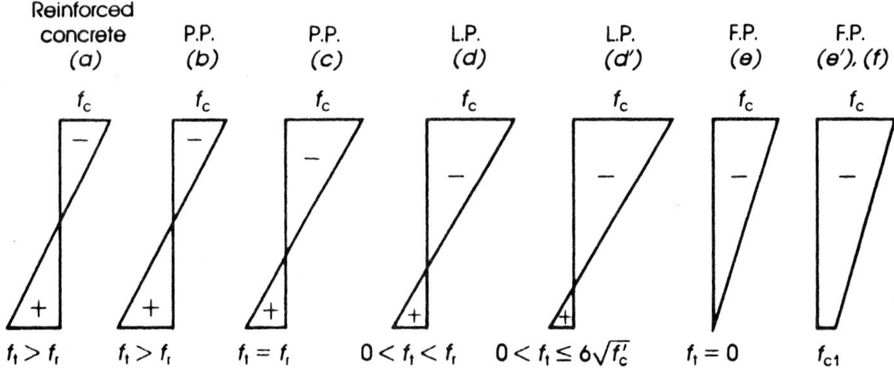

Figure 20.4 Distribution of stresses in beams with varying amounts of prestressed and nonprestressed reinforcement.

Once cracks develop, the effective moment of inertia of the critical section is reduced and a greater deflection is expected. However, partial prestressing has been used with satisfactory results, and its practical application is increasing.

20.2 MATERIALS AND ALLOWABLE STRESSES

20.2.1 Concrete

The physical properties of concrete were discussed in Chapter 2. Although reinforced concrete members are frequently made of concrete with a compressive strength of 3 to 5 Ksi (21 to 35 MPa), prestressed concrete members are made of higher-strength material, usually from 4 to 8 Ksi (28 to 56 MPa). High-strength concrete may be adopted for precast prestressed concrete members where components are prepared under optimum control of mixing concrete, placing, vibrating, and curing.

The allowable stresses in concrete according to the ACI Code, Section 18.4, are as follows.

1. Stresses after prestress transfer and before prestress losses:
   a. Maximum compressive stress of $0.6f_{ci}$
   b. Maximum tensile stress (except as permitted below) of $3\sqrt{f_{ci}}$
   c. Maximum tensile stress at the ends of simply supported members of $6\sqrt{f_{ci}}$, where $\sqrt{f_{ci}}$ is the ultimate strength of concrete at transfer

   If tensile stresses are exceeded, reinforcement must be provided in the tensile zone to resist the total tensile force in concrete (based on uncracked gross section).
2. Stresses at service loads after all losses:
   a. Maximum compressive stress of $0.45f'_c$ due to prestresses plus sustained loads, and of $0.6f'_c$ due to prestress plus total load.
   b. Maximum tensile stress in the precompressed zone of $6\sqrt{f'_c}$
   c. Maximum tensile stress where analysis based on transformed cracked sections shows that the deflections are within the ACI Code limitations of $12\sqrt{f'_c}$
3. These stresses may be exceeded if shown by tests or analysis that performance is satisfactory.

20.2.2 Prestressing Steel

The most common type of steel tendons used in prestressed concrete are strands (or cables) made with several wires, usually 7 or 19. Wires and bars are also used. The strands and wires are manufactured according to ASTM Standard A421 for uncoated stress-relieved wires and A416 for uncoated 7-wire stress-relieved strands. Properties of prestressing steel are given in Table 20.1.

Prestressing steel used in prestressed concrete must be of high-strength quality, usually of ultimate strength, f_{pu}, of 250 Ksi to 270 Ksi (1730–1860 MPa). High-strength steel is necessary to permit high elongation and to maintain a permanent sufficient prestress in the concrete after the inelastic shortening of the concrete.

The allowable stresses in prestressing steel according to the ACI Code, Section 18.5, are as follows:

1. Maximum stress due to tendon jacking force must not exceed the smaller of $0.8f_{pu}$ or $0.94f_{py}$. (The smaller value must not exceed that stress recommended by the manufacturer of tendons or anchorages.)
2. Maximum stress in pretensioned tendons immediately after transfer must not exceed the smaller of $0.74f_{pu}$ or $0.82f_{py}$.
3. Maximum stress in posttensioning tendons after tendon is anchored is $0.70f_{pu}$.

Table 20.1 Properties of Prestressing Steel, Nominal Diameters, Areas, and Weights

Type	Diameter (in.)	Area (in.^{2})	Weight (lb/ft)	Diameter (mm)	Area (mm^2)	Mass (kg/m)
Seven-wire strand (grade 250)	$\frac{1}{4}$ (0.250)	0.036	0.12	6.350	23.2	0.179
	$\frac{5}{16}$ (0.313)	0.058	0.20	7.950	37.4	0.298
	$\frac{3}{8}$ (0.375)	0.080	0.27	9.525	51.6	0.402
	$\frac{7}{16}$ (0.438)	0.108	0.37	11.125	69.7	0.551
	$\frac{1}{2}$ (0.500)	0.144	0.49	12.700	92.9	0.729
	(0.600)	0.216	0.74	15.240	139.4	1.101
Seven-wire strand (grade 270)	$\frac{3}{8}$ (0.375)	0.085	0.29	9.525	54.8	0.432
	$\frac{7}{16}$ (0.438)	0.115	0.40	11.125	74.2	0.595
	$\frac{1}{2}$ (0.500)	0.153	0.53	12.700	98.7	0.789
	(0.600)	0.215	0.74	15.250	138.7	1.101
Prestressing wire grades (250)	0.192	0.029	0.10	4.877	18.7	0.146
(250)	0.196	0.030	0.10	4.978	19.4	0.149
(240)	0.250	0.049	0.17	6.350	31.6	0.253
(235)	0.276	0.060	0.20	7.010	38.7	0.298
Prestressing bars (smooth)	$\frac{3}{4}$ (0.750)	0.44	1.50	19.050	283.9	2.232
(grade 145 or 160)	$\frac{7}{8}$ (0.875)	0.60	2.04	22.225	387.1	3.036
	1 (1.000)	0.78	2.67	25.400	503.2	3.973
	$1\frac{1}{8}$ (1.125)	0.99	3.38	28.575	638.7	5.030
	$1\frac{1}{4}$ (1.250)	1.23	4.17	31.750	793.5	6.206
	$1\frac{3}{8}$ (1.385)	1.48	5.05	34.925	954.8	7.515
Prestressing bars (deformed)	$\frac{5}{8}$ (0.625)	0.28	0.98	15.875	180.6	1.458
(grade 150–160)	$\frac{3}{4}$ (0.750)	0.42	1.49	19.050	271.0	2.218
	1 (1.000)	0.85	3.01	25.400	548.4	4.480
	$1\frac{1}{4}$ (1.250)	1.25	4.39	31.750	806.5	6.535
	$1\frac{3}{8}$ (1.385)	1.58	5.56	34.925	1006	8.274

Seven-wire prestressing strands (shipped in coils as shown).

20.2.3 Reinforcing Steel

Nonprestressed reinforcing steel is commonly used in prestressed concrete structural members, mainly in the prestressed precast concrete construction. The reinforcing steel is used as shear reinforcement, as supplementary reinforcement for transportation and handling the precast elements, and in combination with the prestressing steel in partially prestressed concrete members. The types and allowable stresses of reinforcing bars were discussed in Chapters 2 and 5.

20.3 LOSS OF PRESTRESS

20.3.1 Lump-Sum Losses

Following the transfer of the prestressing force from the jack to the concrete member, a continuous loss in the prestressing force occurs; the total loss of prestress is the reduction in the prestressing force during the lifespan of the structure. The amount of loss in tendon stress varies between 15% and 30% of the initial stress, because it depends on many factors. For most normal-weight concrete structures constructed by standard methods, the tendon stress loss due to elastic shortening, shrinkage, creep, and relaxation of steel is about 35 Ksi (241 MPa) for pretensioned members and 25 Ksi (172 MPa) for posttensioned members. Friction and anchorage slip are not included.

Two current recommendations for estimating the total loss in prestressed concrete members are presented by AASHTO and the Posttensioning Institute (PTI). AASHTO [23] recommends a total loss (excluding friction loss) of 45 Ksi (310 MPa) for pretensioned strands and 33 Ksi (228 MPa) for posttensioned strands and wires when a concrete strength $f'_c = 5$ Ksi is used. The PTI [24] recommends a total lump-sum prestress loss for posttensioned members of 35 Ksi (241 MPa) for beams and 30 Ksi (207 MPa) for slabs (excluding friction loss). These values can be used unless a better estimate of the prestress loss by each individual source is made, as is explained shortly.

In general, the sources of prestress loss are

- Elastic shortening of concrete
- Shrinkage of concrete
- Creep of concrete
- Relaxation of steel tendons
- Friction
- Anchorage set

20.3.2 Loss Due To Elastic Shortening of Concrete

In pretensioned members, estimating loss proceeds as follows. Consider a pretensioned concrete member of constant section and stressed uniformly along its centroidal axis by a force F_o. After the transfer of the prestressing force, the concrete beam and the prestressing tendon shorten by an equal amount, because of the bond between the two materials. Consequently, the starting prestressing force F_o drops to F_i and the loss in the prestressing force is $F_o - F_i$. Also, the strain in the concrete, ε_c, must be equal to the change in the tendon strain, $\Delta\varepsilon_s$. Therefore, $\varepsilon_c = \Delta\varepsilon_s$, or $(f_c/E_c) = (\Delta f_s/E_s)$, and the stress loss due to the elastic shortening is

$$\Delta f_s = \frac{E_s}{E_c} \times f_c = nf_c = \frac{nF_i}{A_c} \approx \frac{nF_o}{A_c} \tag{20.1}$$

where

A_c = the area of the concrete section
$n = E_s/E_c$ = modular ratio
f_c = the stress in the concrete at the centroid of the prestressing steel

Multiply the stress by the area of the prestressing steel, A_{sp}, to get the total force; then the elastic loss is

$$ES = F_o - F_i = \Delta f_s A_{sp} = (nf_c)A_{sp} \approx \left(\frac{nF_o}{A_c}\right)A_{sp} \tag{20.2}$$

$$F_i = F_o - (nf_c)A_{sp} \tag{20.3}$$

For practical design, the loss in the prestressing force, Δf_s per unit A_{sp}, may be taken to be approximately nF_o/A_c. If the force F_o acts at an eccentricity e, then the elastic loss due to the presence of F_o and the applied dead load at transfer is

$$\begin{aligned} ES &= -(nf_c)A_{sp}\,(\text{due to prestress}) + (nf_c)A_{sp}\,(\text{dead load}) \\ ES &= F_o - F_i = -\left(\frac{F_i}{A} + \frac{F_i e^2}{I}\right)nA_{sp} + \left(\frac{M_D e}{I}\right)nA_{sp} \end{aligned} \tag{20.4}$$

An approximate value of $F_i = (0.63f_{pu})A_{sp}$ may be used in the above equation.

$$\begin{aligned} F_o + f_c\,(\text{D.L.})nA_{sp} &= F_i\left[1 + nA_{sp}\left(\frac{1}{A} + \frac{e^2}{I}\right)\right] \\ F_i &= \frac{F_o + (nA_{sp})f_c\,(\text{D.L.})}{1 + (nA_{sp})\left(\frac{1}{A} + \frac{e^2}{I}\right)} \end{aligned} \tag{20.5}$$

For posttensioned members where the tendons or individual strands are not stressed simultaneously, the loss of the prestress can be taken as half the value ES for pretensioned members.

Also, it is practical to consider the elastic shortening loss in slabs equal to one-quarter of the equivalent pretensioned value, because stretching of one tendon will have little effect on the stressing of the other tendons.

20.3.3 Loss Due to Shrinkage

The loss of prestress due to shrinkage is time-dependent. It may be estimated as follows:

$$SH = \Delta f_s\,(\text{shrinkage}) = \varepsilon_{sh} E_s \tag{20.6}$$

where $E_s = 29 \times 10^6$ psi and ε_{sh} = shrinkage strain in concrete.

The average strain due to shrinkage may be assumed to have the following values: for pretensioned members, $\varepsilon_{sh1} = 0.0003$; for posttensioned members, $\varepsilon_{sh2} = 0.0002$. If posttensioning is carried out within 5 to 7 days after concreting, the shrinkage strain can be taken to be $0.8\varepsilon_{sh1}$. If posttensioning is carried out between 1 and 2 weeks, $\varepsilon_{sh} = 0.7\varepsilon_{sh1}$ can be used, and if it occurs more than 2 weeks later, $\varepsilon_{sh} = \varepsilon_{sh2}$ can be adopted. Shrinkage loss, SH, can also be estimated as follows [28]:

$$SH = 8.2 \times 10^{-6} K_{sh} E_s\left(1 - \frac{0.06V}{S}\right)(100 - RH)$$

where V/S = volume-to-surface ratio and RH = average relative humidity. K_{sh} is 1.0 for pretensioned members and is 0.8, 0.73, 0.64, and 0.58 for posttensioned members if posttensioning is carried out after 5, 10, 20, and 30 days, respectively.

20.3.4 Loss Due to Creep of Concrete

Creep is a time-dependent deformation that occurs in concrete under sustained loads. The developed deformation causes a loss of prestress from 5% to 7% of the applied force.

The creep strain varies with the magnitude of the initial stress in the concrete, the relative humidity, and time. The loss in strcss due to creep can be expressed as follows:

$$CR = \Delta f_s \text{ (creep)} = C_c(nf_c) = C_c(\varepsilon_{cr} E_s) \tag{20.7}$$

where

$$C_c = \text{creep coefficient} = \frac{\text{creep strain, } \varepsilon_{cp}}{\text{initial elastic strain, } \varepsilon_i}$$

The value of C_c may be taken as follows [22].

Concrete strength	$f'_c \leq 4$ Ksi		$f'_c > 4$ Ksi	
Relative humidity	100%	50%	100%	50%
C_c	1–2	2–4	0.7–1.5	1.5–3

Linear interpolation can be made between these values. Considering that half the creep takes place in the first $1\frac{3}{4}$ months of the first 6 months after transfer and under normal humidity conditions, the creep strain can be assumed for practical design as follows:

1. For pretensioned members, $\varepsilon_{cr} = 48 \times 10^{-5} \times$ stress in concrete (Ksi).
2. For posttensioned members, $\varepsilon_{cr} = 36 \times 10^{-5} \times$ stress in concrete (Ksi). This value is used when posttensioning is made within 2 to 3 weeks. For earlier posttensioning, an intermediate value may be used.

These values apply when the strength of concrete at transfer is $f'_{ci} \geq 4$ Ksi. When $f'_{ci} < 4$ Ksi, the creep strain should increase in the ratio of (4/actual strength).

$$\text{Total loss of prestress due to creep} = \varepsilon_{cr} E_s \tag{20.8}$$

20.3.5 Loss Due to Relaxation of Steel

Relaxation of steel causes a time-dependent loss in the initial prestressing force, similar to creep in concrete. The loss due to relaxation varies for different types of steel; its magnitude is usually furnished by the steel manufacturers. The loss is generally assumed to be 3% of the initial steel stress for posttensioned members and 2% to 3% for pretensioned members. If test information is not available, the loss percentages for relaxation at 1000 h can be assumed as follows:

1. In low-relaxation strands, when the initial prestress is $0.7f_{pu}$ and $0.8f_{pu}$, relaxation (RE) is 2.5% and 3.5%, respectively.
2. In stress-relieved strands or wire, when the initial prestress is $0.7f_{pu}$ or $0.8f_{pu}$, relaxation (RE) is 8% and 12%, respectively.

20.3.6 Loss Due to Friction

With pretensioned steel, friction loss occurs when wires or strands are deflected through a diaphragm. This loss is usually small and can be neglected. When the strands are deflected to follow a concordant profile, the friction loss may be considerable. In such cases, accurate load measuring devices are commonly used to determine the force in the tendon.

With posttensioned steel, the effect of friction is considerable because of two main factors: the curvature of the tendon and the lack of alignment (wobble) of the duct. The curvature effect may be visualized if a belt around a fixed cylinder is tensioned on one end with a force P_2; then the force, P_1, at the other end to initiate slippage in the direction of P_1 is

$$P_1 = P_2 e^{\mu\alpha} \tag{20.9}$$

where μ = the coefficient of static angular friction and α = the angle between P_1 and P_2. It is a general practice to treat the wobbling effect similarly:

$$\begin{aligned} P_x &= P_s e^{-(\mu\alpha + K l_x)} \\ P_s &= P_x e^{(\mu\alpha + K l_x)} \end{aligned} \tag{20.10}$$

where

P_x = the prestressing tendon force at any point x
P_s = the prestressing tendon force at the jacking end
μ = curvature friction coefficient
α = total angular change of prestressing tendon profile, in radians, from tendon jacking end to any point x
$= \dfrac{\text{length of curve}}{\text{radius of curvature}}$
K = wobble friction coefficient per foot of the prestressing tendon

As an approximation, the ACI Code gives the following expression:

$$P_s = P_x(1 + \mu\alpha + K l_x) \quad \text{(ACI Code, Equation 18.2)} \tag{20.11}$$

provided that $(\mu\alpha + K l_x) \leq 0.30$.

The frictional coefficients α and K depend on the type of prestressing strands or wires, type of duct, and the surface conditions. Some approximate values for μ and K are given in the ACI Commentary, Section 18.6 [1] and in Table 20.2.

Friction loss in the jack is variable and depends on many factors, including the length of travel of the arm over a given load range. The use of accurate load cells to measure directly the force in the tendon is recommended. The use of pressure gauges may lead to inaccuracies unless they are calibrated against a known force in the tendon.

The friction loss in the anchorage is dependent mainly upon the type of anchorage and the amount of deviation of the tendon as it passes through the anchorage. This loss is usually small and may be neglected. Guidance in particular cases should be obtained from the manufacturers.

Table 20.2 Friction Coefficients For Posttensioned Tendons

Type of Tendon	Wobble Coefficient *K* Per Foot ($\times 10^{-3}$)	Curvature Coefficient μ
Tendon in flexible metal sheathing (grouted)		
Wire tendons	1.0–1.5	0.15–0.25
Seven-wire strand	0.5–2.0	0.15–0.25
High-strength bars	0.1–0.6	0.08–0.30
Pregreased unbonded tendon		
Wire tendons and seven-wire strand	0.3–2.0	0.05–0.15
Mastic-coated unbonded tendons		
Wire tendons and seven-wire strand	1.0–2.0	0.05–0.15

20.3.7 Loss Due to Anchor Set

When the force in a tendon is transferred from the jack to the anchorage unit, a small inward movement of the tendon takes place due to the seating of the gripping device or wedges. The slippage causes a shortening of the tendon, which results in a loss in the prestressing force. The magnitude of slippage varies between 0.1 and 0.25 in. (2.5 and 6 mm) and is usually specified by the manufacturer. The loss due to the anchor set may be calculated as follows:

$$\Delta f_s = \Delta \varepsilon E_s = \frac{\Delta L}{L} \times E_s \tag{20.12}$$

where

$\Delta\varepsilon$ = magnitude of the anchor slippage
$E_s = 29 \times 10^6$ psi
L = length of the tendon

Because the loss in stress is inversely proportional to the length of the tendon (or approximately half the length of the tendon if it is stressed from both ends simultaneously), the percentage loss in steel stress decreases as the length of the tendon increases. If the tendon is elongated by $\Delta\varepsilon$ at transfer, the loss in prestress due to slippage is neglected.

Example 20.2

A 36-ft-span pretensioned simply supported beam has a rectangular cross section with $b = 18$ in. and $h = 32$ in. Calculate the elastic loss and all time-dependent losses. Given: prestressing force at transfer is $F_i = 435$ K, area of prestressing steel is $A_{PS} = 3.0$ in.^{2}, $f'_c = 5$ Ksi, $E_c = 5000$ Ksi, $E_s = 29{,}000$ Ksi; profile of tendon is parabolic, eccentricity at midspan = 6.0 in., and eccentricity at ends = 0.

Solution

1. Elastic shortening: Stress due to the prestressing force at transfer is

$$\frac{F_i}{A_{ps}} = \frac{435}{3} = 145 \text{ Ksi}$$

$$\text{Strain in prestressing steel} = \frac{f_s}{E_s} = \frac{145}{29{,}000} = 0.005$$

Using equation (20.1),

$$n = \frac{E_s}{E_c} = \frac{29{,}000}{5000} = 5.8, \quad \text{or} \quad 6$$

$$\Delta f_s = \frac{nF_i}{A_c} = \frac{6 \times 435}{32 \times 18} = 4.5 \text{ Ksi}$$

Considering the variation in the eccentricity along the beam,

$$\text{Strain at end of section} = \frac{F_i}{A_c E_c} = \frac{435}{(18 \times 32) \times 5000} = 0.151 \times 10^{-3}$$

$$\text{Strain at midspan} = \frac{F_i}{A_c E_c} + \frac{F_i e^2}{I E_c}$$

$$I = \frac{bh^3}{12} = \frac{18(32)^3}{12} = 49{,}152 \text{ in.}^4$$

$$\text{Strain} = 0.151 \times 10^{-3} + \frac{435(6)^2}{49{,}152(5000)} = 0.215 \times 10^{-3}$$

$$\text{Average strain} = \frac{1}{2}(0.151 + 0.215) \times 10^{-3} = 0.183 \times 10^{-3}$$

$$\text{Prestress loss} = \text{strain} \times E_s = 0.183 \times 10^{-3} \times 29{,}000 = 5.3 \text{ Ksi}$$

$$\text{Percent loss} = \frac{5.3}{145} = 3.66\%$$

2. Loss due to shrinkage:

$$\text{Shrinkage strain} = 0.0003$$

$$\Delta f_s = \varepsilon_{\text{sh}} E_s = 0.0003 \times 29{,}000 = 8.7 \text{ Ksi}$$

$$\text{Percent loss} = \frac{8.7}{145} = 6\%$$

3. Loss due to creep of concrete: Assuming $C_c = 2.0$, then

$$\Delta f_s = C_c(\varepsilon_{\text{cr}} E_s)$$

$$\text{Elastic strain} = \frac{F_i}{A_c E_c} = 0.151 \times 10^{-3}$$

$$\Delta f_s = 2(0.151 \times 10^{-3} \times 29{,}000) = 8.8 \text{ Ksi}$$

$$\text{Percent loss} = \frac{8.8}{145} = 6.1\%$$

Or, approximately, $\varepsilon_{\text{cr}} = 48 \times 10^{-5} \times$ stress in the concrete (Ksi):

$$\varepsilon_{\text{cr}} = 48 \times 10^{-5}\left(\frac{435}{32 \times 18}\right) = 36 \times 10^{-5}$$

$$\Delta f_s = \varepsilon_{\text{cr}} E_s = 36 \times 10^{-5} \times 29{,}000 = 10.4 \text{ Ksi}$$

$$\text{Percent loss} = \frac{10.4}{145} = 7.2\%$$

This is a conservative value, and the same ratio is obtained if $C_c = 2.38$ is adopted in the preceding calculations.

4. Loss due to relaxation of steel: For low-relaxation strands, the loss is assumed to be 2.5%.

$$\Delta f_s = 0.025 \times 145 = 3.6 \text{ Ksi}$$

5. Assume the losses due to bending, friction of cable spacers, and the end block of the pretensioning system are 2%.

$$\Delta f_s = 0.02 \times 145 = 2.9 \text{ Ksi}$$

6. Loss due to friction in tendon is 0.

7. Total losses are as follows.

Elastic shortening loss	5.3 Ksi	3.6%
Shrinkage loss	8.7 Ksi	6.0%
Creep of concrete loss	8.8 Ksi	6.1%
Relaxation of steel loss	3.6 Ksi	2.5%
Other losses	2.9 Ksi	2.0%
Total losses	29.3 Ksi	20.2%

$$\text{Effective prestress} = 145 - 24 = 121 \text{ Ksi}$$

$$\text{Effective prestressing force } F = 121 \times 3 \text{ in.}^2 = 363 \text{ Ksi}$$

$$F = (1 - 0.166)F_i = 0.834F_i$$

For $F = \eta F_i$, $\eta = 0.834$.

Example 20.3

Calculate all losses of a 120-ft-span posttensioned beam that has an I-section with the following details. Area of concrete section, $A_c = 760$ in.^{2}; moment of inertia, $I_g = 1.64 \times 10^5$ in.^{4}; prestressing force at transfer, $F_i = 1110$ K; area of prestressing steel, $A_{ps} = 7.5$ in.^{2}; $f'_c = 5$ Ksi, $E_c = 5000$ Ksi, and $E_s = 29{,}000$ Ksi; profile of tendon is parabolic; eccentricity at midspan = 20 in.; and eccentricity at ends = 0.

Solution

1. Loss due to elastic shortening:

$$\text{Steel stress at transfer} = \frac{F_i}{A_{ps}} = \frac{1110}{7.5} = 148 \text{ Ksi}$$

$$\text{Stress in concrete at end section} = \frac{1110}{760} = 1.46 \text{ Ksi}$$

$$\text{Stress in concrete at midspan} = \frac{F_i}{A_c} + \frac{F_i e^2}{I} - \frac{M_D e}{I}$$

$$\text{Weight of beam} = \frac{760}{144} \times 150 = 790 \text{ lb/ft}$$

$$M_D = 0.79\frac{(120)^2}{8} = 1422 \text{ K}\cdot\text{ft}$$

$$\text{Stress at midspan} = \frac{1110}{760} + \frac{1110(20)^2}{164{,}000} - \frac{(1422 \times 12)(20)}{164{,}000}$$

$$= 1.46 + 2.71 - 2.08 = 2.09 \text{ Ksi}$$

$$\text{Average stress} = \frac{1.46 + 2.09}{2} = 1.78 \text{ Ksi}$$

$$\text{Average strain} = \frac{1.78}{E_c} = \frac{1.78}{5000} = 0.356 \times 10^{-3}$$

Elastic loss is $\Delta f_s = \varepsilon_c E_s = 0.356 \times 10^{-3} \times 29{,}000 = 10.3$ Ksi, assuming that the tendons are tensioned two at a time. The first pair will have the greatest loss, whereas the last pair will have zero loss. Therefore, average $\Delta f_s = 10.3/2 = 5.15$ Ksi.

$$\text{Percent loss} = \frac{5.15}{148} = 3.5\%$$

2. Loss due to shrinkage of concrete:

$$\Delta f_s\,(\text{shrinkage}) = 0.0002E_s = 0.0002 \times 29{,}000 = 5.8 \text{ Ksi}$$

$$\text{Percent loss} = \frac{5.8}{148} = 3.9\%$$

3. Loss due to creep of concrete: Assume $C_c = 1.5$.

$$\text{Elastic strain} = \frac{F_i}{A_c E_c} = \frac{1110}{760 \times 5000} = 0.92 \times 10^{-3}$$

$$\Delta f_s\,(\text{creep}) = C_c(\varepsilon_{cr} E_s)$$

$$= 1.5(0.292 \times 10^{-3} \times 29{,}000) = 12.7 \text{ Ksi}$$

$$\text{Percent loss} = \frac{12.7}{148} = 8.6\%$$

4. Loss due to relaxation of steel: For low-relaxation strands, the loss is 2.5%.

$$\Delta f_s = 0.025 \times 148 = 3.7 \text{ Ksi}$$

5. Slip in anchorage: For tensioning from one end only, assume a slippage of 0.15 in. The length of the cable is 120 × 12 = 1440 in.

$$\Delta f_s = \frac{\Delta L}{L} \times E_s = \frac{0.15}{1440} \times 29{,}000 = 3 \text{ Ksi} \tag{20.12}$$

To allow for anchorage slip, set the tensioned force to 148 + 3 = 151 Ksi on the pressure gauge to leave a net stress of 148 Ksi in the tendons.

6. Loss due to friction: The equation of parabolic profile is

$$e_x = \frac{4e}{L^2}(Lx - x^2)$$

where e_x = the eccentricity at a distance x measured from the support and e = eccentricity at midspan.

$$\frac{d(e_x)}{dx} = \frac{4e}{L^2}(L - 2x)$$

is the slope of the tendon at any point. At the support, $x = 0$ and the slope

$$\frac{d(e_x)}{dx} = \frac{4e}{L} = \frac{4 \times 20}{120 \times 12} = 0.056$$

The slope at midspan is 0; therefore, $\alpha = 0.056$. Using flexible metallic sheath, $\mu = 0.5$ and $K = 0.001$. At midspan, $x = 60$ ft. Check if $(\mu\alpha + Kl_x) \le 0.30$:

$$\mu\alpha + Kl_x = 0.5 \times 0.056 + 0.001 \times 60 = 0.088 < 0.3$$

$$P_s = P_x(1 + \mu\alpha + Kl_x)$$

$$= P_x(1 + 0.088) = 1.088P_x \tag{20.11}$$

$$= 1.088 \times 148 = 161 \text{ K} \quad \text{(force at jacking end)}$$

$$\Delta f_s = 161 - 148 = 13 \text{ Ksi}$$

$$\text{Percent loss} = \frac{13}{148} = 8.8\%$$

7. Total losses:

Elastic shortening loss	5.2 Ksi	3.5%
Shrinkage loss	5.8 Ksi	3.9%
Creep of concrete loss	12.7 Ksi	8.6%
Relaxation of steel loss	3.7 Ksi	2.5%
Friction loss	13.0 Ksi	8.8%
Total losses	40.4 Ksi	27.3%

$$\text{Effective prestress} = 148 - 35.2 = 112.8 \text{ Ksi}$$

$$\text{Effective prestressing force } F = (1 - 0.238)F_i = 0.762F_i$$

$$F = 0.762 \times 1110 = 846 \text{ K}$$

For $F = \eta F_i$, $\eta = 0.762$.

20.4 ELASTIC ANALYSIS OF FLEXURAL MEMBERS

20.4.1 Stresses Due to Loaded and Unloaded Conditions

In the analysis of prestressed concrete beams, two extreme loadings are generally critical. The first occurs at transfer, when the beam is subjected to the prestressing force F_i and the weight of the beam or the applied dead load at the time of transfer of the prestressing force. No live load or additional dead loads are considered. In this unloaded condition, the stresses at the top and bottom fibers of the critical section must not exceed the allowable stresses at transfers, f_{ci} and f_{ti}, for the compressive and tensile stresses in concrete, respectively.

The second case of loading occurs when the beam is subjected to the prestressing force after all losses F and all dead and live loads. In this loaded condition, the stresses at the top and bottom fibers of the critical section must not exceed the allowable stresses, f_c and f_t, for the compressive and tensile stresses in concrete, respectively.

These conditions can be expressed mathematically as follows.

1. For the unloaded condition (at transfer):
   - At top fibers,

$$\sigma_{ti} = -\frac{F_i}{A} + \frac{(F_i e)y_t}{I} - \frac{M_D y_t}{I} \leq f_{ti} \tag{20.13}$$

   - At bottom fibers,

$$\sigma_{bi} = -\frac{F_i}{A} - \frac{(F_i e)y_b}{I} + \frac{M_D y_b}{I} \geq -f_{ci} \tag{20.14}$$

2. For the loaded condition (all loads are applied after all losses):
   - At top fibers,

$$\sigma_t = -\frac{F}{A} + \frac{(Fe)y_t}{I} - \frac{M_D y_t}{I} - \frac{M_L y_t}{I} \geq -f_c \tag{20.15}$$

   - At bottom fibers,

$$\sigma_b = -\frac{F}{A} - \frac{(Fe)y_b}{I} + \frac{M_d y_b}{I} + \frac{M_L y_b}{I} \leq f_t \tag{20.16}$$

where

F_i and F = the prestressing force at transfer and after all losses
f_{ti} and f_t = allowable tensile stress in concrete at transfer and after all losses
f_{ci} and f_c = allowable compressive stress in concrete at transfer and after all losses
M_D and M_L = moments due to dead load and live load
y_t and y_b = distances from the neutral axis to the top and bottom fibers

In this analysis, it is assumed that the materials behave elastically within the working range of stresses applied.

20.4.2 Kern Limits

If the prestressing force is applied at the centroid of the cross section, uniform stresses will develop. If the prestressing force is applied at an eccentricity e below the centroid such that the stress at the top fibers is equal to zero, that prestressing force is considered acting at the lower Kern point (Figure 20.5). In this case e is denoted by K_b, and the stress distribution is triangular, with maximum compressive stress at the extreme bottom fibers. The stress at the top fibers is

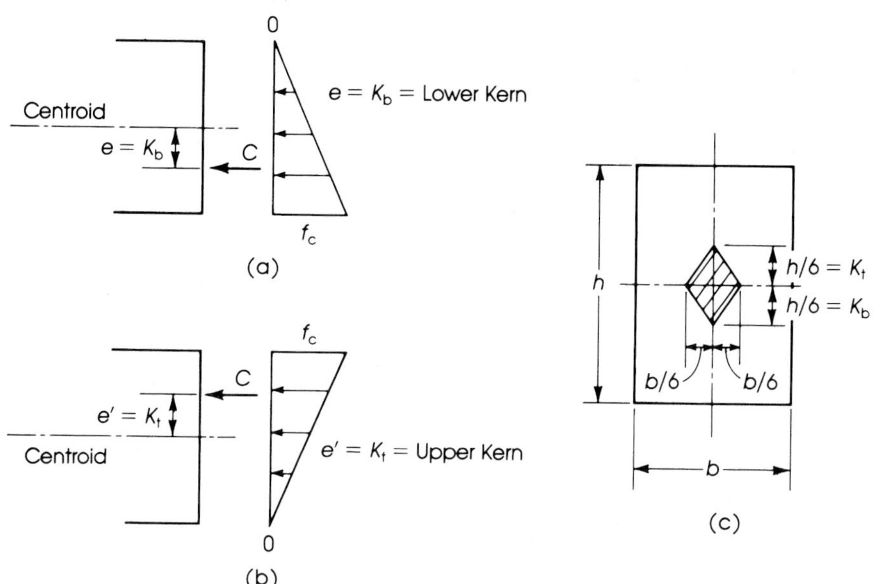

Figure 20.5 Kern points: (a) lower, (b) upper, and (c) central.

$$\sigma_t = -\frac{F_i}{A} + \frac{(F_i e) y_t}{I} = 0$$
$$e = K_b = \text{lower Kern} = \frac{I}{A y_t} \tag{20.17}$$

Similarly, if the prestressing force is applied at an eccentricity e' above the centroid such that the stress at the bottom fibers is equal to zero, that prestressing force is considered acting at the upper Kern point (Figure 20.5). In this case the eccentricity e' is denoted by K_t, and the stress distribution is triangular, with maximum compressive stress at the extreme top fibers. The stress at the bottom fibers is

$$\sigma_b = -\frac{F_i}{A} + \frac{(F_i e') y_b}{I} = 0$$
$$e' = K_t = \text{upper Kern} = \frac{I}{A y_b} \tag{20.18}$$

The Kern limits of a rectangular section are shown in Figure 20.5.

20.4.3 Limiting Values of Eccentricity

The four stress equations, (20.13) through (20.16), can be written as a function of the eccentricity e for the various loading conditions. For example, equation (20.13) can be rewritten as follows:

$$\sigma_{ti} = -\frac{F_i}{A} + \frac{(F_i e) y_t}{I} - \frac{M_D y_t}{I} \leq f_{ti}$$

$$\frac{(f_i e) y_t}{I} \leq f_{ti} + \frac{F_i}{A} + \frac{M_D y_t}{I}$$

$$e \leq \frac{I}{F_i y_t}\left(\frac{F_i}{A} + \frac{M_D y_t}{I} + f_{ti}\right) \tag{20.19}$$

If the lower Kern limit $K_b = I/Ay_t$ is used, then

$$e \leq K_b + \frac{M_D}{F_i} + \frac{f_{ti}AK_b}{F_i} \tag{20.20}$$

This value of e represents the maximum eccentricity based on the top fibers, unloaded condition.

Similarly, from equation (20.14),

$$e \leq \frac{I}{F_i y_b}\left(-\frac{F_i}{A} + \frac{M_D y_b}{I} + f_{ci}\right) \tag{20.21}$$

$$e \leq -K_t + \frac{M_D}{F_i} + \frac{f_{ci}AK_t}{F_i} \tag{20.22}$$

This value of e represents the maximum eccentricity based on the bottom fibers, unloaded condition. The two maximum values of e should be calculated from the preceding equations and the smaller value used.

From equation (20.15),

$$e \geq \frac{I}{F y_t}\left(\frac{F}{A} + \frac{M_T y_t}{I} - f_c\right) \tag{20.23}$$

$$e \geq K_b + \frac{M_T}{F} - \frac{f_c AK_b}{F} \tag{20.24}$$

where M_T = moment due to dead and live loads $= (M_D + M_L)$. This value of e represents the minimum eccentricity based on the top fibers, loaded condition. From equation (20.16),

$$e \geq \frac{I}{F y_b}\left(-\frac{F}{A} + \frac{M_T y_b}{I} - f_t\right) \tag{20.25}$$

$$e \geq K_t + \frac{M_T}{F} - \frac{f_t AK_t}{F} \tag{20.26}$$

This value of e represents the minimum eccentricity based on the bottom fibers, loaded condition. The two minimum values of e should be calculated from the preceding equations and the larger of the two minimum eccentricities used.

20.4.4 Limiting Values of the Prestressing Force at Transfer F_i

Considering that $F = \eta F_i$, where η represents the ratio of the net prestressing force after all losses, and for the different cases of loading, equations (20.20), (20.22), (20.24), and (20.26) can be rewritten as follows:

$$(e - K_b)F_i \leq M_D + f_{ti}AK_b \tag{20.27}$$

$$(e + K_t)F_i \leq M_D + f_{ci}AK_t \tag{20.28}$$

$$(e - K_b)F_i \geq \frac{M_D}{\eta} + \frac{M_L}{\eta} - \frac{1}{\eta}(f_c AK_t) \tag{20.29}$$

$$(e + K_t)F_i \geq \frac{M_D}{\eta} + \frac{M_L}{\eta} - \frac{1}{\eta}(f_t AK_t) \tag{20.30}$$

Subtract equation (20.27) from equation (20.30) to get

$$F_i(K_b + K_t) \geq M_D\left(\frac{1}{\eta} - 1\right) + \frac{M_L}{\eta} - \frac{f_t AK_t}{\eta} - f_{ti}AK_t$$

or

$$F_i \ge \frac{1}{(K_b + K_t)}\left[\left(\frac{1}{\eta} - 1\right)M_D + \frac{M_L}{\eta} - \left(\frac{f_t A K_t}{\eta}\right) - (f_{ti} A K_b)\right] \tag{20.31}$$

This value of F_i represents the *minimum* value of the prestressing force at transfer without exceeding the allowable stresses under the loaded and unloaded conditions. Subtract equation (20.29) from equation (20.28) to get

$$F_i \le \frac{1}{(K_b + K_t)}\left[\left(1 - \frac{1}{\eta}\right)M_D - \frac{M_L}{\eta} + \left(\frac{f_c A K_b}{\eta}\right) + (f_{ci} A K_t)\right] \tag{20.32}$$

This value of F_i represents the *maximum* value of the prestressing force at transfer without exceeding the allowable stresses under the loaded and unloaded conditions. Subtracting equation (20.31) from equation (20.32), then

$$\left(1 - \frac{1}{\eta}\right)2M_D - \frac{2M_L}{\eta} + \left(f_{ti} + \frac{f_c}{\eta}\right)AK_b + \left(f_{ci} + \frac{f_t}{\eta}\right)AK_t \ge 0 \tag{20.33}$$

This equation indicates that $(\text{maximum } F_i) - (\text{minimum } F_i) \ge 0$. If this equation is checked for any given section and proved to be satisfactory, then the section is adequate.

Precast prestressed concrete sections: (a) single T-, (b) double T-, and (c) U-sections.

Example 20.4

A pretensioned simply supported beam of the section shown in Figure 20.6(a) is to be used on a span of 48 ft. The beam must carry a dead load of 900 lb/ft (excluding its own weight), which will be applied at a later stage, and a live load of 1100 lb/ft. Assuming that prestressing steel is made of 20 tendons that are $\frac{7}{16}$ in. in diameter, with $E_s = 29 \times 10^6$ psi, $F_o = 175$ Ksi, and ultimate strength $f_{pu} = 250$ Ksi, it is required to do the following:

1. Determine the location of the upper and lower limits of the tendon profile (centroid of the prestressing steel) for the section at midspan and for three other sections between the midspan section and the beam end.
2. Locate the tendon to satisfy these limits by harping some of the tendons at one-third points of the span. Check the limiting values of the prestressing force at transfer.
3. Revise the prestress losses, taking into consideration the chosen profile of the tendons and the variation of the eccentricity, e.

Use f_{ci} (at transfer) $= 4$ Ksi, $f'_c = 5$ Ksi, $E_c = 4000$ Ksi, and $E_{ci} = 3600$ Ksi.

Solution

1. Determine the properties of the section:

$$\text{Area} = 18 \times 6 + 24 \times 6 + 12 \times 10 = 372 \text{ in.}^2$$

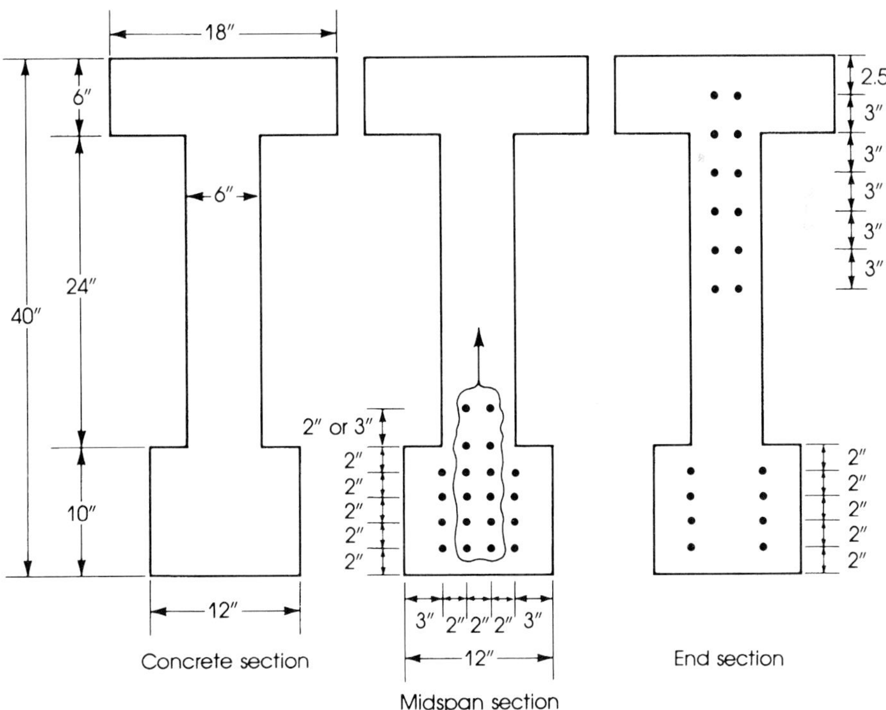

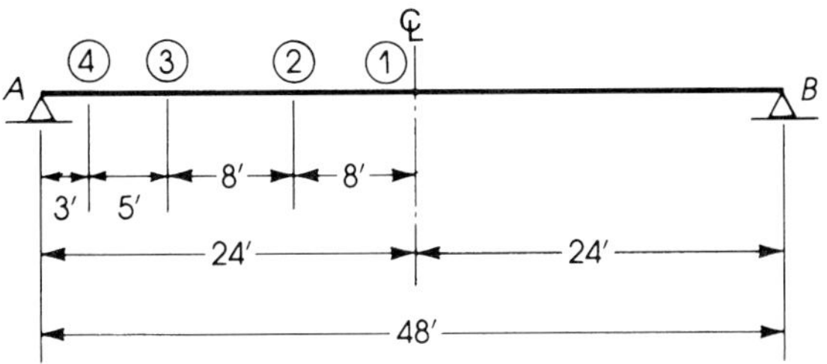

Figure 20.6 (a) Example 20.4.

Determine the centroid of the section by taking moments about the base line.

$$y_b = \frac{1}{372}(120 \times 5 + 144 \times 22 + 108 \times 37) = 20.8 \text{ in.}$$

$$y_t = 40 - 20.8 = 19.2 \text{ in.}$$

Calculate the gross moment of inertia I_g:

$$I_g = \left[\frac{18(6)^3}{12} + 108(16.2)^2\right] + \left[\frac{6(24)^3}{12} + 144(1.2)^2\right] + \left[\frac{12(10)^3}{12} + 120(15.8)^2\right]$$

$$= 66{,}862 \text{ in.}^4$$

$$K_b = \frac{I}{Ay_t} = \frac{66{,}862}{372 \times 19.2} = 9.4 \text{ in.}$$

$$K_t = \frac{I}{Ay_b} = \frac{66{,}862}{372 \times 20.8} = 8.6 \text{ in.}$$

2. Estimate prestress losses, given $F_o = 175$ Ksi.
a. Assume elastic loss is 4%, or $0.04 \times 175 = 7$ Ksi.
b. Loss due to shrinkage is $0.0003E_s = 0.0003 \times 29{,}000 = 8.7$ Ksi.
c. Loss due to creep of concrete: A good first estimate of creep loss is 1.67 times the elastic loss.

$$1.67 \times 7 = 11.7 \text{ Ksi}$$

d. Loss due to relaxation of steel is 4%:

$$0.04 \times 175 = 7 \text{ Ksi}$$

Time-dependent losses are $8.7 + 11.7 + 7 = 27.4$ Ksi.

$$\text{Percentage} = \frac{27.4}{175} = 15.7\%$$

e. The total loss is $27.4 + 7$ (elastic loss) $= 34.4$ Ksi. The percentage of total loss is $34.4/175 = 19.7\%$.
f. Prestress stresses are

$$F_i = 175 - 7 = 168 \text{ Ksi} \quad \text{(at transfer)}$$

$$F = 175 - 34.4 = 140.6 \text{ Ksi}$$

$$F = \eta F_i$$

$$\eta = 1 - \text{time-dependent losses ratio}$$

$$= \frac{140.6}{168} = 0.837$$

3. Limits of the eccentricity, e, at midspan section: Calculate the allowable stresses and moments. At transfer, $f'_{ci} = 4000$ psi, $f_{ci} = 0.6 \times 4000 = 2400$ psi, and $f_{ti} = 3\sqrt{f'_{ci}} = 190$ psi. At service load, $f'_c = 5000$ psi, $f_c = 0.45f'_c = 2250$ psi, and $f_t = 6\sqrt{f'_c} = 424$ psi.

$$\text{Self-weight of beam} = \frac{372}{144} \times 150 = 388 \text{ lb/ft}$$

$$M_D \text{ (self-weight)} = \frac{0.388}{8}(48)^2 \times 12 = 1341 \text{ K}\cdot\text{in.}$$

$$M_a \text{ (additional load and live load)} = \frac{w_a L^2}{8}$$

$$= \frac{(0.9 + 1.1)}{8}(48)^2 \times 12 = 6912 \text{ K}\cdot\text{in.}$$

Total moment $M_T = M_D + M_a = 8253 \text{ K}\cdot\text{in.}$

F_i = stress at transfer × area of prestressing steel

The area of 20 tendons, $\frac{7}{16}$ in. in diameter, is $20 \times 0.1089 = 2.178 \text{ in.}^2$

$$F_i = 2.178 \times 168 = 365.9 \text{ K}$$

$$F = 2.178 \times 140.6 = 306.2 \text{ K}$$

a. Consider the section at midspan.

Top fibers, *unloaded* condition:

$$e \le K_b + \frac{M_D}{F_i} + \frac{f_{ti} A K_b}{F_i} \qquad (20.20)$$

$$\le 9.4 + \frac{1341}{365.9} + \frac{0.190(372)(9.4)}{365.9} \le 14.9 \text{ in.}$$

Bottom fibers, *unloaded* condition:

$$e \le -K_t + \frac{M_D}{F_i} + \frac{f_{ci} A K_t}{F_i} \qquad (20.22)$$

$$\le -8.6 + \frac{1341}{365.9} + \frac{2.4(372)(8.6)}{365.9} \le 16.1 \text{ in.}$$

Maximum $e = 14.9$ in. controls.

Top fibers, *loaded* condition:

$$e \ge K_b + \frac{M_T}{F} - \frac{f_c A K_b}{F} \qquad (20.24)$$

$$\ge 9.4 + \frac{8253}{306.2} - \frac{0.424(372)(8.6)}{306.2} \ge 10.7 \text{ in.}$$

Bottom fibers, *loaded* condition:

$$e \ge -K_t + \frac{M_T}{F} - \frac{f_t A K_t}{F} \qquad (20.26)$$

$$\ge -8.6 + \frac{8253}{306.2} - \frac{0.424(372)(8.6)}{306.2} \ge 13.9 \text{ in.}$$

Minimum $e = 13.9$ in. controls.

b. Consider a section 8 ft from the midspan (section 2, Figure 20.6(a)):

$$M_D \text{ (self-weight)} = R_A(16) - \frac{w_D}{2} \times (16)^2$$

$$= 0.388(24)(16) - \frac{0.388}{2}(16)^2 = 99.3 \text{ K}\cdot\text{ft} = 1192 \text{ K}\cdot\text{in.}$$

$$M_a = 2(24)(16) - \frac{2}{2}(16)^2 = 512 \text{ K}\cdot\text{ft} = 6144 \text{ K}\cdot\text{in.}$$

$$M_T = 6144 + 1192 = 7336 \text{ K}\cdot\text{in.}$$

Top fibers, *unloaded* condition:

$$e \le 9.4 + \frac{1192}{365.9} + \frac{0.190(372)(9.4)}{365.9} \le 14.5 \text{ in.}$$

Bottom fibers, *unloaded* condition:

$$e \le -8.6 + \frac{1192}{365.9} + \frac{2.4(372)(8.6)}{365.9} \le 15.6 \text{ in.}$$

Maximum $e = 14.5$ in. controls.

Top fibers, *loaded* condition:

$$e \geq 9.4 + \frac{7336}{306.2} - \frac{2.25(372)(9.4)}{306.2} \geq 7.7 \text{ in.}$$

Bottom fibers, *loaded* condition:

$$e \geq -8.6 + \frac{7336}{306.2} - \frac{0.424(372)(8.6)}{306.2} \geq 11.0 \text{ in.}$$

Minimum $e = 11.0$ in. controls.

c. Consider a section 16 ft from midspan (section 3, Figure 20.6(a)): M_D (self-weight) $= 745$ K·in., $M_a = 3840$ K·in., and $M_T = 4585$ K·in.
- Top fibers, unloaded condition, $e \leq 13.3$ in. (max) controls.
- Bottom fibers, unloaded condition, $e \leq 14.4$ in.
- Top fibers, loaded condition, $e \geq -1.3$ in.
- Bottom fibers, loaded condition, $e \geq 1.9$ in. (min) controls.

d. Consider a section 3 ft from the end (anchorage length): $M_D = 314$ K·in., $M_a = 1620$ K·in., and $M_T = 1934$ K·in.
- Top fibers, unloaded condition, $e \leq 12.1$ in. (max) controls.
- Bottom fibers, unloaded condition, $e \leq 13.3$ in.
- Top fibers, loaded condition, $e \geq -10$ in.
- Bottom fibers, loaded condition, $e \geq -6.7$ in. (min) controls.

4. The tendon profile is shown in Figure 20.6(b). The eccentricity chosen at midspan is $e = 14.5$ in., which is adequate for section B at 8 ft from midspan. The centroid of the prestressing steel is horizontal between A and B and then harped linearly between B and the end section at E. The eccentricities at sections C and D are calculated by establishing the slope of line BE, which is $14.5/16 = 0.91$ in./ft. The eccentricity at C is 7.25 in. and at D it is 2.72 in. The tendon profile chosen satisfies the upper and lower limits of the eccentricity at all sections.

Harping of tendons is performed as follows:

a. Place the 20 tendons ($\frac{7}{16}$ in. diameter) within the middle third of the beam at spacings of 2 in., as shown in Figure 20.6(a). To calculate the actual eccentricity at midspan section, take moments for the number of tendons about the base line of the section:

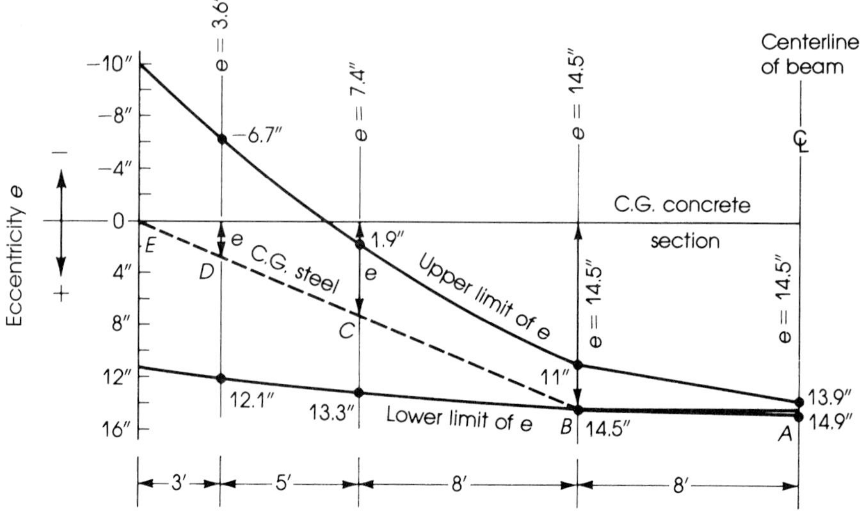

Figure 20.6 (b) Tendon profile, Example 20.4.

$$\text{Distance from base} = \tfrac{1}{20}(16 \times 5 + 4 \times 11) = 6.2 \text{ in.}$$

$$e\ (\text{midspan}) = y_b - 6.2 \text{ in.}$$

$$= 20.8 - 6.2 = 14.6 \text{ in.}$$

which is close to the 14.5 in. assumed. If the top two tendons are placed at 3 in. from the row below them, then the distance from the base becomes

$$\tfrac{1}{20}(16 \times 5 + 2 \times 10 + 2 \times 13) = 6.3 \text{ in.}$$

The eccentricity becomes $20.8 - 6.3 = 14.5$ in., which is equal to the assumed eccentricity. Practically, all tendons may be left at 2 in. spacing by neglecting the difference of 0.1 in.

b. Harp the central 12 tendons only. The distribution of tendons at the end section is shown in Figure 20.6(a). To check the eccentricity of tendons, take moments about the centroid of the concrete section for the 12 tendons at top and the 8 tendons left at bottom:

$$e = \tfrac{1}{20}(8 \times 14.5 - 12 \times 9.2) = 0.28 \text{ in.}$$

This value of e is small and adequate. The actual eccentricity at 3 ft from the end section is

$$e = \tfrac{3}{16}(14.5 - 0.28) + 0.28 = 2.95 \text{ in.} \qquad (3 \text{ in.})$$

The actual eccentricity at 8 ft from the end section is

$$e = \tfrac{1}{2}(14.5 - 0.28) + 0.28 = 7.4 \text{ in.}$$

5. Limited values of F_i: The value of F_i used in the preceding calculations is $F_i = 365.9$ K. Check minimum F_i by equation (20.31):

$$\text{Min. } F_i = \frac{1}{(K_b + K_t)}\left[\left(\frac{1}{\eta} - 1\right)M_D + \frac{M_L}{\eta} - \frac{(f_t A K_t)}{\eta} - (f_{ti} A K_b)\right]$$

$$= \frac{1}{(9.4 + 8.6)}\left[\left(\frac{1}{0.843} - 1\right)1341 + \frac{6912}{0.843}\right.$$

$$\left. - \frac{(0.424 \times 372 \times 8.6)}{0.843} - (0.19 \times 372 \times 9.4)\right] = 343.1 \text{ K}$$

which is less than the F_i used.

Check maximum F_i using equation (20.32):

$$\text{Max. } F_i = \frac{1}{(K_b + K_t)}\left[\left(1 - \frac{1}{\eta}\right)M_D - \frac{M_L}{\eta} + \frac{(f_c A K_b)}{\eta} + (f_{ci} A K_t)\right]$$

$$= \frac{1}{18}\left[\left(1 - \frac{1}{0.843}\right)1341 - \frac{6912}{0.843} + \frac{(2.25 \times 3.72 \times 9.4)}{0.843}\right.$$

$$\left. + (2.4 \times 372 \times 8.6)\right] = 475.7 \text{ K}$$

which is greater than the F_i used. Therefore, the critical section at midspan is adequate.

6. Check prestress losses, recalling that $F_o = 175$ Ksi and $A_{ps} = 2.178 \text{ in.}^2$

$$\text{Total } F_o = 2.178 \times 175 = 381 \text{ K}$$

$$E_c = 4000 \text{ Ksi}$$

$$n = \frac{E_s}{E_c} = \frac{29}{4.0} = 7.25$$

n can be assumed to be 7.

$$M_D \text{ at midspan} = 1341 \text{ K}\cdot\text{in.}$$

$$F_i = \frac{F_o + nA_{ps}f_c\,(\text{D.L.}) \times \frac{2}{3}}{1 + (nA_{ps})\left(\frac{1}{A} + \frac{e^2}{I}\right)} \tag{20.5}$$

The value of f_c due to the distributed dead load is multiplied by $\frac{2}{3}$ to reflect the parabolic variation of the dead load stress along the span, giving a better approximation of F_i.

a. Determine the average value of e^2, as adopted in the beam. The curve representing e^2 is shown in Figure 20.6(c):

$$\begin{aligned}\text{Average } e^2 &= \tfrac{1}{24}\left[\left(\tfrac{1}{3} \times 3 \times 9\right) + (9 \times 13)\right.\\ &\quad \left.+ \left(\tfrac{1}{3} \times 13 \times 201\right) + (210 \times 8)\right]\\ &= 111.5 \text{ in.}^2\\ e &= 10.56 \text{ in.}\end{aligned}$$

The area of a parabola is one-third of area of its rectangle.

b. Stress due to dead load at the level of the tendons is

$$f_c\,(\text{D.L.}) = \frac{1341 \times 10.56}{66{,}862} = 0.212 \text{ Ksi}$$

Therefore,

$$F_i = \frac{381 + 7(2.178) \times 0.212 \times 2/3}{1 + (7 \times 2.178)(1/372 + 111.5/66{,}862)} = 358 \text{ K}$$

Elastic loss is $381 - 358 = 23 \text{ K} = 6.1\%$. This value is greater than the assumed elastic loss of 4%.

$$\text{Elastic loss per unit steel area} = \frac{23}{2.178} = 10.6 \text{ Ksi}$$

$$F_i \text{ per unit steel area} = \frac{358}{2.178} = 164.4 \text{ Ksi}$$

c. Time-dependent losses:

$$\text{Loss due to shrinkage} = 8.7 \text{ Ksi (as before)}$$

Loss due to creep:

$$\text{Elastic strain} = \frac{F_i}{A_cE_c} = \frac{358}{372 \times 4000} = 0.240 \times 10^{-3}$$

$$\Delta f_s = C_c(\varepsilon_{cr}E_s)$$

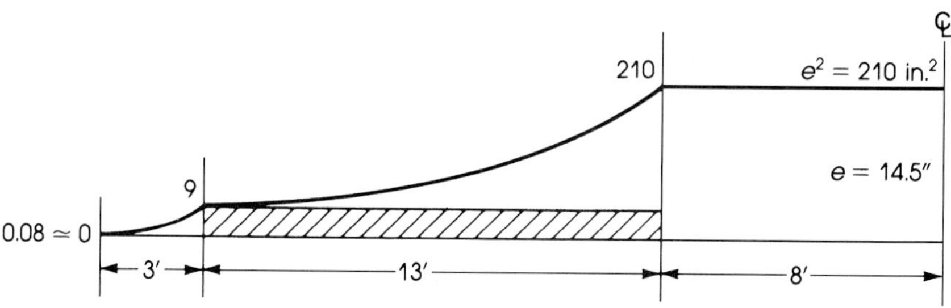

Figure 20.6 (c) Average e^2, Example 20.4.

Let $C_c = 1.5$. Then

$$\Delta f_s = 1.5(0.24 \times 10^{-3} \times 29{,}000) = 10.4 \text{ Ksi}$$

$$\text{Percent loss} = \frac{10.4}{164.4} = 6.3\%$$

Loss due to relaxation of steel is 7 Ksi (as before). Time-dependent losses equal $8.7 + 10.4 + 7 = 26.1$ Ksi, for a percentage loss of $26.1/164.4 = 15.8\%$, which is very close to the previously estimated value of 15.7%.

$$F = \eta F_i = (1 - 0.158)F_i = 0.842F_i$$

$$\eta = 0.842$$

20.5 STRENGTH DESIGN OF FLEXURAL MEMBERS

20.5.1 General

The previous section emphasized that the stresses at the top and bottom fibers of the critical sections of a prestressed concrete member must not exceed the allowable stresses for all cases or stages of loading. In addition to these conditions, a prestressed concrete member must be designed with an adequate factor of safety against failure. The ACI Code requires that the moment due to the factored service loads, M_u, must not exceed ϕM_n, the flexural strength of the designed cross section.

For the case of an under-reinforced prestressed concrete beam, failure begins when the steel stress exceeds the yield strength of steel used in the concrete section. The high-tensile prestressing steel will not exhibit a definite yield point, such as that of the ordinary mild steel bars used in reinforced concrete. But under additional increments of load, the strain in the steel increases at an accelerated rate, and failure occurs when the maximum compressive strain in the concrete reaches a value of 0.003 (Figure 20.7).

20.5.2 Rectangular Sections

The nominal ultimate moment capacity of a rectangular section may be determined as follows (refer to Figure 20.7):

$$M_n = C\left(d - \frac{a}{2}\right) = T\left(d - \frac{a}{2}\right) \tag{20.34}$$

where $T = A_{ps}f_s$ and $C = 0.85f'_c ab$. For $C = T$,

$$a = \frac{A_{ps}f_{ps}}{0.85f'_c b} = \frac{\rho_p f_{ps}}{0.85f'_c}d \tag{20.35}$$

where the prestressing steel ratio is $\rho_p = A_{ps}/bd$, and A_{ps} and f_{ps} refer to the area and tensile stress of the prestressing steel. Let

$$\omega_p = \rho_p\left(\frac{f_{ps}}{f'_c}\right) \le 0.36\beta_1$$

Then

$$a = \frac{\omega_p}{0.85}d \tag{20.36}$$

The quantity ω_p is a direct measure of the force in the tendon. To ensure an under-reinforced behavior, the ACI Code, Section 18.8.1, specifies that ω_p must not exceed $0.36\beta_1$, where

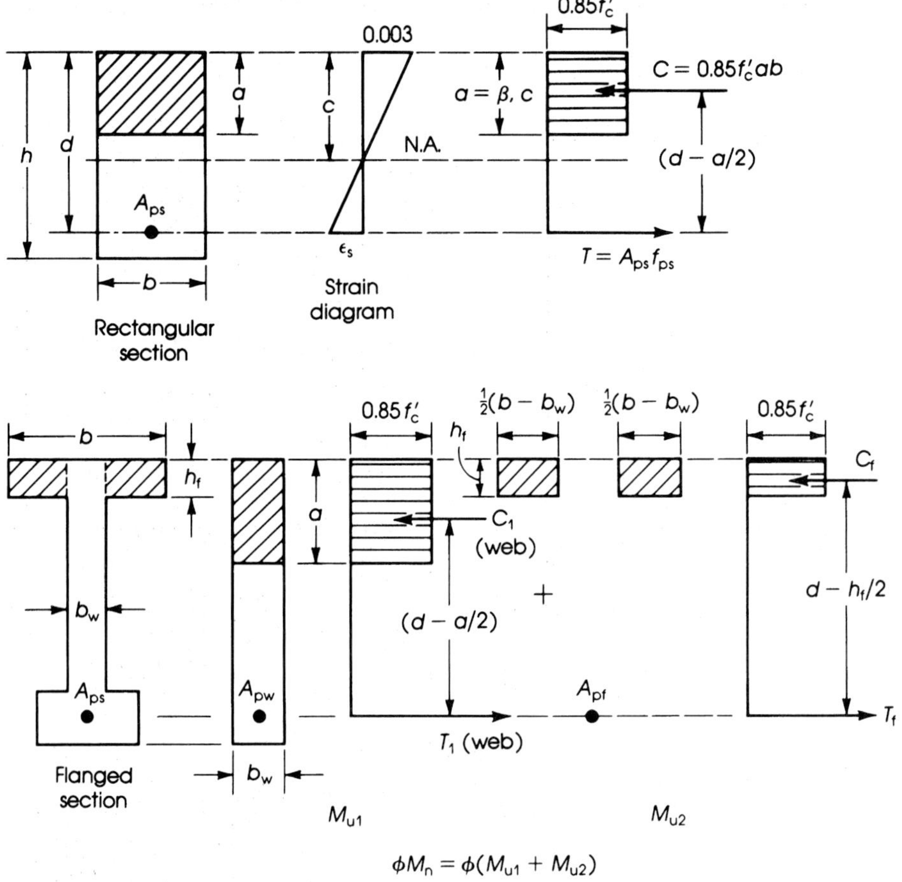

Figure 20.7 Ultimate moment capacity of prestressed concrete beams.

$\beta_1 = 0.85$ for $f'_c \leq 4$ Ksi and reduces by 0.05 for each 1 Ksi greater than 4 Ksi (ACI Code, Section 10.2.7.3). M_n can also be written as follows:

$$M_n = A_{ps} f_{ps}\left(d - \frac{a}{2}\right)$$

$$M_n = A_{ps} f_{ps} d\left(1 - \frac{\rho_p f_{ps}}{1.7 f'_c}\right) \tag{20.37}$$

$$M_n = A_{ps} f_{ps} d\left(1 - \frac{\omega_p}{1.7}\right) \tag{20.38}$$

and $M_u = \phi M_n$.

In the preceding equations, f_{ps} indicates the stress in the prestressing steel at failure. The actual value of f_{ps} may not be easily determined. Therefore, the ACI Code, Section 18.7.2, permits f_{ps} to be evaluated as follows (all stresses are in psi). For *bonded* tendons,

$$f_{ps} = f_{pu}\left[1 - \frac{\gamma_p}{\beta_1}\left(\rho_p \times \frac{f_{pu}}{f'_c}\right)\right] \tag{20.39}$$

For *unbonded* tendons in members with a span-to-depth ratio less than or equal to 35,

$$f_{ps} = \left(f_{se} + 10{,}000 + \frac{f'_c}{100\rho_p}\right) \leq f_{py} \tag{20.40}$$

provided that $f_{se} \geq 0.5f_{pu}$ and that f_{ps} for unbonded tendons must not exceed either f_{py} or $f_{se} + 60{,}000$ psi. For *unbonded* tendons in members with a span-to-depth ratio greater than 35,

$$f_{ps} = \left(f_{se} + 10{,}000 + \frac{f'_c}{300\rho_p}\right) \tag{20.41}$$

but not greater than f_{py} or $f_{se} + 30{,}000$ psi, where

γ_p = factor for the type of prestressing tendon
= 0.55 for f_{py}/f_{pu} not less than 0.8
= 0.4 for f_{py}/f_{pu} not less than 0.85
= 0.28 for f_{py}/f_{pu} not less than 0.9
f_{pu} = ultimate tensile strength of prestressing steel
f_{se} = effective stress in prestressing steel after all losses
f_{py} = specified yield strength of prestressing steel

In the event that $\omega_p > 0.36\beta_1$, an over-reinforced prestressed concrete beam may develop. To ensure a ductile failure, the ACI Code limits ω_p to a maximum value of $0.36\beta_1$. For $\omega_p = 0.36\beta_1$, $a = 0.424\beta_1 d$ (from equation (20.36)). Substituting this value of a in equation (20.38),

$$\begin{aligned} M_n &= A_{ps}f_{ps}d\left(1 - \frac{0.36\beta_1}{1.7}\right) \\ &= (\rho_p bd)f_{ps}d(1 - 0.21\beta_1) \\ &= \omega_p f'_c(1 - 0.21\beta_1)bd^2 \\ &= (0.36\beta_1 - 0.0756\beta_1^2)f'_c bd^2 \end{aligned} \tag{20.42}$$

for $f'_c = 5$ Ksi, $\beta_1 = 0.8$. Then

$$M_n = 0.24f'_c bd^2 = 1.2bd^2$$

Similarly, for $f'_c = 4$ Ksi, $M_n = bd^2$, and for $f'_c = 6$ Ksi, $M_n = 1.365bd^2$.

20.5.3 Flanged Sections

For flanged sections (T- or I-sections), if the stress block depth a lies within the flange, it will be treated as a rectangular section. If a lies within the web, then the web may be treated as a rectangular section using the web width, b_w, and the excess flange width $(b - b_w)$ will be treated similarly to that of reinforced concrete T-sections discussed in Chapters 3 and 4. The ultimate moment capacity of a flanged section can be calculated as follows (see Figure 20.7):

$$M_n = M_{n1} \text{ (moment capacity of the web)} + M_{n2} \text{ (moment capacity of excess flange)}$$

$$M_n = A_{pw}f_{ps}\left(d_p - \frac{a}{2}\right) + A_{pf}f_{ps}\left(d_p - \frac{h_f}{2}\right) \tag{20.43}$$

$$M_u = \phi M_n \quad \text{and} \quad a = \frac{A_{pw}f_{ps}}{0.85f'_c b_w}$$

where $A_{pw} = A_{ps} - A_{pf}$
$A_{pf} = [0.85f'_c(b - b_w)h_f]/f_{ps}$
h_f = thickness of the flange

Note that the total prestressed steel, A_{ps} is divided into two parts, A_{pw} and A_{pf} developing the web and flange moment capacity. For flanged sections, the reinforcement index, ω_{pw}, must not exceed $0.36\beta_1$, where

$$\omega_{pw} = \left(\frac{A_{pw}}{b_w d}\right)\left(\frac{f_{ps}}{f'_c}\right) = \text{prestressed web steel ratio} \times \left(\frac{f_{ps}}{f'_c}\right)$$

20.5.4 Nonprestressed Reinforcement

In some cases, nonprestressed reinforcing bars (A_s) are placed in the tension zone of a prestressed concrete flexural member together with the prestressing steel (A_{ps}) to increase the ultimate moment capacity of the beam. In this case, the total steel $(A_{ps}$ and $A_s)$ is considered in the moment analysis. For rectangular sections containing prestressed and nonprestressed steel, the design moment strength, ϕM_n, may be computed as follows ($\phi = 0.9$):

$$M_n = A_{ps}f_{ps}\left(d_p - \frac{a}{2}\right) + A_s f_y\left(d - \frac{a}{2}\right) \tag{20.44}$$

where

$$a = \frac{A_{ps}f_{ps} + A_s f_y}{0.85 f'_c b}$$

Also, d_p and d are the distances from extreme compression fibers to the centroid of the prestressed and nonprestressed steels, respectively. For flanged sections,

$$M_n = A_{pw}f_{ps}\left(d_p - \frac{a}{2}\right) + A_s f_y\left(d - \frac{a}{2}\right) + A_{pf}f_{ps}\left(d_p - \frac{h_f}{2}\right) \tag{20.45}$$

where

$$A_{pw} = A_{ps} - A_{pf}$$

$$a = \frac{A_{ps}f_{ps} + A_s f_y}{0.85 f'_c b_w}$$

For rectangular sections with compression reinforcement, and taking moments about the force C,

$$M_n = A_{ps}f_{ps}\left(d_p - \frac{a}{2}\right) + A_s f_y\left(d - \frac{a}{2}\right) + A'_s f_y\left(\frac{a}{2} - d'\right) \tag{20.46}$$

where

$$a = \frac{A_{ps}f_{ps} + A_s f_y - A'_s f_y}{0.85 f'_c b}$$

Prestressed concrete beds for slabs and wall panels.

This equation is valid only if compression steel yields. The condition for compression steel to yield is

$$\left(\frac{A_{ps}f_{ps} + A_s f_y - A'_s f_y}{bd}\right) \geq 0.85\beta_1 \frac{f'_c d'}{d}\left(\frac{87}{87 - f_y}\right)$$

If this condition is not met, then compression steel does not yield. In this case, A'_s may be neglected (let $A'_s = 0$), or alternatively, the stress in A'_s may be determined by general analysis, as explained in Chapter 3.

When prestressed and nonprestressed reinforcement are used in the same section, equation (20.39) should read as follows:

$$f_{ps} = f_{pu}\left[1 - \frac{\gamma_p}{\beta_1}\left(\rho_p \frac{f_{pu}}{f'_c} + \frac{d}{d_p}(\omega - \omega')\right)\right] \tag{20.47}$$

(ACI Code, equation (18.3)). If any compression reinforcement is taken into account when calculating f_{ps}, the term

$$\rho_p \frac{f_{pu}}{f'_c} + \frac{d}{d_p}(\omega - \omega')$$

must be greater than or equal to 0.17 and d' must be less than or equal to $0.15d_p$, where d, d', and d_p are the distances from the extreme compression fibers to the centroid of the nonprestressed tension steel, compression steel, and prestressed reinforcement, respectively,

γ_p = factor for type of prestressing tendon

$= 0.55$ for $\frac{f_{py}}{f_{pu}}$ not less than 0.8

$= 0.40$ for $\frac{f_{py}}{f_{pu}}$ not less than 0.85

$= 0.28$ for $\frac{f_{py}}{f_{pu}}$ not less than 0.90

$\beta_1 = 0.85$ for $f'_c \leq 4$ Ksi less 0.05 for each 1 Ksi increase in f'_c, but $\beta_1 \geq 0.65$

1. For rectangular sections, the ACI Code, Section 18.8, limits the reinforcement ratio as follows:

$$\omega_p + \frac{d}{d_p}\omega \leq 0.36\beta_1$$

where

$$\omega_p = \rho_p\left(\frac{f_{ps}}{f'_c}\right) \quad \text{and} \quad \rho_p = \frac{A_{ps}}{bd} \quad \text{(prestressed steel)}$$

$$\omega = \rho\left(\frac{f_y}{f'_c}\right) \quad \text{and} \quad \rho = \frac{A_s}{bd} \quad \text{(nonprestressed steel)}$$

2. If ordinary reinforcing bars A'_s are used in the compression zone, then the condition becomes

$$\omega_p + \frac{d}{d_p}(\omega - \omega') \leq 0.36\beta_1$$

where $\omega' = \rho'(f_y/f'_c)$ and $\rho' = A'_s/bd$. This reinforcement limitation is necessary to ensure a plastic failure of under-reinforced concrete beams.

3. For flanged sections, the steel area required to develop the strength of the web (A_{pw}) is used to check the reinforcement index.

$$\omega_{pw}\,(\text{web}) = \rho_{pw}\left(\frac{f_{ps}}{f'_c}\right) \le 0.36\beta_1$$

where

$$\rho_{pw} = \frac{A_{pw}}{b_w d_b}$$

If nonprestressed reinforcement is used, then the reinforcement limitations are

$$\omega_{pw} + \frac{d}{d_p}(\omega_w - \omega'_w) \le 0.36\beta_1$$

where

$$\omega_w \quad \text{and} \quad \omega'_w = \frac{A_s}{b_w d}\left(\frac{f_y}{f'_c}\right) \quad \text{and} \quad \frac{A'_s}{b_w d}\left(\frac{f_y}{f'_c}\right)$$

respectively. When compression steel A'_s is not used, then $\omega'_w = 0$. The preceding reinforcement conditions must be met in the analysis and design of partially prestressed concrete members.

20.6 CRACKING MOMENT

Cracks may develop in a prestressed concrete beam when the tensile stress at the extreme fibers of the critical section equals or exceeds the modulus of rupture of concrete, f_r. The value of f_r for normal-weight concrete may be assumed to be equal to $7.5\sqrt{f'_c}$. The stress at the bottom fibers of a simply supported beam produced by the prestressing force and the cracking moment is

$$\sigma_b = -\frac{F}{A} - \frac{(Fe)y_b}{I} + \frac{M_{cr}\,y_b}{I}$$

When $\sigma_b = f_r = 7.5\sqrt{f'_c}$, then the cracking moment is

$$M_{cr} = \frac{I}{y_b}\left(7.5\sqrt{f'_c} + \frac{F}{A} + \frac{(Fe)y_b}{I}\right) \tag{20.48}$$

The maximum tensile stress allowed by the ACI Code after all losses is $6\sqrt{f'_c}$, which represents $0.8f_r$. In this case, prestressed concrete beams may remain uncracked at service loads. To ensure adequate strength against cracking, the ACI Code, Section 18.8.3, requires that the ultimate moment capacity of the member ϕM_n be at least 1.2 times the cracking moment, M_{cr}.

Example 20.5

For the beam of Example 20.4, check the ultimate strength and cracking moment against the ACI Code requirements.

Solution

1. Check if the stress block depth a lies within the flange.

$$a = \frac{A_{ps}f_{ps}}{0.85 f'_c b} \tag{20.35}$$

$$A_{ps}\left(\text{of 20 tendons } \tfrac{7}{16} \text{ in. in diameter}\right) = 2.178 \text{ in.}^2$$

Let $f_{py}/f_{pu} = 0.85$, $\gamma_p = 0.4$, and $\gamma_p/\beta_1 = 0.4/0.8 = 0.5$. For bonded tensions,

$$f_{ps} = f_{pu}\left(1 - \frac{\gamma_p}{\beta_1}\rho_p \times \frac{f_{pu}}{f'_c}\right) \qquad (20.39)$$

$$d = 40 - 6.3 = 33.7 \text{ in.}$$

$$\rho_p = \frac{A_{ps}}{bd} = \frac{2.178}{18 \times 33.7} = 0.00359$$

Given $f_{pu} = 250$ Ksi,

$$f_{ps} = 250\left[1 - 0.5(0.00359) \times \frac{250}{5}\right] = 228 \text{ Ksi}$$

$$a = \frac{2.178 \times 228}{0.85 \times 5 \times 18} = 6.5 \text{ in.}$$

which is greater than 6 in. Therefore, the section acts as a flanged section.

2. For flanged sections,

$$M_n = A_{pw}f_{ps}\left(d - \frac{a}{2}\right) + A_{pf}f_{ps}\left(d - \frac{h_f}{2}\right)$$

where

$$A_{pw}\text{ (web)} = A_{ps} - A_{pf}\text{ (flange)}$$

$$A_{pf} = \frac{1}{f_{ps}}\left[0.85f'_c(b - b_w)h_f\right]$$

$$= \frac{1}{228}\left[0.85 \times 5(18 - 6)6\right] = 1.342 \text{ in.}^2$$

$$A_{pw} = 2.178 - 1.342 = 0.836 \text{ in.}^2$$

$$a = \frac{A_{pw}f_{ps}}{0.85f'_c b_w} = \frac{0.836(228)}{0.85 \times 5 \times 6} = 7.5 \text{ in.}$$

$$M_n = 0.836(228)\left(33.7 - \frac{7.5}{2}\right) + 1.342 \times 228\left(33.7 - \frac{6}{2}\right)$$

$$= 15{,}102 \text{ K}\cdot\text{in.} = 1258.5 \text{ K}\cdot\text{ft}$$

$$\phi M_n = 0.9(1258.5) = 1132.7 \text{ K}\cdot\text{ft}$$

Check the reinforcement index for the flanged section:

$$\rho_{pw}\text{ (web)} = \frac{A_{pw}}{b_w d} = \frac{0.836}{6 \times 33.7} = 0.00413$$

$$\omega_{pw}\text{ (web)} = \rho_{pw}\frac{f_{ps}}{f'_c} \le 0.36\beta_1 = 0.36 \times 0.8 = 0.288$$

($\beta_1 = 0.8$ for $f'_c = 5$ Ksi.)

$$\omega_{pw} = 0.00413\frac{(228)}{5} = 0.188 < 0.288$$

3. Calculate the external ultimate moment due to dead and live loads.

$$\text{Dead load} = \text{self-weight} + \text{additional dead load}$$

$$= 0.388 + 0.9 = 1.29 \text{ K/ft}$$

$$\text{Live load} = 1.1 \text{ K/ft}$$

$$U = 1.4D + 1.7L$$

$$M_u = \frac{(48)^2}{8}[1.4(1.29) + 1.7(1.1)] = 1058.7 \text{ K}\cdot\text{ft}$$

This external moment is less than the ultimate moment capacity of the section of 1132.7 K · ft; therefore, the section is adequate.

4. The cracking moment (equation (20.48)) is

$$M_{cr} = \frac{I}{y_b}\left(7.5\sqrt{f'_c} + \frac{F}{A} + (Fe)\frac{y_b}{I}\right)$$

From Example 20.4, $F = 306.2$ K, $A = 372$ in.^{2}, $e = 14.5$ in., $y_b = 20.8$ in., $I = 66{,}862$ in.^{4}, $f'_c = 5$ Ksi, and $7.5\sqrt{f'_c} = 7.5\sqrt{5000} = 530$ psi.

$$M_{cr} = \frac{66{,}862}{20.8}\left[0.53 + \frac{306.2}{372} + \frac{(306.2)(14.5)(20.8)}{66{,}862}\right]$$

$$= 8790 \text{ K}\cdot\text{in.} = 732.5 \text{ K}\cdot\text{ft}$$

Check that $1.2M_{cr} \le \phi M_n$.

$$1.2M_{cr} = 1.2(732.5) = 879 \text{ K}\cdot\text{ft}$$

This value is less than $\phi M_n = 1132.7$ K · ft; thus, the beam is adequate against cracking.

20.7 DEFLECTION

Deflection of a point in a beam is the total movement of the point, either downward or upward, due to the application of load on that beam. In a simply supported prestressed concrete beam, the prestressing force is usually applied below the centroid of the section, causing an upward deflection called *camber.* The self-weight of the beam and any external gravity loads acting on the beam will cause a downward deflection. The net deflection will be the algebraic sum of both deflections.

In computing deflections, it is important to consider both the short-term, or immediate, deflection and the long-term deflection. To ensure that the structure remains serviceable, the maximum short- and long-term deflections at all critical stages of loading must not exceed the limiting values specified by the ACI Code (see Section 6.3 in this text).

The deflection of a prestressed concrete member may be calculated by standard deflection equations or by the conventional methods given in books on structural analysis. For example, the midspan deflection of a simply supported beam subjected to a uniform gravity load w is equal to $(5wL^4/384EI)$. The modulus of elasticity of concrete is $E_c = 33\omega^{1.5}\sqrt{f'_c} = 57{,}000\sqrt{f'_c}$ for normal-weight concrete.

The moment of inertia of the concrete section I is calculated based on the properties of the gross section for an uncracked beam. This case is appropriate when the maximum tensile stress in the concrete extreme fibers does not exceed the modulus of rupture of concrete, $f_r = 7.5\sqrt{f'_c}$, or more conservatively, $6\sqrt{f'_c}$, as recommended by the ACI Code. When the maximum tensile stress based on the properties of the gross section exceeds $6\sqrt{f'_c}$, the effective moment of inertia, I_e, based on the cracked and uncracked sections must be used (as explained in Chapter 6). Typical midspan deflections for simply supported beams due to gravity loads and prestressing forces are shown in Table 20.3.

Upward deflection (camber) in double-T prestressed concrete beams.

Example 20.6

For the beam of Example 20.4, calculate the camber at transfer and then calculate the final anticipated immediate deflection at service load.

Solution

1. Deflection at transfer:
   a. Calculate the downward deflection due to dead load at transfer, self-weight in this case. For a simply supported beam subjected to a uniform load,

$$\Delta_D \text{ (dead load)} = \frac{5wL_4}{384EI}$$

From Example 20.4, $w_D = 388$ lb/ft, $L = 48$ ft, $E_{ci} = 3600$ Ksi, and $I = 66{,}862$ in.^{4}

$$\Delta_D = \frac{5(0.388/12)(48 \times 12)^4}{384(3600)(66{,}862)} = 0.192 \text{ in.} \qquad \text{(downward)}$$

   b. Calculate the camber due to the prestressing force. For a simply supported beam harped at one-third points with the eccentricity $e_1 = 14.5$ in. at the middle third and $e_2 = 0$ at the ends,

$$\Delta_p = \frac{23(F_i e_1)L^2}{216E_{ci}I} \qquad \text{(Table 20.3)}$$

$$= \frac{23(365.9 \times 14.5)(48 \times 12)^2}{216(3600)(66{,}862)} = -0.779 \text{ in.} \qquad \text{(upward)}$$

   c. Final camber at transfer is $-0.779 + 0.192 = -0.587$ in. (upward).
2. Deflection at service load: The total uniform service load is $W_T = 0.388 + 0.9 + 1.1 = 2.388$ K/ft, and $E_c = 4000$ Ksi. The downward deflection due to W_T is

$$\Delta_w = \frac{5W_T L^4}{384E_c I} = \frac{5(2.388/12)(48 \times 12)^4}{384(4000)(66{,}862)} = +1.067 \text{ in.} \qquad \text{(downward)}$$

Table 20.3 Midspan Deflections of Simply Supported Beams

Schematic	*Deflection equations*	
	Camber due to prestressing force	
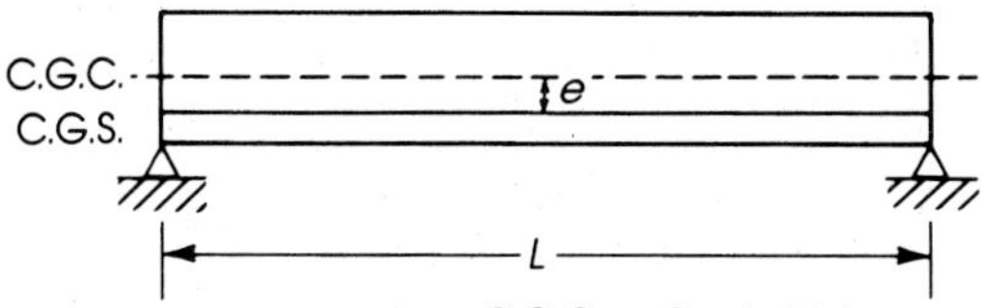 C.G.C. = Centroid (concrete); C.G.S. = Centroid (steel); eccentricity e, span L	$\Delta = \dfrac{(Fe)L^2}{8EI}$ (Horizontal tendons)	(1)
$+e_2$, e_1, Parabolic	$\Delta = \dfrac{FL^2}{8EI}\left[\dfrac{5}{6}e_1 + \dfrac{1}{6}e_2\right]$	(2)
	When $e_2 = 0$: $\Delta = \dfrac{5(Fe_1)L^2}{48EI}$	(3)
$+e_2$, e_1; a — $(L-2a)$ — a	$\Delta = \dfrac{FL^2}{8EI}\left[e_1 + \dfrac{4}{3}\left(\dfrac{a}{L}\right)^2(e_2 - e_1)\right]$	(4)
	When $a = \dfrac{L}{3}$: $\Delta = \dfrac{FL^2}{8EI}\left[e_1 + \dfrac{4}{27}(e_2 - e_1)\right]$	(5)
	When $a = \dfrac{L}{3}$ and $e_2 = 0$: $\Delta = \dfrac{23(Fe_1)L^2}{216EI}$	(6)
$+e_2$, e_1; $L/2$ — $L/2$	$\Delta = \dfrac{FL^2}{24EI}[2e_1 + e_2]$	(7)
	When $e_2 = 0$: $\Delta = \dfrac{(Fe_1)L^2}{12EI}$	(8)
	Deflection due to gravity loads	
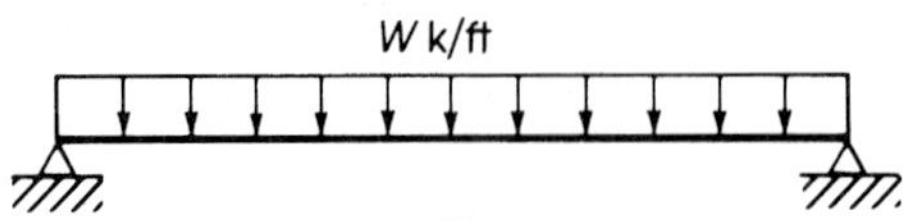 W k/ft	$\Delta = \dfrac{5wL^4}{384EI}$	(9)
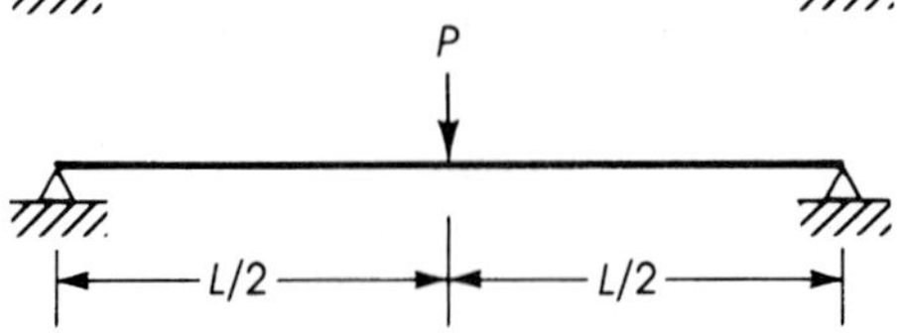 P; $L/2$ — $L/2$	$\Delta = \dfrac{PL^3}{48EI}$	(10)

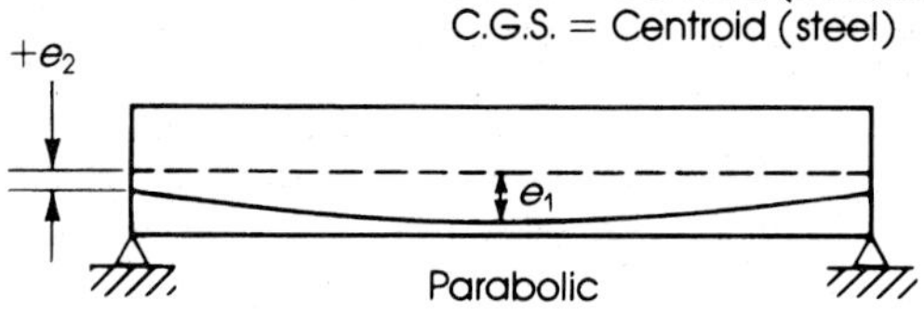

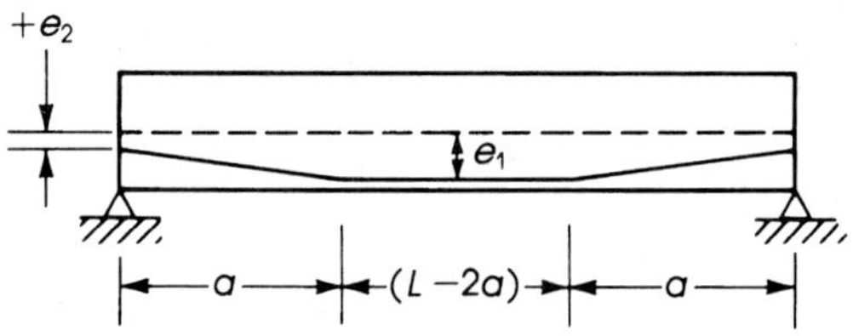

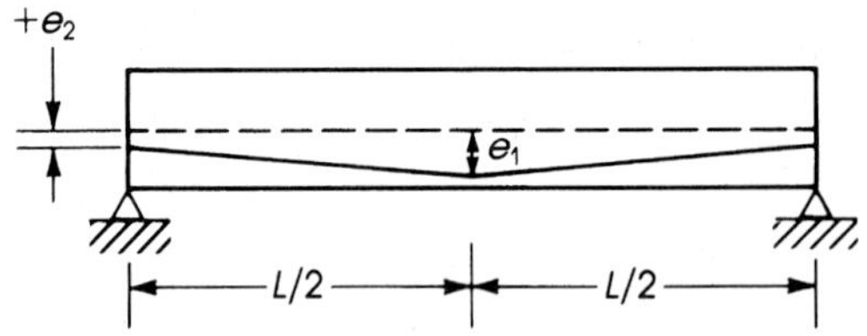

The camber due to prestressing force $F = 306.2$ K and $E_c = 4000$ Ksi is

$$\Delta_p = \frac{23(306.2 \times 14.5)(48 \times 12)^2}{216(4000)(66{,}862)} = -0.587 \text{ in.} \qquad \text{(upward)}$$

The final immediate deflection at service load is

$$\Delta = \Delta_w - \Delta_p = 1.067 - 0.587 = +0.48 \text{ in.} \qquad \text{(downward)}$$

20.8 DESIGN FOR SHEAR

The design approach to determine the shear reinforcement in a prestressed concrete beam is almost identical to that used for reinforced concrete beams. Shear cracks are assumed to develop at 45° measured from the axis of the beam. In general, two types of shear-related cracks form. One type is due to a combined effect of flexure and shear: The cracks start as flexural cracks and then deviate and propagate at an inclined direction due to the effect of diagonal tension. The second type, web-shear cracking, occurs in beams with narrow webs when the magnitude of principal tensile stress is high in comparison to flexural stress. Stirrups must be used to resist the principal tensile stresses in both cases. The ACI design criteria for shear will be adopted here.

20.8.1 Basic Approach

The ACI design approach is based on ultimate strength requirements using the load factors mentioned in Chapter 3. When the factored shear force, V_u, exceeds half the nominal shear strength $(\phi V_c/2)$, shear reinforcement must be provided. The required design shear force, V_u, at each section must not exceed the nominal design strength, ϕV_n, of the cross section based on the combined nominal shear capacity of concrete and web reinforcement:

$$V_u \leq \phi V_n \leq \phi(V_c + V_s) \qquad (20.49)$$

where

V_c = nominal shear capacity of concrete
V_s = nominal shear capacity of reinforcement
ϕ = capacity reduction factor = 0.85

When the factored shear force, V_u, is less than $\frac{1}{2}\phi V_c$, minimum shear reinforcement is required.

20.8.2 Shear Strength Provided by Concrete

The ACI Code, Section 11.4, presents a simple empirical expression to estimate the nominal ultimate shear capacity of a prestressed concrete member in which the tendons have an effective prestress f_{se} at least 40% of the ultimate tensile strength, f_{pu}:

$$V_c = \left(0.6\sqrt{f'_c} + 700\frac{V_u d}{M_u}\right)b_w d \qquad \text{(ACI Code, equation(11.9))} \qquad (20.50)$$

where V_u and M_u = factored shear and moment at the section under consideration

b_w = width of web

$d\left(\text{in the term } \frac{V_u d}{M_u}\right)$ = the distance from the compression fibers to the centroid of the prestressing steel

d(in V_{ci} or V_{cw} equations) = the larger of the above d or $0.8h$ (ACI Code, Section 11.4.2)

The use of equation (20.50) is limited to the following conditions:

1. The quantity $\frac{V_u d}{M_u} \leq 1.0$ (to account for small values of V_u and M_u)
2. $V_c \geq (2\sqrt{f'_c})b_w d$ (minimum V_c)
3. $V_c \leq (5\sqrt{f'_c})b_w d$ (maximum V_c)

The variation of the concrete shear capacity for a simply supported prestressed concrete beam subjected to a uniform load is shown in Figure 20.8. Note that the maximum shear reinforcement may be required near the supports and near one-fourth of the span where ϕV_s reaches maximum values. In contrast, similar reinforced concrete beams require maximum shear reinforcement (or minimum spacings) only near the support where maximum ϕV_s develops.

The values of V_c calculated by equation (20.50) may be conservative sometimes; therefore, the ACI Code, Section 11.4.2, gives an alternative approach for calculating V_c that takes into consideration the additional strength of concrete in the section. In this approach, V_c is taken as the smaller of two calculated values of the concrete shear strength V_{ci} and V_{cw} (Figure 20.8). Both are explained next.

The shear strength, V_{ci}, is based on the assumption that flexural-shear cracking occurs near the interior extremity of a flexural crack at an approximate distance of $d/2$ from the load point in the direction of decreasing moment. The ACI Code specifies that V_{ci} be computed as follows:

$$V_{ci} = (0.6\sqrt{f'_c})b_w d + \left(V_d + \frac{V_i M_{cr}}{M_{max}}\right) \tag{20.51}$$

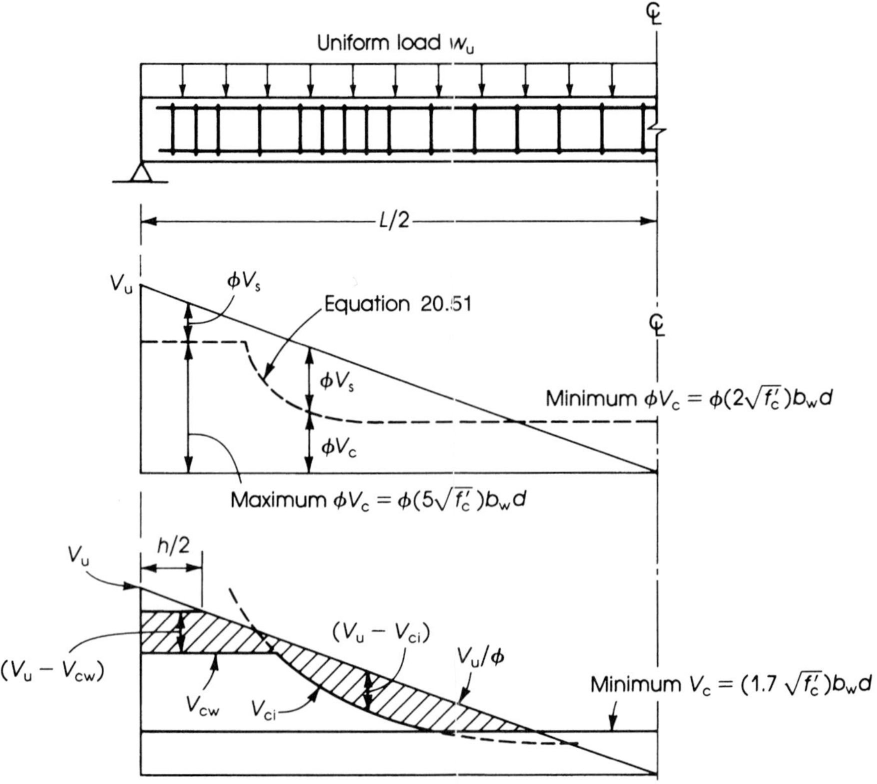

Figure 20.8 Distribution of shear forces along span. The middle diagram shows shear capacity of a simply supported prestressed concrete beam. The bottom diagram shows ACI analysis (stirrups are required for shaded areas).

but it is not less than $(1.7\sqrt{f'_c})b_w d$, where

V_d = shear force at section due to unfactored dead load

V_i = factored shear force at section due to externally applied loads occurring simultaneously with M_{max}

M_{max} = maximum factored moment at the section due to externally applied loads

M_{cr} = cracking moment

The cracking moment can be determined from the following expression:

$$M_{cr} = \frac{I}{y_t}(6\sqrt{f'_c} + f_{pe} - f_d) \qquad \text{(ACI Code, equation (11.11))} \tag{20.52}$$

where

I = moment of inertia of the section resisting external factored loads

y_t = distance from the centroidal axis of the gross section neglecting reinforcement to the extreme fiber in tension

f_{pe} = compressive strength at the extreme fibers of the concrete section due to the effective prestress force after all losses

f_d = stress due to the unfactored dead load at the extreme fiber, where tensile stress is caused by external loads

The web-shear strength, V_{cw}, is based on shear cracking in a beam that has not cracked by flexure. Such cracks develop near the supports of beams with narrow webs. The ACI Code, Section 11.4.2, specifies that V_{cw} be computed as follows:

$$V_{cw} = (3.5\sqrt{f'_c} + 0.3f_{pc})b_w d + V_p \tag{20.53}$$

where

V_p = vertical component of the effective prestress force at the section considered

f_{pc} = compressive stress (psi) in the concrete (after allowance for prestress losses) at the centroid of the section resisting the applied loads or at the junction of the web and flange when the centroid lies within the flange

Alternatively, V_{cw} may be determined as the shear force that produces a principal tensile stress of $4\sqrt{f'_c}$ at the centroidal axis of the member or at the intersection of the flange and web when the centroid lies within the flange. The equation for the principal stresses may be expressed as follows:

$$f_t = 4\sqrt{f'_c} = \sqrt{v_{cw}^2 + \left(\frac{f_{pc}}{2}\right)^2} - \frac{f_{pc}}{2}$$

or

$$V_{cw} = f_t\left(\sqrt{1 + \frac{f_{pc}}{f_t}}\right)b_w d \tag{20.54}$$

where $f_t = 4\sqrt{f'_c}$. When applying equations (20.51) and (20.53) or (20.54), the value of d is taken as the distance between the compression fibers and the centroid of the prestressing tendons but is not less than $0.8h$.

The critical section for maximum shear is to be taken at $h/2$ from the face of the support. The same shear reinforcement must be used at sections between the support and the section at $h/2$.

20.8.3 Shear Reinforcement

The value of V_s must be calculated to determine the required area of shear reinforcement.

$$V_u = \phi(V_c + V_s) \tag{20.49}$$

$$V_s = \frac{1}{\phi}(V_u - \phi V_c) \tag{20.55}$$

For vertical stirrups,

$$V_s = \frac{A_v f_y d}{s}, \quad \text{and} \tag{20.56}$$

$$A_v = \frac{V_s s}{f_y d} \quad \text{or} \quad s = \frac{A_v f_y d}{V_s} \tag{20.57}$$

where A_v = area of vertical stirrups and s = spacing of stirrups. Equations for inclined stirrups are the same as those discussed in Chapter 8.

20.8.4 Limitations

1. Maximum spacing, s_{max}, of the stirrups must not exceed $3h/4$ or 24 in. If V_s exceeds $(4\sqrt{f'_c})b_w d$, the maximum spacing must be reduced to half the preceding values (ACI Code, Section 11.5.4).
2. Maximum shear V_s must not exceed $(8\sqrt{f'_c})b_w d$; otherwise increase the dimensions of the section (ACI Code, Section 11.5.6).
3. The minimum shear reinforcement, A_v, required by the ACI Code is

$$A_v\,(\text{min}) = \frac{50 b_w s}{f_y} \tag{20.58}$$

When the effective prestress, f_{pe}, is greater than or equal to $0.4f_{pu}$, the minimum A_v is

$$A_v = \frac{A_{ps}}{80} \times \frac{f_{pu}}{f_y} \times \frac{s}{d} \times \sqrt{\frac{d}{b_w}} \tag{20.59}$$

The effective depth d need not be taken less than $0.8h$. Generally, equation (20.59) requires greater minimum shear reinforcement than equation (20.58).

Example 20.7

For the beam of Example 20.4, determine the nominal shear strength and the necessary shear reinforcement. Check the sections at $h/2$ and 10 ft from the end of the beam. Use f_y = 40 Ksi for the shear reinforcement.

Solution

1. For the section at $h/2$:

$$\frac{h}{2} = \frac{40}{2} = 20 \text{ in.} = 1.67 \text{ ft from the end}$$

2. The ultimate uniform load on beam is

$$W_u = 1.4(0.388 + 0.9) + 1.7 \times 1.1 = 3.68 \text{ K/ft}$$

$$V_u \text{ at a distance } \frac{h}{2} = 3.68(24 - 1.67) = 82.2 \text{ K}$$

Using the simplified ACI method (equation (20.50)), determine M_u at section $h/2$.

$$M_u = (3.68 \times 21) \times 1.67 - 3.68\frac{(1.67)^2}{2} = 142.4 \text{ K}\cdot\text{ft} = 1708 \text{ K}\cdot\text{in.}$$

The value of d at section $h/2$ from the end (Figure 20.6(b)) is

$$d = 33.7 \text{ (at midspan)} - \frac{(16 - 1.67)}{16} \times 14.5 = 20.7 \text{ in.}$$

$$\frac{V_u d}{M_u} = \frac{82.2 \times 20.7}{1708} = 0.966 \leq 1.0$$

as required by the ACI Code.

$$V_c = \left(0.6\sqrt{f'_c} + 700\frac{V_u d}{M_u}\right)b_w d$$

$$= (0.6\sqrt{5000} + 700 \times 0.996)6 \times 20.7 = 91{,}800 \text{ lb} = 91.8 \text{ K}$$

$$\text{Minimum } V_c = 2\sqrt{f'_c}\,b_w d = 2\sqrt{5000} \times 6 \times 20.7 = 17.6 \text{ K}$$

$$\text{Maximum } V_c = 5\sqrt{f'_c}\,b_w d = 43.9 \text{ K}$$

The maximum V_c of 43.9 K controls.

3. The alternative approach presented by the ACI Code is that V_c may be taken as the smaller value of V_{ci} and V_{cw}.

a. Based on the flexural-shear cracking strength,

$$V_{ci} = (0.6\sqrt{f'_c})b_w d + \left(V_d + \frac{V_i M_{cr}}{M_{max}}\right) \tag{20.51}$$

Calculate each item separately:

$$(0.6\sqrt{f'_c})b_w d = 0.6\sqrt{5000} \times 6 \times 20.7 = 5.3 \text{ K}$$

$$V_d = \text{unfactored dead load shear} = 1.288(24 - 1.67) = 28.8 \text{ K}$$

$$M_{max} = \text{maximum factored moment at section (except for weight of beam)}$$

$$\text{Factored load} = 1.4 \times 0.9 + 1.7 \times 1.1 = 3.13 \text{ K/ft}$$

$$M_{max} = 3.13\left[24 \times 1.67 - \frac{(1.67)^2}{2}\right] = 121 \text{ K/ft} = 1453 \text{ K}\cdot\text{in.}$$

$$V_i = 3.13(24 - 1.67) = 69.9 \text{ K}$$

$$M_{cr} = \frac{I}{y_t}(6\sqrt{f'_c} + f_{pe} - f_d)$$

$$I = 66{,}862 \text{ in.}^4, \qquad y_b = 20.8 \text{ in.}$$

$$f_e = \text{compressive stress due to prestressing force}$$

$$= \frac{F}{A} + \frac{Fey_b}{I}$$

$$= \frac{306.2}{372} + \frac{306.2(1.5)(20.8)}{66{,}862} = 0.966 \text{ Ksi}$$

$$f_d = \text{dead load stress} = \frac{M_D y_b}{I}$$

$$M_D = (1.288)\left[24 \times 1.67 - \frac{(1.67)^2}{2}\right] = 49.8 \text{ K}\cdot\text{ft} = 598 \text{ K}\cdot\text{in.}$$

$$f_d = \frac{598 \times 20.8}{66{,}862} = 0.186 \text{ Ksi}$$

$$M_{cr} = \frac{66{,}862}{20.8}(6\sqrt{5000} + 966 + 186) = 3871 \text{ K}\cdot\text{in.}$$

Therefore,

$$V_{ci} = (5.3) + (28.8) + 69.9\left(\frac{3871}{1453}\right) = 220.3 \text{ K}$$

V_{ci} must not be less than $(1.7\sqrt{f'_c})b_w d = (1.7\sqrt{5000}) \times 6 \times 20.7 = 14.9$ K

b. Shear strength based on web-shear cracking is

$$V_{cw} = (3.5\sqrt{f'_c} + 0.3f_{pc})b_w d + V_p$$

$$f_{pc} = \frac{306.2}{372} = 0.823 \text{ Ksi} \tag{20.53}$$

$$d = 20.7 \text{ in.} \quad \text{or} \quad 0.8h = 0.8 \times 40 = 32 \text{ in.}$$

Use $d = 32$ in.

$$V_p = 306.2 \times \frac{1}{13.2} = 23.2 \text{ K}$$

where $1/13.2$ = slope of tendon profile = 14.5 in./(16 × 12).

$$3.5\sqrt{f'_c} = 3.5\sqrt{5000} = 248 \text{ psi}$$

Therefore,

$$V_{cw} = (0.248 + 0.3 \times 0.823) \times 6 \times 32 + 23.2 = 118.2 \text{ K}$$

c. Because $V_{cw} < V_{ci}$, the value $V_{cw} = 118.2$ K represents the nominal shear strength at section $h/2$ from the end of the beam. In most cases, V_{cw} controls at $h/2$ from the support.

4. Web reinforcement:

$$V_u = 82.3 \text{ K} \qquad \phi V_{cw} = 0.85 \times 118.2 = 100.5 \text{ K}$$

Because $V_u < \phi V_{cw}$, $V_s = 0$; therefore, use minimum stirrups. Use no. 3 stirrups. $A_v = 2 \times 0.11 = 0.22$ in.^{2} Maximum spacing is the least of

$$s_1 = \frac{3}{4}h = \frac{3}{4} \times 40 = 30 \text{ in.} \qquad s_2 = 24 \text{ in.}$$

Calculate s_3 from the equation of minimum web reinforcement:

$$\text{Min. } A_v = \frac{A_{ps}}{80} \times \frac{f_{pu}}{f_y} \times \frac{s}{d} \times \sqrt{\frac{d}{b_w}}$$

$$0.22 = \frac{2.178}{80} \times \frac{250}{40} \times \frac{s_3}{20.7}\sqrt{\frac{20.7}{6}} \tag{20.59}$$

$$s_3 = 14.4 \text{ in.} \quad (14 \text{ in.})$$

Also,

$$\text{Min. } A_v = \frac{50b_w s}{f_y}$$

$$s_4 = \frac{A_v f_y}{50b_w} = \frac{0.22 \times 40{,}000}{50 \times 6} = 29 \text{ in.}$$

$s_{max} = s_3 = 14$ in. controls. Thus, use no. 3 stirrups spaced at 14 in.

5. For the section at 10 ft from the end, the calculation procedure is similar to that for the section at $h/2$. Using the ACI simplified method,

$$V_u = 3.68(24 - 10) = 51.5 \text{ K}$$

$$M_u = 3.68\left[24 \times 10 - \frac{(10)^2}{2}\right] = 699.2 \text{ K}\cdot\text{ft} = 8390 \text{ K}\cdot\text{in.}$$

$$d = 33.7 \text{ (at midspan)} - \frac{(16 \times 10)}{16} \times 14.5 = 28.3 \text{ in.}$$

$$\frac{V_u d}{M_u} = \frac{515. \times 28.3}{8390} = 0.174 < 1.0$$

$$V_c = (0.6\sqrt{5000} + 0.174 \times 700)6 \times 28.3 = 27{,}886 \text{ lb} = 27.9 \text{ K (controls)}$$

Minimum $V_c = 17.6$ K and maximum $V_c = 43.9$ K.

6. Using the ACI Code equations to compute V_{ci} and V_{cw}, calculate V_{ci} first (which controls at this section):

$$0.6\sqrt{f'_c}\,b_w d = 0.6\sqrt{5000} \times 6 \times 28.3 = 7.2 \text{ K}$$

$$V_d = 1.288(24 - 10) = 18 \text{ K}$$

$$M_{max} = 3.13\left[24 \times 10 - \frac{(10)^2}{2}\right] = 594.7 \text{ K}\cdot\text{ft} = 7136 \text{ K}\cdot\text{in.}$$

$$V_i = 3.13(24 - 10) = 43.8 \text{ K}$$

$$f_{pe} = \frac{306.2}{372} + \frac{306.2(9.1)(20.8)}{66{,}862} = 1.69 \text{ Ksi}$$

$$M_D = 1.288\left[24 \times 10 - \frac{(10)^2}{2}\right] = 244.7 \text{ K}\cdot\text{ft} = 2937 \text{ K}\cdot\text{in.}$$

$$f_d = \frac{2937 \times 20.8}{66{,}862} = 0.914 \text{ Ksi} \qquad M_{cr} = 3858 \text{ K}\cdot\text{in.}$$

Therefore,

$$V_{ci} = 7.2 + 18 + \frac{43.8(3858)}{7136} = 48.9 \text{ K}$$

$$V_{ci\,min} = (1.7\sqrt{5000})6 \times 28.3 = 20.4 \text{ K}$$

Thus the minimum is met. Then calculate V_{cw}:

$$f_{pc} = 0.893 \text{ Ksi} \qquad V_p = 23.2 \text{ K} \qquad \text{(as before)}$$

$$d = 28.3 \text{ in.} \quad \text{or} \quad 0.8h = 32 \text{ in.}$$

Use $d = 32$ in.

$$V_{cw} = (3.5\sqrt{f'_c} + 0.3f_{pc})b_w d + V_p$$

$$= (0.248 + 0.3 \times 0.823)6 \times 32 + 23.2 = 118.2 \text{ K}$$

This value of V_{cw} is not critical. At about span/4, the critical shear value is V_{ci} (Figure 20.8).

7. To calculate web reinforcement,

$$V_u = 51.5 \text{ K} \qquad \phi V_{ci} = 0.85 \times 48.9 = 41.5 \text{ K}$$

$$V_u = \phi(V_c + V_s)$$

$$V_s = \frac{1}{0.85}(51.5 - 41.5) = 10 \text{ K}$$

Use no. 3 stirrups; $A_v = 0.22$ in^2. Check maximum spacing: $s_{max} = 14$ in. (as before).

$$\text{Required } A_v = \frac{V_s s}{f_y d} = \frac{10 \times 14}{40 \times 28.3} = 0.124 \text{ in.}^2$$

A_v used is 0.22 in.^{2} > 0.124 in^2. Therefore, use no. 3 stirrups spaced at 14 in.

20.9 PRELIMINARY DESIGN OF PRESTRESSED CONCRETE FLEXURAL MEMBERS

20.9.1 Shapes and Dimensions

The detailed design of prestressed concrete members often involves a considerable amount of computation. A good guess at the dimensions of the section can result in a savings of time and effort. Hence it is important to ensure, by preliminary design, that the dimensions are reasonable before starting the detailed design.

At the preliminary design stage, some data are usually available to help choose proper dimensions. For example, the bending moments due to the applied external loads, the permissible stresses, and the data for assessing the losses are already known or calculated.

The shape of the cross section of a prestressed concrete member may be a rectangular, T-, I-, or box section. The total depth of the section, h, may be limited by headroom considerations or may not be specified. Given the freedom of selection, an empirical practical choice of dimensions for a preliminary design is as follows (Figure 20.9):

1. Total depth of section is $h = \frac{1}{20}$ to $\frac{1}{30}$ of the span L; for heavy loading $h = L/20$ and for light loading $h = L/30$ or $h = 2\sqrt{M_D + M_L}$, where M is in K·ft.
2. The depth of top flange is $h_f = h/8$ to $h/6$.
3. The width of top flange is $b \geq 2h/5$.
4. The thickness of the web is $b_w \geq 4$ in. Usually b_w is taken as $h/30 + 4$ in.
5. b_w and t are chosen to accommodate and uniformly distribute the prestressing tendons, keeping appropriate concrete cover protection.
6. The approximate area of the concrete section required is

$$A_c\,(\text{ft}^2) = \frac{M_D + M_L}{30h}$$

where $M_D + M_L$ are in K·ft and h is in ft. In SI units,

$$A_c\,(\text{m}^2) = \frac{M_D + M_L}{1450h} \qquad (M_D + M_L \text{ in kN·m and } h \text{ in m})$$

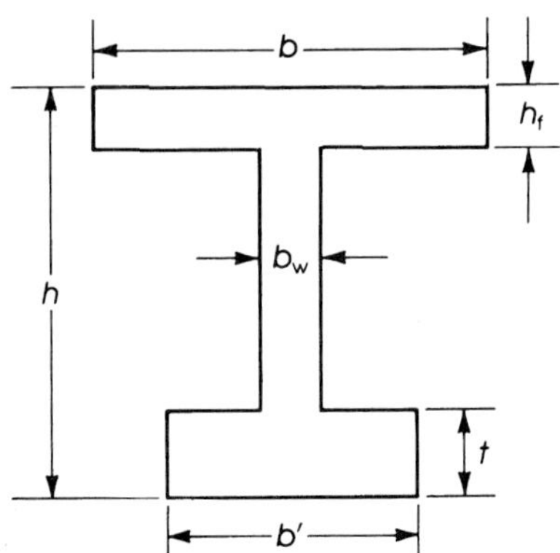

Figure 20.9 Proportioning prestressed concrete sections.

Table 20.4 AASHTO Girders, Normal Weight Concrete [25]

Designation	A (*in.*^{2})	I (*in.*^{4})	y_b (*in.*)	Z_b (*in.*^{3})	Z_t (*in.*^{3})	*Weight* (*lb/ft*)
Type II	369	50,979	15.83	3,220	2527	384
Type III	560	125,390	20.27	6,186	5070	593
Type IV	789	260,741	24.73	10,544	8908	822

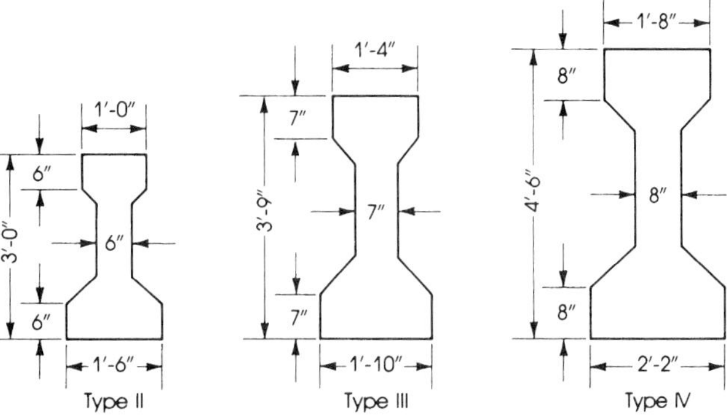

For practical and economical design of prestressed concrete beams and floor slabs, the precast concrete industry has introduced a large number of standardized shapes and dimensions from which the designer can choose an adequate member. Tables of standard sections are available in the PCI Design Handbook [3]. AASHTO [23] has also presented standard girders to be used in bridge construction (Table 20.4).

20.9.2 Prestressing Force and Steel Area

Once the shape, depth, and other dimensions of the cross section have been selected, approximate values of the prestressing force and the area of the prestressing steel, A_{ps}, can be determined. From the internal couple concept, the total moment M_T due to the service dead and live loads is equal to the tension force, T, times the moment arm, jd.

$$M_T = T(jd) = C(jd)$$

$$M_T = A_{ps} f_{se}(jd) \qquad A_{ps} = \frac{M_T}{f_{se}(jd)}$$

where A_{ps} is the area of the prestressing steel and f_{se} is the effective prestressing stress after all losses. The value of the moment arm, jd, varies from $0.4h$ to $0.8h$, with a practical range of $0.6h$ to $0.7h$. An average value of 0.65 may be used. Therefore,

$$A_{ps} = \frac{M_T}{(0.65h) f_{se}} \tag{20.60}$$

and the prestressing force is

$$F = T = A_{ps} f_{se} = \frac{M_T}{0.65h} \tag{20.61}$$

The prestressing force at transfer is $F_i = F/\eta$, where η is the factor of time-dependent losses.

The compressive force C on the section is equal to the tension force, T:

$$C = T = A_{ps} f_{se}$$

In terms of stresses,

$$\frac{C}{A_c} = \frac{A_{ps}f_{se}}{A_c} = f_{c1}$$

where f_{c1} is an assumed uniform stress on the section.

For preliminary design, a triangular stress distribution is assumed with maximum allowable compressive stress, f_{ca}, on one extreme fiber; therefore, the average stress is $0.5f_{ca} = f_{c1}$. The allowable compressive stress in concrete is $f_{ca} = 0.45f'_c$. Thus, the required concrete area, A_c, can be estimated from the force, T, as follows:

$$A_c = \frac{T}{f_{c1}} = \frac{A_{ps}f_{se}}{f_{c1}} = \frac{A_{ps}f_{se}}{0.5f_{ca}} = \frac{A_{ps}f_{se}}{0.225f'_c} \tag{20.62}$$

$$A_c = \frac{T}{0.5f_{ca}} = \frac{M_T}{(0.65h)(0.5f_{ca})} = \frac{M_T}{0.33f_{ca}} = \frac{M_T}{0.15f'_c} \tag{20.63}$$

This analysis is based on the design for service loads and not for the factored loads. The eccentricity, e, is measured from the centroid of the section to the centroid of the prestressing steel and can be estimated approximately as follows:

$$e = K_b + \frac{M_D}{F_i} \tag{20.64}$$

where K_b is the lower Kern limit and M_D is the moment due to the service dead load.

20.10 END-BLOCK STRESSES

20.10.1 Pretensioned Members

Much as a specific development length is required in every bar of a reinforced concrete beam, the prestressing force in a prestressed concrete beam must be transferred to the concrete by embedment or end anchorage or a combination thereof. In pretensioned members, the distance over which the effective prestressing force is transferred to the concrete is called the transfer length, l_t. After transfer, the stress in the tendons at the extreme end of the member is equal to zero, whereas the stress at a distance l_t from the end is equal to the effective prestress, f_{pe}. The transfer length, l_t, depends on the size and type of the tendon, surface condition, concrete strength, f'_c, stress, and method of force transfer. A practical estimation of l_t ranges between 50 and 100 times the tendon diameter. For strands, a practical value of l_t is equal to 50 tendon diameters, whereas for single wires, l_t is equal to 100 wire diameters.

In order that the tension in the prestressing steel develops full ultimate flexural strength, a bond length is required. The purpose is to prevent general slip before the failure of the beam at its full design strength. The development length, l_d, is equal to the bond length plus the transfer length l_t. Based on established tests, the ACI Code, Section 12.9.1, gives the following expression for computing the development length of three- or seven-wire pretensioning strands:

$$l_d \text{ (in.)} = \left(f_{ps} - \frac{2}{3}f_{se}\right)d_b \tag{20.65}$$

where

f_{ps} = stresses in prestressed reinforcement at nominal strength (Ksi)
f_{se} = effective stress in prestressed reinforcement after all losses (Ksi)
d_b = nominal diameter of wire or strand (in.)

In pretensioned members, high-tensile stresses exist at the end zones, for which special reinforcement must be provided. Such reinforcement in the form of vertical stirrups is uniformly distributed within a distance $h/5$ measured from the end of the beam. The first stirrup is usually placed at 1 to 3 in. from the beam end or as close as possible. It is a common practice to add nominal reinforcement for a distance d measured from the end of the beam. The area of the vertical stirrups A_v to be used at the end zone can be calculated approximately from the following expression:

$$A_v = 0.021\,\frac{F_i h}{f_{sa} l_t} \tag{20.66}$$

where f_{sa} = allowable stress in the stirrups (usually 20 Ksi) and l_t = 50 tendon diameters.

Example 20.8

Determine the necessary stirrup reinforcement required at the end zone of the beam given in Example 20.4.

Solution

$$F_i = 365.9\text{ K},\quad h = 40\text{ in.},\quad f_s = 20\text{ Ksi},\quad \text{and}\quad l_t = 50 \times \frac{7}{16} = 22\text{ in.}$$

Therefore,

$$A_v = 0.021 \times \frac{365.9 \times 40}{20 \times 22} = 0.7\text{ in.}^2$$

$$\frac{h}{2} = \frac{40}{5} = 8\text{ in.}$$

Use four no. 3 closed stirrups within the first 8 in. distance from the support. A_v (provided) = $4 \times 0.22 = 0.88\text{ in.}^2$.

20.10.2 Posttensioned Members

In posttensioned concrete members, the prestressing force is transferred from the tendons to the concrete, for both bonded and unbonded tendons, at the ends of the member by special anchorage devices. Within an anchorage zone at the end of the member, very high compressive stresses and transverse tensile stresses develop, as shown in Figure 20.10. In practice, it is found that the length of the anchorage zone does not exceed the depth of the end of the member; nevertheless, the state of stress within this zone is extremely complex.

The stress distribution due to one tendon within the anchorage zone is shown in Figure 20.11. At a distance h from the end section, the stress distribution is assumed uniform all over the section. Considering the lines of force (trajectories) as individual elements acting as curved struts, the trajectories tend to deflect laterally toward the centerline of the beam in zone A, inducing compressive stresses. In zone B, the curvature is reversed in direction and the struts deflect outward, inducing tensile stresses. In zone C, struts are approximately straight, inducing uniform stress distribution.

The reinforcement required for the end anchorage zones of posttensioned members generally consists of a closely spaced grid of vertical and horizontal bars throughout the length of the end block to resist the bursting and tensile stresses. It is a common practice to space the bars not more than 3 in. in each direction and to place the bars not more than 1.5 in. from the inside face of the bearing plate. Approximate design methods are presented in [24] to [27].

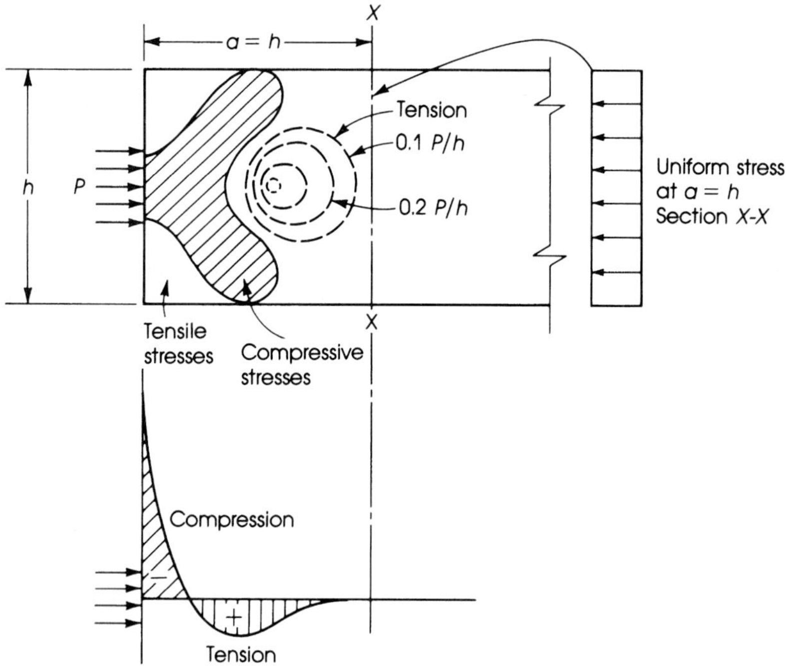

Figure 20.10 Tension and compression zones in a posttensioned member.

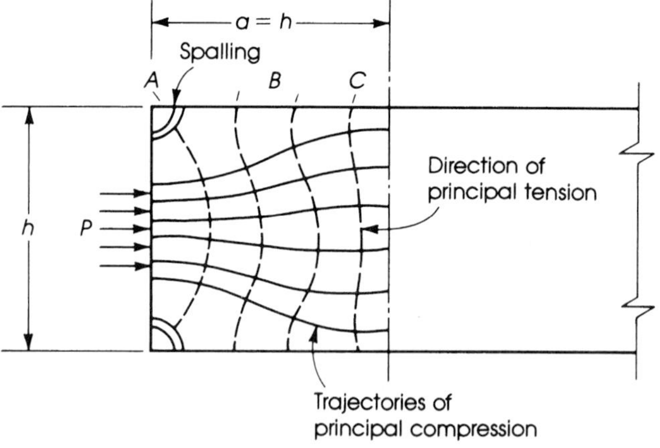

Figure 20.11 Tension and compression trajectories in a posttensioned member.

SUMMARY

Section 20.1

The main objective of prestressing is to offset or counteract all or most of the tensile stresses in a structural member produced by external loadings, thus giving some advantages over a reinforced concrete member. A concrete member may be pretensioned or posttensioned. Nonprestressed reinforcement may also be added to the concrete member to increase its ultimate strength.

Section 20.2

1. The allowable stresses in concrete at transfer are

$$\text{Maximum compressive stress} = 0.6f_{ci}$$

$$\text{Maximum tensile stress} = 3\sqrt{f_{ci}}$$

The allowable stresses after all losses are $0.45f'_c$ for compression and $6f'_c$ for tension.

2. The allowable stress in a pretensioned tendon at transfer is the smaller of $0.74f_{pu}$ or $0.82f_{py}$. The maximum stress due to the tendon jacking force must not exceed $0.85f_{pu}$ or $0.94f_{py}$; and the maximum stress in a posttensioned tendon after the tendon is anchored is $0.70f_{pu}$.

Section 20.3

The sources of prestress loss are the elastic shortening, shrinkage, and creep of concrete; relaxation of steel tendons; and friction. An approximate lump sum loss is 35 Ksi for pretensioned members and 25 Ksi for posttensioned members (friction is not included).

$$\text{Loss due to elastic shortening} = \frac{nF_i}{A_c} \tag{20.1}$$

$$\text{Loss due to shrinkage} = \varepsilon_{sh} E_s \tag{20.6}$$

$$\text{Loss due to creep} = C_c(\varepsilon_c E_s) \tag{20.7}$$

Loss due to relaxation of steel varies between 2.5% and 12%. Loss due to friction in posttensioned members stems from the curvature and wobbling of the tendon.

$$P_s = P_x e^{(\mu\alpha + Klx)} \tag{20.10}$$

$$P_s = P_x(1 + \mu\alpha + Klx) \tag{20.11}$$

Section 20.4

Elastic stresses in a flexural member due to loaded and unloaded conditions are given by equations (20.13) through (20.16). The limiting values of the eccentricity e are given by equations (20.20), (20.22), (20.24), and (20.26). The minimum and maximum values of F_i are given by equations (20.31) and (20.32), respectively.

Section 20.5

The nominal ultimate strength of a rectangular prestressed concrete member is

$$M_n = T\left(d - \frac{a}{2}\right) = A_{ps}f_{ps}d\left(1 - \frac{\rho_p f_{ps}}{1.7f'_c}\right) \tag{20.37}$$

The values of f_{ps} are given by equations (20.39) and (20.40). For flanged sections,

$$M_n = A_{pw}f_{ps}\left(d - \frac{a}{2}\right) + A_{pf}f_{ps}\left(d - \frac{h_f}{2}\right) \tag{20.43}$$

If nonprestressed reinforcement is used in the flexural member, then

$$M_n = A_{ps}f_{ps}\left(d_p - \frac{a}{2}\right) + A_s f_y\left(d - \frac{a}{2}\right) \tag{20.44}$$

where $a = (A_{ps}f_{ps} + A_s f_y)/0.85f'_c b$. For M_n of flanged and rectangular sections with compression reinforcement, refer to equations (20.46) and (20.47), respectively.

Sections 20.6–20.7

1. The cracking moment is

$$M_{cr} = \frac{I}{y_b}\left[7.5\sqrt{f'_c} + \frac{F}{A} + \frac{(Fe)y_b}{I}\right] \tag{20.48}$$

2. Midspan deflections of simply supported beams are summarized in Table 20.3.

Section 20.8

$$\text{Shear strength of concrete } V_c = \left(0.6\sqrt{f'_c} + 700\frac{V_u d}{M_u}\right)b_w d$$

$$\text{Minimum } V_c = 2\sqrt{f'_c}\,b_w d \tag{20.50}$$

$$\text{Maximum } V_c = 5\sqrt{f'_c}\,b_w d$$

The shear strength, V_{ci}, based on flexural shear, is given by equation (20.47), and the web-shear strength V_{cw} is given by equation (20.53):

$$V_s = \frac{1}{\phi}(V_u - \phi V_c) \quad \text{and} \quad A_v = \frac{A_{ps}}{80} \times \frac{f_{pu}}{f_y} \times \frac{S}{d} \times \sqrt{\frac{d}{b_w}} \tag{20.59}$$

Section 20.9

Empirical practical dimensions for the preliminary design of prestressed concrete members are suggested in this section.

Section 20.10

The development length of three- to seven-wire strands is

$$l_d = \left(f_{ps} - \frac{2}{3}f_{se}\right)d_b \tag{20.65}$$

The area of stirrups in an end block is

$$A_v = 0.021\frac{F_i h}{f_{sa} l_t} \tag{20.66}$$

REFERENCES

1. American Concrete Institute. *Building Code Requirements for Structural Concrete* (318–99). Detroit, 1999.
2. P. W. Abeles. "Design of Partially Prestressed Concrete Beams." *ACI Journal* 64 (October 1967).
3. Prestressed Concrete Institute, PCI Handbook, PCI, Chicago, 1999.
4. E. Freyssinet. "A Revolution in the Technique of Utilization of Concrete." *Structural Engineer* 14 (May 1936).
5. H. Straub. *History of Civil Engineering.* London: Leonard Hill, 1952.
6. P. W. Abeles and L. Dzuprynski. "Partial Prestressing, Its History, Research, Application and Future Development." Convention on Partial Prestressing, Brussels (October 1965). *Annales des Travaux Publics de Belgique* 2 (April 1966).
7. P. W. Abeles. "Fully and Partly Prestressed Reinforced Concrete." *ACI Journal* 41 (January 1945).
8. F. Leonherdt. "Prestressed Concrete Design and Construction." Berlin: Wilhelm Ernest and Son, 1964.
9. P. W. Abeles. *Introduction to Prestressed Concrete,* vols. I and II. London: Concrete Publications, 1966.
10. S.T.U.P. *Gazette.* "The Death of Mr. Freyssinet," vol. 11. 1962.
11. P. W. Abeles. "Partial Prestressing and Its Possibilities for Its Practical Application." *PCI Journal* 4, no. 1 (June 1959).

12. W. H. Hewett. "A New Method of Constructing Reinforced Concrete Tanks." *ACI Journal* 19 (1932).
13. R. E. Dill. "Some Experience with Prestressed Steel in Small Concrete Units." *ACI Journal* 38 (November 1941).
14. F. W. Dodge. *The Structures of Eduardo Torroja.* New York: McGraw-Hill, 1959.
15. E. Freyssinet. "The Birth of Prestressing." *Cement and Concrete Association Translation,* no. 59. London, 1956.
16. G. Mangel. *Prestressed Concrete.* London: Concrete Publications, 1954.
17. W. Zerna. "Partially Prestressed Concrete" (in German). *Beton Stahlbeton,* no. 12 (1956).
18. R. H. Evans. "Research and Development in Prestressing." *Journal of the Institution of Civil Engineers* 35 (February 1951).
19. E. Freyssinet. "Prestressed Concrete, Principles and Applications." *Journal of the Institution of Civil Engineers* 4 (February 1950).
20. P. J. Verna. "Economics of Prestressed Concrete in Relation to Regulations, Safety, Partial Prestressing ... in the U.S.A." *Fourth Congress of the FIP. Rome-Naples,* Theme III, paper 1, 1962.
21. S. Chaikes. "The Reinforced Prestressed Concrete." *Fourth Congress of the FIP. Rome-Naples,* Theme III, paper 2, 1962.
22. ACI Committee 435. "Deflection of Prestressed Concrete Members." *ACI Journal* 60 (December 1963).
23. American Association of Highway and Transportation Officials. *AASHTO Specifications for Bridges.* Washington, D.C., 1975.
24. Posttensioning Institute. *Posttensioning Manual.* Phoenix, 1976.
25. Prestressed Concrete Institute. *PCI Design Handbook.* Chicago, 1980.
26. P. Gergley and M. A. Sozen. "Design of Anchorage Zone Reinforcement in Prestressed Concrete Beams." *PCI Journal* 12 (April 1967).
27. Y. Guyon. *Prestressed Concrete.* New York: Wiley and Sons, 1960.
28. P. Zia, H. K. Peterson, N. L. Scott, and E. B. Workman. "Estimating Prestress Losses." *Concrete International* 1 (June 1979).

PROBLEMS

20.1 A 60-ft-span simply supported pretensioned beam has the section shown in Figure 20.12. The beam is prestressed by a force $F_i = 395$ K at transfer (after the elastic loss). The prestress force after all losses is $F = 320$, f'_{ci} (compressive stress at transfer) $= 4$ Ksi and f'_c (compressive stress after all losses) $= 6$ Ksi. For the midspan section and using the ACI Code allowable stresses, (a) calculate the extreme fiber stresses due to the prestressing force plus dead load and (b) calculate the allowable uniform live load on the beam.

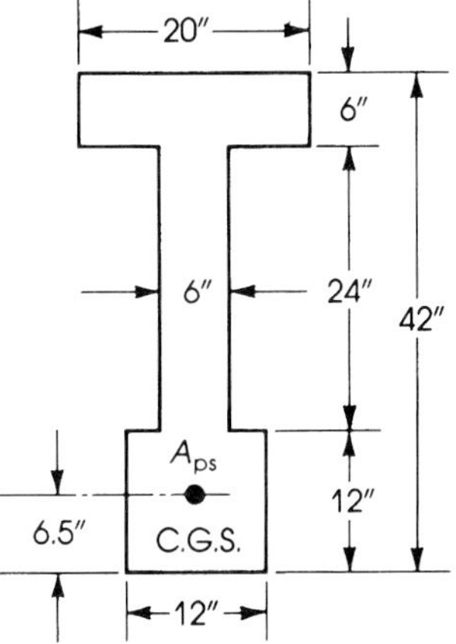

Figure 20.12 Problem 20.1.

20.2 For the beam of Problem 20.1 (Figure 20.12), calculate the elastic loss and all time-dependent losses using the following data: $F_i = 405$ K, $A_{ps} = 2.39$ in.^{2} located at 6.5 in. from the base, $f'_{ci} = 4$ Ksi, and $f'_c = 6$ Ksi. $E_c = 57{,}000\sqrt{f'_c}$, and $E_s = 28{,}000$ Ksi. The profile of the tendon is parabolic, and the eccentricity at the supports is 0.

20.3 The cross section of a 56-ft-span simply supported posttensioned girder that is prestressed by 30 cables $\frac{7}{16}$ in. in diameter (area of one cable is 0.1089 in.^{2}) is shown in Figure 20.13. The cables are made of seven-wire stress-relieved strands. The profile of the cables is parabolic with the centroid of the prestressing steel (C.G.S.) located at 9.6 in. from the base at the midspan section and located at the centroid of the concrete section ($e = 0$) at the ends. Calculate the elastic loss of prestress and all other losses. Given: $f'_{ci} = 4$ Ksi, $f'_c = 6$ Ksi, $E_c = 57{,}000\sqrt{f'_c}$, $E_s = 28{,}000$ Ksi, $f_{pu} = 250$ Ksi, $F_o = 175$ Ksi, D.L. $= 1.0$ K/ft (excluding self-weight), and L.L. $= 1.6$ K/ft.

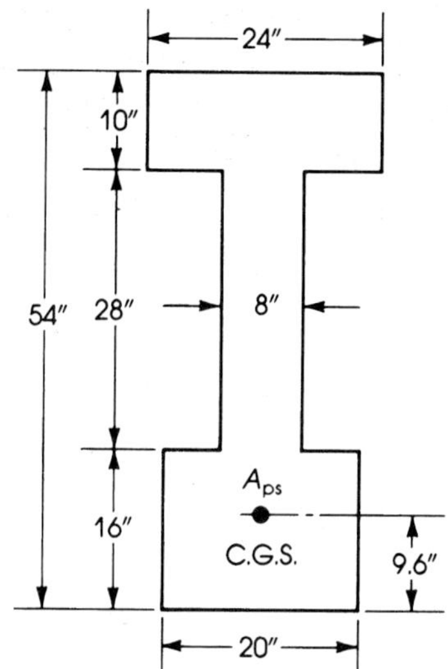

Figure 20.13 Problem 20.3.

20.4 For the girder of Problem 20.3,

a. Determine the location of the upper and lower limits of the tendon profile for the section at midspan and for at least two other sections between midspan and support. (Choose sections at 12 ft, 18 ft, and 25 ft from support.)

b. Check if the parabolic profile satisfies these limits.

20.5 For the girder of Problem 20.3, check the limiting values of the prestressing force at transfer F_i.

20.6 A 64-span simply supported pretensioned girder has the section shown in Figure 20.14. The loads on the girder consist of a dead load $= 1.2$ K/ft (excluding its own weight) that will be applied at a later stage and a live load of 0.6 K/ft. The prestressing steel consists of 24 cables $\frac{1}{2}$ in. in diameter (area of one cable $= 0.114$ in.^{2}), with $E_s = 28{,}000$ Ksi, $F_o = 175$ Ksi, and $f_{pu} = 250$ Ksi. The strands are made of seven-wire stress-relieved steel. The concrete compressive strength at transfer is $f_{ci} = 4$ Ksi, and at 28 days $f'_c = 5$ Ksi. The modulus of elasticity is $E_c = 57{,}000\sqrt{f'_c}$. For the beam just described,

a. Determine the upper and lower limits of the tendon profile for the section at midspan and three other sections between the midspan section and the support. (Choose sections at 3 ft, 11 ft, and 22 ft from the support.)

b. Locate the tendons to satisfy these limits using straight horizontal tendons within the middle third of the span.

c. Check the limiting values of the prestressing force at transfer.

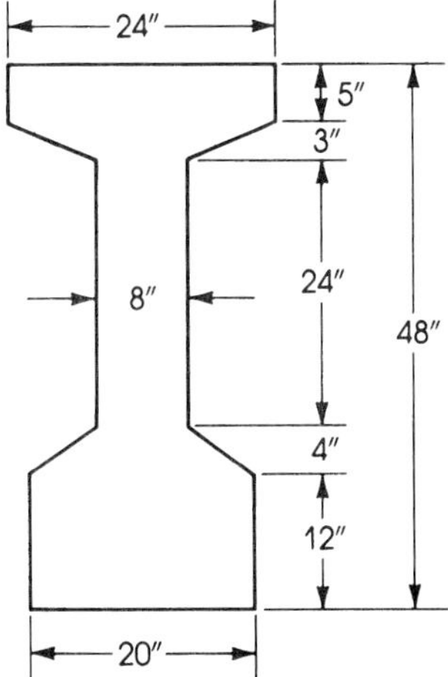

Figure 20.14 Problem 20.6.

20.7 For the girder of Problem 20.6,

a. Harp some of the tendons at one-third points, and draw sections at midspan and at the end of the beam showing the distribution of tendons.

b. Revise the prestress losses, taking into consideration the variation of the eccentricity, e, along the beam.

c. Check the ultimate moment capacity of the section at midspan.

d. Determine the cracking moment.

20.8 For the girder of Problem 20.6,

a. Calculate the camber at transfer.

b. Calculate the immediate deflection at service load.

20.9 For the girder of Problem 20.6, determine the shear capacity of the section and calculate the necessary web reinforcement.

20.10 Determine the normal ultimate moment capacity, M_n, of a pretensioned concrete beam that has the cross section shown in Figure 20.15. Given: $f'_c = 5$ Ksi, $f_{pu} = 270$ psi, $f_{se} = 160$ Ksi, and $A_{sp} = 2.88$ in^2.

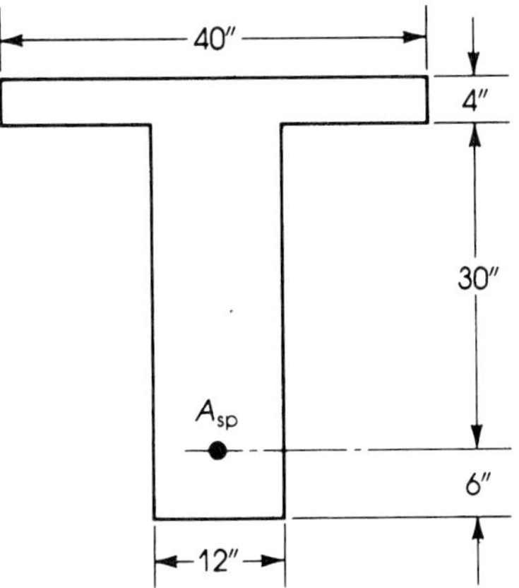

Figure 20.15 Problem 20.10.

21

Reinforced Concrete Institutional Building
Waterloo University, Waterloo, Canada.

COMPUTER PROGRAMS AND FLOWCHARTS

21.1 INTRODUCTION

Using computer programs for the design of reinforced concrete structural members is becoming essential for both the practicing engineer and the student. Such programs speed up calculations and ensure that designers consider all necessary aspects of design limitations. A number of program packages, ranging in sophistication, are available commercially. The programs presented here are short and easy to apply. The output includes calculated values at different steps as well as the final answer. These data will help the user to detect the location of any mistakes made in hand calculations.

The programs are written in Q Basic. Versions using customary units and SI units are available. The analysis and design procedure and limitations are based on the ACI Code. The code contains a number of mathematical expressions that can be incorporated in flowcharts and then in simple computer programs. The programs presented here are as accurate as possible, but they should be considered as a guide only and no warranty is given.

Seven programs using customary units and seven similar programs using SI units are available with this textbook. They cover the analysis and design of singly reinforced, doubly reinforced, and T-sections due to flexure. They also cover the design for shear, torsion, rectangular and circular columns, and footings. All these topics are normally covered in a first course on reinforced concrete design. Other programs, which deal with advanced topics, can be ordered through the author.

ANALYSIS OF ECCENTRICALLY LOADED RECTANGULAR SHORT COLUMN

1-INPUT DATA :		
ECCENTRICITY,	e	(in.) = 6
COLUMN WIDTH,	b	(in.) = 16
COLUMN DEPTH,	h	(in.) = 24
EFFECTIVE DEPTH,	d	(in.) = 21.5
COMP. STEEL DEPTH,	d′	(in.) = 2.5
TENSION STEEL,	AS	(in2) = 5
COMPRESSION STEEL,	AS′	(in2) = 5
CONCRETE STRENGTH,	F'_c	(Ksi) = 4
STEEL YIELD STRESS,	f_y	(Ksi) = 60
2-RESULTS:		
1-) BALANCED SECTION:		
COMP. STRESS DEPTH.	ab	(in.) = 10.81582
BALANCED LOAD,	Pb	(kips) = 571.3805
BALANCED MOMENT,	Mb	(K · ft) = 784.7631
	Phi.Pb	(k) = 399.9663
	Phi.Mb	(K · ft) = 549.3342
BALANCED ECCENT.	eb	(in.) = 16.48141
2-) e (in.) = 6 :		
	Compression controls	
TOTAL STEEL, PERCENT:	Rho	(%) = 2.604167
PLASTIC CENTROID,	YP	(in.) = 12
COMP. STRESS DEPTH,	a	(in.) = 16.20326
NOMINAL LOAD,	Pn	(kip) = 1108.854
NOMINAL MOMENT,	Mn	(K · ft) = 554.427
	phi	= .7
	PHI* Pn	(kip) = 776.1978****
	PHI* Mn	(K · ft) = 388.0989
3-) AXIAL LOAD :		
NOMINAL AXIAL LOAD,	Pno	(kip) = 1871.6
	0.7*Pno	(kip) = 1310.12
	(0.7* 0.8) Pno	(kip) = 1048.096**

Figure 21.1 Computer program output for an eccentrically loaded column.

Typical computer input and output results for the analysis of rectangular columns in customary units and SI units are shown in Figures 21.1 and 21.2.

21.2 COMPUTER PROGRAMS

The following computer programs are available:

1. ANALYSIS: The program deals with the analysis of
   a. Singly reinforced rectangular sections;
   b. Doubly reinforced rectangular sections;
   c. T- and L-sections.

ANALYSIS OF ECCENTRICALLY LOADED SHORT RECTANGULAR COLUMN WITH SIDE BARS

1- INPUT DATA :

ECCENTRICITY	e (mm) = 150
COLUMN WIDTH	b (mm) = 400
COLUMN DEPTH	h (mm) = 600
EFFECTIVE DEPTH	d (mm) = 535
COMP. STEEL DEPTH	d′ (mm) = 65
TENSION STEEL	AS (mm2) = 3125
COMPRESSION STEEL	AS′ (mm2) = 3125
CONCRETE STRENGTH	f'_c (MPa) = 30
STEEL YIELD STRESS	f_y (MPa) = 400

2- RESULTS:

1) BALANCED SECTION:

COMP. STRESS DEPTH	ab (mm) = 272.85
BALANCED LOAD	Pb (kN) = 2703.383
BALANCED MOMENT	Mb(kN · m) = 1024.014
	Phi.Pb(kN) = 1892.368
	Phi.Mb(K · ft) = 716.8099
BALANCED ECCENT.	eb (mm) = 378.7899

2) e (mm) = 150 :

	Compression controls
TOTAL STEEL, PERCENT:	Rho (%) = 2.604167
PLASTIC CENTROID	YP (mm) = 300
COMP. STRESS DEPTH	a (mm) = 398.7457
NOMINAL LOAD	Pn (kN) = 4974.541
NOMINAL MOMENT	Mn (kN·m) = 746.1811
	phi = .7
	Pu (kN) = 3482.179****
	Mu (kN·m) = 522.3268

3) AXIAL LOAD :

NOMINAL AXIAL LOAD	Pno (kN) = 8460.625
	0.7*Pno (kN) = 5922.438
	0.7*0.8 Pno (kN) = 4737.95 **

Figure 21.2 Computer program output for an eccentrically loaded column with sidebars, SI units.

2. DESIGN: The program deals with the design of the same sections with the options of changing ρ or h any time during the running of the program to adopt the most adequate section. Moreover, the program contains tables that allow one to choose bars and to check the minimum beam width to fit the chosen bars.
3. SHEAR: The program deals with shear design using U-stirrups.
4. TORSION: The program deals with the design of rectangular, T-, and L-sections due to shear, torsion, or both shear and torsion.
5. RCOL: This program deals with the analysis of rectangular columns with equal or unequal tension and compression steel on opposite sides.

6. CCOL: This program analyzes circular columns.
7. RFOOT: This program deals with the design of square and rectangular footings.

21.3 FLOWCHARTS

To help the design engineer or the student prepare their own programs, summaries of the analysis and design procedures are given in most chapters. However, flowcharts are given here as well.

1. Flowcharts 21.1, 21.2, and 21.3 explain the analysis of singly reinforced, doubly reinforced, and T-sections (covered in Chapter 3).
2. Flowcharts 21.4, 21.5, and 21.6 explain the design of singly reinforced, doubly reinforced, and T-sections (Chapter 4).
3. Flowchart 21.7 explains the steps in shear design (Chapter 8).
4. Flowchart 21.8 explains the calculation of a development length (Chapter 7).
5. Flowchart 21.9 explains the analysis of rectangular columns, balanced condition (Chapter 11).
6. Flowchart 21.10 explains the analysis of rectangular columns (Chapter 11).
7. Flowchart 21.11 explains the design of single footings (Chapter 13).
8. Flowchart 21.12 explains the design for shear and torsion (Chapter 15).

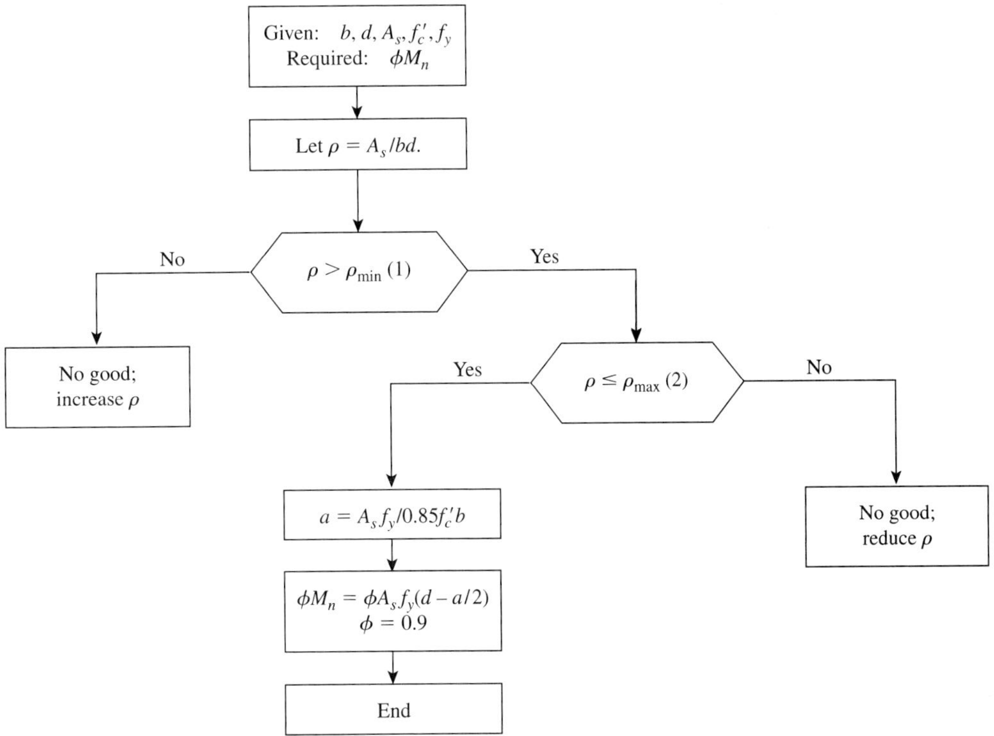

Flowchart 21.1 Analysis of singly reinforced rectangular section.

(1) $\rho_{min} = 3\sqrt{f'_c}/f_y \geq 200/f_y$. (2) $\rho_{max} = 0.75\,(0.85\beta_1)(f'_c/f_y)\left(\dfrac{87}{87 + f_y}\right)$.

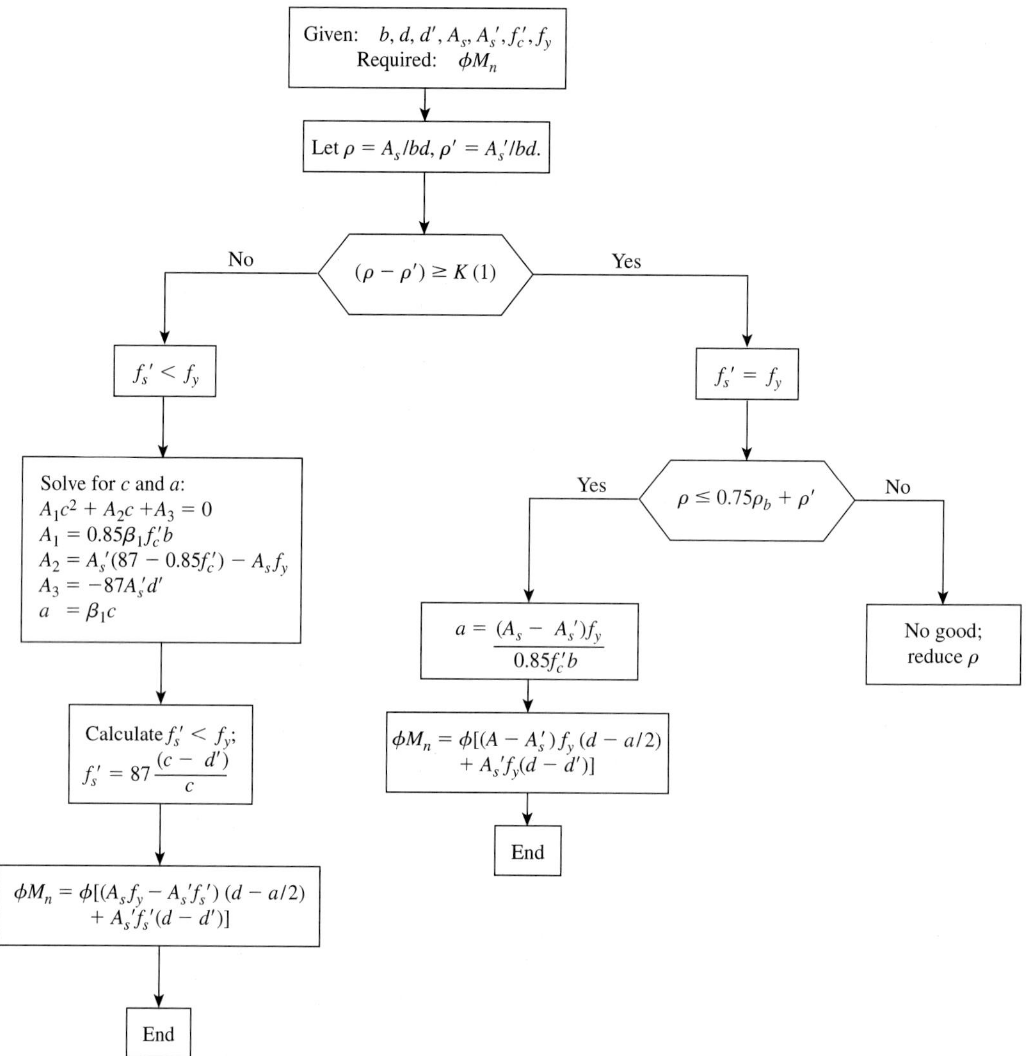

Flowchart 21.2 Analysis of doubly reinforced rectangular sections.

(1) $K = (0.85\beta_1)\left(\dfrac{f_c'}{f_y}\right)\left(\dfrac{d'}{d}\right)\left(\dfrac{87}{87 - f_y}\right)$.

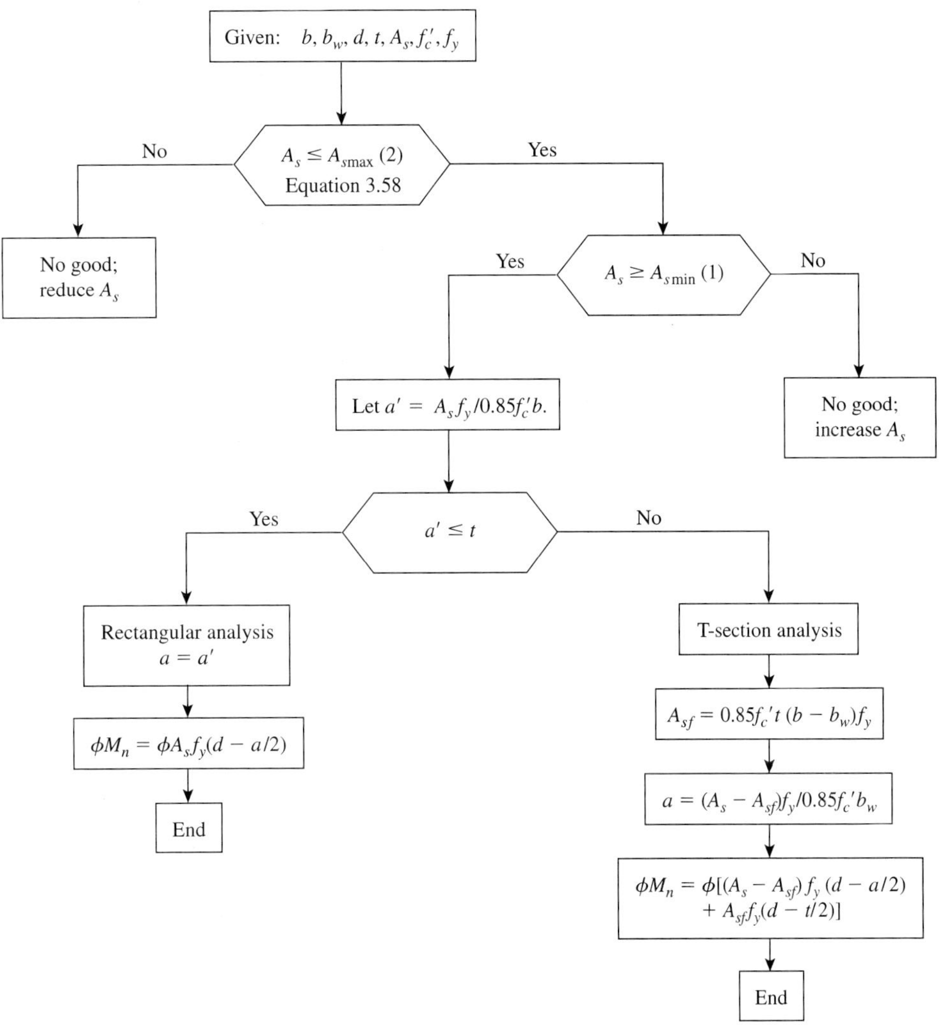

Flowchart 21.3 Analysis of T- and L-sections.

(1) $A_{s\,min} = \left(\dfrac{3\sqrt{f_c'}}{f_y}\right) b_w\, d \le \left(\dfrac{200}{f_y}\right) b_w\, d.$

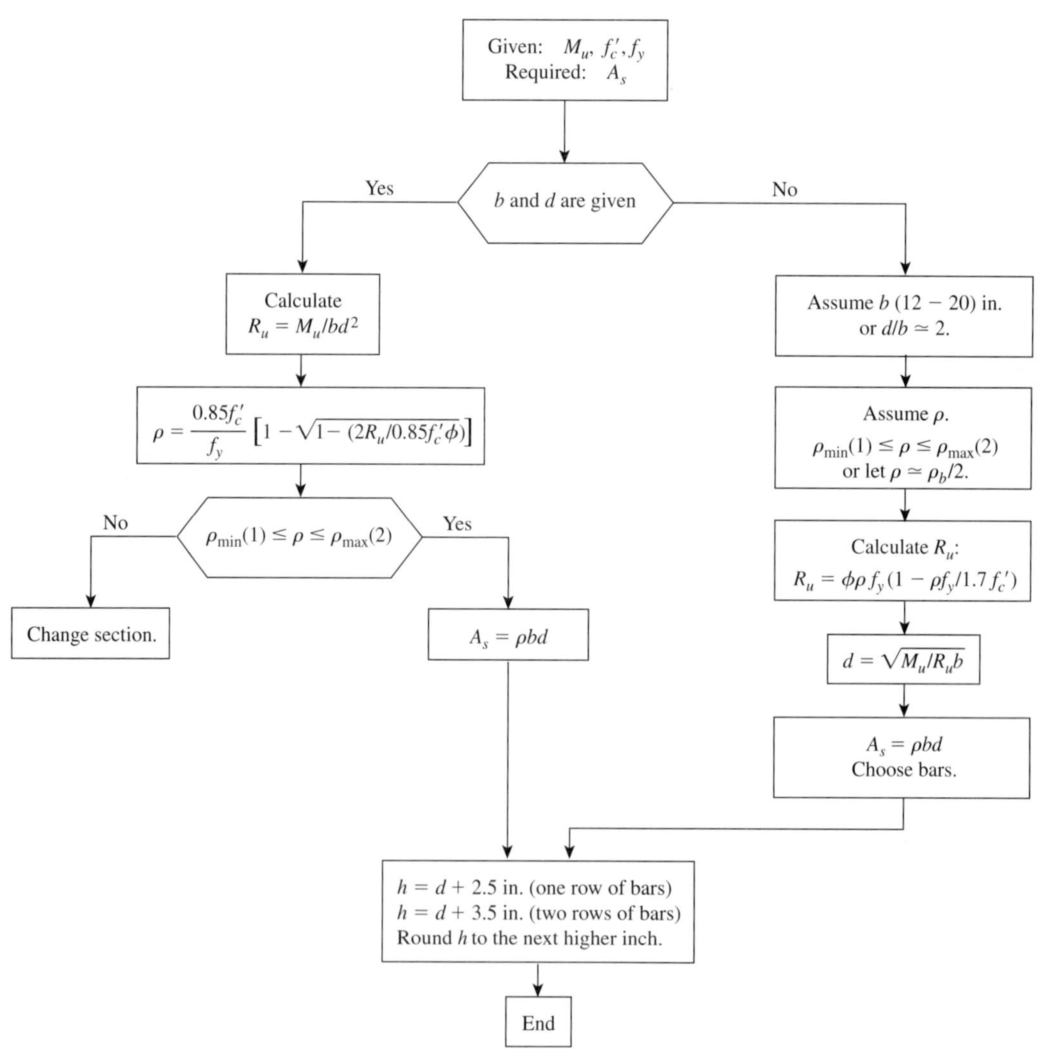

Flowchart 21.4 Design of singly reinforced rectangular sections. For $\rho_{min(1)}$ and $\rho_{max(2)}$, refer to flowchart 21.1.

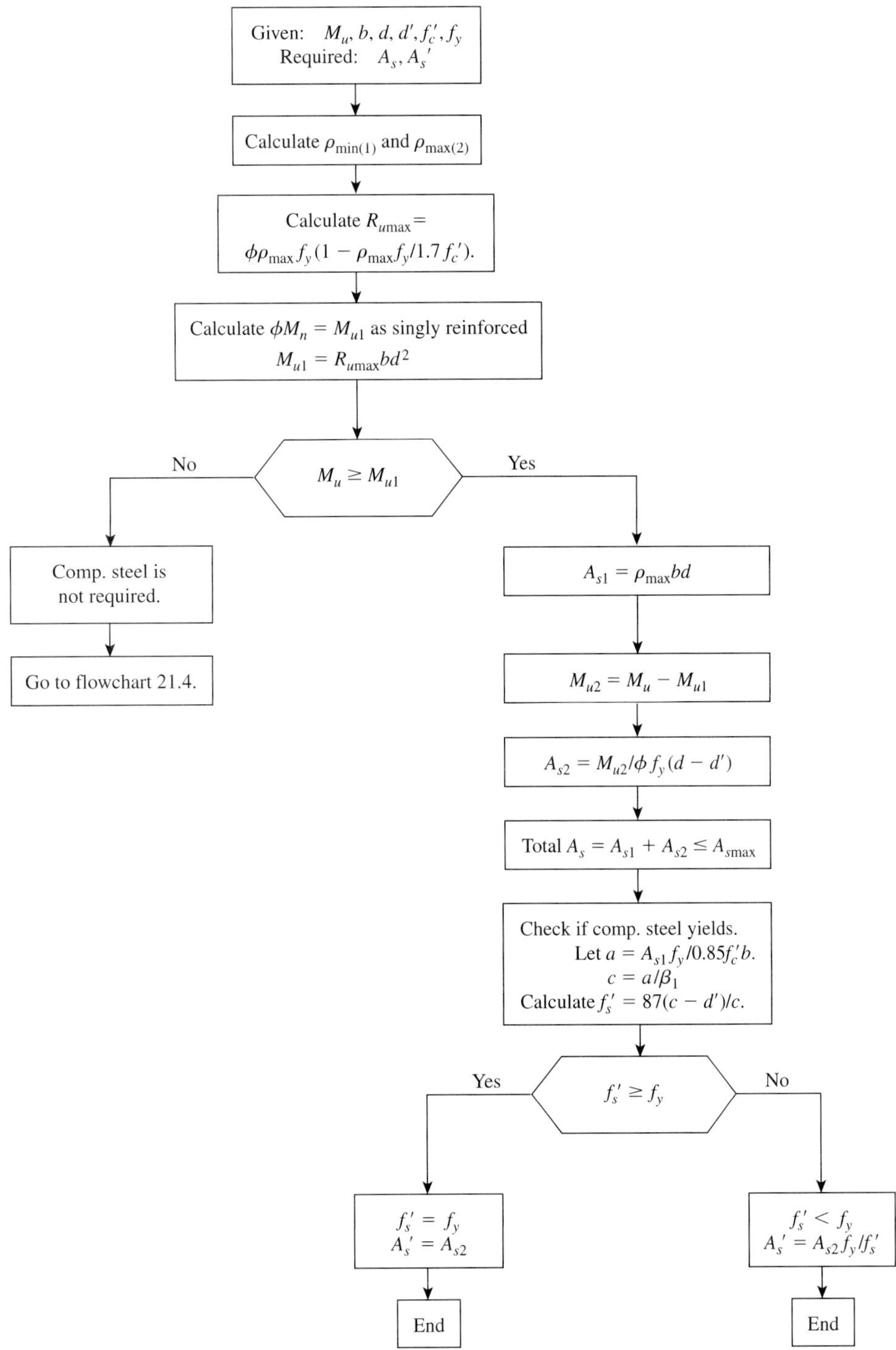

Flowchart 21.5 Design of rectangular sections with compression steel. For $\rho_{\min(1)}$ and $\rho_{\max(2)}$, refer to flowchart 21.1.

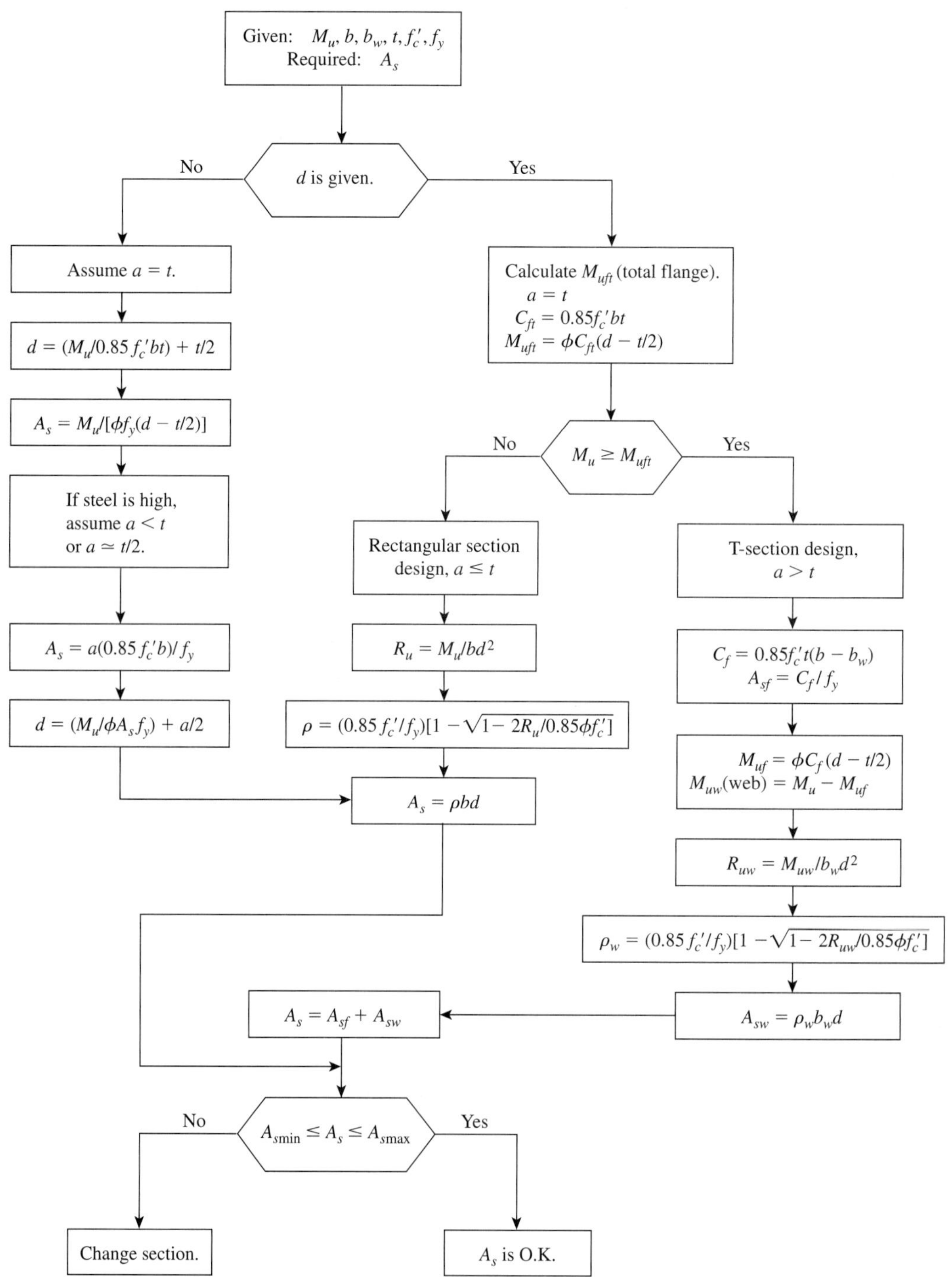

Flowchart 21.6 Design of T- or L-sections.

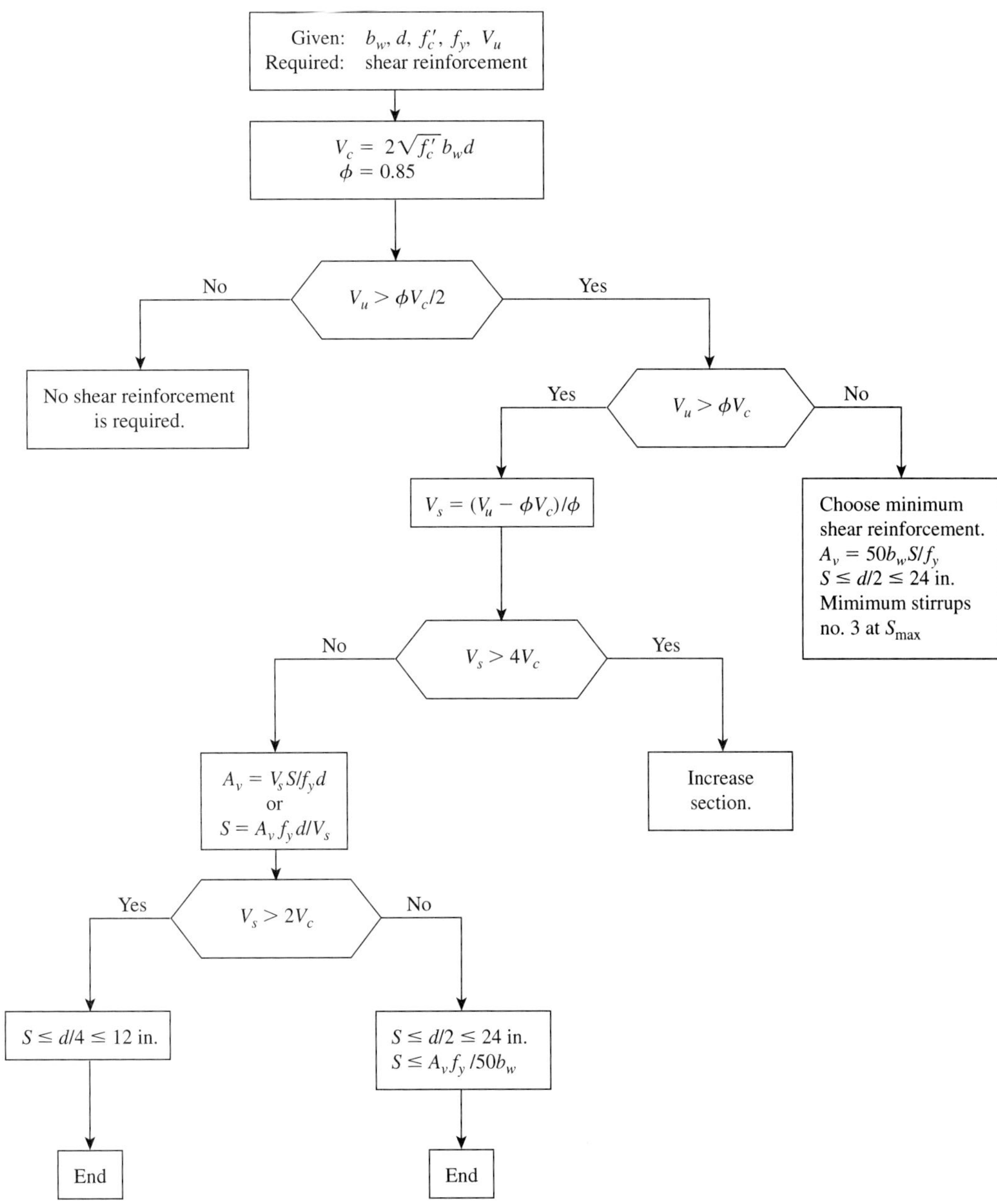

Flowchart 21.7 Shear reinforcement.

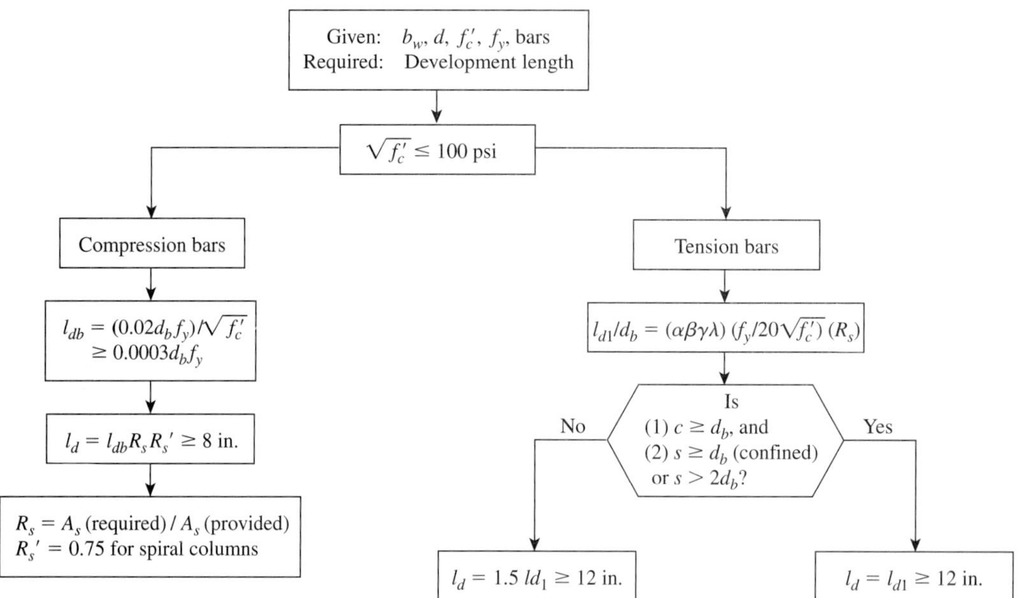

Flowchart 21.8 Development length.

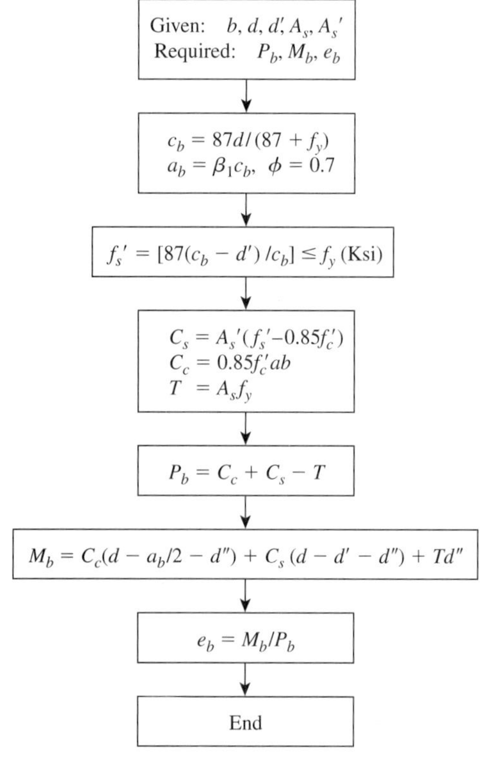

Flowchart 21.9 Balanced load, moment, and eccentricity for rectangular column sections (use Ksi for f'_c and f_y; d'' = distance from the plastic centroid to A_s).

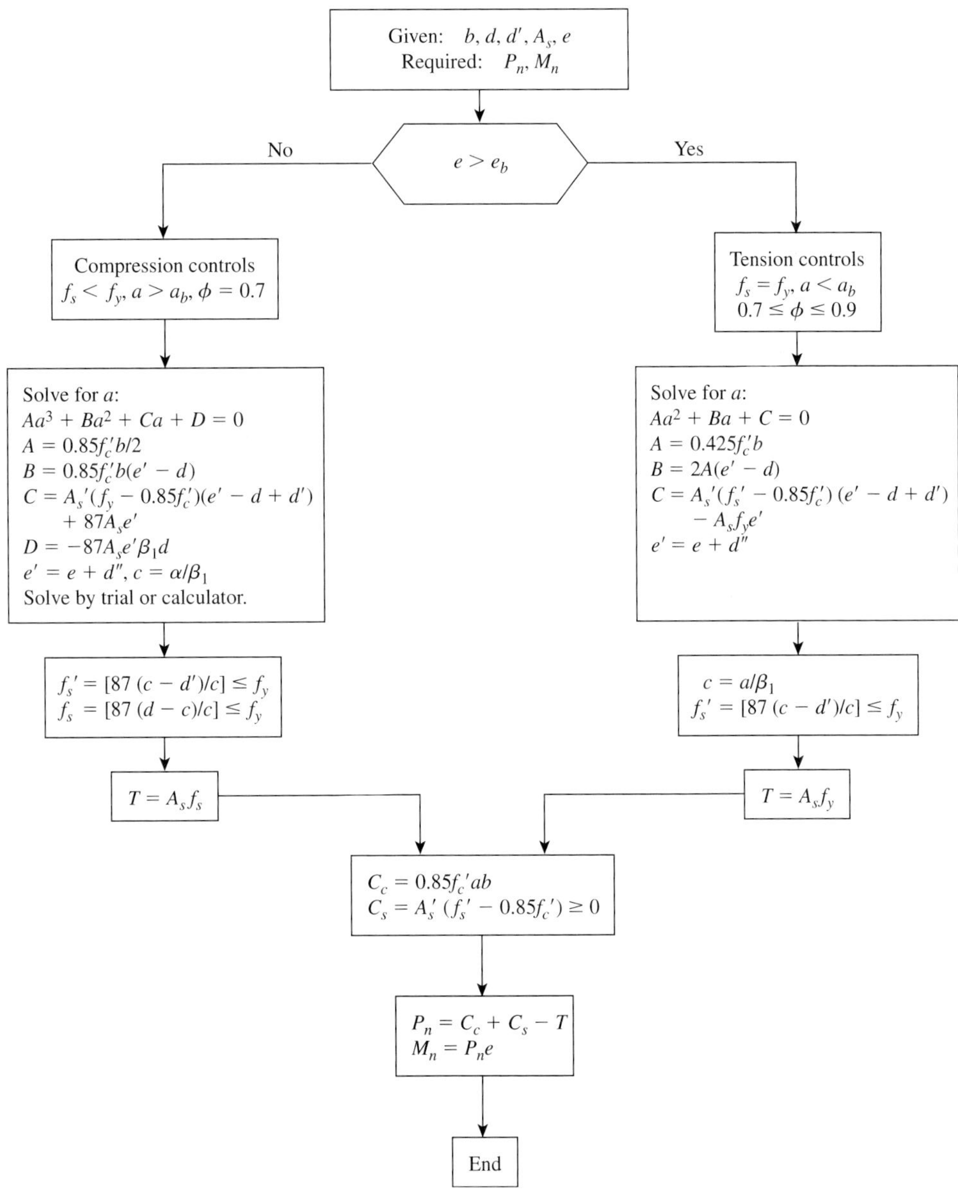

Flowchart 21.10 Analysis of rectangular columns (use Ksi for f_c' and f_y).

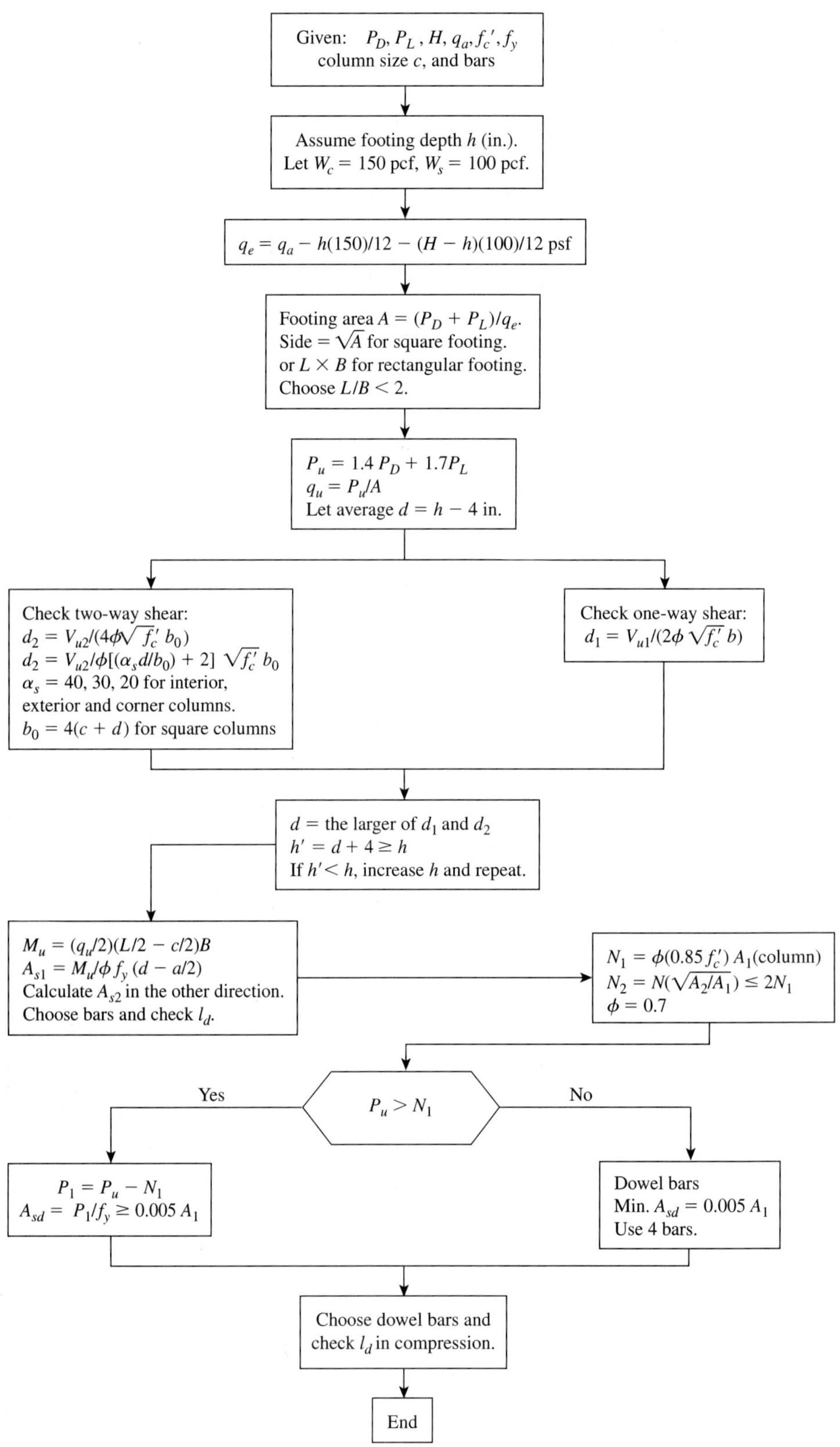

Flowchart 21.11 Design of single footings.

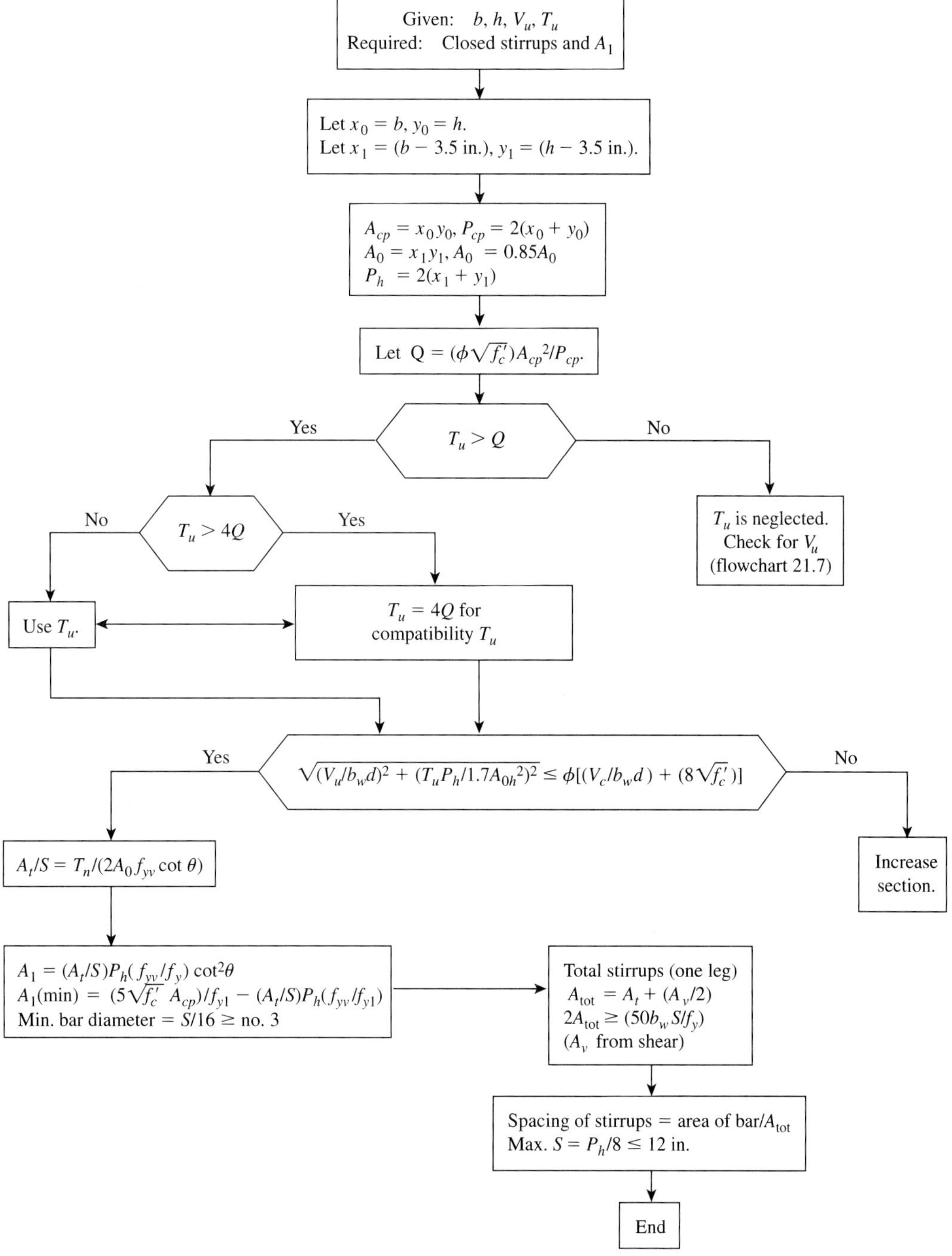

Flowchart 21.12 Design for shear and torsion.

SUMMARY

Sections 21.1–21.2

A list of the computer programs that are provided with this textbook is presented.

Section 21.3

Twelve flowcharts that describe the procedures for the analysis and design of beams, columns, and footings are presented in this section.

22

Concrete Residential Building, Waterloo, Canada.

UNIFIED DESIGN METHOD

22.1 INTRODUCTION

The design of a structure may be regarded as the process of selecting the proper materials and proportioning the different elements of the structure according to state-of-the-art engineering science and technology. In order to fulfill its purpose, the structure must meet the conditions of safety, serviceability, economy, and functionality. Meeting these goals can be achieved using the strength design method (SDM), which was explained in the previous chapters, or the alternate design method (ADM), based on the provisions of Appendix A of the ACI Code, which was explained in Chapter 5.

A third approach for the design of reinforced and prestressed concrete flexural and compression members is called the unified design method (UDM). This new approach introduces substantial changes in design for flexure and axial loads. Provisions for this method were introduced in the ACI Code, Appendix B. The reason for introducing this approach relates to the fact that different design methods were developed for reinforced and prestressed concrete beams and columns. Also, design procedures for prestressed concrete differed from those of reinforced concrete. The purpose of the code in the UDM approach is to simplify and unify the design requirements for reinforced and prestressed flexural and compression members. When this method (UDM) is used in design, the designer shall adhere to all sections of Appendix B, and these requirements shall be substituted for the corresponding sections in the code. Reinforcement limits, strength-reduction factor, ϕ, and moment redistribution are affected. The provisions of the unified design method satisfy the code and are equally acceptable. Specific reference is made to Appendix B in the main body of the ACI Code, Section 8.1.3, to make it a legal part of the code.

The unified design provisions apply to reinforced, prestressed, and partially prestressed concrete flexural members and compression members. They also apply to composite cross sections with more than one type of concrete and to cross sections of any shape with steel located in one location or at various depths within the section.

As in the strength design method, members designed by the unified design provisions must have a design strength greater than or equal to the required strength, $\phi M_n \geq M_u$, and the amount of steel reinforcement must not exceed the limits specified by the code. The strength-reduction factor, ϕ, was explained in Section 3.6, where $\phi = 0.9$ for flexure and axial tension, whereas $\phi = 0.85$ for shear and torsion. Also, $\phi = 0.7$ for axial compression (members without spirals), and $\phi = 0.75$ for spirally reinforced concrete columns. Other values are given in Section 3.6.

The reason the ACI Code limits the maximum steel reinforcement to be used in the concrete flexural members, either reinforced or prestressed, is to ensure ductile behavior of the members and to provide ample warning before their complete failure. The same limitations discussed in Chapters 3, 4, and 11 for reinforced concrete flexural members and Chapter 20 for prestressed concrete flexural members also apply in this chapter.

22.2 DEFINITIONS

The unified design method unifies the design provisions for both reinforced and prestressed concrete members by developing two concepts: compression-controlled sections and tension-controlled sections (ACI Code, Section B10.3.3). Both are defined in terms of ε_t, the *net tensile strain* (NTS) in the extreme tension steel at nominal strength, exclusive of prestress strain.

Moreover, two other conditions may develop: One is the balanced strain condition and the second is the transition region condition. These four conditions are defined as follows:

1. *Compression-controlled sections* are all those sections in which the net tensile strain, ε_t, in the extreme tension steel at nominal strength is equal to or less than the compression-controlled strain limit at the time the concrete in compression reaches its assumed strain limit of 0.003 $(\varepsilon_c = 0.003)$. For grade 60 steel $(f_y = 60 \text{ Ksi})$, the compression-controlled strain limit may be taken as a net tensile strain, ε_t, of 0.002. See Figure 22.1(a).
2. *Tension-controlled sections* are those sections in which the net tensile strain, ε_t, is equal to or greater than 0.005 just as the concrete in compression reaches its assumed strain limit of 0.003 (Figure 22.1(c)).
3. Sections with net tensile strain (NTS), ε_t, in the extreme tension steel between the compression-controlled strain limit $(0.002 \text{ for } f_y = 60 \text{ Ksi})$ and 0.005 constitute the *transition region* between compression-controlled and tension-controlled sections (Figure 22.1(b)).
4. The *balanced strain condition* develops in the section when the tension reinforcement (at its centroidal area) reaches a strain corresponding to its specified yield strength, f_y, or $\varepsilon_s = f_y/E_s$, just as the maximum strain in concrete at the extreme compression fibers reaches 0.003 $(\varepsilon_c = 0.003)$. This case is similar to that given in Section 3.4 for the strength design method (Figure 22.2).

These four cases are summarized in Table 22.1.

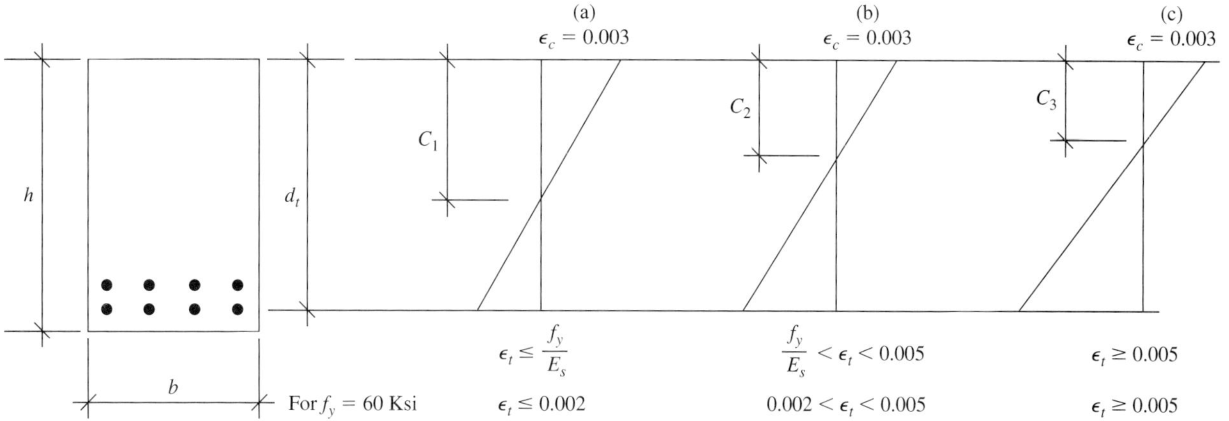

Figure 22.1 Strain limit distribution, $c_1 > c_2 > c_3$: (a) compression-controlled section, (b) transition region, and (c) tension-controlled section.

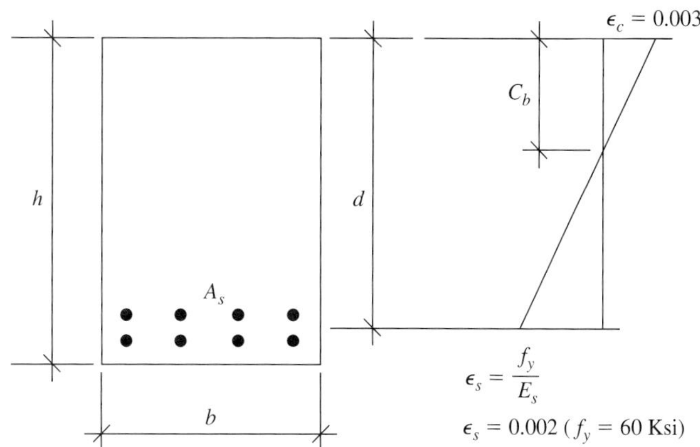

Figure 22.2 Balanced strain section (d is located at the centroid of tension steel, A_s).

Table 22.1 Section Behavior (Figure 22.1)

Section Condition	Concrete Strain	Steel Strain	Notes (f_y = 60 K)
1. Compression-controlled	0.003	$\varepsilon_t \le f_y/E_s$	$\varepsilon_t = 0.002$
2. Tension-controlled	0.003	$\varepsilon_t \ge 0.005$	$\varepsilon_t \ge 0.005$
3. Transition region	0.003	$f_y/E_s < \varepsilon_t < 0.005$	$0.002 < \varepsilon_t < 0.005$
4. Balanced strain	0.003	$\varepsilon_s = f_y/E_s$	$\varepsilon_s = 0.002$

22.3 STRENGTH-REDUCTION FACTOR ϕ

The nominal strength of a section, say, M_n, for flexural members, calculated in accordance with the requirements of the unified design method or the strength design method, must be multiplied by the strength-reduction factor, ϕ, which is always less than 1. The strength-reduction factor ϕ has several purposes:

1. To allow for the probability of under-strength sections due to variations in dimensions, material properties, and inaccuracies in the design equations.
2. To reflect the importance of the member in the structure.
3. To reflect the degree of ductility and required reliability under the applied loads.

A higher ϕ-factor is used for tension-controlled sections than for compression-controlled sections, because the latter sections have less ductility and they are more sensitive to variations in concrete strength. Also, spirally reinforced compression members have $\phi = 0.75$, as compared to $\phi = 0.7$ for tied compression members; these variations reflect the greater ductility behavior of spirally reinforced members under the applied loads. In the unified design method, the ϕ-factor is based on the behavior of the cross section at nominal strength, (P_n, M_n), defined in terms of the net tensile strain (NTS), ε_t, in the extreme tensile strains, as given in Table 22.1. For tension-controlled members, $\phi = 0.9$. For compression-controlled members, $\phi = 0.75$ (with spiral reinforcement) and $\phi = 0.7$ (with other reinforcement).

For the transition region, ϕ may be determined by linear interpolation between 0.7 (or 0.75) and 0.9. Figure 22.3 shows the variation of ϕ for grade 60 steel. The linear equations are as follows:

$$\phi = 0.56 + 68\varepsilon_t \quad \text{(for tied sections)} \tag{22.1}$$

$$\phi = 0.65 + 50\varepsilon_t \quad \text{(for spiral sections)} \tag{22.2}$$

Alternatively, ϕ in the transition region can be determined as a function of (d_t/c) for grade 60 steel as follows:

$$\phi = 0.356 + 0.204(d_t/c) \quad \text{(for tied sections)} \tag{22.3}$$

$$\phi = 0.50 + 0.15(d_t/c) \quad \text{(for spiral sections)} \tag{22.4}$$

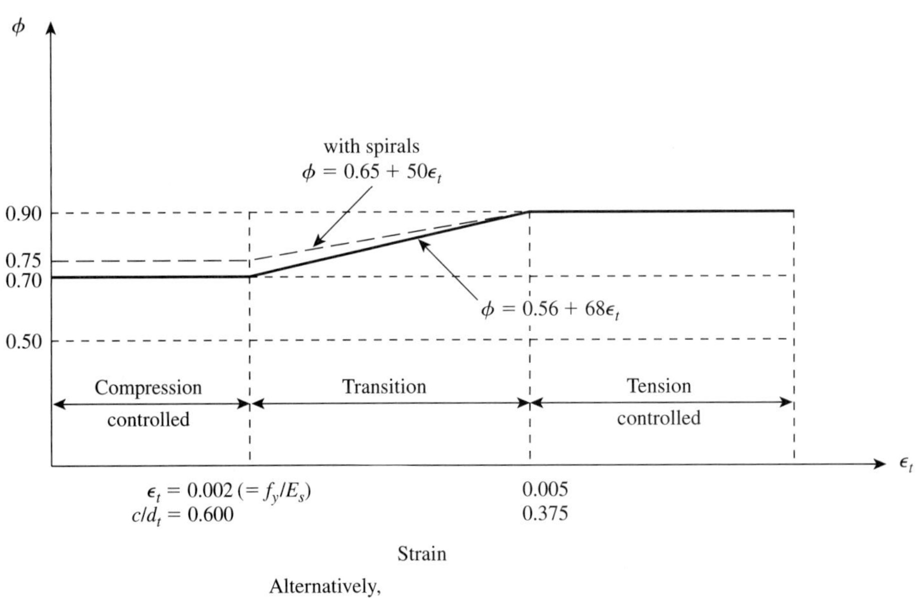

Figure 22.3 Variation of ϕ with NTS for grade 60 and for prestressed steel [1].

where c is the depth of the neutral axis at nominal strength (c_2 in Figure 22.1). At the limit strain of $\varepsilon_t = 0.002$ for grade 60 steel and from the triangles of Figure 22.1(a), $c/d_t = 0.003/(0.002 + 0.003) = 0.6$. Similarly, for $\varepsilon_t = 0.005$, $c/d_t = 0.003/(0.005 + 0.003) = \frac{3}{8} = 0.375$. Both values are shown in Figure 22.3.

22.4 MOMENT REDISTRIBUTION OF NEGATIVE MOMENTS IN CONTINUOUS FLEXURAL MEMBERS

22.4.1 ACI Code, Section 8.4, Limits

Moment redistribution is dependent on adequate ductility in plastic hinge regions, as discussed in Chapter 16. Moment redistribution based on the ACI Code, Section 8.4, for nonprestressed flexural members was also discussed in Section 16.15. It stated that the maximum negative moments calculated by the elastic theory at the supports of continuous flexural members may be increased or decreased by not more than q:

$$q = 20\left(1 - \frac{\rho - \rho'}{\rho_b}\right) \tag{22.5}$$

where

$$\rho = \frac{A_s}{bd}$$

$$\rho' = \frac{A'_s}{bd}$$

$$\rho_b = 0.85\beta_1 \frac{f'_c}{f_y}\left(\frac{87}{87 + f_y}\right)$$

Equation (22.5) must not be applied to the moments calculated by approximate methods or the alternate design method of Appendix A of the ACI Code. The code limits the steel ratio ρ (or $\rho - \rho'$) at the section where the moment is reduced to a maximum ratio of $0.5\rho_b$. The minimum steel ratio allowed in the section, for flexural design, is $\rho_{min} = (3\sqrt{f'_c}/f_y) \geq 200/f_y$. If the extreme values, $0.5\rho_b$ and ρ_{min}, are substituted in equation (22.5), the maximum and minimum redistribution percentages will be as shown in Table 22.2.

In practice, a minimum steel percentage, ρ_{min}, is not adopted for a good design, and therefore a maximum redistribution percentage will not practically be used. A steel percentage of $0.5\rho_b$ or $0.5\rho_{max}$ is very common in the design of sections; thus, a redistribution percentage of 10% to 15% may be expected.

Whatever percentage of moment redistribution is used, it is essential to ensure that no section is likely to suffer local damage or excessive cracking at service loads and that adequate rotation capacity is maintained at every critical section in the structure. The redistribution of moments in a statically indeterminate structure will result in a reduction in the negative moments at the supports and in the positive moments within the spans. This reduction will not imply that the safety of the structure has been reduced or jeopardized as compared with determinate structures. In fact, continuity in structures provides additional strength, stability, and economy in the design.

Table 22.2 Maximum and Minimum q (ACI Code)

f'_c (Ksi)	f_y (Ksi)	ρ_b	ρ_{min}	q_{max} (%) (for ρ_{min})	q_{min} (%) (for $0.5\rho_b$)
3	60	0.0215	0.0033	16.9	10
4	60	0.0285	0.0033	17.7	10
5	60	0.0339	0.0035	17.9	10

22.4.2 ACI Code, Appendix B, Limits

In the unified design method, moment redistribution of negative moments in continuous flexural members is based on NTS, the net tensile strain (ε_t) for both reinforced and prestressed concrete members. Figure 22.4 shows the permissible limits on moment redistribution. It indicates that the percentage increase or decrease in the negative moments at supports of continuous flexural members, q', calculated by the elastic theory, must not exceed $1000\varepsilon_t$ percent, with a maximum of 20%. Moment redistribution is allowed only when $\varepsilon_t \geq 0.0075$, indicating adequate ductility is available at the section at which moment is reduced. When $\varepsilon_t < 0.0075$, no moment redistribution is allowed. The modified negative moments must be used to calculate the modified positive moments within the span, as was explained in Example 16.8. Moment redistribution does not apply to members designed by the alternate design method (Appendix A of the ACI Code) or the direct design method for slab systems (refer to Chapter 17).

In summary, the percentage of increase or decrease in negative moments is as follows:

1. When $\varepsilon_t < 0.0075$, no moment redistribution is allowed.
2. When $\varepsilon_t = 0.0075$, the percentage of moment redistribution is 7.5%.
3. When $\varepsilon_t \geq 0.020$, the percentage of moment redistribution is 20%.
4. When $0.0075 < \varepsilon_t < 0.020$, the percentage of moment redistribution is

$$q' = 1000\varepsilon_t \tag{22.6}$$

For example, if $\varepsilon_t = 0.010$, then the percentage of moment redistribution is 10%.

The relationship between the steel percentage, ρ, in the section and the net tensile strain, ε_t, is as follows:

$$\varepsilon_t = \left(\frac{0.003 + f_y/E_s}{\rho/\rho_b}\right) - 0.003 \tag{22.7a}$$

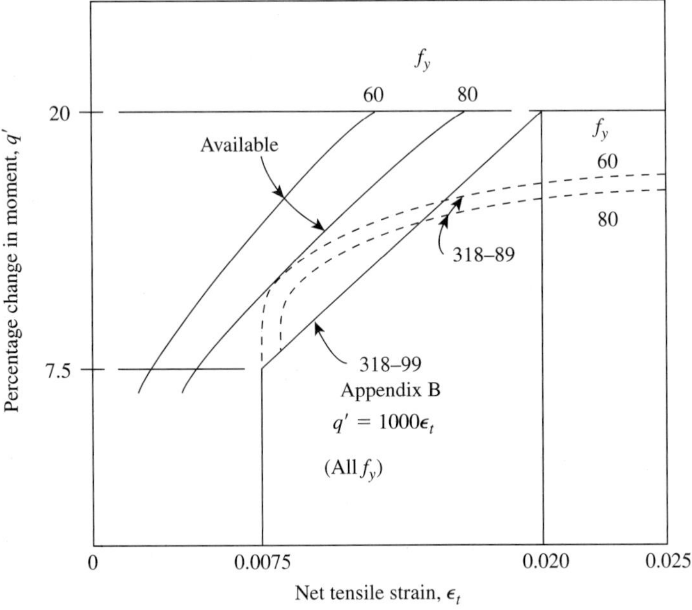

Figure 22.4 Permissible moment redistribution for minimum rotation capacity [1].

For $f_y = 60$ Ksi,

$$\varepsilon_t = \left(\frac{0.005}{\rho/\rho_b}\right) - 0.003 \tag{22.7b}$$

These expressions are obtained by referring to Figure 22.5. For a balanced section,

$$c_b = \frac{a_b}{\beta_1} = \frac{A_{sb}f_y}{0.85f'_c b\beta_1} = \frac{\rho_b f_y d}{0.85f'_c\beta_1}$$

Similarly, for any steel ratio ρ:

$$c = \frac{\rho f_y d}{0.85f'_c\beta_1} \quad \text{and} \quad \frac{c}{c_b} = \frac{\rho}{\rho_b}$$

Divide both sides by d to get

$$\frac{c}{d} = \left(\frac{\rho}{\rho_b}\right)\left(\frac{c_b}{d}\right) \tag{22.8}$$

From the triangles of the strain diagrams,

$$\frac{c}{d} = \frac{0.003}{0.003 + \varepsilon_t}$$

$$\varepsilon_t = \frac{0.003}{(c/d)} - 0.003 \tag{22.9}$$

Similarly,

$$\frac{c_b}{d} = \frac{0.003}{0.003 + f_y/E_s} \tag{22.10}$$

From equations (22.8) and (22.10),

$$\frac{c}{d} = \left(\frac{\rho}{\rho_b}\right)\left(\frac{c_b}{d}\right) = \left(\frac{\rho}{\rho_b}\right)\left(\frac{0.003}{0.003 + f_y/E_s}\right)$$

Substitute this value in equation (22.9) to get

$$\varepsilon_t = \frac{0.003}{(c/d)} - 0.003 = \left[\frac{0.003 + f_y/E_s}{\rho/\rho_b}\right] - 0.003 \tag{22.7a}$$

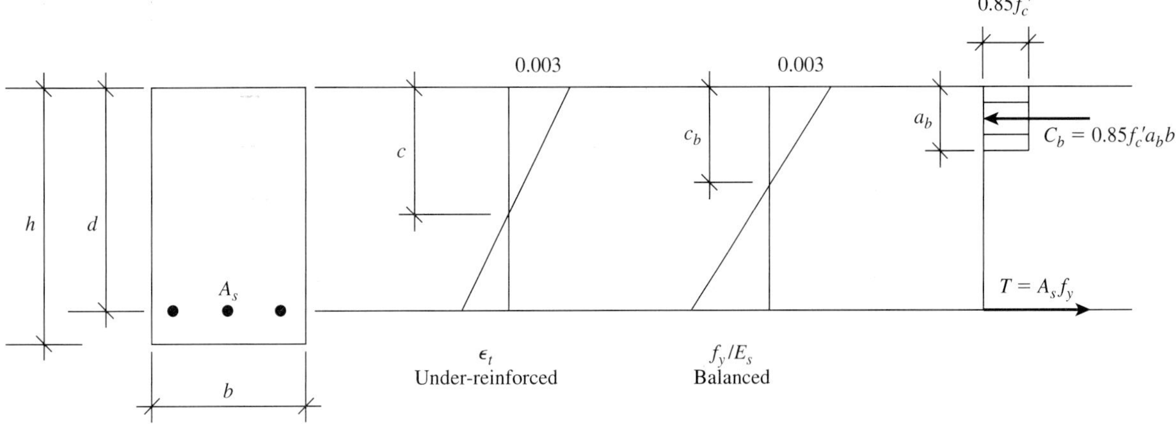

Figure 22.5 Strains in under-reinforced and balanced conditions.

Table 22.3 Percent Change in Moment Redistribution (q'), $f_y = 60$ Ksi

ε_t	0.0075	0.0100	0.0125	0.0150	0.0175	0.0200	0.0225
ρ/ρ_b	0.476	0.385	0.323	0.278	0.244	0.217	0.196
q'%	7.5	10.0	12.5	15.0	17.5	20.0	20.0

Table 22.4 Percent Change in Moment Redistribution (q') for a Given ρ/ρ_b Ratio

ρ/ρ_b	0.48	0.45	0.40	0.35	0.30	0.25	0.20
ε_t	0.0074	0.0081	0.0095	0.0113	0.0137	0.017	0.022
q'%	0	8.1	9.5	11.3	13.7	17.0	20.0

For grade 60 steel, $f_y = 60$ Ksi and $E_s = 29{,}000$ Ksi. Assuming $f_y/E_s = 0.002$, then

$$\varepsilon_t = \left(\frac{0.005}{\rho/\rho_b}\right) - 0.003 \tag{22.7b}$$

The percentage change in moment redistribution according to these limitations and for $f_y = 60$ Ksi is given in Tables 22.3 and 22.4.

22.5 DESIGN OF CONCRETE SECTIONS BY THE UNIFIED DESIGN METHOD

Concrete sections subjected to flexure or axial load and bending moment may lie in compression-controlled, transition, or tension-controlled zones, as shown in Figure 22.3. When $\varepsilon_t \le 0.002$ (or $c/d_t \ge 0.6$), compression controls, whereas when $\varepsilon_t \ge 0.005$ (or $c/d_t \le 0.375$), tension controls. The transition zone occurs when $0.002 < \varepsilon_t < 0.005$ or $0.6 > c/d_t > 0.375$.

For members subjected to flexure, the relationship between the steel ratio ρ or $(\rho - \rho')$, for sections with compression steel was given in equation (22.7):

$$\varepsilon_t + 0.003 = \frac{0.003 + f_y/E_s}{\rho/\rho_b} \tag{22.7a}$$

or

$$\frac{\rho}{\rho_b} = \frac{0.003 + f_y/E_s}{0.003 + \varepsilon_t} \tag{22.11}$$

For $f_y = 60$ Ksi and $E_s = 29{,}000$ Ksi, f_y/E_s may be assumed to be 0.002:

$$\frac{\rho}{\rho_b} = \frac{0.005}{0.003 + \varepsilon_t} \tag{22.12}$$

The limit for tension to control is $\varepsilon_t \ge 0.005$. For $\varepsilon_t = 0.005$, equation (22.12) becomes

$$\frac{\rho}{\rho_b} = \frac{0.005}{0.008} = \frac{5}{8} = 0.625$$

or $\rho \le 0.625\rho_b$ for tension-controlled sections. For example, if $f'_c = 4$ Ksi and $f_y = 60$ Ksi, $\rho_b = 0.0285$ and $\rho \le 0.0178$ for tension to control (as in the case of flexural members). The ϕ-factor in this case is 0.9. This value is less than $\rho_{max} = 0.75\rho_b = 0.0214$ allowed by the ACI Code for flexural members, when $\phi = 0.9$ can be used.

Design of beams and other flexural members can be simplified using the formulas given in Chapters 3 and 4 with a limit of $\varepsilon_t = 0.005$.

$$\frac{\rho}{\rho_b} = \frac{0.003 + f_y/E_s}{0.008} \tag{22.13}$$

Table 22.5 Values of ρ and $R_u = M_u/bd^2$ for Flexural Tension-Controlled Sections with One Row of Bars $(\varepsilon_t = 0.005)$

f'_c (Ksi)	f_y (Ksi)	ρ_b	ρ (limit)	R_u (psi)
3	60	0.0215	0.0134	610
4	60	0.0285	0.0178	810
5	60	0.0339	0.0212	975
6	60	0.0377	0.0236	1100

And for $f_y = 60$ Ksi,

$$\frac{\rho}{\rho_b} = 0.625 \qquad \left(\rho = \frac{5\rho_b}{8}\right) \tag{22.14}$$

It was established in Chapter 3 that $M_u = R_u bd^2$ (equation (3.18)), where $R_u = \phi\rho f_y(1 - \rho f_y/1.7f'_c)$ (equation (3.19)). Once f'_c and f_y are known, then ρ_b, ρ, R_u, and bd^2 can be calculated. For example, for $f'_c = 4$ Ksi, $f_y = 60$ Ksi, $\phi = 0.9$, $\varepsilon_t = 0.005$, and one row of bars in the section,

$$\rho_b = 0.0285, \qquad \rho = 0.0178, \quad \text{and} \quad R_u = 810 \text{ psi}$$

Note that for one row of bars in the section, it can be assumed that $d = d_t = h - 2.5$ in., whereas for two rows of bars, then $d = h - 3.5$ in. and $d_t = h - 2.5$ in.; refer to Figures 22.1 and 22.2.

Table 22.5 gives the values of ρ, ρ_b, and $R_u = M_u/bd^2$ for flexural tension-controlled sections with one row of bars. Note that in cases of sections with compression steel, use $\rho_1 = (\rho - \rho')$.

Example 22.1

Determine the ultimate moment capacity of a rectangular concrete section reinforced with four no. 9 bars in one row (Figure 22.6) by

1. The ACI Code;
2. The unified design method, Appendix B.

Given: $b = 12$ in., $d = 16$ in., $h = 18.5$ in., $f'_c = 4$ Ksi, and $f_y = 60$ Ksi.

Solution

1. By the ACI Code limitations:
   a. For $f'_c = 4$ Ksi and $f_y = 60$ Ksi,

$$\rho_b = 0.0285, \qquad \rho_{\max} = 0.0214, \qquad \phi = 0.9$$

   b. To check, $\rho = A_s/b_d = 4/(12 \times 16) = 0.02083 < \rho_{\max}$.

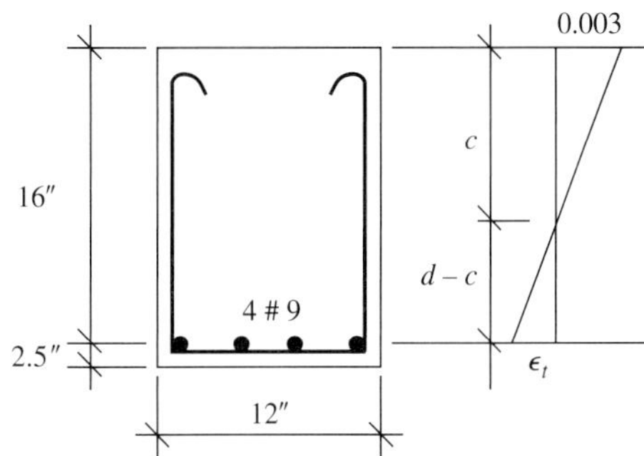

Figure 22.6 Example 22.1 $(d = d_t)$.

c. $\phi M_n = \phi A_s f_y(d - a/2)$:

$$a = \frac{A_s f_y}{0.85 f'_c b} = \frac{4(60)}{0.85(4)(12)} = 5.882 \text{ in.}$$

$$\phi M_n = \frac{0.9}{12}(4)(60)\left(16 - \frac{5.882}{2}\right) = 235 \text{ K} \cdot \text{ft}$$

2. By UDM:

a. Calculate ε_t.

$$a = 5.882, \qquad c = \frac{a}{\beta_1} = \frac{a}{0.85} = 6.92 \text{ in.}$$

From the strain diagram, $d_t = d = 16$ in.

$$\frac{0.003}{c} = \frac{\varepsilon_t}{d_t - c} \quad \text{or} \quad \frac{0.003}{6.92} = \frac{\varepsilon_t}{16 - 6.92}$$

$$\varepsilon_t = 0.003936 \quad \text{and} \quad \varepsilon_t = \frac{0.003(d_t - c)}{c}$$

Because $0.002 < \varepsilon_t < 0.005$, section behavior lies within the transition zone.

b. Calculate $\phi = 0.56 + 68\varepsilon_t = 0.828$.

c. $\phi M_n = (0.828/0.9)(235) = 216.1 \text{ K} \cdot \text{ft}$.

Discussion:

The provisions of Appendix B give a smaller ϕM_n when ε_t lies within the transition zone.

$$\text{Percentage of reduction} = \frac{235 - 216.1}{235} = 8\%$$

It is obvious from Table 22.5 that the limit of ρ for tension to control is $0.0178 < \rho = 0.02083$. It is recommended that $\varepsilon_t = 0.005$ be used in flexural members, with maximum ρ as indicated in Table 22.5 or equation (22.13).

Example 22.2

Determine the ultimate moment capacity of a rectangular concrete section reinforced with six no. 9 bars in two rows (Figure 22.7) by

1. The ACI Code;
2. The unified design method, Appendix B.

Given: $b = 12$ in., $d = 23.5$ in., $h = 27$ in., $d_t = 24.5$ in., $f'_c = 4$ Ksi, and $f_y = 60$ Ksi.

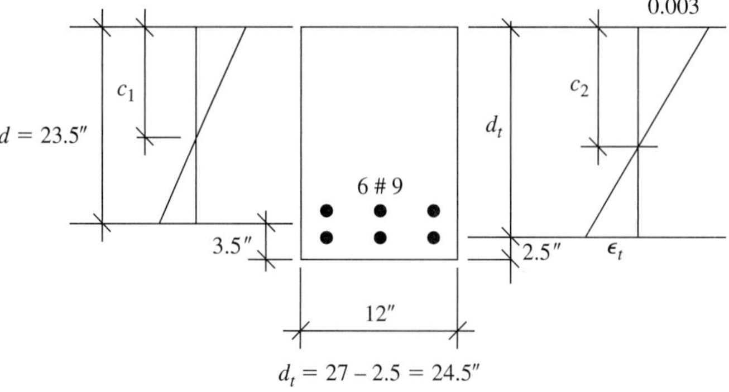

Figure 22.7 Example 22.2.

Solution

1. By the ACI Code limitations:
   a. For $f'_c = 4$ Ksi and $f_y = 60$ Ksi,

$$\rho_b = 0.0285, \qquad \rho_{\max} = 0.0214, \qquad \phi = 0.9$$

   b. $\rho = A_s/bd = 6/(12 \times 23.5) = 0.02128 < \rho_{\max}$.
   c. $\phi M_n = \phi A_s f_y(d - a/2)$

$$a = \frac{A_s f_y}{0.85 f'_c b} = \frac{6(60)}{0.85(4)(12)} = 8.824 \text{ in.}$$

$$\phi M_n = \frac{0.9}{12}(6)(60)(23.5 - 8.824/2) = 515.4 \text{ K}\cdot\text{ft}$$

2. By UDM:
   a. Calculate ε_t.

$$a = 8.824, \qquad c = \frac{a}{\beta_1} = \frac{a}{0.85} = 10.38 \text{ in.}$$

From the strain diagram, $d_t = 27 - 2.5 = 24.5$ in.

$$\varepsilon_t = \frac{0.003(d_t - c)}{c} = \frac{0.003(24.5 - 10.38)}{10.38} = 0.00408$$

Because $0.002 < \varepsilon_t < 0.005$, section behavior lies within the transition zone.

   b. Calculate $\phi = 0.56 + 68\varepsilon_t = 0.837$.
   c. $\phi M_n = (0.837/0.9)(515.4) = 479.3$ K · ft, which is less than 515.4 K · ft. (A reduction of 7% from part 1.)

Example 22.3

Design a rectangular concrete section to support an ultimate moment $M_u = 288$ K · ft based on

1. The ACI Code;
2. The unified design method, Appendix B.

Given: $b = 13$ in., $f'_c = 5$ Ksi, and $f_y = 60$ Ksi.

Solution

1. Based on the ACI Code requirements:
   a. For $f'_c = 5$ Ksi, and $f_y = 60$ Ksi and from tables in Appendix A:

$$\rho_{\max} = 0.0252, \qquad \rho_b = 0.0339$$

$$R_{u\max} = 1110 \text{ psi} \quad \text{and} \quad \phi = 0.9$$

   b. $M_u = R_u bd^2$, or $288 \times 12 = 1.11(13)d^2$, so $d = 15.47$ in. $A_s = 0.0252(13)(15.47) = 5.06$ in.^{2} Use four no. 10 bars in one row ($A_s = 5.06$ in.^{2}). See Figure 22.8. Also, $h = 15.47 + 2.50 = 17.97$, so use 18 in. Actual d is 15.5 in.
   c. $a = A_s f_y/0.85 f'_c b = 5.06(60)/0.85(5)(13) = 5.495$ in.

$$\phi M_n = \phi A_s f_y\left(d - \frac{a}{2}\right) = \frac{0.9}{12}(5.06)(60)\left(15.5 - \frac{5.495}{2}\right) = 290 \text{ K}\cdot\text{ft}$$

2. Based on UDM:
   a. Check the section designed in Step 1.

$$d_t = d = 15.5 \text{ in.}, \qquad a = 5.495 \text{ in.}$$

$$c = \frac{a}{\beta_1} = \frac{a}{0.8} = 6.87 \text{ in.} \qquad (\beta_1 = 0.8 \text{ for } f'_c = 5 \text{ Ksi})$$

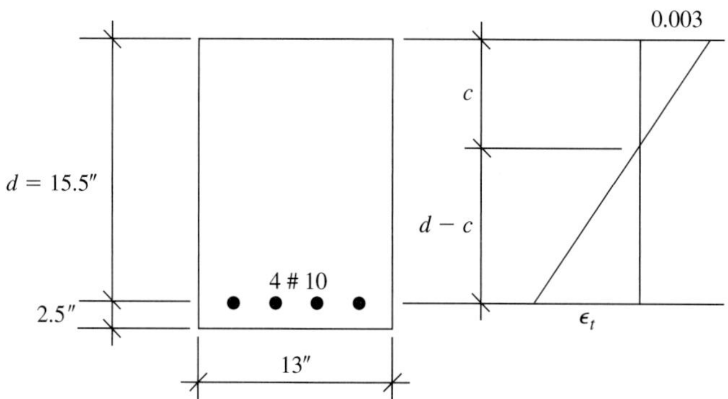

Figure 22.8 Example 22.3.

From the strain diagram (Figure 22.8),

$$\varepsilon_t = \left(\frac{d_t - c}{c}\right)0.003 = \left(\frac{15.5 - 6.87}{6.87}\right)0.003 = 0.00377$$

Behavior of this beam lies in the transition zone.

$$\phi = 0.56 + 68\varepsilon_t = 0.56 + 68(0.00377) = 0.816$$

$$\phi M_n = \frac{0.816}{0.9}(290) = 262.9 \text{ K}\cdot\text{ft} < 288 \text{ K}\cdot\text{ft}$$

Therefore, the section is not adequate based on the UDM.

b. Redesign the beam to have $\varepsilon_t = 0.005$ and $\phi = 0.9$. For $f_y = 60$ Ksi, $\rho/\rho_b = 0.625$ (equation (22.14)). $\rho = 0.625(0.0339) = 0.0212$.

$$R_u = \phi\rho f_y\left(1 - \frac{\rho f_y}{1.7f'_c}\right) = 0.9(60)(0.0212)\left(1 - \frac{0.0212 \times 60}{1.7 \times 5}\right) = 975 \text{ psi}$$

$$M_u = R_u bd^2$$

$$288(12) = 0.975(13)d^2, \qquad d = 16.5 \text{ in.}$$

$$A_s = 0.0212(13)(16.5) = 4.54 \text{ in.}^2$$

Use two no. 10 and two no. 9 bars, $A_s = 4.53$ in.^{2}, and $h = 19$ in. Check ϕM_n: $a = 4.53(60)/(0.85 \times 5 \times 13 = 4.92$ in.)

$$\phi M_n = \left(\frac{0.9}{12}\right)(60)(4.53)\left(16.5 - \frac{4.92}{2}\right) = 286.2 \text{ K}\cdot\text{ft}$$

which is close to 288 K·ft. Practically, four no. 10 bars (or $h = 19$ in.) are recommended, giving $\phi M_n > M_u$. It can be observed that the provisions of Appendix B reduce the design flexural moment strength of the member as long as $\varepsilon_t < 0.005$.

Example 22.4

Design a rectangular section to support $M_u = 780$ K·ft based on

1. The ACI Code;
2. The unified design method, Appendix B.

Given: $b = 12$ in., $h = 31$ in., $f'_c = 4$ Ksi, and $f_y = 60$ Ksi.

Solution

1. Based on the ACI Code:
   a. For $f'_c = 4$ Ksi, and $f_y = 60$ Ksi, $\rho_{\max} = 0.0214$, $\rho_b = 0.0285$, $\phi = 0.9$, and $R_{u\max} = 936$ psi.

b. Check $\phi M_{n1\,max}$ as singly reinforced assuming $d = 31 - 3.5 = 27.5$ in. (two rows of bars).

$$\phi M_{n1} = R_{u\,max} bd^2 = \frac{0.936(12)(27.5)^2}{12} = 707.85 \text{ K}\cdot\text{ft}$$

The section needs compression reinforcement. Let $d' = 2.5$ in.

$$A_{s1} = 0.0214(12)(27.5) = 7.05 \text{ in.}^2$$

c. $M_{u2} = 780 - 707.85 = 72.15$ K·ft

$M_{u2} = \phi A_{s2} f_y (d - d') = \phi A'_s f'_s (d - d')$

$72.15 \times 12 = 0.9A_{s2}(60)(27.5 - 2.5)$ and so $A_{s2} = 0.64$ in.$^2 = A'_s$. Total A_s is $A_{s1} + A_{s2} = 7.69$ in.^{2}; choose six no. 10 bars in two rows $(A_s = 7.62 \text{ in.}^2)$. Choose two no. 6 bars (0.88 in.^{2}) for A'_s.

d. Check if compression steel yields (refer to Example 4.5). Compression steel yields if

$$(\rho - \rho') \geq K = 0.85\beta_1 \frac{f'_c}{f_y}\left(\frac{d'}{d}\right)\left(\frac{87}{87 - f_y}\right)$$

$$K = (0.85)(0.85)\left(\frac{4}{60}\right)\left(\frac{2.5}{27.5}\right)\left(\frac{87}{27}\right) = 0.0141$$

$$\rho - \rho' = \frac{(7.62 - 0.88)}{12(27.5)} = 0.0204 > K$$

Therefore, compression steel yields. The final section is shown in Figure 22.9(a).

e. Check $\phi M_n = \phi\left[(A_s - A_{s1}) f_y (d - a/2) + A'_s f_y (d - d')\right]$.

$$a = \frac{(A_s - A'_s) f_y}{0.85 f'_c b} = \frac{(7.62 - 0.88)60}{0.85(4)(12)} = 9.91 \text{ in.}$$

$$\phi M_n = 782.6 \text{ K}\cdot\text{ft} > M_u = 780 \text{ K}\cdot\text{ft}$$

2. Based on UDM:

a. Check net tensile strain, NTS (Figure 22.9(b)): $h = 31$ in., $d = 27.5$ in., $d_t = 31 - 2.5 = 28.5$ in., $(\rho - \rho') = 0.0204$. Also, $c = a/\beta_1 = 9.91/0.85 = 11.66$ in.

$$\varepsilon_t = \frac{(d_t - c)}{c}(0.003) = \left(\frac{28.5 - 11.16}{11.16}\right)(0.003) = 0.00433$$

Therefore, the section lies in the transition zone.

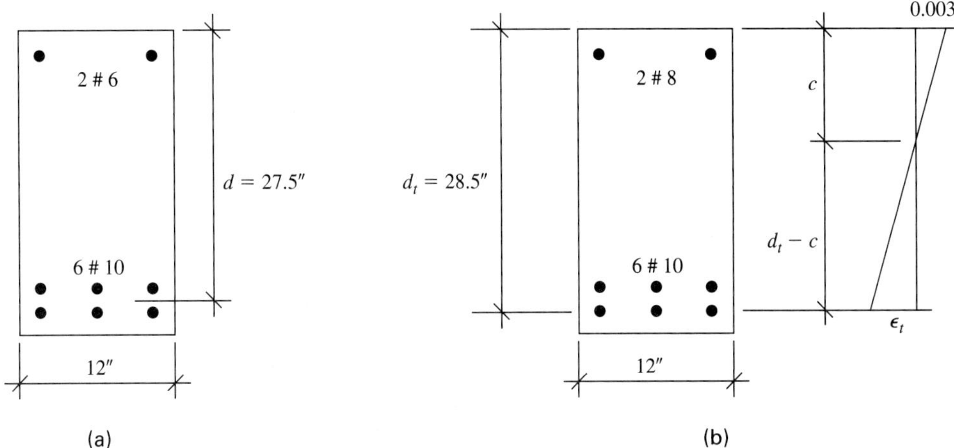

Figure 22.9 Example 22.4.

b. Check if compression steel yields:

$$(\rho - \rho') = 0.0204 > K = 0.0141 \qquad \text{(checked above)}$$

Therefore, compression steel yields.

c. Determine $\phi = 0.56 + 68\varepsilon_t = 0.855$.

$$\phi M_n = \frac{0.855}{0.9}(782.6) = 743.4 \text{ K}\cdot\text{ft} < 780 \text{ K}\cdot\text{ft}$$

Redesign the beam.

d. Increase A'_s so that $\rho_1 = (\rho - \rho')$ for the singly reinforced basic section is reduced. Let $A'_s = 1.57$ in.^{2} (two no. 8 bars).

$$\text{New } a = \frac{(A_s - A'_s)f_y}{0.85 f'_c b} = \frac{(7.62 - 1.57)60}{0.85(4)(12)} = 8.9 \text{ in.}$$

$$c = \frac{a}{0.85} = 10.5 \text{ in.}$$

$$\varepsilon_t = \left(\frac{28.5 - 10.5}{10.5}\right)(0.003) = 0.00514 > 0.005$$

Therefore, tension controls and $\phi = 0.9$.

e. $$\varepsilon'_s = \frac{0.003(c - d')}{c}$$

$$= \left(\frac{10.5 - 2.5}{10.5}\right)0.003 = 0.00228 > \varepsilon_y = 0.002$$

Therefore, compression steel yields.

$$\phi M_n = \left[(7.62 - 1.57)(60)\left(27.5 - \frac{8.9}{2}\right) + 1.57(60)(27.5 - 2.5)\right]\frac{0.9}{12}$$

$$= 804.2 \text{ K}\cdot\text{ft} > M_u = 780 \text{ K}\cdot\text{ft}$$

Example 22.5

For the rectangular concrete column section shown in Figure 22.10, determine P_n, M_n, and e using the UDM.

1. Corresponding to compression-controlled behavior;

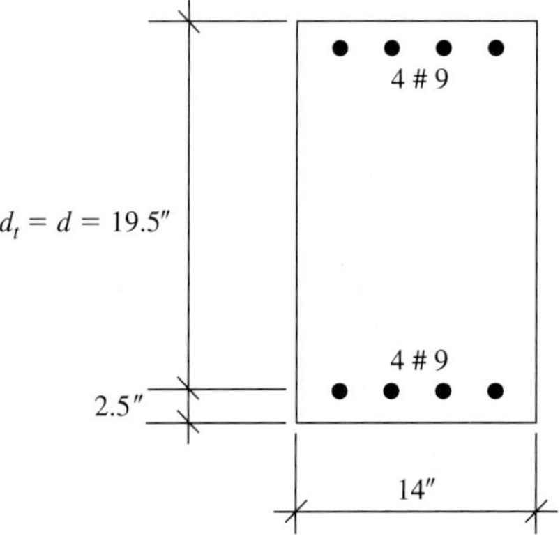

Figure 22.10 Example 22.5.

2. Corresponding to tension-controlled behavior.
3. Compare results with Examples 11.2, 11.3, and 11.4, which use the same section.

Given: $b = 14$ in., $h = 22$ in., $d' = 2.5$ in., $A_s = A'_s =$ four no. 9 bars, $f'_c = 4$ Ksi, and $f_y = 60$ Ksi.

Solution

(Refer to Chapter 11 for column analysis and design.)

1. For compression behavior, $\varepsilon_t = 0.002$ or $c/d_t = 0.6$; refer to Figure 22.3.
   a. For $f_y = 60$ Ksi, $\varepsilon_t = f_y/\varepsilon_s = 60/29{,}000 = 0.002$. From the strain diagram, $d_t = 22 - 2.5 = 19.5$ in., and

$$\frac{c}{d_t} = \frac{0.003}{0.003 + \varepsilon_t} = \frac{3}{5} = 0.6$$

   Also, $c = 0.6(19.5) = 11.7$ in. and $a = 0.85c = 9.945$ in.

   b. Check stress in compression steel:

$$\varepsilon'_s = \left(\frac{c - d'}{c}\right)0.003 = \left(\frac{11.7 - 2.5}{11.7}\right)0.003 = 0.00236$$

   Because $\varepsilon'_s > \varepsilon_y = 0.002$, compression steel yields; $f'_s = 60$ Ksi.

   c. Referring to Sections 11.5 and 11.6,

$$P_n = C_c + C_s - T \tag{22.15}$$

   Taking moments about the plastic centroid of the section (refer to Figure 11.7 and equations (11.9) and (11.10)),

$$M_n = P_n e = C_c\left(d - \frac{a}{2} - d''\right) + C_s(d - d' - d'') + Td'' \tag{22.16}$$

   When $A_s = A'_s$, then

$$M_n = P_n e = C_c\left(\frac{h}{2} - \frac{a}{2}\right) + C_s\left(\frac{h}{2} - d'\right) + T\left(d - \frac{h}{2}\right) \tag{22.17}$$

   where $C_c = 0.85 f'_c ab$, $C_s = A'_s f'_s$, and $T = A_s f_s$. In this example, $A_s = A'_s = 4$ in.^{2}, $f'_s = 60$ Ksi, and $f_s = f_y = 60$ Ksi.

$$C_c = 0.85(4)(9.945)(14) = 473.4 \text{ K}, \qquad C_s = T = 4 \times 60 = 240 \text{ K}$$

   Therefore, $P_n = 473.4 + (4 \times 60) - (4 \times 60) = 473.4$ K.

$$M_n = 473.4\left(\frac{22}{2} - \frac{9.945}{2}\right) + 240\left(\frac{22}{2} - 2.5\right) + 240(19.5 - 11)$$

$$= 6933.3 \text{ K}\cdot\text{in.} = 577.7 \text{ K}\cdot\text{ft}$$

$$e = \frac{M_n}{P_n} = 14.7 \text{ in.}$$

   For compression-controlled behavior, $\phi = 0.7$ and $\phi P_n = 331.4$ K.

2. For tension behavior, $\varepsilon_t = 0.005$.
   a. $\dfrac{c}{d_t} = \dfrac{0.003}{0.003 + 0.005} = 0.375$ (Figure 22.3)

   $d_t = d = 19.5$ in.

   $c = 0.375(19.5) = 7.31$ in., $\quad a = 0.85c = 6.22$ in.

b. Check stress in compression steel, ε_s':

$$\varepsilon_s' = \left(\frac{c - d'}{c}\right)0.003 = \left(\frac{7.31 - 2.5}{7.31}\right)0.003 = 0.001974$$

$$f_s' = 29{,}000\varepsilon_s' = 57.25 \text{ Ksi} < 60 \text{ Ksi}, \qquad f_s = f_y = 60 \text{ Ksi}$$

c. $C_c = 0.85f_c'ab = 0.85(4)(6.22)(14) = 296.1 \text{ K}$

$$C_s = 4(57.25) = 229 \text{ K}, \qquad T = 4(60) = 240 \text{ K}$$

$$P_n = C_c + C_s - T = 296.1 + 229 - 240 = 285.1 \text{ K}$$

$$M_n = P_n e = 296.1(11 - 6.22/2) + 229(11 - 2.5)$$
$$+ 240(19.5 - 11) = 6322.7 \text{ K}\cdot\text{in.} = 526.9 \text{ K}\cdot\text{ft}$$

$$e = \frac{M_n}{P_n} = 22.2 \text{ in.}$$

For tension-controlled behavior, $\phi = 0.9$ and $\phi P_n = 256.6$ K.

3. From Chapter 11, Examples 11.2, 11.3, and 11.4 cover the solution for balanced, tension-controlled, and compression-controlled conditions based on the ACI Code limitations. The results are summarized in Table 11.1.

a. For a balanced condition (Example 11.2): $P_b = 453.2$ K, $M_b = 567.6$ K·ft, $e_b = 15$ in., $\phi = 0.7$, $\phi P_b = 317.4$ K, and $\phi M_b = 397.3$ K·ft.

b. For $e = 20$ in., tension controls (Example 11.3):

$$P_n = 324.4 \text{ K}, \qquad M_n = 540.67 \text{ K}\cdot\text{ft}, \qquad \phi = 0.7$$

$$\phi P_n = 227.1 \text{ K}, \qquad \phi M_n = 378.5 \text{ K}\cdot\text{ft}$$

c. For $e = 10$ in., compression controls (Example 11.4):

$$P_n = 612.9 \text{ K}, \qquad M_n = 510.8 \text{ K}\cdot\text{ft}, \qquad \phi = 0.7$$

$$\phi P_n = 429 \text{ K}, \qquad \phi M_n = 357.5 \text{ K}\cdot\text{ft}$$

From Example 22.5 based on the unified design method, Appendix B, results are different:

i. When $e \le 14.7$ in., compression-controlled behavior develops with $\phi = 0.7$ (or $e = 15$ in., as in (a)).

ii. When $e \ge 22.2$ in., tension-controlled behavior develops with $\phi = 0.9$.

iii. When $14.7 < e < 22.2$ in., a transition case develops, and ϕ varies between 0.7 and 0.9.

When $e = 20$ in. < 22.2 in. (from Example 11.3 and (b)), the behavior lies in the transition zone and $\phi > 0.7$, which can be determined as follows: From Example 11.3, $c = 8.35$ in., $d_t = d = 19.5$ in., and $\phi = 0.356 + 0.204(d_t/c) = 0.356 + 0.204(19.5/8.35) = 0.8324$. $\phi P_n = 0.8324(324.4) = 270$ K, which is greater than 227.1 K (as allowed by the Code), giving an increase of about 19% in this example.

There are no changes in (a) and (c) based on Examples 11.2 and 11.4, because (a) lies at the boundaries of compression control, $\varepsilon_t = 0.002$; in (c), $e = 10$ in. < 14.47 in. when compression-controlled behavior occurs and $\phi = 0.7$.

Example 22.6

Determine the permissible redistribution of negative moments at supports B, C, D, and E of the continuous beam ABCDEF shown in Figure 22.11. The beam has a rectangular section, $b = 12$ in., $h = 22$ in., and $d = 19.5$ in., and it is reinforced as shown in the following table ($f_c' = 4$ Ksi and $f_y = 60$ Ksi).

1. Use the ACI Code limitations.
2. Use the UDM, Appendix B.

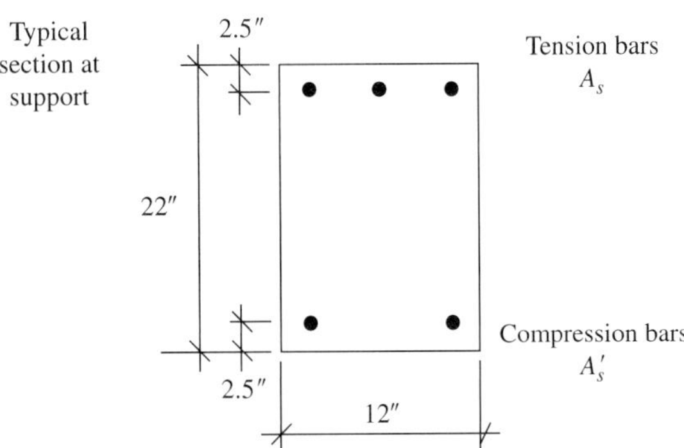

Figure 22.11 Example 22.6.

Solution

1. For $f'_c = 4$ Ksi and $f_y = 60$ Ksi, $\rho_b = 0.0285$. The ACI Code redistribution factor was given as follows:

$$q = 20\left[1 - \frac{\rho - \rho'}{\rho_b}\right] \tag{22.5}$$

2. The UDM redistribution factor is a function of the net tensile strain, ε_t, and varies between 7.5% and 20%, as shown in Figure 22.4.

$$q' = 1000\varepsilon_t \tag{22.6}$$

$$\varepsilon_t = \frac{0.003 + f_y/E_s}{\rho/\rho_b} - 0.003$$

$$\text{and } \varepsilon_t = \frac{0.005}{\rho/\rho_b} - 0.003 \qquad (\text{for } f_y = 60 \text{ Ksi}) \tag{22.7}$$

The table shows the values of q and q', which are not compatible.

Support	Tension Bars (A_s)	ρ	Compression Bars (A'_s)	ρ'	$\frac{\rho - \rho'}{\rho_b}$	q% Code	ε_t	q' UDM
B	3 no. 9	0.01282	0	0	0.45	11.0	0.0113	11.3
C	3 no. 10	0.0160	0	0	0.56	8.8	0.006	0
D	3 no. 6	0.00564	0	0	0.198	16.0	0.0226	20
E	4 no. 8	0.01342	3 no. 6	0.0056	0.273	14.5	0.0153	15.3

SUMMARY

Sections 22.1–22.2

A new approach for the design of reinforced and prestressed concrete members is introduced as the unified design approach. This approach is covered in the ACI Code, Appendix B. Four different conditions may develop under load: compression controlled, tension controlled, balanced, and a transition region condition.

Section 22.3

For compression controlled condition, the strength reduction factor ϕ is equal to 0.7. For tension controlled condition, ϕ is equal to 0.9. For the transition region, ϕ varies between 0.7 (or 0.75) and 0.9. The value of ϕ can be determined according to equations (22.1) through (22.4).

Section 22.4

The moment redistribution coefficient q' based on the new approach, Figure 22.4, is different from the ACI Code coefficient q, equation (22.6) and was explained in Chapter 16. Example 22.6 gives the variation in application results.

Section 22.5

Six different examples are presented to explain the applications of the new approach to beams, columns and moment redistribution.

REFERENCES

1. American Concrete Institute. "Building Code Requirements for Structural Concrete." ACI Code 318-99, American Concrete Institute, Detroit, 1999.
2. Mast, R. F. "Unified Design Provisions for Reinforced and Prestressed Concrete Flexural and Compression Members." ACI *Structural Journal* 89, no. 2 (March–April 1992): 185–199.

PROBLEMS

22.1 Determine the ultimate moment capacity of a rectangular concrete section reinforced with five no. 9 bars placed in one row by (a) the ACI Code, (b) the unified design method. Given: $b = 16$ in., $h = 24$ in., $d = 21.5$ in., $f'_c = 4$ Ksi, and $f_y = 60$ Ksi.

22.2 Repeat Problem 22.1 if the tension steel is seven no. 9 placed in two rows.

22.3 Design a rectangular concrete section to support an ultimate moment of 315 K·ft by (a) the ACI Code using maximum steel ratio, (b) the unified design method. Given: $b = 14$ in., $f'_c = 5$ Ksi, and $f_y = 60$ Ksi.

22.4 Design a rectangular concrete section to support an ultimate moment of 690 K·ft by (a) the ACI Code, (b) the unified design method. Given: $b = 14$ in., $h = 30$ in., $f'_c = 5$ Ksi and $f_y = 60$ Ksi.

22.5 Calculate the nominal load capacity P_n and the nominal moment capacity M_n for a rectangular concrete column section by the unified design method using $b = 16$ in., $h = 24$ in., $A_s = A'_s = 5$ no. 9 bars, $f'_c = 4$ Ksi and $f_y = 60$ Ksi for the following two cases:

a. corresponding to the compression controlled behavior

b. corresponding to the tension controlled behavior

22.6 Repeat Problem 22.5 using the ACI Code provisions when the eccentricity e is (a) 6 in., (b) 20 in., (c) the balanced e.

APPENDIX A

DESIGN TABLES (U.S. CUSTOMARY UNITS)

Table A.1 Values of R_u and a/d for $f'_c = 3000$ psi

100 ρ	$f_y = 40$ Ksi R_u	$f_y = 40$ Ksi a/d	$f_y = 50$ Ksi R_u	$f_y = 50$ Ksi a/d	$f_y = 60$ Ksi R_u	$f_y = 60$ Ksi a/d	$f_y = 75$ Ksi R_u	$f_y = 75$ Ksi a/d
0.2	71	0.031	88	0.039	106	0.047	131	0.059
0.3	105	0.047	131	0.059	156	0.071	192	0.089
0.4	140	0.062	173	0.078	206	0.094	254	0.118
0.5	173	0.078	214	0.098	254	0.118	310	0.148
0.6	206	0.094	254	0.118	301	0.141	368	0.177
0.7	238	0.110	293	0.138	347	0.165	421	0.207
0.8	270	0.126	332	0.157	391	0.189	475	0.236
0.9	301	0.142	369	0.177	434	0.213	524	0.266
1.0	332	0.157	406	0.196	476	0.238	572	0.295
1.1	362	0.173	441	0.216	517	0.260	620	0.325
1.2	390	0.188	476	0.235	556	0.282	648	0.340
1.3	420	0.204	510	0.255	594	0.306	$(\rho_{max} = 1.16)$	
1.4	450	0.220	543	0.274	631	0.330		
1.5	476	0.236	575	0.294	667	0.353		
1.6	504	0.252	607	0.314	700	0.376		
1.7	530	0.267	637	0.334	702	0.378		
1.8	556	0.282	667	0.353	$(\rho_{max} = 1.61)$			
1.9	582	0.298	695	0.373				
2.0	607	0.314	723	0.392				
2.1	630	0.330						
2.2	655	0.345	740	0.405				
2.3	677	0.360	$(\rho_{max} = 2.06)$					
2.4	700	0.376						
2.5	723	0.392						
2.6	745	0.408						
2.7	767	0.424						
2.8	783	0.436						
	$(\rho_{max} = 2.78)$							

Note: Last values are the maximum allowed by the ACI Code.

Table A.2 Values of R_u and a/d for $f'_c = 4000$ psi

100 ρ	$f_y = 40$ Ksi		$f_y = 50$ Ksi		$f_y = 60$ Ksi		$f_y = 75$ Ksi	
	R_u	a/d	R_u	a/d	R_u	a/d	R_u	a/d
0.2	71	0.024	89	0.029	106	0.035	132	0.044
0.3	106	0.036	132	0.044	158	0.053	194	0.066
0.4	140	0.047	175	0.059	208	0.071	257	0.088
0.5	175	0.059	217	0.074	258	0.089	317	0.110
0.6	208	0.071	260	0.088	307	0.106	378	0.132
0.7	242	0.083	300	0.103	355	0.123	434	0.154
0.8	274	0.094	340	0.118	400	0.141	490	0.176
0.9	307	0.106	378	0.132	447	0.158	545	0.198
1.0	340	0.118	419	0.147	492	0.176	600	0.220
1.1	370	0.130	455	0.161	536	0.194	650	0.242
1.2	400	0.141	492	0.176	580	0.212	702	0.264
1.3	432	0.153	530	0.191	620	0.230	752	0.286
1.4	462	0.165	565	0.206	662	0.247	801	0.308
1.5	492	0.177	600	0.221	700	0.265	846	0.330
1.6	522	0.188	635	0.236	742	0.282	867	0.342
							($\rho_{max} = 1.55$)	
1.7	550	0.200	670	0.250	780	0.300		
1.8	580	0.212	702	0.265	818	0.318		
1.9	607	0.224	735	0.280	853	0.335		
2.0	635	0.236	768	0.294	890	0.353		
2.1	662	0.248	800	0.309	924	0.370		
2.2	690	0.260	830	0.323	936	0.378		
2.3	717	0.271	860	0.338	($\rho_{max} = 2.14$)			
2.4	742	0.282	890	0.352				
2.5	767	0.294	918	0.367				
2.6	792	0.306	946	0.381				
2.7	817	0.318	974	0.396				
2.8	842	0.330	988	0.405				
2.9	866	0.342	($\rho_{max} = 2.75$)					
3.0	890	0.354						
3.1	193	0.365						
3.2	935	0.376						
3.3	958	0.388						
3.4	980	0.400						
3.5	1000	0.412						
3.6	1020	0.424						
3.7	1040	0.436						
	1044	0.438						
	($\rho_{max} = 3.72$)							

Note: Last values are the maximum allowed by the ACI Code.

Table A.3 Values of R_u and a/d for f'_c = 5000 psi

100	f_y = 40 Ksi		f_y = 50 Ksi		f_y = 60 Ksi		f_y = 75 Ksi	
ρ	R_u	a/d	R_u	a/d	R_u	a/d	R_u	a/d
0.2	71	0.019	89	0.024	106	0.028	132	0.035
0.3	106	0.029	133	0.036	159	0.042	196	0.052
0.4	141	0.038	176	0.047	210	0.056	260	0.070
0.5	176	0.047	218	0.060	260	0.070	322	0.088
0.6	210	0.056	260	0.071	310	0.085	384	0.106
0.7	244	0.066	302	0.083	360	0.100	442	0.123
0.8	277	0.075	343	0.094	408	0.113	500	0.141
0.9	310	0.085	383	0.106	455	0.127	556	0.159
1.0	343	0.094	424	0.118	502	0.141	612	0.177
1.1	375	0.104	463	0.130	550	0.155	667	0.195
1.2	408	0.113	500	0.141	593	0.169	722	0.212
1.3	440	0.123	540	0.153	637	0.183	776	0.230
1.4	470	0.132	578	0.165	681	0.198	830	0.247
1.5	502	0.141	615	0.177	724	0.212	875	0.265
1.6	532	0.150	652	0.188	766	0.226	920	0.282
1.7	563	0.160	688	0.200	808	0.240	970	0.300
1.8	593	0.169	724	0.212	848	0.254	1020	0.318
1.9	623	0.179	760	0.224	890	0.268	1033	0.322
2.0	652	0.188	794	0.235	927	0.282	(ρ_{max} = 1.83)	
2.1	681	0.198	830	0.247	965	0.292		
2.2	710	0.207	862	0.259	1003	0.311		
2.3	738	0.217	894	0.271	1040	0.325		
2.4	766	0.226	927	0.282	1076	0.339		
2.5	794	0.235	958	0.294	1116	0.358		
2.6	821	0.244	990	0.306	1120	0.364		
					(ρ_{max} = 2.52)			
2.7	848	0.254	1021	0.318				
2.8	875	0.263	1052	0.329				
2.9	900	0.272	1082	0.341				
3.0	127	0.282	1111	0.353				
3.1	952	0.292	1140	0.365				
3.2	978	0.301	1168	0.376				
3.3	1003	0.311						
3.4	1028	0.320	1179	0.381				
3.5	1052	0.329	(ρ_{max} = 3.24)					
3.6	1076	0.338						
3.7	1100	0.348						
3.8	1122	0.357						
3.9	1145	0.367						
4.0	1168	0.376						
4.1	1190	0.385						
4.2	1212	0.395						
4.3	1235	0.404						
	1248	0.410						
	(ρ_{max} = 4.37)							

Note: Last values are the maximum allowed by the ACI Code.

Table A.4 Values of ρ_{max}, R_{umax}, ρ_b, and ρ_{min}

$$\rho_{max} = 0.64\beta_1 \frac{f'_c}{f_y} \times \frac{87}{87 + f_y}, \qquad R_u = \phi\rho f_y\left(1 - \frac{\rho f_y}{1.7f'_c}\right)$$

	$f_y = 40$ Ksi				$f_y = 50$ Ksi			
f'_c psi	$100\ \rho_{max}$	$R_{u\,max}$ (psi)	$100\ \rho_b$	$100\ \rho_{min}$	$100\ \rho_{max}$	$R_{u\,max}$ (psi)	$100\ \rho_b$	$100\ \rho_{min}$
3000	2.78	783	3.71	0.5	2.06	740	2.75	0.4
4000	3.72	1044	4.96	0.5	2.75	987	3.67	0.4
5000	4.36	1244	5.81	0.53	3.24	1180	4.32	0.42
6000	4.90	1424	6.53	0.58	3.64	1364	4.85	0.47
	$f_y = 60$ Ksi				$f_y = 75$ Ksi			
f'_c psi	$100\ \rho_{max}$	$R_{u\,max}$ (psi)	$100\ \rho_b$	$100\ \rho_{min}$	$100\ \rho_{max}$	$R_{u\,max}$ (psi)	$100\ \rho_b$	$100\ \rho_{min}$
3000	1.61	704	2.15	0.33	1.16	648	1.55	0.27
4000	2.14	937	2.85	0.33	1.55	867	2.07	0.27
5000	2.52	1120	3.35	0.35	1.83	1033	2.44	0.28
6000	2.83	1275	3.77	0.39	2.06	1180	2.75	0.31

Table A.5 Suggested Design Steel Ratios, ρ_s, and Comparison With Other Steel Ratios

f'_c (psi)	f_y (Ksi)	$100\ \rho_b$	$100\ \rho_{max}$	$100\rho_s$ (suggested)	R_u for ρ_s (psi)	$100\rho_b$ (elastic)	Ratio ρ_s/ρ_b	Ratio ρ_s/ρ_{max}	Weight Of ρ_s (lb/ft^3 of concrete)
3000	40	3.71	2.78	1.4	450	1.29	0.377	0.500	7
	50	2.75	2.06	1.2	476	1.11	0.436	0.583	6
	60	2.15	1.61	1.2	556	0.96	0.558	0.745	6
4000	40	4.96	3.72	1.4	462	1.88	0.282	0.376	7
	50	3.67	2.75	1.4	565	1.62	0.381	0.510	7
	60	2.85	2.14	1.4	662	1.41	0.491	0.654	7
5000	40	5.81	4.36	1.4	470	2.50	0.241	0.321	7
	50	4.32	3.24	1.4	578	2.15	0.324	0.432	7
	60	3.35	2.52	1.4	681	1.87	0.418	0.551	7

Table A.6 Minimum Thickness of Beams and One-Way Slabs

Member	Yield Strength f_y (Ksi)	Simply Supported	One End Continuous	Both Ends Continuous	Cantilever
Solid one-way slabs	40	$L/25$	$L/30$	$L/35$	$L/12.5$
	50	$L/22$	$L/27$	$L/31$	$L/11$
	60	$L/20$	$L/24$	$L/28$	$L/10$
Beams or ribbed one-way slabs	40	$L/20$	$L/23$	$L/26$	$L/10$
	50	$L/18$	$L/20.5$	$L/23.5$	$L/9$
	60	$L/16$	$L/18.5$	$L/21$	$L/8$

Table A.7 Minimum Beam Width (in.) (Using Stirrups)

Size of Bars	Number of Bars in Single Layer of Reinforcement							Add For Each Added Bar (in.)
	2	*3*	*4*	*5*	*6*	*7*	*8*	
No. 4	6.1	7.6	9.1	10.6	12.1	13.6	15.1	1.50
No. 5	6.3	7.9	9.6	11.2	12.8	14.4	16.1	1.63
No. 6	6.5	8.3	10.0	11.8	13.5	15.3	17.0	1.75
No. 7	6.7	8.6	10.5	12.4	14.2	16.1	18.0	1.88
No. 8	6.9	8.9	10.9	12.9	14.9	16.9	18.9	2.00
No. 9	7.3	9.5	11.8	14.0	16.3	18.6	20.8	2.26
No. 10	7.7	10.2	12.8	15.3	17.8	20.4	22.9	2.54
No. 11	8.0	10.8	13.7	16.5	19.3	22.1	24.9	2.82
No. 14	8.9	12.3	15.6	19.0	22.4	25.8	29.2	3.39
No. 18	10.5	15.0	19.5	24.0	28.6	33.1	37.6	4.51

Table A.8 Values of bd^2 (in.^{3}) $bd^2 = \left[\frac{M_u}{R_u}\left(\frac{\text{lb}\cdot\text{in.}}{\text{psi}}\right)\right]$

d (in.)	Values of b (in.) 6	7	8	9	10	11	12	13	14	15	16	20
4	96	112	128	144	160	176	192	208	224	240	256	320
4.5	122	142	162	182	202	223	244	264	284	305	325	405
5	150	175	200	225	250	275	300	325	350	375	400	500
5.5	182	212	242	273	303	333	364	394	424	455	485	605
6	216	252	288	324	360	396	432	468	504	540	576	720
6.5	255	297	340	382	425	467	510	552	595	637	680	850
7	294	343	392	441	490	539	588	637	686	735	784	980
8	384	448	512	576	640	704	768	832	896	960	1,024	1,280
9	486	567	648	729	810	891	972	1,053	1,134	1,215	1,296	1,620
10	600	700	800	900	1,000	1,100	1,200	1,300	1,400	1,500	1,600	2,000
11	726	847	968	1,089	1,210	1,331	1,452	1,573	1,694	1,815	1,936	2,420
12	864	1,008	1,152	1,296	1,440	1,584	1,728	1,872	2,016	2,160	2,304	2,880
13	1,014	1,183	1,352	1,521	1,690	1,859	2,028	2,197	2,366	2,535	2,704	3,380
14	1,176	1,372	1,568	1,764	1,960	2,156	2,352	2,548	2,744	2,940	3,136	3,920
15	1,350	1,575	1,800	2,025	2,250	2,475	2,700	2,925	3,150	3,375	3,600	4,500
16	1,536	1,792	2,048	2,304	2,560	2,816	3,072	3,328	3,584	3,840	4,096	5,120
17	1,734	2,023	2,312	2,601	2,890	3,179	3,468	3,757	4,046	4,335	4,624	5,780
18	1,944	2,268	2,592	2,916	3,240	3,564	3,888	4,212	4,536	4,860	5,184	6,480
19	2,166	2,527	2,888	3,249	3,610	3,971	4,332	4,693	5,054	5,415	5,776	7,220
20	2,400	2,800	3,200	3,600	4,000	4,400	4,800	5,200	5,600	6,000	6,400	8,000
21	2,646	3,087	3,528	3,969	4,410	4,851	5,292	5,733	6,174	6,615	7,056	8,820
22	2,904	3,388	3,872	4,356	4,840	5,324	5,808	6,292	6,776	7,260	7,744	9,680
23	3,174	3,703	4,232	4,761	5,290	5,819	6,348	6,877	7,406	7,935	8,464	10,580
24	3,456	4,032	4,608	5,184	5,760	6,336	6,912	7,488	8,064	8,640	9,216	11,520
28	4,704	5,488	6,272	7,056	7,840	8,624	9,408	10,192	10,976	11,760	12,544	15,680
30	5,400	6,300	7,200	8,100	9,000	9,900	10,800	11,700	12,600	13,500	14,400	18,000
34	6,936	8,092	9,248	10,404	11,560	12,716	13,872	15,028	16,184	17,340	18,496	23,120
40	9,600	11,200	12,800	14,400	16,000	17,600	19,200	20,800	22,400	24,000	25,600	32,000

Table A.9 Rectangular Sections With Compression Steel
Minimum Steel Percentage $100(\rho - \rho')$ for Compression Steel to Yield

$$(\rho - \rho') \geq 0.85\beta_1 \frac{f'_c}{f_y} \times \frac{d'}{d} \times \frac{87}{(87 - f_y)} \; (f'_c \text{ and } f_y \text{ in Ksi})$$

f'_c (psi)	β_1	d'/d	f_y 40 Ksi	50 Ksi	60 Ksi	75 Ksi
3000	0.85	0.10	1.00	1.02	1.16	2.09
4000	0.85	0.10	1.33	1.35	1.55	2.78
5000	0.80	0.10	1.57	1.59	1.81	3.27
6000	0.75	0.10	1.78	1.81	2.06	3.71
3000	0.85	0.12	1.20	1.22	1.39	2.51
4000	0.85	0.12	1.60	1.62	1.86	3.34
5000	0.80	0.12	1.88	1.91	2.17	3.92
6000	0.75	0.12	2.14	2.17	2.47	4.45
3000	0.85	0.15	1.50	1.53	1.74	3.14
4000	0.85	0.15	2.00	2.03	2.33	4.17
5000	0.80	0.15	2.36	2.39	2.72	4.91
6000	0.75	0.15	2.67	2.72	3.09	5.57

Note: Minimum $(\rho - \rho')$ for any value of $d'/d = 10 \times (d'/d) \times$ value shown in table with $d'/d = 0.10$.

Table A.10 Modulus of Elasticity of Concrete, E_c (Ksi)

Concrete Cylinder Strength (f'_c)	Unit Weight of Concrete (psi) 90	100	110	125	145
3000	1540	1800	2080	2520	3150
4000	1780	2090	2410	2920	3640
5000	1990	2330	2690	3260	4060
6000	2185	2560	2950	3580	4500
7000	2360	2760	3190	3870	4800
8000	2520	2950	3410	4130	5200

Note: $E_c = 33\, W^{1.5}\sqrt{f'_c}$
$E_c = 57{,}000\sqrt{f'_c}$ for $W = 145$ pcf (normal-weight concrete)

Table A.11(a) Values of ℓ_d/d_b for Various Values of f'_c and f_y (Tension Bars)

	$f_y = 40$ Ksi				$f_y = 60$ Ksi			
	≤ No. 6 Bars		≥ No. 7 Bars		≤ No. 6 Bars		≥ No. 7 Bars	
f'_c (Ksi)	Conditions Met	Others	Conditions Met	Others	Conditions Met	Others	Conditions Met	Others
3	29.3	43.9	36.6	54.8	43.9	65.8	54.8	82.2
4	25.3	38.0	31.7	47.5	38.0	57.0	47.5	71.2
5	22.7	34.0	28.3	42.5	34.0	51.0	42.5	63.7
6	20.7	31.0	25.9	38.8	31.0	46.5	38.8	58.1

Table A.11(b) Development Length ℓ_d for Tension Bars and $f_y = 60$ Ksi ($\alpha = \beta = \lambda = 1.0$)

		Development Length ℓ_d (in.)—Tension Bars			
		$f'_c = 3$ Ksi		$f'_c = 4$ Ksi	
Bar Number	Bar Diameter (in.)	Conditions Met	Others	Conditions Met	Others
3	0.375	17	25	15	21
4	0.500	22	33	19	29
5	0.625	28	41	24	36
6	0.750	33	50	29	43
7	0.875	48	72	42	63
8	1.000	55	83	48	72
9	1.128	62	93	54	81
10	1.270	70	105	61	92
11	1.410	78	116	68	102

Table A.12 Designations, Areas, Perimeters, and Weights of Standard U.S. Bars

Bar No.	Diameter (in.)	Cross-Sectional Area (in.^{2})	Perimeter (in.)	Unit Weight Per Foot (lb)	Diameter (mm)	Area (mm^2)
2	$\frac{1}{4}$ = 0.250	0.05	0.79	0.167	6.4	32
3	$\frac{3}{8}$ = 0.375	0.11	1.18	0.376	9.5	71
4	$\frac{1}{2}$ = 0.500	0.20	1.57	0.668	12.7	129
5	$\frac{5}{8}$ = 0.625	0.31	1.96	1.043	15.9	200
6	$\frac{3}{4}$ = 0.750	0.44	2.36	1.502	19.1	284
7	$\frac{7}{8}$ = 0.875	0.60	2.75	2.044	22.2	387
8	1 = 1.000	0.79	3.14	2.670	25.4	510
9	$1\frac{1}{8}$ = 1.128	1.00	3.54	3.400	28.7	645
10	$1\frac{1}{4}$ = 1.270	1.27	3.99	4.303	32.3	820
11	$1\frac{3}{8}$ = 1.410	1.56	4.43	5.313	35.8	1010
14	$1\frac{3}{4}$ = 1.693	2.25	5.32	7.650	43.0	1450
18	$2\frac{1}{4}$ = 2.257	4.00	7.09	13.600	57.3	2580

Table A.13 Areas of Groups of Standard U.S. Bars in Square Inches

Bar Number	Number of Bars											
	1	*2*	*3*	*4*	*5*	*6*	*7*	*8*	*9*	*10*	*11*	*12*
3	0.11	0.22	0.33	0.44	0.55	0.66	0.77	0.88	1.00	1.10	1.21	1.32
4	0.20	0.39	0.58	0.78	0.98	1.18	1.37	1.57	1.77	1.96	2.16	2.36
5	0.31	0.61	0.91	1.23	1.53	1.84	2.15	2.45	2.76	3.07	3.37	3.68
6	0.44	0.88	1.32	1.77	2.21	2.65	3.09	3.53	3.98	4.42	4.84	5.30
7	0.60	1.20	1.80	2.41	3.01	3.61	4.21	4.81	5.41	6.01	6.61	7.22
8	0.79	1.57	2.35	3.14	3.93	4.71	5.50	6.28	7.07	7.85	8.64	9.43
9	1.00	2.00	3.00	4.00	5.00	6.00	7.00	8.00	9.00	10.00	11.00	12.00
10	1.27	2.53	3.79	5.06	6.33	7.59	8.86	10.12	11.39	12.66	13.92	15.19
11	1.56	3.12	4.68	6.25	7.81	9.37	10.94	12.50	14.06	15.62	17.19	18.75
14	2.25	4.50	6.75	9.00	11.25	13.50	15.75	18.00	20.25	22.50	24.75	27.00
18	4.00	8.00	12.00	16.00	20.00	24.00	28.00	32.00	36.00	40.00	44.00	48.00

Table A.14 Areas of Bars in Slabs (square inches per foot)

Spacing (in.)	Bar Number								
	3	*4*	*5*	*6*	*7*	*8*	*9*	*10*	*11*
3	0.44	0.78	1.23	1.77	2.40	3.14	4.20	5.06	6.25
$3\frac{1}{2}$	0.38	0.67	1.05	1.51	2.06	2.69	3.43	4.34	5.36
4	0.33	0.59	0.92	1.32	1.80	2.36	3.00	3.80	4.68
$4\frac{1}{2}$	0.29	0.52	0.82	1.18	1.60	2.09	2.67	3.37	4.17
5	0.26	0.47	0.74	1.06	1.44	1.88	2.40	3.04	3.75
$5\frac{1}{2}$	0.24	0.43	0.67	0.96	1.31	1.71	2.18	2.76	3.41
6	0.22	0.39	0.61	0.88	1.20	1.57	2.00	2.53	3.12
$6\frac{1}{2}$	0.20	0.36	0.57	0.82	1.11	1.45	1.85	2.34	2.89
7	0.19	0.34	0.53	0.76	1.03	1.35	1.71	2.17	2.68
$7\frac{1}{2}$	0.18	0.31	0.49	0.71	0.96	1.26	1.60	2.02	2.50
8	0.17	0.29	0.46	0.66	0.90	1.18	1.50	1.89	2.34
9	0.15	0.26	0.41	0.59	0.80	1.05	1.33	1.69	2.08
10	0.13	0.24	0.37	0.53	0.72	0.94	1.20	1.52	1.87
12	0.11	0.20	0.31	0.44	0.60	0.79	1.00	1.27	1.56

Table A.15 Common Styles of Welded Wire Fabric

Style Designation	Steel Area (in.^{2}/ft)		Weight Approx. lb/100 ft^2
	Longitudinal	*Transverse*	
6 × 6—W1.4 × W1.4	0.03	0.03	21
6 × 6—W2 × W2	0.04	0.04	29
6 × 6—W2.9 × W2.9	0.06	0.06	42
6 × 6—W4 × W4	0.08	0.08	58
6 × 6—W5.5 × W5.5	0.11	0.11	80
4 × 4—W1.4 × W1.4	0.04	0.04	31
4 × 4—W2 × W2	0.06	0.06	43
4 × 4—W2.9 × W2.9	0.09	0.09	62
4 × 4—W4 × W4	0.12	0.12	86

Table A.16 Size and Pitch of Spirals

f_y (Ksi)	Diameter of Column (in.)	Outside to Outside of Spiral (in.)	f'_c (psi)		
			3000	*4000*	*5000*
40	14, 15	11, 12	$\frac{3}{8} - 1\frac{3}{4}$	$\frac{1}{2} - 2\frac{1}{2}$	$\frac{1}{2} - 1\frac{3}{4}$
	16	13	$\frac{3}{8} - 1\frac{3}{4}$	$\frac{1}{2} - 2\frac{1}{2}$	$\frac{1}{2} - 2$
	17–19	14–16	$\frac{3}{8} - 1\frac{3}{4}$	$\frac{1}{2} - 2\frac{1}{2}$	$\frac{1}{2} - 2$
	20–23	17–20	$\frac{3}{8} - 1\frac{3}{4}$	$\frac{1}{2} - 2\frac{1}{2}$	$\frac{1}{2} - 2$
	24–30	21–27	$\frac{3}{8} - 2$	$\frac{1}{2} - 2\frac{1}{2}$	$\frac{1}{2} - 2$
60	14, 15	11, 12	$\frac{3}{8} - 2\frac{3}{4}$	$\frac{3}{8} - 2$	$\frac{1}{2} - 2\frac{3}{4}$
	16–23	13–20	$\frac{3}{8} - 2\frac{3}{4}$	$\frac{3}{8} - 2$	$\frac{1}{2} - 2\frac{3}{4}$
	24–29	21–26	$\frac{3}{8} - 3$	$\frac{3}{8} - 2\frac{1}{4}$	$\frac{1}{2} - 3$
	30	27	$\frac{3}{8} - 3$	$\frac{3}{8} - 2\frac{1}{4}$	$\frac{1}{2} - 3\frac{1}{4}$

APPENDIX B

DESIGN TABLES (SI UNITS)

Table B.1 Values of R_u and a/d for $f'_c = 21$ MPa (R_u in MPa)

100 ρ	$f_y = 280$ MPa R_u	$f_y = 280$ MPa a/d	$f_y = 350$ MPa R_u	$f_y = 350$ MPa a/d	$f_y = 420$ MPa R_u	$f_y = 420$ MPa a/d	$f_y = 520$ MPa R_u	$f_y = 520$ MPa a/d
0.2	0.50	0.031	0.62	0.039	0.75	0.047	0.92	0.059
0.3	0.74	0.046	0.92	0.059	1.10	0.071	1.35	0.089
0.4	0.98	0.062	1.22	0.078	1.45	0.094	1.79	0.118
0.5	1.21	0.078	1.50	0.098	1.79	0.118	2.18	0.148
0.6	1.45	0.094	1.79	0.118	2.12	0.141	2.59	0.177
0.7	1.68	0.110	2.06	0.138	2.44	0.165	2.96	0.207
0.8	1.90	0.126	2.33	0.157	2.75	0.189	3.34	0.236
0.9	2.12	0.142	2.59	0.177	3.05	0.213	3.68	0.266
1.0	2.33	0.157	2.84	0.196	3.35	0.238	4.02	0.295
1.1	2.55	0.173	3.10	0.216	3.64	0.260	4.36	0.325
1.2	2.74	0.188	3.35	0.235	3.91	0.280	4.56	0.340
1.3	2.95	0.204	3.59	0.255	4.18	0.306	($\rho_{max} = 1.16$)	
1.4	3.16	0.220	3.82	0.274	4.44	0.330		
1.5	3.35	0.236	4.04	0.294	4.69	0.353		
1.6	3.54	0.252	4.27	0.314	4.92	0.376		
1.7	3.73	0.267	4.48	0.334	4.94	0.378		
1.8	3.91	0.282	4.69	0.353	($\rho_{max} = 1.61$)			
1.9	4.09	0.298	4.89	0.373				
2.0	4.27	0.314	5.08	0.392				
2.1	4.43	0.330	5.20	0.405				
			($\rho_{max} = 2.06$)					
2.2	4.61	0.345						
2.3	4.76	0.360						
2.4	4.92	0.376						
2.5	5.08	0.392						
2.6	5.24	0.408						
2.7	5.39	0.424						
2.8	5.50	0.436						
2.9	($\rho_{max} = 2.78$)							

Note: Last values are the maximum allowed by the ACI Code.

Table B.2 Values of R_u and a/d for $f'_c = 28$ MPa (R_u in MPa)

100 ρ	$f_y = 280$ MPa R_u	$f_y = 280$ MPa a/d	$f_y = 350$ MPa R_u	$f_y = 350$ MPa a/d	$f_y = 420$ MPa R_u	$f_y = 420$ MPa a/d	$f_y = 520$ MPa R_u	$f_y = 520$ MPa a/d
0.2	0.50	0.024	0.63	0.029	0.75	0.025	0.93	0.044
0.3	0.74	0.036	0.93	0.044	1.11	0.053	1.36	0.066
0.4	0.98	0.047	1.23	0.059	1.46	0.071	1.81	0.088
0.5	1.23	0.059	1.53	0.074	1.81	0.089	2.23	0.110
0.6	1.46	0.071	1.83	0.088	2.16	0.106	2.66	0.132
0.7	1.70	0.083	2.11	0.103	2.50	0.123	3.05	0.154
0.8	1.93	0.094	2.39	0.118	2.81	0.141	3.45	0.176
0.9	2.16	0.106	2.66	0.132	2.14	0.158	3.83	0.198
1.0	2.39	0.118	2.95	0.147	3.46	0.176	4.22	0.220
1.1	2.60	0.130	3.20	0.161	3.77	0.194	4.57	0.242
1.2	2.81	0.141	3.46	0.176	4.08	0.212	4.94	0.264
1.3	3.04	0.153	3.73	0.191	4.36	0.230	5.29	0.286
1.4	3.25	0.165	3.97	0.206	4.65	0.247	5.63	0.308
1.5	3.46	0.177	4.22	0.221	4.92	0.265	5.95	0.330
1.6	3.67	0.188	4.46	0.236	5.22	0.282	6.10	0.342
							($\rho_{max} = 1.55$)	
1.7	3.87	0.200	4.71	0.250	5.48	0.300		
1.8	4.08	0.212	4.94	0.265	5.75	0.318		
1.9	4.27	0.224	5.17	0.280	6.00	0.335		
2.0	4.46	0.236	5.40	0.294	6.26	0.353		
2.1	4.65	0.248	5.62	0.309	6.50	0.370		
2.2	4.85	0.260	5.84	0.323	6.58	0.378		
2.3	5.04	0.271	6.05	0.338	($\rho_{max} = 2.14$)			
2.4	5.22	0.282	6.26	0.352				
2.5	5.39	0.294	6.43	0.367				
2.6	5.57	0.306	6.65	0.381				
2.7	5.74	0.318	6.85	0.396				
2.8	5.92	0.330	6.95	0.405				
2.9	6.09	0.342	($\rho_{max} = 2.75$)					
3.0	6.26	0.354						
3.1	6.42	0.365						
3.2	6.57	0.376						
3.3	6.74	0.388						
3.4	6.89	0.400						
3.5	7.03	0.412						
3.6	7.17	0.424						
3.7	7.31	0.436						
	7.34	0.438						
	($\rho_{max} = 3.72$)							

Note: Last values are the maximum allowed by the ACI Code.

Table B.3 Values of R_u and a/d for $f'_c = 35$ MPa (R_u in MPa)

100 ρ	$f_y = 350$ MPa R_u	$f_y = 350$ MPa a/d	$f_y = 420$ MPa R_u	$f_y = 420$ MPa a/d	$f_y = 520$ MPa R_u	$f_y = 520$ MPa a/d
0.2	0.63	0.024	0.75	0.028	0.93	0.035
0.3	0.93	0.036	1.12	0.042	1.38	0.052
0.4	1.24	0.047	1.48	0.056	1.83	0.070
0.5	1.53	0.060	1.83	0.070	2.26	0.088
0.6	1.83	0.071	2.18	0.085	2.70	0.106
0.7	2.12	0.083	2.53	0.100	3.11	0.123
0.8	2.41	0.094	2.87	0.113	3.52	0.141
0.9	2.69	0.106	3.20	0.127	3.91	0.159
1.0	2.98	0.118	3.53	0.141	4.30	0.177
1.1	3.26	0.130	3.87	0.155	4.69	0.195
1.2	3.52	0.141	4.17	0.169	5.08	0.212
1.3	3.80	0.153	4.48	0.183	5.46	0.230
1.4	4.06	0.165	4.79	0.198	5.84	0.247
1.5	4.32	0.177	5.09	0.212	6.15	0.265
1.6	4.58	0.188	5.39	0.226	6.47	0.282
1.7	4.84	0.200	5.68	0.240	6.82	0.300
1.8	5.09	0.212	5.96	0.254	7.17	0.318
1.9	5.34	0.224	6.26	0.268	7.26	0.322
2.0	5.58	0.235	6.52	0.282	($\rho_{max} = 1.83$)	
2.1	5.84	0.247	6.78	0.296		
2.2	6.06	0.259	7.05	0.311		
2.3	6.29	0.271	7.31	0.325		
2.4	6.52	0.282	7.56	0.339		
2.5	6.74	0.294	7.85	0.352		
2.6	6.96	0.306	7.87	0.354		
			($\rho_{max} = 2.52$)			
2.7	7.18	0.318				
2.8	7.40	0.329				
2.9	7.61	0.341				
3.0	7.81	0.353				
3.1	8.01	0.365				
3.2	8.21	0.376				
	8.29	0.381				
	($\rho_{max} = 3.24$)					

Note: Last values are the maximum allowed by the ACI Code.

Table B.4 Values of ρ_{max}, $R_{u\,max}$, ρ_b, and ρ_{min}

$$\rho_{max} = \left(0.64\beta_1 \frac{f'_c}{f_y} \times \frac{600}{600 + f_y}\right), \quad R_u = \phi\rho f_y\left(1 - \frac{\rho f_y}{1.7 f'_c}\right) \quad (R_u \text{ in MPa})$$

f'_c (MPa)	$f_y = 280$ MPa				$f_y = 350$ MPa			
	100 ρ_{max}	$R_{u\,max}$ (psi)	100 ρ_b	100 ρ_{min}	100 ρ_{max}	$R_{u\,max}$ (psi)	100 ρ_b	100 ρ_{min}
21.0	2.78	5.50	3.71	0.5	2.06	5.20	2.75	0.4
28.0	3.72	7.34	4.96	0.5	2.75	6.95	3.67	0.4
35.0	4.37	8.77	5.81	0.53	3.24	8.29	4.32	0.42
42.0	4.9	10.0	6.53	0.58	3.64	9.46	4.85	0.47

f'_c (MPa)	$f_y = 420$ MPa				$f_y = 520$ MPa			
	100 ρ_{max}	$R_{u\,max}$ (psi)	100 ρ_b	100 ρ_{min}	100 ρ_{max}	$R_{u\,max}$ (psi)	100 ρ_b	100 ρ_{min}
21.0	1.61	4.94	2.15	0.33	1.16	4.56	1.55	0.267
28.0	2.14	6.58	2.85	0.33	1.55	6.10	2.07	0.267
35.0	2.52	7.87	3.39	0.35	1.83	7.26	2.44	0.283
42.0	2.83	9.00	3.77	0.39	2.06	8.3	2.75	0.310

Table B.5 Suggested Design Steel Ratios, ρ_s, and Comparison with Other Steel Ratios

f'_c (MPa)	f_y (MPa)	100 ρ_b	100 ρ_{max}	100ρ_s (suggested)	R_u For ρ_s (MPa)	100ρ_b (elastic)	Ratio ρ_s/ρ_b	Ratio ρ_s/ρ_{max}	Weight of ρ_s (kg/m^3 of concrete)
21.0	280	3.71	2.78	1.4	3.16	1.29	0.377	0.50	112
	350	2.75	2.06	1.2	3.35	1.11	0.436	0.58	96
	420	2.15	1.61	1.2	3.91	0.96	0.558	0.74	96
28.0	280	4.96	3.72	1.4	3.25	1.58	0.282	0.38	112
	350	3.67	2.75	1.4	3.97	1.62	0.381	0.51	112
	420	2.85	2.52	1.4	4.65	1.41	0.491	0.65	112
35.0	280	5.81	4.37	1.4	3.30	2.50	0.241	0.32	112
	350	4.32	3.24	1.4	4.06	2.15	0.324	0.43	112
	420	3.35	2.52	1.4	4.79	1.87	0.418	0.55	112

Table B.6 Minimum Thickness of Beams and One-Way Slabs

Member	Yield Strength f_y (MPa)	Simply Supported	One End Continuous	Both Ends Continuous	Cantilever
Solid one-way slabs	280	$L/25$	$L/30$	$L/35$	$L/12.5$
	350	$L/22$	$L/27$	$L/31$	$L/11$
	420	$L/20$	$L/24$	$L/28$	$L/10$
Beams or ribbed one-way slabs	280	$L/20$	$L/23$	$L/26$	$L/10$
	350	$L/18$	$L/20.5$	$L/23.5$	$L/9$
	420	$L/16$	$L/18.5$	$L/21$	$L/8$

Table B.7 Rectangular Sections with Compression Steel. Minimum Steel Percentage $100(\rho - \rho')$ for Compression Steel to Yield

$$(\rho - \rho') \geq 0.85\beta_1 \left(\frac{f'_c}{f_y}\right) \times \left(\frac{d'}{d}\right) \times \frac{600}{(600 - f_y)}, \qquad (f_y, f'_c \text{ in MPa})$$

f'_c MPa	β_1	d'/d	f_y 300 MPa	f_y 400 MPa	f_y 500 MPa
21	0.85	0.10	1.20	1.35	2.16
28	0.85	0.10	1.45	1.63	2.60
35	0.80	0.10	1.59	1.80	2.85
42	0.75	0.10	1.70	1.91	3.06
21	0.85	0.12	1.45	1.63	2.60
28	0.85	0.12	1.73	1.95	3.12
35	0.80	0.12	2.02	2.27	3.64
42	0.75	0.12	2.04	2.29	3.67
21	0.85	0.15	1.81	2.03	3.25
28	0.85	0.15	2.17	2.44	3.90
35	0.80	0.15	2.38	2.68	4.28
42	0.75	0.15	2.55	2.87	4.59

Note: Minimum $(\rho - \rho')$ for any value of $d'/d = 10 \times (d'/d) \times$ value shown in table with $d'/d = 0.10$.

Table B.8 Modulus of Elasticity of Normal-weight Concrete

General: $E_c = 0.043 W^{1.5}\sqrt{f_c}$ **MPa**

For normal-wieght concrete,
$W = 2350$ **kg/m**^{3}**:**

$$E_c = 4730\sqrt{f'_c}\ MPa$$

f'_c MPa	E_c (kN/mm^2)
17.5	20.0
21.0	22.5
28.0	25.0
35.0	29.0
42.0	32.0
49.0	33.5
56.0	36.5

Table B.9(a) Values of ℓ_d/d_b for Various Values of f'_c and f_y (Tension Bars)

	$f_y = 300$ MPa				$f_y = 400$ MPa			
	≤ *20 M Bars*		≥ *25 M Bars*		≤ *20 M Bars*		≥ *25 M Bars*	
f'_c MPa	*Conditions Met*	*Others*	*Conditions Met*	*Others*	*Conditions Met*	*Others*	*Conditions Met*	*Others*
20	34.0	50.5	42.0	63.0	45.0	67.0	56.0	84.0
30	27.5	41.5	34.5	51.5	36.5	55.0	46.0	68.5
35	25.5	38.5	32.0	47.5	34.0	51.0	42.5	63.5
40	23.5	35.5	29.5	44.5	31.5	47.5	39.5	59.5

Table B.9(b) Development Length ℓ_d for Tension Bars and $f_y = 400$ MPa ($\alpha = \beta = \lambda = 1.0$)

		Development Length ℓ_d (mm)—Tension Bars			
		$f'_c = 20$ MPa		$f'_c = 30$ MPa	
Bar Number	Bar Diameter (mm)	*Conditions Met*	*Others*	*Conditions Met*	*Others*
10M	11.3	510	765	415	620
15M	16.0	720	1080	585	875
20M	19.5	880	1320	710	1070
25M	25.2	1410	2120	1160	1740
30M	29.9	1675	2510	1375	2065
35M	35.7	2000	3000	1640	2465

Table B.10 Designations, Areas, and Mass of Bars

Bar No.	Nominal Dimensions: *Diameter* (mm)	*Area* (mm^2)	*Mass* (kg/m)
#10	9.5	71	0.560
#13	12.7	129	0.994
#16	15.9	199	1.552
#19	19.1	284	2.235
#22	22.2	387	3.042
#25	25.4	510	3.973
#29	28.7	645	5.060
#32	32.3	819	6.404
#36	35.8	1006	7.907
#43	43.0	1452	11.38
#57	57.3	2581	20.24

ASTM A615M Grade 300 is limited to sizes #10 through #19, otherwise, grades are 400 or 500 MPa (these bars are soft conversion of #3 to #18 in US customary units).

Table B.11 ASTM Standard Metric Reinforcing Bars

Bar-size Designation (number)	Nominal Dimensions: *Diameter* (mm)	*Area* (mm^2)	*Mass* (kg/m)
10M	11.3	100	0.785
15M	16.0	200	1.570
20M	19.5	300	2.355
25M	25.2	500	3.925
30M	29.9	700	5.495
35M	35.7	1000	7.850
45M	43.7	1500	11.775
55M	56.4	2500	19.625

ASTM A615M grade 300 is limited to size 10M through no. 20M; otherwise grades are 400 or 500 MPa for all other sizes.

Table B.12 Areas of Group of Bars (mm^2)—Metric

Bar No. Metric	Number of Bars: *1*	*2*	*3*	*4*	*5*	*6*	*7*	*8*	*9*	*10*
#10	71	142	213	384	355	426	497	568	639	710
#13	129	258	387	516	645	774	903	1032	1161	1290
#16	199	398	597	796	995	1194	1393	1592	1791	1990
#19	284	568	852	1136	1420	1704	1988	2272	2556	2840
#22	387	774	1161	1548	1935	2322	2709	3096	3483	3870
#25	510	1020	1530	2040	2550	3060	3570	4080	4590	5100
#29	645	1290	1935	2580	3225	3870	4515	5160	5805	6450
#32	819	1638	2457	3276	4095	4914	5733	6552	7371	8190
#36	1006	2012	3018	4024	5030	6036	7042	8048	9054	10060

APPENDIX C

STRUCTURAL AIDS

Note: S.D. stands for shearing force diagram. B.D. stands for bending moment diagram. Bending moments are drawn on the tension sides of beams.
Moments, shearing forces and deflections, for any combination of loadings, are obtained by superposition.

Table C.1. Simple Beams (w = load/unit length)

1. Uniform load

W = total load = wL

$$R_A = R_B = V_A = V_B = \frac{W}{2}$$

$$M_x = \frac{Wx}{2}\left(1 - \frac{x}{L}\right)$$

$$M_{\max} = \frac{WL}{8} \quad \text{(at center)}$$

$$\Delta_{\max} = \frac{5}{384} \times \frac{WL^3}{EI} \quad \text{(at center)}$$

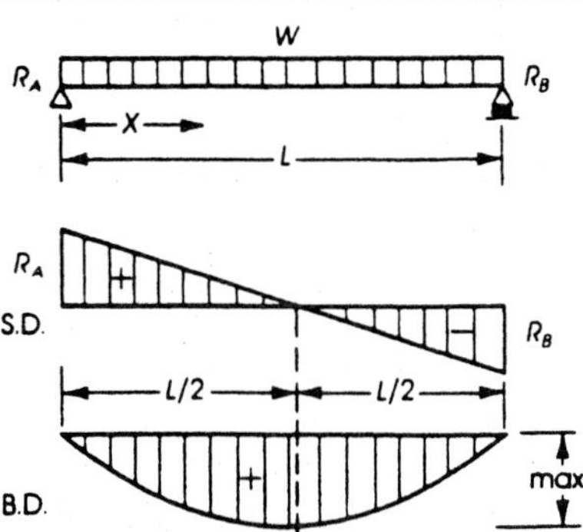

2. Uniform partial load

W = total load = wb

$$R_A = V_A = \frac{W}{L}\left(\frac{b}{2} + c\right)$$

$$R_B = V_B = \frac{W}{L}\left(\frac{b}{2} + a\right)$$

$$M_{\max} = \frac{W}{2b}(x^2 - a^2) \quad \text{when } x = a + \frac{R_A b}{W}$$

$$\Delta_{\max} = \frac{W}{384EI}(8L^3 - 4Lb^2 + b^3) \quad \text{when } a = c$$

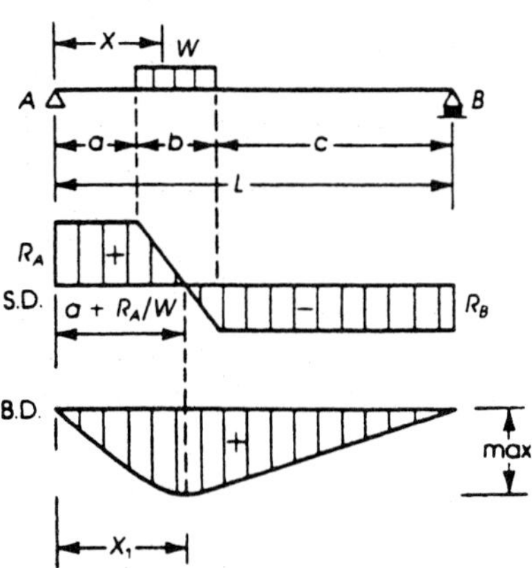

3. Uniform partial load at one end

W = total load = wa

$$R_A = V_A = W\left(1 - \frac{a}{2L}\right)$$

$$R_B = V_B = \frac{Wa}{2L}$$

$$M_{\max} = \frac{Wa}{2}\left(1 - \frac{a}{2L}\right)^2 \quad \text{when } x = a\left(1 - \frac{a}{2L}\right)$$

$$\Delta = \frac{WL^4}{24aEI}n^2[2m^3 - 6m^2 + m(4 + n^2) - n^2]$$

when $x \geq a$

$$\Delta = \frac{WL^4 m}{24aEI}[n^2(2 - n)^2 - 2nm^2(2 - n) + m^3] \quad \text{when } x < a$$

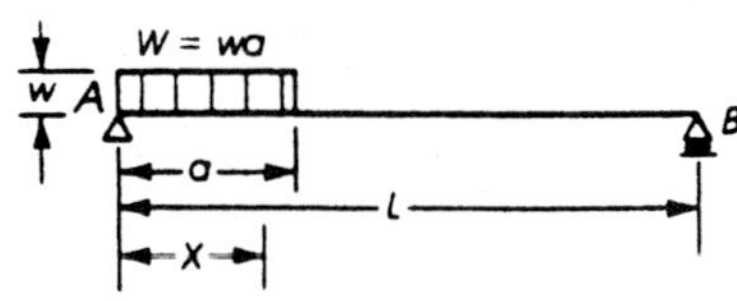

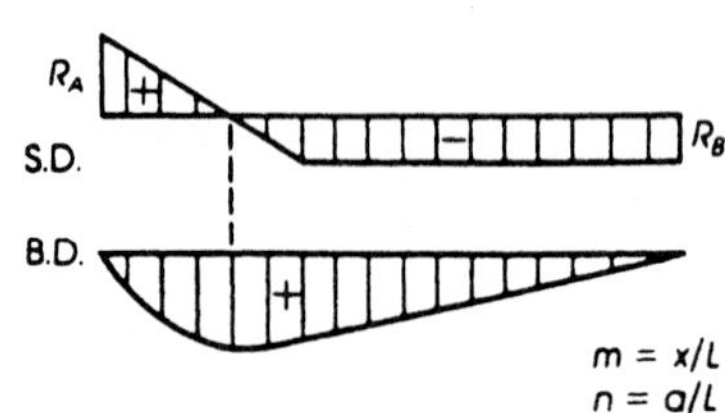

(continued)

Table C.1. *(continued)*

4. Triangular load on span with maximum value at one end

$$W = \text{total load} = \frac{wL}{2}$$

$$R_A = V_A = \frac{W}{3}$$

$$R_B = V_B = \frac{2W}{3}$$

$$M_x = \frac{Wx}{3}\left(1 - \frac{x^2}{L^2}\right)$$

$$M_{max} = 0.128WL \quad \text{when } x = 0.5774L$$

$$\Delta_{max} = \frac{0.01304WL^3}{EI} \quad \text{when } x = 0.5193L$$

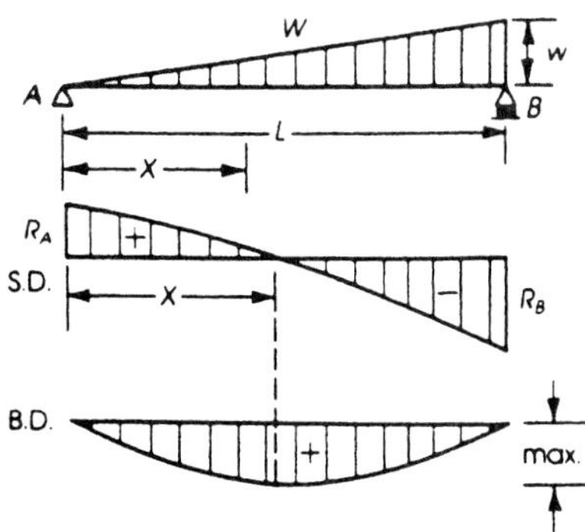

5. Triangular load with maximum value at midspan

$$W = \text{total load} = \frac{wL}{2}$$

$$R_A = R_B = V_A = V_B = \frac{W}{2}$$

$$M_x = Wx\left(\frac{1}{2} - \frac{2x^2}{3L^2}\right)$$

$$M_{max} = \frac{WL}{6} \quad \text{(at midspan)}$$

$$\Delta_{max} = \frac{WL^3}{60EI} \quad \text{(at midspan)}$$

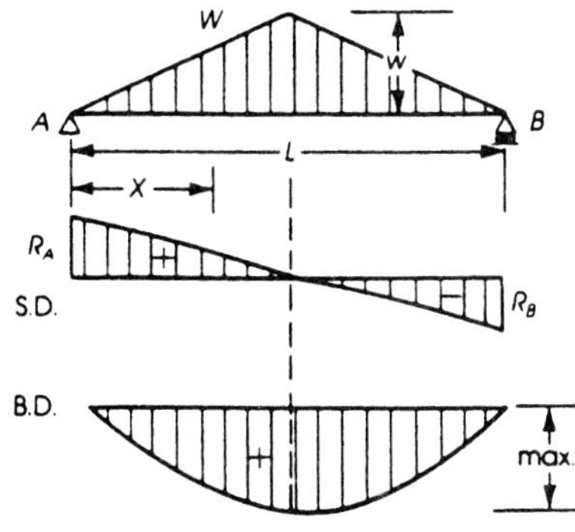

6. Moments at ends

$$R_A = R_B = V_A = V_B = \frac{M_A - M_B}{L}$$

$$\Delta_{max} \text{ (at midspan)} = \frac{ML^2}{8EI} \quad \text{when } M_A = M_B$$

$$\Delta \text{ (at midspan)} = \frac{M_A L^2}{16EI} \quad \text{when } M_B = 0$$

$$\Delta \text{ (at midspan)} = \frac{M_B L^2}{16EI} \quad \text{when } M_A = 0$$

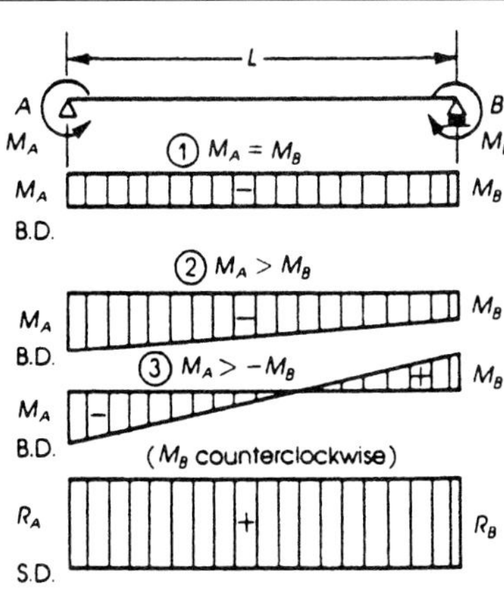

(continued)

Table C.1. *(continued)*

7. External moment at any point

$$R_A = -R_B = V_A = V_B = \frac{M}{L}$$

$$M_{CA} = \frac{Ma}{L}$$

$$M_{CB} = \frac{Mb}{L}$$

$$\Delta_c = \frac{-Mab}{3EIL}(a - b)$$

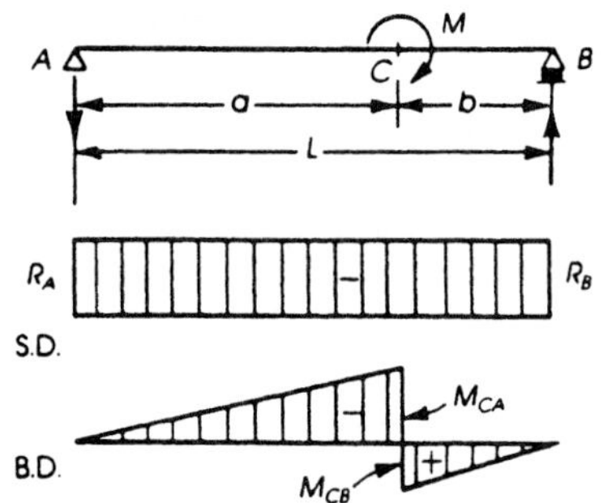

8. Concentrated load at midspan

$$R_A = R_B = V_A = V_B = \frac{P}{2}$$

$$M_{max} = \frac{PL}{4} \quad \text{(at midspan)}$$

$$\Delta_{max} = \frac{PL^3}{48EI} \quad \text{(at midspan)}$$

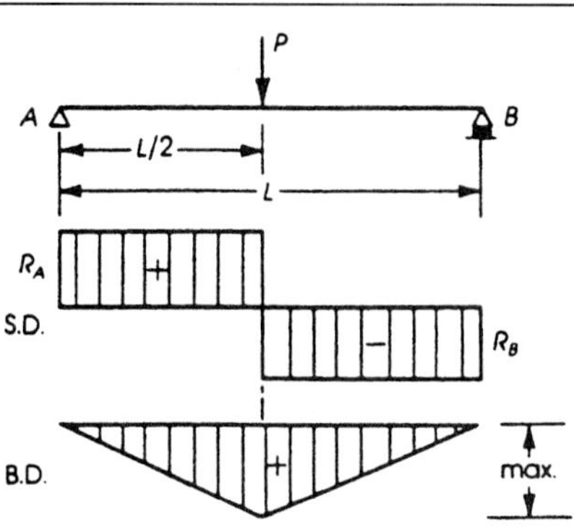

9. Concentrated load at any point

$$R_A = V_A = \frac{Pb}{L}$$

$$R_B = V_B = \frac{Pa}{L}$$

$$M_{max} = \frac{Pab}{L} \quad \text{(at point load)}$$

$$\Delta_C = \frac{Pa^2b^2}{3EIL} \quad \text{(at point load)}$$

$$\Delta_{max} = \frac{PL^3}{48EI}\left[\frac{3a}{L} - 4\left(\frac{a}{L}\right)^3\right] \quad \text{(when } a \geq b\text{)}$$

at $x = \sqrt{a(b + L)/3}$

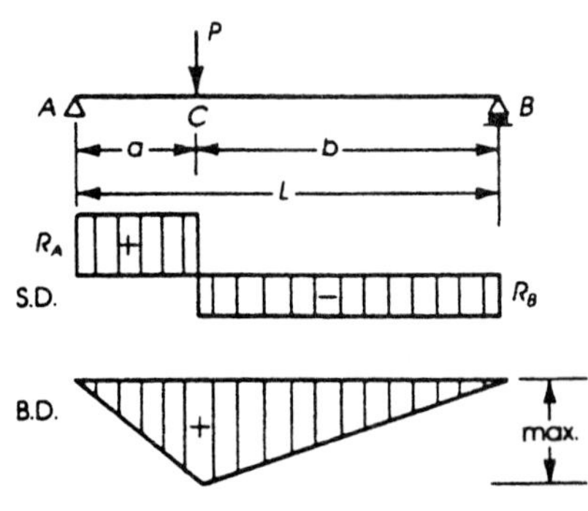

(continued)

Table C.1. *(continued)*

10. Two symmetrical concentrated loads

$$R_A = R_B = V_A = V_B = P$$

$$M_{\max} = Pa$$

$$\Delta_{\max} = \frac{PL^3}{6EI}\left[\frac{3a}{4L} - \left(\frac{a}{L}\right)^3\right] \quad \text{(at midspan)}$$

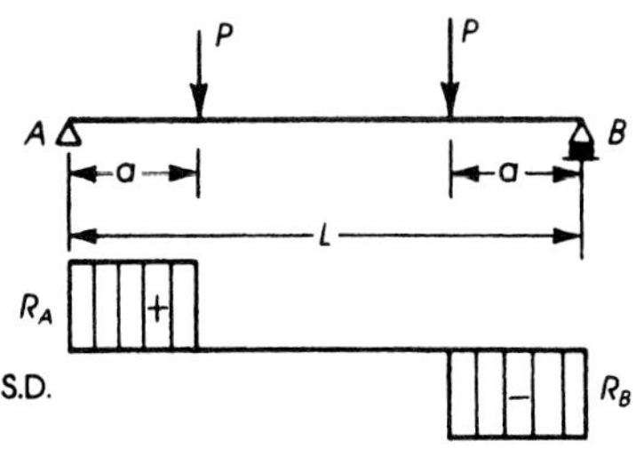

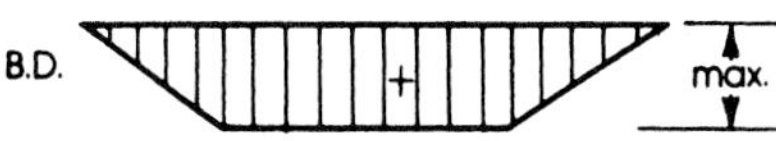

11. Two concentrated loads

$$R_A = V_A = \frac{P(b + 2c)}{L}$$

$$R_B = V_B = \frac{P(b + 2a)}{L}$$

$$M_C = \frac{Pa(b + 2c)}{L}$$

$$M_D = \frac{Pc(b + 2a)}{L}$$

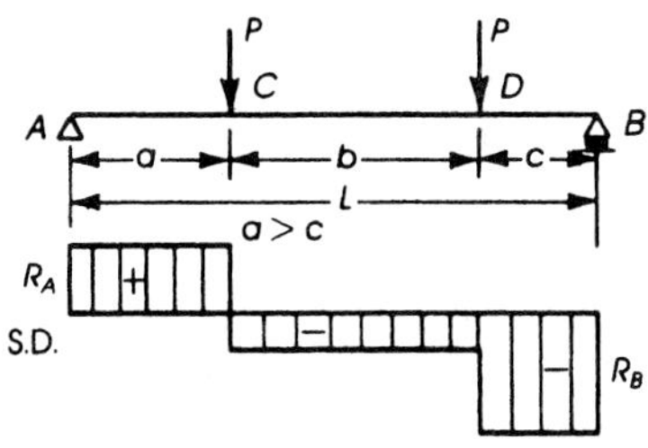

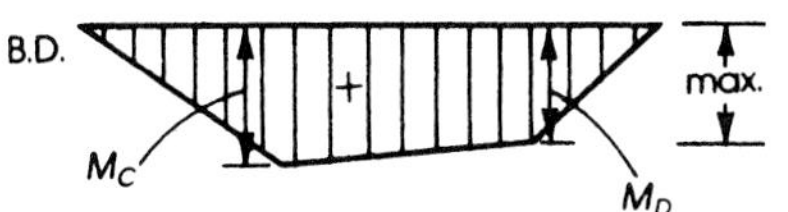

12. Two concentrated loads at one-third points

$$R_A = R_B = V_A = V_B = P$$

$$M_{\max} = \frac{PL}{3}$$

$$\Delta_{\max} = \frac{23PL^3}{648EI} \quad \text{(at midspan)}$$

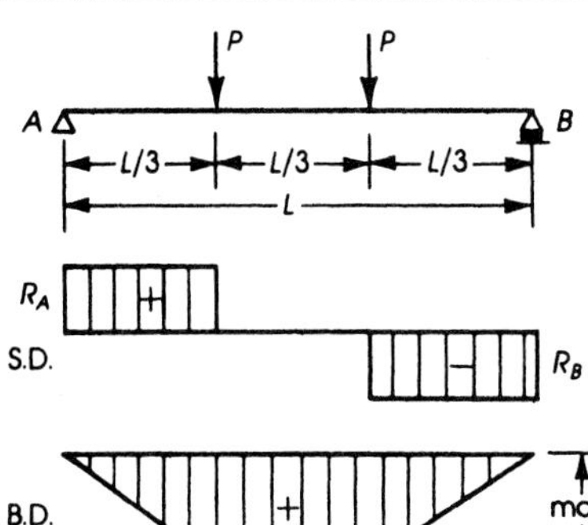

(continued)

Table C.1. *(continued)*

13. Three concentrated loads at one-fourth points

$$R_A = R_B = V_A = V_B = \frac{3P}{2}$$

$$M_C = M_E = \frac{3PL}{8}$$

$$M_D = \frac{PL}{2}$$

$$\Delta_{max} = \frac{19PL^3}{384EI} \quad \text{(at midspan)}$$

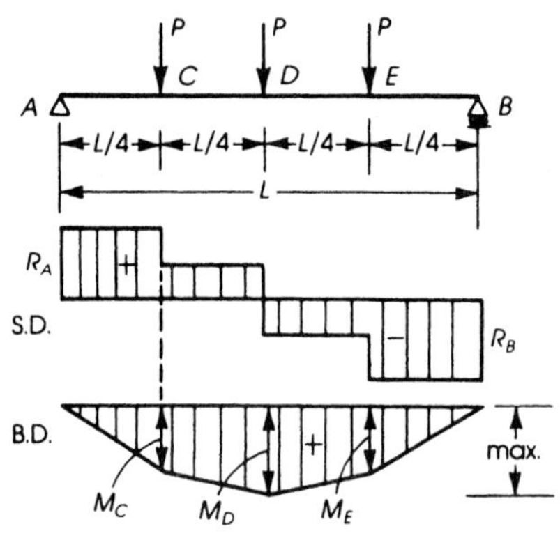

14. Three concentrated loads as shown

$$R_A = R_B = V_A = V_B = \frac{3P}{2}$$

$$M_C = M_E = \frac{PL}{4}$$

$$M_D = \frac{5PL}{12}$$

$$\Delta_{max} = \frac{53PL^3}{1296EI} \quad \text{(at midspan)}$$

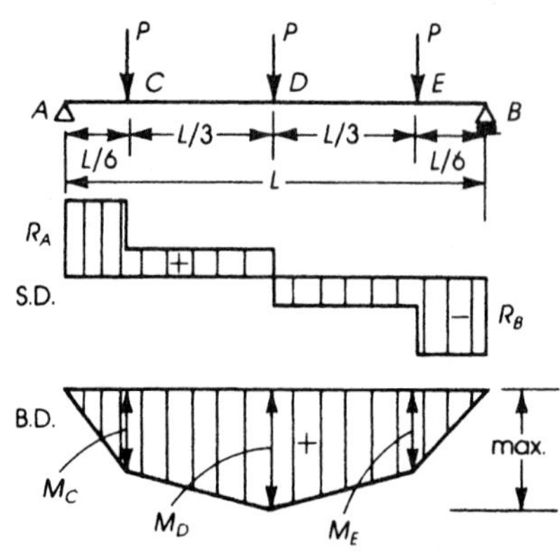

15. Uniformly distributed load and variable end moments

$$W = \text{total load} = wL$$

$$R_A = V_A = \frac{W}{2} + \frac{M_1 - M_2}{L}$$

$$R_B = V_B = \frac{W}{2} - \frac{M_1 - M_2}{L}$$

$$M_3 = \frac{WL}{8} - \frac{M_1 + M_2}{2} + \frac{(M_1 - M_2)^2}{2WL}$$

$$\text{at } x = \frac{L}{2} + \frac{M_1 - M_2}{W}$$

$$\Delta_x = \frac{Wx}{24EIL}\left[x^3 - \left(2L + \frac{4M_1}{W} - \frac{4M_2}{W}\right)x^2 + \frac{12M_1L}{W}x + L^3 - \frac{8M_1L^2}{W} - \frac{4M_2L^2}{W}\right]$$

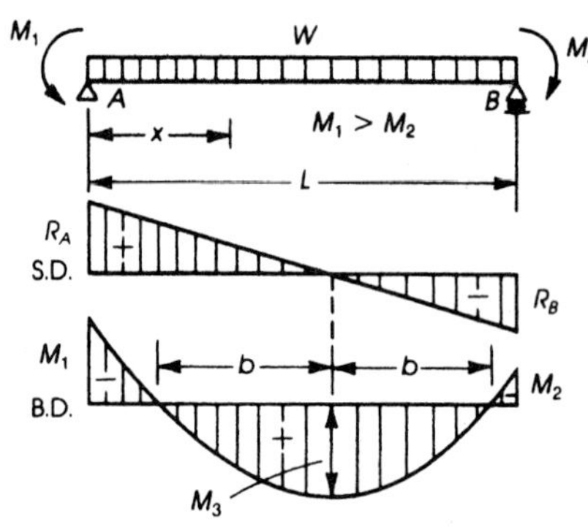

(continued)

Table C.1. *(continued)*

16. Concentrated load at center and variable end moments

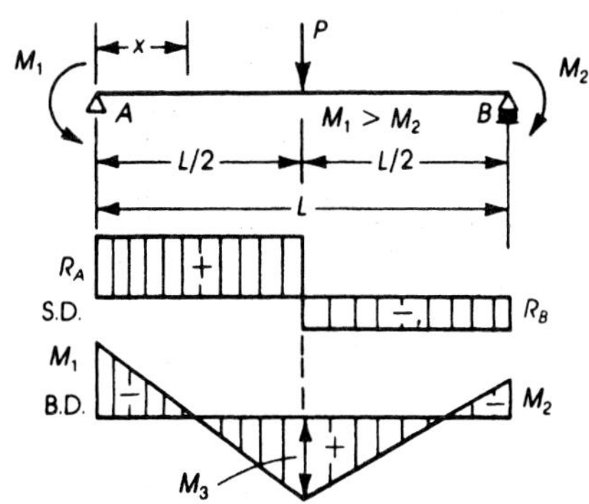

$$R_A = V_A = \frac{P}{2} + \frac{M_1 - M_2}{L}$$

$$R_B = V_B = \frac{P}{2} - \frac{M_1 - M_2}{L}$$

$$M_3 = \frac{PL}{4} - \frac{M_1 + M_2}{2} \quad \text{(at midspan)}$$

$$M_x = \left(\frac{P}{2} + \frac{M_1 - M_2}{L}\right)x - M_1 \quad \text{when } x < \frac{L}{2}$$

$$M_x = \frac{P}{2}(L - x) + \frac{(M_1 - M_2)}{L}x - M_1 \quad \text{when } x > \frac{L}{2}$$

$$\Delta_x = \frac{Px}{48EI}\left[3L^2 - 4x^2 - \frac{8(L - x)}{PL}\{M_1(2L - x) + M_2(L + x)\}\right] \quad \text{when } x < \frac{L}{2}$$

17. One concentrated moving load

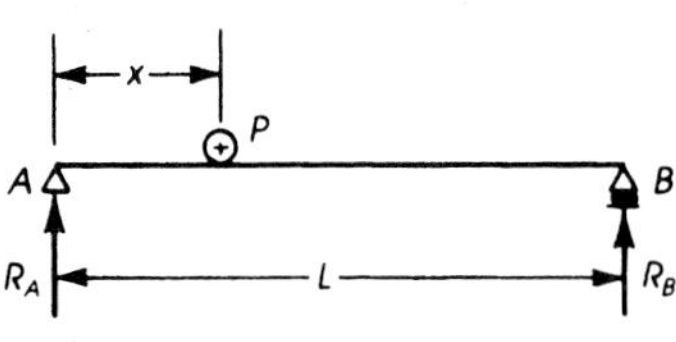

$$R_A\text{max} = V_A\text{max} = P \quad \text{at } x = 0$$

$$R_B\text{max} = V_A\text{max} = P \quad \text{at } x = L$$

$$M_{\text{max}} = \frac{PL}{4} \quad \text{at } x = \frac{L}{2}$$

$$M_x = \frac{P}{L}(L - x)x$$

18. Two equal concentrated moving loads

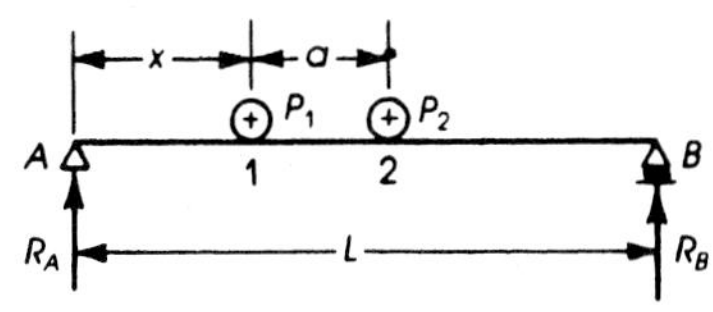

$$R_A\text{max} = V_A\text{max} = P\left(2 - \frac{a}{L}\right) \quad \text{at } x = 0$$

$$M_{\text{max}} = \frac{P}{2L}\left(L - \frac{a}{2}\right)^2$$

when $a < 0.586L$ under load 1 at $x = \frac{1}{2}\left(L - \frac{a}{2}\right)$

$$M_{\text{max}} = \frac{PL}{4} \quad \text{when } a > 0.5L \text{ with one load at midspan}$$

(continued)

Table C.1. *(continued)*

19. Two unequal concentrated moving loads

$$R_A\text{max} = V_A\text{max} = P_1 + P_2\left(\frac{L-a}{L}\right) \quad \text{at } x = 0$$

$$M_{\text{max}} = (P_1 + P_2)\frac{x^2}{L}$$

$$\text{under load } P_1 \text{ at } x = \frac{1}{2}\left(L - \frac{P_2 a}{P_1 + P_2}\right)$$

$$M_{\text{max}} = \frac{P_1 L}{4}$$ may occur with larger load at center of span and other load off span

$P_1 > P_2$

20. General rules for simple beams carrying moving concentrated loads

V_{max} occurs at one support and other loads on span (trial method)

For M_{max}: place center line of beam midway between center of gravity of loads and nearest concentrated load. M_{max} occurs under this load (here P_1)

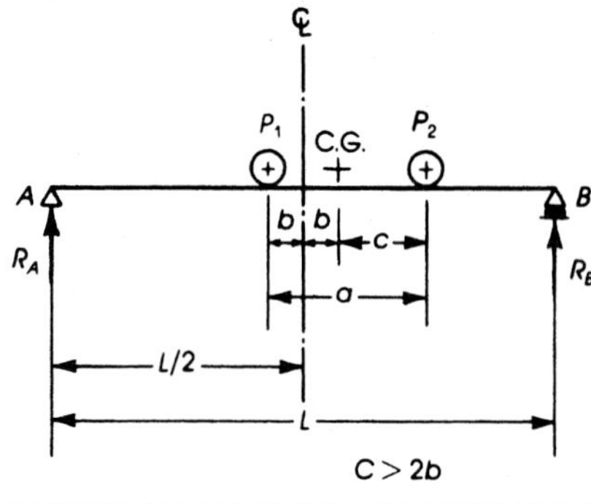

Table C.2. Cantilever Beams

21. Uniform load

W = total load = wL

$R_A = V_A = W$

$$M_A = \frac{WL}{2} \quad \text{(at support A)}$$

$$M_x = \frac{Wx^2}{2L}$$

$$\Delta_B\text{max} = \frac{WL^3}{8EI}$$

$$\Delta_x = \frac{W}{24EIL}(x^4 - 4L^3x + 3L^4)$$

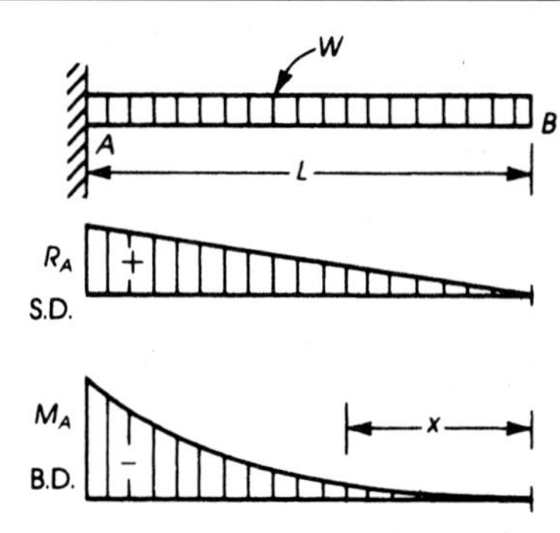

(continued)

Table C.2. *(continued)*

22. Partial uniform load starting from support

W = total load = wa

$R_A = V_A = W$

$M_A = \dfrac{Wa}{2}$ (at support A)

$M_x = \dfrac{Wx^2}{2a}$

$\Delta_C = \dfrac{Wa^3}{8EI}$

$\Delta_B \max = \dfrac{Wa^3}{8EI}\left(1 + \dfrac{4b}{3a}\right)$

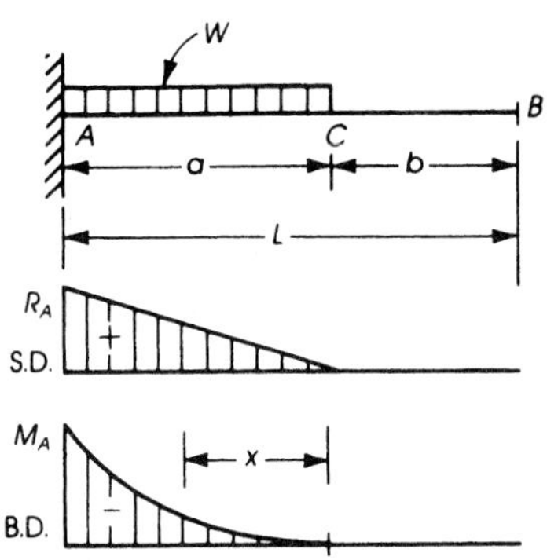

23. Concentrated load

$R_A = V_A = P$

$M_{max} = Pa$ (at support A)

$M_x = Px$

$\Delta_C = Pa^3/3EI$

$\Delta_B \max = \dfrac{Pa^3}{3EI}\left(1 + \dfrac{3b}{2a}\right)$ (at free end)

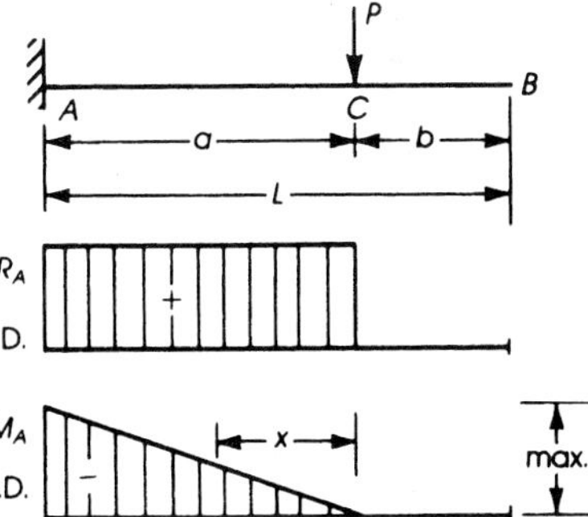

24. Concentrated load at free end

$R_A = V_A = P$

$M_{max} = PL$ (at A)

$M_x = Px$

$\Delta_B \max = \dfrac{PL^3}{3EI}$

$\Delta_x = \dfrac{P}{6EI}(2L^3 - 3L^2x + x^3)$

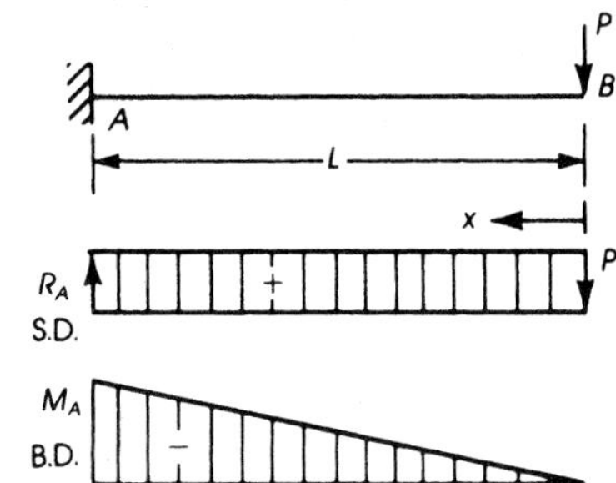

(continued)

Table C.3. Propped Beams

25. Uniform load

W = total load = wL

$$R_A = V_A = \frac{5W}{8} \qquad R_B = V_B = \frac{3W}{8}$$

$$M_A = -\frac{WL}{8} \qquad M_C = \frac{9WL}{128} \quad \left(\text{at } x = \frac{3}{8}L\right)$$

$$\Delta_x = \frac{WL^3}{48EI}(m - 3m^3 + 2m^4) \quad \text{where } m = \frac{x}{L}$$

$$\Delta_{max} = \frac{WL^3}{185EI} \quad \text{at a distance } x = 0.4215L \text{ (from support B)}$$

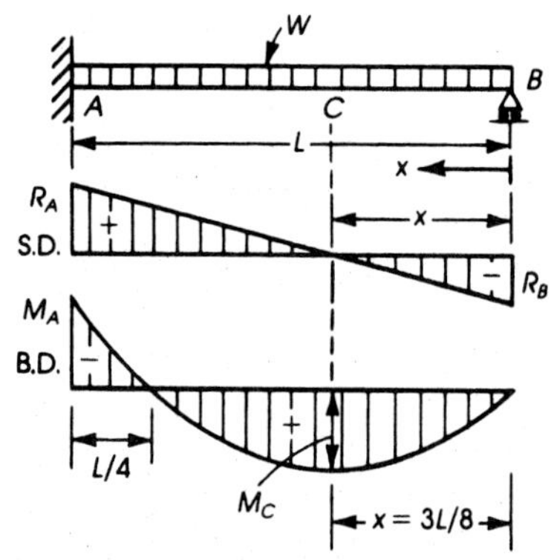

26. Partial uniform load starting from hinged support

$$W = wb \qquad n = \frac{b}{L}$$

$$R_A = V_A = \frac{Wn}{8}(6 - n^2)$$

$$R_B = V_B = \frac{W}{8}(n^3 - 6n + 8)$$

$$M_A = -\frac{Wb}{8}(2 - n^2) \qquad M_C = \frac{Wb}{8}(6n - n^3 - 4)$$

$$\Delta_x = \frac{WbL^2}{48EI}\left[(n^2 - 6)m^3 - (3n^2 - 6)m^2\right] \quad \text{when } x \le a$$

$$\Delta_x = \frac{WL^4}{48bEI}\left[2P^4 - P^3n(n^3 - 6n + 8) + Pn^2(3n^2 - 8n + 6)\right] \quad \text{when } x \ge a \text{ and } P = \frac{L - x}{L}$$

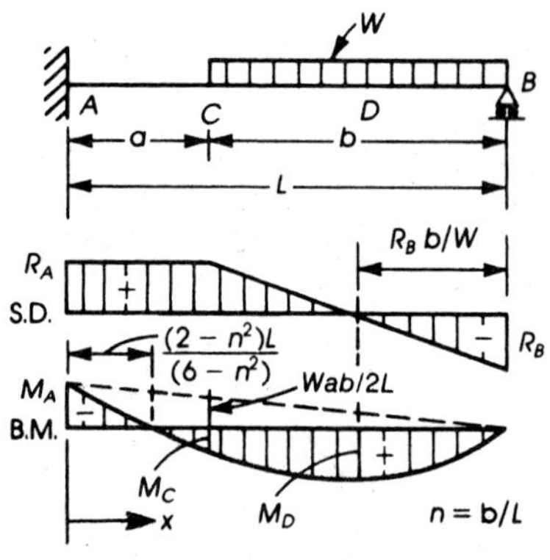

27. Partial uniform load starting from fixed end

$$W = wa \qquad n = \frac{a}{L}$$

$$R_A = V_A = \frac{W}{8}\left[8 - n^2(4 - n)\right]$$

$$R_B = V_B = \frac{Wn^2}{8}(4 - n) \qquad Y = b + an^2(4 - n)$$

$$M_A = -\frac{Wa}{8}(2 - n)^2$$

$$M_{max} = \frac{Wa}{8}\left\{-\frac{\left[8 - n^2(4 - n)\right]^2}{16} + 4 - n(4 - n)\right\}$$

$$\Delta_C = \frac{Wa^3}{48EI}(6 - 12n + 7n^2 - n^3)$$

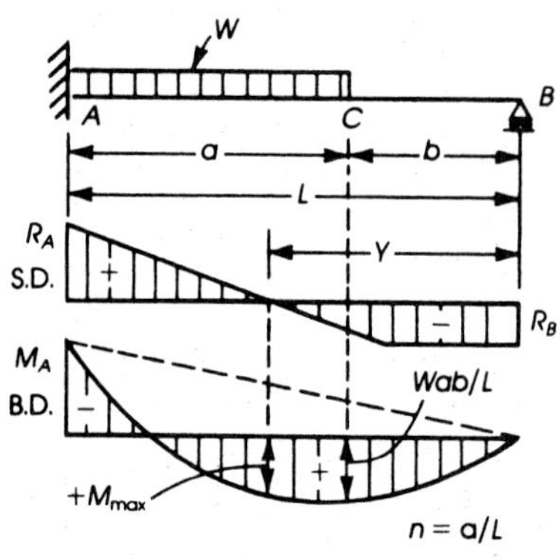

(continued)

Table C.3. *(continued)*

28. Triangular load on all span L

$$W = \text{total load} = \frac{wL}{2}$$

$$R_A = V_A = \frac{4}{5}W \qquad R_B = \frac{W}{5} = V_B$$

$$M_A = -\frac{2}{15}WL$$

$$M_C = +\frac{3}{50}WL$$

$$\Delta_{max} = \frac{WL^3}{212EI} \quad (\text{at } x = 0.447L)$$

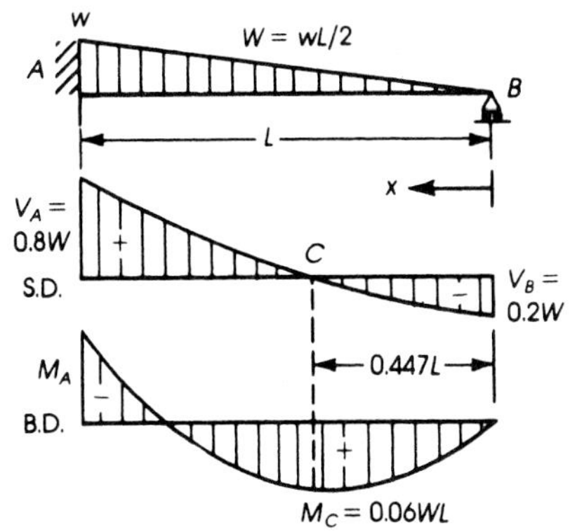

29. Triangular load on part of the span

$$W = \frac{wa}{2}$$

$$R_B = V_B = \frac{Wa^2}{20L^3}(5L - a)$$

$$R_A = W - R_B$$

$$M_A = \frac{Wa}{60L^2}(3a^2 - 15aL + 20L^2)$$

Maximum positive moment at

$$S = b + \frac{a^2}{2L}\sqrt{1 - \frac{a}{5L}}$$

M_{max} (positive) at D:

$$M_D = R_B S - \frac{WL}{3a^3}(-b + S)^3$$

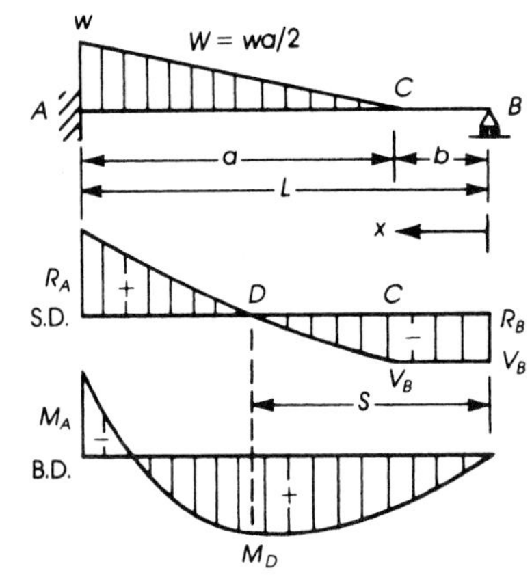

30. Concentrated load at midspan

$$R_A = V_A = \frac{11P}{16}$$

$$R_B = V_B = \frac{5P}{16}$$

$$M_A = -\frac{3PL}{16}$$

$$M_C = \frac{5PL}{32}$$

$$\Delta_C = \frac{7PL^3}{768EI}$$

$$\Delta_{max} = \frac{PL^3}{107EI} \quad (\text{at } x = 0.447L \text{ from B})$$

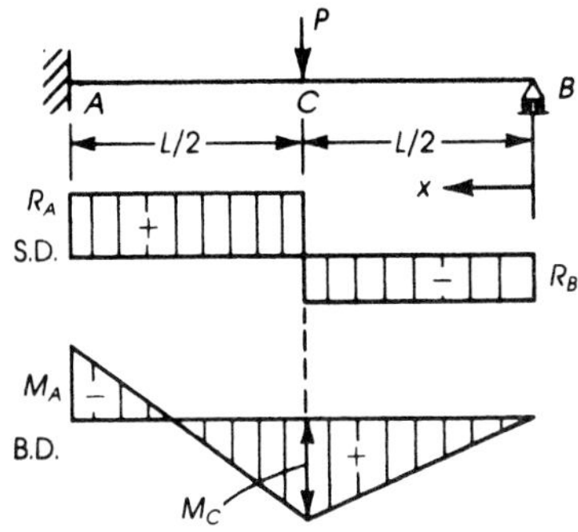

(continued)

Table C.3. *(continued)*

31. Concentrated load at any point

$$R_A = V_A = P - R_B \qquad R_B = V_B = \frac{Pa^2}{2L^3}(b + 2L)$$

$$M_A = -\frac{Pb(L^2 - b^2)}{2L^2}$$

M_A max $= 0.193PL$ when $b = 0.577L$

$$M_C = \frac{Pb}{2}\left(2 - \frac{3b}{L} + \frac{b^3}{L^3}\right)$$

M_C max $= 0.174PL$ when $b = 0.366L$

$$\Delta_C = \frac{Pa^3b^2}{12EIL^3}(4L - a)$$

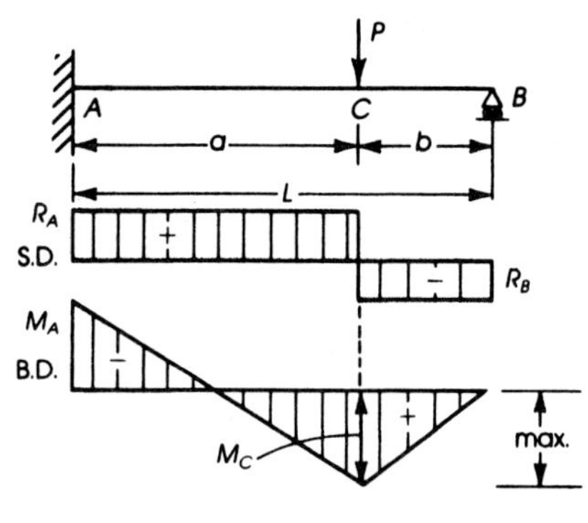

32. Two concentrated loads at one–third points

$$R_A = V_A = \frac{4P}{3}$$

$$R_B = V_B = \frac{2P}{3}$$

$$M_A = -\frac{PL}{3}$$

$$M_C = \frac{PL}{9} \qquad M_D = \frac{2PL}{9}$$

$$\Delta_{max} = \frac{PL^3}{65.8EI}$$

occurs at point $= 0.423L$ from support B

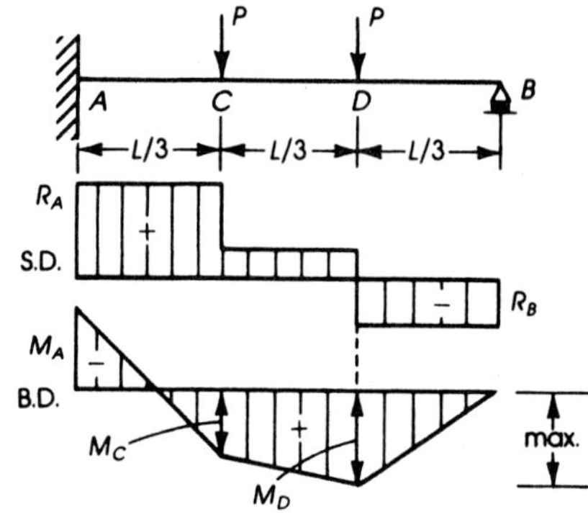

Table C.4. Fixed End Beams

33. Uniform load

W = total load = wL

$$R_A = V_A = R_B = V_B = \frac{W}{2}$$

$$M_A = M_B = -\frac{WL}{12} \quad \text{(at support)}$$

$$M_C \max = \frac{WL}{24} \quad \text{(at midspan)}$$

$$\Delta_{\max} = \frac{WL^3}{384EI} \quad \text{(at midspan)}$$

$$\Delta_x = \frac{Wx^3}{24EIL}(L - x)^2 \quad \text{(from } A \text{ or } B\text{)}$$

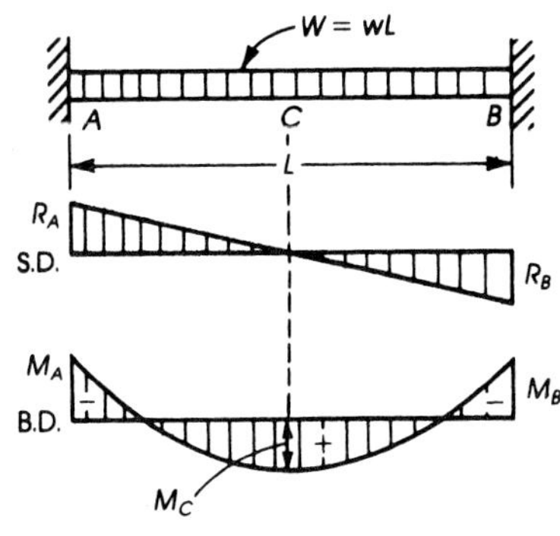

34. Uniform partial load at one end

W = total load = wa $\quad m = \frac{a}{L}$

$$R_A = V_A = \frac{W(m^3 - 2m^2 + 2)}{2}$$

$$R_B = V_B = \frac{Wm^2(2 - m)}{2} = W - R_A$$

$$M_A = \frac{WLm}{12}(3m^2 - 8m + 6)$$

$$M_B = \frac{WLm^2}{12}(4 - 3m)$$

$$M_{\max} = \frac{WLm^2}{12}\left(-\frac{3}{2}m^5 + 6m^4 - 6m^3 - 6m^2 + 15m - 8\right)$$

when $x = \frac{a}{2}(m^3 - 2m^2 + 2)$

$$\Delta_{\max} = \frac{WL^3}{333EI}$$

$$\Delta_C = \frac{WL^3}{384EI}$$

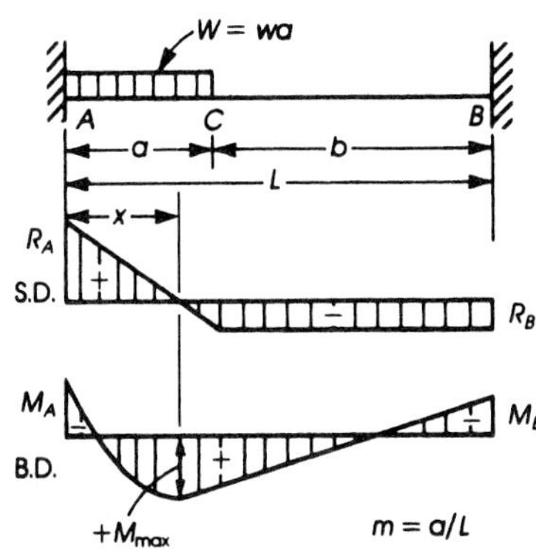

(continued)

Table C.4. *(continued)*

35. Triangular load

$$W = \frac{wL}{2}$$

$$R_A = V_A = 0.7W$$

$$R_B = V_B = 0.3W$$

$$M_A = \frac{WL}{10} \qquad M_B = \frac{WL}{15}$$

$$\Delta_{\max} = \frac{WL^3}{382EI} \quad \text{(at } x = 0.55L \text{ from B)}$$

$$M_C \text{ (maximum positive moment)} = +\frac{WL}{23.3} \quad \text{(at } 0.55L \text{ from B)}$$

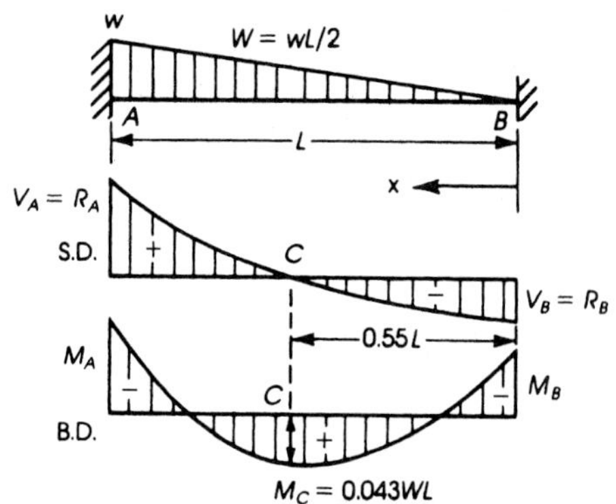

36. Triangular load on part of the span

$$W = \frac{wa}{2}$$

$$R_B = V_B = \frac{Wa^2}{10L^3}(5L - 2a)$$

$$R_A = W - R_B$$

$$M_A = \frac{Wa}{30L^2}(3a^2 + 10bL)$$

$$M_B = \frac{Wa}{30L^2}(-3a^2 + 5aL)$$

Maximum positive moment at $S = b + \frac{a^2}{3.16L}\sqrt{5 - \frac{2a}{L}}$

$$M_D = R_B S - \frac{WL}{3a^3}(a + S - L)^3 - M_B$$

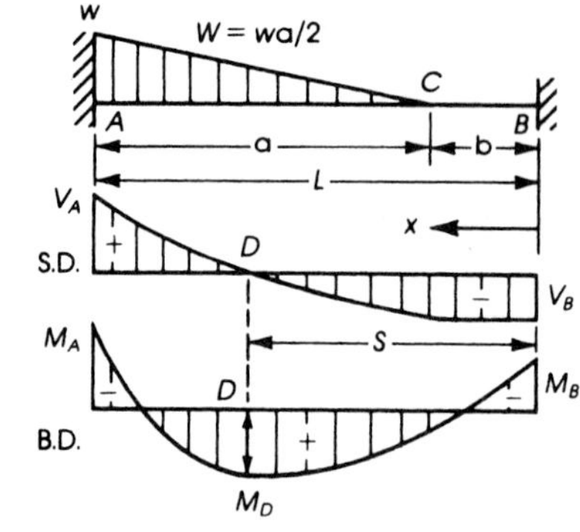

37. Triangular load, maximum intensity at midspan

$$W = \text{total load} = \frac{wL}{2}$$

$$R_A = R_B = \frac{W}{2}$$

$$M_A = M_B = -\frac{5}{48}WL$$

$$M_C \text{ (maximum positive)} = \frac{WL}{16}$$

$$\Delta_{\max} = \frac{1.4WL^3}{384EI} \quad \text{(at midspan)}$$

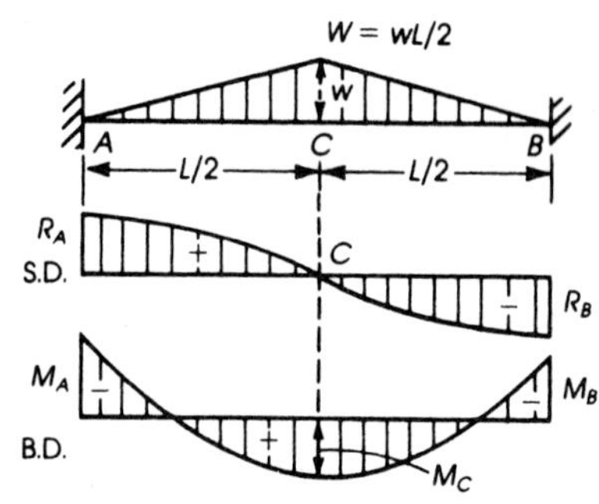

(continued)

Table C.4. *(continued)*

38. Concentrated load at midspan

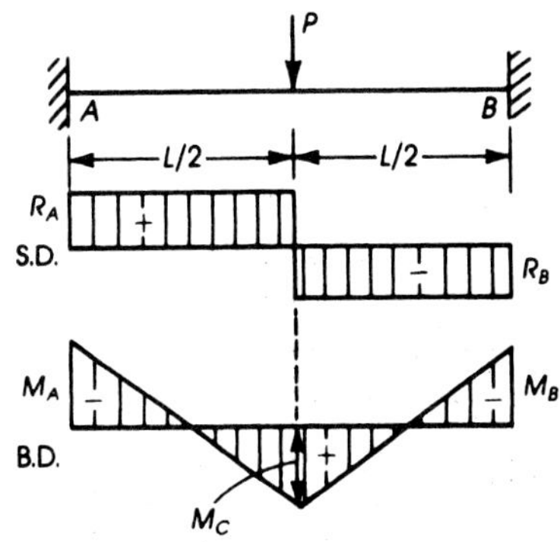

$$R_A = V_A = R_B = V_B = \frac{P}{2}$$

$$M_A = M_B = M_C = -\frac{PL}{8}$$

$$\Delta_{\max} = \frac{PL^3}{192EI} \quad \text{(at midspan)}$$

$$\Delta_x = \frac{Px^2}{48EI}(3L - 4x) \quad \left(x < \frac{L}{2}\right)$$

39. Two symmetrical concentrated loads

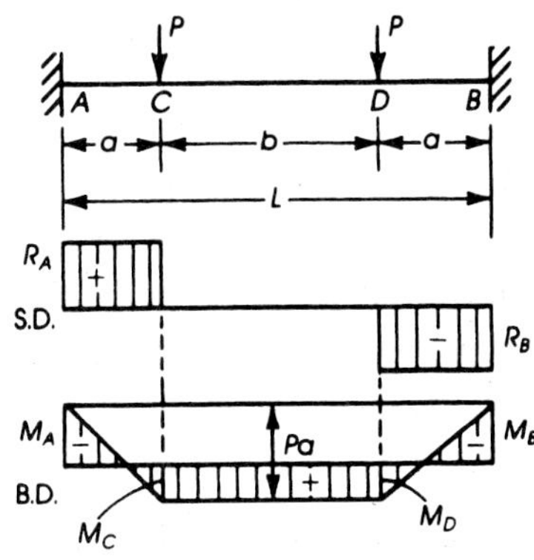

$$R_A = V_A = R_B = V_B = P$$

$$M_A = M_B = -\frac{Pa(L - a)}{L}$$

$$M_C = M_D = \frac{Pa^2}{L}$$

$$\Delta_{\max} = \frac{PL^3}{6EI}\left[\frac{3a^2}{4L^2} - \left(\frac{a}{L}\right)^3\right] \quad \text{(at midspan)}$$

If $a = \dfrac{L}{3}$,

$$M_A = M_B = \frac{2}{9}PL$$

$$\Delta_{\max} = \frac{5PL^3}{648EI} \quad \text{(at centerline)}$$

If $a = \dfrac{L}{4}$,

$$M_A = M_B = \frac{3}{16}PL$$

$$\Delta_{\max} = \frac{PL^3}{192EI} \quad \text{(at centerline)}$$

(continued)

Table C.4. *(continued)*

40. Concentrated load at any point

$$R_A = V_A = P\left(\frac{b}{L}\right)^2\left(1 + \frac{2a}{L}\right)$$

$$R_B = V_B = P\left(\frac{a}{L}\right)^2\left(1 + \frac{2b}{L}\right)$$

$$M_A = -\frac{Pab^2}{L^2} \quad M_B = -\frac{Pba^2}{L^2} \quad M_C = \frac{2Pa^2b^2}{L^3}$$

$$\Delta_C = \frac{Pa^3b^3}{3EIL^3} \text{ (at point C)}$$

$$\Delta_{\max} = \frac{2Pa^2b^3}{3EI(3L - 2a)^2} \text{ when } x = \frac{2bL}{3L - 2a} \text{ and } b > a$$

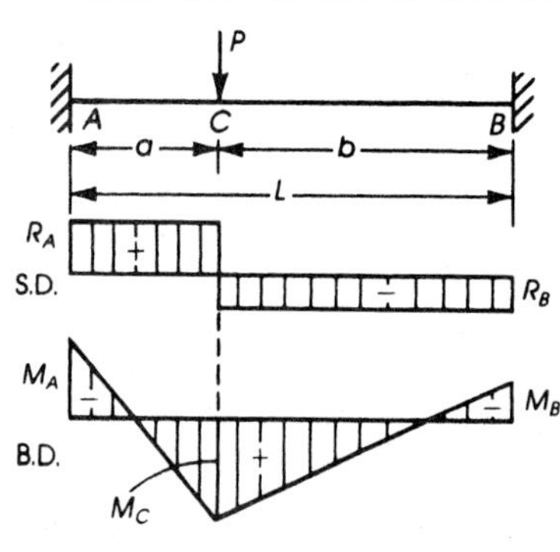

Table C.5. Moments in Two Unequal Spans and Values of the Coefficient K (w = unit load/unit length)

1. Load on short span

$$M_B = \frac{wL_2^3}{8(L_1 + L_2)} = \frac{wL_2^2}{K}$$

L_2/L_1	0.20	0.25	0.30	0.40	0.50	0.60	0.70	0.80	0.90	1.00
K	46.0	40.0	34.7	28.0	24.0	21.4	19.5	18.0	16.9	15.9

2. Load on long span

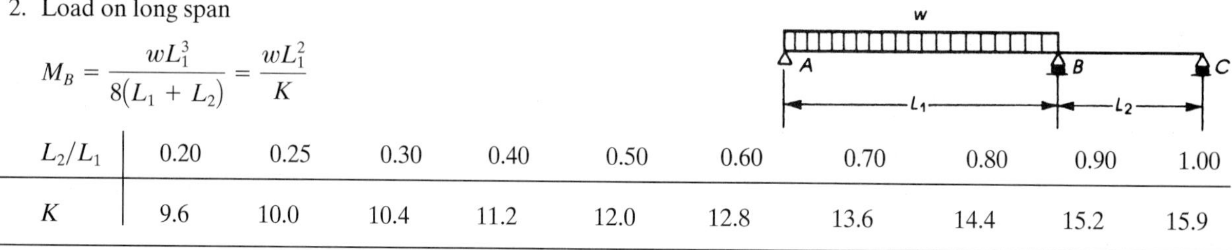

$$M_B = \frac{wL_1^3}{8(L_1 + L_2)} = \frac{wL_1^2}{K}$$

L_2/L_1	0.20	0.25	0.30	0.40	0.50	0.60	0.70	0.80	0.90	1.00
K	9.6	10.0	10.4	11.2	12.0	12.8	13.6	14.4	15.2	15.9

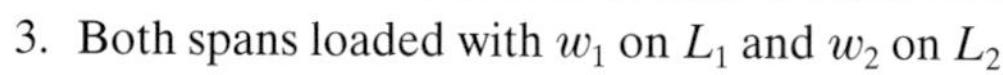

3. Both spans loaded with w_1 on L_1 and w_2 on L_2

$$M_B = \frac{w_1L_1^3 + w_2L_2^3}{8(L_1 + L_2)}$$

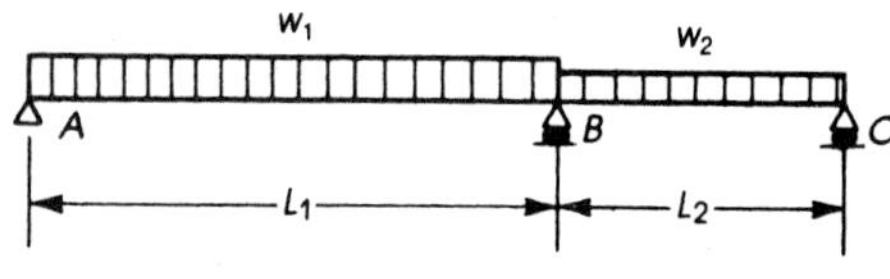

Table C.6. Moments in Three Unequal Spans and Values of the Coefficient K (w = load/unit length)

4. Load on span CD

L_2/L_3	$M_B = \frac{wL_3^2}{K}$ (*positive*)	$M_C = \frac{wL_3^2}{K}$ (*negative*)
0.25	100.0	9.9
0.30	90.9	10.3
0.40	76.3	11.0
0.50	70.4	11.7
0.60	65.8	12.3
0.70	62.9	13.0
0.80	61.7	13.7
1.00	59.9	14.9

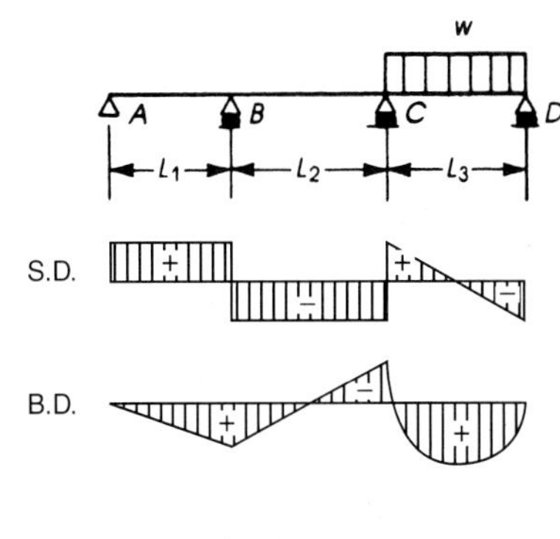

5. Load on middle span

L_2/L_1	$M_B = M_C = \frac{wL_2^2}{K}$ (*negative*)
0.25	43.5
0.30	38.5
0.40	32.3
0.50	27.8
0.60	25.6
0.70	23.3
0.80	22.2
1.00	20.0

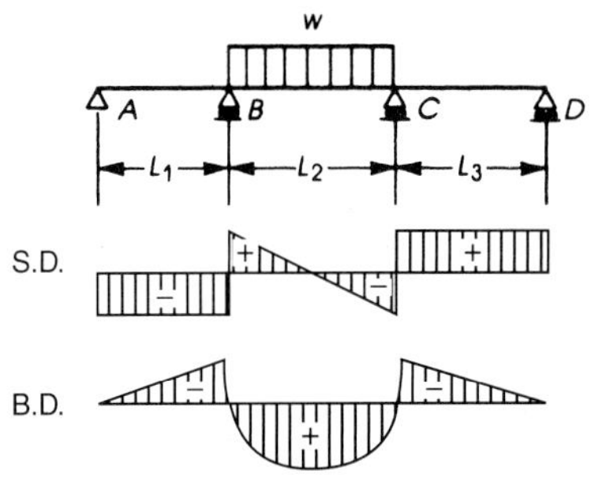

6. Load on span AB

L_2/L_1	$M_B = \frac{wL_1^2}{K}$ (*negative*)	$M_C = \frac{wL_1^2}{K}$ (*positive*)
0.25	9.9	100.0
0.30	10.3	90.9
0.40	11.0	76.3
0.50	11.7	70.4
0.60	12.3	65.8
0.70	13.0	62.9
0.80	13.7	61.7
1.00	14.9	59.9

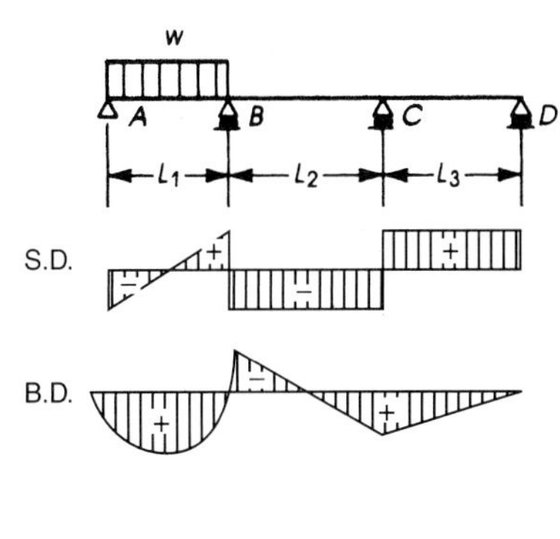

Table C.7. Maximum and Minimum Moments in Equal Spans Continuous Beams

7. Uniform loads

$$M = \frac{wL^2}{K}$$

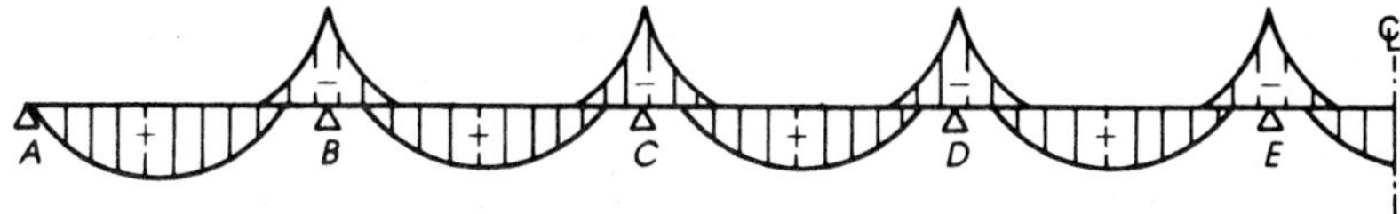

where w = (D.L. + L.L.) per unit length D.L. = Uniform dead load L.L. = Uniform live load

Values of coefficient K

Ratio	*First span AB (positive moment)*				*Second support B (negative moment)*			
	Number of spans				*Number of spans*			
D.L./w	**2**	**3**	**4**	**5**	**2**	**3**	**4**	**5**
0.0	10.5	10.0	10.2	10.1	8.0	8.6	8.3	8.3
0.1	10.8	10.2	10.4	10.3	8.0	8.7	8.4	8.5
0.2	11.1	10.4	10.6	10.6	8.0	8.8	8.5	8.6
0.3	11.4	10.6	10.9	10.8	8.0	9.0	8.6	8.7
0.4	11.8	10.9	11.1	11.0	8.0	9.1	8.6	8.8
0.5	12.1	11.1	11.4	11.3	8.0	9.2	8.8	8.9
0.6	12.5	11.4	11.7	11.6	8.0	9.4	8.9	9.0
0.7	12.9	11.6	12.0	11.9	8.0	9.5	9.0	9.1
0.8	13.3	11.9	12.3	12.2	8.0	9.7	9.1	9.2
0.9	12.8	12.2	12.6	12.5	8.0	9.8	9.2	9.4
1.0	14.3	12.5	13.0	12.8	8.0	9.9	9.3	9.5

Ratio D.L./w	*Second span BC (positive moment)*			*Third support C (negative moment)*		*Third span CD (positive moment)*	*Interior span (positive moment)*	*Interior support (negative moment)*
	Number of spans			*Spans*		*Span*		
	3	**4**	**5**	**4**	**5**	**5**		
0.0	13.4	12.4	12.7	9.3	9.0	11.7	12.0	8.8
0.1	14.3	13.2	13.5	9.7	9.3	12.3	12.6	9.1
0.2	15.4	14.0	14.3	10.0	9.6	12.9	13.3	9.8
0.3	16.7	14.9	15.3	10.4	9.9	13.6	14.1	9.5
0.4	18.2	16.0	16.5	10.8	10.2	14.3	15.0	9.9
0.5	20.0	17.2	17.9	11.5	10.5	15.2	16.0	10.1
0.6	22.2	18.7	19.5	11.7	10.9	16.2	17.2	10.5
0.7	25.0	20.4	21.4	12.2	11.3	17.3	18.4	10.8
0.8	28.6	22.4	23.8	12.7	11.7	18.5	20.0	11.2
0.9	33.3	24.9	26.6	13.3	12.2	20.0	21.8	11.6
1.0	40.0	28.3	30.0	14.0	12.7	21.7	24.0	12.0

(continued)

Table C.7. *(continued)*

Example: K values

1. $\frac{\text{D.L.}}{w} = 0.4$

2. $\frac{\text{D.L.}}{w} = 1.0$

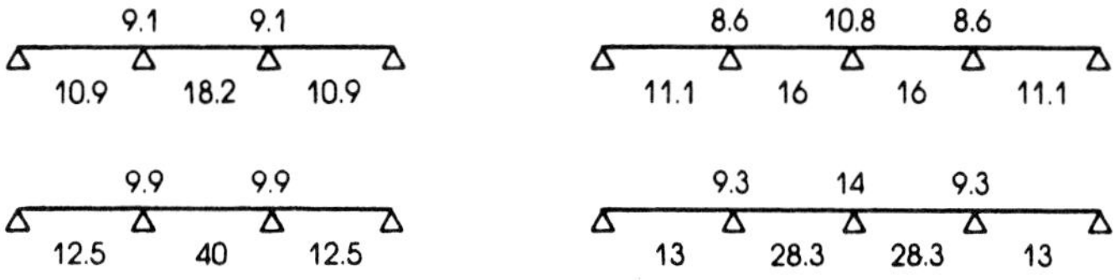

8. Concentrated loads

P' = concentrated dead load

P'' = concentrated live load

$$M = \left(\frac{P'}{K_1} + \frac{P''}{K_2}\right)L$$

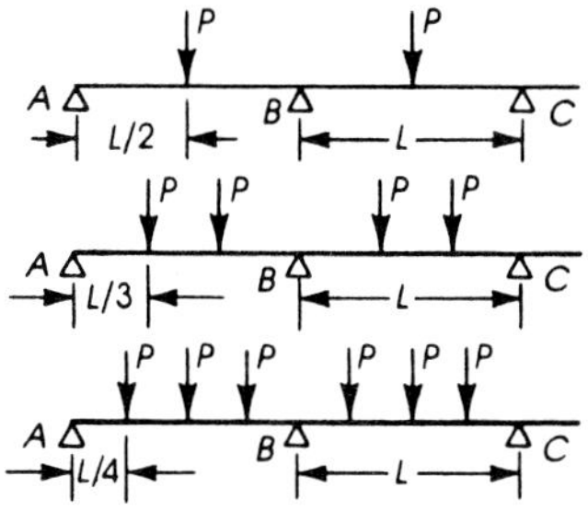

Number of spans	First span AB						Second support B					
	K_1 (D.L.)			K_2 (L.L.)			K_1 (D.L.)			K_2 (L.L.)		
	2	3	4	2	3	4	2	3	4	2	3	4
Central load	6.40	5.71	5.89	4.92	4.70	4.76	5.35	6.67	6.22	5.33	5.71	5.53
One-third-point loads	4.50	4.09	4.20	3.60	3.46	3.50	3.00	3.75	3.50	3.00	3.21	3.11
One-fourth-point loads	3.67	3.20	3.34	2.61	2.46	2.50	2.13	2.67	2.49	2.13	2.28	2.21

Number of spans	Second span BC				Third support C	
	K_1		K_2		K_1	K_2
	3	4	3	4	4	4
Central load	10.00	8.61	5.71	5.46	9.33	6.22
One-third-point loads	15.00	9.00	5.00	4.50	5.25	3.50
One-fourth-point loads	8.00	6.05	3.20	3.01	3.72	2.49

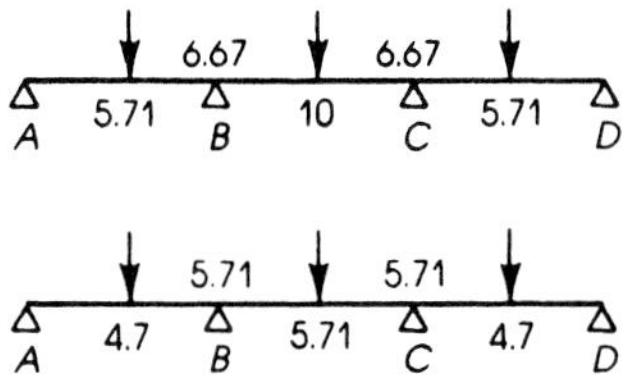

Example: K values

K_1 (dead load) $\quad M_{AB}(\max) = \left(\frac{P'}{5.71} + \frac{P''}{4.7}\right)L$

$$-M_B(\max) = \left(\frac{P'}{6.67} + \frac{P''}{5.71}\right)L$$

K_2 (live load) $\quad M_{BC}(\max) = \left(\frac{P'}{10} + \frac{P''}{5.71}\right)L$

Table C.8. Moments in Unequal Spans Continuous Beams Subjected to Unequal Loads.

9. Unequal spans and unequal loads. For approximate bending moments in continuous beams, use

$L' = 0.8L$ for spans continuous at both ends

$L' = L$ for spans continuous at only one end

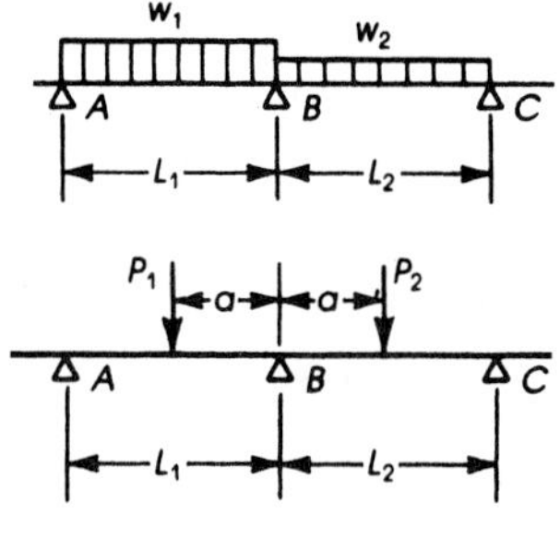

1. Uniform loads (load on two adjacent spans):

$$M_B = \frac{w_1 L_1'^3 + w_2 L_2'^3}{8.5(L_1' + L_2')}$$

2. Concentrated loads:

$$M_B = \frac{K P_1 L_1'^2}{L_1' + L_2'} \quad \text{due to load } P_1$$

$$M_B = \frac{K P_2 L_2'^2}{L_1' + L_2'} \quad \text{due to load } P_2$$

a/L	0.1	0.2	0.3	0.4	0.5	0.6	0.7	0.8	0.9	1.0
K	0.080	0.136	0.168	0.182	0.176	0.158	0.128	0.090	0.050	0.000

3. Moments within span:
   - Maximum positive moment is obtained by superposing B.M. due to D.L. + L.L. and the negative moments at supports due to D.L. only.
   - Maximum negative moment is obtained by superposing B.M. due to D.L. only and the negative moments at supports due to D.L. + L.L.

Example:

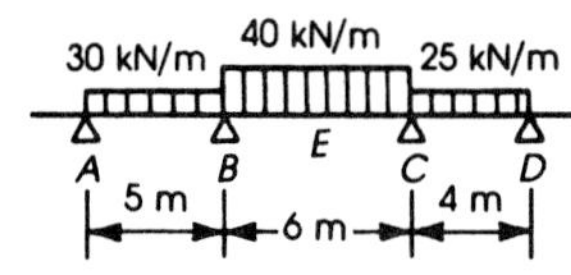

$$M_B = -\frac{30(0.8 \times 5)^3 + 40(0.8 \times 6)^3}{8.5(0.8 \times 5 + 0.8 \times 6)} = -84.8\ \text{kN}\cdot\text{m}$$

$$M_C = -\frac{40(0.8 \times 6)^3 + 25(0.8 \times 4)^3}{8.5(0.8 \times 4 + 0.8 \times 6)} = -77.1\ \text{kN}\cdot\text{m}$$

$$M_E = \text{(at centerline of BC)} = +\frac{wL^2}{8} + \frac{1}{2}(M_B + M_C) = \frac{40 \times 36}{8} + \frac{1}{2}(-84.8 - 77.1) = +99\ \text{kN}\cdot\text{m}$$

ANSWERS TO SELECTED PROBLEMS

CHAPTER 2: **2.3** $E_c = 4067$ Ksi, $G_c = 1723$ Ksi; **2.9** 4723 Ksi, 3644 Ksi, 2306 Ksi, 29,910 MPa, 25,979 MPa, 20,690 Mpa; **2.11** **a.** $f'_c = 3820$ psi at strain $= 0.003$; **b.** Initial $E_c = 4760$ Ksi, $E_c = 3131$ Ksi; **c.** $E_c = 3523$ Ksi greater than 3130 Ksi

CHAPTER 3: **3.1** **a.** 441.2; **c.** 370.1; **e.** 600; **g.** 200.5 K · ft; **3.3** **a.** 374.8; **c.** 769.9; **e.** 660.1; **g.** 415 K · ft; **3.5** **a.** 272.3 KN · m; **c.** N.G., use $\rho_{\max}$(352.6 KN · m); **e.** 283.1 K · ft; **3.7** 3.628 K/ft; **3.9** **a.** 277.8 K · ft; **b.** 157.1 K · ft; **3.11** 6.724 K/ft; **3.13** 236.4 K · ft; **3.15** 466.8 K · ft;

CHAPTER 4: **4.1** **a.** $A_s = 3.15$ in.^{2}; **c.** $d = 28$ in. and $A_s = 7.62$ in.^{2}; **e.** $d = 24.5$ in. and $A_s = 6.0$ in.^{2}; **g.** $d = 20.5$ in. and $A_s = 6.0$ in.^{2}; **i.** $d = 18.5$ in. and $A_s = 4.91$ in.^{2}; **k.** $b = 12$ in., $d = 30$ in., and $A_s = 6.82$ in.^{2}; **4.3** **a.** $A_s = 5.06$ in.^{2}; **c.** 3.96 in.^{2}; **e.** 7.59 in.^{2}; **g.** 6.0 in.^{2}; **i.** 8.0 in.^{2}; **k.** 6.0 in.^{2}; **m.** 6.43 in.^{2}; **o.** 6.0 in.^{2}; **4.5** $A_s = 3.5$ in.^{2}; **4.7** **a.** Using max ρ:$b = 12$ in., $h = 15$ in., $A_s = 3.1$ in.^{2}; **b.** Using $\rho_s = 0.014$: $b = 12$, $h = 17$ in.,$A_s = 2.38$ in.^{2}; **4.9** For $\rho = 0.014$: $b = 14$ in., $h = 23$ in., $A_s = 4.0$ in.^{2}; **4.11** $A_s = 4.43$ in.^{2}, $A'_s = 0.76$ in.^{2}; **4.13** For $a = t$, $d = 26$ in. and $A_s = 5.05$ in.^{2}.

CHAPTER 5: **5.1** **a.** $I_g = 13{,}824$ in.^{4}, $I_{cr} = 6609$ in.^{4}, $M_{cr} = 45.5$ K · ft, $f_c = 1522$ psi, $f_s = 2217$ psi, $M_a = 119.1$ K · ft; **g.** $I_g = 10{,}648$ in.^{4}, $I_{cr} = 8236$ in.^{4}, $M_{cr} = 38.24$ K · ft, $f_c = 1466$ psi, $f_s = 16.3$ Ksi, $f'_s = 15.88$ Ksi, $M_a = 159.6$ K · ft; **5.3** $M_a = 64.6$ K · ft, $f_c = 1107$ psi, $f_s = 13.94$ Ksi; **5.5** $f_c = 2086$ psi $> f_{ac} = 1800$ psi, $f_s = 23.7$ Ksi $< f_{sa} = 24$ Ksi; **5.7** $w = 5.856$ K/ft; **5.9** $f_c = 464$ psi, $f_s = 10.37$ Ksi, $M_a = 167.7$ K · ft; **5.11** $M = 144.3$ K · ft $= M_s$; **5.13** $M = 68.3$ K · ft $= M_c$

CHAPTER 6: **6.1** **a.** $\Delta_I = 0.667$ in., Δ (long-time) $= 1.46$ in.; **e.** $\Delta_I = 0.55$ in., Δ (long-time) $= 1.07$ in.; **6.3** $\Delta_I = 1.01$ in.; **6.5** **a.** $z = 132.6$ K/in.; **b.** $z = 123.1$ K/in.; **c.** $z = 112.6$ K/in.; **d.** $z = 121$ K/in.; All can be used as exterior and interior members.

CHAPTER 7: **7.1** **a.** 28 in.; **c.** 56 in.; **f.** 54 in.; **i.** 48 in.; **7.3** **a.** 70 in.; **b.** 22 in.; **c.** 22 in.; **7.5** **a.** 54 in.; **b.** 25 in.; **c.** 25 in.; **7.7** **a.** $l_{st} = l_{db} = 54$ in.; For b., c., and d.: $l_{st} = 71$ in.; **7.9** 34 in.; **7.11** 54 in.

CHAPTER 8: **8.1** **a.** Use stirrups #3 spaced at 8.5 in.; **b.** Use stirrups #3 at 5 in.; **c.** Use stirrups #3 at 3.5 in.; **8.3** **a.** Use stirrups #3 at 13.5 in.; **b.** Use stirrups #3 at 4.5 in.; **c.** Use stirrups #3 at 3 in. (or #4 at 5.5 in.) **8.5** Use stirrups #3 at 5.5 in.; **8.7** $b = 12$ in., $d = 17.5$ in., $h = 20$ in., $A_s = 3$ no. 9 bars. Use stirrups #3 at 6.0 in.; **8.9** Section 16×32 in. $(b \times d)$ or 16×31.5 in. $(b \times d)$ gives stirrups #3 at 11 in.; **8.11** Use stirrups #3 at 3 in. or #4 at 4 in.

CHAPTER 9: **9.11** Answers are in table of Problem 9.1.; **9.3** W_L (allowable) = 186 psf; **9.5** $A_s = 0.36$ in.^{2} (no. 4 spaced at 6 in.); Shrinkage bars, 0.144 in.^{2} (no. 3 spaced at 9 in.); **9.7** Reinforcement area A_s at A, B, C, D and E is: 0.21, 0.38, 0.54, 0.49 and 0.33 in.^{2}; **9.9** Reinforcement area A_s at A, B, C, D and E is: 0.16, 0.24, 0.36, 0.33 and 0.22 in.^{2}; **9.11** Total $h = 12$ in., slab thickness = 2 in. Use 2 no. 5 bars per rib.

CHAPTER 10: **10.1 a.** 741 K; **c.** 473 K; **e.** 778 K; **g.** 1340 K; **10.3 a.** (14×14) and 8 no. 9 bars; **c.** (12×12) and 8 no. 10 bars; **e.** (12×18) and 6 no. 8 bars; **g.** (12×16) and 6 no. 9 bars; **i.** (Diameter = 16 in.) and 8 no. 9 bars; **k.** (Diameter = 18 in.) and 10 no. 10 bars

CHAPTER 11: **11.1 a.** 400 K, $e_b = 17.4$ in.; **c.** 593 K, $e_b = 20.1$ in.; **e.** 212 K, $e_b = 15.2$ in.; **g.** 273 K, $e_b = 15.2$ in.; **i.** 334 K, $e_b = 13.4$ in.; **11.3 a.** 275 K; **c.** 482 K; **e.** 124 K; **g.** 159 K; **i.** 154 K; **k.** 329 K; **11.5** 708 K; **11.7** 822 K; **11.9** (288 K), (1066 K), (496 K), (20 in. and 5 no. 10 bars), (24 in. and 6 no. 10 bars), (24 in. and 6 no. 10 bars); **11.11** 183 K, 160 K, 10.4 in.; **11.13** 498 K, 435 K · ft; and 10.5 in.

CHAPTER 12: **12.1** $\delta_{ns} = 1.16$, $M_c = 535.7$ K · ft, $P_n = 493$ K; **12.3** $\delta_{ns} = 1.127$, $M_c = 521.6$ K · ft, $P_n = 533$ K; **12.5** $\delta_{ns} < 1.0$, use $\delta_{ns} = 1.0$, $P_n = 1436$ K; **12.7** $\delta_s = 1.06$, $M_c = 418$ K · ft, $P_n = 1539$ K; **12.9** $\delta_s = 1.11$, $M_c = 240.6$ K · ft, $P_n = 964.5$ K

CHAPTER 13: **13.1, 13.2** Answers are in Tables 13.1 and 13.2; **13.3 a.** $B \times L = 6 \times 10.5$ ft, $h = 23$ in.; **c.** $B \times L = 8 \times 10$ ft, $h = 23$ in.; **e.** $B \times L = 7 \times 10$ ft, $h = 21$ in.; **13.7** Areas (sq ft) = 60.5, 98, 72, 80.1, 90.7, 103.2; **13.9** $B \times L = 8 \times 16$ ft, $h = 28$ in.

CHAPTER 14: **14.1** $q_1 = 2.19$ Ksf, $q_2 = 0.19$ Ksf, wall is adequate; **14.3 a.** $F. S.$ (overturning) = 3.73, sliding = 1.76; **c.** $F. S.$ (overturning) = 3.15, sliding = 1.64; **e.** $F. S.$ (overturning) = 2.95, sliding = 1.63; **14.5 c.** $F. S.$ (overturning) = 2.35, sliding < 1.5, use a key; **14.7** $L = 12$ ft, $H = 20$ ft, $h = 18$ ft, stem 1.0 ft; top and 2.0 ft bottom, $F. S.$ (overturning) = 2.3 > 2, $F. S.$ (sliding) = 1.3, (use key 1.5×1.5 ft); **14.9** $F. S.$ (overturning) = 2.82 > 2, $F. S.$ (overturning) = 1.69 > 1.5 with 1.5×1.5 ft key; A_s (stem) = no. 8 at 7 in.; **14.11** Wall thickness = 8 in., use no. 5 at 12 in. vertical bars and no. 4 at 12 in. as horizontal bars.

CHAPTER 15: **15.1** 396.4 K · in., 99.1 K · in.; **15.3** 244.3 K · in., 61.1 K · in.; **15.5** 319.4 K · in., 87.36 K · in.; **15.7** No. 4 bars spaced at 7 in., $A_l = 0.76$ in.^{2}; **15.9** No. 4 bars spaced at 3 in., $A_l = 1.19$ in.^{2}; **15.11** $A_s = 4$ no. 9 bars, no. 3 spaced at 6 in., $A_l = 1.9$ in.^{2}; **15.13** $A_s = 4$ no. 9 bars, no. 3 spaced at 5 in., $A_l = 2.0$ in.^{2}; **15.15** $A_s = 4$ no. 9 bars, no. 3 spaced at 6 in., $A_l = 1.6$ in.^{2}

CHAPTER 16: **16.1** $h = 9.5$ in., $A_s = 2.12$ in.^{2}, 3.77 in.^{2}, and 3.34 in.^{2}, at A, B and C respectively; **16.3** At AB: $M_u = 416.2$ K · ft, $A_s = 5.06$ in.^{2}; At BC: $M_u = 249.1$ K · ft, $A_s = 2.83$ in.^{2}; At B: $M_u = -619.5$ K · ft, $A_s = 6.44$ in.^{2}, $A'_s = 1.34$ in.^{2}; **16.5** Use 16×34 in. beam; At B and C: $M_u = -1006.6$ K · ft, $A_s = 8.31$ in.^{2}; At BC: $M_u = +1087.5$ K · ft, $A_s = 9.13$ in.^{2}; Column: 16×30 in., $A_s = 10.46$ in.^{2}, $A'_s = 5.21$ in.^{2}; **16.7** $M_p = 494.2$ K · ft; **16.9** $h = 19$ in., $A_s = 3$ no. 9 bars

CHAPTER 17: **17.1** Slab thickness: h_1 = exterior, h_2 = interior; **a.** $h_1 = 7.2$ in., $h_2 = 6.6$ *in.*; **b.** $h_1 = 8.8$ in., $h_2 = 8.0$ in.; **c.** $h_1 = 9.6$ in., $h_2 = 8.73$ *in.*; **e.** $h_1 = 8.8$ in., $h_2 = 8.0$ *in.*; **g.** $h_1 = 11.2$ in., $h_2 = 10.2$ in.; **17.3 a.** Let $h = 8$ in.; Column strip: M_u (ext.) = −80, +96, −161 (K · ft); Middle strip: M_u (ext.) = 0, +64, −54 (K · ft); **d.** Let $h = 8$ in.; Column strip: M_u (ext.) = −64, +77, −129 (K · ft); Middle strip: M_u (ext.) = 0, +51, −43 (K · ft); **17.5 a.** Let h (slab) = 7.5 in., h (drop panel) = 9 in. (Interior panel); Column strip: M_u: −149, +64.2 (K · ft); Middle strip: M_u: −50, +43 (K · ft); **d.** Let h (slab) = 7.5 in., h (drop panel) = 9 in. (Interior panel); Column strip: M_u = −119.5, +51.5 (K · ft); Middle strip: M_u = −40, +34.5 (K · ft)

CHAPTER 18: **18.1** One solution: Slab thickness = 8 in. reinforced with #6 bars spaced at 7 in. Transverse bars #4 at 12.

CHAPTER 19: **19.1** Let beam $b \times h = 14 \times 20$ in., $d = 17.5$ in., $-M_u = 187$ K · ft, $+M_u = 94.5$ K · ft., $T_u = 18.9$ K · ft

CHAPTER 20: **20.1** At transfer: $f_{\text{top}} = -21$ psi, $f_{\text{bot}} = -2016$ psi; After all losses: $f_{\text{top}} = -124$ psi, $f_{\text{bot}} = -1514\ psi$, $w_L = 1.35$ K/ft; **20.3** $F_i = 166.5$ Ksi (per unit area), $F_i = 543.9$ K, $F = 415$ K; **20.5** F_i (max) $= 1616.6$ K, F_i (min) $= 349.5$ K; **20.7** **a.** harp 12 tendons; **b.** $F = 482.5$ K (139.6 Ksi); **c.** $\phi M_n = 2332.5$ K · ft; **d.** $M_{\text{cr}} = 1521.3$ K · ft; **20.9** At $h/2$ distance from support: $V_c = 75.5$, $V_u = 109.7$ K, use stirrups no. 3 spaced at 16 in.

CHAPTER 21: No problems.

CHAPTER 22: **22.1** **a.** $\phi M_n = 421.7$ K · ft; **b.** $\phi M_n = 412.7$ K · ft; **22.3** **a.** $h = 18$ in., $A_s = 5.5$ in.^{2}; **b.** $h = 19$ in., $A_s = 5.0$ in.^{2}; **22.5** **a.** $P_n = 597$ K, $M_n = 823$ K · ft, $e = 16.55$ in.; **b.** $P_n = 465.8$ K, $M_n = 782.8$ K · ft, $e = 20.2$ in.

INDEX